D0894583

Thermal Insulation Handbook

THERMAL INSULATION HANDBOOK

by

William C. Turner, E.E., M.E., P.E.

and

John F. Malloy, M.E., P.E.

ROBERT E. KRIEGER PUBLISHING COMPANY • **McGRAW-HILL BOOK COMPANY**

Malabar, Florida **New York, New York**

1981

Original Edition 1981
(Based upon a previous book "Thermal Insulation")

Printed and Published by
ROBERT E. KRIEGER PUBLISHING COMPANY, INC.
Krieger Drive
Malabar, Florida 32950

Joint edition with
McGRAW-HILL BOOK COMPANY

Material from Original book
Copyright © 1976 in the name of John Malloy
Copyright © 1981 (new material) by
ROBERT E. KRIEGER PUBLISHING COMPANY, INC.

All rights reserved. No part of this book may be reproduced in any form or by any electronic or mechanical means including information storage and retrieval systems without permission in writing from the publisher.

Printed in the United States of America

Library of Congress Cataloging in Publication Data

Turner, William C., 1913-
Thermal insulation handbook.

Revision of: Thermal insulation/by John F. Malloy.
New York: Van Nostrand Reinhold, 1969.
1. Insulation (Heat)—Handbooks, manuals, etc.
I. Malloy, John F., 1901- II. Title.
[TH1715.T89 1981] 693.8'32 76-52962
ISBN 0-88275-510-2 (Krieger) AACR2
ISBN 0-07-039805-4 (McGraw-Hill)

AUTHORS' DEDICATION

"To our wives Catherine and JoAnn, whose whole-hearted support and encouragement, and endless patience with the constant use of our spare hours, this book would not have been possible, we affectionately dedicate it."

Preface

Thermal insulation has one major function—that is to conserve energy. The importance of energy conservation has not yet been fully appreciated by business or individuals. As America developed, energy was plentiful in respect to needed uses. Our development has caused us to use more and more energy per person. Only in the past few years has there been an energy crisis.

Due to the past abundance of energy, most business management individuals have very little idea as to the money cost of energy. Because of low cost, little was done to reduce energy loss—especially as heat energy is invisible to the human eye. To illustrate this point most management of petrochemical companies, refineries, or power plants are still under the impression that their companies are being operated very efficiently. Whereas if they would make an energy survey, as to the amount of energy which goes into their product and that which is wasted, they find in most cases that 90% of all fuel purchased is wasted by inefficient production methods and poorly thermally insulated plants.

When energy was cheap and plentiful, there was little economic need to conserve it. Because it was inexpensive, as compared to raw materials and labor, it had relatively little effect on the monetary cost of the product. This has now changed as the cost of energy has multiplied 10 to 15 times its cost 30 years ago. From this time on, energy is going to be a major factor in the cost of merchandise. However, most business executives have yet to recognize this change.

One of the major difficulties involved in the understanding of the economics of energy is that it is purchased as fuel then converted to usable form. The fuel is purchased as gallons of oil or gasoline, cubic feet of gas, tons of coal or kilowatts of electricity. These are frequently converted to heat energy, mechanical, or chemical energy. These various forms of energy are used to produce the product. During these transfer processes, most of the losses are heat losses and can not be seen; thus they are not recorded as waste. Although heat loss is a major factor as to earnings of a company, this loss is seldom shown in the accounting statements.

The previous statements are given to illustrate the need for thermal insulation. The economic justification is presented in "Handbook of Thermal Insulation Design Economics for Pipes and Equipment". Having determined the economic need and understanding that our very existence depends on correct use of energy, "Thermal Insulation Handbook" presents the manner in which thermal insulation can be properly designed and used to conserve heat energy.

The control of heat is essential to man's existence. Control of heat, either external or internal or both, is necessary to assure the proper functioning of processes, and the proper heat balance and operation of control systems. The control of heat allows transportation of hot or cold materials, not only in industrial processes, but also in transportation of foods. This book provides basic information as to material and accessory properties and requirements of process and environment to guide the designer as to selections which will serve satisfactorily.

Since the performance of an insulation system depends upon a proper matching of material properties to the system requirement, this book provides a comprehensive list of commercially available insulating materials.

This book provides the mathematical formula, in both British and Metric S.I. Units for the calculation of heat transfer and temperature gradients. It covers the field of industrial thermal insulation in depth, including complete instructions for design of insulation systems for process pipe and equipment. Included are examples covering the design of high temperature applications, low temperature applications, cyclic application, heat traced applications, underground applications, maximum heat gain or loss services and the use of thermal insulation for fire protective services.

Specific recommendations are presented as to the design of insulation systems to correctly provide for expansion and contraction of piping and equipment. Equations and tables are provided for design of heat traced systems. This includes the use of heat transfer cements and precast heat transfer elements.

Material and application specifications are need to correctly obtain installation of systems as designed. The suggested methods of preparing such specifications are presented. This includes references to the ASTM Test Method for evaluation and testing of materials. A sample of a correctly prepared material and application specification is given.

Another essential, presented in this book, is a set of guide lines to use in the preparation of an insulation contract. This protects both parties to the contract, i.e. the consumer (owner) and the installer (contractor). This is written in layman's language; not as a legal document. The purpose was that the check list presented could be used to make certain that no important considerations are overlooked in the preparation of the contract.

Another phase of importance to insulation efficiency is "inspection". Proper inspection prevents errors and assures that important details receive the attention they should have. Good inspection assures that full energy and money savings designed into the system is realized. A list of potential application trouble spots is included for the inspector's use. Good inspection can save both the contractor and the owner money.

For any subject to be understood, the definitions of the terms used in discussing the subject must be understood. Especially as they are used in respect to the subject under consideration. For this reason a Glossary of nearly 500 words as they are used in relation to the subject of thermal insulation has been included as part of the Appendix. The definitions given bear the connotations of their use in the insulation industry, thus in some cases do not strictly follow ordinary dictionary definitions.

Of great convenience to the engineer, designer student or contractor is a collection of tables which are most used in the process of insulation design, calculations, cost estimating and money values of energy. Over 50 of these tables are in the Appendix.

This book, with its theory-example combinations, presented in both British and Metric S.I. Units, will be of interest to students and professors in the fields of Mechanical, Chemical, Environment, and Electrical Engineering. Architects will also find the information usefull in the design of energy-efficient buildings. It will also serve as a reference for construction, maintenance, purchasing and contract departments in industry. Insulation contractors, applicators and manufacturers will find it to be an essential reference book.

To summarize, this book presents the fundamentals of heat transfer, the installation requirements, and the properties of materials so that proper selection of materials is possible. It also presents how to design insulation systems based on properly selected materials and their correct application. It presents information as how to prepare contracts and points out correct inspection procedures. Thus it is the basic book for the manufacturer, applicator, and user of thermal insulation.

William C. Turner, P.E., M.E., E.E.

January 1981

Contents

1 Fundamentals of Heat Transfer

Basic Terms and Laws

As a large part of the misunderstanding in this field is due to terminology, it is necessary to begin with the definition of words and how they apply to this science.

The three basic words of this science are those which apply to most of man's problems. These words are force, resistance and energy. The dictionary gives many meanings to the word *force*, such as strength for war, to do violence; however, the meaning used herein is the one related to physics: "the potential for movement, or relative movement, of physical mass or temperature." The word *resistance* is used in the sense of the ability to oppose movements, or relative movements, of physical mass or temperature. *Energy* is a result of force and may be in either stored or transient form. The definitions for the forms of energy appear below.

Stored energy is commonly classified into three forms.

Mechanical potential energy Energy possessed by a body by virtue of its vertical distance above a horizontal plane.

Mechanical kinetic energy Energy possessed by a body by virtue of the relative motion between it and other parts of a system.

Internal energy Energy within a body such as a gas, liquid, or solid. Energy stored within any material and which can be released by chemical reaction comes under this classification.

Stored energy can be transformed into *transient energy,* which has two classifications.

Work is energy in transient form. Force acting through distance transforms mechanical energy into a flow of energy.

Heat is a result of molecular motion and interacting forces and is energy in transient form. A temperature difference existing between two bodies causes energy to flow, or transfer, from the body at the higher temperature to the body at the lower temperature.

Thus is determined the definition of *heat*; the more exact wording being:

Heat is energy in transient form. It is energy in transition, or transfer, from one body to another by virtue of a temperature difference existing between the bodies. Conversion of mechanical energy to heat energy can be by friction or by chemical reaction.

Friction converts mechanical energy into heat energy by changing the temperature of the body upon which the mechanical energy is acting.

Chemical reaction, or combustion, converts internal energy into heat energy by changing the temperature of the gas, liquid, or solid.

(*Note*: These definitions, as presented, would not be acceptable to a physicist, as they are too general and not sufficiently detailed or restrictive. They are given to establish the relationship of terms used to present the fundamentals of heat transfer.)

Not only can energy be converted from one form to another, but it can also be transmitted from one body to another. For example, electrical energy is conducted from one point to another by wire, or heat energy is transmitted from one body to another by radiation through space.

The basic law of the flow of energy is defined as follows:

A steady flow of energy through any medium of transmission is directly proportional to the force causing the flow and inversely proportional to the resistance to that force.

In simple terms:

$$\text{energy} = \frac{\text{force}}{\text{resistance}}$$

This simple equation is the base upon which the insulation industry is built. From this basic law it is now possible to expand or restrict each factor pertinent to this branch of engineering.

To start: *Heat* is a form of energy
Temperature difference is a form of force

Thus:

$$\text{Heat} = \frac{\text{Temperature difference}}{\text{Resistance}}$$

There remains the problem of *resistance.* Resistance to what? Of course, the answer is *resistance to heat flow*, or

$$\text{Heat} = \frac{\text{Temperature difference}}{\text{Resistance to heat flow}}$$

If heat flows it must have a path or means of transfer or transmission, of which there are two. One is *conduction* and the other is *radiation.*

Conduction is the transfer of energy within a body or between two bodies in physical contact. The transfer is from a higher temperature region to a lower temperature region by tangible contact.

Radiation is the transfer of energy from a higher temperature body, through space, to another lower temperature body or bodies some distance away. True radiation is the transfer of heat between these bodies which does not raise the temperature of the medium through which the heat passes.

A third natural phenomenon is often considered as a means of heat transfer. This is a movement of bodies called *Convection.*

Convection is the movement of a mass, with its associated energy, from one location to another.

Liquids or gases in contact with a body of a higher temperature have energy transmitted to them by conduction and radiation. This energy increases the temperature of the liquid or gas, which in turn changes density. This change in density causes a movement of the gas or liquid. As the liquids or gases of this new density move away from the higher temperature body, lower temperature masses of the gas or liquid move in. Thus the process of heating the close molecules by conduction and radiation and their subsequent movement away from the higher temperature body continues. This movement is called *Natural Convection.* The rate of this movement may be increased by some outside influence such as wind or by a fan. This is called *Forced Convection.*

The process of convection is so closely related to heat transmission that it has been accepted as a means of heat transfer. Even if not true, this definition is convenient. For this reason it must be understood that from this point on, although not strictly correct, reference will be made to "heat transfer by convection."

The other term in the equation, used frequently even before it was defined, is *Temperature.*

Temperature is an indicated level by which temperature differences, which are forces, may be measured. It is the thermal state of a body in reference to its ability to impart heat to other bodies. Temperature is a measurement of level; it is not a measurement of amount.

To illustrate this principle in terms of liquids: the relation of temperature to heat is the same as the number of feet of head is to water pressure. Just as head is no measure of quantity (cubic feet) of water, neither is temperature a measure of quantity (Btu) of heat.

When energy is used, or transmitted, at a certain rate, then the element of time becomes a factor.

Power is the time rate of doing work, or of expenditure of energy.

Units of Measurement

Heat

Any form of energy may be expressed quantitatively in any unit of energy. These units are derived from the standards of mass, length, and time. The fundamental energy unit in the English system is the foot-pound. The fundamental unit of heat energy in the English system is the British thermal unit (Btu).

British thermal unit (Btu) was defined in the early days of science as the quantity of energy (heat) required to raise the temperature of 1 lb of water 1 ° F, from and at 32 ° F. As the boundaries of knowledge expanded, the definition was changed to 1/180 of the quantity of energy (heat) required to change 1 lb of water from the ice point to the steam point at standard atmospheric pressure. The difficulty of setting up standards for such a definition, taking into account the now known variations of specific heat with both pressure and temperature, was so great that no standard was ever generally accepted. To relate the measurement of heat with other energies and to arrive at an acceptable standard for a Btu, the International Steam Table Conference in 1926 recommended that the International Calorie be set at 1/860 of an International Watt-hour. By calculation, this established the Btu as 778.26 ft-lb.

A ***Joule*** is the work done when the point of application of a force of one of one Newton is displaced a distance of one metre in the direction of the force.

A ***Watt*** is the power which gives rise to the production of energy at the rate of one joule per second.

A ***Calorie*** (International Table) = 4.1868 Joule (J).

conversion

Btu (British thermal unit) = 1055.06 J (Joule)

J (Joule) = 9.478133 *E − 4 Btu (British thermal unit)

* E = Exponent

Temperature

The purpose of a temperature scale is to assign a number to every level of thermal state. In the English system two scales are used. The most common scale is the *Fahrenheit (F)* scale and next the *Rankine (R)* scale. In the Metric system two scales are also used, these are *Celsius (C)*—in the past known as *Centigrade*—and *Kelvin (K)*.

Three basic states are considered in all temperature scales. These are: (1) The theoretically lowest thermal state—sometimes referred to as absolute zero. (2) The ice point of water. (3) The steam point of water.

Fahrenheit (F) scale subdivides the temperature interval between ice point and the steam point into 180 parts. The ice point is assigned the value of 32 degrees F so the steam point has a temperature of 212 degrees F.

Rankine (R) scale subdivides the temperature interval between ice point and the steam point into 180 parts. The theoretically lowest thermal state was assigned a value of 0 degrees R so the ice point of water then has a value of 491.7 degrees R and the steam point of water has a value of 671.7 degrees R.

Celsius (C) scale subdivides the temperature interval between ice point of water and the steam point of water into 100 parts and assigns the value of 0 degrees C to the freezing point of water, and 100 degrees C to the steam point of water. The lowest thermal state then has a value of − 273.16 degrees C.

Kelvin (K) scale subdivides the temperature interval between ice point of water and the steam point of water into 100 parts and assigns the value of lowest thermal state as 0 degrees K. The ice point then has a value of 273.2 degrees K and the steam point has a value of 373.2 degrees K.

Thermal resistance

Resistance is that property of a material that opposes the passage of energy (heat) through the material. In heat flow the force causing the flow is the temperature difference between two bodies or from one region of a body to another region of the same body. If a homogeneous body is at a higher temperature on one side than it is on the other, then the amount of heat energy passing from the higher to the lower side is determined by the resistance of the body to heat flow. The thermal resistance of a homogeneous body of uniform cross section varies directly as its cross section.

Scales or units for thermal resistance have neither been named nor established. However, the resistance is the reciprocal of conductance.

The conductance of a homogeneous material is the amount of heat transmitted through a unit area of the material in a unit time, through its *total* thickness, and with a unit of temperature difference between the surfaces of the two opposite sides.

Thus, for a given thickness of a homogeneous body the resistance is

$$\text{Thermal resistance} = \frac{\text{Thickness of homogeneous body}}{\text{Conductivity of that body}}$$

Heat transfer equation for single flat homogeneous body (insulation)

From the basic law of energy, force, and resistance, the first equation of heat transfer becomes:

$$\text{Heat flow} = \frac{\text{Temperature difference}}{\text{Thermal resistance}} \tag{1}$$

but:

$$\text{Thermal resistance} = \frac{\text{Thickness of homogeneous body}}{\text{Conductivity of that body}}$$

thus:

$$\text{Heat flow} = \frac{\text{Temperature difference}}{\left(\dfrac{\text{Thickness of homogeneous body}}{\text{Conductivity of that body}}\right)} \tag{2}$$

In our considerations the homogeneous body is mass thermal insulation, so:

$$\text{Heat flow} = \frac{\text{Temperature difference}}{\left(\dfrac{\text{Thickness of insulation}}{\text{Conductivity of insulation}}\right)} \tag{3}$$

The units of measurement used have absolutely no effect on the fundamental equation of heat transfer. However all equations must be solved using measurement units of a single system. The units of the English and Metric systems must not be intermingled.

In both systems there are many units of measurement for length, area, volume, temperature, energy, power etc. With these numerous units conversion from English system to Metric system (or reverse) becomes complex. To assist in handling this complex conversion, the most commonly used units in heat transfer problems are provided, and conversion factors between the English and Metric systems are given.

For simplification, the English units given in this book will be as indicated, such as temperature t, t_1, t_2 and length ℓ, ℓ_1, ℓ_2. The Metric units will be given as t_{m1}, t_{m2}, t_{m3} for temperature and ℓ_m, ℓ_{m1}, ℓ_{m2} for length. Note that the Metric symbols have a small "m" subscript.

English Units	− ***conversion***	= ***Metric Units***	− ***conversion***	= ***English Units***
t °F	(t °F − 32) ÷ 1.8	= t_m °C	(t_m °C × 1.8) + 32	= t °F
t °F	(t °F + 459.67) ÷ 1.8	= t_m °K	(t_m °K × 1.8) − 459.67	= t °F
t °R	t °R ÷ 1.8	= t_m °K	× 1.8	= t °R
t °R	(t °R ÷ 1.8) − 273.16	= t_m °C	(t °C × 1.8) + 491.67	= t °R
Δt °F	t °F ÷ 1.8	= Δt_m °C or Δt_m °K	× 1.8	= Δt °F or Δt °R
ℓ" (inches)	× 0.0254	= ℓ_m in metres	× 39.37	= ℓ" (inches)
ℓ' (feet)	× 0.3048	= ℓ_m in metres	× 3.280839	= ℓ' (feet)
ft^2 (sq ft)	× 0.092903	= m^2 (sq metres)	× 10.76391	= ft^2 (sq ft)
Btu	× 1055.056	= J (joule)	× 0.0009478	= Btu
Btu/hr	× 0.2930711	= W (watt)	× 3.412141	= Btu/hr
Q (Btu/ft^2, hr)	× 3.152591	= Q_m (W/m^2)	× 0.317199	= Q (Btu/ft^2, hr)
U (Btu/ℓft, hr)	× 0.9609097	= $U_m \ell_m$ (W/ln m)	× 1.04068	= $U_{\ell f}$ (Btu/ℓft, hr)
f (Btu/ft^2, hr, °F)	× 5.67826	= f_m (W/m^2 K)	× 0.17611	= f (Btu/ft^2, hr, °F)
k (Btu in./ft, hr, °F)	× 0.1442	= k_m (W/mK)	× 6.933	= k (Btu in./ft^2, hr, °F)
R (°F, ft, hr, Btu)	× 0.1761	= R_m (Km^2W)	× 5.6782	= R (°F, ft^2, hr, Btu)
V (ft/min)	× 0.00508	= V_m (m/s)	× 196.85039	= V (ft/min)

(*Note*: All energy conversions are based on International Table Units.)

When: *In English Units*

Q = Heat flow in Btu/ft^2, hr

Δt = Temperature difference between higher temperature and lower temperature surfaces in degrees F.

ℓ = length of heat path—thickness of insulation in inches.

k = conductivity in Btu's through 1 inch thickness of insulation, per one square ft of area, per one hour, per one degree of temperature difference in °F. (Btu in./ft^2 °F). Value of k is at the mean temperature of the insulation [°F higher temperature, (t_1) surface plus °F lower temperature, (t_2) divided by 2] or $\left(\frac{t_1 + t_2}{2}\right)$

Then:

$$Q = \frac{\Delta t}{\left(\frac{\ell}{k}\right)} \qquad (4)$$

However $\Delta t = t_1 - t_2$

Where t_1 = °F of surface at higher temperature
t_2 = °F of surface at lower temperature

Thus: *(English Units)*

$$Q = \frac{t_1 - t_2}{\left(\frac{\ell}{k}\right)} \qquad (5)$$

Which is the equation for heat transfer through a single insulation. In Figure 1 these factors are shown schematically.

In Metric units the form of the equation does not change. The equation in Metric units is:

(Metric Units)

$$Q_m = \frac{t_{m1} - t_{m2}}{\frac{\ell_m}{k_m}} \qquad (5a)$$

When Q_m = Heat flow in W/m^2
t_{m1} = °C or °K of surface at higher temperature
t_{m2} = °C or °K of surface at lower temperature
ℓ_m = length of heat path in metres
k_m = conductivity in W/mK

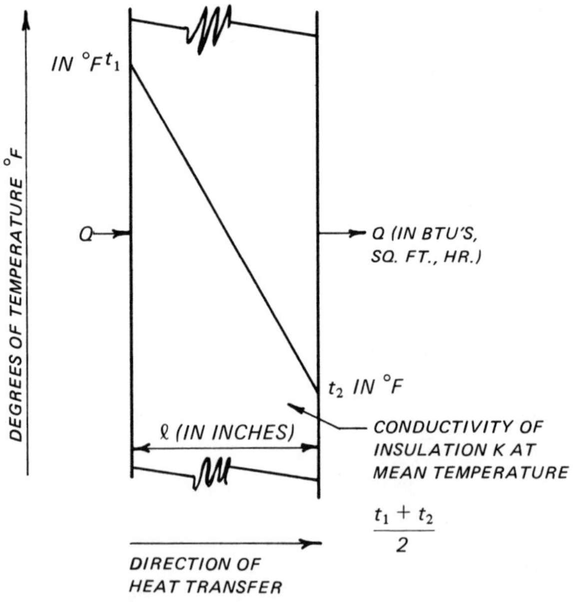

EQUATION 5

$$Q = \frac{t_1 - t_2}{\frac{\ell}{K}}$$

Figure 1 Schematic diagram of heat transfer through a single flat material.

Of the terms in equations 5 and 5a only conductivity (k or k_m) is troublesome. The term conductivity is somewhat misleading. As shown in Figure 2 page 6, the heat flow through mass insulation is transmitted by more than one mode of heat transfer. However these are all measured by test methods* in a single operation and the result labeled conductivity. For any mass insulation conductivity varies with mean temperature and temperature differential between higher and lower temperature surfaces. Thus, true conductivity should be determined at the mathematical mean at full temperature difference, rather than a simple average mean of lower and higher temperatures of the two surfaces.

In most instances, unless the curve of conductivity is a pronounced curve the simple average is sufficiently close to solve most practical problems. However with light weight or translucent materials it is very important that conductivity be determined by tests which have been performed with the temperature difference from higher to lower surface being approximately the same as expected in service. At full temperature difference the conductivity, using average mean, will be two or three percent low. In the case of light weight material, or translucent material where conductivity is determined by very small temperature differences test methods will determine the conductivity 30 to 50% too low. Where heat loss is critical this should be taken into consideration.

*See ASTM Test Methods C-177, C-335, C-513.

■ EXAMPLE 1

If the temperature of insulation is 300° F (t_1) on one surface and 70° F (t_2) on the other, and insulation is 2″ thick and its conductivity (k) at 185° F is 43, what is the amount of heat (Btu) transferred for 1 sq ft of surface in one hour?

$$Q = \frac{t_1 + t_2}{\left(\frac{\ell}{k}\right)}$$

As stated $t_1 = 300°$ F

$t_2 = 70°$ F

$\ell = 2''$

$k = .43$

$$Q = \frac{300 - 70}{\left(\frac{2}{.43}\right)} = \frac{230}{4.65} = 49.4 \text{ Btu/sq ft, hr}$$

■

■ EXAMPLE 1a

Metric Units

English Values (as given)	conversion	Metric Values
$t_1 = 300°$ F	$(300 - 32) \div 1.8$	$= t_{m1}$ is 148.89° C*
$t_2 = 70°$ F	$(70 - 32) \div 1.8$	$= t_{m2}$ 21.1° C*
$\ell = 2''$	× 0.0254	$= \ell_{m2}$ is 0.0508_m
k = 0.43 Btu in/ft^2, hr, °F	× 0.1442	$= k_m$ is 0.062 W/mK

$$Q_m = \frac{148.89 - 21.1}{\left(\frac{0.0508}{0.062}\right)} = \frac{127.79}{0.8193} = 155.96 \text{ W/m}^2$$

Check

$$Q = 49.4 \times 3.152591 = Q_m = 155.7 \text{ W/m}^2$$

In some instances the maximum allowable heat transfer may be known, and it is desired to determine the thickness of insulation necessary to provide sufficient resistance to control the heat flow. The formula can be changed to determine this.

$$Q = \frac{t_1 - t_2}{\frac{\ell}{k}}, \quad \frac{\ell}{k} Q = t_1 - t_2.$$

$$\frac{\ell}{k} = \frac{t_1 - t_2}{Q}, \quad \ell = k\left(\frac{t_1 - t_2}{Q}\right) \tag{6}$$

In Metric Units

$$\frac{\ell_m}{k_m} = \frac{t_{m1} - t_{m2}}{Q_m} \tag{6a}$$

■

■ EXAMPLE 2

All factors same as given in Example 1 except that heat transfer (Q) should not be over 30 Btu. What thickness of insulation (ℓ) in inches is required?

$$\ell = k\left(\frac{t_1 - t_2}{Q}\right) = .43\left(\frac{300 - 70}{30}\right) = .43 \frac{230}{30}$$

$$= .43(7.66) = 3.29'' \text{ of insulation required}$$

■

■ EXAMPLE 2a

Metric Units

Heat transfer Q_m should not be over

$$(30 \text{ Btu} \times 3.152591) = 94.577 \text{ W/m}^2$$

$$\ell_m = k_m\left(\frac{t_{m1} - t_{m2}}{Q_m}\right) = 0.062\left(\frac{148.89 - 21.1}{94.577}\right)$$

$$= 0.062\left(\frac{127.79}{94.577}\right) = 0.0838\text{m}$$

Check $0.0838 \times 39.37 - 3.298'' = \ell$

■

Equation for Heat Transfer for Two or More Thicknesses of Flat Insulation

All resistances in the path of heat flow are additive. Thus: If t_1 = °F higher temperature surface and t_s lower temperature surface, then,

$$Q_t = \frac{t_1 - t_s}{R_1 + R_2 + R_3 \text{ etc.}} \tag{7}$$

When *In English Units*

ℓ_1 = Thickness in inches of first insulation material
ℓ_2 = Thickness in inches of second insulation material
ℓ_3 = Thickness in inches of third insulation material
k_1 = Conductivity in Btu in./sq ft, hr, °F of first insulation
k_2 = Conductivity in Btu in./sq ft, hr, °F of second insulation.
k_3 = Conductivity in Btu in./sq ft, hr, °F of third insulation.

then the R_1, R_2, and R_3 resistance of each material is $\frac{\ell_1}{k_1}, \frac{\ell_2}{k_2}, \frac{\ell_3}{k_3}$, respectively, then equation 7 is

$$Q = \frac{t_1 - t_s}{\left[\left(\frac{\ell_1}{k_1}\right) + \left(\frac{\ell_2}{k_2}\right) + \left(\frac{\ell_3}{k_3}\right)\right]} \quad (7)$$

A schematic diagram is shown in Figure 2.

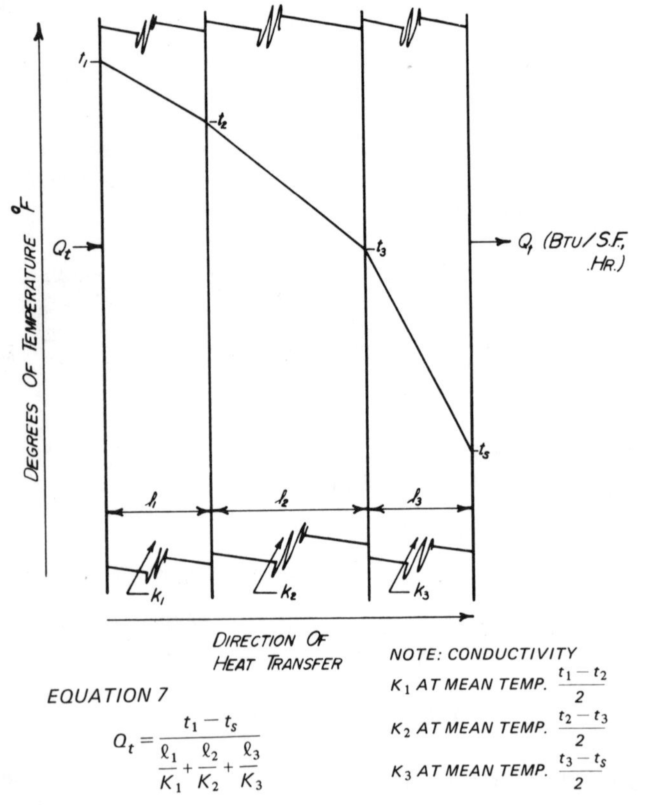

Figure 2 Schematic diagram of heat transfer through three flat materials.

In Metric Units

$$Q_{m1} = \frac{t_{1m} - t_{sm}}{\frac{\ell_{1m}}{k_{1m}} + \frac{\ell_{2m}}{k_{2m}} + \frac{\ell_{3m}}{k_{3m}}} = W/m^2 \quad (7a)$$

Care must be taken as to direction of heat flow, as the temperature gradient across the insulations will affect the mean temperature which determines the conductivity of each insulation.

As the temperature gradients are not known until the problem is solve, though they are necessary to determine the various conductivities, this presents a set of unknowns in the problem which has no finite method of calculation. Therefore, the only solution is to assume a mean temperature for each material, use the conductivity for that mean temperature for solving the problem, and after completion to calculate the temperature gradient and compare the mean temperatures with those used in the solution.

■ EXAMPLE 3

If the surface temperature of the first insulation is 2000° F (t_1) and the surface temperature of the third insulation is 50° F (t_s), the thickness of the first insulation is 1″ (ℓ_1), the thickness of the second insulation is 3″ (ℓ_2), the thickness of the third insulation is 2″ (ℓ_3), and the conductivities of each are as shown below, what would the heat transfer (Q_t) be?

Assume mean temperature of first insulation to be 1500° F, then $k_1 = 2.5$
Assume mean temperature of second insulation to be 1100° F, then $k_2 = 1.2$
Assume mean temperature of third insulation to be 150° F, then $k_3 = 0.4$

As shown in Example 3, thus,

$$Q_t = \frac{t_1 - t_s}{\left(\frac{\ell_1}{k_1}\right) + \left(\frac{\ell_2}{k_2}\right) + \left(\frac{\ell_3}{k_3}\right)}$$

$$= \frac{2000 - 50}{\left(\frac{1}{2.5}\right) + \left(\frac{3}{1.2}\right) + \left(\frac{2}{0.4}\right)} = \frac{1950}{0.4 + 2.5 + 5.0}$$

$$= \frac{1950}{7.9} = 247 \text{ Btu/sq ft, hr}$$

However, with this information an approximate temperature gradient can be established, as the temperature change through a series of insulations is directly proportioned to their resistance (as shown in Figure 3A) or:

when

t_2 = Temperature °F at juncture of first and second insulations
t_3 = Temperature °F at juncture of second and third insulations

Starting at

$t_1 = 2000°$ F

$$t_2 = t_1 - Q_t\left(\frac{\ell_1}{k_1}\right)$$

$$t_3 = t_2 - Q_t\left(\frac{\ell_2}{k_2}\right)$$

$t_s = t_3 - Q_t\left(\frac{\ell_3}{k_3}\right)$, also given as 50° F

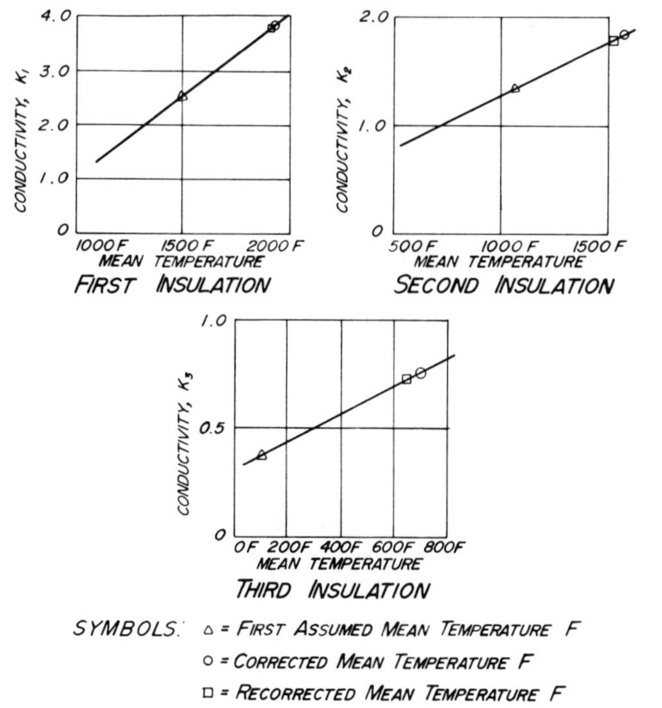

Example 3 Conductivity curves for insulations for Example 3.

Thus the temperature gradient as calculated

$t_2 = 2000 - (247 \times 0.4) = 2000 - 99 = 1901\,°\,F$

$t_3 = 1901 - (247 \times 2.5) = 1901 - 618 = 1283\,°\,F$

$t_s = 1283 - (247 \times 5.0) = 1283 - 1233 = 50\,°\,F$

Rechecking the first assumptions on mean temperature shows that the assumed mean temperature values were too low.

Corrected values of mean temperatures follows:

Mean temperature °F of first insulation

$$\frac{2000 + 1900}{2} \text{ or approximately } 1950 \quad k_1 = 3.9$$

Mean temperature °F of second insulation

$$\frac{1900 + 1300}{2} \text{ or approximately } 1600 \quad k_2 = 1.8$$

Mean temperature °F of third insulation

$$\frac{1300 + 50}{2} \text{ or approximately } 675 \quad k_3 = 0.75$$

Corrected calculation then is:

$$Q_t = \frac{2000 - 50}{\left(\frac{1}{3.9}\right) + \left(\frac{3}{1.8}\right) + \left(\frac{2}{0.75}\right)} = \frac{1950}{0.26 + 1.66 + 2.67}$$

$$= \frac{1950}{4.59} = 425 \text{ Btu/sq ft, hr}$$

Rechecking the temperature gradients:

$t_1 = 2000$

$t_2 = 2000 - (425 \times 0.26)$

$= 2000 - 110 = 1890\,°\,F$

$t_3 = 1890 - (425 \times 1.66)$

$= 1890 - 706 = 1184\,°\,F$

$t_s = 1184 - (425 \times 2.67)$

$= 1184 - 1134 = 50\,°\,F$

Check on problem:

Mean temperature of first material

$$= \frac{2000 + 1890}{2} = \frac{3890}{2} = 1945\,°\,F$$

Mean temperature of second material

$$= \frac{1890 + 1184}{2} = \frac{3074}{2} = 1537\,°\,F$$

Mean temperature of third material

$$= \frac{1184 + 50}{2} = \frac{1234}{2} = 617\,°\,F$$

In this case the rechecked conductivities were sufficiently close to make the calculations close enough to be valid. A diagram of this result is shown in Figure 3B.

The special case of boiler or hopper insulation with an inner layer of refractory material is not as difficult as it might seem at first glance. The refractory is an insulation and all that is necessary is to add it to the chain of resistances shown in Eq. 7, inserting the proper values for ℓ and k. Values of k for refractories are shown in Appendix B. ■

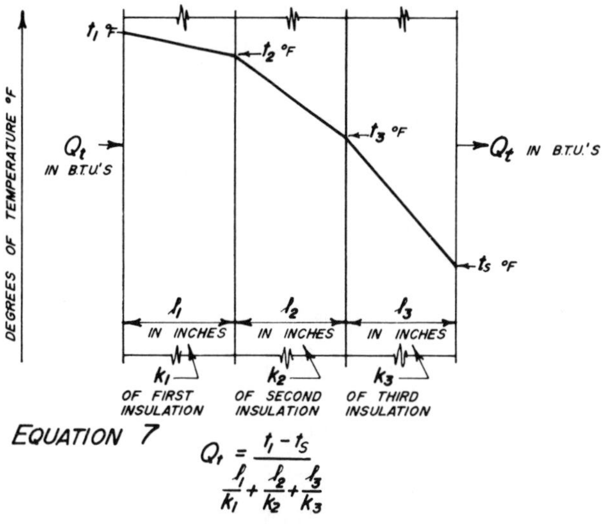

Figure 3A Heat transfer through flat insulations.

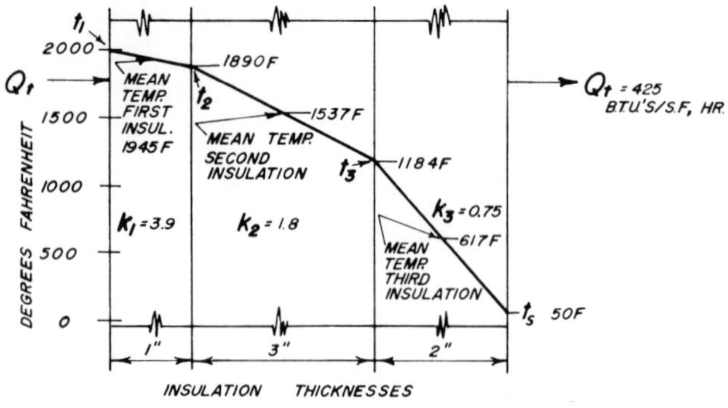

Figure 3B Diagram of results as calculated—

■ EXAMPLE 3a

Metric Units

English Values (as given)	***conversion***	***Metric Values***	
t_1 = 2000° F	(2000 − 32) ÷ 1.8	= 1093.33° C	= t_{m1}
t_s = 50° F	(50 − 32) ÷ 1.8	= 10.° C	= t_{ms}
ℓ_1 = 1"	× 0.0254	= 0.0254m	= ℓ_{m1}
ℓ_2 = 3"	× 0.0254	= 0.0762m	= ℓ_{m2}
ℓ_3 = 2"	× 0.0254	= 0.0508m	= ℓ_{m3}
k_1 = 3.9 Btu in/ft^2, hr, °F	× 0.1442	= 0.56238 W/mK	= k_{m1}
k_2 = 1.8 Btu in/ft^2, hr, °F	× 0.1442	= 0.25956 W/mK	= k_{m2}
k_3 = 0.75 Btu in/ft^2, hr, °F	× 0.1442	= 0.10815 W/mK	= k_{m3}

$$Q_m = \frac{t_{m1} - t_{m2}}{\left[\left(\frac{\ell_{m1}}{k_{m1}}\right) + \left(\frac{\ell_{m2}}{k_{m2}}\right) + \left(\frac{\ell_{m3}}{k_{m3}}\right)\right]}$$

$$= \frac{1093.33 - 10}{\left[\left(\frac{0.0254}{0.5624}\right) + \left(\frac{0.0762}{0.2596}\right) + \left(\frac{0.0508}{0.1082}\right)\right]}$$

$$= \frac{1083.33}{0.045 + 0.294 + 0.470} = \frac{1083.33}{0.809}$$

$$= 1339.09 \text{ W/m}^2$$

Check: 1339.09 × 0.317199 = 424.8 Btu/ft^2, hr = Q_t

TO DETERMINE TEMPERATURE GRADIENTS

When t_{m2} = Temperature °C at junction of first and second insulations.

t_{m3} = Temperature °C at junction of second and third insulations.

Then when t_{m1} is 1093.33° C (given)

$$t_{m2} = t_{m1} - Q_{mt}\left(\frac{\ell_{m1}}{k_{m1}}\right)$$

$$t_{m3} = t_{m2} - Q_{mt}\left(\frac{\ell_{m2}}{k_{m2}}\right)$$

$$t_{ms} = t_{m3} - Q_{mt}\left(\frac{\ell_{m3}}{k_{m3}}\right)$$

t_{ms} is given as 10 °C

t_{m2} = 1093.33 − (1339 × 0.045) = 1093.33 − 60.25 = 1033.08 °C

t_{m3} = 1033.06 − (1339 × 0.294) = 1033.08 − 393.66 = 639.42 °C

t_{ms} = 639.42 − (1339 × 0.470) = 639.42 − 629.33 = 10.09 °C

The mean temperature of the insulations is:

Mean temperature of first material is:

$$\frac{t_{m1} + t_{m2}}{2} = \frac{1093.33 + 1033.08}{2}$$

$$= \frac{2126.41}{2} = 1063.2 \text{ °C}$$

Mean temperature of second material is:

$$\frac{t_{m2} + t_{m3}}{2} = \frac{1033.08 + 639.42}{2}$$

$$= \frac{1672.5}{2} = 836.25 \text{ °C}$$

Mean temperature of third material is:

$$\frac{t_{m3} + t_{ms}}{2} = \frac{639.42 + 10}{2}$$

$$= \frac{649.42}{2} = 324.7 \text{ °C}$$

Substituting Metric Units for English Units

In Figure 3B these are:

Temperature °C	***Mean Temperature °C***	***Conductivity W/mK***	***Thickness in metres***
t_{m1} = 1093.33	First Insulation = 1063.2	K_{m1} = 0.562	ℓ_1 = 0.0254
t_{m2} = 1033.08	Second Insulation = 836.25	K_{m2} = 0.260	ℓ_2 = 0.0762
t_{m3} = 639.42	Third Insulation = 324.7	K_{m3} = 0.108	ℓ_3 = 0.0508
t_{ms} = 10.0			

Q_m (Heat transfer through all three insulations) = 1339 W/m^2

In Metric Units

When Q_m = W/m^2

t_{1m} = temperature in °C

t_{2m} = temperature in °C

r_{1m} = radius in metres

r_{2m} = radius in metres

k_m = conductivity in W/mK

$$Q_m = \frac{t_{1m} - t_{2m}}{\left[\dfrac{r_{2m} \log_e \left(\dfrac{r_{2m}}{r_{1m}}\right)}{k_m}\right]}$$

■

Equation for Heat Transfer Through One Thickness of Insulation

As heat flows through a cylindrical insulation the difference in inner and outer areas within a given length affects the amount of heat flow per unit of area. Solutions to problems involving areas of insulation in cylindrical form can be accomplished by using actual thicknesses and solving the problems based on the mean area involved. Many engineers use this method of solution, but it does not lend itself to precalculated tables for easy reference.

A more common method of calculating heat transfer is to base all heat transfer on outer area of the insulation and to use *"equivalent thickness"* in the heat transfer formulas to arrive at resistances.

Equivalent Thickness is when

r_1 = Inner radius of a single layer of cylindrical insulation
r_2 = Outer radius of a single layer of cylindrical insulation

$$\text{Equivalent thickness} = r_2 \log_e \frac{r_2}{r_1}\ .^{*}$$

**Note:* Logarithm is to base e.

A schematic diagram is shown in Figure 4.

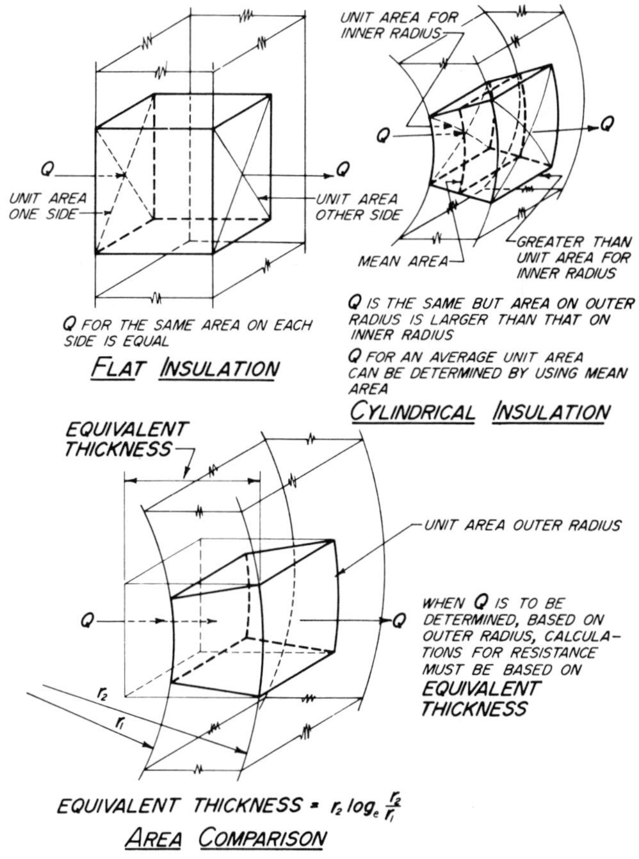

Figure 4 Equivalent thickness of cylindrical insulation.

Thus the equation for heat transfer through one thickness of cylindrical insulation is:

English Units

$$Q = \frac{t_1 - t_2}{\left[\dfrac{r_2 \log_e \left(\dfrac{r_2}{r_1}\right)}{k}\right]} \tag{8}$$

In Metric Units

$$Q_m = \frac{t_{m3} - t_{m2}}{\left[\dfrac{r_{m2} \log_e \left(\dfrac{r_{m2}}{r_{m1}}\right)}{k_m}\right]} \tag{8a}$$

■ EXAMPLE 4

English Units

A cylindrical insulation 6″ inside diameter (r_1 = 3″) is 4″ thick. The temperature of the inner surface is 900° F (t_1), and the temperature of the outer surface is 70° F (t_2). The conductivity (k) of the insulation $\left(\frac{900 + 70}{2}\right)$, which is 985° F mean temperature, is 0.6 Btu, in./sq ft, hr, °F. What is the heat transfer?

$r_1 = 3''$

$r_2 = 3 + 4 = 7''$

$$Q = \frac{900 - 70}{\left[\dfrac{7 \log_e \left(\dfrac{7}{3}\right)}{0.6}\right]} = \frac{830}{\left[\dfrac{7 \log_e 2.33}{0.6}\right]}$$

from the table of natural logarithms

$$\log_e 2.33 = 0.8459$$

$$Q = \frac{830}{\left[\dfrac{7 \times 0.8459}{0.6}\right]} = \frac{830}{\left[\dfrac{5.93}{0.6}\right]} = \frac{830}{9.9}$$

= 83.5 Btu/sq ft (outer radius surface), hr ■

With cylindrical surfaces the heat transfer must be based on one of the surfaces. As mentioned previously, the transfer, using equivalent thickness, is based on the surface at the outer radius. Thus, if the transfer is to be found for a linear foot of cylinder, this transfer will be:

When $q_{\ell f}$ = Heat transfer for a linear foot of cylinder

$$q_{\ell f} = Q \times \frac{2 \times r_2 \times \pi}{12}$$

An example of this would be:

$$U = 83.5 \times \frac{2 \times 7 \times 3.1416}{12}$$

$$= 83.5 \times 3.65 = 304 \text{ Btu/lin, ft, hr}$$

■ EXAMPLE 4a

Metric Units

English Values (as given)	***conversion***	***Metric Values***	
t_1 = 900° F	(900 − 32) ÷ 1.8	= 482.2° C	= t_{m1}
t_2 = 70° F	(70 − 32) ÷ 1.8	= 21.1° C	= t_{m2}
r_1 = 3″	× 0.0254	= 0.0762m	= r_{m1}
r_2 = 7″	× 0.0254	= 0.1778m	= r_{m2}
k = 0.6 Btu in./ft², hr, °F	× 0.1442	= 0.08652 W/mK	= k_m

$$Q_m = \frac{t_{m1} - t_{m2}}{\left[\dfrac{r_{m2}\log_e\left(\dfrac{r_{m2}}{r_{m1}}\right)}{k_m}\right]} = \frac{482.2 - 21.1}{\left[\dfrac{0.1778\log_e\left(\dfrac{0.1778}{0.0762}\right)}{0.08652}\right]}$$

$$= \frac{461.1}{\left[\dfrac{0.1778\log_e 2.333}{0.08652}\right]} = \frac{461.1}{\left[\dfrac{0.1778 \times 0.8459}{0.08652}\right]}$$

$$= \frac{461.1}{\left[\dfrac{0.1504}{0.08652}\right]} = \frac{461.1}{1.7383} = 265.25 \text{ W/m}^2$$

Check:

$$Q_m = 265.25 \text{ W/m}^2 \times 0.317199 = 84.1 \text{ Btu/ft}^2\text{, hr} = Q$$

To determine heat loss per linear metre.

$$U_{m\ell m} = Q_m \times 2r_{m2}\pi = 265.25\,(2 \times 0.1778\pi)$$

$$= 265.25 \times 0.3556\pi = 265.25 \times 1.11715$$

$$= 296.32 \text{ W/lin m.}$$

Check:

$$U_{m\ell m} = 296.32 \times 1.04068\pi = 308.37 \text{ Btu/lin ft, hr} = U_{\ell f}$$ ■

■ EXAMPLE 5

English Units

If a 2″ NPS* pipe is insulated with 2-1/2″ nominal dimensional standard insulation and the surface of the pipe is 700° F, the temperature of the outer surface of the insulation is 100° F, and the conductivity of the insulation is 0.5 at 400° F $\left(\dfrac{700 + 100}{2}\right)$ mean temperature, what would the heat flow be?

$$r_1 = \frac{2.375}{2}$$

The thickness of 2-1/2″ nominal dimensional standard insulation is approximately 2-5/8″ = 2.625″

$$r_2 = \frac{2.375}{2} + 2.625$$

$$r_2 = 1.1875 + 2.625 = 3.8125$$

*IPS (iron pipe size) has been renamed NPS (nominal pipe size).

$$Q = \frac{700 - 100}{\left[\dfrac{3.8125 \times \log_e\left(\dfrac{3.8125}{1.1875}\right)}{0.5}\right]}$$

The factor $3.8125 \times \log_e \dfrac{3.8125}{1.1875}$ can be calculated. However, for convenience, it has been precalculated in Table B6, Appendix B. From this table of values of $r_2 \log_e \dfrac{r_2}{r_1}$ the value is found to be 4.445.

$$Q = \frac{600}{\left[\dfrac{4.445}{0.5}\right]} = \frac{600}{8.89} = 67.6 \text{ Btu/sq ft, hr}$$

and U for linear foot of pipe

$$U = \frac{3.8125 \times 2 \times \pi}{12} \times 67.6$$

$$U = 1.99 \times 67.6 = 135 \text{ Btu/lin ft, hr}$$ ■

■ EXAMPLE 5a

Metric Units

English Values (as given)	***conversion***	***Metric Values***	
t_1 = 700° F	(700 − 32) ÷ 1.8	= 371.1° C	= t_{m1}
t_2 = 100° F	(100 − 32) ÷ 1.8	= 37.7° C	= t_{m2}
r_1 = 1.1875″	× 0.0254	= 0.0301m	= r_{m1}
r_2 = 3.8125″	× 0.0254	= 0.0968m	= r_{m2}
k = 0.5 Btu in./ft², hr, °F	× 0.1442	= 0.0721 W/mK	= k_m

$$Q_m = \frac{t_{m1} - t_{m2}}{\left[\dfrac{r_{m2}\log_e\left(\dfrac{r_{m2}}{r_{m1}}\right)}{k_m}\right]} = \frac{371.1 - 37.7}{\left[\dfrac{0.0968\log_e\left(\dfrac{0.0968}{0.0301}\right)}{0.0721}\right]}$$

$$= \frac{333.4}{\left[\dfrac{0.0968\log_e 3.216}{0.0721}\right]} = \frac{333.4}{\left[\dfrac{0.0968 \times 1.1663}{0.0721}\right]}$$

$$= \frac{333.4}{\left[\dfrac{0.1129}{0.0721}\right]} = \frac{333.4}{1.566} = 212.9 \text{ W/m}^2$$

Check:

$$Q_m = 212.9 \text{ W/m}^2 \times 0.317199 = 67.54 \text{ Btu/ft}^2\text{, hr.}$$

To determine heat loss per linear metre

$$U_{m\ell m} = Q_m \times 2r_{m2}\,\pi = 212.9\,(2 \times 0.0968\pi)$$

$$= 212.9 \times 0.1936\pi = 212.9 \times 0.6082 = 129.49 \text{ W/lin m.}$$

Check:

$$U_{m\ell m} = 129.49 \times 1.04068 = 134.76 \text{ Btu/lin, ft, hr} = U_{\ell f}$$ ■

Equation for Heat Transfer Through Two or More Thicknesses of Cylindrical Insulation

The same form of equation presented to calculate the heat flow through one thickness of cylindrical insulation should also be used where the total thickness is composed of two or more thicknesses of different k values. In all cases, however, the multiplier for the $\log_e$ factor is the outermost radius.

Thus when:

In English Units

r_1 = Inner radius of innermost insulation in inches
r_2 = Outer radius of innermost insulation and inner radius of the outermost insulation in inches
r_s = Outer radius of the outermost insulation in inches
t_1 = Temperature °F innermost surface
t_2 = Temperature °F at radius between innermost and outermost insulations
t_s = Temperature °F of outer surface
k_1 = Conductivity of innermost insulation in Btu, in./ft², hr, °F
k_2 = Conductivity of outermost insulation in Btu, in./ft², hr, °F

Heat transfer Q through two different types of cylindrical insulations is:

$$Q = \frac{t_1 - t_s}{\left[\dfrac{r_s \log_e\left(\dfrac{r_2}{r_1}\right)}{k_1}\right] + \left[\dfrac{r_s \log_e\left(\dfrac{r_s}{r_2}\right)}{k_2}\right]} \quad (9)$$

In Metric Units

When r_{1m} = Inner radius of innermost insulation in metres
r_{2m} = Outer radius of innermost insulation and inner radius of outermost insulation, in metres
r_{sm} = Outer radius of outermost insulation in metres
t_{1m} = Temperature °C (or °K) of innermost surface
t_{2m} = Temperature °C (or °K) of radius between inner and outermost insulation
t_{sm} = Temperature °C (or °K) of outer surface
k_{1m} = Conductivity of innermost insulation in W/mK
k_{2m} = Conductivity of outermost insulation in W/mK

Heat transfer Q_m through two different types of cylindrical insulation in W/m² is:

$$Q_m = \frac{t_{1m} - t_{sm}}{\left[\dfrac{r_{sm} \log_e\left(\dfrac{r_{2m}}{r_{1m}}\right)}{k_{1m}}\right] + \left[\dfrac{r_{sm} \log_e\left(\dfrac{r_{sm}}{r_{2m}}\right)}{k_{2m}}\right]} \text{ in W/m}^2 \quad (9a)$$

Heat transfer Q through three different types of cylindrical insulations is:

In English Units

$$Q = \frac{t_1 - t_s}{\left[\dfrac{r_s \log_e\left(\dfrac{r_2}{r_1}\right)}{k_1}\right] + \left[\dfrac{r_s \log_e\left(\dfrac{r_3}{r_2}\right)}{k_2}\right] + \left[\dfrac{r_s \log_e\left(\dfrac{r_s}{r_3}\right)}{k_3}\right]} \quad (10)$$

A schematic diagram is shown in Figure 5.

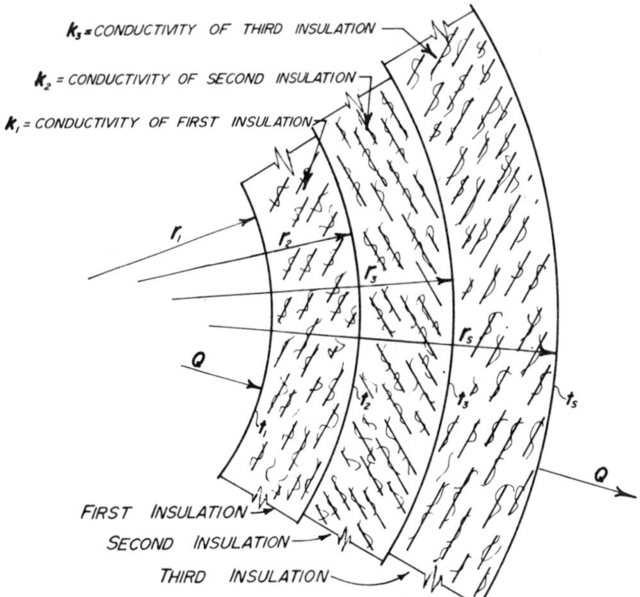

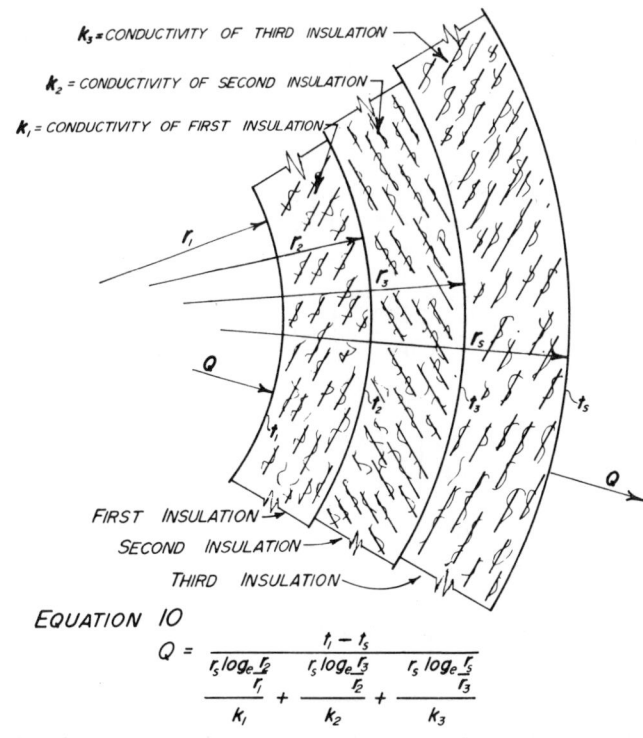

$$Q = \frac{t_1 - t_s}{\dfrac{r_s \log_e \frac{r_2}{r_1}}{k_1} + \dfrac{r_s \log_e \frac{r_3}{r_2}}{k_2} + \dfrac{r_s \log_e \frac{r_s}{r_3}}{k_3}}$$

Figure 5 Schematic diagram of heat transfer through three different types of cylindrical insulation materials.

In Metric Units (10a)

$$Q_m = \frac{t_{m1} - t_{ms}}{\left[\dfrac{r_{ms} \log_e\left(\dfrac{r_{m2}}{r_{m1}}\right)}{k_{m1}}\right] + \left[\dfrac{r_{ms} \log_e\left(\dfrac{r_{m3}}{r_{m2}}\right)}{k_{m2}}\right] + \left[\dfrac{r_{ms} \log_e\left(\dfrac{r_{ms}}{r_{m3}}\right)}{k_{m3}}\right]}$$

■ EXAMPLE 6

English Units

If an inside insulation with a 20″ inside radius is 2 1/2″ thick and the outer insulation is 4″ thick, the temperature of the inner surface is 1100° F, and the outer surface of outer insulation is 150° F, with conductivities of each material as shown below, what is the heat flow Q in Btu/sq ft?

Assume the mean temperature of the inner insulation to be 1000° F; then its conductivity (k_1) is 0.95. Assume the mean temperature of the outer insulation to be 600° F; then its conductivity (k_2) is 0.51. Then: r_1 is given as 20″, r_2 = 20″ + 2 1/2″ = 22 1/2″, r_s = 22 1/2″ + 4″ = 26 1/2″.

$$Q = \frac{t_1 - t_s}{\left[\dfrac{r_s \log_e\left(\dfrac{r_2}{r_1}\right)}{k_1}\right] + \left[\dfrac{r_s \log_e\left(\dfrac{r_s}{r_2}\right)}{k_2}\right]}$$

$$= \frac{1100 - 150}{\left[\dfrac{26.5 \log_e\left(\dfrac{22.5}{20}\right)}{0.95}\right] + \left[\dfrac{26.5 \log_e\left(\dfrac{26.5}{22.5}\right)}{0.51}\right]}$$ ■

$$Q = \frac{950}{\left[\dfrac{26.5 \log_e 1.12}{0.95}\right] + \left[\dfrac{26.5 \log_e 1.18}{0.51}\right]}$$

$$= \frac{950}{\left[\dfrac{26.5 \times 0.1122}{0.95}\right] + \left[\dfrac{26.5 \times 0.1655}{0.51}\right]}$$

$$= \frac{950}{\left[\dfrac{3.01}{0.95}\right] + \left[\dfrac{4.50}{0.51}\right]} = \frac{950}{3.17 + 8.60}$$

$$= \frac{950}{11.77}$$

= 81.0 Btu/sq ft (outer surface)

Calculating the temperature gradient to determine if assumed mean temperatures were approximately correct:

$t_2 = 1100 - (3.17 \times 81.0)$

$= 1100 - 256 = 844$

mean temperature inner insulation

$$= \frac{1100 + 844}{2} = \frac{1944}{2} = 972$$

$t_s = 844 - (8.60 \times 81.0)$

$= 844 - 696 = 148$ checks within 2° F

mean temperature outer layer

$$= \frac{844 + 150}{2} = \frac{944}{2} = 497$$

Plotting on conductivity charts it is evident that there is a slight difference in the k values as selected, based on assumed temperature gradients. Although close, the problem could be recalculated for correct values of $k_1 = 0.94$ and $k_2 = 0.49$ by using calculated temperature gradients.

$$Q = \frac{950}{\left[\dfrac{3.01}{0.94}\right] + \left[\dfrac{4.40}{0.49}\right]} = \frac{950}{3.2 + 9.0} = \frac{950}{12.2}$$

= 78.0 Btu/sq ft, hr (outer surface)

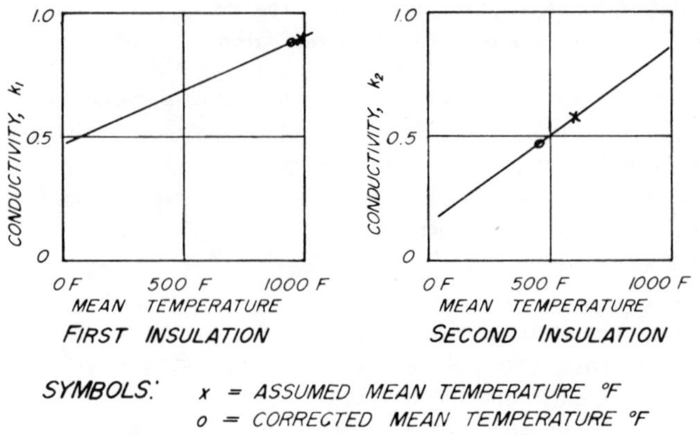

Example 6 Conductivity curves for insulations.

■ EXAMPLE 6a

Metric Units

English Values (as given)			conversion		Metric Values		
t_1	=	1100° F	(1100 − 32) ÷ 1.8	=	593.3° C	=	t_{m1}
t_2	=	150° F	(150 − 32) ÷ 1.8	=	65.6° C	=	t_{m2}
r_1	=	20″	× 0.0254	=	0.508m	=	r_{m1}
r_2	=	22.5″	× 0.0254	=	0.572m	=	r_{m2}
r_s	=	26.5″	× 0.0254	=	0.673m	=	r_{ms}
k_1	=	0.94 Btu in./ft², hr, °F	× 0.1442	=	0.1355 W/mK	=	k_{m1}
k_2	=	0.49 Btu in./ft², hr, °F	× 0.1442	=	0.0707 W/mK	=	k_{m2}

$$Q_m = \frac{t_{m1} - t_{m2}}{\left[\dfrac{r_{ms} \log_e\left(\dfrac{r_{m2}}{r_{m1}}\right)}{k_{m1}}\right] + \left[\dfrac{r_{ms} \log_e\left(\dfrac{r_{ms}}{r_{m2}}\right)}{k_{m2}}\right]}$$

$$= \frac{593.3 - 65.6}{\left[\dfrac{0.673 \log_e\left(\dfrac{0.572}{0.508}\right)}{0.1355}\right] + \left[\dfrac{0.673 \log_e\left(\dfrac{0.673}{0.572}\right)}{0.0707}\right]}$$

$$= \frac{527.7}{\left[\dfrac{0.673 \times 0.1183}{0.1355}\right] + \left[\dfrac{0.673 \times 0.1620}{0.0707}\right]}$$

$$= \frac{527.7}{0.5876 + 1.5421} = \frac{527.7}{2.1309} = 247.6 \text{ W/m}^2$$

Check:

$Q_m = 247.6 \text{ W/m}^2 \times 0.317199 = 78.6 \text{ Btu/ft}^2\text{, hr} = Q$ ■

Equation for Heat Transfer Through One Flat Insulation and One Surface Resistance

Surface resistance located on lower temperature surface

The basic law that energy flows in inverse ratio to the resistance still applies under this condition. If the "film conductance" is f, then the surface resistance equals $\frac{1}{f}$. The temperature difference is no longer that between surface and surface, however, but between surface and air.

When: *In English Units*

t_1 = Temperature of higher temperature surface °F
t_s = Temperature of lower temperature surface °F
t_a = Temperature of air °F
ℓ = Thickness of insulation in inches
k = Conductivity of insulation in Btu, in./ft², hr, °F
f = Film conductance in Btu/ft², hr, °F
Q = Heat flow in Btu/sq ft, hr, °F

then

$$Q = \frac{t_1 - t_a}{\left[\left(\frac{\ell}{k}\right) + \left(\frac{1}{f}\right)\right]} \tag{11}$$

A schematic diagram for this equation is shown in Figure 6.

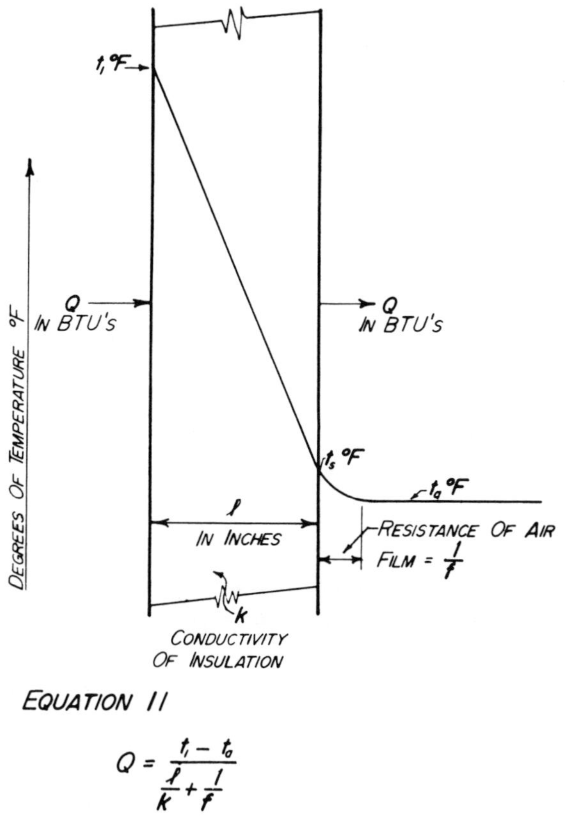

Figure 6 Schematic diagram of heat transfer through one insulation and one surface resistance.

When: *In Metric Units*

t_{m1} = Temperature of higher temperature surface in °C
t_{m2} = Temperature of lower temperature surface in °C
t_{ma} = Temperature of air in °C
ℓ_m = Thickness of insulation in metres
k_m = Conductivity of insulation in W/mK
f_m = Film conductance in W/m²
Q_m = Heat flow in W/m²

Then:

$$Q_m = \frac{t_{m1} - t_{m2}}{\left[\left(\frac{\ell_{m1}}{k_{mi}}\right) + \left(\frac{1}{f_{mi}}\right)\right]} \tag{11a}$$

■ EXAMPLE 7

English Units

If a flat insulation is 1 1/2″ thick (ℓ), the temperature of the higher surface is 400° F (t_1), the temperature of the air is 70° F (t_a), the insulation conductivity is 0.35 (k), and the conductance of the air film is 1.7 (f), what would be the heat flow Q?

$$Q = \frac{400 - 70}{\left[\frac{1.5}{0.35}\right] + \left[\frac{1}{1.7}\right]} = \frac{330}{4.29 + 59} = \frac{330}{4.88}$$

$$= 67.7 \text{ Btu/sq ft, hr}$$

If the temperature of the surface of the insulation on the lower side is required, it can be determined as

$$t_1 - \left(Q \times \frac{\ell}{k}\right) = t_s \text{ and } t_s - \left(Q \times \frac{1}{f}\right) = t_a$$

$$400 - (67.7 \times 4.29) = 400 - 290 = 110° \text{ F} = t_s$$

$$110 - (67.7 \times .59) = 110 - 40 = 70° \text{ F} = t_a \text{ (as given)}$$

■

■ EXAMPLE 7a

Metric Units

English Values (as given)	conversion		Metric Values	
t_1 = 400° F	(400 − 32) ÷ 1.8	=	204.4° C	= t_{m1}
t_a = 70° F	(70 − 32) ÷ 1.8	=	21.1° C	= t_{m2}
ℓ = 1.5″	× 0.0254	=	0.0381m	= ℓ_m
k = 0.35 Btu in./ft², hr, °F	× 0.1442	=	0.0505 W/mK	= k_m
f = 1.7 Btu in./ft², hr, °F	× 5.6782	=	9.6529 W/m²K	= f_m

$$Q_m = \frac{t_{m1} - t_{m2}}{\left[\left(\frac{\ell_m}{k_m}\right) + \left(\frac{1}{f_m}\right)\right]} = \frac{204.4 - 21.1}{\left[\left(\frac{0.0281}{0.0505}\right) + \left(\frac{1}{9.6529}\right)\right]}$$

$$= \frac{183.3}{0.7544 + 0.1036} = \frac{183.3}{0.858}$$

$$= 213.6 \text{ W/m}^2$$

Check:

$$Q_m = 213.6\ W/m^2 \times 0.317199 = 67.7\ Btu/sq, ft, hr.$$

To determine the surface temperature on lower side,

$$t_{m1} - \left(Q_m \times \frac{\ell}{k_m}\right) = t_{ms} \text{ and } t_{ms} - \left(Q_m \times \frac{1}{f_m}\right) = t_{ma}$$

$$204.4 - (213.6 \times 0.7544) = 204.4 - 161.1 = 43.3° C\ t_{ms} \text{ and}$$

$$43.3 - (213.6 \times 0.1036) = 43.3 - 22.1 = 21.2° C = t_{m2}$$

Check:

$$(21.2° \times 1.8) + 32 = 38.1 + 32 = 70.1° F = t_s$$ ■

Film resistance located on higher temperature surface

The equation for these conditions is the same as Eq. 11. The difference is that due to the direction of heat flow the answer will bear a minus sign. This merely indicates direction, as heat is always a positive factor.

■ **EXAMPLE 8**

English Units

If a flat insulation is 4" thick (ℓ), and the temperature of the lower temperature surface is −10° F (t_1), and the temperature of the air is 70° F (t_a), the insulation conductivity is 0.33 (k), and the conductance of the air film is 1.5 (f), what would be the heat flow Q?

$$Q = \frac{-10 - 70}{\left[\left(\frac{4}{0.33}\right) + \left(\frac{1}{1.5}\right)\right]} = \frac{-80}{12.1 + .66} = \frac{-80}{12.76}$$

$$= -6.3\ Btu/sq\ ft, hr$$

If the temperature of the surface of the insulation on the higher side is required, it is determined as

$$t_1 - \left(Q \times \frac{\ell}{k}\right) = t_s$$

$$-10 - (-6.3 \times 12.1) = t_s$$

$$= -10 - (-76) = -10 + 76 = +66° F$$

$$t_a = 66 - (-6.3 \times .66) = 66 - (-4)$$

$$= 66 + 4 = 70 \text{ (as given)}$$ ■

■ **EXAMPLE 8a**

Metric Units

English Values (as given)	***conversion***	***Metric Values***	
t_1 = −10° F	(−10 − 32) ÷ 1.8	= −23.2° C	= t_{m1}
t_a = 70° F	(70 − 32) ÷ 1.8	= 21.1° C	= t_{ma}
ℓ = 4"	× 0.0254	= 0.1016m	= ℓ_m
k = 0.33 Btu in./sq, ft, hr, °F	× 0.1442	= 0.047586 W/mK	= k_m
f = 1.5 Btu/sq, ft, hr, °F	× 5.6782	= 8.5173 W/m²K	= f_m

$$Q_m = \frac{t_{m1} - t_{ma}}{\left[\left(\frac{\ell_m}{k_m}\right) + \left(\frac{1}{f_m}\right)\right]} = \frac{23.3 - (21.1)}{\left[\left(\frac{0.1016}{0.047586}\right) + \left(\frac{1}{8.5173}\right)\right]}$$

$$= \frac{-44.4}{2.135 + 0.1174} = \frac{-44.4}{2.2524}$$

$$= 19.71\ W/m^2$$

Check:

$$Q_m = -19.71 \times 0.317199 = -6.25\ Btu/sq, ft, hr = Q$$

To determine the surface temperature on warmer side,

$$t_{m1} - \left(Q_m \times \frac{\ell_m}{k_m}\right) = t_{ms} \text{ and } t_{ms} - \left(Q_m \times \frac{1}{f_m}\right) = t_{ma}$$

$$-23.3 - (-19.71 \times 2.135) = -23.2 - (-42) = 18.7° C = t_{ms}$$

$$18.7 - (-19.71 \times 0.1174) = 18.7 - (-2.3) = 21° C = t_{ma}$$

Check:

$$t_{ma} = 21° C + (21 \times 1.8) + 32 = 37.8 + 32 = 69.8° F = t_a$$ ■

Equation for Heat Transfer Through Two Different Types of Insulations and One Surface Resistance

This follows the same form as Eq. 11 with the additional resistance of another type of insulation added to the equation.

When,

In English Units

t_1 = Temperature of surface, first insulation
t_2 = Temperature of surface of first insulation in contact with second insulation in °F
t_s = Temperature of surface of second insulation next to air in °F
t_a = Temperature of ambient air in °F
ℓ_1 = Thickness of first insulation in inches
ℓ_2 = Thickness of second insulation in inches
k_1 = Conductivity of first insulation in Btu, in./ft², hr, °F

k_2 = Conductivity of second insulation in Btu, in./ft², hr, °F
f = Film conductance in Btu/ft², hr
Q = Heat flow in Btu/ft², hr

then

$$Q = \frac{t_1 - t_a}{\left[\left(\frac{\ell_1}{k_1}\right) + \left(\frac{\ell_2}{k_2}\right) + \left(\frac{1}{f}\right)\right]} \qquad (12)$$

A schematic diagram for Eq. 12 is shown in Figure 7.

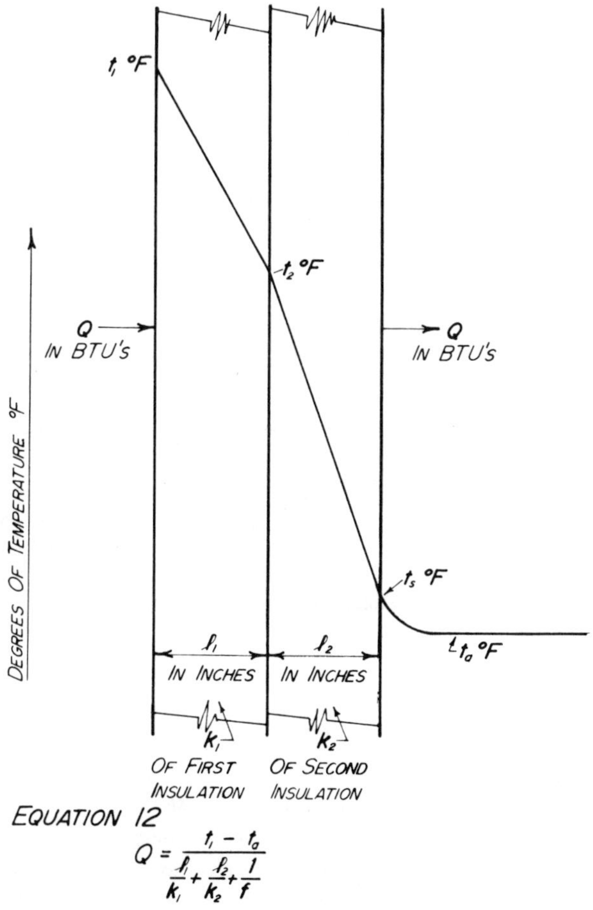

Figure 7 Schematic diagram of heat transfer through two different types of insulation and one surface resistance.

When,

In Metric Units

t_{m1} = Temperature of surface, first insulation in °C (or °K)
t_{m2} = Temperature of surface, or first insulation in contact with second insulation in °C (or °K)
t_{ms} = Temperature of surface of second insulation in °C (or °K)
t_{ma} = Temperature of ambient air in °C (or °K)
k_{m1} = Conductivity of first insulation in W/mK
k_{m2} = Conductivity of second insulation in W/mK
ℓ_{m1} = Thickness of first insulation in metres
ℓ_{m2} = Thickness of second insulation in metres
f_m = Film conductance in W/m²
Q_m = Heat flow in W/m²

Then

$$Q_m = \frac{t_{m1} - t_{m2}}{\left[\left(\frac{\ell_{m1}}{k_{m1}}\right) + \left(\frac{\ell_{m2}}{k_{m2}}\right) + \left(\frac{1}{f_m}\right)\right]} \qquad (12a)$$

■ EXAMPLE 9

If two different types of flat insulations, the first of which is 2″ thick and the second 4″ thick, are used to resist a temperature of 1500 °F on the inner surface, and the air temperature is 90° F, what is the heat transfer? The conductivities of the two insulations are given below. The conductance of the film is 1.8.

Assume the mean temperature of the first material to be 1350° F; then its conductivity (k_1) is 1.1. Assume the mean temperature of the second material to be 750° F; then its conductivity is 0.70.

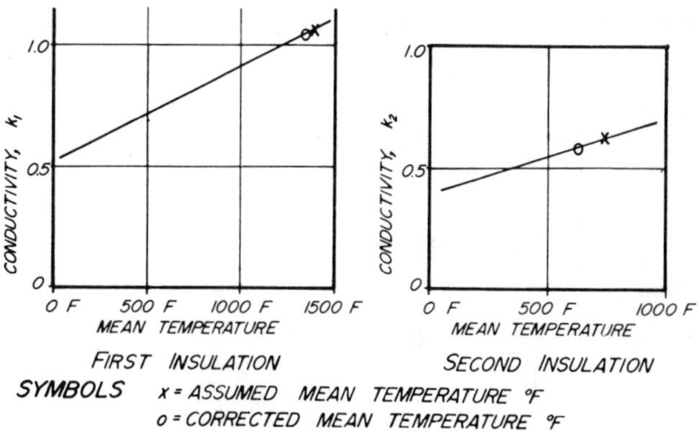

Example 9 Conductivity curves for insulations.

Then:

$$Q = \frac{1500 - 90}{\left[\left(\frac{2}{1.1}\right) + \left(\frac{4}{0.70}\right) + \left(\frac{1}{1.8}\right)\right]} = \frac{1410}{1.82 + 5.71 + 0.55}$$

$$= \frac{1410}{8.08} = 175 \text{ Btu/sq ft, hr}$$

Calculating the temperature gradient to determine if the assumed mean temperatures were approximately correct:

$$t_2 = t_1 - \left(Q \times \frac{\ell_1}{k_1}\right) = 1500 - (175 \times 1.82)$$

$$= 1500 - 319 = 1181$$

thus the mean temperature of the first insulation

$$\frac{1500 + 1181}{2} = \frac{2681}{2} = 1340° \text{ F}$$

$$t_s = t_2 - \left(Q \times \frac{\ell_2}{k_2}\right) = 1181 - (175 \times 5.71)$$

$$= 1181 - 999 = 182$$

thus the mean temperature of the second insulation

$$= \frac{1181 + 182}{2} = \frac{1363}{2} = 681°\text{ F}$$

Check:

$$t_a = t_s - \left(Q \times \frac{1}{f}\right) = 182 - (175 \times 0.55)$$

$$= 182 - 96 = 86 \text{ (within 4\% of given)}$$

Examination shows the assumed value of the first insulation to be close to correct and the second slightly high. Using new values of conductivities $k_1 = 1.1$ and $k_2 = 0.65$

$$Q = \frac{1500 - 90}{\left[\left(\frac{2}{1.1}\right)+\left(\frac{4}{0.65}\right)+\left(\frac{1}{1.8}\right)\right]} = \frac{1410}{1.82 + 6.17 + 0.55}$$

$$= \frac{1410}{8.54} = 165 \text{ Btu/sq ft, hr}$$

Then the temperature gradient would be

$$t_2 = t_1 - (165 \times 1.82)$$

$$= 1500 - 300 = 1200°\text{ F}$$

thus the mean temperature of the first material

$$= \frac{1500 + 1200}{2} = \frac{2700}{2} = 1350°\text{ F}$$

$$t_s = 1200 - (165 \times 6.17)$$

$$= 1200 - 1020 = 180°\text{ F}$$

mean temperature of the second material

$$= \frac{1200 + 180}{2} = \frac{1380}{2} = 690°\text{ F}$$

Check:

$$t_a = 180 - (165 \times 0.55) - 180 - 91$$

$$= 89°\text{ F (or within 1 °F of that given)}$$

A diagram of these results is shown in Figure 8. ■

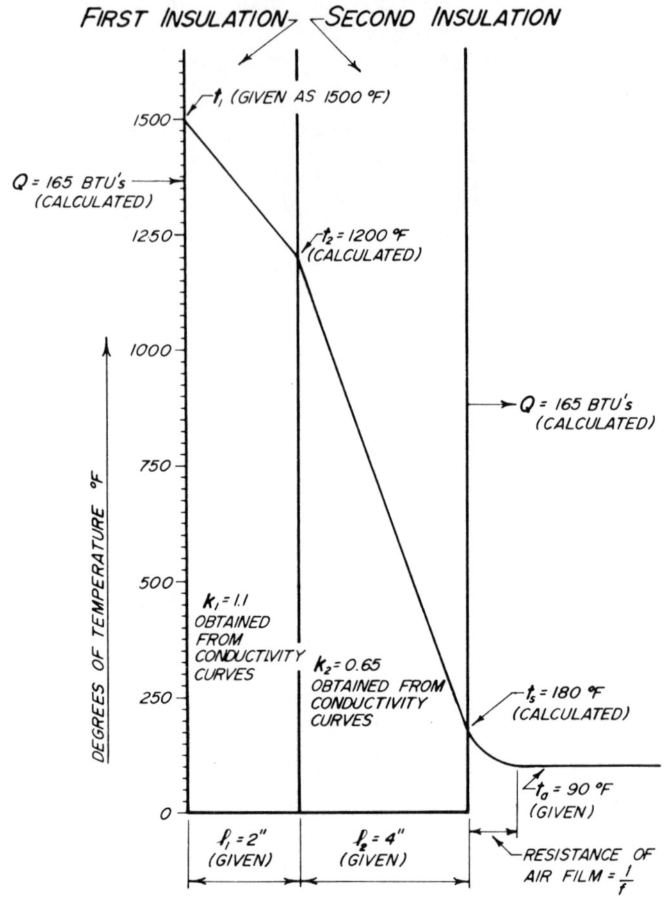

Figure 8 Diagram of results Example 9.

■ EXAMPLE 9a

Metric Units

English Values (as given)	***conversion***		***Metric Values***	
t_1 = 1500° F	(1500 − 32) ÷ 1.8	=	815.55° C	= t_{m1}
t_a = 90° F	(90 − 32) ÷ 1.8	=	32.22° C	= t_{ma}
ℓ_1 = 2"	× 0.0254	=	0.0508m	= ℓ_{m1}
ℓ_2 = 4"	× 0.0.254	=	0.1016m	= ℓ_{m2}
k_1 = 1.1 Btu in./ft^2, hr, °F	× 0.1442	=	0.1586 W/mK	= k_{m1}
k_2 = 0.65 Btu in./ft^2, hr, °F	× 0.1442	=	0.0937 W/mK	= k_{m2}
f = 1.8 Btu/ft^2, hr, °F	× 5.6782	=	10.2208 W/m^{2}K	= f_m

$$Q_m = \frac{t_{m1} - t_{ma}}{\left[\left(\frac{\ell_{m1}}{k_{m1}}\right)+\left(\frac{\ell_{m2}}{k_{m2}}\right)+\left(\frac{1}{f_m}\right)\right]}$$

$$= \frac{815.55 - 32.22}{\left[\left(\frac{0.0508}{0.1586}\right)+\left(\frac{0.1016}{0.0937}\right)+\left(\frac{1}{10.2208}\right)\right]}$$

$$= \frac{783.33}{0.320 + 1.084 + 0.098}$$

$$= \frac{783.33}{1.502} = 521.52 \text{ W/m}^2$$

Check:

$$Q_m = 521.52 \text{ w/m}^2 \times 0.317199 = 165.4 \text{ Btu/ft}^2\text{, hr.}$$

To determine the temperature gradients

First insulation and second insulation junction

$$t_{m2} = t_{m1} - \left(Q_m \times \frac{\ell_{m1}}{k_{m1}}\right) = 815.55 - (521.52 \times 0.32)$$

$$= 815.55 - (166.9) = 648.65° \text{ C } (1200° \text{ F})$$

Second insulation, outer surface

$$t_{ms} = t_{m2} - \left(Q_m \times \frac{\ell_{m2}}{k_{m2}}\right) = 648.65 - (521.52 \times 1.084)$$

$$= 648.65 - (565.33) = 83.32° \text{ C } (182° \text{ F})$$

Thus the mean temperature of first material

$$= \frac{815.55 + 648.65}{2} = 732° \text{ C } (1350° \text{ F})$$

The mean temperature of the second material is:

$$= \frac{648.65 + 83.32}{2} = \frac{731.97}{2} = 365.99° \text{ C } (690° \text{ F})$$

Check:

$$t_{ma} = 83.32 - (521.52 \times 0.098) = 83.32 - 51.11 = 32.21° \text{ C}$$ ■

Heat Transfer through One Flat Insulation with Surface Resistances on Each Side

Following the same procedure, using as a basis the temperature difference divided by the total resistance, the temperature difference will be from the ambient air on one side to the ambient air on the other side and the total resistance must include the resistances on each of the sides plus the resistance of the insulation.

When

In English Units

t_{a1} = Temperature of the ambient air, one side °F
t_1 = Temperature of the insulation, same side as t_{a1} °F
t_s = Temperature of the insulation, other side °F
t_{as} = Temperature of the ambient air on side t_s °F
ℓ = Thickness of the insulation in inches
k = Conductivity of the insulation in Btu, in./ft², hr, °F
f_1 = Conductance of the air film on side t_{a1} in Btu/hr, °F
f_2 = Conductance of the air film on side t_s in Btu/hr, °F
Q = Heat transfer in Btu/sq ft, hr

then

$$Q = \frac{t_{a1} - t_{as}}{\left[\left(\frac{1}{f_1}\right) + \left(\frac{\ell}{k}\right) + \left(\frac{1}{f_2}\right)\right]} \quad (13)$$

A schematic diagram is shown in Figure 9.

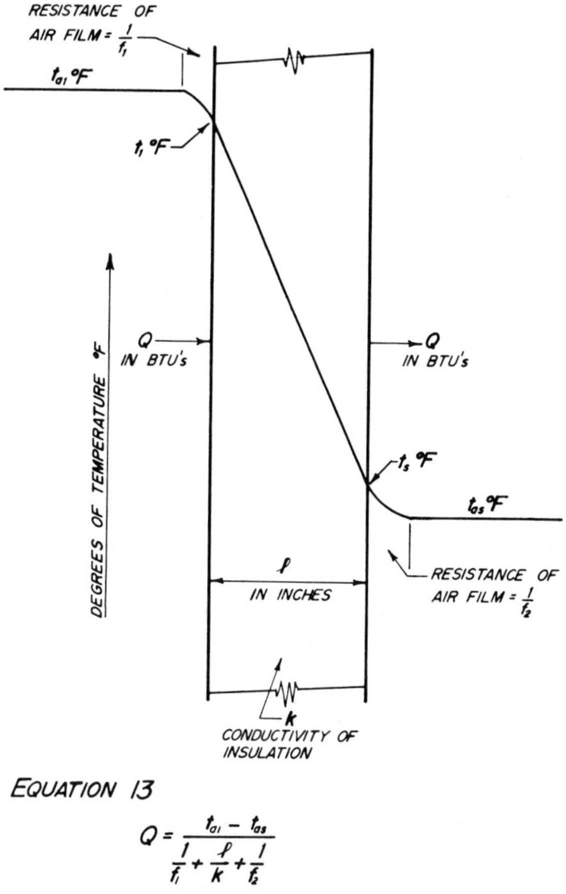

Figure 9 Schematic diagram of heat transfer through one insulation and surface resistance on each side.

When

In Metric Units

t_{ma1} = Temperature of ambient air, one side in °C
t_{m1} = Temperature of the insulation same side as t_{ma1} in °C
t_{ms} = Temperature of the insulation, other side in °C
t_{mas} = Temperature of ambient air on side t_{ms} in °C
ℓ_m = Thickness of insulation in metres
k_m = Conductivity of insulation in W/mK
f_{m1} = Conductance of air film on side t_{ma1} in W/m²K
f_{m2} = Conductance of air film on side t_{ms} in W/m²K
Q_m = Heat transfer in W/m²

then

$$Q_m = \frac{t_{ma2} - t_{mas}}{\left[\left(\frac{1}{f_{m1}}\right) + \left(\frac{\ell_m}{k}\right) + \left(\frac{1}{f_{m2}}\right)\right]} \quad (13a)$$

■ EXAMPLE 10

English Units

If an insulation 3″ thick has a conductivity of 0.48, the ambient air on the hotter side is 400° F, the ambient air on the cooler side is 50° F, the conductance of the air film is 1.9 on the hotter side and 1.4 on the cooler side, what is the heat transfer in Btu and what are its surface temperatures?

t_{a1} = 400° F ambient air temperature
t_1 = Temperature on one surface—to be calculated
t_s = Temperature on other surface—to be calculated
t_{as} = 50° F ambient air temperature
ℓ = Thickness of insulation = 3″
k = Conductivity of insulation = 0.48
Q = Heat transfer in Btu—to be calculated
f_1 = Conductance of air film—hot side = 1.9
f_2 = Conductance of air film—cold side = 1.4

Then:

$$Q = \frac{400 - 50}{\frac{1}{1.9} + \frac{3}{0.48} + \frac{1}{1.4}} = \frac{350}{0.52 + 6.25 + .71}$$

$$= \frac{350}{7.48} = 46.8 \text{ Btu/sq ft, hr}$$

To determine t_1 and t_s

$$t_1 = t_{a1} - \left(Q \times \frac{1}{f_1}\right) = 400 - \left(46.8 \times \frac{1}{1.9}\right)$$

$$= 400 - 25 = 375° \text{ F}$$

$$t_s = t_1 - \left(Q \times \frac{\ell}{k}\right) = 375 - \left(46.8 \times \frac{3}{0.48}\right)$$

$$= 375 - 292 = 83° \text{ F}$$

Check:

$$t_{as} = t_s - \left(Q \times \frac{1}{f_2}\right) = 83 - \left(46.8 \times \frac{1}{1.4}\right)$$

$$= 83 - 33 = 50° \text{ F (as given)}$$

■

■ EXAMPLE 10a

Metric Units

English Values (as given)	***conversion***	***Metric Values***	
t_{a1} = 400° F	(400 − 32) ÷ 1.8	= 204.4° C	= t_{ma1}
t_{as} = 50° F	(50 − 32) ÷ 1.8	= 10° C	= t_{mas}
ℓ = 3″	× 0.0254	= 0.0762m	= l_m
k = 0.48 Btu in./ft², hr, °F	× 0.1442	= 0.0692 W/mK	= k_m
f_1 = 1.9 Btu/ft², hr, °F	× 5.6782	= 10.7886 W/m²K	= f_{m1}
f_2 = 1.4 Btu/ft², hr, °F	× 5.6782	= 7.9495 W/m²K	= f_{m2}

To determine t_{m1} and t_{ms}, Q_m must be calculated first.

$$Q_m = \frac{t_{ma1} - t_{mas}}{\left[\left(\frac{1}{f_{m1}}\right) + \left(\frac{\ell_m}{k_m}\right) + \left(\frac{1}{f_{m2}}\right)\right]}$$

$$= \frac{204.4 - 10}{\left[\left(\frac{1}{10.7886}\right) + \left(\frac{0.0762}{0.0692}\right) + \left(\frac{1}{7.9495}\right)\right]}$$

$$= \frac{194.4}{0.0927 + 1.1011 + 0.1258}$$

$$= \frac{194.4}{1.3196} = 147.3 \text{ W/m}^2$$

Check:

$$Q_m = 147.3 \text{ W/m}^2 \times 0.317199 = 46.7 \text{ Btu/ft}^2\text{, hr}$$

To determine temperature gradients

$$t_{m1} = t_{a1} - \left(Q_m \times \frac{\ell}{f_{m1}}\right) = 204.4 - (147.3 \times 0.0927)$$

$$= 204.4 - 13.7 = 190.7° \text{ C } (375° \text{ F})$$

$$t_{m1} = t_{m1} - \left(Q \times \frac{\ell_m}{k_m}\right) = 190.7 - (147.3 \times 1.1011)$$

$$= 190.7 - 162.2 = 28.5° \text{ C } (83° \text{ F})$$

Check:

$$t_{mas} = t_{ms} - \left(Q_m \times \frac{1}{f_{m2}}\right) = 28.5 - (147.3 \times 0.1258)$$

$$= 28.5 - 18.5 = 10° \text{ C } (50° \text{ F})$$

■

Heat Transfer through Two Different Types of Flat Insulations with Surface Resistance on Both Hotter and Colder Sides

This equation is the same as the one previous, with one additional resistance for an additional different type of insulation. The number of insulation resistances that can be added in series is unlimited.

When

English Units

t_{a1} = Temperature of the ambient air on side t_{a1} in °F
t_1 = Temperature of the surface of first insulation in °F
t_2 = Temperature of surfaces of first and second insulations in contact in °F
t_s = Temperature of the second insulation in °F

t_{as} = Temperature of the ambient air on side t_s in °F
ℓ_1 = Thickness of the first insulation in inches
ℓ_2 = Thickness of the second insulation in inches
k_1 = Conductivity of the first insulation in Btu, in./ft^2, hr, °F
k_2 = Conductivity of the second insulation in Btu, in./ft^2, hr, °F
f_1 = Conductance of the air film on side t_{a1} in Btu/ft^2, hr, °F
f_2 = Conductance of the air film on side t_{as} in Btu/ft^2, hr, °F
Q = Heat transfer in Btu/sq ft, hr

then:

$$Q = \frac{t_{a1} - t_{as}}{\left[\left(\frac{1}{f_1}\right)+\left(\frac{\ell_1}{k_1}\right)+\left(\frac{\ell_2}{k_2}\right)+\left(\frac{1}{f_2}\right)\right]} \qquad (14)$$

In similar manner if a third different type of flat insulation is in the series of insulations and its thickness is ℓ_3 and conductivity is k_3

then:

$$Q = \frac{t_{a1} - t_{as}}{\left[\left(\frac{1}{f_1}\right)+\left(\frac{\ell_1}{k_1}\right)+\left(\frac{\ell_2}{k_2}\right)+\left(\frac{\ell_3}{k_3}\right)+\left(\frac{1}{f_2}\right)\right]} \qquad (15)$$

Any additional insulation, or materials having thermal resistance can be added to the equation as needed.

$$Q = \frac{t_{a1} - t_{as}}{\left[\left(\frac{1}{f_1}\right)+\left(\frac{\ell_1}{k_1}\right)+\left(\frac{\ell_2}{k_2}\right)+\left(\frac{\ell_3}{k_3}\right)+\left(\frac{\ell_4}{k_4}\right)\text{etc.} +\left(\frac{1}{f_2}\right)\right]} \qquad (16)$$

The schematic diagram of this equation is the same as that shown in Figure 9, with the proper number of insulations added to the diagram. Care must be taken, however, that each insulation is put in proper sequence in relation to the temperature drop, or heat flow, otherwise the mean temperature which determines the conductivity of the material may be in error.

In Metric Units

t_{ma1} = Temperature of the ambient air on side t_{a1} in °C
t_{m1} = Temperature of the surface of the first insulation in °C
t_{m2} = Temperature of the surfaces of the first and second insulation in contact in °C
t_{m3} = Temperature of the ambient air on side t_s in °C
ℓ_{m1} = Thickness of first insulation in metres
ℓ_{m2} = Thickness of second insulation in metres
k_{m1} = Conductivity of the first insulation in W/mK
k_{m2} = Conductivity of the second insulation in W/mK
f_{m1} = Conductance of air film on side t_{ma1} in W/m^{2}K
f_{m2} = Conductance of air film on side t_{ma2} in W/m^{2}K
Q_m = Heat transfer in W/m^2 hr

then

$$Q_m = \frac{t_{ma1} - t_{ma2}}{\left[\left(\frac{1}{f_{m1}}\right)+\left(\frac{\ell_{m1}}{k_{m1}}\right)+\left(\frac{\ell_{m2}}{k_{m2}}\right)+\left(\frac{1}{f_{m2}}\right)\right]} \qquad (14a)$$

In similar manner, if a third different type of flat insulation is in the series of insulations and its thickness is ℓ_{m3} and conductivity is k_{m3}

then:

$$Q_m = \frac{t_{ma1} - t_{ma2}}{\left[\left(\frac{1}{f_{m3}}\right)+\left(\frac{\ell_{m1}}{k_{m1}}\right)+\left(\frac{\ell_{m2}}{k_{m2}}\right)+\left(\frac{\ell_{m3}}{k_{m3}}\right)+\left(\frac{1}{f_{m2}}\right)\right]} \qquad (15a)$$

Any additional insulation or materials having thermal resistance can be added to the equation as needed.

$$Q_m = \frac{t_{ma1} - t_{ma3}}{\left[\left(\frac{1}{f_{m1}}\right)+\left(\frac{\ell_{m1}}{k_{m1}}\right)+\left(\frac{\ell_{m2}}{k_{m2}}\right)+\left(\frac{\ell_{m3}}{k_{m3}}\right)+\left(\frac{\ell_{m4}}{k_{m4}}\right)\text{etc.} +\left(\frac{1}{f_{m2}}\right)\right]} \qquad (16a)$$

■ EXAMPLE 11

English Units

Given the arrangement shown in Figure 10 for four insulations, with conditions as illustrated, find heat transfer Q and temperatures at t_1, t_2, t_3, t_4, and t_s.

First, the mean temperature of each different material must be assumed to determine a conductivity for each for a first determination of approximate heat transfer.

Assume mean temperature first insulation to be 1900° F.

Assume mean temperature second insulation to be 1600° F.

Assume mean temperature third insulation to be 750° F.

Assume mean temperature fourth insulation to be 150° F.

Then:

k_1 is approximately 1.9 Btu, in./ft^2, hr, °F
k_2 is approximately 1.2 Btu, in./ft^2, hr, °F
k_3 is approximately .85 Btu, in./ft^2, hr, °F
k_4 is approximately .55 Btu, in./ft^2, hr, °

$$Q = \frac{2000 - 90}{\frac{1}{4.5}+\frac{2}{1.9}+\frac{3}{1.2}+\frac{5}{0.85}+\frac{1}{0.55}+\frac{1}{1.8}}$$

$$= \frac{1910}{0.22 + 1.05 + 2.5 + 5.88 + 1.82 + 0.55}$$

$$= \frac{1910}{12.02} = 159 \text{ Btu/sq ft, hr}$$

Calculation for approximate values of t_1, t_2, t_3, t_4, and t_s:

$$t_1 = t_{a1} - \left(Q \times \frac{1}{f_1}\right) = 2000 - (159 \times 0.22)$$

$$= 2000 - 35 = 1965° \text{ F}$$

$$t_2 = t_1 - \left(Q \times \frac{\ell_1}{k_1}\right) = 1965 - (159 \times 1.05)$$

$$= 1965 - 167 = 1798°\ F$$

$$t_3 = t_2 - \left(Q \times \frac{\ell_2}{k_2}\right) = 1798 - (159 \times 2.5)$$

$$= 1798 - 397 = 1401°\ F$$

$$t_4 = t_3 - \left(Q \times \frac{\ell_3}{k_3}\right) = 1401 - (159 \times 5.88)$$

$$= 1401 - 934 = 467°\ F$$

$$t_s = t_4 - \left(Q \times \frac{\ell_4}{k_4}\right) = 467 - (159 \times 1.82)$$

$$= 467 - 288 = 179°\ F$$

$$t_{as} = t_s - \left(Q \times \frac{1}{f_2}\right) = 179 - (159 \times 0.55)$$

$$= 179 - 88 = 91°\ F \text{ (given as } 90°\ F)$$

Re-established mean temperatures for each different insulation:

$$\text{Insulation 1} = \frac{t_1 + t_2}{2} = \frac{1965 + 1798}{2}$$

$$= \frac{3763}{2} = 1881°\ F$$

$$\text{Insulation 2} = \frac{t_2 + t_3}{2} = \frac{1798 + 1401}{2}$$

$$= \frac{3199}{2} = 1600°\ F$$

$$\text{Insulation 3} = \frac{t_3 + t_4}{2} = \frac{1401 + 467}{2}$$

$$= \frac{1868}{2} = 934°\ F$$

$$\text{Insulation 4} = \frac{t_4 + t_s}{2} = \frac{467 + 179}{2}$$

$$= \frac{646}{2} = 323°\ F$$

Corrected values of conductivities:

Insulation 1 so close it is approximately same
$k_1 = 1.9$

Insulation 2 so close it is approximately same
$k_2 = 1.2$

Insulation 3 should be corrected to be
$k_3 = 0.95$

Insulation 4 should be corrected to be
$k_4 = 0.63$

Thus:

$$Q = \frac{2000 - 90}{\left[\left(\frac{1}{4.5}\right) + \left(\frac{2}{1.9}\right) + \left(\frac{4}{1.2}\right) + \left(\frac{5}{0.95}\right) + \left(\frac{1}{0.63}\right) + \left(\frac{1}{1.8}\right)\right]}$$

$$= \frac{1910}{[0.22 + 1.05 + 2.50 + 5.27 + 1.59 + 0.55]}$$

$$= \frac{1910}{11.18} = 170.5 \text{ Btu/sq ft, hr—Answer}$$

To determine t_1, t_2, t_3, t_4, and t_s:

$$t_1 = t_{a1} - \left(Q - \frac{1}{f_1}\right)$$

$$= 2000 - (170.5 \times 0.22)$$

$$= 2000 - 37 = 1963°\ F$$

$$t_2 = t_1 - \left(Q - \frac{\ell_1}{k_1}\right)$$

$$= 1963 - (170.5 \times 1.05)$$

$$= 1963 - 180 = 1783°\ F$$

$$t_3 = t_2 - \left(Q - \frac{\ell_2}{k_2}\right)$$

$$= 1783 - (170.5 \times 2.50)$$

$$= 1783 - 427 = 1356°\ F$$

$$t_4 = t_3 - \left(Q - \frac{\ell_3}{k_3}\right)$$

$$= 1356 - (170.5 \times 5.27)$$

$$= 1356 - 900 = 456°\ F$$

$$t_s = t_4 - \left(Q - \frac{\ell_4}{k_4}\right)$$

$$= 456 - (170.5 \times 1.59)$$

$$= 456 - 272 = 184°\ F$$

$$t_{as} = t_s - \left(Q - \frac{1}{f_2}\right)$$

Answer

$$= 184 - (170.5 \times 0.55)$$

$$= 184 - 94 = 90°\ F \text{ (as given)}$$

■

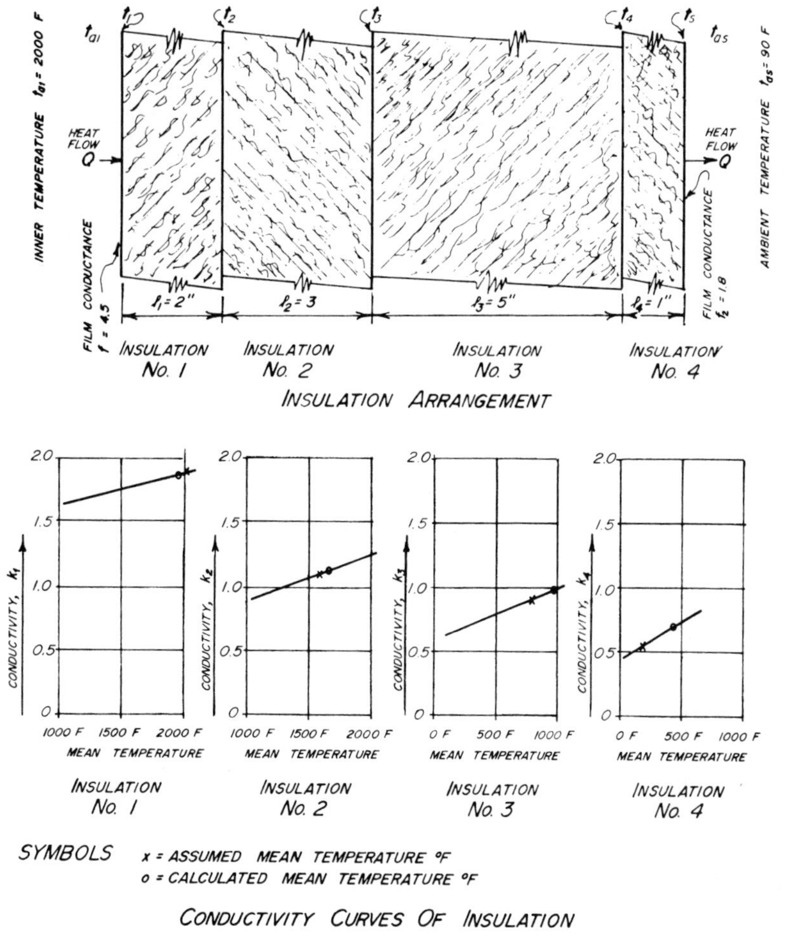

Figure 10 Given conditions and conductivity curves for Example 11.

■ EXAMPLE 11a

Metric Units

English Values (as given)	***conversion***		***Metric Values***	
t_{a1} = 2000° F	(2000 − 32) ÷ 1.8	=	1093.3° C	= t_{ma1}
t_{as} = 90° F	(90 − 32) ÷ 1.8	=	32.2° C	= t_{mas}
ℓ_1 = 2″	× 0.0254	=	0.0508	= ℓ_{m1}
ℓ_2 = 3″	× 0.0254	=	0.0762m	= ℓ_{m2}
ℓ_3 = 5″	× 0.0254	=	0.127m	= ℓ_{m3}
ℓ_4 = 1″	× 0.0254	=	0.0254m	= ℓ_{m4}
k_1 = 1.9 Btu in./ft^2, hr, °F	× 0.1442	=	0.274 W/mK	= k_{m1}
k_2 = 1.2 Btu in./ft^2, hr, °F	× 0.1442	=	0.173 W/mK	= k_{m2}
k_3 = 0.95 Btu in./ft^2, hr, °F	× 0.1442	=	0.137 W/mK	= k_{m3}
k_4 = 0.63 Btu in./ft^2, hr, °F	× 0.1442	=	0.0908 W/mK	= k_{m4}
f_1 = 4.5 Btu/ft^2, hr	× 5.6782	=	25.552 W/m^{2}K	= f_{m1}
f_2 = 1.8 Btu/ft^2, hr	× 5.6782	=	10.221 W/m^{2}K	= f_{m2}

Determine Q_m, t_{m1}, t_{m2}, t_{m3}, t_{ma} and t_{ms}

$$Q_m = \frac{t_{ma1} - t_{mas}}{\left[\left(\frac{1}{f_{m1}}\right)+\left(\frac{\ell_{m1}}{k_{m1}}\right)+\left(\frac{\ell_{m2}}{k_{m2}}\right)+\left(\frac{\ell_{m3}}{k_{m3}}\right)+\left(\frac{\ell_{m4}}{k_{m4}}\right)+\left(\frac{1}{f_{m2}}\right)\right]}$$

$$= \frac{1093.3 - 32.2}{\left[\left(\frac{1}{25.552}\right)+\left(\frac{0.0508}{0.274}\right)+\left(\frac{0.0762}{0.173}\right)+\left(\frac{0.127}{0.137}\right)+\left(\frac{0.0254}{0.0908}\right)+\left(\frac{1}{10.221}\right)\right]}$$

$$= \frac{1061.1}{[0.0391 + 0.1854 + 0.4405 + 0.927 + 0.280 + 0.0978]}$$

$$= \frac{1061.1}{1.97} = 538.7\ \text{W/m}^2$$

Check:

$$Q_m = 538.7\ \text{W/m}^2 \times 0.317199 = 170.8\ \text{Btu/ft}^2\text{, hr} = Q$$

Determine t_{m1}, t_{m2}, t_{m3}, t_{m4}, t_{ms}

$$t_{m1} = t_{ma1} - \left(Q_m \times \frac{1}{f_{m1}}\right) = 1093.3 - (538.7 \times 0.0391)$$

$$= 1093.3 - 21.1 = 1072.2°\ \text{C}\ (1962°\ \text{F})$$

$$t_{m2} = t_{m1} - \left(Q_m \times \frac{\ell_{m1}}{k_{m1}}\right) = 1072.2 - (538.7 \times 0.1854)$$

$$= 1072.2 - 99.9 = 972.3° \text{ C } (1783° \text{ F})$$

$$t_{m3} = t_{m2} - \left(Q_m \times \frac{\ell_{m2}}{k_{m2}}\right) = 972.3 - (538.7 \times 0.4405)$$

$$= 972.3 - 237.3 = 735° \text{ C } (1355° \text{ F})$$

$$t_{m4} = t_{m3} - \left(Q_m \times \frac{\ell_{m3}}{k_{m3}}\right) = 735 - (538.7 \times 0.927)$$

$$= 735 - 499.4 = 235.6° \text{ C } (457° \text{ F})$$

$$t_{ms} = t_{m4} - \left(Q_m \times \frac{\ell_{m4}}{k_{m4}}\right) = 235.6 - (538.7 \times 0.28)$$

$$= 235.6 - 150.8 = 84.8° \text{ C } (184° \text{ F})$$

Check:

$$t_{mas} = t_{ms} - \left(Q \times \frac{1}{10.221}\right) = 84.8 - (538.7 \times 0.0978)$$

$$= 84.8 - 52.6 = 32.2° \text{ C } (0° \text{ F}) \ (32° \text{ C given})$$

■

Equation for Heat Transfer through One Cylindrical Insulation and One Surface Resistance

Surface resistance located on outermost radius

In an identical manner as shown for flat insulations, the surface resistance can be added to a series of resistances in the basic equation of heat transfer:

When,

In English Units

t_1 = Temperature of inner radius surfaces, °F
t_s = Temperature of outer radius surface, °F
t_a = Temperature of the ambient air °F
r_1 = Inner radius of insulation in inches
r_2 = Outer radius of insulation in inches
k = Conductivity of insulation in Btu, in./ft^2, hr, °F
f = Film conductance in Btu/ft^2, hr °F
$q_{\ell f}$ = Total heat transfer in Btu/lin ft, hr
Q = Heat transfer in Btu/sq ft, hr—(outer surface)

then:

$$Q = \frac{t_1 - t_a}{\left[\left(\dfrac{r_2 \log_e \dfrac{r_2}{r_1}}{k}\right) + \left(\dfrac{1}{f}\right)\right]} \tag{17}$$

When,

In Metric Units

t_{m1} = Temperature of inner radius surface in °C (or °K)
t_{ms} = Temperature of outer surface in °C (or °K)
t_{ma} = Temperature of ambient air in °C (or °K)
r_{m1} = Inner radius of insulation in metres
r_{m2} = Outer radius of insulation in metres
k_m = Conductivity of insulation in W/mK
f_m = Film conductance in W/m^2
$q_{\ell m}$ = Heat transfer in W/lin m.
Q_m = Heat transfer in W/m^2 (outer surface)

then:

$$Q_m = \frac{t_{m1} - t_{ma}}{\left[\left(\dfrac{r_{m2} \log_e \dfrac{r_{m2}}{r_{m1}}}{k_m}\right) + \left(\dfrac{1}{f_m}\right)\right]} \tag{17a}$$

Note: Direction of flow will be indicated by plus or minus sign. Heat transfer to the air will have a plus sign, and heat transfer from surrounding ambient air will have a minus sign.

■ EXAMPLE 12

English Units

The surface temperature of the inner radius of the insulation is 300° F (t_1). The inner radius of the insulation is 2 1/4″ (r_1); the outer radius is 3 3/4″ (r_2). The conductivity (k) of the insulation is 0.4. The air temperature is 90° F. Film conductance is 2.2 (f). What is heat transfer Q in sq ft and lin. ft and the outer surface temperature of the insulation (t_s)?

$$Q = \frac{300 - 90}{\left[\left(\dfrac{3\tfrac{3}{4} \log_e \dfrac{3\tfrac{3}{4}}{2\tfrac{1}{4}}}{0.4}\right) + \left(\dfrac{1}{2.2}\right)\right]}$$

$$= \frac{210}{\left[\left(\dfrac{3.75 \times \log_e 1.66}{0.4}\right) + \left(\dfrac{1}{2.2}\right)\right]}$$

$$= \frac{210}{\left[\left(\dfrac{(3.75 \times 0.5068)}{0.4}\right) + 0.45\right]}$$

$$= \frac{210}{\left[\left(\dfrac{1.9}{0.4}\right) + 0.45\right]} = \frac{210}{\left[4.75 + .45\right]}$$

(from $\log_e$ tables; $\log_e 1.66 = 0.5068$)

$$= \frac{210}{5.2} = 40.3 \text{ Btu/sq ft, hr (outer surface)}$$

Temperature gradients can be determined starting from either direction from a known temperature:

$$t_s = t_a + \left(Q \frac{1}{f}\right) = 90 + 18 = 108°\text{ F}$$

to obtain loss per linear foot.

Heat transfer $q_{\ell f}$ per linear foot of insulation:

$$q_{\ell f} = Q\left(\frac{2r_2\pi}{12}\right)$$

$$q_{\ell f} = 40.3\left(\frac{3.75 \times 2 \times \pi}{12}\right)$$

$$= 40.3 \times 1.96 = 79 \text{ Btu/lin. ft, hr}$$

■

■ EXAMPLE 12a

Metric Units

English Values (as given)	conversion	Metric Values	
t_1 = 300° F	(300 − 32) ÷ 1.8	= 149.9° C	= t_{m1}
t_a = 90° F	(90 − 32) ÷ 1.8	= 32.2° C	= t_{ma}
r_1 = 2.25″	× 0.0254	= 0.0571m	= r_{m1}
r_2 = 3.75″	× 0.0254	= 0.0952m	= r_{m2}
k = 0.4 Btu in./ft², hr, °F	× 0.1442	= 0.05768 W/mK	= k_m
f_1 = 2.2 Btu/ft², hr, °F	× 5.6782	= 12.492 W/m²K	= f_{m1}

To determine t_{mas} outer surface temperature, it is first necessary to determine Q_m the heat transfer per square meter, per hr.

$$Q_m = \frac{t_{m1} - t_{ma}}{\left[\left(\dfrac{r_{m2}\log_e \dfrac{r_{m2}}{r_{m1}}}{k_{m2}}\right) + \left(\dfrac{1}{f_m}\right)\right]}$$

$$= \frac{148.9 - 32.2}{\left[\left(\dfrac{0.0952 \log_e \dfrac{0.0952}{0.0571}}{0.05768}\right) + \left(\dfrac{1}{12.492}\right)\right]}$$

$$= \frac{116.7}{\left[\left(\dfrac{0.0952 \times 0.5068}{0.05768}\right) + 0.08\right]}$$

$$= \frac{116.7}{\left[\left(\dfrac{0.0482}{0.05768}\right) + 0.08\right]}$$

$$= \frac{116.7}{[0.8364 + 0.08]} = \frac{116.7}{0.9165} = 127.3 \text{ W/m}^2$$

Check:

$$Q_m = 127.3 \text{ W/m}^2 \times 0.317199 = 40.4 \text{ Btu/ft}^2\text{, hr}$$

Determine outer surface temperature of insulation t_{ms}

$$t_{ms} = t_{ma} + \left(Q_m \times \frac{1}{f_m}\right) = 32.2 + (127.3 \times 0.08)$$

$$= 32.2 + 10.2 = 42.4°\text{ C } (108°\text{ F})$$

■

■ EXAMPLE 13

English Units

With all factors the same, except the inner surface temperature previously given as 300° F, assume this to be a problem where inner surface is refrigerated and inner surface temperature is 30° F.

Then Q, the heat loss per square foot of outer surface, is

$$Q = \frac{30 - 90}{\dfrac{3\tfrac{3}{4} \log_e \dfrac{3\tfrac{3}{4}}{2\tfrac{1}{4}}}{0.4} + \dfrac{1}{2.2}} = \frac{-60}{5.2}$$

$$= -11.5 \text{ Btu/sq ft, hr}$$

(The minus sign indicates that the heat transfer is from the air to the insulation.)

$$t_s = t_a + \left(Q \times \frac{1}{f}\right) = 90 + (-11.5 \times 0.45)$$

$$= 90 + (-5.2) = 84.8°\text{ F}$$

$q_{\ell f}$ = heat loss per lin. ft

$$q_{\ell f} = -11.5 \times (1.96) = -22.5 \text{ Btu/lin. ft, hr}$$

■

■ EXAMPLE 13a

Metric Units

English Values (as given)	conversion	Metric Values	
All same as Example 12 & 13 — Metric Units except			
t_1 = 30° F	(30 − 32) ÷ 1.8	= −1.1° C	= t_{m1}

$$Q_m = \frac{-1.1 - 32.2}{\left(\dfrac{0.0952 \times 0.5068}{0.05768}\right) + 0.08} = \frac{-33.3}{0.9165} = -36.3 \text{ W/m}^2$$

Check:

$$Q_m = -36.3 \text{ W/m}^2 \times 0.317199 = -11.5 \text{ Btu/ft}^2\text{, hr.}$$

(Minus sign indicates that the heat transfer is from air to insulation.)

$$t_{ms} = t_{ma} + \left(Q_m \times \frac{1}{f_{m1}}\right) = 32.2 + (-36.3 \times 0.08)$$

$$= 32.2 - 2.9 = 29.3°\text{ C } (84.8°\text{ F})$$

■

Equation for Heat Transfer through Two Different Types of Cylindrical Insulations and One Surface Resistance

Surface resistance located on outermost surface

When

t_1 = Temperature of inner radius surface, °F
t_2 = Temperature of junction of first insulation and second insulation, °F
t_s = Temperature of outer radius surface, °F
t_a = Temperature of ambient air, °F
r_1 = Inner radius of first insulation
r_2 = Outer radius of first insulation and inner radius of second insulation
r_3 = Outer radius of second insulation
k_1 = Conductivity of first insulation
k_2 = Conductivity of second insulation
f = Film conductance
$q_{\ell f}$ = Total heat transfer in Btu/lin. ft, hr
Q = Heat transfer in Btu/sq ft, hr

then:

$$Q = \frac{t_1 - t_a}{\left[\left(\frac{r_3 \log_e \frac{r_2}{r_1}}{k_1}\right) + \left(\frac{r_3 \log_e \frac{r_3}{r_2}}{k_2}\right) + \left(\frac{1}{f}\right)\right]} \tag{18}$$

Note: Additional layers can be included, as for three layers.

$$Q = \frac{t_1 - t_a}{\left[\left(\frac{r_4 \log_e \frac{r_2}{r_1}}{k_1}\right) + \left(\frac{r_4 \log_e \frac{r_3}{r_2}}{k_2}\right) + \left(\frac{r_4 \log_e \frac{r_4}{r_3}}{k_3}\right) + \left(\frac{1}{f}\right)\right]} \tag{19}$$

Notice multiplier of $\log_e$ factor in all resistances is outermost radius. In this case r_4.

Where the loss is to be calculated on basis of linear foot then when,

In English Units

$Q_{\ell f}$ = Btu per linear foot per hour

$$Q_{\ell f} = \frac{t_1 - t_a}{\left[\left(\frac{r_4 \log_e \frac{r_2}{r_1}}{k_1}\right) + \left(\frac{r_4 \log_e \frac{r_3}{r_2}}{k_2}\right) + \left(\frac{r_4 \log_e \frac{r_4}{r_3}}{k_3}\right) + \left(\frac{1}{f}\right)\right]} \times \left(\frac{2r_4\pi}{12}\right)$$

When,

In Metric Units

t_{m1} = Temperature of inner radius surface in °C (or °K)
t_{m2} = Temperature of juncture of first and second insulation in °C (or °K)
t_{ms} = Temperature of outer surface of outer insulation in °C (or °K)
t_{ma} = Temperature of ambient air °C (or °K)
r_{m1} = Inner radius of first insulation in metres
r_{m2} = Outer radius of first insulation and inner radius of second insulation in metres
r_{m3} = Outer radius of second insulation in metres
k_{m1} = Conductivity of first insulation in W/mK
k_{m2} = Conductivity of second insulation in W/mK
f_m = Film conductance in W/m^2
$q_{\ell m}$ = Total heat transfer in W/lin. metre
Q_m = Heat transfer in W/m^2 (outer surface)

then,

$$Q_m = \frac{t_{m1} - t_{ma}}{\left[\left(\frac{r_{m3} \log_e \frac{r_{m2}}{r_{m1}}}{k_{m1}}\right) + \left(\frac{r_{ms} \log_e \frac{r_{m3}}{r_{m2}}}{k_{m2}}\right) + \left(\frac{1}{f_m}\right)\right]} \tag{18a}$$

and

$$q_{\ell m} = \frac{t_{m1} - t_{ma}}{\left[\left(\frac{r_{ms} \log_e \frac{r_{m2}}{r_{m1}}}{k_{m1}}\right) + \left(\frac{r_{ms} \log_e \frac{r_{m3}}{r_{m2}}}{k_{m2}}\right) + \left(\frac{1}{f_m}\right)\right]} \times (2r_{m3}\,\pi)$$

following Equation 19

In Metric Units

$$Q_m = \frac{t_{m1} - t_{ma}}{\left[\left(\frac{r_{m4} \log_e \frac{r_{m2}}{r_{m1}}}{k_{m1}}\right) + \left(\frac{r_{m4} \log_e \frac{r_{m3}}{r_{m2}}}{k_{m2}}\right) + \left(\frac{r_{m4} \log_e \frac{r_{m4}}{r_{m3}}}{k_{m3}}\right) + \left(\frac{1}{f_m}\right)\right]} \tag{19a}$$

and

$$q_{\ell m} = \frac{t_{m1} - t_{ma}}{\left[\left(\frac{r_{m4} \log_e \frac{r_{m2}}{r_{m1}}}{k_{m1}}\right) + \left(\frac{r_{m4} \log_e \frac{r_{m3}}{r_{m2}}}{k_{m2}}\right) + \left(\frac{r_{m4} \log_e \frac{r_{m4}}{r_{m3}}}{k_{m3}}\right) + \left(\frac{1}{f_m}\right)\right]} \times (2r_{m4}\,\pi)$$

■ EXAMPLE 14

English Units

Given the following:

t_1 = Temperature of inner surface, 1775° F
t_a = Ambient air temperature, 50° F

r_1 = Inner radius first insulation, 4″
r_2 = Outer radius first insulation and inner radius second insulation, 6″
r_3 = Outer radius second insulation, 10″
f = Film conductance, 2.5
k_1 = Conductivity of first insulation, given below
k_2 = Conductivity of second insulation, given below

Find: Q, q_ℓ, t_2, and t_s

Assume mean temperature of the first insulation to be 1500° F.

Assume mean temperature of the second insulation to be 800° F

Then $k_1 = 1.2$ and $k_2 = 0.8$

$$Q = \frac{1775 - 50}{\left[\left(\dfrac{10 \log_e \frac{6}{4}}{1.2}\right) + \left(\dfrac{10 \log_e \frac{10}{6}}{0.8}\right) + \left(\dfrac{1}{2.5}\right)\right]}$$

$$= \frac{1725}{\left[\left(\dfrac{10 \log_e 1.5}{1.2}\right) + \left(\dfrac{10 \log_e 1.66}{0.8}\right) + \left(\dfrac{1}{2.5}\right)\right]}$$

$(\log_e 1.5 = 0.4055; \log_e 1.66 = 0.5068)$

$$Q = \frac{1725}{\left[\left(\dfrac{10 \times 0.405}{1.2}\right) + \left(\dfrac{10 \times 0.507}{0.8}\right) + \left(\dfrac{1}{2.5}\right)\right]}$$

$$= \frac{1725}{\left[\left(\dfrac{4.05}{1.2}\right) + \left(\dfrac{5.07}{0.8}\right) + \left(\dfrac{1}{2.5}\right)\right]}$$

$$= \frac{1725}{[3.38 + 6.32 + 0.40]} = \frac{1725}{10.10}$$

= 171 Btu/sq ft, hr (approximate)

Calculation for approximate values of t_2, t_s:

$$t_2 = t_1 - \left(Q \times \frac{r_3 \log_e \frac{r_2}{r_1}}{k_1}\right)$$

$= 1775 - (171 \times 3.38)$

$= 1775 - 578 = 1197°$ F

$$t_s = t_2 - \left(Q \times \frac{r_3 \log_e \frac{r_3}{r_2}}{k_2}\right)$$

$= 1197 - (171 \times 6.32)$

$= 1197 - 1080 = 117°$ F

$$t_a = t_s - \left(Q \times \frac{1}{f}\right)$$

$= 117 - (171 \times 0.4)$

$= 117 - 67 = 50°$ F (as given)

Checking assumed mean temperatures:

$$\text{First insulation} = \frac{t_1 + t_2}{2} = \frac{1775 + 1197}{2}$$

$$= \frac{2972}{2} = 1486$$

$$\text{Second insulation} = \frac{t_2 + t_s}{2} + \frac{1197 + 117}{2}$$

$$= \frac{1314}{2} = 657$$

In this case the assumed mean temperatures proved to be close for the first insulation but slightly off for the second. Thus, the conductivity k_1 for the first material remains the same, but the conductivity k_2 must be changed from 0.8 to 0.7.

Recalculation for heat transfer:

$$Q = \frac{1725}{\dfrac{4.05}{1.2} + \dfrac{5.07}{0.7} + \dfrac{1}{2.5}}$$

$$= \frac{1725}{3.38 + 7.23 + 0.40}$$

$$= \frac{1725}{11.01} = 157 \text{ Btu/sq ft, hr}$$

Recalculation for values of t_2 and t_s:

$t_2 = 1775 - (157 \times 3.38)$

$= 1775 - 530 = 1245°$ F

$t_s = 1245 - (157 \times 7.23)$

$= 1245 - 1135 = 110°$ F

$t_a = 110 - (157 \times 0.40)$

$= 110 - 63 = 47°$ F

To find heat transfer per lin. ft:

$$q_{\ell f} = Q\left(\frac{2 r_3 \pi}{12}\right) = 157 \times 5.22$$

= 820 Btu/lin. ft, hr

■

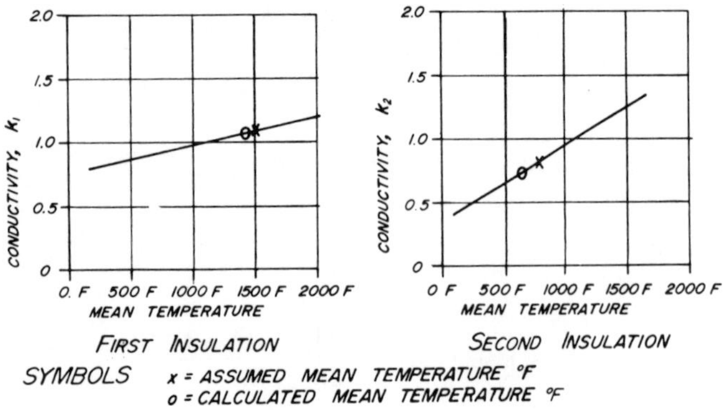

Example 14 Conductivity curves for insulations.

■ EXAMPLE 14a

Metric Units

English Values (as given)	***conversion***		***Metric Values***	
t_1 = 1775° F	(1775 − 32) ÷ 1.8	=	968.3° C	= t_{m1}
t_a = 50° F	(50 − 32) ÷ 1.8	=	10° C	= t_{ma}
r_1 = 4″	× 0.0254	=	0.1016m	= r_{m1}
r_2 = 6″	× 0.0254	=	0.1524m	= r_{m2}
r_3 = 10″	× 0.0254	=	0.2540m	= r_{m3}
k_1 = 1.2 Btu in./ft², hr, °F	× 0.1442	=	0.173 W/mK	= k_{m1}
k_2 = 0.7 Btu in./ft², hr, °F	× 0.1442	=	0.101 W/mK	= k_{m2}
f = 2.5 Btu/ft², hr, °F	× 5.6782	=	14.195 W/m²K	= f_m

Determine Q_m, $q_{m\ell f}$, t_{m2} and t_{ms}

$$Q_m = \frac{t_{m1} - t_{ma}}{\left[\left(\dfrac{r_{m3}\log_e \dfrac{r_{m2}}{r_{m1}}}{k_{m1}}\right) + \left(\dfrac{r_{m3}\log_e \dfrac{r_{m3}}{r_{m2}}}{k_{m2}}\right) + \left(\dfrac{1}{f_m}\right)\right]}$$

$$= \frac{968.3 - 10}{\left[\left(\dfrac{0.254\log_e \dfrac{0.1524}{0.1016}}{0.173}\right) + \left(\dfrac{0.254\log_e \dfrac{0.254}{0.1524}}{0.101}\right) + \left(\dfrac{1}{14.195}\right)\right]}$$

$$= \frac{958.3}{\left[\left(\dfrac{0.254 \times 0.405}{0.173}\right) + \left(\dfrac{0.254 \times 0.512}{0.101}\right) + 0.07\right]}$$

$$= \frac{958.3}{[0.595 + 1.288 + 0.07]}$$

$$= \frac{958.3}{1.95} = 491 \text{ W/m}^2$$

Check:

$$Q_m = 491 \text{ W/m}^2 \times 0.317199 = 155.7 \text{ Btu/sq, ft, hr}$$

$$Q_{m\ell} = Q_m(2r_{m3}\pi) = 491\ (2 \times 0.254\pi)$$

$$= 491\ (1.596) = 783.6 \text{ W/lin. metre}$$

Check:

$$q_{m\ell f} = 783.6 \text{ W/lin. metre} \times 1.04068$$

$$= 815.4 \text{ Btu/lin. ft, hr} = Q_{\ell f}$$

$$t_{m2} = t_{m1} - \left(Q_m \times \frac{r_{m3}\log_e \dfrac{r_{m2}}{r_{m1}}}{k_{m1}}\right) = 968.3 - (491 \times .595)$$

$$= 968.3 - 292.1 = 676.2° \text{ C } (1245° \text{ F})$$

$$t_{ms} = t_{m2} - \left(Q_m \times \frac{r_{m3}\log_e \dfrac{r_{m3}}{r_{m2}}}{k_{m2}}\right) = 676.2 - (491 \times 1.275)$$

$$= 674.4 - 626.0 = 48.4° \text{ C } (119° \text{ F})$$

Check:

$$t_{ma} = t_{ms} - \left(Q_m \times \frac{1}{f_m}\right) = 48.4 - (491 \times 0.07)$$

$$= 48.4 - 34.4 = 14° \text{ C } (57° \text{ F})$$

■

Heat Transfer through One Cylindrical Insulation with Surface Resistances on Each Side

In this case the temperature difference is from an ambient air temperature on the inner side to the ambient air temperature on the outer surface. Therefore, the total resistance must include the resistance of the insulation material.

When,

In English Units

t_{a1} = Temperature of air on one side °F
t_1 = Temperature of insulation, same side as t_{a1} °F
t_s = Temperature of insulation, same side as t_{as} °F
t_{as} = Temperature of ambient air on side next to t_s °F
r_1 = Inner radius of insulation in inches
r_2 = Outer radius of insulation in inches
f_1 = Conductance of air film—side t_{a1} in Btu/ft², hr, °F
f_2 = Conductance of air film—side t_s in Btu/ft², hr, °F
k = Conductivity of the insulation material in Btu, in./ft², hr, °F
$q_{\ell f}$ = Total heat transfer in Btu/lin. ft, hr

then:

$$Q = \frac{t_{a1} - t_{as}}{\left[\left(\dfrac{1}{f_1}\right) + \left(\dfrac{r_2\log_e \dfrac{r_2}{r_1}}{k}\right) + \left(\dfrac{1}{f_2}\right)\right]} \tag{20}$$

and

$$Q_{lf} = \frac{t_{a1} - t_{as}}{\left[\left(\frac{1}{f_1}\right) + \left(\frac{r_2 \log_e \frac{r_2}{r_3}}{k}\right) + \left(\frac{1}{f_2}\right)\right]} \times \left(\frac{2r_2 \pi}{12}\right)$$

In Metric Units

t_{ma1} = Temperature of air one side, in °C (or °K)
t_{m1} = Temperature of insulation surface, same side as t_{ma1}, in °C (or °K)
t_{ms} = Temperature of insulation surface, same side as t_{mas}, in °C (or °K)
t_{mas} = Temperature of ambient air on side next to t_{mas}, in °C (or °K)
r_{m1} = Inner radius of insulation, in metres
r_{m2} = Outer radius of insulation, in metres
f_{m1} = Conductance of air film—side t_{ma1}, in W/m²
f_{m2} = Conductance of air film—side t_{ms}, in W/m²
k_m = Conductivity of insulation material, in W/mK
q_{lm} = Total heat transfer, in W/lin. metre
Q_m = Heat transfer, in W/m² (outer surface)

then

$$Q_m = \frac{t_{ma1} - t_{mas}}{\left[\left(\frac{1}{f_{m1}}\right) + \left(\frac{r_{m2} \log_e \frac{r_{m2}}{r_{m1}}}{k_m}\right) + \left(\frac{1}{f_{m2}}\right)\right]} \qquad (20a)$$

and

$$q_{lf} = \frac{t_{ma1} - t_{mas}}{\left[\left(\frac{1}{f_{m1}}\right) + \left(\frac{r_{m2} \log_e \frac{r_{m2}}{r_{m1}}}{k}\right) + \left(\frac{1}{f_{m2}}\right)\right]} \times (2r_{m2} \pi)$$

■ EXAMPLE 15

English Units

The air temperature on the inside of a cylindrical insulation is 200 °F. The inner radius of the insulation is 9″ and the insulation is 1-1/2″ thick. The conductivity of the insulation is 0.25 Btu/sq ft, hr, in., °F. The film conductance is 3.6 on the inside and 1.75 on the outside. The outside air temperature is 0° F. What is the heat loss and the inner and outer surface temperatures of the insulation? Also, what is its loss per linear foot?

The heat loss

$$Q = \frac{t_{a1} - t_{as}}{\left[\left(\frac{1}{f_1}\right) + \left(\frac{r_2 \log_e \frac{r_2}{r_1}}{k}\right) + \left(\frac{1}{f_2}\right)\right]}$$

$$= \frac{200 - 0}{\left[\left(\frac{1}{3.6}\right) + \left(\frac{10\tfrac{1}{2} \log_e \frac{10\tfrac{1}{2}}{9}}{0.25}\right) + \left(\frac{1}{1.75}\right)\right]}$$

$$= \frac{200}{\left[(0.28) + \left(\frac{10\tfrac{1}{2} \log_e 1.17}{0.25}\right) + (0.57)\right]}$$

(from $\log_e$ tables—$\log_e$ 1.17 = 0.1570)

$$= \frac{200}{\left[0.28 + \left(\frac{10.5 \times 0.1570}{0.25}\right) + 0.57\right]}$$

$$= \frac{200}{\left[0.28 + \left(\frac{1.64}{0.25}\right) + 0.57\right]}$$

$$= \frac{200}{[0.28 + 6.60 + 0.57]} = \frac{200}{7.45}$$

= 26.8 Btu/sq ft, hr

The temperature of the inner surface of the insulation is:

$$t_1 = t_{a1} - \left(Q \times \frac{1}{f_1}\right) = 200 - (26.8 \times .28)$$

$$= 200 - 7.5 = 192.5° \text{ F}$$

The temperature of the outer surface of the insulation is:

$$t_s = t_1 - \left(Q \times \frac{r_2 \log_e \frac{r_2}{r_1}}{k}\right)$$

$$= 192.5 - (26.8 \times 6.60)$$

$$= 192.5 - 177 = 15.5° \text{ F}$$

or this could be calculated:

$$t_s = t_a - \left(Q \times \frac{1}{f_2}\right) = 0 + (26.8 \times 0.57)$$

= 15.4° F
(slight difference is due to dropping off of figures after second decimal point)

The loss per linear foot is:

$$q_{lf} = Q \times \left(\frac{2r_2 \pi}{12}\right) = 26.8 \times \left(\frac{2 \times 10.5 \times \pi}{12}\right)$$

$$= 26.8 \left(\frac{21\pi}{12}\right)$$

= 26.8 × 5.51 = 148 Btu/lin. ft, hr ■

■ **EXAMPLE 15a**

Metric Units

English Values (as given)	***conversion***	***Metric Values***	
t_1 = 200° F	(200 − 32) ÷ 1.8	= 93.3° C	= t_{m1}
t_a = 0° F	(0 − 32) ÷ 1.8	= −17.7° C	= t_{ma}
r_1 = 9″	× 0.0254	= 0.2286m	= r_{m1}
r_2 = 10.5″	× 0.0254	= 0.2667m	= r_{m2}
k = 0.25 Btu in./ft², hr, °F	× 0.1442	= 0.036 W/mK	= k_m
f_1 = 3.6 Btu/ft², hr, °F	× 5.6782	= 20.442 W/m²K	= f_{m1}
f_2 = 1.75 Btu/ft², hr, °F	× 5.6782	= 9.937 W/m²K	= f_{m2}

Determine Q_m, $Q_{m\ell}$, t_{ma1} and t_{ms}

$$Q_m = \frac{t_{m1} - t_{ma}}{\left[\left(\frac{1}{f_{m1}}\right) + \left(\frac{r_{m2}\log_e \frac{r_{m2}}{r_{m1}}}{k_m}\right) + \left(\frac{1}{f_{ma}}\right)\right]}$$

$$= \frac{93.3 - (-17.7)}{\left[\left(\frac{1}{20.442}\right) + \left(\frac{0.2667\log_e \frac{0.2667}{0.2286}}{0.03605}\right) + \left(\frac{1}{9.937}\right)\right]}$$

$$= \frac{111.0}{\left[0.049 + \left(\frac{0.2667 \times 0.157}{0.036}\right) + 0.101\right]}$$

$$= \frac{111}{[0.049 + 1.161 + 0.101]} = \frac{111}{1.311}$$

$$= 84.7\ \text{W/m}^2$$

Check:

$$Q_m = 84.7\ \text{W/m}^2 \times 0.317199 = 26.8\ \text{Btu/ft}^2\text{, hr} = Q$$

$$t_{ma1} = t_{m1} - \left(Q_m \times \frac{1}{f_{m1}}\right) = 93.3 - (84.7 \times 0.049)$$

$$= 93.3 - 4.1 = 89.2°\ \text{C}\ (192°\ \text{C})$$

$$t_{ms} = t_{ma1} - \left(Q_m \times \frac{r_{m2}\log_e \frac{r_{m2}}{r_{m1}}}{k_m}\right)$$

$$= 89.2 - (84.9 \times 1.161)$$

$$= 89.2 - 98.3 = -9.1°\ \text{C}\ (15°\ \text{F})$$

Check:

$$t_{ma} = t_{ms} - \left(Q_m \times \frac{1}{f_{ma}}\right) = -9.1 - (84.7 \times 0.101)$$

$$= -9.1 - (8.6) = -17.7°\ \text{C}\ (0°\ \text{F})$$ ■

Heat Transfer through Two (or more) Different Materials and Surface Resistances on Innermost and Outermost Surfaces

Adding additional layers of materials into the line of resistance to the heat transfer takes the same form as presented in Eqs. 18 and 19. Thus:

$$Q = \frac{t_{a1} - t_a}{\left[\left(\frac{1}{f_1}\right) + \left(\frac{r_3\log_e \frac{r_2}{r_1}}{k_1}\right) + \left(\frac{r_3\log_e \frac{r_3}{r_2}}{k_2}\right) + \left(\frac{1}{f_2}\right)\right]} \tag{21}$$

Btu/sq ft, hr (outer surface)

$$q_{\ell f} = Q\left(\frac{2r_3\,\pi}{12}\right)\ \text{Btu/lin. ft, hr}$$

This is illustrated in Figure 11.

More than two materials can be added into the equation—as indicated below.

$$Q = \frac{t_{a1} - t_a}{\left[\left(\frac{1}{f_1}\right) + \left(\frac{r_4\log_e \frac{r_2}{r_1}}{k_1}\right) + \left(\frac{r_4\log_e \frac{r_3}{r_2}}{k_2}\right) + \left(\frac{r_4\log_e \frac{r_4}{r_3}}{k_3}\right) + \left(\frac{1}{f_2}\right)\right]}$$

The r multiplier of each $\log_e$ is always the outermost radius. In the above case it is r_4.

Then:

In Metric Units

$$Q_m = \frac{t_{ma1} - t_{ma}}{\left[\left(\frac{1}{f_{m1}}\right) + \left(\frac{r_{m3}\log_e \frac{r_{m2}}{r_{m1}}}{k_{m1}}\right) + \left(\frac{r_{m3}\log_e \frac{r_3}{r_2}}{k_{m2}}\right) + \left(\frac{1}{f_{m2}}\right)\right]} \tag{21a}$$

and

$$q_{\ell m} = Q_m\,(2r_{m3}\,\pi)$$

■ **EXAMPLE 16**

English Units

Given:

Inner air temperature, t_{a1} = 1600° F
Outer air temperature, t_a = 90° F
Inner radius of first insulation, r_1 = 12″
Thickness of first insulation = 2″
Thickness of second insulation = 3-1/2″
Inner film conductance = 5.0 Btu/sq ft, hr, °F

Outer film conductance = 3.0 Btu/sq ft, hr, °F
Conductivity of insulations = k_1 and k_2 in Btu/sq ft, hr, in., °F in curves above

Assume a mean temperature of 1400° F for first material and 700° F for second material. Then $k_1 = 0.95$ and $k_2 = 0.65$.

Putting known factors and assumed factors for conductivities in Eq. 21 to find Q = t_1, t_2, t_s, and $q_{\ell f}$:

$$r_1 = 12$$

$$r_2 = 12 + 2 = 14$$

$$r_3 = 12 + 2 + 3\tfrac{1}{2} = 17\tfrac{1}{2}$$

$$Q = \frac{1600 - 90}{\left[\left(\frac{1}{5.0}\right) + \left(\frac{17.5 \log_3 \frac{14}{12}}{0.95}\right) + \left(\frac{17.5 \log_e \frac{17}{14}}{0.65}\right) + \left(\frac{1}{3.0}\right)\right]}$$

$$= \frac{1510}{\left[0.2 + \left(\frac{17.5 \log_e 1.16}{0.95}\right) + \left(\frac{17.5 \log_e 1.25}{0.65}\right) + 0.33\right]}$$

(from $\log_e$ table: $\log_e = 0.1484$, $\log_e 1.25 = 0.2231$)

$$Q = \frac{1510}{\left[0.2 + \left(\frac{17.5 \times 0.1484}{0.95}\right) + \left(\frac{17.5 \times 0.2231}{0.65}\right) + 0.33\right]}$$

$$= \frac{1510}{\left[0.2 + \left(\frac{2.6}{0.95}\right) + \left(\frac{3.9}{0.65}\right) + 0.33\right]}$$

$$= \frac{1510}{[0.2 + 2.73 + 6.00 + 0.33]}$$

$$Q = \frac{1510}{9.26} = 164 \text{ Btu/sq ft, hr (outer surface)}$$

Calculating to check assumed mean temperatures:

$$t_1 = t_{a1} - \left(Q \times \frac{1}{f_1}\right) = 1600 - (164 \times 0.2)$$

$$= 1600 - 32.8 = 1567° \text{ F}$$

$$t_2 = t_1 - \left(Q \times \frac{r_3 \log_e \frac{r_2}{r_1}}{k_1}\right)$$

$$= 1567 - (164 \times 2.73)$$

$$= 1567 - 446 = 1121° \text{ F}$$

$$t_s = t_2 - \left(Q \times \frac{r_3 \log_e \frac{r_3}{r_2}}{k_2}\right)$$

$$= 1121 - (164 \times 6.00)$$

$$= 1121 - 980 = 141° \text{ F}$$

$$t_a = t_s - \left(Q \times \frac{1}{f_2}\right) = 141 - (164 \times .33)$$

$$= 141 - 54 = 87° \text{ F (given as } 90° \text{ F)}$$

(difference due to leaving off figures after decimal point)

Checking assumed mean temperatures:

$$\text{First insulation mean temperature} = \frac{t_1 + t_2}{2}$$

$$= \frac{1567 + 1121}{2} = \frac{2688}{2} = 1344° \text{ F}$$

■

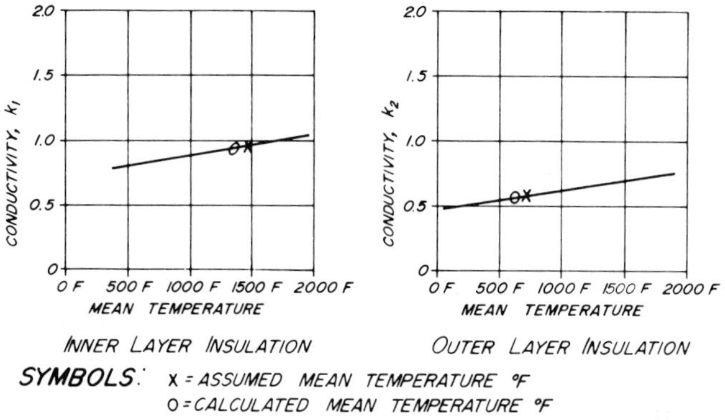

Example 16 Conductivity curves for insulations.

■ EXAMPLE 16a

Metric Units

English Values (as given)	***conversion***	***Metric Values***	
t_{a1} = 1600° F	(1600 − 32) ÷ 1.8	= 871.1° C	= t_{ma1}
t_a = 90° F	(90 − 32) ÷ 1.8	= 32.2° C	= t_{ma}
r_1 = 12″	× 0.0254	= 0.3048m	= r_{m1}
r_2 = 14″	× 0.0254	= 0.3556m	= r_{m2}
r_3 = 17.5″	× 0.0254	= 0.4445m	= r_{m3}
k_1 = 0.95 Btu in./ft^2, hr, °F	× 0.1442	= 0.137 W/mK	= k_{m1}
k_2 = 0.65 Btu in./ft^2, hr, °F	× 0.1442	= 0.0937 W/mK	= k_{m2}
f_1 = 5.0 Btu/ft^2, hr, °F	× 5.6782	= 28.391 W/m^2K	= f_{m1}
f_a = 3.0 Btu/ft^2, hr, °F	× 5.6782	= 17.035 W/m^2K	= f_{m2}

Determine Q_m, $Q_{m\ell}$, t_{m1}, t_{m2}, and t_{ms}

$$Q_m = \frac{t_{ma1} - t_{ma}}{\left[\left(\frac{1}{f_{m1}}\right) + \left(\frac{r_{m3} \log_e \frac{r_{m2}}{r_{m1}}}{k_{m1}}\right) + \left(\frac{r_{m3} \log_e \frac{r_{m3}}{r_{m2}}}{k_{m2}}\right) + \left(\frac{1}{f_{ma}}\right)\right]} \qquad (21a)$$

$$= \frac{871.1 - 32.2}{\left[\left(\frac{1}{28.391}\right) + \left(\frac{0.4445 \log_e \frac{0.3556}{0.3048}}{0.137}\right) + \left(\frac{0.0445 \log_e \frac{0.4445}{0.3556}}{0.094}\right) + \left(\frac{1}{17.035}\right)\right]}$$

$$= \frac{838.9}{\left[0.035 + \left(\frac{0.4445 \times 0.1484}{0.137}\right) + \left(\frac{0.4445 \times 0.2231}{0.094}\right) + 0.059\right]}$$

$$= \frac{838.9}{0.035 + 0.481 + 1.055 + 0.059} = \frac{838.9}{1.63} = 514.7 \text{ W/m}^2$$

Check:

$$Q_m = 514.7 \text{ W/m}^2 \times 0.317199 = 163.3 \text{ Btu/sq, ft, hr} = Q$$

$$t_{m1} = t_{ma1} - \left(Q_m \times \frac{1}{f_{m1}}\right) = 871.1 - (514.7 \times 0.035)$$

$$= 871.1 - 18 = 853.1\,°\text{C}\ (1567\,°\text{F})$$

$$t_{m2} = t_{m1} - \left(Q_m \times \frac{r_{m3} \log_e \frac{r_{ms}}{r_{m1}}}{k_{m1}}\right) = 853.1 - (514.7 \times 0.481)$$

$$= 853.1 - 247.6 = 605.5\,°\text{C}\ (1121\,°\text{F})$$

$$t_{m3} = t_{m2} - \left(Q_m \times \frac{r_{m3} \log_e \frac{r_{m3}}{r_{m2}}}{k_{m2}}\right) = 605.5 - (514.7 \times 1.055)$$

$$= 605.5 - 543 = 62.5\,°\text{C}\ (144\,°\text{F})$$

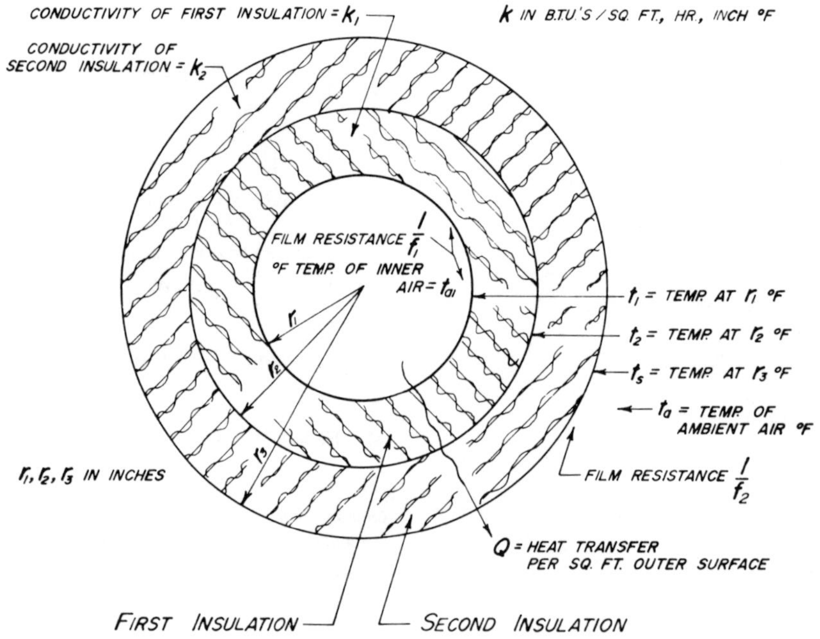

EQUATION 21

$$Q = \frac{t_{a1} - t_a}{\frac{1}{f_1} + \frac{r_3 \log_e \frac{r_2}{r_1}}{k_1} + \frac{r_3 \log_e \frac{r_3}{r_2}}{k_2} + \frac{1}{f_2}} = \text{BTU'S/SQ. FT., HR.}$$

AND

q_{lf} = HEAT TRANSFER PER LINEAR FT.

$$q_{lf} = \left(\frac{t_{a1} - t_a}{\frac{1}{f_1} + \frac{r_3 \log_e \frac{r_2}{r_1}}{k_1} + \frac{r_3 \log_e \frac{r_3}{r_2}}{k_2} + \frac{1}{f_2}}\right)\left(\frac{2 r_3 \pi}{12}\right)$$

Figure 11 Schematic diagram of heat transfer through two cylindrical insulations and two surface resistances.

Check:

$$t_{ma} = t_{m3} - \left(Q_m \times \frac{1}{f_{ma}}\right) = 62.5 - (514 \times 0.059)$$

$$= 62.5 - 32.2° \text{ C } (32.2° \text{ C given as } t_{ma})$$

$$\text{Second insulation mean temperature} = \frac{t_2 + t_s}{2}$$

$$= \frac{1121 + 141}{2} = \frac{1262}{2} = 631° \text{ F}$$

Assumed mean temperatures are sufficiently close that conductivities used in calculation are essentially correct.

Thus the answer for heat transfer in square feet or linear feet is:

Heat transfer Q = 164 Btu/sq ft, hr (outer surface) or Heat transfer $q_{\ell f}$ per linear foot of cylinder

$$q_{\ell f} = Q\left(\frac{2r_3\pi}{12}\right) = 164\left(\frac{2 \times 17.5 \times \pi}{12}\right)$$

$$= 164\,(9.16) = 1502 \text{ Btu/lin. ft, hr}$$

■

Special Considerations

The previous equations are the basic equations for the calculation of heat transfer. Examples of their use have been given. The examples, however, only illustrate one manner in which the equations may be used. In all previous examples the temperatures and resistances were given, then the heat transfer was determined. By rearrangement, if the maximum (or minimum) heat transfer is known, it is possible to determine the insulation thickness required.

Prior to illustrating the above, one item that should be clarified is that in most calculations of heat transfer the resistance through the pipe wall is ignored. To illustrate why this is done, the following example will be calculated with and without consideration of pipe wall in the equation.

■ EXAMPLE 17

English Units

Given: A 4″ NPS pipe operating at 600° F is insulated with 3″ calcium silicate insulation. The ambient temperature is 70° F, still air conditions. The fluid flow through the pipe is such that film conductance from liquid to inner pipe wall is approximately 200. The conductivity of the steel at 600° F is approximately 370 (k_1). Other factors are:

Inside diameter of pipe is 4″, thus $r_1 = 2″$
Outside diameter of pipe is 4 1/2″, thus $r_2 = 2\ 1/4″$
Inside diameter of pipe insulation is 4 1/2″, $r_2 = 2\ 1/4″$
Outside diameter of pipe insulation is 10 3/4″, $r_3 = 5\ 3/8″$

Inside water to steel conductance, $f_1 = 200$
Outer air film conductance, $f_2 = 1.9$

$$\text{Mean temperature of insulation} = \frac{600 + 70}{2} = 335, \text{ from tables on}$$

conductivity of calcium silicate $k_2 = 0.44$

$$Q = \frac{t_{a1} - t_a}{\left[\left(\frac{1}{f_1}\right) + \left(\frac{r_3 \log_e \frac{r_2}{r_1}}{k_1}\right) + \left(\frac{r_3 \log_e \frac{r_3}{r_2}}{k_2}\right) + \left(\frac{1}{f_2}\right)\right]} \tag{21}$$

$$Q = \frac{600 - 70}{\left[\left(\frac{1}{200}\right) + \left(\frac{5\ 3/8 \log_e \frac{2\ 1/4}{2}}{370}\right) + \left(\frac{5\ 3/8 \log_e \frac{5\ 3/8}{2\ 1/4}}{0.44}\right) + \left(\frac{1}{1.9}\right)\right]}$$

$$= \frac{530}{\left[0.005 + \left(\frac{5.375 \log_e 1.12}{370}\right) + \left(\frac{5.375 \log_e 2.38}{0.44}\right) + 0.33\right]}$$

from $\log_e$ tables: $\log_e 1.12 = 0.1133$, $\log_e 2.38 = 0.8671$)

$$Q = \frac{530}{\left[0.005 + \left(\frac{5.375 \times 0.1133}{370}\right) + \left(\frac{5.375 \times 0.8671}{0.44}\right) + 0.53\right]}$$

$$= \frac{530}{\left[0.005 + \left(\frac{0.62}{370}\right) + \left(\frac{4.65}{0.44}\right) + 0.53\right]}$$

$$= \frac{530}{[0.005 + .0016 + 10.55 + 0.53]}$$

$$= \frac{530}{11.0866} = 47.8$$

If the first two terms, the resistance through the fluid to the pipe and the resistance of the pipe wall, were dropped from the equation, then:

$$Q = \frac{530}{\left[\left(\frac{r_3 \log_e \frac{r_3}{r_2}}{k_2}\right) + \left(\frac{1}{f_2}\right)\right]} = \frac{530}{[10.55 + 0.53]} = \frac{530}{11.08}$$

$$= 47.8$$

Dropping these two terms changes the resistance from 11.0866 to 11.08, which changes the answer in the third decimal place and is not worth consideration.

For this reason, in heat transfer calculations covering the losses from insulated pipes and equipment the resistance of the steel is ignored. Of course, on uninsulated pipe and equipment these become significant factors. ■

■ EXAMPLE 17a

Metric Units

English Values (as given)	conversion		Metric Values	
t_{a1} = 600° F	(600 − 32) ÷ 1.8	=	315.6° C	= t_{ma1}
t_a = 70° F	(70 − 32) ÷ 1.8	=	21.1° C	= t_{ma}
r_1 = 2″	× 0.0254	=	0.0508m	= r_{m1}
r_2 = 2.25″	× 0.0254	=	0.0572m	= r_{m2}
r_3 = 5.375″	× 0.0254	=	0.1365m	= r_{m3}
k_1 = 370 Btu in./ft², hr, °F	× 0.1442	=	53.352 W/mK	= k_{m1}
k_2 = 0.44 Btu in./ft², hr, °F	× 0.1442	=	0.0634 W/mK	= k_{m2}
f_1 = 200 Btu/ft², hr, °F	× 5.6782	=	1135.64 W/m²K	= f_{m2}
f_2 = 1.9 Btu/ft², hr, °F	× 5.6782	=	10.789 W/m²K	= f_{m2}

Determine Q_m, the heat transfer if inner metal wall and inner film factor is included in the equation and eliminated from equation Q_m with inner film factor f_{m1} and conductivity of metal k_{m1} included in the equation:

(21a)

$$Q_m = \frac{t_{ma1} - t_{ma}}{\left[\left(\frac{1}{f_{m1}}\right) + \left(\frac{r_{m3}\log_e \frac{r_{m2}}{r_{m1}}}{k_{m1}}\right) + \left(\frac{r_{m3}\log_e \frac{r_{m3}}{r_{m2}}}{k_{m2}}\right) + \left(\frac{1}{f_{m2}}\right)\right]}$$

$$= \frac{315.6 - 21.1}{\left[\left(\frac{1}{1135.64}\right) + \left(\frac{0.1365\log_e \frac{0.0572}{0.0508}}{53.354}\right) + \left(\frac{0.1365\log_e \frac{0.1365}{0.0572}}{0.063}\right) + \left(\frac{1}{10.789}\right)\right]}$$

$$= \frac{294.5}{\left[0.008 + \left(\frac{0.1365 \times 0.1133}{53.354}\right) + \left(\frac{0.1365 \times 0.8671}{0.063}\right) + 0.093\right]}$$

$$= \frac{294.5}{[0.0008 + 0.003 + 1.880 + 0.093]}$$

$$= \frac{294.5}{1.977} + 149 \text{ W/m}^2$$

Q_m with inner film factor and metal of pipe not in equation

$$Q_m = \frac{t_{ma1} - t_{ma}}{\left[\left(\frac{r_{m3}\log_e \frac{r_{m3}}{r_{m2}}}{k_{m2}}\right) + \left(\frac{1}{f_{m2}}\right)\right]}$$

$$= \frac{315.6 - 21.1}{\left[\left(\frac{0.1365 \times 0.8671}{0.0634}\right) + \left(\frac{1}{10.789}\right)\right]}$$

$$= \frac{294.5}{[1.879 + 0.093]} = \frac{294.5}{1.972} = 149.3 \text{ W/m}^2$$

A difference of 0.3 W/m² ■

When Maximum Heat Transfer is Given

Elevated temperature, single flat layer of insulation

In English Units

Equation 6 gives $\ell = k\left(\frac{t_1 - t_2}{Q}\right)$. Therefore, if the thickness of insulation required to provide not over a given heat transfer is wanted, it can be determined.

In Metric Units

$$\ell_m = k_m\left(\frac{t_{m1} - t_{m2}}{Q_m}\right)$$

■ EXAMPLE 18

English Units

Temperature of higher side of insulation, = t_1 = 450
Temperature of lower side of insulation, = t_2 = 20
Maximum heat transfer (Q) shall be less than 75 Btu/sq ft, hr
Conductivity (k) of insulation = 0.42
What shall thickness (ℓ) be?

$$\ell = \left(\frac{450 - 20}{75}\right) 0.42 = \left(\frac{430}{75}\right) 0.42$$

$$= 5.72 \times .42 = 2.41''$$

However, as most insulation does not come in thicknesses such as this, 2.5″ thickness would be required. To determine the heat transfer through the 2.5″ selected:

$$Q = \frac{450 - 20}{\frac{2.5}{0.52}} = \frac{430}{5.95} = 72 \text{ Btu/sq ft, hr}$$

which is less than maximum stated to be allowed. ■

■ **EXAMPLE 18a**

Metric Units

English Values (as given)	*conversion*	*Metric Values*	
t_1 = 450° F	(450 − 32) ÷ 1.8	= 232.2° C	= t_{m1}
t_2 = 20° F	(20 − 32) ÷ 1.8	= −6.6° C	= t_{m2}
k = 0.42 Btu in./ft², hr, °F	× 0.1442	= 0.0606 W/mK	= k_m

Maximum Heat Transfer Q is 75 Btu/ft², hr,
or 75 × 3.152591 = 236.4 W/m² = Q_m

find ℓ_m the necessary thickness in metres

$$\ell_m = k_m \left(\frac{t_{m1} - t_{m2}}{Q_m}\right) = 0.0606 \left(\frac{232.2 - (-6.6)}{236.4}\right)$$

$$= 0.0606 \left(\frac{238.8}{236.4}\right) = 0.0606 \times 1.0102 = 0.0612\text{m}$$

Check:

$$\ell_m = 0.0612\text{m} \times 39.37 = 2.4'' = \ell$$ ■

Elevated temperature, two flat layers of different insulation

English Units

$$\text{Equation 7 gives } Q = \frac{t_1 - t_s}{\left[\left(\frac{\ell_1}{k_1}\right) + \left(\frac{\ell_2}{k_2}\right)\right]}$$

$$\left[\left(\frac{\ell_1}{k_1}\right) + \left(\frac{\ell_2}{k_2}\right)\right] = \frac{t_1 - t_a}{Q}$$

however

$$\left[\left(\frac{\ell_1}{k_1}\right) + \left(\frac{\ell_2}{k_2}\right)\right] = \text{total resistance to heat flow, } R_T$$

or

$$R_T = \frac{t_1 - t_s}{Q}$$

Metric Units

$$R_{mT} = \frac{t_{m1} - t_{ms}}{Q_m}$$

■ **EXAMPLE 19**

English Units

Find the thickness of two different insulations which will provide resistance so that the heat transfer will be less than 100 Btu per sq ft per hour when the temperature on one side is 1400° F (t_1) and the temperature on the other side is 32° F (t_s). Conductivity of one insulation is 0.75 and that of the other is 0.50. Maximum temperature of the insulation with the lower conductivity is 1200 °F (t_2).

$$R_T = \frac{t_1 - t_s}{Q} = \frac{1400 - 32}{100} = \frac{1368}{100} = 13.68$$

As $R_T = \frac{\ell_1}{k_1} + \frac{\ell_2}{k_2}$, to obtain the least possible insulation thickness the one with the higher k should be the one of least thickness, provided that the temperature drop through it is more than 200° F, so that the inner face temperature of second insulation is less than 1200° F.

As it is known that $t_1 - t_2$ must be 200 and Q is > 100 then:

resistance of the insulation of higher conductivity

$$R = \frac{t_1 - t_2}{Q}, \text{ or } \frac{\ell}{k} = \frac{t_1 - t_2}{Q}, \text{ or}$$

$$\ell_1 = \left(\frac{t_1 - t_2}{Q}\right) k = \left(\frac{200}{100}\right) .75 = 2 \times .75 = 1.5$$

$$\text{R of first insulation} = \frac{1.5}{.75} = 2.00$$

R of second insulation = 13.68 − 2.00 = 11.68

and

$$\ell_2 = (11.68)k_2 = 11.68 \times 0.50 = 5.83$$

However, as most insulations come in ½″ thickness increments, 6″ of insulation is required. If 6″ is used, however, then the 1.5″ thickness of the first insulation is insufficient to obtaiin the needed 200° F temperature drop. Thus, the answer must be a first insulation thickness of 2″ and a second thickness of 6″ . Rechecking:

$$Q = \frac{1400 - 32}{\frac{2}{0.75} + \frac{6}{0.5}} = \frac{1368}{2.66 + 12.0} = \frac{1368}{14.66}$$

$$= 93.3 \text{ Btu/ft}^2\text{, hr}$$ ■

■ **EXAMPLE 19a**

Metric Units

English Values (as given)	*conversion*	*Metric Values*	
t_1 = 1400° F	(1400 − 32) ÷ 1.8	= 760° C	= t_{m1}
t_2 = 1200° F	(1200 − 32) ÷ 1.8	= 649° C	= t_{m2}
t_s = 32° F	(32 − 32) ÷ 1.8	= 0° C	= t_{ms}
k_1 = 0.75 Btu in./ft², hr, °F	× 0.1442	= 0.108 W/mK	= k_{m1}
k_2 = 0.5 Btu in./ft², hr, °F	× 0.1442	= 0.072 W/mK	= k_{m2}
Q = 100 Btu/ft², hr	× 3.152591	= 315.26 W/m	= Q_m (max.)

Determine ℓ_{m1}, the thickness of insulation with conductivity of

0.108 W/mK so that temperature t_{m2} is no greater than 649° C.

$$R_{mT} = \frac{\ell_{m1}}{k_{m1}} + \frac{\ell_{m2}}{k_{m2}} \text{ or } R_{mT} = \frac{t_{m1} - t_{ms}}{Q_m} = \frac{760 - 0}{315.26} = 2.41$$

As it is stated that $t_{m1} - t_{m2}$ must be 760 − 649 = 111Δ° C or less and Q_m is stated to be no more than 315.26 W/m²

$$\ell_{m1} = \left(\frac{t_{m1} - t_{m2}}{Q_m}\right) k_{m1} = \left(\frac{760 - 649}{315.26}\right) 0.108$$

$$= \left(\frac{111}{315.26}\right) 0.108 = 0.038\text{m}$$

Check:

$$\ell_m = 0.038\text{m} \times 39.37 = 1.497'' \ \ell$$

Resistance of first insulation $R_{m1} = \frac{0.038}{0.108} = 0.352$ Km²W

Resistance of second insulation $R_{m2} = R_{mt} - R_{m1} = 2.41 - 0.352$ = 2.058 Km²W

$$\ell_{m2} = \left(\frac{t_{m2} - t_{ms}}{Q_m}\right) k_{m2} = \left(\frac{649 - 0}{315.26}\right) .072$$

$$= 2.058 \times 0.072 = 0.148\text{m}$$

Check:

$$\ell_{m2} = 0.148\text{m} \times 39.37 = 5.84'' \ (6'' = 0.152\text{m})$$ ■

Elevated temperature, one flat insulation and one film resistance to find thickness when maximum heat transfer is given

■ EXAMPLE 20

English Units

The temperature of higher side of insulation, t_1 = 325° F
The temperature of ambient air, t_a = 70° F
The conductivity of the insulation, k = 0.45 Btu in./sq ft, hr, °F
The conductance of the air film, f = 2.1
The heat transfer Q should not exceed 40 Btu/sq ft, hr

Find thickness of insulation ℓ in inches:
From Eq. 11:

$$Q = \frac{t_1 - t_a}{\left(\frac{\ell}{k} + \frac{1}{f}\right)}$$

$$\frac{\ell}{k} + \frac{1}{f} = \frac{t_1 - t_a}{Q}$$

however,

$$\frac{\ell}{k} + \frac{1}{f} = \text{total resistance } R_T \text{ to heat flow}$$

This total R_T can be easily determined, in this case

$$R_T = \frac{325 - 70}{40} = \frac{255}{40} = 6.37$$

thus:

$$\frac{\ell}{k} + \frac{1}{f} = 6.37 \text{ and } \frac{1}{f} = \frac{1}{2.1} = 0.47$$

then:

$$\frac{\ell}{k} = 6.37 - 0.47 = 5.9$$

With k given as 0.45, ℓ = k (5.9) = 0.45 × 5.9 = 2.66″, which is a mathematical answer but not a practial one, as most insulations are in ½″ increments. For that reason 3″ of insulation must be used. Using this selected thickness in Eq. 11 to determine the actual heat transfer Q:

$$Q = \frac{325 - 70}{\frac{3}{.45} + \frac{1}{2.1}} = \frac{255}{6.66 + .47} = \frac{255}{7.13}$$

$$= 35.7 \text{ Btu/sq ft, hr}$$

If the temperature of the surface of the insulation facing the air, t_s, is desired, then this can be found as

$$t_s = t_a + \left(Q \times \frac{1}{f}\right) = 70 + (35.7 \times 0.47)$$

$$= 70 + 17 = 87° \text{ F}$$

Using the same equations it is possible to find the thickness of the insulation if its surface temperature is the critical factor. This, of course, is based on the assumption that the *film conductance is known.* ■

■ EXAMPLE 20a

Metric Units

English Values (as given)	***conversion***	***Metric Values***	
t_1 = 325° F	(325 − 32) ÷ 1.8 =	162.7° C	= t_{m1}
t_a = 70° F	(70 − 32) ÷ 1.8 =	21.1° C	= t_{ma}
k = 0.45 Btu in./ft², hr, °F	× 0.1442 =	0.06489 W/mK	= k_m
f = 2.1 Btu/ft², hr, °F	× 5.67826 =	11.9243 W/m²K	= f_m
Q = 40 Btu/ft², hr. (max.)	× 3.152591 =	126.10 W/m²	= Q_m

Determine required thickness of insulation in metres, ℓ_m.

First determine resistance R_{mT} required to maintain not over 126.1 W/m^2 heat transfer Q_m.

$$R_{mT} = \frac{t_{m1} - t_{ma}}{Q_m} = \frac{162.7 - 21.1}{126.1} = \frac{141.6}{126.1} = 1.123$$

$$\frac{\ell_m}{k_m} + \frac{1}{f_m} = R_{mT} \text{ or } \ell_m = \left(R_{mT} - \frac{1}{f_m}\right) k_m$$

$$\ell_m = \left(1.123 - \frac{1}{11.9243}\right) 0.06489$$

$$= (1.123 - 0.084)\ 0.06489$$

$$= 1.039 \times 0.06489 = 0.06742\text{m, or} \times 39.37 = 2.65''$$

However the next available thickness of conventional insulation is 3″ or $3 \times 0.0254 = 0.0762\text{m} = \ell_m$

$$Q_m = \frac{t_{m1} - t_{ma}}{\dfrac{\ell_m}{k_m} + \dfrac{1}{f_m}} = \frac{141.6}{\dfrac{0.0762}{0.06489} + \dfrac{1}{11.9243}}$$

$$= \frac{141.6}{1.174 + 0.084} = \frac{141.6}{1.258} = 112.6\ W/m^2$$

Check:

$$Q_m = 112.6\ W/m^2 \times 0.317199 = 35.7\ \text{Btu/ft}^2\text{, hr.}$$

$$t_{ms} = t_{ma} + \left(Q_m + \frac{1}{f_m}\right) = 21.1 + (112.6 \times 0.084)$$

$$= 21.1 + 9.5 = 30.6°\text{C}\ (87°\text{F})$$

■

■ EXAMPLE 21

English Units

The temperature of the hotter surface, $t_1 = 500°$ F

The temperature of the air, $t_a = 90°$ F

The conductivity of the insulation, k = 0.45 Btu, in./ft², hr, °F

The conductance of the air film, f = 2.2 Btu/ft², hr, °F

The surface temperature of the insulation t_s should not exceed 130°F

To find insulation thickness, ℓ, in inches:

$$t_s = 130°\text{ F max.}$$

$$t_s = t_a + \left(Q \times \frac{1}{f}\right),\ \text{as } t_s = 130 \text{ and } t_a = 90°\text{ F}$$

$$t_s - t_a = Q \times \frac{1}{f},\ \text{or } 130 - 90 = Q \times \frac{1}{2.2}$$

$$Q = \frac{40}{\dfrac{1}{2.2}} = \frac{40}{.455} = 88\ \text{Btu/sq ft, hr}$$

However,

$$Q = \frac{t_1 - t_a}{\left(\dfrac{\ell}{k} + \dfrac{1}{f}\right)} \text{ or } \frac{\ell}{k} + \frac{1}{f} = \frac{t_1 - t_a}{Q}$$

Again,

$$\frac{\ell}{k} + \frac{1}{f} = R_T$$

$$R_T = \frac{500 - 90}{88} = \frac{410}{88} = 4.66$$

$$\frac{\ell}{k} = R_T - \left(\frac{1}{f}\right) = 4.66 - 0.455 = 4.205$$

or $\ell = 4.205 \times k = 4.205 \times .45 = 1.89''$, which is the mathematical answer.

Rechecking, using 2″ as thickness of insulation selected, using Eq. 11:

$$Q = \frac{t_1 - t_a}{\left(\dfrac{\ell}{k} + \dfrac{1}{f}\right)} = \frac{500 - 90}{\dfrac{2.0}{0.45} + \dfrac{1}{2.2}} = \frac{410}{4.43 + 0.455}$$

$$= \frac{410}{4.885} = 84\ \text{Btu/sq ft, hr}$$

$$t_s = t_a + \left(Q \times \frac{1}{f}\right) = 90 + (84 \times 0.455)$$

$$= 90 + 38 = 128°\text{ F}$$

(or slightly below the upper limit of 130° F given) ■

■ EXAMPLE 21a

Metric Units

English Values (as given)			**conversion**		**Metric Values**		
t_1	=	500° F	(500 − 32) ÷ 1.8	=	260° C	=	t_{m1}
t_a	=	90° F	(90 − 32) ÷ 1.8	=	32.2° C	=	t_{ma}
k	=	0.45 Btu in./ft², hr, °F	× 0.1442	=	0.06489 W/mK	=	k_m
f	=	2.2 Btu/ft², hr, °F	× 5.67826	=	12.492 W/m²K	=	f_m
t_s	=	130° F (maximum)	(130 − 32) ÷ 1.8	=	54.4° C	=	t_{ms}

Determine maximum allowable Q_m—based on film factor f_m, then calculate to establish minimum ℓ_m which would control heat transfer to be less than maximum allowable heat transfer.

$$\text{max. } Q_m = \frac{t_s - t_a}{\frac{1}{f_m}} = \frac{54.4 - 32.2}{\frac{1}{12.492}} = \frac{22.2}{0.08} = 277.5 \text{ W/m}^2$$

Check:

$$Q_m = 277.5 \text{ W/m}^2 \times 0.317199 = 88 \text{ Btu/ft}^2\text{, hr.}$$

Resistance R_{mt} must be no less than

$$\frac{t_{m1} - t_{ma}}{Q_m} = \frac{260 - 32.2}{277.5} = 0.8210 \; k_{mt}$$

minimum to control heat transfer.

Resistance through insulation must be not less than:

$$R_m - \frac{1}{f} = 0.821 - 0.08 = 0.741 = R_{m1}$$

$$\text{minimum } \ell_m = R_{mi} \times k_m = 0.741 \times 0.06489 = 0.0481\text{m}$$

Check:

$$\ell_m = 0.0481\text{m} \times 39.37 = 1.89''$$

Using 2″ thickness of insulation which is 2 × 0.0254 = 0.0508m = ℓ_m, then:

$$Q_m = \frac{t_{m1} - t_{m2}}{\frac{\ell_m}{k_m} + \frac{1}{f_m}} = \frac{260 - 32.2}{\frac{0.0508}{0.0649} + \frac{1}{12.492}}$$

$$= \frac{227.8}{0.782 + 0.08} = \frac{227.8}{0.862} = 264 \text{ W/m}^2$$

264 W/m² is less than the maximum 277.5 W/m²

The surface temperature t_{ms} is:

$$t_{ms} = t_{ma} + \left(Q_m \times \frac{1}{f}\right) = 32.2 + (264 \times 0.08)$$

$$= 32.2 + 21.1 = 53.3° \text{ C}$$

which is less than the 54.4° C given as maximum. ■

Elevated temperature, one cylindrical insulation.

To find thickness:

■ EXAMPLE 22

English Units

The temperature of the higher side, t_1 = 450° F
The temperature of lower (outer) side, t_2 = 100° F
The conductivity of the insulation = 0.40 Btu in./ft² hr, °F
The inner radius, r_1 = 1 1/2″
The heat transfer, Q, shall not exceed 70 Btu/sq ft, hr (outer surface)

From Eq. 8:

$$Q = \frac{t_1 - t_2}{\frac{r_2 \log_e \frac{r_2}{r_1}}{k}}$$

$$r_2 \log_e \frac{r_2}{r_1} = \left(\frac{t_1 - t_2}{Q}\right) k$$

$$r_2 \log_e \frac{r_2}{r_1} = \left(\frac{450 - 100}{70}\right) 0.40$$

$$= \left(\frac{350}{70}\right) 0.40 = 2.0$$

As r_2 is the unknown and appears twice in the remaining factor, it is difficult to compute directly. However, by looking in Table B-6a "Values of $r_2 \log_e \frac{r_2}{r_1}$," it can be determined that the value listed for 3″ diameter with 1-1/2″ of insulation is 2.08. As this is slightly higher than the required 2.00, the answer is 1-1/2″ thick insulation and r_2 = 3″. ■

■ EXAMPLE 22a

Metric Units

English Values (as given)	***conversion***	***Metric Values***	
t_1 = 450° F	(450 − 32) ÷ 1.8	= 232.2° C	= t_{m1}
t_2 = 100° F	(100 − 32) ÷ 1.8	= 37.7° C	= t_{m2}
r_1 = 1.5″	× 0.0254	= 0.0381m	= r_{m2}
k = 0.40 Btu in./ft², hr, °F	× 0.1442	= 0.0577	= k_m
Q = 70 Btu/ft², hr, °F (maximum)	× 3.152591	= 220.7 W/m²K	= Q_m

Determine ℓ_m the thickness, and r_{m2} outer diameter of insulation.

$$r_{m2} \log_e \frac{r_{m2}}{r_{m1}}\text{, however, } \log_e \frac{r_2}{r_1} \text{ is identical to } \log_e \frac{r_{m2}}{r_{m1}}$$

$$\text{Thus } 37.39 \times r_{m2}\left(\log_e \frac{r_{m2}}{r_{m1}}\right) \text{ in metres} = r_2 \log_e \frac{r_2}{r_1} \text{ in inches}$$

Using 2″ thickness of insulation which is 2 × 0.0254 = 0.0508m = ℓ_m, then:

$$Q_m = \frac{t_{m1} - t_{m2}}{\left(\frac{\ell_m}{k_m} + \frac{1}{f_m}\right)} = \frac{260 - 32.2}{\left(\frac{0.0508}{0.0649} + \frac{1}{12.492}\right)}$$

$$= \frac{227.8}{[0.782 + 0.08]} = \frac{227.8}{0.862} = 264 \text{ W/m}^2$$

264 W/m² is less than the maximum 277.5 W/m²

The surface temperature t_{ms} is:

$$t_{ms} = t_{ma} + \left(Q_m \times \frac{1}{f}\right) = 32.2 + (264 \times 0.08)$$

$$= 32.2 + 21.1 = 53.3° \text{ C}$$

which is less than the 54.4° C given as maximum.

$$Q_m = \frac{t_{m1} - t_{m2}}{\dfrac{r_{m2} \log_e \dfrac{r_{m2}}{r_{m1}}}{k_m}} = 220.7 \text{ W/m}^2 \text{ given as maximum}$$

also

$$r_{m2} \log_e \frac{r_{m2}}{r_{m1}} = \left(\frac{t_{m1} - t_{m2}}{Q_m}\right) k_m = \left(\frac{232.2 - 37.7}{220.7}\right) 0.0577$$

$$= \left(\frac{194.5}{220.7}\right) \times 0.0577 = 0.881 \times 0.0577$$

$$= 0.051 \text{ equivalent metres}$$

Table B-6 is equivalent for values of cylindrical inches. Thus to obtain equivalent inches, multiply 0.051 × 39.37 = 2.01 equivalent inches.

From Table B-6 pick the closest value to r = 3.0″ (or r_{m2} = 3 × 0.0254 = 0.0762 metres) then ℓ_1 (in inches) = 1.5. Convert to metric 1.5 × 0.254 = 0.038 metres = ℓ_m ■

Elevated temperature, one cylindrical insulation and one surface film to find thickness when maximum heat transfer is given

■ **EXAMPLE 23**

English Units

The temperature of the higher side, t_1 = 650° F
The temperature of the ambient air, t_a = 30° F
The conductivity of the insulation, k = 0.49 Btu, in./ft², hr, °F
The conductance of film, f = 2.5 Btu/sq ft, hr (outer surface)
The insulation fits a 2″ NPS pipe (r_1 = 1.8″)
The heat transfer shall not exceed 80 Btu/sq ft, hr

From Eq. 17:

$$Q = \frac{t_1 - t_a}{\dfrac{r_2 \log_e \dfrac{r_2}{r_1}}{k} + \dfrac{1}{f}}$$

$$\frac{r_2 \log_e \dfrac{r_2}{r_1}}{k} + \frac{1}{f} = \frac{t_1 - t_a}{Q}$$

$$\frac{r_2 \log_e \dfrac{r_2}{r_1}}{k} = \frac{t_1 - t_a}{Q} - \frac{1}{f}$$

$$r_2 \log_e \frac{r_2}{r_1} = k\left(\frac{t_1 - t_a}{Q}\right) - k\left(\frac{1}{f}\right)$$

or

$$r_2 \log_e \frac{r_2}{r_1} = .49\left(\frac{650 - 30}{80}\right) - .49\left(\frac{1}{2.5}\right)$$

$$= .49\,(7.75) - .49\,(0.4)$$

$$= 3.80 - .20 = 3.6$$

From Table B-6a for $r_2 \log_e \frac{r_2}{r_1}$, values for 2″ NPS pipe, 2″ thickness equals 3.16 and 2-1/2″ thickness equals 4.17. Thus, the thickness must be nominal 2-1/2″ and r_2 = 3.81.

$$Q = \frac{650 - 30}{\left[\left(\dfrac{r_2 \log_e \dfrac{r_2}{r_1}}{0.49}\right) + \left(\dfrac{1}{2.5}\right)\right]} = \frac{620}{\left[\left(\dfrac{4.17}{0.49}\right) + .4\right]} = \frac{620}{[8.5 + .4]}$$

$$= \frac{620}{8.9} = 70 \text{ Btu/sq ft, hr}$$

which is less than the 80 Btu/sq ft, hr as required.

If the surface temperature, t_s, of the insulation is desired:

$$t_s = t_a + \left(Q \times \frac{1}{f}\right) = 30 + (70 \times .4)$$

$$= 30 + 28 = 58° \text{ F}$$

When more than one cylindrical material is involved, the problem is solved to determine total insulation resistance required. Then by trial and error the combination of insulations which, with their respective conductances, thicknesses, and radii, will provide the needed total resistance. ■

■ **EXAMPLE 23a**

Metric Units

English Values (as given)	conversion	Metric Values	
t_1 = 650° F	(650 − 32) ÷ 1.8	= 343.3° C	= t_{m1}
t_a = 30° F	(30 − 32) ÷ 1.8	= −1.1° C	= t_{ma}
r_1 = 1.18″ (radius of 2″ pipe	× 0.0254	= 0.030m	= r_{m1}
k = 0.49 Btu in./ft², hr, °F	× 0.1442	= 0.071 W/mK	= k_m
f = 2.5 Btu/ft², hr, °F	× 5.67826	= 14.196 W/m²K	= f_m
Q = 80 Btu/ft², hr. (maximum)	× 3.152591	= 252.2 W/m²	= Q_m

Determine thickness of insulation ℓ_m required so that Q_{m1} heat transfer per square metre of outer surface is less than 252.2

W/m^2. The inside radius of the insulation is approx. 1.24″ or 0.031m = r_{mn}.

The value of $r_{m2} \log_e \frac{r_{m2}}{r_{mn}}$ must be determined to establish r_{m2} so as to calculate ℓ_m which is $r_{m2} - r_{mn}$.

$$r_{m2} \log_e \frac{r_{m2}}{r_{m1}} = k_m \left(\frac{t_{m1} - t_{ma}}{Q_m}\right) - k_m \frac{1}{f_m}$$

$$= 0.071 \left(\frac{343.3 - (-1.1)}{252.2}\right) - 0.071 \left(\frac{1}{14.196}\right)$$

$$= 0.071\,(1.3656) - 0.071\,(0.07) = 0.0969 - 0.005$$

$$= 0.092\text{m}$$

Convert to inches so as to use Table B-6

$$0.092 \times 39.37 = 3.62''$$

From Table B-6 for 2″ NPS pipe the next higher value above 3.62″ is 4.17″ equivalent thickness and this requires 2-1/2″ nominal thickness insulation.

From Table B-18 the outer diameter of insulation 2-1/2″ thick for 2″ NPS pipe is 7.62″. Radius would be 3.81″. In metres this is 0.0254 × 3.81 = 0.096m, ℓ_1 = 3.81 − 1.24 = 2.57 or 0.096m − 0.031m = 0.065m = ℓ_m.

The heat transfer Q_m with this thickness of insulation is:

$$Q_m = \frac{t_{m1} - t_{ma}}{\left[\left(\dfrac{r_{m2} \log_e \frac{r_{m2}}{r_{mn}}}{k_m}\right) + \left(\dfrac{1}{f_m}\right)\right]}$$

$$= \frac{343.3 - (-1.1)}{\left[\left(\dfrac{0.096 \log_e \frac{0.096}{0.031}}{0.071}\right) + \left(\dfrac{1}{f_m}\right)\right]}$$

$$= \frac{344.4}{\left[\left(\dfrac{0.096 \times \log_e 3.09}{0.071}\right) + \left(\dfrac{1}{14.196}\right)\right]}$$

$$= \frac{344.4}{\left[\left(\dfrac{0.096 \times 1.1282}{0.071}\right) + 0.07\right]}$$

$$= \frac{344.4}{[1.525 + 0.07]} = \frac{344.4}{1.595} = 215.9 \text{ W/m}^2$$

Check:

$$Q_m = 215.9 \text{ W/m}^2 \times 0.317199 = 68.5 \text{ Btu/ft}^2\text{, hr.}$$

If surface temperature, t_{ms}, surface of insulation is desired, then:

$$t_{ms} = t_{m1} + \left(Q_m \times \frac{1}{f_m}\right) = 1.1 + (215.9 \times 0.07)$$

$$= -1.1 + 15.1$$

$$= 14.0°\text{C}\ (57.2°\text{F})$$

■

Theory of Heat Flow, from or to a Surface

In the previous discussion of heat transfer the equations were all based upon the conductance of heat through a material. Even the air films on one or both sides were treated as a mass which had a given conductance to heat. Unfortunately, heat transfer from a surface to air and surrounding bodies is not as simple.

Heat is transferred from (or to) a surface to (or from) surrounding air and bodies by radiation, convection, and conduction. In most instances the amount of heat transferred by conduction through air is negligible. This is due to the fact that air, without movement, has a conductivity of approximately 0.16 Btu/sq ft, hr, in., °F. Thus, the resistance to heat flow from one surface to another just three feet away would be 36 (inches) divided by 0.16, or 225. This illustrates that absolutely still air would be a good insulator if it were not for radiation and for the fact that air never remains still when temperature differences occur. For most cases, then, the conductivity of air may be eliminated when calculating heat transfer from surfaces, as still air never occurs in practice.

Radiation

For the bases of heat transfer from surfaces we must return to the work of Langmuir and Stefan-Boltzman. According to the Stefan-Boltzman law on heat transfer by radiation:

English Units

When Q_r = heat transfer by radiation in Btu/ft², hr, °F
ε = surface emittance
T_s = absolute temperature of hot surface °R (t_s + 459.6)
T_a = absolute temperature of ambient air and bodies °R (t_a + 459.6)
T° R = t° F + 459.6)

$$Q_r = 0.174\,\varepsilon \left[\left(\frac{T_s}{100}\right)^4 - \left(\frac{T_a}{100}\right)^4\right] \quad (22)$$

(Stefan-Boltzman Law)

Surface emittance is the ability of an opaque material to emit radiant energy as a result of its temperature. It is measured by the ratio of the rate of radiant emission of the material to the corresponding emission of a thermally black body at the same temperature (1.0).

In Metric Units

When Q_{mr} = heat transfer by radiation in W/m²
ε = surface emittance
T_{ms} = absolute temperature of hot surface in °K
T_{mn} = absolute temperature of air and bodies in °K

$$Q_{mr} = 0.548\,\varepsilon \left[\left(\frac{T_{ms}}{55.55}\right)^4 - \left(\frac{T_{ma}}{55.55}\right)^4\right] \quad (22a)$$

■ **EXAMPLE 24**

The temperature of the surface, $t_s = 150°$ F
The temperature of the ambient air, $t_a = 70°$ F
The surface emittance is 0.85

What is the heat transfer by radiation, Q_r?

$$T_s = t_s + 459.6 = 150 + 459.6 = 609.6° \text{ R}$$

$$T_a = t_a + 459.6 = 70 + 459.6 = 529.6° \text{ R}$$

$$Q_r = 0.174 \times 0.085 \left[\left(\frac{609.6}{100}\right)^4 - \left(\frac{529.6}{100}\right)^4\right]$$

using (22)

$$Q_r = 0.174 \times 0.85\,[6.09^4 - 5.29^4]$$

$$= 0.174 \times 0.85\,[1375.5 - 783.1]$$

$$= 0.174 \times 0.85\,[592.4] = 0.174 \times 505$$

$$= 88 \text{ Btu/sq ft, hr}$$

(Use tables on fourth power of numbers.) ■

A simpler method of solution is provided in this manual by use of radiation charts B-1 through B-10, Appendix B.

From table:

heat transfer by radiation of t_s at 150° F = 242
heat transfer by radiation of t_a at 70° F = 138
Subtract 104
Multiply by ε of .85
104 × .85 = 88 Btu/sq ft, hr

It must be pointed out that an ambient condition where air and all surrounding objects are a single given temperature just does not exist in nature. For example, the calculations presented do not take into account solar radiation. In the previous example if the surface under consideration were located so that it faced the sun directly, it would not be emitting heat from its surface but would be receiving heat, even in 70° ambient air. At night the opposite is true. On clear nights a horizontal surface pointing upward will radiate much greater quantities of heat than calculations based upon ambient air temperatures would indicate. (This is the reason we must scrape frost off the modern car windshields, which face almost upward, at air temperatures up to 40° F, but we have little trouble with side windows which are in the vertical position.)

■ **EXAMPLE 24a**

Metric Units

		English Values (as given)		**conversion**		**Metric Values**		
T_s	=	609.6° R	÷	1.8	=	338.7K	=	T_{ms}
T_a	=	529.6° R	÷	1.8	=	294.2K	=	T_{ma}
ε	=	0.85	×	1	=	0.85	=	ε*

Determine Q_{mr} the heat transfer by radiation

$$Q_{mr} = 0.548 \times 0.85 \left[\left(\frac{338.7}{55.5}\right)^4 - \left(\frac{294.2}{55.5}\right)^4\right]$$

$$= 0.4658\left[(6.10)^4 - (5.3)^4\right]$$

$$= 0.4658\,[1385 - 789] = .4658 \times 596 = 278 \text{ W/m}^2$$

Check:

$$Q_{mr} = 278 \text{ W/m}^2 \times 0.317199 = 88.1 \text{ Btu/ft}^2\text{, hr.}$$

*ε is a ratio, thus has same value in English and Metric Units. ■

Convection

As stated in the definition, convection is a movement of a mass of gas or liquid due to temperature difference, and physical contact of the gas or liquid is the actual method of heat transfer. Thus, the speed at which this gas or liquid passes over a surface will change the rate of heat transfer. As convection is not a basic physical law but a combination of physical phenomena, it cannot be arranged into a nice clean-cut mathematical formula. For this reason the equations used by some will differ from those used by others. Also, the equations must be applied within the particular limits and under the conditions set forth by the investigators.

Because of these facts, statements must be made as to what the conditions are that apply to each equation.

The formula for heat transfer from a surface, at relatively moderate temperatures, to air moved by "natural convection" was developed by Langmuir as being

$$Q_c = 0.296\,(t_s - t_a)^{5/4}$$

(Langmuir's Equation) (23)

When Q_c = Heat transferred by natural convection in Btu/sq ft, hr
t_s = Temperature of surface, °F
t_a = temperature of ambient air, °F

Note: Natural convection is the movement of air caused by variation in its density as its temperature changes. As air is heated its density becomes less and it tends to rise. As the heated air rises it is replaced by cooler air, which in its turn is heated, rises, and is replaced in a continuous, recurring process.

When air has movement (caused by an external force) other than the natural movement caused by any change of its temperature, this movement will increase the rate of heat transfer. The effect of this movement on heat transfer was determined by Langmuir to be:

$$\sqrt{\frac{V + 68.9}{68.9}}$$ where V is the velocity of the air, ft/min

Thus, the equation for heat transfer by convection by air at a forced velocity is:

In English Units

$$Q_{cv} = 0.296\,(t_s - t_a)^{5/4} \sqrt{\frac{V + 68.9}{68.9}} \qquad (24)$$

(Langmuir)

Where Q_{cv} = heat transfer by forced convection in Btu/sq ft, hr
V = air velocity, feet/min

Note: There are two different elements of time in the single equation. Although confusing, the equation is simpler to solve as shown than by having air velocity expressed in units per hour and having much larger compensating constants in the square root factor.

In Metric Units

Langmuir's equation is:

$$Q_{mc} = 1.957\,(T_{ms} - T_{ma})^{5/4}$$

(Langmuir) (23a)

When Q_{mc} = Heat transferred by natural convection, in W/m²
T_{ms} = Temperature of surface, °C (or °K)*
T_{ma} = Temperature of ambient air, °C (or °K)*

**Note:* T_{ms} and T_{ma} must be on same temperature scale as C or K

In Metric Units

Q_{mcv} = heat transfer by forced convection in W/m²
V_m = air velocity, metre per second, m/s

$$Q_{mcv} = 1.957\,(T_{ms} - T_{ma})^{5/4} \sqrt{\frac{196.85\,V_m + 68.9}{68.9}} \qquad (24a)$$

Additional note: Heilman contributed later work on the effect of location and size on conduction. This will be discussed later in the manual.

■ EXAMPLE 25

English Units

The temperature of the surface, t_s = 150° F
The temperature of the ambient air, t_a = 70° F
Still air (natural convection)

What is the heat transfer by convection, Q_c?

$$Q_c = 0.296\,(150 - 70)^{5/4} = 0.296\,(80)^{5/4}$$

(from table of 5/4 powers, $80^{5/4} = 239.26$)

$$Q_c = 0.296\,(239.26) = 71.0 \text{ Btu/sq ft, hr}$$ ■

■ EXAMPLE 25a

Metric Units

English Values (as given)			conversion		Metric Values		
t_s	=	150° F	×	(150 − 32) ÷ 1.8	=	65.5° C	= T_{ms}
t_a	=	70° F	×	(70 − 32) ÷ 1.8	=	21.1° C	= T_{ma}

$$Q_{mc} = 1.957\,(T_{ms} - T_{ma})^{5/4} = 1.957\,(65.5 - 21.1)^{5/4}$$

$$= 1.957\,(44.4)^{5/4} = 1.957 \times 114 = 223 \text{ W/m}^2$$

Check:

$$Q_{mc} = 223 \text{ W/m}^2 \times 0.317199 = 70.8 \text{ Btu/ft}^2\text{, hr.}$$ ■

A surface transmits heat to (or from) the ambient air, as previously stated, by both radiation and convection. Thus, the total heat transfer $Q_t = Q_r + Q_c$ or $Q_r + Q_{cv}$, depending upon conditions. Substituting the equations for Q_r, Q_c, and Q_{cv}, then:

In English Units

Under still air (natural convection) conditions, total heat transfer from surface to ambient air by radiation and convection is

$$Q_t = 0.174\varepsilon \left[\left(\frac{T_s}{100}\right)^4 - \left(\frac{T_a}{100}\right)^4\right] + 0.296\,(t_s - t_a)^{5/4} \qquad (25)$$

Total heat transfer from surface to ambient air by radiation and convection, with forced air movement over surface is

$$Q_t = 0.174\,\varepsilon \left[\left(\frac{T_s}{100}\right)^4 - \left(\frac{T_a}{100}\right)^4\right] + 0.296\,(t_s - t_a)^{5/4} \sqrt{\frac{V + 68.9}{68.9}} \qquad (26)$$

In Metric Units, the total heat transfer from surface to ambient air by radiation and convection is:

$$Q_{mc} = 0.548 \left[\left(\frac{T_{ms}}{55.55}\right)^4 - \left(\frac{T_{ma}}{55.55}\right)^4\right] + 1.957\,(T_{ms} - T_{ma})^{5/4} \qquad (25a)$$

In Metric Units the total heat transfer from surface to ambient air by radiation, and convection, with forced air movement is:

$$Q_{mt} = 0.548 \left[\left(\frac{T_{ma}}{55.55}\right)^4 - \left(\frac{T_{mn}}{55.55}\right)^4\right] + 1.957\,(T_{ms} - T_{ma})^{5/4} \sqrt{\frac{196.85\,V_m + 68.9}{68.9}} \qquad (26a)$$

Note: T_{ms} and T_{ma} are in °K

■ EXAMPLE 26

English Units

With all other factors the same as above, except the air conditions, the air is moving at a rate of 5 mph due to wind:

$$Q_{cv} = 0.296\,(150 - 70)^{5/4} \sqrt{\frac{V + 68.9}{68.9}}$$

$$\left(\text{V is in feet per minute; a wind at 5 mph}\right.$$

$$\left. = \frac{5280 \times 5}{60} = 440 \text{ ft/min}\right)$$

$$= 71.0 \sqrt{\frac{440 + 68.9}{68.9}} = 71.0 \sqrt{\frac{508.9}{68.9}}$$

$$= 71.0 \sqrt{7.4} = 71.0\,(2.72)$$

$$= 193 \text{ Btu/sq ft, hr}$$ ■

■ EXAMPLE 26a

Metric Units

English Values (as given)		conversion		Metric Values		
V =	440 Ft/min.	×	0.00508	=	2.2352 m/sec.	= V_m

$$Q_{mc} = 1.957\,(T_{ms} - T_{ma})^{5/4} \sqrt{\frac{196.85\,V_m + 68.9}{68.9}}$$

$$= 1.957\,(114) \sqrt{\frac{(196.85 \times 2.2352) + 68.9}{68.9}}$$

$$= 223 \sqrt{\frac{440 + 68.9}{68.9}}$$

$$= 223 \sqrt{7.386} = 223 \times 2.71 = 606 \text{ W/m}^2$$

Check:

$$Q_m = 606 \text{ W/m}^2 \times 0.317199 = 192.2 \text{ Btu/ft}^2\text{, hr.}$$ ■

■ EXAMPLE 27

English Units

The temperature of the surface, $t_s = 212°$ F
The temperature of the air, $t_a = 50°$ F
The emittance of the surface, $\varepsilon = 0.9$

When there is no forced air movement (natural convection), what is the total heat transfer, Q_t?

$$Q_t = 0.174 \times 0.9 \left[\left(\frac{212 + 459.6}{100}\right)^4 - \left(\frac{50 + 459.6}{100}\right)^4\right] + 0.296\,(212 - 50)^{5/4} \tag{25}$$

$$= 0.174 \times 0.9 \left[\left(\frac{671.6}{100}\right)^4 - \left(\frac{509.6}{100}\right)^4\right] + 0.296\,(162)^{5/4}$$

$$= 0.174 \times 0.9 \left[(6.7)^4 - (5.1)^4\right] - 0.296\,(577.95)$$

$$= 0.174 \times 0.9 \left[2015 - 676\right] + 171$$

$$= 0.174 \times 0.9 \left[1339\right] + 171$$

$$= 0.174 \times 1206 + 171 = 210 + 171$$

$$= 381 \text{ Btu/sq ft, hr.}$$ ■

■ EXAMPLE 27a

Metric Units

English Values (as given)	conversion	Metric Values
$t_s = 212°$ F	T_s = 212 + 459.6 = 671.6° R ÷ 1.8 =	373.1° K = T_{ms}
$t_a = 50°$ F	T_a = 50 + 459.6 = 509.6° R ÷ 1.8 =	283.1° K = t_{ma}
$\varepsilon = 0.9$	× l	= 0.9

$$Q_{mt} = 0.548 \times 0.9 \left[\left(\frac{373.1}{55.55}\right)^4 - \left(\frac{283.1}{55.55}\right)^4\right] + 1.957\,(373.1 - 283.1)^{5/4}$$

$$= 0.4932 \left[2035 - 674\right] + 1.957\,(90)^{5/4}$$

$$= 0.4932 \left[1361\right] + 1.957\,(277.2)$$

$$= 671 + 542 = 1213 \text{ W/m}^2$$

Check:

$$Q_{mt} = 1213 \text{ W/m} \times 0.317199 = 385 \text{ Btu/sq, ft, hr.}$$ ■

■ EXAMPLE 28

English Units

With all factors remaining the same, except the air. If the air had a 10 mph wind velocity, what would the total heat transfer be?

$$Q_t = 0.174 \times 0.9 \left[\left(\frac{212 + 459.6}{100}\right)^4 - \left(\frac{50 + 459.6}{100}\right)^4\right] + 0.296\,(212 - 50)^{5/4} \sqrt{\frac{V + 68.9}{68.9}}$$

$$V = \frac{5280 \times 10}{60} = 880 \text{ ft/min}$$

$$Q_t = 210 + 171 \sqrt{\frac{880 + 68.9}{68.9}}$$

$$= 210 + 171 \sqrt{\frac{949.9}{68.9}} = 210 + 171 \sqrt{13.8}$$

$$= 210 + 171 \times 3.71 = 210 + 634$$

$$= 844 \text{ Btu/sq ft, hr}$$

■

■ EXAMPLE 28a

Metric Units

English Values (as given)	***conversion***	***Metric Values***
V = 880 ft/m	× 0.0508	= 4.47 m/sec.

$$Q_{mt} = 0.548 \times 0.9 \left[\left(\frac{373.1}{55.55}\right)^4 - \left(\frac{283.1}{55.55}\right)^4\right]$$

$$+ 1.957\,(373.1 - 283.1)^{5/4} \sqrt{\frac{196.35\,(4.47) + 68.9}{68.9}}$$

$$Q_{mt} = 0.4932\left[2035 - 674\right] + 1.957\,(90)^{5/4} \sqrt{\frac{948.9}{68.9}}$$

$$= 667 + 542 \times \sqrt{13.8}$$

$$= 671 + (542 \times 3.71) = 671 + 2011 = 2682 \text{ W/m}^2$$

Check:

$$Q_{mt} = 2682 \text{ W/m}^2 \times 0.317199 = 851 \text{ Btu/sq, ft, hr.}$$

■

The previous problems were selected to illustrate the rapid rate of heat transfer on a surface from boiling point temperature to an ordinary ambient air temperature; the problems also show the amount of heat transfer which would occur with just a moderate wind velocity. This illustrates the need for thermal insulation for piping operating at very moderately elevated temperatures.

Solution of Heat Transfer Insulation Problems Which Have an Unknown Film Conductance

Returning to Eq. 11 "Heat Transfer Through One Insulation and One Surface Film":

$$Q = \frac{t_1 - t_a}{\left[\left(\frac{\ell}{k}\right) + \left(\frac{1}{f}\right)\right]}$$

In most instances the operating temperature, t_1, the air temperature, t_a, the thickness of the insulation, ℓ, and its conductivity, k, are known. Past practice has been to assume a value for $\frac{1}{f}$, generally about 0.5. As can be seen by examination of Eqs. 26 and 27, the emittance ε and velocity V completely influence the amount of heat a surface at a given temperature will transmit; and the amount of heat transmitted will influence the surface temperature, t_s, when it is dependent upon the heat conducted through the insulation. Even knowing the emittance of the surface and the velocity of the air, V, it is still impossible to set up an equation which may be solved directly, as the surface temperature, t_s, is unknown. However, *under static conditions the heat transfer through the insulation to the surface must equal the heat transfer from that surface.* Thus Q, or in this case, Q_t:

$$Q_t = \frac{t_1 - t_a}{\left[\left(\frac{\ell}{k}\right) + \left(\frac{1}{f}\right)\right]}$$

but if the surface temperature is t_s, then

$$Q_t = \frac{t_1 - t_s}{\left(\frac{\ell}{k}\right)} = \frac{t_s - t_a}{\left(\frac{1}{f}\right)}$$

However, the value of Q_t was established by Eqs. 25 and 26. Thus:

Heat transfer through one flat insulation and one air film, under still conditions, is

In English Units

$$\frac{t_1 - t_s}{\frac{\ell}{k}} = 0.174\,\varepsilon \left[\left(\frac{t_s + 459.6}{100}\right)^4 - \left(\frac{t_a + 459.6}{100}\right)^4\right] + 0.296\,(t_s - t_a)^{5/4} \tag{27}$$

In Metric Units

$$\frac{T_{m1} - T_{ms}}{\frac{\ell_m}{k_m}} = 0.548\,\varepsilon \left[\left(\frac{T_{ms}}{55.55}\right)^4 - \left(\frac{T_{ma}}{55.55}\right)^4\right] + 1.957\,(t_{ms} - t_{ma})^{5/4} \tag{27a}$$

Note: T_{m1} and T_{ms} in °K

and with Forced Convection, V

In English Units

$$\frac{t_1 - t_s}{\frac{\ell}{k}} = 0.174\,\varepsilon \left[\left(\frac{t_s + 459.6}{100}\right)^4 - \left(\frac{t_a + 459.6}{100}\right)^4\right] + 0.296\,(t_s - t_a)^{5/4} \times \sqrt{\frac{V + 68.9}{68.9}} \tag{28}$$

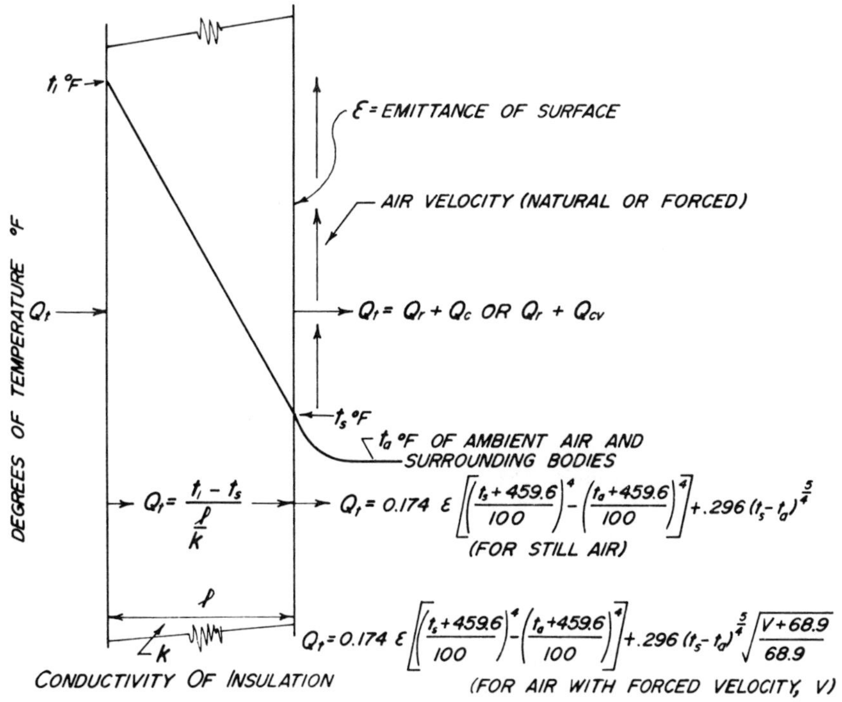

EQUATION 27——FOR NATURAL CONVECTION

$$\frac{t_1 - t_s}{\frac{\ell}{k}} = 0.174\,\varepsilon\left[\left(\frac{t_s+459.6}{100}\right)^4-\left(\frac{t_a+459.6}{100}\right)^4\right] + 0.296\,(t_s - t_a)^{\frac{5}{4}}$$

EQUATION 28——FOR FORCED CONVECTION

$$\frac{t_1 - t_s}{\frac{\ell}{k}} = 0.174\,\varepsilon\left[\left(\frac{t_s+459.6}{100}\right)^4-\left(\frac{t_a+459.6}{100}\right)^4\right] + 0.296\,(t_s - t_a)^{\frac{5}{4}}\sqrt{\frac{V+68.9}{68.9}}$$

Figure 12 Schematic diagram of heat transfer through one flat insulation and one surface air film. Equations in English Units. For Equations in Metric Units, see Equations 27a and 28a.

In Metric Units

$$\frac{T_{m1} - T_{ms}}{\frac{\ell_m}{k_m}} = 0.548\,\varepsilon\left[\left(\frac{T_{ms}}{55.55}\right)^4 - \left(\frac{T_{ma}}{55.55}\right)^4\right] + 1.957\,(t_{ms} - t_{ma})^{5/4}\sqrt{\frac{196.85\,V_m + 68.9}{68.9}} \qquad (28a)$$

Note: T_{m1} and T_{ms} in °K. This is illustrated in Figure 12.

Heat transfer through one cylindrical insulation and one air film under still air conditions, is

In English Units

$$\frac{t_1 - t_s}{\left(\frac{r_2 \log_e \frac{r_2}{r_1}}{k}\right)} = 0.174\,\varepsilon\left[\left(\frac{t_s + 459.6}{100}\right)^4 - \left(\frac{t_a + 459.6}{100}\right)^4\right] + 0.296\,(t_s - t_a)^{5/4} \qquad (29)$$

In Metric Units

$$\frac{T_{m1} - T_{ms}}{\left(\frac{r_{m2} \log_e \frac{r_{m2}}{r_{m1}}}{k_m}\right)} = 0.548\,\varepsilon\left[\left(\frac{T_{ms}}{55.55}\right)^4 - \left(\frac{T_{ma}}{55.55}\right)^4\right] + 1.957\,(t_{ms} - t_{ma})^{5/4} \qquad (29a)$$

Note: T_{m1} and T_{ms} in °K

and with Forced Convection, V

In English Units

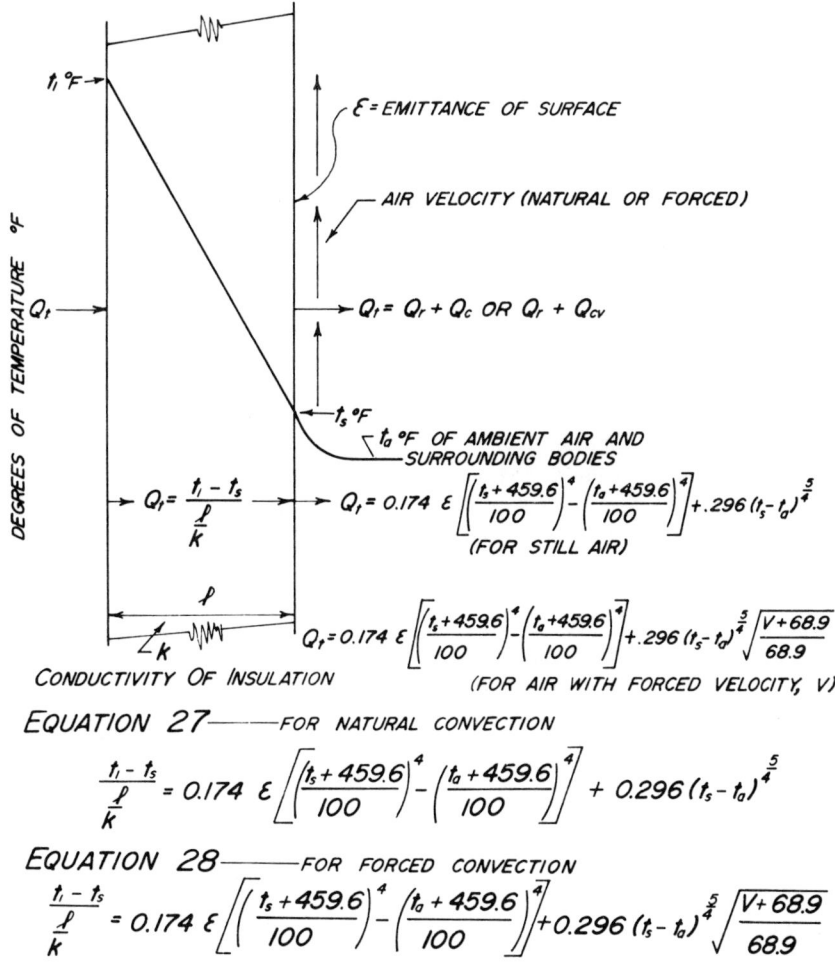

$$\frac{t_1 - t_s}{\left(\frac{r_2 \log_e \frac{r_2}{r_1}}{k}\right)} = 0.174\,\varepsilon\left[\left(\frac{t_s + 459.6}{100}\right)^4 - \left(\frac{t_a + 459.6}{100}\right)^4\right] + 0.296\,(t_s - t_a)^{5/4}\sqrt{\frac{V + 68.9}{68.9}} \qquad (30)$$

In Metric Units

$$\frac{T_{m1} - T_{ms}}{\left(\dfrac{r_{m2} \log_e \dfrac{r_{m2}}{r_{m1}}}{k_m}\right)} = 0.548\,\varepsilon \left[\left(\frac{T_{m1}}{55.55}\right)^4 - \left(\frac{T_{ms}}{55.55}\right)^4\right]$$

$$+ 1.957\,(t_{m1} - t_{ms})^{5/4} \sqrt{\frac{196.85\,V_m + 68.9}{68.9}} \qquad (30a)$$

Note: T_{m1} and T_{ms} in °K

Although presented as heat flow from a surface, the four previous equations are also true for heat transfer to a surface. In this case the result would have a minus sign, indicating flow of heat from the ambient air.

As developed, the equations cannot be solved directly, nor do they lend themselves to simultaneous solution, due to the fourth power and the 5/4 power factors. The solution developed by Turner is a graphic one and is shown in Figure 13; it can be programmed and solved relatively simply on a digital computer by using t_s as a variable until both sides of the equation balance. However, if no computer is available, the graphic method presented below can be used to obtain the answer.

■ EXAMPLE 29

English Units

The operating temperature of a 4″ pipe = 700° F
The ambient temperature = 90° F, still air conditions
The conductivity of the insulation = 0.5 in Btu, in./ft² hr, °F
The thickness of the insulation = 3″
The weather barrier is a mastic with $\varepsilon = 0.85$

or

t_1 = 700, $t_a = 90$, $r_1 = 2.25$ (radius of 4″ NPS pipe)
r_2 = 5.375 (radius of 4″ NPS insulation nominally 3″ thick)
k = 0.5, $\varepsilon = 0.85$
Assume various values of t_s

Find the heat transfer and the insulation surface temperature

Solving the left side of equation:

If $t_s = 120$,

$$Q_t = \frac{t_1 - t_s}{\left(\dfrac{r_2 \log_e \dfrac{r_2}{r_1}}{k}\right)} = \frac{700 - 120}{\left(\dfrac{5.375 \log_e \dfrac{5.375}{2.25}}{k}\right)}$$

$$= \frac{580}{\left(\dfrac{5.375 \log_e 2.38}{0.5}\right)} = \frac{580}{\left(\dfrac{5.375 \times 0.8671}{0.5}\right)}$$

$$= \frac{580}{\left(\dfrac{4.65}{0.5}\right)} = \frac{580}{9.3} = 62 \text{ Btu/ft}^2\text{, hr.}$$

If

$$t_s = 140, \quad Q_t = \frac{560}{9.3} = 60$$

$$t_s = 160, \quad Q_t = \frac{540}{9.3} = 58$$

$$t_s = 180, \quad Q_t = \frac{520}{9.3} = 56$$

Plot on graph as shown in Figure 14

Solving the right side of equation:

If $t_s = 120$,

$$Q_t = 0.174\,\varepsilon \left[\left(\frac{t_s + 459.6}{100}\right)^4 - \left(\frac{t_a + 459.6}{100}\right)^4\right]$$

$$+ 0.296\,(t_s - t_a)^{5/4}$$

$$Q_t = 0.174 \times 0.85 \left[\left(\frac{120 + 459.6}{100}\right)^4\right.$$

$$\left. - \left(\frac{90 + 459.6}{100}\right)^4\right] + 0.296\,(120 - 90)^{5/4}$$

$$Q_t = 0.148\left[(5.79)^4 - (5.49)^4\right] + 0.296\,(30)^{5/4}$$

$$Q_t = 0.148\left[1123 - 908\right] + 0.296\,(70.2)$$

$$Q_t = 0.148\left[215\right] + 21 = 32 + 21$$

$$= 53 \text{ Btu/sq ft, hr}$$

If $t_s = 140$,

$$Q_t = 0.174\,\varepsilon \left[\left(\frac{140 + 459.6}{100}\right)^4 - \left(\frac{90 + 459.6}{100}\right)^4\right]$$

$$+ 0.296\,(140 - 90)^{5/4}$$

$$= 0.174\,\varepsilon\left[(5.99)^4 - 908\right] + 0.296\,(50)^{5/4}$$

$$= 0.174\,\varepsilon\left[1287 - 908\right] + 0.296\,(133)$$

$$= 0.148\left[379\right] + 0.296\,(133) = 56 + 39$$

$$= 95 \text{ Btu}$$

Plot in Figure 14. Where the two curves cross shows that the surface temperature, t_s, is approximately 125 and that Q_t = approximately 62 Btu/sq ft, hr.

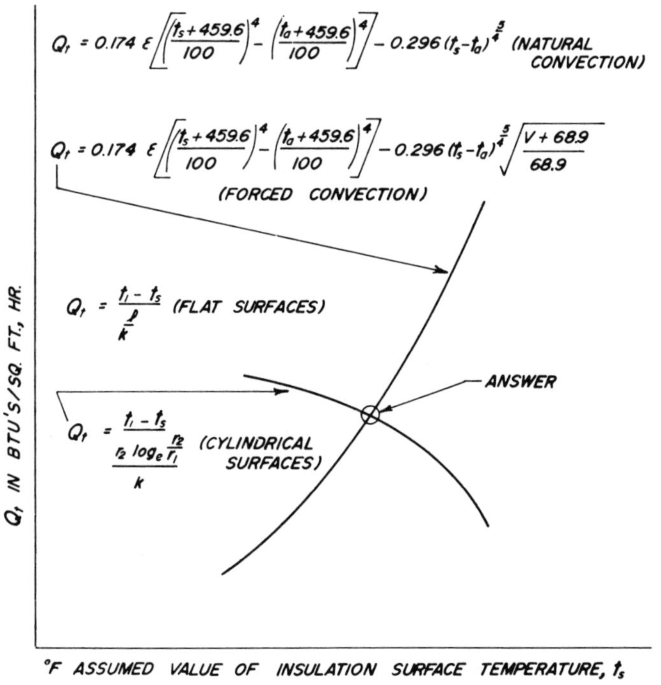

Q1 = 0.174 ε [((ts+459.6)/100)^4 − ((ta+459.6)/100)^4] − 0.296 (ts−ta)^5/4 (NATURAL CONVECTION)
Q1 = 0.174 ε [((ts+459.6)/100)^4 − ((ta+459.6)/100)^4] − 0.296 (ts−ta)^5/4 √((V+68.9)/68.9)
(FORCED CONVECTION)
Q1 = (t1 − ts)/(l/k) (FLAT SURFACES)
ANSWER
Q1 = (t1 − ts)/((r2 loge r2/r1)/k) (CYLINDRICAL SURFACES)
Q1 IN BTU'S/SQ. FT., HR.
°F ASSUMED VALUE OF INSULATION SURFACE TEMPERATURE, ts

Figure 13 Graphic solution for determination of heat transfer and surface temperature of insulation. Above equations are in English Units. Metric equations provide identical curves.*

**Note:* For Equations in Metric see Equations 25a, 25b, 5a, and 7a.

■

■ EXAMPLE 29a

Metric Units

English Values (as given)	***conversion***	***Metric Values***	
t_1 = 700° F	(700 − 32) ÷ 1.8	= 371.1° F	= t_{m1}*
t_a = 90° F	(90 − 32) ÷ 1.8	= 32.2° C	= t_{ma}**
r_1 = 2.25″	× 0.0254	= 0.0571 m	= r_{m1}
r_2 = 5.375″	× 0.0254	= 0.1365 m	= r_{m2}
k = 0.5 Btu, in./ft², hr, °F	× 0.1442	= 0.072 W/mK	= k_m
ε = 0.85	× 1	= 0.85	= ε

*t_{m1} + 273.2 = 644.3°K T_{m1}

**t_{ma} + 273.2 = 305.4°K T_{ma}

Find Q_m and t_{ms}

Solution is by solving for various values of t_{ms}

$$\frac{T_{m1} - T_{ms}}{\left(\dfrac{r_{m2} \log_e \dfrac{r_{m2}}{r_{m1}}}{k_m}\right)} = Q_{mt} = 0.548\,\varepsilon \left[\left(\frac{T_{ms}}{55.55}\right)^4 - \left(\frac{T_{ma}}{55.55}\right)^4\right] + 1.957\,(t_{ms} - t_{ma})^{5/4}$$

Solving for Q_m, using left hand side of equation = Q_{mt}, shown on Figure 14a

Assuming T_{ms} = 322.2° K (49° C)

$$Q_{mt} = \frac{644.3 - 322.2}{\left(\dfrac{0.1365 \log_e \dfrac{0.1365}{0.057}}{0.072}\right)} = \frac{322.1}{\left(\dfrac{0.1365 \log_e 2.39}{0.072}\right)}$$

$$= \frac{322.1}{\left(\dfrac{0.1365 \times 0.8713}{0.072}\right)} = \frac{322.1}{1.651} = 195 \text{ W/m}^2$$

Assuming T_{ms} = 333.2° K (60° C)

$$Q_{mt} = \frac{644.3 - 333.2}{1.651} = \frac{311.1}{1.651} = 188.4 \text{ W/m}^2$$

Assuming T_{ms} = 344.2° K (71° C)

$$Q_{mt} = \frac{644.3 - 344.2}{1.651} = \frac{300.1}{1.651} = 181.7 \text{ W/m}^2$$

Assuming T_{ms} = 355.2° K (82° C)

$$Q_{mt} = \frac{644.3 - 355.2}{1.651} = \frac{289.1}{1.651} = 175 \text{ W/m}^2$$

These are plotted on graph as shown in Figure 14a

Solving for Q_m using left hand side of equation = Q_{mt}

Assuming T_{ms} = 322.2° K (49° C), T_{ma} = 305.4° K (32.2° C)

$$Q_{mt} = 0.548 \times 0.85 \left[\left(\frac{322.2}{55.55}\right)^4 - \left(\frac{305.4}{55.55}\right)^4\right] + 1.957\,(322.2 - 305.4)^{5/4}$$

$$Q_{mt} = 0.4658\left[(5.8)^4 - (5.5)^4\right] + 1.957\,(16.7)^{5/4}$$

$$= 0.4658\left[1131 - 915\right] + 1.957\,(33.8)$$

$$= 0.4658\left[216\right] + 66.1$$

$$= 100.6 + 66.1 = 166.7 \text{ W/m}^2$$

Assume T_{ms} = 333.2° K (60° C), T_{ma} still remains 305° K (32.2° C)

$$Q_m = 0.4658 \left[\left(\frac{333.2}{55.55}\right)^4 - \left(\frac{305.4}{55.55}\right)^4\right] + 1.957\,(333.2 - 305.4)^{5/4}$$

$$= 0.4658\left[12.94 - 913.6\right] + 1.957\,(27.7)^{5/4}$$

$$= 0.4658\left[382.4\right] + 124.5 = 178.1 + 124.5$$

$$= 302.6 \text{ W/m}^2$$

Assume T_{ms} = 344.2° K (71° C) T_{ma} still remains 305° K (32.2° C)

$$Q_m = 0.4658\left[\left(\frac{344.2}{55.55}\right)^4 - \left(\frac{305.4}{55.55}\right)^4\right]$$

$$+ 1.957\,(344.2 - 305.4)^{5/4}$$

$$= 0.4658\left[1474 - 913.6\right] + 1.957\,(38.8)^{5/4}$$

$$= 0.4658\left[560.4\right] + 1.957\,(96.5) = 261.0 + 188.9$$

$$= 449.9\ \text{W/m}^2$$

Plot in Figure 14a where the two curves are approximately 195 W/m² and T_{ms} = approximately 326° K (53° C) ■

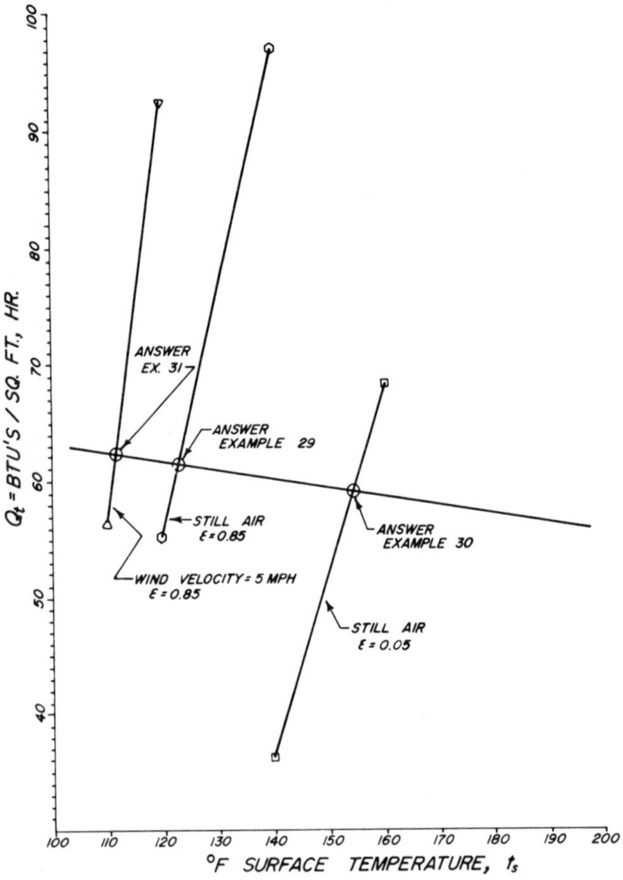

Figure 14 Solution of Examples 29, 30, and 31.

■ EXAMPLE 30

English Units

To show the effect of surface emittance on heat transfer, Q_t, and surface temperature, t_s, with all factors remaining as given in Example 29, change surface emittance ε to 0.05, as would be the case if aluminum jacket had been used.

The solution for the left side of equation remains the same, thus the curve plotted on Figure 13 is identical.

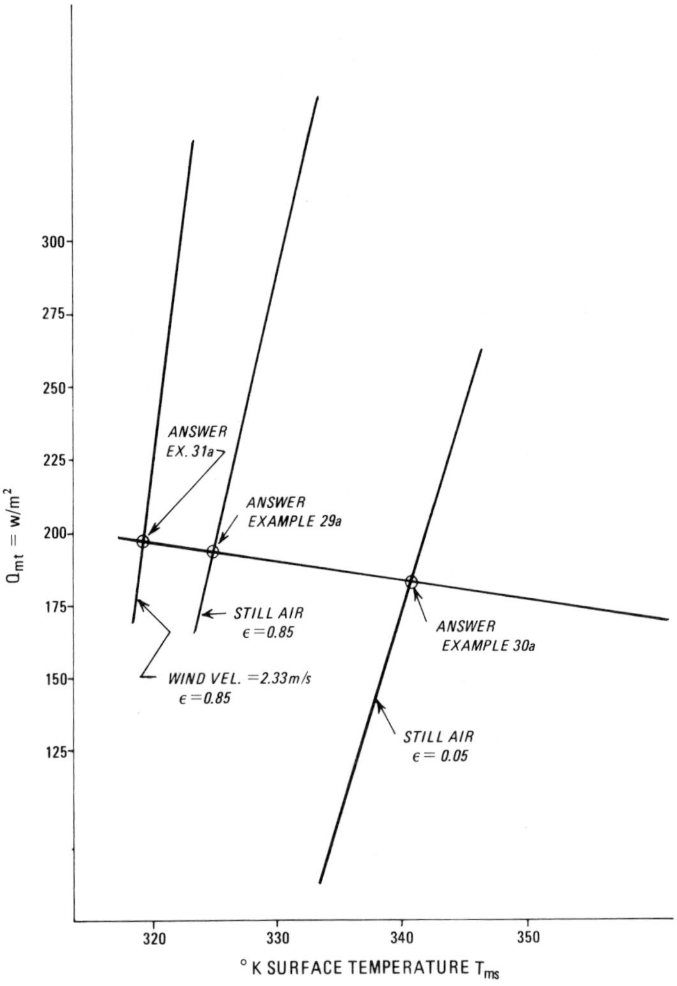

Figure 14a Solution of Examples 29a, 30a, and 31a.

Solving the right side of equation, the only change will be ε. Thus, if t_s is assumed to be 120° F, then:

$$Q_t = 0.174\,\varepsilon\,\left[215\right] + 21$$

$$= 0.174 \times 0.05\left[215\right] + 21$$

$$= 0.0087\left[215\right] + 21 = 2 + 21$$

$$= 23\ \text{Btu/sq ft, hr}$$

(too small heat transfer to plot in graph)

If t_s = 140, then:

$$Q_t = 0.0087\left[379\right] + 30 = 3 + 30$$

$$= 33\ \text{Btu/sq ft, hr}$$

Plotted on Figure 14

As can be seen by comparison, surface temperature must be higher than 140 since 33 Btu/sq ft, hr is too small to balance the left side of equation. Thus, if t_s = 160, then:

$$Q_t = 0.0087\left[\left(\frac{160 + 459.6}{100}\right)^4 - 908\right]$$

$$+ 0.296\,(160 - 90)^{5/4}$$

$$= 0.0087\left[(6.19)^4 - 908\right] + 0.296\,(70)^{5/4}$$

$$= 0.0087\left[1468 - 908\right] + 0.296\,(202)$$

$$= 0.0087\left[560\right] + 0.296\,(202)$$

$$= 5 + 60 = 65 \text{ Btu/sq ft, hr}$$

Plotted on Figure 14. Where the two curves cross shows that surface temperature $t_s = 156°$ F and that heat transfer $Q_t = 58$ Btu/sq ft, hr. ■

■ EXAMPLE 30a

Metric Units

English Values (as given)	***conversion***	***Metric Values***
ε = 0.5 (aluminum outer surface)	×1	ε = 0.05

(All other values same as in Example 29a)

The solution to determine Q_{mt} on left side of equation is identical to Example 30a. Curve on Figure 14a for calculations of equation on left is identical for Example 30a.

Assuming T_{ms} = 322.2° K (49° C)

$$Q_{mt} = 0.548 \times 0.05\left[\left(\frac{322.2}{55.55}\right)^4 - \left(\frac{305.4}{55.55}\right)^4\right]$$

$$+ 1.957\,(322.2 - 305.4)^{5/4}$$

$$= 0.0274\left[(5.8)^4 - (5.5)^4\right] + 1.957\,(16.8)^{5/4}$$

$$= 0.02074\left[1131 - 913.5\right] + 1.957\,(34.0)$$

$$= 0.0274\left[217.5\right] + 66.1 = 6.0 + 66.5 = 72.5 \text{ W/m}^2$$

Assuming T_{ms} = 333.2° K (60° C)

$$Q_{mt} = 0.0274\left[\left(\frac{333.2}{55.55}\right)^4 - \left(\frac{305.4}{55.55}\right)^4\right]$$

$$+ 1.957\,(333.2 - 305.4)^{5/4}$$

$$= 0.0274\left[1294 + 913.5\right] + 1.957\,(27.8)^{5/4}$$

$$= 0.0274\left[308.5\right] + 124.9 = 10.4 + 124.9$$

$$= 135.8 \text{ W/m}^2$$

Assuming T_{ms} = 344.2° K (71°)

$$Q_{mt} = 0.0274\left[\left(\frac{344.2}{55.55}\right)^4 - \left(\frac{305.4}{55.55}\right)^4\right]$$

$$+ 1957\,(344.2 - 305.4)^{5/4}$$

$$= 0.0274\left[1474 - 913.5\right] + 1.957\,(38.8)^{5/4}$$

$$= 0.0274\left[560.5\right] + 1.957\,(96.8) = 15.4 + 189.4$$

$$= 204.8 \text{ W/m}^2$$

The curve of this Q_{mt} based on assumed surface temperatures T_{ms} is plotted on Figure 14a. Where it crosses curve plotted by left hand side equation is the answer Q_{mt} = 182 W/m² and T_{ms} = 341° K (68° C) or 154° F.

Check:

$$Q_m = 182 \text{ W/m}^2 \times 0.317199 = 58 \text{ Btu/ft}^2\text{, hr.}$$ ■

■ EXAMPLE 31

English Units

To illustrate the effect of air velocity on heat transfer.

Returning to Example 29 all factors remain the same except that instead of still air conditions, in this case there is a 5 mph wind, or

$$V = \frac{5280 \times 5}{60} = 440 \text{ ft/min}$$

Again the solution on the left side of the equation is identical.

Solving the right side of the equation:

If t_s = 110° F

$$Q_t = 0.174 \times 0.85\left[\left(\frac{110 + 459.6}{100}\right)^4 - \left(\frac{90 + 459.6}{100}\right)^4\right]$$

$$+ 0.296\,(110 - 90)^{5/4} \times \sqrt{\frac{V + 68.9}{68.9}}$$

$$= 0.148\left[(5.69)^4 - 908\right]$$

$$+ 0.296\,(20)^{5/4}\sqrt{\frac{440 + 68.9}{68.9}}$$

$$= 0.148\left[1048 - 908\right] + 0.296\,(42)$$

$$\times \sqrt{\frac{508.9}{68.9}}$$

$$= 0.148\left[140\right] + (12)\sqrt{7.4}$$

$$= 21 + (12)\,2.73 = 21 + 33$$

$$= 54 \text{ Btu/sq ft, hr}$$

If t_s = 120,

$$Q_t = 0.148\left[215\right] + (21)\,2.73$$

$$= 32 + 57 = 89 \text{ Btu/sq ft, hr}$$

Plotted on Figure 14, (Δ), the point where the two curves cross shows that the surface temperature t_s = approximately 112° F and that the heat transfer is 63 Btu/sq ft, hr. ■

■ EXAMPLE 31a

Metric Units

English Values (as given)	**conversion**	**Metric Values**
V (air velocity) = 440 ft/mi	×0.00508	= 2.2352 m/s

All other factors same as given in Example 29a.

Other than still air, wind is blowing over surface at the rate of 2.2351 m/s.

Calculate right hand side of equation, with various values of assumed surface temperature to determine the heat transfer Q_{mt} and surface temperature T_{ms} as influenced by wind.

Assume T_{ms} = 322° K (49° C), T_{ma} still remains 305.4° K (32.2° C)

$$Q_{mt} = 0.548 \times 0.85 \left[\left(\frac{322.4}{55.55}\right)^4 - \left(\frac{305.4}{55.55}\right)^4\right]$$

$$+ 1.957\,(322.2 - 305.4)^{5/4} \sqrt{\frac{(196.85 \times 2.2352) + 68.9}{68.9}}$$

$$= 0.4658\left[1131 - 913.5\right] + 1.956\,(16.8)^{5/4} \sqrt{7.38}$$

$$= 0.4658\left[216\right] + 66.5 \times 2.717 = 10.3 + 180.7$$

$$= 282.0 \text{ W/m}^2$$

Assume T_{ms} 333.2° K (60° C), T_{ma} still remains 305° K (32.2° C)

$$Q_{mt} = 0.548 \times 0.85 \left[\left(\frac{333.2}{55.55}\right)^4 - \left(\frac{305.4}{55.55}\right)^4\right]$$

$$+ 1.957\,(333.2 - 305.4)^{5/4} \sqrt{\frac{(196.85 \times 2.2352) + 68.9}{68.9}}$$

$$= 0.4658\left[1294 - 913.5\right] + (124.8 \times 2.717)$$

$$= 177.2 + 339.1 = 516.3 \text{ W/m}^2$$

as this value of T_{ms} was too high a value of T_{ms} lower must be used for plotting on chart.

Assume T_{ms} = 318.2° K (45° C), T_{ma} still remains 305.4° K (32.2° C).

$$Q_{mt} = 0.548 \times 0.85 \left[\left(\frac{318.2}{55.55}\right)^4 - \left(\frac{305.4}{55.55}\right)^4\right]$$

$$+ 1.957\,(318.2 - 305.5)^{5/4} \sqrt{\frac{(196.85 \times 2.2352) + 68.9}{68.9}}$$

$$= 0.4658\left[1076 - 913.5\right] + (48.4 \times 2.717)$$

$$= 75.6 + 131.5 = 207.2 \text{ W/m}^2$$

These calculated Q_{mt} based on assumed values of T_{ms} are plotted on Figure 14a, where this curve crosses curve based on left hand equation is the answer. Q_{mt} = 197 W/m² and T_{ms} is approximately 320° K (47° C) (116° F).

Check:

$$Q_{mt} = 207.2 \text{ W/m} \times 0.317199 = 65.7 \text{ Btu/sq, ft, hr.}$$ ■

These examples not only illustrate the method of solution. They also show that outer surface emittance on well-insulated surfaces has only a relatively small effect on total heat transfer but an extreme effect on surface temperature. They show that unless all factors are known, measurement of surface temperature is no indicator of heat transfer. To further illustrate this fact, if this same surface was wet with 50° F rain, the surface resistance would be practically zero and the temperature of the surface, t_s = 50° F, then:

$$Q_t = \frac{t_1 - t_s}{\left(\dfrac{r_2 \log_e \dfrac{r_2}{r_1}}{k}\right)} = \frac{700 - 50}{\left(\dfrac{5.375 \log_e \dfrac{5.375}{2.25}}{0.5}\right)}$$

$$= \frac{650}{9.3} = 71 \text{ Btu}$$

So, as conditions were changed in this example, the lowest surface temperature provided the conditions of highest heat loss. This again illustrates that the measurement of surface temperature, by itself, is no measure of heat transfer.

These same equations can be used to determine insulation thickness to maintain not over a certain surface temperature. In this case the surface temperature, t_s, is known and the equations can be solved without using a graphical solution.

When Insulation Surface Temperature is Given, to Find Insulation Thickness

■ EXAMPLE 32

English Units

This problem has the same factors as Example 29, except that ℓ, the thickness of insulation, is unknown, and the surface temperature to be maintained is given. The factors are:

4″ NPS pipe operating at 700° F
ambient air = 90° F, still air
conductivity of insulation = 0.5
emittance, ε, of surface = 0.85
surface temperature not to exceed 125° F

Thus:

$$Q_t = 0.174 \times 0.85 \left[\left(\frac{125 + 459.6}{100}\right)^4 - \left(\frac{90 + 459.6}{100}\right)^4\right]$$

$$+ 0.296\,(125 - 90)^{5/4}$$

$$Q_t = 0.174 \times 0.85 \left[(5.84)^4 - (5.49)^4\right] + 0.296\,(35)^{5/4}$$

$$Q_t = 0.174 \times 0.85 \left[1163 - 908\right] + 0.296\,(85.1)$$

$$Q_t = 0.174 \times 0.85 \left[255\right] + 25.3$$

$$Q_t = 0.174 \times 217 + 25.3$$

$$Q_t = 37.8 + 25.3$$

$$= 63.1 \text{ Btu/sq ft, hr max Btu}$$

if surface temperature is to be maintained as given.

Using the left side of equation we know that $Q_t > 63.1$ or

$$Q_t = \frac{t_1 - t_s}{\dfrac{r_2 \log_e \dfrac{r_2}{r_1}}{k}}$$

or

$$\frac{r_2 \log_e \dfrac{r_2}{r_1}}{k} = \frac{t_1 - t_s}{Q_t}$$

or

$$r_2 \log_e \frac{r_2}{r_1} = k\left(\frac{t_1 - t_s}{Q_t}\right)$$

Knowing Q_t, t_1, t_s, and k:

$$r_2 \log_e \frac{r_2}{r_1} = 0.5\left(\frac{700 - 125}{63.1}\right) = 0.5\left(\frac{575}{63.1}\right)$$

$$= 0.5 \times 9.1 = 4.55$$

From the table of values of $r_2 \log_e \frac{r_2}{r_1}$ NPS pipe, the value of 3″ thickness is 4.64, which is all right. Thus, 3″ thickness of insulation is required. ■

■ EXAMPLE 32a

Metric Units

English Values (as given)	**conversion**	**Metric Values**	
$t_1 = 700°$ F	$(700 - 32) \div 1.8 = 371.1°$ C t_{m1}, 644°K		$= T_{m1}$
$t_a = 90°$ F	$(90 - 32) \div 1.8 = 32.2°$ C t_{ma}, 305.4°K		$= T_{ma}$
k = 0.5 Btu in./ft², hr °F	0.1442	= 0.0721 W/mK	$= K_m$
4″ NPS pipe r_1 = 2.3 (insul. I.D.)	0.0254	= 0.0584 m	$= r_{mi}$
$\varepsilon = 0.85$	× 1	= 0.85	$= \varepsilon$
	Still air conditions		
t_s maximum given as 125° F	$(125 - 32) \div 1.8 = 51.6°$ C t_{ms}, 325° K T_{ms}		

Find ℓ_{m2} and r_{m2}

$$Q_{mt} = 0.548\,\varepsilon\left[\left(\frac{T_{ms}}{55.55}\right)^4 - \left(\frac{T_{ma}}{55.55}\right)^4\right]$$

$$+ 1.957\,(T_{ms} - T_{ma})^{5/4}$$

$$= 0.548 \times 0.85\left[\left(\frac{325}{55.55}\right)^4 - \left(\frac{305.4}{55.55}\right)^4\right]$$

$$+ 1.957\,(325 - 305.4)^{5/4}$$

$$= 0.4658\left[1171 - 913.5\right] + 1.957\,(42.3)$$

$$= 119.7 + 81.2 = 200.9 \text{ W/m}^2$$

Having calculated maximum Q_{mt} and knowing t_{mi}, t_{ms} and k_m it is then possible to determine minimum required equivalent thickness.

$$r_{m2} \log_e \frac{r_{m2}}{r_{m1}} = k_m\left(\frac{T_{mi} - T_{ms}}{Q_{mt}}\right) = 0.0721\left(\frac{644 - 325}{200.5}\right)$$

$$= 0.0721\,(1.591) = 0.1147$$

Convert to inches to be able to use Table B-6.

0.115 × 39.37 = 4.50 minimum equivalent thickness required. For 4″ NPS 3″ thickness gives an equivalent thickness of 4.46″, thus 3″ thickness is required. 4″ NPS insulation 3″ thickness has outside diameter of 10.75″.

In Metric Units

= 3 × 0.0254 = 0.0762 metres
= 5.375 × 0.0254 = 0.1365 metres

■

■ EXAMPLE 33

English Units

Assume that all factors in Example 32 remain identical, except that a low emittance jacket, such as aluminum with emittance of 0.08, is used on the outer surface. What thickness of insulation would then be required to maintain not over 125° F surface temperature? Q_t would be:

$$Q_t = 0.174 \times 0.08\left[255\right] + 25.3$$

$$= 0.174 \times 20.4 + 25.3 = 3.5 + 25.3$$

$$= 28.8 \text{ Btu/sq ft, hr}$$

and

$$r_2 \log_e \frac{r_2}{r_1} = k\left(\frac{575}{28.8}\right) = 0.5 \times (20.0) = 10.0$$

From the table of values of $r_2 \log_e \frac{r_2}{r_1}$ for NPS pipe, the value of 5″ thickness is 8.98, which is too thin. Thus, the answer is that *5½″ of thickness, which has a value of 10.09, is required.* ■

EXAMPLE 33a

Metric Units

English Units (as given)	**conversion**	**Metric Units**
Same as Example 32a except = 0.08		Same as Example 32a except = 0.08

$$Q_{mt} = 0.548\,\varepsilon \left[\left(\frac{T_{ms}}{55.55}\right)^4 - \left(\frac{T_{ma}}{55.55}\right)^4\right]$$

$$+ 1.957\,(T_{ms} - T_{ma})^{5/4}$$

$$= 0.548 \times 0.08 \left[\left(\frac{325}{55.55}\right)^4 - \left(\frac{305.4}{55.55}\right)^4\right]$$

$$+ 1.957\,(325 - 305.4)^{5/4}$$

$$= 0.04384\,[258] + 80.6 = 11.3 + 80.6$$

$$= 91.9 \text{ W/m}^2$$

Check:

$$Q_{mt} = 91.9 \text{ W/m}^2 \times 0.317199 = 29.2 \text{ Btu/ft}^2\text{, hr, °F}$$

and

$$r_m \log_e \frac{r_{m2}}{r_{m1}} = k_m \left(\frac{T_{mi} - T_{ms}}{Q_{mt}}\right)$$

$$= 0.0721 \left(\frac{319}{91.9}\right) = 0.250\text{m} \times 39.37 = 9.85''$$

From Table B-6 5½″ thickness = required to provide equivalent thickness of 9.61″.

5½″ ×0.0254 = 0.1397m of insulation thickness required. ■

The solutions presented with the previous equations are, except in rare instances, sufficiently close for solving problems related to pipes and equipment operating at hot temperatures. They ignore, however, the effect of shape, and of direction of heat flow. Where surface temperatures are of paramount importance and temperature differences are small, such as the calculation for dew point surface temperatures for low temperature applications, then more accuracy is required.

Additional Equations for Free Convection from a Surface

The Langmuir equation for convection reduced a complex problem to one condition in steady state. Rice, however, carried the development further by assuming that the film thickness will depend upon temperature difference, the coefficient of density change per degree of temperature change, the acceleration of gravity, and the density and viscosity. From this work, and additional work by Heilman, the following equations was developed:

English Units

$$Q_{cv} = C \times \left(\frac{1}{d}\right)^{0.2} \times \left(\frac{1}{T_{ave}}\right)^{0.181} \quad (31)$$

$$\times \Delta t^{1.266} \text{ Btu/sq ft, hr}$$

where: C = Constant values for various conditions given in Table below

d = Cylinder diameter in inches, with d to be taken as 24 for flat and diameters greater than 24″

T_{ave} = Average of cylinder and ambient temperature, °R (which equals degrees Fahrenheit absolute)

Δt = Temperature difference F between surface and air.

In Metric Units

$$Q_{mcv} = 3.152591 \left[C \left(\frac{39.37}{d_m}\right)^{0.2} \times \left(\frac{0.55}{T_{mave}}\right)^{0.181} \times (1.8\,\Delta t_m)\right]^{1.266} \text{ W/m}^2 \quad (31a)$$

Where: Q_{mcv} = heat transfer by convection in W/m²

d_m = cylinder diameter in metres, with d_m to be taken as 0.6 for flat or diameters over 0.6m

T_{mave} = average of cylinder and ambient temperature, °K

ΔT_m = temperature difference °K (or °C) between surface temperature and air temperature

C = constant—values for various conditions given in the Table below.

Also developed by Heilman is a table of values for C for various shapes and direction of heat flow.* This table follows.

Free convection, values for constant, C, in Eq. 31

Shape and Condition	Value of C
Horizontal cylinders	1.016
Long vertical cylinders	1.235
Vertical plates	1.394
Horizontal plates, warmer than air, facing upward	1.79
Horizontal plates, warmer than air, facing downward	0.89
Horizontal plates, cooler than air, facing upward	0.89
Horizontal plates, cooler than air, facing downward	1.79

Equation 31 and the Correction Factors above provide the means for calculating the thickness of insulation required to prevent sweating on the surface of low-temperature installations.

*Presented in "Transactions, Soc. of Mechanical Engineers" Vol. 1, Part 1 FSP-51-91, pp. 289-301 (1929) and "Industrial and Engineering Chemistry" Vol. 28 No. 9, pp. 782-986 (1936).

Determination of Insulation Thickness Required to Prevent Condensation

When dry, mass insulations are effective: when wet, they are ineffective. The conductivity of wet insulation is 16-20 times greater than dry insulation. For this reason insulations must be kept dry.

Insulation must be protected from two forms of moisture: (1) in liquid state and (2) in vapor state. Insulation, then, must be protected by a vapor-barrier as well as by a weather-barrier. Many experiments have shown that when the surface of a vapor-barrier is wet with liquid water its transmission of moisture is approximately four times that than when the surface is dry. This more rapid transmission of moisture shortens the effective life of insulation. Thus the importance of designing low-temperature insulation with surface temperatures above those at which water will condense out of the atmosphere onto the surfaces.

No thickness of insulation, no matter how great, installed on a low-temperature surface can prevent condensation on the outer weather-barrier surface when the ambient relative humidity reaches 100%. However, periods of 100% relative humidity are generally short; therefore, insulation thickness to prevent excessive condensation can be based on a more reasonable percentage. Selection of design conditions of air temperature and relative humidity should be such that they are not expected to be exceeded for any extended length of time.

Due to variables in weather conditions, locations, and surrounding conditions of a given installation, no general recommendations of "design" conditions are possible. Good judgment must be exercised in selecting realistic conditions. Selections which are too severe result in an excess of insulation, and selection of conditions too mild results in insufficient insulation, which allows surface condensation too large a part of the time. The results completely depend upon design selection of dry-bulb and wet-bulb temperatures (or dry-bulb temperature and relative humidity).

After selection of design conditions and determination of a surface temperature which will satisfy these design conditions, determination of insulation thickness which can maintain this surface temperature is possible. The previous method of solution was to assume an air-film resistance. Unfortunately, using the assumed air-film resistance method resulted in errors of insulation thickness ranging from 50% more than needed to 66% less than required. Using Heilman's equations, Dorsey and Turner developed a method of determining insulation thickness required to prevent condensation. The following is taken from the manual "Determination of Insulation Thickness to Prevent Condensation."*

The mathematical method of solving the problem is based on calculating the heat gain from the ambient air to the lower temperature surface. This eliminates the need for guessing at the air-film resistance. Solution to this problem is divided into three basic steps:

STEP 1: Determination of design surface temperature
STEP 2: Determination of surface heat gain
STEP 3: Determination of insulation thickness

*One manual developed by Joint Program on Design of Low Temperature Insulation by Union Carbide Corp., Olefins Division, and National Insulation Manufacturers Association, 1965-1966.

Step 1 is no different from any other method of determining insulation thickness to prevent condensation. The surface temperature to which the installation must be designed depends upon the conditions to which the installation is subjected. Tables 1, 2 and 2a are provided in this manual to eliminate the need for a psychometric chart. To prevent condensation the insulation surface temperature must be slightly above dew point.

Step 2 is the key to the solution of the problem. The heat gain from ambient air to the lower temperature surface by radiation and convection determines the maximum allowable heat flow through the insulation if the design surface temperature is to be maintained. The radiation heat gain and convection heat gain must be found separately, then added together to obtain total heat gain. When:

Q_a = Total heat gain to surface
Q_{ra} = Heat gain due to radiation
Q_{ca} = Heat gain due to convection
$Q_a = Q_{ra} + Q_{ca}$

Tables 3 and 3a give the values of radiation heat transfer rates in reference to absolute zero. It is based on Stefan-Boltzmann Formula:

English Units

When: Q_r = Radiation heat transfer rate from a body at temperature to a body at absolute zero
ε = Absorptance which is 1.00 (black body conditions)
t = Temperature of body, °F
$Q_{ra} = 0.174 \times 10^{-8}\, \varepsilon\, (t + 459.6)^4$

Since we are concerned with radiation between the surface and ambient air and absorptance other than 1.00, when:

Q_{ra} = Radiation heat transfer rate from ambient air and surrounding bodies at temperature t_a to the insulation surface at a temperature t_2
ε = Absorptance of the insulation surface or its outer barrier

then:

$$Q_{ra} = [0.174 \times 10^{-8} (t_a + 459.6)^4] - [0.174 \times 10^{-8} (t_s + 459.6)^4]\ \varepsilon$$

The values of the expression $0.174 \times 10^{-8} (t + 459.6)^4$ have been calculated for different values of t_a and are tabulated in Table 3. Using the temperature of $t_a = t$, the value of the expression $0.174 \times 10^{-8} (t_a + 459.6)^4$ can be determined. Similarly, the value of the expression $0.174 = 10^{-8} (t_s + 459.6)^4$ can be determined by using temperature $t_s = t$. Multiplying the difference of these values by the correct value of ε will determine Q_{ra}. This multiplication has been precalculated in Table 4.

Because surface heat gain due to solar radiation lessens the possibility of condensation, the heat gain to insulated vessels and pipes located outdoors where they are exposed to solar heat is not included in this manual. If for reasons other than condensation control the maximum heat gain rate must be known, solar radiation must be taken into consideration. Neither has the cooling effect of radiation from the insulation surface into space during darkness been considered, as it is not practical to attempt to insulate for this condition.

Table 5 gives values of convection heat transfer rates for various conditions of shape, size, and position of surfaces. The proper table and correction factor, dictated by installation conditions, must be selected. In case the installation has several conditions and it is impractical to have various insulation thicknesses, then the most severe condition is the controlling factor. These tables are based on the Heilman Formula, given below.

When:

Q_{ca} = Convection heat gain for Δt
C = A constant depending on shape of surface (horizontal cylinders, C = 1.016, etc.)
d = Outside diameter of cylindrical surface or height of vertical flat surface. (For anything 24″ or more use values listed for 24.)
T = Average absolute temperature; for this study was taken as 540° R (80° F)
Note: A change to 500° R would produce an error of slightly more than one percent.
Δt = Dry-bulb ambient air temperature (t_a) less design insulation surface temperature (t_s)

then:

$$Q_{ca} = C\left(\frac{1}{d}\right)^{0.2}\left(\frac{1}{T}\right)^{0.181}(\Delta t)^{1.266}$$

The effect of a forced air velocity has not been considered, since, due to faster evaporation, air velocity most often will tend to retard condensation. If for reasons other than condensation control the maximum heat rate must be known, then air velocity must be considered. To obtain convection loss for a certain forced air velocity, all that is necessary is to multiply Q_{ca} by

$$\sqrt{\frac{V + 68.9}{68.9}}$$

when V equals air velocity in feet per minute.

The insulation (or its outer barrier) surface heat gain rate, Q_a, is found by:

$$Q_a = Q_{ra} + Q_{ca}$$

This Q_a is the key factor for design of insulation to prevent condensation. It is the maximum allowable surface heat gain rate if the design surface temperature is to be maintained. Under static conditions the heat gain to a surface must equal the heat loss from that surface. Thus, all that remains is to determine the thickness of insulation which provides sufficient resistance so that the heat flow from the outer surface of "design" temperature t_s to the inner surface of temperature t_1 (operating) is equal to or less than Q_a.

Step 3. The required resistance to heat flow is determined. Table 6 gives the values of resistance R. This table is based on the formula given below.

When:

Q_a = Heat gain to the surface of the insulation and through the insulation
Δt_s = Design surface temperature t_s less operating temperature t_1 of pipe or equipment
R = Thermal resistance

then:

$$Q_a = \frac{\Delta t_s}{R}$$

Table 7 gives the flat surface (or equivalent) thickness. This table is based on the formula below.

When:

ℓ = insulation thickness, flat surface (or equivalent L)
k = conductivity of insulation, Btu/sq ft, hr, in., °F

then:

$$\ell = Rk, \text{ or } L = Rk$$

The thickness ℓ is the minimum thickness of insulation on flat surfaces to prevent condensation. However, L is equivalent thickness of insulation to prevent condensation on curved surfaces. Table B-6 provides the conversion from equivalent thickness to nominal thicknesses of pipe covering; this table also gives actual insulation thickness for small diameter equipment. Table B-6 is based on the formula:

$$L = r_2 \log_e \frac{r_2}{r_1}$$

when: r_1 = Inner radius of insulation, in inches
r_2 = Outer radius of insulation, in inches

The above r_1 and r_2 were used to calculate Table B-6 but do not appear in the tabulation. The tabulations presented in NPS pipe diameters and nominal thicknesses are those manufactured for certain combinations of $r_2 - r_1$.

The curved thickness is selected so that its equivalent thickness L is equal to or more than the required thickness ℓ for flat surface.

A diagram of this method for selection of insulation thickness to prevent condensation is shown in Figure 15.

Solution by Use of Tables and Work Sheet

Although the mathematical solution appears relatively complex, the actual solution to the problem is easy. Selection of proper design wet and dry-bulb temperatures is based on installation requirements. After this selection is made, determination of insula-

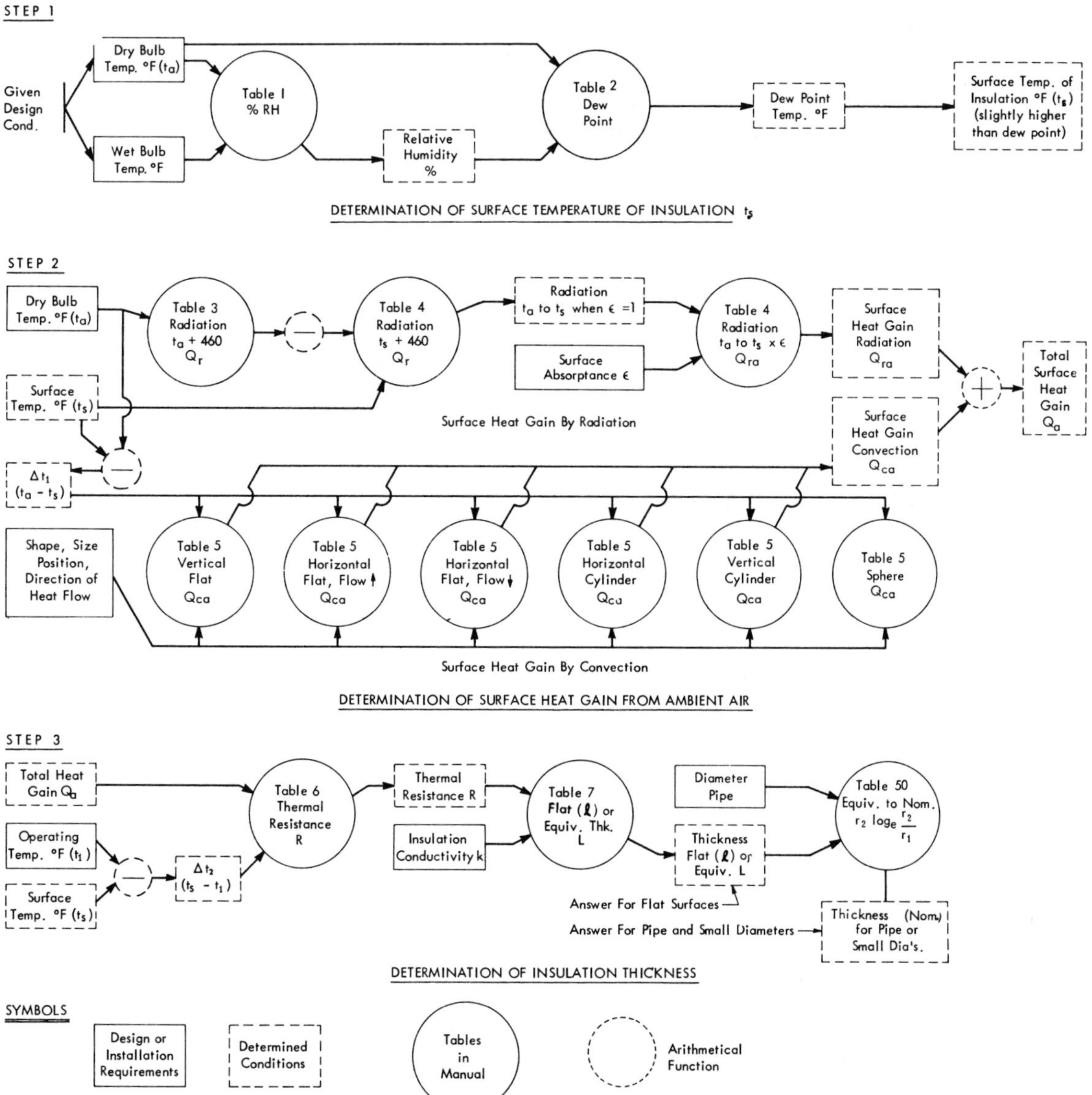

Note: The following method and tables 6 and 7 are based on English Units. These will not be presented in Metric Units as, at present, all pipe, tube and insulation thicknesses are in inches. However if units were in Metric, the method of conversion to English and back to Metric Units are given in examples 34a, 35a and 36a.

Figure 15 Diagram of solution.

tion thickness to prevent condensation can be made by using the tables in this manual. The tables used are Tables 1, 2, 2a, 3, 3a, 4, 5, 5a, 6, 7 and B-6.

Step 1—determination of design surface temperature

Using the selected dry-bulb (t_a) and wet-bulb temperatures, the percent relative humidity is found in Table 1.

Using the selected dry-bulb (t_a) and determined relative humidity, the dew-point temperature is found in Table 2 English Units, and 2a Metric Units.

The design surface temperature (t_s) must be slightly higher than dew point. Add 1 °F to the dew-point temperature to obtain design surface temperature.

Step 2—determination of surface heat gain

The total heat gain by radiation (Q_{ra}) and convection (Q_{ca}) to the surface at temperature t_s from ambient air at temperature t_a must be determined.

Radiation gain is found by determining the difference in radiation levels of surface temperature t_s and ambient temperature t_a in reference to absolute zero.

Determine radiation level of ambient temperature t_a using Table 3 English Units and 3a for Metric Units. From this, subtract the radiation level of surface temperature t_s. The result is radiation

Table 1

DIFFERENCE BETWEEN READINGS OF WET-BULB AND DRY-BULB TEMPERATURE °F — ENGLISH UNITS

PER CENT RELATIVE HUMIDITY

Eng. Units Dry-Bulb Temp °F	1	2	3	4	5	6	7	8	9	10	11	12	13	14	15	16	17	18	19	20	21	22	23	24	25	26	27	28	29	30	31	32	33	34	35	36	37	38	39	40	41	42	43	44	45	46	47	48	49	50	Metric Units Dry-Bulb Temp °C
0	70	40	10																																																-17.8
5	75	50	26	4																																															-15.0
10	80	60	40	22	4																																														-12.3
15	84	68	52	34	20	5																																													-3.4
20	86	72	59	46	33	20	7																																												-6.7
21	86	73	60	48	34	22	11																																												-6.1
22	87	74	61	50	36	25	13	2																																											-5.6
23	87	75	62	51	38	27	16	5																																											-5.0
24	88	76	64	53	41	30	19	8																																											-4.4
25	88	77	65	54	42	32	22	11																																											-3.9
26	88	77	66	55	44	34	24	14	4																																										-3.3
27	88	78	67	56	46	36	26	16	6																																										-2.8
28	89	78	68	57	48	37	28	18	9																																										-2.2
29	89	78	69	58	49	39	30	20	12	3																																									-1.7
30	89	80	70	60	51	41	32	23	14	6																																									-1.1
31	90	80	71	61	52	43	34	26	17	8																																									-0.6
32	90	81	72	62	53	44	36	27	19	10	3																																								0
33	90	82	73	63	55	46	38	29	21	13	5																																								0.6
34	90	82	74	65	56	48	40	32	24	16	8																																								1.1
35	90	82	73	66	58	50	42	34	26	18	10	4																																							1.7
36	91	82	74	65	59	51	44	36	28	20	13	6																																							2.2
37	91	83	75	66	58	53	45	38	30	22	16	8																																							2.8
38	91	83	75	67	59	51	47	39	32	24	17	10	4																																						3.5
39	91	83	76	68	60	53	46	41	34	27	20	13	6																																						3.9
40	92	84	77	68	61	53	47	39	36	28	22	15	8																																						4.4
41	92	84	77	69	62	54	47	40	33	31	24	17	11	5																																					3.0
42	92	84	77	70	62	55	48	41	34	28	26	19	13	6																																					5.6
43	93	84	78	70	63	56	49	43	36	29	23	22	15	9	3																																				6.1
44	93	85	78	71	64	57	50	44	37	31	24	18	17	12	5																																				6.7
45	93	86	78	72	65	58	52	45	39	32	26	20	14	13	8	3																																			7.2
46	93	86	79	72	65	58	53	46	40	33	27	22	16	11	11	5																																			7.8
47	93	86	79	72	66	59	54	47	41	35	29	23	17	12	7	7	2																																		8.3
48	93	86	80	73	67	60	54	48	42	36	30	24	19	14	8	4	4																																		8.9
49	93	86	80	74	67	61	55	49	43	38	32	26	20	16	10	5	2																																		9.4
50	93	87	81	75	68	62	56	50	45	39	33	28	23	18	13	7	3																																		10.0
51	94	87	81	75	68	62	56	50	45	39	34	29	23	19	14	8	4																																		10.6
52	94	87	81	75	69	63	57	51	46	40	35	30	24	20	15	10	5	2																																	11.1
53	94	87	81	75	69	63	58	52	47	41	36	31	25	21	17	12	8	3																																	11.7
54	94	88	82	76	70	64	58	53	47	42	37	32	27	23	18	14	9	4																																	12.2
55	94	88	82	76	70	65	59	54	49	44	39	34	29	24	20	15	11	5	2																																12.8
56	94	88	82	76	71	66	60	55	50	45	40	35	30	25	21	17	13	8	4																																13.3
57	94	88	83	77	71	66	61	56	51	46	41	36	31	27	23	18	14	9	6	2																															13.9
58	94	88	83	77	72	67	62	57	52	47	42	37	32	28	24	20	15	11	7	4																															14.4
59	94	89	84	77	73	68	62	57	53	48	43	38	34	29	25	21	16	13	8	5	2																														15.0
60	94	89	84	78	73	68	63	58	53	48	44	39	35	31	27	22	18	14	10	6	3																														15.6
61	94	89	84	79	74	69	63	59	54	49	45	41	36	32	28	24	20	15	12	8	4																														16.1
62	94	89	84	79	74	69	64	59	54	50	46	42	37	33	29	25	21	17	13	9	6	3																													16.7
63	95	90	84	79	74	70	65	60	55	51	47	43	38	34	30	26	22	19	14	11	8	4																													17.2
64	95	90	85	80	75	70	65	61	56	52	48	44	39	34	31	27	23	20	16	12	9	6	2																												17.8
65	95	90	85	80	75	71	65	61	57	53	49	45	40	36	32	28	25	21	17	13	11	7	4																												18.3
66	95	90	85	80	75	71	66	62	57	53	49	46	41	37	33	30	26	22	19	15	12	8	5	2																											18.9
67	95	90	85	80	76	71	67	63	58	55	50	47	42	39	34	30	27	23	20	17	13	10	7	4																											19.4
68	95	90	86	81	76	72	67	63	58	55	51	48	43	39	35	31	28	24	21	18	14	11	8	5	2																										20.0
69	95	90	86	81	76	72	68	64	59	56	52	48	44	40	36	32	29	25	22	19	15	12	9	6	3																										20.6
70	95	90	86	81	76	72	68	64	60	56	52	48	45	41	37	33	30	27	23	20	17	14	10	7	5	2																									21.1
71	95	90	86	81	76	72	69	64	60	56	53	49	45	41	38	34	31	28	24	21	18	15	12	8	6	3																									21.7
72	95	91	86	82	77	73	69	65	61	57	53	50	46	42	39	35	32	28	25	22	19	16	13	10	7	4	2																								22.6
73	95	91	86	82	77	73	69	65	61	58	54	50	46	43	39	36	32	29	26	23	20	17	14	11	8	5	3																								22.8
74	95	91	86	82	78	74	70	65	62	58	54	51	47	43	40	36	33	30	27	24	21	18	15	12	9	6	4	2																							23.3
75	95	91	87	82	78	74	70	66	62	58	55	52	48	45	41	37	34	31	28	25	22	19	16	13	11	8	6	3																							23.9
76	96	91	87	83	78	75	70	66	63	59	55	52	49	45	42	38	35	32	29	26	23	20	17	15	12	9	7	4	2																						24.4
77	96	91	87	83	78	75	71	67	64	59	56	53	49	46	42	39	36	33	30	27	24	21	18	16	13	11	8	5	3																						25.4
78	96	91	87	83	79	75	71	67	64	60	57	53	50	46	43	40	37	34	31	27	25	22	19	17	14	12	9	6	4																						25.6
79	96	91	87	83	79	75	71	67	64	60	57	54	50	47	44	40	37	34	32	28	26	23	20	18	16	13	10	7	5	3																					26.1
80	96	91	87	83	79	75	72	68	64	61	57	54	50	47	44	41	38	35	32	29	27	24	21	19	17	14	11	8	6	5	2																				26.7
82	96	92	88	84	80	76	72	69	65	62	58	55	51	48	45	42	39	37	34	31	28	25	23	21	18	16	13	11	9	7	4																				27.8
84	96	92	88	84	80	76	73	69	66	63	59	57	53	50	47	44	41	38	35	32	30	28	25	22	20	17	15	12	11	8	5	4	2																		28.4
86	96	92	88	84	80	77	73	70	66	63	60	57	54	51	48	45	42	39	37	34	32	29	26	24	22	19	17	14	13	11	8	6	4	2																	30.0
88	96	92	88	85	81	78	74	70	67	64	61	58	55	52	49	46	43	41	38	35	33	31	28	26	24	21	18	17	15	13	11	8	6	4	2																31.1
90	96	92	89	85	81	78	74	71	68	65	61	59	55	53	50	47	44	42	39	37	34	32	29	27	25	22	20	18	16	14	12	10	8	6	4	2															32.2
92	96	92	89	85	82	79	75	72	68	65	62	59	56	54	51	48	46	43	41	38	36	33	31	28	26	24	22	20	18	16	14	12	10	8	6	4	2														33.3
94	96	93	89	85	82	79	75	72	69	66	63	60	57	55	52	49	47	44	42	39	37	34	32	30	28	26	24	21	19	17	15	13	12	10	8	7	5	3													34.4
96	96	93	89	86	82	79	76	73	70	67	64	61	58	56	53	50	48	45	43	40	38	35	33	31	29	27	25	23	21	19	17	15	14	12	10	8	6	4	3												35.6
98	96	93	89	86	92	79	77	73	70	67	64	62	59	57	54	51	49	46	44	41	39	36	34	32	30	28	26	24	22	20	18	16	15	13	11	9	8	6	4	3											36.7
100	96	93	89	86	83	80	77	74	71	68	65	62	60	57	54	51	49	47	44	42	40	37	35	33	31	29	27	25	23	22	19	18	16	14	13	11	9	8	6	5	3										37.8
102	96	93	90	86	83	80	77	74	71	68	65	63	60	58	55	52	50	48	45	42	41	38	36	34	32	30	28	26	24	23	20	19	17	15	14	13	10	10	7	7	5	2									44.4
104	97	93	90	87	83	81	77	74	71	69	66	63	61	58	55	53	51	48	46	43	42	39	37	35	33	31	29	27	25	24	21	20	18	16	16	14	12	11	8	8	7	4	2								45.6
106	97	93	90	87	84	81	78	75	72	69	66	63	62	59	56	53	51	49	46	44	43	40	38	36	34	32	30	28	26	25	22	21	19	17	17	16	13	12	10	10	8	5	4	3							46.1
108	97	93	90	87	84	81	78	75	72	70	67	64	62	60	57	54	52	50	47	45	44	41	39	37	35	33	31	29	27	26	23	22	20	18	18	17	14	14	11	11	10	7	5	4	3						42.8
110	97	93	90	87	84	81	78	75	73	70	67	64	63	60	57	55	53	50	48	45	45	42	40	38	36	34	32	30	28	27	24	23	21	19	19	18	16	15	12	12	11	8	7	6	4	3					43.9
112	97	94	90	87	84	81	79	76	73	70	68	65	63	61	58	56	53	51	49	46	45	43	41	39	37	35	33	31	30	28	26	24	23	20	20	19	18	16	14	13	12	9	8	7	6	4	3	2			38.9
114	97	94	91	88	85	82	79	76	74	71	68	65	64	61	58	56	54	52	49	47	46	44	42	40	38	36	34	32	31	29	27	26	24	22	21	20	19	17	16	14	13	12	9	8	7	6	5	3	2		40.0
116	97	94	91	88	85	82	79	76	74	71	69	66	64	62	59	57	54	52	50	48	47	45	43	41	39	37	35	33	32	30	28	27	25	23	22	21	20	18	17	15	14	13	12	10	9	7	6	6	4	2	41.1
118	97	94	91	88	85	82	79	77	74	72	o9	67	65	62	59	58	55	53	51	49	48	46	44	42	40	38	36	34	33	31	29	28	26	24	23	22	21	19	18	16	15	14	13	12	10	9	8	7	5	4	42.2
120	97	94	91	88	85	82	80	77	74	72	69	67	65	62	60	58	55	53	51	50	48	46	44	42	41	39	37	35	34	32	30	29	27	26	24	23	22	20	19	18	16	15	14	13	11	10	9	8	7	6	43.3
Eng. Units Dry-Bulb Temp °F	0.5	1.1	1.6	2.2	2.5	3.3	3.8	4.4	5.0	5.5	6.1	6.6	7.2	7.7	8.3	8.8	9.4	10.0	10.5	11.1	11.6	12.2	12.7	13.3	13.8	14.4	15.0	15.5	16.1	16.6	17.2	17.7	18.3	18.8	19.4	20.0	20.5	21.1	21.6	22.2	22.7	23.3	23.8	24.4	25.4	25.5	26.1	26.6	27.2	29.7	Metric Units Dry-Bulb Temp °C

DIFFERENCE BETWEEN READINGS OF WET BULB AND DRY BULB TEMPERATURE CELSIUS °C — METRIC UNITS

NOTE:
Values above heavy line are for ice, while those below line are for water.

Table 2 Dew-point temperature.

Dry-bulb temp. °F	Percent relative humidity 10	15	20	25	30	35	40	45	50	55	60	65	70	75	80	85	90	95	100
5	−35	−30	−25	−21	−17	−14	−12	−10	−8	−6	−5	−4	−2	−1	1	2	3	4	5
10	−31	−25	−20	−16	−13	−10	−7	−5	−3	−2	0	2	3	4	5	7	8	9	10
15	−28	−21	−16	−12	−8	−5	−3	−1	1	3	5	6	8	9	10	12	13	14	15
20	−24	−16	−11	−8	−4	−2	2	4	6	8	10	11	13	14	15	16	18	19	20
25	−20	−15	−8	−4	0	3	6	8	10	12	15	16	18	19	20	21	23	24	25
30	−15	−9	−3	2	5	8	11	13	15	17	20	22	23	24	25	27	28	29	30
35	−12	−5	1	5	9	12	15	18	20	22	24	26	27	28	30	32	33	34	35
40	−7	0	5	9	14	16	19	22	24	26	28	29	31	33	35	36	38	39	40
45	−4	3	9	13	17	20	23	25	28	30	32	34	36	38	39	41	43	44	45
50	−1	7	13	17	21	24	27	30	32	34	37	39	41	42	44	45	47	49	50
55	3	11	16	21	25	28	32	34	37	39	41	43	45	47	49	50	52	53	55
60	6	14	20	25	29	32	35	39	42	44	46	48	50	52	54	55	57	59	60
65	10	18	24	28	33	38	40	43	46	49	51	53	55	57	59	60	62	63	65
70	13	21	28	33	37	41	45	48	50	53	55	57	60	62	64	65	67	68	70
75	17	25	32	37	42	46	49	52	55	57	60	62	64	66	69	70	72	74	75
80	20	29	35	41	46	50	54	57	60	62	65	67	69	72	74	75	77	78	80
85	23	32	40	45	50	54	58	61	64	67	69	72	74	76	78	80	82	83	85
90	27	36	44	49	54	58	62	66	69	72	74	77	79	81	83	85	87	89	90
95	30	40	48	54	59	63	67	70	73	76	79	82	84	86	88	90	91	93	95
100	34	44	52	58	63	68	71	75	78	81	84	86	88	91	92	94	96	98	100
105	38	48	56	62	67	72	76	79	82	85	88	90	93	95	97	99	101	103	105
110	41	52	60	66	71	77	80	84	87	90	92	95	98	100	102	104	106	108	110
115	45	56	64	70	75	80	84	88	91	94	97	100	102	105	107	109	111	113	115
120	48	60	68	74	79	85	88	92	96	99	102	105	107	109	112	114	116	118	120
125	52	63	72	78	84	89	93	97	100	104	107	109	111	114	117	119	121	123	125

Table 2a Dew-point temperature °C Metric Units

Dry-bulb temp. °C	Percent relative humidity 10	15	20	25	30	35	40	45	50	55	60	65	70	75	80	85	90	95	100
−15.0	−37.2	−34.4	−31.7	−29.4	−27.2	−26.5	−24.4	−23.3	−22.2	−21.1	−20.6	−20.0	−18.9	−18.3	−17.2	−16.7	−16.1	−15.6	−15.0
−12.2	−35.0	−31.7	−28.9	−26.7	−25.0	−23.3	−21.7	−20.6	−19.4	−18.9	−17.8	−16.7	−16.1	−15.6	−15.0	−13.9	−13.3	−12.8	−12.2
−9.4	−33.3	−29.4	−26.7	−24.4	−22.2	−20.6	−19.4	−18.3	−17.2	−16.1	−15.0	−14.4	−13.3	−12.8	−12.2	−11.1	−10.6	−10.6	−9.4
−6.7	−31.1	−26.7	−23.9	−22.2	−20.0	−18.9	−16.7	−15.6	−14.4	−13.3	−12.2	−11.7	−10.6	−10.0	−9.4	−8.9	−7.8	−7.2	−6.7
−3.9	−28.9	−26.1	−22.2	−20.0	−17.8	−16.1	−14.4	−13.3	−12.2	−11.1	−9.4	−8.9	−7.8	−7.2	−6.7	−6.1	−5.0	−4.4	−3.9
−1.1	−26.1	−22.8	−19.4	−16.7	−15.0	−13.3	−11.7	−10.6	−9.4	−8.3	−6.7	−5.6	−5.0	−4.4	−3.9	−2.8	−2.2	−1.7	−1.1
1.7	−24.4	−20.6	−17.2	−15.0	−12.8	−11.1	−9.9	−7.8	−6.7	−5.6	−4.4	−3.3	−2.8	−2.2	−1.1	0	0.6	1.1	1.7
4.4	−21.7	−17.8	−15.0	−12.8	−10.0	−8.9	−7.2	−5.6	−4.4	−3.3	−2.2	−1.7	−0.6	0.6	1.7	2.2	3.3	3.9	4.4
7.2	−20.0	−16.1	−12.8	−10.6	−8.3	−6.7	−5.0	−3.9	−2.2	−1.1	0	1.1	2.2	3.2	4.1	5.0	5.8	6.7	7.2
10.0	−18.3	−13.9	−11.1	−8.3	−6.1	−4.4	−2.8	−1.1	0	1.0	2.0	3.0	4.0	5.0	6.0	7.0	8.0	9.0	10.0
12.8	−16.1	−11.7	−8.9	−6.1	−3.9	−2.2	0	1.1	2.8	3.9	5.0	6.1	7.2	8.3	9.4	10.0	11.1	11.7	12.8
15.6	−14.4	−10.0	−6.7	−3.9	−1.7	0	1.7	3.9	5.6	6.7	7.8	8.9	10.0	11.1	12.2	18.3	13.9	15.0	15.6
18.3	−12.2	−7.8	−4.4	−2.2	0.6	3.3	4.4	6.1	7.8	9.4	10.6	11.7	12.8	13.9	15.0	15.6	16.7	17.2	18.3
21.1	−10.6	−6.1	−2.2	0.6	2.8	5.0	7.2	8.9	10.0	11.7	12.8	13.9	15.6	16.7	17.8	18.3	19.4	20.0	21.1
23.9	−8.3	−3.9	0	2.8	5.6	7.8	9.4	11.1	12.8	13.9	15.6	16.7	17.8	18.9	20.6	21.1	22.2	23.3	23.9
26.7	−6.7	−1.7	1.7	5.0	7.8	10.0	12.2	13.9	15.6	16.7	18.3	19.4	20.6	22.2	23.3	24.2	25.0	25.6	26.7
29.4	−5.0	0	4.4	7.2	10.0	12.2	14.4	16.1	17.8	19.4	20.6	22.2	23.3	24.4	25.6	26.7	27.8	28.3	32.4
32.2	−2.8	2.2	6.7	9.4	12.2	14.4	16.7	18.9	20.6	22.2	23.3	25.0	26.1	27.2	28.3	29.4	30.6	31.7	32.2
35.0	−1.1	4.4	8.9	12.4	15.0	17.2	19.4	21.1	22.8	24.4	26.1	27.8	28.9	30.0	31.1	32.2	32.8	33.9	35.0
37.8	1.1	6.7	11.1	14.4	17.2	20.0	21.7	23.9	25.6	27.2	28.9	30.0	31.1	32.8	33.3	34.4	35.6	36.7	37.8
40.6	3.3	8.9	13.3	16.7	19.4	22.2	24.4	26.1	27.8	29.4	31.1	32.2	33.9	35.0	36.1	37.2	38.3	39.4	40.6
43.3	5.0	11.1	15.6	18.9	21.7	25.0	26.7	28.9	30.6	32.2	33.3	35.0	36.7	37.8	38.9	40.0	41.1	42.2	43.3
46.1	7.2	13.3	17.8	21.1	23.9	26.7	28.9	31.1	32.8	34.4	36.1	37.8	38.9	40.6	41.7	42.8	43.9	45.0	46.1
48.9	8.9	15.6	20.0	23.3	26.1	29.4	31.1	33.3	35.6	37.2	38.9	40.6	41.7	42.8	44.4	45.6	46.7	47.8	48.9
51.7	11.1	17.2	22.2	25.6	28.9	31.7	33.9	36.1	37.8	40.0	41.7	42.8	43.9	45.6	47.2	48.3	49.4	50.6	51.7

Table 3 Radiation heat transfer rate for black body conditions (to absolute zero)

VALUES Of $Q_r = 0.174\,\epsilon\left(\frac{t + 459.6}{100}\right)^4$, WHEN $\epsilon = 1.0$

t Temp. °F	Q_r, Btu/hr, sq ft	t Temp. °F	Q_r, Btu/hr, sq ft	t Temp. °F	Q_r, Btu/hr, sq ft	t Temp. °F	Q_r, Btu/hr, sq ft	t Temp. °F	Q_r, Btu/hr, sq ft
−35	56.77	−3	75.90	29	99.49	61	128.2	93	162.7
−34	57.30	−2	76.56	30	100.3	62	129.2	94	163.9
−33	57.84	−1	77.23	31	101.1	63	130.2	95	165.1
−32	58.39	0	77.91	32	102.0	64	131.2	96	166.3
−31	58.94	1	78.59	33	102.8	65	132.2	97	167.5
−30	59.49	2	79.27	34	103.6	66	133.2	98	168.7
−29	60.04	3	79.96	35	104.5	67	134.2	99	169.9
−28	60.60	4	80.65	36	105.3	68	135.2	100	171.1
−27	61.17	5	81.35	37	106.2	69	136.3	101	172.4
−26	61.73	6	82.05	38	107.0	70	137.3	102	173.6
−25	62.30	7	82.76	39	107.9	71	138.3	103	174.8
−24	62.88	8	83.47	40	108.8	72	139.4	104	176.1
−23	63.46	9	84.19	41	109.6	73	140.4	105	177.3
−22	64.04	10	84.91	42	110.5	74	141.5	106	178.6
−21	64.63	11	85.63	43	111.4	75	142.6	107	179.8
−20	65.22	12	86.36	44	112.3	76	143.6	108	181.1
−19	65.81	13	87.10	45	113.2	77	144.7	109	182.4
−18	66.41	14	87.83	46	114.1	78	145.8	110	183.7
−17	67.01	15	88.58	47	115.0	79	146.9	111	185.0
−16	67.62	16	89.33	48	115.9	80	148.0	112	186.3
−15	68.23	17	90.08	49	116.8	81	149.1	113	187.6
−14	68.85	18	90.84	50	117.7	82	150.2	114	188.9
−13	69.47	19	91.60	51	118.6	83	151.3	115	190.2
−12	70.09	20	92.37	52	119.6	84	152.4	116	191.5
−11	70.72	21	93.14	53	120.5	85	153.5	117	192.9
−10	71.35	22	93.92	54	121.5	86	154.6	118	194.2
−9	71.99	23	94.70	55	122.4	87	155.8	119	195.6
−8	72.63	24	95.48	56	123.4	88	156.9	120	196.9
−7	73.27	25	96.28	57	124.3	89	158.1		
−6	73.92	26	97.07	58	125.3	90	159.2		
−5	74.58	27	97.87	59	126.3	91	160.4		
−4	75.23	28	98.68	60	127.2	92	161.6		

from t_a to t_s if the surface has black body absorptance (ε)* equal to one. Should surface absorptance be a ratio less than one, the previously determined result is multiplied by the proper absorptance ratio ε *. This multiplication is found in Table 4. The final result is the heat gain by radiation Q_{ra}.

Convection heat gain (Q_{ca}) is found using the proper column in Table 5. The heat gain by convection (Q_{ca}) is found opposite the proper temperature difference, Δt_1, between air and surface ($\Delta t_1 = t_a - t_s$). In some instances this result must be multiplied by a correction factor, as shown at bottom of table.

Add radiation heat gain (Q_{ra}) and convection heat gain (Q_{ca}) to obtain total heat gain (Q_a). *This is also maximum allowable heat transfer through the insulation (Q).*

*Note that absorptance and emittance ratio are being used interchangeably here. This is true in the temperature ranges and wave lengths normally encountered in thermal insulation.

Step 3—determination of insulation thickness

Using maximum allowable heat transfer through the insulation Q and the temperature difference across the insulation Δt_s (which equals the operating temperature t_1 subtracted from the outer surface temperature t_s), the required thermal resistance (R) of the insulation is found in Table 6.

Using required thermal resistance R and the conductivity of insulation (k), the flat surface (or equivalent) thickness of insulation (ℓ or L) is found in Table 7. This thickness table gives thickness ℓ or L in decimals. The actual thickness of the insulation must be that which is commercially available and is equal to or slightly larger than the figure given. *This is the thermal insulation thickness required to prevent condensation for flat surfaces.*

Using the equivalent thickness L for various pipe sizes, the nominal thickness required can be found in Table B-6. *This is the*

Table 3a Radiation heat transfer rate for black body conditions [to absolute zero 0°K, (−273.2°C)] Metric Units

VALUES Of $Q_{mr} = 0.598\ \epsilon\ (K/55.5)^4$, WHEN $\epsilon = 1.0$

Temp. °K	Temp. °C	Q_{mr} = W/m²	Temp. °K	Temp. °C	Q_{mr} = W/m²	Temp °K	Temp °C	Q_{mr} = W/m²
236.9	−37.2	191.59	263.0	−10.2	300.46	290.0	16.8	444.18
237.0	−36.2	198.13	264.0	− 9.2	305.06	291.0	17.8	450.34
238.0	−35.2	201.49	265.0	− 8.2	309.71	292.0	18.8	456.56
239.0	−34.2	204.91	266.0	− 7.2	314.41	293.0	19.8	462.84
240.0	−33.2	208.36	267.0	− 6.2	319.16	294.0	20.8	469.20
241.0	−32.2	211.85	268.0	− 5.2	323.97	295.0	21.8	475.61
242.0	−31.2	215.39	269.0	− 4.2	328.83	296.0	22.8	482.09
243.0	−30.2	218.37	270.0	− 3.2	333.75	297.0	23.8	488.64
244.0	−29.2	222.60	271.0	− 2.2	338.72	298.0	24.8	495.26
245.0	−28.2	226.27	272.0	− 1.2	343.75	299.0	25.8	501.94
246.0	−27.2	229.99	273.0	− .2	348.83	300.0	26.8	508.69
247.0	−26.2	233.75	274.0	0.8	253.97	301.0	27.8	515.50
248.0	−25.2	237.56	275.0	1.8	359.17	302.0	28.8	522.39
249.0	−24.2	241.41	276.0	2.8	364.42	303.0	29.8	529.34
250.0	−23.2	245.32	277.0	3.8	369.73	304.0	30.8	536.36
251.0	−22.2	249.26	278.0	4.8	375.10	305.0	31.8	543.46
252.0	−21.2	253.26	279.0	5.8	380.52	306.0	32.8	550.62
253.0	−20.2	257.30	280.0	6.8	386.01	307.0	33.8	557.85
254.0	−19.2	261.40	281.0	7.8	391.55	308.0	34.8	565.16
255.0	−18.2	265.50	282.0	8.8	397.16	309.0	35.8	572.53
256.0	−17.2	269.73	283.0	9.8	402.82	310.0	36.8	579.98
257.0	−16.2	273.97	284.0	10.8	408.54	311.0	37.8	587.50
258.0	−15.2	278.26	285.0	11.8	414.33	312.0	38.8	595.09
259.0	−14.2	282.59	286.0	12.8	420.17	313.0	39.8	602.76
260.0	−13.2	296.98	287.0	13.8	426.08	314.0	40.8	610.50
261.0	−12.2	291.43	288.0	14.8	432.05	315.0	41.8	618.31
262.0	−11.2	295.92	289.0	15.8	438.08	316.0	42.8	626.20

thermal insulation thickness for pipe, or curved surfaces, required to prevent condensation.

To assist in the use of these tables, a work sheet has been prepared. Several illustrative examples are given.

Illustrative examples

Three design problems applying different ambient air conditions, different installation requirements, and different insulations are presented to illustrate the use of the tables to determine the insulation thickness needed to prevent condensation.

■ EXAMPLE 34

English Units

Data

Bring lines 2″, 4″, and 6″ NPS in both horizontal and vertical positions.

Ambient air, high mean average is 90° F dry-bulb, 80° F wet-bulb
Operating temperature −20° F
Black outer weather-vapor surface, absorptance = .9.
Insulation—cellular glass, conductivity k = .35.

Enter data on work sheet, as shown in Figure 16.

Description:—Work Order—Figure 16

Pipe or vessel—Pipe.
Description of surface—cylindrical, vertical and horizontal 2″, 4″, and 6″ NPS.
Heat flow does not apply.

Dry-bulb temperature in column a	90° F
Wet-bulb temperature in column b	80° F
Surface absorptance in column i	ε = .9
Operating temperature in column q	−20° F
Conductivity of insulation in column t	k = 0.35

Table 4 Radiation heat transfer to surface with absorptance ϵ listed.

$$0.174\left(\frac{t_s + 459.6}{100}\right)^4 - 0.174\left(\frac{t_a + 459.6}{100}\right)^4$$ English Units t_a, t_g in °F

Note: In cases of radiation of low level with a decimal in the term such as 8.3 use 83 in table and shift answer decimal point one figure to left.

	ABSORPTANCE ε																			
	0.05	0.10	0.15	0.20	0.25	0.30	0.35	0.40	0.45	0.50	0.55	0.60	0.65	0.70	0.75	0.80	0.85	0.90	0.95	1.00
1	0.05	0.10	0.15	0.20	0.25	0.30	0.35	0.40	0.45	0.50	0.55	0.60	0.65	0.70	0.75	0.80	0.85	0.90	0.95	1.00
2	0.10	0.20	0.30	0.40	0.50	0.60	0.70	0.80	0.90	1.00	1.10	1.20	1.30	1.40	1.50	1.60	1.70	1.80	1.90	2.00
3	0.15	0.30	0.45	0.60	0.75	0.90	1.05	1.20	1.35	1.50	1.65	1.80	1.95	2.10	2.25	2.40	2.55	2.70	2.85	3.00
4	0.20	0.40	0.60	0.80	1.00	1.20	1.40	1.60	1.80	2.00	2.20	2.40	2.60	2.80	3.00	3.20	3.40	3.60	3.80	4.00
5	0.25	0.50	0.75	1.00	1.25	1.50	1.75	2.00	2.25	2.50	2.75	3.00	3.25	3.50	3.75	4.00	4.25	4.50	4.75	5.00
6	0.30	0.60	0.90	1.20	1.50	1.80	2.10	2.40	2.70	3.00	3.30	3.60	3.90	4.20	4.50	4.80	5.10	5.40	5.70	6.00
7	0.35	0.70	1.05	1.40	1.75	2.10	2.45	2.80	3.15	3.50	3.85	4.20	4.55	4.90	5.25	5.60	5.95	6.30	6.65	7.00
8	0.40	0.80	1.20	1.60	2.00	2.40	2.80	3.20	3.60	4.00	4.40	4.80	5.20	5.60	6.00	6.40	6.80	7.20	7.60	8.00
9	0.45	0.90	1.35	1.80	2.25	2.70	3.15	3.60	4.05	4.50	4.95	5.40	5.85	6.30	6.75	7.20	7.65	8.10	8.55	9.00
10	0.50	1.00	1.50	2.00	2.50	3.00	3.50	4.00	4.50	5.00	5.50	6.00	6.50	7.00	7.50	8.00	8.50	9.00	9.50	10.00
11	0.55	1.10	1.65	2.20	2.75	3.30	3.85	4.40	4.95	5.50	6.05	6.60	7.15	7.70	8.25	8.80	9.35	9.90	10.45	11.00
12	0.60	1.20	1.80	2.40	3.00	3.60	4.20	4.80	5.40	6.00	6.60	7.20	7.80	8.40	9.00	9.60	10.20	10.80	11.40	12.00
13	0.65	1.30	1.95	2.60	3.25	3.90	4.55	5.20	5.85	6.50	7.15	7.80	8.45	9.10	9.75	10.40	11.05	11.70	12.35	13.00
14	0.70	1.40	2.10	2.80	3.50	4.20	4.90	5.60	6.30	7.00	7.70	8.40	9.10	9.80	10.50	11.20	11.90	12.60	13.30	14.00
15	0.75	1.50	2.25	3.00	3.75	4.50	5.25	6.00	6.75	7.50	8.25	9.00	9.75	10.50	11.25	12.00	12.75	13.50	14.25	15.00
16	0.80	1.60	2.40	3.20	4.00	4.80	5.60	6.40	7.20	8.00	8.80	9.60	10.40	11.20	12.00	12.80	13.60	14.40	15.20	16.00
17	0.85	1.70	2.55	3.40	4.25	5.10	5.95	6.80	7.65	8.50	9.35	10.20	11.05	11.90	12.75	13.60	14.45	15.30	16.15	17.00
18	0.90	1.80	2.70	3.60	4.50	5.40	6.30	7.20	8.10	9.00	9.90	10.80	11.70	12.60	13.50	14.40	15.30	16.20	17.10	18.00
19	0.95	1.90	2.85	3.80	4.75	5.70	6.65	7.60	8.55	9.50	10.45	11.40	12.35	13.30	14.25	15.20	16.15	17.10	18.05	19.00
20	1.00	2.00	3.00	4.00	5.00	6.00	7.00	8.00	9.00	10.00	11.00	12.00	13.00	14.00	15.00	16.00	17.00	18.00	19.00	20.00
21	1.05	2.10	3.15	4.20	5.25	6.30	7.35	8.40	9.45	10.50	11.55	12.60	13.65	14.70	15.75	16.80	17.85	18.90	19.95	21.00
22	1.10	2.20	3.30	4.40	5.50	6.60	7.70	8.80	9.90	11.00	12.10	13.20	14.30	15.40	16.50	17.60	18.70	19.80	20.90	22.00
23	1.15	2.30	3.45	4.60	5.75	6.90	8.05	9.20	10.35	11.50	12.65	13.80	14.95	16.10	17.25	18.40	19.55	20.70	21.85	23.00
24	1.20	2.40	3.60	4.80	6.00	7.20	8.40	9.60	10.80	12.00	13.20	14.40	15.60	16.80	18.00	19.20	20.40	21.60	22.80	24.00
25	1.25	2.50	3.75	5.00	6.25	7.50	8.75	10.00	11.25	12.50	13.75	15.00	16.25	17.50	18.75	20.00	21.25	22.50	23.75	25.00
26	1.30	2.60	3.90	5.20	6.50	7.80	9.10	10.40	11.70	13.00	14.30	15.60	16.90	18.20	19.50	20.80	22.10	23.40	24.70	26.00
27	1.35	2.70	4.05	5.40	6.75	8.10	9.45	10.80	12.15	13.50	14.85	16.20	17.55	18.90	20.25	21.60	22.95	24.30	25.65	27.00
28	1.40	2.80	4.20	5.60	7.00	8.40	9.80	11.20	12.60	14.00	15.40	16.80	18.20	19.60	21.00	22.40	23.80	25.20	26.60	28.00
29	1.45	2.90	4.35	5.80	7.25	8.70	10.15	11.60	13.05	14.50	15.95	17.40	18.85	20.30	21.75	23.20	24.65	26.10	27.55	29.00
30	1.50	3.00	4.50	6.00	7.50	9.00	10.50	12.00	13.50	15.00	16.50	18.00	19.50	21.00	22.50	24.00	25.50	27.00	28.50	30.00
31	1.55	3.10	4.65	6.20	7.75	9.30	10.85	12.40	13.95	15.50	17.05	18.60	20.15	21.70	23.25	24.80	26.35	27.90	29.45	31.00
32	1.60	3.20	4.80	6.40	8.00	9.60	11.20	12.80	14.40	16.00	17.60	19.20	20.80	22.40	24.00	25.60	27.20	28.80	30.40	32.00
33	1.65	3.30	4.95	6.60	8.25	9.90	11.55	13.20	14.85	16.50	18.15	19.80	21.45	23.10	24.75	26.40	28.05	29.70	31.35	33.00
34	1.70	3.40	5.10	6.80	8.50	10.20	11.90	13.60	15.30	17.00	18.70	20.40	22.10	23.80	25.50	27.20	28.90	30.60	32.30	34.00
35	1.75	3.50	5.25	7.00	8.75	10.50	12.25	14.00	15.75	17.50	19.25	21.00	22.75	24.50	26.25	28.00	29.75	31.50	33.25	35.00
36	1.80	3.60	5.40	7.20	9.00	10.80	12.60	14.40	16.20	18.00	19.80	21.60	23.40	25.20	27.00	28.80	30.60	32.40	34.20	36.00
37	1.85	3.70	5.55	7.40	9.25	11.10	12.95	14.80	16.65	18.50	20.35	22.20	24.05	25.90	27.75	29.60	31.45	33.30	35.15	37.00
38	1.90	3.80	5.70	7.60	9.50	11.40	13.30	15.20	17.10	19.00	20.90	22.80	24.70	26.60	28.50	30.40	32.30	34.20	36.10	38.00
39	1.95	3.90	5.85	7.80	9.75	11.70	13.65	15.60	17.55	19.50	21.45	23.40	25.35	27.30	29.25	31.20	33.15	35.10	37.05	39.00
40	2.00	4.00	6.00	8.00	10.00	12.00	14.00	16.00	18.00	20.00	22.00	24.00	26.00	28.00	30.00	32.00	34.00	36.00	38.00	40.00
41	2.05	4.10	6.15	8.20	10.25	12.30	14.35	16.40	18.45	20.50	22.55	24.60	26.65	28.70	30.75	32.80	34.85	36.90	38.95	41.00
42	2.10	4.20	6.30	8.40	10.50	12.60	14.70	16.80	18.90	21.00	23.10	25.20	27.30	29.40	31.50	33.60	35.70	37.80	39.90	42.00
43	2.15	4.30	6.45	8.60	10.75	12.90	15.05	17.20	19.35	21.50	23.65	25.80	27.95	30.10	32.25	34.40	36.55	38.70	40.85	43.00
44	2.20	4.40	6.60	8.80	11.00	13.20	15.40	17.60	19.80	22.00	24.20	26.40	28.60	30.80	33.00	35.20	37.40	39.60	41.80	44.00
45	2.25	4.50	6.75	9.00	11.25	13.50	15.75	18.00	20.25	22.50	24.75	27.00	29.25	31.50	33.75	36.00	38.25	40.50	42.75	45.00
46	2.30	4.60	6.90	9.20	11.50	13.80	16.10	18.40	20.70	23.00	25.30	27.60	29.90	32.20	34.50	36.80	39.10	41.40	43.70	46.00
47	2.35	4.70	7.05	9.40	11.75	14.10	16.45	18.80	21.15	23.50	25.85	28.20	30.55	32.90	35.25	37.60	39.95	42.30	44.65	47.00
48	2.40	4.80	7.20	9.60	12.00	14.40	16.80	19.20	21.60	24.00	26.40	28.80	31.20	33.60	36.00	38.40	40.80	43.20	45.60	48.00
49	2.45	4.90	7.35	9.80	12.25	14.70	17.15	19.60	22.05	24.50	26.95	29.40	31.85	34.30	36.75	39.20	41.65	44.10	46.55	49.00
50	2.50	5.00	7.50	10.00	12.50	15.00	17.50	20.00	22.50	25.00	27.50	30.00	32.50	35.00	37.50	40.00	42.50	45.00	47.50	50.00
51	2.55	5.10	7.65	10.20	12.75	15.30	17.85	20.40	22.95	25.50	28.05	30.60	33.15	35.70	38.25	40.80	43.35	45.90	48.45	51.00
52	2.60	5.20	7.80	10.40	13.00	15.60	18.20	20.80	23.40	26.00	28.60	31.20	33.80	36.40	39.00	41.60	44.20	46.80	49.40	52.00
53	2.65	5.30	7.95	10.60	13.25	15.90	18.55	21.20	23.85	26.50	29.15	31.80	34.45	37.10	39.75	42.40	45.05	47.70	50.35	53.00
54	2.70	5.40	8.10	10.80	13.50	16.20	18.90	21.60	24.30	27.00	29.70	32.40	35.10	37.80	40.50	43.20	45.90	48.60	51.30	54.00
55	2.75	5.50	8.25	11.00	13.75	16.50	19.25	22.00	24.75	27.50	30.25	33.00	35.75	38.50	41.25	44.00	46.75	49.50	52.25	55.00
56	2.80	5.60	8.40	11.20	14.00	16.80	19.60	22.40	25.20	28.00	30.80	33.60	36.40	39.20	42.00	44.80	47.60	50.40	53.20	56.00
57	2.85	5.70	8.55	11.40	14.25	17.10	19.95	22.80	25.65	28.50	31.35	34.20	37.05	39.90	42.75	45.60	48.45	51.30	54.15	57.00
58	2.90	5.80	8.70	11.60	14.50	17.40	20.30	23.20	26.10	29.00	31.90	34.80	37.70	40.60	43.50	46.40	49.30	52.20	55.10	58.00
59	2.95	5.90	8.85	11.80	14.75	17.70	20.65	23.60	26.55	29.50	32.45	35.40	38.35	41.30	44.25	47.20	50.15	53.10	56.05	59.00
60	3.00	6.00	9.00	12.00	15.00	18.00	21.00	24.00	27.00	30.00	33.00	36.00	39.00	42.00	45.00	48.00	51.00	54.00	57.00	60.00
61	3.05	6.10	9.15	12.20	15.25	18.30	21.35	24.40	27.45	30.50	33.55	36.60	39.65	42.70	45.75	48.80	51.85	54.90	57.95	61.00
62	3.10	6.20	9.30	12.40	15.50	18.60	21.70	24.80	27.90	31.00	34.10	37.20	40.30	43.40	46.50	49.60	52.70	55.80	58.90	62.00
63	3.15	6.30	9.45	12.60	15.75	18.90	22.05	25.20	28.35	31.50	34.65	37.80	40.95	44.10	47.25	50.40	53.55	56.70	59.85	63.00
64	3.20	6.40	9.60	12.80	16.00	17.20	22.40	25.60	28.80	32.00	35.20	38.40	41.60	44.80	48.00	51.20	54.40	57.60	60.80	64.00
65	3.25	6.50	9.75	13.00	16.25	17.50	22.75	26.00	29.25	32.50	35.75	39.00	42.25	45.50	48.75	52.00	55.25	58.50	61.75	65.00
66	3.30	6.60	9.90	13.20	16.50	17.80	23.10	26.40	29.70	33.00	36.30	39.60	42.90	46.20	49.50	52.80	56.10	59.40	62.70	66.00
67	3.35	6.70	10.05	13.40	16.75	18.10	23.45	26.80	30.15	33.50	36.85	40.20	43.55	46.90	50.25	53.60	56.95	60.30	63.65	67.00
68	3.40	6.80	10.20	13.60	17.00	18.40	23.80	27.20	30.60	34.00	37.40	40.80	44.20	47.60	51.00	54.40	57.80	61.20	64.60	68.00
69	3.45	6.90	10.35	13.80	17.25	18.70	24.15	27.60	31.05	34.50	37.95	41.40	44.85	48.30	51.75	55.20	58.65	62.10	65.55	69.00
70	3.50	7.00	10.50	14.00	17.50	21.00	24.50	28.00	31.50	35.00	38.50	42.00	45.50	49.00	52.50	56.00	59.50	63.00	66.50	70.00
71	3.55	7.10	10.65	14.20	17.75	21.30	24.85	28.40	31.95	35.50	39.05	42.60	46.15	49.70	53.25	56.80	60.35	63.90	67.45	71.00
72	3.60	7.20	10.80	14.40	18.00	21.60	25.20	28.80	32.40	36.00	39.60	43.20	46.80	50.40	54.00	57.60	61.20	64.80	68.40	72.00
73	3.65	7.30	10.95	14.60	18.25	21.90	25.55	29.20	32.85	36.50	40.15	43.80	47.45	51.10	54.75	58.40	62.05	65.70	69.35	73.00
74	3.70	7.40	11.10	14.80	18.50	22.20	25.90	29.60	33.30	37.00	40.70	44.40	48.10	51.80	55.50	59.20	62.90	66.60	70.30	74.00
75	3.75	7.50	11.25	15.00	18.75	22.50	26.25	30.00	33.75	37.50	41.25	45.00	48.75	52.50	56.25	60.00	63.75	67.50	71.25	75.00
76	3.80	7.60	11.40	15.20	19.00	22.80	26.60	30.40	34.20	38.00	41.80	45.60	49.40	53.20	57.00	60.80	64.60	68.40	72.20	76.00
77	3.85	7.70	11.55	15.40	19.25	23.10	26.95	30.80	34.65	38.50	42.35	46.20	50.05	53.90	57.75	61.60	65.45	69.30	73.15	77.00
78	3.90	7.80	11.70	15.60	19.50	23.40	27.30	31.20	35.10	39.00	42.90	46.80	50.70	54.60	58.50	62.40	66.30	70.20	74.10	78.00
79	3.95	7.90	11.85	15.80	19.75	23.70	27.65	31.60	35.55	39.50	43.45	47.40	51.35	55.30	59.25	63.20	67.15	71.10	75.05	79.00
80	4.00	8.00	12.00	16.00	20.00	24.00	28.00	32.00	36.00	40.00	44.00	48.00	52.00	56.00	60.00	64.00	68.00	72.00	76.00	80.00
81	4.05	8.10	12.15	16.20	20.25	24.30	28.35	32.40	36.45	40.50	44.55	48.60	52.65	56.70	60.75	64.80	68.85	72.90	76.95	81.00
82	4.10	8.20	12.30	16.40	20.50	24.60	28.70	32.80	36.90	41.00	45.10	49.20	53.30	57.40	61.50	65.60	69.70	73.80	77.90	82.00
83	4.15	8.30	12.45	16.60	20.75	24.90	29.05	33.20	37.35	41.50	45.65	49.80	53.95	58.10	62.25	66.40	70.55	74.70	78.85	83.00
84	4.20	8.40	12.60	16.80	21.00	25.20	29.40	33.60	37.80	42.00	46.20	50.40	54.60	58.80	63.00	67.20	71.40	75.60	79.80	84.00
85	4.25	8.50	12.75	17.00	21.25	25.50	29.75	34.00	38.25	42.50	46.75	51.00	55.25	59.50	63.75	68.00	72.25	76.50	80.75	85.00
86	4.30	8.60	12.90	17.20	21.50	25.80	30.10	34.40	38.70	43.00	47.30	51.60	55.90	60.20	64.50	68.80	73.10	77.40	81.70	86.00
87	4.35	8.70	13.05	17.40	21.75	26.10	30.45	34.80	39.15	43.50	47.85	52.20	56.55	60.90	65.25	69.60	73.95	78.30	82.65	87.00
88	4.40	8.80	13.20	17.60	22.00	26.40	30.80	35.20	39.60	44.00	48.40	52.80	57.20	61.60	66.00	70.40	74.80	79.20	83.60	88.00
89	4.45	8.90	13.35	17.80	22.25	26.70	31.15	35.60	40.05	44.50	48.95	53.40	57.85	62.30	66.75	71.20	75.65	80.10	84.55	89.00
90	4.50	9.00	13.50	18.00	22.50	27.00	31.50	36.00	40.50	45.00	49.50	54.00	58.50	63.00	67.50	72.00	76.50	81.00	85.50	90.00
91	4.55	9.10	13.65	18.20	22.75	27.30	31.85	36.40	40.95	45.50	50.05	54.60	59.15	63.70	68.25	72.80	77.35	81.90	86.45	91.00
92	4.60	9.20	13.80	18.40	23.00	27.60	32.20	36.80	41.40	46.00	50.60	55.20	59.80	64.40	69.00	73.60	78.20	82.80	87.40	92.00
93	4.65	9.30	13.95	18.60	23.25	27.90	32.55	37.20	41.85	46.50	51.15	55.80	60.45	65.10	69.75	74.40	79.05	83.70	88.35	93.00
94	4.70	9.40	14.10	18.80	23.50	28.20	32.90	37.60	42.30	47.00	51.70	56.40	61.10	65.80	70.50	75.20	79.90	84.60	89.30	94.00
95	4.75	9.50	14.25	19.00	23.75	28.50	33.25	38.00	42.75	47.50	52.25	57.00	61.75	66.50	71.25	76.00	80.75	85.50	90.25	95.00
96	4.80	9.60	14.40	19.20	24.00	28.80	33.60	38.40	43.20	48.00	52.80	57.60	62.40	67.20	72.00	76.80	81.60	86.40	91.20	96.00
97	4.85	9.70	14.55	19.40	24.25	29.10	33.95	38.80	43.65	48.50	53.35	58.20	63.05	67.90	72.75	77.60	82.45	87.30	92.15	97.00
98	4.90	9.80	14.70	19.60	24.50	29.40	34.30	39.20	44.10	49.00	53.90	58.80	63.70	68.60	73.50	78.40	83.30	88.20	93.10	98.00
99	4.95	9.90	14.85	19.80	24.75	29.70	34.65	39.60	44.55	49.50	54.45	59.40	64.35	69.30	74.25	79.20	84.15	89.10	94.05	99.00
100	5.00	10.00	15.00	20.00	25.00	30.00	35.00	40.00	45.00	50.00	55.00	60.00	65.00	70.00	75.00	80.00	85.00	90.00	95.00	100.00

$$0.548\left(\frac{T_{ms}}{55.55}\right)^4 - 0.548\left(\frac{T_{ma}}{55.5}\right)^4$$ Metric Units T_{ms} + T_{ma} in °K

Note: Results are in units of heat transfer same as units used for temperature.

Table 5 Convection heat gain rate (Q_{ac})

FREE CONVECTION

Heat gain rate, Btu/hr, sq ft

Temp. diff. Δt_1	Vertical flat surfaces[1]	Horizontal flat surfaces, facing upward[2]	Horizontal flat surfaces, facing downward[2]	Horizontal cylinders and pipes[1]	Vertical cylinders and pipes[1,3]	Spheres
1	0.24	0.15	0.30	0.17	0.21	0.31
2	0.57	0.36	0.73	0.41	0.50	0.74
3	0.95	0.61	1.22	0.69	0.84	1.24
4	1.34	0.87	1.76	1.00	1.21	1.79
5	1.81	1.16	2.33	1.32	1.61	2.37
6	2.29	1.46	2.93	1.67	2.02	2.98
7	2.78	1.77	3.56	2.02	2.46	3.63
8	3.29	2.10	4.22	2.40	2.91	4.29
9	3.82	2.44	4.90	2.78	3.38	4.98
10	4.36	2.79	5.60	3.18	3.86	5.69
11	4.92	3.14	6.31	3.58	4.36	6.43
12	5.49	3.51	7.06	4.00	4.87	7.17
13	6.08	3.88	7.80	4.43	5.39	7.94
14	6.68	4.26	8.57	4.86	5.92	8.72
15	7.29	4.65	9.36	5.31	6.46	9.52
16	7.91	5.50	10.2	5.76	7.01	10.3
17	8.54	5.45	11.0	6.22	7.57	11.1
18	9.18	5.86	11.8	6.69	8.13	12.0
19	9.83	6.28	12.6	7.17	8.71	12.8
20	10.4	6.70	13.5	7.65	9.29	13.7
21	11.2	7.12	14.3	8.13	9.89	14.6
22	11.8	7.56	15.2	8.63	10.5	15.5
23	12.5	7.99	16.1	9.13	11.1	16.4
24	13.2	8.44	17.0	9.63	11.7	17.3
25	13.9	8.88	17.9	10.1	12.3	18.2
26	14.6	9.34	18.8	10.7	13.0	19.1
27	15.3	9.79	19.7	11.2	13.6	20.1
28	16.1	10.3	20.6	11.7	14.2	21.0
29	16.8	10.7	21.6	12.2	14.9	21.9
30	17.5	11.2	22.5	12.8	15.5	22.9
31	18.3	11.7	23.5	13.3	16.2	23.9
32	19.0	12.1	24.4	13.9	16.8	24.8
33	19.8	12.6	25.4	14.4	17.5	25.8
34	20.5	13.1	26.4	15.0	18.2	26.8
35	21.3	13.6	27.4	15.5	18.9	27.8
36	22.1	14.1	28.4	16.1	19.6	28.8
37	22.9	14.6	29.4	16.7	20.3	29.8
38	23.6	15.1	30.4	17.2	20.9	30.9
39	24.4	15.6	31.4	17.8	21.7	31.9
40	25.2	16.1	32.4	18.4	22.4	32.9
41	26.0	16.6	33.4	19.0	23.1	34.0
42	26.8	17.1	34.5	19.6	23.8	35.0
43	27.7	17.7	35.5	20.2	24.5	36.1
44	28.5	18.2	36.6	20.7	25.2	37.2
45	29.3	18.7	37.6	21.3	25.9	38.2
46	30.1	19.2	38.7	22.0	26.7	39.3
47	30.9	19.7	39.7	22.6	27.4	40.4
48	31.8	20.3	40.8	23.2	28.2	41.5
49	32.6	20.8	41.9	23.8	28.9	42.6
50	33.5	21.4	43.0	24.4	29.7	43.7
51	34.3	21.9	44.1	25.0	30.4	44.8
52	35.2	22.5	45.2	25.6	31.2	45.9
53	36.0	23.0	46.3	26.3	31.9	47.0
54	36.9	23.5	47.4	26.9	32.7	48.2
55	37.8	24.1	48.5	27.5	33.5	49.3
56	38.6	24.7	49.6	28.2	34.2	50.4
57	39.5	25.2	50.7	28.8	35.0	51.6
58	40.4	25.8	51.9	29.4	35.8	52.7
59	41.3	26.3	53.0	30.1	36.6	53.9
60	42.2	26.9	54.1	30.7	37.3	55.0
61	43.0	27.5	55.3	31.4	38.1	56.2
62	43.9	28.1	56.4	32.0	38.9	57.4
63	44.8	28.6	57.6	32.7	39.7	58.5
64	45.7	29.2	58.7	33.3	40.5	59.7
65	46.7	29.8	59.9	34.0	41.3	60.9
66	47.6	30.4	61.1	34.7	42.1	62.1
67	48.5	31.0	62.2	35.3	42.9	63.3
68	49.4	31.5	63.4	36.0	43.8	64.5
69	50.3	32.1	64.6	36.7	44.6	65.7
70	51.2	32.7	65.8	37.3	45.4	66.9
71	52.2	33.3	67.0	38.0	46.2	68.1
72	53.1	33.9	68.2	38.7	47.0	69.3

CORRECTION FACTORS FOR SURFACE HEIGHT OR DIAMETER OF INSULATION

Approx. Height or OD, Inches	1	2	3	4	5	6	7	8	9	10	12	14	16	18	20	22
Correction Factor (multiply)	1.89	1.64	1.52	1.43	1.37	1.32	1.28	1.25	1.22	1.19	1.15	1.11	1.08	1.06	1.04	1.02

No correction needed for heights or diameters 24 inches and greater.

Notes 1, 2, and 3 –

1. See table for correction factors for surface height or diameter.
2. Figures in this column are more accurate for surfaces having dimensions 24 inches or greater.
3. Figures in this column are more accurate for vertical heights over 24 inches.

Table 5a Convection heat gain rate (Q_{acm}) – Metric Units

FREE CONVECTION

Heat gain – W/m^2

Temp. diff. °K or °C	Vertical flat surfaces[1]	Horizontal flat surfaces, facing upward[2]	Horizontal flat surfaces, facing downward[3]	Horizontal cylinders and pipes[1]	Vertical cylinders and pipes[1,3]	Spheres
1	1.36	0.85	1.70	0.96	1.19	1.76
2	3.17	2.04	4.07	2.31	2.80	4.14
3	6.16	3.94	7.93	4.49	5.48	8.06
4	8.94	5.69	11.45	6.50	7.91	11.67
5	12.04	7.69	15.45	8.76	10.66	15.70
6	14.85	9.50	19.07	10.82	13.14	19.37
7	18.17	11.61	23.37	13.24	16.12	23.73
8	21.66	13.81	27.78	15.72	19.20	28.27
9	25.25	16.11	32.56	18.38	22.38	32.88
10	28.94	18.47	37.20	21.09	25.63	37.83
11	32.46	20.91	42.13	23.88	28.99	42.76
12	36.32	23.09	46.37	26.36	32.06	47.34
13	40.09	25.63	51.63	29.28	35.60	52.60
14	44.17	28.22	56.88	32.10	39.09	57.84
15	48.23	30.86	62.11	35.30	42.88	63.37
16	52.21	33.40	66.80	37.94	46.05	68.10
17	56.27	36.02	72.35	41.16	49.84	73.64
18	60.65	38.62	77.88	49.37	53.63	79.16
19	65.01	41.54	83.72	47.57	57.71	84.99
20	69.67	44.45	39.53	50.76	61.79	90.79
21	73.75	47.02	94.69	53.79	65.38	95.97
22	78.10	49.93	100.51	56.93	69.46	102.11
23	82.76	52.84	106.32	60.48	73.54	108.23
24	87.73	56.06	112.44	63.98	77.60	114.34
25	92.31	58.95	118.54	67.15	81.65	120.42
26	96.54	61.58	124.13	70.56	85.64	126.05
27	101.51	64.80	130.23	74.05	90.01	132.46
28	106.46	68.01	136.65	77.54	94.38	138.87
29	111.40	71.21	143.05	81.02	98.74	145.26
30	116.33	74.09	149.43	84.80	103.09	151.95
31	120.90	77.08	155.12	87.96	107.15	157.68
32	125.83	80.28	161.52	91.75	111.50	164.39
33	131.08	83.48	168.22	95.54	116.17	171.08
34	136.01	86.98	174.91	99.32	120.51	177.76
35	141.24	90.16	181.59	103.09	125.16	184.43
36	145.87	93.21	187.37	106.29	129.28	190.56
37	151.43	96.71	194.37	110.39	133.93	197.56
38	156.65	99.89	201.05	114.16	138.90	204.54
39	161.87	103.38	208.03	117.93	143.54	211.51
40	167.40	106.87	215.01	122.01	147.17	218.47

Correction factors for surface height or diameter of insulation is same as given at bottom of Table 5.

References 1, 2, and 3 same as for Table 5 except that reference to 24 inches is 0.61 metre.

Table 6 Values of resistance R(R = $\Delta t_s Q_a$) — English Units

Note: $Q_a = Q_{cr} + Q_{ca}$

Δt_s °F	SURFACE HEAT GAIN Q_a in Btu/sf hr																			
	1	2	3	4	5	6	7	8	9	10	11	12	13	14	15	16	17	18	19	20
5	5.0	2.5	1.67	1.25	1.0	0.83	0.71	0.63	0.56	0.5	0.45	0.42	0.39	0.36	0.33	0.31	0.29	0.28	0.26	0.25
10	10.0	5.0	3.33	2.50	2.0	1.67	1.43	1.25	1.11	1.0	0.91	0.83	0.77	0.71	0.66	0.63	0.59	0.56	0.53	0.50
15	15.0	7.5	5.00	3.75	3.0	2.50	2.14	1.88	1.67	1.5	1.36	1.25	1.15	1.07	1.00	0.94	0.88	0.83	0.79	0.75
20	20.0	10.0	6.67	5.00	4.0	3.33	2.86	2.50	2.22	2.0	1.82	1.67	1.54	1.43	1.33	1.25	1.18	1.11	1.05	1.00
25	25.0	12.5	8.33	6.25	5.0	4.17	3.57	3.13	2.78	2.5	2.27	2.08	1.92	1.79	1.66	1.56	1.47	1.39	1.32	1.25
30	30.0	15.0	10.00	7.50	6.0	5.00	4.29	3.75	3.33	3.0	2.73	2.50	2.31	2.14	2.00	1.88	1.76	1.67	1.57	1.50
35	35.0	17.5	11.67	8.75	7.0	5.83	5.00	4.38	3.89	3.5	3.18	2.92	2.69	2.50	2.33	2.19	2.06	1.94	1.84	1.75
40	40.0	20.0	13.33	10.00	8.0	6.67	5.71	5.00	4.44	4.0	3.64	3.33	3.08	2.86	2.66	2.50	2.35	2.22	2.11	2.00
45	45.0	22.5	15.00	11.25	9.0	7.50	6.43	5.63	5.00	4.5	4.09	3.75	3.46	3.21	3.00	2.81	2.65	2.50	2.37	2.25
50	50.0	25.0	16.67	12.50	10.0	8.33	7.14	6.25	5.56	5.0	4.55	4.17	3.85	3.57	3.33	3.13	2.94	2.78	2.63	2.50
55	55.0	27.5	18.33	13.75	11.0	9.17	7.86	6.88	6.11	5.5	5.00	4.58	4.23	3.93	3.66	3.44	3.24	3.06	2.89	2.75
60	60.0	30.0	20.00	15.00	12.0	10.00	8.57	7.50	6.67	6.0	5.46	5.00	4.62	4.29	4.00	3.75	3.53	3.33	3.16	3.00
65	65.0	32.5	21.67	16.25	13.0	10.83	9.29	8.13	7.22	6.5	5.91	5.42	5.00	4.64	4.33	4.06	3.82	3.61	3.42	3.25
70	70.0	35.0	23.33	17.50	14.0	11.67	10.00	8.75	7.78	7.0	6.37	5.83	5.39	5.00	4.66	4.38	4.12	3.89	3.68	3.50
75	75.0	37.5	25.00	18.75	15.0	12.00	10.71	9.38	8.33	7.5	6.82	6.25	5.77	5.36	5.00	4.69	4.41	4.17	3.95	3.75
80	80.0	40.0	26.67	20.00	16.0	13.33	11.43	10.00	8.89	8.0	7.27	6.67	6.15	5.71	5.33	5.00	4.71	4.44	4.21	4.00
85	85.0	42.5	28.33	21.25	17.0	14.17	12.14	10.63	9.44	8.5	7.73	7.08	6.54	6.07	5.66	5.31	5.00	4.72	4.47	4.25
90	90.0	45.0	30.00	22.50	18.0	15.00	12.86	11.25	10.00	9.0	8.19	7.50	6.92	6.43	6.00	5.63	5.29	5.00	4.74	4.50
95	95.0	47.5	31.67	23.75	19.0	15.83	13.57	11.88	10.56	9.5	8.64	7.92	7.31	6.79	6.33	5.94	5.59	5.23	5.00	4.75
100	100.0	50.0	33.33	25.00	20.0	16.67	14.29	12.50	11.11	10.0	9.09	8.33	7.69	7.14	6.66	6.25	5.88	5.56	5.26	5.00
110	110.0	55.0	36.67	27.50	22.0	18.33	15.71	13.75	12.22	11.0	10.00	9.17	8.46	7.86	7.33	6.88	6.47	6.11	5.79	5.50
120	120.0	60.0	40.00	30.00	24.0	20.00	17.14	15.00	13.33	12.0	10.91	10.00	9.23	8.57	8.00	7.50	7.06	6.67	6.32	6.00
130	130.0	65.0	43.33	32.50	26.0	21.67	18.57	16.25	14.44	13.0	11.82	10.83	10.00	9.29	8.66	8.13	7.65	7.22	6.84	6.50
140	140.0	70.0	46.67	35.00	28.0	23.33	20.00	17.50	15.56	14.0	12.73	11.67	10.77	10.00	9.30	8.75	8.24	7.78	7.37	7.00
150	150.0	75.0	50.00	37.50	30.0	25.00	21.43	18.75	16.67	15.0	13.64	12.50	11.54	10.71	10.00	9.38	8.82	8.33	7.89	7.50
160	160.0	80.0	53.33	40.00	32.0	26.67	22.86	20.00	17.78	16.0	14.55	13.33	12.31	11.42	10.66	10.00	9.41	8.89	8.42	8.00
170	170.0	85.0	56.67	42.50	34.0	28.33	24.29	21.25	18.89	17.0	15.46	14.17	13.08	12.14	11.33	10.63	10.00	9.44	8.95	8.50
180	180.0	90.0	60.00	45.00	36.0	30.00	25.71	22.50	20.00	18.0	16.37	15.00	13.85	12.86	12.00	11.25	10.59	10.00	9.47	9.00
190	190.0	95.0	63.33	47.50	38.0	31.67	27.14	23.75	21.11	19.0	17.27	15.83	14.62	13.57	12.66	11.88	11.18	10.56	10.00	9.50
200	200.0	100.0	66.67	50.00	40.0	33.33	28.57	25.00	22.22	20.0	18.19	16.67	15.39	14.29	13.33	12.50	11.76	11.11	10.53	10.00
210	210.0	105.0	70.00	52.50	42.0	35.00	30.00	26.25	23.33	21.0	19.09	17.50	16.15	15.00	14.00	13.13	12.35	11.67	11.05	10.50
220	220.0	110.0	73.33	55.00	44.0	36.67	31.43	27.50	24.44	22.0	20.00	18.33	16.92	15.71	14.66	13.75	12.94	12.22	11.57	11.00
230	230.0	115.0	76.67	57.50	46.0	38.33	32.86	28.75	25.56	23.0	20.91	19.17	17.69	16.43	15.33	14.38	13.53	12.78	12.11	11.50
240	240.0	120.0	80.00	60.00	48.0	40.00	34.29	30.00	26.67	24.0	21.82	20.00	18.46	17.14	16.00	15.00	14.12	13.33	12.63	12.00
250	250.0	125.0	83.33	62.50	50.0	41.67	35.71	31.25	27.78	25.0	22.73	20.83	19.23	17.86	16.66	15.63	14.71	13.89	13.16	12.50
260	260.0	130.0	86.67	65.00	52.0	43.33	37.14	32.50	28.89	26.0	23.64	21.67	20.00	18.57	17.33	16.25	15.29	14.44	13.68	13.00
270	270.0	135.0	90.00	67.50	54.0	45.00	38.57	33.75	30.00	27.0	24.55	22.50	20.77	19.29	18.00	16.88	15.88	15.00	14.21	13.50
280	280.0	140.0	93.33	70.00	56.0	46.67	40.00	35.00	31.11	28.0	25.46	23.33	21.54	20.00	18.66	17.50	16.47	15.56	14.74	14.00
290	290.0	145.0	96.67	72.50	58.0	48.33	41.43	36.25	32.22	29.0	26.37	24.17	22.31	20.71	19.33	18.13	17.06	16.11	15.26	14.50
300	300.0	150.0	100.00	75.00	60.0	50.00	42.86	37.50	33.33	30.0	27.27	25.00	23.08	21.43	20.00	18.75	17.65	16.67	15.79	15.00

Δt_s °F	SURFACE HEAT GAIN Q_a in Btu/sf hr																			
	21	22	23	24	25	26	27	28	29	30	31	32	33	34	35	36	37	38	39	40
5	0.24	0.23	0.22	0.21	0.20	0.20	0.19	0.18	0.17	0.17	0.16	0.16	0.15	0.15	0.14	0.14	0.14	0.13	0.13	0.13
10	0.48	0.46	0.44	0.42	0.40	0.39	0.37	0.36	0.35	0.33	0.32	0.31	0.30	0.30	0.29	0.28	0.27	0.26	0.26	0.25
15	0.71	0.68	0.65	0.63	0.60	0.58	0.56	0.54	0.52	0.50	0.48	0.47	0.46	0.44	0.43	0.42	0.41	0.40	0.39	0.38
20	0.95	0.91	0.87	0.83	0.80	0.77	0.74	0.71	0.69	0.67	0.65	0.63	0.61	0.59	0.57	0.56	0.54	0.53	0.51	0.50
25	1.19	1.14	1.09	1.04	1.00	0.96	0.93	0.89	0.86	0.83	0.81	0.78	0.76	0.74	0.72	0.70	0.68	0.66	0.64	0.63
30	1.43	1.36	1.30	1.25	1.20	1.15	1.11	1.07	1.04	1.00	0.97	0.94	0.91	0.88	0.86	0.83	0.81	0.79	0.77	0.75
35	1.67	1.59	1.52	1.46	1.40	1.35	1.30	1.25	1.21	1.17	1.13	1.09	1.06	1.03	1.00	0.97	0.95	0.92	0.90	0.88
40	1.91	1.82	1.74	1.67	1.60	1.54	1.48	1.43	1.38	1.33	1.29	1.25	1.21	1.18	1.14	1.11	1.08	1.05	1.03	1.00
45	2.14	2.05	1.96	1.88	1.80	1.73	1.67	1.61	1.55	1.50	1.45	1.41	1.36	1.32	1.29	1.25	1.22	1.18	1.15	1.13
50	2.38	2.27	2.17	2.08	2.00	1.92	1.85	1.79	1.72	1.67	1.61	1.56	1.52	1.47	1.43	1.39	1.35	1.32	1.28	1.25
55	2.62	2.50	2.39	2.29	2.20	2.12	2.04	1.96	1.90	1.83	1.77	1.72	1.67	1.62	1.57	1.53	1.49	1.45	1.41	1.38
60	2.86	2.73	2.61	2.50	2.40	2.31	2.22	2.14	2.07	2.00	1.94	1.88	1.82	1.77	1.71	1.67	1.62	1.58	1.54	1.50
65	3.10	2.96	2.83	2.71	2.60	2.50	2.41	2.32	2.24	2.17	2.10	2.03	1.97	1.91	1.86	1.81	1.76	1.71	1.67	1.63
70	3.33	3.18	3.04	2.92	2.80	2.69	2.59	2.50	2.41	2.33	2.26	2.19	2.12	2.06	2.00	1.94	1.89	1.84	1.80	1.75
75	3.57	3.41	3.26	3.13	3.00	2.89	2.78	2.68	2.59	2.50	2.42	2.34	2.27	2.21	2.14	2.08	2.03	1.97	1.92	1.88
80	3.81	3.64	3.48	3.33	3.20	3.08	2.96	2.86	2.76	2.67	2.58	2.50	2.42	2.35	2.29	2.22	2.16	2.11	2.11	2.00
85	4.05	3.86	3.70	3.54	3.40	3.27	3.15	3.04	2.93	2.83	2.74	2.66	2.58	2.50	2.43	2.36	2.29	2.24	2.18	2.13
90	4.29	4.09	3.91	3.75	3.60	3.46	3.33	3.21	3.10	3.00	2.90	2.81	2.73	2.65	2.57	2.50	2.43	2.37	2.31	2.25
95	4.52	4.32	4.13	3.96	3.80	3.65	3.52	3.39	3.28	3.17	3.07	2.97	2.88	2.79	2.71	2.64	2.57	2.50	2.44	2.38
100	4.76	4.55	4.35	4.17	4.00	3.85	3.70	3.57	3.45	3.33	3.23	3.13	3.03	2.94	2.86	2.78	2.70	2.63	2.56	2.50
110	5.24	5.00	4.78	4.58	4.40	4.23	4.07	3.93	3.79	3.67	3.55	3.44	3.33	3.24	3.14	3.06	2.97	2.90	2.82	2.75
120	5.71	5.46	5.22	5.00	4.80	4.62	4.44	4.29	4.14	4.00	3.87	3.75	3.64	3.53	3.43	3.33	3.24	3.16	3.08	3.00
130	6.19	5.91	5.65	5.42	5.20	5.00	4.82	4.64	4.48	4.33	4.19	4.06	3.94	3.82	3.71	3.61	3.51	3.42	3.33	3.25
140	6.67	6.36	6.09	5.83	5.60	5.39	5.19	5.00	4.83	4.67	4.52	4.38	4.24	4.12	4.00	3.89	3.78	3.68	3.59	3.50
150	7.14	6.82	6.52	6.25	6.00	5.77	5.56	5.36	5.17	5.00	4.84	4.69	4.55	4.41	4.29	4.17	4.05	3.95	3.85	3.75
160	7.62	7.27	6.96	6.67	6.40	6.15	5.93	5.71	5.52	5.33	5.16	5.00	4.85	4.71	4.57	4.44	4.32	4.21	4.10	4.00
170	8.10	7.73	7.39	7.08	6.80	6.54	6.30	6.07	5.86	5.66	5.48	5.31	5.15	5.00	4.86	4.72	4.60	4.47	4.36	4.25
180	8.57	8.18	7.83	7.50	7.20	6.92	6.67	6.43	6.21	6.00	5.81	5.63	5.46	5.29	5.14	5.00	4.87	4.74	4.62	4.50
190	9.05	8.64	8.26	7.92	7.60	7.31	7.04	6.79	6.55	6.33	6.13	5.94	5.76	5.59	5.43	5.28	5.14	5.00	4.87	4.75
200	9.52	9.09	8.70	8.33	8.00	7.69	7.41	7.14	6.90	6.67	6.45	6.25	6.06	5.88	5.71	5.56	5.41	5.26	5.13	5.00
210	10.00	9.55	9.13	8.75	8.40	8.08	7.78	7.50	7.24	7.00	6.77	6.56	6.36	6.18	6.00	5.83	5.68	5.53	5.39	5.25
220	10.48	10.00	9.57	9.17	8.80	8.46	8.15	7.86	7.59	7.33	7.10	6.88	6.67	6.47	6.29	6.11	5.95	5.79	5.64	5.50
230	10.95	10.46	10.00	9.58	9.20	8.85	8.52	8.21	7.93	7.67	7.42	7.19	6.97	6.77	6.57	6.39	6.22	6.05	5.90	5.75
240	11.43	10.91	10.44	10.00	9.60	9.23	8.89	8.57	8.28	8.00	7.74	7.50	7.27	7.06	6.86	6.67	6.49	6.32	6.15	6.00
250	11.91	11.36	10.87	10.42	10.00	9.62	9.26	8.93	8.62	8.33	8.07	7.82	7.58	7.36	7.14	6.94	6.76	6.58	6.41	6.25
260	12.38	11.82	11.30	10.83	10.40	10.00	9.63	9.29	8.97	8.67	8.39	8.13	7.88	7.65	7.43	7.22	7.03	6.84	6.67	6.50
270	12.86	12.27	11.74	11.25	10.80	10.39	10.00	9.64	9.31	9.00	8.71	8.44	8.18	7.94	7.71	7.50	7.30	7.11	6.92	6.75
280	13.33	12.73	12.17	11.67	11.20	10.77	10.37	10.00	9.66	9.33	9.03	8.75	8.49	8.24	8.00	7.78	7.57	7.37	7.18	7.00
290	13.81	13.18	12.61	12.08	11.60	11.15	10.74	10.36	10.00	9.67	9.36	9.06	8.79	8.53	8.29	8.06	7.84	7.63	7.44	7.25
300	14.29	13.64	13.04	12.50	12.00	11.54	11.11	10.71	10.35	10.00	9.68	9.38	9.09	8.82	8.57	8.33	8.11	7.90	7.69	7.50

Note: Table 6 is for the use of work sheet for "Selection of Insulation to Prevent Condensation" as illustrated in Figure, thus equivalent Metric Units is not provided as in other tables.

Table 7 Flat ℓ (or equivalent L) thickness of insulation (ℓ or L = Rk) — English Units

R	k OF INSULATION IN Btu/sf, hr, °F, INCH																			
	0.10	0.12	0.14	0.16	0.18	0.20	0.22	0.24	0.26	0.28	0.30	0.35	0.40	0.45	0.50	0.60	0.70	0.80	0.90	1.00
0.5	0.05	0.06	0.07	0.08	0.09	0.10	0.11	0.12	0.13	0.14	0.15	0.18	0.20	0.23	0.25	0.30	0.35	0.40	0.45	0.50
1.0	0.10	0.12	0.14	0.16	0.18	0.20	0.22	0.24	0.26	0.28	0.30	0.35	0.40	0.45	0.50	0.60	0.70	0.80	0.90	1.00
1.5	0.15	0.18	0.21	0.24	0.27	0.30	0.33	0.36	0.39	0.42	0.45	0.53	0.60	0.68	0.75	0.90	1.05	1.20	1.35	1.50
2.0	0.20	0.24	0.28	0.32	0.36	0.40	0.44	0.48	0.52	0.56	0.60	0.70	0.80	0.90	1.00	1.20	1.40	1.60	1.80	2.00
2.5	0.25	0.30	0.35	0.40	0.45	0.50	0.55	0.60	0.65	0.70	0.75	0.88	1.00	1.13	1.25	1.50	1.75	2.00	2.25	2.50
3.0	0.30	0.36	0.42	0.48	0.54	0.60	0.66	0.72	0.78	0.84	0.90	1.05	1.20	1.35	1.50	1.80	2.10	2.40	2.70	3.00
3.5	0.35	0.42	0.49	0.56	0.63	0.70	0.77	0.84	0.91	0.98	1.05	1.23	1.40	1.58	1.75	2.10	2.45	2.80	3.15	3.50
4.0	0.40	0.48	0.56	0.64	0.72	0.80	0.88	0.96	1.04	1.12	1.20	1.40	1.60	1.80	2.00	2.40	2.80	3.20	3.60	4.00
4.5	0.45	0.54	0.63	0.72	0.81	0.90	0.99	1.08	1.17	1.26	1.35	1.58	1.80	2.03	2.25	2.70	3.15	3.60	4.05	4.50
5.0	0.50	0.60	0.70	0.80	0.90	1.00	1.10	1.20	1.30	1.40	1.50	1.75	2.00	2.25	2.50	3.00	3.50	4.00	4.50	5.00
5.5	0.55	0.66	0.77	0.88	0.99	1.10	1.21	1.32	1.43	1.54	1.65	1.93	2.20	2.48	2.75	3.30	3.85	4.40	4.95	5.50
6.0	0.60	0.72	0.84	0.96	1.08	1.20	1.32	1.44	1.56	1.68	1.80	2.10	2.40	2.70	3.00	3.60	4.20	4.80	5.40	6.00
6.5	0.65	0.78	0.91	1.04	1.17	1.30	1.43	1.56	1.69	1.82	1.95	2.28	2.60	2.93	3.25	3.90	4.55	5.20	5.85	6.50
7.0	0.70	0.84	0.98	1.12	1.26	1.40	1.54	1.68	1.82	1.96	2.10	2.45	2.80	3.15	3.50	4.20	4.90	5.60	6.30	7.00
7.5	0.75	0.90	1.05	1.20	1.35	1.50	1.65	1.80	1.95	2.10	2.25	2.63	3.00	3.38	3.75	4.50	5.25	6.00	6.75	7.50
8.0	0.80	0.96	1.12	1.28	1.44	1.60	1.76	1.92	2.08	2.24	2.40	2.80	3.20	3.60	4.00	4.80	5.60	6.40	7.20	8.00
8.5	0.85	1.02	1.19	1.36	1.53	1.70	1.87	2.04	2.21	2.38	2.55	2.98	3.40	3.83	4.25	5.10	5.95	6.80	7.65	8.50
9.0	0.90	1.08	1.26	1.44	1.62	1.80	1.98	2.16	2.34	2.52	2.70	3.15	3.60	4.05	4.50	5.40	6.30	7.20	8.10	9.00
9.5	0.95	1.14	1.33	1.52	1.71	1.90	2.09	2.28	2.47	2.66	2.85	3.33	3.80	4.28	4.75	5.70	6.65	7.60	8.55	9.50
10.0	1.00	1.20	1.40	1.60	1.80	2.00	2.20	2.40	2.60	2.80	3.00	3.50	4.00	4.50	5.00	6.00	7.00	8.00	9.00	10.00
11.0	1.10	1.32	1.54	1.76	1.98	2.20	2.42	2.64	2.86	3.08	3.30	3.85	4.40	4.95	5.50	6.60	7.70	8.80	9.90	11.00
12.0	1.20	1.44	1.68	1.92	2.16	2.40	2.64	2.88	3.12	3.36	3.60	4.20	4.80	5.40	6.00	7.20	8.40	9.60	10.80	12.00
13.0	1.30	1.56	1.82	2.08	2.34	2.60	2.86	3.12	3.38	3.64	3.90	4.55	5.20	5.85	6.50	7.80	9.10	10.40	11.70	13.00
14.0	1.40	1.68	1.96	2.24	2.52	2.80	3.08	3.36	3.64	3.92	4.20	4.90	5.60	6.30	7.00	8.40	9.80	11.20	12.60	14.00
15.0	1.50	1.80	2.10	2.40	2.70	3.00	3.30	3.60	3.90	4.20	4.50	5.25	6.00	6.75	7.50	9.00	10.50	12.00	13.50	15.00
16.0	1.60	1.92	2.24	2.56	2.88	3.20	3.52	3.84	4.16	4.48	4.80	5.60	6.40	7.20	8.00	9.60	11.20	12.80	14.40	16.00
17.0	1.70	2.04	2.38	2.72	3.06	3.40	3.74	4.08	4.42	4.76	5.10	5.95	6.80	7.65	8.50	10.20	11.90	13.60	15.30	17.00
18.0	1.80	2.16	2.52	2.88	3.24	3.60	3.96	4.32	4.68	5.04	5.40	6.30	7.20	8.10	9.00	10.80	12.60	14.40	16.20	18.00
19.0	1.90	2.28	2.66	3.04	3.42	3.80	4.18	4.56	4.94	5.32	5.70	6.65	7.60	8.55	9.50	11.40	13.30	15.20	17.10	19.00
20.0	2.00	2.40	2.80	3.20	3.60	4.00	4.40	4.80	5.20	5.60	6.00	7.00	8.00	9.00	10.00	12.00	14.00	16.00	18.00	20.00
21.0	2.10	2.52	2.94	3.36	3.78	4.20	4.62	5.04	5.46	5.88	6.30	7.35	8.40	9.45	10.50	12.60	14.70	16.80	18.90	21.00
22.0	2.20	2.64	3.08	3.52	3.96	4.40	4.84	5.28	5.72	6.16	6.60	7.70	8.80	9.90	11.00	13.20	15.40	17.60	19.80	22.00
23.0	2.30	2.76	3.22	3.68	4.14	4.60	5.06	5.52	5.98	6.44	6.90	8.05	9.20	10.35	11.50	13.80	16.10	18.40	20.70	23.00
24.0	2.40	2.88	3.36	3.84	4.32	4.80	5.28	5.76	6.24	6.72	7.20	8.40	9.60	10.80	12.00	14.40	16.80	19.20	21.60	24.00
25.0	2.50	3.00	3.50	4.00	4.50	5.00	5.50	6.00	6.50	7.00	7.50	8.75	10.00	11.25	12.50	15.00	17.50	20.00	22.50	25.00
26.0	2.60	3.12	3.64	4.16	4.68	5.20	5.72	6.24	6.76	7.28	7.80	9.10	10.40	11.70	13.00	15.60	18.20	20.80	23.40	26.00
27.0	2.70	3.24	3.78	4.32	4.86	5.40	5.94	6.48	7.02	7.56	8.10	9.45	10.80	12.15	13.50	16.20	18.90	21.60	24.30	27.00
28.0	2.80	3.36	3.92	4.48	5.04	5.60	6.16	6.72	7.28	7.84	8.40	9.80	11.20	12.60	14.00	16.80	19.60	22.40	25.20	28.00
29.0	2.90	3.48	4.06	4.64	5.22	5.80	6.38	6.96	7.54	8.12	8.70	10.15	11.60	13.05	14.50	17.40	20.30	23.20	26.10	29.00
30.0	3.00	3.60	4.20	4.80	5.40	6.00	6.60	7.20	7.80	8.40	9.00	10.50	12.00	13.50	15.00	18.00	21.00	24.00	27.00	30.00
31.0	3.10	3.72	4.34	4.96	5.58	6.20	6.82	7.44	8.06	8.68	9.30	10.85	12.40	13.95	15.50	18.60	21.70	24.80	27.90	31.00
32.0	3.20	3.84	4.48	5.12	5.76	6.40	7.04	7.68	8.32	8.96	9.60	11.20	12.80	14.40	16.00	19.20	22.40	25.60	28.80	32.00
33.0	3.30	3.96	4.62	5.28	5.94	6.60	7.26	7.92	8.58	9.24	9.90	11.55	13.20	14.85	16.50	19.80	23.10	26.40	29.70	33.00
34.0	3.40	4.08	4.76	5.44	6.12	6.80	7.48	8.16	8.84	9.52	10.20	11.90	13.60	15.30	17.00	20.40	23.80	27.20	30.60	34.00
35.0	3.50	4.20	4.90	5.60	6.30	7.00	7.70	8.40	9.10	9.80	10.50	12.25	14.00	15.75	17.50	21.00	24.50	28.00	31.50	35.00
36.0	3.60	4.32	5.04	5.76	6.48	7.20	7.92	8.64	9.36	10.08	10.80	12.60	14.40	16.20	13.00	21.60	25.20	28.80	32.40	36.00
37.0	3.70	4.44	5.18	5.92	6.66	7.40	8.14	8.88	9.62	10.36	11.10	12.95	14.80	16.65	13.50	22.20	25.90	29.60	33.30	37.00
38.0	3.80	4.56	5.32	6.08	6.84	7.60	8.36	9.12	9.88	10.64	11.40	13.30	15.20	17.10	19.00	22.80	26.60	30.40	34.20	38.00
39.0	3.90	4.68	5.46	6.24	7.02	7.80	8.58	9.36	10.14	10.92	11.70	13.65	15.60	17.55	19.50	23.40	27.30	31.20	35.10	39.00
40.0	4.00	4.80	5.60	6.40	7.20	8.00	8.80	9.60	10.40	11.20	12.00	14.00	16.00	18.00	20.00	24.00	28.00	32.00	36.00	40.00
41.0	4.10	4.92	5.74	6.56	7.38	8.20	9.02	9.84	10.66	11.48	12.30	14.35	16.40	18.45	20.50	24.60	28.70	32.80	36.90	41.00
42.0	4.20	5.04	5.88	6.72	7.56	8.40	9.24	10.08	10.92	11.76	12.60	14.70	16.80	18.90	21.00	25.20	29.40	33.60	37.80	42.00
43.0	4.30	5.16	6.02	6.88	7.74	8.60	9.46	10.32	11.18	12.04	12.90	15.05	17.20	19.35	21.50	25.80	30.10	34.40	38.70	43.00
44.0	4.40	5.28	6.16	7.04	7.92	8.80	9.68	10.56	11.44	12.32	13.20	15.40	17.60	19.80	22.00	26.40	30.80	35.20	39.60	44.00
45.0	4.50	5.40	6.30	7.20	8.10	9.00	9.90	10.80	11.70	12.60	13.50	15.75	18.00	20.25	22.50	27.00	31.50	36.00	40.50	45.00
46.0	4.60	5.52	6.44	7.36	8.28	9.20	10.12	11.04	11.96	12.88	13.80	16.10	18.40	20.70	23.00	27.60	32.20	36.80	41.40	46.00
47.0	4.70	5.64	6.58	7.52	8.46	9.40	10.34	11.28	12.22	13.16	14.10	16.45	18.80	21.15	23.50	28.20	32.90	37.60	42.30	47.00
48.0	4.80	5.76	6.72	7.68	8.64	9.60	10.56	11.52	12.48	13.44	14.30	16.80	19.20	21.60	24.00	28.80	33.60	38.40	43.20	48.00
49.0	4.90	5.88	6.86	7.84	8.82	9.80	10.78	11.76	12.74	13.72	14.70	17.15	19.60	22.05	24.50	29.40	34.70	39.20	44.10	49.00
50.0	5.00	6.00	7.00	8.00	9.00	10.00	11.00	12.00	13.00	14.00	15.00	17.50	20.00	22.50	25.00	30.00	35.00	40.00	45.00	50.00
52.0	5.20	6.24	7.28	8.32	9.36	10.40	11.44	12.48	13.52	14.56	15.60	18.20	20.80	23.40	26.00	31.20	36.40	41.60	46.80	52.00
54.0	5.40	6.48	7.56	8.64	9.72	10.80	11.88	12.96	14.04	15.12	16.20	18.90	21.60	24.30	27.00	32.40	37.80	43.20	48.60	54.00
56.0	5.60	6.72	7.84	8.96	10.04	11.20	12.32	13.44	14.56	15.68	16.80	19.60	22.40	25.20	28.00	33.60	39.20	44.80	50.40	56.00
58.0	5.80	6.96	8.12	9.28	10.44	11.60	12.76	13.92	15.08	16.24	17.40	20.30	23.20	26.10	29.00	34.80	40.60	46.40	52.20	58.00
60.0	6.00	7.20	8.40	9.60	10.80	12.00	13.20	14.40	15.60	16.80	18.00	21.00	24.00	27.00	30.00	36.00	42.00	48.00	54.00	60.00
62.0	6.20	7.44	8.68	9.92	11.16	12.40	13.64	14.88	16.12	17.36	18.60	21.70	24.80	27.90	31.00	37.20	43.40	49.60	55.80	62.00
64.0	6.40	7.68	8.96	10.24	11.52	12.80	14.08	15.36	16.64	17.92	19.20	22.40	25.60	28.80	32.00	38.40	44.80	51.40	57.60	64.00
66.0	6.60	7.92	9.24	10.56	11.88	13.20	14.52	15.84	17.16	18.48	19.80	23.10	26.40	29.70	33.00	39.60	46.20	53.20	59.40	66.00
68.0	6.80	8.16	9.52	10.88	12.24	13.60	14.96	16.32	17.68	19.04	20.40	23.80	27.20	30.60	34.00	40.80	47.60	54.80	61.20	68.00
70.0	7.00	8.40	9.80	11.20	12.60	14.00	15.40	16.80	18.20	19.60	21.00	24.50	28.00	31.50	35.00	42.00	49.00	56.00	63.00	70.00
75.0	7.50	9.00	10.50	12.00	13.50	15.00	16.50	18.00	19.50	21.00	22.50	26.25	30.00	33.75	37.50	45.00	52.50	60.00	67.50	75.00
80.0	8.00	9.60	11.20	12.80	14.40	16.00	17.60	19.20	20.80	22.40	24.00	28.00	32.00	36.00	40.00	48.00	56.00	64.00	72.00	80.00
85.0	8.50	10.20	11.90	13.60	15.30	17.00	18.70	20.40	22.10	23.80	25.50	29.75	34.00	38.25	42.50	51.00	59.50	68.00	76.50	85.00
90.0	9.00	10.80	12.60	14.40	16.20	18.00	19.80	21.60	23.40	25.20	27.00	31.50	36.00	40.50	45.00	54.00	63.00	72.00	81.00	90.00
95.0	9.50	11.40	13.30	15.20	17.10	19.00	20.90	22.80	24.20	26.60	28.50	33.25	38.00	42.75	47.50	57.00	66.50	76.00	85.50	95.00
100.0	10.00	12.00	14.00	16.00	18.00	20.00	22.00	24.00	26.00	28.00	30.00	35.00	40.00	45.00	50.00	60.00	70.00	80.00	90.00	100.00

Note: Equivalent thicknesses are tabulated for use in method for "Selection of Insulation Thickness to Prevent Condensation", thus an equivalent Table in Metric Units is not provided.

STEP 1

Determine relative humidity from Table 1, 90° F dry-bulb, 80° F wet-bulb = 65% RH.
Enter in column c.

Determine dew point from Table 2, 90° F dry-bulb — 65% RH = 77° F dew point
Enter in column d.

Add 1° F to dew-point temperature: 77 + 1 = 78° F, the surface temperature above and at which condensation will not occur.
Enter in column e.

STEP 2

From Table 3 determine radiation from absolute zero to ambient dry-bulb temperature t_a of 90° F; this equals 159.2 Btu.
Enter in column f.

From Table 3 determine radiation from absolute zero to surface temperature t_s of 78° F, this equals 145.8 Btu.
Enter in column g.

Subtract amount of Btu in column g from that in column f; this equals 13.4.
Enter in column h.

Multiply amount in column h by column i (given in Table 4).
Enter in column j.

Enter Δt_1, which is t_a (Column a) less t_s (Column e) and which is 12° F, in column k.

Selection of proper convection from Table 5 is necessary to obtain Btu heat gain for column 1.

In this case, as the problem is with piping, the answer will be in column 5 or 6 of Table 5. Also, in this case the pipes are installed both in horizontal and vertical position, and as it is impractical to change insulation thickness for various positions, the most severe conditions must be used. Therefore, column 6 for vertical cylinders and pipes is selected. For 12° F Δt_1, this equals 4.87.

At the bottom of Table 5 are correction factors for outside diameters of insulation. An assumed thickness of the insulation must be made at this point. Assuming the insulation thickness to be about 2″, then the outside diameter of the 2″ NPS by 2″ thick insulation would be approximately 6″ OD. The 4″ × 2″ would be approximately 8″ OD and the 6″ × 2″ would be approximately 10″ OD. These figures are entered in column m.

The correction factors of the OD of insulation are entered in column n. These are 1.32, 1.25, and 1.22. As one can see, an error of 1″ thickness in assumed to actual thickness of insulation makes a relatively small difference in the correction factor.

Total convection heat gain is obtained by the multiplication of Column l by n. These results, which are 6.43, 6.08, and 5.93, are entered in Column o.

Total heat gain to surface is obtained by adding Btu by radiation Q_{ra} (Column j) to Btu by convection Q_{ca} (Column o). These, which are 18.62, 18.28, and 18.13, are entered in Column p.

STEP 3

Operating temperature t_1, in Column q is given. The Δt_s is surface temperature t_s Column e minus operating temperature t_1 or 78 − (−20) = 98. Enter in Column r.

Insulating resistance is found in Table 6. Using 18 Btu as heat flow to surface at 90 Δt_s, resistance R would be approximately 5.45. Enter in Column s. Note total differences in Column p are so small in this case that they have little effect on thickness.

Flat thickness ℓ, or equivalent thickness L is found in Table 7. Using resistance of 5.5 with insulation k of .35, L or ℓ equals 1.925.

From Table B-6 the nominal pipe insulation thickness equivalent to a flat thickness of 1.925. Two-inch NPS pipe requires 1-1/2″ nominal thickness, 4″ IPS pipe requires 1-1/2″ nominal thickness, and 6″ IPS pipe requires 2″ nominal thickness of pipe insulation.

As shown in the Example 34 illustration. ■

■ EXAMPLE 34a

Metric Units

Given Metric Units	***conversion***		***English Values***
Pipe No. 1 0.060325 m dia. pipe	39.37	=	2.375″ dia. (2″ NPS)
Pipe No. 2 0.1143 m dia. pipe	39.37	=	4.5″ dia. (4″ NPS)
Pipe No. 3 0.168275 m dia. pipe	39.37	=	6.625″ dia. (6″ NPS)
Dry Bulb Temp. of Ambient Air 32.2° C (305.4° K)		=	90° F (549.7° R)
Wet Bulb Temp. of Ambient Air 26.7° C (299.8° K)		=	80° F (539.7° R)
Operating Temp. of pipe −28.9° C (244.3° K)		=	−20° F (439.7° R)

Metric Units	***conversion***		***English Values***
Conductivity of insulation (Cellular glass) 0.0505 W/mK	× 6.983	=	0.35 Btu in./ft^2, hr °F
ε = 0.9 for Black outer weather barrier	× 1.0	=	0.9 = ε

Work steps 1, 2, and 3 using English Units are as shown previously (00). The results obtained by these steps are:

In English Values	***conversion***		***Metric Values***
ℓ = 1.925″ (min. equivalent thickness)	0.2054	=	0.0489 m ℓ_m
From Table B-6, then			
for 2″ NPS pipe min ℓ = 1.5″	0.0254	=	0.0381 m insul. thk.
for 4″ NPS pipe min ℓ = 1.5″	0.0254	=	0.0381 m insul. thk.
for 6″ NPS pipe min ℓ = 2.0″	0.0254	=	0.0381 m insul. thk.

■

■ EXAMPLE 35

English Units

Less explanation is provided in solving this problem.

Data

Large duct 48″ × 48″ passing through high humidity space.
Ambient air temperature dry bulb = 70 °F, relative humidity = 90%.
Operating temperature of duct 40° F.
Outer coating on duct insulation—light gray ε = .7.
Insulation, glass fiber, conductivity (published) = .25 + 20% = 0.3.
(20% safety factor to allow for some moisture in the insulation)
Enter data on work sheet, as shown in Figure 17.

Description:	Example 35 Duct Flat surfaces: one horizontal—heat upward, one horizontal—heat downward, one vertical
column a	Dry-bulb temperature = 70° F
columb b	Not needed—RH given.
column c	90% RH
column i	Surface absorptance ε = .7
column q	Operating temperature t_1 = 40° F
column t	Conductivity of insulation k = 0.3

STEP 1

column d (from Table 2)	Dew point = 67° F
column e	Surface temperature = 68° F

STEP 2

column f (from Table 3)	Absolute 0 to 70 = 137.3 Btu
column g (from Table 3)	Absolute 0 to 68 = 135.2 Btu
column h	137.3 − 135.2 = 2.1
column j	2.1 × .7 = 1.47
column k	$t_a - t_s$, 70 − 68 = 2
column l	Horizontal flat, facing down, Table 5—0.73. Horizontal flat, facing up, Table 5—0.36 Vertical flat, Table 5—0.57
column m	Horizontal flat, facing down, no correction factor
column n	Horizontal flat, facing up, no correction factor
column o	Vertical flat, above 24″, no correction required
column p	Columns j and l = 2.20, 1.83, and 2.04
column q	Given as 40° F
column r	$t_s - t_1$ = 68 − 40 = 28° F
column s	In this case of small differences it is more accurate to divide results in column r by results in column p, 28/2.20 = 12.3, 28/1.83 = 15.3, 28/2.04 = 13.7
column t	k = 0.3 given
column u (from Table 7)	ℓ or L = 3.7 ℓ or L = 4.6 ℓ or L = 4.0 Or in practical thickness: horizontal flat, surface facing down, requires 4″ thickness horizontal flat, surface facing up, requires 5″ thickness vertical flat, requires 4″ thickness Example 35 is entered on work as shown in Figure 17, Example 11, page ~~00~~.

Note: With a relatively small temperature difference between air and operating temperatures, in a very humid atmosphere, heavy thicknesses of insulation are required to prevent condensation. ■

■ EXAMPLE 35a

Metric Units

Given Metric Units	***conversion***		***English Values***
Duct 1.22m × 1.22m	× 39.37	=	48″ × 48″
Ambient air, dry bulb temp. 21.1° C (294.3° K)		=	70° F
Relative humidity = 90%	× 1	=	90% = Rel. humidity
Duct operating temp. = 4.4° C (277.6° K)		=	40° F
Conductivity of insulation = 0.0432 W/mK	× 6.933	=	0.3 Btu in./ft^2, hr °F
Surface emittance—light gray ε = 0.7	× 1	=	0.7 = ε
Dew Point temp. = 19.4° C (292.6° K)		=	67° F

Using English Unit values enter on work sheet as shown in Figure 17 and by steps 1 and 2 as indicated in Example 35 determine the thicknesses required for the various positions of heat transfer direction.

Horizontal, flat, heat flow up ℓ = 3.7″ min. × 0.0254 = 0.094m (min.)
Horizonal, flat, heat flow down ℓ = 4.6″ min. × 0.0254 = 0.117m (min.)
Vertical, flat, heat flow horizontal ℓ = 4.0″ min. × 0.0254 = 0.102m (min.)

■

■ EXAMPLE 36

English Units

Data

Heat exchanger, horizontal, 3′ 0″ diameter with dished heads.
Ambient air, high mean average of 100° F dry bulb, 85° F wet bulb.
Operating temperature −100° F.
White outer weather barrier surface, absorptance ε = .5.
Insulation—cellular glass, conductivity k = .33.
Enter data on work sheet, as shown in Example 36.

Body is horizontal cylinder.
Ends being over 24″ will be considered as vertical flat.
Surface facing up or down does not apply.

Column	Given	Found
a	100	
b	85	
c		From Table 1 = 50
d		From Table 2 = 78
e		78 + 1 = 79
f		From Table 3 = 171.1
g		From Table 3 = 146.9
h		171.1 − 146.9 = 24.2
i	.5	
j		24.2 × .5 = 12.1
k		100 − 79 = 21
l		Hor-cyl (Table 5) = 8.13, Vert-flat (Table 5) = 11.2
m		Greater than 24″
n		None (above 24″)
o		8.13, 11.2
p		12.1 + 8.13 = 20.23, 12.1 + 11.2 = 23.3
q	−100	
r		79 − (−100) = 179
s		179/20.3 = 8.33, 179/23.3 = 7.68
t	.33	
u		8.83 × .33 = 2.91, 7.68 × .33 = 2.54

Therefore, 3″ thickness would be required, as shown in Figure 18.

Note: The resultant thickness of only 3″ is the direct result of the basic ambient condition that 50% RH would be highest relative humidity. ■

Had the surface been black instead of white, the film resistances would have been:

$$R_s = \frac{21}{32.33} = 0.65$$

$$R_s = \frac{21}{35.5} = 0.59$$

From just these three examples the surface resistances range from 0.59 to 1.04, which shows that a single assumed surface resistance cannot be used to determine insulation thickness to prevent condensation. As shown in the following example, using a single assumed surface resistance can lead to considerable error: If Q_a is determined by $\frac{t_a - t_s}{R_s}$ using an assumed R_s, then R for insulation is determined by $R = \frac{t_2 - t_1}{Q_{ca}}$. The R is no more accurate than the assumed R_s. Thus, in assuming R_s, the R which determines insulation thickness basically is also assumed, unless the values of R_s are known for each particular condition of wet-bulb temperature, dry-bulb temperature, surface shape, size, position, and emittance. As such tabulation for values of R_s does not exist, the use of R_s to determine insulation thickness to prevent condensation is not recommended.

■ EXAMPLE 36a

Metric Units

Given Metric Units	*conversion*		*English Values*
Heat Exchanger, horizontal 0.9144m dia.	× 39.37	=	36″ dia.
Ambient air, dry bulb temp. = 37.8° C (311° K)		=	100° F ambient dry bulb
Ambient air, wet bulb temp. = 29.4° C (302.6° K)		=	85° F ambient wet bulb
Operating temperature = −73.3° C (199.9° K)		=	−100° F oper. temp.
Conductivity of insulation = 0.0476 W/mK	× 6.933	=	0.33 Btu in./ft^2, hr, °F
Outer surface is white ε = 0.5	× 1	=	0.5 = ε

Using English Unit values enter on work sheet as shown in Figure 18 and as indicated in Example 36, determine the insulation thickness to prevent condensation. ■

Discussion on illustrative examples

After using this method of determining insulation thickness to prevent condensation, it becomes apparent that accurate answers can be obtained in just a few minutes. As can be observed, the thickness is directly affected by the ambient conditions to which the installation is exposed. The three illustrative examples show this.

In Example 34, where dry-bulb and wet-bulb temperatures dictate a maximum relative humidity of only 65%, only 1-1/2 and 2″ of insulation with a conductivity of .35 are required to prevent condensation at an operating temperature of −20° F.

In Example 35, where the relative humidity is 90% at 70° F dry bulb, 4 and 5″ of insulation are required to prevent condensation with an insulation conductivity equal to .3 and operating temperature of only 40° F.

In Example 36, with a very dry condition of 50% maximum relative humidity, 3″ of insulation will prevent condensation even when the operating temperature is −100° F.

The above shows the importance of an accurate analysis of the conditions to which the installation will be subjected. In most instances design conditions should be selected which are seldom exceeded for more than a few hours per year.

Film Resistance

In this method of computation, resistance to heat flow across the air film is not used even though past practices used an assumed

SELECTION OF INSULATION THICKNESS TO PREVENT CONDENSATION – WORK SHEET

Figure 16

EXAMPLE 34

DATE 1-15-64
WORK ORDER 936-4901
CALCULATED BY HENSON

			STEP 1 SELECT DESIGN SURFACE TEMPERATURE					STEP 2 DETERMINE SURFACE HEAT GAIN RATE											STEP 3 DETERMINATION OF THICKNESS OF INSULATION						
			ATMOSPHERIC DESIGN CONDITIONS					HEAT GAIN RATE DUE TO RADIATION																	
PIPE OR VESSEL NUMBER	DESCRIPTION OF SURFACE: HORIZONTAL, VERTICAL, FLAT, CYLINDER, SPHERE – SIZE, LENGTH	HEAT FLOW	DRY BULB TEMP t_a	WET BULB TEMP	RELATIVE HUMIDITY TABLE 1	DEW POINT TABLE 2	DESIGN SURFACE TEMP °F t_s	DRY BULB TO ZERO TABLE 3	SURFACE TEMP TO ZERO TABLE 3	NET GAIN COLUMN f−g	ABSORPTANCE ϵ	TOTAL RADIATION HEAT GAIN q_{ra} TABLE 4	Δt_1 $(t_a - t_s)$	HEAT GAIN TABLE 5	SIZE	CORRECTION FACTOR TABLE 5	TOTAL CONVECTION HEAT GAIN q_{ca}	TOTAL RAD AND CONV HEAT GAIN $q_a = q_{ra}+q_{ca}$	OPERATING TEMP t_1	Δt_2 $(t_s - t_1)$	THERMAL RESISTANCE R TABLE 6	INSULATION CONDUCTIVITY k	THK REQD TO PREVENT CONDENSATION ON FLAT SURFACE L or L TABLE 7	PIPE SIZE OR DIA OF SMALL VESSELS	THK REQD TO PREVENT CONDENSATION ON PIPE OR CURVED SURFACE TABLE 54
			a	b	c	d	e	f	g	h	i	j	k	l	m	n	o	p	q	r	s	t	u	v	w
P-107	V&H CYL 2" NPS	—	90	80	65	77	78	159.2	145.8	13.4	.9	12.2	12	4.87	6"OD	1.32	6.43	18.62	-20	98	5.45	.35	1.925	2"	1½"
P-110	" " 4" "	—	"	"	"	"	"	"	"	"	"	"	"	"	8"OD	1.25	6.08	18.28	"	"	5.45	"	1.925	4"	1½"
P-116	" " 6" "	—	"	"	"	"	"	"	"	"	"	"	"	"	10"OD	1.22	5.93	18.13	"	"	5.45	"	1.925	6"	2"

SELECTION OF INSULATION THICKNESS TO PREVENT CONDENSATION – WORK SHEET

Figure 17

EXAMPLE 35

DATE 1-15-64
WORK ORDER 936-4891
CALCULATED BY JONES

			STEP 1 SELECT DESIGN SURFACE TEMPERATURE					STEP 2 DETERMINE SURFACE HEAT GAIN RATE											STEP 3 DETERMINATION OF THICKNESS OF INSULATION						
			ATMOSPHERIC DESIGN CONDITIONS					HEAT GAIN RATE DUE TO RADIATION																	
PIPE OR VESSEL NUMBER	DESCRIPTION OF SURFACE: HORIZONTAL, VERTICAL, FLAT, CYLINDER, SPHERE – SIZE, LENGTH	HEAT FLOW	DRY BULB TEMP t_a	WET BULB TEMP	RELATIVE HUMIDITY TABLE 1	DEW POINT TABLE 2	DESIGN SURFACE TEMP °F t_s	DRY BULB TO ZERO TABLE 3	SURFACE TEMP TO ZERO TABLE 3	NET GAIN COLUMN f−g	ABSORPTANCE ϵ	TOTAL RADIATION HEAT GAIN q_{ra} TABLE 4	Δt_1 $(t_a - t_s)$	HEAT GAIN TABLE 5	SIZE	CORRECTION FACTOR TABLE 5	TOTAL CONVECTION HEAT GAIN q_{ca}	TOTAL RAD AND CONV HEAT GAIN $q_a = q_{ra}+q_{ca}$	OPERATING TEMP t_1	Δt_2 $(t_s - t_1)$	THERMAL RESISTANCE R TABLE 6	INSULATION CONDUCTIVITY k	THK REQD TO PREVENT CONDENSATION ON FLAT SURFACE L or L TABLE 7	PIPE SIZE OR DIA OF SMALL VESSELS	THK REQD TO PREVENT CONDENSATION ON PIPE OR CURVED SURFACE TABLE 54
			a	b	c	d	e	f	g	h	i	j	k	l	m	n	o	p	q	r	s	t	u	v	w
DUCT	HOR. FLAT	UP	70		90	67	68	137.3	135.2	2.1	.7	1.47	2	0.73	—	—	0.73	2.20	40	28	12.3	0.30	3.7	FLAT	
"	HOR. FLAT	DN	70		90	67	68	137.3	135.2	2.1	.7	1.47	2	0.36	—	—	0.36	1.83	40	28	15.3	0.30	4.6	"	
"	VERT. FLAT	—	70		90	67	68	137.3	135.2	2.1	.7	1.47	2	0.57	OVER 24"	—	0.57	2.04	40	28	13.7	0.30	4.0	"	

SELECTION OF INSULATION THICKNESS TO PREVENT CONDENSATION – WORK SHEET

Figure 18

EXAMPLE 36

DATE 1-15-64
WORK ORDER 936-4632
CALCULATED BY CARSON

			STEP 1 SELECT DESIGN SURFACE TEMPERATURE					STEP 2 DETERMINE SURFACE HEAT GAIN RATE											STEP 3 DETERMINATION OF THICKNESS OF INSULATION						
			ATMOSPHERIC DESIGN CONDITIONS					HEAT GAIN RATE DUE TO RADIATION																	
PIPE OR VESSEL NUMBER	DESCRIPTION OF SURFACE: HORIZONTAL, VERTICAL, FLAT, CYLINDER, SPHERE – SIZE, LENGTH	HEAT FLOW	DRY BULB TEMP t_a	WET BULB TEMP	RELATIVE HUMIDITY TABLE 1	DEW POINT TABLE 2	DESIGN SURFACE TEMP °F t_s	DRY BULB TO ZERO TABLE 3	SURFACE TEMP TO ZERO TABLE 3	NET GAIN COLUMN f−g	ABSORPTANCE ϵ	TOTAL RADIATION HEAT GAIN q_{ra} TABLE 4	Δt_1 $(t_a - t_s)$	HEAT GAIN TABLE 5	SIZE	CORRECTION FACTOR TABLE 5	TOTAL CONVECTION HEAT GAIN q_{ca}	TOTAL RAD AND CONV HEAT GAIN $q_a = q_{ra}+q_{ca}$	OPERATING TEMP t_1	Δt_2 $(t_s - t_1)$	THERMAL RESISTANCE R TABLE 6	INSULATION CONDUCTIVITY k	THK REQD TO PREVENT CONDENSATION ON FLAT SURFACE L or L TABLE 7	PIPE SIZE OR DIA OF SMALL VESSELS	THK REQD TO PREVENT CONDENSATION ON PIPE OR CURVED SURFACE TABLE 54
			a	b	c	d	e	f	g	h	i	j	k	l	m	n	o	p	q	r	s	t	u	v	w
E-101	HOR. 3'-0"OD CYL	—	100	85	50	78	79	171.1	146.9	24.2	.5	12.1	21	8.13	3'-0"	—	8.13	20.23	-100	179	8.83	.33	2.91		
"	ENDS VERT. OVER 3' (FLAT)	—	100	85	50	78	79	171.1	146.9	24.2	.5	12.1	21	11.2	3'-0"	—	11.2	23.3	-100	179	7.68	.33	2.54		

Figures 16, 17, and 18.

value of film resistance to determine thickness needed to prevent condensation. If it is desirable to determine this air-film resistance R_s, it can be found after the heat flow to the surface has been established.

$$Q_a = \frac{t_a - t_s}{R_s} \text{ or } R_s = \frac{t_a - t_s}{Q_a}$$

For Example 34 the film resistance for the three different surfaces would be:

$$R_s = \frac{12}{18.62} = 0.644$$

$$R_s = \frac{12}{18.28} = 0.657$$

$$R_s = \frac{12}{18.13} = 0.662$$

For Example 35 the film resistances for the three different surfaces would be:

$$R_s = \frac{2}{2.20} = 0.91$$

$$R_s = \frac{2}{1.83} = 1.09$$

$$R_s = \frac{2}{2.04} = 0.97$$

For Example 36 the film resistances for the two different surfaces would be:

$$R_s = \frac{21}{20.3} = 1.09$$

$$R_s = \frac{21}{23.3} = 0.90$$

Heating and Cooling of Piping, Vessels, and Equipment

There are a large number of plant processes involving transfer and storage of materials where the addition or removal of heat is not required. Many products can be successfully handled in the lowest winter ambient temperature as well as in the highest summer ambient temperatures.

Under certain operating conditions, the use of thermal insulation by itself, or combined with flow rates, may retard the loss or gain of heat so that a material can be transferred from one point to another without difficulty. There may be other cases where, due to extreme temperature differences, slow or no flow conditions, or the temperature sensitivity of the materials itself, that addition or removal of heat is required to successfully transfer or store the material.

There are many reasons for adding heat to materials being stored or transported. The viscosity of most liquids decreases with increasing temperature. Thus, pump requirements for transferring a product may be reduced by either maintaining or heating up a product to an elevated temperature. Also many products may freeze at some temperature in winter. Pipes, vessels and pumps containing such materials must be heated to temperatures above their freezing points. In some processes it is required that materials be held above their vaporization temperature. This must be achieved by the addition of heat.

In a completely opposite direction some materials may require cooling for transportation or storage. It is well known that temperature rise increases the speed with which a chemical reaction will occur. There is often a need to keep this reaction under control. For example, an epoxy might polymerize if not kept cool. Other products require cooling to keep pressure in storage vessels and pipes within design limitations. Still other products require cooling to retard corrosive effects on equipment containing these products.

The two basic media for adding (or removing) heat to process piping and equipment are fluid systems (vapor or liquid) and electrical systems.

Typical fluid heating systems are shown in Figure 19 and 20. Refrigeration cooling systems are shown in Figure 21.

From an insulation point of view, the proper selection of the heating or cooling system must accompany the proper selection and installation of the insulation. That is; improper selection of the heating or cooling system can be as disastrous as the improper selection and installation of the insulation. In the sections which follow, information concerning the proper design and installation of the heating and cooling systems will be presented. Energy users to be considered are shown as outlines in Figure 23.

Electrical systems utilize a number of methods to convert electrical energy to thermal energy. One is by resistance heating and the other by inductance heating. These basic methods are shown in Figure 22.

Internal heating and cooling systems

In this system the products inside storage vessels are heated or cooled by coils located in the interior. Piping is heated or cooled by a small tube or pipe located inside the product pipe.

Internal tracing in piping systems requires stuffing boxes and packing boxes at every cross or elbow in order to allow for physical placements in the piping system and to allow for differential expansion between the tracer and the process line. Stuffing boxes are expensive to install and constitute a maintenance problem. Cross contamination hazards are also a problem. Where internal coils are used in vessels, the same physical disadvantages exist. Internally heated and cooled systems are shown in Figures 19 and 20.

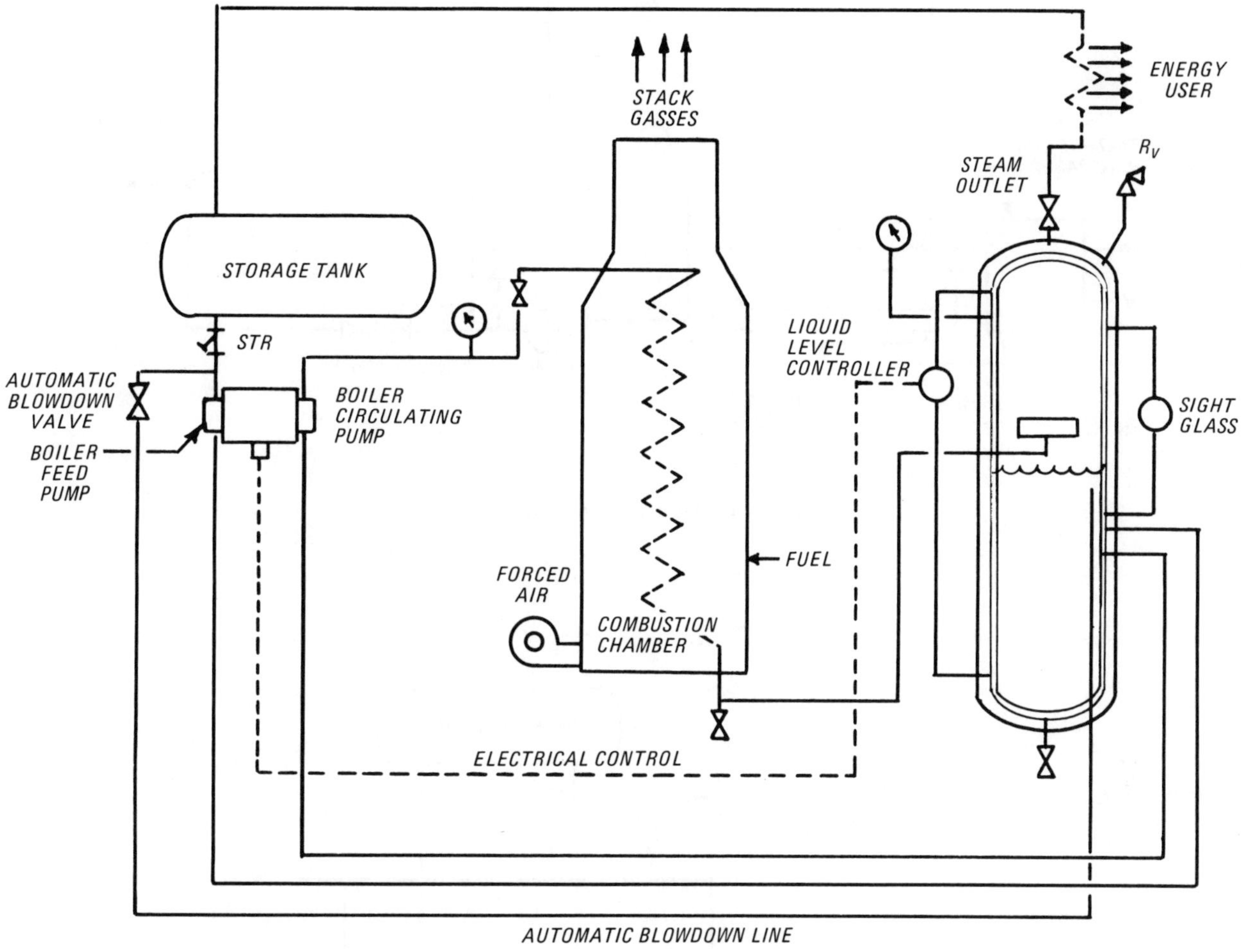

Figure 19 Direct steam heating system. (Typical vapor heating system.)

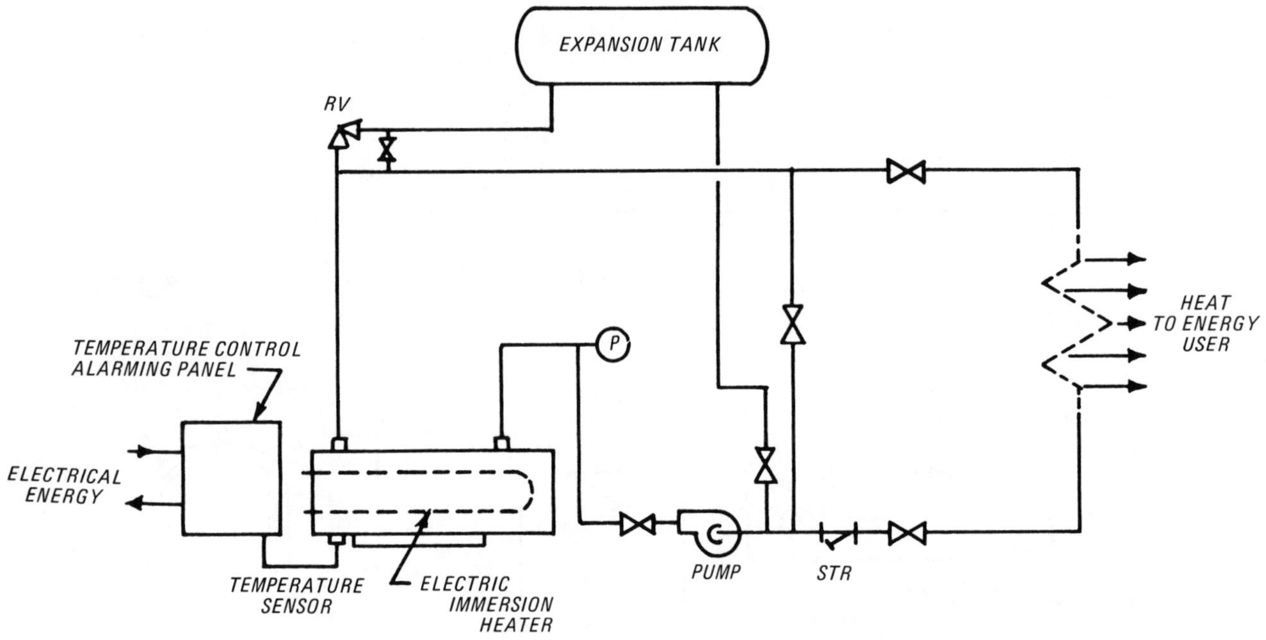

Electrically heated liquid circulation system.

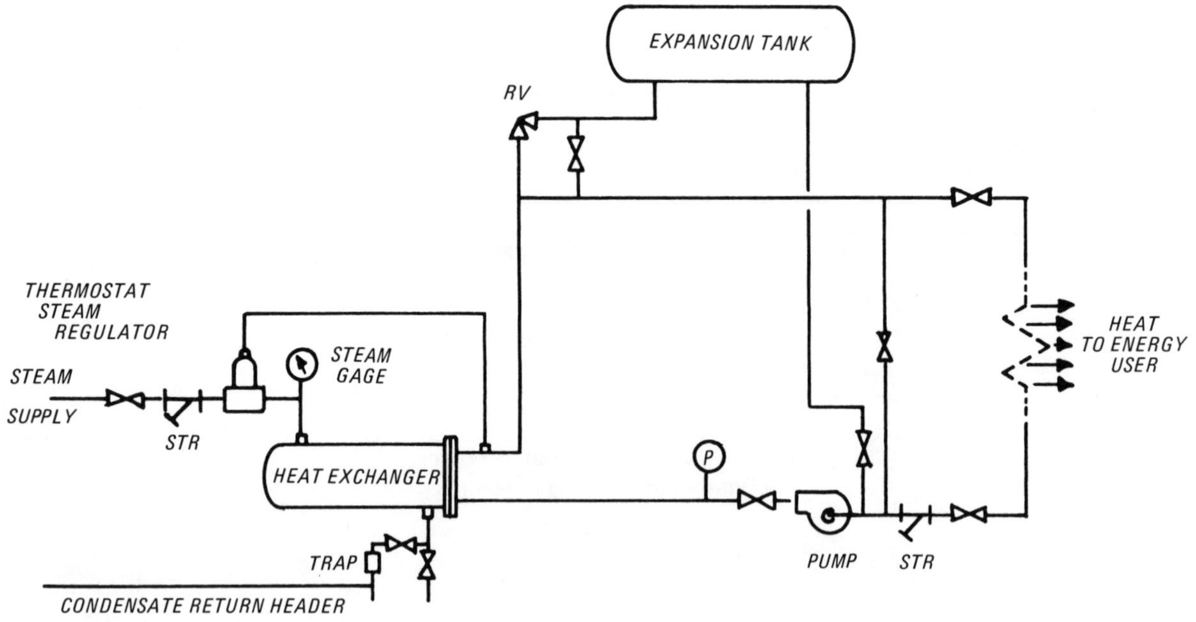

Steam heated liquid circulation system.
(Typical liquid heating system.)

Figure 20

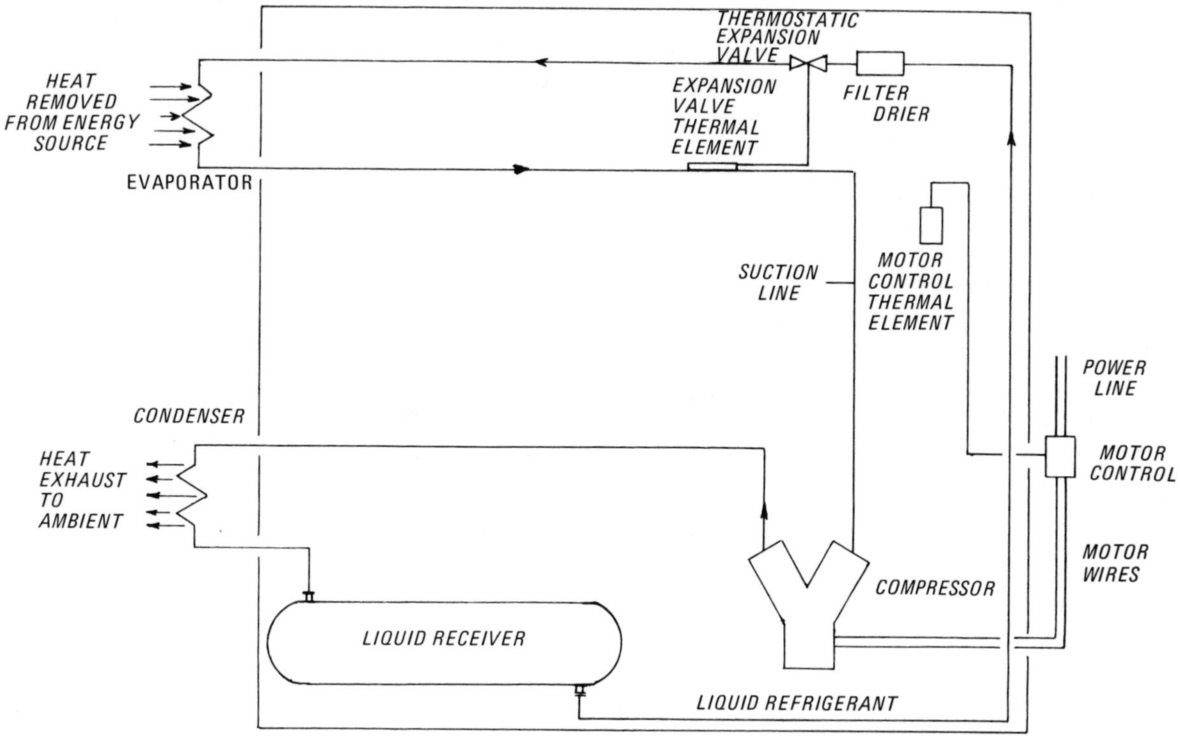

Basic refrigeration system.

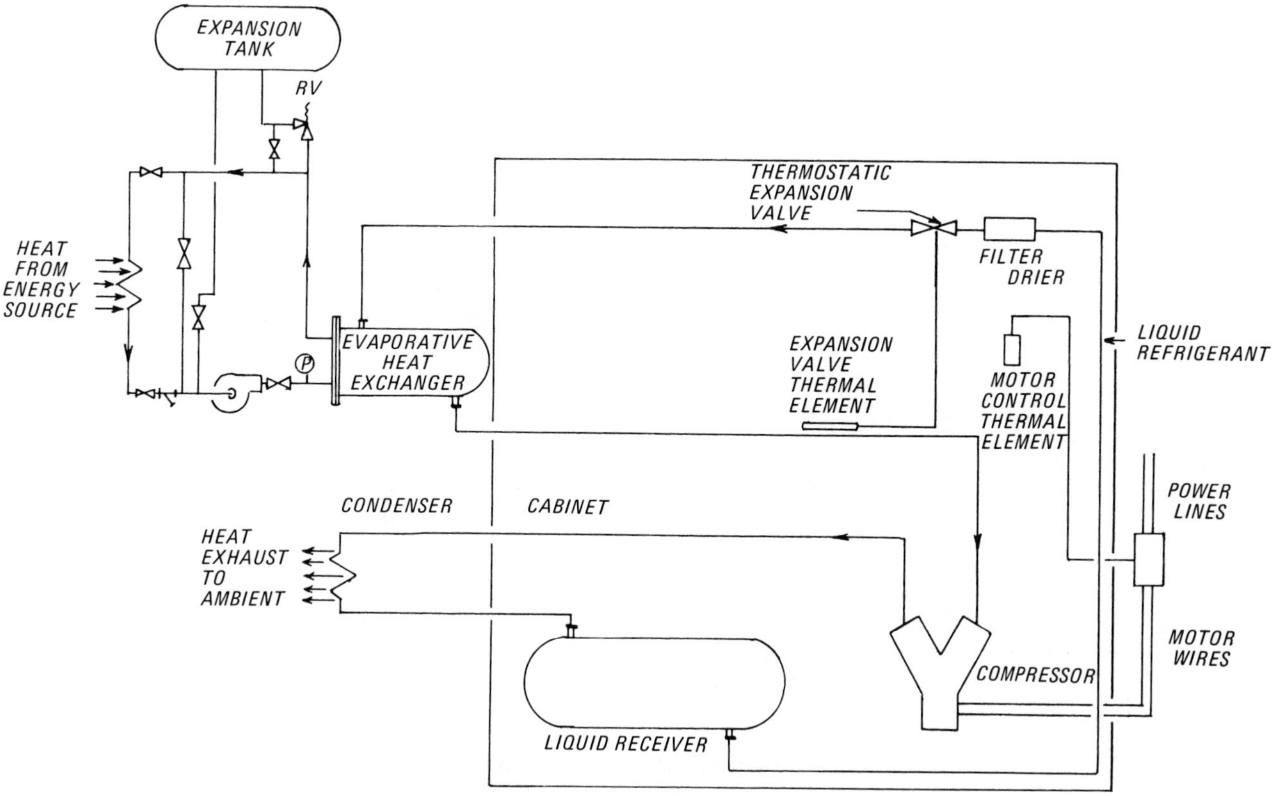

Basic refrigeration system with secondary liquid circulation loop.

Figure 21

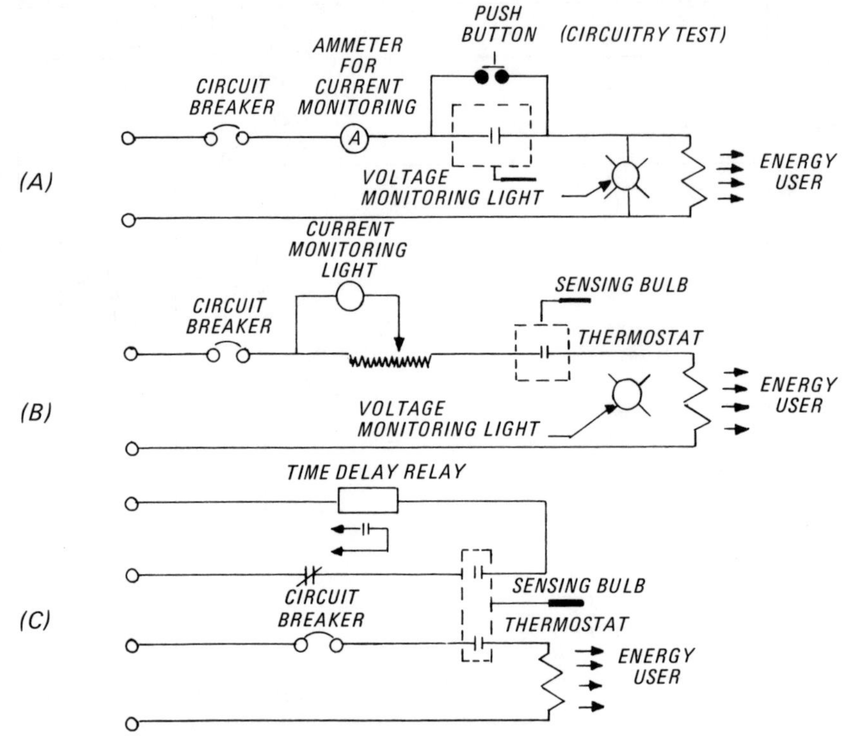

Typical resistance heating system for piping.

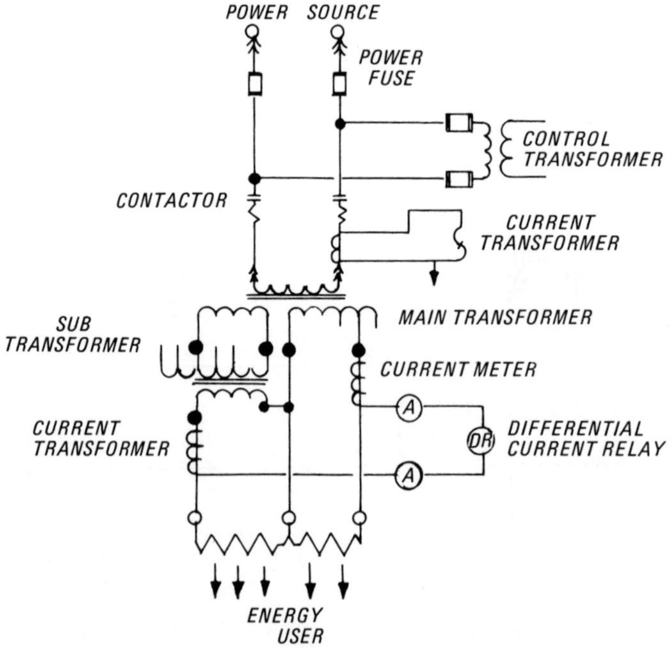

Typical inductance heating system for piping.

Figure 22

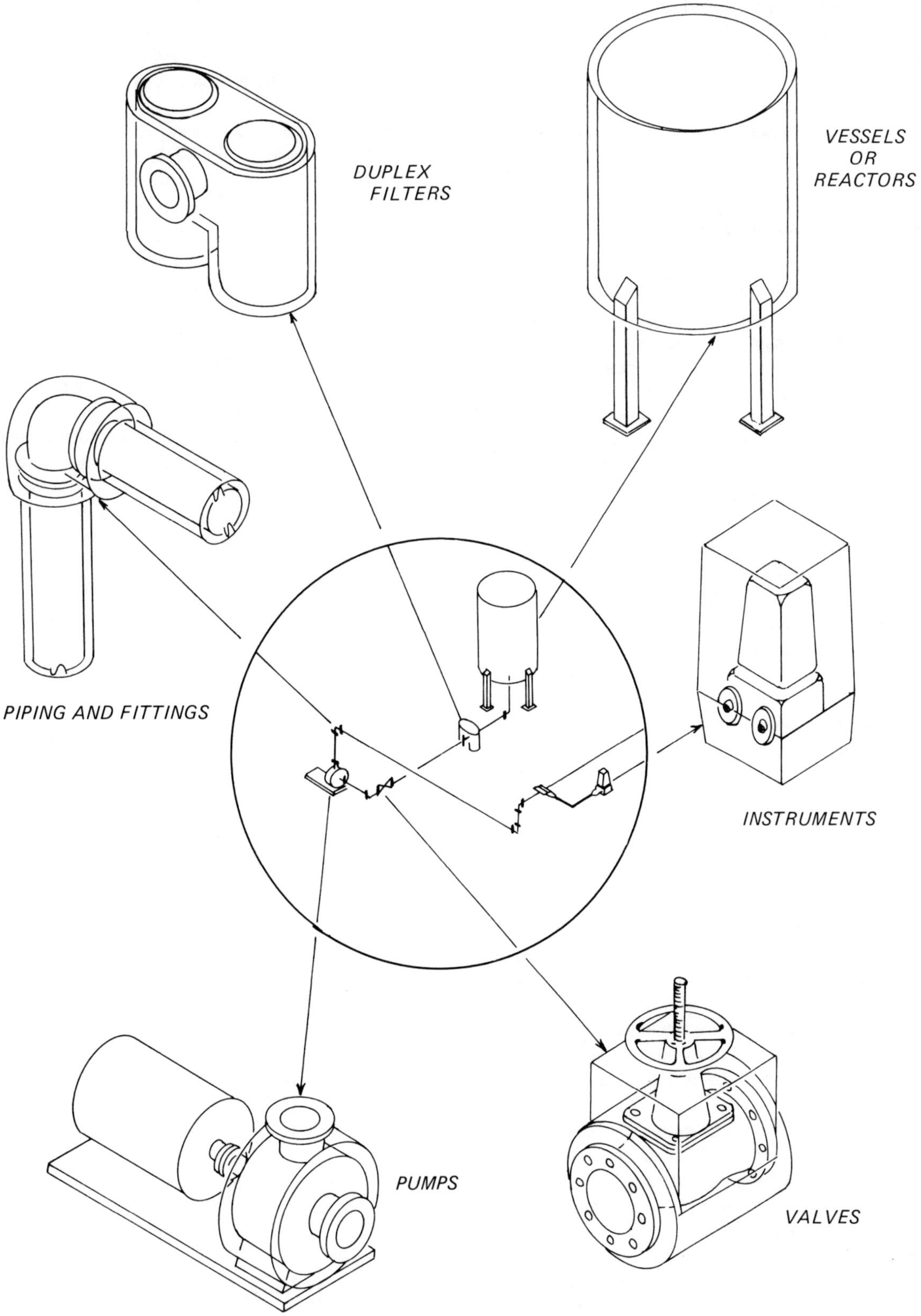

Typical items which require heating or cooling.

Figure 23

The heat losses are calculated based on the temperature of material in the pipe or vessel. For vessels the heat transfer equations used to calculate the heat transfer through the vessel insulation would be Equations 13, 13a, 14, or 14a depending on the particular conditions of interest. For piping the equations used would be 27, 27a, 28, 28a, 29, or 29a depending upon the particular conditions of interest.

Surface heating and cooling systems

Where fluid heating (or cooling) is used the outer surface of the pipe or vessel can be jacketed. The inner pipe or vessel contains the product while the annular region between the two walls carries the steam or other heat transfer fluid. Jacketing may also be specially fabricated to fit pumps, valves, etc. Since the only thermal resistances present between the fluid media and the energy user are the fluid media film resistance, the energy user wall, and the product film resistance, the efficiency of a jacketed system is extremely high. Jacketed systems typically maintain the product at very near the heating media temperature (assuming the pipe, vessel, or equipment has little or no product flow). On the negative side, the construction cost of such a system is extremely high in both materials and labor. The system is extremely expensive to maintain. Leakage or failure of the system is difficult to locate. Where failure does occur, it is frequently leakage of the inner wall such that the heating media passes into the process material and vice versa. The removal of product which plugs the jacket is an extremely difficult repair job. The usual solution is a complete replacement of the piping system. In addition, the possibility of cross contamination could cause complete loss of the product. In some processes, this cross contamination may be a critical hazard.

The heat transfer losses of jacketed pipe and equipment are based on the outer surface temperature of the jacket. On insulated pipe and equipment, this outer metal surface temperature would be approximately the same as the heating (or cooling) media. The heat transfer through the vessel insulation would be calculated by Equations 11, 11a, 13, 13a, 14 or 14a depending on the particular conditions. For piping, the equations used would be 27, 27a, 28, 28a, 29, or 29a depending upon conditions.

Surface heating can be accomplished by the use of electricity as the heating media. Three major means are available: (1) Resistance heating, (2) Induction heating and (3) Skin effect heating.

Straight resistance heating is accomplished by using the pipe as an electrical conductor and the heat is generated by its resistance to current flow. In the case of a pipe one end is connected to the minus side of an electrical circuit and the other end is connected to the positive side of the circuit. The heating is produced by the flow of electrical current from one end of the pipe to the other. Although simple, this system is seldom used because of the hazard. A break in the pipe would result in full voltage potential from one side of the break to the other. Such a system should never be used with any combustible product lines.

Induction heating is accomplished by impressing alternating flux across the ferromagnetic metal of the pipe. The pipe itself or even the outside of the thermal insulation is wrapped with coils of electrical wire which transmit alternating electrical currents. As in the case of a pipe, the induced current concentrates at the surface of the metallic object and will result in Joulian heating of the object. Typically, frequencies can range from 1,000 hz to 2,000,000 hz.

Skin effect induction heating is very effective where piping runs are a mile or more in length. When ferromagnetic material such as steel is used as an electrical conductor, skin effect heating will occur when transmitting an alternating current. That is, when an alternating current is passed through a steel tube, it does not flow uniformly through the cross section of the tube but concentrates near the inner surface of the tube. This restriction of the current to the surface of the conductor substantially increases the resistance of the steel tube and as a result increases the Joulian Heating (I^2R) of the conductor. This skin effect tracer installed on a pipe is shown in Figure 24. The heat tube is typically made of carbon steel and the internal electrical copper cable is electrically insulated with Teflon, PVC, polyethylene or silicone rubber. The commerical frequency voltage source typically ranges from 300 to 700 volts per kilometer of pipeline.

The induction and skin effect heating systems cause the temperature of the metal pipe or equipment to rise. The temperature of the pipe or equipment provides the temperature differential across the insulation to the ambient air. Thus Equations 11, 11a, 13, 13a, 14, 14a, 27, 27a, 28, 28a, 29 or 29a may be used to determine the heat loss. However, the heat supplied to the pipe and equipment is in electrical units, watts per unit length.

From Equation 11

When t_1 = temperature of inner surface °F (vessel or pipe temperature)
t_a = temperature of ambient air °F
r_1 = inner radius of insulation in inches
r_2 = outer radius of insulation in inches
k = conductivity of insulation Btu in./ft², hr, °F
f = conductance of air film Btu/ft², hr, °F
Q = heat transfer in Btu/ft², hr (based on outer surface area)

Then:

$$Q = \frac{t_1 - t_a}{\left[\left(\dfrac{r_2 \log_e \dfrac{r_2}{r_1}}{k}\right) + \left(\dfrac{1}{f}\right)\right]} \quad \text{in Btu/ft}^2\text{, hr}$$

or when $Q_{\ell f}$ = heat transfer per linear foot

$$Q_{\ell f} = \frac{t_1 - t_a}{\left[\left(\dfrac{r_2 \log_e \dfrac{r_2}{r_1}}{k}\right) + \left(\dfrac{1}{f}\right)\right]} \times \frac{2 r_2 \pi}{12} \quad \text{Btu/lin ft, hr}$$

Under static conditions these losses are supplied by electrical energy, which heats the metal pipe or equipment. Thus it is desirable to express the energy units in electrical units needed per

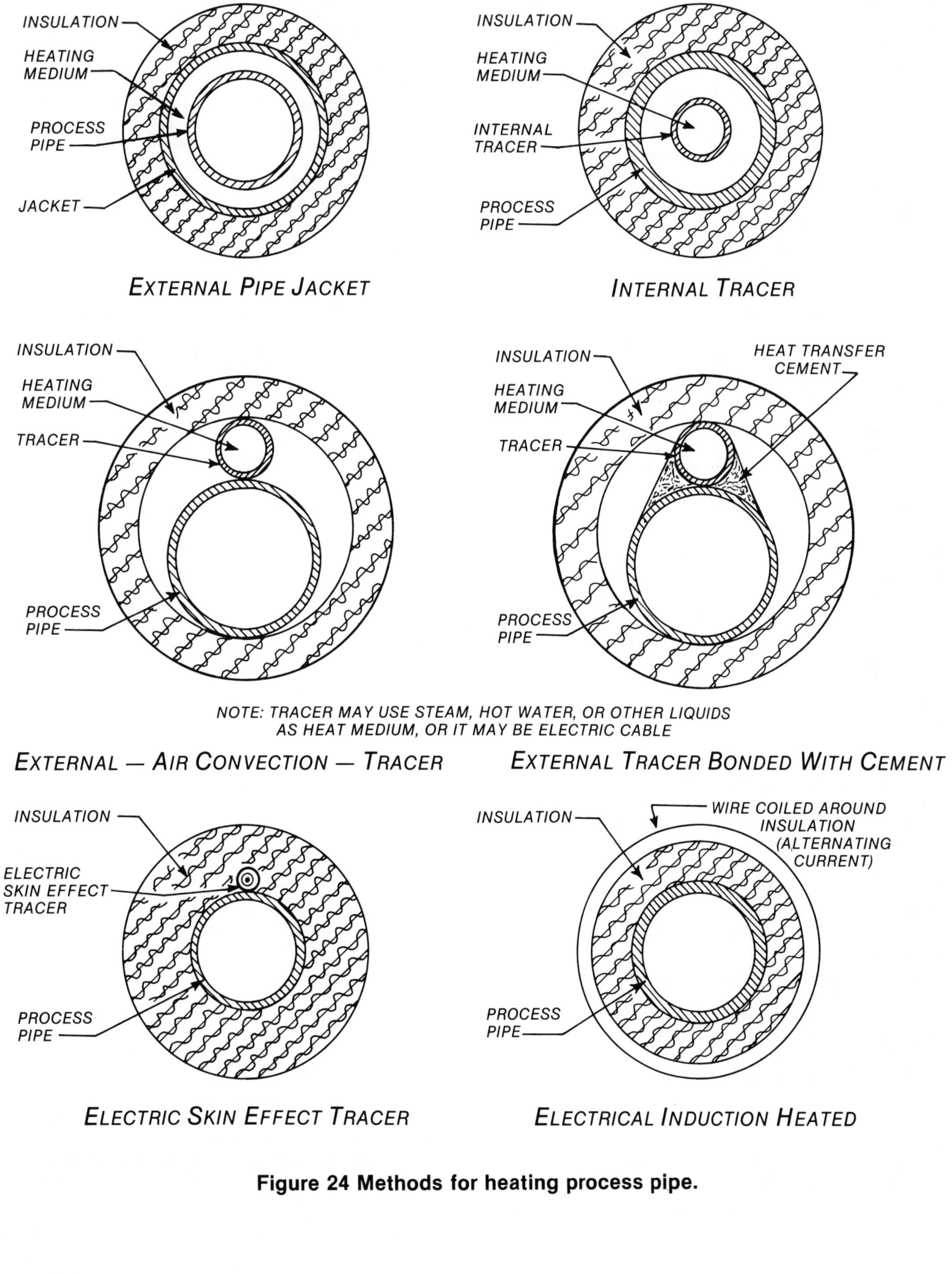

Figure 24 Methods for heating process pipe.

linear foot of pipe.

$$Q_{lf} = \frac{t_1 - t_a}{\left(\dfrac{r_2 \log_e \dfrac{r_2}{r_1}}{k}\right)} \times \frac{2r_2\pi}{12} \times \frac{1}{3.413}$$

When dimensions and heat units are in the metric system:

t_{m1} = temperature of inner surface °C (pipe or vessel)
t_{ma} = temperature of ambient air °C
r_{m1} = inner radius of insulation in metres
r_{m2} = outer radius of insulation in metres
k_m = conductivity of insulation W/m °K
f_m = conductance of air film in W/f² °K
Q_m = heat transfer in W/m²

$$Q_m = \frac{t_{m1} - t_{ma}}{\left[\left(\dfrac{r_{m2} \log_e \dfrac{r_{m2}}{r_{mi}}}{k_m}\right) + \left(\dfrac{1}{f_m}\right)\right]} \text{ in W/m}^2$$

When Q_{mlm} = heat transfer in W/lin m

$$Q_{mlm} = Q_m \times 2r_{m2}\pi$$

The Watts per linear metre is:

$$Q_{mlm} = \frac{t_{mi} - t_{ma}}{\left[\left(\dfrac{r_{m2} \log_e \dfrac{r_{m2}}{r_{m1}}}{k_m}\right) + \left(\dfrac{1}{f_m}\right)\right]} \times 2r_{m2}\pi \qquad (32a)$$

in Watts per linear metre

■ EXAMPLE 37

English Units

Determine the electrical energy needed per linear foot of pipe under following conditions:

The temperature of the pipe, t_1, is 200° F.
The temperature of ambient air, t_a, is 0° F.
The pipe size is 4 NPS, r_p = 2.25 in.
The inner radius of insulation, r_1 = 2.3 in.
The outside radius of insulation = 4.31 in.
The conductivity of the insulation k = 0.5 Btu in./ft², hr, °F at 100° F mean.
The conductance of the outer air film = 2 Btu/ft², hr, °F (typical for outdoors and sheltered from wind).

Find the Watt loss per linear foot of pipe:

$$Q_{lf} = \frac{200 - 0}{\dfrac{4.31 \log_e \dfrac{4.31}{2.3}}{0.5} + \dfrac{1}{2}} \times \frac{2 \times 4.31 \times \pi}{12} \times \frac{1}{3.413}$$

$$= \frac{200}{\dfrac{2.71}{0.5} + \dfrac{1}{2}} \times 2.26 \times \frac{1}{3.413}$$

$$= \left(\frac{200}{5.92} \times 2.26\right) \div 3.413 = 22.4 \text{ W/lin. ft.}$$

■

■ EXAMPLE 37a

Metric Units

English Values (as given)	**conversion**		**Metric Values**	
t_1 = 200° F	(200 − 32) ÷ 1.8	=	93.3° C	= t_{mi}
t_{ai} = 0° F	(0 − 32) ÷ 1.8	=	−17.8° C	= t_{mai}
r_1 = 2.3″	× 0.0254	=	0.05842m	= r_1
r_2 = 4.31	× 0.0254	=	0.10947m	= r_2
k = 0.5 Btu, in./ft², hr, °F	× 0.1442	=	0.0721 W/mK	= k_m
f = 2 Btu/ft², hr, °F	× 5.67826	=	11.3565	= f_m

Find the Watt loss per linear metre of pipe:

$$Q_{lf} = \frac{93.3 - (-17.8)}{\dfrac{0.10947 \log_e \dfrac{0.10947}{0.0384}}{0.0721} + \dfrac{1}{11.3565}}$$

$$\times 2 \times 0.10947 \times \pi$$

$$= \frac{111.1}{\dfrac{0.10947 \times 0.6283}{0.0721} + 0.088} \times 0.6878$$

$$= \frac{111.1}{0.954 + 0.008} \times 0.6878 = \frac{111.1}{1.092} \times 0.6878$$

$$= 73.3 \text{ W/lin m}$$

Check:

W/lin. m = 73.3 ÷ 3.280839 = 22.4 W/lin. ft

■

Heat added to process material

When it is necessary to add heat to the process from an external source of energy, this additional heat must be calculated and be added to the pipe loss. As such calculations must be for specific pipe conditions, the length of the pipe must be one of the factors.

The equation for calculating the heat loss, in watts per unit length, of the pipe remains the same as given in Equation 32 with the change of multiplication of heat loss per unit length by length L.

The heat to be added to raise the process material Δt is calculated by $\frac{W_p \times S_p \times \Delta t}{3.413}$ and is added to loss of pipe.

To obtain the heat energy required for total length of pipe, the heat required per unit length must be multiplied by total length of pipe (L). Thus total energy, Q, required for total pipe and to raise temperature of process material is, when:

S_p = specific heat of material (Btu/lb, °F)
W_p = weight of material in lbs per hour being added to system
Δtp = degrees of temperature, °F, material is to be increased
L = length of pipe in feet
t_{ap} = in this case, the average temperature, °F, over the length of the pipe
t_a = the ambient air temperature, °F
r_1 = inner radius of insulation in inches
r_2 = outer radius of insulation in inches
k = conductivity of insulation in Btu, in./ft², hr, °F
f = conductance of air film in Btu/ft², hr, °F

Then:

$$Q = \frac{t_{ap} - t_a}{\left[\left(\frac{r_2 \log_e \frac{r_2}{r_1}}{k}\right) + \left(\frac{1}{f}\right)\right]} \times \frac{2r_2\pi}{12} \times L \times \frac{1}{3.413} + \frac{W_p \times S_p \times \Delta t_p}{3.413} \text{ Btu/hr} \quad (33)$$

In Metric Units, when:

W_{mp} = weight of material in Kg per hour being added to system
S_{mp} = specific heat of material (Joule/kg °C)
Δt_{mp} = degrees of temperature °C (or °K) material is to be increased
L_m = length of pipe in metres
t_{map} = in this case, the average temperature, °C, over the length of the pipe
t_{ma} = the ambient air temperature, °C
r_{m1} = inner radius of insulation in metres
r_{m2} = outer radius of insulation in metres
k_m = conductivity of insulation in W/mK
f_m = conductance of air film in W/m²K

Then:

$$Q_m = \frac{t_{map} - t_{ma}}{\left[\left(\frac{r_{m2} \log_e \frac{r_{m2}}{r_{m1}}}{k_m}\right) + \left(\frac{1}{f_m}\right)\right]} \times 2r_{m2}\pi \times L_m + \frac{W_{mp} \times S_{mp} \times \Delta t_{mp}}{3600} \text{ Watts} \quad (33a)$$

■ **EXAMPLE 38**

English Units

With all values the same as given Example 37. Find the heat required if the pipe was 300′-0″ in length and it was necessary to raise the temperature of 1500 lbs of material 100° F every hour. The specific heat of the material is 0.95 Btu/lb °F.

$$Q = \frac{200 - 0}{\left[\left(\frac{4.31 \log_e \frac{4.31}{23}}{0.5}\right) + \left(\frac{1}{f}\right)\right]} \times \frac{2 \times 4.31 \times \pi}{12} \times \frac{1}{3.413} \times 300 + \frac{1500 \times 0.95 \times 100}{3.413}$$

$$= 22.37 \times 300 + \frac{142500}{3.413} = 6711 + 41752 = 48450 \text{ Watts}$$

■

■ **EXAMPLE 38a**

Metric Units

English Values	***conversion***		***Metric Values***	
L = 300′	×	0.3048 =	91.44m	$= L_m$
W_p = 1500 lbs	×	0.4535 =	680.25 Kg	$= W_{mp}$
S_p = 0.95	× 4187		= 3977.6 J/Kg	$= S_{mp}$
Δt = 100° F	÷	1.8	= 55.55° C	$= \Delta t_m$

$$Q_m = \frac{93.3 - (-17.8)}{\left[\left(\frac{0.10947 \log_e \frac{0.10947}{0.0584}}{0.0721}\right) + \left(\frac{1}{11.3565}\right)\right]} \times 2 \times 0.10947 \times \pi \times 91.44 + \frac{680.25 \times 3977.6 \times 55.5}{3.609}$$

$$= 6711 + 41752 = 48450 \text{ Watts}$$

■

External heating and cooling systems

External heating or cooling can be accomplished by running a small line (pipe or more commonly tubing) containing the heating media either parallel or spirally around the pipe or vessel. Heating can be accomplished in a similar manner by use of electrical heating cables. The first part of this section is directed to the fluid media of external heat tracing.

Fluid systems of external tracing

The external companion (tracer) tube or pipe may be a single or multiple parallel tracer system, or occasionally, tubing is spirally wrapped around the process pipe. The parallel tracers may depend upon air for convection movement of heat around the pipe or they may be thermally attached. Spiral tracers are the least efficient of all tracer systems. Because of the tortuous path, thermal expansion prevents contact with the pipe and also makes thermal banding ineffective. Being in the spiral configuration, they add considerable friction resistance to fluid flow or, when heated by steam, must be trapped approximately four times more frequently than straight run tracers. For these reasons, no effort was made to develop mathematical means of calculating heat transfer of spirally wrapped pipe systems.

The parallel systems can be installed in two basic systems. One depends on the convection of air to transmit the heat (or cooling) from the media to the process pipe and the other depends on the conduction of heat through a thermal bonding material to or from the process pipe. For very moderate requirements, the first may be satisfactory. The air convection system has the tracer attached to the pipe by stainless steel wire (or banding) or by stitch welding or brazing. These attachments are too small in area to be effective as thermal conductors. At most, the contact between pipe and tracer is a line contact. Thus, most of the heat is transferred by air convection currents moving in the annulus between the insulation and pipe. As the tracer is most often very small as compared to the pipe a relatively small amount of heat is transferred by radiation. However, where tracers are used for such applications as preventing the freezing of water lines, high heat delivery capability is not required. Table 8 provides dimensions of insulation for these systems.

Table 8 Insulation sizes for heat traced systems.

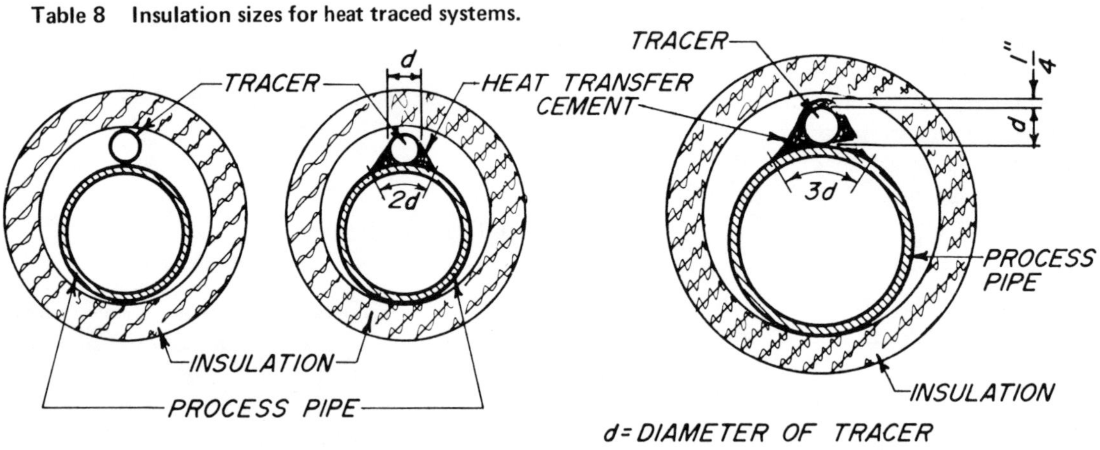

Table 8A

Process line size	Tracer size OD	Nominal pipe size insul	Process line size	Tracer size OD	Nominal pipe size insul
3/8 1/2 OD	1/4, 3/8 1/4, 3/8	1/2 1/2	2	1/4, 3/8 1/2, 5/8	2 1/2
1/2	1/4, 3/8 1/2, 5/8	3/4 1	2 1/8 OD	1/4, 3/8 1/2, 5/8	2 2 1/2
5/8 OD	1/4, 3/8 1/2, 5/8	3/4 1	2 1/2 OD	1/4, 3/8 1/2, 5/8	2 1/2
3/4 OD	1/4, 3/8 1/2, 5/8	1	2 1/2	1/4, 3/8 1/2, 5/8	3
3/4	1/4, 3/8 1/2, 5/8	1 1 1/2	3 OD	1/4, 3/8 1/2, 5/8	3 1/2
7/8 OD	1/4, 3/8, 1/2 5/8	1 1 1/2	3 NPS & 3 1/2 OD	1/4, 3/8 1/2, 5/8	3 1/2
1 OD	1/4, 3/8 1/2, 5/8	1 1 1/2	4 OD	1/4, 3/8 1/2, 5/8	4
1	1/4, 3/8 1/2, 5/8	1 1/2	4	1/4, 3/8, 1/2 5/8	4 1/2 5
1 1/8 OD	1/4, 3/8 1/2, 5/8	1 1/2	5 OD	1/4, 3/8 1/2, 5/8	5
1 1/4 OD	1/4, 3/8 1/2, 5/8	1 1/2	6 OD	1/4, 3/8 1/2, 5/8	6
1 1/4	1/4, 3/8, 1/2 5/8	1 1/2 2	6	1/4, 3/8 1/2, 5/8	7
1 3/8 OD	1/4, 3/8 1/2, 5/8	1 1/2	8 OD	1/4, 3/4 1/2, 5/8	8
1 1/2 OD	1/4, 3/8, 1/2 5/8	1 1/2 2	8	1/4, 3/4 1/2, 5/8	9
1 1/2	1/4, 3/8 1/2, 5/8	2	10 OD	1/4, 3/4 1/2, 5/8	10
1 5/8 OD	1/4, 3/8 1/2, 5/8	1 1/2 2	10	1/4, 3/4 1/2, 5/8	11
2 OD	1/4, 3/8 1/2, 5/8	2 2 1/2	12	1/4, 3/4 1/2, 5/8	14

Table 8B

Process line size	Tracer size OD	Nominal pipe size insul	Process line size	Tracer size OD	Nominal pipe size insul
3/8 1/2 OD	1/4, 3/8 1/4, 3/8	1 1	2	1/4, 3/8 1/2, 5/8	3
1/2	1/4, 3/8 1/2, 5/8	1 1/2	2 1/8 OD	1/4, 3/8 1/2, 5/8	3
5/8 OD	1/4, 3/8 1/2, 5/8	1 1/2	2 1/2 OD	1/4, 3/8 1/2, 5/8	3
3/4 OD	1/4, 3/8 1/2, 5/8	1 1/2	2 1/2	1/4, 3/8 1/2, 5/8	3 1/2
3/4	1/4, 3/8 1/2, 5/8	1 1/2	3 OD	1/4, 3/8 1/2, 5/8	4
7/8 OD	1/4, 3/8 1/2, 5/8	1 1/2	3 NPS & 3 1/2 OD	1/4, 3/8 1/2, 5/8	4
1 OD	1/4, 3/8 1/2, 5/8	1 1/2	4 OD	1/4, 3/8 1/2, 5/8	5
1	1/4, 3/8 1/2, 5/8	2	4	1/4, 3/8 1/2, 5/8	5
1 1/8 OD	1/4, 3/8 1/2, 5/8	2	5 OD	1/4, 3/8 1/2, 5/8	6
1 1/4 OD	1/4, 3/8 1/2, 5/8	2	6 OD	1/4, 3/8 1/2, 5/8	7
1 1/4	1/4, 3/8 1/2, 5/8	2 1/2	6	1/4, 3/8 1/2, 5/8	7
1 3/8 OD	1/4, 3/8 1/2, 5/8	2	8 OD	1/4, 3/8 1/2, 5/8	8 9
1 1/2 OD	1/4, 3/8 1/2, 5/8	2	8	1/4, 3/8 1/2, 5/8	9
1 1/2	1/4, 3/8 1/2, 5/8	2 1/2	10 OD	1/4, 3/8 1/2, 5/8	10 11
1 5/8 OD	1/4, 3/8 1/2, 5/8	2 1/2	10	1/4, 3/8 1/2, 5/8	11
2 OD	1/4, 3/8 1/2, 5/8	2 1/2 3	12	1/4, 3/8 1/2, 5/8	14

Note: To obtain English Dimensions for these Nominal Pipe Sizes see on Pages 477 and 478 in Appendix B, Tables B-6 and B-6a.
To obtain Metric Dimensions for these Nominal Pipe Sizes see on Pages 479 and 480 in Appendix B, Tables B-6b and B-6c.

Fluid-air convection systems of external tracing

Because there is considerable amount of piping that is traced with parallel tracers depending upon air convection, Estep and Turner developed empirical equations for estimating design data. The diagram and identification of symbols are shown in Figure 25. The inside film resistance indicated was taken to be equal to 0.5. The outside film resistance indicated was taken to be equal to .166 which is typical for 25 mph wind conditions. Although it has been previously shown that film resistances do change under changes in conditions, without making this assumption the problem is unsolvable because of the many unknown variables. As this system is restricted to mean temperatures across the insulation of less than 200° F (93° C) the thermal conductivity of the rigid mass type insulation was taken as 0.45 Btu in./ft^2, hr °F. Should the conductivity of the insulation used be different than the assumed value, the insulation thickness requirement can be corrected in the same ratio as the true conductivity bears to the assumed conductivity.

Derivation of this following formula to determine temperature which could be maintained in the product material under no flow conditions is, if:

Q_{ta} = heat transfer from tracer to annulus space
Q_{ap} = heat transfer from the annulus space to pipe
Q_{ia} = heat transfer from annulus space through insulation to air.

Then:

$Q_{ta} = Q_{ap} + Q_{ia}$ as illustrated in Figure 25.

However, under static conditions where material temperature in process line is only being maintained and not added to or reduced $Q_{ap} = 0$. Therefore $Q_{ta} = Q_{ia}$. Also under static conditions t_m must equal t_{ap}.

When d_t = outside diameter of tracer, in inches
d_p = outside diameter of pipe, in inches
d_i = inside diameter of insulation, in inches
d_o = outside diameter of insulation, in inches
t_{st} = temperature of steam in tracer, in °F
t_{ap} = temperature of annulus space, in °F
t_a = temperature of ambient air, in °F
t_m = temperature of material in process line, in °F
$\frac{1}{f_t}$ = film resistance of tracer = 0.5 Btu/ft^2, hr, °F
$\frac{1}{f_p}$ = film resistance of process pipe = .05 Btu/ft^2, hr, °F
$\frac{1}{f_i}$ = film resistance of inside of insulation = .05 Btu/ft^2, hr, °F
$\frac{1}{f_o}$ = film resistance of outside of insulation = 0.166 Btu/ft^2, hr, °F (includes wind factor)
k = conductivity of insulation, in Btu, in./ft^2, hr, °F

Then:

$$Q_{ta} = \left(\frac{t_{st} - t_{ap}}{\frac{1}{f_t}}\right) \times \frac{\pi d_t}{12}$$

and

$$Q_{ia} \left(\frac{t_{ap} - t_a}{\frac{d_o \div 2 \times \log_e \frac{d_o}{d_i}}{k} + \frac{1}{f_i}\left(\frac{d_o}{d_i}\right) + \frac{1}{f_o}}\right) \times \frac{\pi d_o}{12}$$

substituting t_m for t_{ap}

$$\left(\frac{t_m - t_a}{\frac{d_o \div 2 \log_e \frac{d_o}{d_i}}{k} + \frac{1\, d_o}{f_i\, d_i} + \frac{1}{f_o}}\right) \times \frac{\pi d_o}{12}$$

$$= \left(\frac{t_{st} - t_m}{\frac{1}{f_t}}\right) \times \frac{\pi d_t}{12}$$

solving for t_m: For one tracer:

$$t_m = \frac{t_{st}\, d_t \left(2.22\, d_o \log_e \frac{d_o}{d_i} + \frac{d_o}{d_i} + 0.332\right) + d_o\, t_a}{d_o + d_t \left(2.22\, d_o \log_e \frac{d_o}{d_i} + \frac{d_o}{d_i} + 0.332\right)} \tag{34}$$

If more than one tracer is installed when n = number of tracers:

$$t_m = \frac{n\, t_{st}\, d_t \left(2.22\, d_o \log_e \frac{d_o}{d_i} + \frac{d_o}{d_i} + 0.332\right) + d_o\, t_a}{d_o + n\, d_t \left(2.22\, d_o \log_e \frac{d_o}{d_i} + \frac{d_o}{d_i} + 0.332\right)} \tag{35}$$

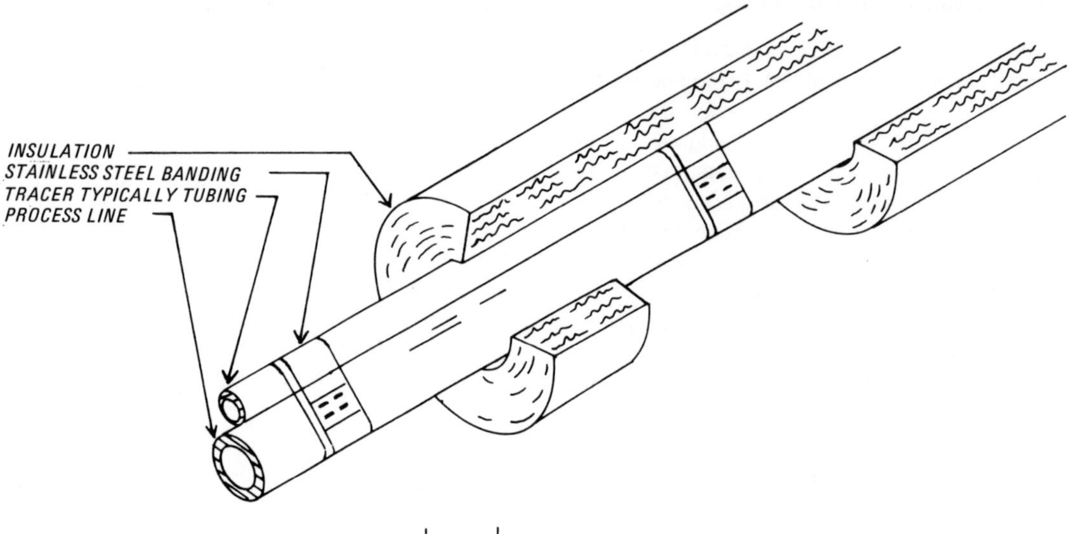

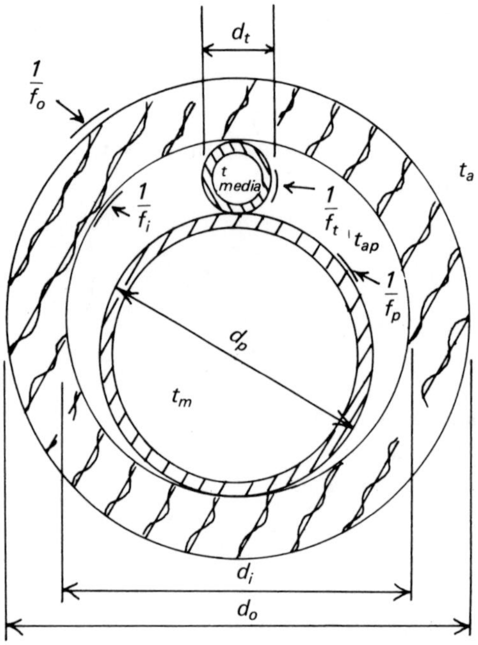

d_i and d_o obtained from nominal insulation sizes given on Table 8

ENGLISH		METRIC
d_t =	OUTSIDE DIAMETER OF TRACER (IN., MM)	d_{mt}
d_p =	OUTSIDE DIAMETER OF PIPE (IN., MM)	d_{mp}
d_i =	INSIDE DIAMETER OF INSULATION (IN., MM)	d_{mi}
d_o =	OUTSIDE DIAMETER OF INSULATION (IN., MM)	d_{mo}
t_{media} =	MEDIA TEMPERATURE OF TRACER (°F, °C)	t_{media}
t_{ap} =	TEMPERATURE OF ANNULUS SPACE (°F, °C)	t_{map}
t_a =	TEMPERATURE OF AMBIENT AIR (°F, °C)	t_{ma}
t_m =	TEMPERATURE OF PROCESS LINE (°F, °C)	t_{mm}
$\frac{1}{f_t}$ =	FILM RESISTANCE OF TRACER SURFACE	$1/f_{mt}$
$\frac{1}{f_p}$ =	FILM RESISTANCE OF PROCESS PIPE	$1/f_{mp}$
$\frac{1}{f_i}$ =	FILM RESISTANCE OF INSIDE OF INSULATION	$1/f_{mi}$
$\frac{1}{f_o}$ =	FILM RESISTANCE OF OUTSIDE OF INSULATION	$1/f_{mo}$
n =	NUMBER OF TRACERS	n

Units of the film resistances ($1/f_{mt}$, $1/f_{mp}$, $1/f_{mi}$, $1/f_{mo}$): $\left\{ \frac{hr \cdot ft^2 \cdot °F}{BTU}, \frac{M^2\ °C}{W} \right\}$

Note: Subscript m added to symbols when used with metric units.

Figure 25 Schematic diagram of external tracer air convection system.

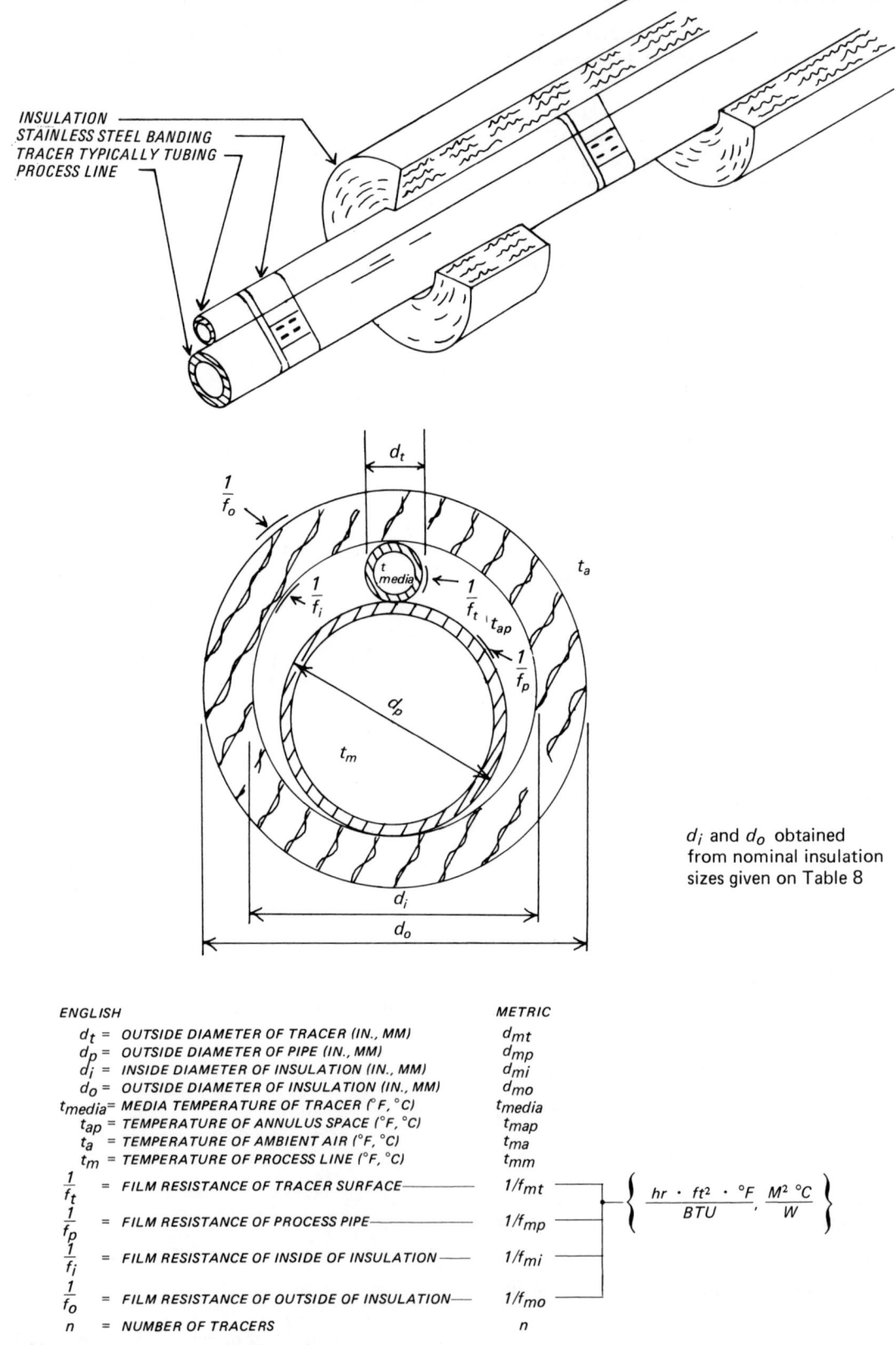

In Metric Units, with dimensions in mm, temperatures in °C, and heat transfer in W/m^2:

In Metric Units (subscript m added to symbols): For one tracer:

(34a)

$$t_{mm} = \frac{t_{mst}\, d_{mt} \left(0.087\, d_{mo} \log_e \frac{d_{mo}}{d_{mi}} + \frac{d_{mo}}{d_{mi}} + 0.332\right) + t_{ma}\, d_{mo}}{d_{mo} + \left(0.087\, d_{mo} \log_e \frac{d_{mo}}{d_{mi}} + \frac{d_{mo}}{d_{mi}} + 0.332\right) d_{mt}}$$

For more than one tracer when number of tracers = n:

(35a)

$$t_{mm} = \frac{n\, t_{mst}\, d_{mt} \left(0.087\, d_{mo} \log_e \frac{d_{mo}}{d_{mi}} + \frac{d_{mo}}{d_{mi}} + 0.332\right) + t_{ma}\, d_{mo}}{d_{mo} + \left(0.087\, d_{mo} \log_e \frac{d_{mo}}{d_i} + \frac{d_{mo}}{d_i} + 0.332\right) n\, d_{mt}}$$

While the previously presented equations in this section are quite complex; these equations still required certain simplifying assumptions be made in order that a direct solution might be made. These equations made the following simplifying assumptions:

A. The heat transfer coefficient of the air convection tracer is a constant 2.0 Btu/hr-ft²-°F (11.36 Watt/m²-°C).

B. The heat transfer coefficient of the air convection and radiation to the inside and from the outside of the insulation is 2.0 Btu/hr-ft²-°F (11.36 W/m²-°C) and 6.0 Btu/hr-ft²-°F (34.08 W/m²-°C)

C. The thermal conductivity of the insulation is a constant value of .45 Btu-in./hr-ft²-°F (.065 Watt/m²-°C).

Assumption A is not strictly correct. That is, the heat transfer coefficient of an air convection tracer has been found to range from a low of 2.0 Btu/hr-ft²-°F (11.36 Watt/m²-°C) to as high as 5.0 Btu/hr-ft²-°F (28.39 Watt/m²-°C) depending on the method of tracer attachment. Assumption B is likewise not strictly correct. The air film coefficient on either side of the insulation is clearly a function of such parameters as the pipe temperatures maintained, wind velocity, insulation thickness, etc., and can not be considered a constant. Assumption C is an oversimplification since thermal insulations increase in conductivity with temperature.

In order to consider all of these parameters as variables, an iterative solution of simultaneous equations is required. The iterative solution of necessity must be performed by a high speed digital computer. Tables 9 and 9a list steam tracing data for a number of different design conditions and are the results of an iterative solution developed by the Thermon Manufacturing Company.

For easy determination Table 9 was calculated for 0° F 25 mph wind ambient air conditions. These were selected as being the probable extreme design conditions. The pipe size ranges from 1-1/2" NPS to 30" NPS. Insulation thicknesses nominal 1-1/2", 2", 2-1/2" and 3" were used. The temperature of the steam varies from 250° F to 406° F. The results presented give temperature °F maintained in process pipe, lbs of steam condensed per hour per foot of tracer and heat loss in Btu per hour linear foot.

In Metric Table 9a the units are ambient air −17.8° C 40 Kph wind or 37.8° C still air. Pipe sizes considered are 43 mm to 762 mm. Insulation thickness include 38, 51, 64 and 76 mm. The temperature of the steam ranges from 121° C to 208° C. The results presented include temperature °C maintained in process pipe, GM/sec per linear metre of steam condensation; and heat loss in Watts per linear metre.

The conductivities of the insulation used in the calculations are as given below:

Insulation Design Conductivities

English Units Mean Temp.	0°F	50°F	100°F	150°F	200°F	250°F	300°F	350°F
Btu in./ft², hr, °F	0.34	0.37	0.41	0.45	0.48	0.52	0.56	0.59

Metric Units Mean Temp.	−20°C	10°C	40°C	70°C	100°C	130°C	150°C	180°C
W/mk	0.475	0.0541	0.0593	0.0650	0.0707	0.0752	0.0797	0.0842

Tables 9, 9a, 12 and 12a were computer calculated by Eneron Company in cooperation with Therman Company, San Marcus Texas, and are published by their courtesy.

Table 9 Steam tracing data (English units)

Design Conditions:

Tracer: One 1/2" O.D. Air convection tubing tracer
Insulation Thermal Conductivity: See Chart, page 79
Ambient: 0° F, 25 MPH wind

Nomenclature:

TS = Steam Temperature (°F)
TP = Pipe Temperature (°F)
#/HR-FT = Tracer condensate rate per linear foot
Q3 = Pipe Heat Losses in Btu/hr, lin ft

N.P.S.	Steam	INSULATION THICKNESS											
	Temp °F	1.5" Nom.			2.0" Nom			2.5" Nom			3.0" Nom		
	TS	TP	#/HR-FT	Q3	TP	#/HR-FT	Q3	TP	#/HR-FT	Q3	TP	#/HR-FT	Q3
1.5"	250	135.1	.03638	34.5	143.5	.03360	31.8	149.6	.03140	29.7	155.0	.02957	28.1
	274	148.0	.04115	38.3	157.5	.03804	35.4	164.4	.03558	33.1	169.9	.03355	31.3
	298	161.7	.04651	42.5	171.6	.04286	39.3	179.2	.04018	36.7	185.2	.03799	34.7
	320	173.2	.05153	46.1	183.8	.04752	42.6	191.9	.04448	39.9	198.4	.04201	37.7
	338	181.9	.05542	49.0	198.1	.05113	45.2	201.7	.04787	42.3	208.5	.04532	40.0
	366	195.1	.06193	53.3	207.2	.05713	49.2	216.5	.05362	46.1	223.8	.05071	43.7
	388	206.0	.06775	56.9	218.8	.06256	52.6	228.6	.05865	49.3	236.4	.05561	46.7
	406	214.7	.07274	59.9	228.1	.06727	55.4	238.4	.06310	52.0	246.5	.05979	49.2
2"	250	123.4	.04201	39.8	133.8	.03796	36.1	141.7	.03527	33.4	148.0	.03319	31.4
	274	135.1	.04739	44.4	146.6	.04314	40.2	155.7	.03994	37.3	162.4	.03759	35.0
	298	147.1	.05353	48.8	160.0	.04855	44.4	169.6	.04511	41.3	177.0	.04235	38.7
	320	157.6	.05898	52.9	171.4	.05368	48.2	181.7	.04999	44.8	189.6	.04689	42.1
	338	165.7	.06351	56.1	180.0	.05792	51.1	190.9	.05378	47.5	199.3	.05053	44.6
	360	177.8	.07089	61.0	193.1	.06460	55.6	204.8	.06013	51.7	213.8	.05648	48.6
	388	187.7	.07739	65.1	203.9	.07066	59.4	216.4	.06583	55.3	225.9	.06185	52.0
	406	195.7	.08305	68.5	212.6	.07577	62.5	225.5	.07067	58.2	235.5	.06651	54.8
3"	250	117.1	.04614	43.9	128.2	.04217	40.0	136.5	.03886	37.0	143.6	.03629	34.4
	274	128.3	.05220	48.7	140.4	.04757	44.4	149.8	.04415	41.1	157.7	.04105	38.2
	298	139.7	.05876	53.8	153.3	.05375	49.0	163.2	.04967	45.4	171.9	.04625	42.3
	320	149.6	.06488	58.2	164.2	.05921	53.1	174.9	.05490	49.3	184.4	.05142	46.1
	338	157.4	.06984	61.7	172.4	.06388	56.3	183.7	.05922	52.2	193.7	.05530	48.8
	366	168.9	.07779	67.0	185.0	.07118	61.2	197.1	.06606	56.8	207.9	.06180	53.2
	388	178.4	.08508	71.5	195.4	.07770	65.4	208.2	.07225	60.7	219.6	.06763	56.8
	406	185.9	.09122	75.1	203.7	.08348	68.7	217.1	.07747	63.9	229.0	.07262	59.8
4"	250	109.4	.05091	48.2	120.7	.04583	43.6	130.5	.04229	40.1	137.3	.03951	37.6
	274	119.9	.05756	53.5	132.2	.05185	48.4	143.0	.04771	44.5	150.8	.04488	41.8
	298	130.5	.06465	59.0	144.1	.05850	53.4	156.1	.05389	49.2	164.3	.05048	46.2
	320	139.8	.07118	63.9	154.6	.06446	57.9	167.2	.05939	53.3	176.0	.05579	50.1
	338	146.9	.07657	67.7	162.4	.06952	61.3	175.7	.06405	56.5	184.9	.06016	53.1
	366	157.7	.08506	73.3	174.2	.07730	66.6	188.5	.07138	61.4	198.5	.05711	57.7
	388	166.5	.09298	78.1	184.0	.08455	71.1	199.1	.07793	65.6	209.7	.07342	61.7
	406	173.0	.09953	82.1	191.8	.09060	74.7	207.6	.08374	69.0	218.6	.07870	64.9
6"	250	97.8	.05780	54.3	109.8	.05245	49.7	118.7	.04802	45.5	127.9	.04442	42.1
	274	107.1	.06459	60.1	120.1	.05901	54.9	130.1	.05414	50.5	140.4	.05007	46.7
	298	116.5	.07226	65.9	130.9	.06630	60.5	142.0	.06097	55.8	153.0	.05654	51.6
	320	124.8	.07957	71.2	140.2	.07303	65.5	152.2	.06730	60.4	163.9	.06230	55.9
	338	131.0	.08525	75.4	147.6	.07850	69.6	159.7	.07210	63.8	172.2	.06717	59.3
	366	140.6	.09479	81.7	158.1	.08716	75.1	171.4	.08050	69.3	184.8	.07485	64.4
	388	148.7	.10338	87.0	167.0	.09520	80.0	181.1	.08809	74.0	195.2	.08180	68.8
	406	155.0	.11088	91.3	174.1	.10207	84.0		.09444	77.9	203.6	.08766	72.3
8"	250	88.8	.06175	58.5	103.2	.05575	52.8	112.3	.05213	49.4	119.3	.04859	46.1
	274	97.1	.06947	64.7	113.1	.06286	58.5	123.1	.05893	54.8	130.5	.05457	50.9
	298	105.8	.07786	71.2	123.2	.07060	64.5	133.9	.06570	60.2	142.5	.06150	56.2
	320	113.3	.08588	76.9	132.0	.07769	69.7	143.7	.07265	65.2	152.8	.06791	61.0
	338	118.8	.09162	81.0	138.5	.08313	73.5	151.0	.07811	69.0	160.6	.07311	64.6
	366	127.5	10196	87.7	148.8	.09263	79.7	162.0	.08702	74.9	172.2	.08130	70.0
	388	134.6	.11097	93.4	157.2	.10100	84.9	170.9	.09477	79.7	181.9	.08883	74.7
	406	140.4	.11892	97.9	163.9	.10817	89.1	178.2	.10156	83.7	189.9	.09852	78.8

10″	274	85.5	.07454	69.6	97.2	.06983	65.3	107.4	.06564	61.1	116.1	.06214	57.8
	298	93.2	.08368	76.5	105.9	.07853	71.8	117.1	.07356	67.3	126.5	.06985	63.7
	320	99.8	.09232	82.6	113.5	.08662	77.0	125.4	.08121	72.7	135.6	.07682	68.9
	338	104.9	.09884	87.3	119.0	.09241	81.7	131.7	.08701	76.9	142.7	.08253	73.0
	366	112.5	.10975	94.5	127.7	.10284	88.5	141.6	.09678	83.4	153.1	.09200	79.1
	388	118.9	.11952	100.6	134.8	.11191	94.2	149.6	.10549	88.8	161.5	.09995	84.0
	406	124.0	.12802	105.4	140.6	.11992	98.8	155.9	.11313	93.2	168.3	.10708	88.2
12″	250	70.4	.07024	66.5	82.5	.06547	62.9	92.3	.06118	57.9	100.6	.05794	54.9
	274	77.1	.07869	73.5	90.2	.07297	68.2	101.1	.06884	64.1	110.0	.06496	60.5
	298	84.0	.08885	80.8	98.3	.08193	74.9	110.1	.07714	70.6	119.8	.07282	66.6
	320	90.0	.09718	87.2	105.3	.09016	80.9	118.0	.08508	76.2	128.4	.08040	72.0
	338	94.0	.10428	92.1	110.6	.09680	85.5	124.0	.09120	70.6	134.9	.08615	76.1
	366	101.5	.11509	99.7	118.7	.10754	92.6	132.8	.10096	80.6	145.0	.09580	82.5
	388	107.0	.12556	105.5	125.4	.11712	98.5	140.5	.11003	92.6	153.1	.10448	87.9
	406	111.6	.13402	110.6	130.8	.12547	103.3	146.5	.11792	97.1	159.7	.11201	92.3
14″	250	64.5	.07294	69.1	76.3	.06784	64.3	86.2	.06401	60.6	94.3	.06048	57.3
	274	70.9	.08208	76.4	83.7	.07638	71.0	94.5	.07178	67.1	103.3	.06807	63.4
	298	77.3	.09171	83.9	91.2	.08534	78.1	102.7	.08023	73.4	112.6	.07628	69.8
	320	82.6	.10041	89.9	97.7	.09387	84.3	110.0	.08827	79.3	120.6	.08416	75.4
	338	86.8	.10778	95.0	102.6	.10078	89.1	115.6	.09482	83.8	126.7	.09020	79.7
	366	93.1	.11927	102.8	110.1	.11187	96.4	124.0	.10554	90.7	136.0	.10045	86.4
	388	98.4	.12975	109.2	116.4	.12202	102.5	131.0	.11478	96.0	143.9	.10929	92.0
	406	102.7	.13882	114.5	121.3	.13048	107.5	136.6	.12296	101.3	149.8	.11672	96.1
16″	250	60.1	.07522	71.3	71.6	.07015	66.5	81.1	.06606	62.6	89.3	.06291	59.6
	274	65.9	.08425	78.7	78.5	.07858	73.5	88.8	.07409	69.2	97.8	.07054	65.9
	298	71.9	.09385	85.8	85.5	.08823	80.7	96.8	.08314	76.1	106.6	.07928	72.5
	320	77.0	.10340	92.6	91.7	.09701	87.1	103.7	.09151	82.1	114.0	.08704	78.0
	338	81.0	.11088	97.8	96.3	.10416	92.0	109.0	.09822	86.8	119.8	.09327	82.4
	366	86.9	.12262	105.7	103.4	.11549	99.5	116.9	.10908	94.0	138.5	.10382	89.3
	388	91.8	.13371	112.3	109.2	.12598	105.9	123.6	.11882	100.0	135.8	.11292	95.0
	406	95.8	14276	117.7	113,7	.13390	110.5	128.8	.12725	104.8	141.8	.12103	99.7
18″	250	56.2	.07656	72.6	67.2	.07219	68.4	76.7	.06815	64.6	84.8	.06512	61.7
	274	61.6	.08569	80.1	74.0	.08082	75.6	84.0	.07670	71.3	92.7	.07261	67.8
	298	67.1	.09681	88.0	80.6	.09078	83.0	91.0	.08513	78.4	101.0	.08155	74.6
	320	72.2	.10597	94.8	86.4	.10002	89.6	98.1	.09455	84.7	108.2	.08974	80.6
	338	75.9	.11357	100.2	90.8	.10736	94.6	103.1	.10125	89.5	113.7	.09633	85.1
	366	81.5	.12565	108.3	97.2	.11810	101.8	110.7	.11239	96.8	122.0	.10703	92.2
	388	86.1	.13679	115.1	102.7	.12877	108.2	116.9	.12260	103.0	128.9	.11658	98.1
	406	89.8	.14616	120.5	107.1	.13752	113.4	121.9	.13108	108.0	134.4	.12488	102.9
20″	250	52.8	.07776	74.1	63.5	.07403	70.2	72.7	.07003	66.3	80.5	.06667	63.2
	274	57.9	.08786	81.8	69.9	.08298	77.5	79.7	.07842	73.3	88.2	.07478	69.8
	298	63.1	.09851	89.8	76.2	.09306	85.1	86.9	.08808	80.6	96.1	.08391	76.8
	320	67.7	.10795	96.9	81.4	.10189	91.2	93.1	.09686	87.0	103.0	.09252	82.8
	338	71.4	.11592	102.3	85.6	.10930	96.4	97.9	.10393	91.8	108.3	.09909	87.6
	366	76.7	.12830	110.6	91.9	.12092	104.2	105.0	.11531	99.3	116.2	.11004	94.8
	388	81.1	.13984	117.5	97.1	.13166	110.8	111.0	.12580	105.7	122.8	.11986	100.8
	406	84.5	.14928	123.1	101.3	.14074	116.1	115.7	.13427	110.8	128.0	.12836	105.7
24″	250	49.1	.07975	76.0	58.9	.07643	72.4	67.1	.07267	68.9	74.7	.06944	65.8
	274	53.6	.09010	83.9	64.6	.08556	80.0	73.9	.08168	76.0	81.9	.07807	72.7
	298	58.7	.10094	92.0	70.4	.09532	78.2	80.5	.09134	83.5	89.3	.08733	79.9
	320	62.9	.11060	99.3	75.5	.10505	94.0	86.3	.10064	90.1	95.6	.09608	86.2
	338	66.2	.11851	104.8	79.3	.11259	99.3	90.7	.10801	95.2	100.5	.10307	91.1
	366	71.3	.13132	113.2	85.2	.12455	107.3	97.2	.11882	102.4	107.9	.11439	98.5
	388	75.4	.14327	120.3	90.0	.13559	114.1	102.7	.13933	108.9	114.0	.12477	104.8
	406	78.7	.15286	126.0	93.9	.14493	119.5	107.1	.13831	114.1	118.9	.13320	109.9
30″	250	42.5	.08322	79.2	51.2	.07907	75.4	59.1	.07659	72.6	65.9	.07348	69.6
	274	46.6	.09395	87.4	56.1	.08929	83.1	64.8	.08574	80.2	72.6	.08236	76.9
	298	50.8	.10516	95.9	61.2	.10010	91.3	70.7	.09551	87.4	79.1	.09235	84.5
	320	54.5	.11520	103.4	65.6	.10966	98.4	75.8	.10527	94.2	84.8	.10174	91.1
	338	57.3	.12383	109.1	69.3	.11772	103.9	79.7	.11276	99.4	89.1	.10915	96.2
	366	61.0	.13672	117.8	74.4	.13027	112.3	85.5	.12481	107.6	95.4	.12008	103.5
	388	65.1	.14879	125.2	78.7	.14198	119.3	90.4	.13601	114.3	100.9	.13070	110.0
	406	67.7	.15803	130.1	82.1	.15152	124.9	94.3	.14524	119.8	105.2	.13978	115.3

Table 9 Steam tracing data (English units continued)

Design Conditions:

Tracer: One 1/2" O.D. Air convection tubing tracer
Insulation Thermal Conductivity: See Chart, page 79
Ambient: 100° F, no wind

Nomenclature:

TS = Steam Temperature (° F)
TP = Pipe Temperature (° F)
#/HR-FT = Tracer condensate rate per linear foot
Q3 = Pipe Heat Losses in Btu/hr, lin ft

N.P.S.	Steam Temp °F	INSULATION THICKNESS 1.5" Nom.			2.0" Nom			2.5" Nom			3.0" Nom		
	TS	TP	#/HR-FT	Q3	TP	#/HR-FT	Q3	TP	#/HR-FT	Q3	TP	#/HR-FT	Q3
1.5"	250	179.8	.02129	20.2	183.9	.01986	18.9	187.1	.01873	17.8	189.7	.01783	16.9
	274	192.9	.02571	24.0	197.7	.02397	22.4	201.5	.02263	21.1	204.8	.02165	20.2
	298	206.4	.03055	27.9	212.0	.02855	26.1	216.6	.02708	24.7	220.1	.02580	23.6
	320	218.1	.03509	31.4	224.5	.03290	29.5	229.5	.03106	27.8	233.4	.02960	26.5
	338	227.0	.03869	34.2	233.9	.03632	32.1	239.2	.03430	30.3	243.5	.03269	28.8
	366	240.7	.04477	38.5	248.4	.04186	36.0	254.3	.03956	34.0	259.1	.03770	32.4
	388	252.1	.05009	42.1	260.2	.04682	39.4	266.6	.04423	37.2	271.9	.04217	35.5
	406	260.9	.05464	45.0	269.6	.05108	42.1	276.4	.04829	39.8	282.0	.04602	37.9
2"	250	173.6	.02413	22.9	178.9	.02240	21.3	182.8	.02097	19.9	186.0	.01982	18.8
	274	185.9	.02926	27.3	191.9	.02701	25.2	196.5	.02531	23.6	200.2	.02393	22.3
	298	198.3	.03481	31.8	205.2	.03210	29.3	210.6	.03009	27.5	214.9	.02850	26.0
	320	209.0	.03985	35.8	216.8	.03681	33.0	222.8	.03453	30.9	227.6	.03271	29.3
	338	217.2	.04395	38.8	225.6	.04004	35.9	232.0	.03812	33.6	237.5	.03622	32.0
	366	229.6	.05087	43.5	238.8	.04683	40.3	246.5	.04406	37.9	252.3	.04175	35.9
	388	239.8	.05558	47.6	250.1	.05239	44.0	258.2	.04930	41.4	264.5	.04673	39.3
	406	248.2	.06170	50.8	258.9	.05710	47.0	267.5	.05377	44.3	274.2	.05094	42.0
3"	250	170.0	.02654	25.2	175.5	.02462	23.4	179.9	.02314	22.0	183.6	.02174	20.7
	274	181.8	.03217	30.0	188.0	.02967	27.7	193.1	.02791	26.0	197.5	.02623	24.5
	298	193.6	.03824	34.9	200.8	.03520	32.2	206.7	.03320	30.3	211.7	.03115	28.5
	320	203.9	.04374	39.3	212.1	.04056	36.4	218.4	.03798	34.1	224.1	.03575	32.0
	338	211.7	.04821	42.6	220.5	.04470	39.5	227.3	.04191	37.0	233.5	.03940	34.8
	366	223.4	.05549	47.8	233.3	.05152	44.3	241.1	.04829	41.5	248.0	.04541	39.1
	388	233.2	.06199	52.2	244.2	.05753	48.4	252.4	.05396	45.4	259.8	.05080	42.7
	406	241.3	.06764	55.7	252.6	.06273	51.7	261.3	.05886	48.5	269.4	.05557	45.8
4"	250	165.4	.02898	27.5	171.3	.02674	25.4	176.3	.02493	23.7	179.8	.02354	22.4
	274	176.4	.03496	32.5	183.0	.03222	30.1	188.9	.03005	28.0	193.3	.02855	26.6
	298	187.7	.04167	38.0	195.4	.03849	35.1	201.9	.03573	32.6	206.8	.03394	31.0
	320	197.6	.04789	43.0	205.8	.04403	39.5	213.1	.04088	36.7	218.6	.03888	34.9
	338	205.1	.05304	46.9	213.8	.04852	42.9	221.7	.04506	39.8	227.6	.04285	37.9
	366	216.1	.06096	52.4	225.6	.05563	47.9	234.8	.05210	44.9	241.4	.04937	42.5
	388	225.2	.06790	57.1	235.8	.06240	52.5	245.8	.05823	48.9	252.7	.05520	46.4
	406	232.6	.07402	61.0	244.0	.06801	56.0	254.4	.06349	52.3	261.7	.06016	49.6
6"	250	159.4	.03517	31.4	165.3	.03092	29.3	189.5	.02823	26.8	174.2	.02633	25.0
	274	169.3	.03977	37.1	176.1	.03704	34.5	181.1	.03411	31.7	186.8	.03193	29.8
	298	179.3	.04710	43.1	187.2	.04383	40.1	192.9	.04045	36.9	199.4	.03792	34.6
	320	188.0	.05388	48.3	196.6	.04996	44.8	203.2	.04629	41.5	210.4	.04336	38.9
	338	194.3	.05907	52.1	203.9	.05499	48.6	211.0	.05101	45.1	218.7	.04778	42.2
	366	204.2	.06778	58.3	214.8	.06328	54.4	222.7	.05853	50.4	231.4	.05499	47.4
	388	212.4	.07558	63.6	223.7	.07034	59.1	232.6	.06561	55.1	242.2	.06137	51.6
	406	218.9	.08227	67.8	230.9	.07853	63.1	240.6	.07137	58.8	250.5	.06696	55.2
8"	250	154.1	.03587	34.0	161.4	.03307	31.3	165.7	.03101	29.4	169.2	.02870	27.3
	274	162.9	.04309	40.1	171.5	.03947	36.8	176.9	.03732	34.7	180.8	.03478	32.4
	298	172.3	.05081	46.5	181.8	.04661	42.6	187.9	.04389	40.1	192.4	.04110	37.5
	320	180.2	.05804	52.1	190.8	.05332	47.8	197.6	.05033	45.1	202.9	.04715	42.3
	338	185.8	.06352	56.1	197.7	.05875	51.8	204.9	.05536	48.9	210.8	.05218	46.1
	366	194.9	.07292	62.7	207.7	.06709	57.7	216.1	.06367	54.8	223.0	.06032	51.9
	388	202.3	.08120	68.3	216.2	.07476	62.9	225.1	.07071	59.5	232.7	.06733	56.0
	406	208.3	.08829	72.8	223.0	.08137	67.1	232.4	.07696	63.5	240.9	.07349	60.5

10″	274	155.7	.04837	43.3	161.8	.04396	40.9	167.6	.04174	38.9	172.1	.03955	36.8
	298	163.6	.05493	50.1	170.9	.05192	47.5	177.2	.04892	44.7	182.7	.04664	42.7
	320	170.8	.06255	56.1	178.4	.05889	52.9	185.6	.05591	50.2	191.8	.05336	47.9
	338	176.1	.06894	60.8	184.3	.06483	57.3	192.1	.06154	54.3	198.6	.05886	51.9
	366	184.1	.07884	67.9	193.2	.07426	63.9	201.8	.07062	60.8	208.9	.06715	57.8
	388	190.5	.08737	73.4	200.6	.08284	69.6	209.8	.07868	66.2	217.5	.07486	63.0
	406	195.7	.09491	78.2	206.4	.09000	74.2	216.3	.08567	70.8	224.5	.08144	67.1
12″	250	143.9	.04066	38.7	149.8	.03831	36.5	154.5	.03619	34.3	158.8	.03450	32.7
	274	151.0	.04891	45.6	157.7	.04568	42.6	163.5	.04347	40.5	168.7	.04151	38.7
	298	158.3	.05790	52.9	166.2	.05414	49.4	172.9	.05125	46.9	178.5	.04861	44.5
	320	164.4	.06539	58.7	173.5	.06150	55.2	180.9	.05854	52.5	187.1	.05556	49.9
	338	169.5	.07193	63.5	179.0	.06785	59.9	186.7	.06416	56.6	193.6	.06125	54.0
	366	176.8	.08235	70.9	187.3	.07762	66.9	196.0	.07345	63.2	203.6	.07018	60.4
	388	182.8	.09183	77.2	194.2	.08663	72.8	203.5	.08190	68.8	211.8	.07821	65.8
	406	187.6	.09973	82.2	199.7	.09408	77.6	209.6	.08904	73.3	218.3	.08516	70.1
14″	250	140.7	.04246	40.3	146.2	.03956	37.6	151.1	.03779	36.0	155.1	.03604	34.1
	274	147.0	.05039	47.0	153.7	.04756	44.4	159.5	.04538	42.3	164.2	.04322	40.2
	298	153.8	.05932	54.5	161.5	.05638	51.5	168.1	.05334	48.7	173.8	.05098	46.6
	320	159.6	.06789	60.9	168.4	.06429	57.7	175.5	.06073	54.5	181.9	.05823	52.3
	338	164.0	.07462	65.9	173.6	.07064	62.4	181.2	.06695	59.1	188.1	.06410	56.5
	366	171.0	.08538	73.5	181.3	.08087	69.7	189.7	.07661	65.0	197.1	.07311	62.9
	388	176.6	.09515	80.0	187.4	.08966	75.4	196.9	.08539	71.8	204.8	.08151	68.5
	406	181.1	.10384	85.3	192.5	.09731	80.2	202.5	.09295	76.5	211.0	.08854	73.0
16″	250	138.0	.04856	41.3	143.5	.04130	39.1	148.3	.03934	37.4	152.2	.03753	35.6
	274	144.1	.05206	48.6	150.6	.04936	46.1	156.3	.04709	43.9	160.9	.04490	41.9
	298	150.4	.06148	56.2	157.9	.05841	53.3	164.1	.05533	50.5	170.0	.05315	48.5
	320	155.8	.07017	63.0	164.2	.06648	59.7	171.3	.06300	56.5	177.4	.06013	54.0
	338	160.1	.07696	68.0	168.9	.07256	64.1	176.7	.06945	61.2	183.1	.06614	58.5
	366	166.6	.08825	76.0	176.2	.08309	71.6	184.7	.07940	68.4	192.0	.07575	65.3
	388	171.5	.09759	82.0	162.2	.09265	77.9	191.5	.08846	74.4	199.3	.08455	71.1
	406	175.7	.10588	87.2	187.0	.10046	82.9	196.6	.09563	78.8	205.2	.09189	75.7
18″	250	135.7	.04491	42.5	141.1	.04255	40.4	145.5	.04027	38.3	149.7	.03868	36.8
	274	141.5	.05359	50.0	147.8	.05091	47.5	153.0	.04832	45.1	157.8	.04651	43.4
	298	147.4	.06323	57.8	154.4	.05954	54.5	160.7	.05707	52.1	166.2	.05457	49.8
	320	152.6	.07202	64.6	160.4	.06774	60.8	167.5	.06507	58.4	173.5	.06216	55.8
	338	156.3	.07845	69.3	164.8	.07465	66.0	172.7	.07149	63.2	179.0	.06842	60.3
	366	162.2	.08988	77.3	171.9	.08548	73.6	180.1	.08135	70.1	187.4	.07825	67.4
	388	167.3	.10000	84.0	177.7	.09506	79.9	186.4	.09072	76.2	194.3	.08732	73.0
	406	171.2	.10864	89.5	182.2	.10342	83.2	191.5	.09842	81.2	199.8	.09479	78.1
20″	250	133.8	.04558	43.4	138.9	.04374	41.6	143.3	.04155	39.4	147.3	.04001	36.0
	274	139.2	.05492	51.3	145.3	.05225	48.8	150.4	.04968	46.4	154.9	.04742	44.3
	298	144.8	.06477	59.2	151.5	.06106	55.8	157.6	.05863	53.7	162.8	.05619	51.3
	320	149.7	.07370	66.1	157.1	.06969	62.5	164.0	.06682	60.0	169.9	.06396	57.4
	338	153.1	.08021	70.9	161.4	.07648	67.6	168.7	.07294	64.4	175.2	.07039	62.2
	366	158.6	.09201	79.1	168.1	.08767	75.5	176.0	.08354	72.0	183.2	.08052	69.4
	388	163.3	.10214	85.9	173.5	.09752	82.0	182.0	.09313	78.3	189.8	.08982	75.5
	406	167.3	.11104	91.5	177.8	.10603	87.3	186.8	.10098	83.3	195.1	.09746	80.4
24″	250	131.0	.04714	44.9	136.0	.04528	43.1	140.1	.04302	40.9	143.6	.04151	39.3
	274	136.3	.05638	52.6	141.0	.05356	50.0	146.0	.05161	48.2	150.8	.04960	46.3
	298	141.5	.06071	60.9	147.6	.06323	57.8	153.1	.06042	55.3	158.2	.05856	53.5
	320	145.8	.07516	67.4	152.8	.07196	64.6	158.0	.06880	61.8	164.6	.06667	59.8
	338	149.1	.08244	72.9	156.8	.07914	70.0	163.3	.07570	66.9	169.7	.07335	64.8
	366	154.3	.09440	81.2	162.7	.09067	78.0	170.3	.08669	74.6	177.1	.08385	72.2
	388	158.5	.10502	88.4	167.7	.10014	84.2	175.8	.09660	81.2	183.0	.09288	78.1
	406	162.0	.11409	94.0	172.0	.10937	90.1	180.3	.10487	85.4	187.8	.10083	83.2
30″	250	127.1	.04956	47.0	131.1	.04701	44.8	135.3	.04571	43.5	138.7	.04386	41.7
	274	131.2	.05875	54.7	136.5	.05642	52.7	140.9	.05402	50.4	145.1	.05246	49.0
	298	135.9	.06912	63.1	141.7	.06665	60.9	146.8	.06373	58.3	151.6	.06213	56.7
	320	139.8	.07867	70.6	146.3	.07565	67.9	151.9	.07235	64.9	157.0	.06989	52.7
	338	142.8	.08619	76.2	149.5	.08236	72.8	155.8	.07976	70.5	161.3	.07690	68.0
	366	147.3	.09878	85.0	154.8	.09415	81.0	161.7	.09118	78.4	168.1	.08791	75.7
	388	151.0	.10933	92.0	159.0	.10476	88.2	166.8	.10168	85.4	173.5	.09793	82.3
	406	153.7	.11817	97.3	162.6	.11383	93.8	170.8	.11034	99.9	177.8	.10649	87.7

Table 9 Steam tracing data (English units continued)

Design Conditions:

Tracer: One 3/4" O.D. Air convection tubing tracer
Insulation Thermal Conductivity: See Chart, page 79
Ambient: 0° F, 25 MPH wind

Nomenclature:

TS = Steam Temperature (°F)
TP = Pipe Temperature (°F)
#/HR-FT = Tracer condensate rate per linear foot
Q3 = Pipe Heat Losses in Btu/hr, lin ft

N.P.S.	Steam Temp °F	INSULATION THICKNESS											
		1.5" Nom			2" Nom			2.5" Nom			3" Nom		
	TS	TP	#/HR-FT	Q3	TP	#/HR-FT	Q3	TP	#/HR-FT	Q3	TP	#/HR-FT	Q3
1.5"	250	152.7	.04189	39.8	160.6	.03819	36.2	166.4	.03538	33.6	170.9	.03322	31.5
	274	167.6	.04758	44.4	176.1	.04335	40.4	182.5	.04015	37.5	187.5	.03771	35.2
	298	182.6	.05381	49.2	192.0	.04907	44.8	199.6	.04553	41.6	204.7	.04289	39.2
	320	195.7	.05962	53.4	205.7	.05435	48.8	213.3	.05060	45.4	219.5	.04760	42.7
	338	203.6	.06431	56.8	216.2	.05871	51.8	224.4	.05467	48.3	230.7	.05141	45.4
	366	220.7	.07196	61.9	232.3	.06590	56.7	240.9	.06125	52.7	247.7	.05762	49.6
	388	233.1	.07885	66.3	245.5	.07226	60.7	254.7	.06720	56.5	261.9	.06523	53.2
	406	243.1	.08475	69.8	256.0	.07770	64.0	265.6	.07228	59.6	273.2	.06805	56.1
2"	250	142.1	.04935	.46.9	152.2	.04418	42.0	159.8	.04045	38.4	165.4	.03767	35.7
	274	155.7	.05599	52.7	167.0	.05017	46.8	175.2	.04596	42.8	181.4	.04277	39.9
	298	169.9	.06317	57.8	182.1	.05665	51.8	191.0	.05199	47.5	197.8	.04841	44.2
	320	182.0	.06996	62.7	195.1	.06280	56.3	204.7	.05762	51.7	212.1	.05364	48.1
	338	191.1	.07540	66.5	205.0	.06767	59.8	215.1	.06218	54.9	222.9	.05795	51.1
	366	205.1	.08424	72.5	220.0	.07576	65.2	230.9	.06960	59.9	239.5	.06505	56.0
	388	216.6	.09217	77.5	232.4	.08300	69.8	244.0	.07629	64.1	253.2	.07134	60.0
	406	225.8	.09905	81.6	242.3	.08916	73.5	254.5	.08202	67.6	264.1	.07672	63.2
3"	250	136.9	.05519	52.4	147.8	.04957	47.1	155.9	.04535	43.1	162.6	.04178	39.7
	274	150.0	.06248	58.3	162.3	.05624	52.4	171.0	.05149	48.0	178.4	.04745	44.2
	298	163.7	.07051	64.5	176.8	.06346	58.0	186.4	.05813	53.2	194.5	.05368	49.0
	320	175.6	.07828	70.2	189.5	.07030	63.0	199.7	.06445	57.8	208.5	.05948	53.3
	338	184.4	.08433	74.4	199.1	.07576	66.9	209.9	.06944	61.3	219.2	.06417	56.6
	366	197.9	.09405	81.0	213.7	.08465	72.8	225.3	.07773	66.9	235.8	.07182	61.8
	388	209.0	.10290	86.5	225.7	.09263	77.9	238.1	.08515	71.6	248.7	.07671	66.2
	406	217.9	.11047	91.0	235.4	.09955	82.0	248.3	.09151	75.4	259.4	.08461	69.7
4"	250	129.5	.06125	58.2	141.0	.05477	52.1	150.9	.04974	47.3	157.3	.04620	43.9
	274	141.9	.06938	64.7	154.8	.06201	57.9	165.4	.05644	52.6	172.5	.05245	48.9
	298	154.8	.07819	71.4	168.7	.07000	64.0	180.3	.06369	58.2	188.0	.05922	54.2
	320	165.9	.08632	77.4	180.9	.07773	69.7	193.2	.07054	63.2	201.6	.06564	58.8
	338	174.2	.09299	82.1	190.1	.07773	73.9	203.0	.07603	67.1	211.8	.07073	62.5
	366	186.9	.10371	89.2	204.0	.09340	80.4	217.9	.08495	73.1	222.4	.07915	68.1
	388	197.5	.11336	95.3	215.0	.10221	86.0	230.3	.09297	78.2	240.4	.08666	72.9
	406	205.8	.12156	100.2	224.7	.10973	90.4	240.1	.09992	82.3	250.7	.09312	76.8
6"	250	118.0	.07005	66.4	130.2	.06320	60.0	139.9	.05773	54.9	148.9	.05264	50.0
	274	129.3	.07919	73.7	143.0	.07146	66.7	153.5	.06538	61.0	163.5	.05981	55.8
	298	140.9	.08912	81.4	156.1	.08063	73.7	167.4	.07380	67.4	178.2	.06756	61.8
	320	151.2	.09842	88.2	167.2	.08909	79.8	179.3	.08150	73.1	191.0	.07480	67.0
	338	158.9	.10585	93.4	175.1	.09587	84.6	188.4	.08788	77.5	200.8	.08060	71.1
	366	170.5	.11797	101.5	188.5	.10683	92.0	202.2	.09801	84.4	215.5	.09002	77.5
	388	180.1	.12878	108.3	199.1	.11683	98.2	213.7	.10726	90.2	227.7	.09847	82.8
	406	187.8	.13824	113.9	207.6	.12527	103.3	222.8	.11506	94.8	237.5	.10581	87.2
8"	250	109.5	.07791	73.9	124.2	.06816	64.6	133.4	.06284	59.7	141.0	.05847	55.6
	274	120.0	.08788	82.0	136.1	.07700	71.7	146.4	.07105	66.3	154.5	.06630	61.8
	298	130.5	.09854	90.1	148.6	.08670	79.2	159.6	.08028	73.3	168.5	.07473	68.3
	320	139.8	.10882	97.5	159.2	.09576	85.8	171.0	.08862	79.4	180.5	.08254	74.0
	338	147.1	.11695	103.3	167.3	.10299	90.9	179.6	.09538	84.2	189.7	.08857	78.5
	366	157.9	.13033	112.1	179.5	.11479	98.8	192.8	.10628	91.5	203.6	.09929	85.4
	388	166.6	.14169	119.2	189.6	.12538	105.4	203.7	.11625	97.7	215.1	.10859	91,3
	406	173.6	.15193	125.2	197.7	.13442	110.8	212.4	.12465	102.7	224.3	.11649	96.0

10″	274	107.8	.09661	90.2	120.3	.08863	82.7	130.7	.08127	75.8	139.8	.07587	70.8
	298	117.2	.10812	98.9	131.0	.09974	91.2	142.7	.09147	83.7	152.3	.08559	78.2
	320	125.6	.11924	106.9	140.4	.11015	98.7	152.9	.10103	90.6	163.2	.09454	84.7
	338	131.9	.12794	113.0	147.8	.11835	104.5	160.4	.10832	95.7	171.5	.10170	89.8
	366	141.5	.14245	122.5	158.3	.13138	113.0	172.2	.12094	104.0	184.0	.11337	97.5
	388	149.7	.15530	130.5	167.2	.13138	120.5	182.0	.13210	111.1	194.4	.12385	104.1
	406	156.1	.16621	137.0	174.3	.15363	126.6	189.8	.14182	116.8	202.7	.13281	109.5
12″	250	89.8	.09151	86.9	103.2	.08847	79.2	113.8	.07715	73.1	121.8	.07106	67.4
	274	98.4	.10306	96.2	113.1	.09398	87.7	124.6	.08672	89.9	133.5	.08019	74.8
	298	107.2	.11594	105.9	123.2	.10570	96.6	135.8	.09770	89.4	145.7	.09041	82.6
	320	114.8	.12765	114.4	132.0	.11663	104.5	145.6	.10745	96.4	156.2	.09974	89.5
	338	120.7	.13693	121.0	138.4	.12467	110.2	153.6	.11560	102.1	164.1	.10728	94.8
	366	129.2	.15162	130.4	148.8	.13886	119.5	164.2	.12888	110.9	176.1	.11909	103.0
	388	136.5	.16527	138.9	157.1	.15141	127.3	173.4	.14068	118.3	186.1	.13067	109.9
	406	142.3	.17676	145.7	163.8	.16207	133.6	180.8	.15084	124.3	194.0	.14025	115.5
14″	250	83.5	.09508	91.3	96.6	.08774	83.2	107.5	.08139	77.4	115.6	.07529	71.6
	274	91.3	.10758	100.4	105.9	.09863	92.1	117.5	.09145	85.4	127.0	.08555	79.8
	298	99.4	.12098	110.5	115.3	.11103	101.4	128.0	.10285	94.0	138.7	.09634	88.1
	320	106.5	.13310	119.3	123.5	.12225	109.6	137.2	.11352	101.7	148.5	.10612	95.2
	338	111.9	.14268	126.1	129.8	.13131	115.9	144.3	.12202	107.7	156.1	.11422	100.9
	366	120.0	.15864	136.5	139.2	.14599	125.6	154.9	.13577	116.8	167.6	.12747	109.7
	388	126.8	.17288	145.3	147.3	.15915	133.8	163.6	.14800	124.5	176.8	.13861	116.6
	406	132.2	.18485	152.4	153.6	.17028	140.3	170.3	.15818	130.3	184.4	.14872	122.5
16″	250	78.3	.09924	94.3	91.5	.09138	86.8	102.1	.08517	80.8	110.8	.07951	75.6
	274	85.8	.11171	104.2	100.2	.10293	96.1	111.9	.09584	89.5	121.4	.08978	83.8
	298	93.5	.12532	114.6	109.2	.11576	105.7	121.7	.10731	98.1	132.3	.10106	92.4
	320	100.2	.13809	123.8	117.0	.12747	114.2	130.3	.11837	106.1	141.8	.11116	99.6
	338	105.2	.14821	130.8	122.9	.13676	120.8	136.9	.12720	112.2	149.0	.11953	105.4
	366	112.9	.16453	141.6	131.6	.15148	130.3	147.1	.14143	121.7	159.8	.13299	114.4
	388	119.3	.17924	150.6	139.0	.16509	138.8	155.5	.15419	129.6	168.9	.14503	122.0
	406	124.4	.19166	157.9	145.2	.17657	145.6	162.1	.16503	136.0	176.1	.15548	128.1
18″	250	74.0	.10251	97.4	86.9	.09482	90.1	97.2	.08829	83.7	106.0	.08293	78.7
	274	81.0	.11540	107.7	95.2	.10678	99.7	106.5	.09925	92.6	116.1	.09340	87.2
	298	88.3	.12904	118.3	103.5	.11929	109.1	116.0	.11172	102.0	126.5	.10506	96.1
	320	94.6	.14230	127.7	110.9	.13146	117.8	124.3	.12301	110.2	135.5	.11591	103.9
	338	99.4	.15291	134.9	116.5	.14097	124.6	130.6	.13213	116.6	142.7	.12443	109.9
	366	106.7	.16952	146.0	124.9	.15678	134.9	140.4	.14688	126.4	152.8	.13809	188.8
	388	112.7	.18481	155.3	132.0	.17080	143.6	148.3	.16011	134.6	161.4	.15059	126.6
	406	117.5	.19769	162.9	137.6	.18282	150.6	154.6	.17147	141.2	168.3	.16119	132.9
20″	250	70.1	.10543	100.2	82.7	.09792	93.1	92.9	.09121	86.7	101.7	.08621	81.7
	274	76.8	.11868	110.7	90.4	.10957	102.3	101.8	.10274	94.9	111.4	.09699	90.5
	298	83.7	.13302	121.6	98.5	.12320	112.5	110.9	.11556	105.5	121.1	.10856	99.2
	320	89.6	.14631	131.3	105.5	.13555	121.5	118.8	.12725	114.4	129.8	.11973	107.3
	338	94.2	.15718	138.7	110.8	.14544	128.4	124.8	.13649	120.6	136.4	.12855	113.5
	366	109.8	.17336	149.3	118.9	.16151	139.0	133.7	.15124	130.1	146.6	.14301	123.0
	388	106.5	.18893	158.8	125.6	.17600	147.9	141.5	.16483	138.5	154.8	.15593	131.1
	406	111.1	.20202	165.4	131.0	.18814	155.1	147.5	.17629	145.3	161.4	.16686	137.5
24″	250	65.4	.10892	103.5	77.1	.10129	96.3	87.0	.09501	90.9	95.2	.09046	85.7
	274	71.9	.12286	114.4	84.5	.11402	106.4	95.3	.10767	100.5	104.3	.10163	94.9
	298	78.4	.13733	125.6	92.1	.12817	117.0	103.6	.12048	110.0	113.7	.11429	104.4
	320	83.8	.15015	134.7	98.7	.14073	126.3	111.0	.13253	118.8	121.6	.12588	112.8
	338	88.1	.16126	142.3	103.7	.15124	133.5	116.7	.14207	125.5	128.0	.13506	119.3
	366	94.5	.17872	153.9	111.3	.16762	144.4	125.2	.15800	135.9	137.6	.15025	129.3
	388	99.9	.19470	163.6	117.6	.18282	153.6	132.3	.17215	144.7	145.3	.16377	137.6
	406	104.2	.20821	171.5	122.6	.19555	161.1	138.2	.18411	151.8	151.3	.17467	143.9
30″	250	57.4	.11514	109.4	68.3	.10770	103.4	77.5	.10162	96.6	85.7	.09699	92.2
	274	62.5	.12857	119.7	74.9	.12121	113.1	84.9	.11441	106.8	93.7	.10864	101.4
	298	68.3	.14377	131.5	81.6	.13584	124.2	92.6	.12849	117.3	103.1	.12191	111.5
	320	73.5	.15806	141.8	87.5	.14937	134.0	99.2	.14108	126.6	109.4	.13434	120.4
	338	77.2	.16968	149.7	91.9	.16044	141.6	104.2	.15165	133.8	115.0	.14403	127.3
	366	82.9	.18819	161.9	98.4	.17687	152.3	111.9	.16816	144.8	123.4	.16014	137.8
	388	87.6	.20475	172.1	104.0	.19272	162.0	118.2	.18330	154.0	130.4	.17450	146.7
	406	91.4	.21881	180.2	108.5	.20605	169.7	123.3	.19609	161.5	135.9	.18655	153.8

Table 9 Steam tracing data (English units continued)

Design Conditions:

Tracer: One 3/4'' O.D. Air convection tubing tracer
Insulation Thermal Conductivity: See Chart, page 79
Ambient: 100° F, no wind

Nomenclature:

TS = Steam Temperature (°F)
TP = Pipe Temperature (°F)
#/HR-FT = Tracer condensate rate per linear foot
Q3 = Pipe Heat Losses in Btu/hr, lin ft

N.P.S.	Steam Temp °F	INSULATION THICKNESS 1.5'' Nom			2.0'' Nom			2.5'' Nom			3.0'' Nom		
	TS	TP	#/HR-FT	Q3	TP	#/HR-FT	Q3	TP	#/HR-FT	Q3	TP	#/HR-FT	Q3
1.5''	250	190.4	.02452	23.3	194.2	.02265	21.5	197.2	.02121	20.1	199.5	.02008	19.1
	274	205.3	.02966	27.7	209.9	.02742	25.6	213.4	.02571	24.0	216.2	.02428	22.7
	298	220.7	.03532	32.3	225.9	.03267	29.8	230.0	.03061	28.0	233.3	.02900	26.5
	320	234.0	.04061	36.4	239.9	.03754	33.7	244.5	.03518	31.6	248.4	.03331	29.9
	338	244.1	.04488	39.6	250.7	.04151	36.7	255.7	.03892	34.4	259.7	.03686	32.5
	366	259.7	.05177	44.6	266.9	.04793	41.3	272.4	.04498	38.7	276.9	.04259	36.7
	388	272.5	.05801	48.8	280.2	.05373	45.2	286.5	.05057	42.5	291.3	.04793	40.3
	406	282.6	.06337	52.2	290.9	.05872	48.4	297.5	.05525	45.5	303.7	.05238	43.2
2''	250	184.8	.02829	26.9	189.9	.02595	24.6	193.6	.02410	22.9	196.5	.02259	21.5
	274	199.0	.03435	32.0	204.8	.03135	29.2	209.2	.02913	27.1	212.7	.02736	25.5
	298	213.4	.04091	37.4	220.1	.03736	34.1	225.2	.03467	31.7	229.2	.03259	29.8
	320	225.8	.04699	42.1	233.4	.04288	38.5	239.1	.03988	35.7	243.5	.03745	33.6
	338	235.3	.05189	45.8	243.4	.04741	41.8	249.9	.04404	38.9	254.7	.04138	36.5
	366	249.9	.05983	51.5	259.0	.05470	47.1	265.9	.05083	43.8	271.3	.04782	41.1
	388	261.8	.06700	56.3	271.7	.06130	51.5	279.2	.05696	47.9	285.1	.05357	45.1
	406	271.3	.07309	60.2	281.8	.06688	59.0	289.8	.06222	51.3	296.1	.05854	48.2
3''	250	182.1	.03171	30.1	187.3	.02889	27.4	191.3	.02680	25.4	194.7	.02496	23.7
	274	195.6	.03829	35.7	201.8	.03490	32.6	206.7	.03252	30.3	210.8	.03030	28.3
	298	209.5	.04556	41.6	216.9	.04175	38.1	222.3	.03871	35.4	227.1	.03607	33.0
	320	221.5	.05227	46.9	229.8	.04792	43.0	235.9	.04445	39.8	241.5	.04144	37.1
	338	230.9	.05795	51.1	239.6	.05291	46.7	246.4	.04911	43.3	252.2	.04581	40.4
	366	245.1	.06674	57.4	254.7	.06099	52.5	262.1	.05664	48.7	268.5	.05280	45.5
	388	256.6	.07467	62.8	267.0	.06830	57.4	275.0	.06344	53.3	282.1	.05919	49.8
	406	265.7	.08146	67.1	276.9	.07451	61.4	285.4	.06923	57.1	292.9	.06462	63.3
4''	250	178.0	.03514	88.3	183.7	.03193	30.3	188.4	.02981	27.9	191.7	.02751	26.1
	274	190.9	.04236	39.5	197.5	.03858	36.0	203.1	.03541	33.0	207.2	.03337	31.1
	298	204.1	.05034	46.0	211.7	.04590	41.9	218.4	.04234	38.7	222.9	.03975	36.3
	320	215.5	.05773	51.7	224.0	.03260	47.2	231.5	.04857	43.6	236.6	.04558	40.9
	338	224.1	.06367	56.2	233.6	.05832	51.5	241.7	.05365	47.4	247.2	.05037	44.5
	366	237.5	.07336	63.1	248.0	.06723	57.8	256.8	.06185	53.2	263.0	.05809	50.0
	388	248.4	.08200	68.9	259.8	.07520	63.2	269.3	.06924	58.2	276.0	.06506	54.7
	406	257.1	.08937	73.7	269.2	.08197	67.6	279.3	.07550	62.2	286.5	.07098	58.5
6''	250	171.4	.03995	37.9	177.4	.03668	34.8	182.4	.03404	32.3	187.0	.03145	29.8
	274	183.4	.04846	45.2	190.3	.04419	41.2	196.1	.04110	38.3	201.4	.03790	35.4
	298	195.5	.05742	52.5	203.6	.05273	48.2	210.1	.04882	44.6	216.3	.04511	41.2
	320	205.9	.06586	59.0	215.0	.06045	54.2	222.2	.05596	50.2	229.2	.05167	46.3
	338	213.8	.07258	64.1	223.7	.06666	58.8	231.5	.06174	54.5	239.2	.05706	50.4
	366	225.8	.08359	71.9	237.0	.07670	66.0	245.7	.07115	61.2	254.3	.06599	56.8
	388	236.1	.09333	78.5	247.9	.08568	72.1	257.3	.07951	66.8	266.7	.07384	62.1
	406	244.0	.10177	83.8	256.6	.09342	77.0	266.6	.08668	71.5	276.6	.08049	66.3
8''	250	166.3	.04412	41.9	173.8	.03980	37.7	178.3	.03695	35.1	182.4	.03489	33.1
	274	177.5	.05360	50.0	186.0	.04790	44.6	191.3	.04452	41.5	196.0	.04206	39.2
	298	189.0	.06387	58.4	198.5	.05675	51.9	204.7	.05285	48.3	210.0	.04996	45.6
	320	198.7	.07312	65.5	209.3	.06510	58.4	216.4	.06088	54.6	222.3	.05720	51.3
	338	205.9	.08038	70.9	217.6	.07121	63.3	225.2	.06713	59.3	231.5	.06310	55.7
	366	217.1	.09245	79.6	230.1	.08252	71.0	238.8	.07724	66.5	245.8	.07265	62.5
	388	226.2	.10291	86.5	240.7	.09216	77.5	249.8	.08629	72.6	257.4	.08123	68.3
	406	233.9	.11212	92.4	248.9	.10049	82.8	258.7	.09400	77.5	266.7	.08349	72.9

10″	274	170.1	.05946	55.5	176.7	.05538	51.6	181.9	.05071	47.3	186.7	.04780	44.6
	298	180.3	.07038	64.4	187.9	.06564	60.0	193.9	.06014	55.0	199.6	.05697	52.1
	320	189.0	.08047	72.2	197.3	.07486	67.1	204.2	.06903	61.9	210.7	.06524	58.5
	338	195.4	.08829	77.9	204.6	.08247	72.8	212.1	.07616	67.2	219.0	.07191	63.5
	366	205.4	.10127	87.2	215.7	.09467	81.5	224.1	.08764	75.4	231.7	.08269	71.1
	388	213.7	.11311	95.1	224.6	.10536	88.6	233.8	.09758	82.1	242.4	.09235	72.7
	406	220.4	.12293	101.3	231.9	.11467	94.5	242.1	.10650	87.7	250.8	.10069	83.0
12″	250	155.6	.05259	50.0	162.2	.04882	46.4	167.6	.04524	43.0	171.3	.04175	39.7
	274	164.7	.06320	59.0	172.8	.05868	54.8	178.7	.05458	50.9	183.1	.05038	47.0
	298	174.2	.07494	68.5	182.9	.06913	63.2	190.3	.06481	59.3	195.3	.05972	54.6
	320	182.3	.08557	76.8	192.0	.07907	71.0	200.0	.07395	66.3	205.9	.06852	61.4
	338	188.4	.09427	83.2	199.0	.08705	76.8	207.6	.08147	71.9	213.9	.07553	66.7
	366	197.4	.10745	82.5	209.4	.09999	86.0	218.7	.09321	80.3	226.1	.08691	74.8
	388	205.1	.11977	100.7	217.8	.11111	93.4	228.2	.10411	87.6	236.4	.09711	81.7
	406	211.3	.13026	107.4	224.7	.12073	99.5	236.0	.11342	93.5	244.5	.10589	87.2
14″	250	152.1	.05532	52.6	158.4	.05134	48.8	163.7	.04792	45.5	168.1	.04483	42.5
	274	160.5	.06654	62.0	168.3	.06173	57.6	174.5	.05770	53.8	179.6	.05433	50.6
	298	169.2	.07818	71.4	177.9	.07254	66.3	185.3	.06827	62.4	191.1	.06425	58.7
	320	176.7	.08922	79.9	186.4	.06294	74.4	194.4	.07774	69.7	201.1	.07333	65.7
	338	182.4	.09822	88.7	192.8	.09130	80.6	201.4	.08566	75.6	208.7	.08089	71.4
	366	191.0	.11267	96.9	203.6	.10482	90.2	212.2	.09833	84.6	220.3	.09296	80.0
	388	198.2	.12542	105.5	210.7	.11686	98.2	221.1	.10972	92.2	229.7	.10340	87.0
	406	203.9	.13663	112.5	217.2	.12708	104.7	227.9	.11889	98.0	237.6	.11258	92.8
16″	250	149.2	.05783	55.0	155.3	.05335	50.7	160.6	.05007	47.6	165.2	.04755	45.2
	274	150.9	.06888	64.2	164.4	.06402	59.8	170.9	.06014	56.1	176.0	.05687	53.0
	298	165.3	.08142	74.4	173.9	.07572	69.2	181.1	.07120	65.1	187.1	.06731	61.5
	320	172.4	.09285	83.3	182.0	.08666	72.7	190.0	.08143	73.1	196.7	.07711	69.1
	338	177.8	.10224	90.2	188.2	.09532	84.1	196.5	.08926	78.8	203.8	.08444	74.5
	366	185.7	.11648	100.2	197.2	.10863	93.5	206.7	.10230	88.1	214.7	.09693	83.4
	388	192.4	.12969	109.1	204.8	.12122	101.9	215.2	.11418	96.0	223.9	.10817	90.9
	406	197.9	.14092	116.1	210.9	.13165	108.5	221.9	.12418	102.3	231.2	.11771	97.0
18″	250	146.4	.05936	56.4	152.7	.05548	52.7	157.9	.05225	49.6	162.4	.04949	47.0
	274	154.0	.07124	66.5	161.3	.06668	62.1	167.4	.06234	58.2	172.7	.05010	55.2
	298	161.7	.08416	76.9	170.4	.07879	71.9	177.2	.07382	67.4	183.3	.06997	64.0
	320	168.7	.09590	86.0	177.8	.08943	80.1	185.7	.08431	75.6	192.4	.08003	71.8
	338	173.8	.10553	93.1	183.6	.09824	86.7	192.1	.09276	81.9	199.4	.08814	77.8
	366	181.2	.12019	103.4	192.5	.11263	96.9	201.9	.10637	91.6	209.6	.10068	86.6
	388	187.6	.13368	112.4	199.7	.12547	105.6	209.9	.11872	99.8	218.4	.11224	94.3
	406	192.7	.14559	119.9	205.5	.13656	112.5	216.1	.12843	105.9	225.3	.12206	100.6
20″	250	144.2	.06124	58.2	150.3	.05739	54.5	155.2	.05376	51.1	159.7	.05116	48.6
	274	151.3	.07358	68.6	158.2	.06843	63.8	164.3	.06455	60.2	169.8	.06143	57.3
	298	158.4	.08595	78.5	166.9	.08071	73.7	173.9	.07649	69.9	179.7	.07225	66.1
	320	164.7	.09832	88.1	174.2	.09209	82.6	181.9	.08733	78.3	188.4	.08261	74.1
	338	169.8	.10793	95.4	179.7	.10145	89.5	188.1	.09603	84.7	195.1	.09093	80.2
	366	177.2	.12375	106.5	188.1	.11627	100.0	197.2	.10944	94.2	205.2	.10433	89.8
	388	183.2	.13779	115.8	195.0	.12950	108.9	204.8	.12213	102.7	213.5	.11633	97.8
	406	188.1	.14952	123.2	200.6	.14087	116.0	211.0	.13269	109.4	220.2	.12652	104.3
24″	250	141.3	.06358	60.4	146.7	.05954	56.6	151.6	.05662	53.8	155.8	.05369	51.0
	274	148.0	.07626	71.1	154.3	.07144	66.7	160.1	.06811	63.5	165.0	.06443	60.1
	298	154.5	.08922	81.5	162.2	.08444	77.1	169.1	.08024	73.3	174.7	.07626	69.7
	320	160.4	.10184	91.4	169.2	.09625	86.4	176.3	.09115	81.7	182.9	.08705	78.1
	338	165.2	.11192	98.9	174.4	.10593	93.5	182.1	.10013	88.4	189.1	.09589	84.6
	366	172.0	.12823	110.3	181.9	.12060	103.8	190.8	.11479	98.8	198.6	.10971	94.5
	388	177.7	.14272	120.0	188.4	.13430	112.9	197.9	.12784	107.5	206.1	.12181	102.4
	406	182.2	.15516	127.8	193.6	.14594	120.2	203.7	.13903	114.5	212.3	.13285	109.1
30″	250	136.2	.06700	63.5	141.5	.06365	60.5	146.0	.06024	57.2	150.3	.05790	55.0
	274	142.1	.08011	74.7	148.3	.07640	71.2	153.5	.07229	67.4	158.5	.06960	64.9
	298	148.1	.09452	86.5	155.3	.08983	82.1	161.4	.08527	77.9	167.0	.08142	74.4
	320	153.3	.10788	96.8	161.0	.10171	91.2	168.3	.09748	87.4	174.3	.09275	83.2
	338	157.3	.11833	104.5	165.8	.11150	98.5	173.4	.10697	94.4	179.9	.10223	90.2
	366	162.9	.13457	115.8	172.7	.12808	110.2	181.2	.12256	105.4	188.4	.11716	100.8
	388	168.2	.14924	125.6	178.5	.14257	119.8	187.3	.13570	114.1	195.5	.13029	109.6
	406	172.2	.16231	133.8	183.1	.15479	127.5	192.5	.14735	121.4	201.0	.14185	116.8

Table 9a Steam tracing data (Metric units)

Design Conditions:

Tracer: One 13 MM O.D. Air convection tubing tracer
Insulation Thermal Conductivity: See Chart, page 79
Ambient: 37.8 °C, no wind

Nomenclature:

TS = Steam Temperature (°C)
TP = Pipe Temperature (°C)
GM/SEC-M = Tracer condensate rate per linear metre
Q3 = Pipe Heat Losses in Watts/lin m

PIPE SIZE	Steam Temp °C	INSULATION THICKNESS											
		38 MM Nom			51 MM Nom			64 MM Nom			76 MM Nom		
	TS	TP	GM/SEC-M	Q3	TP	GM/SEC-M	Q3	TP	GM/SEC-M	Q3	TP	GM/SEC-M	Q3
48 MM	121	82.1	.00881	19.4	84.4	.00821	18.1	86.1	.00775	17.1	87.6	.00738	16.3
	134	89.4	.01064	23.1	92.1	.00992	21.5	94.2	.00936	20.3	96.0	.00895	19.4
	148	96.9	.01264	26.9	100.0	.01181	25.1	103.5	.01120	23.8	104.5	.01067	22.6
	160	103.4	.01451	30.2	107.0	.01381	28.3	109.7	.01285	26.8	111.9	.01225	25.5
	170	108.3	.01601	32.8	112.3	.01502	30.8	115.1	.01419	29.1	117.5	.01352	27.7
	186	115.9	.01852	37.0	120.2	.01732	34.6	123.5	.01636	32.7	126.2	.01560	31.2
	198	122.3	.02072	40.5	126.8	.01937	37.8	130.4	.01829	35.8	133.3	.01744	34.1
	208	127.2	.02260	43.3	132.0	.02113	40.5	135.8	.01998	38.3	138.9	.01904	36.5
60 MM	121	78.6	.00996	22.0	61.0	.00926	20.5	83.8	.00867	19.2	85.5	.00820	18.1
	134	85.5	.01210	26.2	88.8	.01117	24.2	91.4	.01047	22.7	93.5	.00990	21.5
	148	92.4	.01440	30.5	96.2	.01328	28.2	99.2	.01245	26.5	101.6	.01179	25.0
	160	98.4	.01648	34.4	102.7	.01528	31.8	106.0	.01428	29.7	108.6	.01353	28.2
	170	102.9	.01818	37.3	107.5	.01681	34.5	111.1	.01577	32.3	114.1	.01498	30.7
	186	109.8	.02092	41.9	114.9	.01937	38.7	119.2	.01823	36.4	122.4	.01727	34.5
	198	115.5	.02340	45.7	121.2	.02167	42.3	125.7	.02039	39.8	129.2	.01933	37.8
	208	120.1	.02552	48.8	120.1	.02362	45.2	130.8	.02224	42.6	134.6	.02107	40.4
89 MM	121	96.7	.01098	24.2	79.7	.01018	22.5	82.2	.00957	21.1	84.2	.00899	19.9
	134	83.2	.01331	28.8	86.7	.01227	28.6	89.5	.01154	25.0	91.9	.01085	23.5
	148	89.6	.01582	33.6	93.6	.01456	30.9	97.0	.01373	29.1	99.8	.01288	27.4
	160	95.5	.01809	37.7	100.0	.01678	35.0	103.6	.01571	32.8	106.7	.01479	30.8
	170	99.8	.01994	40.5	104.7	.01849	38.0	108.5	.01733	35.6	111.9	.01630	33.4
	186	106.4	.02295	45.9	111.8	.02131	42.6	116.2	.01987	39.9	120.0	.01878	37.5
	198	111.8	.02564	50.1	117.9	.02380	46.5	122.4	.02232	43.6	126.3	.02101	41.0
	208	116.3	.02798	53.5	122.5	.02595	49.7	127.4	.02435	46.6	131.9	.02299	44.0
114 MM	121	74.1	.01198	26.5	77.4	.01106	24.4	80.2	.01031	22.8	82.1	.00974	21.5
	134	80.2	.01448	31.3	83.9	.01333	28.9	87.2	.01243	26.9	89.6	.01181	25.6
	148	86.5	.01724	36.6	98.8	.01592	33.8	94.4	.01478	31.3	97.1	.01404	29.8
	160	92.0	.01981	41.3	96.6	.01821	38.0	100.6	.01691	35.3	103.7	.01606	33.5
	170	96.1	.02194	45.0	101.0	.02007	41.2	105.4	.01864	38.3	108.7	.01773	36.4
	186	102.3	.02522	50.4	107.5	.02301	46.0	112.7	.02155	43.1	116.4	.02042	40.8
	198	107.3	.02809	54.8	113.2	.02581	50.4	118.8	.02408	47.0	122.6	.02283	44.6
	208	111.4	.03062	58.6	117.8	.02813	53.8	123.5	.02626	50.3	127.6	.02489	47.6
168 MM	121	70.6	.01372	30.2	74.1	.01279	28.1	76.4	.01168	25.8	79.0	.01089	24.0
	134	76.3	.01645	35.7	80.1	.01532	35.1	82.8	.01411	30.5	86.0	.01321	28.6
	148	81.8	.01948	41.4	88.2	.01813	38.5	89.4	.01673	35.5	93.0	.01568	33.3
	160	86.6	.02229	40.5	91.4	.02066	43.0	95.1	.01915	39.9	99.1	.01793	37.4
	170	90.2	.02443	50.1	95.5	.02274	46.7	99.5	.02110	43.3	103.7	.01976	40.6
	180	95.7	.02804	56.0	101.6	.02618	52.3	105.9	.02421	48.4	110.8	.02275	45.9
	198	100.2	.03126	61.1	106.5	.02910	56.8	111.4	.02714	53.0	116.8	.02539	49.6
	208	103.9	.03403	65.2	110.5	.03166	60.8	115.9	.02952	56.5	121.4	.02770	53.0
219 MM	121	67.8	.01484	32.7	71.9	.01368	30.1	74.3	.01283	28.2	76.2	.01187	26.2
	134	72.7	.01782	38.5	77.5	.01633	35.3	80.5	.01544	33.4	82.6	.01439	31.1
	148	77.9	.02102	44.7	83.2	.01928	41.0	86.6	.01815	38.6	89.1	.01700	36.1
	160	82.3	.02401	50.1	88.2	.02206	46.0	92.0	.02082	43.3	94.9	.01950	40.6
	170	85.5	.02027	53.9	92.0	.02430	49.8	96.1	.02290	47.0	99.3	.02159	44.3
	186	90.5	.03016	60.3	97.6	.02775	55.5	102.3	.02638	52.6	106.1	.02495	49.9
	198	94.6	.03359	65.6	102.1	.03092	60.4	107.3	.02925	57.1	111.5	.02785	54.4
	208	98.0	.03652	70.0	106.1	.03366	64.5	111.3	.03184	61.0	116.1	.03040	58.2

273 MM	134	68.7	.01918	41.6	72.1	.01818	39.3	75.4	.01727	37.4	77.9	.01636	35.4
	148	73.1	.02272	48.2	77.2	.02148	45.6	80.6	.02024	43.0	83.7	.01929	41.0
	160	77.1	.02587	54.0	81.3	.02436	50.8	85.3	.02313	48.2	88.8	.02207	46.0
	170	80.0	.02852	58.4	84.6	.02682	55.0	88.9	.02545	52.2	92.6	.02435	49.9
	186	84.5	.03261	65.3	89.6	.03072	61.4	94.3	.02921	58.4	98.3	.02778	55.5
	198	88.0	.03614	70.6	93.7	.03427	66.9	98.8	.03255	63.6	103.1	.03097	60.5
	208	91.0	.03926	75.2	96.9	.03723	71.3	102.4	.03544	67.8	106.9	.03369	64.5
324 MM	121	62.2	.01682	37.1	65.4	.01585	35.8	68.1	.01497	32.9	70.4	.01429	31.5
	134	66.1	.02023	43.9	69.8	.01890	41.0	73.0	.01788	38.9	75.9	.01717	37.2
	148	70.2	.02395	50.8	74.6	.02239	47.5	78.3	.02120	45.0	81.4	.02011	42.7
	160	73.6	.02705	56.4	78.6	.02544	53.1	82.7	.02421	50.5	86.2	.02298	47.9
	170	76.4	.02975	61.1	81.7	.02807	57.5	86.0	.02654	54.4	89.8	.02534	50.9
	186	80.4	.03406	68.2	88.3	.03211	64.3	91.1	.03038	60.7	95.3	.02903	58.0
	198	83.8	.03799	74.2	94.1	.03584	70.0	95.3	.03388	66.2	99.9	.03235	63.3
	208	86.4	.04125	79.0	93.1	.03892	74.5	98.7	.03683	70.5	103.5	.03522	67.4
356 MM	121	60.4	.01756	38.7	63.4	.01636	36.2	66.1	.01563	34.6	66.4	.01491	32.8
	134	63.9	.02084	45.2	67.6	.01967	42.7	70.8	.01877	40.7	73.4	.01788	38.7
	148	67.6	.02462	52.3	71.9	.02332	49.5	75.6	.02207	46.8	78.8	.02109	44.8
	160	70.9	.02808	58.6	75.8	.02659	55.5	79.7	.02512	52.4	83.3	.02409	50.2
	170	73.4	.03087	63.4	78.6	.02922	68.0	82.9	.02769	56.8	86.7	.02652	54.3
	186	77.2	.03537	70.7	82.9	.03345	68.9	87.6	.03169	63.4	91.7	.03024	60.4
	198	80.3	.03936	76.9	85.3	.03709	72.4	91.6	.03532	69.0	96.0	.03371	65.8
	208	82.8	.04283	82.0	88.2	.04025	77.1	94.7	.03845	73.6	99.4	.03662	70.1
406 MM	121	58.9	.01802	39.7	61.9	.01708	37.6	64.0	.01627	36.0	66.8	.01552	34.2
	134	62.3	.03154	46.7	65.9	.02042	44.3	69.0	.01948	42.3	71.6	.01861	40.2
	148	65.8	.02543	54.0	69.9	.02416	51.3	73.4	.02289	48.5	76.6	.02199	46.6
	160	68.8	.02903	60.5	73.4	.02750	57.3	77.4	.02606	54.3	80.8	.02467	51.9
	170	71.1	.03184	65.4	76.1	.03002	61.6	80.4	.02873	58.8	84.0	.02736	56.2
	186	74.8	.03650	73.0	80.1	.03437	68.8	84.9	.03284	65.7	88.9	.03132	62.7
	198	77.5	.04037	78.8	83.4	.03832	74.8	88.6	.03659	71.5	92.9	.03497	68.3
	208	79.8	.04380	83.8	86.1	.04155	79.6	91.4	.03956	95.8	96.2	.03801	72.7
457 MM	121	57.6	.01858	40.9	60.6	.01760	38.8	63.1	.01666	36.8	65.4	.01600	35.4
	134	60.8	.02217	48.1	64.3	.02106	45.7	67.2	.01997	43.3	69.9	.01924	41.7
	148	64.1	.02613	55.6	68.0	.02403	52.4	71.5	.02361	50.1	74.6	.02257	41.9
	160	67.0	.02919	62.1	71.3	.02807	58.4	75.5	.02691	56.1	78.6	.02571	53.6
	170	69.0	.03248	66.6	73.6	.03088	63.4	78.1	.02957	60.7	81.7	.02830	58.0
	186	72.3	.03716	74.3	77.7	.03536	70.7	82.3	.03365	67.3	86.3	.03237	64.8
	198	75.3	.04136	80.8	80.9	.03932	76.8	85.8	.03753	73.3	90.2	.03612	76.5
	208	77.3	.04496	88.0	83.4	.04278	81.9	83.6	.04071	78.8	93.2	.03921	75.1
508 MM	121	56.6	.01885	41.7	59.4	.01809	39.9	61.8	.01718	37.8	64.1	.01655	36.4
	134	59.6	.02272	49.3	63.0	.02161	46.9	65.8	.02055	44.6	68.3	.01962	42.5
	148	62.7	.02679	56.9	66.4	.02526	53.7	69.8	.02425	51.6	72.7	.02324	49.3
	160	65.4	.03048	63.6	69.5	.02883	60.1	73.3	.02764	57.6	76.6	.02847	55.2
	170	67.3	.03318	68.1	71.9	.03164	65.0	76.0	.03017	61.9	79.5	.02912	59.8
	186	70.4	.03800	76.0	75.6	.03626	72.6	80.0	.03456	69.1	84.0	.03331	66.7
	198	73.0	.04225	82.5	78.6	.04034	78.8	83.3	.03852	75.2	87.7	.03716	72.6
	208	75.2	.04593	87.9	81.0	.04386	83.9	86.0	.04177	80.0	90.8	.04031	77.2
610 MM	121	55.0	.01950	43.1	57.8	.01873	41.4	60.0	.01780	39.3	62.0	.01717	37.8
	134	58.0	.02332	50.6	60.9	.02215	48.0	63.7	.02135	46.3	66.0	.02052	44.5
	148	60.9	.02759	58.5	64.2	.02615	55.6	67.3	.02499	53.1	70.1	.02422	51.4
	160	63.2	.03109	64.8	67.1	.02976	62.1	70.5	.02846	59.3	73.7	.02758	57.5
	170	65.1	.03410	70.0	69.3	.03274	67.2	72.9	.03131	64.3	76.5	.03033	62.3
	186	68.6	.03905	78.0	72.6	.03750	74.9	76.8	.03588	71.7	80.6	.03468	69.4
	198	70.3	.04344	85.0	75.4	.04142	80.9	79.9	.03996	78.0	83.9	.03842	75.0
	208	73.4	.04719	90.3	77.8	.04524	86.6	82.4	.04338	83.0	86.5	.04171	79.9
762 MM	121	52.8	.02050	45.2	55.1	.01945	43.0	57.4	.01891	41.8	59.3	.01814	40.1
	134	55.1	.02430	52.5	58.0	.02334	50.6	60.5	.02234	48.4	62.8	.02175	47.1
	148	57.7	.02859	60.6	60.9	.02757	58.5	63.8	.02636	56.0	68.5	.02570	54.5
	160	59.0	.03254	67.8	63.5	.03129	65.2	66.6	.02993	62.4	69.4	.02891	60.3
	170	61.6	.03565	73.2	65.3	.03407	70.0	88.8	.03299	67.7	71.8	.03181	65.3
	186	64.0	.04086	81.7	68.2	.03895	77.8	72.1	.03772	75.4	75.6	.03636	72.7
	198	66.1	.04523	88.4	70.6	.04333	84.7	74.9	.04265	82.1	78.6	.04051	79.1
	208	67.6	.04888	93.5	72.5	.04709	90.1	77.1	.04564	87.3	81.0	.04405	84.3

Table 9a Steam tracing data (Metric units continued)

Design Conditions:

Tracer: One 13 MM O.D. Air convection tubing tracer
Insulation Thermal Conductivity: See Chart, page 79
Ambient: −17.8° C, 40 KPH wind

Nomenclature:

TS = Steam Temperature (°C)
TP = Pipe Temperature (°C)
GM/SEC-M = Tracer condensate rate per linear metre
Q3 = Pipe Heat Losses in Watts/lin m

PIPE SIZE	Steam Temp °C	INSULATION THICKNESS											
		38 MM Nom			51 MM Nom			64 MM Nom			76 MM Nom		
	TS	TP	GM/SEC-M	Q3	TP	GM/SEC-M	Q3	TP	GM/SEC-M	Q3	TP	GM/SEC-M	Q3
43 MM	121	57.3	.01565	33.1	61.9	.01396	30.6	65.4	.01299	28.6	66.3	.01223	27.8
	134	64.4	.01762	36.8	69.7	.01574	34.0	73.6	.01472	31.8	76.6	.01388	30.1
	146	72.0	.01924	40.9	74.5	.01773	37.7	81.8	.01667	35.3	85.1	.01571	38.3
	160	78.4	.02132	44.4	84.3	.01966	40.9	88.9	.01848	38.3	92.5	.01738	36.2
	170	83.3	.02292	47.0	89.5	.02115	43.4	94.3	.01986	40.7	98.1	.01875	38.5
	186	90.6	.02562	51.2	97.3	.02363	47.3	102.4	.02218	44.3	106.5	.02098	42.0
	198	96.7	.02803	54.7	103.8	.02588	50.6	109.2	.02426	47.8	113.6	.02300	44.9
	208	101.5	.03009	57.6	108.9	.02783	53.2	114.6	.02610	49.9	119.2	.02473	47.3
60 MM	121	50.8	.01738	38.3	56.6	.01570	34.7	60.9	.01459	32.1	64.4	.01373	30.2
	134	57.3	.01968	42.5	63.7	.01785	38.6	68.7	.01652	35.8	72.5	.01555	33.6
	148	64.0	.02214	46.9	71.1	.02008	42.7	76.5	.01866	39.7	80.6	.01752	37.2
	160	69.8	.02439	50.9	77.4	.02220	46.3	83.2	.02068	43.0	87.6	.01946	40.4
	170	74.3	.02627	53.9	82.2	.02396	49.1	88.3	.02225	45.7	92.9	.02090	42.9
	186	81.0	.02932	58.6	89.5	.02672	53.4	96.0	.02487	49.7	101.0	.02336	46.7
	198	86.5	.03281	62.6	95.5	.02923	57.1	102.4	.02723	53.2	107.7	.02558	50.6
	208	90.9	.03435	65.8	100.3	.03134	60.0	107.5	.02923	55.9	113.1	.02751	52.7
89 MM	121	47.3	.01968	42.2	53.4	.01744	38.4	58.0	.01607	35.5	62.0	.01501	33.0
	134	53.5	.02159	46.8	60.2	.01968	42.7	65.4	.01826	39.5	69.8	.01698	36.7
	148	59.8	.03430	51.7	67.4	.02223	47.1	72.9	.02054	43.7	77.7	.01913	40.7
	160	65.4	.02684	56.0	73.4	.02449	51.1	79.4	.02271	47.3	84.7	.02127	44.3
	170	69.7	.02889	59.3	78.0	.02642	54.1	84.3	.02449	50.2	89.9	.02287	46.9
	186	76.0	.03218	64.4	85.0	.02944	58.8	91.7	.02733	54.6	97.7	.02556	51.1
	198	81.3	.03519	68.7	90.8	.03214	62.8	97.9	.02988	58.4	104.2	.02798	54.0
	208	85.5	.03773	72.2	95.4	.03454	66.1	102.8	.03201	61.4	109.5	.03004	57.5
114 MM	121	43.0	.02106	46.4	49.3	.01896	41.9	54.7	.01749	38.5	58.5	.01634	36.1
	134	48.8	.02381	54.5	55.7	.02145	46.5	61.7	.01973	42.8	66.0	.01857	40.1
	148	54.7	.02674	58.7	62.3	.02420	51.3	69.0	.02229	47.3	73.5	.02088	44.4
	160	59.9	.02944	61.4	68.1	.02686	55.6	75.1	.02456	51.2	80.0	.02308	48.1
	170	63.8	.03187	65.0	72.4	.02876	58.9	79.8	.02649	54.3	85.0	.02489	51.0
	186	69.8	.03519	70.4	79.0	.03198	64.0	87.0	.02953	59.0	92.5	.02776	55.5
	198	74.7	.03846	75.1	84.4	.03497	68.3	92.9	.03224	63.0	98.7	.03037	59.3
	208	78.7	.04117	78.9	88.8	.03748	71.8	97.6	.03464	66.3	103.7	.03255	62.4
168 MM	121	36.6	.02376	52.2	43.2	.02176	47.8	48.2	.01986	43.8	53.3	.01837	40.3
	134	41.7	.02672	57.8	49.0	.02441	52.8	54.5	.02240	48.6	60.2	.02071	44.9
	148	46.9	.02969	63.3	54.9	.02742	58.2	61.1	.02522	53.6	61.2	.02339	49.6
	160	51.5	.03291	68.5	60.1	.03021	63.0	66.8	.02784	58.1	73.3	.02577	53.7
	170	55.0	.03526	72.4	64.2	.03247	66.7	70.9	.02986	61.3	77.9	.02778	57.0
	186	60.3	.03921	78.5	70.1	.03605	72.1	77.4	.03330	66.6	84.9	.03096	61.9
	198	64.8	.04276	83.6	75.0	.03938	76.9	82.9	.03644	71.2	90.7	.03384	66.1
	208	68.4	.04586	87.8	79.0	.04220	80.8	87.2	.03966	74.8	95.3	.03626	69.5
219 MM	121	31.5	.02554	56.2	39.6	.02306	50.8	44.6	.02156	47.5	48.5	.02010	44.3
	134	36.2	.02874	62.2	45.1	.02600	50.2	50.6	.02438	52.7	54.7	.02257	49.9
	148	41.0	.03220	68.4	50.7	.02923	61.9	58.6	.02722	57.9	61.4	.02544	54.0
	160	45.2	.03552	73.9	55.6	.03214	67.0	62.0	.03005	62.7	67.1	.02809	58.6
	170	48.2	.03790	77.8	59.1	.03439	70.6	66.1	.03231	66.3	71.4	.03024	62.1
	186	53.0	.04217	84.3	64.9	.03832	76.6	72.2	.03599	72.0	77.9	.03363	67.3
	198	57.0	.04590	89.7	69.5	.04178	81.6	77.2	.03920	76.5	83.6	.03675	71.8
	208	60.2	.04919	94.1	73.3	.04474	85.6	81.2	.04201	80.4	87.7	.03951	75.7

273 MM	134	29.7	[illegible]	66.9	36.2	.02888	62.7	41.9	.02715	58.7	46.7	.02371	55.6
	148	34.0	.03461	73.6	41.1	.03248	69.0	47.3	.03043	64.6	52.5	.02889	61.2
	160	37.7	.03819	79.4	45.3	.03583	74.6	51.9	.03359	69.9	57.5	.03178	66.2
	170	40.5	.04088	83.9	48.3	.03822	78.5	55.4	.03599	75.9	61.5	.03414	70.1
	186	44.7	.04590	90.8	53.1	.04254	85.0	60.9	.04003	80.1	67.2	.03806	76.0
	198	48.3	.04944	96.6	57.1	.03629	90.5	65.3	.04361	85.3	71.9	.04124	80.7
	208	51.1	.05296	101.3	60.3	.04961	94.9	68.8	.04680	89.6	75.7	.04430	84.8
324 MM	121	21.3	.02965	64.0	28.1	.02708	59.6	33.5	.02531	55.7	38.1	.02397	52.7
	134	25.1	.03255	70.7	32.3	.03018	65.5	38.4	.02848	61.6	43.3	.02687	58.1
	148	28.9	.03655	77.7	36.8	.03389	72.0	43.4	.03191	68.8	48.6	.03012	64.0
	160	32.2	.04018	83.6	40.7	.03730	77.8	47.8	.03519	73.2	53.5	.03326	69.2
	170	34.6	.04314	88.5	43.7	.04004	82.2	51.1	.03703	77.5	57.1	.03563	73.2
	186	38.6	.03786	95.8	48.2	.04448	89.0	56.0	.04176	83.6	62.8	.03903	79.3
	198	41.7	.05194	101.4	51.9	.04844	94.7	60.3	.04551	89.0	67.3	.04321	84.5
	208	44.2	.05544	106.2	54.9	.05199	99.3	63.6	.04878	93.3	70.9	.04633	88.7
356 MM	121	18.0	.03017	66.4	24.6	.02806	61.8	30.1	.02648	58.3	34.6	.02502	55.0
	134	21.6	.03395	73.4	28.7	.03159	68.3	34.7	.02989	64.5	39.6	.02816	60.9
	148	25.2	.03794	80.6	32.9	.03530	75.0	39.3	.03319	70.5	44.8	.03155	67.1
	160	28.1	.04153	86.4	36.5	.03883	81.0	43.3	.03651	76.2	49.2	.03481	72.4
	170	30.4	.04458	91.3	39.2	.04169	85.6	46.4	.03922	80.5	52.6	.03731	76.6
	186	34.0	.04934	98.8	43.4	.04628	92.6	51.1	.04365	87.2	57.8	.04155	83.0
	198	36.9	.05367	105.0	46.9	.05447	98.5	55.0	.04748	92.8	62.2	.04521	88.4
	208	39.3	.05742	110.0	49.8	.03397	103.3	58.1	.05086	97.3	65.4	.04828	92.4
406 MM	121	15.6	.03111	68.5	22.0	.02902	63.9	27.3	.02737	60.1	31.8	.02602	57.3
	134	18.8	.03485	75.7	25.8	.03250	70.6	31.6	.03065	66.5	36.6	.02918	63.4
	148	22.1	.03882	82.5	29.7	.03650	77.6	36.0	.03439	73.1	41.5	.03279	69.7
	160	25.0	.04277	89.0	33.1	.04013	83.7	39.9	.03785	78.9	45.6	.03600	74.9
	170	27.2	.04586	94.0	35.7	.04306	88.4	42.8	.04063	83.4	40.8	.03858	79.2
	186	30.5	.05072	101.5	39.7	.04777	95.6	47.2	.04512	90.3	53.6	.04295	85.8
	198	33.2	.05531	108.0	42.9	.05211	101.7	50.9	.04915	96.1	57.7	.04671	91.3
	208	35.4	.05905	113.1	45.4	.05539	106.2	53.8	.05264	100.7	61.0	.05007	95.8
457 MM	121	13.4	.03167	69.7	19.6	.02986	65.7	24.8	.02819	62.0	29.3	.02694	59.3
	134	16.4	.03544	77.0	23.3	.03343	72.6	28.9	.03173	68.6	33.7	.03003	65.2
	148	19.5	.03992	84.5	27.0	.03755	79.8	33.1	.03546	75.4	38.3	.03373	71.7
	160	22.3	.04383	91.1	30.2	.04137	86.1	36.7	.03911	81.4	42.3	.03712	77.4
	170	24.4	.04696	96.3	32.6	.04441	90.9	39.5	.04188	86.0	45.4	.03985	81.8
	186	27.5	.03198	104.1	36.2	.04085	97.8	43.7	.04649	93.0	50.0	.04427	88.6
	198	30.1	.05658	110.6	39.3	.05326	104.0	47.2	.05071	99.0	53.9	.04822	94.3
	208	32.1	.06046	115.8	41.7	.05688	109.0	50.0	.05422	103.8	56.9	.05166	98.8
503 MM	121	11.6	.03216	71.2	17.5	.03062	67.4	22.6	.02897	63.8	28.9	.02758	60.7
	134	14.4	.03634	78.6	21.1	.03432	74.5	26.5	.03244	70.5	31.2	.03033	67.1
	148	17.3	.04075	86.3	24.6	.03850	81.8	30.5	.03643	77.4	35.6	.03471	73.6
	160	19.8	.04465	93.1	27.5	.04215	87.7	34.0	.04007	83.6	39.5	.03827	79.6
	170	21.9	.04795	98.3	29.8	.04521	92.6	36.0	.04299	88.3	42.4	.04099	84.2
	186	24.8	.05307	108.3	33.3	.05002	100.1	40.0	.04770	95.5	48.8	.04552	91.1
	198	27.3	.05784	112.9	36.2	.05446	106.5	43.9	.05204	101.6	50.4	.04958	96.9
	208	29.2	.06175	118.3	38.5	.05821	111.5	46.5	.05554	106.4	53.3	.05309	101.6
610 MM	121	9.5	.03299	73.0	14.9	.03162	69.6	19.5	.03006	66.2	23.7	.02877	63.2
	134	12.1	.03727	80.8	18.1	.03539	76.9	23.3	.03379	73.1	27.7	.03229	69.9
	148	14.8	.04175	88.5	21.3	.03943	83.8	27.0	.03728	80.3	31.8	.03612	76.8
	160	17.2	.04575	95.4	24.1	.04345	90.4	30.2	.04163	86.6	35.4	.03974	82.9
	170	19.0	.04902	100.7	28.3	.04657	95.4	32.6	.04468	91.5	38.1	.04264	87.5
	186	21.5	.05432	108.8	29.5	.05152	103.1	36.2	.04915	98.4	43.3	.04731	94.7
	198	24.1	.05926	115.6	32.1	.05609	109.6	39.3	.05350	104.6	45.6	.05161	100.7
	208	25.9	.06373	121.1	34.4	.03995	114.9	41.7	.05721	109.6	48.3	.05510	105.6
762 MM	121	5.8	.03442	70.2	10.6	.03271	72.4	15.1	.03168	69.8	18.8	.03040	66.9
	134	8.1	.03886	84.0	13.4	.03693	79.9	18.2	.03547	77.1	22.5	.03407	73.9
	148	10.4	.04350	92.2	16.2	.04141	87.7	21.5	.03951	84.0	26.2	.03820	81.2
	160	12.5	.04765	99.4	18.7	.04536	94.6	24.3	.04354	90.5	29.3	.04208	87.0
	170	14.0	.05122	104.9	20.7	.04869	99.8	26.5	.04644	95.6	31.7	.04515	92.5
	186	16.4	.05655	113.2	23.8	.05389	107.9	29.7	.05163	103.4	35.2	.04967	99.4
	198	18.4	.06155	120.3	25.9	.03873	114.0	32.5	.05626	109.8	38.3	.05406	105.7
	208	19.8	.06537	125.0	27.8	.06268	120.1	34.6	.06008	115.1	40.7	.05782	110.8

Table 9a Steam tracing data (Metric units continued)

Design Conditions:

Tracer: One 22 MM O.D. Air convection tubing tracer
Insulation Thermal Conductivity: See Chart, page 79
Ambient: 37.8° C, no wind

Nomenclature:

TS = Steam Temperature (°C)
TP = Pipe Temperature (°C)
GM/SEC-M = Tracer condensate rate per linear metre
Q3 = Pipe Heat Losses in Watts/lin m

PIPE SIZE	Steam Temp °C	INSULATION THICKNESS											
		38 MM Nom			51 MM Nom			64 MM Nom			76 MM Nom		
	TS	TP	GH/SEC-M	Q3	TP	GM/SEC-M	Q3	TP	GM/SEC-M	Q3	TP	GM/SEC-M	Q3
48 MM	121	88.0	.01014	22.4	90.1	.00937	20.7	91.8	.00877	19.4	93.1	.00831	18.3
	134	96.3	.01227	26.6	98.8	.01134	24.6	100.8	.01063	23.0	102.3	.01004	21.8
	148	104.8	.01461	31.0	107.7	.01351	28.7	110.0	.01266	26.9	111.8	.01199	25.5
	160	112.2	.01680	35.0	115.5	.01553	32.4	113.0	.01455	30.3	120.2	.01378	28.7
	170	117.9	.01856	38.1	121.5	.01717	35.2	124.3	.01610	33.0	126.5	.01525	31.3
	186	126.5	.02141	42.8	130.5	.01923	39.7	133.6	.01860	37.2	136.0	.01762	35.2
	198	133.6	.02430	46.9	137.9	.02223	43.4	141.4	.02092	40.9	144.1	.01983	38.7
	208	139.2	.02621	50.2	143.8	.02429	46.5	147.5	.02285	43.8	150.4	.02167	41.5
60 MM	121	84.9	.01170	25.8	87.7	.01074	23.7	89.8	.00997	22.0	91.4	.00935	20.6
	134	92.8	.01421	30 8	98.0	.01297	28.1	98.4	.01205	26.1	100.4	.01132	24.5
	148	100.8	.01692	35.9	104.5	.01545	32.8	107.3	.01434	30.4	109.5	.01348	28.6
	160	107.7	.01944	40.5	111.9	.01774	37.0	115.0	.01650	34.4	117.5	.01549	32.3
	170	112.9	.02147	44.0	117.5	.01961	40.2	121.0	.01822	37.4	123.7	.01712	35.1
	186	121.1	.02475	49.5	126.1	.02262	45.2	129.9	.02102	42.0	133.0	.01978	39.5
	198	127.7	.02771	54.1	133.1	.02536	49.9	137.3	.02356	46.0	140.6	.02216	43.3
	208	132.9	.03623	57.9	138.8	.02767	53.0	143.2	.02574	49.3	146.7	.02422	46.4
89 MM	121	83.4	.01312	28.9	86.3	.01195	26.4	88.5	.01108	24.4	90.4	.01033	22.8
	134	90.9	.01584	34.3	94.3	.01444	31.3	97.1	.01345	29.2	99.3	.01253	27.2
	148	98.6	.01884	40.0	102.7	.01727	36.7	105.7	.01601	34.0	108.4	.01492	31.7
	160	105.3	.02162	45.1	109.9	.01982	41.3	113.3	.01839	38.3	116.4	.01714	35.7
	170	110.5	.02397	49.2	115.3	.02189	44.9	119.1	.02031	41.7	122.3	.01895	38.9
	186	118.4	.02761	55.2	123.7	.02523	50.4	127.8	.02343	46.8	131.4	.02184	43.7
	198	124.8	.03069	60.3	130.6	.02825	55.2	135.0	.02624	51.3	138.9	.02448	47.3
	208	129.9	.03369	64.5	136.0	.03082	59.0	140.8	.02864	54.8	144.9	.02673	51.2
114 MM	121	81.1	.01453	32.0	84.3	.01321	29.2	86.9	.01213	26.8	88.7	.01138	25.1
	134	88.3	.01752	37.9	91.9	.01596	34.6	95.0	.01465	31.7	97.3	.01380	29.9
	148	95.6	.02087	44.2	99.8	.01899	40.3	103.5	.01752	37.2	106.1	.01644	34.9
	160	101.9	.02388	49.7	106.7	.02176	45.3	110.6	.02009	41.9	113.7	.01885	39.3
	170	106.7	.02634	54.0	112.0	.02412	49.5	116.5	.02219	45.5	119.6	.02084	42.7
	186	114.2	.03084	60.7	120.0	.02781	55.6	124.9	.02558	51.1	128.3	.02403	48.0
	198	120.2	.03392	66.3	126.6	.03110	60.8	131.8	.02864	55.9	135.6	.02691	52.6
	208	125.1	.03697	70.8	131.8	.03391	64.9	137.4	.03123	59.8	141.4	.02936	56.2
168 MM	121	77.4	.01652	36.4	80.8	.01517	33.4	83.6	.01408	31.1	86.1	.01301	28.7
	134	84.1	.02005	43.4	87.9	.01828	39.6	91.1	.01700	36.8	94.1	.01568	34.0
	148	90.8	.02375	50.5	95.4	.02181	46.3	98.9	.02019	42.9	102.4	.01866	39.6
	160	96.6	.02724	56.7	101.7	.02506	52.1	105.7	.02315	48.2	109.6	.02137	44.5
	170	101.0	.03002	61.6	106.5	.02757	56.5	110.8	.02554	52.4	115.1	.02360	48.4
	186	107.7	.03457	69.1	113.9	.03173	63.4	118.7	.02943	58.8	123.5	.02730	54.6
	198	113.4	.03661	75.4	120.0	.03544	69.3	125.2	.03289	64.2	130.4	.03054	59.7
	208	117.8	.04210	80.6	124.8	.03864	74.0	130.3	.03586	68.7	135.5	.03330	63.6
219 MM	121	74.6	.01825	40.3	78.8	.01646	36.3	81.3	.01529	33.7	83.5	.01448	31.8
	134	80.8	.02217	48.0	85.6	.01981	42.9	88.5	.01842	39.9	91.1	.01740	37.7
	148	87.2	.02642	56.1	92.5	.02348	49.9	95.9	.02186	46.4	98.9	.02066	43.9
	160	92.6	.03024	63.0	98.5	.02693	56.1	102.5	.02518	52.4	105.7	.02366	49.3
	170	96.6	.03325	68.2	103.1	.02966	60.9	107.3	.02777	56.9	110.8	.02610	53.5
	186	102.8	.03824	76.5	110.0	.03414	68.2	114.9	.03195	63.9	118.8	.03005	60.1
	198	107.9	.04257	83.2	115.9	.03812	74.5	121.0	.08560	69.7	125.2	.03369	65.6
	208	112.1	.04638	88.8	120.5	.04157	79.6	125.9	.03868	74.5	130.4	.03661	70.1

273 MM	134	76.7	.02459	53.3	80.4	.02291	49.6	83.3	.02098	45.4	86.0	.01977	42.8
	148	82.4	.02911	61.9	86.6	.02715	57.7	89.9	.02488	52.8	93.1	.02356	50.1
	160	87.2	.03328	69.4	91.8	.03096	64.5	95.7	.02855	59.5	99.3	.02699	56.2
	170	90.8	.03652	74.9	95.9	.03411	69.9	100.0	.03150	64.6	103.9	.02975	61.0
	186	96.3	.04189	83.8	102.1	.03916	78.3	106.7	.03625	72.5	110.9	.03421	68.4
	198	100.9	.04679	91.4	107.0	.04358	85.2	112.1	.04037	78.9	116.9	.03820	74.6
	208	104.6	.05085	97.4	111.1	.04743	90.8	116.7	.04405	84.3	122.6	.04165	79.7
324 MM	121	68.7	.02075	48.0	72.3	.02019	44.6	75.3	.01871	41.3	77.4	.01727	36.1
	134	73.7	.02614	56.7	78.2	.02427	52.6	81.5	.02258	48.9	84.0	.02084	45.1
	148	79.6	.03100	65.8	83.9	.02859	60.7	87.9	.02681	57.0	90.7	.02470	52.5
	160	83.5	.03539	73.8	88.9	.03271	68.2	93.3	.03059	63.7	96.6	.02831	59.0
	170	86.9	.03899	80.0	92.8	.03601	73.6	97.5	.03370	69.1	101.0	.03124	64.1
	186	91.9	.04445	88.9	98.6	.04136	82.7	103.7	.03855	77.1	107.8	.03595	71.9
	198	96.2	.04954	96.8	103.2	.04596	89.8	109.0	.04307	84.1	113.6	.04017	78.5
	208	99.6	.05388	103.2	107.0	.04994	95.6	113.3	.04692	89.8	118.1	.04380	83.8
356 MM	121	66.7	.02286	50.5	70.2	.02124	46.9	78.1	.01982	43.8	75.6	.01854	40.9
	134	71.4	.02753	59.6	75.7	.02554	55.4	79.2	.02387	51.7	82.0	.02247	48.7
	148	76.2	.03234	68.6	81.0	.03000	63.8	85.2	.02824	60.0	88.4	.02658	58.5
	160	80.4	.03690	76.8	85.8	.03431	71.5	90.2	.03216	67.0	93.9	.03033	65.2
	170	83.6	.04063	83.3	89.3	.03777	77.4	94.1	.03543	72.6	98.2	.03346	68.6
	186	88.3	.04661	93.2	94.8	.04336	86.7	100.1	.04067	81.3	104.6	.03845	76.9
	198	92.3	.05188	101.4	99.3	.03834	94.4	105.1	.04538	88.6	109.8	.04277	83.6
	208	95.5	.05652	108.2	102.9	.05257	100.7	108.9	.04918	94.2	114.2	.04657	89.2
406 MM	121	65.1	.02392	52.6	68.5	.02207	48.7	71.4	.02071	45.7	74.0	.01967	43.4
	134	69.4	.02849	61.7	73.5	.02648	57.4	77.1	.02488	53.9	80.0	.02352	50.9
	148	74.1	.03388	71.5	78.9	.03132	66.5	82.8	.02945	62.6	86.2	.02784	59.1
	160	78.0	.03841	80.1	83.3	.03585	74.7	87.8	.03368	70.2	91.5	.03196	66.4
	170	81.0	.04229	86.7	86.8	.03943	80.8	91.4	.03692	75.7	95.4	.03493	71.6
	186	85.4	.04818	96.3	91.8	.04493	89.9	97.0	.04232	84.6	101.5	.04010	80.2
	198	69.1	.05365	104.8	96.0	.05014	97.9	101.8	.04723	92.2	106.6	.04475	87.4
	208	92.1	.05829	111.6	99.4	.05445	104.3	105.5	.05137	98.4	110.6	.04869	93.2
457 MM	121	63.6	.02456	54.2	67.0	.02295	50.7	70.0	.02161	47.7	72.5	.02047	45.2
	134	67.8	.02947	63.9	71.5	.02758	59.7	75.2	.02479	55.9	78.2	.02444	53.0
	148	72.1	.03481	73.9	76.9	.03256	69.1	80.7	.03054	64.8	84.6	.02891	61.5
	160	76.0	.03967	82.7	81.0	.03699	77.0	85.4	.03487	72.7	89.1	.03310	69.0
	170	78.8	.04365	89.5	84.2	.04064	83.3	89.0	.03837	78.7	93.0	.03646	74.7
	186	82.9	.04971	99.4	89.1	.04659	93.1	94.4	.04400	88.0	98.7	.04165	83.3
	198	86.5	.05530	108.1	93.2	.05190	101.4	98.8	.04911	95.9	103.6	.04643	90.7
	208	89.3	.06022	115.3	96.4	.05649	108.1	102.3	.05312	101.7	107.4	.05049	96.7
508 MM	121	62.3	.02533	55.9	65.7	.02374	52.4	68.5	.02224	49.1	70.9	.02116	46.7
	134	66.3	.03043	65.9	70.1	.07831	61.3	73.5	.02670	57.9	76.6	.02541	55.1
	148	70.2	.03555	75.4	74.9	.03338	70.8	78.8	.03164	67.2	82.1	.02989	63.5
	160	73.7	.04067	84.7	79.0	.03809	79.4	83.3	.03612	75.3	86.9	.03417	71.2
	170	76.8	.04464	91.6	82.1	.04196	86.0	86.7	.03972	81.4	90.6	.03761	77.1
	186	80.6	.05115	102.3	86.7	.04809	96.1	91.8	.04527	90.6	96.2	.04316	86.3
	198	84.0	.05700	111.3	90.6	.05357	104.7	96.0	.05052	98.7	100.8	.04812	94.0
	208	86.7	.06199	118.4	93.6	.05827	111.5	99.4	.05489	105.1	104.5	.05234	100.2
610 MM	121	60.7	.02630	58.1	63.7	.02463	54.4	66.5	.02347	51.7	68.8	.02221	49.0
	134	64.4	.03155	68.3	67.9	.02955	64.1	71.2	.02817	61.0	73.9	.02665	58.8
	148	68.1	.03691	78.3	72.3	.03493	74.1	76.2	.03319	70.4	79.3	.03154	66.9
	160	71.4	.04213	87.8	76.2	.03981	83.0	80.2	.03770	78.5	83.8	.03601	75.0
	170	74.0	.04629	95.0	79.1	.04382	89.9	83.4	.04142	84.9	87.3	.03966	81.3
	186	77.8	.05304	106.0	83.3	.04989	99.7	88.2	.04748	94.9	92.5	.04538	90.8
	198	80.9	.05904	115.3	86.9	.05558	108.5	92.2	.05288	103.4	96.7	.05039	98.4
	208	83.4	.06418	122.8	89.0	.06037	115.5	95.4	.05751	110.1	100.2	.05475	104.8
762 MM	121	57.9	.02771	61.0	50.8	.02633	58.1	63.3	.02492	55.0	65.7	.02395	52.8
	134	61.1	.03314	71.8	64.6	.03160	68.4	67.5	.02990	64.8	70.3	.02879	62.3
	148	64.5	.03910	83.1	68.5	.03716	78.9	71.9	.03527	74.8	75.0	.03368	71.5
	160	67.4	.04463	93.0	71.7	.04207	87.7	75.7	.04032	84.0	79.1	.03830	80.0
	170	69.6	.04895	100.4	74.4	.04612	94.7	78.6	.04425	90.7	82.2	.04239	86.7
	186	72.7	.05566	111.3	78.2	.05298	105.9	82.9	.05070	101.3	86.9	.04846	96.9
	198	75.7	.06173	120.7	81.4	.05897	115.2	86.3	.05613	109.6	90.2	.03396	105.3
	208	77.9	.06714	128.6	84.0	.06403	122.6	89.2	.06095	116.6	93.5	.05868	112.3

Table 9a Steam tracing data (Metric units continued)

Design Conditions:

Tracer: One 22 MM O.D. Air convection tubing tracer
Insulation Thermal Conductivity: See Chart, page 79
Ambient: −17.8° C, 40 KPH wind

Nomenclature:

TS = Steam Temperature (°C)
TP = Pipe Temperature (°C)
GM/SEC-M = Tracer condensate rate per linear metre
Q3 = Pipe Heat Losses in Watts/lin m

PIPE SIZE	Steam Temp °C	INSULATION THICKNESS											
		38 MM Nom			51 MM Nom			64 MM Nom			76 MM Nom		
	TS	TP	GH/SEC-M	Q3	TP	GM/SEC-M	Q3	TP	GM/SEC-M	Q3	TP	GM/SEC-M	Q3
48 MM	121	67.0	.01733	33.2	71.5	.01580	34.8	74.7	.01403	32.3	77.2	.01374	30.3
	134	75.3	.01968	42.6	80.0	.01793	38.8	83.6	.01661	36.0	86.4	.01560	33.8
	148	83.7	.02226	47.2	83.9	.02030	43.1	92.8	.01883	40.0	95.9	.01714	37.7
	160	90.9	.02466	57.4	96.5	.02248	46.9	100.8	.02093	43.6	104.1	.01969	41.0
	170	96.5	.02660	54.6	102.3	.02429	49.8	106.9	.02262	46.4	110.4	.02127	43.6
	186	104.8	.02977	59.5	111.3	.02726	54.5	116.1	.02534	50.6	119.9	.02383	47.6
	198	111.7	.03261	63.7	118.6	.02989	58.4	123.7	.02780	54.3	127.7	.02616	51.1
	208	117.3	.03505	67.1	124.4	.03214	61.5	129.8	.02990	57.3	134.0	.02815	53.9
60 MM	121	61.1	.02041	45.1	68.8	.01827	40.3	71.0	.01673	36.9	74.1	.01558	34.3
	134	68.7	.02316	50.1	75.0	.02075	44.9	79.5	.01901	41.2	83.0	.01769	38.3
	148	76.6	.02613	55.5	83.4	.02343	49.8	88.3	.02151	45.6	92.1	.02002	42.5
	160	83.3	.02894	60.3	90.6	.02598	54.1	95.9	.02384	49.6	100.0	.02219	46.2
	170	88.4	.03119	64.0	96.1	.02799	57.5	101.7	.02572	52.7	106.0	.02397	49.2
	186	96.1	.03485	69.7	104.4	.03134	62.6	110.5	.02879	57.6	115.3	.02691	53.8
	198	102.6	.03613	74.5	111.3	.03433	67.0	117.8	.03156	61.6	122.9	.02951	57.6
	208	107.7	.04097	78.4	116.9	.03688	70.6	123.6	.03393	65.0	128.9	.03173	60.8
89 MM	121	58.3	.02283	50.4	64.3	.02057	45.3	68.9	.01876	41.4	72.6	.01728	36.4
	134	65.9	.02585	56.0	72.4	.02327	50.4	77.2	.02130	46.1	81.3	.01963	42.5
	148	73.2	.02917	62.0	80.5	.02625	55.8	85.8	.02405	51.1	90.3	.02220	47.1
	160	79.8	.03238	67.4	87.5	.02908	60.6	93.2	.02666	55.5	98.1	.02460	51.2
	170	84.7	.03488	71.5	92.8	.03134	64.3	98.8	.02872	58.9	104.6	.02655	54.4
	186	92.1	.03890	77.8	100.9	.03502	70.0	107.4	.03215	64.3	113.0	.02971	59.4
	198	98.3	.04256	83.2	107.6	.03832	74.9	114.5	.03522	68.8	120.4	.03256	63.6
	208	103.3	.04569	87.5	113.0	.04118	78.8	120.2	.03785	72.4	126.4	.03500	67.0
114 MM	121	54.2	.02534	55.9	60.5	.02266	50.0	66.0	.02057	45.4	69.6	.01911	42.2
	134	61.0	.02870	62.1	68.2	.02565	55.6	74.1	.02335	50.6	78.0	.02170	47.0
	148	68.2	.03234	68.6	75.9	.02896	61.5	82.4	.02634	56.0	86.7	.02450	52.0
	160	74.4	.03571	74.4	82.7	.03215	67.0	89.5	.02918	60.8	94.2	.02715	56.5
	170	79.0	.03846	78.9	87.8	.03464	71.0	95.0	.03145	64.5	99.9	.02926	60.0
	186	86.1	.04290	85.7	95.6	.03863	77.3	103.3	.03514	70.2	108.6	.03274	65.4
	198	91.9	.04689	91.6	102.0	.04228	82.6	110.1	.03846	75.1	115.8	.03585	70.0
	208	96.6	.05028	96.3	107.1	.04539	86.9	115.6	.04133	79.1	121.5	.03852	73.8
168 MM	121	47.8	.02898	63.9	54.7	.02614	57.7	59.9	.02388	52.7	65.0	.02177	48.1
	134	54.1	.03275	70.9	61.7	.02956	64.1	67.5	.02705	58.6	73.1	.02474	53.7
	148	60.5	.03687	78.2	68.9	.03335	70.8	75.2	.03053	64.8	81.2	.02795	59.4
	160	66.2	.04071	84.7	75.1	.03685	76.7	81.8	.03371	70.3	88.3	.03894	64.4
	170	70.5	.04378	89.8	79.8	.03960	81.3	86.9	.03635	74.5	93.8	.03334	68.4
	186	76.9	.04880	97.5	80.9	.04419	88.4	94.6	.04054	81.1	101.9	.03724	74.4
	198	82.3	.05327	104.1	92.8	.04833	94.4	100.9	.04437	86.6	108.7	.04073	79.6
	208	86.5	.05718	109.4	97.6	.05182	99.2	106.0	.04760	91.1	114.2	.04377	88.8
219 MM	121	43.0	.03223	71.0	51.2	.02819	62.1	56.3	.02600	57.4	60.6	.02418	53.4
	134	48.9	.03635	78.8	57.8	.03185	69.0	63.0	.02939	63.7	68.1	.02743	59.3
	148	54.7	.04076	86.6	64.8	.03586	76.1	70.9	.03317	70.4	75.8	.03091	65.6
	160	59.9	.04501	93.7	70.7	.03961	82.5	77.2	.03666	76.3	82.5	.03414	71.2
	170	64.0	.04838	99.3	75.1	.04260	87.4	82.0	.03945	80.9	87.6	.03680	75.5
	186	69.9	.05391	107.8	81.9	.04748	94.9	89.3	.04396	87.9	95.3	.04107	82.1
	198	74.8	.05861	114.5	87.6	.05186	101.3	95.4	.04808	93.9	101.7	.04492	87.7
	208	78.7	.06285	120.3	92.0	.05560	106.5	100.2	.05156	98.7	105.9	.04819	92.3

273 MM	134	42.1	.03996	86.7	49.0	.03666	79.5	54.8	.03362	72.9	59.9	.03138	66.1
	148	47.4	.04473	95.0	55.0	.04126	87.7	61.5	.03784	80.4	66.8	.03540	75.1
	160	52.0	.04932	102.7	60.2	.04556	94.9	67.2	.04179	87.1	72.9	.03910	81.4
	170	55.5	.05292	108.6	64.3	.04895	100.5	71.3	.04481	92.0	77.5	.04207	86.3
	186	60.8	.05892	117.6	70.2	.05435	108.6	77.9	.05002	100.0	84.5	.04689	93.7
	198	65.4	.06424	125.5	75.1	.05931	115.8	83.3	.05464	106.8	90.2	.05123	100.1
	208	68.9	.06875	131.7	79.1	.06355	121.6	87.7	.05866	112.3	94.8	.05494	105.2
324 MM	121	32.1	.03785	83.6	39.6	.03453	76.1	45.4	.03191	70.3	49.9	.02940	64.8
	134	36.9	.04263	92.4	45.0	.03887	84.3	51.4	.03597	77.8	56.4	.03317	71.9
	148	41.8	.04796	101.8	50.7	.04372	92.9	57.7	.04041	85.9	63.2	.05740	79.4
	160	46.0	.05280	109.9	55.5	.04824	100.5	63.1	.04444	92.7	69.0	.04126	86.0
	170	49.3	.05664	116.2	59.1	.05157	105.9	67.2	.04782	98.1	73.4	.04438	91.1
	186	54.0	.06272	125.4	64.9	.05744	114.8	73.4	.05331	106.5	80.1	.04951	99.0
	198	58.0	.06836	133.5	69.5	.06267	122.3	78.6	.05819	113.7	85.6	.05405	105.6
	208	61.3	.07311	140.0	73.2	.06704	128.4	82.7	.06239	119.4	90.0	.05801	114.0
356 MM	121	28.6	.03974	87.8	35.9	.03629	79.9	41.9	.03367	74.4	46.5	.03114	68.8
	134	32.9	.04450	96.5	41.0	.04080	88.5	47.5	.03783	62.6	52.8	.03539	76.7
	148	37.4	.05004	106.2	46.3	.04593	97.4	53.3	.04254	90.4	59.3	.03985	84.7
	160	41.4	.05505	114.6	50.9	.05057	105.3	58.4	.04696	97.8	64.7	.04389	91.5
	170	44.4	.05902	121.2	54.3	.05431	111.4	62.4	.05047	103.5	68.9	.04725	97.0
	186	48.9	.06562	131.2	59.6	.06039	120.7	68.3	.05616	112.2	75.3	.05273	105.4
	198	52.7	.07151	139.6	64.1	.06583	128.6	73.1	.06122	119.7	80.5	.05734	112.1
	208	55.7	.07646	146.4	67.5	.07043	134.9	76.9	.06543	125.2	84.7	.06152	117.7
406 MM	121	25.7	.04105	90.6	33.0	.03780	83.4	38.9	.03523	77.6	43.8	.03289	72.7
	134	29.9	.04621	100.2	37.9	.04258	92.3	44.4	.03964	86.0	49.7	.03714	80.5
	148	34.2	.05192	110.2	42.9	.04788	101.6	49.8	.04439	94.3	55.7	.04180	88.8
	160	37.9	.05712	118.9	47.2	.05273	109.8	54.6	.04896	102.0	61.0	.04598	95.7
	170	40.7	.06130	125.7	50.5	.05657	116.1	58.3	.05261	107.8	65.0	.04944	101.3
	186	45.0	.06806	136.1	55.3	.06266	125.2	64.0	.05850	116.9	71.0	.05501	110.0
	198	48.5	.07414	144.8	59.5	.06829	133.4	68.6	.06378	124.6	76.0	.05990	117.2
	208	51.3	.07928	151.7	62.9	.07304	139.9	72.3	.06826	130.7	88.8	.06431	123.1
457 MM	121	23.3	.04240	93.6	30.5	.03922	86.6	36.2	.03652	80.4	41.1	.03430	75.6
	134	27.2	.04773	103.5	35.1	.04417	95.8	41.4	.04105	89.0	46.7	.03863	83.8
	148	31.3	.05363	113.7	39.7	.04934	104.8	46.7	.04621	98.0	52.5	.04346	92.3
	160	34.8	.05886	122.7	43.8	.05438	113.2	51.3	.05088	106.0	57.5	.04795	99.8
	170	37.4	.06325	129.7	46.9	.05831	119.7	54.8	.05465	112.0	61.5	.05147	105.7
	186	41.5	.07012	140.3	51.6	.06485	129.6	60.2	.06075	121.4	67.1	.05712	114.2
	198	44.8	.07644	149.3	55.6	.07065	138.0	64.6	.06623	129.3	71.9	.06229	121.7
	208	47.5	.08177	156.5	58.7	.07562	144.7	68.1	.07093	135.7	75.7	.06667	127.7
508 MM	121	21.1	.04361	96.3	28.2	.04050	89.4	33.8	.03773	83.3	38.7	.03566	78.6
	134	24.9	.04909	106.4	32.4	.04533	98.3	38.8	.04250	92.2	44.1	.04012	87.0
	148	28.7	.05502	116.9	36.9	.05096	108.1	43.8	.04780	101.4	49.5	.04490	95.4
	160	32.0	.06052	126.2	40.8	.05607	116.8	48.2	.05264	109.6	54.3	.04957	103.1
	170	34.5	.06502	133.3	43.8	.06016	123.4	51.6	.05646	115.9	58.0	.05317	109.1
	186	38.2	.07171	143.5	48.3	.06681	133.6	56.5	.06256	125.0	63.6	.05915	118.2
	198	41.4	.07815	152.6	52.0	.07280	142.2	60.8	.06818	133.1	68.2	.06450	126.0
	208	43.9	.08357	159.9	55.0	.07782	149.1	64.2	.07292	139.7	71.9	.06902	132.2
610 MM	121	18.6	.04505	99.5	25.1	.04190	92.5	30.6	.03955	87.3	35.1	.03742	82.4
	134	22.2	.05082	109.9	29.2	.04716	102.3	35.2	.04454	96.6	40.2	.04204	91.2
	148	25.8	.05680	120.7	33.4	.05302	112.5	39.8	.04584	105.7	45.4	.04728	100.3
	160	28.8	.06211	129.5	37.0	.05821	121.4	43.9	.05482	114.1	49.9	.05207	108.4
	170	31.1	.06670	136.8	39.8	.06256	128.3	47.0	.05877	120.6	53.3	.05587	114.6
	186	34.7	.07393	147.9	44.0	.06934	138.7	51.6	.06536	130.6	58.6	.06215	124.2
	198	37.7	.08054	157.3	47.5	.07562	147.7	55.7	.07121	139.0	63.0	.06774	132.3
	208	40.1	.08613	164.8	50.3	.08089	154.8	59.0	.07616	145.8	66.3	.07225	138.3
762 MM	121	14.1	.04763	105.2	20.2	.04455	98.4	25.3	.04203	92.8	29.8	.04012	88.6
	134	17.0	.05318	115.1	23.8	.05014	108.7	29.4	.04733	102.6	34.4	.04494	97.4
	148	20.2	.05947	126.4	27.6	.05619	119.4	33.7	.05315	112.7	39.0	.05043	107.1
	160	23.0	.06538	136.3	30.8	.06178	128.8	37.3	.05836	121.7	43.0	.05537	115.7
	170	25.1	.07018	143.9	33.3	.06636	136.1	40.1	.06273	128.6	46.1	.05958	122.3
	186	25.3	.07784	155.6	36.9	.07316	146.4	44.4	.06956	139.2	50.8	.06624	132.4
	198	30.9	.08469	165.4	40.0	.07972	155.7	47.9	.07582	148.0	54.6	.07218	140.9
	208	33.0	.09051	172.2	42.5	.08923	163.1	50.7	.08111	155.2	57.7	.07716	147.8

Electrical heater external tracing systems

Piping and equipment may be heated to desired temperature by a number of types of resistance heating cables or tapes. These cables are designed to have specific resistances which when connected to commercially available voltages will yield outputs of heat energy suitable for most heating requirements. Figure 26-A illustrates a schematic diagram of series resistance cable and Figure 26-B is a schematic diagram of parallel resistance cable. The cable conductor is most often nickel or copper alloy metal. Where temperature of cable is moderate they may be electrically insulated with any of a variety of the thermoplastics such as PVC, Teflon, Thermoplastic rubber or polyethylene. Each of these plastics have an upper temperature limit which must not be exceeded. Another type cable is mineral insulated cable. The conductor (or conductors) are jacketed in a solid metal sheath and electrically insulated with magnesium oxide. Heating tape is made with bus wires or strips with a conductive matrix formed between them. This is shown schematically in Figure 26-C.

The conversion of electrical energy into heat energy is caused by the resistance of the electrical conductor to the passage of electrons, i.e. current. The series resistance heater cable, in its most basic form is simply one continuous wire which when connected to voltage begins to heat. Resistance of a wire is mathematically expressed, when:

ϱ = resistivity of the conductor wire (R/ft or R/ε m)
A = cross sectional area of wire (ft^2 or ε m^2)
L = length of wire (ft or ε m)
R = resistance (r)

Then $R = \varrho \dfrac{L}{A}$

Since the Watt per unit length output of a cable is, when:

V = voltage applied to cable

$$\text{Watt per unit length of cable} = \frac{V^2}{R \times L}$$

Combining equations

$$\frac{V^2 A}{\varrho \Lambda^{II}} = \text{Watt per unit length} \qquad (36)$$

From Equation 32 (and 32a) it is possible, knowing the thermal conditions, to determine the Watts loss per unit length. Thus with above equation it is possible to select cable length, conductor resistance, and conductor area to yield the desired cable output.

The cable selected must be such as to provide sufficient heat energy to fulfill the thermal requirements of the process pipe. Various types of series cables are available and they can be installed on pipe as parallel tracers, with or without being thermally bonded to the pipe. Various types of series installed cable are shown in Figure 27-A, -C, and -D, without thermal bonding.

Parallel resistance cables and tapes may also be used. These may be described as a succession of resistance elements con-

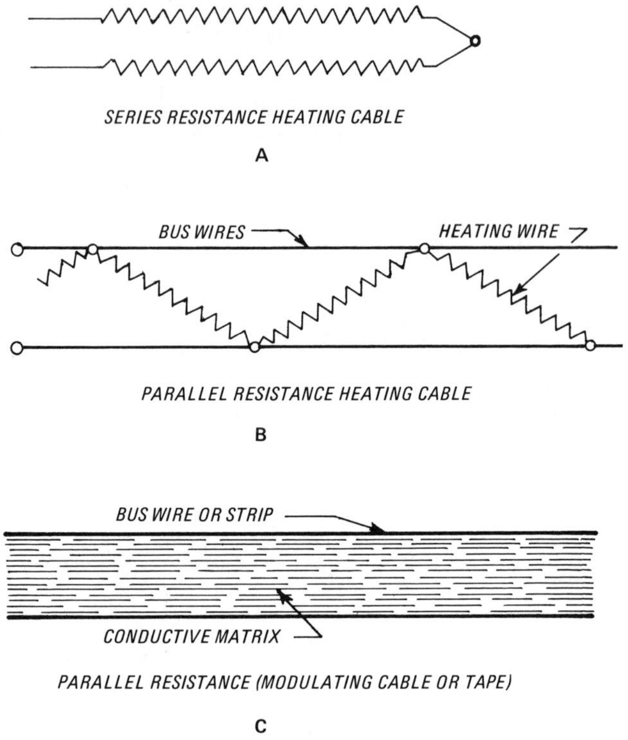

Figure 26 Electric heating cables and tapes.

nected in parallel. Unlike the series resistance cable, the watt per unit length output of the parallel cable is basically unchanged regardless of cable length. The output over the cable length will only be affected by the small voltage drop that occurs along the main buss wires. It is recommended that the maximum length of cable used be no greater than that which will permit the watts per unit length at the last heating zone to be no less than 90% of the initial value. The principal advantage of parallel cable is its relative insensitivity to length. Identical to series resistance cables if the thermal conditions are known, it is possible to determine watts loss per unit length from Equation 32 (or 32a). Knowing the loss in watts per unit length, proper cable can be easily chosen since the parallel resistance cable is specified by its manufacturer as to watts per unit length output.

A third type of heating cable and tape is similar to previously discussed parallel heating cables, however it has modulating capability. It contains a graphite matrix core, the electrical resistance of which increases with temperature. The output of this cable, like the parallel cable, is only affected by the small voltage drop that occurs along the main buss wires. Thus its watt output is relatively unaffected by length. Its output is, however, affected by the temperature environment surrounding the cable.

This characteristic is both an advantage and a disadvantage. It is an advantage as the increase in resistance with increase in temperature serves as a high temperature limit. However, this characteristic makes determination of heat output difficult since output will vary with temperature of pipe (or equipment), ambient air conditions, thickness and conductivity of insulation and method of attachment. The schematic illustration of this type heater is in Figure 26-C.

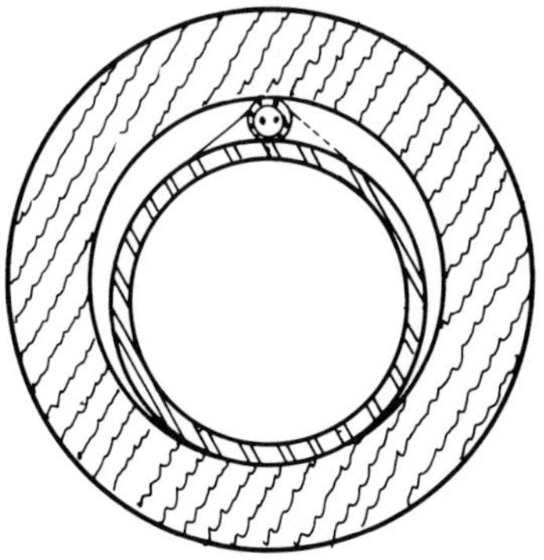

MINERAL INSULATED CABLE (COPPER OR NICKEL ALLOY RESISTANCE CONDUCTOR) WITHOUT HEAT TRANSFER CEMENT

(A)

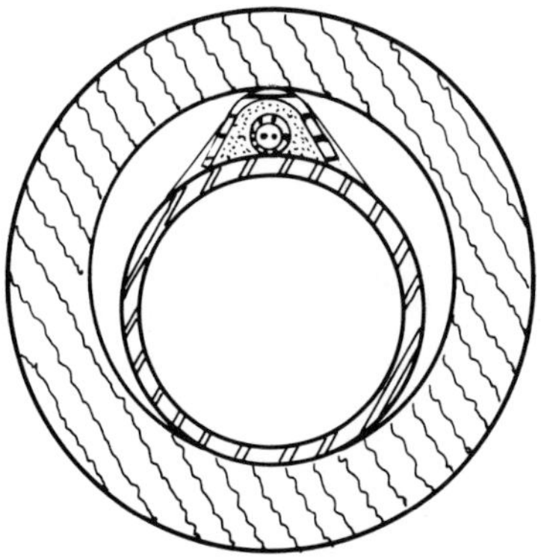

MINERAL INSULATED CABLE (COPPER OR NICKEL ALLOY RESISTANCE CONDUCTOR) WITH HEAT TRANSFER CEMENT AND CHANNEL

(B)

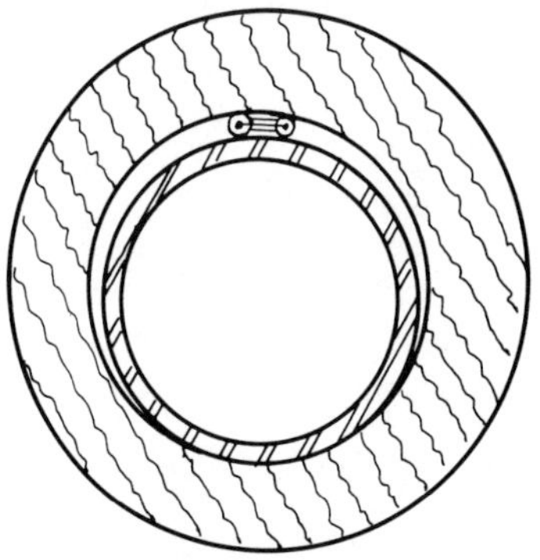

POLYETHYLENE INSULATED PARALLEL RESISTANCE GRAPHITE MATRIX MODULATING CONDUCTOR WITHOUT HEAT TRANSFER CEMENT

(C)

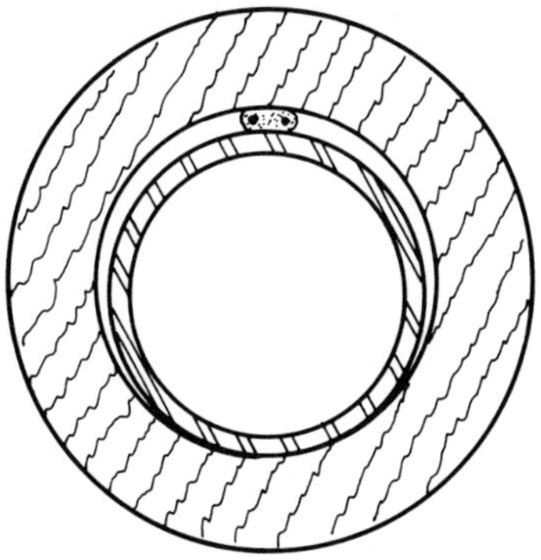

TEFLON, PVC, OR THERMOPLASTIC RUBBER INSULATED PARALLEL RESISTANCE NICHROME CONDUCTOR WITHOUT HEAT TRANSFER CEMENT

(D)

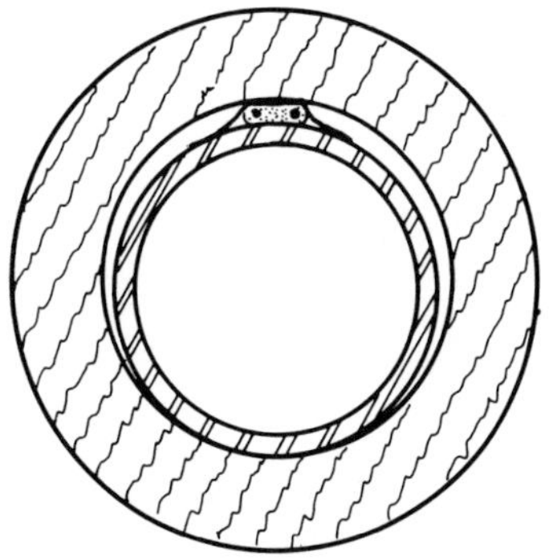

TEFLON, PVC, OR THERMOPLASTIC RUBBER INSULATED PARALLEL RESISTANCE NICHROME CONDUCTOR COVERED WITH ALUMINUM TAPE

(E)

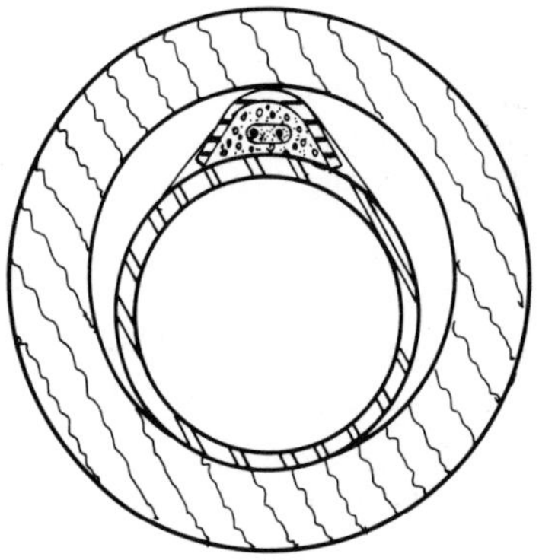

TEFLON, PVC, OR THERMOPLASTIC RUBBER INSULATED PARALLEL RESISTANCE NICHROME CONDUCTOR WITH HEAT TRANSFER CEMENT

(F)

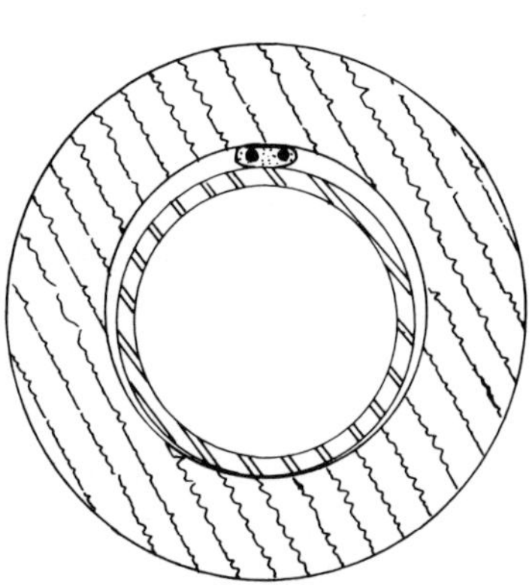

TEFLON INSULATED CABLE (COPPER OR NICKEL ALLOY SERIES RESISTANCE CONDUCTOR) WITHOUT HEAT TRANSFER CEMENT

(G)

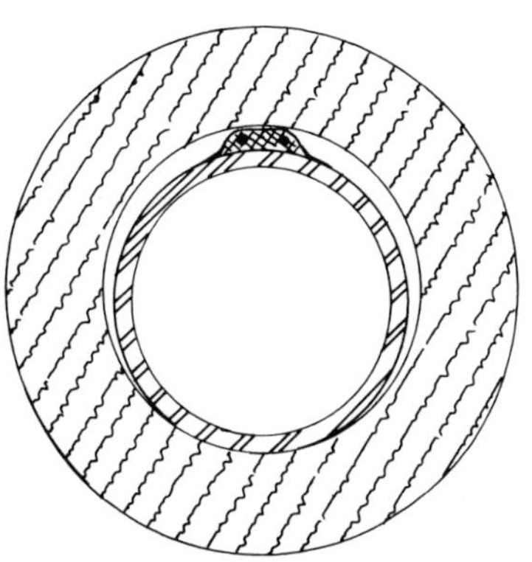

TEFLON INSULATED CABLE (COPPER OR NICKEL ALLOY SERIES RESISTANCE CONDUCTOR) COVERED WITH ALUMINUM TAPE

(H)

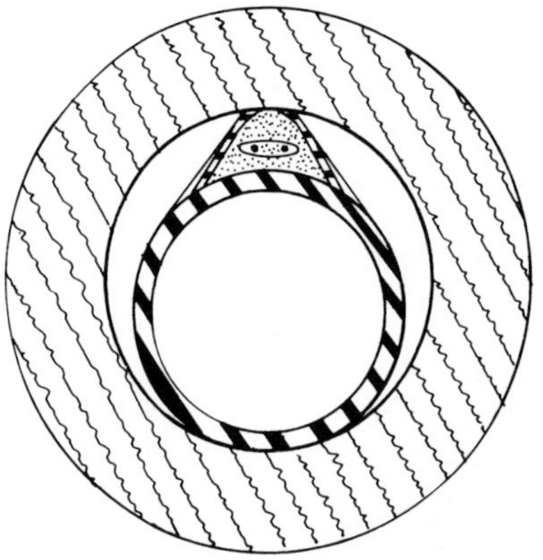

TEFLON INSULATED CABLE (COPPER OR NICKEL ALLOY SERIES RESISTANCE CONDUCTOR) WITH HEAT TRANSFER CEMENT AND CHANNEL

(I)

Figure 27 Methods for heating process pipe.

As in fluid heat tracing, the method of attachment is very important to its functional characteristics. In Figure 27-A, 27-C, 27-D, and 27-E the various cables are shown installed on a pipe by simple wiring or strapping. In these conditions the heat transfer is mainly accomplished by air convection. Figures 27-E, and H indicate that aluminum tape was placed over the tracer to obtain some reflectivity of heat from the tape, however as tape restricts the movement of air convection currents around the tracer, not all reflective advantage is obtained.

Tracer thermally bonded to pipe — fluid systems

To increase the effectiveness of external tracing, thermal bonding of the tracer to the process pipe is recommended. The basic concept of thermal attachment of the tracer tube is shown in Figure 28. In 1956, Turner and Estep developed a formula for calculation of t_m and q_t for such a traced system of the configuration noted and for "trowelled on" highly conductive cement.

As new methods of thermally bonding tracers to pipe (such as channel and fillet type installation) were developed, as shown in Figure 29, it was necessary to modify the original equations. These modifications done in 1978 by Barth and McDonald, as before, required the following assumptions:

(1) f_t, f_p, f_i were equal to 2 Btu/ft^2, hr, ° (11.86 W/m^2, °C)
(2) f_o was equal to 6 Btu/ft^2, hr, °F (34.08 W/m^2, °C)
(3) k of insulation was .45 Btu, in./ft^2, hr, °F (.065 W/m^2, °C)

With these assumptions and using the nomenclature in Figure 28, it is possible to solve directly for t_m and q_t.

When A_{cc} = cement contact area ft^2/ft (or m^2/m in metric) and A_{cp} = cement or channel perimeter ft^2/ft (or m^2/m in metric), the heat transfer from the tracer to the process pipe q_{tp} through cement is:

$$q_{tp} = n\, q_t\, A_{cc}\, (t_{media} - t_m) \tag{37}$$

where q_t is given in Table 11 and n is the number of tracers.

The heat transfer from process pipe to annulus space is:

$$Q_{pa} = 2\left(\frac{d\,p\,\pi}{12} - n\,A_{cc}\right)(t_m - t_{ap}) \tag{38}$$

The heat transfer from tracer to the annulus space is:

$$Q_{ta} = 2\,(A_{cp})\,(t_{media} - t_{ap}) \tag{39}$$

The heat transfer through the insulation

$$Q_{ia} = \left(\frac{\pi\, d_o}{12}\right) \frac{t_{ap} - t_a}{\left[\left(\dfrac{d_o}{f_i\, d_i}\right) + \left(\dfrac{\dfrac{d_o}{2}\log_e \dfrac{d_o}{d_i}}{k}\right) + \left(\dfrac{1}{f_o}\right)\right]} \tag{40}$$

or

$$Q_{ia} = \frac{\pi\, d_o\, [t_{ap} - t_a]}{\left[6\,\dfrac{d_o}{d_i} + \left(13.32\, d_o \log_e \dfrac{d_o}{d_i}\right) + 2\right]} \tag{41}$$

Under equilibrium conditions:

$$Q_{pa} + Q_{ta} = Q_{ia} \tag{42}$$

Also

$$Q_{tp} = Q_{pa} \tag{43}$$

Substituting for Q_{tp} and for Q_{pa} in Equation 43:

$$q_t\, n\, A_{cc}\, (t_{media} - t_m) = 2\left(\frac{d_p\pi}{12} - n\,A_{cc}\right)(t_m - t_{ap}) \tag{44}$$

Substituting for Q_{pa}, Q_{ta}, and Q_{ia} in Equation 42

$$2\left(\frac{d_p\pi}{12} - n\,A_{cc}\right)(t_m - t_{ap}) + 2\,A_{cp}\,(t_{media} - t_{ap}) = \frac{\pi\, d_o\,(t_{ap} - t_a)}{6\,\dfrac{d_o}{d_i} + 13.32\, d_o \log_e \dfrac{d_o}{d_i} + 2} \tag{45}$$

Solving Equation 44 for t_{ap}

$$t_{ap} = \frac{t_m - \dfrac{q_t}{2}\, n\, A_{cc}\,(t_{media} - t_m)}{\dfrac{d_p\pi}{12} - n\,A_{cc}}$$

Solving for t_m and substituting for t in Equation 45:

Then in English Units,

$$t_m = \frac{t_{media}\left[q_t\, n\, A_{cc} + 2A_{cp} + \left(\dfrac{n\,A_{op}\,A_{cc}\,q_t}{\dfrac{d_o\,\pi}{12} - n\,A_{cc}}\right) \div \left(\dfrac{\dfrac{\pi\, d_o}{2}\, q_t\, n\, A_{cc}}{\left(\dfrac{d_p\,\pi}{12} - n\,A_{cc}\right)\left(\dfrac{6d_o}{d_i} + 13.32\, d_o \log_e \dfrac{d_o}{d_i} + 2\right)}\right)\right] + \left[\left(\dfrac{\pi\, d_o}{\dfrac{6\,d_o}{d_i} + 13.32\, d_o \log_e \dfrac{d_o}{d_i} + 2}\right) t_a\right]}{q_t\, n\, A_{cc} + 2A_{cp} + \left(\dfrac{n\,A_{op}\,A_{cc}\,q_t}{\dfrac{d_p\,\pi}{12} - n\,A_{cc}}\right) \div \left(\dfrac{\pi\, d_o}{\dfrac{6d_o}{d_i} + 13.32\, d_o \log_e \dfrac{d_o}{d_i} + 2}\right) + \left[\dfrac{\dfrac{\pi\, d_o}{2}\, q_t\, n\, A_{cc}}{\left(\dfrac{d_p\,\pi}{12} - n\,A_{cc}\right)\left(\dfrac{6\,d_o}{d_i} + 13.32\, d_o \log_e \dfrac{d_o}{d_i} + 2\right)}\right]} \tag{46}$$

ENGLISH		*METRIC*
d_o =	outside diameter of insulation (in., mm)	d_{mo}
d_i =	inside diameter of insulation (in., mm)	d_{mi}
d_p =	outside diameter of pipe (in., mm)	d_{mp}
d_t =	outside diameter of tracer (in., mm)	d_{mt}
n =	number of tracers	n
t_m =	process line temperature (°F, °C)	t_{mm}
t_{media} =	media temperature (°F, °C)	$t_{m\ media}$
t_{ap} =	temperature of annulus space (°F, °C)	t_{map}
t_a =	air temperature (°F, °C)	t_{ma}
f_t =	air film coefficient, conductive cement surface $\frac{\text{Btu}}{\text{hr-ft}^2\text{-°F}}, \frac{\text{Watt}}{\text{m}^2\text{-°C}}$	f_{mt}
f_p =	air film coefficient, process pipe surface $\frac{\text{Btu}}{\text{hr-ft}^2\text{-°F}}, \frac{\text{Watt}}{\text{m}^2\text{-°C}}$	f_{mp}
f_i =	air film coefficient, inside surface of insulation $\frac{\text{Btu}}{\text{hr-ft}^2\text{-°F}}, \frac{\text{Watt}}{\text{m}^2\text{-°C}}$	f_{mi}
f_o =	air film coefficient, outside surface of insulation $\frac{\text{Btu}}{\text{hr-ft}^2\text{-°F}}, \frac{\text{Watt}}{\text{m2-°C}}$	f_{mo}
k =	conductivity of insulation, (Btu-in./hr-ft^2-°F, watt/m^2-°C)	k_m
q_t =	overall heat transmittance from tracer, throught cement to process pipe (Btu, hr-ft^2-°F, watt/m^2-°C)	q_{mt}
d_{ti} =	inside diameter of tracer (in., mm)	d_{mti}
A_{cc} =	cement contact area (ft^2/ft, m^2/m)	A_{mcc}
A_{cp} =	cement or channel perimeter (ft^2/ft, m^2/m)	A_{mcp}

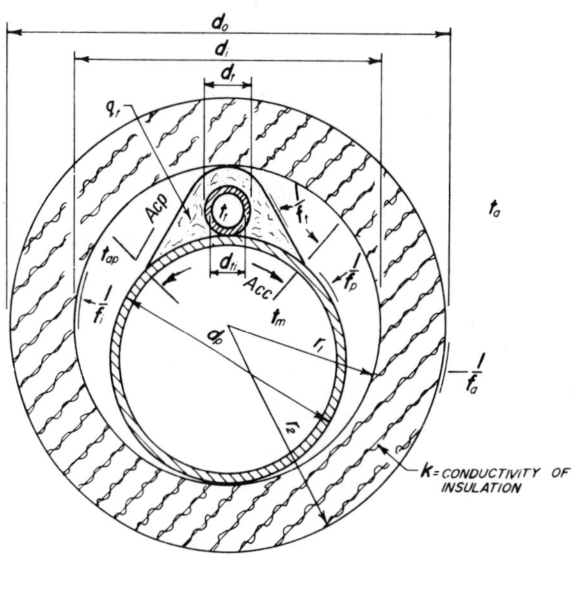

Figure 28 Schematic diagram—external tracer bonded to process pipe with conductive cement.

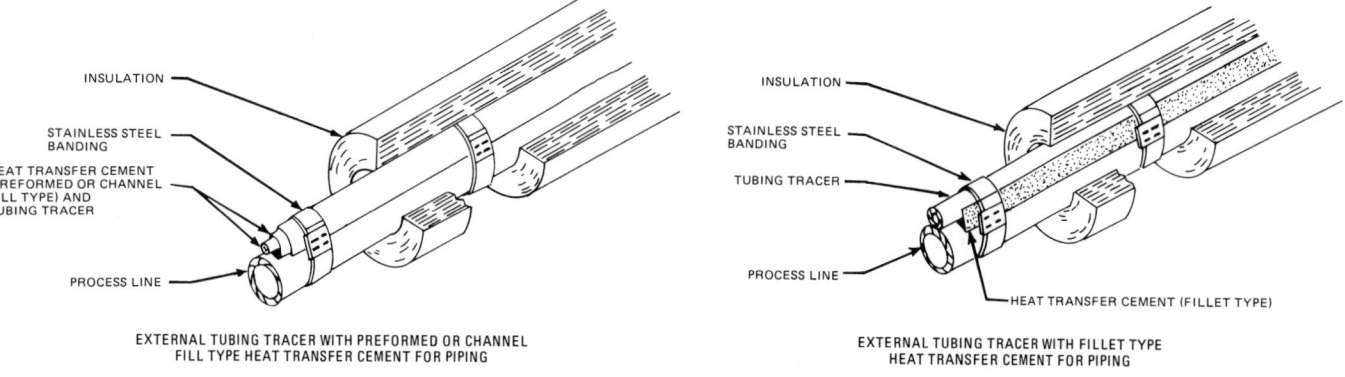

EXTERNAL TUBING TRACER WITH PREFORMED OR CHANNEL FILL TYPE HEAT TRANSFER CEMENT FOR PIPING

EXTERNAL TUBING TRACER WITH FILLET TYPE HEAT TRANSFER CEMENT FOR PIPING

Figure 29 Channel and fillet thermal bonding of tracer to process pipe.

$$t_{mm} = \frac{t_{m\,media}\left[q_{mt}\, n\, A_{mcc} + 11.36A_{mqt} + \left(\dfrac{n\, A_{moq}\, A_{mcc}\, q_{mt}}{\dfrac{d_{mo}\,\pi}{1000} - n\, A_{mcc}}\right)\right.}{q_{mt}\, n\, A_{mcc} + 11.36A_{mcp} + \left(\dfrac{n\, A_{mcp}\, A_{mcc}\, q_{mt}}{\dfrac{d_{mp}\,\pi}{1000} - n\, A_{mcc}}\right)}$$

$$\frac{\left. + \left(\dfrac{\dfrac{\pi\, d_{mo}}{11.36}\, q_{mt}\, n\, A_{mcc}}{\left(\dfrac{d_{mp}\,\pi}{1000} - n\, A_{mcc}\right)\left(\dfrac{88d_{mo}}{d_{mi}} + 7.69\, d_{mo}\log_e\dfrac{d_{mo}}{d_{mi}} + 29\right)}\right)\right]}{+\left(\dfrac{\pi\, d_{mo}}{\dfrac{88d_{mo}}{d_{mi}} + 7.69\, d_{mo}\log_e\dfrac{d_{mo}}{d_{mi}} + 29.3}\right)}$$

$$\frac{+\left[\left(\dfrac{\pi\, d_{mo}}{\dfrac{88\, d_{mo}}{d_{mi}} + 7.69\, d_{mo}\log_e\dfrac{d_{mo}}{d_{mi}} + 29.3}\right) t_{ma}\right]}{+\left[\dfrac{\dfrac{\pi\, d_{mo}}{11.36}\, q_{mt}\, n\, A_{mcc}}{\left(\dfrac{d_{mp}\,\pi}{1000} - n\, A_{mcc}\right)\left(\dfrac{88\, d_{mo}}{d_{mi}} + 7.69\, d_{mo}\log_e\dfrac{d_{mo}}{d_{mi}} + 29.3\right)}\right]} \qquad (46a)$$

Once t_m (or t_{mm}) has been determined, the heat delivery from the tracer (or heat loss from the insulation) may be obtained as follows:

In *English Units,*

$$Q_t = q_t\, n\, A_{cc}\,(t_{media} - t_m) \qquad (47)$$

In *Metric Units,*

$$Q_{mt} = q_{mt}\, n\, A_{mcc}\,(t_{m\,media} - t_{mm}) \qquad (47a)$$

To solve these equations, values for contact areas A_{cc} and A_{cp} must be determined. For the commonly used methods of installing heat transfer cement, the following values can be used. Table 10 gives English unit values and Table 10a presents Metric unit values:

Table 10 and 10a Type of installation of heat transfer cement.

Type of installation of heat transfer cement	Table 10 — English Units		Table 10a — Metric Units	
	Values of		Values of	
	Acc ft^2/ft	Acp ft^2/ft	Amcc m^2/m	Amop m^2/m
Tracer bonded with fillet applied heat transfer cement 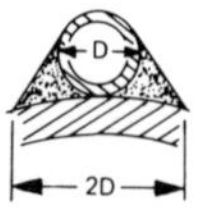	$0.01666d_t$	$0.2357d_t$	$0.002d_{mt}$	$0.00283d_{mt}$
Hand trowel ½″ over top of tracer 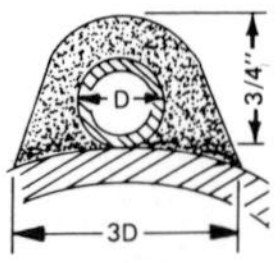	$0.25d_t$	$\sqrt{0.0908d_t^2 + 0.01388d_t + 0.00174}$	$0.003d_{mt}$	$\sqrt{0.0065d_{mt}^2 + 0.0254d_{mt} + 0.0806}$
Channel-small filled with heat transfer cement 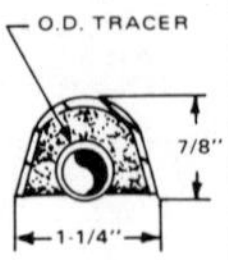	0.1167	0.2083	0.0356	0.0635
Channel-large filled with heat transfer cement 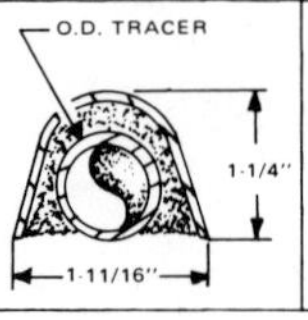	0.169	0.2708	0.0515	0.0826

d_t is in inches
d_{mt} is in millimeters

Table 11 Values of q_t or q_{mt} for Equation 37.

English & Metric Units

NOMINAL PIPE SIZE	q_t*	q_{mt}*
	Btu/Ft²-°F-Hr	Watt/m²-°C
1.5″ (48 mm)	34.3	194.8
2.0″ (60 mm)	32.6	185.1
3.0″ (89 mm)	29.1	165.2
4.0″ (114 mm)	26.9	152.7
6.0″ (168 mm)	23.8	135.1
8.0″ (219 mm)	21.5	122.1
10.0″ (273 mm)	18.4	104.5
12.0″ (324 mm)	14.6	82.9
14.0″ (356 mm)	12.2	69.3
16.0″ (406 mm)	9.8	55.6
18.0″ (457 mm)	9.8	55.6
20.0″ (508 mm)	9.8	55.6
24.0″ (610 mm)	9.8	55.6

*For multiple tracers multiply q_t by .9.

As was the case with the Estep-Turner equation for convection tracers, these equations are based on simplifying assumptions which were necessary to obtain a direct solution. The assumption of constant air film coefficients and a constant thermal conductivity are not necessarily true. In order to consider all of these parameters as variables, an iterative solution of simultaneous equations is required. Tables 11 and 11a are based on an iterative solution developed by Thermon Mfg. Company and have been processed on high speed digital computer.

Table 12 was calculated for 1/2″ and 3/4″ OD tracers being installed on pipe with channels filled with heat transfer cement, thermally bonding the tracers to the process pipe. The generally used extremes of design of ambient air temperature at 0° F, 25 mph wind and 100° F with no wind were chosen. For a particular insulation thickness and steam temperature of the process pipe, pounds of steam being condensed per linear ft, per hour and the heat loss in Btu per linear ft per hour are tabulated.

Table 12a was tabulated in metric units. The tracer tube sizes are 13 and 22 mm OD installed on pipe with channels filled with heat transfer cement. The ambient air temperature was −17.8° C, 40 kph wind and 37.8° C no wind. Temperature of steam 121° C to 208° C. Nominal insulation thicknesses were 33mm, 51mm, 64mm, and 76mm. The tabulated results provide the temperature of the process pipe in °C. The condensate rate of steam in gm/sec per linear metre. The heat loss per linear metre of length of pipe is in Watts/metre.

By the use of Tables 9 and 9a, the tabulation for tracers not thermally bonded to process pipe, and Tables 12 and 12a, direct comparison of the two systems are possible.

Table 12 Steam tracing data (English units)

Design Conditions:

Tracer: One 1/2" O.D. tracer thermally bonded to pipe with a TFK-4 channel and heat transfer cement.
Insulation Thermal Conductivity: See Chart, page 79
Ambient: 0° F, 25 MPH wind

Nomenclature:

TS = Steam Temperature (°F)
TP = Pipe Temperature (°F)
#/HR-FT = Tracer condensate rate per linear foot
Q3 = Pipe Heat Losses in Btu/hr, lin ft

N.P.S.	Steam Temp °F	INSULATION THICKNESS											
		1.5" Nom			2.0" Nom			2.5" Nom			3.0" Nom		
	TS	TP	#/HR-FT	Q3	TP	#/HR-FT	Q3	TP	#/HR-FT	Q3	TP	#/HR-FT	Q3
1.5"	250	232.0	.05814	55.1	233.4	.05126	48.6	234.4	.04623	44.0	235.1	.04261	40.6
	274	255.5	.06639	61.8	257.1	.05865	54.6	258.1	.05297	49.4	259.0	.04885	45.6
	298	279.9	.07548	69.0	281.6	.06676	61.0	283.0	.06052	55.4	283.9	.05595	51.1
	320	301.4	.08450	75.7	303.3	.07451	66.9	304.6	.06753	60.6	305.6	.06231	55.9
	338	317.7	.09140	80.8	319.7	.08077	71.4	321.1	.07326	64.7	322.2	.06772	59.7
	366	342.6	.10321	88.8	344.8	.09125	78.5	346.4	.08265	71.2	347.5	.07626	65.7
	388	363.5	.11363	95.6	365.8	.10061	84.6	367.5	.09129	76.7	368.7	.08428	70.8
	406	380.2	.12290	101.2	382.7	.10873	89.6	384.4	.09856	81.3	385.7	.09100	75.0
2"	250	227.9	.07261	68.8	230.2	.06223	58.9	231.7	.05523	52.3	232.8	.05014	47.5
	274	250.9	.08267	77.2	253.5	.07078	66.2	255.2	.06311	58.7	256.4	.05739	53.4
	298	274.8	.09414	86.1	277.6	.08074	73.9	279.5	.07194	65.6	280.8	.06519	59.6
	320	295.6	.10518	94.1	298.9	.09020	81.0	301.0	.08037	72.0	302.5	.07291	65.4
	338	311.8	.11410	100.6	315.1	.09778	86.4	317.3	.08690	76.8	318.9	.07905	69.9
	366	336.2	.12819	110.5	339.7	.11019	94.9	342.1	.09818	84.4	343.9	.08934	76.8
	388	356.5	.14153	118.9	360.3	.12170	102.2	362.9	.10806	90.9	364.8	.09848	82.6
	406	372.8	.15258	125.8	376.9	.13137	108.2	379.6	.11693	96.3	381.6	.10643	87.7
3"	250	222.3	.08435	80.3	225.4	.07223	68.4	227.5	.06383	60.5	229.2	.05712	54.1
	274	244.6	.09674	90.0	248.1	.08258	76.8	250.5	.07263	68.8	252.4	.06523	60.7
	298	267.7	.10994	100.3	271.6	.09363	85.6	274.3	.08278	75.7	276.4	.07414	68.8
	320	287.9	.12208	109.6	292.2	.10459	93.6	295.1	.09250	82.8	297.6	.08302	74.3
	338	303.4	.13243	116.8	307.9	.11324	99.9	311.2	.10022	88.6	313.7	.08978	79.4
	366	327.2	.14902	128.4	332.2	.12745	109.8	335.5	.11289	97.3	338.2	.10120	87.2
	388	346.9	.16402	138.1	352.2	.14070	118.2	355.8	.12465	104.7	358.7	.11160	93.9
	406	362.7	.17732	146.0	368.3	.15169	125.1	372.1	.13437	110.8	375.1	.12072	99.4
4"	250	216.3	.09795	92.7	220.5	.08253	78.6	223.6	.07193	68.2	225.5	.06496	61.5
	274	238.0	.11166	103.8	242.6	.09468	88.1	246.1	.08218	76.4	248.3	.07391	69.0
	298	260.4	.12678	115.6	265.5	.10764	98.2	269.3	.09322	85.3	271.8	.08421	77.0
	320	279.9	.14062	126.2	285.5	.11952	107.3	289.7	.10411	93.2	292.3	.09383	84.2
	338	294.9	.15216	134.5	300.8	.12942	114.4	305.2	.11273	99.4	308.0	.10171	89.9
	366	317.6	.17143	147.4	324.1	.14560	125.4	329.2	.12680	109.3	332.3	.11477	98.9
	388	336.7	.18812	158.4	343.8	.16059	135.2	349.0	.14003	117.7	352.3	.12669	106.5
	406	352.2	.20324	167.7	359.5	.17337	143.0	364.9	.15097	124.5	368.4	.13658	112.7
6"	250	205.5	.11884	112.6	211.4	.10068	95.4	215.4	.08772	83.5	218.9	.07712	73.1
	274	226.0	.13529	125.9	232.4	.11483	106.8	236.9	.10051	93.5	240.9	.08801	81.9
	298	247.1	.15355	140.0	254.2	.12990	118.9	259.2	.11412	104.2	263.5	.10019	91.4
	320	265.5	.17041	152.6	273.2	.14453	129.7	278.6	.12669	113.7	283.4	.11122	99.8
	338	279.5	.18381	162.5	287.7	.15632	138.2	293.5	.13748	121.2	298.5	.12038	106.4
	366	300.9	.20681	177.8	309.9	.17601	151.3	316.1	.15423	132.9	321.6	.13547	116.7
	388	318.7	.22731	190.8	328.3	.19315	162.6	335.0	.16966	142.8	340.9	.14923	125.6
	406	333.0	.24420	201.5	343.1	.20823	171.8	350.1	.18314	151,9	356,6	.16157	133.1
8"	250	195.3	.13558	128.4	204.0	.11041	105.3	208.6	.09816	93.0	212.1	.08806	83.8
	274	214.6	.15331	143.4	224.3	.12657	117.7	229.4	.11198	104.1	233.3	.10084	93.8
	298	234.5	.17477	159.2	245.2	.14349	130.9	250.8	.12707	115.9	255.1	.11444	104.5
	320	251.8	.19384	173.4	263.4	.15950	142.8	269.5	.14087	126.4	274.2	.12702	114.0
	338	265.1	.20932	184.4	277.3	.17194	152.0	283.8	.15238	134.7	288.8	.13782	121.5
	366	285.2	.23390	201.5	298.5	.19351	166.3	305.6	.17158	147.5	311.0	.15455	133.1
	388	302.0	.25740	216.1	316.1	.21217	178.6	323.7	.18822	158.5	329.5	.16996	143.1
	406	315.4	.27647	228.0	330.3	.22850	188.5	338.2	.20292	167.4	344.3	.18342	151.2

10″	[illegible]	[illegible]	[illegible]	[illegible]	[illegible]	[illegible]	[illegible]	[illegible]	[illegible]	[illegible]	[illegible]	[illegible]	[illegible]
	274	196.8	.17246	160.4	205.9	.14944	139.8	213.4	.13323	124,0	218.9	.12015	111.7
	298	214.9	.19523	177.9	225.2	.17080	155.6	233.2	.15099	137.7	239.2	.13570	124.2
	320	230.6	.21609	193.4	241.8	.18921	169.4	250.4	.16750	150.0	256.9	.15131	135.4
	338	242.6	.23212	205.5	254.4	.20439	180.1	263.5	.18050	159.5	270.4	.16299	144.1
	366	261.1	.26069	224.7	273.6	.22823	196.6	283.5	.20282	174.4	291.0	.18283	157.6
	388	276.3	.28661	240.6	289.6	.25033	210.8	300.1	.22232	187.1	308.1	.20093	169.1
	406	288.4	.30788	253.5	302.4	.26944	222.2	313.4	.23922	197.4	321.8	.21639	178.5
12″	250	158.9	.15529	148.3	170.4	.13506	127.9	178.8	.11957	113.3	185.0	.10730	102.2
	274	174.3	.17647	164.9	187.1	.15231	142.5	196.3	.13572	126.3	203.2	.12264	114.0
	298	190.2	.20019	182.4	204.2	.17318	157.8	214.3	.15358	140.0	221.9	.13818	126.6
	320	204.0	.22126	198.0	219.1	.19073	171.4	230.0	.17003	152.3	238.4	.15434	138.1
	338	214.5	.23738	210.0	230.4	.20659	182.0	241.9	.18304	161.8	250.9	.16607	146.8
	366	230.5	.26530	228.6	247.6	.23025	198.4	260.0	.20535	176.5	269.8	.18607	160.3
	388	243.8	.29039	244.4	262.0	.25223	212.3	275.4	.22573	189.5	285.5	.20409	171.8
	406	254.4	.31154	257.1	273.4	.27111	223.6	287.5	.24204	199.7	298.0	.21954	181.1
14″	250	141.6	.15436	147.4	154.3	.13587	128.7	163.9	.12177	115.3	171.2	.10991	104.8
	274	155.3	.17504	163.8	169.3	.15297	143.1	179.9	.13797	128.4	187.9	.12550	116.7
	298	169.4	.19846	180.8	184.9	.17419	158.7	196.3	.15590	142.1	205.0	.14170	129.3
	320	181.6	.21898	196.0	198.3	.19170	172.2	210.5	.17167	154.4	220.0	.15705	140.5
	338	190.9	.23467	207.7	208.5	.20729	182.6	221.4	.18532	163.8	231.4	.16883	149.2
	366	205.0	.26194	225.7	223.9	.23065	198.7	237.8	.20761	178.4	248.6	.18935	162.7
	388	216.8	.28715	241.0	236.8	.25235	212.4	251.6	.22733	190.9	263.0	.20692	174.2
	406	226.2	.30772	253.4	247.1	.27094	223.5	262.5	.24355	200.9	274.5	.22233	183.5
16″	250	123.1	.14790	140.1	136.5	.13126	124.3	146.8	.11886	112.6	154.8	.10821	103.1
	274	134.9	.16718	155.3	149.7	.14746	138.0	161.0	.13450	125.1	169.8	.12330	114.7
	298	147.5	.18908	172.3	163.2	.16725	152.4	175.6	.15165	138.3	185.2	.13845	126.8
	320	158.5	.20932	187.4	175.0	.18396	165.1	188.3	.16746	150.0	198.9	.15428	138.1
	338	166.5	.22480	198.2	184.0	.19869	175.0	198.0	.17988	159.0	209.1	.16569	146.5
	366	178.6	.24925	214.8	197.6	.22074	190.2	212.6	.20109	172.9	224.6	.18497	159.4
	388	188.9	.27299	229.3	208.9	.24165	203.0	224.8	.21952	184.7	237.5	.20241	170.4
	406	196.9	.29203	240.3	217.9	.25868	213.4	234.5	.23549	194.3	247.8	.21727	179.3
18″	250	118.1	.15506	148.1	131.2	.13818	130.9	141.7	.12540	118.8	149.9	.11417	109.0
	274	129.5	.17557	164.2	143.9	.15534	145.2	155.4	.14170	131.9	164.4	.13019	121.1
	298	141.2	.19868	181.0	156.9	.17591	160.3	169.4	.15984	145.7	179.3	.14677	133.9
	320	151.4	.21897	196.0	168.2	.19398	173.6	181.7	.17573	157.9	192.3	.16223	145.2
	338	158.8	.23358	206.7	176.8	.20874	183.9	191.0	.18938	167.4	202.1	.17424	154.0
	366	170.5	.26029	224.3	189.9	.23185	199.8	205.1	.21180	182.0	217.1	.19435	167.6
	388	180.3	.28489	239.1	200.5	.25256	212.6	216.8	.23137	194.3	229.5	.21275	179.1
	406	188.0	.30504	251.1	209.1	.27094	223.5	226.1	.24763	204.3	239.5	.22827	188.3
20″	250	113.0	.16147	154.1	126.2	.14430	136.7	136.7	.13099	124.0	145.3	.12077	114.4
	274	123.9	.18226	170.8	138.2	.16166	151.0	149.9	.14713	137.7	159.4	.13662	127.1
	298	135.1	.20643	188.1	150.8	.18309	166.8	163.5	.16690	152.1	173.8	.15408	140.5
	320	144.8	.22750	203.5	161.7	.20128	180.8	175.3	.18359	164.6	186.4	.17007	152.3
	338	152.2	.24359	215.4	170.0	.21738	191.6	184.3	.19831	174.7	195.9	.18268	161.5
	366	163.2	.27007	232.7	182.4	.24095	207.6	197.9	.22034	189.8	210.4	.20425	175.6
	388	172.4	.29476	248.1	192.9	.26422	222.0	209.2	.24124	202.7	222.4	.22290	187.6
	406	179.8	.31560	260.4	201.3	.28292	233.3	218.2	.25825	213.0	232.0	.23909	197.3
24″	250	107.0	.16998	162.0	120.0	.15273	145.9	129.7	.13935	132.0	138.2	.12906	122.2
	274	117.1	.19097	178.5	131.6	.17284	161.9	142.3	.15667	146.5	151.6	.14497	135.7
	298	127.6	.21561	196.6	143.3	.19536	178.0	155.1	.17750	161.7	165.3	.16448	149.9
	320	136.8	.23766	212.5	153.6	.21527	192.7	166.3	.19509	175.2	177.2	.18090	162.5
	338	143.8	.25493	224.8	161.5	.23136	204.0	174.9	.21067	185.6	186.5	.19544	172.7
	366	154.3	.28277	243.7	173.4	.25702	221.5	187.8	.23405	201.7	200.3	.21790	187.7
	388	163.1	.30839	259.6	183.1	.28046	235.5	198.5	.25642	215.3	211.7	.23857	200.4
	406	170.1	.33036	272.4	190.9	.30049	247.3	207.1	.27435	226.3	220.8	.25537	210.7
30″	250	96.0	.18472	174.8	109.0	.16704	159.3	119.5	.15359	146.7	127.6	.14240	134.9
	274	105.2	.20639	193.4	119.5	.18887	176.5	131.1	.17393	162.8	139.9	.16018	149.7
	298	114.7	.23333	212.8	130.1	.21229	193.5	142.7	.19632	178.9	152.6	.18133	165.2
	320	122.7	.25603	228.9	139.4	.23399	209.3	153.0	.21631	193.6	163.6	.19933	179.0
	338	129.0	.27337	242.0	146.5	.25130	221.4	160.8	.23163	204.1	171.9	.21509	189.6
	366	138.5	.30525	262.1	157.3	.27847	240.0	172.6	.25824	222.6	184.4	.23833	205.4
	388	146.3	.33133	279.0	166.2	.30376	255.7	182.2	.28175	236.6	195.0	.26050	219.3
	406	152.6	.35564	292.6	173.3	.32611	268.4	190.0	.30184	248.5	203.4	.27965	230.6

Table 12 Steam tracing data (English units continued)

Design Conditions:

Tracer: One 1/2″ O.D. tracer thermally bonded to pipe with a TFK-4 channel and heat transfer cement.
Insulation Thermal Conductivity: See Chart, page 79
Ambient: 100° F, no wind

Nomenclature:

TS = Steam Temperature (°F)
TP = Pipe Temperature (°F)
#/HR-FT = Tracer condensate rate per linear foot
Q3 = Pipe Heat Losses in Btu/hr, lin ft

N.P.S.	Steam Temp °F	INSULATION THICKNESS 1.5″ Nom			2.0″ Nom			2.5″ Nom			3.0″ Nom		
	TS	TP	#/HR-FT	Q3	TP	#/HR-FT	Q3	TP	#/HR-FT	Q3	TP	#/HR-FT	Q3
1.5″	250	237.1	.03341	31.0	237.8	.03008	28.5	238.3	.20753	26.2	236.7	.02504	24.4
	274	260.8	.04067	37.9	261.6	.03661	34.1	202.2	.03354	31.3	262.6	.03125	29.2
	298	285.2	.04873	44.0	266.2	.04362	40.1	288.9	.04031	36.8	287.4	.03746	34.3
	320	306.6	.05637	50.8	307.7	.05081	45.5	308.5	.04658	41.8	309.1	.04337	38.9
	338	323.0	.06259	55.3	324.2	.05634	49.8	325.0	.05170	45.7	325.7	.04817	42.6
	366	348.0	.07287	62.7	349.3	.06559	56.4	350.3	.06021	51.8	351.1	.05603	48.3
	388	368.9	.08216	69.1	370.4	.07390	62.2	371.5	.06791	57.1	372.3	.06321	53.2
	406	386.0	.09037	74.4	387.5	.08128	67.0	386.7	.07470	61.5	389.6	.06960	57.3
2″	250	235.0	.04071	38.7	236.1	.03591	34.0	236.8	.03244	30.7	237.4	.02984	26.3
	274	258.2	.04961	46.3	259.5	.04375	40.7	260.4	.03951	36.8	261.0	.03635	33.8
	298	282.2	.05951	54.5	283.7	.05246	47.9	284.6	.04729	43.2	285.6	.04350	39.8
	320	303.2	.06884	61.8	304.9	.06052	54.3	306.1	.05466	49.1	307.0	.05038	45.1
	338	319.2	.07640	67.5	321.1	.06730	59.4	322.4	.06080	53.0	323.4	.05587	49.4
	366	343.7	.08900	76.5	345.9	.07811	67.3	347.4	.07068	60.8	348.5	.06504	56.0
	388	364.2	.10021	84.2	366.6	.08817	74.1	368.2	.07959	67.0	389.4	.07333	61.6
	406	380.9	.11014	90.7	383.4	.09687	79.9	385.2	.08754	72.2	388.5	.08059	66.4
3″	250	231.6	.04702	44.8	233.1	.04137	39.2	234.2	.03718	35.2	235.0	.03376	32.0
	274	254.4	.05776	53.7	256.2	.05041	47.0	257.5	.04535	42.3	256.5	.04122	38.4
	298	277.8	.06922	63.2	279.9	.06047	55.3	261.4	.05448	49.7	262.6	.04934	45.1
	320	298.2	.07999	71.6	300.6	.06987	62.7	302.3	.06284	56.4	303.7	.05705	51.2
	338	313.5	.08853	78.3	316.4	.07752	68.5	318.3	.06916	61.6	319.8	.06352	55.9
	366	337.6	.10302	88.6	340.6	.09027	77.6	342.6	.08107	69.8	344.4	.07370	63.4
	388	357.5	.11585	97.5	360.7	.10160	85.4	363.0	.09143	76.8	364.9	.08304	69.8
	406	373.4	.12718	104.8	376.9	.11146	91.8	379.4	.10020	82.8	381.4	.09098	75.0
4″	250	228.3	.05431	51.4	230.3	.04705	44.8	231.9	.04182	39.6	232.9	.03810	36.3
	274	250.5	.06627	61.7	252.9	.05780	53.8	254.7	.05096	47.0	255.9	.04661	43.5
	298	273.2	.07921	72.5	276.0	.06916	63.1	278.2	.06107	55.9	279.6	.05595	51.1
	320	293.1	.09152	82.1	296.2	.07995	71.6	298.6	.07057	63.3	300.2	.06455	57.9
	338	308.2	.10151	89.7	311.6	.08843	78.2	314.3	.07832	69.2	318.0	.07162	63.3
	366	331.3	.11778	101.5	335.1	.10289	88.5	338.2	.09109	78.3	340.1	.08324	71.7
	388	350.6	.13260	111.6	354.8	.11569	97.3	358.1	.10255	88.2	360.3	.09384	78.9
	406	366.0	.14531	119.8	370.5	.12693	104.5	374.1	.11248	92.6	376.4	.10290	84.8
6″	250	222.4	.06581	62.3	225.2	.05722	54.2	227.2	.05098	48.3	229.0	.04521	43.0
	274	243.2	.07995	74.3	246.7	.06972	64.9	249.2	.06218	57.9	251.3	.05520	51.5
	298	264.9	.09567	87.5	268.8	.08330	76.2	271.6	.07428	67.9	274.2	.06602	60.4
	320	283.7	.11044	99.1	288.1	.09614	86.3	291.2	.08586	76.9	294.1	.07634	68.5
	338	298.0	.12232	108.1	302.8	.10657	94.1	306.2	.09500	84.0	309.3	.08465	74.8
	366	319.8	.14173	122.1	325.2	.12348	106.4	329.1	..11028	95.0	332.6	.09839	84.6
	388	338.0	.15938	134.2	343.9	.13919	117.0	348.1	.12424	104.4	352.0	.11058	93.1
	406	352.6	.17453	144.0	358.8	.15226	125.5	363.4	.13591	112.1	367.5	.12139	100.0
8″	250	216.7	.07499	71.1	220.9	.06338	60.0	223.2	.05693	53.9	225.0	.05186	49.1
	274	236.4	.09107	84.8	241.4	.07695	71.6	244.2	.06908	64.3	246.3	.06302	58.7
	298	257.0	.10916	99.6	262.9	.09207	84.2	266.1	.08272	75.7	268.6	.07546	69.0
	320	274.8	.12548	112.6	281.4	.10647	95.3	285.0	.09563	85.6	287.9	.08722	78.1
	338	288.4	.13924	122.7	295.5	.11781	103.9	299.4	.10580	93.5	302.8	.09645	85.2
	366	309.0	.16085	138.5	317.0	.13625	117.4	321.4	.12255	105.6	324.9	.11188	96.4
	388	325.1	.18059	152.0	334.9	.15317	128.9	339.8	.13780	116.0	343.6	.12600	105.9
	406	339.9	.19763	163.0	349.2	.16775	138.3	354.5	.15096	124.5	358.6	.13782	113.7

10″	274	225.8	.10285	95.6	230.5	.09132	85.0	234.4	.08204	76.3	237.4	.07452	69.6
	298	244.4	.12213	111.6	249.9	.10896	99.4	254.5	.09790	89.3	258.0	.08909	81.5
	320	260.6	.14059	126.2	266.8	.12507	112.3	272.2	.11285	101.3	276.1	.10326	92.4
	338	273.0	.15544	137.4	280.0	.13923	122.7	285.5	.12508	110.3	289.8	.11420	100.7
	366	292.0	.18055	155.3	299.6	.16061	138.3	305.8	.14442	124.4	310.6	.13193	113.7
	388	307.6	.20215	170.2	315.9	.18018	151.7	322.7	.16219	136.5	328.0	.14820	124.7
	406	320.1	.22086	182.2	329.0	.19699	162.5	336.2	.17771	146.3	341.8	.16247	133.6
12″	250	196.2	.08631	82.1	202.0	.07680	72.8	206.1	.06923	65.6	209.4	.06318	59.8
	274	212.5	.10542	98.0	219.1	.09291	86.5	224.0	.08345	77.9	227.8	.07652	71.2
	298	229.0	.12550	114.4	236.6	.11069	101.0	242.4	.09978	91.0	246.8	.09095	83.2
	320	243.4	.14348	128.8	251.9	.12683	113.9	258.3	.11434	102.6	263.3	.10492	93.9
	338	254.3	.15856	140.2	263.6	.14056	123.9	270.5	.12642	111.7	275.9	.11594	102.3
	366	270.9	.18270	157.5	281.2	.16189	139.4	289.0	.14602	125.8	295.1	.13374	115.2
	388	284.7	.20464	172.2	296.0	.18123	152.6	304.4	.16373	137.8	311.3	.15044	126.6
	406	295.7	.22334	181.3	307.7	.19805	163.4	317.0	.17981	148.0	324.0	.16447	135.7
14″	250	186.4	.08659	82.4	192.5	.07701	73.4	197.2	.07024	66.5	200.9	.06453	61.1
	274	200.8	.10501	97.6	208.0	.09352	87.0	213.8	.08489	79.3	218.2	.07834	73.0
	298	215.5	.12432	113.7	223.8	.11112	101.4	230.5	.10147	92.5	235.6	.09308	85.1
	320	228.3	.14261	128.0	237.9	.12780	114.7	245.1	.11625	104.3	250.8	.10695	96.0
	338	238.0	.15727	139.0	248.4	.14148	124.7	256.2	.12853	113.3	262.4	.11804	104.4
	366	252.7	18105	156.1	264.3	.16272	140.1	273.0	.14788	127.4	279.9	.13618	117.4
	388	265.0	.20270	170.6	277.5	.18198	153.2	287.0	.16558	139.4	294.5	.15264	128.5
	406	274.8	.22087	182.2	288.1	.19860	163.8	298.1	.18124	149.2	306.2	.16672	137.5
16″	250	175.6	.08272	78.8	182.0	.07509	71.2	187.4	.06902	65.4	191.4	.06395	60.6
	274	188.1	.10035	93.3	195.7	.09047	84.2	201.9	.08308	77.6	206.7	.07715	71.8
	298	200.7	.11809	108.1	209.7	.10748	98.1	216.8	.09904	90.3	222.4	.09142	83.6
	320	211.9	.13554	121.7	221.8	.12288	110.3	229.8	.11315	101.6	236.1	.10506	94.0
	338	220.4	.14951	132.1	231.3	.13639	120.3	239.6	.12507	110.3	246.4	.11594	102.3
	366	233.3	.17256	148.3	245.3	.15667	135.0	254.6	.14366	123.8	262.2	.13331	114.9
	388	244.1	.19249	162.1	257.0	.17501	147.3	267.0	.16066	135.2	275.5	.14962	125.9
	406	252.4	.20918	172.6	266.3	.19090	157.4	277.0	.17555	144.6	285.9	.16347	134.6
18″	250	172.4	.08664	82.4	179.0	.07866	75.0	184.1	.07245	68.6	188.5	.06758	64.0
	274	184.6	.10571	98.3	192.2	.09532	88.7	198.4	.08774	81.7	203.3	.08145	75.8
	298	197.3	.12635	115.2	205.6	.11316	103.3	212.8	.10422	95.1	218.5	.09684	88.3
	320	208.0	.14423	129.4	217.3	.12941	116.2	225.3	.11914	106.9	231.7	.11066	99.3
	338	216.3	.15914	140.7	226.2	.14293	125.9	234.8	.13129	116.1	241.7	.12207	107.9
	366	228.5	.18241	157.2	239.7	.16443	141.3	249.2	.15120	130.3	256.9	.14055	121.1
	388	238.5	.20333	171.1	250.9	.18319	154.2	261.2	.16916	142.4	269.5	15722	132.3
	406	246.7	.22150	182.8	259.8	.19971	164.7	270.8	.18437	152.0	279.8	.17167	141.4
20″	250	170.2	.09248	87.5	176.1	.08217	78.3	181.4	.07603	72.1	185.6	.07052	66.8
	274	181.8	.11186	104.0	188.7	.09953	92.6	194.9	.09163	85.3	200.2	.08515	79.6
	298	193.4	.13142	120.1	201.6	.11771	107.8	208.8	.10881	99.3	214.9	.10158	92.6
	320	203.7	.15075	134.9	212.9	.13504	121.2	220.9	.12429	111.6	227.6	.11613	104.2
	338	211.6	.16570	146.5	221.5	.14890	131.6	230.0	.13749	121.2	237.3	.12824	113.1
	366	223.1	.19000	163.3	234.5	.17159	147.5	243.9	.15781	135.9	252.0	.14741	127.0
	388	232.9	.21175	178.2	245.0	.19075	160.6	255.5	.17629	148.4	264.3	16469	138.6
	406	240.7	.23051	190.2	253.7	.20786	171.5	264.7	.19228	158.6	274.0	.18001	148.2
24″	250	166.4	.09818	93.0	172.1	.08802	83.7	177.1	.08070	76.9	181.6	.07584	71.9
	274	177.0	.11737	109.1	184.4	.10709	99.6	189.9	.09777	90.9	195.1	.09154	85.2
	298	188.2	.13809	126.5	196.7	.12715	116.0	203.0	.11562	105.9	209.0	.10861	99.1
	320	197.8	.15902	142.3	207.5	.14608	130.7	214.5	.13264	119.1	221.2	.12397	111.3
	338	205.0	.17335	153.2	215.4	.15970	141.2	223.2	.14618	129.2	230.3	.13713	120.9
	366	216.0	.19919	171.7	227.8	.18366	158.3	236.4	.16853	144.8	244.3	.15742	135.6
	388	225.3	.22234	187.1	237.8	.20450	172.1	247.4	.18782	158.1	255.9	.17585	148.1
	406	232.6	.24208	199.7	246.0	.22257	183.6	256.1	.20475	168.9	265.2	.19167	158.1
30″	250	159.6	.10546	100.4	166.2	.09766	93.1	171.3	.09111	86.2	175.0	.08300	79.0
	274	169.4	.12741	118.5	176.8	.11748	109.2	183.2	.10984	102.1	187.5	.10056	93.5
	298	179.4	.15076	137.5	188.0	.13833	126.7	195.2	.12889	118.1	200.4	.11883	108.8
	320	187.9	.17104	153.2	197.7	.15891	142.2	205.7	.14841	132.8	211.6	.13638	122.4
	338	194.4	.18783	166.0	204.8	.17363	153.5	213.5	.16208	143.3	220.1	.15046	133.0
	366	204.4	.21549	185.6	215.9	.19925	171.8	225.6	.18629	160.5	233.0	.17343	149.1
	388	212.7	.24051	202.0	225.2	.22227	187.0	235.6	.20819	175.2	243.5	.19272	162.2
	406	219.4	.26114	215.4	232.5	.24200	199.7	243.4	.22564	186.2	252.1	.21036	173.6

Table 12 Steam tracing data (English units continued)

Design Conditions:

Tracer: One 3/4'' O.D. tracer thermally bonded to pipe with a TFK-7 channel and heat transfer cement.
Insulation Thermal Conductivity: See Chart, page 79
Ambient: 0° F, 25 MPH wind

Nomenclature:

TS = Steam Temperature (° F)
TP = Pipe Temperature (° F)
#/HR-FT = Tracer condensate rate per linear foot
Q3 = Pipe Heat Losses in Btu/hr, lin ft

N.P.S.	Steam	INSULATION THICKNESS											
	Temp °F	1.5'' Nom			2.0'' Nom			3.5'' Nom			3.0'' Nom		
	TS	TP	#/HR-FT	Q3	TP	#/HR-FT	Q3	TP	#/HR-FT	Q3	TP	#/HR-FT	Q3
4''	250	220.7	.11858	112.7	225.0	.09675	91.9	227.3	.08458	80.2	229.1	.07537	71.6
	274	242.9	.13517	126.3	247.6	.11050	103.1	250.3	.09643	90.0	252.3	.08617	80.4
	298	265.9	.15379	140.7	271.1	.12595	115.0	274.0	.11005	100.5	276.2	.09817	89.8
	320	285.9	.17147	153.7	291.6	.14009	125.7	294.8	.12258	109.9	297.4	.10982	98.4
	338	301.2	.18558	163.8	307.3	.15196	134.1	311.0	.13315	117.5	313.5	.11896	105.7
	366	324.8	.20888	179.9	331.4	.17123	147.5	335.2	.14993	129.0	338.0	.13417	115.4
	388	344.4	.22986	193.4	351.4	.18884	158.7	355.5	.16524	138.9	358.4	.14778	124.3
	406	360.0	.24828	204.5	367.5	.20383	167.9	371.7	.17845	147.0	374.9	.15974	131.6
6''	250	213.2	.13739	130.3	218.2	.11465	109.0	222.2	.09675	91.9	224.4	.08705	82.5
	274	234.5	.15656	145.8	240.1	.13106	122.1	244.5	.11047	103.1	247.0	.09917	92.6
	298	256.6	.17742	162.3	262.7	.14871	136.0	267.6	.12590	115.0	270.4	.11312	103.3
	320	275.8	.19739	177.1	282.5	.16586	148.6	287.8	.14001	125.7	290.8	.12600	112.9
	338	290.5	.21353	188.6	297.6	.17944	158.4	303.2	.15186	134.0	306.4	.13637	120.5
	366	312.8	.24017	206.6	320.6	.20185	173.6	327.0	.17108	147.3	330.5	.15403	132.5
	388	331.5	.26389	222.0	340.0	.22250	187.0	346.7	.18864	158.5	350.4	.16972	142.6
	406	346.5	.28409	234.5	355.4	.24008	197.7	362.4	.20360	167.7	366.4	.18326	150.9
8''	250	205.8	.15342	145.9	211.6	.13013	123.3	215.8	.11285	107.3	218.5	.10201	97.0
	274	226.2	.17477	163.1	232.7	.14795	138.0	237.4	.12901	120.1	240.4	.11646	108.7
	298	247.3	.19839	181.4	254.5	.16800	158.7	259.7	.14633	133.9	263.0	.13248	121.1
	320	265.8	.22032	197.8	273.6	.18717	167.7	279.2	.16296	146.2	282.8	.14746	132.3
	338	279.8	.23829	210.5	288.1	.20221	178.7	294.1	.17654	155.8	297.9	15987	141.1
	366	301.2	.26782	230.4	310.3	.22749	195.7	316.7	.19832	170.8	320.9	.17972	154.7
	388	319.1	.29387	247.3	328.7	.25029	210.3	335.7	.21845	183.6	340.1	.19804	166.4
	406	333.4	.31678	261.1	343.5	.26976	222.2	350.8	.23569	194.1	355.8	.21409	176.3
10''	250	194.9	.16766	158.8	201.7	.14351	136.6	207.0	.12613	119.5	210.9	.11244	106.9
	274	214.1	.18999	177.4	221.7	.16413	152.7	227.6	.14356	133.7	231.9	.12846	119.6
	298	234.0	.21576	197.0	242.3	.18566	169.8	248.8	.16264	148.8	253.6	.14565	133.2
	320	251.3	.23892	214.5	260.3	.20624	185.1	267.3	.18110	162.3	272.5	.16208	145.4
	338	264.5	.25867	228.1	274.1	.22336	197.0	281.5	.19558	172.8	287.0	.17553	154.9
	366	284.6	.28976	249.3	295.0	.25050	215.5	303.0	.21995	189.2	309.0	.19702	169.7
	388	301.3	.31805	267.3	312.4	.27527	231.3	321.0	.24161	203.3	327.4	.21697	182.4
	406	314.7	.34202	282.0	326.3	.29611	244.2	335.4	.26066	214.7	342.1	.23400	192.7
12''	250	177.2	.17717	168.0	187.2	.14985	142.5	194.2	.13158	124.7	199.7	.11770	111.9
	274	194.6	.20058	187.2	205.7	.17041	159.0	213.7	.14963	139.7	219.4	.13420	125.0
	298	212.5	.22679	207.5	224.9	.19352	177.0	233.5	.16965	155.2	239.8	.15196	139.0
	320	228.0	.25126	225.6	241.4	.21457	192.6	250.7	.18866	169.0	257.6	.16911	151.5
	338	239.9	.27112	239.6	254.1	.23203	204.7	263.9	.20351	179.8	271.2	.18273	161.3
	366	257.9	.30326	261.3	273.2	.25990	223.6	283.8	.22847	196.6	291.8	.20486	176.4
	388	273.2	.33357	280.3	289.2	.28477	239.6	300.5	.25090	210.8	309.0	.22523	189.3
	406	285.2	.35821	295.3	302.0	.30646	252.7	313.8	.27005	222.4	322.7	.24268	199.9

14″	250	161.8	.18103	171.6	173.4	.15487	147.2	181.6	.13689	129.8	187.9	.12314	116.8
	274	177.6	.20458	191.0	190.3	.17574	164.0	199.4	.15541	144.7	206.3	.13950	130.2
	298	193.8	.23102	211.3	207.7	.19874	181.7	217.8	.17550	160.5	225.6	.15842	144.9
	320	207.8	.25549	229.4	222.9	.22002	197.5	233.7	.19466	174.6	242.2	.17578	157.8
	338	218.6	.27536	243.3	234.5	.23738	209.8	245.9	.21006	185.6	254.8	.18981	167.3
	366	234.9	.30758	265.0	252.0	.26586	228.7	264.6	.23602	203.0	274.1	.21274	183.2
	388	248.4	.33709	283.3	266.6	.29090	244.6	280.0	.25878	217.5	290.1	.23344	198.4
	406	259.2	.36159	298.2	278.3	.31280	257.9	292.3	.27834	229.3	302.9	.25153	207.2
16″	250	144.3	.17670	167.6	157.0	.15323	145.7	166.6	.13731	130.2	173.6	.12447	117.9
	274	158.3	.19940	186.1	172.5	.17423	162.6	182.8	.15566	145.0	190.8	.14079	131.4
	298	172.6	.22473	205.7	188.2	.19666	179.8	199.5	.17546	160.5	208.3	.15915	145.6
	320	185.1	.24829	222.9	201.8	.21741	195.2	214.0	.19434	174.4	223.5	.17638	158.3
	338	194.6	.26740	236.3	212.2	.23461	207.0	225.1	.20947	185.1	235.0	.19022	168.1
	366	209.0	.29814	256.9	227.9	.26198	225.4	241.8	.23442	201.7	252.6	.21287	183.3
	388	221.0	.32640	274.3	241.1	.28631	240.9	255.8	.25669	215.7	267.2	.23328	196.3
	406	230.6	.35028	288.4	251.5	.30748	253.5	266.9	.27580	227.2	278.9	.25110	206.6
18″	250	138.7	.18591	176.2	152.0	.16246	154.4	161.9	.14525	136.3	169.4	.13235	125.4
	274	152.4	.21046	196.4	166.7	.18384	171.6	177.6	.16490	153.9	185.9	.14964	139.7
	298	166.2	.23706	216.9	181.8	.20791	189.7	193.6	.18617	170.3	202.9	.16917	154.6
	320	178.2	.26176	235.0	195.0	.22929	205.8	207.8	.20600	164.9	217.7	.18703	166.1
	338	187.3	.28177	249.0	205.0	.24751	218.3	218.6	.22243	196.2	228.9	.26205	176.5
	366	201.1	.31412	270.6	220.5	.27662	238.1	234.8	.24836	213.7	246.0	.22619	194.6
	388	212.6	.34369	288.8	233.1	.30274	254.5	248.5	.27786	228.5	260.2	.24761	208.3
	406	221.8	.36817	303.6	243.2	.32502	267.7	259.1	.29169	240.5	271.5	.26633	219.4
20″	250	133.8	.19532	185.0	147.2	.17083	162.5	157.2	.15288	145.4	165.2	.13978	132.5
	274	146.7	.22001	205.3	161.5	.19345	180.6	172.7	.17382	162.2	181.3	.15849	147.6
	298	160.0	.24829	226.7	176.1	.21860	199.6	188.4	.19624	179.5	197.6	.17860	163.4
	320	171.6	.27358	245.6	188.9	.24105	216.4	202.1	.21696	194.8	212.2	.19777	177.4
	338	180.4	.29435	260.2	198.6	.25976	229.5	212.4	.23415	206.6	223.2	.21320	186.3
	366	193.7	.32819	282.6	213.3	.29007	249.5	228.2	.26150	225.0	239.8	.23855	205.4
	388	204.8	.35893	301.6	225.5	.31713	266.5	241.3	.28579	240.5	253.6	.26093	219.6
	406	213.6	.38490	317.0	235.2	.34034	280.3	251.8	.30695	253.1	264.0	.28065	231.2
24″	250	127.6	.20639	195.4	140.8	.18328	173.7	150.7	.16460	156.5	158.6	.15030	142.9
	274	139.9	.23314	216.9	154.4	.20668	192.9	165.3	.18641	174.0	174.3	.17100	159.6
	298	152.6	.26241	239.4	168.4	.23351	213.1	180.3	.21078	192.4	190.1	.19308	176.6
	320	168.6	.28893	259.3	180.5	.25724	230.9	193.4	.23242	208.7	203.9	.21354	191.7
	338	172.0	.31067	274.7	189.8	.27690	244.7	203.3	.25079	221.3	214.4	.23048	203.3
	366	184.5	.34586	297.4	203.8	.30882	266.0	218.4	.27986	240.0	230.3	.25741	221.4
	388	195.1	.37786	317.6	215.4	.33794	284.0	230.9	.30610	257.3	243.5	.28141	236.7
	406	203.6	.40541	333.9	224.7	.36214	298.6	240.9	.32856	270.0	254.1	.30220	249.2
30″	250	116.9	.22924	217.0	129.6	.20229	191.6	140.3	.18429	174.7	148.6	.16852	160.2
	274	128.2	.25847	240.7	142.1	.22859	212.6	153.8	.20779	194.0	163.0	.19078	178.1
	298	139.8	.29009	265.4	155.0	.25725	234.7	167.7	.23472	214.2	177.6	.21564	196.9
	320	149.6	.31935	286.0	166.2	.28334	254.3	179.8	.25863	232.2	190.6	.23780	213.5
	338	157.3	.34245	302.7	174.7	.30476	269.4	189.0	.27844	246.1	200.4	.25666	228.4
	366	168.8	.38182	328.5	187.6	.34046	292.7	203.0	.31050	267.5	215.2	.28626	246.3
	388	178.5	.41601	350.1	198.3	.37175	312.4	214.6	.33979	285.6	227.5	.31309	263.1
	406	186.1	.44644	367.7	206.9	.39871	328.4	223.6	.36411	300.2	237.4	.33614	276.8

Table 12 Steam tracing data (English units continued)

Design Conditions:

Tracer: One 3/4″ O.D. tracer thermally bonded to pipe with a TFK-7 channel and heat transfer cement.
Insulation Thermal Conductivity: See Chart, page 79
Ambient: 100° F, no wind

Nomenclature:

TS = Steam Temperature (°F)
TP = Pipe Temperature (°F)
#/HR-FT = Tracer condensate rate per linear foot
Q3 = Pipe Heat Losses in Btu/hr, lin ft

N.P.S.	Steam Temp °F	INSULATION THICKNESS											
		1.5″ Nom			2″ Nom			2.5″ Nom			3″ Nom		
	TS	TP	#/HR-FT	Q3	TP	#/HR-FT	Q3	TP	#/HR-FT	Q3	TP	#/HR-FT	Q3
4″	250	231.0	.06514	61.7	232.9	.05483	52.1	234.1	.04879	46.4	235.0	.04418	42.0
	274	253.7	.07967	74.2	256.0	.06718	62.6	257.5	.05960	55.6	258.5	.05408	50.4
	298	276.9	.09531	87.2	279.7	.08047	73.5	281.3	.07148	65.4	282.6	.00476	59.2
	320	297.2	.11025	98.8	300.3	.09297	83.4	302.2	.08267	74.2	303.7	.07497	67.2
	338	312.7	.12236	108.0	316.1	.10326	91.1	318.2	.09189	81.1	319.6	.08323	73.4
	366	336.4	.14209	122.2	340.2	.12001	103.2	342.6	.10670	91.9	344.3	.09667	83.2
	388	356.1	.15996	134.5	360.4	.13521	113.6	363.0	.12031	101.1	364.9	.10904	91.6
	406	372.0	.17537	144.5	376.6	.14823	122.1	379.3	.13189	108.7	381.4	.11954	96.5
6″	250	226.6	.07528	71.6	229.1	.06494	61.6	231.1	.05596	53.2	232.3	.05086	46.3
	274	248.7	.09201	85.9	251.5	.07930	73.9	253.9	.06844	63.8	255.2	.06210	58.0
	298	271.0	.11043	100.8	274.4	.09484	86.7	277.1	.08204	74.9	278.7	.07459	68.1
	320	290.6	.12739	114.2	294.3	.10968	98.3	297.5	.09469	85.0	299.3	.08607	77.2
	338	305.5	.14109	124.7	309.6	.12171	107.4	313.0	.10510	92.8	315.0	.09562	84.4
	366	328.3	.16367	141.0	332.8	.14123	121.5	336.7	.12211	105.1	338.9	.11104	95.6
	388	347.2	.18448	155.0	352.3	.15896	133.6	356.5	.13744	115.5	359.0	.12509	105.2
	406	362.5	.20201	166.4	367.8	.17427	143.6	372.4	.15074	124.2	375.0	.13719	113.0
8″	250	222.6	.08476	80.4	225.4	.07336	69.8	227.5	.06521	61.8	228.9	.05946	56.5
	274	243.7	.10325	98.4	247.0	.08959	83.6	249.5	.07951	74.0	251.2	.07266	67.8
	298	265.2	.12361	113.0	269.1	.10733	98.1	272.0	.09505	88.9	274.0	.08707	79.5
	320	284.0	.14263	128.0	288.4	.12397	111.1	291.7	.10989	98.5	293.9	.10048	90.1
	338	298.3	.15821	139.6	303.1	.13726	121.2	306.7	.12179	107.5	309.1	.11136	98.4
	366	320.2	.16343	157.8	325.6	.15928	137.0	329.6	.14135	121.6	332.3	.12943	111.3
	388	338.4	.20622	173.3	344.3	.17928	150.7	348.8	.15906	133.7	351.7	.14552	122.4
	406	353.6	.22585	186.0	359.3	.19630	161.7	364.1	.17425	143.5	367.2	.15962	131.5
10″	250	216.5	.09230	87.7	219.8	.08120	77.2	222.4	.07243	68.7	224.4	.06565	62.2
	274	236.2	.11207	104.6	240.1	.09860	92.0	243.2	.08784	82.0	245.6	.07976	74.3
	298	256.8	.13442	122.9	261.3	.11837	108.3	265.6	.10543	96.4	267.8	.09554	87.4
	320	274.5	.15499	139.1	279.7	.13662	122.4	283.8	.12172	109.1	266.9	.11038	98.9
	338	288.1	.17181	151.6	293.6	.15134	133.6	298.1	.13478	119.0	301.5	.12238	108.0
	366	308.6	.19862	171.1	314.9	.17516	150.8	319.9	.15637	134.5	323.8	.14181	122.0
	388	325.8	.22324	187.6	332.6	.19701	165.6	338.1	.17574	147.7	342.3	.15955	134.1
	406	339.5	.24430	201.2	346.8	.21567	177.6	352.7	.19242	158.5	357.2	.17470	143.9
12″	250	207.0	.09746	92.6	211.8	.08501	80.6	215.3	.07550	71.8	217.9	.06862	65.1
	274	225.0	.11813	110.2	230.6	.10278	95.9	234.8	.09156	85.5	237.9	.08312	77.6
	298	243.5	.14079	128.8	250.0	.12273	112.2	254.9	.10960	100.0	258.5	.09924	90.8
	320	259.6	.16211	145.4	267.0	.14127	126.8	272.6	.12658	113.4	276.8	.11489	103.0
	338	272.1	.18000	158.8	280.1	.15698	138.5	286.0	.13997	123.7	290.5	.12726	112.3
	366	290.7	.20790	179.0	299.7	.18146	156.2	306.3	.16215	139.5	311.4	.14723	126.1
	388	306.2	.23312	196.1	316.1	.20376	171.3	323.5	.18209	153.0	326.8	.16544	139.1
	406	318.7	.25502	210.0	329.1	.22291	183.6	336.8	.19917	164.1	342.7	.18082	149.1

14″	250	198.3	.09995	95.0	203.8	.08808	83.5	207.9	.07879	74.9	211.1	.07186	68.2
	274	214.7	.12103	112.9	221.2	.10626	99.2	226.1	.09536	89.0	229.9	.08683	81.0
	298	231.5	.14397	131.7	239.1	.12682	115.8	244.8	.11394	104.0	249.2	.10366	94.8
	320	246.2	.16577	148.5	254.7	.14564	130.7	261.1	.13100	117.4	266.0	.11939	107.0
	338	257.3	.18315	161.6	266.6	.16119	142.2	273.5	.14454	127.7	278.9	.13201	116.5
	366	274.3	.21096	181.7	284.6	.18613	160.1	292.3	.16714	144.0	298.6	.15308	131.7
	388	288.4	.23641	198.9	299.6	.20861	175.3	308.3	.18819	156.2	314.8	.17183	144.4
	406	299.6	.25829	212.7	311.8	.22867	188.3	320.8	.20566	169.4	327.8	.18763	154.7
16″	250	188.1	.09811	93.2	194.2	.08734	82.8	198.9	.07873	74.8	202.6	.07238	68.7
	274	202.7	.11859	110.6	209.9	.10531	98.3	215.7	.09583	89.5	220.1	.08786	82.0
	298	217.8	.14084	128.8	226.3	.12612	115.2	232.8	.11414	104.2	237.9	.10459	95.6
	320	230.8	.16184	145.2	240.4	.14449	129.7	247.6	.13104	117.4	253.3	.12030	107.8
	338	240.7	.17866	157.7	251.1	.15973	141.0	258.9	.14455	127.7	265.1	.13299	117.3
	366	255.8	.20567	177.1	267.3	.18428	158.5	276.0	.16695	143.8	282.9	.15346	132.0
	388	266.6	.23076	194.2	280.8	.20624	173.3	290.3	.18705	157.2	297.8	.17189	144.5
	406	278.6	.25201	207.6	291.6	.22494	185.3	301.7	.20427	166.3	309.7	.18764	154.7
18″	250	185.1	.10371	98.6	191.4	.09233	87.7	196.2	.08365	79.9	200.0	.07708	73.1
	274	199.2	.12555	116.9	206.5	.11162	104.2	212.3	.10109	94.4	217.6	.09339	87.2
	298	213.7	.14867	136.0	222.2	.13271	121.3	229.1	.12108	110.6	234.3	.11132	101.6
	320	226.2	.17101	153.2	235.7	.15232	136.7	243.4	.13903	124.6	249.4	.12779	114.5
	338	235.8	.18846	166.3	246.4	.16900	149.1	254.4	.15352	135.5	260.8	.14101	124.6
	366	250.3	.21876	186.7	262.1	.19472	167.5	271.0	.17684	152.3	278.2	.16286	140.1
	388	262.3	.24244	204.0	275.1	.21779	183.1	284.8	.19820	166.6	282.6	.18251	153.4
	406	272.0	.28462	218.0	285.5	.23769	195.8	295.9	.21630	178.2	304.2	.19917	164.1
20″	250	182.0	.10827	102.9	188.6	.09730	92.5	193.6	.08860	84.0	197.6	.08122	77.2
	274	195.6	.13097	122.0	203.3	.11759	109.7	209.2	.10689	99.8	214.2	.09863	92.1
	298	209.5	.15541	142.2	218.5	.13968	127.8	225.3	.12719	116.1	231.0	.11736	107.4
	320	221.9	.17942	160.7	231.6	.16033	143.8	239.5	.14645	131.4	245.6	.13494	120.9
	338	231.2	.19740	174.4	241.6	.17706	156.2	250.1	.16186	142.8	256.7	.14902	131.5
	366	245.2	.22754	195.8	256.8	.20390	175.6	266.2	.18654	160.5	273.6	.17168	147.9
	388	256.8	.25456	213.9	269.4	.22796	191.8	279.6	.20877	175.5	287.7	.19250	161.8
	406	266.0	.27749	228.6	279.5	.24695	205.0	290.4	.22771	187.5	298.9	.21003	173.0
24″	250	178.0	.11506	109.4	184.4	.10394	98.8	189.7	.09550	90.8	193.7	.08850	83.9
	274	190.9	.13890	129.7	198.7	.12656	117.8	204.6	.11542	107.7	209.4	.10655	99.4
	298	204.5	.16572	151.6	213.2	.14988	137.1	219.9	.13714	125.4	225.5	.12682	115.8
	320	216.0	.19053	170.7	225.7	.17216	154.3	233.3	.15729	141.1	239.8	.14588	130.9
	338	224.8	.20957	185.1	235.2	.18962	167.5	243.4	.17373	153.3	250.4	.16125	142.3
	366	238.1	.24143	207.7	249.7	.21835	188.1	258.8	.20008	172.1	266.6	.18586	159.9
	388	249.1	.26947	226.7	261.7	.24419	205.4	271.9	.22456	188.8	280.1	.20799	174.8
	406	258.0	.29365	242.1	271.3	.26656	219.6	282.1	.24506	201.8	290.8	.22700	187.0
30″	250	172.0	.12936	122.6	177.9	.11544	109.7	183.3	.10631	101.0	187.6	.09896	94.1
	274	184.2	.15697	146.2	190.9	.13907	129.8	197.1	.12842	119.6	202.5	.11939	111.4
	298	196.3	.18565	169.8	204.1	.16537	151.3	211.7	.15286	139.8	217.5	.14177	129.7
	320	206.7	.21178	190.1	216.0	.19067	170.8	224.0	.17527	157.3	230.5	.16272	140.0
	338	214.8	.23404	206.5	224.7	.20979	185.3	233.4	.19347	171.0	240.5	.17970	158.6
	366	226.8	.26772	230.3	238.0	.24168	207.9	247.7	.22285	191.7	255.5	.20666	178.0
	388	236.9	.29933	251.6	249.9	.26909	226.1	259.5	.25889	209.3	268.1	.23112	194.5
	406	245.0	.32618	268.7	257.7	.29291	241.5	269.0	.27137	223.5	278.1	.25221	207.7

Table 12a Steam tracing data (Metric units)

Design Conditions:

Tracer: One 13 MM O.D. tracer thermally bonded to pipe with a TFK-4 channel and heat transfer cement.
Insulation Thermal Conductivity: See Chart, page 79
Ambient: −17.8° C, 40 KPH wind

Nomenclature:

TS = Steam Temperature (°C)
TP = Pipe Temperature (°C)
GM/SEC-M = Tracer condensate rate per linear metre
Q3 = Pipe Heat Losses in Watts/lin m

PIPE SIZE	Steam Temp °F	INSULATION THICKNESS											
		38 MM Nom			51 MM Nom			64 MM Nom			76 MM Nom		
	TS	TP	GM/SEC-M	Q3	TP	GM/SEC-M	Q3	TP	GM/SEC-M	Q3	TP	GM/SEC-M	Q3
48 MM	121	111.1	.02405	52.9	111.9	.02120	46.7	112.4	.01912	42.3	112.8	.01763	39.0
	134	124.2	.02746	59.4	125.0	.02426	52.4	125.6	.02191	47.5	126.1	.02021	43.8
	148	137.7	.03122	66.3	138.7	.02762	58.6	139.5	.02503	53.2	140.0	.02314	49.1
	160	149.7	.03495	72.7	150.7	.03082	64.3	151.5	.02793	58.3	152.0	.02577	53.8
	170	158.7	.03761	77.6	159.9	.03341	68.6	160.6	.03030	62.2	161.2	.02801	57.4
	186	172.6	.04269	85.3	173.8	.03775	75.4	174.7	.03419	68.4	175.3	.03155	63.1
	198	184.2	.04700	91.9	185.5	.04162	81.3	186.4	.03776	73.7	187.1	.03486	68.1
	208	193.4	.05083	97.3	194.8	.04498	86.1	195.8	.04077	78.1	196.5	.03764	72.1
60 MM	121	108.8	.03004	66.1	110.1	.02574	56.6	110.9	.02284	50.3	111.0	.02074	45.7
	134	121.6	.03420	74.4	123.0	.02928	63.6	124.0	.02611	56.5	124.7	.02374	51.3
	148	134.9	.03894	82.8	136.4	.03340	71.0	137.5	.02976	63.1	138.2	.02696	57.3
	160	146.5	.04351	90.5	148.3	.03731	77.8	149.4	.03324	69.2	150.3	.03016	62.9
	170	155.5	.04720	96.7	157.3	.04045	83.0	158.5	.03595	73.8	159.4	.03270	67.1
	186	169.0	.05302	106.2	171.0	.04558	91.2	172.3	.04061	81.1	173.3	.03695	73.8
	198	180.3	.05854	114.3	182.4	.05034	98.2	183.8	.04470	87.4	184.9	.04074	79.5
	208	189.4	.06311	120.9	191.6	.05434	104.0	193.1	.04837	92.6	194.2	.04403	84.3
89 MM	121	105.7	.03489	77.2	107.5	.02988	65.8	108.6	.02640	58.1	109.6	.02363	52.0
	134	118.1	.04002	86.5	120.1	.03416	73.8	121.4	.03004	65.2	122.5	.02698	58.4
	148	131.0	.04548	96.4	133.1	.03873	82.3	134.6	.03424	72.8	135.8	.03067	65.2
	160	142.2	.05050	105.3	144.6	.04326	90.0	146.2	.03826	79.6	147.6	.03434	71.5
	170	150.8	.05478	112.3	153.3	.04684	96.0	155.1	.04145	85.1	156.5	.03714	76.3
	186	164.0	.06164	123.4	166.8	.05272	105.5	168.6	.04670	93.5	170.1	.04186	83.8
	198	174.9	.06785	132.7	177.9	.05820	113.6	179.9	.05156	100.6	181.5	.04616	90.3
	208	183.7	.07335	140.3	186.8	.06275	120.2	188.9	.05558	106.5	190.6	.04993	95.6
114 MM	121	102.4	.04052	89.1	104.7	.03414	75.5	106.4	.02975	65.5	107.5	.02687	59.1
	134	114.4	.04619	99.8	117.6	.03916	84.7	118.9	.03399	73.5	120.1	.03057	66.3
	148	126.9	.05244	111.1	129.7	.04453	94.4	131.9	.03856	81.9	133.2	.03483	74.0
	160	137.7	.05816	121.3	140.8	.04954	103.1	143.2	.04307	89.6	144.6	.03881	81.0
	170	146.0	.06294	129.2	149.3	.05353	109.9	151.8	.04663	95.5	153.4	.04207	86.4
	186	158.7	.07091	141.6	162.3	.06023	120.6	165.1	.05248	105.1	166.8	.04748	95.0
	198	169.3	.07781	152.2	173.2	.06643	129.9	176.1	.05792	113.1	178.0	.05240	102.3
	208	177.9	.08407	161.1	181.9	.07172	137.4	185.0	.06245	119.6	186.9	.05650	108.3
168 MM	121	96.4	.04916	108.2	99.6	.04165	91.7	101.9	.03629	80.2	103.8	.03190	70.3
	134	107.8	.05596	121.0	111.4	.04750	102.6	113.9	.04158	89.8	116.0	.03640	78.8
	148	119.5	.06351	134.5	123.5	.05373	114.2	126.2	.04720	100.1	128.6	.04144	87.8
	160	129.7	.07049	146.7	134.0	.05978	124.7	137.0	.05241	109.3	139.6	.04600	95.9
	170	137.5	.07603	156.1	142.1	.06466	132.8	145.3	.05687	116.5	148.1	.04979	102.3
	186	149.4	.08555	170.8	154.4	.07281	145.4	157.8	.06380	127.7	160.9	.05604	112.2
	198	159.3	.09402	183.4	164.6	.07990	156.3	168.3	.07018	137.3	171.6	.06173	120.7
	208	167.2	.10101	193.6	172.8	.08613	165.1	176.7	.07575	145.1	180.3	.06683	127.9
219 MM	121	90.7	.05608	123.4	95.6	.04567	101.2	98.1	.04061	89.4	100.1	.03642	80.5
	134	101.4	.06341	137.8	106.8	.05235	113.1	109.7	.04632	100.1	111.8	.04171	90.1
	148	112.5	.07229	153.0	118.4	.05935	125.8	121.6	.05256	111.4	124.0	.04734	100.4
	160	122.1	.08018	166.7	128.5	.06597	137.2	132.0	.05827	121.5	134.6	.05254	109.6
	170	129.5	.08659	177.2	138.3	.07112	146.1	139.9	.06303	129.4	142.7	.05701	116.8
	186	140.7	.09675	193.7	148.0	.08004	159.8	152.0	.07097	141.7	155.0	.06393	127.9
	198	150.0	.10647	207.7	157.8	.08776	171.6	162.1	.07785	152.3	165.3	.07030	137.5
	208	157.4	.11436	219.1	165.7	.09452	181.2	170.1	.08394	160.9	173.5	.07587	145.3

273 MM	121	81.8	.06232	138.3	86.4	.05474	120.4	90.0	.04795	106.3	92.9	.04344	96.1
	134	91.6	.07134	154.2	96.6	.06182	134.4	100.8	.05511	119.1	103.8	.04970	107.4
	148	101.6	.08076	171.0	107.3	.07065	149.6	111.8	.06246	132.3	115.1	.05613	119.4
	160	110.3	.08939	185.9	118.5	.07827	162.8	121.3	.06928	144.1	124.9	.06259	130.1
	170	117.0	.09602	197.5	123.6	.08455	173.0	128.6	.07466	153.3	132.5	.06742	138.5
	186	127.3	.10783	215.9	134.2	.09441	189.0	139.7	.08389	167.6	143.9	.07563	151.4
	198	135.7	.11856	231.2	143.1	.10355	202.5	149.0	.09196	179.8	153.4	.08311	162.5
	208	142.5	.12735	243.6	150.2	.11145	213.6	156.3	.09895	189.7	161.0	.08951	171.6
324 MM	121	70.5	.06424	142.5	76.9	.05587	122.9	81.5	.04946	108.9	85.0	.04438	98.2
	134	79.1	.07300	153.5	86.1	.06300	136.9	91.3	.05614	121.4	95.1	.05073	109.6
	148	87.9	.08281	175.3	95.7	.07164	151.6	101.3	.06353	134.6	105.5	.05716	121.6
	160	95.8	.09152	190.3	103.9	.07889	164.8	110.0	.07033	146.3	114.7	.06384	132.7
	170	101.4	.09819	201.8	110.2	.08546	174.9	116.6	.07571	155.5	121.6	.06869	141.1
	186	110.3	.01974	219.7	119.8	.09524	190.7	126.7	.08494	169.7	132.1	.07696	154.1
	198	117.6	.12012	234.9	127.8	.10433	204.0	135.2	.09337	182.1	140.8	.08422	165.1
	208	123.5	.12887	247.1	134.1	.11215	214.9	141.9	.10012	191.9	147.8	.09081	174.1
356 MM	121	60.9	.06385	141.7	67.9	.05620	123.6	73.3	.05037	110.8	77.3	.04546	100.7
	134	68.5	.07240	157.4	76.3	.06328	137.5	82.1	.05707	123.4	86.6	.05191	112.2
	148	76.3	.08209	173.8	84.9	.07205	152.5	91.3	.06449	136.6	96.1	.05861	124.3
	160	83.1	.09058	188.3	92.4	.07929	165.5	99.2	.07101	148.3	104.4	.06496	135.1
	170	88.3	.09707	199.6	98.0	.08574	175.5	105.2	.07666	157.4	110.8	.06984	143.4
	186	96.1	.10835	216.9	106.6	.09541	191.0	114.4	.08588	171.5	120.4	.07832	156.4
	198	102.7	.11878	231.6	113.8	.10439	204.1	122.0	.09403	183.4	128.3	.08559	167.4
	208	107.9	.12729	243.5	119.5	.11207	214.8	128.0	.10074	193.1	134.7	.09197	176.3
406 MM	121	50.6	.06118	134.6	58.0	.05430	119.5	63.8	.04916	108.2	68.2	.04476	99.1
	134	57.2	.06915	149.2	65.4	.06099	132.6	71.7	.05563	120.3	76.6	.05100	110.2
	148	64.2	.07821	165.6	72.9	.06918	146.5	79.8	.06273	132.9	85.1	.05727	121.9
	160	70.3	.08658	180.1	79.4	.07609	158.7	86.8	.06927	144.1	92.7	.06382	132.7
	170	74.7	.09299	190.5	84.4	.08219	168.2	92.2	.07441	152.8	98.4	.06654	140.8
	186	81.5	.10310	206.4	92.0	.09131	182.8	100.3	.08318	166.2	107.0	.07651	153.2
	198	87.2	.11292	220.4	98.3	.09996	195.1	107.1	.09080	177.5	114.2	.08373	163.7
	208	91.6	.12079	230.9	103.3	.10700	205.1	112.5	.09741	186.7	119.9	.08967	172.3
457 MM	121	47.8	.06414	142.3	55.1	.05716	125.8	60.9	.05187	114.1	65.5	.04723	104.7
	134	54.2	.07262	157.8	62.2	.06426	139.6	68.5	.05861	126.7	73.6	.05385	116.4
	148	60.7	.08218	174.0	69.4	.07277	154.0	76.4	.06612	140.0	61.8	.06071	128.7
	160	66.3	.09057	188.3	75.7	.08024	166.9	83.1	.07269	151.8	89.0	.06710	139.6
	170	70.5	.09662	198.7	80.4	.08634	176.7	88.3	.07834	160.9	94.5	.07207	148.0
	186	77.0	.10767	215.6	87.7	.09590	192.0	96.1	.08761	174.9	102.8	.08039	161.0
	198	82.4	.11784	229.8	93.6	.10447	204.3	102.7	.09571	186.7	109.7	.08800	172.1
	208	86.7	.12618	241.3	98.4	.11207	214.8	107.8	.10243	196.3	115.3	.09442	181.0
508 MM	121	45.0	.06679	148.1	52.4	.05969	131.3	58.2	.05418	119.2	63.0	.04996	110.0
	134	51.1	.07539	164.1	59.0	.06687	145.1	65.5	.06086	132.3	70.8	.05651	122.2
	148	57.3	.08539	180.8	66.0	.07573	160.3	73.1	.06904	146.2	78.8	.06373	135.0
	160	62.7	.09410	195.6	72.0	.08326	173.8	79.6	.07594	158.4	85.8	.07035	146.4
	170	66.8	.10076	207.0	76.7	.08992	184.1	84.6	.08203	167.8	91.1	.07556	155.2
	186	72.9	.11171	223.7	83.5	.09967	199.5	92.1	.09114	182.4	99.1	.08449	168.8
	198	78.0	.12193	238.4	89.4	.10929	213.3	98.4	.09979	194.8	105.8	.09220	180.3
	208	82.1	.13055	250.3	94.0	.11703	224.2	103.4	.10682	204.7	111.1	.09890	189.6
610 MM	121	41.7	.07031	155.7	48.9	.06317	140.2	54.3	.05764	126.8	59.0	.05338	117.5
	134	47.3	.07899	171.5	55.3	.07150	155.6	61.3	.06481	140.8	66.4	.05997	130.4
	148	53.1	.08919	188.9	61.8	.08081	171.1	68.4	.07342	155.4	74.0	.06804	144.0
	160	58.2	.09831	204.2	67.6	.08904	185.2	74.6	.08070	168.4	80.7	.07483	156.1
	170	62.1	.10545	216.0	71.9	.09570	196.1	79.4	.08714	178.4	85.8	.08084	166.0
	186	68.0	.11696	234.2	78.6	.10632	212.9	86.5	.09682	193.8	93.5	.09013	180.4
	198	72.8	.12756	249.5	83.9	.11601	226.3	92.5	.10606	206.9	99.8	.09868	192.6
	208	76.7	.13665	261.8	88.3	.12430	237.7	97.3	.11348	217.4	104.9	.10563	202.5
762 MM	121	35.6	.07641	168.0	42.8	.06909	153.1	48.6	.06353	141.0	53.1	.05890	129.6
	134	40.7	.08537	185.9	48.6	.07812	169.6	55.0	.07195	156.5	59.9	.06626	143.8
	148	46.0	.09651	204.5	54.5	.08781	186.0	61.5	.08121	171.9	67.0	.07501	158.8
	160	50.4	.10590	219.9	59.7	.09679	201.1	67.2	.08948	186.1	73.1	.08245	172.0
	170	53.9	.11308	232.5	63.6	.10395	212.8	71.5	.09581	197.1	77.7	.08897	182.2
	186	59.1	.12627	251.9	69.6	.11519	230.6	78.1	.10682	213.9	84.7	.09859	197.4
	198	63.5	.13705	268.1	74.6	.12565	245.7	83.5	.11654	227.3	90.6	.10775	210.6
	208	67.0	.14711	281.2	78.5	.13489	257.9	87.8	.12485	238.8	95.2	.11567	221.6

Table 12a Steam tracing data (Metric units continued)

Design Conditions:

Tracer: One 13 MM O.D. tracer thermally bonded to pipe with a TFK-4 channel and heat transfer cement.
Insulation Thermal Conductivity: See Chart, page 79
Ambient: 37.8° C, no wind

Nomenclature:

TS = Steam Temperature (°C)
TP = Pipe Temperature (°C)
GM/SEC-M = Tracer condensate rate per linear metre
Q3 = Pipe Heat Losses in Watts/lin m

PIPE SIZE	Steam Temp °C	INSULATION THICKNESS											
		38 MM Nom			51 MM Nom			64 MM Nom			76 MM Nom		
	TS	TP	GM/SEC-M	Q3	TP	GM/SEC-M	Q3	TP	GM/SEC-M	Q3	TP	GM/SEC-M	Q3
48 MM	121	114.0	.01382	30.4	114.3	.01244	27.4	114.6	.01139	25.1	114.8	.01081	23.4
	134	127.1	.01682	36.4	127.5	.01515	32.8	127.9	.01367	30.1	128.1	.01292	28.0
	148	140.7	.02016	42.8	141.2	.01813	38.5	141.6	.01668	35.4	141.9	.01550	33.0
	160	152.6	.03332	48.6	153.2	.02102	43.7	153.6	.01927	40.2	153.9	.01794	37.4
	170	161.7	.02589	53.2	162.3	.02330	47.8	162.8	.02139	43.9	163.2	.01992	40.9
	186	175.6	.03014	60.3	176.3	.02713	54.2	176.8	.02491	49.8	177.3	.02318	46.4
	198	187.2	.03398	66.4	188.0	.03057	59.7	188.6	.02809	54.9	189.1	.02615	51.1
	208	196.6	.03738	71.5	197.5	.03362	64.4	198.2	.03090	59.1	198.7	.02879	55.1
60 MM	121	112.8	.01684	37.2	113.4	.01485	32.7	113.8	.01342	29.5	114.1	.01234	27.2
	134	125.7	.02052	44.5	128.4	.01810	39.1	126.9	.01634	35.4	127.2	.01504	32.5
	148	139.0	.02462	52.3	139.8	.02170	46.0	140.4	.01956	41.6	140.9	.01800	38.2
	160	150.7	.02847	59.4	151.8	.02503	52.2	152.3	.02262	47.2	152.8	.02084	43.4
	170	159.8	.03160	64.5	160.6	.02784	57.1	161.4	.02515	51.6	161.9	.02311	47.4
	186	173.2	.03681	73.5	174.4	.03231	64.7	175.2	.02924	58.4	175.8	.02690	53.8
	198	184.6	.04145	80.9	185.9	.03647	71.2	186.8	.03292	64.4	187.5	.03033	59.2
	208	193.9	.04556	87.2	195.2	.04007	76.8	196.2	.03621	69.4	197.0	.03333	63.6
89 MM	121	110.9	.01945	43.6	111.7	.01711	37.6	112.3	.01538	33.9	112.8	.01397	30.7
	134	123.6	.02390	51.7	124.6	.02085	45.2	125.3	.01876	40.7	125.9	.01705	36.9
	148	136.5	.02663	60.7	137.7	.02501	53.2	138.5	.02254	47.8	139.2	.02041	43.4
	160	147.9	.03309	68.8	149.2	.02890	60.3	150.2	.02600	54.2	150.9	.02360	49.2
	170	156.6	.03662	75.2	158.0	.03207	65.8	159.0	.02885	59.2	159.9	.02619	53.8
	186	169.8	.04261	85.2	171.4	.03734	74.8	172.6	.03353	67.1	173.5	.03049	60.9
	198	180.8	.04792	93.7	182.6	.04203	82.1	183.9	.03782	73.9	185.0	.03435	67.1
	208	189.7	.05261	100.7	191.6	.04611	88.2	193.0	.04145	79.4	194.1	.03763	72.1
114 MM	121	109.1	.02246	49.4	110.2	.01946	43.1	111.0	.01730	38.1	111.6	.01576	34.8
	134	121.4	.02741	59.3	122.7	.02391	51.7	123.7	.02108	45.7	124.4	.01928	41.8
	148	134.0	.03278	69.6	135.6	.02861	60.7	136.8	.02526	53.7	137.5	.02315	49.1
	160	145.0	.03786	78.9	146.8	.03307	68.8	148.1	.02919	60.9	149.0	.02670	55.7
	170	153.4	.04199	86.2	155.3	.03658	75.1	156.8	.03240	66.5	157.8	.02963	60.8
	186	166.3	.04872	97.5	168.4	.04256	85.0	170.1	.03768	75.3	171.2	.03443	68.9
	198	177.0	.05485	107.2	179.3	.04785	93.5	181.2	.04242	82.8	182.4	.03882	75.8
	208	185.6	.06011	115.2	186.1	.05250	100.5	196.0	.04653	89.0	191.3	.04256	81.5
168 MM	121	105.8	.02722	59.9	107.3	.02367	52.1	108.4	.02109	46.4	109.4	.01870	41.3
	134	117.3	.03307	71.4	119.3	.02884	62.4	120.6	.02572	55.6	121.8	.02283	49.5
	148	129.4	.03956	84.1	131.6	.03446	73.2	133.1	.03073	65.3	134.5	.02731	58.1
	160	139.8	.04566	95.3	142.3	.03977	82.9	144.0	.03552	73.9	145.6	.03158	65.8
	170	147.5	.05060	103.9	150.4	.04408	90.5	152.4	.03930	80.7	154.1	.03502	71.9
	186	159.9	.05862	117.4	162.9	.05108	102.3	165.0	.04562	91.3	167.0	.04070	81.3
	198	170.0	.06593	128.9	173.3	.05758	112.4	175.6	.05139	100.3	177.8	.04574	89.4
	208	178.1	.07219	138.4	181.6	.06298	120.8	184.1	.05622	107.7	186.4	.05021	96.1
219 MM	121	102.6	.03182	68.3	104.9	.02622	57.7	106.2	.02355	51.8	107.2	.02145	47.2
	134	113.6	.03767	81.5	116.3	.03183	68.8	117.9	.02858	61.8	119.1	.02607	56.4
	148	125.0	.04515	95.7	128.3	.03808	80.9	130.1	.03422	72.7	131.5	.03121	66.3
	160	134.9	.05190	108.2	138.5	.04404	91.6	140.6	.03956	82.3	142.1	.03608	75.1
	170	142.4	.05759	117.9	146.4	.04873	99.9	148.6	.04376	89.8	150.3	.03990	81.9
	186	153.9	.06653	133.1	158.3	.05636	112.8	160.8	.05069	101.5	162.7	.04628	92.6
	198	163.4	.07470	146.1	168.3	.06336	123.9	171.0	.05700	111.5	173.1	.05212	101.7
	208	171.0	.08175	156.7	178.2	.06939	132.9	179.2	.06244	119.6	181.4	.05701	109.2

273 MM	121	97.6	.03469	77.2	99.8	.03114	68.6	101.7	.02799	61.6	103.1	.02550	56.1
	134	107.6	.04255	91.9	110.3	.03777	81.7	112.5	.03394	73.3	114.1	.03083	66.9
	148	118.0	.05052	107.4	121.1	.04507	95.5	123.6	.04050	85.8	125.5	.03685	78.3
	160	127.0	.05815	121.3	130.5	.05173	107.9	133.4	.04668	97.3	135.6	.04271	88.5
	170	133.9	.06430	132.0	137.8	.05759	117.9	140.8	.05174	106.0	143.2	.04724	96.8
	186	144.4	.07468	149.2	148.7	.06644	132.9	152.1	.05974	119.6	154.8	.05457	109.3
	198	153.1	.08362	163.5	157.7	.07453	145.8	161.5	.06709	131.2	164.4	.06130	119.6
	208	160.1	.09136	175.1	165.0	.08148	156.2	169.0	.07351	140.6	172.1	.06721	128.6
324 MM	121	91.2	.03578	76.9	94.4	.03177	70.0	96.7	.02864	63.0	98.6	.02613	57.5
	134	100.3	.04361	94.2	103.9	.03843	83.1	106.7	.03452	74.9	108.8	.03165	68.4
	148	109.4	.05191	109.9	113.7	.04579	97.1	116.9	.04127	87.4	119.3	.03762	80.0
	160	117.4	.05935	123.8	122.2	.05246	109.4	125.7	.04730	98.6	128.5	.04340	90.2
	170	123.5	.06559	134.7	128.6	.05814	119.0	132.5	.05229	107.4	135.5	.04796	98.3
	186	132.7	.07557	151.3	138.5	.06696	134.0	142.8	.06040	120.9	146.1	.05532	110.8
	198	140.4	.08405	165.5	146.6	.07497	146.6	151.3	.06773	132.5	155.2	.06223	121.7
	208	146.5	.09238	177.1	153.2	.08192	157.0	158.3	.07438	142.2	162.2	.06803	130.4
356 MM	121	85.8	.03562	79.2	89.2	.03186	70.6	91.8	.02905	64.0	93.9	.02669	58.7
	134	93.8	.04344	93.8	97.8	.03868	83.6	101.6	.03512	76.3	103.4	.03241	70.1
	148	101.9	.05143	109.3	100.6	.04590	97.5	110.3	.04197	88.9	113.1	.03850	81.8
	160	109.0	.05899	123.0	114.4	.05286	110.3	118.4	.04809	100.3	121.6	.04424	92.2
	170	114.4	.06505	133.6	120.2	.05852	119.8	124.5	.05316	108.9	128.0	.04883	100.3
	186	122.6	.07489	150.0	129.0	.06731	134.7	133.9	.06117	122.4	137.7	.05633	112.6
	198	129.5	.08384	164.0	136.4	.07528	147.2	141.7	.06849	134.0	145.8	.06314	123.5
	208	134.9	.09136	175.1	142.3	.08215	157.4	147.9	.07497	143.4	152.3	.06896	132.1
406 MM	121	79.8	.03422	75.7	83.4	.03108	68.4	86.3	.02855	62.8	88.6	.02645	58.2
	134	86.7	.04151	89.7	90.9	.03742	80.9	94.4	.03436	74.6	97.1	.03191	69.0
	148	93.7	.04885	103.9	98.7	.04446	94.2	102.7	.04097	86.8	105.0	.03782	80.4
	160	99.9	.05606	116.9	105.5	.05083	106.0	109.9	.04680	97.6	113.4	.04346	90.4
	170	104.7	.06164	127.0	110.7	.05642	115.6	115.4	.05173	106.0	119.1	.04796	98.3
	186	111.8	.07134	142.6	118.5	.06480	129.7	123.7	.05942	119.0	127.9	.05514	110.4
	198	117.8	.07982	155.7	125.0	.07239	141.6	130.6	.06646	130.0	135.3	.06189	121.0
	208	122.4	.08633	165.9	130.2	.07896	151.3	136.1	.07262	139.0	141.1	.06762	129.4
457 MM	121	78.0	.03584	79.2	81.7	.03254	72.1	84.5	.02997	66.0	67.0	.02795	61.5
	134	84.8	.04373	94.5	89.0	.03943	85.2	92.4	.03629	78.5	95.2	.03369	72.9
	148	91.8	.05226	110.7	96.4	.04661	99.3	100.4	.04311	91.4	103.6	.04006	84.8
	160	97.8	.05900	124.4	102.9	.05353	111.6	107.4	.04928	102.8	110.9	.04577	95.4
	170	102.4	.06583	135.2	107.9	.05912	121.0	112.7	.05431	111.8	116.5	.05049	103.7
	186	109.1	.07545	151.1	115.4	.06802	135.8	120.7	.06254	125.2	124.9	.05814	116.4
	198	114.7	.08411	164.5	121.6	.07578	148.2	127.3	.06997	136.9	132.0	.06503	127.2
	208	119.3	.01962	175.6	126.5	.08261	158.3	132.6	.07626	146.1	137.6	.07101	135.9
508 MM	121	75.8	.03825	84.1	80.0	.03399	75.2	83.0	.03145	69.3	85.3	.02917	64.2
	134	83.2	.04827	100.0	87.1	.04117	89.0	90.5	.03790	82.0	93.4	.03522	76.5
	148	89.7	.05436	115.4	94.2	.04869	103.6	98.2	.04501	95.4	101.6	.04202	89.0
	160	95.4	.06236	129.6	100.5	.05586	116.5	104.9	.05141	107.2	108.7	.04804	100.2
	170	99.8	.06854	140.8	105.3	.06159	126.5	110.0	.05687	116.5	114.1	.05304	108.7
	186	106.2	.07059	156.9	112.5	.07098	141.8	117.7	.06528	130.6	122.2	.06098	122.1
	198	111.6	.05759	171.3	118.3	.07890	154.3	124.2	.07292	142.6	129.0	.06812	133.2
	208	116.0	.09535	182.8	123.1	.08598	164.8	129.3	.07954	152.4	134.4	.07446	142.4
610 MM	121	74.7	.04061	89.4	77.9	.03641	80.5	80.6	.03338	73.9	83.1	.03137	69.1
	134	80.6	.04855	104.9	84.7	.04430	95.7	87.7	.04044	87.4	90.6	.03786	81.9
	148	86.8	.05712	121.6	91.5	.05259	111.5	95.0	.04783	101.8	98.3	.04493	95.2
	160	92.1	.06578	136.8	97.5	.06043	125.6	101.4	.05486	114.4	105.1	.05128	107.0
	170	96.1	.07170	147.3	101.9	.06606	135.7	106.2	.06047	124.1	110.2	.05672	116.2
	186	102.2	.08239	165.0	108.8	.07597	152.1	113.6	.06971	139.2	117.9	.06512	130.3
	198	107.4	.09197	179.8	114.3	.08459	165.4	119.6	.07769	152.0	124.4	.07274	142.3
	208	111.5	.10014	192.0	118.9	.09206	176.5	124.5	.08469	162.3	129.5	.07928	151.9
762 MM	121	70.9	.04362	96.5	74.6	.04040	89.5	77.4	.03769	82.8	79.5	.03433	76.0
	134	76.4	.05278	113.9	80.4	.04859	105.0	84.0	.04544	98.2	80.4	.04160	89.9
	148	81.9	.06236	132.1	86.7	.05722	121.8	90.6	.05331	113.5	93.6	.04915	104.6
	160	86.6	.07075	147.2	92.1	.06573	136.7	96.5	.06139	127.6	99.8	.05641	117.6
	170	90.2	.07769	159.5	96.0	.07182	147.5	100.8	.06704	137.7	104.5	.06224	127.8
	186	95.8	.08914	178.4	102.1	.08242	165.1	107.6	.07706	154.2	111.7	.07174	143.3
	198	100.4	.09946	194.2	107.3	.09194	179.7	113.1	.08612	168.4	117.5	.07972	155.9
	208	104.1	.01802	207.0	111.4	.10010	191.9	117.5	.09333	178.9	122.3	.08701	166.8

Table 12a Steam tracing data (Metric units continued)

Design Conditions:

Tracer: One 22 MM O.D. tracer thermally bonded to pipe with a TFK-7 channel and heat transfer cement.
Insulation Thermal Conductivity: See Chart, page 79
Ambient: −17.8° C, 40 KPH wind

Nomenclature:

TS = Steam Temperature (°C)
TP = Pipe Temperature (°C)
GM/SEC-M = Tracer condensate rate per linear metre
Q3 = Pipe Heat Losses in Watts/lin m

PIPE SIZE	Steam Temp °C	INSULATION THICKNESS											
		38 MM Nom			51 MM Nom			64 MM Nom			76 MM Nom		
	TS	TP	GM/SEC-M	Q3	TP	GM/SEC-M	Q3	TP	GM/SEC-M	Q3	TP	GM/SEC-M	Q3
114 MM	121	104.9	.04905	108.3	107.2	.04002	88.4	108.5	.03499	77.1	109.5	.03118	68.6
	134	117.2	.05591	121.3	119.8	.04571	99.1	121.5	.03989	86.5	122.4	.03565	77.3
	148	129.9	.06361	135.2	132.8	.05210	110.5	134.5	.04552	96.5	135.7	.04061	86.3
	160	141.1	.01093	147.7	144.2	.05795	120.8	146.0	.05070	105.6	147.5	.04543	94.6
	170	149.6	.07677	157.4	152.9	.06286	128.9	155.0	.05508	112.9	156.4	.04921	101.0
	186	162.7	.08640	172.0	168.4	.07083	141.7	168.4	.06202	124.0	124.0	.05550	110.9
	198	173.5	.09506	185.9	177.5	.07811	152.5	179.7	.06835	133.5	181.4	.06113	119.5
	208	182.2	.10226	190.5	186.4	.08431	161.3	188.7	.07381	141.3	190.3	.06608	128.5
168 MM	121	100.7	.05688	125.2	103.4	.04743	104.7	105.7	.04002	85.4	106.9	.03601	79.3
	134	112.5	.06478	140.1	115.6	.05421	117.3	118.1	.04570	99.1	119.4	.04102	89.0
	148	124.8	.07339	156.0	128.2	.06151	130.7	130.9	.05208	110.5	132.4	.04670	99.3
	160	135.4	.08165	170.2	139.1	.06861	142.8	142.1	.05792	120.8	143.8	.05212	108.5
	170	143.6	.08832	181.3	147.5	.07422	152.2	150.7	.06282	128.8	152.4	.05641	115.8
	186	156.0	.09934	198.6	160.3	.08349	166.9	163.9	.07677	141.6	165.8	.06371	127.3
	198	166.4	.10918	213.3	171.1	.09204	179.7	174.8	.07803	152.4	176.9	.07020	137.1
	208	174.7	.11776	225.4	179.7	.09931	198.0	183.6	.08422	161.2	185.8	.07561	145.1
219 MM	121	96.5	.06346	140.2	99.8	.05363	118.5	102.1	.04668	103.1	103.6	.04226	93.2
	134	107.9	.07229	156.6	111.5	.06120	132.7	114.1	.05336	115.5	115.8	.04817	104.4
	148	119.8	.08206	174.3	123.6	.06949	147.7	126.5	.06053	128.6	128.3	.05460	116.4
	160	129.9	.09113	190.1	134.2	.07742	161.2	137.3	.06741	140.5	139.3	.06099	127.2
	170	137.7	.09857	202.3	142.3	.08364	171.7	145.8	.07303	149.7	147.7	.06615	135.5
	186	149.6	.11078	221.4	154.6	.09410	188.1	158.2	.08203	164.1	168.5	.07434	148.7
	198	159.5	.12156	237.7	164.9	.10353	202.1	168.7	.09030	176.5	171.2	.08194	160.0
	208	167.5	.13103	250.9	173.1	.11159	213.5	177.1	.09749	186.5	179.9	.08856	169.5
273 MM	121	90.5	.06935	152.6	94.3	.05936	131.3	97.2	.05217	114.9	99.4	.04651	102.7
	134	101.2	.07859	170.4	105.4	.06789	146.8	108.6	.05936	128.5	111.0	.05314	115.0
	148	112.2	.08925	189.3	116.9	.07680	163.2	120.4	.06728	143.0	123.1	.06025	128.6
	160	121.8	.09883	206.1	126.8	.08531	177.9	130.7	.07491	156.0	133.6	.06764	139.7
	170	129.2	.10700	219.2	134.5	.09239	189.3	138.6	.08096	166.1	141.7	.07261	148.9
	186	140.3	.11987	239.6	146.1	.10362	207.1	150.6	.09098	181.0	155.9	.08149	163.1
	198	149.6	.13156	256.9	155.8	.11386	222.3	160.5	.09994	195.3	164.1	.08975	175.3
	208	157.1	.14147	271.0	163.5	.12249	234.7	188.5	.10782	206.3	172.3	.09679	185.2
324 MM	121	80.7	.07329	161.5	86.2	.06199	137.0	90.1	.05443	119.8	93.7	.04809	107.3
	134	90.3	.08297	179.9	96.5	.07049	152.8	100.0	.06189	134.3	104.1	.05551	120.2
	148	100.3	.09381	199.4	107.2	.08005	170.1	111.9	.07018	149.2	115.5	.06260	133.6
	160	108.9	.10393	216.8	116.4	.08876	185.1	121.5	.07804	162.5	125.3	.06995	145.6
	170	115.5	.11215	230.2	123.4	.09598	196.8	128.8	.08418	172.8	132.9	.07559	155.0
	186	125.5	.12544	251.1	134.0	.10751	214.0	139.9	.09450	188.9	144.3	.08475	169.5
	198	134.0	.13798	269.4	142.9	.11780	230.3	149.2	.10378	202.6	153.9	.09317	182.0
	208	140.7	.14817	283.6	150.0	.12678	242.8	156.6	.11170	213.8	161.5	.10036	192.1

356 MM	121	72.1	.07485	164.9	78.5	.06406	141.5	83.1	.05662	124.7	86.6	.05093	112.1
	134	80.9	.08463	183.5	87.9	.07269	157.6	93.0	.06429	139.1	90.0	.05770	125.1
	148	89.9	.09556	203.4	97.6	.08221	174.7	103.2	.07259	154.3	107.6	.06553	139.3
	160	97.7	.10561	220.4	106.1	.09101	189.8	112,1	.08052	167.6	116.8	.07271	151.6
	170	103.7	.11390	233.8	112.5	.09819	201.6	118.8	.08669	178.4	123.5	.07851	161.2
	186	112.7	.12722	254.6	122.2	.10997	219.6	129.2	.08763	195.1	134.5	.08698	176.1
	198	120.2	.13944	272.3	130.3	.12033	235.3	137.8	.10704	209.0	143.4	.09650	188.8
	208	126.2	.14957	286.5	136.8	.12939	247.8	144.6	.11514	230.3	150.5	.10405	199.1
406 MM	121	62.4	.07309	101.9	69.4	.06338	140.0	74.8	.05680	125.1	78.8	.05149	113.3
	134	70.2	.08248	178.9	78.8	.07207	156.3	83.8	.06459	139.3	88.2	.05824	126.3
	148	78.1	.02969	197.7	86.8	.08135	172.3	93.1	.07258	154.3	97.9	.06588	139.9
	160	85.1	.10270	214.2	94.3	.08993	187.6	101.1	.08039	162.6	106.4	.07290	152.1
	170	90.3	.11061	227.1	100.1	.09704	199.0	107.3	.08664	177.6	112.8	.07866	161.6
	186	98.3	.12332	246.9	108.9	.10837	216.0	116.6	.09896	193.8	122.6	.08805	176.4
	198	105.0	.13501	263.6	116.1	.11843	231.5	124.3	.10618	207.5	130.7	.09650	188.6
	208	110.3	.14489	277.2	122.0	.12719	243.0	130.5	.11408	218.3	137.2	.10347	198.8
457 MM	121	59.3	.07690	169.3	66.0	.06716	148.3	72.3	.06008	132.9	76.3	.05473	120.6
	134	66.9	.08705	188.8	74.8	.07605	164.9	80.9	.06821	147.9	85.5	.06190	134.6
	148	74.5	.09806	208.4	83.2	.08600	182.3	89.5	.07701	163.6	94.9	.06997	148.7
	160	81.2	.10627	225.8	90.6	.09484	197.0	97.7	.08521	177.7	103.1	.07781	161.6
	170	88.3	.11655	239.3	96.1	.10230	209.8	103.6	.09201	185.5	109.4	.08358	171.8
	186	94.0	.12993	260.0	104.7	.11450	226.9	112.7	.10274	205.3	118.9	.09350	187.0
	198	100.4	.14216	277.6	111.7	.12523	244.9	120.2	.11245	219.6	126.8	.10242	200.2
	208	105.5	.15229	291.8	117.3	.13444	257.3	126.2	.12068	231.2	133.1	.11016	210.1
508 MM	121	56.6	.08079	177.8	64.0	.07066	156.2	69.5	.06324	139.7	74.0	.05782	127.4
	134	63.7	.09101	197.3	71.9	.08002	173.5	78.2	.07190	155.9	82.9	.06556	141.8
	148	71.1	.10270	217.9	80.1	.09042	191.8	86.9	.08117	172.5	92.1	.07358	157.0
	160	77.5	.11317	236.0	87.2	.09971	208.0	94.5	.08975	187.2	100.1	.08181	170.5
	170	82.4	.12175	250.0	92.5	.10745	220.5	100.2	.09685	198.6	106.2	.08819	181.0
	186	89.8	.13575	271.6	100.7	.11995	239.6	109.0	.10817	216.2	115.4	.09868	197.2
	198	96.0	.14847	289.9	107.5	.13118	256.1	116.3	.11821	231.1	123.1	.10793	211.0
	208	100.9	.15921	304.7	112.9	.14076	269.4	122.1	.12697	243.2	129.2	.11689	222.2
610 MM	121	53.1	.08537	187.8	60.4	.07881	167.0	66.0	.06809	150.4	70.3	.06217	137.4
	134	60.0	.09644	208.4	68.0	.08549	185.4	74.1	.07711	167.2	79.0	.07074	153.4
	148	67.0	.10854	230.0	75.8	.09659	204.8	82.4	.08719	184.9	87.8	.07956	169.7
	160	73.1	.11952	249.2	82.5	.10641	221.9	89.7	.09614	200.5	95.5	.08533	184.2
	170	77.8	.12851	263.9	87.7	.11454	235.2	95.2	.10374	212.7	101.3	.09534	195.4
	186	84.7	.14306	285.8	95.4	.12774	255.7	103.5	.11576	231.4	110.2	.10647	212.0
	198	90.6	.15630	305.2	101.9	.13979	272.9	110.5	.12662	247.3	117.5	.11646	227.5
	208	95.3	.16770	320.9	107.1	.14980	287.0	116.0	.15591	260.1	123.4	.12500	239.5
762 MM	121	47.2	.09482	208.6	54.2	.08607	184.1	60.1	.07623	167.9	64.5	.06971	154.0
	134	53.5	.10692	231.3	61.2	.09453	204.3	67.7	.08595	186.4	72.6	.07891	171.1
	148	59.9	.11999	255.0	68.3	.10641	225.8	75.4	.09769	205.9	81.0	.08920	189.2
	160	65.3	.13216	274.9	74.5	.11726	244.4	82.1	.10698	223.1	88.1	.09837	205.2
	170	69.6	.14165	290.9	79.5	.12606	258.9	87.2	.11517	236.5	93.5	.10617	217.5
	186	76.0	.15794	315.7	86.4	.14083	281.3	95.0	.12844	257.1	101.8	.11841	236.7
	198	81.4	.17208	336.5	92.4	.15377	300.3	101.4	.14055	274.4	108.0	.12951	252.9
	208	85.6	.18467	353.3	97.2	.16493	315.6	106.6	.15061	288.5	114.1	.13904	266.0

Table 12a Steam tracing data (Metric units continued)

Design Conditions:

Tracer: One 22 MM O.D. tracer thermally bonded to pipe with a TFK-7 channel and heat transfer cement.
Insulation Thermal Conductivity: See Chart, page 79
Ambient: 37.8° C, no wind

Nomenclature:

TS = Steam Temperature (°C)
TP = Pipe Temperature (°C)
GM/SEC-M = Tracer condensate rate per linear metre
Q3 = Pipe Heat Losses in Watts/lin m

PIPE SIZE	Steam Temp °C	INSULATION THICKNESS											
		38 MM Nom			51 MM Nom			64 MM Nom			76 MM Nom		
	TS	TP	GM/SEC-M	Q3	TP	GM/SEC-M	Q3	TP	GM/SEC-M	Q3	TP	GM/SEC-M	Q3
114 MM	121	110.5	.02694	59.3	111.0	.02268	50.1	112.3	.02018	44.5	112.8	.01827	40.3
	134	123.1	.03296	71.3	124.5	.02779	60.1	125.3	.02465	53.5	125.9	.02236	48.4
	148	136.0	.03943	83.8	137.6	.03329	70.6	138.5	.02957	62.8	139.2	.02679	50.9
	160	147.3	.04501	95.0	149.1	.03846	80.2	150.1	.03420	71.3	150.9	.03101	64.6
	170	155.9	.05061	103.8	157.9	.04271	87.6	159.0	.03801	77.9	159.9	.03443	70.6
	186	169.1	.05878	117.5	171.2	.04964	99.2	172.5	.04418	88.3	173.5	.03999	80.0
	198	180.1	.06617	129.2	182.4	.05593	109.2	188.9	.04876	97.2	184.9	.04510	88.1
	208	188.9	.07254	138.8	191.4	.06131	117.3	192.9	.05455	104.5	194.1	.04945	94.7
168 MM	121	108.2	.03114	68.8	109.5	.02686	59.2	110.6	.02315	51.1	111.3	.02104	46.4
	134	120.4	.03806	82.5	121.9	.03280	71.0	123.3	.02831	61.3	124.0	.02569	55.7
	148	132.8	.04588	96.9	134.6	.03923	83.4	136.2	.03394	72.0	137.1	.03085	65.5
	160	143.7	.05269	109.7	145.7	.04537	94.5	147.5	.03917	81.7	148.5	.03560	74.2
	170	152.0	.05836	119.8	154.2	.05035	103.2	156.1	.04348	89.2	157.2	.03955	81.1
	186	164.8	.06778	135.6	167.1	.05842	116.8	169.3	.05051	101.0	170.5	.04593	91.8
	198	175.1	.07631	149.0	177.9	.06575	128.4	180.3	.05685	111.0	181.6	.05174	101.1
	208	183.6	.08356	159.9	185.6	.07208	138.0	189.1	.06235	119.3	190.6	.05675	108.6
219 MM	121	105.9	.03508	77.3	107.4	.03035	67.0	108.0	.02697	59.4	109.4	.02460	54.3
	134	117.6	.04271	92.6	119.5	.03706	80.4	120.8	.03289	71.2	121.8	.03006	65.1
	148	129.6	.05113	108.0	131.7	.04439	94.3	133.4	.03932	83.6	134.4	.03602	76.4
	160	140.0	.05900	123.0	142.4	.05128	106.8	144.3	.04546	94.7	145.5	.04156	86.6
	170	148.0	.06544	134.2	150.6	.05677	116.5	152.6	.05036	103.3	154.0	.04608	94.5
	186	160.1	.07587	151.6	163.1	.06589	131.7	165.4	.05847	116.9	166.8	.05354	107.0
	198	170.2	.08530	166.6	173.5	.07416	144.8	176.0	.06579	128.5	177.6	.06019	117.6
	208	178.3	.09342	178.8	181.9	.08120	155.4	184.5	.07208	137.9	186.2	.06603	126.4
273 MM	121	102.5	.03618	84.3	104.3	.03359	74.2	105.8	.02996	66.0	106.9	.02715	59.8
	134	113.5	.04036	100.5	115.6	.04079	88.4	117.4	.03633	78.6	118.7	.03299	71.4
	148	124.9	.05560	118.1	127.4	.04896	104.0	129.4	.04361	92.6	131.0	.03953	84.0
	160	134.7	.06411	133.6	137.6	.05651	117.7	139.9	.05035	104.8	141.6	.04566	95.1
	170	142.3	.07107	145.7	145.3	.06260	128.4	147.8	.05575	114.4	149.7	.05062	103.8
	186	153.7	.08216	164.4	157.2	.07245	145.0	159.9	.06468	129.3	162.1	.05866	117.2
	198	163.2	.09234	180.3	167.0	.08149	159.1	170.1	.07269	141.9	172.4	.06680	128.9
	208	170.5	.10105	193.4	174.9	.08921	170.7	178.2	.07959	152.3	180.7	.07227	138.3
324 MM	121	97.2	.04032	89.0	99.9	.03516	77.5	101.8	.03123	69.0	103.3	.02839	62.5
	134	107.2	.04886	105.9	110.4	.04251	92.2	112.7	.03787	82.1	114.4	.03438	74.5
	148	117.5	.05824	123.8	121.1	.05078	107.8	123.8	.04534	96.1	125.9	.04105	87.2
	160	126.4	.06705	139.8	130.5	.05843	121.9	133.7	.05236	109.0	136.0	.04752	99.0
	170	133.4	.07446	152.7	137.8	.06493	133.1	141.1	.05790	118.8	143.6	.05265	107.9
	186	143.7	.08600	172.0	148.7	.07506	150.1	152.4	.06707	134.1	155.2	.06090	121.7
	198	152.4	.09643	188.5	157.7	.08428	164.6	161.8	.07532	147.1	164.9	.06843	133.8
	208	159.3	.10549	201.9	165.1	.09220	176.4	169.3	.08239	157.7	172.6	.07480	143.3

356 MM	121	92.4	.04134	91.3	95.4	.03643	80.2	97.7	.03259	72.0	99.5	.02973	65.5
	134	101.5	.05006	108.5	105.1	.04395	95.3	107.8	.03945	85.5	109.9	.03592	77.9
	148	110.9	.05956	126.6	115.1	.05246	111.3	118.2	.04713	100.0	120.7	.04288	91.1
	160	119.0	.06857	142.7	123.7	.06024	125.6	127.3	.05419	112.8	130.0	.04939	102.8
	170	125.2	.07570	155.3	130.6	.06668	136.7	134.2	.05979	122.7	137.1	.05460	111.9
	186	134.6	.08720	174.6	140.3	.07699	153.9	144.6	.06914	138.3	148.1	.06332	126.6
	198	142.4	.09779	191.2	148.7	.08629	168.5	153.5	.07784	152.0	157.1	.07108	138.8
	208	148.7	.10664	204.4	155.5	.09459	181.0	160.5	.08507	162.8	164.4	.07761	148.7
406 MM	121	86.7	.04058	89.6	90.1	.03613	79.6	92.7	.03257	71.9	94.8	.02994	66.0
	134	94.8	.04905	106.3	98.9	.04356	94.5	102.1	.03904	86.0	104.5	.03634	78.8
	148	103.2	.05826	123.8	108.0	.05217	110.7	111.5	.04721	100.2	114.4	.04326	91.9
	160	110.4	.06695	139.5	115.6	.05977	124.6	119.8	.05420	112.9	123.0	.04976	103.6
	170	116.0	.07896	151.5	121.7	.06607	135.5	126.1	.05979	122.6	129.5	.05501	112.8
	186	124.3	.08507	170.2	130.7	.07623	152.3	135.6	.06906	138.2	139.4	.06348	126.9
	198	131.4	.09545	186.0	138.2	.08831	166.6	143.5	.07737	151.1	147.7	.07110	138.8
	208	137.6	.10424	199.5	144.2	.09505	178.0	149.8	.08846	161.7	154.3	.07762	148.7
457 MM	121	85.0	.04296	94.7	88.5	.03819	84.3	91.2	.03469	76.4	93.3	.03188	70.2
	134	92.9	.05193	112.3	97.0	.04617	100.1	100.1	.04182	90.7	102.8	.03863	83.6
	148	100.9	.06150	130.7	105.7	.05496	116.6	109.5	.05009	106.2	112.4	.04605	97.7
	160	107.9	.07078	147.3	113.2	.06301	131.4	117.5	.05751	119.8	120.8	.05286	110.1
	170	113.2	.07791	159.8	119.1	.06991	143.3	123.5	.06350	130.2	127.1	.05833	119.7
	186	121.3	.08967	179.5	127.8	.08054	160.9	132.8	.07315	146.3	138.8	.06737	134.7
	198	125.0	10029	198.0	135.1	.09009	175.9	140.5	.08199	160.1	144.6	.07549	147.4
	208	133.8	.01946	209.5	140.8	.09832	188.1	146.6	.08949	171.3	151.2	.08239	157.7
508 MM	121	83.4	.04476	98.9	87.6	.04025	80.9	89.6	.03605	80.7	92.0	.03360	74.2
	134	90.9	.05417	117.2	95.2	.04864	105.4	98.4	.04422	95.9	101.2	.04080	86.5
	148	98.6	.06426	136.6	103.6	.05776	122.6	107.4	.05261	111.6	110.5	.04655	103.2
	160	105.5	.07423	154.4	110.9	.06632	136.2	115.3	.06058	126.3	118.7	.05582	116.2
	170	110.7	.08165	167.6	116.5	.07324	150.2	121.2	.06695	137.3	124.9	.06164	126.4
	186	118.4	.09412	188.1	124.9	.08434	168.7	130.1	.07716	154.2	134.2	.07102	142.1
	198	124.9	.10530	205.6	131.9	.09429	184.4	137.6	.08636	166.6	142.0	.07965	155.5
	208	130.0	.11478	219.7	137.5	.10298	197.0	143.5	.09419	180.2	145.3	.08688	160.3
610 MM	121	81.1	.04759	105.1	84.7	.04299	94.9	87.6	.03950	87.2	89.0	.03661	80.6
	134	88.3	.05745	124.6	92.6	.05235	113.3	95.9	.04774	103.5	98.5	.04407	95.6
	148	95.8	.06655	145.7	100.6	.06200	131.8	104.4	.05673	120.5	107.5	.05246	111.3
	160	102.2	.07881	164.0	107.6	.07121	148.3	111.8	.06500	135.6	115.4	.06034	125.5
	170	107.1	.08869	177.9	112.9	.07852	161.0	117.5	.07186	147.3	121.4	.06670	136.8
	186	114.5	.09987	199.6	120.9	.09032	180.8	126.0	.08276	165.4	130.3	.07688	153.8
	198	120.6	.11147	217.9	127.6	.10101	197.4	133.3	.09289	181.4	137.8	.08603	168.0
	208	125.5	.12146	232.7	132.9	.11026	211.0	138.9	.10137	194.0	143.8	.09390	179.7
762 MM	121	77.8	.05351	117.8	81.1	.04775	105.4	84.0	.04397	97.1	86.0	.04094	90.4
	134	84.5	.06493	140.5	88.3	.05753	124.7	91.7	.05312	114.9	94.7	.04938	107.1
	148	91.3	.07679	163.2	95.6	.06840	145.4	99.8	.06323	134.4	103.1	.05804	124.6
	160	97.1	.08760	182.7	102.2	.07887	164.1	106.7	.07250	151.2	110.3	.06731	140.3
	170	101.6	.09681	198.5	107.1	.08678	178.1	111.9	.08003	164.3	115.6	.07433	152.4
	186	108.2	.11074	221.3	114.5	.09997	199.6	119.8	.09218	184.3	124.2	.08549	171.0
	198	113.9	.12362	241.6	120.5	.11131	217.3	126.4	.10295	201.2	131.1	.09560	186.9
	208	118.4	.13492	258.2	125.4	.12116	232.1	131.7	.11225	214.8	136.7	.10432	199.6

Table 13 R_t factor for tracers thermally bonded to pipe.

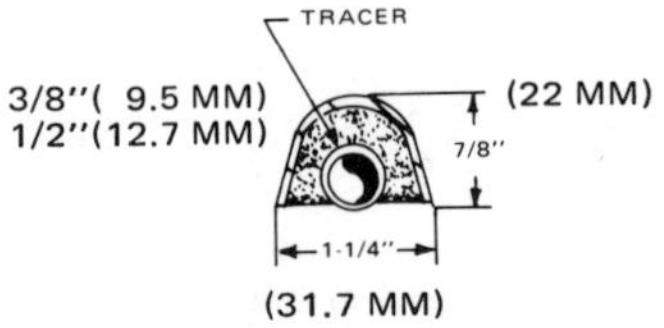

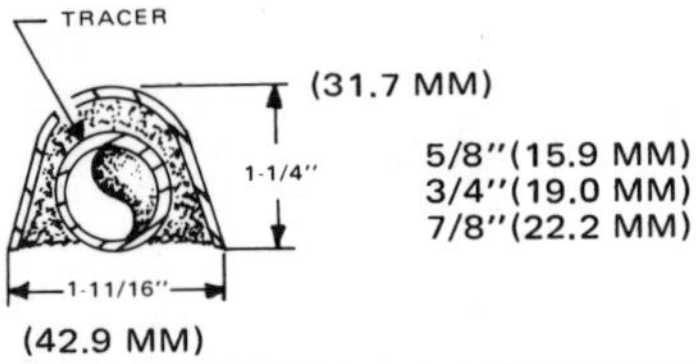

PROCESS PIPE SIZE		THERMAL BONDING OF TRACER (3/8"–1/2" tracer, 1-1/4" (31.7 MM) × 7/8" (22 MM))			THERMAL BONDING OF TRACER (5/8"–7/8" tracer, 1-11/16" (42.9 MM) × 1-1/4" (31.7 MM))		
NOM. INCH	MM	1 TRACER	2 TRACER	3 TRACER	1 TRACER	2 TRACER	3 TRACER
1-1/2	48	13.9 / 15.4	19.8 / 21.6	– / –	– / –	– / –	– / –
2	60	10.8 / 12.3	16.1 / 18.2	21.3 / 23.8	– / –	– / –	– / –
2-1/2	73	8.8 / 10.1	15.2 / 17.3	20.3 / 22.7	– / –	– / –	– / –
3	89	8.2 / 9.5	12.5 / 14.4	17.0 / 19.3	– / –	– / –	– / –
4	114	6.5 / 7.6	10.1 / 12.0	13.8 / 16.4	7.7 / 9.3	12.7 / 14.8	17.2 / 19.7
6	168	4.6 / 5.5	7.7 / 9.1	10.7 / 12.6	5.5 / 6.9	9.7 / 11.9	13.4 / 16.2
8	219	3.5 / 4.4	6.1 / 7.3	8.7 / 10.2	4.6 / 5.5	7.4 / 8.7	10.4 / 12.1
10	273	2.5 / 2.9	4.7 / 5.5	6.6 / 7.8	3.5 / 4.1	6.4 / 7.6	7.9 / 9.5
12	324	1.7 / 2.1	3.2 / 3.4	4.1 / 5.1	2.4 / 2.9	4.5 / 5.5	5.7 / 7.0
14	356	1.3 / 1.6	2.4 / 3.0	3.5 / 4.3	1.8 / 2.2	3.4 / 4.2	4.9 / 5.9
16	406	1.0 / 1.2	1.8 / 2.2	2.6 / 3.2	1.3 / 1.6	2.5 / 3.1	3.6 / 4.5
18	457	.9 / 1.1	1.7 / 2.0	2.4 / 2.9	1.2 / 1.5	2.3 / 2.8	3.3 / 4.1
20	508	.8 / 1.0	1.5 / 1.9	2.2 / 2.7	1.1 / 1.4	2.1 / 2.6	3.1 / 3.8
24	610	.7 / .9	1.4 / 1.7	2.0 / 2.4	1.0 / 1.3	1.9 / 2.4	2.8 / 3.4
30	762	.6 / .7	1.1 / 1.4	1.7 / 2.0	.9 / 1.0	1.6 / 2.0	2.3 / 2.9

Note: The upper figure is based on 1.5" (38 mm) thickness and the lower figure is based on 2" (51 mm) thickness of insulation of thermal conductivity shown in chart. A 25 mph (40 kph) wind was considered in making this table—no heat to process considered. (Cal. by "Thermon Mfg. Co.")

Stagnation method

Equations are quite complex so for ease of rapid determination a method, based on the assumption that no heat was flowing to or from the process material, was developed. In an equilibrium situation the heat loss from the pipe through the insulation must equal the heat delivered from the tracer. Under these conditions, when

Q_t = total loss per unit area of outside pipe insulation
U_t = total heat transference through the annulus air space, insulation, weather barrier and outside air film.
A = unit area of outside surface of insulation
t_m = process pipe temperature
t_a = ambient air temperature

then

$$Q_t = U_t A (t_m - t_a) \tag{48}$$

If f_{si} = conductance per unit area of insulation = U_a
$Q_t = f_{si} (t_m - t_a)$

Likewise, the heat transfer of tracer to pipe, when

Q_d = heat delivered from tracer to pipe
V_{tr} = heat transfer from tracer to pipe
A_{tr} = effective heat transfer area or tracer
t_{media} = media temperature
t_m = process pipe temperature
f_{tr} = conductance of tracer = $U_{tr} \times A_{tr}$

then

$$Q_d = f_{tr} (t_{media} - t_m)$$

However, under equilibrium conditions heat transfer from the tracer equals heat loss from outer surface of insulation.

Then

$$Q_t = Q_d$$

or

$$f_{si} (t_m - t_a) = f_{tr} (t_{media} - t_m)$$

Solving for t_m

$$t_m = t_a + \frac{f_{tr}}{f_{si}} (t_{media} - t_m)$$

and

$$t_m + \frac{f_{tr}}{f_{si}} t_m = t_a + \frac{f_{tr}}{f_{si}} t_{media}$$

$$t_m = \frac{t_a + \frac{f_{tr}}{f_{si}} t_{media}}{1 + \frac{f_{tr}}{f_{si}}} \tag{49}$$

The ratio of tracer conductance to the insulation conductance is designated by symbol R_t

$$\frac{t_m - t_a}{t_{media} - t_a} = \frac{f_{tr}}{f_{si}} = R_t$$

$$t_m = \frac{R_t\, t_{media} + t_a}{R + 1} \tag{50}$$

The values of R_t have been determined based on experimental data and have been tabulated in Table 13. Note the R_t factor is dimensionless and thus may be used for both English and Metric calculations. This table was calculated for two thicknesses of insulation, 1-1/2″ thick (38mm thick) and 2″ thick (51mm), nominal NPS pipe insulation.

Graphical solution — thermally bonded fluid system

The Equations 34 and 34a which were developed in 1956 made a number of assumptions for simplification. These assumptions were (1) the heat transfer coefficient (q_t) was a constant (2) the heat transfer coefficient of the air convection and radiation both inside and outside of the insulation was identical and constant (3) the thermal conductivity of the insulation was set a constant (4) the heat transfer of conduction of tracer was constant.

To consider all the above listed constants as variables, solution of equations require iterative simultaneous calculations. Such iterative solutions must be performed by high speed digital computer. To aid in the design of tracers, the Thermon Mfg. Co. has developed such an iterative technique and has plotted information for both English and Metric units.

As it is impossible to provide graphic solutions for all possible condition, the following graph, pages 120 to 133 was calculated for the most commonly used design conditions of outdoor traced lines. The ambient air temperature was taken to be 0° F (−17.8° C). Wind was assumed to be 25 mph (40 kph) and there was no allowance for any heat being added to the process medium. The thermal conductivity of the insulation was taken to be as listed on page 79. The graphs plot the effect of multiple tracers on the larger pipe sizes. These are shown on Graphs 1 through 14.

1.5" NOMINAL PIPE (48MM ACTUAL O.D.)
INSULATION 1-1/2" NOM. (38MM) THICK
AMBIENT AIR TEMPERATURE 0°F (—17.8°C), 25MPH(40KPH) WIND

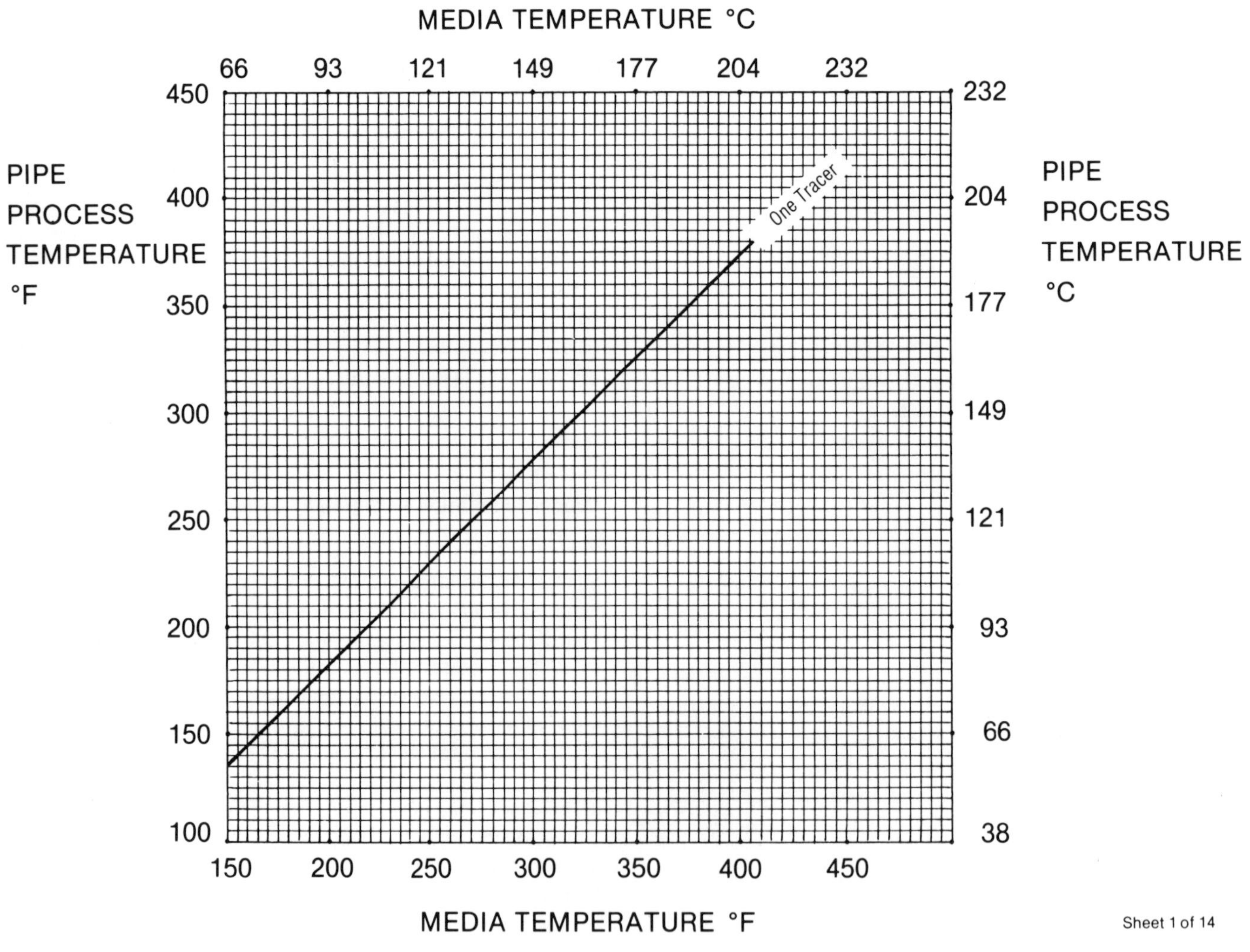

Sheet 1 of 14

CHART 1

2" NOMINAL PIPE (60MM ACTUAL O.D.)
INSULATION 1-1/2" NOM. (38MM) THICK
AMBIENT AIR TEMPERATURE 0°F (—17.8°C), 25MPH(40KPH) WIND

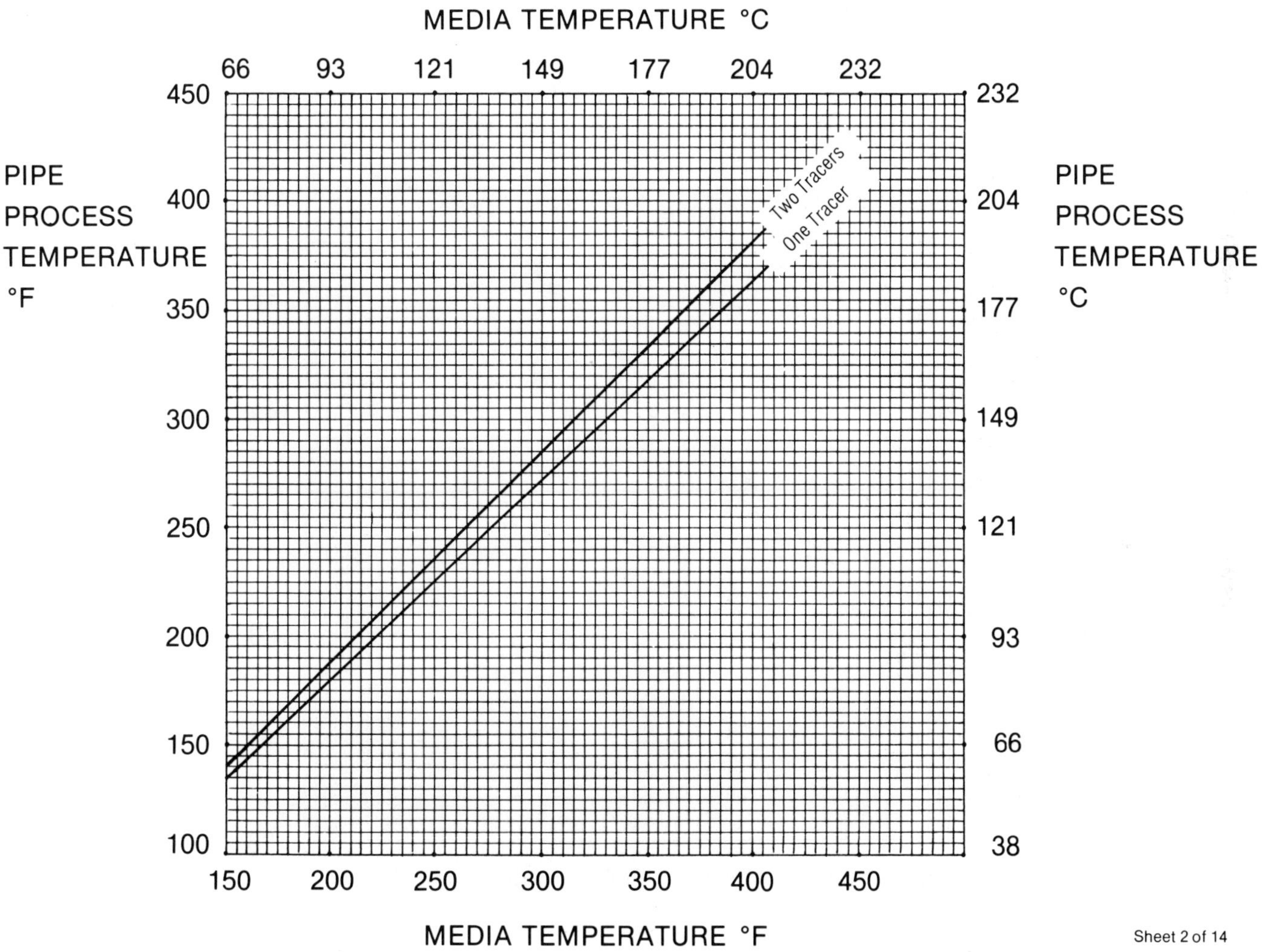

Sheet 2 of 14

CHART 2

3" NOMINAL PIPE (89MM ACTUAL O.D.)
INSULATION 1-1/2" NOM. (38MM) THICK
AMBIENT AIR TEMPERATURE 0 °F (—17.8 °C), 25MPH(40KPH) WIND

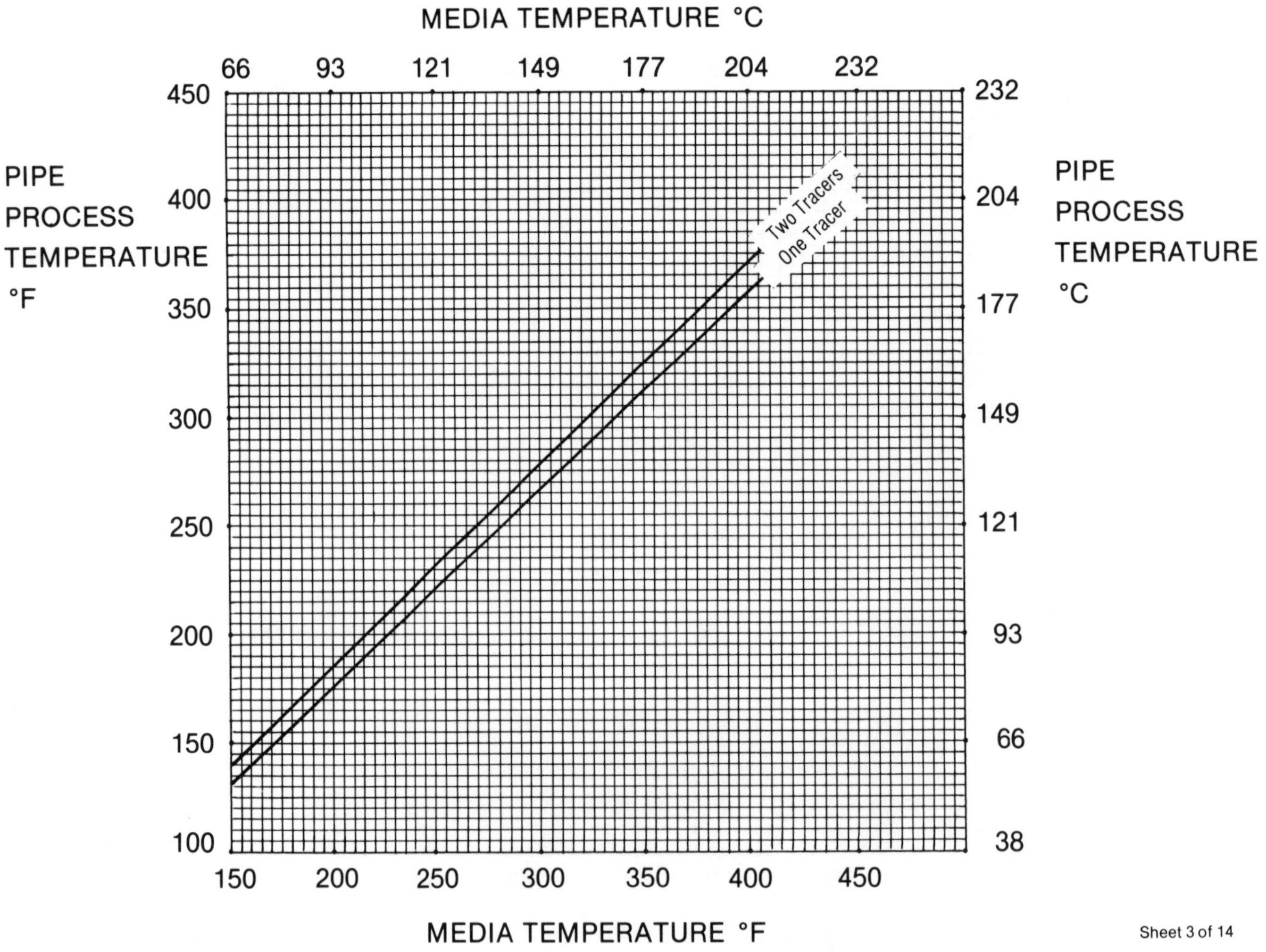

Sheet 3 of 14

CHART 3

4" NOMINAL PIPE (114MM ACTUAL O.D.)
INSULATION 1-1/2" NOM. (38MM) THICK
AMBIENT AIR TEMPERATURE 0°F (—17.8°C), 25MPH(40KPH) WIND

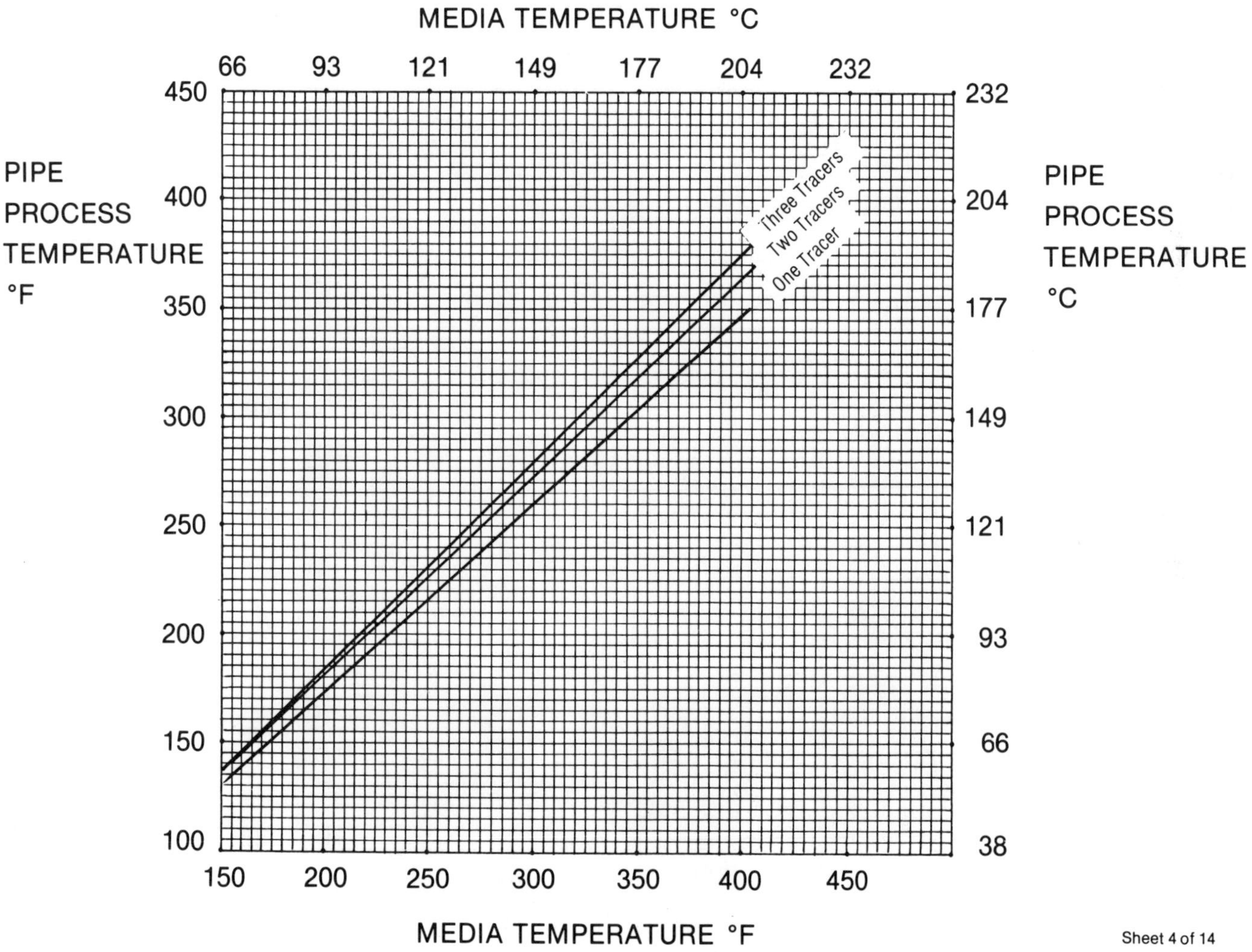

Sheet 4 of 14

CHART 4

6" NOMINAL PIPE (168MM ACTUAL O.D.)
INSULATION 1-1/2" NOM. (38MM) THICK
AMBIENT AIR TEMPERATURE 0°F (—17.8°C), 25MPH(40KPH) WIND

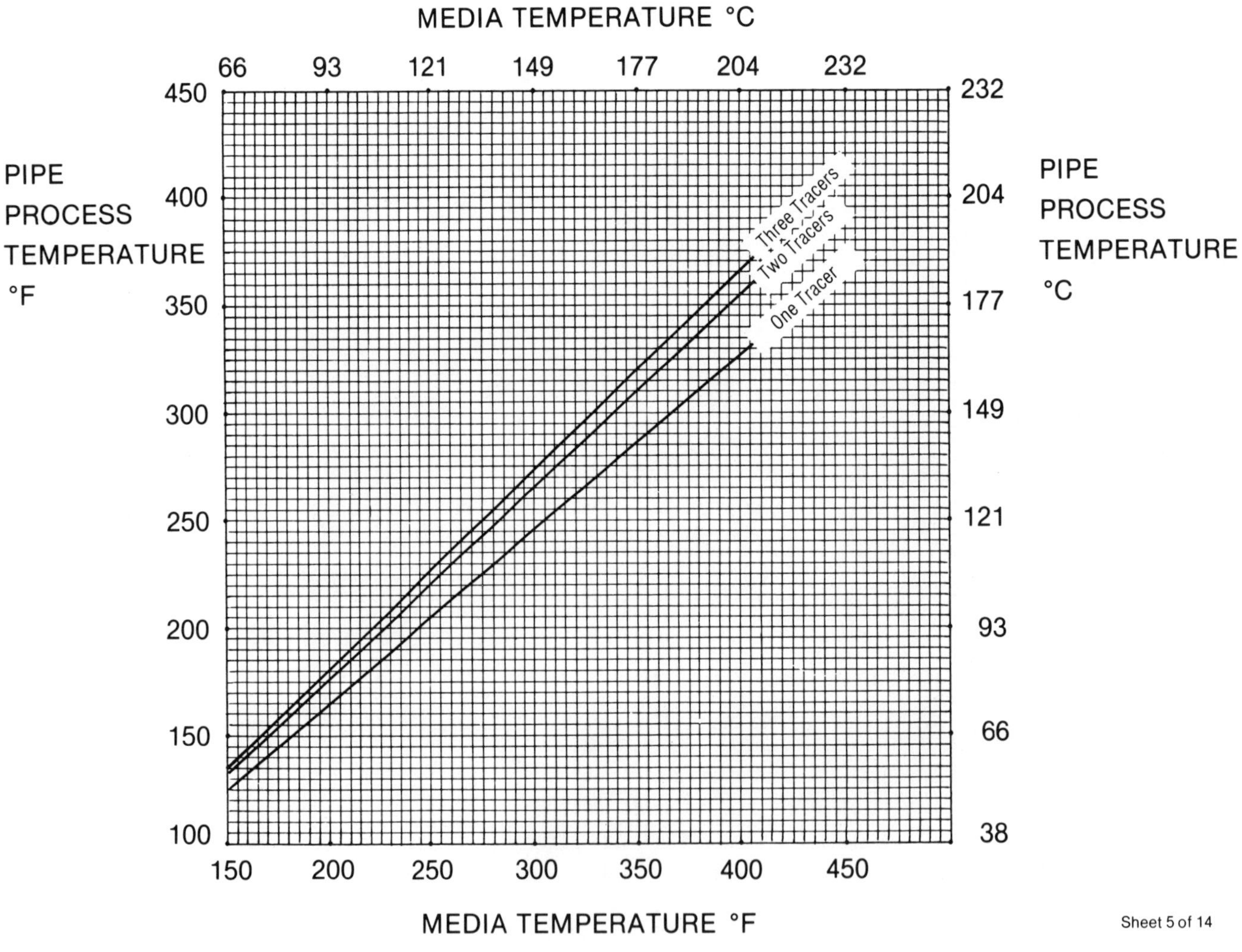

Sheet 5 of 14

CHART 5

8" NOMINAL PIPE (219MM ACTUAL O.D.)
INSULATION 1-1/2" NOM. (38MM) THICK
AMBIENT AIR TEMPERATURE 0°F (—17.8°C), 25MPH(40KPH) WIND

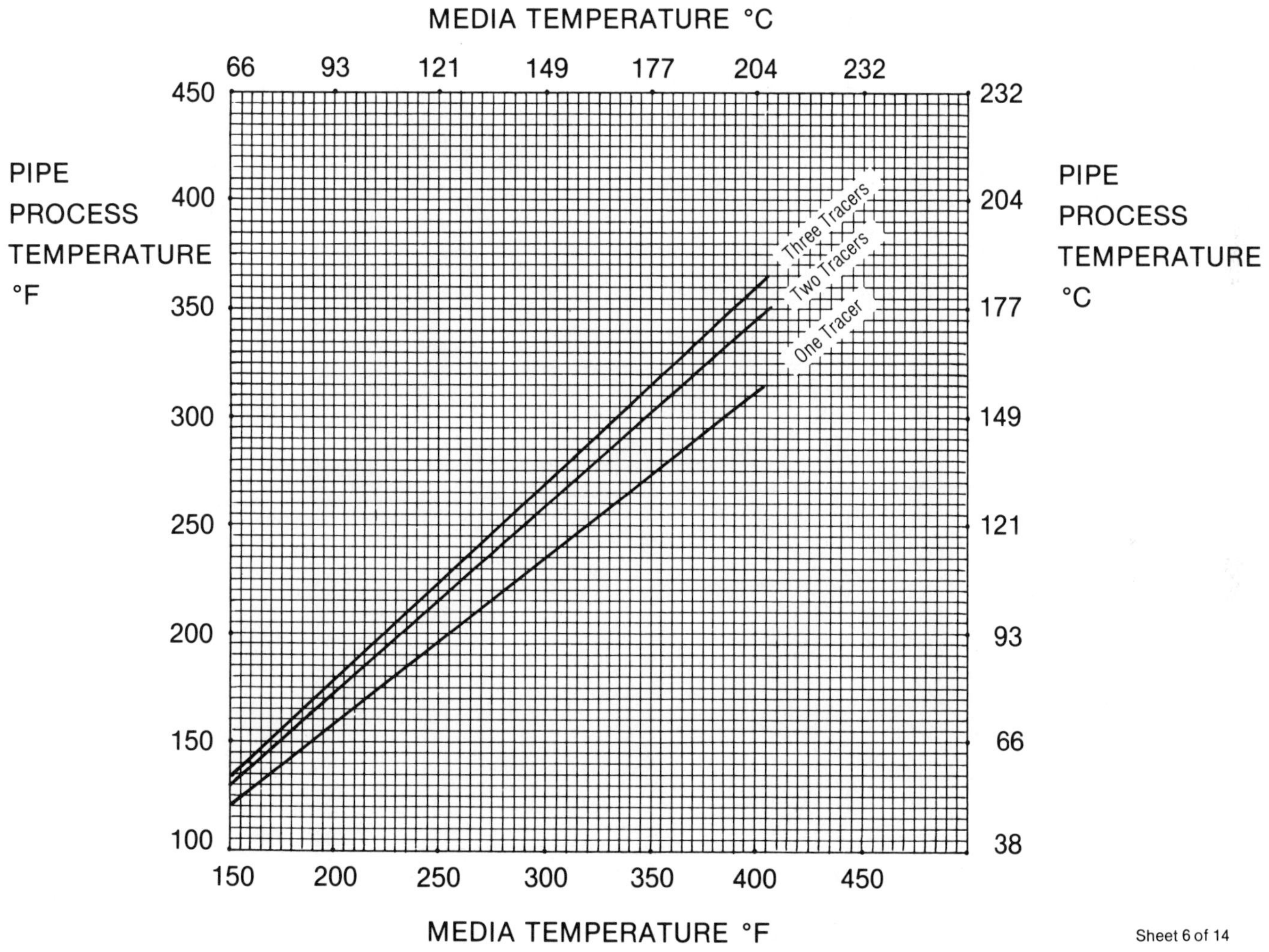

Sheet 6 of 14

CHART 6

10" NOMINAL PIPE (273MM ACTUAL O.D.)
INSULATION 1-1/2" NOM. (38MM) THICK
AMBIENT AIR TEMPERATURE 0°F (—17.8°C), 25MPH(40KPH) WIND

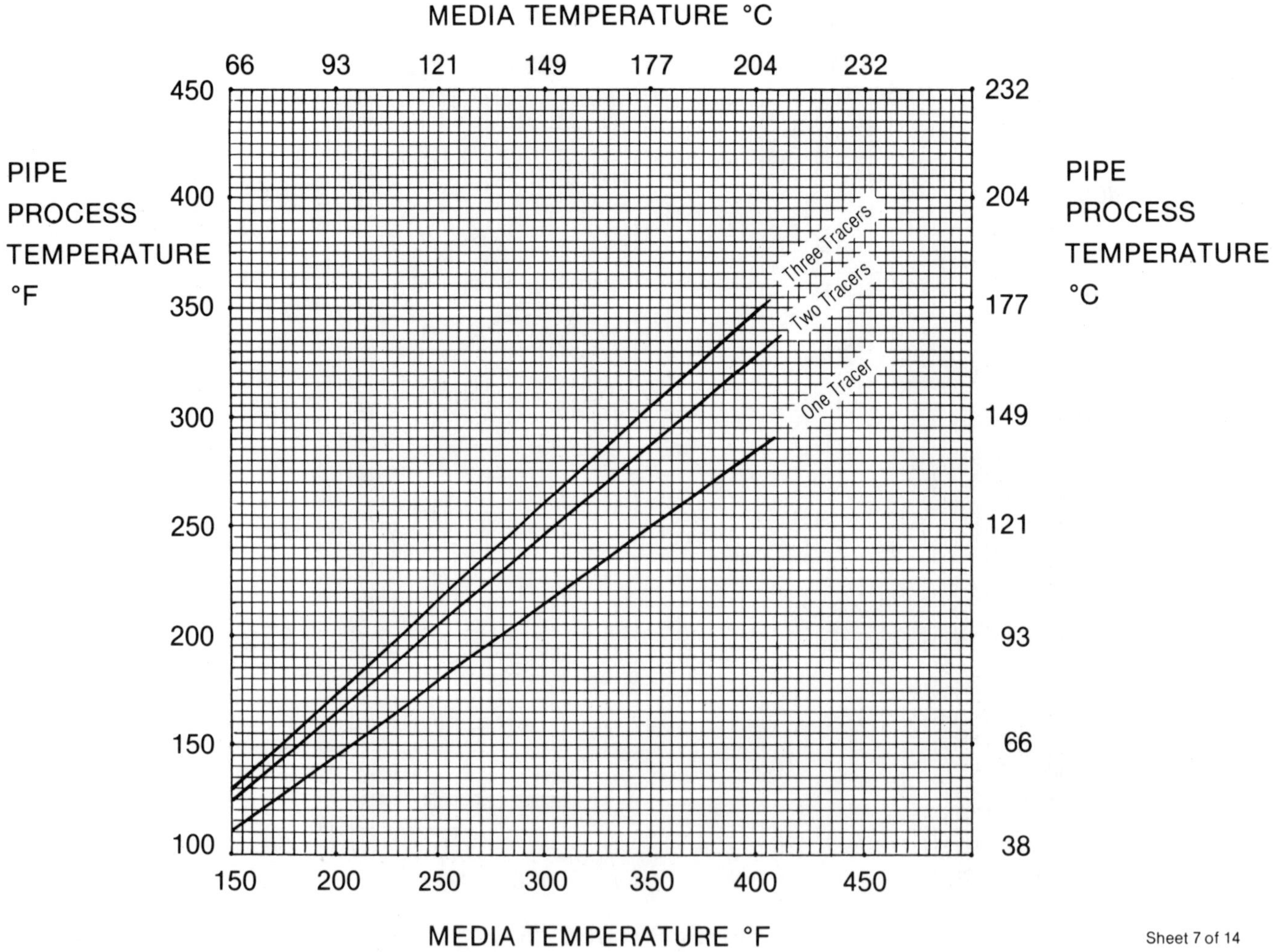

Sheet 7 of 14

CHART 7

12" NOMINAL PIPE (324MM ACTUAL O.D.)
INSULATION 1-1/2" NOM. (38MM) THICK
AMBIENT AIR TEMPERATURE 0°F (—17.8°C), 25MPH(40KPH) WIND

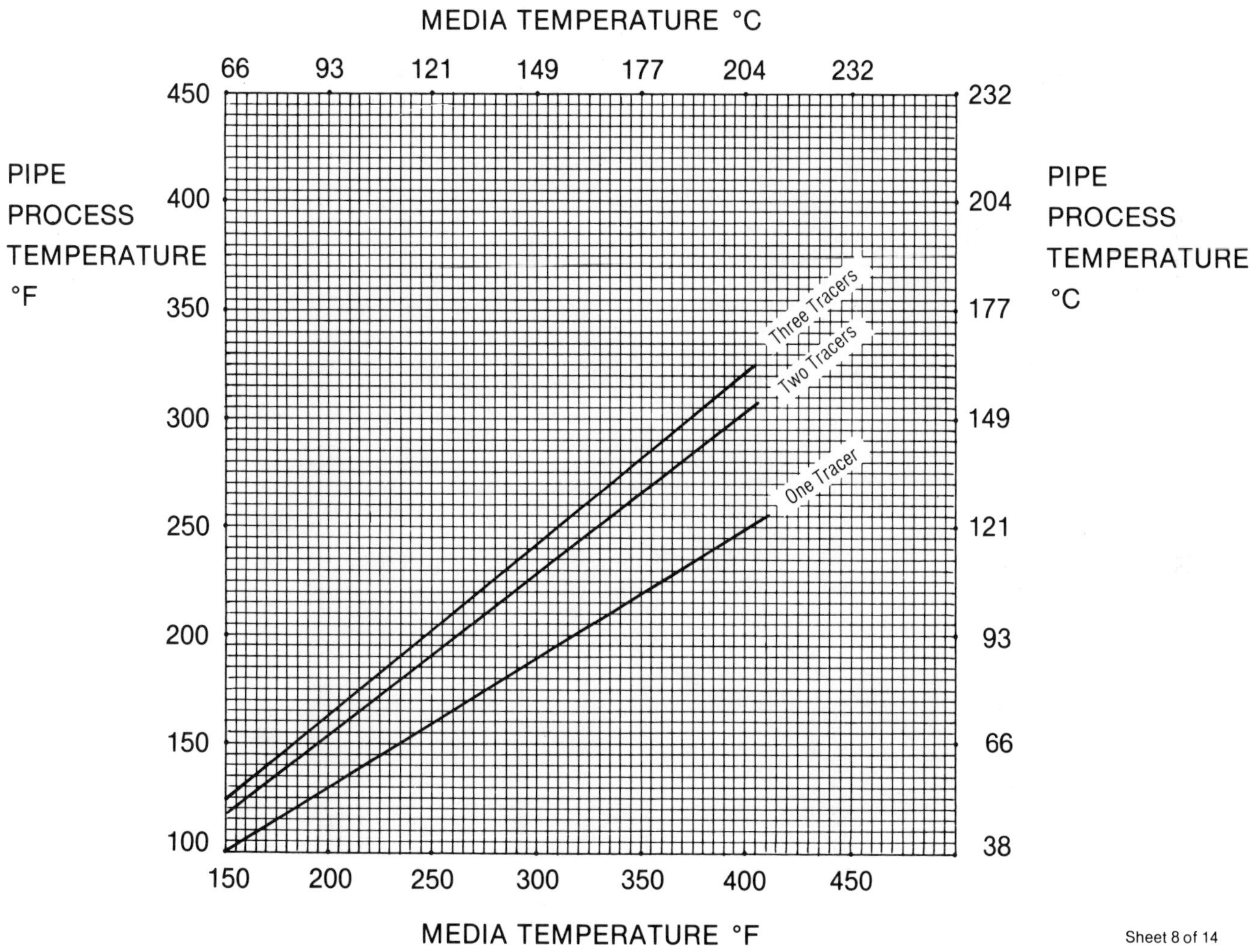

Sheet 8 of 14

CHART 8

14" NOMINAL PIPE (356MM ACTUAL O.D.)
INSULATION 1-1/2" NOM. (38MM) THICK
AMBIENT AIR TEMPERATURE 0°F (—17.8°C), 25MPH(40KPH) WIND

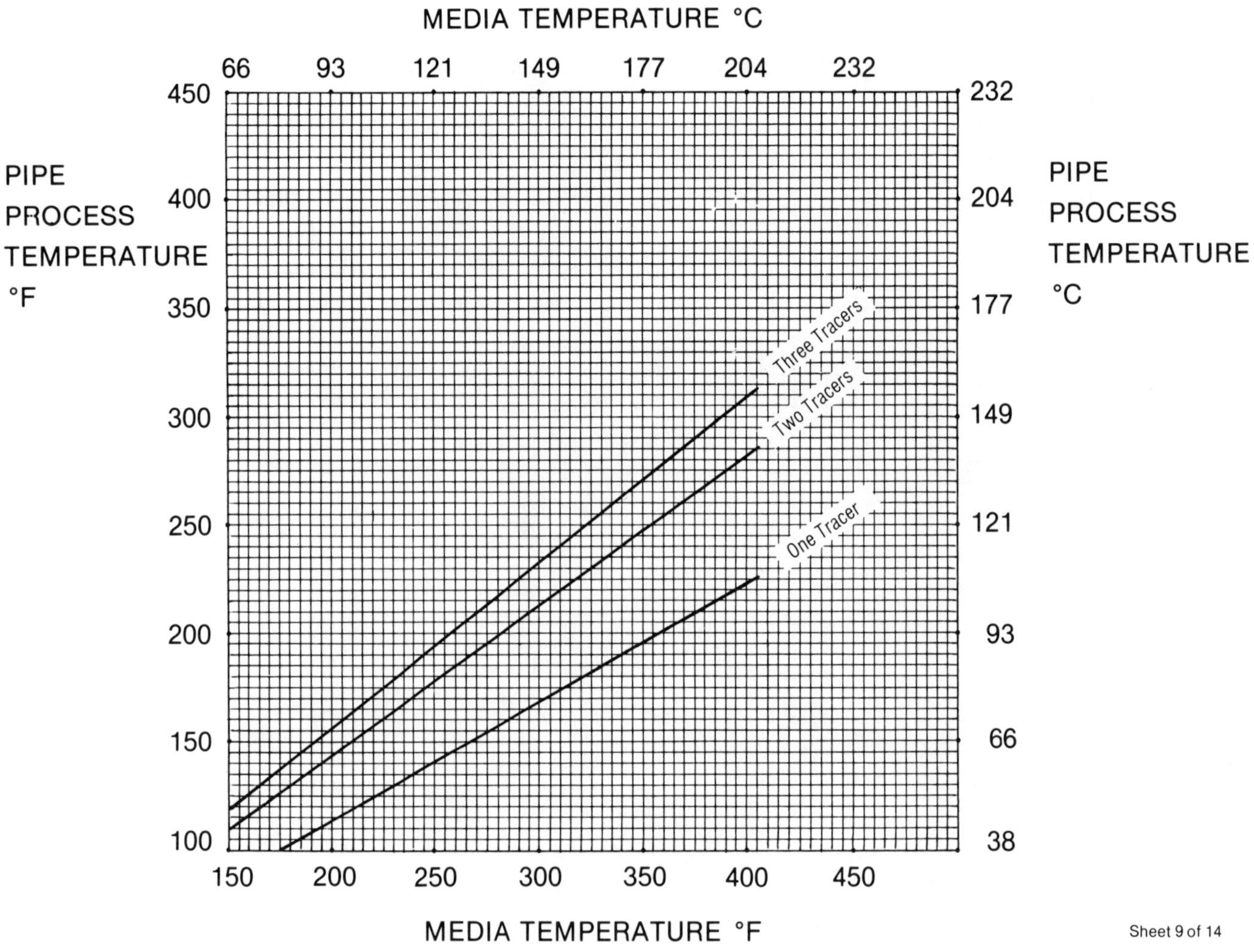

Sheet 9 of 14

CHART 9

16" NOMINAL PIPE (406MM ACTUAL O.D.)
INSULATION 1-1/2" NOM. (38MM) THICK
AMBIENT AIR TEMPERATURE 0°F (—17.8°C), 25MPH(40KPH) WIND

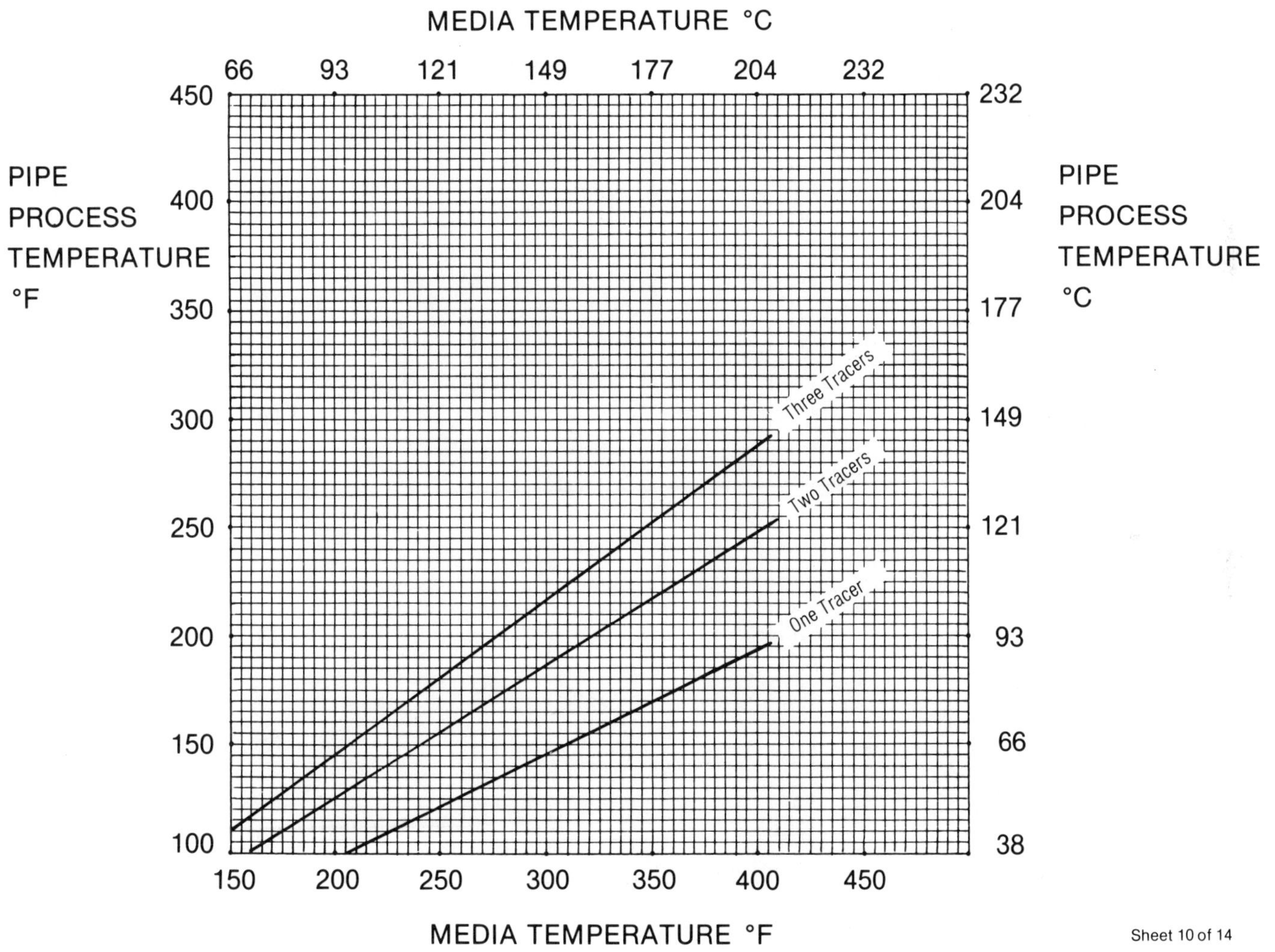

Sheet 10 of 14

CHART 10

18" NOMINAL PIPE (457MM ACTUAL O.D.)
INSULATION 1-1/2" NOM. (38MM) THICK
AMBIENT AIR TEMPERATURE 0°F (—17.8°C), 25MPH(40KPH) WIND

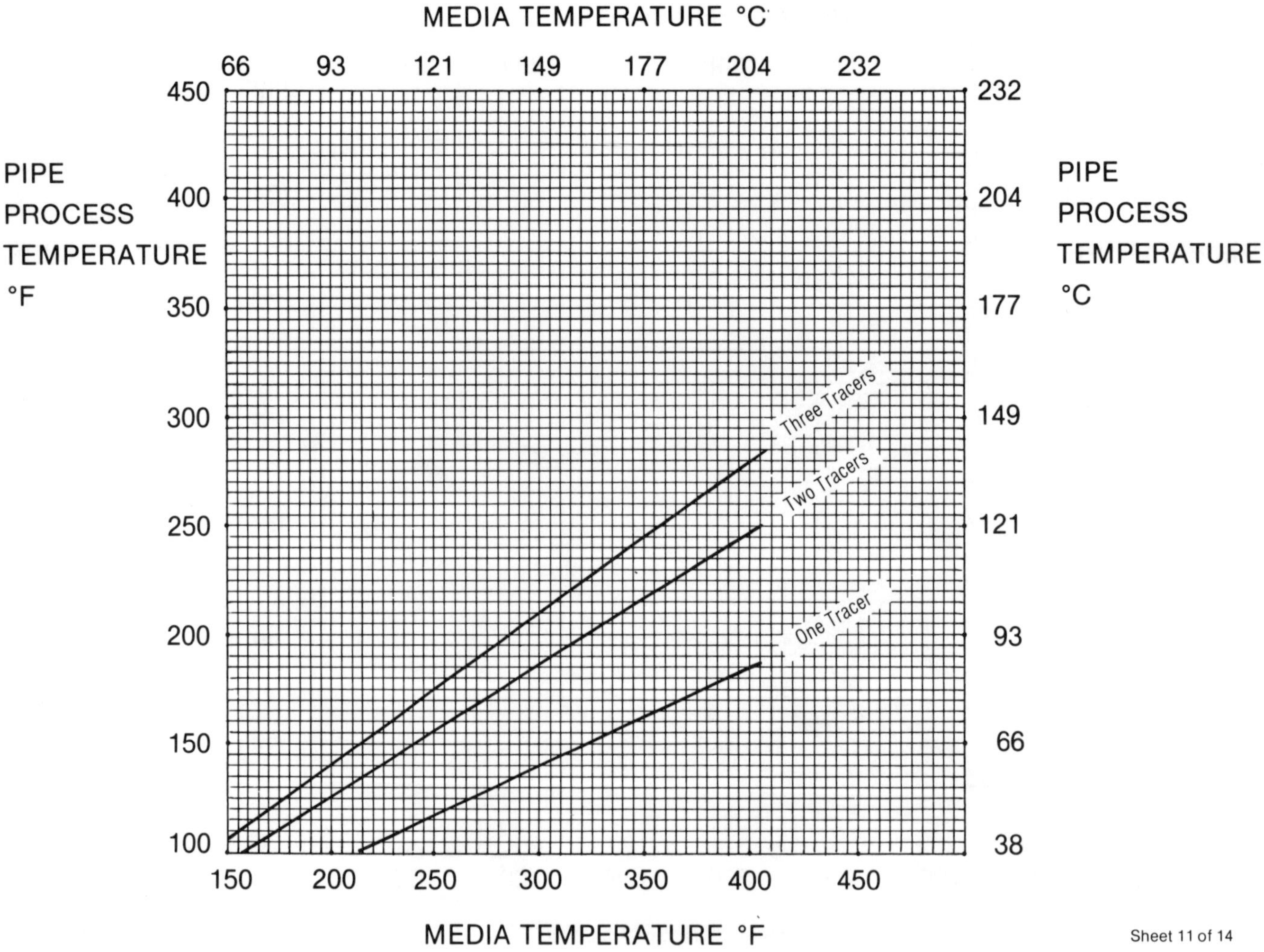

Sheet 11 of 14

CHART 11

20" NOMINAL PIPE (508MM ACTUAL O.D.)
INSULATION 1-1/2" NOM. (38MM) THICK
AMBIENT AIR TEMPERATURE 0°F (—17.8°C), 25MPH(40KPH) WIND

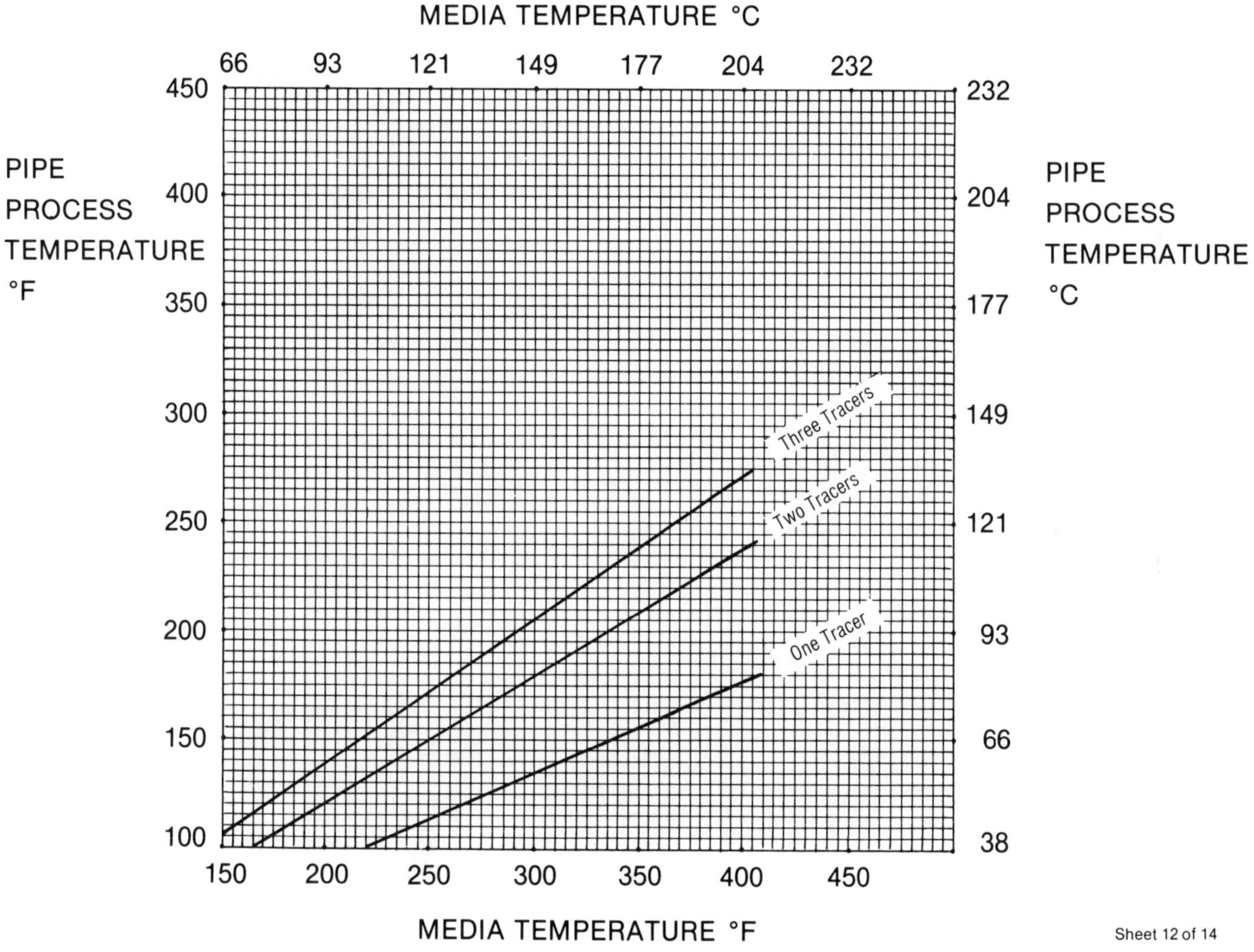

Sheet 12 of 14

CHART 12

24" NOMINAL PIPE (610MM ACTUAL O.D.)
INSULATION 1-1/2" NOM. (38MM) THICK
AMBIENT AIR TEMPERATURE 0 °F (—17.8 °C), 25MPH(40KPH) WIND

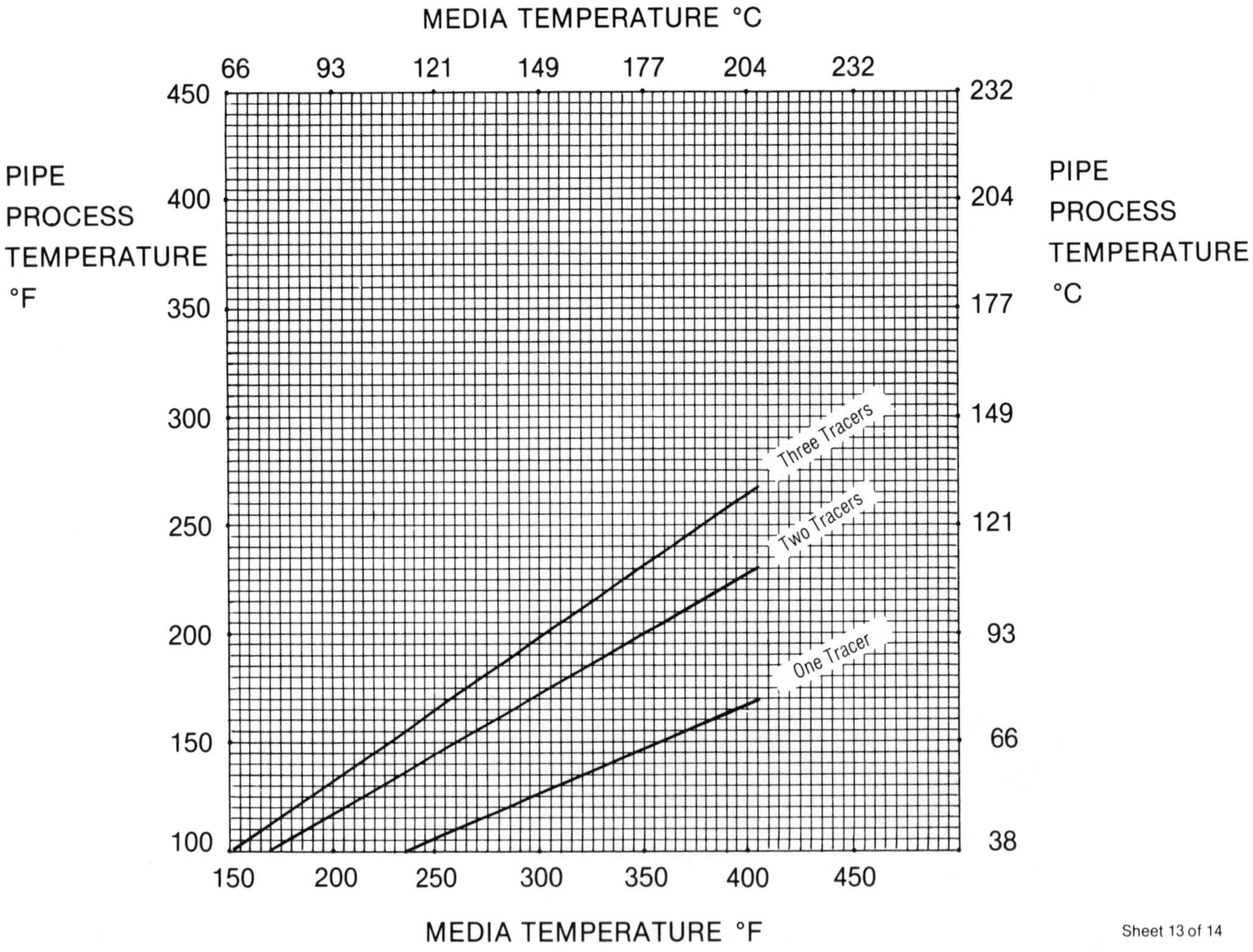

Sheet 13 of 14

CHART 13

30" NOMINAL PIPE (762MM ACTUAL O.D.)
INSULATION 1-1/2" NOM. (38MM) THICK
AMBIENT AIR TEMPERATURE 0 °F (—17.8 °C), 25MPH(40KPH) WIND

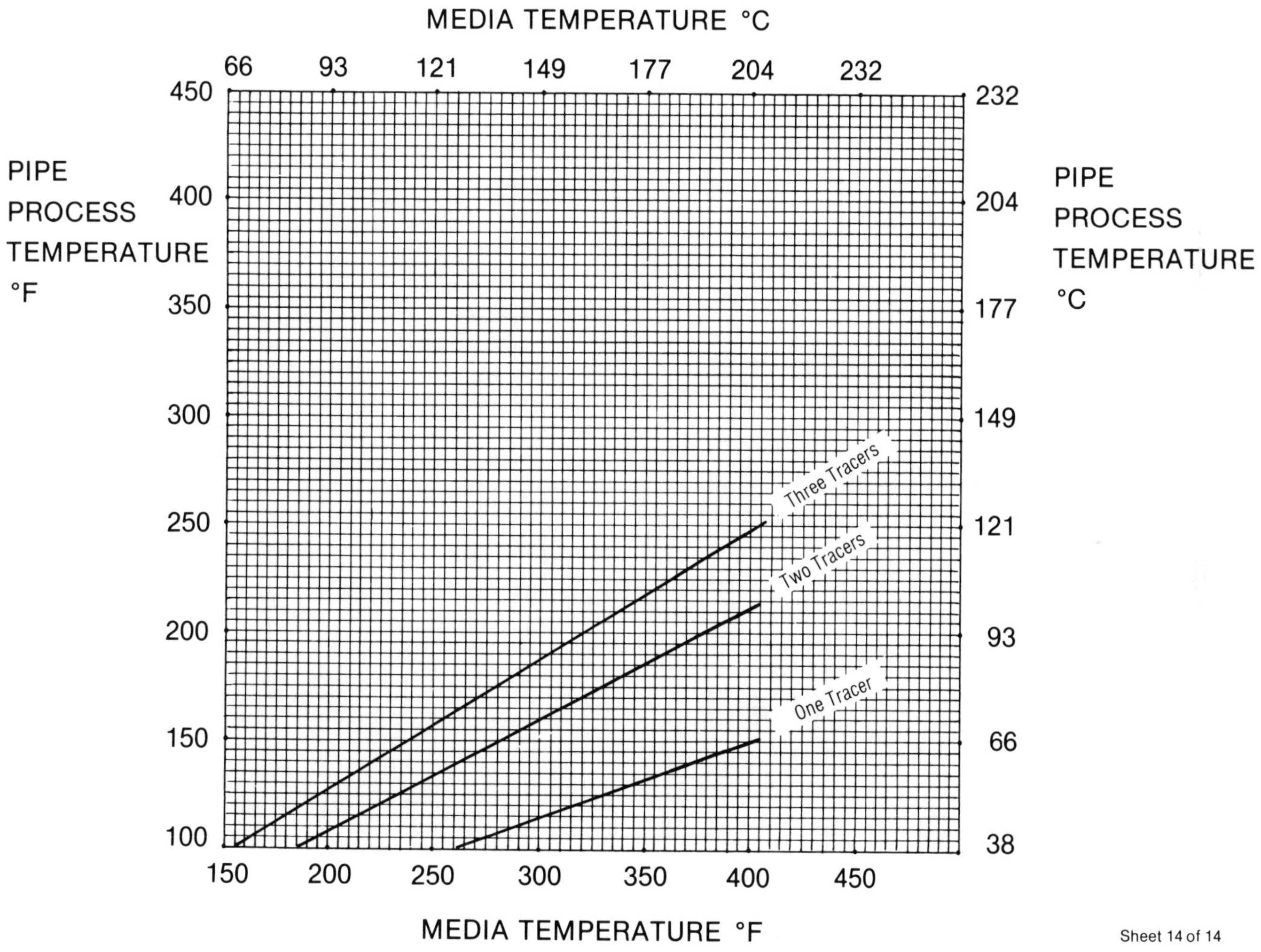

Sheet 14 of 14

CHART 14

■ EXAMPLE 39

English Units

Find the minimum media temperature necessary and number of tracers necessary to maintain 2″ NPS, 4″ NPS, 12″ NPS and 24″ NPS pipes at 220° F. From Graph 1 to maintain 220° F:

Pipe Size ″	Min Temp of Media °F	Number of Tracers
2	243	1
2	232	2
4	255	1
4	238	2
4	233	3
12	350	1
12	288	2
12	272	3
24	305	2
24	332	3

■

■ EXAMPLE 39a

Metric Units

Find the minimum media temperature necessary and number of tracers necessary to maintain 60mm, 114mm, 324mm and 610mm pipes at 104° C. From Graph 1 to maintain 104° C:

Pipe Size mm	Min Temp of Media °C	Number of Tracers
60	117	1
60	111	2
114	123	1
114	114	2
114	112	3
324	176	1
324	142	2
324	133	3
610	196	2
610	166	3

■

Tracers thermally bonded to pipe — electric systems

The various types of electric heating cables have been previously discussed. Of equal importance is the method by which the cables are attached to the process pipe or the equipment. For maximum efficiency in transferring the heat generated by the cable, heat transfer cements are utilized. Various types of heating cable attached to the process pipe with heat transfer cements are shown in Figure 30. In addition to the channel fill, types of heat transfer cements have been developed which are extruded onto the cable by the manufacturer. This pre-extruded cement and cable may simply be unrolled onto the process pipe, or equipment, covered with metal raceway and permanently fastened with either stainless steel banding or specially designed clamps.

A major reason for the use of heat transfer cements on electrical cables is to obtain good heat release from the cable at relatively low temperatures. The Heat Transfer coefficient (U_1) for cables which have been installed with heat transfer cements on metallic piping has been determined to be 30 Btu/hr-ft^2-°F (170 W/m^2 °C) which is approximately a magnitude of ten greater than that of a cable which is simply banded or wired to a metallic pipe. The knowledge of the heat transfer coefficient (U_1) permits prediction of cable operating temperatures as follows:

$$T_{sheath} = \frac{(q_{cable} \times C_1)}{U\,A_{cable}} + T_p \tag{51}$$

English Symbols		Metric Symbols
q_{cable}	= cable output W/ft	$q_{cable\ m}$ W/m
C_1	= Unit conversion 3.413	C_{im} = 1
U_1	= Heat trans. coefficient Btu/hr ft^2 °F	U_{im} = W/m^2
A_{cable}	= Surface area of cable ft^2/lin ft	$A_{cable\ m}$ = m^2/lin m
T_p	= Pipe temperature °F	T_{pm} = °C
T_{sheath}	= Cable sheath temperature °	$T_{sheath\ m}$ = °C

The difference between cable sheath and the actual resistive element (conductor) must also be considered in any design. This difference, however, is dependent upon the conductor material and size, cable type, and thickness and density of cable insulation. Typically, this information should be supplied by the cable manufacturer.

■ EXAMPLE 40

Determine the cable sheath temperature for a cable having a nominal output of 8 W/ft (26.24 W/m) which has a surface area of .065 ft^2/ft (.0198 m^2/m) when it is to operate on a pipe being held at 200° F (93.3° C) by a control thermostat.

English Units

$$T_{sheath} = \frac{8 \times 3.413}{30 \times 0.065} + 200$$

$$= 14 + 200 = 214°\,F$$

■

■ EXAMPLE 40a

Metric Units

$$T_{sheath\ m} = \frac{26.24 \times 1}{170 \times 0.0198} + 93.3$$

$$= 101.1\,°C$$

■

Note that if the U_1 value of 3 Btu/hr ft^2 °F (17 W/m^2 °C) had been used, the T sheath would be 340° F or 171° C. In many cases the

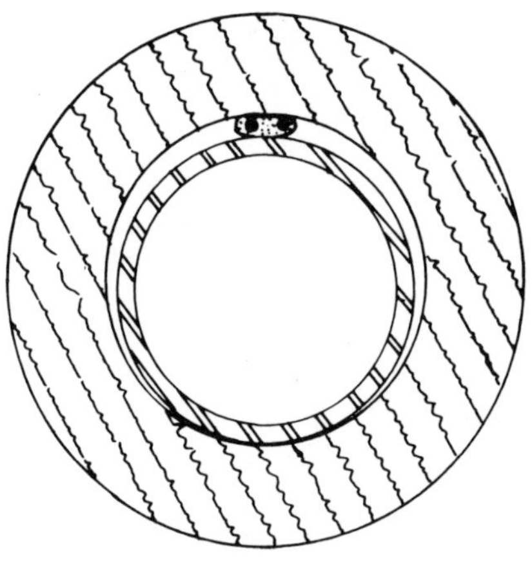

TEFLON INSULATED CABLE (COPPER OR NICKEL ALLOY SERIES RESISTANCE CONDUCTOR) WITHOUT HEAT TRANSFER CEMENT

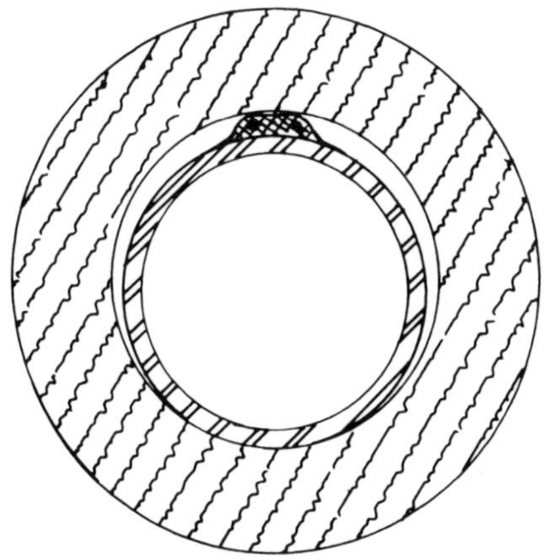

TEFLON INSULATED CABLE (COPPER OR NICKEL ALLOY SERIES RESISTANCE CONDUCTOR) COVERED WITH ALUMINUM TAPE

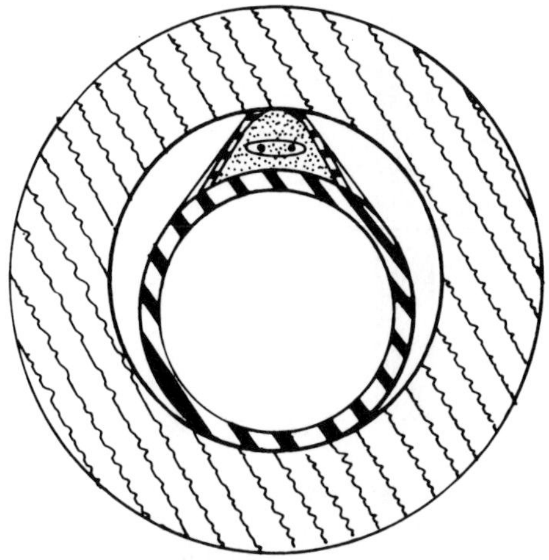

TEFLON INSULATED CABLE (COPPER OR NICKEL ALLOY SERIES RESISTANCE CONDUCTOR) WITH HEAT TRANSFER CEMENT AND CHANNEL

Figure 30 Electrical tracing systems for process pipe.

lack of heat transfer cement can cause the cable to operate at temperatures which exceed the cable's electrical insulation or conductor temperature limits.

Electrically heated pipes and vessels have one major advantage over fluid systems in that they can be controlled so as to obtain the temperature required. It is "on" only when the traced pipe line demands additional heat. The basic control system is "on-off", using thermostats to control the feed power to the tracer cable. However, an inherent disadvantage of "on-off" control is its effect of cycling.

By using a proportioning control system the deleterious effect of cycling can be eliminated. Basically it is an "on-off" control except with an anticipating circuit in addition, it automatically adjusts the ratio of power "on" and "off" over a given time cycle to maintain any desired temperature. This results in a straightening out of the temperature control curve giving a much closer and more accurate control.

Another control system is the stepless temperature control. This system uses a saturable core reactor arrangement and modulates the input power to heating cable according to demand, supplying more or less power to the load, but passing full power only during initial heating period. When control temperature is reached the heating cable receives only enough power to take care of heat loss.

External heating of equipment — fluid systems

Vessels and large pieces of equipment can be heated or cooled by pipe coils, tubing or sheet panels installed on their exterior. The thermal insulation is installed over these exterior heating or cooling mechanisms. Figure 31A illustrates the proper method of installing tubing on a vertical or horizontal vessel. The tubing should be installed in almost horizontal position with expansion loops between straight runs. Fluid flow should be downward. Also illustrated, Figure 31B is panel for fluid heating or cooling. These panels are simply two metal plates which have been either cast or resistance welded so that hollow passages (for fluid media) are present between the two plates. These panels are designed to fit the contour of the vessel. Typically, these panels are constructed of such materials as stainless steel, copper or carbon steel. The method of mounting these panels externally on the vessel is dependent on the pressure rating of the panel.

The principal advantage of panel or plate coils over external tubing tracer is the simplicity of the installation. The advantage of tubing is that it can be so installed that it provides more even distribution of heat over large surfaces. In either case the purpose of the external heater is to provide heat to the shell of vessel or equipment, and the use of heat transfer cements makes the external heaters more effective. For thermal bonding of panels or plate coils a non-hardening type of heat transfer cement is available. Tubing is thermally bonded in same manner as described for piping.

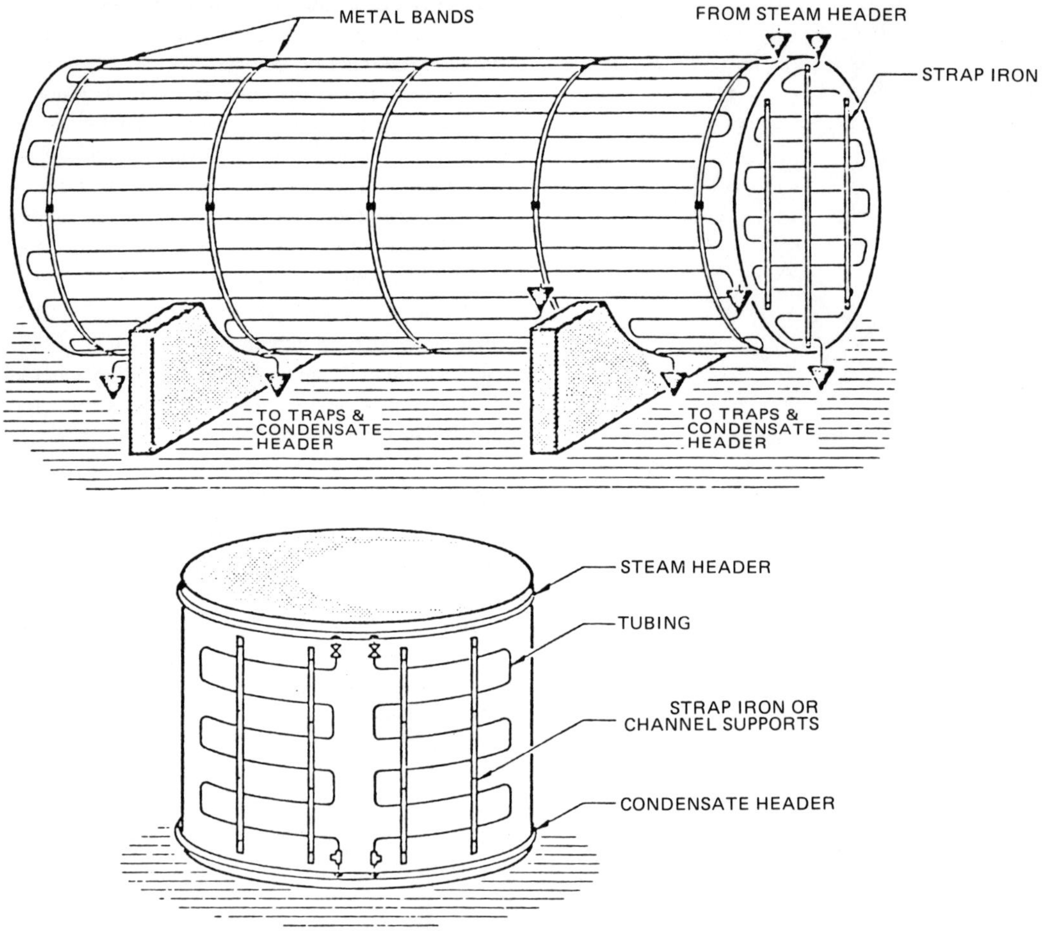

A

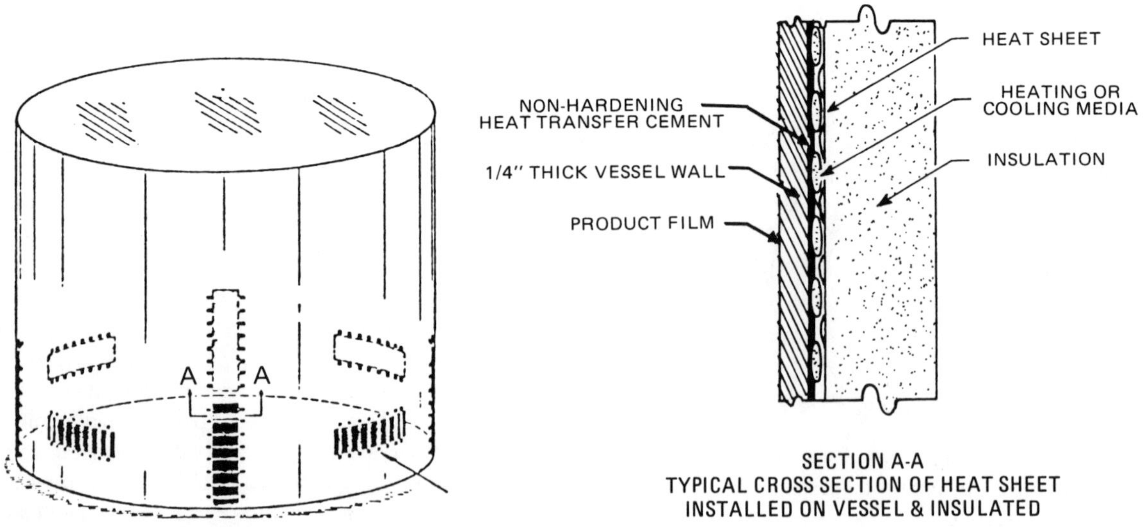

TYPICAL INSTALLATION ON VERTICAL VESSEL OVER 30 FT.

B

Figure 31 Fluid systems for external heating or cooling of vessels.

The method of calculating the heat loss for a vessel or irregular shaped object is to use the desired temperature of the equipment for t_1 and the ambient temperature for t_2. For insulated surfaces, heat loss may be calculated by Equation 11, 11a or equation 12a. Figure 37 is also presented as another means of determining heat loss for an insulated surface. For uninsulated portions of a vessel or irregular surface, Figure 36 may be used. Sizing of the proper heating or cooling system is a matter of providing at minimum the heat loss (or gain) calculated. For a fluid heating or cooling system, determination of the heat delivery or removal capability involves the calculation of the overall heat transfer coefficient. That is, the resistance to heat flow between the heating or cooling unit and the vessel contents must be evaluated. Tables 11 and 13 have been compiled to aid in this determination for a fluid panel applied externally to a vessel with non-hardening heat transfer cement. The basic equation for determining the overall heat transfer coefficient is:

where *General Equation, English or Metric Units*

U_t = overall heat transfer coefficient from media to product Btu/hr, ft^2 °F (W/m^2 °C)
U_1 = heat transfer coefficient from media to vessel wall Btu/hr, ft^2 °F (W/m^2 °C)
U_2 = heat transfer coefficient from outside vessel wall to inside vessel wall Btu/hrm ft^2, °F (W/m^2 °C)
U_3 = heat transfer coefficient from inside vessel wall to product Btu/hr, ft^2, °F, (W/m^2 °C)

$$\frac{1}{U_t} = \frac{1}{U_1} + \frac{1}{U_2} + \frac{1}{U_3} \qquad (52)$$

Having determined the overall heat transfer coefficient (U_t) of the thermally bonded panel, the area of heating or cooling surface may be determined by the following equation.

where *General Equation, English or Metric Units*

$Q_{delivered}$ = heat load required (for temperature maintenance this is simply the heat loss Btu/hr (W)
T_{media} = heating or cooling media temperature °F (°C)
t_1 = process temperature desired °F (°C)
U_t = overall heat transfer coefficient Btu/hr, ft^2, °F (W/m^2 – °C)
A = heat transfer surface area required ft^2 (m^2)

$$A = Q_{delivered}/U_t\,(T_{media} - t_1) \qquad (53)$$

Electrical Systems

Similar to fluid systems, equipment or vessels can be heated by electric cables, as described previously in this chapter, or by panels. The electrical cables should be installed and thermally bonded to vessel as shown in Figure 32A.

Vessels and large pieces of regular shape which require high heat input can use specially designed high watt density heating panels advantageously. These panels generally employ a Nichrome or Inconel resistance wire or ribbon for Joulian heating. The electrical insulation consists of glass, fabric and/or silicone rubber. It is possible to obtain panels which can operate at watt densities as high as 2.0 watts/inch^2. The maximum exposure temperature of the silicone rubber insulation is 450° F (232° C). The principal advantage of panel heaters is the relative ease with which they can be installed. Such is illustrated in Figure 32B.

For this particular example with saturated steam, the steam temperature at inlet would be 298° F (148° C) and be reduced to 293° F (145° C) at outlet when the steam circuit length is based on this criterion. Based on this criteria of not greater than 10% loss in pressure (English Units) the maximum trapping distances, for various sizes of tracers have been plotted on Graph No. 35 (English Units) and Graph No. 35A (Metric Units). These graphs were based on the Darcy equation and include safety factors to compensate for start up loads.

Heating or cooling requirements of vessels

The previous parts of this chapter relating to tracing provided information as to temperature which could be maintained in pipe and effectiveness of tracer or heating sheets or plate coils in adding or removing heat to equipment or vessels. In most instances it was assumed that no heat was removed or added to the process fluid. However, in many instances it may be required to heat or cool the contents of a vessel. This will influence the spacing and number of tracers required.

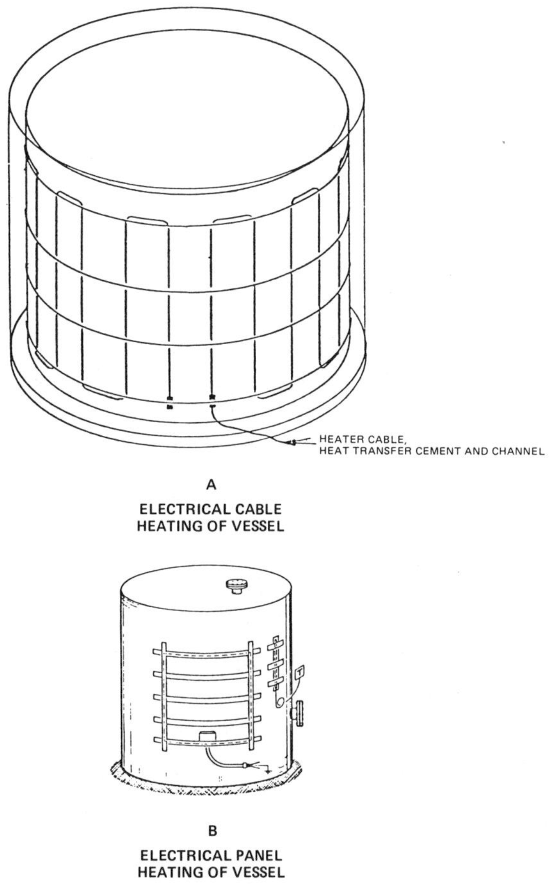

Figure 32 External electrical systems for heating of vessels.

The design of externally heated vessels involving heating or cooling loads is a matter of heat transfer balance. It involves balancing the required heat load against the available heat input from the external heaters.

To calculate the external tracer or panel heating or cooling requirements the following formulas are presented.

English Units

When Q_{tv} = Total heat load in Btu/hr
Q_{1v} = Heat input or removal, required to raise or lower the temperature of the contents of the vessel from initial to final temperature, in Btu/hr
Q_{2v} = Heat input or removal required to raise or lower the temperature of the vessel wall from initial to final temperature in Btu/hr
Q_{3v} = Heat input or removal, required to offset the heat loss or gain to the vessel wall from the ambient atmosphere (either uninsulated or insulated), in Btu/hr

Then

$$Q_{tv} = Q_{1v} + Q_{2v} + Q_{3v} \tag{54}$$

In *Metric Units*

when Q_{mtv}, Q_{m1v}, Q_{2v}, and Q_{m3v} are in watts (W)

$$Q_{mtv} = Q_{m1v} + Q_{m2v} + Q_{m3v} \tag{54a}$$

To determine the input or removal of heat required to raise or lower the temperature of the contents of the vessel it is necessary to know the volume of the process fluid, the specific heat of the fluid, the density of the fluid and the temperature increased or decreased. Thus when:

English Units

V = Volume of process fluid, in cubic feet
S_f = Specific heat of process fluid in Btu/lb, °F
W_f = Density of process fluid in lb per cubic foot
t_{2p} = Final process temperature °F
t_{1p} = Initial process temperature °F
C_y = Cycle time, hr
Q_{1v} = Heat input or removal, required to raise or lower the temperature of the contents of the vessel, in Btu/hr

Then

$$Q_{1v} = \frac{VS_f\, W_f\, (t_{2p} - t_{1p})}{C_y} \text{ in Btu/hr} \tag{55}$$

When:

In Metric Units

V_m = Volume of process fluid in cubic metres
S_{mf} = Specific heat of process fluid in J/Kg °C
W_{mf} = Density of process fluid in kg/m³
t_{m2p} = Final process temperature °C
t_{m1p} = Initial process temperature °C
C_{ym} = Cycle time, sec.
Q_{m1v} = Heat input or removal, required to raise or lower the temperature of the contents of the vessel from initial to final temperature, in W/m²

Then:

$$Q_{m1v} = \frac{V_m\, S_{mf}\, W_{mf}\, (t_{m2p} - t_{m1p})}{C_{ym}} \text{ in W/m}^2 \tag{55a}$$

To determine the heat input or removal to raise or lower the temperature of the vessel from initial to final temperature,

when,

in English Units

V_{st} = Total volume of metal contained in vessel, cubic ft
S_w = Specific heat of vessel metal in Btu/lb °F
W_w = Density of metal of vessel in lb/ft³
Q_{2v} = Heat input or removal, required to raise or lower the temperature of the vessel from initial to final temperature, in Btu/hr.

Then:

$$Q_{2v} = \frac{V_{st} S_w\, W_w\, (t_{2p} - t_{1p})}{C_y} \text{ in Btu/hr} \tag{56}$$

When:

In *Metric Units*

V_{stm} = Total volume of metal contained in vessel, cubic metres
S_{mw} = Specific heat of vessel metal in J/kg °C
W_{mw} = Density of metal in kg/m³
Q_{m2v} = Heat input or removal, required to raise or lower the temperature of the vessel from initial to final temperature, in W/m²

Then:

$$Q_{m2v} = \frac{V_{stm}\, S_{mw}\, W_{mw}\, (t_{m2p} - t_{m1p})}{C_{gm}} \tag{56a}$$

The remaining factor to obtain total heat input or removal is to determine the heat loss or gain from the surface of the vessel to the ambient air. When q_{tv} is the heat transfer per sq ft hr, this factor is multiplied by the outer surface area of insulation (if insulated) or vessel (if uninsulated).

The value q_{tv} (English Units) or q_m (Metric Units) is determined by Equation 11, 11a, 12, 12a, 13, 13a, 17, 17a, 27, 27a, 28, 28a, 29, 29a, 30, 30a, 36 or 37 depending upon installation and ambient conditions. When:

In English Units

D = Diameter of vessel in feet
H = Height of vessel in feet

Then:

$$Q_{3v} = q\left(\frac{2\pi D^2}{4} + \pi DH\right) \text{ in Btu/hr} \quad (57)$$

When:

In Metric Units

D_m = Diameter of vessel in metres
H_m = Height of vessel in metres

Then:

$$Q_m = q_m\left(\frac{2\pi D_m^2}{4} + \pi D_m H_m\right) \text{ in Watts per hr.} \quad (57a)$$

Combining the various parts making up the total heat load per hour.

In English Units

$$Q_{tv} = Q_{1v} + Q_{2v} + Q_{3v}$$

$$Q_{tv} = \frac{VS_f W_f (t_{2p} - t_{1p})}{C_y} + \frac{V_{st}S_w W_w (t_{2p} - t_{1p})}{C_y} + q\left(\frac{2\pi D2}{C} + \pi DH\right) \text{in Btu/hr} \quad (58)$$

In Metric Units

when Q_{mtu} is Watts per hour

$$Q_{mtv} = \frac{V_m S_{mf} W_{mf} (t_{m2p} - t_{m1p})}{C_{ym}} + \frac{V_{stm} S_{mw} W_{mw} (t_{m2p} - t_{m1p})}{C_{ym}} + q_m\left(\frac{2\pi D_m}{4} + \pi D_m H_m\right) \text{ in Watts per hour} \quad (58a)$$

Flat head vessels

Volume of the process fluid in vessel when full, is V

in *English Units*

When D = Diameter of vessel in feet
H = Height of vessel in feet

Then

$$V = \frac{\pi D^2 H}{C} \text{ in cubic feet} \quad (59)$$

V_m in *Metric Units*

When D_m = Diameter of vessel in metres
H_m = Height of vessel in metres

Then

$$V_m = \frac{\pi D_m^2 H_m}{4} \text{ in cubic metres} \quad (59a)$$

The volume of the metal of vessel wall and head, is V_{st}

In English Units

When t_w = Thickness of the vessel wall (and heads) in inches
D = Diameter of vessel in feet
H = Height of vessel in feet

Then

$$V_{st} = t_w\left(2\pi \frac{D^2}{4} + \pi DH\right) \quad (60)$$

V_{stm} in *Metric Units*

When t_{mw} = Thickness of the vessel wall in metres
D_m = Diameter of vessel in metres
H_m = Height of vessel in metres

$$V_{stm} = t_{mw}\left(2\pi \frac{D_m^2}{4} + \pi D_m H_m\right) \quad (60a)$$

The factors V, V_m, v, and v_m are used in Equations 58 and 58a but are listed separately due to the additional length of 58 or 58a if all basic units were listed in these equations.

To assist in calculation of heating or cooling requirements, the following tables give values of various terms used in the calculation of external heating and cooling systems.

Tables 14 and 14a give values of thermal conductivities k (k_m), specific weights W_n (W_{mn}) and specific heats S_w (S_{mw}) for metals:

Table 14 English Units / **Table 14a Metric Units**

	K Btu in./ft², hr, °F	Ww lb/ft³	Sw Btu/lb °F	Km W/mK	Wmn Kg/m³	Smw J/kg K
Mild Steel	312	490	0.10	45.0	7850	418.7
Wrought Iron	400	480	0.11	58.0	7690	460.6
Stainless Steel	93	500	0.12	13.4	8010	502.4
Copper	2600	556	0.10	375.0	8907	418.1
Aluminum	1400	165	0.23	201.9	2643	963.0

Tables 15 and 15a values of inside (of vessel) film coefficients:

Table 15 English Units / **Table 15a Metric Units**

Fluid	Film Coefficient Btu/ft², hr °F q_{3i}	Film Coefficient W/m²K q_{m3i}
Water	100 to 1500	570 to 8520
Oil	60 to 300	340 to 1700
Air	4 to 12	23 to 68

Note: The values given in Tables 15 and 15a are typical film coefficients. Film coefficients vary considerably with the degree of agitation, changing state, impurities, scale deposits, etc. For more exact information for determination of heat transfer for various chemicals under various conditions, see handbook on chemical engineering or heat transfer texts.

Tables 16 and 16a contain outside heating surface areas of tubes and pipes used as external tracers:

Table 16
English Units

Tracer or Tubing OD/in.	Area A ft^2/lin ft	Tracer Pipe NPS/in.	Area A ft^2/lin ft
1/4 (6.3 mm)	0.065	1/4 (13.7 mm)	0.141
3/8 (9.5 mm)	0.098	3/8 (17.1 mm)	0.177
1/2 (12.7 mm)	0.130	1/2 (21.3 mm)	0.220
5/8 (15.8 mm)	0.163	3/4 (26.7 mm)	0.275
3/4 (19.1 mm)	0.196	1 (33.4 mm)	0.344
7/8 (22.2 mm)	0.229	1 1/4 (42.2 mm)	0.435
1 (25.4 mm)	0.261	1 1/2 (48.3 mm)	0.498

Table 16a
Metric Units

Tracer Tubing OD	Area A m^2/lin m	Tracer Pipe NPS	Area A m^2/lin m
6.3 mm (1/4")	0.019	13.7 mm (1/4")	0.043
9.5 mm (3/8")	0.029	17.1 mm (3/8")	0.054
12.7 mm (1/2")	0.039	21.3 mm (1/2")	0.068
15.8 mm (5/8")	0.049	26.7 mm (3/4")	0.08
19.1 mm (3/4")	0.059	33.4 mm (1")	0.105
22.2 mm (7/8")	0.070	42.2 mm (1 1/4")	0.132
25.4 mm (1")	0.079	48.3 mm (1 1/2")	0.152

■ EXAMPLE 41

English Units

Given a low-carbon steel vessel handling a water solution, determine the heat input required per hour when the following conditions exist:

Diameter of vessel D = 8′ – 0″
Initial process temperature t_{ip} = 50° F
Final process temperature t_{fp} = 200° F
Density of water solution W_p = 60 lb/cu ft
Height of vessel H = 9′ – 0″
Ambient air temperature t_a = 20° F
Cycle time C_y = 4.0 hr
Specific heat of process fluids = 0.9 Btu/lb °F
Thickness of vessel wall t_w = 1/4″
Heating media temperature t_t = 298° F
Working level of product in vessel t_{ip} = 6′ – 0″
Insulation 3″ thick
Density of metal wall W_w = 490 lb/cu ft

STEP 1: From Equations 54, 55 and 59

$$Q_{tv} = Q_{1v} + Q_{2v} + Q_{3v}$$

and

$$Q_{1v} = \frac{V_{sf}\,W_f\,(t_{2p} - t_{1p})}{C_y} \text{ in Btu/hr}$$

Note: See Equation 59 for V_{st}.

$$Q_{1v} = \frac{\dfrac{\pi\, 8^2\,(6)}{4} \times 0.9 \times 60\,(200 - 50)}{4} = \frac{2450000}{4}$$

$$= 612{,}500 \text{ Btu/hr}$$

STEP 2: and

$$Q_{2v} = \frac{V\,S_w\,W_w\,(t_{sp} - t_{1p})}{C_y} \text{ in Btu/hr}$$

Note: See Equation 55 for V.

$$Q_{2v} = \frac{\dfrac{1/4}{12}\left[2\pi\,\dfrac{8^2}{4} + \pi \times 8 \times 9\right] 0.10 \times \left[490\,(200 - 50)\right]}{4}$$

$$= 12.500 \text{ Btu/hr}$$

STEP 3: also

$$Q_{3v} = q\left(2\pi\,\frac{D^2}{4} + \pi\,DH\right)$$

Note: Q the heat transfer through the insulation for accuracy should be calculated in accordance with equations previously given in this chapter. However, Charts 15 and 16 are furnished for estimating purposes. From Chart 16 the approximate heat transfer through 3″ thickness of insulation at given temperature difference at 200° F – 20° F = 180° F Δt the heat transfer through the insulation is approximately 23 Btu/ft², hr.

$$Q_{3v} = 44\left(2\pi\,\frac{8^2}{4} + \pi \times 8 \times 9\right) = 14400 \text{ Btu/hr}$$

$$Q_{tv} = Q_{1v} + Q_{2v} + Q_{3v}$$

$$= 612500 + 12500 + 14400 = 639.400 \text{ Btu/hr}$$

Chart 17 is furnished for estimating friction loss if water is used for heating or cooling media. ■

■ EXAMPLE 41a

Metric Units

Given a low-carbon steel vessel handling a water solution, determine the heat input required when the following conditions exist:

Diameter of vessel D_m = 2.438 metres
Initial process temperature t_{mip} = 10° C
Final process temperature t_{msp} = 93.3° C
Density of water solution W_{nf} = 961 kg/m³
Height of vessel H_m = 2.74 metres
Ambient air temperature t_{ma} = –6.7° C
Specific heat of process fluid S_{mf} = 3760 j/kg °K
Thickness of vessel wall t_{mw} = 0.00625 m
Heating media temperature t_{mt} = 148° C
Working level of product in vessel H_{mf} = 1.83 metres
Insulation thickness I_{mi} = 76.2 mm
Specific heat of vessel wall S_{mw} = 418.7 j/kg °K
Heat up time required = 14400 sec (4 hours)

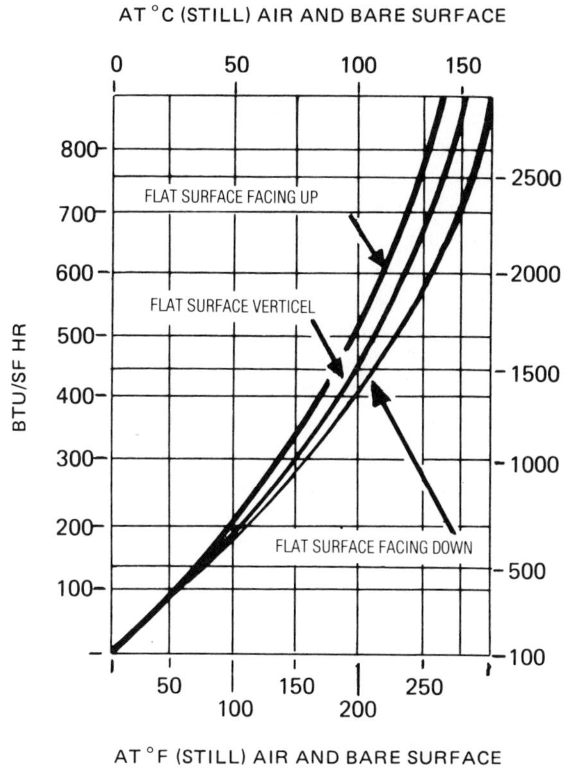

HEAT LOSS — Q
BARE SURFACE TO AIR
Chart 15

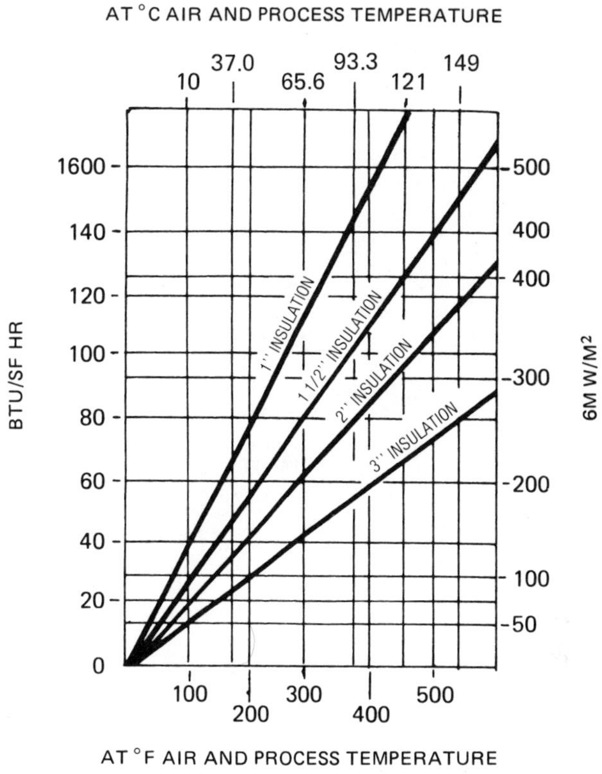

HEAT LOSS — Q
THROUGH INSULATION
Chart 16

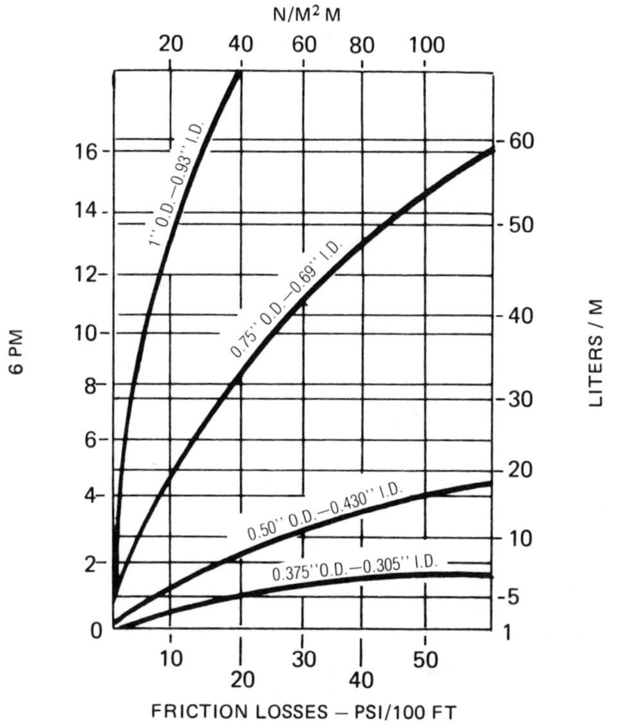

FRICTION LOSSES FOR WATER
FLOW IN COPPER TUBING
Chart 17

NOTE:

INSULATION INCHES	THICKNESS MM
1	25.4
1 1/2	38.1
2	50.8
3	76.2

COPPER TUBING OUTSIDE DIAMETER

Q.D. INCHES	Q.D. MM
0.375	9.5
0.500	12.7
0.750	19.05
1.000	25.4

STEP 1: From Equations 54a, 55a and 59a

$$Q_{mtv} = Q_{m1v} + Q_{m2v} + Q_{m3v}$$

and

$$Q_{m1v} = \frac{V_m\, S_{mf}\, W_{mf}\, (t_{m2p} - t_{m1p})}{C_{ym}} = W/hr$$

$$V_m = \frac{\pi\, D_m^2\, H_m}{4}$$

$$Q_{mv} = \frac{\frac{\pi\, 2.438^2\, (1.83)}{4} \times 3760 \times 961\, (93.3 - 10)}{14400}$$

$$= 178{,}577 \text{ Watts per hour}$$

STEP 2:

$$Q_{mv} = \frac{V_{stm}\, S_{mv}\, W_{mw}\, (t_{m2p} - t_{m1p})}{C_{ym}}$$

$$v_m = t_{mv}\left(2\pi\, \frac{D_m^2}{4} + \pi\, D_m\, H_m\right)$$

$$Q_{m2v} = \frac{t_{mv}\left(2\pi\, \frac{D_m^2}{4} + \pi\, D_m\, H_m\right) S_{mv}\, W_{mv}\, (t_{m2p} - t_{m1p})}{C_{ym}}$$

$$= \frac{0.00625\left(2\pi \frac{2.43^2}{4} + \pi 2.74 \times 2.43\right) \times 413.7 \times 7850 (93.3 - 10)}{144.00}$$

$$= 3{,}585 \text{ Watts per hour}$$

STEP 3:

$$Q_{m3v} = q\left(2\pi\, \frac{D^2}{4} + \pi\, D_m\, H_m\right)$$

$$= 139\,(9.36 + 20) = 4081 \text{ watts}$$

$$Q_{mtv} = 17877 + 3585 + 4081 = 186243 \text{ watts}$$

To check with results obtained in English Units, the conversion factor from W (per hour) to Btu/hr is 3.413.

$$186243 \times 3.413 = 635647 \text{ Btu/hr}$$

Knowing the heating or cooling requirements of vessels which are externally heated or cooled in the heat transfer capability of fluid or electrically heated tracers previously presented the system can be designed. ■

Tracer lengths — temperature or pressure drop fluids (steam)

In all the methods which have been outlined for the calculation of temperature maintained by the tracing systems, it has been assumed that the media temperature was a constant level. This is seldom true. That is, as the heating media gives up its energy to the process pipe and to the ambient surroundings its temperature is reduced. The amount of temperature drop will depend upon the type of media being used and the heat delivered or removed by the tracer. When the heat media is steam or vapor, the amount of pressure drop will be determined by the type of media, the type and size of tracer, and the fluid design flow rate (which is based on the heat delivered or removal rate).

For liquid media heating or cooling systems, heat transfer by the tracer involves the addition or removal of "sensible" heat from the media. Thus, the liquid system temperature drops proportionately as heat is removed. The temperature drop in the liquid system is given in the following relationship:

English Units: when,

t_t = temperature of heating (or cooling) media, in °F
t_{out} = outlet temperature of heating (or cooling) media, in °F
Q_{tp} = heat delivered (or gained) by a single tracer in Btu/ft, hr
L = length of tracer in feet
P_m = density of heating (or cooling) media, lb/ft³
S_{hm} = specific heat of heating (or cooling) media, Btu/lb °F
F_l = flow rate of heating (or cooling) media, in gal per min

then:

$$t_t - t_{out} = \frac{q_{tp} \times L}{8_{pm} \times 5_{hm} \times F_l} \quad (61)$$

Metric Units: when,

t_{mt} = temperature of heating or cooling media in °C
t_{mout} = outlet temperature of heating or cooling media in °C
q_{mtp} = heat delivered (or gained) by a single tracer in W/m
L_m = length of tracer in metres
P_{mm} = density of heating or cooling media in kg/m³
S_{mhm} = specific heat of heating or cooling media in J/kg °C
F_{m2} = flow rate of heating or cooling media in m³/sec

then:

$$t_{mt} - t_{mout} = \frac{q_{mtp} \times L_m}{P_{mm} \times S_{mhm} \times F_{ml}} \quad (61a)$$

Note: The above equations permit the calculation of flow rate of media required to hold the outlet temperature above some minimum value. Once the required flow rate is determined, the pressure drop of the media must be determined to assure that the design is practical. For determination of pressure drop, reference is made to fluid flow textbooks.

For steam media heating systems, it is standard practice to

design the tracer circuit length such that the steam pressure does not drop more than 10% of its initial steam pressure. This criterion would, for example, restrict a 50 psig (3.45×10^5 N/m^2) steam supplied tracer to be trapped and returned before the steam pressure dropped below 45 psig (3.1×10^5 N/m^2). Heat loss vs trap distances are shown in Charts 18 and 18a.

Heat Loss vs. Trap Distance
(English Units)
Trap Distance = $L_t \times C_f$

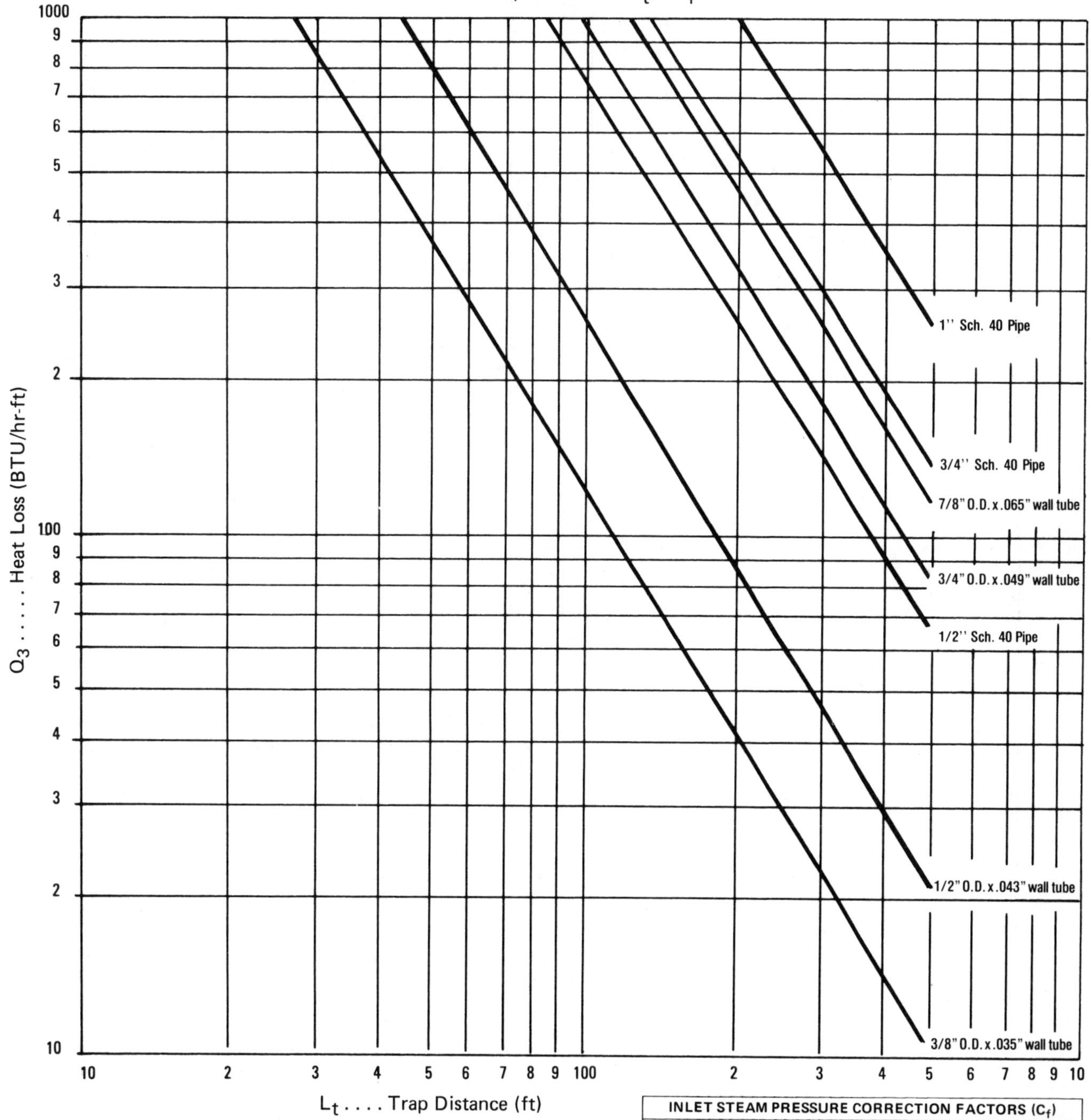

INLET STEAM PRESSURE CORRECTION FACTORS (C_f)		
Steam Pressure	Pipe Tracer	Tube Tracer
50 psig	1	1
100 psig	1.477	1.506
150 psig	1.869	1.93
200 psig	2.21	2.301
250 psig	2.51	2.633

Chart 18

Heat Loss vs. Trap Distance
(Metric Units)
Trap Distance = $L_{mt} \times C_f$

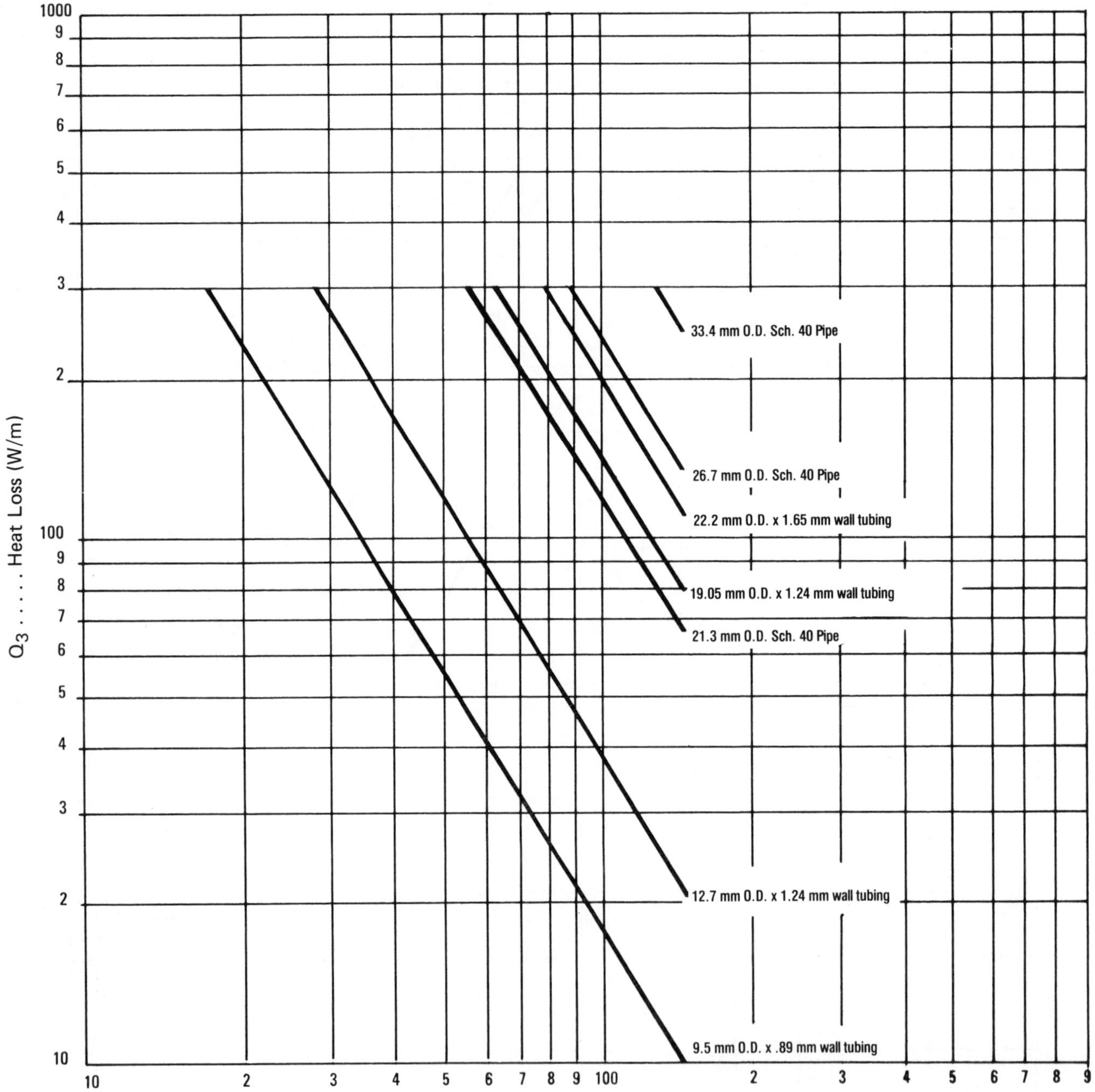

INLET STEAM PRESSURE CORRECTION FACTOR (C_f)		
Steam Pressure (n/m^2 Gauge)	Pipe Tracer	Tube Tracer
3.45×10^5	1	1
6.89×10^5	1.477	1.506
10.34×10^5	1.869	1.93
13.79×10^5	2.21	2.301
17.24×10^5	2.51	2.633

Chart 18a

Types of thermal bonding materials

The installation and service requirements imposed on heat transfer materials are many and varied. Heat transfer requirements may require that heat transfer cement be trowelled as shown in A, Figure 33. This is quite common for tracing of valves, fittings etc. In other instances a Fillet type of resin type heat transfer cement may be preferable. This is shown in B, Figure 33.

Where maximum thermal transfer is required, the heat transfer cement or molded strip may be installed in channels. This is shown in C, Figure 33. One, two, or three tracers may be needed to provide sufficient heating or cooling of process pipe as shown in D, Figure 33.

Plate coils and heater panels require heat transfer cement between their surface and the substrate. This is illustrated in E, Figure 33.

Tracers may be fluid tracers or a number of types of electrical heat tracers. The metal of tracers may be aluminum, copper or stainless steel (carbon steel not recommended) and the process pipe or equipment may be aluminum, copper, stainless steel or carbon steel. The temperature to be maintained and the ambient air temperature can be quite varied.

Because no single material can fulfill all conditions, the heat transfer cement (or strip) must be selected to function for the particular installation. Table 17 is provided as listing of characteristics of available heat transfer materials.

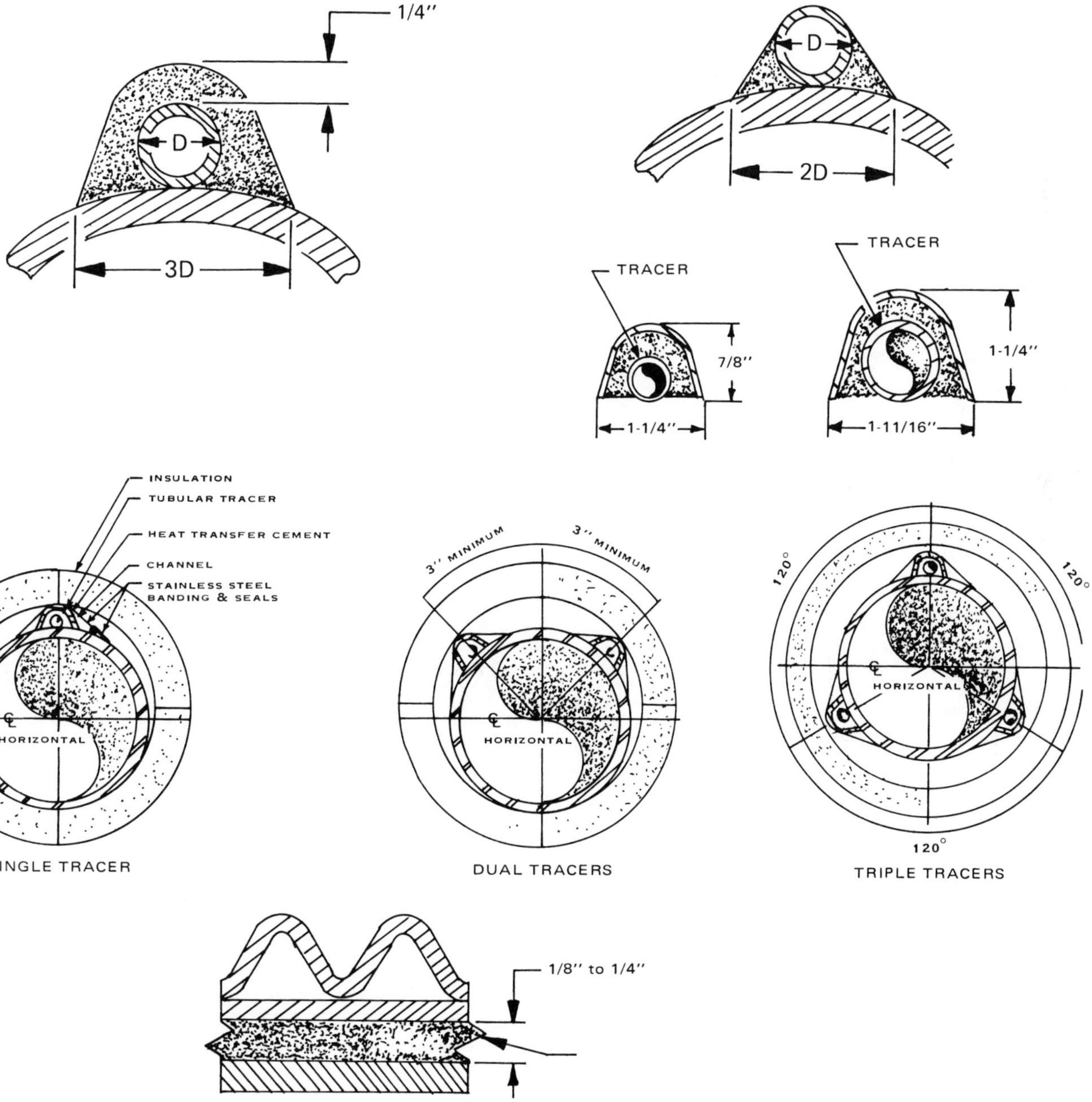

Figure 33 Thermally bonded fluid tracing systems.

Table 17 Characteristic of heat transfer cement and strip.

Properties	Generic Type of Heat Transfer Cement			Generic Type of Heat Trans. Cement Molded Trans. Strip		
Service	Emulsion Mastic	Emulsion Mastic	Emulsion Mastic	Solvent Mastic	Solvent Mastic	Shape to Fit Around Tracer & in Channel
Use	On Fluid & Elect. Tracers	On Fluid & Elect. Tracers	Panels	On Fluid Tracer	On Fluid Tracer 3/8"	1/2" & 3/4" O.D. Tracers
Temperature Max. Limits: Min.	1250°F 675°C −320°F −190°C	700°F 371°C −320°F −190°C	700°F 371°C −320°F −190°C	325°F 162°C −320°F −190°C	375°F 190°C −320°F −190°C	406°F 205°C −100°F −70°C
Thermal Overall Heat Transfer from Tracer to Pipe Wall "u" Heating: Cooling:	English Metric btu/hr w/m²k °F, ft² 20-40 113-227 Not Recommended	English Metric btu/hr w/m²k °F, ft² 20-40 113-227 Not Recommended	English Metric btu/hr w/m²k °F, ft² 20-40 113-227	English Metric btu/hr w/m²k °F, ft 20-25 113-141 10-20 56-113	English Metric btu/hr w/m²k °F, ft 20-25 113-141 10-20 56-113	English Metric btu/hr w/m²k °F, ft 20-40 113-227 10-20 56-113
Shear Strength to Substrate After Curing	psi n/m² 200 to 140000- 250 1750000	psi n/m² 200 to 140000- 250 1750000	psi n/m² 200 to 140000- 250 1750000	psi n/m² 1000- 6.895X10⁶ 1800 1.24X10⁷	psi n/m² 1000- 6.896X10 1800 1.24X10⁷	psi n/m² 100-150 6.895X10⁵ 1.03X10⁶
Electrical Resistance	3.3omhs 0.129omhs /in. /mm	.267ohms .011ohms /in. /mm	.261ohms .011ohms /in. /mm	146ohms 5.7ohms /in. /mm	146ohms 5.7ohms /in. /mm	146ohms 5.7ohms /in. /mm
Resistance to Soluability	Soluable in Most Liquids	Soluable in Most Liquids	Soluable in Most Liquids	Excellent	Excellent	Excellent
Cured Material	Hard, Solid	Hard, Solid	Solid	Elastic Solid	Elastic Solid	Elastic Solid
Application						
Min. Ambient Temperature	33°F 1°C	33°F 1°C	33°F 1°C	0°F(−17.8°C)Amb. 10°F(−12°C) Prod.	32°F(0°C)Amb. 70°F(21°C)Prod.	−10°F(−23°C)
Coverage: Tracer Size Pipe 3/8"O.D. 11/2-21/2" 3/8"O.D. 3 to Flat 1/2"O.D. 11/2-21/2" 1/2"O.D. 3 to Flat 3/4"O.D. 3 to Flat 7/8"O.D. 4 to Flat	lin ft lin m /gal /liter 37-41 3.0-3.3 34-37 2.7-3.0 43-47 3.0-3.8 40-43 3.2-3.5 18-20 1.4-1.6 21-24 1.7-1.9	lin ft lin m /gal /liter 37-41 3.0-3.3 34-37 2.7-3.0 43-47 3.0-3.8 40-43 3.2-3.5 18-20 1.4-1.6 21-24 1.7-1.9	ft²/gal m²/liter 1/4"tk 32 sq ft			
Coverage: (1/10 gal cartridge) Tracer Size Pipe 3/8"O.D. 11/2" to Flat 1/2"O.D. 11/2" to Flat 3/4"O.D. 11/2" to Flat 1" O.D. 11/2" to Flat				lin ft lin m /cart. /cart. 20.0 6.09 14.0 4.26 6.6 2.05 4.2 1.28	lin ft lin m /cart. /cart. 20.0 6.09 14.0 4.26 6.6 2.05 4.2 1.28	lin ft lin m /cart. /cart.
Method of Installation	Hand Trowelled/ Used in Channels	Hand Trowelled/ Used in Channels	Hand Trowelled/ Used in Channels	Manual, Air or Electric Operated Gun	Manual, Air or Electric Operated Gun	Install in Channel-strip, Wire or Clamp
Start Up Technique	Cure 4 to 12hrs at 160°F(71°C) to 211°F(99°C) no Curing Req'd. if Used in Channels	Cure 4 to 12hrs at 160°F(71°C) to 211°F(99°C) no Curing Req'd. if Used in Channels	Cure 4 to 12hrs at 160°F(71°C) to 211°F(99°C) no Curing Req'd. if Used in Channels	No Special Curing Req'd.	No Special Curing Req'd.	No Special Curing Req'd.
Solvents for Clean Up	Soap & Water	Soap & Water	Soap & Water	MEK, Tolvene Turpentine	MEK, Tolvene Turpentine	
Weight	13 lbs/gal	14 lbs/gal	14 lbs/gal	13 lbs/gal	13 lbs/gal	1/3 lb/lin ft 4'0"(1.22M)Sect.
Container Size Available	1 & 5 gal Pail	1 & 5 gal Pail	1 & 5 gal Pail	1/10 gal Cartridge	1/10 gal Cartridge	4'0"(1.22M) Section
Shelf Life	1 Year	1 Year	1 Year	30 to 90 Days Refrigerated - 1 Yr.	30 to 90 Days Refrigerated - 1 Yr	1 Year
Hazards in Application	Alkaline Keep Away From Eyes	Alkaline Keep Away From Eyes	Alkaline Keep Away From Eyes	Sticky-Wear Gloves	Sticky-Wear Gloves	

Table 18 Coefficient of heat transfer I/U for various metals, plastics to walls through walls and various produsts.

Table 18a Coefficient to Tank Wall: $\frac{1}{U_1}$
English Units

Steam	(40)	(0.0250)
Water	(20)	(0.0500)
Oil	(20)	(0.0500)

Notes:

1. These figures are applicable when installation conforms with securement details shown on **Drawing P72-35A Rev-3.**
2. These figures can be used for Glycol, Dowtherm, Therminol, Mobiltherm, etc. Contact our **Customer Service Department** for ($\frac{1}{U_1}$) values of Heat Transfer fluids not shown.

Table 18b Coefficient Thru Tank Wall: $\frac{1}{U_2}$
English Units

Tank Wall Thickness (tw)		1/16"	1/8"	3/16"	1/4"	3/8"
$\frac{1}{U_2}$	M.S.	0.0002	0.0004	0.0006	0.0008	0.0012
	S.S.	0.0007	0.0014	0.0021	0.0028	0.0042
	AL.	0.00004	0.00009	0.00013	0.00017	0.00026
Tank Wall Thickness (tw)		**1/2"**	**5/8"**	**3/4"**	**7/8"**	**1"**
$\frac{1}{U_2}$	M.S.	0.0016	0.0020	0.0024	0.0028	0.0032
	S.S.	0.0056	0.0070	0.0084	0.0098	0.0112
	AL.	0.00035	0.00043	0.00052	0.00061	0.00070

Note: M.S. = Mild Steel: S.S. = Stainless Steel: AL. = Aluminum

Table 18c Plastic Tanks
Coefficient Through Tank Wall: $\frac{1}{U_2}$
English Units

Wall Thickness (tw)	1/4"	3/8"	1/2"	5/8"	3/4"
Polyester or Epoxy	0.1250	0.1875	0.2500	0.3125	0.3750
Additional Coefficient for Tank Wall Lining: $\frac{1}{U_2}$					
Lining Thickness (tw)	1/32"	1/16"	3/32"	1/8"	1/4"
Rubber, Glass or Plastic	0.0325	0.0651	0.0976	0.1302	0.2604

Note: In the case of lined tanks, these values are to be added to the values indicated in **Table IV.**

Example:

Steel Tank Wall 1/4" Thick	0.0008
Rubber Lining 1/16" Thick	+0.0651
Total $\frac{1}{U_2}$ =	0.0659

Table 18d Typical Coefficients to Product
English Units

Product	Media	Process Temp.	Media Temp.	$1/U_3$ No Agitation	$1/U_3$ Mild Agitation
Asphalt, API 10.2	Steam	300	366	.0518	.029
Caustic, 50%	Water	70	160	.0256	.0153
Caustic, 73%	Steam	200	298	.0137	.008
Diethanolamine	Water	150	180	.0793	.0438
Ethanol	Water	140	180	.0269	.0154
Napthalene	Steam	250	338	.0224	.0126
No. 6 Fuel Oil	Steam	150	298	.0952	.0532
Para-Xylene	Steam	175	298	.0203	.0114
Phthalic Anhydride	Steam	300	366	.0377	.0207
Maleic Anhydride	Steam	160	250	.0306	.0169
SAE 30 Oil	Steam	175	298	.054	.0301
SAE 90 Oil	Steam	200	298	.0588	.0327
Styrene	Steam	150	298	.0177	.01
Sulfur	Steam	270	298	.0599	.0325
Sulfuric Acid 98%	Water	100	180	.0283	.0169
Toluene	Water	125	180	.0277	.0158
Tar	Steam	150	250	.0724	.0389
Water	50%-50% Water-Glycol	40	90	.0245	.0148
Isomerose	Water	85	100	.0980	.0531
Lard	Steam	130	250	.0581	.0313
Benzene	Water	100	150	.0261	.0149
Phenol	Steam	170	250	.0271	.0150
Paraffin	Steam	215	298	.1335	.0185

LIST OF SYMBOLS

English Symbols	Definitions and English Units	Metric Symbols	Metric Units
a_{mi}	= Constant (McMillan)	a_{mi}	
a	= Total surface area of vessel in sq ft	a_m	m^2
b	= Cost of installed insulation per sq ft per 1″ thickness per year (McMillan)	b_m	per m^2 per 25 min thk per year
d	= Diameter in inches	d_m	mm
d_i	= Inside diameter of insulation in inches	d_{mi}	mm
d_o	= Outside diameter of insulation in inches	d_{mo}	mm
d_p	= Diameter of pipe in inches	d_{mp}	mm
d_t	= Diameter of tracer in inches	d_{mt}	mm
d_{ti}	= Inside diameter of tracer in inches	d_{mti}	mm
f	= Conductance of surface air film in Btu/sq ft, hr, °F	f_m	W/m^2K
f_i	= Conductance of surface film, inner surface of insulation in Btu/sq ft, hr, °F	f_{mi}	W/m^2K
f_o	= Conductance of surface air film—outside of insulation in Btu/sq ft, hr, °F	f_{mo}	W/m^2K
f_p	= Conductance of surface air film—outer side of process pipe in Btu/sq ft, hr, °F	f_{mp}	W/m^2K
f_t	= Conductance of surfaceair film—outside of tracer (or cemented tracer), Btu/sq ft, hr, °F	f_{mt}	W/m^2K
f_{tr}	= Conductance of surface—outside surface exposed to air, Btu/ft^2, hr °F	f_{mtr}	W/m^2K
f_1	= Conductance of inner film, Btu/sq ft, hr, °F	f_{m1}	W/m^2K
f_2	= Conductance of outer ambient air film, Btu/sq ft, hr, °F	f_{m2}	W/m^2K
h_s	= Heat content of steam, Btu/lb	h_{ms}	W/kg
h_1	= Heat loss or gain through insulation (Fig. 36)	h_{m1}	W/m^2
k	= Conductivity of insulation in Btu/sq ft, hr. in., °F	k_m	W/m^2K
k_1	= Conductivity of 1st insulation in Btu/sq ft. hr. in., °F	k_{mi}	W/m^2K
k_2	= Conductivity of 2nd insulation in Btu/sq ft. hr. in., °F	k_{m2}	W/m^2K
k_3	= Conductivity of 3rd insulation in Btu/sq ft. hr. in., °F	k_{m3}	W/m^2K
k_4	= Conductivity of 4th insulation in Btu/sq ft. hr. in., °F	k_{m4}	W/m^2K
k_w	= Conductivity of vessel wall in Btu/sq ft. hr. in., °F	k_{mw}	W/m^2K
ℓ	= Thickness of insulation in inches	ℓ_m	m
ℓ_1	= Thickness of 1st insulation in inches	ℓ_{m1}	m
ℓ_2	= Thickness of 2nd insulation in inches	ℓ_{m2}	m
ℓ_3	= Thickness of 3rd insulation in inches	ℓ_{m3}	m
ℓ_4	= Thickness of 4th insulation in inches	ℓ_{m4}	m
$\log_e$	= Natural logarithm, to base e	$\log_e$	
m	= Cost of heat loss per linear foot (McMillan)	m_m	Cost of heat per lin m
n	= Number of parallel tracers	n	
n_m	= Cost of insulation per linear foot per year (McMillan)	n_{mm}	Cost of insul. per lin m per yr.
p	= Density, lb/cu ft	p_m	Density kgs/m^2
ϱ	= Resistivity of conductor wire	R/f	R/em
q_c	= Overall heat transfer from tracer, through heat transfer cement, to metal pipe or vessel, Btu/sq ft, hr, °F	q_{mc}	W/m^2
$q_{\ell f}$	= Total heat transfer, through cylindrical insulation(s) and air film(s). Btu/linear ft, hr.	$q_{m\ell f}$	W/m^2
q_t	= Total heat transfer from the heating medium to contents of vessel, Btu/sq ft, hr, °F	q_{mt}	W/m^2
q_{ti}	= Overall heat transfer, from the heating medium in the tracer through the heat transfer cement, through the vessel wall, and to the product in the vessel, Btu/sq ft, hr, °F	q_{mti}	W/m^2
q_{1i}	= Heat transfer from the heating medium in the tracer, through the tracer wall, through the heat transfer cement, to the outside of the vessel wall, Btu/sq ft, hr, °F	q_{m1i}	W/m^2
q_{2i}	= Heat transfer through the wall of the vessel, Btu/sq ft, hr, °F	q_{m2i}	W/m^2

English Symbols	Definitions and English Units	Metric Symbols	Metric Units
q_{3i}	= Heat transfer from the inner wall of the vessel, through the product film, to the product mass, Btu/hr, sq ft, °F	q_{m3i}	W/m^2
r_1	= Inside radius of innermost insulation in inches	r_{m1}	mm
r_2	= Outside radius of innermost insulation, and inside radius second layer of insulation, in inches	r_{m2}	mm
r_3	= Outside radius of second layer of insulation and inside radius of third layer of insulation, in inches	r_{m3}	mm
r_s	= Outermost radius of insulation, in inches	r_{ms}	mm
s_w	= Specific heat per unit weight, Btu/lb. °F		
t	= Temperature °F	t_m	°C
t_a	= Temperature of ambient air, °F	t_{ma}	°C
t_{as}	= Temperature of ambient air beyond air film, °F	t_{mas}	°C
t_{ap}	= Temperature of annulus space, °F	t_{map}	°C
t_{a1}	= Temperature of inner air, gas, or liquid, °F	t_{ma1}	°C
t_{a2}	= Temperature of outer ambient air, °F	t_{ma2}	°C
t_m	= Temperature of process pipe, °F	t_{mm}	°C
t_{mt}	= Mean temperature of process pipe, °F	t_{mt}	°C
t_p	= Temperature of process to be maintained, °F	t_{mp}	°C
t_s	= Temperature of outer surface, °F	t_{ms}	°C
t_{st}	= Temperature of steam in tracer, °F	t_{mst}	°C
t_t	= Temperature of tracer, °F	t_{mt}	°C
t_{in}	= Inlet temperature of heating media, if not constant, °F	t_{min}	°C
t_{out}	= Outlet temperature of heating media, if not constant, °F	t_{mout}	°C
t_{1p}	= Initial process temperature, °F	t_{m1p}	°C
t_{2p}	= Final process temperature, °F	t_{m2p}	°C
t_1	= Temperature of inner surface of 1st insulation (or operating temperature), °F	t_{m1}	°C
t_2	= Temperature of outer surface of 1st insulation (or junction temperature between 1st and 2nd insulations), °F	t_{m2}	°C
t_3	= Temperature of outer surface of 2nd insulation (or junction temperature between 2nd and 3rd insulations), °F	t_{m3}	°C
t_4	= Temperature of outer surface of 3rd insulation (or junction temperature between 3rd and 4th insulations), °F	t_{m4}	°C
t_w''	= Thickness of vessel wall, in inches	t_{mw}''	m
Δt	= Temperature difference, °F	Δ_{mt}	°
Δt_m	= Log mean temperature difference between the heating or cooling media and the average of the initial and final process temperatures, °F	Δ_{mtm}	°C
Δt_p	= Degrees Fahrenheit material is to be raised	Δt_{mp}	°C
Δt_t	= Temperature difference between heating	Δt_{mt}	°C
v	= Volume of metal in vessel wall, cu ft	v_m	m^3
y	= m + n_m (McMillan)	Y_m	$m_m + n_{mm}$
ε	= Emittance ratio in reference to thermal black body, with which it has a value of 1	ε	same as English
ϱ	= Weight of one cubic foot of steam, lb	ϱ	m^3/kg
Δ	= Loss in steam pressure, psi	Δ_m	kgs/m^2
A	= Heat transfer surface of the tracer, sq ft/ft of tracer	A	m^2/m
A_d	= Thermal diffusivity, ft/hr	A_{md}	m/hr
C_y	= Cycle time, hr	C_y	hr
D	= Diameter of vessel, ft	D_m	m
E	= Electromotive force, volts	E	volts
F	= Flow rate of the heating or cooling media, gpm	F_m	liters/sec.
F′	= Flow rate of the process fluid, gpm	F'_m	liters/sec.
H	= Height of vessel, ft	H_m	m
H′	= Working level of product in vessel, ft	H'_m	m
I	= Electric current, amp	I	amp
K	= Conductivity in Btu/sq ft, ft, hr, °F	K	W/m^2K

LIST OF SYMBOLS (Cont.)

English Symbols	Definitions and English Units	Metric Symbols	Metric Units
L	= Equivalent thickness, in inches	L	m
L_t	= Length of tracer in equivalent feet, (maximum trapping distance) in ft	L_{m1}	m
L'	= Length of process pipe, ft	L'_m	m
M	= Value of heat in dollars per million Btu (McMillan)	M	S/kw
P	= Tracer spacing, in.	P_m	m
Q	= Heat transfer in Btu/sq ft, hr	Q_m	W/m^2
Q_a	= Heat transfer thru air film in Btu/sq ft, hr	Q_{ma}	W/m^2
Q_c	= Heat transfer by natural convection, Btu/sq	Q_{m2}	W/m^2
Q_{ca}	= Heat transfer due to convection, Btu/sq ft,	Q_{mca}	W/m^2
Q_{cv}	= Heat transfer by forced convection, Btu/sq ft, hr	Q_{mcv}	W/m^2
Q_{ia}	= Heat transfer from annulus space through insulation to ambient air, Btu/sq ft, hr	Q_{mia}	W/m^2
Q_{pa}	= Heat transfer from process pipe to annulus space, Btu/sq ft, hr	Q_{mpa}	W/m^2
Q_{ap}	= Heat transfer from annulus space to process pipe, Btu/sq ft, hr	Q_{map}	W/m^2
Q_r	= Heat transfer by radiation, Btu/sq ft, hr	Q_{mr}	W/m^2
Q_{ra}	= Heat transfer by radiation at given emittance (or absorptance) ε, Btu/sq ft, hr	Q_{mra}	W/m^2
Q_t	= Total heat transfer, Btu/sq ft, hr	Q_{mt}	W/m^2
Q_{ta}	= Heat transfer from tracer to annulus space, Btu/sq ft, hr	Q_{mta}	W/m^2
Q_{tp}	= Heat transfer from tracer through heat transfer cement to metal surface, Btu/sq ft. hr	Q_{mtp}	W/m^2
Q_{tv}	= Total heat load, Btu/sq ft, hr	Q_{mtv}	W/m^2
Q_{tv}	= Heat input or removal required to raise or lower the temperature of vessel from initial to final temperature, Btu/sq ft, hr	Q_{mtv}	W/m^2
Q_{2v}	= Heat input or removal required to raise or lower the temperature of vessel wall from initial to final temperature, Btu/sq ft, hr	Q_{m2v}	W/m^2

English Symbols	Definitions and English Units	Metric Symbols	Metric Units
Q_{3v}	= Heat input or removal required to offset heat gain or loss from the vessel wall either through insulation, or if uninsulated, to the surrounding ambient air, Btu/sq ft, hr	Q_{m3v}	W/m^2
R	= Thermal resistance ($\ell \div k$)	R_m	$\ell_m \div k_m$
R_e	= Electrical resistance, ohms	R_e	ohms
R'	= Variable constant (Eq. 38)	R'	
R_s	= Air film resistance °F, hr, ft^2/Btu	R_{ms}	km^2/W
S_f	= Specific heat of process fluid, Btu/lb. °F (Water = 1.0)	S_{mf}	J/kg K
S_p	= Specific heat of material (product), Btu/lb. ° (Water = 1.0)	S_{mp}	J/kg K
S_{hm}	= Specific heat of heating media, Btu/lb, °F (Water = 1.0)	S_{mhm}	J/kg K
S_{vol}	= Specific heat of volume undergoing change, Btu/cu ft, °F	S_{mvol}	W/m^2K
S_w	= Specific heat of vessel wall, Btu/lb (cu ft)	S_{mw}	W/kg
T	= Absolute temperature, °F + 459.6	T_m	°C or °K
V	= Volume of the process fluid, cu ft	V_m	m^3
V_{st}	= Volume (total) of metal contained in vessel cu ft	V_{stm}	m^3
W_c	= Weight of steam condensate per hour, lb	W_{mc}	Kg/hr
W_f	= Weight of heating fluid, lb/cu ft	W_{mf}	Kg/m^3
W_{fp}	= Weight of process fluid in one foot of pipe,	W_{mfp}	Kg/lin m
W_{hm}	= Weight of heating media, lb/gal	W_{mhm}	Kg/m^3
W_p	= Weight of material (product) per hour, lb	W_{mp}	Kg/hr
W_s	= Weight of steam per hour, lb	W_{ms}	Kg/hr
W_{st}	= Weight of steam per hour, lb, per given length of pipe	W_{mst}	Kg/hr, lin m
W_w	= Specific weight, lb/cu ft	W_{mw}	Kg/m^3
W_{wv}	= Density of vessel wall, lbs/cu ft	W_{mwv}	Kg/m^3
Y	= Hours of operation per year (McMillan)	Y	hrs/yr

2 Economics of Thermal Insulation

Fundamentally, thermal insulation is a material used to conserve energy. As energy has monetary value, insulation conserves money. Insulation is also used to control temperature and where so used in manufacturing it also helps in the direct earning of money. In many installations it serves in both capacities. In all applications, wherever it is used—in process plants, cold storage, buildings, mechanical heating and cooling systems, transportation trucks, railroad cars, or ships—the insulation either helps produce profits or provides a savings by reducing heat or refrigeration losses.

Unfortunately, energy is invisible; therefore, the amount of savings or earnings by its conservation is not readily discernible. Because of this, the fact that insulation is a method for producing earnings is almost invariably overlooked or minimized by industry. However, make no mistake, the major reason an industrial user buys insulation is to obtain a return on his investment or the insulation is essential to enable him to stay in business.

The latter part of the above statement, because of the energy shortages, now includes many more businesses than previously. In the past, the abundance of energy in this country gave many people the impression that there were no limits to this vital resource. For this reason it was, and still is, wasted. It must be remembered that without energy no life is possible.

Our existence depends upon many forms and sources of energy. Energy may be in potential or transient state, and in various forms. It can be transferred from one form to another. Probably man's first step toward our present level of technology was his discovery of the chemical process necessary to convert potential energy in wood to transient energy in the form of heat. This process is named "fire."

Energy exists in such forms as atomic energy, chemical energy, mechanical energy, electric energy or heat energy. It has numerous units of measurement, such as Btu's, calories, ergs, dynes, foot-lbs., joules, watts, and therms. With all the different terms for measurement units and the fact that energy is invisible, it is no wonder that comprehension of the concept of energy is difficult.

In most instances the available energy is neither in the form nor location that man wishes to use it. For this reason, it may be converted one, or many times, before its final use. For example: Energy in oil, gas or coal may be converted by fire to heat energy, which is used to convert water to steam—the pressure of which is used to produce mechanical energy, which then is converted to electrical energy, which is then transmitted some distance away where it is converted back to mechanical energy or heat energy.

From this, it is evident that availability and accessibility of energy are two major factors which effect its use for a particular purpose. To measure these interrelated factors, the term used is named "money." Food, shelter, transportation and conveniences are all products of energy. Frequently because it is invisible the loss of energy (or "money") goes unnoticed. Thermal insulation is a material to conserve essential energy, in doing so, it conserves money. The manner in which thermal insulation conserves energy (money) differs for various industries. For this reason, the following discussion points out the particular use of thermal insulation for specific industries.

Depending upon its use, insulation can provide a profit on the money invested in it in many ways. As the manner in which insulation earns a profit is different for various industries, the discussion which follows will deal with its use in these industries.

Frozen Food Industry

The frozen food industry depends upon insulation for its very existence.

At the processing plant low-temperature freezers, which consist of well-insulated enclosures, make possible the rapid freezing of foods immediately after they are processed. From this point the frozen foods are shipped in insulated refrigerator railroad cars, trucks, and ships. At distribution points and stores the foods are stored in freezers until they are sold.

Even the ultimate consumer of the frozen foods is dependent upon the existence of well-insulated home refrigerators which allow the housewife to store the food until she is ready to use it.

Without insulation it would be impractical, if not impossible, to keep food at a temperature below its spoiling temperature. This is an industry in which the producer, wholesaler, and dealer depend upon insulation to enable them to process, distribute, and sell the products at a profit.

Transportation Industry

The transportation field includes many other industries, besides the frozen food and refrigerated food industries, where items must be transported within given temperature limits. Some of these other industrial areas are (1) the transportation of hot molten materials, (2) the transportation of cold liquified gases, (3) the transportation of materials which must be kept from freezing, and not of the least importance, (4) the transportation of people within the comfort range of temperature.

When hot molten materials are transported the temperature range may be such that refractory materials may be essential. For materials of lower molten temperatures, regular high temperature insulations may be sufficient.

Transportation of highly volatile liquids also requires insulated cars and trucks to prevent excessive loss due to vaporization. However, it may require only moderate temperature insulation that is efficient in the atmospheric range. On the low-temperature side liquified gases may require super-efficient insulation to reduce heat gain to low quantities because of the high monetary value of the product-loss or the necessity to control pressure. In this instance the super-efficient cryogenic insulations are necessary.

The comfort of people being transported also depends upon insulation. Modern passenger cars are well insulated and air-conditioned. Of course airplanes must be insulated and heated to prevent freezing of the passengers at high altitudes. The problem of providing suitable temperature for human comfort in space travel is even more difficult than with ground, sea, or air travel, and insulation must be used to solve the problem.

The transportation field brings up some very interesting insulation economic problems. One is that any unnecessary weight of insulation reduces the profitable load. For this reason light-weight, effective insulations can be used very advantageously. Most very light-weight insulations, however, are structurally weak and must be protected from mechanical damage by a jacket or covering. When used in refrigeration service the insulations must be protected by a vapor barrier. Thus, good design requires the selection of a number of materials and processes, among which are the insulation and its manner of installation to form an insulated structure with a weather and/or vapor barrier which will provide reasonable life. An economic balance to achieve the best results as to thermal efficiency, effective life, low weight, and reasonable cost is quite difficult. Unfortunately, many of the designers of transportation equipment are not experts in insulation design, thus many errors are made in the selection of insulation materials and their application. Also, the technical people in the insulation industry are not in a position to obtain the economic facts and problems related to the transportation industry, so they are unable to properly assist the user in the proper selection of materials, installation design, and application.

When the insulation industry starts thinking of insulation as a means by which both it and its customer profits, the customer, with increasing frequency, will turn to the industry for advice.

A condensed listing of the previously mentioned uses of insulation and how they related to money savings or profits, follows:

1. Insulation makes it possible to control the temperature of the product so that it can be processed, shipped, and sold at a profit.
2. It conserves energy, either heat or refrigeration, thus reducing power costs.
3. It reduces the required size of heating or refrigeration equipment, thus reducing investment.
4. It provides the safety without which it would be impossible to ship such items as molten materials or liquefied gases.
5. On ships it provides fire protection by reducing the fire hazard, resulting in lower losses and lower insurance rates.

As any insulation costs money to buy, install, and maintain, these costs must be compared with the money it earns or saves. When the capital investment in insulation is established and the profits or savings it provides are determined in dollars per year, it is then possible to determine the return on investment which it provides to its user. In most instances when the facts are determined it will be apparent that the insulation was the best investment in the entire business enterprise.

Building

Insulation is an essential part of all modern buildings; it performs many functions, all of which affect the cost of the building and its operating expenses. The term "building," used here, includes homes, hotels, schools, office buildings, factory buildings, laboratories, textile and other industrial buildings. Insulation here is essential not only in the floors, walls and ceilings of the buildings, but in their mechanical systems such as heating, cooling, plumbing, and air-conditioning, as well. The insulation being only one component in a structure, many architects are not experts in this single phase of their entire profession, and do not realize the importance of insulation in the economics of a building.

The use of properly designed insulation in floors, walls, and roofs can frequently so reduce the heating and cooling load that the consequent smaller sized heating and cooling equipment reduces the overall cost more than the cost of the insulation. In this way a thermally efficient building is obtained for a lesser cost than a poorly insulated one. Other than this initial savings in capital outlay the insulation provides continuing savings each year from the use of less power and fuel. In such instances insulation pro vides a return with no investment or even a reduction in investment.

In addition to the function of insulating the building itself, insulation is used to conserve energy which would be lost from uninsulated pipes, ducts, heating equipment, and refrigeration within the building. These uses either help control the temperatures within the building or conserve energy. In either instance the use of insulation again saves money. If insulation thicknesses or uses are skimped, additional capital costs and increased operating cost result from this false economy.

Other less tangible advantages result when insulation is used correctly in building design. One of these is the maintenance of human comfort. This is a difficult item to express in dollar values, but it is definitely a part of the earning ability of the building. The human comfort, or lack of it, induced by a building influences its market value and its rent level. It also exerts a definite influence on the efficiency of those who work within the confines of the building.

A human tendency is to disregard items which are difficult to evaluate. Thus human comfort in buildings is frequently given little or no economic attention. Yet everyone has been in offices or manufacturing plants where workers are spending non-productive time trying to keep warm or attempting to cool off in the summer. Under such conditions this thermally inefficient building is probably the most wasteful purchase management could possibly make.

To further illustrate, most persons have experienced a day or night in a hotel where it was impossible to obtain comfort conditions. After having experienced this ask yourself whether you ever attempted to obtain reservations at that hotel again? Then decide if human comfort has dollar value.

Human comfort in buildings depends, of course, upon the proper design of heating and cooling equipment. Efficient operation of this equipment, however, depends upon insulation.

Although it is not simple, the services performed by the insulation can be resolved into terms of dollars, and its economic value can be determined.

Process Industries

In the process industries thermal insulation serves several functions: (1) fire protection, (2) conservation of product, (3) control of temperature, (4) production of product, and (5) conservation of energy. As would be expected, these are identical functions to those previously mentioned for other industries. However, in a process industry the use of insulation is so directly and intimately tied into the production processes, and its economics are so important, that they are better known and easier to establish.

Other functions, such as fire protection, can only be estimated. The economics involved in fire protection are first, to prevent or retard accidental fires. Second, if fires do start then the insulation should protect the pipe and equipment until the fire can be extinguished. In this manner insulation protects the plant, reduces the frequency or duration of shutdowns, saves product, protects personnel, and reduces insurance costs. All of these have extremely high value. The difficulty in calculating the monetary value of the insulation is that overall economics cannot be based on accidental occurrences. In a particular set of conditions the insulation may have prevented a small fire from growing into a several million dollar conflagration. On the other hand if the fire never occurred, the insulation provided no savings at all, unless its presence was reflected in reduced insurance cost.

The conservation of product is one of the fastest growing uses of insulation. More and more materials must be stored, transported, and used within a particular temperature range. The temperatures may vary from very low to very high, depending upon the product involved. To illustrate with a few examples: (1) materials, such as water emulsions, which are ruined if frozen; (2) volatile materials which if allowed to exceed a certain temperature will start to react, resulting in ruined material or even an explosion; (3) materials with high melting points, which if allowed to solidify are difficult to remelt; (4) liquefied gases which vaporize very rapidly and must be kept cold to prevent excessive pressures and losses.

When such materials are in storage vessels, heating or cooling coils are frequently provided to transfer heat to or from the material being stored. In such instances the insulation must be of proper thickness so that excessive heating or cooling is not required. Knowing the input of the heater or coil and the area of the vessel, it is relatively simple by the heat flow formulas already presented to calculate the needed insulation thickness.

When materials are shipped, the tank truck or car may or may not have means to heat or cool its contents. In these instances the insulation thickness required must be based on allowable heat transfer in the time between loading and unloading. This allowable transfer can be determined as follows when:

Δt = temperature change allowable.
S_p = specific heat of material.
lb = material
Allowable Btu transfer = $\Delta \times S_p \times$ lb.

Knowing this, the insulation thickness can be determined by the heat loss per hour, by the standard heat transfer formulas. Remember, though, that this loss per hour must be less than the total allowable Btu transfer divided by the number of hours in transit.

In these cases the economics are not related to the value of the heat saved but must be compared to the loss of product which would occur if insulation were not used. This is another instance where insulation is necessary to earn a profit, and heat saving is incidental to the fundamental function.

Similar to previous examples, when materials must be manufactured at some temperature other than ambient, the insulation becomes a part of the production equipment and a necessary part of the production. This economic function of insulation must be considered as a necessary investment to produce materials to be sold at a profit. In many instances insulation so used also conserves heat or refrigeration energy, but in these cases this is a plus value rather than the main function.

The use of insulation to conserve energy is an application where its ability to provide substantial returns to its user is more readily apparent. This does not mean that the calculations are simple or that the facts are easy to determine, but the answers can be determined without affecting the product being produced. A major part of insulation used in industry falls within this category, so it is well worthwhile to investigate the magnitude of savings which are possible by such use. For this purpose the tables presented in economic considerations are first calculated by Btu, converted into steam savings, and then into dollar values. The reason for this procedure is that steam is the most common heat medium for process plants. Using proper units of conversion, the same can be done for hot water heating, Dowtherm heating, or electrical heating. All that is required is that the dollar value of Btu for any heating method be established. As most of these heating mediums are more expensive per Btu than steam, the savings with steam will be greater.

Economic Considerations When Insulation is Used to Conserve Energy

This economic consideration is based on savings of energy—that heat energy has monetary value. Also, that the equipment which converts fuel to heat has monetary value. It follows, that if insulation can reduce the cost of each of these it will earn a saving for the user. Of the two savings, the one in heat-producing equipment is most frequently overlooked by industry.

As steam is the most commonly used medium of heating, a review of the costs of steam boilers as related to insulation is justified. The most common manner of stating boiler capacity is the number of pounds of steam it will produce per hour. Therefore the cost of equipment necessary to produce a pound of steam per hour is the basis of comparison. Depending upon size, location, fuel, and other factors, the cost for boilers and auxiliary equipment ranges from a low of approximately $10.00 per pound of steam per hour to approximately $80.00. (Should it be desirable to convert this to 1000 lb of steam per year capacity, the conversion factor is 0.114, i.e., 8760 ÷ 1000.)

Other than operating temperature, many factors influence the Btu heat loss, such as ambient air temperature, emittance of surface, air velocity, position (such as horizontal or vertical), and shape. Table 19 based on very moderate conditions, is presented to show the magnitude of heat loss from bare pipe. This table is based on 70° F ambient air temperature, still air conditions, ε being 0.9, and pipe in a horizontal position. From this table it can be seen that the heat loss from a 1/2″ pipe operating at 212° F is 82 Btu/lin. ft, hr and a 36″ pipe at the same temperature loses 2660 Btu/lin. ft, hr. The same pipes operating at 1000° F have heat

Table 19 Heat loss—bare surface. (Based on 70° ambient temperature—natural convection.)

NPS Pipe Size Nom. ″	BTU loss/linear foot, hour – English Units — OPERATING TEMPERATURE											Pipe Size Dia. mm
	212° F 100° C	300° F 149° C	400° F 204° C	500° F 260° C	600° F 316° C	700° F 371° C	800° F 427° C	900° F 482° C	1000° F 538° C	1100° F 593° C	1200° F 649° C	
1/2	82	166	287	445	649	901	1218	1602	2075	2644	3317	21.3
3/4	101	205	353	547	801	1113	1508	1984	2576	3282	4122	26.7
1	123	250	434	674	989	1379	1865	2460	3194	4080	5123	33.4
1 1/4	153	312	541	843	1237	1728	2342	3091	4010	5123	6438	42.2
1 1/2	173	352	613	955	1403	1960	2661	3514	4568	5841	7345	48.3
2	213	431	753	1176	1732	2423	3295	4355	5665	7251	9133	60.3
2 1/2	252	516	899	1408	2077	1907	3956	5235	6817	8732	11005	73.0
3	303	620	1083	1695	2505	3511	4784	6337	8259	10532	13344	88.9
3 1/2	343	700	1226	1922	2841	3987	5434	7201	9390	12039	15189	101.6
4	383	784	1370	2149	3179	4467	6094	8079	10545	13513	17049	114.3
5	420	861	1513	2377	3513	4989	6742	8944	11676	14982	18892	141.3
6	544	1122	1976	3105	4604	6479	8858	11769	15366	19740	24909	168.3
8	697	1436	2530	3988	5927	8356	11433	15211	19895	25568	32293	219.1
10	855	1769	3114	4918	7312	10325	14147	18812	24621	31679	40051	273.0
12	1020	2079	3664	5809	8627	12194	16721	22248	29133	37501	47429	323.8
14	1180	2258	3989	6316	9404	13301	18236	24279	31844	40965	51656	355.6
16	1240	2560	4529	7160	10697	15124	20769	27663	36257	46689	59089	406.4
18	1390	2875	5074	8032	12010	16992	23346	31109	40788	52539	66456	457.2
20	1530	3168	5618	8897	13270	18777	25810	34432	45157	58159	73691	508.0
24	1820	3772	6679	10595	15825	22375	30836	41112	53958	69536	88221	609.6
30	2215	4642	8242	13103	19606	27758	38299	51107	67126	86558	109872	762.6
36	2660	5570	9880	15720	23530	33300	45900	61300	80500	104000	131800	914.4
	BTU LOSS per square foot per hour											
Flat	319	654	1145	1806	2666	3765	5126	6799	8885	11400	14394	Flat

Table 19a Heat Loss — Bare Surface — Metric Units
Based on 21.1°C Ambient temperature — Natural Convection

Pipe Size Dia. mm	HEAT LOSS W/linear metre — Metric Units / OPERATING TEMPERATURE 100°C 212°F	149°C 300°F	204°C 400°F	260°C 500°F	316°C 600°F	371°C 700°F	427°C 800°F	482°C 900°F	538°C 1000°F	593°C 1100°F	649°C 1200°F	NPS Pipe Size Nom."
21.3	85	172	298	463	675	937	1267	1667	2159	2752	3451	1/2
26.7	105	213	367	569	833	1158	1569	2064	2680	3415	4289	3/4
33.4	128	260	452	701	1029	1435	1940	2560	3323	4246	5331	1
42.2	159	324	563	877	1287	1798	2437	3217	4173	5331	6699	1 1/4
48.3	180	366	637	994	1460	2040	2769	3657	4753	6079	7643	1 1/2
60.3	221	448	783	1224	1802	2538	3429	4532	5895	7545	9504	2
73.0	262	537	935	1465	2161	3025	4116	5448	7094	9087	11452	2 1/2
88.9	315	645	1127	1763	2606	3654	4978	6595	8594	10960	13887	3
101.6	356	728	1276	2000	2957	4149	5655	7494	9771	12528	15807	3 1/2
114.3	398	816	1426	2236	3308	4649	6341	8408	10973	14062	17742	4
141.3	437	896	1575	2474	3656	5191	7016	9307	12150	15591	19660	5
168.3	566	1168	2056	3231	4791	6743	9218	12247	15991	20543	25922	6
219.1	725	1494	2633	4150	6168	8696	11898	15830	20704	26608	33606	8
273.0	885	1840	3241	5118	7609	10745	14722	19577	25622	32967	41680	10
323.8	1061	2163	3813	6045	8978	12690	17401	23153	30318	39026	49358	12
355.6	1228	2350	4152	6573	9787	13842	18678	25267	33139	42631	53757	14
406.4	1290	2664	4713	7451	11132	15739	21614	28788	37731	48588	61492	16
457.2	1446	2991	5280	8359	12499	17683	24296	32375	42447	54676	69159	18
508.0	1592	3297	5847	9259	13810	19540	26869	35832	46993	60524	76689	20
609.6	1894	3925	6951	11026	16469	23285	32090	42784	56153	72364	91809	24
762.6	2305	4830	8577	12636	20404	28887	39857	53186	69856	90079	114341	30
914.4	2768	5797	10281	16359	24487	34655	47767	63794	83774	108230	137161	36
						HEAT LOSS W/m²						
Flat	1005	2062	3610	5694	8405	11870	16160	21434	28011	35940	45378	Flat

losses of 2075 Btu/lin. ft, hr, and 80,500 Btu/lin. ft, hr, respectively.

Steam at various pressures but at the same temperature does not have the same Btu content. Thus a table converting from Btu loss to steam loss must consider the given pressure and the superheat contained in the steam, to provide the proper content of Btu per pound. Table 20, then, is correct only when used for saturated steam, which has the maximum Btu content at a given temperature. For the purpose of illustrating economies, however, using the maximum Btu/lb of steam is sufficiently close to obtain a basis of comparison of the cost of using insulation and the resultant savings. Table 20 gives the approximate steam loss per linear foot of uninsulated pipe and per square foot of uninsulated flat surface, based on 70° F ambient temperature with natural convection. If windage occurs, these losses will rise considerably. Also, lower ambient air temperatures would increase the loss. However, under the most favorable conditions for minimum loss it can be seen that bare surfaces cause considerable steam loss per hour.

How much savings can be accomplished by insulation? Of course, this depends upon the thickness of the insulation and its conductivity. If we assume the average conductivity of commonly used high-temperature insulations such as expanded silica, calcium silicate, or asbestos fibers, it is still necessary to decide on "how much." This is a difficult problem to answer at this stage in the study of economics because we must be guided by the "economic thickness," which has yet to be discussed.

To get to the next step in the analysis, an "economic thickness" table was made out for *one* set of conditions: Capital Investment per pound of steam—$25.00, cost of steam produced—$1.00 per thousand pounds, cost of $2.80 per linear foot of 1" thickness of insulation installed on a 1" NPS pipe (Table 22). The "economic thickness" for these stated conditions are shown in Table 21. This table is *not* a table of recommended thicknesses, as it is recommended for *one single set of conditions only.*

Recommended Thickness Tables presented by the industry have probably caused more poorly designed insulation than any other factor. Unless the insulation thickness is based upon the economics of the conditions under which it is to be used, it is wrong. Either too little or too much insulation results in a waste of money. Oddly enough, the "Recommended Thickness" given in past published literature of the manufacturers, for most conditions, recommended much less insulation than the consumer should have used to obtain good economics. Although this paragraph is out of context it has been included here as a caution against the blind use of "Recommended Thickness Tables."

From the table of thicknesses established it is possible to calculate the heat loss of the insulated surfaces. For the same conditions used in determining the bare loss, the insulated losses are shown in Table 23. These values are subtracted from the base loss table to determine the heat savings accomplished by the insulation; these are shown in Table 24. Knowing the heat savings, it then is possible to convert to the approximate steam savings per hour. This table of steam savings, pounds per hour, is shown in Table 25, and steam savings in thousands of pounds of steam per year is shown in Table 26.

These tables regarding heat loss, or steam loss, are similar to tables of this type that have been published in one form or another over a period of many years. The figures presented regarding heat loss and steam loss are quite conservative, in that the ambient conditions are very moderate, with no windage and maximum Btu

Table 20 Steam loss—bare pipe.

NPS Pipe Size Nom. "	POUNDS of steam loss/linear foot, hour – English Units — OPERATING TEMPERATURE											Pipe Size Dia. mm
	212°F 100°C	300°F 149°C	400°F 204°C	500°F 260°C	600°F 316°C	700°F 371°C	800°F 427°C	900°F 482°C	1000°F 538°C	1100°F 593°C	1200°F 649°C	
1/2	0.07	0.14	0.23	0.35	0.49	0.65	0.85	1.08	1.35	1.67	2.02	**21.3**
3/4	0.09	0.17	0.28	0.43	0.60	0.81	1.05	1.34	1.68	2.07	2.52	**26.7**
1	0.11	0.21	0.35	0.53	0.74	0.99	1.30	1.66	2.08	2.58	3.13	**33.4**
1 1/4	0.12	0.26	0.44	0.66	0.93	1.25	1.64	2.03	2.61	3.23	3.93	**42.2**
1 1/2	0.15	0.29	0.49	0.74	1.05	1.41	1.85	2.37	2.98	3.68	4.48	**48.3**
2	0.18	0.36	0.61	0.92	1.29	1.75	2.30	2.93	3.68	4.58	5.58	**60.3**
2 1/2	0.22	0.43	0.72	1.10	1.55	2.10	2.76	3.53	4.44	5.50	6.72	**73.0**
3	0.26	0.52	0.87	1.32	1.88	2.54	3.41	4.29	5.38	6.67	8.15	**88.9**
3 1/2	0.29	0.58	0.99	1.49	2.13	2.88	3.79	4.86	6.12	7.60	9.26	**101.6**
4	0.33	0.65	1.10	1.68	2.37	3.22	4.26	5.43	6.37	8.52	10.40	**114.3**
5	0.36	0.72	1.22	1.86	2.63	3.56	4.70	6.03	7.61	9.44	11.53	**141.3**
6	0.47	0.94	1.59	2.42	3.45	4.68	6.18	7.93	10.01	12.43	15.21	**168.3**
8	0.60	1.20	2.04	3.11	4.43	6.03	7.97	10.25	12.96	16.21	19.72	**219.1**
10	0.75	1.48	2.51	3.84	5.48	7.45	9.85	12.66	16.05	20.00	24.45	**273.0**
12	0.88	1.74	2.94	4.53	6.46	8.81	11.67	15.15	18.97	23.65	28.99	**323.8**
14	1.02	1.89	3.22	4.94	7.05	9.62	12.72	16.38	20.75	25.85	31.70	**355.6**
16	1.07	2.14	3.64	5.60	8.01	10.94	14.45	18.64	23.62	29.45	36.10	**406.4**
18	1.20	2.40	4.09	6.25	9.00	12.22	16.24	20.97	26.60	33.14	40.60	**457.2**
20	1.32	2.64	4.53	7.02	9.93	13.52	18.00	23.22	29.42	36.70	45.10	**508.0**
24	1.57	3.11	5.38	8.28	11.85	16.14	21.52	27.70	35.17	43.80	53.80	**609.6**
30	1.91	3.88	6.63	10.23	14.69	20.01	26.70	34.50	43.80	54.60	67.20	**762.6**
36	2.29	4.66	7.96	12.27	17.61	24.50	32.05	41.30	52.50	65.60	80.40	**914.4**
	POUNDS OF STEAM LOSS per square foot per hour											
Flat	0.28	0.55	0.92	1.41	2.00	2.72	3.57	4.58	5.80	7.18	8.78	**Flat**

Table 21 Insulation thickness used to calculate heat and steam savings. (English Units)

NPS Pipe Size	INSULATION THICKNESSES—NOMINAL — OPERATING TEMPERATURE										
	212°F	300°F	400°F	500°F	600°F	700°F	800°F	900°F	1000°F	1100°F	1200°F
1/2	1	1	1	1	1 1/2	1 1/2	1 1/2	2	2	2	2
3/4	1	1	1	1	1 1/2	1 1/2	1 1/2	2	2	2	2 1/2
1	1	1	1	1 1/2	1 1/2	1 1/2	2	2	2	2	2 1/2
1 1/4	1	1	1	1 1/2	1 1/2	2	2	2	2	2	2 1/2
1 1/2	1	1	1 1/2	1 1/2	2	2	2	2 1/2	2 1/2	2 1/2	2 1/2
2	1	1	1 1/2	1 1/2	2	2	2	2 1/2	2 1/2	2 1/2	3
2 1/2	1	1	1 1/2	1 1/2	2	2	2 1/2	2 1/2	2 1/2	3	3
3	1	1	1 1/2	1 1/2	2	2	2 1/2	2 1/2	2 1/2	3	3
3 1/2	1	1 1/2	1 1/2	2	2	2 1/2	2 1/2	2 1/2	3	3	3
4	1	1 1/2	1 1/2	2	2	2 1/2	2 1/2	3	3	3	3 1/2
5	1	1 1/2	1 1/2	2	2	2 1/2	2 1/2	3	3	3	3 1/2
6	1 1/2	1 1/2	2	2	2	2 1/2	3	3	3 1/2	3 1/2	3 1/2
8	1 1/2	1 1/2	2	2	2 1/2	3	3	3 1/2	3 1/2	4	4
10	1 1/2	1 1/2	2	2 1/2	3	3 1/2	3 1/2	3 1/2	4	4	4 1/2
12	1 1/2	2	2 1/2	2 1/2	3	3 1/2	3 1/2	4	4 1/2	4 1/2	5
14	1 1/2	2	2 1/2	3	3 1/2	3 1/2	4	4	4 1/2	4 1/2	5
16	1 1/2	2	2 1/2	3	3 1/2	3 1/2	4	4	4 1/2	4 1/2	5
18	1 1/2	2	2 1/2	3	3 1/2	4	4	4 1/2	4 1/2	5	5
20	1 1/2	2	2 1/2	3	3 1/2	4	4	4 1/2	4 1/2	5	5
24	1 1/2	2	2 1/2	3	3 1/2	4	4 1/2	4 1/2	5	5	5 1/2
30	2	2 1/2	3	3 1/2	4	4 1/2	4 1/2	5	5 1/2	5 1/2	6
36	2	2 1/2	3	3 1/2	4	4 1/2	5	5 1/2	6	6	6 1/2
	INSULATION THICKNESSES—INCHES										
Flat	2	2 1/2	3	3 1/2	4	4 1/2	5	5 1/2	6	6 1/2	6 1/2

THIS TABLE IS NOT A TABLE OF RECOMMENDED THICKNESSES

It represents only a table suitable for a particular type of insulation, ambient condition, surface emmittance and air movement. These Tables 14, 15, 16, 17, 18, 19, 20, 21, and 22 are all based on this single condition, which is not representative of any particular set of conditions for industry. This is only presented to illustrate how economics of insulation should be determined.

Table 21a Insulation thicknesses used to calculate heat, steam, and dollar savings. (Metric Units)

Pipe Size Dia. mm	INSULATION THICKNESS mm – Metric Units / OPERATING TEMPERATURE 100°C 212°F	149°C 300°F	204°C 400°F	260°C 500°F	316°C 600°F	371°C 700°F	427°C 800°F	482°C 900°F	538°C 1000°F	593°C 1100°F	649°C 1200°F	NPS Pipe Size Nom."
21.3	26	26	26	26	40	40	40	53	53	53	53	1/2
26.7	23	23	23	23	37	37	37	50	50	50	50	3/4
33.4	27	27	27	40	40	40	54	54	54	54	67	1
42.2	23	23	23	42	42	49	49	49	49	49	63	1 1/4
48.3	26	26	39	39	60	60	60	72	72	72	72	1 1/2
60.3	26	26	40	40	53	53	53	66	66	66	79	2
73.0	26	26	40	40	53	53	66	66	66	79	79	2 1/2
88.9	26	26	39	39	52	52	65	65	65	77	77	3
101.6	33	46	46	58	58	71	71	71	85	85	85	3 1/2
114.3	26	39	39	52	52	65	65	79	79	79	92	4
141.3	25	38	38	51	51	65	65	78	78	78	90	5
168.3	37	37	51	51	51	64	77	77	93	93	93	6
219.1	39	39	51	51	67	80	80	93	93	105	105	8
273.0	40	40	53	66	80	93	93	93	105	105	118	10
323.8	40	53	66	66	78	91	91	104	117	117	130	12
355.6	37	50	62	75	88	88	101	101	114	114	127	14
406.4	38	51	64	76	89	89	101	101	114	114	127	16
457.2	38	51	64	76	89	101	101	114	114	127	127	18
508.0	38	51	64	76	89	101	101	114	114	127	127	20
609.6	38	51	64	76	89	101	114	114	127	127	140	24
762.6	51	64	76	89	101	114	114	127	140	140	152	30
914.4	51	64	76	89	101	114	127	140	152	152	165	36
mm THICKNESSES for NPS pipe as per ASTEM C-585												
Flat	51	64	76	89	101	114	127	140	152	165	178	Flat

THIS TABLE IS NOT A TABLE OF RECOMMENDED THICKNESSESS

It represents only a table suitable for a particular type of insulation, ambient condition, surface emittance and air movement. It is only presented to illustrate how economics of insulation should be determined.

Table 22 Example of installed costs to compare with capital savings of steam capacity costs. (English Units) (Based on thicknesses shown in Table 21.)

NPS Pipe Size Nom. "	INSTALLED INSULATION COST – dollars/linear foot – English Units / OPERATING TEMPERATURE 212°F 100°C	300°F 149°C	400°F 204°C	500°F 260°C	600°F 316°C	700°F 371°C	800°F 427°C	900°F 482°C	1000°F 538°C	1100°F 593°C	1200°F 649°C	Pipe Size Dia. mm
1/2	$ 2.60	$ 2.60	$ 2.60	$ 2.60	$ 3.60	$ 3.60	$ 3.60	$ 4.70	$ 4.70	$ 4.70	$ 4.70	21.3
3/4	2.70	2.70	2.70	2.70	3.80	3.80	3.80	5.00	5.00	5.00	5.90	26.7
1	2.80	2.80	2.80	4.00	4.00	4.00	4.00	5.20	5.20	5.20	6.20	33.4
1 1/4	2.86	2.86	2.86	4.10	4.10	5.40	5.40	5.40	5.40	5.40	6.40	42.2
1 1/2	3.14	3.14	4.30	4.30	5.50	5.50	5.50	6.60	6.60	6.60	6.60	48.3
2	3.24	3.24	4.40	4.40	5.80	5.80	5.80	7.00	7.00	7.00	9.00	60.3
2 1/2	3.50	3.50	4.60	4.60	6.00	6.00	6.00	7.40	7.40	9.40	9.40	73.0
3	3.70	3.70	4.80	4.80	6.40	6.40	6.40	8.00	8.00	10.00	10.00	88.9
3 1/2	4.00	5.30	5.30	6.70	6.70	8.60	8.60	8.60	10.60	10.60	10.60	101.6
4	4.40	5.50	5.50	7.10	7.10	9.20	9.20	9.20	11.20	12.20	12.20	114.3
5	4.80	6.00	6.00	7.60	7.60	9.80	9.80	9.80	12.00	13.60	13.60	141.3
6	6.50	6.50	8.50	8.50	8.50	10.80	10.80	13.00	15.00	15.00	15.00	168.3
8	7.60	7.60	9.80	9.80	12.00	14.20	14.20	17.00	17.50	19.50	19.50	219.1
10	8.80	8.80	11.40	14.50	16.40	18.50	18.50	18.50	22.00	22.20	26.00	273.0
12	9.70	12.50	14.40	15.40	18.00	21.00	21.00	24.50	29.00	29.00	25.00	323.8
14	11.00	13.80	17.00	20.00	23.50	23.50	28.00	28.00	33.00	33.00	39.00	355.6
16	12.00	15.00	18.40	22.00	26.00	26.00	32.00	32.00	37.00	37.00	43.00	406.4
18	13.20	16.40	20.00	24.00	28.00	34.00	34.00	40.00	40.00	46.00	46.00	457.2
20	14.60	18.00	22.00	26.00	30.50	38.00	38.00	46.00	46.00	54.00	54.00	508.0
24	16.40	21.00	25.00	30.00	34.50	42.00	52.00	52.00	60.00	60.00	70.00	609.6
30	25.00	29.00	34.00	39.00	46.00	60.00	60.00	68.00	78.00	78.00	86.00	762.6
36	30.00	32.50	40.00	43.00	50.00	72.00	80.00	88.00	92.00	92.00	100.00	914.4
INSTALLED INSULATION COST – dollars/square foot												
Flat	$ 3.80	$ 4.40	$ 5.00	$ 5.80	$ 6.80	$ 7.40	$ 8.00	$ 8.60	$ 9.20	$ 9.80	$ 10.40	Flat

Note to convert: $/linear foot to $/linear metre multiply by 3.28.
$/square foot to $/square metre multiply by 10.764.

Table 22a Example of installed costs to compare with capital savings of steam capacity costs. (Metric Units) (Based on thickness shown in Table 21a.)

Pipe Size Dia. mm	INSTALLED INSULATION COSTS – dollars/linear metre – Metric Units											NPS Pipe Size Nom."
	OPERATING TEMPERATURE											
	100°C 212°F	149°C 300°F	204°C 400°F	260°C 500°F	316°C 600°F	371°C 700°F	427°C 800°F	482°C 900°F	538°C 1000°F	593°C 1100°F	649°C 1200°F	
21.3	8.53	8.53	8.53	8.53	11.81	11.81	11.81	15.42	15.42	15.42	15.42	1/2
26.7	8.86	8.86	8.86	8.86	12.46	12.46	12.46	16.40	16.40	16.40	19.35	3/4
33.4	9.18	9.18	9.18	13.12	13.12	13.12	13.12	17.06	17.06	17.06	20.34	1
42.2	9.38	9.38	9.38	12.45	13.45	17.71	17.71	17.71	17.71	17.71	20.00	1 1/4
48.3	10.13	10.30	14.10	14.10	18.04	18.04	18.04	21.65	21.65	21.65	21.65	1 1/2
60.3	10.63	10.63	14.43	14.43	19.02	19.02	19.02	22.96	22.96	22.96	29.52	2
73.0	11.48	11.48	15.09	15.09	19.68	19.68	19.68	24.27	24.27	30.83	30.83	2 1/2
88.9	12.14	12.14	15.74	15.74	20.99	20.99	20.99	26.24	26.24	32.80	32.80	3
101.6	13.12	17.38	27.38	21.98	21.98	28.21	28.21	28.21	34.77	34.77	34.77	3 1/2
114.3	14.42	18.04	18.04	23.29	23.29	30.18	30.18	30.18	40.02	40.02	40.02	4
141.3	15.74	19.68	19.68	24.93	24.93	32.14	32.14	32.14	44.61	44.61	44.61	5
168.3	21.32	21.32	27.88	27.88	27.88	35.42	35.42	35.42	49.20	49.20	49.20	6
219.1	24.93	24.93	32.14	32.14	39.36	46.58	46.58	55.76	57.40	63.96	63.96	8
273.0	28.86	28.86	37.39	47.56	53.79	60.68	60.68	60.68	72.16	72.16	85.28	10
323.8	31.82	41.00	47.23	49.20	59.04	68.88	68.88	80.36	95.12	95.12	114.80	12
355.6	36.08	45.26	55.76	65.60	77.08	77.08	91.84	91.84	108.24	108.24	127.92	14
406.4	39.36	49.20	59.70	72.16	85.28	85.28	104.96	104.96	121.36	121.36	141.04	16
457.2	43.30	53.79	65.60	78.72	91.84	111.52	111.52	131.20	131.20	150.88	150.88	18
508.0	47.89	59.04	72.16	85.28	100.04	124.64	124.64	150.88	150.88	177.12	177.12	20
609.6	53.79	68.88	82.00	98.40	113.16	137.76	170.56	170.56	196.80	196.80	229.60	24
762.6	62.00	95.12	111.52	127.92	150.68	196.80	196.80	233.04	255.84	255.84	282.06	30
914.4	98.40	106.00	131.20	141.04	164.00	236.16	262.40	288.64	301.76	301.76	328.00	36
	INSTALLED INSULATION COST – dollars/square metre											
Flat	40.90	47.36	53.82	62.43	73.20	79.65	86.11	92.57	99.02	105.50	111.95	Flat

Note to convert: $/linear metre to $/linear foot multiply by 0.3048.
$/square metre to $/square foot multiply by 0.0929.

Table 23 Heat loss—Insulated pipe and equipment. (Based on thicknesses shown in Table 21.)

NPS Pipe Size Nom. "	HEAT LOSS BTU/linear foot, hour – English Units											Pipe Size Dia. mm
	OPERATING TEMPERATURE											
	212°F 100°C	300°F 149°C	400°F 204°C	500°F 260°C	600°F 316°C	700°F 371°C	800°F 427°C	900°F 482°C	1000°F 538°C	1100°F 593°C	1200°F 649°C	
1/2	17	29	46	65	70	88	109	116	137	161	185	21.3
3/4	20	35	55	77	81	102	126	132	156	182	180	26.7
1	22	37	57	66	87	111	118	148	169	197	204	33.4
1 1/4	25	47	74	75	98	113	140	170	201	125	239	42.2
1 1/2	27	48	60	84	89	113	140	150	178	207	239	48.3
2	31	55	68	95	106	134	165	177	210	245	257	60.3
2 1/2	36	63	70	100	112	143	156	188	233	239	274	73.0
3	43	75	89	126	133	176	189	228	270	283	326	88.9
3 1/2	44	57	89	108	142	159	197	237	252	294	338	101.6
4	51	69	108	126	166	181	224	239	283	337	347	114.3
5	61	82	129	149	196	211	260	278	330	385	401	141.3
6	55	96	121	170	224	242	263	317	331	386	445	168.3
8	57	97	150	210	277	284	313	378	402	428	494	219.1
10	78	139	175	209	241	272	356	404	435	508	538	273.0
12	82	130	170	286	276	311	384	419	455	531	567	323.8
14	104	146	191	233	271	342	381	459	497	580	613	355.6
16	118	164	214	260	302	382	424	511	553	645	685	406.4
18	132	182	238	288	335	380	468	512	607	653	752	457.2
20	149	200	260	315	366	415	510	560	662	710	818	508.0
24	172	237	307	370	429	485	541	652	709	827	862	609.6
30	169	242	322	398	468	532	655	724	793	924	990	762.6
36	199	291	387	478	562	639	720	803	883	1004	1107	914.4
	HEAT LOSS–BTU/square foot per hour											
Flat	19	29	38	46	53	60	67	73	79	87	98	Flat

Table 23a Heat Loss – Insulated Pipe and Equipment
(Based on thickness shown in Table 21a)

Pipe Size Dia. mm	HEAT LOSS W/linear metre – Metric Units OPERATING TEMPERATURE 100°C 212°F	149°C 300°F	204°C 400°F	260°C 500°F	316°C 600°F	371°C 700°F	427°C 800°F	482°C 900°F	538°C 1000°F	593°C 1100°F	649°C 1200°F	NPS Pipe Size Nom."
21.3	16	28	44	62	67	85	104	111	132	154	178	1/2
26.7	19	33	53	74	78	98	121	127	149	174	172	3/4
33.4	21	35	54	63	84	107	113	142	162	189	196	1
42.2	24	45	71	72	94	109	135	163	193	226	229	1 1/4
48.3	26	26	57	81	86	109	135	144	171	199	229	1 1/2
60.3	29	53	65	91	102	129	159	170	202	235	246	2
73.0	34	61	67	96	108	137	150	180	223	230	263	2 1/2
88.9	41	72	85	121	127	169	182	219	259	271	313	3
101.6	42	55	85	103	136	153	189	228	242	283	324	3 1/2
114.3	49	67	101	121	160	174	215	230	272	324	333	4
141.3	59	79	124	143	188	203	250	267	317	370	385	5
168.3	53	93	117	163	215	233	253	305	318	371	427	6
219.1	55	94	144	202	266	273	301	363	386	411	474	8
273.0	75	134	169	201	232	261	342	388	418	438	516	10
323.8	79	125	163	274	265	299	369	403	437	510	544	12
355.6	101	140	184	224	260	329	366	441	478	557	589	14
406.4	115	158	205	250	290	367	407	491	531	620	658	16
457.2	127	175	228	276	322	365	450	492	583	627	722	18
508.0	143	192	250	303	353	399	490	538	636	682	786	20
609.6	165	228	295	356	412	466	520	626	681	794	828	24
762.6	162	233	309	382	450	511	629	696	762	888	951	30
914.4	191	280	372	459	540	614	691	771	848	964	1063	36
	HEAT LOSS W/square metre											
Flat	60	91	119	145	167	189	211	230	249	274	308	Flat

Table 24 Heat saved per hour. (Based on insulation thickness shown in Table 21.)

NPS Pipe Size Nom. "	HEAT SAVINGS BTU/linear foot, hour – English Units OPERATING TEMPERATURE 212°F 100°C	300°F 149°C	400°F 204°C	500°F 260°C	600°F 316°C	700°F 371°C	800°F 427°C	900°F 482°C	1000°F 538°C	1100°F 593°C	1200°F 649°C	Pipe Size Dia. mm
1/2	65	137	241	380	579	823	1109	1486	1838	2483	3132	21.3
3/4	81	170	298	470	720	1011	1382	1852	2420	3101	3942	26.7
1	101	213	377	608	902	1268	1747	2319	3025	3883	4919	33.4
1 1/4	128	265	467	768	1139	1615	2202	2921	3809	4888	6199	42.2
1 1/2	145	304	573	871	1314	1847	2521	3364	4390	5634	7106	48.3
2	182	376	685	1081	1626	2289	3130	4205	5455	7006	8876	60.3
2 1/2	216	453	829	1308	1965	2864	3800	5027	6584	8499	10781	73.0
3	237	545	994	1469	2366	3335	4595	6105	7989	10249	13018	88.9
3 1/2	299	643	1037	1796	2675	3806	5210	6962	9238	11845	14851	101.6
4	332	715	1262	2023	3013	4286	5870	7840	10262	13182	16702	114.3
5	359	779	1384	2230	3317	4728	6482	8666	11346	14597	18491	141.3
6	448	1036	1855	2935	4380	6232	8395	11452	15035	19354	24464	168.3
8	640	1339	2380	3778	5650	8082	11120	14833	19493	25140	31799	219.1
10	713	1630	2939	4709	7081	10053	13791	18408	24186	31171	39513	273.0
12	848	1949	3494	5623	8351	11883	16337	21829	28678	36970	46862	323.8
14	1076	2012	3798	6183	9133	12959	17855	23720	31347	40385	51238	355.6
16	1122	2394	4315	7900	10395	14742	20345	27152	35704	46044	58404	406.4
18	1258	2683	4736	7744	11675	16612	22878	30587	40181	51886	65704	457.2
20	1381	2968	5358	8582	12604	18362	25300	33872	44495	57448	72873	508.0
24	1648	3545	6672	10215	15394	21890	30295	40460	53249	68609	87339	609.6
30	2046	4400	7920	12705	19137	27226	37644	50383	66333	85634	108882	762.6
36	2461	5279	9403	15242	22968	32661	45180	60497	79617	102996	130693	914.4
	HEAT SAVINGS–BTU/square foot per hour											
Flat	276	625	1107	1760	2613	3705	5059	6727	8806	11313	14296	Flat

To convert from heat units (in Btu's) the values below were used:

Btu/lb. steam at temperature °F listed below										
212	300	400	500	600	700	800	900	1000	1100	1200
1150	1195	1240	1290	1335	1385	1435	1485	1535	1585	1639

From these values of steam in Btu's the heat saved in KW/linear metre was determined and is shown in Table 24a.

Table 24a Heat saved per hour. (Metric Units)
(Based on insulation thickness shown in Table 21a.)

Pipe Size Dia. mm	HEAT SAVINGS IN KW/linear metre – Metric Units OPERATING TEMPERATURE 100°C 212°F	149°C 300°F	204°C 400°F	260°C 500°F	316°C 600°F	371°C 700°F	427°C 800°F	482°C 900°F	538°C 1000°F	593°C 1100°F	649°C 1200°F	NPS Pipe Size Nom."
21.3	62	131	232	365	556	790	1066	1427	1766	2385	3009	1/2
26.7	78	163	286	452	691	971	1328	1779	2325	2979	3787	3/4
33.4	97	204	362	584	866	1218	1679	2219	2906	3731	4727	1
42.2	123	254	448	738	1094	1552	2115	2806	3660	4696	5957	1 1/4
48.3	139	292	550	836	1262	1775	2422	3233	4219	5413	6828	1 1/2
60.3	174	361	658	1038	1562	2199	3007	4040	5241	6732	8529	2
73.0	207	435	796	1257	1888	2752	3651	4830	6326	8166	10369	2 1/2
88.9	227	523	955	1411	2374	3204	4415	5866	7677	9849	12509	3
101.6	287	617	996	1725	2570	3657	5006	6689	8780	11382	14770	3 1/2
114.3	319	687	1212	1943	2895	4118	5640	7534	9861	12667	16049	4
141.3	344	748	1329	2142	3187	4543	6229	8327	10902	14026	17768	5
168.3	430	995	1782	2820	4208	5988	8067	11004	14447	18597	23508	6
219.1	614	1286	2287	3630	5429	7766	10685	14253	18731	24157	30556	8
273.0	685	1566	2824	4524	6804	9660	13252	17688	23240	29953	37968	10
323.8	814	1872	3357	5403	8024	11418	15698	20975	27557	35525	45030	12
355.6	1033	1933	3649	5941	8776	12452	17157	22793	30121	38806	49235	14
406.4	1078	2300	4146	6630	9989	14166	19549	26090	34308	44244	56121	16
457.2	1208	2578	4550	7441	11219	15962	21983	29391	38610	49858	63136	18
508.0	1327	2851	5148	8247	12111	17644	24311	32547	42756	55202	70024	20
609.6	1583	3406	6411	9815	14792	21934	29111	38878	51249	64927	83924	24
762.6	1966	4228	7610	12208	18388	26161	26172	48413	65740	82287	104626	30
914.4	2364	5072	9035	14646	22070	31384	43413	58132	76505	98970	125584	36
	HEAT SAVINGS IN W/square metre											
Flat	870	1970	3490	5549	8238	11680	15949	21204	27762	35665	45069	Flat

Table 25 Steam saved per hour. (Based on insulation thicknesses shown in Table 21.)

NPS Pipe Size Nom."	STEAM SAVED POUND/linear foot, hour – English Units OPERATING TEMPERATURE 212°F 100°C	300°F 149°C	400°F 204°C	500°F 260°C	600°F 316°C	700°F 371°C	800°F 427°C	900°F 482°C	1000°F 538°C	1100°F 593°C	1200°F 649°C	Pipe Size Dia. mm
1/2	0.05	0.11	0.19	0.30	0.43	0.59	0.77	1.00	1.19	1.56	1.91	21.3
3/4	0.07	0.14	0.24	0.37	0.54	0.73	0.98	1.25	1.57	1.95	2.41	26.7
1	0.09	0.18	0.30	0.49	0.68	0.92	1.22	1.56	1.97	2.44	3.00	33.4
1 1/4	0.11	0.22	0.38	0.60	0.85	1.16	1.54	1.97	2.48	3.08	3.78	42.2
1 1/2	0.12	0.25	0.46	0.68	0.98	1.33	1.76	2.26	2.86	3.55	4.35	48.3
2	0.16	0.31	0.55	0.84	1.21	1.65	2.18	2.83	3.45	4.41	5.42	60.3
2 1/2	0.19	0.38	0.67	1.02	1.45	2.07	2.65	3.39	4.29	5.36	6.60	73.0
3	0.20	0.46	0.80	1.14	1.76	2.89	3.20	4.11	5.21	6.47	7.97	88.9
3 1/2	0.26	0.54	0.84	1.40	2.01	2.74	3.63	4.69	5.95	7.48	9.09	101.6
4	0.29	0.60	1.02	1.58	2.23	3.09	4.10	5.28	6.68	8.33	10.21	114.3
5	0.31	0.65	1.11	1.74	2.48	3.42	4.53	5.84	7.40	9.22	11.30	141.3
6	0.39	0.86	1.49	2.28	3.27	4.51	5.86	7.72	9.77	12.22	14.94	168.3
8	0.55	1.12	1.91	2.94	4.22	5.83	7.76	10.01	12.67	15.85	19.40	219.1
10	0.62	1.36	2.36	3.67	5.27	7.26	9.63	12.42	15.73	19.65	24.10	273.0
12	0.73	1.63	2.81	4.38	6.26	8.58	11.40	14.75	19.33	23.30	28.60	323.8
14	0.93	1.68	3.06	4.82	6.85	9.27	12.45	15.98	20.42	25.45	31.30	355.6
16	0.97	2.00	3.49	5.38	7.78	10.65	14.20	18.32	23.25	29.05	35.65	406.4
18	1.07	2.24	3.80	6.03	8.75	12.00	15.96	20.62	26.16	32.70	40.05	457.2
20	1.19	2.48	4.31	6.68	9.44	13.24	17.64	22.85	29.00	36.20	44.50	508.0
24	1.42	2.96	5.21	7.96	11.52	15.82	21.14	27.43	34.70	43.30	53.20	609.6
30	1.77	3.67	5.97	9.78	14.32	19.67	26.25	33.90	43.20	53.90	66.40	762.6
36	2.22	4.40	7.56	11.88	17.12	23.60	31.46	40.70	51.80	64.90	79.80	914.4
	STEAM SAVED–POUNDS/square foot per hour											
Flat	0.24	0.52	0.89	1.37	1.88	2.68	3.53	4.53	5.74	7.12	8.72	Flat

Table 25a **Heat saved per hour. (Metric Units)**
(Based on insulation thickness shown in Table 21a.)

Pipe Size Dia. mm	HEAT SAVINGS IN W/linear metre - hour — Metric Units											NPS Pipe Size Nom. "
	OPERATING TEMPERATURE											
	100°C 212°F	149°C 300°F	204°C 400°F	260°C 500°F	316°C 600°F	371°C 700°F	427°C 800°F	482°C 900°F	538°C 1000°F	593°C 1100°F	649°C 1200°F	
21.3	62	131	232	365	556	790	1066	1427	1766	2385	3009	1/2
26.7	78	163	286	452	691	971	1328	1779	2325	2979	3787	3/4
33.4	97	204	362	584	866	1218	1679	2219	2906	3731	4727	1
42.2	123	254	448	738	1094	1552	2115	2806	3660	4696	5957	1 1/4
48.3	139	292	550	836	1262	1775	2422	3233	4219	5413	6828	1 1/2
60.3	174	361	658	1038	1562	2199	3007	4040	5241	6732	8529	2
73.0	207	435	796	1257	1888	2752	3651	4830	6326	8166	10369	2 1/2
88.9	227	523	955	1411	2374	3204	4415	5866	7677	9849	12509	3
101.6	287	617	996	1725	2570	3657	5006	6689	8780	11382	14770	3 1/2
114.3	319	687	1212	1943	2895	4118	5640	7534	9861	12667	16049	4
141.3	344	748	1329	2142	3187	4543	6229	8327	10902	14026	17768	5
168.3	430	995	1782	2820	4208	5988	8067	11004	14447	18597	23508	6
219.1	614	1286	2287	3630	5429	7766	10685	14253	18731	24157	30556	8
273.0	685	1566	2824	4524	6804	9660	13252	17688	23240	29953	37968	10
323.8	814	1872	3357	5403	8024	11418	15698	20975	27557	35525	45030	12
355.6	1033	1933	3649	5941	8776	12452	17157	22793	30121	38806	49235	14
406.4	1078	2300	4146	6630	9989	14166	19549	26090	34308	44244	56121	16
457.2	1208	2578	4550	7441	11219	15962	21983	29391	38610	49858	63136	18
508.0	1327	2851	5148	8247	12111	17644	24311	32547	42756	55202	70024	20
609.6	1583	3406	6411	9815	14792	21934	29111	38878	51249	64927	83924	24
762.6	1966	4228	7610	12208	18388	26161	26172	48413	65740	82287	104626	30
914.4	2364	5072	9035	14646	22070	31384	43413	58132	76505	98970	125584	36
	HEAT SAVINGS IN W/square metre											
Flat	870	1970	3490	5549	8238	11680	15949	21204	27762	35665	45069	Flat

Table 26 **Thousands of pounds steam saved per year.**
(Based on insulation thicknesses shown in Table 21 and 8760 hours per year operation time.)

NPS Pipe Size Nom. "	STEAM SAVED—M pound/year, linear foot — English Units											Pipe Size Dia. mm
	OPERATING TEMPERATURE											
	212°F 100°C	300°F 149°C	400°F 204°C	500°F 260°C	600°F 316°C	700°F 371°C	800°F 427°C	900°F 482°C	1000°F 538°C	1100°F 593°C	1200°F 649°C	
1/2	0.5	1.0	1.6	2.6	3.7	5.1	6.3	8.7	10.5	13.7	16.7	21.3
3/4	0.6	1.2	2.1	3.2	4.7	6.3	8.4	10.9	13.8	17.1	21.1	26.7
1	0.8	1.5	2.7	4.2	5.9	7.9	10.7	13.7	17.3	21.4	26.3	33.4
1 1/4	1.0	1.9	3.3	5.2	7.5	10.1	13.5	17.3	21.7	27.0	33.2	42.2
1 1/2	1.1	2.2	4.0	5.9	8.6	11.6	15.4	19.9	25.1	31.1	28.1	48.3
2	1.4	2.8	4.8	7.4	10.7	14.3	19.2	24.8	31.2	38.7	47.5	60.3
2 1/2	1.6	3.3	5.9	8.9	12.9	17.9	22.2	29.7	37.6	46.9	57.7	73.0
3	1.8	4.0	7.0	10.0	15.5	20.9	28.1	36.1	45.6	56.6	69.7	88.9
3 1/2	2.4	4.7	7.3	12.3	17.5	23.8	31.3	41.1	52.2	65.5	79.5	101.6
4	2.5	5.2	8.9	13.8	19.7	26.9	35.9	46.3	57.5	72.8	89.5	114.3
5	2.8	5.7	9.8	15.6	21.7	29.6	39.7	51.2	64.8	80.7	99.0	141.3
6	3.4	7.6	13.1	20.0	28.7	39.0	51.4	67.6	85.9	107.0	131.0	168.3
8	4.9	9.8	16.8	25.8	37.0	50.6	68.0	87.5	111.4	139.0	170.1	219.1
10	5.4	11.9	20.7	32.2	46.3	62.9	84.3	108.7	138.0	172.2	211.3	273.0
12	6.4	14.3	24.6	38.4	54.6	74.4	100.0	129.0	163.8	204.5	251.2	323.8
14	8.2	14.7	26.8	42.3	59.8	81.1	109.3	140.2	179.2	223.5	274.5	355.6
16	8.5	17.6	30.4	47.1	68.2	92.3	124.3	160.4	204.0	254.5	312.5	406.4
18	9.6	19.7	33.4	52.9	76.5	104.2	140.1	180.6	229.9	287.0	351.6	457.2
20	10.5	21.8	37.8	60.3	82.7	115.0	154.8	200.5	256.8	317.5	390.0	508.0
24	12.5	26.0	47.1	69.7	101.0	137.1	185.5	238.8	304.1	379.5	467.0	609.6
30	15.5	32.3	55.8	86.8	125.2	170.5	240.7	298.0	372.0	473.3	585.5	762.6
36	18.7	38.8	66.5	100.4	150.4	204.6	276.5	357.0	455.0	518.0	699.0	914.4
	STEAM SAVED—M pounds per year per square foot											
Flat	2.1	4.6	7.8	12.0	15.8	23.2	31.0	39.7	50.3	62.6	76.5	Flat

Table 26a **KW per linear metre saved per year. (Metric Units)**
(Based on insulation thicknesses as shown in Table 21a and 8760 hours per year operation time.)

Pipe Size Dia. mm	HEAT SAVINGS KW per linear metre, per year — Metric Units — OPERATING TEMPERATURE											NPS Pipe Size Nom."
	100°C 212°F	149°C 300°F	204°C 400°F	260°C 500°F	316°C 600°F	371°C 700°F	427°C 800°F	482°C 900°F	538°C 1000°F	593°C 1100°F	649°C 1200°F	
21.3	547	1153	2028	3198	4874	6928	9335	12508	15471	20900	26363	1/2
26.7	682	1431	2503	3956	6060	8510	11633	15589	20371	26102	33182	3/4
33.4	850	1793	3173	5117	7593	10673	14705	19520	25463	32685	41406	1
42.2	1022	2230	3931	6465	9588	13594	18535	24588	32062	41145	52181	1 1/4
48.3	1221	2559	4823	7331	11060	15547	21221	28317	36953	47424	59815	1 1/2
60.3	1532	3165	5766	9099	13685	19268	26347	35396	45918	58973	74714	2
73.0	1818	3813	6978	11010	16540	24107	31986	42315	55421	71541	90750	2 1/2
88.9	1994	4588	8367	12365	19916	28073	38678	51389	67248	86272	109580	3
101.6	2517	5412	8729	15118	22517	32037	43856	58603	76920	99706	125009	3 1/2
114.3	2795	6019	10622	17028	25362	36078	49411	65993	86381	110960	140590	4
141.3	3021	6557	11649	18771	27921	39798	54563	72947	95596	122871	155649	5
168.3	3771	8721	15614	24705	36869	52458	70665	96398	126558	162914	205927	6
219.1	5387	11271	20033	32801	47559	68030	93603	124858	164084	211618	267670	8
273.0	6001	13720	24739	39638	59604	84621	116087	154951	203587	262384	332603	10
323.8	7138	16406	29411	47331	70295	100025	137518	183747	241399	311197	394464	12
355.6	9057	16936	31969	52045	76878	109083	150295	199664	263865	339944	431299	14
406.4	9445	20151	36321	58081	87501	124092	171255	228554	300540	387579	491619	16
457.2	10589	22584	39866	65185	98275	139832	192577	257468	338226	436754	553067	18
508.0	11625	24983	45101	72239	106095	154563	212964	285114	374539	483573	613414	20
609.6	13872	29841	56162	85985	129580	184260	255010	340575	448227	577521	735182	24
762.6	17222	37037	66667	106945	161087	229177	316870	424102	558363	720830	916522	30
914.4	20715	44436	79150	128300	193334	274926	380305	509238	670182	866976	1100117	36
	HEAT SAVINGS KW per square metre per year											
Flat	7622	17260	30571	48605	72162	102319	139173	185750	243192	312427	394808	Flat

values per pound of steam when converting from heat to steam loss. Overall they range 10% to 25% low as compared to the typical conditions which are the basis of most insulation design. This was done to prove that under conditions most unfavorable to remarkable savings, the economic advantages to be gained from the use of insulation are still quite startling. The tables presented next will show this startling economic value of thermal insulation.

The following tables are simply conversions of steam loss to its value in money. As previously stated there is no one single value which can be placed on cost of capital investment of steam-producing equipment or production cost of steam. For this reason a typical value for each was selected. These values were: Capital investment in steam-producing equipment, including all costs (such as water treating, steam boilers, distribution system), is taken as $25.00 lb/hr (or $2.85 per 1000 lb of steam per year). Production cost of steam is taken to be $1.00/1000 lb.

In Metric Units this is approximately $73.75 capital investment per KWH. Production cost, in Metric Units, is taken to be $0.00294 per KWH.

It should be noted that the practice of putting cost figures in reference to steam is poor practice as the heat content of steam changes in respect to pressure, temperature and superheat. For this reason tables in English Units are given in savings in Btu's as well as steam. Steam savings in Metric Units will not be presented. Losses and saving in metric will be in Watts or KW's. The heat content of steam was based on 1 lb Abs. pressure steam superheated to temperatures listed. At 212° F Btu/lb steam = 1150 Btu and at 1200° F = 1639 Btu/lb.

Table 27 gives the amount of investment saved per linear foot of pipe or one square foot of flat surface. Table 27a gives the amount of investment saved per linear metre of pipe or one square metre of flat surface. The same ratio of change of heat content of steam has been used in the calculation of dollar savings in the Metric System.

Both of these tables present savings obtained by the use of thermal insulation. Conversely, if the insulation were not used this same amount would represent the capital investment necessary to supply the additional capacity to make up for the loss incurred by not using the insulation.

Except for pipe sizes less than 2" (60.3 mm) operating at 212° F (100° C) all other sizes show that it costs less to insulate pipes than the capital investment necessary to supply steam for the loss of heat due to no insulation. In other words it cost more in capital investment to provide heat for bare piping than the cost of the insulation. This makes calculation of return on investment in insulation quite difficult—how can one determine the return on a negative investment?

The savings in production costs of steam given in Table 28 (English Units) and Table 28a (Metric Units) provide savings in dollars per year saved by the insulation—which is in addition to the capital savings. These tables are based on operating time of 8760 hours per year.

On larger sizes the savings in capital costs of steam-producing equipment completely outweigh the cost of insulation. For example, the cost to properly insulate a 600° F, 6" NPS steam line is approximately $8.50 per linear foot. Savings in steam capacity is $81.75 per linear foot. Thus we show that it is $73.25 less costly per linear foot to insulate than not to insulate. In addition to this saving in capital cost the insulation will save $28.70 per linear foot in steam costs each year. In other words, the expenditure of $8.50 saves $101.95 the first year and an additional $28.70 each year thereafter.

To illustrate the economic advantages of insulation in regard to the capital investment necessary to supply capacity to take care of bare pipe losses vs the investment required to supply capacity for insulated pipe losses plus the investment in insulation, Table 29 is presented. Comparing the cost of steam-capacity investment with total investment of insulated lines it can quickly be seen that, except for small sizes operating at relatively low temperatures, less total capital investment is required when pipes are insulated than when they are left bare. In other words, the total capital investment is reduced by the use of thermal insulation.

At times the economic correctness of calculating the capital investment of savings in steam capacity, particularly in existing plants where boiler capacity is available, has been questioned. The only reliable answer to this question is this: if no additional steam-producing equipment is expected and the existing unit is never to be replaced, then steam capacity savings cannot be considered. However, if these conditions are true, then there is no point in calculating savings on that plant as it is going out of business anyway.

Assuming however, that there is a particular set of conditions where savings in capital investment of steam capacity cannot be justified, then what are the economics when the investment in insulation must be judged only on its ability to reduce heat (steam) losses. In this case, where no credit is given to savings in production capacity, all the insulation installation cost must be considered as investment. The saving in steam production cost is the return on this investment. These two items, plus percent return on investment, are shown in Table 30.

This table was prepared based on the following assumed values:

The ambient air temperature is 70° F (21° C)
The steam production costs = $3.00/M lbs, or $0.00882 per KWH.
Insulation thickness as listed in Table 21 and 21a.
Insulation cost as listed in Table 22 and 22a.

In addition, although temperatures and pipe sizes are given in Metric Units, the units used in the calculation, are given in $/lin. ft or $/sq ft only. However the result obtained is % return on investment and it remains unchanged by the use of Metric Units.

Examination of this table clearly indicates that return on insulation investment is so outstanding that all steam-heated pipes should be insulated to take advantage of the economics illustrated.

The previously mentioned tables had to be calculated for a given set of conditions, and some issue may be taken that the conditions chosen do not apply because their steam capital investment, steam-production cost, or insulation-installation cost is different. This is true. However, it is recommended that those interested use this system and calculate their own economics. Unless the installed cost of insulation is very much higher, and the steam costs are very, very low as compared to the typical figures shown, the answer will still show that it is profitable to insulate all steam-heated pipes.

Table 27 Dollar investment savings in production equipment due to insulation. Based on steam capacity investment cost of $25.00/lb hour (or $2.85 per M lb per year.) (Based on insulation thickness shown in Table 21.) (English Units)

NPS Pipe Size Nom. "	DOLLARS INVESTMENT SAVINGS per linear foot – English Units											Pipe Size Dia. mm
	OPERATING TEMPERATURE											
	212°F 100°C	300°F 149°C	400°F 204°C	500°F 260°C	600°F 316°C	700°F 371°C	800°F 427°C	900°F 482°C	1000°F 538°C	1100°F 593°C	1200°F 649°C	
1/2	$ 1.25	$ 2.75	$ 4.75	$ 7.50	$ 10.75	$ 14.75	$ 19.00	$ 25.00	$ 29.00	$ 39.00	$ 47.75	21.3
3/4	1.75	3.50	6.00	9.25	13.50	16.25	24.50	31.25	39.25	48.75	60.25	26.7
1	2.25	4.50	7.50	12.25	17.00	23.00	30.50	39.00	49.25	61.00	75.00	33.4
1 1/4	2.75	5.50	9.50	15.00	21.25	29.00	38.50	49.25	62.00	77.00	94.50	42.2
1 1/2	3.00	6.25	11.50	17.00	24.50	33.25	44.00	56.50	71.50	88.75	108.75	48.3
2	4.00	7.75	13.75	21.00	30.25	41.25	54.50	70.75	86.25	110.25	135.50	60.3
2 1/2	4.75	9.50	16.75	25.50	36.75	51.75	66.25	84.75	107.25	134.00	165.00	73.0
3	5.00	11.50	20.00	28.50	44.00	72.25	80.00	102.75	130.25	161.75	199.25	88.9
3 1/2	6.50	13.50	21.00	35.00	50.25	68.50	90.75	117.25	148.75	187.00	227.25	101.6
4	7.25	15.00	25.50	39.50	55.75	77.25	102.50	132.00	167.00	208.25	255.25	114.3
5	7.75	16.25	27.75	43.50	62.00	85.50	113.25	146.00	185.00	230.50	282.50	141.3
6	9.75	21.50	37.25	57.00	81.75	112.75	146.50	193.00	244.25	305.50	373.50	168.3
8	13.75	28.00	47.75	73.50	105.50	145.75	194.00	250.00	316.75	396.25	485.00	219.1
10	15.50	34.00	59.00	91.75	131.75	181.50	240.00	310.50	393.25	491.25	602.50	273.0
12	18.25	40.75	70.25	109.50	156.50	214.50	285.00	368.75	488.25	582.50	715.00	323.8
14	23.25	42.00	76.50	120.50	171.25	231.75	311.25	399.50	510.50	636.25	782.50	355.6
16	24.25	50.00	87.25	134.50	194.50	266.25	355.00	458.00	581.25	726.25	891.65	406.4
18	26.75	56.00	95.00	150.75	218.75	300.00	399.00	515.50	654.00	817.00	1012.50	457.2
20	29.75	62.00	107.75	167.00	236.00	331.00	441.00	571.25	725.50	905.00	1112.50	508.0
24	35.50	74.00	130.25	199.00	288.00	395.00	528.50	685.75	867.50	1082.50	1330.00	609.6
30	44.25	99.25	149.25	244.50	363.00	491.75	656.25	847.50	1080.00	1347.50	1660.00	762.6
36	55.50	110.00	189.00	297.00	428.00	590.00	786.50	1017.50	1295.00	1622.50	1995.00	914.4
	DOLLARS INVESTMENT SAVINGS per square foot (English Units)											
Flat	$ 6.00	$ 13.00	$ 22.25	$ 34.25	$ 47.00	$ 67.00	$ 88.25	$ 113.25	$ 143.50	$ 178.00	$ 218.00	Flat

Table 27a **Dollar investment savings in production equipment due to insulation. Based on steam investment cost $73.75/KWH. (Metric Units) (Based on insulation thicknesses as shown in Table 21a.)**

Pipe Size Dia. mm	DOLLARS INVESTMENT SAVINGS per linear metre – Metric Units											NPS Pipe Size Nom."
	OPERATING TEMPERATURE											
	100°C 212°F	149°C 300°F	204°C 400°F	260°C 500°F	316°C 600°F	371°C 700°F	427°C 800°F	482°C 900°F	538°C 1000°F	593°C 1100°F	649°C 1200°F	
21.3	$ 4.10	$ 9.02	$ 15.60	$ 24.60	$ 35.25	$ 48.40	$ 63.15	$ 82.00	$ 97.60	$ 127.00	$ 156.60	1/2
26.7	5.74	11.48	19.70	30.35	44.30	53.30	80.35	102.50	128.75	159.90	197.60	3/4
33.4	7.38	14.76	24.60	40.20	55.75	75.45	100.05	127.90	161.55	200.05	246.00	1
42.2	0.02	18.04	31.15	49.20	69.70	95.10	126.30	161.55	203.35	252.55	309.95	1 1/4
48.3	9.84	20.50	37.70	55.75	80.35	109.05	144.30	185.30	234.50	291.10	389.50	1 1/2
60.3	13.12	25.42	45.10	68.90	99.20	135.30	178.75	232.05	282.90	361.60	444.40	2
73.0	15.58	31.15	54.95	83.65	120.55	169.75	217.30	278.00	351.25	439.50	541.20	2 1/2
88.9	16.40	37.70	65.60	93.50	144.30	236.70	262.40	337.00	427.20	530.55	653.55	3
101.6	21.32	44.30	68.90	114.80	164.80	224.70	297.65	384.60	487.90	613.35	745.40	3 1/2
114.3	23.78	49.20	83.64	129.55	182.85	253.40	336.20	432.95	598.75	683.05	837.40	4
141.3	25.42	53.30	91.00	142.70	203.35	280.45	371.45	478.90	606.80	756.05	926.60	5
168.3	31.98	70.50	122.15	187.00	268.15	369.80	480.50	633.05	801.15	1002.05	1225.10	6
219.1	45.10	91.85	156.60	241.10	346.05	478.05	633.30	820.25	1038.95	1299.70	1590.10	8
273.0	50.84	111.50	193.50	301.00	432.15	595.30	812.60	1018.45	1289.85	1611.30	1976.20	10
323.8	59.86	133.65	230.40	359.15	513.30	703.55	934.80	1209.50	1585.05	1910.60	2345.20	12
355.6	76.25	137.75	250.90	395.25	561.70	760.15	1020.90	1310.35	1674.45	2086.90	2566.60	14
406.4	79.54	164.00	286.20	441.15	638.00	873.30	1164.40	1502.25	1906.50	2382.10	2924.60	16
457.2	87.74	183.70	311.60	494.45	717.50	984.00	1308.70	1690.85	2145.10	2681.40	3321.00	18
508.0	97.56	203.35	353.42	547.75	774.10	1085.70	1446.50	1873.70	2378.00	2968.40	3649.00	20
609.6	116.44	242.70	427.20	652.70	944.65	1295.50	1733.50	2249.25	2845.40	3550.60	4362.40	24
762.6	145.14	325.55	489.50	802.00	1190.40	1612.50	2152.50	2779.80	3542.40	4419.80	5444.80	30
914.4	182.04	360.80	619.90	974.15	1403.85	1935.70	2579.70	3337.40	4247.60	5321.80	6543.60	36
	DOLLARS INVESTMENT SAVINGS per square metre – Metric Units											
Flat	$ 64.60	$139.95	$239.50	$368.65	$ 505.90	$ 721.20	$ 949.90	$1219.00	$1544.60	$1916.00	$2346.55	Flat

Table 28 **Production cost of steam saved per year-dollars. (Based on insulation thickness shown in Table 21.) Steam production cost is $1.00 per 1000 lbs of steam. (English Units)**

NPS Pipe Size Nom. "	DOLLAR VALUE OF STEAM SAVED/linear foot per year – English Units											Pipe Size Dia. mm
	OPERATING TEMPERATURE											
	212°F 100°C	300°F 149°C	400°F 204°C	500°F 260°C	600°F 316°C	700°F 371°C	800°F 427°C	900°F 482°C	1000°F 538°C	1100°F 593°C	1200°F 649°C	
1/2	$ 0.50	$ 1.00	$ 1.60	$ 2.60	$ 3.70	$ 5.10	$ 6.80	$ 8.70	$ 10.50	$ 13.70	$ 16.70	21.3
3/4	0.60	1.20	2.10	3.20	4.70	6.30	8.40	10.90	13.80	17.10	21.10	26.7
1	0.80	1.50	2.70	4.20	0.40	7.90	10.70	15.70	17.30	21.40	26.30	33.4
1 1/4	1.00	1.90	3.30	5.20	7.50	10.10	12.50	17.30	21.70	27.00	33.20	42.2
1 1/2	1.10	2.20	4.00	5.90	8.60	11.60	15.40	19.90	25.10	31.10	39.00	48.3
2	1.40	2.80	4.80	7.40	10.70	14.30	19.20	24.80	31.20	38.70	47.50	60.3
2 1/2	1.60	3.30	5.90	8.90	12.90	17.90	22.20	29.70	36.60	46.90	56.70	73.0
3	1.80	4.00	7.00	10.00	15.50	20.90	28.10	36.10	45.60	56.60	69.70	88.9
3 1/2	2.40	4.70	7.30	12.30	17.50	23.80	31.40	41.10	52.20	65.50	79.30	101.6
4	2.50	5.20	8.90	13.80	19.70	26.90	35.90	46.30	57.50	72.80	89.50	114.3
5	2.80	5.70	9.80	15.60	21.70	29.60	39.70	50.20	64.80	80.70	99.00	141.3
6	3.40	7.60	13.10	20.00	28.70	39.00	51.40	67.60	85.90	107.00	131.00	168.3
8	4.90	9.80	16.80	25.80	37.00	50.60	68.00	87.50	111.40	139.00	170.10	219.1
10	5.40	10.90	20.70	32.20	26.30	61.90	84.30	108.70	138.00	172.20	211.30	273.0
12	6.40	14.30	24.60	38.40	54.60	74.40	100.00	127.00	163.80	204.50	251.20	323.8
14	8.40	14.70	26.80	42.30	59.80	81.10	109.30	140.20	179.20	223.50	264.50	355.6
16	8.50	17.60	30.40	47.10	68.20	92.30	124.30	160.40	204.00	254.50	312.50	406.4
18	9.60	19.70	33.40	52.90	76.50	104.20	140.10	180.60	229.90	287.00	351.60	457.2
20	10.50	21.80	37.80	60.30	82.70	115.00	154.80	200.50	254.80	317.50	390.00	508.0
24	12.50	26.00	47.10	69.70	101.00	137.10	185.50	238.80	304.10	379.50	437.00	609.6
30	15.50	32.30	55.80	86.80	125.20	170.50	240.70	298.00	379.00	473.30	585.50	762.6
36	19.70	38.80	66.50	100.00	150.40	204.60	277.50	357.00	455.00	518.00	699.00	914.4
	DOLLAR VALUE OF STEAM SAVED per square foot per year											
Flat	$ 2.10	$ 4.60	$ 7.80	$ 12.00	$ 15.80	$ 23.20	$ 31.00	$ 39.70	$ 50.30	$ 62.60	$ 76.50	Flat

Table 28a Production cost of steam saved per year-dollars.
(Based on production cost of $0.00294/KWH.) (Metric Units)

Pipe Size Dia. mm	DOLLAR VALUE OF HEAT SAVED per linear metre per year – Metric Units — OPERATING TEMPERATURE 100°C 212°F	149°C 300°F	204°C 400°F	260°C 500°F	316°C 600°F	371°C 700°F	427°C 800°F	482°C 900°F	538°C 1000°F	593°C 1100°F	649°C 1200°F	NPS Pipe Size Nom."
21.3	$ 1.61	$ 3.39	$ 5.96	$ 9.40	$ 14.33	$ 20.36	$ 27.45	$ 36.70	$ 45.49	$ 61.45	$ 77.51	1/2
26.7	2.00	4.21	7.37	11.63	17.82	25.02	34.20	45.83	59.89	76.74	97.56	3/4
33.4	2.49	5.27	9.33	15.04	22.32	31.38	43.23	57.49	74.86	96.10	121.73	1
42.2	3.71	6.56	11.56	19.01	28.19	39.97	54.49	72.29	94.26	120.97	153.41	1 1/4
48.3	3.58	7.52	14.18	21.56	32.52	45.71	62.39	83.25	108.64	139.42	175.86	1 1/2
60.3	4.50	9.31	16.95	26.79	40.24	56.65	77.46	104.06	135.00	173.38	219.66	2
73.0	5.34	11.21	20.51	32.37	48.63	70.88	94.04	124.41	162.94	210.33	266.80	2 1/2
88.9	5.87	13.49	24.60	36.35	58.55	82.53	113.72	151.08	197.71	253.64	322.16	3
101.6	7.40	15.91	25.66	44.44	66.20	94.19	128.94	172.29	226.14	293.14	367.53	3 1/2
114.3	8.22	17.69	31.23	50.06	74.56	106.07	145.27	194.02	253.96	326.22	413.34	4
141.3	8.88	19.28	34.25	55.19	82.09	117.00	160.41	214.46	280.79	361.24	457.61	5
168.3	11.09	25.64	45.91	72.63	108.39	154.23	207.76	283.41	372.08	478.97	605.43	6
219.1	15.84	33.14	58.90	93.50	139.82	200.01	275.19	367.08	482.41	622.16	786.95	8
273.0	17.65	40.34	72.73	116.54	175.24	248.79	341.29	455.55	598.55	771.41	977.85	10
323.8	20.99	48.23	86.47	139.16	206.67	294.08	404.30	540.22	709.71	914.92	1159.72	12
355.6	26.63	49.70	93.99	153.01	226.02	320.70	441.87	587.01	775.76	999.43	1268.02	14
406.4	27.77	59.25	106.79	170.75	257.25	364.83	503.49	671.95	883.59	1023.21	1445.36	16
457.2	31.13	66.40	117.20	191.64	288.93	411.11	566.18	756.96	994.39	1284.06	1626.02	18
508.0	34.18	73.45	132.60	212.38	311.92	454.42	626.12	338.25	110.15	1421.70	1803.44	20
609.6	40.78	87.73	165.12	252.80	380.97	541.73	749.73	1001.29	1318.79	1697.91	2161.43	24
762.6	50.63	108.00	196.00	314.42	476.82	673.78	931.60	1246.86	1641.59	2119.24	2694.57	30
914.4	60.90	130.00	232.70	377.21	568.40	808.28	111.09	1970.33	1970.33	2548.91	3234.34	36
	DOLLAR VALUE OF HEAT SAVED per square metre per year											
Flat	$22.40	$ 50.75	$ 89.88	$142.90	$212.16	$300.82	$410.76	$ 546.10	$ 714.99	$ 918.54	$1160.73	Flat

Table 29 Capital investment — uninsulated vs. insulated surfaces*

NPS Pipe Size Nom. Inches	CAPITAL INVESTMENT DOLLARS/LINEAR FOOT — OPERATING TEMPERATURE 212°F, 100°C — Non Insul. Steam Capital Invest.	212°F, 100°C Insulated Steam Capital Invest.	Insulated Insul. Invest.	Insulated Total Invest.	300°F, 149°C Non Insul. Steam Capital Invest.	300°F, 149°C Insulated Steam Capital Invest.	Insulated Insul. Invest.	Insulated Total Invest.	400°F, 204°C Non Insul. Steam Capital Invest.	400°F, 204°C Insulated Steam Capital Invest.	Insulated Insul. Invest.	Insulated Total Invest.	Pipe Size Dia. mm
½	$ 1.60	$0.35	$ 2.60	$ 2.95	$ 3.35	$0.60	$ 2.60	$ 3.20	$ 5.70	$0.95	$ 2.60	$ 3.64	21.3
¾	1.92	0.42	2.70	3.12	4.22	0.72	2.70	3.42	7.15	1.15	2.70	3.85	26.7
1	2.72	0.47	2.80	3.27	5.27	0.77	2.80	3.57	8.70	1.20	2.80	4.00	33.4
1¼	3.30	0.55	2.86	3.41	6.47	0.97	2.86	3.93	11.02	1.52	2.86	4.38	42.2
1½	3.60	0.60	3.14	3.74	7.25	1.00	3.14	4.14	12.75	1.25	3.14	4.39	48.2
2	4.65	0.65	3.24	3.89	8.90	1.15	3.24	4.39	15.15	1.40	3.24	4.64	60.3
2½	5.52	0.77	3.50	4.27	10.82	1.32	3.50	4.82	18.20	1.45	3.50	4.95	73.0
3	5.92	0.92	3.70	4.62	13.07	1.57	3.70	5.27	21.85	1.85	3.70	5.55	88.9
3½	7.45	0.95	4.00	4.95	14.70	1.20	5.30	6.50	23.85	1.87	5.30	7.17	101.6
4	8.35	1.10	4.40	5.50	16.45	1.45	5.50	6.95	27.75	2.75	5.50	7.75	114.3
5	9.07	1.16	4.80	5.96	17.97	1.72	6.00	7.72	30.42	2.67	6.00	8.67	141.3
6	10.95	1.20	6.50	7.70	23.50	2.00	6.50	8.50	29.75	2.58	8.50	11.08	168.3
8	15.00	1.25	7.60	7.85	30.02	2.27	7.60	9.87	50.60	3.10	9.80	12.10	219.1
10	17.20	1.70	8.80	10.50	36.60	2.90	8.80	11.70	62.62	3.65	11.40	15.05	273.0
12	20.02	1.77	9.70	11.47	43.47	2.72	12.50	15.22	64.77	3.72	15.40	19.12	328.0
14	26.00	2.25	11.00	12.15	45.05	3.05	13.80	16.85	80.45	3.95	17.00	20.95	355.0
16	26.82	2.57	12.00	14.57	53.42	3.45	15.00	18.45	91.65	4.42	18.40	22.82	406.4
18	29.62	2.87	13.20	16.07	59.80	3.80	16.40	20.20	99.92	4.92	20.00	24.92	457.2
20	33.00	3.25	14.60	17.85	66.17	4.17	18.00	22.17	13.12	5.37	22.00	27.37	508.0
24	39.25	3.75	16.40	20.15	78.95	4.95	21.00	25.95	36.62	6.37	25.00	31.37	609.6
30	47.92	4.17	25.00	29.17	96.82	5.07	29.00	34.07	155.87	6.62	34.00	40.62	762.6
36	55.82	4.32	30.00	34.32	115.10	6.10	35.00	41.10	197.02	8.02	40.00	48.02	914.4
	CAPITAL INVESTMENT DOLLARS/SQUARE FOOT												
Flat	$ 6.40	$0.40	$ 3.80	$ 4.20	$ 13.60	$0.60	$ 4.40	$ 5.00	$ 23.05	$0.80	$ 5.00	$ 5.80	Flat

To convert to metric units \$/linear foot × 3.28 = \$/linear metre, \$/square foot × 10.764 = \$/m^2

*Based on steam capital investment of $25.00 per pound/hour.
Insulation thickness as shown in Table 21 and cost of installed insulation as shown in Table 22.

(Sheet 1 of 4)

Table 29 Capital investment – uninsulated vs. insulated surfaces*

NPS Pipe Size Nom. Inches	CAPITAL INVESTMENT DOLLARS/LINEAR FOOT – OPERATING TEMPERATURE												Pipe Size Dia. mm
	500°F, 260°C				600°F, 316°C				700°F, 371°C				
	Non Insul.	Insulated			Non Insul.	Insulated			Non Insul.	Insulated			
	Steam Capital Invest.	Steam Capital Invest.	Insul. Invest.	Total Invest.	Steam Capital Invest.	Steam Capital Invest.	Insul. Invest.	Total Invest.	Steam Capital Invest.	Steam Capital Invest.	Insul. Invest.	Total Invest.	
½	$ 8.77	$1.27	$ 2.60	$ 1.87	$ 12.07	$ 1.32	$ 3.60	$ 4.92	$ 16.00	$ 1.60	$ 3.60	$ 5.20	21.3
¾	10.77	1.50	2.70	4.20	15.02	1.52	3.80	5.32	20.10	1.85	3.80	5.65	26.7
1	13.52	1.27	4.00	5.27	18.70	1.70	4.00	5.70	25.00	2.00	4.00	6.00	33.4
1¼	16.45	1.47	4.10	5.57	23.15	1.90	4.10	6.00	31.05	2.05	5.40	7.45	42.2
1½	17.37	1.62	4.30	5.92	26.15	1.65	5.50	7.15	36.05	2.05	5.50	7.55	48.2
2	22.85	1.85	4.40	6.25	32.22	1.97	5.80	7.77	43.67	2.42	5.80	8.22	60.3
2½	27.45	1.95	4.60	6.55	38.92	2.17	6.00	8.17	54.32	2.57	6.00	8.57	73.0
3	30.96	2.45	4.80	7.25	46.50	2.50	6.40	8.90	62.92	3.17	6.40	9.57	88.9
3½	37.10	2.10	6.70	8.80	52.90	2.65	6.70	9.35	71.37	2.87	8.60	11.47	101.6
4	41.95	2.45	7.10	9.55	58.85	3.10	7.10	10.20	80.52	3.27	9.20	12.47	114.3
5	46.40	2.90	7.60	10.50	65.67	3.67	7.60	11.27	89.32	3.82	9.80	13.62	141.3
6	60.30	3.30	8.50	11.80	85.95	4.20	8.50	12.70	117.12	4.37	10.80	15.17	168.3
8	77.57	4.07	9.80	13.81	110.67	5.17	12.00	17.17	150.87	5.12	14.20	19.32	219.1
10	95.82	4.07	14.50	18.57	136.25	4.50	16.40	20.90	186.42	4.92	18.50	23.42	273.0
12	115.05	5.55	15.40	20.95	161.65	5.15	18.00	23.15	220.12	5.62	21.00	26.62	328.0
14	125.02	4.52	20.00	24.52	176.32	5.07	23.50	28.57	237.92	6.17	23.50	29.67	355.0
16	139.55	5.05	22.00	27.05	200.15	5.65	26.00	31.65	273.15	6.90	26.00	32.90	406.4
18	156.35	5.60	24.00	29.60	225.02	6.27	28.00	34.27	306.87	6.87	34.00	40.87	457.2
20	173.12	6.12	26.00	32.12	242.85	6.85	30.50	37.35	338.50	7.50	38.00	45.50	508.0
24	206.17	7.17	30.00	37.17	296.05	8.05	34.50	42.55	404.25	8.75	42.00	50.75	609.6
30	252.25	7.75	39.00	46.75	366.77	8.77	46.00	54.77	501.35	9.60	60.00	69.60	762.6
36	306.30	9.30	43.00	52.30	438.40	10.50	50.00	60.50	605.55	11.55	72.00	83.55	914.4
	CAPITAL INVESTMENT DOLLARS/SQUARE FOOT												
Flat	$ 35.15	$0.90	$ 5.80	$ 6.70	$ 48.00	$ 1.00	$ 6.80	$ 7.80	$ 68.10	$ 1.10	$ 7.40	$ 8.50	Flat

To convert to metric units \$/linear foot X 3.28 = \$/linear metre, \$/square foot X 10.764 = \$/m²

(Sheet 2 of 4)

Table 29 Capital investment – uninsulated vs. insulated surfaces*

NPS Pipe Size Nom. Inches	CAPITAL INVESTMENT DOLLARS/LINEAR FOOT – OPERATING TEMPERATURE												Pipe Size Dia. mm
	800°F, 427°C				900°F, 482°C				1000°F, 538°C				
	Non Insul.	Insulated			Non Insul.	Insulated			Non Insul.	Insulated			
	Steam Capital Invest.	Steam Capital Invest.	Insul. Invest.	Total Invest.	Steam Capital Invest.	Steam Capital Invest.	Insul. Invest.	Total Invest.	Steam Capital Invest.	Steam Capital Invest.	Insul. Invest.	Total Invest.	
½	$ 21.22	$ 1.90	$ 3.60	$ 5.50	$ 26.95	$ 1.95	$ 4.70	$ 6.65	$ 31.97	$ 2.22	$ 4.70	$ 6.90	21.3
¾	26.70	2.20	3.80	6.00	33.47	2.22	5.00	7.22	41.80	2.55	5.00	7.50	26.7
1	32.55	2.05	4.00	6.05	41.50	2.50	5.20	7.70	52.00	2.75	5.20	7.90	33.4
1¼	40.95	2.45	5.40	7.85	52.10	2.85	5.40	8.25	65.25	3.27	5.40	8.60	42.2
1½	46.45	2.45	5.50	7.95	59.02	2.53	6.60	9.13	74.30	2.80	6.60	9.40	48.2
2	57.37	2.87	5.80	8.67	61.22	2.97	7.00	9.97	89.55	3.30	7.00	10.30	60.3
2½	68.97	2.72	6.00	8.72	87.92	3.17	7.90	10.57	110.92	3.67	7.40	11.00	73.0
3	83.30	3.30	6.40	9.70	106.60	3.85	8.00	11.85	134.80	4.55	8.00	12.50	88.9
3½	94.17	3.42	8.60	12.02	121.25	4.00	8.60	12.60	152.72	3.97	10.60	14.50	101.6
4	106.40	3.90	9.20	13.10	136.02	4.02	11.20	15.20	156.45	4.45	11.20	15.60	114.3
5	117.77	4.52	9.80	14.32	151.02	5.02	12.00	17.02	190.20	5.20	12.00	17.20	141.3
6	150.07	4.57	10.80	15.37	198.35	5.35	13.00	18.35	249.45	5.20	15.00	20.00	168.3
8	199.20	5.20	14.20	19.40	256.77	6.52	17.00	23.52	323.10	6.35	17.00	23.35	219.1
10	246.95	6.12	18.50	24.62	317.30	6.80	18.50	25.30	400.10	6.85	22.00	28.85	273.0
12	291.70	6.70	21.00	27.70	375.80	7.05	24.50	31.55	490.42	7.17	29.00	36.17	328.0
14	317.87	6.62	28.00	34.62	407.25	7.75	28.00	35.25	518.57	7.82	33.00	40.82	355.0
16	364.40	7.40	32.00	39.40	466.60	8.60	32.00	40.60	587.50	8.75	37.00	45.75	406.4
18	405.92	8.17	34.00	42.17	524.12	8.62	40.00	48.62	663.57	9.57	40.00	49.57	457.2
20	449.90	8.90	38.00	46.90	565.70	9.45	46.00	55.45	735.45	10.45	46.00	56.45	508.0
24	537.82	9.45	52.00	61.45	696.57	10.95	52.00	62.95	878.67	11.17	60.00	71.17	609.6
30	667.67	11.42	60.00	71.42	853.95	12.07	68.00	80.07	1092.50	12.50	78.00	90.50	762.6
36	799.05	12.55	80.00	92.55	1031.02	13.52	88.00	101.52	1308.92	13.92	92.00	105.92	914.4
	CAPITAL INVESTMENT DOLLARS/SQUARE FOOT												
Flat	$ 89.42	$ 1.17	$ 8.00	$ 9.17	$114.47	$ 1.22	$ 8.60	$ 9.82	$144.75	$ 1.25	$ 9.20	$ 10.45	Flat

To convert to metric units \$/linear foot X 3.28 = \$/linear metre, \$/square foot X 10.764 = \$/m²

*Based on steam capital investment of $25.00 per pound/hour.
Insulation thickness as shown in Table 21 and cost of installed insulation as shown in Table 22.

(Sheet 3 of 4)

Table 29 Capital investment – uninsulated vs. insulated surfaces*

NPS Pipe Size Nom. Inches	CAPITAL INVESTMENT DOLLARS/LINEAR FOOT — OPERATING TEMPERATURE — 1100°F, 593°C — Non Insul. Steam Capital Invest.	1100°F, 593°C — Insulated — Steam Capital Invest.	1100°F, 593°C — Insulated — Insul. Invest.	1100°F, 593°C — Insulated — Total Invest.	1200°F, 649°C — Non Insul. Steam Capital Invest	1200°F, 649°C — Insulated — Steam Capital Invest.	1200°F, 649°C — Insulated — Insul. Invest.	1200°F, 649°C — Insulated — Total Invest.	Pipe Size Dia. mm
½	$ 41.55	$ 2.55	$ 4.70	$ 7.25	$ 50.60	$ 2.85	$ 4.70	$ 7.55	21.3
¾	51.62	2.87	5.00	7.87	63.00	2.75	5.90	8.65	26.7
1	64.10	3.10	5.20	8.30	78.12	3.12	6.20	9.32	33.4
1¼	60.72	3.72	5.40	9.12	98.17	3.67	6.40	10.07	42.2
1½	97.02	3.27	6.60	9.87	112.42	3.67	6.60	10.27	48.2
2	116.66	3.87	7.00	10.87	139.45	3.95	9.00	12.95	60.3
2½	137.77	3.77	9.40	13.17	169.20	4.20	9.40	13.60	73.0
3	166.25	4.50	10.00	14.50	204.25	5.00	10.00	15.00	88.9
3½	191.65	4.65	10.60	15.25	232.42	5.17	10.60	15.77	101.6
4	213.57	5.32	11.20	16.52	260.50	5.25	12.20	17.45	114.3
5	236.57	6.07	12.00	18.07	288.65	6.15	13.60	19.75	141.3
6	311.62	6.12	15.00	21.12	380.32	6.82	15.00	21.82	168.3
8	403.00	6.75	19.50	26.25	492.57	7.57	19.50	27.07	219.1
10	499.27	8.02	22.00	30.02	610.60	8.10	26.00	34.10	273.0
12	590.90	8.40	29.00	37.40	723.67	8.67	35.00	43.67	328.0
14	645.43	9.17	33.00	42.17	791.97	9.47	39.00	48.47	355.0
16	736.45	10.20	37.00	47.20	351.75	10.50	43.00	53.50	406.4
18	827.82	10.32	47.00	56.32	1012.80	11.55	46.00	57.55	457.2
20	916.22	11.22	54.00	65.22	1125.05	12.55	54.00	66.55	508.0
24	1070.00	13.07	60.00	73.07	1393.52	13.52	70.00	83.52	609.6
30	1364.60	14.60	78.00	92.60	1675.15	15.15	86.00	101.15	762.6
36	1638.62	16.12	92.00	108.12	2011.95	16.95	100.00	116.95	914.4
	CAPITAL INVESTMENT DOLLARS/SQUARE FOOT								
Flat	$178.67	$ 1.37	$ 9.80	$ 11.17	$219.50	$ 1.50	$ 10.40	$ 11.90	Flat

To convert to metric units \$/linear foot × 3.28 = \$/linear metre, \$/square foot × 10.764 = $\$/m^2$

*Based on steam capital investment of \$25.00 per pound/hour.
Insulation thickness as shown in Table 21 and cost of installed insulation as shown in Table 22.

(Sheet 4 of 4)

Table 30 Cost of insulation, steam savings linear foot/year and % return on insulation investment*
(No credit given for savings obtained in steam capacity investment.)**

NPS Pipe Size Nom. Inches	DOLLAR INSULATION COST, STEAM SAVINGS/LINEAR FOOT AND % RETURN ON INVESTMENT — OPERATING TEMPERATURE — 212°F, 100°C — Insul. Invest. \$/lin.ft.	212°F, 100°C — Steam Saving \$/year	212°F, 100°C — % Return on Invest.	300°F, 149°C — Insul. Invest. \$/lin.ft.	300°F, 149°C — Steam Saving \$/year	300°F, 149°C — % Return on Invest.	400°F, 204°C — Insul. Invest. \$/lin.ft.	400°F, 204°C — Steam Saving \$/year	400°F, 204°C — % Return on Invest.	500°F, 260°C — Insul. Invest. \$/lin.ft.	500°F, 260°C — Steam Saving \$/year	500°F, 260°C — % Return on Invest.	Pipe Size Dia. mm
½	2.60	1.50	58	2.60	3.00	115	2.60	4.80	184	2.60	7.80	300	21.3
¾	2.70	1.80	67	2.70	3.60	133	2.70	6.30	233	2.70	9.60	355	26.7
1	2.80	2.40	86	2.80	4.00	142	2.80	8.10	289	4.00	12.60	315	33.4
1¼	2.86	3.00	105	2.86	5.70	199	2.86	9.90	346	4.10	15.60	380	42.2
1½	3.14	3.30	105	3.14	6.60	210	3.14	12.00	382	4.30	17.70	411	48.2
2	3.24	4.20	129	3.24	8.40	259	3.24	14.90	448	4.40	22.20	504	60.3
2½	3.50	4.80	137	3.50	9.90	282	3.50	17.70	505	4.60	26.70	580	73.0
3	3.70	5.40	146	3.70	12.00	324	3.70	21.00	567	4.80	30.00	625	88.9
3½	4.00	5.76	144	5.30	14.10	266	5.30	21.90	413	6.70	36.90	550	101.6
4	4.40	7.50	170	5.50	15.60	283	5.50	26.70	485	7.10	41.40	583	114.3
5	4.50	8.40	175	6.00	17.10	285	6.00	29.40	490	7.60	46.80	615	141.3
6	6.50	10.20	157	6.50	22.80	350	8.50	39.30	462	8.50	60.00	705	168.3
8	7.60	14.70	193	7.60	29.40	387	9.80	50.40	514	9.80	77.40	789	219.1
10	8.80	16.20	184	8.80	35.70	405	11.40	62.10	544	14.50	96.60	666	273.0
12	9.70	19.20	197	12.50	42.90	343	15.40	73.80	479	15.40	115.20	754	328.0
14	11.00	25.20	229	13.80	44.10	319	17.00	80.40	472	20.00	126.90	634	355.0
16	12.00	25.50	212	15.00	52.80	352	18.00	91.20	506	22.00	141.50	643	406.4
18	13.20	28.80	218	16.40	59.10	360	20.00	100.20	501	24.00	158.70	661	457.2
20	14.60	31.50	215	18.00	65.40	363	22.00	113.40	515	26.00	180.90	696	508.0
24	16.40	37.50	228	31.00	78.00	371	25.00	141.30	565	30.00	209.10	697	609.6
30	25.00	46.50	186	29.00	96.90	334	34.00	167.40	492	39.00	260.40	667	762.6
36	30.00	59.10	197	35.00	116.40	332	40.00	199.50	498	43.00	301.20	700	914.4
	DOLLAR INSULATION COST, STEAM SAVINGS/SQUARE FOOT AND % RETURN ON INVESTMENT												
Flat	3.80	6.30	166	4.40	13.80	313	5.00	23.40	468	5.88	36.00	620	Flat

* Based on insulation thicknesses given in Table 13.
Ambient air temperature 70°F (21°C), steam costs at \$3.00/M lbs. and 8760 hrs. operation per year.

** In most instances cost of insulation is less than cost of steam capacity saved.
Thus if credit is taken, insulation cost is negative investment.

(Sheet 1 of 3)

Table 30 Cost of insulation, steam savings linear foot/year and % return on insulation investment*
(No credit given for savings obtained in steam capacity investment.)**

NPS Pipe Size Nom. Inches	DOLLAR INSULATION COST, STEAM SAVINGS/LINEAR FOOT AND % RETURN ON INVESTMENT												Pipe Size Dia. mm
	OPERATING TEMPERATURE												
	600°F, 316°C			700°F, 371°C			800°F, 427°C			900°F, 402°C			
	Insul. Invest. $/lin.ft.	Steam Saving $/year	% Return on Invest.	Insul. Invest. $/lin.ft.	Steam Saving $/year	% Return on Invest.	Insul. Invest. $/lin.ft.	Steam Saving $/year	% Return on Invest.	Insul. Invest. $/lin.ft.	Steam Saving $/year	% Return on Invest.	
½	3.60	11.10	308	3.60	15.30	425	3.60	20.40	566	4.70	26.10	555	21.3
¾	3.80	14.10	371	3.80	18.90	497	3.80	25.20	663	5.00	32.70	654	26.7
1	4.00	17.70	442	4.00	23.70	592	4.00	32.10	775	5.20	47.10	905	33.4
1¼	4.10	22.50	548	5.40	30.30	561	5.40	40.50	750	5.40	51.90	961	42.2
1½	5.50	25.80	469	5.50	34.80	632	5.50	46.20	840	6.60	59.70	904	48.2
2	5.80	32.10	553	5.80	42.90	739	5.80	57.60	993	7.00	74.40	1062	60.3
2½	6.00	38.70	645	6.00	53.70	895	6.00	66.60	1110	7.40	89.10	1204	73.0
3	6.40	46.50	726	6.40	62.70	979	6.40	84.30	1317	8.00	108.30	1354	88.9
3½	6.70	52.50	783	8.60	71.40	830	8.60	95.40	1109	8.60	123.30	1433	101.6
4	7.10	59.10	832	9.20	80.70	877	9.20	107.70	1170	11.20	138.90	1240	114.3
5	7.60	65.10	856	9.80	88.80	906	9.80	119.10	1215	12.00	150.60	1255	141.3
6	8.50	86.10	1012	10.80	117.00	1083	10.80	154.20	1427	13.00	202.80	1560	168.3
8	12.00	111.00	925	14.20	151.00	1069	14.20	204.00	1436	17.00	262.50	1544	219.1
10	16.40	138.90	846	18.50	188.00	1020	18.50	252.90	1367	18.50	326.10	1762	273.0
12	18.00	163.80	910	21.00	223.00	1062	21.00	300.00	1428	24.50	381.00	1555	328.0
14	23.50	179.40	763	23.50	243.00	1035	28.00	327.90	1171	28.00	420.60	1502	355.0
16	26.00	204.50	786	26.00	276.00	1065	32.00	372.90	1165	32.00	481.20	1503	406.4
18	28.00	229.50	819	34.00	312.00	919	34.00	420.30	1236	40.00	541.80	1354	457.2
20	30.50	248.10	813	38.00	345.00	907	38.00	464.40	1222	46.00	601.50	1307	508.0
24	34.50	303.00	878	42.00	411.00	979	52.00	556.50	1070	52.00	716.40	1377	609.6
30	46.00	375.60	816	60.00	511.00	852	60.00	722.35	1204	68.00	894.00	1314	762.6
36	50.00	451.20	902	72.00	613.00	852	80.00	832.50	1040	88.00	1071.00	1217	914.4
	DOLLAR INSULATION COST, STEAM SAVINGS/SQUARE FOOT AND % RETURN ON INVESTMENT												
Flat	6.80	47.40	697	7.40	69.60	940	8.00	93.00	1162	8.60	119.10	1384	Flat

(Sheet 2 of 3)

Table 30 Cost of insulation, steam savings linear foot/year and % return on insulation investment*
(No credit given for savings obtained in steam capacity investment.)**

NPS Pipe Size Nom. Inches	DOLLAR INSULATION COST, STEAM SAVINGS/LINEAR FOOT AND % RETURN ON INVESTMENT									Pipe Size Dia. mm
	OPERATING TEMPERATURE									
	1000°F, 538°C			1100°F, 593°C			1200°F, 649°C			
	Insul. Invest. $/lin.ft.	Steam Saving $/year	% Return on Invest.	Insul. Invest. $/lin.ft.	Steam Saving $/year	% Return on Invest.	Insul. Invest. $/lin.ft.	Steam Saving $/year	% Return on Invest.	
½	4.70	31.50	670	4.70	41.10	874	4.70	50.10	1066	21.3
¾	5.00	41.40	828	5.00	51.30	1026	5.90	63.30	1072	26.7
1	5.20	51.90	998	5.20	64.20	1235	6.20	78.90	1272	33.4
1¼	5.40	65.10	1205	5.40	81.00	1500	6.40	99.60	1556	42.2
1½	6.60	75.30	1141	6.60	93.30	1413	6.60	117.00	1772	48.2
2	7.00	93.60	1337	7.00	116.10	1658	9.00	142.50	1583	60.3
2½	7.40	112.80	1524	9.40	140.70	1497	9.40	173.10	1841	73.0
3	8.00	136.80	1710	10.00	169.80	1698	10.00	209.10	2091	88.9
3½	10.60	156.60	1477	10.60	196.50	1853	10.60	238.50	2250	101.6
4	11.20	172.50	1540	11.20	218.40	1950	12.20	268.50	2200	114.3
5	12.00	194.40	1620	12.00	242.10	2017	13.60	297.00	2183	141.3
6	15.00	257.70	1718	15.00	321.00	2140	15.00	393.00	2620	168.3
8	17.00	334.20	1965	19.50	417.00	2138	19.50	510.30	2617	219.1
10	22.00	414.00	1882	22.00	516.60	2348	26.00	633.90	2438	273.0
12	29.00	491.40	1694	29.00	613.50	2115	35.00	753.60	2153	328.0
14	33.00	537.60	1629	33.00	670.50	2031	39.00	793.00	2034	355.0
16	37.00	612.00	1654	37.00	763.50	2063	43.00	937.50	2180	406.4
18	40.00	689.70	1724	46.00	861.00	1872	46.00	1054.80	2293	457.2
20	46.00	764.40	1661	54.00	952.50	1763	54.00	1170.00	2166	508.0
24	60.00	912.30	1520	60.00	1138.50	1898	70.00	1401.00	2001	609.6
30	78.00	1137.00	1457	78.00	1419.90	1820	86.00	1756.50	2042	762.6
36	92.00	1365.00	1483	92.00	1554.00	1689	100.00	2097.00	2097	914.4
	DOLLAR INSULATION COST, STEAM SAVINGS/SQUARE FOOT AND % RETURN ON INVESTMENT									
Flat	9.20	150.90	1640	9.80	187.80	1916	10.40	229.50	2206	Flat

* Based on insulation thicknesses given in Table 13.
Ambient air temperature 70°F (21°C), steam costs at $3.00/M lbs. and 8760 hrs. operation per year.
** In most instances cost of insulation is less than cost of steam capacity saved.
Thus if credit is taken, insulation cost is negative investment.

(Sheet 3 of 3)

Economic Consideration of Insulating Flanges and Valves

Because it appears to cost more to insulate flanges and valves than straight pipe, it has become the practice, for the sake of economy (?), to leave the flanges and valves uninsulated. This is an error that causes considerable waste. Some large plant-design contracting companies have recommended that flanges and valves not be insulated if operating temperatures will be less than 400° F or even 500° F. Such recommendations show a complete lack of knowledge of the economics of insulation.

Using the same procedure as presented for bare pipe it is possible to calculate the heat transfer from bare surfaces and insulated surfaces, then converting these losses to dollar values. The following tables were not calculated in as extensive a manner as those for pipe, for it becomes apparent that the larger the pipe size and the higher the operating temperature the more important it is that flanges be insulated. For that reason these tables are for flanges up to 12 NPS pipe and temperatures up to 600° F.

Table 31 gives the Btu heat loss of bare and insulated pipe flanges and this same loss in terms of pounds of steam per hour. Table 32 shows the amount of steam saved per hour and per year, then converts this saving into dollar values of steam capacity and steam production per year saved by the insulation flange cover. Table 33 provides the comparison of capital investment necessary with and without insulation flange covers. Table 34 shows that even not considering that insulation provides a savings in capital investment, that considering it only in regard to its steam savings, insulation still provides substantial returns on investment.

These tables were calculated on the typical conditions as previously stated for the economics of bare pipe. If there is disagreement as to the basic factors such as installed-insulation cost or steam costs, then it is recommended that new tables be calculated for the set of conditions under consideration.

Tables for economics in respect to other fittings such as valves or elbows have not been calculated, as their bare area compared to insulation cost is greater than flange bare area is in relation to its installed-insulation cost. Thus, it must follow that if it pays to insulate flanges, it pays to insulate all pipe fittings.

The economics prove that insulation is one of the best investments possible and that on today's market it is one of the most undersold products. Unfortunately the underselling of insulation not only harms the insulation manufacturers, distributors, and contractors, but it also causes the potential consumer to lose money. There is no question as to whether hot steam lines or steam-heated equipment should be insulated; the question is how much insulation should be used to take full advantage of its ability to save.

Table 31 Heat and steam losses — bare flanges compared to insulated flanges

HEAT LOSSES

Flange Pipe Size Nom. Inches	HEAT LOSS BTU/HR UNINSULATED — OPERATING TEMPERATURE 212°F 100°C	300°F 149°C	400°F 204°C	500°F 260°C	600°F 316°C	HEAT LOSS BTU/HR INSULATED — OPERATING TEMPERATURE 212°F 100°C	300°F 149°C	400°F 204°C	500°F 260°C	600°F 316°C	Flange Pipe Size mm
1	310	600	980	1390	1970	49	69	103	139	113	33.4
1¼	373	730	1200	1700	2400	52	78	115	155	139	42.2
1½	465	920	1490	2120	2990	52	78	115	106	139	48.2
2	642	1260	2060	2930	4130	67	104	155	142	185	60.3
2½	836	1640	2670	3800	5360	82	120	128	175	228	73.0
3	971	1890	3080	4370	6180	82	120	128	175	228	88.9
3½	1136	2220	3620	5140	7280	87	127	145	186	255	101.6
4	1344	2600	4240	6030	8510	92	134	155	210	275	114.3
5	1665	3240	5280	7520	10640	99	178	214	189	324	141.3
6	1953	3800	6200	8810	12460	105	186	226	153	328	168.3
8	2580	5020	8180	11640	16410	126	211	266	184	372	219.1
10	3510	6830	11400	15850	22380	156	263	256	347	453	273.0
12	4320	8450	13800	19610	27770	166	296	284	384	502	355.0

STEAM LOSSES

Flange Pipe Size Nom. Inches	STEAM LOSS LBS/HR UNINSULATED — OPERATING TEMPERATURE 212°F 100°C	300°F 149°C	400°F 204°C	500°F 260°C	600°F 316°C	STEAM LOSS LBS/HR INSULATED — OPERATING TEMPERATURE 212°F 100°C	300°F 149°C	400°F 204°C	500°F 260°C	600°F 316°C	Flange Pipe Size mm
1	0.27	0.50	0.79	1.08	1.54	0.04	0.06	0.08	0.11	0.09	33.4
1¼	0.32	0.61	0.97	1.37	1.87	0.05	0.07	0.09	0.12	0.11	42.2
1½	0.40	0.77	1.20	1.65	2.33	0.05	0.07	0.09	0.09	0.11	48.2
2	0.55	1.05	1.65	2.28	3.21	0.06	0.09	0.13	0.11	0.14	60.3
2½	0.72	1.37	2.15	2.96	4.18	0.07	0.10	0.11	0.14	0.18	73.0
3	0.84	1.58	2.49	3.41	4.81	0.07	0.10	0.11	0.14	0.18	88.9
3½	0.98	1.86	2.91	4.01	5.68	0.08	0.11	0.12	0.15	0.20	101.6
4	1.15	2.17	3.41	4.70	6.63	0.08	0.11	0.13	0.16	0.21	114.3
5	1.43	2.71	4.25	5.86	8.29	0.09	0.14	0.17	0.23	0.25	141.3
6	1.68	3.18	4.99	6.86	9.72	0.09	0.16	0.18	0.20	0.25	168.3
8	2.22	4.20	6.59	0.09	12.81	0.11	0.18	0.21	0.22	0.29	219.1
10	3.03	5.71	9.18	12.35	17.42	0.13	0.22	0.21	0.27	0.35	273.0
12	3.75	7.08	11.10	15.30	21.56	0.14	0.25	0.23	0.30	0.39	355.0

Table 31a Heat Losses — in watts — bare flanges compared to insulated flanges

Flange Pipe Size mm	HEAT LOSS W UNINSULATED OPERATING TEMPERATURE					HEAT LOSS W INSULATED OPERATING TEMPERATURE					Flange Pipe Size Nom. Inches
	100°C 212°F	149°C 300°F	204°C 400°F	260°C 500°F	316°C 600°F	100°C 212°F	149°C 300°F	204°C 400°F	260°C 500°F	316°C 600°F	
33.4	90	176	287	407	577	14	20	30	41	33	1
42.2	109	214	351	498	703	15	23	34	45	41	1¼
48.2	136	273	437	621	876	15	23	34	31	41	1½
60.3	188	369	604	859	1210	17	30	45	42	54	2
73.0	245	481	782	1114	1571	24	35	38	51	67	2½
88.9	285	554	903	1281	1811	24	35	38	51	67	3
101.6	333	650	1061	1506	2134	25	37	42	55	75	3½
114.3	294	762	1243	1767	2494	27	39	45	62	81	4
141.3	488	950	1547	2204	3118	29	52	63	85	95	5
168.3	572	1114	1817	2582	3652	31	54	66	74	96	6
219.1	756	1471	2397	3411	4809	40	62	78	83	109	8
273.0	1029	2002	3341	4645	6559	46	77	75	102	133	10
355.0	1266	2476	4044	5747	8139	49	87	83	113	147	12

Economic Thickness of Thermal Insulation Installed on Hot Surfaces

It is evident from the preceding that insulation installed to save energy also saves money at a rate that is essential to efficient plant operation. As more insulation is used, more heat is saved, but as the last unit of insulation saves less than the first, the law of diminishing returns sets in and there comes a point above which more insulation begins to cost money rather than save it. To illustrate:

Assume a flat surface operating at 600° F
The bare loss per sq ft is 2666 Btu/hr
One in. of insulation will reduce the loss to 212 Btu/hr
Two in. of insulation will reduce the loss to 106 Btu/hr
Four in. of insulation will reduce the loss to 53 Btu/hr
Eight in. of insulation will reduce the loss to 27 Btu/hr
Thus the first 4″ of insulation saves 2613 Btu and the last 4″ (of 8″ total thickness) saves only 26 Btu/hr.

Because of this effect of diminishing returns, the question becomes one of not whether to insulate, but of how much?

Insulation, by limiting the wasteful dissipation of heat, makes possible the economical transportation of heat. Insulation can be compared to a leaky wall of a pipe used to transport water. The term "leaky" is used advisedly because no insulation can completely stop the escape of heat. All the water that leaks from a hole in a pipe is subtracted from the amount put in, the difference is the usable amount flowing from the delivery end. The analogy is similar for insulated lines: the heat that is wasted cannot be included in the usable amount. The greater the thickness of proper insulation, the smaller the loss of heat through the pipe or vessel walls.

In many instances pipe is insulated but pipe fittings are left bare. This practice was started because the cost of insulating a fitting per unit of area is greater than for pipe or vessels. Leaving flanges bare became almost a universal practice, based on the mistaken idea that economically their being insulated could not be justified and that they should be left bare for the repair and replacement of leaky gaskets. One of the major reasons for the necessity of repairing leaky gaskets was that due to thermal shock and metal expansion-contraction caused by rain, snow and wind the gasket seal was broken. Proper insulation over valves, fittings and flanges protects them and reduces maintenance of these items.

The savings in heat, and money, by the proper insulation of piping fittings is extensive. The larger the pipe size and higher the temperature, the greater the savings in energy and money. However, the savings possible in small and moderate size pipe at relatively moderate temperatures is considerable. To illustrate this, Tables 31, 32, 33 and 34 have been prepared. The production investment of steam is taken to be $25.00/lb/hr ($71.10/KW) and $2.00 per 1000 lbs steam (0.0059 per KW hr) production cost. The thicknesses used for the flange covers are given below. The savings were based on 8760 hours of operation per year. The cost of the installed insulation on each flange is shown on Tables 33 and 34.

THICKNESS OF FLANGE INSULATION											
ENGLISH UNITS — Inches (Nom.)						METRIC UNITS — mm (Nom.)					
NPS Pipe Size	TEMPERATURE °F					Pipe Dia. mm	TEMPERATURE °C				
	212	300	400	500	600		100	149	204	260	316
1	1	1	1½	1½	1½	33.4	25	25	38	38	38
1¼	1	1	1½	1½	1½	42.2	25	25	38	38	38
1½	1	1	1½	1½	1½	48.2	25	25	38	38	38
2	1	1	1½	1½	1½	60.3	25	25	38	38	38
2½	1½	1½	1½	1½	1½	73.0	38	38	38	38	38
3	1½	1½	1½	1½	1½	88.9	38	38	38	38	38
3½	1½	1½	1½	1½	1½	101.6	38	38	38	38	38
4	1½	1½	1½	1½	1½	114.3	38	38	38	38	38
5	1½	1½	1½	1¼	2	141.3	38	38	38	38	50
6	1½	1½	1½	2	2	168.3	38	38	38	50	50
8	1½	1½	1½	2	2	219.1	38	38	38	50	50
10	1½	1½	2	2	2	273.0	38	38	50	50	50
12	1½	1½	2	2	2	355.0	38	38	50	50	50

It should be noted that the thicknesses used are very minimal.

Table 32 Savings accomplished by flange insulation covers

Flange Pipe Size Nom. Inches	STEAM SAVED										Flange Pipe Size mm
	LBS OF STEAM CAPACITY SAVED					M LBS OF STEAM SAVED PER YEAR					
	OPERATING TEMPERATURE					OPERATING TEMPERATURE					
	212°F 100°C	300°F 149°C	400°F 204°C	500°F 260°C	600°F 316°C	212°F 100°C	300°F 149°C	400°F 204°C	500°F 260°C	600°F 316°C	
1	0.23	0.46	0.71	0.97	1.45	2.04	4.03	6.22	8.49	12.70	33.4
1¼	0.27	0.54	0.88	1.28	1.76	2.39	4.73	7.71	11.21	15.42	42.2
1½	0.35	0.70	1.11	1.56	2.22	3.10	6.12	9.42	13.65	19.42	48.2
2	0.49	0.96	1.52	2.17	3.07	4.34	8.41	13.32	19.00	26.85	60.3
2½	0.65	1.27	2.14	2.82	4.00	5.69	11.13	18.75	24.70	35.05	73.0
3	0.74	1.48	2.38	3.27	4.63	6.48	12.96	20.85	28.65	40.50	88.9
3½	0.87	1.75	2.79	3.86	5.48	7.62	15.33	24.41	33.80	48.00	101.6
4	1.07	2.07	3.28	4.54	6.42	9.37	18.13	28.70	39.75	56.30	114.3
5	1.34	2.57	4.08	5.63	8.04	11.74	22.51	35.75	49.30	70.40	141.3
6	1.59	3.02	4.83	6.66	9.47	14.08	26.41	42.30	58.30	82.90	168.3
8	2.11	4.02	6.38	8.87	12.52	18.70	35.20	55.80	77.70	109.70	219.1
10	2.90	5.49	8.97	12.08	17.07	25.65	48.10	78.60	105.70	149.20	273.0
12	3.58	6.83	10.85	15.00	21.17	31.70	59.80	95.10	131.40	185.00	355.0

Flange Pipe Size Nom. Inches	DOLLAR VALUE SAVED										Flange Pipe Size mm
	DOLLAR VALUE OF STEAM CAPACITY SAVED					DOLLAR VALUE OF STEAM SAVED PER YEAR					
	OPERATING TEMPERATURE					OPERATING TEMPERATURE					
	212°F 100°C	300°F 149°C	400°F 204°C	500°F 260°C	600°F 316°C	212°F 100°C	300°F 149°C	400°F 204°C	500°F 260°C	600°F 316°C	
1	$5.75	$11.50	$17.75	$24.25	$36.25	$4.03	$8.06	$12.44	$16.98	$25.40	33.4
1¼	6.75	13.50	22.00	32.00	44.00	4.78	9.46	15.42	22.42	30.82	42.2
1½	8.75	17.50	27.75	39.00	55.50	6.20	12.24	18.84	27.30	38.84	48.2
2	12.25	24.00	38.00	54.25	76.75	8.68	16.82	26.64	38.00	53.70	60.3
2½	16.25	31.75	53.50	70.50	100.00	9.90	22.26	37.50	49.40	70.10	73.0
3	18.50	37.00	59.50	81.75	115.75	12.96	25.92	41.70	57.30	81.00	88.9
3½	21.75	43.75	69.75	96.50	146.00	15.24	30.66	48.82	87.60	96.00	101.6
4	26.75	51.75	82.00	113.50	160.50	18.74	36.26	57.40	79.50	112.60	114.3
5	33.50	64.25	102.00	140.75	201.00	23.48	45.02	71.50	98.60	140.80	141.3
6	39.75	75.20	120.75	166.50	236.75	28.16	52.82	84.60	116.60	165.80	168.3
8	52.75	100.50	159.50	221.75	313.00	37.40	70.40	111.60	155.40	219.40	219.1
10	72.50	137.25	224.25	302.00	426.75	51.30	96.20	157.20	211.40	298.40	273.0
12	89.50	170.75	271.25	375.00	529.25	63.40	119.60	190.20	262.80	370.00	355.0

This was done to establish that even very minimal thicknesses of insulation save appreciable amounts of energy and money.

Flange covers can be fabricated out of preformed pipe insulation as shown in ASTM Standard C-450. So fabricated it can be removed for inspection and repair of flanges, then the two halves replaced. Where required a leak detector tube should be installed prior to application of the insulation.

The heat and steam losses from bare and insulated flanges are given in Table 31—heat, steam, and dollar savings due to flange insulation are presented in Table 32. Capital investment uninsulated vs insulated flanges are given in Table 33 and Table 34 presents the return on investment in flange covers.

Following the present practice of calculating costs involving heat losses in terms of steam capacity and steam saved, Table 32 gives the savings obtained by insulation of flanges in terms of lbs of steam. This is then converted to dollars saved in production capacity required and savings in production cost per year.

The same procedure could have been used by the calculation of savings necessary in Watts production and savings in KW per year and converting their cost into dollar savings.

No matter which system is used the dollar savings per unit will be the same if identical dollar value for heat is used for both English and Metric systems.

Heat energy has monetary value. However, there is no direct conversion from Btu into dollars as the cost depends upon many factors, such as the form, location, and use of the heat. For example: Btu obtainable from gasoline are generally more expensive than those obtainable from coal. Also, Btu obtainable from coal are less expensive at the mine than at some distant point to which coal must be transported to be used. Likewise, the cost of installed insulation varies with materials used, location, labor rates, and other factors. The simple statement that "the greater the thickness of insulation the smaller the loss of heat" may be restated as "the greater the cost of insulation the smaller the cost of heat loss." Depending upon the cost of insulation and the cost of the heat, a certain definite thickness of insulation will provide the lowest total cost.

This basic truth has been recognized for years. At a meeting of the American Society of Mechanical Engineers in New York City on December 6, 1926, Mr. L. B. McMillan presented a paper title "Heat Transfer Through Insulation," in which he presented the formulas for calculating the economic thickness of insulation.

Table 33 Capital investment for production of wasted steam vs. capital investment for flange insulation

Flange Pipe Size Nom. Inches	CAPITAL INVESTMENT PER FLANGE – DOLLARS — OPERATING TEMPERATURE 212°F, 100°C				300°F, 149°C				400°F, 204°C				500°F, 260°C				600°F, 316°C				Flange Pipe Size mm
	No Insul.	Insulated Flange			No Insul.	Insulated Flange			No Insul.	Insulated Flange			No Insul.	Insulated Flange			No Insul.	Insulated Flange			
	Steam Capital Invest.	Steam Capital Invest.	Insul. Invest.	Total Invest.	Steam Capital Invest.	Steam Capital Invest.	Insul. Invest.	Total Invest.	Steam Capital Invest.	Steam Capital Invest.	Insul. Invest.	Total Invest.	Steam Capital Invest.	Steam Capital Invest.	Insul. Invest.	Total Invest.	Steam Capital Invest.	Steam Capital Invest.	Insul. Invest.	Total Invest.	
1	$6.75	$1.00	$15.25	$17.25	$12.50	$1.50	$16.25	$17.75	$19.75	$2.00	$17.50	$19.50	$27.00	$2.75	$17.50	$20.25	$38.50	$2.75	$17.50	$20.25	33.4
1¼	8.00	1.25	17.50	18.75	15.25	1.75	17.50	19.25	24.25	2.25	18.50	20.75	34.25	2.75	18.50	21.25	46.75	2.75	18.50	21.25	42.2
1½	10.00	1.25	18.50	19.75	19.25	1.75	18.50	20.25	30.00	2.25	19.25	21.50	41.25	2.75	19.25	22.00	58.25	2.75	19.25	22.00	48.2
2	13.75	1.50	19.50	21.00	26.25	2.25	19.50	21.75	41.25	3.25	20.25	23.50	57.00	3.00	20.25	23.25	80.25	3.50	20.25	23.75	60.3
2½	18.00	1.75	22.25	24.00	34.25	2.50	22.25	24.75	53.75	2.75	22.25	25.00	74.00	3.50	22.25	25.75	104.50	4.50	22.25	26.75	73.0
3	21.00	1.75	28.00	29.75	39.50	2.50	28.00	30.50	62.25	2.75	28.00	30.75	84.25	3.50	28.00	31.50	120.25	4.50	28.00	32.50	88.9
3½	24.50	2.00	30.00	32.00	46.50	2.75	30.00	32.75	72.75	3.00	30.00	33.00	100.35	3.75	30.00	33.75	142.00	5.00	30.00	35.00	101.6
4	28.75	2.00	32.50	34.50	54.25	2.75	32.50	35.25	85.25	3.25	32.50	35.75	117.50	4.00	32.50	36.50	165.75	5.25	32.50	37.75	114.3
5	35.75	2.25	35.00	37.25	67.75	3.50	35.00	38.50	106.25	4.25	35.00	39.25	146.50	5.75	35.00	40.75	219.75	6.25	35.00	41.25	141.3
6	42.00	2.25	37.50	39.75	79.50	4.00	37.50	41.50	124.75	4.50	37.50	42.00	171.50	5.75	40.00	45.75	243.00	6.25	40.00	46.25	168.3
8	55.50	2.75	43.75	46.50	105.00	4.50	43.75	48.25	164.75	5.25	43.75	49.00	227.25	6.00	46.25	52.25	320.25	7.25	46.25	53.50	219.1
10	75.75	3.25	50.00	53.25	142.75	5.50	50.00	55.50	229.50	5.50	65.00	60.50	313.75	6.75	65.00	71.75	435.50	8.75	65.00	73.75	273.0
12	95.75	3.50	57.50	61.00	177.00	6.25	57.50	63.75	277.50	6.25	72.50	78.75	382.50	7.50	72.50	80.00	539.00	9.75	72.50	80.25	355.0

Table 34 Return on investment dollar/year heat savings by insulation investment (No credit given for savings of capital investment of wasted steam.)

Flange Pipe Size Nom. Inches	INSULATION COST DOLLAR – HEAT SAVINGS DOLLAR/YEAR – % RETURN — OPERATING TEMPERATURE 212°F, 100°C			300°F, 149°C			400°F, 204°C			500°F, 260°C			600°F, 316°C			Flange Pipe Size mm
	Insul. Invest. $	Steam Saving $/yr	% Return on Invest.	Insul. Invest. $	Steam Saving $/yr	% Return on Invest.	Insul. Invest. $	Steam Saving $/yr	% Return on Invest.	Insul. Invest. $	Steam Saving $/yr	% Return on Invest.	Insul. Invest. $	Steam Saving $/yr	% Return on Invest.	
1	$16.25	$4.08	25	$16.25	$8.06	50	$17.50	$12.44	71	$17.50	$16.98	97	$17.50	$25.40	145	33.4
1¼	17.50	4.78	27	17.50	9.46	54	18.50	15.42	83	18.50	22.42	121	18.50	30.82	167	42.2
1½	18.50	6.20	36	18.50	12.24	66	19.25	18.84	98	19.25	27.30	142	19.25	38.84	202	48.2
2	19.50	8.68	46	19.50	16.82	86	20.25	26.64	131	20.25	38.00	187	20.25	53.70	265	60.3
2½	22.25	9.90	45	22.25	22.26	100	22.25	37.50	168	22.25	49.40	222	22.25	70.10	315	73.0
3	27.00	12.96	48	27.00	25.92	96	27.00	41.70	158	27.00	57.30	212	27.00	81.00	300	88.9
3½	30.00	15.24	51	30.00	30.66	102	30.00	48.82	163	30.00	67.60	225	30.00	96.00	320	101.6
4	32.50	18.74	58	32.00	36.26	111	32.50	57.40	177	32.50	79.50	245	32.50	112.60	346	114.3
5	35.00	23.48	67	35.00	45.02	129	35.00	71.50	204	35.00	98.60	282	35.00	140.80	402	141.3
6	37.50	28.16	75	31.50	52.82	141	37.50	84.60	226	40.00	116.60	292	40.00	165.80	414	168.3
8	43.75	37.40	85	43.75	70.40	161	43.75	111.60	255	46.25	155.40	336	46.25	219.40	474	219.1
10	50.00	51.30	103	50.00	96.20	192	65.00	157.20	241	65.00	211.40	325	65.00	298.40	459	273.0
12	57.50	63.40	110	57.50	119.00	208	73.50	190.20	262	72.50	262.80	362	72.50	370.00	510	355.0

McMillan's derivation for determining the most economical thickness for flat surfaces is as follows:

Let the cost of heat loss per sq ft per year $= m = \dfrac{a_m}{\dfrac{\ell}{k} + R}$

$$\text{When } a_m = \frac{Y(t_1 - t_a)M}{1{,}000{,}000}$$

- Y = hours of operation per year
- t_1 = operating temperature of inner surface insulation, °F
- t_a = ambient air temperature, °F
- M = value of heat in dollars per million Btu
- R = resistance (thermal)

Let the cost of insulation per sq ft per year $= n = b\ell + C$ when

- $n = b\ell + C$ when
- b = cost of installed insulation per sq ft per 1″ thickness per year.
- ℓ = thickness of insulation in inches
- C = constant (used only in derivation)

Let y = total cost per sq ft per year = m + n

$$= \frac{a_m}{\dfrac{\ell}{k} + R} + b\ell + C$$

$$= \frac{a_m k}{\ell + Rk} + C + b\ell$$

Differentiating and equating the first derivative to zero:

$$\frac{dy}{dx} = \frac{-a_m k}{(\ell + Rk)^2} + b = 0$$

$$= -a_m k + b\,(\ell + Rk)^2 = 0, \text{ or}$$

$$a_m k = b\,(\ell + Rk)^2$$

Dividing through by b and extracting the square root of both terms:

$$\sqrt{\frac{a_m k}{b}} = \ell + Rk$$

Therefore, the most economical thickness for flat surfaces is:

$$\ell = \sqrt{\frac{a_m k}{b}} + Rk \qquad (62f)$$

McMillan's derivation for determining the most economical thickness of one cylindrical insulation is as follows:

Let the cost of heat loss per sq ft of insulation surface per year

$$= \frac{a_m}{\dfrac{r_2 \log_e \dfrac{r_2}{r_1}}{k} + R_s}$$

When R_s = surface resistance

Then the cost of heat loss per lin. ft:

$$m' = \frac{2\pi}{12}\left(\frac{r_2\, a'}{\dfrac{r_2 \log_e \dfrac{r_2}{r_1}}{k} + R_s}\right)$$

$$= \frac{\pi}{6}\left(\frac{r_2\, a_{m'}\, k}{r_2 \log_e \dfrac{r_2}{r_1} + R_s k}\right)$$

The cost of insulation per lin. ft per year n′:

$$n' = \frac{2\pi}{12}\, r_2\,(r_2 - r_1)\, b + C$$

$$= \frac{\pi}{6}\, r_2\,(r_2 - r_1)\, b + C$$

$$= \frac{\pi}{6}\,(r_2^2 - r_1 r_2)\, b + C$$

Then the total cost per lin. ft per year y′

$$y' = m' + n' = \frac{\pi}{6} \times \left[\frac{r_2\, a_m\, k}{\left(r_2 \log_e \dfrac{r_2}{r_1} + R_s k\right)}\right]$$

$$+ \left[\frac{\pi}{6}\,(r_2^2 - r_1\, r_2)\, b\right] + C$$

The most economical thickness is a function of r_2 and may be found by differentiating the expression for y′ with respect to r_2 and equating the first derivative to zero:

$$\frac{dy}{dr_2} = \frac{\pi}{6} \times$$

$$\frac{\left(r_2 \log_e \dfrac{r_2}{r_1} + R_s\, k\right) a_{m'}\, k - r_2\, a_{m'}\, k \left(1 + \log_e \dfrac{r_2}{r_1}\right)}{\left(r_2 \log_e \dfrac{r_2}{r_1} + R_s\, k\right)^{\cdot 2}}$$

Clearing of fractions and multiplying through by $\frac{6}{\pi b}$:

$$\left(r_2 \log_e \frac{r_2}{r_1} + R_s k\right) \frac{a_{m'} k}{b} -$$

$$\frac{r a_{m'} k}{b} \left(1 + \log_e \frac{r_2}{r_1}\right) +$$

$$\left(r_2 \log_e \frac{r_2}{r_1} + R_s k\right)^2 (2r_2 - r_1) = 0$$

$$r_2 \frac{a_{m'} k}{b} \log_e \frac{r_2}{r_1} + \frac{a_{m'} k}{b} R_s k -$$

$$r_2 \frac{a_{m'} k}{b} - r_2 \frac{a_{m'} k}{b} \log_e \frac{r_2}{r_1} +$$

$$(2r_2 - r_1)\left(r_2 \log_e \frac{r_2}{r_1} + R_s k\right)^2 = 0$$

Simplifying:

$$r_2 \frac{a_{m'} k}{b} - \frac{a_{m'} k}{b} R_s k$$

$$= (2r_2 - r_1)\left(r_2 \log_e \frac{r_2}{r_1} + R_s k\right)^2$$

$$\frac{a_{m'} k}{b} (r_2 - R_s k)$$

$$= (2r_2 - r_1) \quad r_2 \log_e \frac{r_2}{r_1} + R_s k \quad ^2$$

Dividing through by ($r_2 = R_s k$) and extracting the square root of both sides, the most economical thickness for one cylindrical insulation is:

$$\sqrt{\frac{2r_2 - r_1}{r_2 - r_s k}} \left(r_2 \log_e \frac{r_2}{r_1} + R_s k\right) = \sqrt{\frac{a_{m'} k}{b}} \qquad (62c)$$

Substituting in for value of a:

$$\left(r_2 \log_e \frac{r_2}{r_1} + R_s k\right) \sqrt{\frac{2r_2 - r_1}{r_2 - R_s k}}$$

$$= \sqrt{\frac{\left[\frac{Y(t_o - t_a) M}{1{,}000{,}000}\right] k}{b}}$$

Although the formula is complicated, the physical laws upon which it is based are quite simple and can best be understood from graphs.

To illustrate the statement "the greater the thickness of insulation, the lower the heat loss," see Figure 34. Notice that on a bare pipe or piece of equipment the heat loss curve runs completely off the chart. Also notice that after insulation is installed it becomes necessary to double the insulation thickness to cut the remaining heat loss in half.

When:

- b = cost of insulation per year: includes capital investment, cost of money, interest, depreciation, maintenance.
- k = conductivity of insulation
- r_1 = inside radius of insulation
- r_2 = outside radius of insulation
- t_a = temperature of air
- t_o = operating temperature of pipe
- M = value of heat per 1,000,000 Btu, including consideration of fuel cost, capital investment, cost of money, interest, depreciation, maintenance.
- R_s = sum of resistance, including surface resistance.
- Y = hours of operation per year

Then:

$$r_2 \log_e \frac{r_2}{r_1} + R_s k \quad \sqrt{\frac{2r_2 - r_1}{r_2 - R_s k}}$$

$$= \sqrt{\frac{\left[\frac{Y(t_o - t_a) M}{1{,}000{,}000}\right] k}{b}} \qquad (62r)$$

This equation is to be solved to determine r_2.

As previously stated, Btu have a monetary value: therefore, the scale showing "Btu" can be replaced by dollars per year. This in no way affects the curve, as the vertical scale can always be converted to proper units.

Several factors enter into the cost of heat on an annual basis. All must be expressed in terms of the same unit of heat energy. These factors are:

1. Cost of fuel
2. Capital investment in heat-producing equipment and the heat distribution to point of use
3. Cost of money for capital investment
4. Interest on investment
5. Depreciation period
6. Maintenance
7. Number of hours of operation per year.

As the left-hand scale is in dollars per year, insulation costs can be plotted on the same chart. So, in proper units (such as lin. ft) the cost of insulation per year, per inch of thickness is also plotted. This is shown in Figure 35.

Most of the cost factors of heat also apply to insulation. These are:

1. Capital investment of installed insulation

2. Cost of money for capital investment
3. Interest on investment
4. Depreciation period
5. Maintenance cost

The total cost per year, then, becomes cost of lost heat per year added to insulation cost per year. This curve is shown in Figure 35.

It should be noticed that the total cost is quite high when insufficient insulation thickness is used. The cost drops to a minimum when the optimum thickness is used, then rises again when an uneconomical increased thickness is chosen. The low point on the combined curve determines the economic thickness of insulation.

These curves on insulation economics have been misinterpreted since the time they were presented by Mr. McMillan. The basic cause of the misinterpretation was the extremely high order of savings inherent in the use of any insulation. Depending upon temperature and cost factors, the first inch of insulation provides a return on investment of several hundred percent per year. Upon examination of the related costs where temperature differences and all other factors dictated thicker insulation, the last one-half inch added, though much less rewarding than the first, will still provide a return on investment of 20 to 50% per year or more. These increments on investment are shown on Figure 36.

This graph, while typical, is only one of a *family* of curves. Its vertical scale of cost and its curve of heat costs will be somewhat different for each combination of values. The number of different cost factors has already been mentioned. In addition to the cost factors, the following physical factors must also enter into the final determination of curves:

1. Temperature difference
2. Shape
   a. Flat
   b. Curved
      (1) Inside radius
      (2) Outside radius
3. Conductivity of insulation
   a. Mean temperature

The economic thickness could be obtained by having a curve for each set of variable conditions of cost, shape, conductivity of insulation, and temperature difference. Some items of cost, such as return on investment, which stems from *both* cost of insulation and cost of equipment, can be entered into the equations as constants, in accordance with good business practice. Taxes, as they affect capital investment in the steam-producing equipment and insulation when the equations are in balance, were found to have cancelled themselves out. For this reason they do not affect the results and can be ignored.

Many of the other variables cover a wide range of values. Therefore, a part of the problem in arriving at a solution of economic insulation thickness is to establish practical limits as a guide in establishing the magnitude of the calculations required.

A committee under the leadership of W. C. Turner was established in 1959 by Union Carbide Corporation, in cooperation with West Virginia University, to solve this problem. Their first task was to establish the variables and limits and then to decide on practical units of divisions. Following is this list of Variables and the limits of these variables.

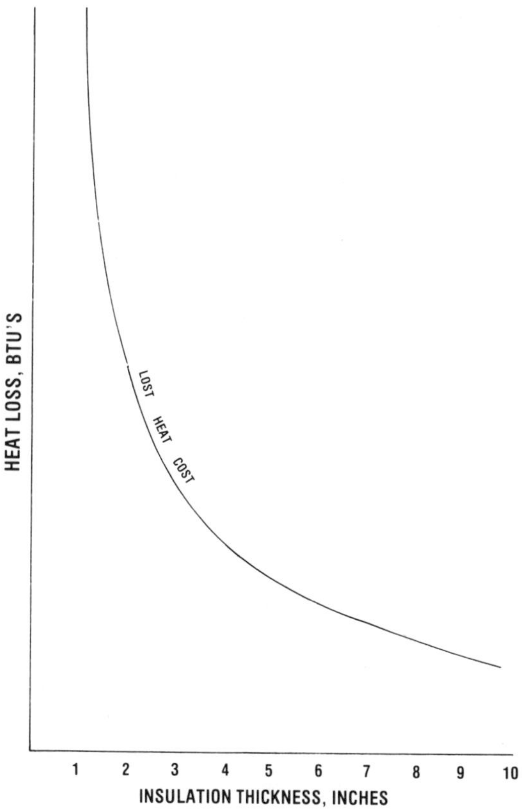

Figure 34 Economic thickness of insulation.

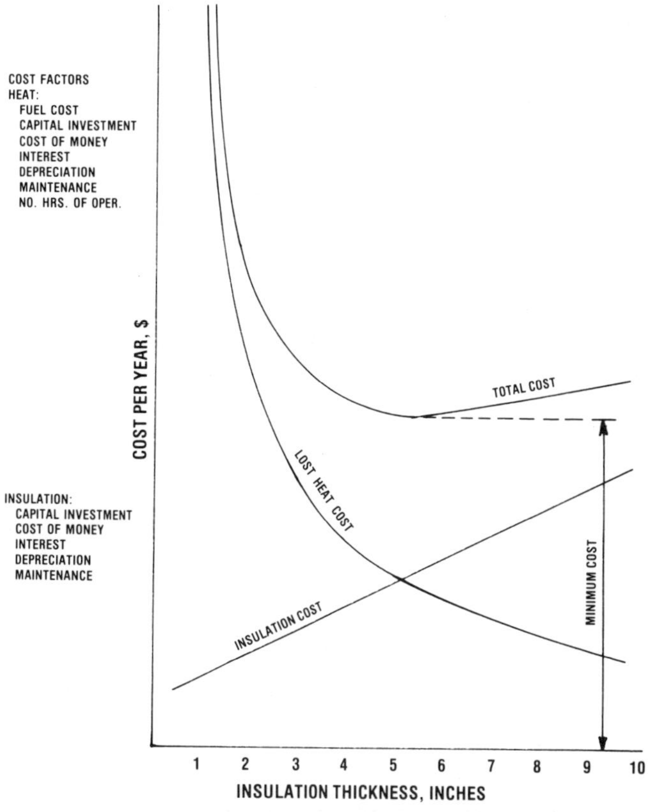

Figure 35 Economic thickness of insulation.

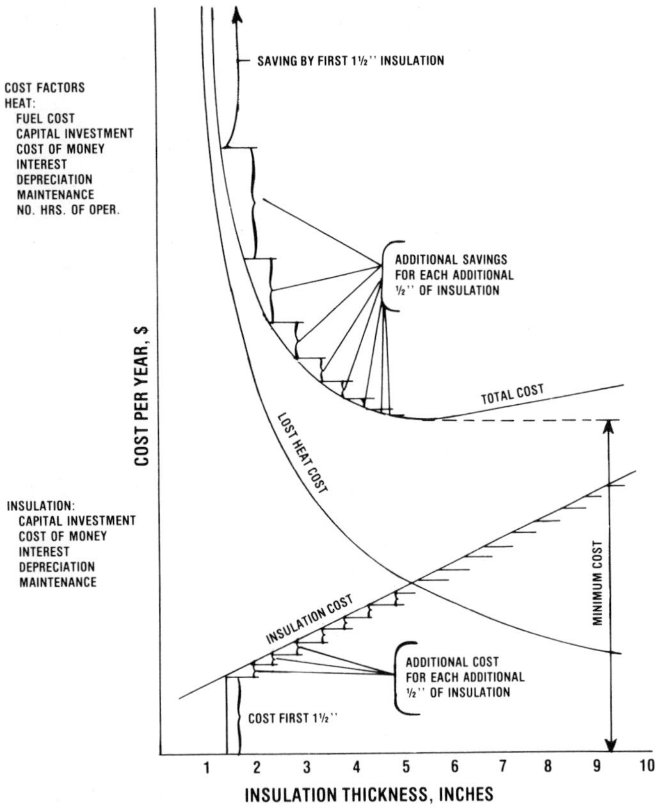

Figure 36 Economic thickness of insulation.

Limits:

Variable	Number of increments	Range
Heat		
Capital costs	12	\$2.00—\$50.00/lb. hr
Depreciation rate	5	5-40 yr
Production cost of heat	90	\$0.40—\$4.00/M lb
Interest rates (fixed as constant)		
Insulation costs		
Capital cost	21	\$0.50—\$10.00/in. thickness, lin. ft
Depreciation rate	4	5—0 yr
Interest rates (fixed as constant)		
Shape		
Flat and curved	22	½″ NPS—36″ NPS and flat
Thermal conditions		
Conductivity of insulation	15	0.1—1.5 Btu/sq ft. in., °F
Temperature differences	35	90—1314° F
Shape		
Flat and curved	22	½″ NPS—36″ NPS and flat

(Flat refers to any surface with a radius greater than 18″.)

The number of curves required for all possible sets of conditions would be all of the variable increments sums multiplied together, or:

$$12 \times 5 \times 90 \times 21 \times 4 \times 22 \times 15 \times 35 \times 22$$

which equals 115,259,760,000

It is understandable that no one would attempt to calculate this number of curves and plot them as charts.

Even by machine calculation this myriad of figures would still be present and some means would have to be devised to break the impasse. The particular device that has been successful is a separation of cost factors from thermal factors. Mr. William Hollenbeck discovered that the cost factors could be resolved into a single variable which could determine the curves in proper relationship to the thermal equations.

Another observation, the use of which has simplified calculations, is that plotted insulation costs fall into a family of relatively straight line curves having a common origin. The tangent angle and location of any curve on the graph may be determined by the common origin and a single point on the curve.

The application of these two observations made possible the calculation of an economic-thickness manual. This manual, produced by the teams of Union Carbide Corporation and West Virginia University, was completed in 1960. In 1961 it was released to the National Insulation Manufacturers Association for publication.

After this manual was out of print, it was updated by a publication EGON-1; however, due to the energy crisis this publication was also sold out of print and other publications for presenting methods of determining insulation thickness, or additional thickness, were published. These are:

R-ECON—A METHOD FOR DETERMINING ECONOMIC THICKNESS OR ADD-ON THERMAL INSULATION—Published by Thermal Insulation Manufacturers Assoc.

ETI—ECONOMIC THICKNESS FOR INDUSTRIAL INSULATION—obtainable from Federal Energy Administration.

The ETI publication is a method of determining insulation thickness by using a series of nomographs.

A number of manufacturers have set up computer programs to assist in the design of insulation based on capital investment, fuel cost, and other cost factors.

As a design method using precalculated charts is still needed the authors updated the original economic thickness manual to present day costs. This updated manual is titled "Economics of Thermal Insulation Design for Pipes and Equipment Above Ambient Temperature" by William C. Turner and John F. Malloy and is published by Robert E. Krieger Publishing Company Inc., Huntington, New York 11743.

This manual provides tabulated information as to the heat loss of bare pipe and equipment, heat loss of insulated pipe and equipment and the dollar cost of bare and insulated pipe and equipment. It also contains information as to the economic design of externally heated pipe and equipment. All of these effect the design thickness of thermal insulation.

In this revised manual the cost of heat, insulation and energy had to be revised to higher figures than presented in the original manual. In addition, with the greater thicknesses of insulation it was necessary to prepare a separate section for flanges and flanged fittings economic thickness design.

The design of thermal insulation as influenced by cost factors is essential. Unfortunately due to their volue all the tables in the "Economics of Thermal Insulation Design Manual" cannot be reproduced in this book which is primarily directed to the thermal, physical and chemical factors influencing the design of thermal insulation. However, even if it is essential to have the Manual "Economics of Thermal Insulation Design for Pipes and Equipment Above Ambient Temperature" to solve the thickness problems, an explanation of the use of that manual is warranted here.

In the previous part of this chapter the value of insulation to conserve energy and money was discussed. It is evident that had different thicknesses of insulation been used, other than the thickness selected, different results would have been obtained. This brings up the question of what is the correct thickness for a specific set of conditions. In some instances the thickness of insulation is set by the maximum allowable heat loss, or gain, or maximum or minimum allowable surface temperature. However, in many instances when these thermal conditions have been met, the question remains, what is the correct thickness based on economics? The economic-thickness method was devised so that the proper thickness could be selected. After the basic thickness of insulation (½" or 1") the economic-thickness evaluates each additional ½" increase in thickness to determine if it is a good investment.

As can be observed on Figure 36 the calculated curve of total cost may indicate that minimum total cost is obtained when insulation thickness is 5.237". Of course in such instance 5" of insulation would be used. However, further investigation may show that the investment in the last ½"—in this case 4½" to 5"—may only give a return of 6% on investment. Thus, on today's economy 4½" thickness would be the correct selection not 5". The equations in the Economic Thickness Manual were set up so that the investment in the *last* ½" gives a minimum return of 20% on investment. Due to the fact that ½" increments are relatively large increments to total thickness, analysis has shown that the last ½" of the recommended thickness provides an average return of 30 to 50%. The total installed insulation almost always provides a return on the investment of at least 250%.

The cost of heat and insulation provided in the manual is completely independent of the accounting methods used to arrive at these costs. The costs are set up as variables, the values of which to be used, must be determined by the user. The manual makes no allowance for energy escalation cost. However, it is well to consider the advisability of using cost of energy factors based on the expected cost of energy over the life expectancy of the installation.

Tables and charts (reproduced from the manual) illustrate the simplicity for the determination of economic thickness of insulation from precalculated data. The tables and charts were selected to solve an example for pipe insulation in both English and Metric Units. Figures 37 (English Units) and 37a (Metric Units) show the method and steps required.

The cost factors and thermal factors are so arranged as to be able to obtain answers in both English and Metric systems. However, as all thermal insulations for NPS pipes are based on nominal inch dimensions which are in accordance with ASTM C-585-76 "Inner and Outer Diameters of Rigid Thermal Insulation for Nominal Sizes of Pipe and Tubing (NPS System)," Table No. 35 is provided to convert nominal inch thicknesses to approximate mm thicknesses. For large diameters (over 14") and flat surfaces the thicknesses are in even ½" increments. To convert to mm simply multiply the thickness in inches by 25.4.

As indicated on the schematic flow chart, the first requirement is to establish costs. Using proper accounting practices the following must be established:

1. *The total cost of labor and material to install the insulation system (C")* required for the particular conditions involved. Although many pipe sizes and thicknesses may be required, in most instances, accurate cost estimates may be determined by using the cost of 1" nominal thickness of insulation installed on 1" NPS pipe. The tables in the Economic Thickness of Insulation for Pipe were based on plotting variable cost of many sizes and many thicknesses and summarizing these to the basic one inch thickness on one inch NPS pipe factor.

2. *The depreciation period of insulation.* This is the expected life of the insulation system without excessive maintenance. The calculations do have a maintenance allowance of 10% per year of capital investment to keep the insulation system in good repair. The life factor "N_n" is on Table No. 35 in terms of 5, 10, 15, 20, and 30 years.

3. *The depreciation period of the plant, unit, or building* onto which the insulation is being installed. This factor "N_h" selects the correct graph for determination of YM × 10^{-6} factor. These graphs are plotted for depreciation periods. N_n = 5, 10, 15, 20, and 30 years (Figure 38).

4. *Cost of Energy.* This cost is the total cost of energy to the point of use. It is not the cost of fuel alone. To illustrate, the cost of fuel may be $0.50 for a million Btu, but by the time it is converted to steam and is transported to the point of use, only 20% of its original energy is available for intended use. Thus, the energy cost at the point of use would be $2.50.

5. *The capital investment, per unit of energy, in equipment to produce the energy from fuel.*

6. *The hours of operation per year.*

To simplify the tabulation and to make the precalculation of tables of economic thickness practical it was necessary to resolve all these factors into a single cost factor. This factor was named "D". Once established, the "D" factor can be used to determine the proper table of economic thickness for a particular pipe size or flat surface.

Schematic Chart for Use of Manual

(English Units)

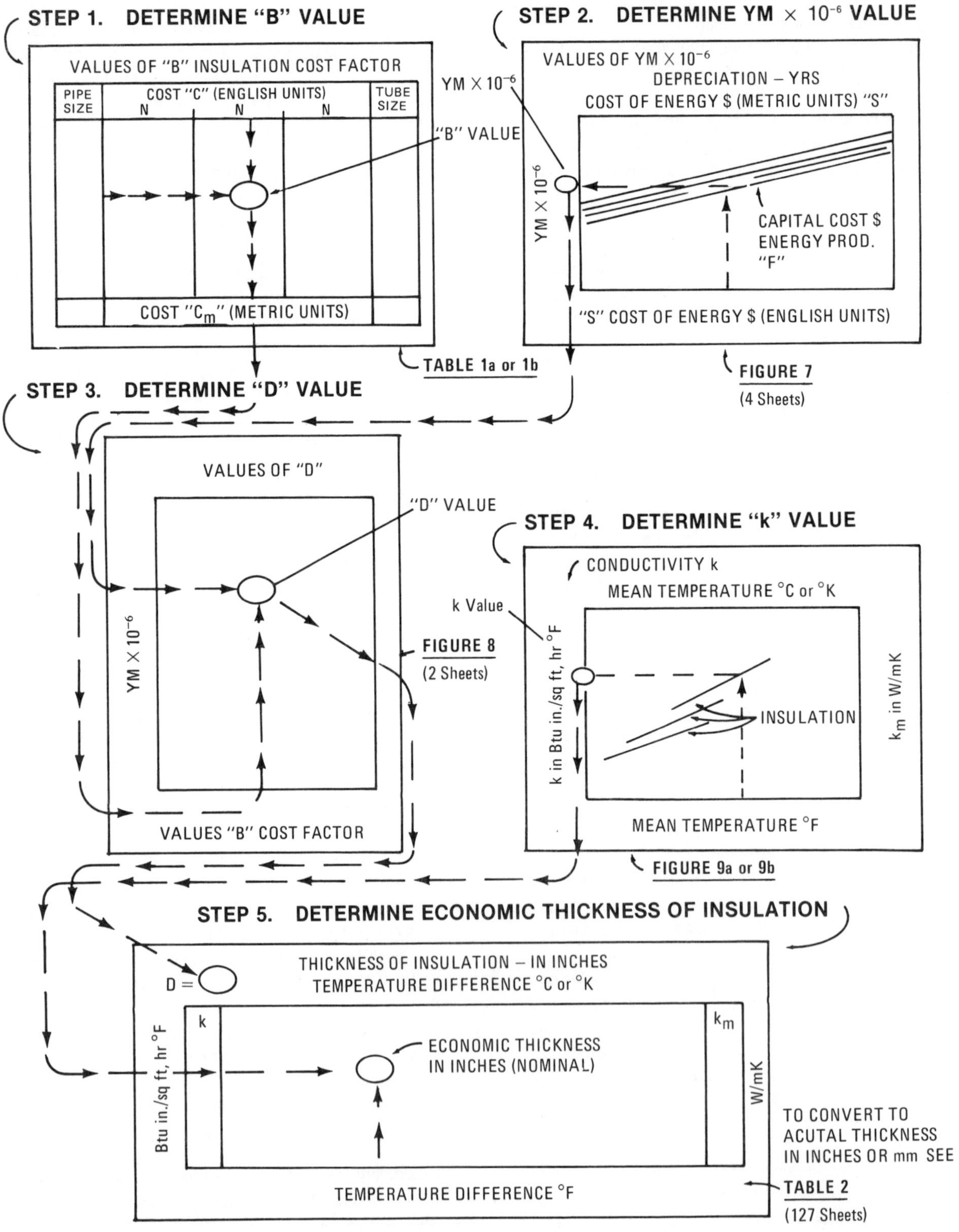

Figure 37

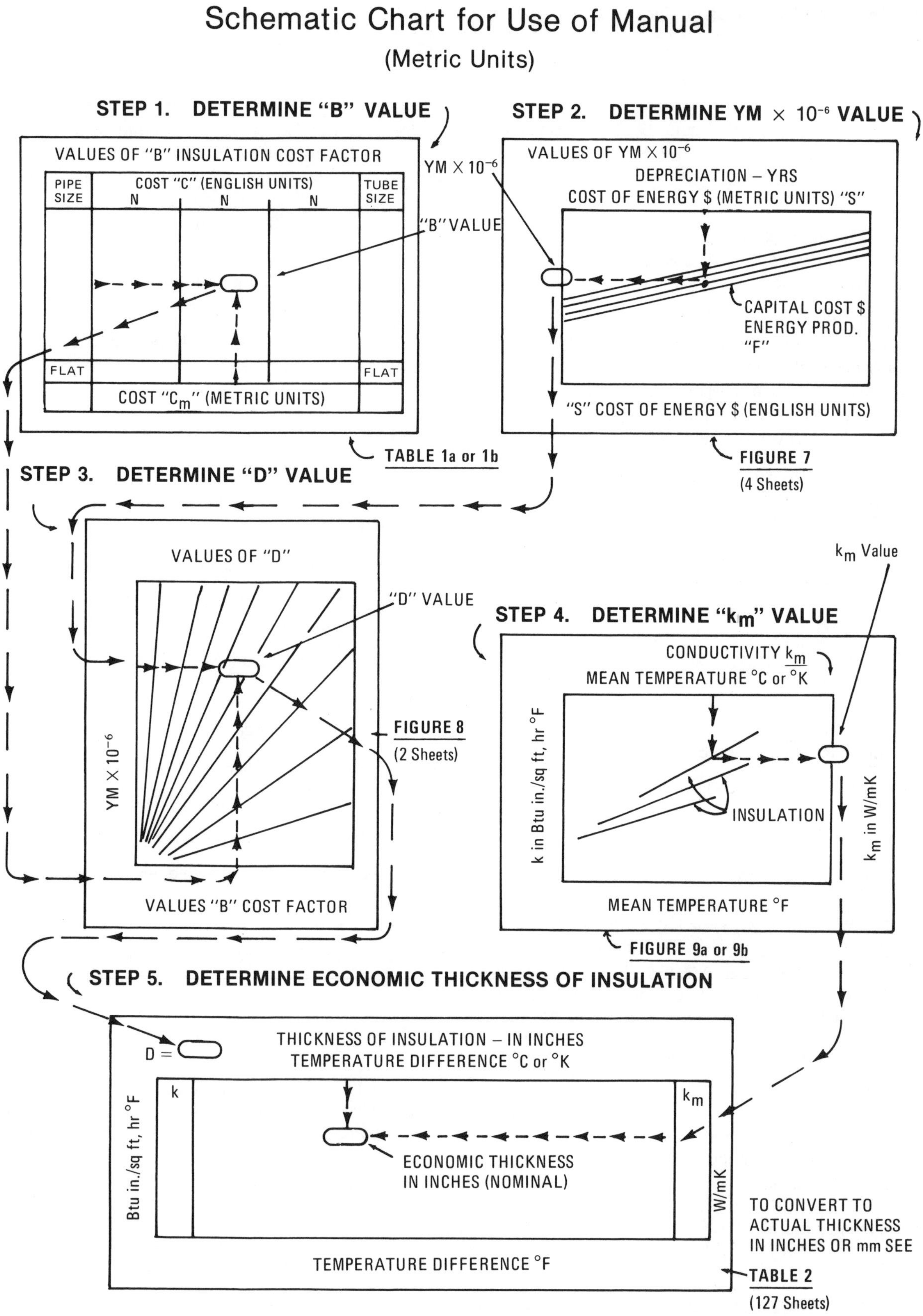
Schematic Chart for Use of Manual
(Metric Units)
STEP 1. DETERMINE "B" VALUE
VALUES OF "B" INSULATION COST FACTOR
PIPE SIZE
COST "C" (ENGLISH UNITS)
N N N
TUBE SIZE
FLAT
FLAT
COST "Cm" (METRIC UNITS)
YM X 10-6
"B" VALUE
TABLE 1a or 1b
STEP 2. DETERMINE YM × 10-6 VALUE
VALUES OF YM X 10-6
DEPRECIATION – YRS
COST OF ENERGY $ (METRIC UNITS) "S"
CAPITAL COST $
ENERGY PROD.
"F"
"S" COST OF ENERGY $ (ENGLISH UNITS)
FIGURE 7
(4 Sheets)
STEP 3. DETERMINE "D" VALUE
VALUES OF "D"
"D" VALUE
YM X 10-6
VALUES "B" COST FACTOR
FIGURE 8
(2 Sheets)
km Value
STEP 4. DETERMINE "km" VALUE
CONDUCTIVITY km
MEAN TEMPERATURE °C or °K
k in Btu in./sq ft, hr °F
INSULATION
km in W/mK
MEAN TEMPERATURE °F
FIGURE 9a or 9b
STEP 5. DETERMINE ECONOMIC THICKNESS OF INSULATION
D =
THICKNESS OF INSULATION – IN INCHES
TEMPERATURE DIFFERENCE °C or °K
k
km
Btu in./sq ft, hr °F
ECONOMIC THICKNESS
IN INCHES (NOMINAL)
W/mK
TEMPERATURE DIFFERENCE °F
TO CONVERT TO
ACTUAL THICKNESS
IN INCHES OR mm SEE
TABLE 2
(127 Sheets)

Figure 37a

Method of Determining Economics of Thickness of Insulation

Determine the cost factor D.

Following the schematic chart, Figure 37 (English Units) or Figure 37a (Metric Units) through steps 1 to 5:

STEP 1

Determine "B" value from established cost of installed insulation, depreciation period and pipe size (Table 35).

STEP 2

Select the proper graph for established plant depreciation "N_n". Using established cost of steam, or energy, follow vertical line to line representing the established capital investment in steam, or energy. From this intersection proceed horizontally to determine value of $YM \times 10^{-6}$ (Figure 38).

The value of $YM \times 10^{-6}$ so determined is based on full time operation of 8760 hours per year. Should the operation time be less than 8760 hours, the value of $YM \times 10^{-6}$ is modified for the hours of operation per year.

$$YM \times 10^{-6} \text{ (modified)} = YM \times 10^{-6} \text{ (graph value)}$$

$$\times \frac{\text{hour per year operation}}{8760}$$

STEP 3

The "D" value is established by one of two methods.

(1) Established $YM \times 10^{-6}$ divided by "B" = "D"

$$\frac{YM \times 10^{-6}}{B}$$

(2) Using proper chart of "B" values, enter at the "B" value on the chart and proceed vertically. At the value of $YM \times 10^{-6}$ proceed horizontally. The point of the intersection of these two lines determines the "D" values as referenced from the straight line curves on the chart (Figure 39).

Using the "D" value and thermal factors—to find economic thickness:

STEP 4

Determine the thermal conductivity at the mean temperature "t_m" across the insulation.

$$t_m = \frac{\text{Operating temperature "}t_n\text{" + ambient temperature "}t_{av}\text{"}}{2}$$

Enter the Figure 40 at value t_m, arrived at above, and proceed vertically to where it intersects the line representing the selected insulation material, then read horizontally to the conductivity for that mean temperature.

STEP 5

The temperature differential Δt is the operating t_o minus the average ambient air temperature t_{av} $\Delta t = t_a - t_{av}$

Using the proper sheet of Table 36, based on "D"* for the individual pipe size (or flat surface), the answer is determined in Step 3 and go horizontally to the column with the Δt determined in Step 4. The values tabulated in the columns are economic thickness of insulation in inches (nominal for NPS pipe in accordance with dimensions in ASTM C-585).

To convert to mm for answers in actual inch measurements, as is necessary in large diameters of piping, equipment and flat surfaces use the table below for conversion.

*Should the determined "D" value differ from that at the top of each table then the sheet of next higher "D" value should be used.

■ EXAMPLE 42

English Units — Cost Data

Capital investment in equipment to convert fuel to steam \$30.00/lb, hr "F".

Capital investment of insulation 1″ thickness on 1″ NPS pipe = \$2.80/lin ft, "C".

Depreciation period of insulation N_n is 15 years

Depreciation period of plant N_h is 15 years

Production cost of steam (to point of use) = \$2.60/1000 lbs.

Physical and Thermal Data

Nominal pipe size is 6″ NPS

Operating temperature of pipe is 600° F "t_o"

Average ambient air temperature is 50° F "t_{av}"

Type of insulation is rigid expanded perlite pipe insulation

STEP 1

From Table 35 at "C" capital investment in insulation = \$2.80 under depreciation period N_n = 20 years for 6″ NPS pipe B = 1.64.

STEP 2

From Figure 38 at the cost of \$2.60/1000 lb read vertically to line of capital investment for steam production equipment which is \$30.00 per lb/hr. Then read horizontally to determine $YM \times 10^{-6}$.

Which, under these conditions is 0.025.

STEP 3

$$\text{"D"} = \frac{YM \times 10^{-6}}{B} = \frac{0.025}{1.62} = 0.015$$

"D" can also be determined from Figure 39. Reading vertically "B" = 1.62 to horizontal line $YM \times 10^{-6}$ = 0.025 "D" curve = 0.015.

Table 35 Values of "B" — Insulation Cost Factor
"C" is cost of labor and material to insulate 1 inch NPS pipe with 1 inch nom. thick insulation, 1 linear foot.
"N_n" is depreciation period of insulation system, in years.

NPS Pipe Size	Diameter Inch	Diameter cm	C = \$2.60, N_n = years: 5	10	15	20	30	C = \$2.80, N_n = years: 5	10	15	20	30	C = \$3.00, N_n = years: 5	10	15	20	30	Tube Size Inches
½	0.840	2.133	1.34	0.90	0.78	0.68	0.64	1.44	0.98	0.84	0.74	0.70	1.54	1.04	0.90	0.78	0.74	¼ to ¾
¾	1.030	2.616	1.40	0.94	0.72	0.72	0.68	1.50	1.02	0.86	0.76	0.72	1.62	1.10	0.94	0.81	0.78	1
1	1.315	3.340	1.56	1.06	0.90	0.80	0.76	1.68	1.14	0.96	0.86	0.82	1.80	1.22	10.4	0.92	0.83	1¼
1¼	1.660	4.216	1.80	1.22	1.04	0.92	0.88	1.93	1.30	1.12	0.98	0.94	2.06	1.40	1.20	1.06	1.01	1½
1½	1.900	4.826	1.46	0.98	0.84	0.74	0.70	1.58	1.06	0.90	0.80	0.76	1.68	1.14	0.98	0.86	0.82	2
2	2.375	6.033	2.02	1.36	1.16	1.02	0.98	2.16	1.46	1.26	1.10	1.06	2.32	1.58	1.24	1.18	1.14	2½
2½	2.875	7.303	2.40	1.64	1.40	1.24	1.20	2.60	1.76	1.50	1.32	1.28	2.78	1.84	1.60	1.42	1.38	3
3	3.500	8.890	2.22	1.50	1.28	1.14	1.10	2.38	1.62	1.38	1.22	1.18	2.56	1.74	1.48	1.30	1.26	
3½	4.000	10.160	2.28	1.56	1.32	1.18	1.14	2.46	1.68	1.42	1.26	1.22	2.64	1.80	1.52	1.34	1.30	4
4	4.500	11.430	2.36	1.60	1.36	1.20	1.16	2.54	1.72	1.48	1.30	1.26	2.72	1.86	1.58	1.40	1.36	
Example 42 → 6	6.625	16.827	1.78	1.52	1.34	1.34	1.30	2.84	1.92	*1.64	1.46	1.42	3.04	2.06	1.76	1.54	1.48	6
8	8.625	21.908	2.94	2.00	1.70	1.50	1.46	3.16	2.14	1.82	1.62	1.56	3.38	2.30	1.96	1.74	1.68	
10	10.75	27.305	3.26	2.20	1.88	1.66	1.60	3.50	2.38	2.02	1.80	1.74	3.76	2.56	2.18	1.92	1.92	10
12	12.75	32.385	3.54	2.40	2.06	1.82	1.76	3.82	2.60	2.22	1.96	1.88	4.10	2.78	2.38	2.10	2.02	12
14	14.00	35.560	3.58	2.42	2.08	1.84	1.78	3.86	2.62	2.24	1.98	1.90	4.14	2.80	2.40	2.12	2.04	14
16	16.00	40.640	3.82	2.60	2.22	2.00	1.92	4.12	2.80	2.38	2.10	2.00	4.42	3.00	2.56	2.26	2.16	16
18	18.00	45.720	4.06	2.76	2.34	2.08	2.00	4.36	2.96	2.52	2.24	2.14	4.68	3.18	2.70	2.40	2.30	18
20	20.00	50.800	4.40	2.98	2.54	2.24	2.14	4.74	3.20	2.74	2.42	2.30	5.06	3.44	2.94	2.58	2.43	20
24	24.00	60.050	4.94	3.34	2.86	2.52	2.38	5.30	3.60	3.08	2.72	2.58	5.68	3.86	3.30	2.90	2.74	24
30	30.00	76.200	5.60	3.80	3.24	2.86	2.66	6.04	4.10	3.50	3.08	2.86	6.46	4.38	3.74	3.30	3.14	
36	36.00	91.440	6.66	5.12	3.86	3.40	3.12	7.18	4.86	4.14	3.66	3.36	7.68	5.22	4.44	3.92	3.62	
Flat	over 36	over 91	0.30	0.20	0.18	0.16	0.14	0.32	0.22	0.18	0.16	0.14	0.34	0.24	0.20	0.18	0.16	Flat
		↑ × 10 for mm	C_m = \$8.53					C_m = \$9.18					C_m = \$9.84					

"C_m" is cost of labor and material insulate 1 metre length of 3.34 cm diameter pipe with 25 mm thick insulation.

*Indicates Figure for Example 42.

STEP 4

$$\text{Mean temperature is } \frac{t_o + t_{av}}{2} = \frac{600 + 50}{2} = 325°\text{ F}$$

From Figure 40. At this mean temperature the thermal conductivity = 0.5 Btu in/ft hr °F

STEP 5

Temperature difference = $t_o - t_{av}$ = 600 − 50 = 550° F Δt. From Table 36, with k = 0.5, Δt = 550, D = 0.015 for 6″ NPS pipe the economic thickness is given as 4.0″.

■

■ EXAMPLE 42a

Metric Units — Cost Data

Capital investment in equipment to convert fuel to steam \$7.00/KW hr "F".

Capital investment of insulation, 25.4 mm thickness on 33.4 mm NPS pipe is \$9.18 for linear metre "C_m".

Depreciation period of insulation N_n is 15 years.

Depreciation period of plant N_h is 15 years.

Production cost of energy is \$0.61/million watt-hr.

Physical and Thermal Data

Nominal pipe size is 168.3 mm

Operating temperature of pipe is 315.5 °C "t_{mo}"

Average ambient air temperature is 10° C "t_{mav}"

Type of insulation is rigid expanded silica pipe insulation

STEP 1

From Table 35. At C_m = \$9.18 for linear metre under depreciation period equals 20 years, for 16.83 cm (168.3 mm) pipe "B" = 1.62.

STEP 2

From Figure 38. At the cost of \$0.61 per million watt-hrs vertically downward to line of capital investment of \$7.00 per KWH, from intersection read horizontal to determine YM × 10^{-6}, which under these conditions is 0.025.

STEP 3

Read off Figure 39 for YM × 10^{-6} = 0.025 and B = 1.62 "D" curve is 0.015.

STEP 4

$$\text{Mean temperature is } \frac{t_{ma} + t_{mav}}{2} = \frac{315.5 + 10}{2} = 162\,°\text{C}$$

From Figure 40. Read vertical downward to intersection of line for rigid expanded perlite then horizontal to conductivity k_m = 0.0721.

STEP 5

The temperature difference = $t_{mo} - t_{mav}$ = 315.4 − 10 = 305.5° C.

From Table 35 with k_m = 0.0721 w/mK, Δt_m = 305.5° C for 168.3 mm pipe the economic thickness is shown to be 4.0″. 4.0 × 25.4 = *101.6 mm* thickness of insulation required.

The Tables 35 and 36, Figures 38, 39, and 40 shown are only the particular sheet or page of these tables or figures necessary to illustrate this one problem. The total of each of these for the calculation of most pipe sizes, flat surfaces and conditions are in the "Economic Thickness Design Manual" previously mentioned.

■

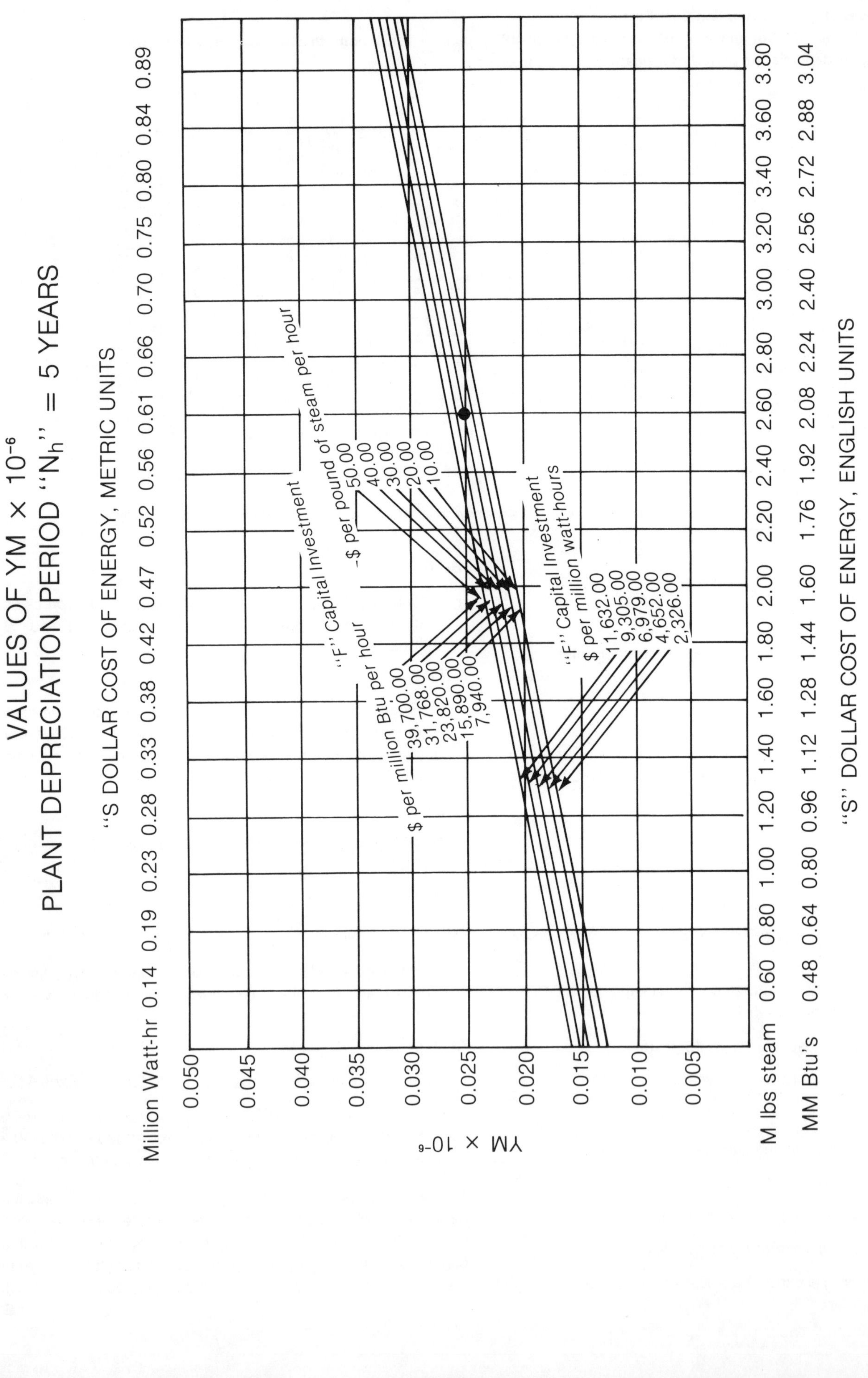
VALUES OF YM × 10⁻⁶
PLANT DEPRECIATION PERIOD "Nh" = 5 YEARS
"S DOLLAR COST OF ENERGY, METRIC UNITS
Million Watt-hr 0.14 0.19 0.23 0.28 0.33 0.38 0.42 0.47 0.52 0.56 0.61 0.66 0.70 0.75 0.80 0.84 0.89
YM × 10⁻⁶
0.050 0.045 0.040 0.035 0.030 0.025 0.020 0.015 0.010 0.005
"F" Capital Investment
$ per pound of steam per hour
50.00
40.00
30.00
20.00
10.00
"F" Capital Investment
$ per million Btu per hour
39,700.00
31,768.00
23,820.00
15,890.00
7,940.00
"F" Capital Investment
$ per million watt-hours
11,632.00
9,305.00
6,979.00
4,652.00
2,326.00
M lbs steam 0.60 0.80 1.00 1.20 1.40 1.60 1.80 2.00 2.20 2.40 2.60 2.80 3.00 3.20 3.40 3.60 3.80
MM Btu's 0.48 0.64 0.80 0.96 1.12 1.28 1.44 1.60 1.76 1.92 2.08 2.24 2.40 2.56 2.72 2.88 3.04
"S" DOLLAR COST OF ENERGY, ENGLISH UNITS

Figure 38

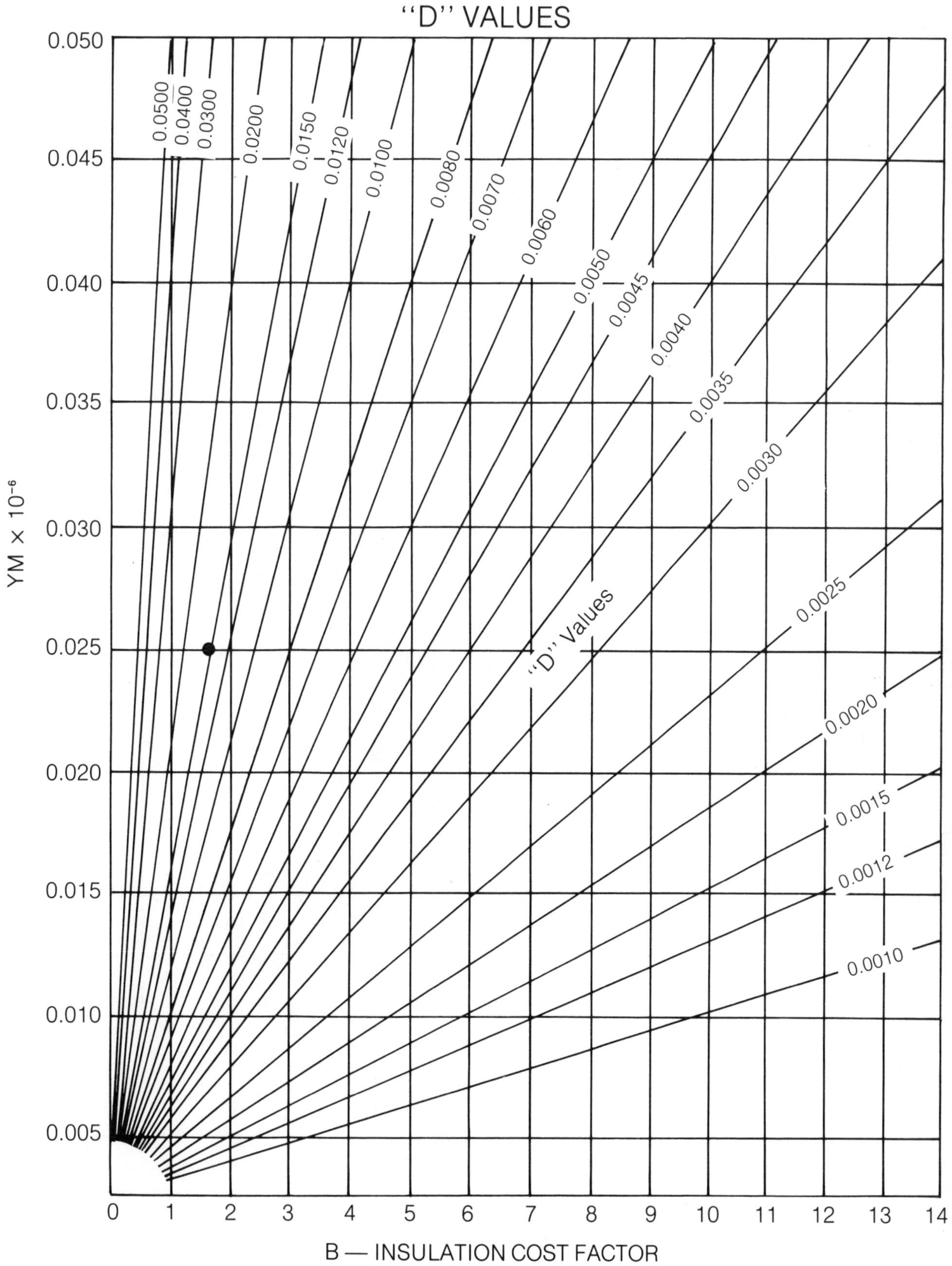
"D" VALUES
0.0500
0.0400
0.0300
0.0200
0.0150
0.0120
0.0100
0.0080
0.0070
0.0060
0.0050
0.0045
0.0040
0.0035
0.0030
0.0025
0.0020
0.0015
0.0012
0.0010
"D" Values
YM × 10⁻⁶
0.050
0.045
0.040
0.035
0.030
0.025
0.020
0.015
0.010
0.005
0 1 2 3 4 5 6 7 8 9 10 11 12 13 14
B — INSULATION COST FACTOR

Figure 39

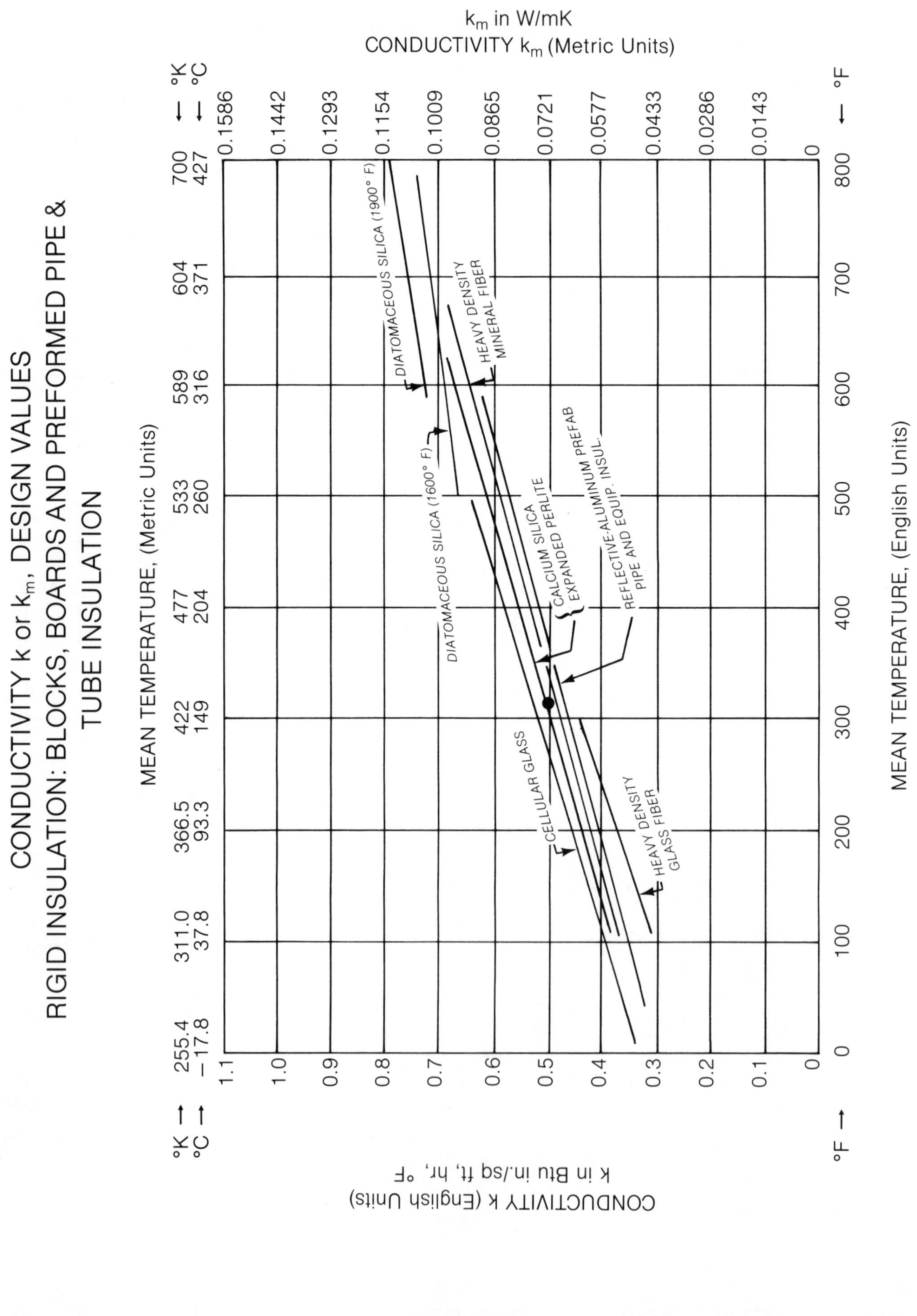
CONDUCTIVITY k or k_m, DESIGN VALUES
RIGID INSULATION: BLOCKS, BOARDS AND PREFORMED PIPE & TUBE INSULATION
k_m in W/mK
CONDUCTIVITY k_m (Metric Units)
MEAN TEMPERATURE, (Metric Units)
°K
°C
255.4
−17.8
311.0
37.8
366.5
93.3
422
149
477
204
533
260
589
316
604
371
700
427
0.1586
0.1442
0.1293
0.1154
0.1009
0.0865
0.0721
0.0577
0.0433
0.0286
0.0143
0
1.1
1.0
0.9
0.8
0.7
0.6
0.5
0.4
0.3
0.2
0.1
0
DIATOMACEOUS SILICA (1900° F)
HEAVY DENSITY MINERAL FIBER
DIATOMACEOUS SILICA (1600° F)
CALCIUM SILICA
EXPANDED PERLITE
REFLECTIVE-ALUMINUM PREFAB PIPE AND EQUIP. INSUL.
CELLULAR GLASS
HEAVY DENSITY GLASS FIBER
CONDUCTIVITY k (English Units)
k in Btu in./sq ft, hr, °F
°F
0
100
200
300
400
500
600
700
800
MEAN TEMPERATURE, (English Units)

Figure 40

Table 36

ECONOMICAL INSULATION THICKNESS TABLE

IN INCHES
(nominal)

PIPE SIZE: 6'' NPS (Tube Size 6-1/2'' OD)
168.3 mm

At the intersection of the ''Δt'' column with the ''k'' row read the economical thickness.
(•—*exceeds 12'' thickness*)

D = 0.0120

Δt Celsius °C or Kelvin °K

Btu, in./sq ft, hr °F

k	50	70	90	110	130	150	170	190	210	230	250	270	290	310	330	350	370	390	410	430	450	470	490	510	530	550	570	590	610	630	650	670	690	710	730	km
0.1	1.0	1.0	1.0	1.0	1.0	1.0	1.5	1.5	1.5	1.5	1.5	1.5	1.5	1.5	2.0	2.0	2.0	2.0	2.0	2.0	2.0	2.0	2.0	2.0	2.0	2.0	2.5	2.5	2.5	2.5	2.5	2.5	2.5	2.5	2.5	.0144
0.2	1.0	1.0	1.5	1.5	1.5	1.5	1.5	2.0	2.0	2.0	2.0	2.0	2.0	2.5	2.5	2.5	2.5	2.5	2.5	2.5	3.0	3.0	3.0	3.0	3.0	3.0	3.0	3.0	3.0	3.0	3.5	3.5	3.5	3.5	3.5	.0288
0.3	1.0	1.5	1.5	1.5	2.0	2.0	2.0	2.0	2.5	2.5	2.5	2.5	2.5	2.5	3.0	3.0	3.0	3.0	3.0	3.0	3.5	3.5	3.5	3.5	3.5	3.5	3.5	4.0	4.0	4.0	4.0	4.0	4.0	4.0	4.0	.0433
0.4	1.5	1.5	1.5	2.0	2.0	2.0	2.5	2.5	2.5	2.5	3.0	3.0	3.0	3.0	3.0	3.5	3.5	3.5	3.5	3.5	4.0	4.0	4.0	4.0	4.0	4.0	4.5	4.5	4.5	4.5	4.5	4.5	4.5	5.0	5.0	.0577
0.5	1.5	1.5	2.0	2.0	2.0	2.5	2.5	2.5	3.0	3.0	3.0	3.0	3.5	3.5	3.5	3.5	4.0	4.0	4.0	4.0	4.0	4.5	4.5	4.5	4.5	4.5	4.5	5.0	5.0	5.0	5.0	5.0	5.0	5.5	5.5	.0721
0.6	1.5	1.5	2.0	2.0	2.5	2.5	2.5	3.0	3.0	3.0	3.5	3.5	3.5	3.5	4.0	4.0	4.0	4.0	4.5	4.5	4.5	4.5	4.5	5.0	5.0	5.0	5.0	5.0	5.5	5.5	5.5	5.5	5.5	6.0	6.0	.0865
0.7	1.5	2.0	2.0	2.5	2.5	2.5	3.0	3.0	3.0	3.5	3.5	3.5	4.0	4.0	4.0	4.5	4.5	4.5	4.5	4.5	5.0	5.0	5.0	5.0	5.5	5.5	5.5	5.5	5.5	6.0	6.0	6.0	6.0	6.0	6.5	.1009
0.8	1.5	2.0	2.0	2.5	2.5	3.0	3.0	3.0	3.5	3.5	4.0	4.0	4.0	4.0	4.5	4.5	4.5	5.0	5.0	5.0	5.0	5.5	5.5	5.5	5.5	6.0	6.0	6.0	6.0	6.0	6.5	6.5	6.5	6.5	6.5	.1154
0.9	1.5	2.0	2.0	2.5	3.0	3.0	3.0	3.5	3.5	4.0	4.0	4.0	4.5	4.5	4.5	5.0	5.0	5.0	5.0	5.5	5.5	5.5	5.5	6.0	6.0	6.0	6.0	6.5	6.5	6.5	6.5	7.0	7.0	7.0	7.0	.1298
1.0	1.5	2.0	2.5	2.5	3.0	3.0	3.5	3.5	4.0	4.0	4.0	4.5	4.5	4.5	5.0	5.0	5.0	5.5	5.5	5.5	5.5	6.0	6.0	6.0	6.5	6.5	6.5	6.5	7.0	7.0	7.0	7.0	7.5	7.5	7.5	.1442
1.1	1.5	2.0	2.5	2.5	3.0	3.0	3.5	3.5	4.0	4.0	4.5	4.5	4.5	5.0	5.0	5.0	5.5	5.5	5.5	6.0	6.0	6.0	6.5	6.5	6.5	6.5	7.0	7.0	7.0	7.5	7.5	7.5	7.5	7.5	8.0	.1586
1.2	1.5	2.0	2.5	3.0	3.0	3.5	3.5	4.0	4.0	4.5	4.5	4.5	5.0	5.0	5.0	5.5	5.5	5.5	6.0	6.0	6.5	6.5	6.5	6.5	7.0	7.0	7.0	7.5	7.5	7.5	7.5	8.0	8.0	8.0	8.0	.1730
1.3	1.5	2.0	2.5	3.0	3.0	3.5	3.5	4.0	4.0	4.5	4.5	5.0	5.0	5.0	5.5	5.5	6.0	6.0	6.0	6.5	6.5	6.5	7.0	7.0	7.0	7.5	7.5	7.5	7.5	8.0	8.0	8.0	8.0	8.5	8.5	.1875
1.4	2.0	2.0	2.5	3.0	3.0	3.5	4.0	4.0	4.5	4.5	5.0	5.0	5.0	5.5	5.5	6.0	6.0	6.0	6.5	6.5	6.5	7.0	7.0	7.0	7.5	7.5	7.5	8.0	8.0	8.0	8.0	8.5	8.5	8.5	9.0	.2019
1.5	2.0	2.0	2.5	3.0	3.5	3.5	4.0	4.0	4.5	4.5	5.0	5.0	5.5	5.5	6.0	6.0	6.0	6.5	6.5	6.5	7.0	7.0	7.5	7.5	7.5	8.0	8.0	8.0	8.0	8.5	8.5	8.5	9.0	9.0	9.0	.2163
k	90	126	162	198	234	270	306	342	378	414	450	486	522	558	594	630	666	702	738	774	810	846	882	918	954	990	1026	1062	1098	1134	1170	1206	1242	1278	1314	km

Δt Fahrenheit

W/mk

Ex. 42

D = 0.0150

Δt Celsius °C or Kelvin °K

CONDUCTIVITY
Btu, in./sq ft, hr °F

k	50	70	90	110	130	150	170	190	210	230	250	270	290	310	330	350	370	390	410	430	450	470	490	510	530	550	570	590	610	630	650	670	690	710	730	km
0.1	1.0	1.0	1.0	1.0	1.5	1.5	1.5	1.5	1.5	1.5	1.5	2.0	2.0	2.0	2.0	2.0	2.0	2.0	2.0	2.0	2.5	2.5	2.5	2.5	2.5	2.5	2.5	2.5	2.5	2.5	2.5	2.5	2.5	3.0	3.0	.0144
0.2	1.0	1.5	1.5	1.5	1.5	2.0	2.0	2.0	2.0	2.0	2.5	2.5	2.5	2.5	2.5	2.5	3.0	3.0	3.0	3.0	3.0	3.0	3.0	3.0	3.5	3.5	3.5	3.5	3.5	3.5	3.5	3.5	4.0	4.0	4.0	.0288
0.3	1.5	1.5	1.5	2.0	2.0	2.0	2.5	2.5	2.5	2.5	3.0	3.0	3.0	3.0	3.0	3.0	3.5	3.5	3.5	3.5	3.5	4.0	4.0	4.0	4.0	4.0	4.0	4.5	4.5	4.5	4.5	4.5	4.5	4.5	4.5	.0433
0.4	1.5	1.5	2.0	2.0	2.0	2.5	2.5	3.0	3.0	3.0	3.0	3.0	3.5	3.5	3.5	3.5	4.0	4.0	4.0	4.0	4.0	4.5	4.5	4.5	4.5	4.5	5.0	5.0	5.0	5.0	5.0	5.0	5.5	5.5	5.5	.0577
0.5	1.5	2.0	2.0	2.5	2.5	2.5	3.0	3.0	3.0	3.5	3.5	3.5	3.5	4.0★	4.0	4.0	4.0	4.5	4.5	4.5	4.5	5.0	5.0	5.0	5.0	5.0	5.5	5.5	5.5	5.5	5.5	6.0	6.0	6.0	6.0	.0721
0.6	1.5	2.0	2.0	2.5	2.5	3.0	3.0	3.0	3.5	3.5	4.0	4.0	4.0	4.0	4.5	4.5	4.5	4.5	5.0	5.0	5.0	5.0	5.5	5.5	5.5	5.5	6.0	6.0	6.0	6.0	6.0	6.5	6.5	6.5	6.5	.0865
0.7	1.5	2.0	2.5	2.5	3.0	3.0	3.0	3.5	3.5	4.0	4.0	4.0	4.5	4.5	4.5	5.0	5.0	5.0	5.0	5.5	5.5	5.5	5.5	6.0	6.0	6.0	6.0	6.5	6.5	6.5	6.5	7.0	7.0	7.0	7.0	.1009
0.8	2.0	2.0	2.5	2.5	3.0	3.0	3.5	3.5	4.0	4.0	4.5	4.5	4.5	5.0	5.0	5.0	5.5	5.5	5.5	5.5	6.0	6.0	6.0	6.5	6.5	6.5	6.5	6.5	7.0	7.0	7.0	7.5	7.5	7.5	7.5	.1154
0.9	2.0	2.0	2.5	3.0	3.0	3.5	3.5	4.0	4.0	4.5	4.5	4.5	5.0	5.0	5.0	5.5	5.5	5.5	6.0	6.0	6.0	6.5	6.5	6.5	6.5	7.0	7.0	7.0	7.5	7.5	7.5	7.5	8.0	8.0	8.0	.1298
1.0	2.0	2.5	2.5	3.0	3.0	3.5	4.0	4.0	4.5	4.5	4.5	5.0	5.0	5.5	5.5	5.5	6.0	6.0	6.0	6.5	6.5	6.5	7.0	7.0	7.0	7.5	7.5	7.5	7.5	8.0	8.0	8.0	8.0	8.5	8.5	.1442
1.1	2.0	2.5	2.5	3.0	3.5	3.5	4.0	4.0	4.5	4.5	5.0	5.0	5.5	5.5	5.5	6.0	6.0	6.5	6.5	6.5	7.0	7.0	7.0	7.5	7.5	7.5	8.0	8.0	8.0	8.0	8.5	8.5	8.5	9.0	9.0	.1586
1.2	2.0	2.5	3.0	3.0	3.5	4.0	4.0	4.5	4.5	5.0	5.0	5.5	5.5	5.5	6.0	6.0	6.5	6.5	6.5	7.0	7.0	7.5	7.5	7.5	8.0	8.0	8.0	8.0	8.5	8.5	8.5	9.0	9.0	9.0	9.5	.1730
1.3	2.0	2.5	3.0	3.0	3.5	4.0	4.0	4.5	5.0	5.0	5.5	5.5	5.5	6.0	6.0	6.5	6.5	6.5	7.0	7.0	7.5	7.5	7.5	8.0	8.0	8.0	8.5	8.5	8.5	9.0	9.0	9.0	9.5	9.5	9.5	.1875
1.4	2.0	2.5	3.0	3.5	3.5	4.0	4.5	4.5	5.0	5.0	5.5	5.5	6.0	6.0	6.5	6.5	7.0	7.0	7.0	7.5	7.5	8.0	8.0	8.0	8.5	8.5	8.5	9.0	9.0	9.0	9.5	9.5	9.5	10.0	10.0	.2019
1.5	2.0	2.5	3.0	3.5	4.0	4.0	4.5	5.0	5.0	5.5	5.5	6.0	6.0	6.5	6.5	7.0	7.0	7.5	7.5	7.5	8.0	8.0	8.0	8.5	8.5	9.0	9.0	9.0	9.5	9.5	9.5	10.0	10.0	10.0	10.5	.2163
k	90	126	162	198	234	270	306	342	378	414	450	486	522	558	594	630	666	702	738	774	810	846	882	918	954	990	1026	1062	1098	1134	1170	1206	1242	1278	1314	km

Ex. 42

Δt Fahrenheit

W/mk

D = 0.0200

Δt Celsius °C or Kelvin °K

Btu, in./sq ft, hr °F

k	50	70	90	110	130	150	170	190	210	230	250	270	290	310	330	350	370	390	410	430	450	470	490	510	530	550	570	590	610	630	650	670	690	710	730	km
0.1	1.0	1.0	1.0	1.5	1.5	1.5	1.5	1.5	2.0	2.0	2.0	2.0	2.0	2.0	2.0	2.5	2.5	2.5	2.5	2.5	2.5	2.5	2.5	2.5	3.0	3.0	3.0	3.0	3.0	3.0	3.0	3.0	3.0	3.0	3.0	.0144
0.2	1.5	1.5	1.5	2.0	2.0	2.0	2.0	2.5	2.5	2.5	2.5	3.0	3.0	3.0	3.0	3.0	3.0	3.5	3.5	3.5	3.5	3.5	3.5	4.0	4.0	4.0	4.0	4.0	4.0	4.0	4.5	4.5	4.5	4.5	4.5	.0288
0.3	1.5	1.5	2.0	2.0	2.5	2.5	2.5	3.0	3.0	3.0	3.0	3.5	3.5	3.5	3.5	4.0	4.0	4.0	4.0	4.0	4.5	4.5	4.5	4.5	4.5	4.5	5.0	5.0	5.0	5.0	5.0	5.0	5.5	5.5	5.5	.0433
0.4	1.5	2.0	2.0	2.5	2.5	3.0	3.0	3.0	3.5	3.5	3.5	4.0	4.0	4.0	4.0	4.5	4.5	4.5	4.5	5.0	5.0	5.0	5.0	5.0	5.5	5.5	5.5	5.5	5.5	6.0	6.0	6.0	6.0	6.0	6.5	.0577
0.5	2.0	2.0	2.5	2.5	3.0	3.0	3.0	3.5	3.5	4.0	4.0	4.0	4.5	4.5	4.5	5.0	5.0	5.0	5.0	5.5	5.5	5.5	5.5	6.0	6.0	6.0	6.0	6.5	6.5	6.5	6.5	6.5	7.0	7.0	7.0	.0721
0.6	2.0	2.0	2.5	3.0	3.0	3.5	3.5	4.0	4.0	4.0	4.5	4.5	4.5	5.0	5.0	5.0	5.5	5.5	5.5	6.0	6.0	6.0	6.0	6.5	6.5	6.5	6.5	7.0	7.0	7.0	7.5	7.5	7.5	7.5	7.5	.0865
0.7	2.0	2.5	2.5	3.0	3.5	3.5	4.0	4.0	4.5	4.5	4.5	5.0	5.0	5.0	5.5	5.5	5.5	6.0	6.0	6.5	6.5	6.5	6.5	7.0	7.0	7.0	7.5	7.5	7.5	7.5	8.0	8.0	8.0	8.0	8.5	.1009
0.8	2.0	2.5	3.0	3.0	3.5	4.0	4.0	4.5	4.5	4.5	5.0	5.0	5.5	5.5	6.0	6.0	6.0	6.5	6.5	6.5	7.0	7.0	7.0	7.5	7.5	7.5	8.0	8.0	8.0	8.0	8.5	8.5	8.5	9.0	9.0	.1154
0.9	2.0	2.5	3.0	3.5	3.5	4.0	4.5	4.5	5.0	5.0	5.0	5.5	5.5	6.0	6.0	6.5	6.5	6.5	7.0	7.0	7.5	7.5	7.5	8.0	8.0	8.0	8.0	8.5	8.5	8.5	9.0	9.0	9.0	9.5	9.5	.1298
1.0	2.0	2.5	3.0	3.5	4.0	4.0	4.5	4.5	5.0	5.0	5.5	5.5	6.0	6.0	6.5	6.5	7.0	7.0	7.5	7.5	7.5	8.0	8.0	8.0	8.5	8.5	8.5	9.0	9.0	9.0	9.5	9.5	9.5	10.0	10.0	.1442
1.1	2.5	3.0	3.0	3.5	4.0	4.5	4.5	5.0	5.0	5.5	5.5	6.0	6.5	6.5	6.5	7.0	7.0	7.5	7.5	8.0	8.0	8.0	8.5	8.5	8.5	9.0	9.0	9.5	9.5	9.5	10.0	10.0	10.5	10.5	10.5	.1586
1.2	2.5	3.0	3.5	4.0	4.0	4.5	5.0	5.0	5.5	5.5	6.0	6.5	6.5	6.5	7.0	7.0	7.5	7.5	8.0	8.0	8.5	8.5	8.5	9.0	9.0	9.5	9.5	9.5	10.0	10.0	10.0	10.5	10.5	11.0	11.0	.1730
1.3	2.5	3.0	3.5	4.0	4.5	4.5	5.0	5.5	5.5	6.0	6.0	6.5	6.5	7.0	7.5	7.5	7.5	8.0	8.0	8.5	8.5	9.0	9.0	9.5	9.5	9.5	10.0	10.0	10.5	10.5	10.5	11.0	11.0	11.5	11.5	.1875
1.4	2.5	3.0	3.5	4.0	4.5	5.0	5.0	5.5	6.0	6.0	6.5	6.5	7.0	7.5	7.5	8.0	8.0	8.0	8.5	8.5	9.0	9.0	9.5	9.5	10.0	10.0	10.5	10.5	11.0	11.0	11.0	11.5	11.5	12.0	12.0	.2019
1.5	2.5	3.0	3.5	4.0	4.5	5.0	5.5	5.5	6.0	6.5	6.5	7.0	7.0	7.5	8.0	8.0	8.5	8.5	9.0	9.0	9.5	9.5	9.5	10.0	10.0	10.5	10.5	11.0	11.5	11.5	11.5	12.0	12.0	•	•	.2163
k	90	126	162	198	234	270	306	342	378	414	450	486	522	558	594	630	666	702	738	774	810	846	882	918	954	990	1026	1062	1098	1134	1170	1206	1242	1278	1314	km

Δt Fahrenheit

W/mk

TEMPERATURE DIFFERENCE

★ Indicates answer to Example 42

EXAMPLE 43

Knowing the economic thickness, it becomes of interest to determine what the economics are for the insulation installation. Under the conditions given in the preceding problem what are the dollar savings?

From Table 11, the bare loss of an uninsulated 6″ NPS pipe operating at 600° F, with 70° F ambient temperature, is 4604 Btu/lin ft, hr. Adjusting to 50° F ambient temperature the heat loss is 4777 Btu/lin ft, hr.

The insulated heat loss is calculated as follows:

$$Q = \frac{t_o - t_{av}}{\left(\frac{7.5 \log_e \frac{7.5}{3.35}}{k}\right) + \frac{1}{f}} = \text{Btu/ft}^2\text{, hr} \qquad (17)$$

$$Q = \frac{600 - 50}{\left(\frac{7.5 \log_e \frac{7.5}{3.35}}{0.5}\right) + 0.52} = \frac{550}{\left(\frac{6.03}{0.5}\right) + 0.52}$$

$$= \frac{550}{12.06 + 0.52} = \frac{550}{12.58} = 43.7 \text{ Btu/ft}^2\text{, hr*}$$

The outer surface area of 6″ NPS pipe insulation 4″ thick is 3.93 sq ft/lin ft.

The heat loss per lin ft of insulated pipe then is:

$$43.7 \times 3.93 = 172 \text{ Btu/lin ft, hr.}$$

The heat savings per lin ft due to insulation is then:

4777 (bare loss) − 172 (insulated loss) = 4605 Btu/lin ft, hr.

The Btu/lb of 300 psi steam superheated to 600° F is 1315 Btu. The pounds of steam per hour saved by the insulation for each lin foot is then:

$$\frac{4605}{1315} = 3.50 \text{ lbs/lin ft, hr steam savings.}$$

The example stated the capital investment of steam producing equipment was $30.00 per lb/hr. Therefore, the savings in necessary capital investment for steam production-equipment is:

$30.00 × 3.50 = $105.00 per lin foot.

or

3.048 × $105.00 = $320.00 per lin metre.

Also based on the curves of installed cost (based $2.60 for 1″ thickness on 1″ NPS) for 4″ thickness on 6″ NPS pipe the cost is $21.25 per lin foot. In other words, the investment of $21.25 in insulation saves $105.00 in required steam production capacity cost.

Note: Some individuals do not agree that the insulation savings should be applied to steam production capital investment, on the basis that the steam production equipment is already in existence. However, consider what happens if insulation is not used. One thousand feet of uninsulated line would require equipment producing 3470 lbs of steam/hr just to supply losses which would not have occurred if pipe had been properly insulated. It may be that the fact that it cost less to insulate than not to insulate upsets the common practice of calculating return on investment, as the question arises how do you calculate a return on a negative investment?

As mentioned above, to be able to calculate return on investment in this example, the savings in capital investment of steam producing equipment must be ignored.

The approximate installed cost of 4″ thick insulation on a 6″ NPS pipe is $21.25 per lin ft. This saves the cost of producing 3.50 lbs of steam for every lin ft of pipe every hour. Based on operation of 8760 hours per year, the savings in steam not produced would be 3.5 × 8760 or 30630 lbs per year per lin ft of insulated pipe.

It was stated in the example that the production cost of steam was $2.60 per 1000 lbs. The savings in production cost—for steam not produced—is 30.63 × $2.60 = $79.72 per lin foot of pipe.

Thus, an investment of $21.25 (not considering capital savings of production equipment) saves $79.72 of energy cost per year. Even calculated on this basis, the return on investment in insulation is 375%.

Other than units to measure length of pipe, the use of Metric Units would not influence the answers in dollars.

If the savings per linear foot of pipe is $79.72, the savings per linear metre would be $79.72 × 3.048 = $242.99/lin m.

Likewise the cost of installed insulation per linear foot at 21.25 would be $21.25 × 3.048 = $64.77/lin m.

The return on investment being a % would be $242.99 ÷ 69.77 × 100 = 348.

As insulation is furnished in Increments of ½″ thickness, the question remains: Does it pay to install the last ½″ of thickness?

The additional cost to install 4″ thick on 6 NPS pipe over the cost to install 3½″ thickness is $1.35 per linear foot. Even if, again, the savings of capital investment is ignored, what is the return on investment on this $1.65?

If 3½″ thickness had been used, the heat loss per sq ft/hr "Q" would be:

$$Q = \frac{t_o - t_{av}}{\left(\frac{r_2 \log_e \frac{r_2}{r_1}}{k}\right) + \frac{1}{f}} = \frac{600 - 50}{\left(\frac{7.0 \log_e \frac{7.0}{3.35}}{k}\right) + 0.52}$$

$$= \frac{550}{\left(\frac{5.16}{0.5}\right) + 0.52} = \frac{550}{10.32 + 0.52}$$

*Table 37 provides a quick check of the heat loss per hour per sq ft knowing the resistance R.

$$= \frac{550}{10.84} = 50.7 \text{ Btu/ft}^2\text{, hr.}$$

As outer surface area of 6″ NPS insulation 3½″ = 3.67 sq ft/lin ft

$$Q_l = 50.7 \times 3.67 = 186.07 \text{ Btu/lin ft hr,}$$

$$Q_l \text{ with 4″ thick insulation} = 172 \text{ Btu/lin ft.}$$

The savings per lin ft in Btu = 186.07 − 172 = 14.07 Btu lin ft/hr in 4″ insulation as compared to 3½″ insulation.

In one year this is 8764 × 14.07 = 123,253 Btu/lin ft, year
Converting to steam 123,253 ÷ 1315 = 93.93 lbs of steam
As the production cost of steam was given at $2.60/m lbs
Savings per lin ft = 0.094 × $2.60 = $0.24 per lin ft/year

The additional cost to install the last ½″ of 4″ thickness was given as $1.35 per lin ft.

Thus: % return on $1.35 is (0.24 ÷ 1.35) × 100 = 17.7%.

The economic thickness was calculated as 4″ thickness and even the last ½″ (3½″ to 4″) provided a return of 17.7% on investment. However, it must be remembered the total installation gave a return of investment of 375%. In addition the insulation saved 490% of cost on needed production capacity, as compared to bare pipe loss.

Again, regardless of the astounding savings shown for the *total* insulation installation, the basis of the economic thickness study was that the *final* ½″ thickness had to provide an excellent return on investment. To attempt to save money by using less than the economic thickness of insulation proves to be very costly.

The proper selection of insulation thicknesses for pipes and equipment will result in savings of millions of dollars per year for the large user.

Insulation Economics in Buildings

The solution of the insulation requirements for buildings is not as straightforward as the calculation of economic thickness of insulation for pipes and vessels. One of the reasons is that most building materials provide at least some resistance to heat flow, and the insulation is part of a system used in the construction. Therefore, the solution must be based on comparison of the systems. Other factors enter into the economics of building insulation, such as location (with corresponding extremes of weather and anticipated yearly heat or/and cooling load), proportion of glassed areas, and whether the building is only heated or heated and cooled.

For these reasons design of insulation of buildings becomes an integral part of the design of the buildings' floors, walls, ceilings, and roofs. Because of this, in calculating heat transfer problems related to buildings, the conductivity of insulation by itself is seldom used. In most instances the conductance "U" value of the floor systems, wall systems, and ceiling or roof systems is the basis of the calculation. Of course the conductivity of the insulation, and its thickness, affects the "U" value, which is the reciprocal of the total resistances.

$$U = \frac{1}{R}$$

The values of U for various constructions of walls, floors, ceilings, and roofs, with and without insulation, is listed in ASHRAE Guide. It is recommended that for conventional constructions their U values be obtained from this source. However, if heat transfer is needed for special constructions not listed, they can be calculated as previously shown in Eq. 16.

To establish the order of magnitude, in general, walls in buildings which are heated should have a U value of less than 0.15, and if cooled, the U value should be less than 0.10. In northern climates, or where heating cost is high, lower U values are justified for heating. Of course the more southern climates and where refrigeration costs are high justify lower U values than those mentioned.

The popularity of large areas of glass in a building increases the difficulty of insulating such a building to be thermally efficient. However, as the glass areas cause excessive heat losses (or gains in cooled buildings), it becomes increasingly important that the other areas be well insulated to offset these inefficiencies.

Regardless of the basic building design, the insulation industry must do its part to assist the designers and the owners to understand the economic advantage of a well-insulated structure. "Dividend Engineering for Buildings," published by the Owens-Corning Corporation, is an excellent approach to this problem. The use of this economic engineering method can prove again that dollar savings, or return on investment, obtained by properly installed insulation is the best investment the owner can make in his building budget.

The discussion on economics of insulation in buildings has so far been based on dollars which can be saved in fuel and refrigeration costs. Frequently, in such discussions, the fundamental purpose of the building is lost sight of. This fundamental purpose is to provide a suitable environment in which man can work and live. The walls, floors, ceilings, and roofs are the enclosure necessary to provide this comfortable environment. A number of years ago the John B. Pierce Laboratory of Hygiene performed a series of tests and demonstrated the physiological relationship between heat-production and heat-loss of humans and the various environmental factors of air temperature, radiant energy, humidity, and air movement. From these experiments the following equation was arrived at to express what happens between the human body and its environment when the heat that is dissipated to the atmosphere and the heat taken from it is in equilibrium. When:

M = metabolism
E = evaporation (sweating and breathing)
Q_{cc} = heat gained or lost through convection and conduction
Q_R = heat gained or lost through radiation

then

$$M = E \pm Q_{cc} \pm Q_R = 0 \qquad (63)$$

Table 37

HEAT LOSS "Q" $\left(Q = \dfrac{\Delta t}{R + R_s}\right)$

Q in Btu, sq ft, hr when $R + R_s$ is in English Units and Δt is °F.
Q in Wm² when $R + R_s$ is in Metric Units and Δt is °C or °K.

VALUE	Δt TEMPERATURE DIFFERENCE (Pipe or Equipment Surface Temperature Minus Ambient Air Temperature)											
$R + R_s$	100	200	300	400	500	600	700	800	900	1000	1100	1200
.1	1000	2000	3000	4000	5000	6000	7000	8000	9000	10000	11000	12000
.2	500	1000	1500	2000	2500	3000	3500	4000	4500	5000	5500	6000
.3	333	667	1000	1333	1667	2000	2333	2667	3000	3333	3667	4000
.4	250	500	750	1000	1250	1500	1750	2000	2250	2500	2750	3000
.5	200	400	600	800	1000	1200	1400	1600	1800	2000	2200	2400
.6	166	333	500	666	833	1000	1166	1333	1500	1666	1833	2000
.7	143	286	428	571	714	857	1000	1142	1285	1428	1571	1714
.8	125	250	375	500	625	750	875	1000	1125	1250	1375	1500
.9	111	222	333	444	556	667	778	889	1000	1111	1222	1333
1.0	100	200	300	400	500	600	700	800	900	1000	1100	1200
1.1	91	182	273	364	455	545	636	727	818	909	1000	1090
1.2	83	166	249	332	415	500	583	666	750	833	917	1000
1.3	77	154	231	308	383	461	588	615	692	769	846	923
1.4	71	142	213	284	355	426	500	571	642	714	785	857
1.5	66	133	200	266	333	400	467	533	600	666	733	800
1.6	63	125	188	250	313	375	438	500	563	625	687	750
1.7	59	118	176	235	294	353	412	470	529	588	697	706
1.8	56	111	166	222	277	333	388	444	500	555	611	666
1.9	53	105	157	210	263	315	368	421	473	526	578	631
2.0	50	100	150	200	250	300	350	400	450	500	550	600
2.1	48	95	142	190	238	286	333	380	428	476	524	571
2.2	45	91	136	182	227	272	318	364	409	454	500	545
2.3	43	87	130	174	217	261	304	347	391	434	478	522
2.4	42	83	125	166	208	250	291	333	375	416	458	500
2.5	40	80	120	160	200	240	280	320	360	400	440	480
2.6	38	77	115	154	192	231	269	308	346	384	423	462
2.7	37	74	111	148	185	222	259	296	333	370	407	444
2.8	35	71	107	142	178	214	250	285	321	357	392	428
2.9	34	69	103	137	172	206	241	275	310	344	379	413
3.0	33	67	100	133	167	200	233	267	300	333	367	400
3.1	32	65	97	129	161	193	225	258	290	322	354	387
3.2	31	62	94	125	156	187	218	250	281	312	343	375
3.3	30	61	91	121	151	181	212	242	272	303	333	363
3.4	29	59	88	118	147	176	205	235	264	294	323	352
3.5	29	57	86	114	143	171	200	228	257	285	314	343
3.6	28	55	83	111	138	167	194	222	250	278	306	333
3.7	27	54	81	108	135	162	189	216	243	270	297	324
3.8	26	53	79	105	131	157	184	210	236	263	289	315
3.9	26	51	77	102	128	153	179	205	230	256	282	307
4.0	25	50	75	100	125	150	175	200	225	250	275	300
4.1	24	48	73	98	121	146	171	195	219	243	268	293
4.2	24	47	71	95	119	142	166	190	214	238	261	285
4.3	23	46	70	93	116	139	162	186	209	232	256	279
4.4	23	45	68	91	114	136	159	182	204	227	250	272
4.5	22	44	66	89	111	133	156	178	200	222	244	267
4.6	22	43	65	87	109	130	152	174	196	217	239	260
4.7	21	43	64	85	106	127	149	170	191	213	234	255
4.8	21	42	62	83	104	125	146	167	188	208	229	250
4.9	20	41	61	81	102	122	142	163	184	204	224	245
5.0	20	40	60	80	100	120	140	160	180	200	220	240
5.1	20	39	59	78	98	118	137	157	176	196	216	235
5.2	19	38	58	77	96	115	134	154	173	192	211	231
5.3	19	38	57	75	94	113	132	150	170	189	207	226
5.4	19	37	56	74	93	111	130	148	167	185	203	222
5.5	18	36	55	73	91	109	127	145	164	181	200	218
5.6	18	36	54	71	89	107	125	142	160	179	196	214
5.7	18	35	53	70	88	105	122	140	158	175	192	210
5.8	17	34	52	69	86	103	120	138	155	172	188	207
5.9	17	34	51	68	85	102	119	136	153	169	186	203
6.0	17	33	50	67	83	100	117	133	150	167	183	200
6.1	16	33	49	66	82	98	115	131	148	164	180	197
6.2	16	32	48	65	81	97	113	129	145	161	177	193
6.3	16	32	48	63	79	95	111	127	142	159	174	190
6.4	16	31	47	63	78	94	109	125	140	156	172	187
6.5	15	31	46	62	77	92	108	123	138	154	169	185
6.6	15	30	45	61	76	91	106	121	136	152	167	182
6.7	15	30	45	60	75	90	104	119	134	150	164	179
6.8	15	29	44	59	74	88	102	118	132	147	162	176
6.9	14	29	43	58	72	87	101	116	130	145	159	174
7.0	14	29	43	57	71	86	100	114	129	143	157	171

VALUE	Δt TEMPERATURE DIFFERENCE (Pipe or Equipment Surface Temperature Minus Ambient Air Temperature)											
$R + R_s$	100	200	300	400	500	600	700	800	900	1000	1100	1200
7.1	14.0	28.2	42.2	56.3	70.4	84.5	98.6	112.7	126.8	140.8	154.9	169.0
7.2	13.9	27.7	41.6	55.5	69.4	83.3	97.2	111.1	125.0	138.8	152.8	166.6
7.3	13.7	27.4	41.1	54.8	68.5	82.2	95.9	109.6	123.3	137.0	150.7	164.4
7.4	13.5	27.0	40.5	54.1	67.6	81.1	94.6	108.1	121.6	135.1	148.6	162.2
7.5	13.3	26.6	39.9	53.3	66.7	80.0	93.3	106.6	120.0	133.3	146.6	160.0
7.6	13.2	26.3	39.4	52.6	65.8	78.9	92.1	105.2	118.4	131.6	144.7	157.9
7.7	13.0	26.0	39.0	51.9	64.9	77.9	90.9	104.0	116.9	129.9	142.9	155.8
7.8	12.8	25.6	38.5	51.3	64.1	76.9	89.7	102.6	115.4	128.2	141.0	153.8
7.9	12.7	25.3	38.0	50.6	63.3	75.9	88.6	101.3	118.9	126.6	139.2	151.9
8.0	12.5	25.0	37.5	50.0	62.5	75.0	87.5	100.0	112.5	125.5	137.5	150.0
8.1	12.3	24.7	37.0	49.4	61.7	74.1	86.4	98.8	111.1	123.5	135.8	148.1
8.2	12.2	24.4	36.6	48.8	60.9	73.2	85.4	97.6	109.8	123.0	134.1	146.3
8.3	12.0	24.1	36.1	48.1	60.2	72.3	84.3	96.3	108.4	120.5	132.5	144.6
8.4	11.9	23.8	35.7	47.6	59.5	71.4	83.3	95.2	107.1	119.0	131.1	142.9
8.5	11.8	23.5	35.3	47.0	58.8	70.9	82.3	94.1	105.8	117.6	129.4	141.2
8.6	11.6	23.3	34.9	46.5	58.1	69.8	82.0	93.0	104.7	116.3	127.9	139.5
8.7	11.5	23.0	34.5	46.0	57.5	69.0	80.5	92.0	103.4	114.9	126.4	137.9
8.8	11.4	22.7	34.1	45.5	56.8	68.2	79.5	90.9	102.2	113.6	125.0	136.4
8.9	11.2	22.4	33.7	44.9	56.2	67.4	78.7	89.9	101.1	112.4	123.6	134.8
9.0	11.1	22.2	33.3	44.4	55.5	66.7	77.8	88.9	100.0	111.1	122.2	133.3
9.1	11.0	22.0	33.0	44.0	55.0	66.0	77.0	88.0	99.0	110.0	120.9	131.9
9.2	10.9	21.7	32.6	43.4	54.3	65.2	76.1	87.0	97.8	108.7	119.6	130.4
9.3	10.8	21.5	32.3	43.0	53.8	64.5	75.3	86.0	96.8	107.5	118.3	129.0
9.4	10.6	21.3	31.9	42.6	53.2	63.8	74.5	85.1	95.7	106.4	117.0	127.7
9.5	10.5	21.0	31.6	42.1	52.6	63.2	73.7	84.2	94.7	105.3	115.8	126.3
9.6	10.4	20.8	31.3	41.7	52.1	62.5	72.9	83.3	93.8	104.2	114.6	125.0
9.7	10.3	20.6	30.9	41.2	51.5	61.9	72.2	82.5	92.8	103.1	113.4	123.7
9.8	10.2	20.4	30.6	40.8	51.0	61.2	71.4	81.6	91.8	102.0	112.2	122.4
9.9	10.1	20.2	30.3	40.4	50.5	60.6	70.7	80.8	90.9	101.0	111.1	121.2
10.0	10.0	20.0	30.0	40.0	50.0	60.0	70.0	80.0	90.0	100.0	110.0	120.0
10.2	9.8	19.6	29.4	39.2	49.0	58.8	66.6	78.4	88.2	98.0	107.8	117.6
10.4	9.6	19.2	28.8	38.5	48.1	57.7	67.3	76.9	86.5	96.2	105.8	115.4
10.6	9.4	18.9	28.3	37.7	47.2	56.6	66.0	75.5	84.9	94.3	103.8	113.2
10.8	9.2	18.5	27.8	37.0	46.3	55.6	64.8	74.1	83.3	92.6	101.9	111.1
11.0	9.1	18.2	27.3	36.4	45.5	54.5	63.6	72.7	81.8	90.9	100.0	109.0
11.2	8.9	17.9	26.8	35.7	44.6	53.6	62.5	71.4	80.4	89.3	98.2	107.1
11.4	8.8	17.5	26.3	35.0	43.9	52.6	61.4	70.2	78.9	87.7	96.5	105.3
11.6	8.6	17.2	25.9	34.5	43.1	51.7	60.3	69.0	77.6	86.2	94.8	108.4
11.8	8.5	16.9	25.4	33.9	42.4	50.8	59.3	67.8	76.3	84.7	93.2	101.7
12.0	8.3	16.6	24.9	33.2	41.5	50.0	58.3	66.6	75.0	83.3	91.7	100.0
12.2	8.2	16.4	24.6	32.8	41.0	49.2	57.4	65.6	73.8	81.7	90.2	98.4
12.4	8.1	16.1	24.2	32.3	40.3	48.4	56.5	64.5	72.6	80.6	88.7	96.8
12.6	7.9	15.9	23.8	31.7	39.7	47.6	55.6	63.5	71.4	79.4	87.3	95.2
12.8	7.8	15.6	23.4	31.3	39.1	46.9	54.7	62.5	70.3	78.1	85.9	93.8
13.0	7.7	15.4	23.1	30.8	38.5	46.1	53.8	61.5	69.2	76.9	84.6	92.3
13.2	7.6	15.2	22.7	30.3	37.9	45.5	53.0	60.6	68.2	75.8	83.3	90.9
13.4	7.5	14.9	22.4	29.9	37.3	44.8	52.2	59.7	67.2	74.6	82.1	89.6
13.6	7.4	14.7	22.0	29.4	36.8	44.1	51.5	58.8	66.2	73.5	80.1	88.2
13.8	7.2	14.5	21.7	29.0	36.2	43.4	50.7	58.0	65.2	72.5	79.7	87.0
14.0	7.1	14.2	21.3	28.4	35.5	42.6	50.0	57.1	64.2	71.4	78.5	85.7
14.2	7.0	14.1	21.1	28.2	35.2	42.2	49.3	53.3	63.4	70.4	77.5	84.5
14.4	6.9	13.9	20.8	27.8	34.7	41.7	48.6	55.6	62.5	69.4	76.4	83.3
14.6	6.8	13.7	20.5	27.4	34.2	41.0	47.9	54.8	61.6	68.5	75.3	82.2
14.8	6.7	13.5	20.3	27.0	33.8	40.5	47.3	54.1	60.8	67.6	74.3	81.1
15.0	6.6	13.3	20.0	26.6	33.3	40.0	46.7	53.3	60.0	66.6	73.3	80.0
15.2	6.6	13.2	19.7	26.3	32.8	39.5	46.1	52.6	59.2	65.7	72.4	78.9
15.4	6.5	13.0	19.5	25.9	32.5	39.0	45.4	51.9	58.4	64.9	71.4	77.9
15.6	6.4	12.8	19.2	25.6	32.0	38.5	44.9	51.3	57.7	64.1	70.5	76.9
15.8	6.3	12.7	19.0	25.3	31.6	38.0	44.3	50.6	57.0	63.2	69.6	75.9
16.0	6.3	12.5	18.8	25.0	31.3	37.5	43.8	50.0	56.3	62.5	68.7	75.0
16.2	6.2	12.3	18.5	24.7	30.9	37.0	43.2	49.4	55.6	61.7	67.9	74.1
16.4	6.1	12.2	18.3	24.3	30.5	36.6	42.7	48.8	54.9	61.0	67.0	73.1
16.6	6.0	12.0	18.0	24.1	30.1	36.1	42.2	48.2	54.2	60.2	66.3	72.3
16.8	6.0	11.9	17.8	23.8	29.8	35.7	41.7	47.6	53.6	59.5	65.5	71.4
17.0	5.9	11.8	17.6	23.5	29.4	35.3	41.2	47.0	52.9	58.8	64.7	70.6
17.2	5.8	11.6	17.4	23.3	29.0	34.9	40.6	46.5	52.3	58.1	64.0	69.8
17.4	5.7	11.5	17.2	23.0	28.7	34.5	40.2	46.0	51.7	57.5	63.2	69.0
17.6	5.7	11.4	17.0	22.7	28.4	34.1	40.0	45.5	51.1	56.8	62.5	68.2
17.8	5.6	11.2	16.8	22.5	28.0	33.7	39.3	44.9	50.6	56.2	61.8	67.4
18.0	5.6	11.1	16.6	22.2	27.7	33.3	38.8	44.4	50.0	55.5	61.1	66.6

Table 37 (Continued)

HEAT LOSS "Q" ($Q = \frac{\Delta t}{R + R_s}$)

Q in Btu, sq ft, hr when $R + R_s$ is in English Units and Δt is °F.
Q in Wm² when $R + R_s$ is in Metric Units and Δt is °C or °K.

VALUE	Δt TEMPERATURE DIFFERENCE (Pipe or Equipment Surface Temperature Minus Ambient Air Temperature)											
$R + R_s$	100	200	300	400	500	600	700	800	900	1000	1100	1200
18.2	5.5	11.0	16.5	22.0	27.5	33.0	38.5	44.0	49.5	54.9	60.4	65.9
18.4	5.4	10.9	16.3	21.7	27.2	32.6	38.0	43.5	48.9	54.3	60.0	65.2
18.6	5.4	10.8	16.1	21.5	26.9	32.3	37.6	43.0	48.4	53.8	59.1	64.5
18.8	5.3	10.6	15.9	21.2	26.6	31.9	37.2	42.6	47.9	53.2	58.5	63.8
19.0	5.3	10.5	15.7	21.0	26.3	31.5	36.8	42.1	47.3	52.6	57.3	62.5
19.2	5.2	10.4	15.6	20.8	26.0	31.3	36.5	41.7	46.9	52.1	57.3	62.5
19.4	5.1	10.3	15.5	20.6	25.8	30.9	36.1	41.2	46.4	51.5	56.7	61.8
19.6	5.1	10.2	15.3	20.4	25.5	30.6	35.7	40.8	45.9	51.0	56.1	61.2
19.8	5.1	10.1	15.2	20.2	25.3	30.3	35.4	40.4	45.5	50.5	55.6	60.6
20.0	5.0	10.0	15.0	20.0	25.0	30.0	35.0	40.0	45.0	50.0	55.0	60.0
20.5	4.9	9.8	14.6	19.5	24.4	29.3	34.1	39.0	43.9	48.8	53.7	58.5
21.0	4.8	9.5	14.2	19.0	23.8	28.6	33.3	38.0	42.8	47.6	52.4	57.1
21.5	4.7	9.3	13.9	18.6	23.3	27.9	32.6	37.2	41.8	46.5	51.2	55.8
22.0	4.5	9.1	13.6	18.2	22.7	27.2	31.8	36.4	40.9	45.4	50.0	54.5
22.5	4.4	8.8	13.3	17.8	22.2	26.7	31.1	35.6	40.0	44.4	48.9	53.3
23.0	4.3	8.7	13.0	17.4	21.7	26.1	30.4	34.7	39.1	43.4	47.8	52.2
23.5	4.3	8.5	12.8	17.0	21.2	25.5	29.8	34.0	38.3	42.6	46.8	51.1
24.0	4.2	8.3	12.5	16.6	20.8	25.0	29.1	33.3	37.5	41.6	45.0	50.0
24.5	4.1	8.2	12.2	16.3	20.4	24.5	28.6	32.7	36.7	40.8	44.9	49.0
25.0	4.0	8.0	12.0	16.0	20.0	24.0	28.0	32.0	36.0	40.0	44.0	48.0
26	3.8	7.7	11.5	15.4	19.2	23.1	26.9	30.8	34.6	38.4	42.3	46.2
27	3.7	7.4	11.1	14.8	18.5	22.2	25.9	29.6	33.3	37.0	40.7	44.4
28	3.5	7.1	10.7	14.2	17.8	21.4	25.0	28.5	32.1	35.7	39.2	42.8
29	3.4	6.9	10.3	13.7	17.2	20.6	24.1	27.5	31.0	34.4	37.9	41.3
30	3.3	6.7	10.0	13.3	16.7	20.0	23.3	26.7	30.0	33.3	36.7	40.0
31	3.2	6.5	9.7	12.9	16.1	19.3	22.5	25.8	29.0	32.2	35.4	38.7
32	3.1	6.2	9.4	12.3	15.6	18.7	21.8	25.0	28.1	31.2	34.3	37.5
33	3.0	6.1	9.1	12.1	15.1	18.1	21.2	24.2	27.2	30.3	33.3	36.3
34	2.9	5.9	8.8	11.8	14.7	17.6	20.5	23.5	26.4	29.4	32.3	35.2
35	2.9	5.7	8.6	11.4	14.3	17.1	20.0	22.8	25.7	28.5	31.4	34.3
36	2.8	5.5	8.3	11.1	13.8	16.7	19.4	22.2	25.0	27.8	30.6	33.3
37	2.7	5.4	8.1	10.8	13.5	16.2	18.9	21.6	24.3	27.0	29.7	32.4
38	2.6	5.3	7.9	10.5	13.1	15.7	18.4	21.0	23.6	26.3	28.9	31.5
39	2.6	5.1	7.7	10.2	12.8	15.3	17.9	20.5	23.0	25.6	28.2	30.7
40	2.5	5.0	7.5	10.0	12.5	15.0	17.5	20.0	22.5	25.0	27.5	30.0
41	2.4	4.8	7.3	9.8	12.1	14.6	17.1	19.5	21.9	24.3	26.8	29.3
42	2.4	4.7	7.1	9.5	11.9	14.2	16.6	19.0	21.4	23.8	26.1	28.5
43	2.3	4.6	7.0	9.3	11.6	13.9	16.2	18.6	20.9	23.2	25.6	27.6
44	2.3	4.5	6.8	9.1	11.4	13.6	15.9	18.2	20.4	22.7	25.0	27.2
45	2.2	4.4	6.6	8.9	11.1	13.3	15.6	17.8	20.0	22.2	24.4	26.7
46	2.2	4.3	6.5	8.7	10.9	13.0	15.2	17.4	19.6	21.7	23.9	26.0
47	2.1	4.3	6.4	8.5	10.6	12.7	14.9	17.0	19.1	21.3	23.4	25.5
48	2.1	4.2	6.2	8.3	10.4	12.5	14.6	16.7	18.8	20.8	22.9	25.0
49	2.0	4.1	6.1	8.1	10.2	12.2	14.2	16.3	18.4	20.4	22.4	24.5
50	2.0	4.0	6.0	8.0	10.0	12.0	14.0	16.0	18.0	20.0	22.0	24.0
51	2.0	3.9	5.9	7.8	9.8	11.8	13.7	15.7	17.6	19.6	21.6	23.5
52	1.9	3.8	5.8	7.7	9.6	11.5	13.4	15.4	17.3	19.2	21.1	23.1
53	1.9	3.8	5.7	7.5	9.4	11.3	13.2	15.0	17.0	18.9	20.7	22.6
54	1.9	3.7	5.6	7.4	9.3	11.1	13.0	14.8	16.7	18.5	20.3	22.2
55	1.8	3.6	5.5	7.3	9.1	10.9	12.7	14.5	16.4	18.1	20.0	21.8
56	1.8	3.6	5.4	7.1	8.9	10.7	12.5	14.2	16.0	17.9	19.6	21.4
57	1.8	3.5	5.3	7.0	8.8	10.5	12.2	14.0	15.8	17.5	19.2	21.0
58	1.7	3.4	5.2	6.9	8.6	10.3	12.0	13.8	15.5	17.2	18.8	20.7
59	1.7	3.4	5.1	6.8	8.5	10.2	11.9	13.6	15.3	16.9	18.6	20.3
60	1.7	3.3	5.0	6.7	8.3	10.0	11.7	13.3	15.0	16.7	18.3	20.0
61	1.6	3.3	4.9	6.6	8.2	9.8	11.5	13.1	14.8	16.4	18.0	19.7
62	1.6	3.2	4.8	6.5	8.1	9.7	11.3	12.9	14.5	16.1	17.7	19.3
63	1.6	3.2	4.8	6.3	7.9	9.5	11.1	12.7	14.2	15.9	17.4	19.0
64	1.6	3.1	4.7	6.3	7.8	9.4	10.9	12.5	14.0	15.6	17.2	18.7
65	1.5	3.1	4.6	6.2	7.7	9.2	10.8	12.3	13.8	15.4	16.9	18.5
66	1.5	3.0	4.5	6.1	7.6	9.1	10.6	12.1	13.6	115.2	16.7	18.2
67	1.5	3.0	4.5	6.0	7.5	9.0	10.4	11.9	13.4	15.0	16.4	17.9
68	1.5	2.9	4.4	5.9	7.4	8.8	10.2	11.8	13.2	14.7	16.2	17.6
69	1.4	2.9	4.3	5.8	7.2	8.7	10.1	1.6	13.0	14.5	15.9	17.3
70	1.4	2.9	4.3	5.7	7.1	8.6	10.0	11.4	12.9	14.3	15.7	17.1

VALUE	Δt TEMPERATURE DIFFERENCE (Pipe or Equipment Surface Temperature Minus Ambient Air Temperature)											
$R + R_s$	100	200	300	400	500	600	700	800	900	1000	1100	1200
71	1.40	2.82	4.22	5.63	7.04	8.45	9.86	11.27	12.68	14.08	15.49	16.90
72	1.39	2.77	4.16	5.55	6.94	8.33	9.72	11.11	12.50	13.88	15.28	16.66
73	1.37	2.74	4.11	5.48	6.85	8.22	9.59	10.96	12.33	13.70	15.07	16.44
74	1.35	2.70	4.05	5.41	6.76	8.11	9.46	10.81	12.16	13.51	14.86	16.22
75	1.33	2.66	3.99	5.33	6.67	8.00	9.33	10.66	12.00	13.33	14.66	16.00
76	1.32	2.63	3.94	5.26	6.58	7.89	9.21	10.52	11.84	13.16	14.47	15.79
77	1.30	2.60	3.90	5.19	6.49	7.79	9.09	10.40	11.69	12.99	14.29	15.58
78	1.28	2.56	3.85	5.13	6.41	7.69	8.97	10.26	11.54	12.82	14.10	15.38
79	1.27	2.53	3.80	5.06	6.33	7.59	8.86	10.13	11.39	12.66	13.92	15.19
80	1.25	2.50	3.75	5.00	6.25	7.50	8.75	10.00	11.25	12.55	13.75	15.00
82	1.22	2.44	3.66	4.88	6.09	7.32	8.54	9.76	10.98	12.20	13.41	14.63
84	1.19	2.38	3.57	4.76	5.95	7.14	8.33	9.52	10.71	11.90	13.11	14.29
86	1.16	2.33	3.49	4.65	5.81	6.98	8.20	9.30	10.47	11.63	12.79	13.95
88	1.14	2.27	3.41	4.55	5.68	6.82	7.95	9.09	10.22	11.36	12.50	13.64
90	1.11	2.22	3.33	4.44	5.56	6.67	7.78	8.89	10.00	11.11	12.22	13.33
92	1.09	2.17	3.26	4.34	5.43	6.52	7.61	8.70	9.78	10.87	11.96	13.04
94	1.06	2.13	3.19	4.26	5.32	6.38	7.45	8.51	9.57	10.64	11.70	12.77
96	1.04	2.08	3.13	4.17	5.21	6.25	7.29	8.33	9.38	10.42	11.46	12.50
98	1.02	2.04	3.06	4.08	5.10	6.12	7.14	8.16	9.18	10.20	11.22	12.24
100	1.00	2.00	3.00	4.00	5.00	6.00	7.00	8.00	9.00	10.00	11.00	12.00

If, due to body metabolism or environment, the equation becomes greatly out of balance, humans must make some adjustment to induce a more equitable balance. The heat-transfer to or from a human body is by the three methods previously discussed—radiation, conduction, and convection—plus the additional factor of evaporation. Human comfort is thus effected by the environment which affects those methods of heat transfer. Affecting environmental factors are air temperature, radiant temperature of the surroundings, air movement, and relative humidity. The problem of maintaining comfortable conditions is further complicated by the fact that human activity generates heat, and the greater the physical energy we exert the greater the need to transfer this resultant excessive heat from the body to a cooler surrounding. Conversely, the lesser the human activity the smaller heat-transfer from the body. For average human activity, however, it has been found that when air temperature is between 68° F and 75° F, relative humidity moderate, and wall, ceiling, and floor temperatures 60° F to 80° F, comfort is obtained.

However, even the ranges given above are found to be incorrect if both air temperature and enclosure (wall, floor, and ceiling) are both on the high or low side. As the temperatures of the enclosure go down, the air temperature must go up to compensate for additional radiation from the body to the ambient surfaces. Should the walls, ceilings, and floor go either much higher than 80° F or lower than 60° F, a person will not be comfortable—regardless of the fact that temperature and relative humidity of the ambient air are at so-called comfort range. *Therefore, sufficient area of the walls, floors, and ceilings of buildings must be insulated properly so that considerable surface facing the human occupants is between 60° F and 80° F. If not, human comfort is impossible.*

Without proper insulation a building cannot be comfortable, and a building where insulation is cheated upon is not modern, but obsolete at the time it is built. *This is very poor economics.*

Slab floors are a particularly flagrant example of incorrect building design. Ground temperatures in most sections of the United States range from 35° F to 50° F under slab floors. The relatively small difference between these temperatures and the ambient temperatures in the room above the slab indicate that little heating cost could be saved if insulation were installed below the slab. Thus insulation is not justified—it is assumed.

A common incorrect assumption is that concrete floors *must* be cold, so people who must walk on these floors purchase heavy shoes and socks. Where conditions are excessively bad in some plants the management buys insulating rubber pads or even electrically heated pads to provide some degree of comfort for their workers. Other than manufacturing plants, some of the new houses built in the more northern climates are also being constructed on uninsulated slabs. Of course the most common users of this construction are the motels. A recent experience of a friend of the writer brought this fact unpleasantly to attention.

On a relatively moderate winter day (approximately 25° F), after checking into a motel for the evening and settling back to watch television, the friend soon found that he had to keep his feet propped up on a chair so that his feet and ankles would be warm enough to be comfortable, even though the floor was carpeted. Upon awakening the next morning, and not being accustomed to wearing shoes while taking a bath, he was subjected to a sudden shock when he walked into the bathroom. Regardless of whether or not the insulation could be justified by heat savings, the economics are still very poor, as the attitude of this guest, as with many others, is "I shall not return."

Why, with today's knowledge, do such conditions exist? One is that the insulation industry has not shown the courage to make a definite statement, and back it up, that insulation should be used under slab floors of heated buildings. Yet such a stand would benefit both them and their customers.

Economic Thickness of Thermal Insulation on Cold Surfaces

The economic thicknesses for thermal insulation used on cold surfaces was investigated by the same team which made the study for insulation on high-temperature surfaces. The cost of capital investment in refrigeration equipment, and the cost of power, as related to cost of insulation, was electronically calculated and tabulated. The results, however, showed that except in rare instances, the thickness necessary to prevent condensation was greater than the economic thickness. With the perfection of exceptionally low-conductivity insulations and the resultant installation savings, it appears to be only a matter of time before this study becomes of significant economic interest.

The study, which was a joint effort by National Insulation Manufacturers Association, Union Carbide Corporation, and West Virginia University was completed in 1962 and distributed to the member companies.

The summary of this chapter is simple—Thermal Insulation is an excellent investment.

3 Functions of Thermal Insulation

In the previous chapters the theory of heat transfer through insulations and the economics involved were discussed. Before these heat transfer and economic equations can be put to practical use the actual conditions which affect a material functioning as an insulation must be understood.

Use by Early Man

In his limited capacity early man understood many of the functions of insulation and the physical requirements of installing it to insure the realization of its intended use. To illustrate, he knew that when he became cold he wrapped himself with more clothes (or animal skins), thereby finding that there was a direct relation between the quantity of heat flow and the thickness of insulation. He also found that wet insulation was less effective than dry insulation. When he got caught in the rain and his clothes became wet he knew he had to dry them before they again became effective in keeping him warm. In his use of clothes he was using insulation in two of its functions: (1) the conservation of energy, and (2) the control of temperature.

As time passed he learned to build shelters to control his environment. He found that these structures, lean-tos or huts made of wood, leaves, or reeds, were effective as thermal barriers when not allowed to get too hot from his fire. He learned that rock, clay, and sand, although less efficient insulations, were effective in the control of heat at high temperatures, and he used these materials around his fire to control its spread. So he used the wooden shelter to conserve heat and control temperature and the rock and clay for fire protection.

Early man even knew a considerable amount about thermal diffusivity. He found that if he had rocks around his fire and they had sufficient mass and were heated to a high temperature, they would stay warm for a long period of time and provide him with comfort long after his fire had died down.

Other than to keep himself warm, early man found that his shelters could be constructed to help keep him cool by shielding him from solar radiation. Thus, by using insulating materials man began to build an environment for his comfort.

Within his scope, early man realized that insulation served him in many ways. Of course he did not call it insulation, nor did he select or set down the ways it served him or designate them as technical functions. However, he did use insulation to accomplish the following:

1. Conserve energy
2. Control heat transfer
3. Control temperature
4. Retard freezing
5. Protection from burns
6. Control fire

These functions are the same today. In almost every case where insulation is used today it is for one or more of the same functions.

As the use of primitive thermal insulation to accomplish these functions was one of man's earliest achievements, he soon found out that he had to select his insulating materials with care. If a material were combustible, he found it would not serve when in direct contact with fire. When wood or leaves were used as a shelter he soon learned to keep them at a "safe" distance from the fire. In other words, he learned the temperature limitations of his insulations. In addition, he soon learned to choose materials that had water resistance, so they would not become water-soaked or wash away. Thus he became selective as to what materials he used, depending upon the requirements of the application. He also soon realized that one single material was not the best insulation for all applications; each material had to be used only within its own limitations.

What does this mean to us now? It simply means we must try to be as smart as early man. We must understand the functions for which insulations are used, the requirements of each installation, and the properties of materials so as to make the proper selection. One of the most prevalent ills of the insulation industry today is this lack of understanding of principles, *including the limitations of materials.* Thus we are faced with the problem of materials being sold and used for applications with little thought as to whether they are suitable for the use intended.

Mechanism of Thermal Insulation

An understanding of the mechanism of thermal insulation (or *how* it works) is essential to its proper selection and use.

With the exception of the reflective types, thermal insulation depends upon totally enclosed, very small pockets or bubbles in the material, containing air or gas. These contained pockets of air or gas retard the flow of heat. These small pockets are formed by the flakes, fibers, or nodules of solids, or by cells in the material itself. Each pocket or bubble must be sufficiently small to cause considerable resistance to air (or gas) flow so that little heat is transferred by convection from one side of the cell to the other. The path which heat follows through the solid matter must be a long, circuitous one to limit the amount of heat transferred by conduction through the solid portion of the mass. The solid material must also be sufficiently opaque (or reflective) to reduce heat transmitted by radiation. Thus, when a temperature difference exists between the surfaces of an insulation, all methods of heat transfer combine to transmit the heat through the insulating material. A schematic diagram of the mechanism is shown in Figure 41.

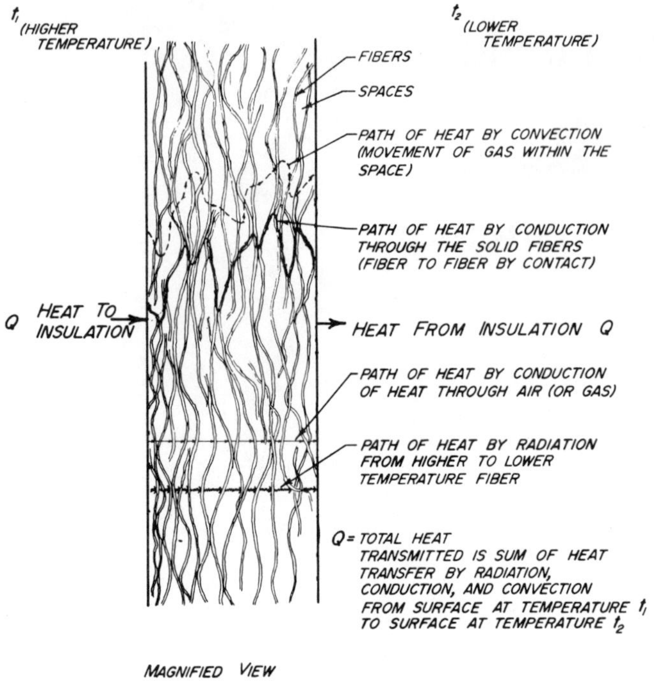

Figure 41 Schematic heat flow through the mass of the insulation.

Reflective insulation is insulation which obtains its major resistance to heat transfer by sheet material of low emittance and absorptance. These sheet barriers must be so located and spaced as to minimize convection of the air (or gas) and must be supported in such a manner that the supports provide little heat transmission by direct conduction. Because of its necessary construction, reflective insulation is a system of material assembly rather than homogeneous mass. However, the passage of heat through a reflective system is identical to that described for mass insulation in that all modes of heat transfer combine to transmit the heat from the higher temperature surface to the lower temperature surface. A schematic diagram of heat flow through reflective insulation is shown in Figure 42.

Variations of both the mass insulation and the reflective insulation are the vacuum insulations. This type of insulation is so constructed that either the mass insulation or reflective shields are located in a space which is sufficiently tight and strong that the air in the space can be evacuated. By this means the transfer of that part of the heat which normally is transmitted through air is reduced to a very low level. As the conductivity of still air at 70° F is approximately 0.16, the reduction of this conducted heat to almost 0 by air is a major factor in the control of heat flow. Although this principle is only recently being adapted into industrial insulation, its basis is as long and well-known as the Dewar flask, named after Sir James Dewar who developed it in the early part of this century.

To reduce the weight of construction of sheets which can withstand the pressures caused by an internal vacuum, another development has been to replace the air in the space by a gas of lower conductivity than air. In most instances this gas is inert and is cycled and dried to keep the internal insulation free of moisture and to control the rusting of the steel sheets.

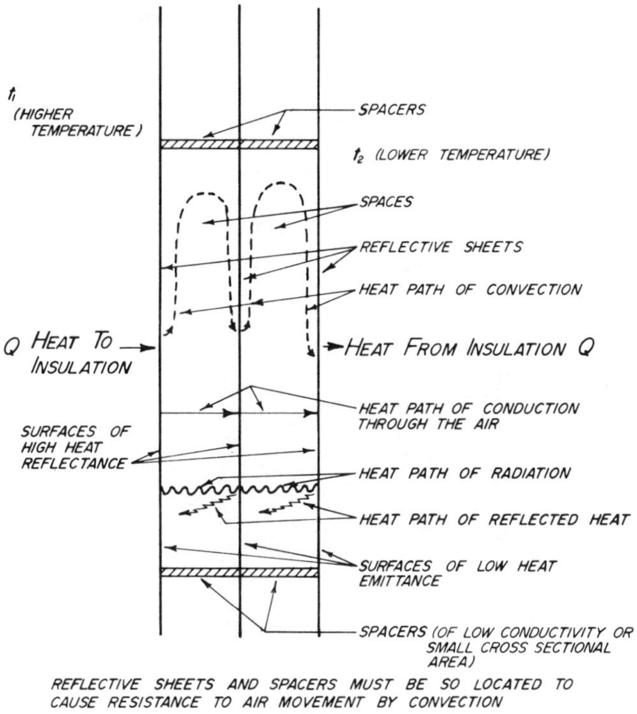

Figure 42 Schematic heat flow through reflective insulation.

Types of Insulation

Insulation may be separated into these general types:

1. Mass insulation (atmospheric air)
2. Mass insulation in space where air is evacuated
3. Mass insulation in space with low conductivity gas
4. Reflective insulation (atmospheric air)
5. Reflective insulation in space where air is evacuated
6. Reflective insulation in space with low conductivity gas

Because of space considerations, all the types of insulation listed above, other than mass insulations, must be fabricated into specific shape or form to be effective. Thus, in most instances, the types must be considered as *systems* of insulation. For this reason their effectiveness as an insulation must be measured as a system and not as a single material. Although this does add to the complexity of calculations related to their design and use, these insulation systems do utilize the basic principles to obtain their effectiveness, and they can be designed to provide a lower heat transfer for a given space than is possible with ordinary mass insulations.

Mass insulations can be produced in volume without specific design for each end use, so they are manufactured in greater quantity than all the other types combined. They are made from many different materials to form the small air spaces upon which their resistance to heat transfer depends. These materials produce insulation of various forms which can be classified into the following:

Rigid block, board, sheets, or formed shapes
Semi-rigid block, sheets, or formed shapes
Blankets
Felts
Batts
Cements
Loose fill
Sprayable solids
Sprayable foams
Liquid foams
Intumescent coatings

The physical characteristics of these materials range from very strong structural materials such as concrete, which is a relatively poor insulation but quite strong, to liquid foam, which has practically no structural strength but very low conductivity and is effective in control of accidental fires.

From the number of types of insulation, the number of forms in which it can be obtained, and the wide range in physical properties available, it is evident that each has certain advantages and limitations and was developed to serve certain uses. Later, the requirements, as determined by uses, will be discussed, but one division of use is so basic that it must be considered as a fundamental function. This division relates to the temperature range at which the insulation is to serve, i.e., whether said temperature is below or above ambient air temperature. In other words, is the insulation to conserve refrigeration or to conserve heat? This basic division is necessary because the direction of vapor transmission into or out of the insulation determines many of the insulation's characteristics necessary for successful operation. With the exception of

Table 38 Vapor pressures at 100% relative humidity
English Units

Temp °F	Vapor Pressure		Temp °F	Vapor Pressure	
	In. of Hg	Lb/sq ft		In. of Hg	Lb/sq ft
60	.522	36.91	94	1.61	113.8
61	.541	38.24	95	1.66	117.4
62	.560	39.62	96	1.71	121.1
63	.580	41.04	97	1.76	124.8
64	.601	42.50	98	1.82	128.7
65	.622	44.01	99	1.87	132.6
66	.642	45.56	100	1.93	136.7
67	.667	47.17	105	2.28	160.9
68	.690	48.83	110	2.60	183.9
69	.714	50.53	115	2.99	211.8
70	.739	52.28	120	3.45	244.0
71	.765	54.04	125	3.95	279.6
72	.791	55.96	130	4.53	320.4
73	.818	57.88	135	5.16	365.4
74	.846	59.31	140	5.88	415.9
75	.875	61.89	145	6.66	472.5
76	.906	63.98	150	7.57	535.4
77	.935	66.15	155	8.56	605.5
78	.967	68.37	160	9.65	683.5
79	.99	70.65	165	10.86	705.2
80	1.03	72.99	170	12.20	862.9
81	1.06	75.40	175	13.67	949.6
82	1.10	77.94	180	15.3	1081
83	1.14	80.49	185	17.0	1207
84	1.18	83.11	190	19.0	1344
85	1.21	85.76	195	21.1	1496
86	1.25	88.63	200	23.5	1660
87	1.29	91.45	205	25.9	1839
88	1.34	94.45	210	28.7	2034
89	1.38	97.47	215	31.9	2245
90	1.42	99.58	220	34.9	2414
91	1.47	103.8			
92	1.51	107.0			
93	1.56	110.4			

Table 38a Vapor pressure at 100% relative humidity
Metric Units

Temp °C	Vapor Pressure Pa	Temp °C	Vapor Pressure Pa
18	2160	36	6027
19	2276	37	6265
20	2337	38	6603
21	2492	39	6874
22	2618	40	7375
23	2790	45	9344
24	2977	50	12326
25	3167	55	15747
26	3319	60	19912
27	3556	65	25330
28	3827	70	31155
29	4013	75	37856
30	4243	80	47341
31	4555	85	57568
32	4843	90	70038
33	5147	100	101320
34	5300	105	120892
35	5621		
°C	N/m^2	°C	N/m^2

the liquid foams and intumescent coating insulation, all of the types of insulations and the various forms of mass types of insulations listed have been used both as refrigeration insulation and heat insulation. An insulation applied to a refrigerated surface requires completely different methods of application and accessory materials than the same insulation applied to a hot surface.

The vapor pressure in ambient air changes with temperature and relative humidity. The vapor is moisture in its gaseous state, and like all gases it can be condensed into a liquid by lowering its temperature below its condensation point. Table 38 is included to illustrate the rate of change of vapor pressurre with change of temperature.

Insulation Applied to Hot Surfaces

A hot gas tends to expand and attempts to escape into a lower pressure area. Most insulations contain moisture, so when the insulation is attached to a hot surface the pressure of the vapor is increased and it attempts to escape to the lower pressure ambient air. This vapor pressure can be quite high. For example the vapor pressure of air at 100% relative humidity at 220° F is 2414.48 lb/sq ft (104.4° C is 118,185 Pu). Of course, at this same temperature, as the vapor pressure finds an outlet and the moisture content of the surrounding air is less, the pressure drops until it is in equilibrium with the ambient conditions. In simple terms, a heated surface dries out the insulation with which it is in contact. However, moisture is always present, the amount depending upon the temperature and the vapor pressure of the surrounding air. As the vapor pressure of ambient air is constantly changing, due to time of day, time of year, and weather conditions, this amount of moisture in insulation and surrounding the hot surface is also constantly changing. For convenience this is called the "breathing" action which takes place in insulation.

Under all practical conditions of application, mass insulations other than hermetically sealed cell types contain some moisture when they are applied. When a pipe or vessel is heated, that moisture's vapor pressure rises and the moisture is driven outward. Should this high pressure, moist vapor be trapped by a very tight weather-barrier jacket or coating, the moisture will condense on the inner surface of the jacket or coating. This explains why most metal securements, wire, or netting made of carbon steel will rust rapidly under the weather-barriers on high-temperature insulation. A schematic diagram of these conditions is shown in Figure 43. This diagram also illustrates why, on insulations installed on high temperatures, the outer jacket or weather-barrier should *not* be a good vapor-barrier. The outer weather-barrier should protect the insulation from entry of liquid water, but it should allow water vapor to pass through to the ambient air.

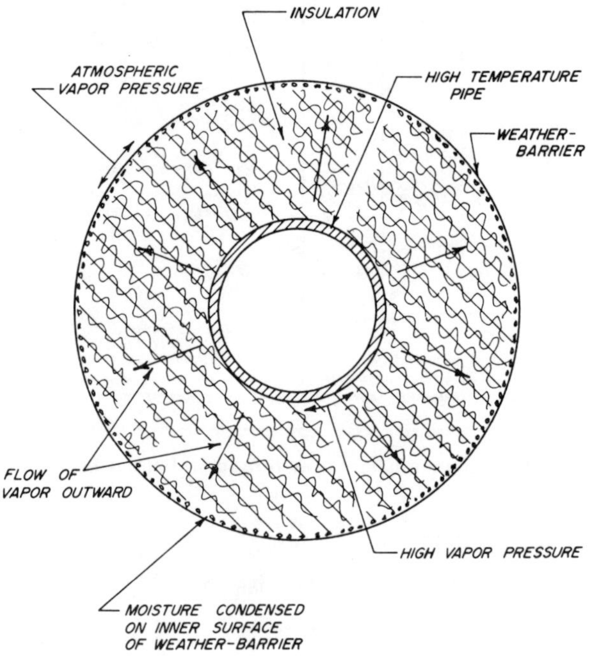

Figure 43 Schematic diagram of vapor flow insulated high temperature pipe.

Insulation on Heat Traced Piping

A variation of insulation on hot pipes is insulation installed on process lines externally heated by a heat "tracer." The best tracerr may be tubing supplied by steam, other hot vapor or liquid, or by electrically heated cable or tape. During operation the process line itself may supply sufficient heat for its satisfactory operation. It could operate at a considerable temperature above the temperature which the tracer could produce. For this reason there are frequently two operating temperatures involved in the design of a traced line. One, the temperature during operation, and the other, the temperature at which the tracer will maintain the process pipe during periods of low flow or no flow.

A schematic diagram of a heat-traced process pipe is shown in Figure 44. As shown, the tracer is not thermally connected to the process pipe with heat transfer cement. Instead, it keeps the pipe hot by heating the air space around it. If the tracer had been bonded to the pipe, the heat flow would have been from the tracer, through the heat transfer cement to the metal pipe, then from the pipe to the annulus space. This direction of heat flow will keep the

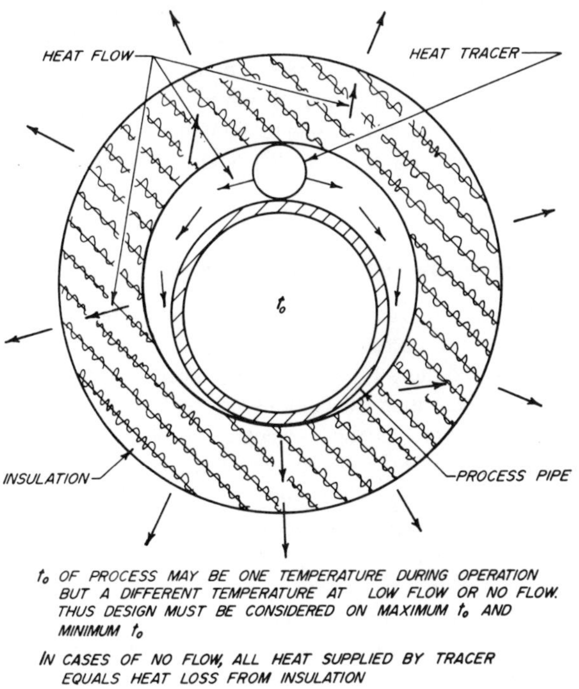
HEAT FLOW
HEAT TRACER
INSULATION
PROCESS PIPE
t_o
t_o OF PROCESS MAY BE ONE TEMPERATURE DURING OPERATION BUT A DIFFERENT TEMPERATURE AT LOW FLOW OR NO FLOW. THUS DESIGN MUST BE CONSIDERED ON MAXIMUM t_o AND MINIMUM t_o
IN CASES OF NO FLOW, ALL HEAT SUPPLIED BY TRACER EQUALS HEAT LOSS FROM INSULATION

Figure 44 Schematic diagram of heat traced process pipe.

process pipe at a higher temperature than the air convection, as shown in the figure, and the annulus space will be somewhat cooler. As the heat loss of the system is dependent upon the temperature difference between the annulus space and the outside air, the heat loss will be less, even though the process pipe is maintained at a higher temperature.

Determination of insulation thickness for such a system may be either by the temperature of the process while in operation or by the insulation thickness necessary to provide the proper temperature of process pipe during no flow. Whichever requires the greater thickness will be the determining factor.

A major difference exists between the use of steam tracing as compared to electric tracing. The temperature of the steam can be assumed to be almost constant, and for this reason its rate of transmitting heat energy will change as conditions change. With electric tracing, however, the heat energy output is practically constant and the temperature of the heater cable changes with conditions, unless the system is temperature controlled.

The temperature of steam used in a tracer system is always above 212° F (100° C) and is constant for the pressure at which it is used. As the temperature of the process pipe will bear definite relation to the temperature gradient of the system, when the temperature of the ambient air changes, the temperature of the process pipe will change in accordance with that ratio. For these reasons it is difficult to design a steam tracer system for a given temperature and maintain it within reasonable limits over a wide range of ambient conditions.

These limitations of steam tracing have made the use of electric tracing systems very important in the last few years since close temperature control of process pipes has become important. This in turn has made the design of the insulation correspondingly important. Efficient electric tracing requires that its power output be carefully designed. As this energy output can only be controlled within relatively narrow limits, it does not have the ability to make up large losses due to faulty insulation. The insulation must be complete, with no voids, and its conductivity must be relatively constant. Poorly installed weather-barriers which allow entry of water into the systtem can make an electrically traced system completely inoperative. Because of this limited heat output, it is almost impossible for an electric tracer to dry out insulation if it becomes wet. For this reason only moisture resistant or water-repellent insulation should be used on these systems.

Although not specifically mentioned, process equipment and vessels are frequently heat traced, or heated, with external heating coils. The factors presented for heat-traced process piping are identical for these.

Insulations Used to Conserve Heat of Stored Materials

Frequently insulation is used to conserve the heat of a material in storage vessels, in pipes, or in transit. When the vessel, pipe, or storage tank, together with the material, has considerable mass, conventional insulation materials can be used and the time-rate of temperature drop can be determined by the rate of heat transfer through the insulation to the outside atmosphere in relation to the amount of heat held by the original mass. Such use of insulation is becoming increasingly common as materials are loaded into tank cars and hauled to the users, and with no additional heat having been added, they arrive at their destinations within 5 to 15° F (2.8 to 8.3° C) of their temperature at loading.

The success of these systems depends upon having a considerable amount of heat sink in the mass. For this reason liquids of fairly high specific gravity and high specific heat can lose a fair amount of heat (Btu) with little reduction in temperature. The same, in reverse, is true on the cold side with liquefied gases.

This is not true with stored gases or vapors. The low specific gravities and specific heats of these materials prevent formation of a "heat sink" and insure that small amounts of heat gain or loss make considerable difference in their temperatures. For this reason the use of insulation to reduce the solar effect and heat-up of the vapor in a storage vessel is completely ineffective. To be effective the insulation would have to be the heat sink itself, with sufficient mass and low diffusivity so that any solar heat radiated to its outer surface would be retarded from reaching the inside for approximately 12 hours, at which time the heat flow would reverse and dissipate to the cooler ambient evening or night air. Depending on the mass of the insulation and its diffusivity, the thickness required generally calculates to be 12″ to 20″, to reduce the effect of atmospheric cycles.

Of course if the gas or space is refrigerated, then insulation is effective. This is because the problem is no longer one of leveling out atmospheric cycles but becomes one of reducing heat gain to a refrigerated space.

Insulation Used to Control Surface Temperature

On hot pipes, equipment, and vessels insulation is also used to control surface temperature to prevent personnel from being burned by coming into contact with the surface and to prevent surfaces being so hot that spilled materials might ignite. The function of the insulation is the same as that described for the insulation of hot surfaces, and the surface temperature can be calculated by the proper equation as given in Chapter 1.

The emittance of a surface has a direct relation to its surface temperature. To illustrate this relationship Dr. Jerome F. Parmer, Professor at West Virginia University, calculated the following table of surface temperatures. This table is based on 4″ pipe operating 600° F in still air ambient environment. Using the conductivity of standard high-temperature insulations such as calcium silicate, expanded silica, or asbestos fibers, the surface temperatures in respect to insulation thickness and surface emittance are:

Table 38b *English Units*

Insulation thickness, in.	Surface Temperature °F — When surface emittance is .05	.10	.90	.95
1	217	212	165	163
1½	185	181	143	142
2	167	163	131	130
3	154	151	123	122

Table 38c *Metric Units*

Insulation thickness, mm	Surface Temperature °C — When surface emittance ε is .05	.10	.90	.95
25	102.7	100.0	73.9	72.8
38	85.0	82.8	61.7	61.1
51	75.0	72.8	55.0	54.4
76	67.8	66.1	50.6	50.0

Thus, highly reflective (low emittance) surface materials will have much higher surface temperatures than low reflective surfaces, when such surfaces encase an insulated hot operating pipe or vessel. For this reason highly reflective metallic surfaces, on insulated hot pipe or equipment, can be a personnel hazard if insufficient insulation is used to reduce surface temperature to safe limits.

There is another factor which is often not considered, i.e., that the surface itself can make the difference as to whether an individual is burned or not, even at the same surface temperature. A mastic, felt, or canvas surface at a temperature of 160°F (71.1°C) is warm but will not blister the skin. Bare metal at this same temperature, however, will cause a burn and blistering. This is because the mastic or felts have a relatively low conductivity of approximately 5.5 Btu, in./ft^2, hr °F (0.7931 w/mK). Thus, when the skin is in contact with such a surface the heat is transmitted to the skin relatively slowly and at a rate at which the skin can absorb and dissipate it. In contrast, steel, with a conductivity of 350 Btu in./ft^2, hr °F (50.97 w/mK), and aluminum, with a conductivity of 1230 Btu in./ft^2, hr °F (177.36 w/mK), rapidly conduct heat to any body with which they are in direct contact. Metals, then, will transmit heat to the skin at such a rate that the skin temperature itself will almost immediately rise to the temperature of the metal. Thus, surface temperature by itself is not the only measure of safe surfaces.

Insulations on Low-Temperature Surfaces

When thermal insulation is applied to equipment or piping which operates at temperatures less than the ambient air temperature, the lower temperature surface causes a lower vapor pressure area adjacent to it than exists in the ambient air. Moisture in the vapor state in the ambient air, seeking to equalize the pressure difference, attempts to flow to the low-temperature surface. Whenever a barrier exists between a high and low pressure area, the vapor will seek by all means to reach the lower pressure by going through the material itself, through cracks in it, through joints, or around ends. Vapor will pass through almost all materials—some very quickly and others very slowly.

As the moisture vapor does get through the outer vapor-barrier, cracks, or joiints in its path toward the cold surface, it encounters lower temperatures. At some point it will reach its dew-point temperature, where the vapor will condense to a liquid. This water will replace some of the air in the crack or the insulation itself. Still further inward, if the low-temperature surface is below 32° F, (0° C), this liquid water will freeze into ice. As this process continues, the insulation becomes filled with water and ice and

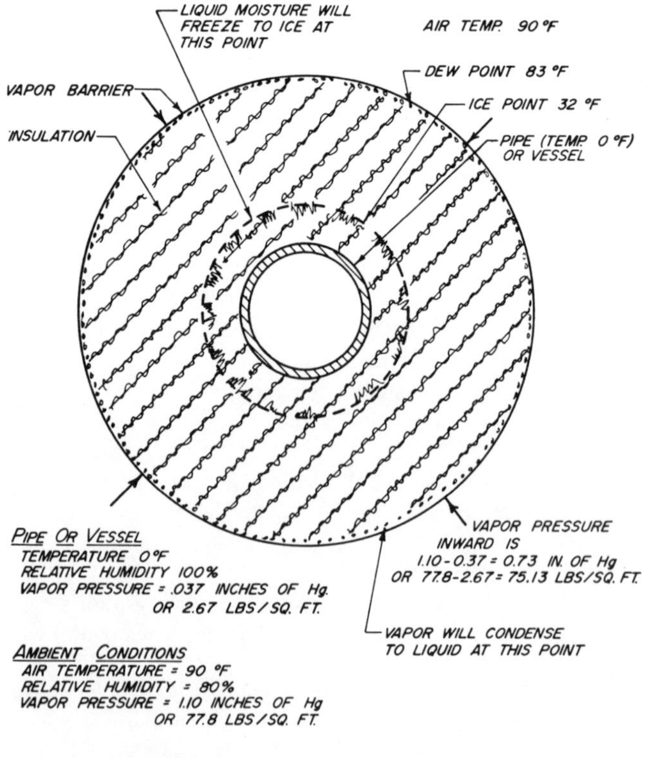

Figure 45 Vapor-retarder diagram.

loses most of its insulating ability. The schematic diagram in Figure 45 illustrates the conditions existing when insulation is installed on a 0° F (−17.8° C) pipe or vessel and the ambient conditions are at 90° F (32.2° C) with a relative humidity of 80%.

Effect of Moisture on Insulations

Thermal insulations depend upon being relatively dry to function efficiently. When insulation becomes wet, or partly so, all or part of the air or gas spaces become filled with water. Heat transmission through the water-filled spaces then approaches the rate of conductivity of water instead of that of the air or gas. The conductivity of water at 70° F (21.1° C) mean temperature is 4.1 Btu, in./sq ft, hr, °F (0.591 W/mK), as compared to 0.17 Btu, in./sq ft, hr °F (0.0245 W/mK) for air. Thus, the transmission of heat across each space is approximately 24 times as great when it is filled with water as compared to air. A completely dry insulation, therefore, having a conductivity of from 0.16 to 0.38 Btu, in./sq ft, hr, °F (0.023 to 0.0598 W/mK) will have a conductivity of 5.0 to 5.5 Btu/sq ft, hr, in., °F when wet. The moisture which is frozen causes an even greater rise in the conductivity as the conductivity of ice at 32° F (0° C) is approximately 15.5 Btu, in./sq ft, hr, °F (2.235 W/mK). Thus, the conductivity of an insulation can change from 0.15 to 0.35 Btu in./sq ft, hr, °F (0.0216 to 0.0505 W/mK) to greater than 15.5 Btu in./sq ft, hr, °F (2.239 W/mK)—a rise of 50 to 100 times its original rate of heat transmission.

Considerable effort was made by Professors E. R. Queer and F. A. Joy of Pennsylvania State University to measure the effect of moisture content in insulation. This is quite difficult, as the very act of measuring heat conductivity changes the placement and amount of moisture in the insulation. Several methods of measurement were used in these tests, i.e., the guarded hot-plate method and the D. Eustachio probe-line heat source method. The results of these tests clearly indicate that moisture does have considerable effect on the conductivity of the materials. These tests also showed that the measurements include two kinds of heat transfer: sensible heat and latent heat. Also shown was the fact that moisture location and arrangement of moisture in the insulation structure has considerable effect on the measured heat transfer. Because of the complexity, there are no tables that can state that for a given percentage of moisture in insulation there will be a certain percentage increase in its conductivity. The only certain statement which can be made is that moisture in the liquid phase will increase the conductivity and that when the insulation becomes completely saturated with liquid water its conductivity will be something greater than that of water by itself.

In insulation composed of cellular material, whose permeability will allow vapor migration, the temperature of the vapor eventually reaches its dew point, at which location it will condense. As the resultant water travels on toward the low temperature side it will reach the freezing point and solidify into ice. These ice crystals will soon rupture the cells, with the consequent breakdown of the insulating value of the material. For this reason the life and proper functioning of the insulation is totally dependent upon a vapor-barrier on its warm side which will prevent penetration of any water vapor over its entire surface. Poor vapor-barriers, such as those with pinholes, tears, and ruptures, will rapidly accelerate the deterioration of the insulation. This action is illustrated in Figure 45.

Cellular glass insulation has such a low rate of vapor permeability that its permeability has never been measured. Its vapor resistance is much greater than any vapor-barrier that can be practically installed on it. At first thought this would make it appear that no special care would be required in using cellular glass as a low-temperature insulation. However, when it is used at operating temperatures it must be installed so as to prevent its eventual breakdown due to freezing and thawing. This freeze and thaw action occurs in the joints between the pieces of insulation. Moisture vapor, upon entering the joint, condenses to liquid, then freezes at the point where it reaches 32° F (0° C). This freezing breaks open a few of the glass cells at this point. Should the temperature remain completely constant, little damage would result and the ice would probably act as a seal to prevent further damage. However, when there is temperature change, such as is always true outdoors, the ice point shifts in its location and some ice is melted, then refrozen. Each temperature cycle causes this freeze and thaw action which can lead to a complete breakdown in the material. This is illustrated in Figure 46.

Fortunately this action can be retarded by sealing the joints

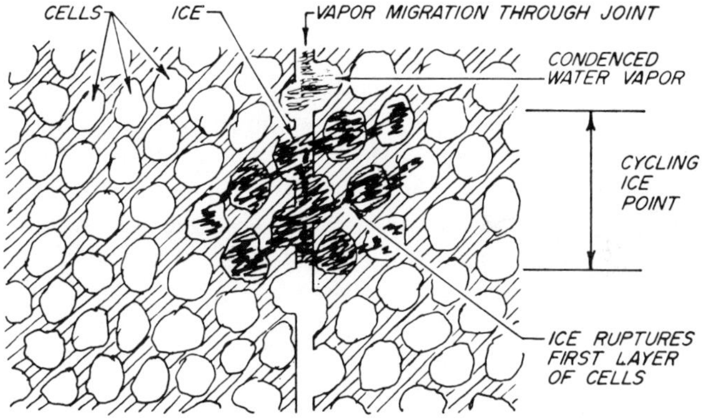

Figure 46 Microscopic section of cellular glass at ice point in joint.

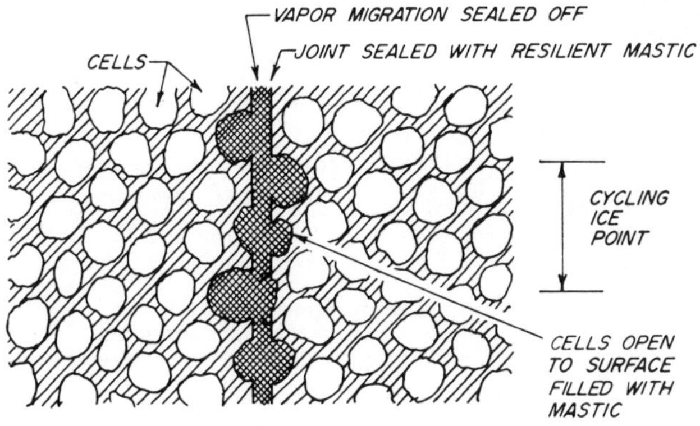

Figure 47 Microscopic section of cellular glass at ice point in joint.

with a resilient mastic which fills the cut cells and prevents accumulation of water and ice and which also retards the vapor flow inward. The mastic chosen, however, must be one which remains resilient at the low temperature to which it is subjected and does not harden and shrink to cause the same damage it is to prevent. A joint filled with mastic is illustrated in Figure 47. By vapor-sealing the vapor resistant insulation to itself in a complete envelope, the insulation and its sealed joints together become the vapor-barrier required to prevent the migration inward to the low pressure area.

Other than moisture in the vapor phase, insulations must be protected from liquid moisture. This is accomplished by an outer weather-barrier or jacket. The weather-barriers or jackets also serve another purpose—that of protecting the insulation from physical and mechanical damage. Because of the latter, coatings or jackets are used indoors as well as outdoors.

Insulation on Cyclic Pipes and Equipment

Pipes and equipment which cycle from refrigerated to heated service are the most difficult of all to insulate properly. The moisture which may enter the system when the equipment is cold can cause very high vapor pressure when the equipment is hot. If the operating temperature is above 212° F (0° C) and the vapor-barrier is quite tight, the liquid moisture can be converted into steam.

Such internal pressures will rupture most vapor-barriers. The only solution is to use an insulation or combination of insulations which are as vapor and water-resistant as possible. In many instances it becomes necessary to use a vapor-resistant insulation as an outer layer and an insulation suitable for the high temperature as an inner layer. Although the vapor-barrier, or sealed cellular glass construction, is fundamentally wrong when the equipment is operating on the high-temperature cycle, it must be used to prevent moisture from entering on the low-temperature cycle. Here, hopefully, it must be attempted to keep the insulation so dry during the refrigeration cycle that the amount of moisture trapped between the vapor-barrier and the steel will be insufficient to cause excessive vapor pressure or steam pressure.

On this application the use of an outer welded steel container over the insulation still does not solve the problem, for if it were hermetically sealed, excessive pressures would be built up on the heating cycle and a partial vacuum would be obtained during the refrigeration cycle. Simple venting will not solve the problem either, as excessive moisture would be drawn in when the pipe or equipment is cold. Where double-wall vessels or pipe are used on extremely cyclic service they must be connected to a dry gas system, which controls the pressure and includes driers to remove moisture absorbed by the gas.

Insulation on Low Temperature Spaces

The action of moisture vapor entering insulation installed in the walls, ceiling, and floor of a refrigerated space is different than insulation installed on cold equipment or pipe. A vapor-barrier (or vapor-sealed, vapor-resistant insulation) must be installed to retard excessive amounts of water vapor from entering the ther-

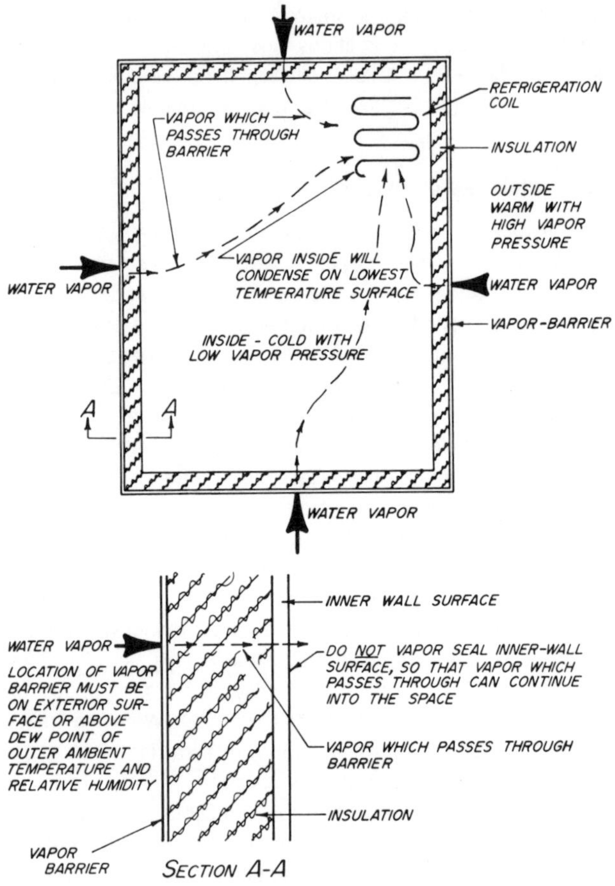

Figure 48 Schematic diagram of insulated refrigerated space.

mal barrier. However, if the innermost surfaces are constructed to have little vapor resistance, these small amounts can continue through the insulation and will condense on the cold refrigerator coils or on other cold surfaces within the room. This is illustrated in Figure 48.

As shown, an ambient vapor pressure completely surrounds the enclosed refrigerated space. The inner area, due to its low temperature, has low vapor pressure. Thus, the outer vapor tries to migrate to the low pressure area. This migration is resisted by the outer vapor-barrier, which should be so located within the construction that it is above the dew-point temperature of the outer ambient air. If it is not above this temperature, condensation will occur on its outer surface. However, even the best of vapor-barriers do allow some moisture to pass. Any moisture which passes through the vapor-barrier should not be impeded in its attempt to condense on the lowest temperature surface within the room, which in most cases is the refrigeration coil. Although many factors can affect the over-all design, a rule of thumb which has proved to be satisfactory for most applications is that the outer vapor-barrier should be ten times more resistant to the passage of vapor than the total of all the other resistances in its path between the barriers and the interior of the room. *In no case* should the interior be vapor-sealed, as the resultant build-up of water and ice will rapidly ruin the insulation.

Insulation Applied in Buildings

Heated buildings

Most insulations installed in buildings are installed at the time of the original construction. For this reason the insulations used are a part of the system of construction of the building. The number of construction materials and the combinations used are quite numerous. Specific applications, then, cannot be discussed here; only the fundamentals that the construction is to achieve will be presented.

Most building walls are constructed with outer and inner building materials, with the insulation placed between the two. For effective service the insulation itself must be highly vapor-resistant, or a vapor-barrier must be installed in the wall construction. The importance of the vapor-barrier is frequently overlooked, but the fundamental laws of vapor migration—presented earlier in this chapter still apply.

In the case of a building which is heated for winter comfort, the vapor migration which affects the insulation will be from the inside to the outside. The conditions are shown in Figure 49.

Assuming the inside air temperature to be 80° F (26.7° C) and the relative humidity 50% then the vapor pressure is 0.5172″ of Hg., or approximately 36 lb/sq ft (1751 Pu). If the outdoor temperature was 35° F with 100% relative humidity, the vapor pressure of the ambient air would be 0.2035″ of Hg., or approximately 14.4 lb/sq ft, (689 Pu). Thus, there is a vapor pressure difference from inside to outside of 36–14.4 or a force of 21.6 lb/sq ft (1062 Pu), tending to drive the water vapor toward the outside. The dew-point temperature of the vapor is 60° F, so the vapor at that moisture content must be restricted from entering the wall or liquid water will condense in the wall where the temperature is 60° F (15.6° C) or less. Of course, the vapor-barrier must be located in the wall where it is above the 60° F (15.6° C) dew-point temperature, otherwise condensate will collect on its inner surface. Beyond the vapor-barrier there should be no barrier to restrict the vapor passage outward.

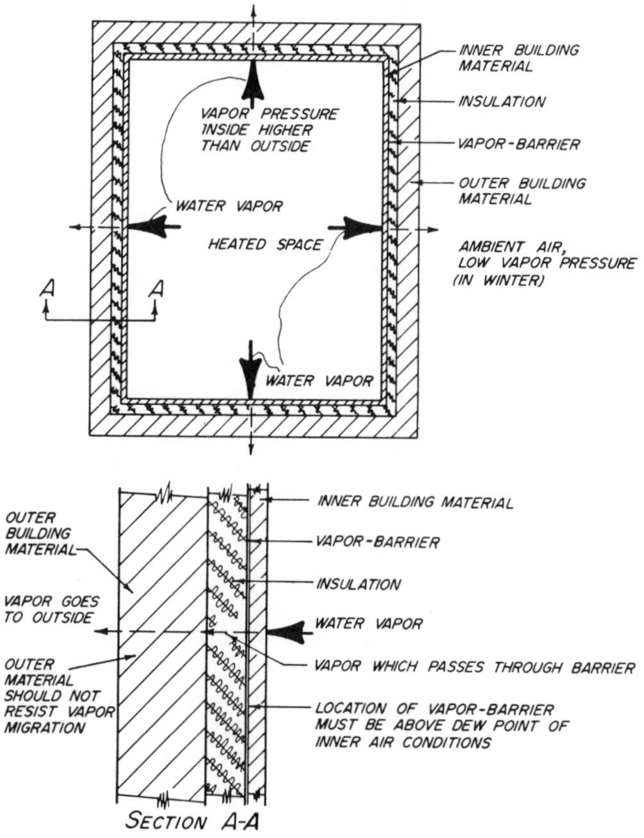

Figure 49 Schematic diagram of insulated wall in heated building.

These basic laws of psychrometry are frequently ignored in building construction. On frame houses the major cause of paint blistering on the outside siding is that no vapor-barrier was installed on the inner part of the wall and the paint on the outside was the greatest barrier to vapor in the system. Therefore, condensation occurs under the paint film, and in its attempt to get to the atmosphere, it pushes the paint off the wood. Another bad practice started a number of years ago was to use asphalt paper under the siding. With no inside vapor-barrier this paper caused condensation in the wall cavity, causing early wood rot. About the only factor that saved these houses and buildings from complete decay within a few years was the fact that the conditions presented above could never be obtained. The simple window glass condenses moisture, and if the outside temperature is 35° F (1.7° C), relative humidity cannot exceed 43%. However, if storm windows and a humidifier are added to houses incorrectly constructed, then rapid destruction is assured. (See Table 2)

More recently, siding is manufactured with aluminum facings and with the need for conservation of energy the cavity between the studs is being filled with fibrous mass insulation. This construction assures that the temperature of the air at the junction of fibrous insulation will be lower than dew point at inner surface of the aluminum faced siding and that condensation will occur. The condensate will wet the fibrous insulation so that it is ineffective (as discussed in previous parts of this chapter). If the outer sheeting is combustible organic foam or cellulose fiber, the aluminum surfacing does protect them from becoming wet and thus maintains their fire hazard properties. This dangerous type of construction is increasing in magnitude. Where combustible insulation boards covered with aluminum foil are used on the interior of a heated building, the vapor problem is corrected but the fire hazard is increased.

Cooled buildings

Buildings in the southern climates where the heating season is very short are mainly insulated to prevent the ambient heat from entering. In this case the vapor-barrier must be outside the insulation to prevent moisture from entering and condensing in the insulation. In simple terms, the vapor-barrier *must* always be on the *hotter* side of insulation when the insulation separates two spaces of air which are at different temperatures. Although the space inside may not be as cold as a refrigerated storage room, the vapor migration is in the same direction, as illustrated in Figure 48.

Heated and cooled buildings

Buildings which are heated in the winter and cooled in the summer present a difficult problem in placement of vapor-barriers. Theoretically the vapor-barrier should be on one side for heating and the other side for cooling. In the cases of heated and cooled buildings the only answer, if non-vapor-resistant construction is used, is to have a vapor-barrier on each side. In each instance the vapor-barrier should be located where it will be above the dew-point temperature for the air temperature and relative humidity from which it is to protect the insulation. This is illustrated in Figure 50.

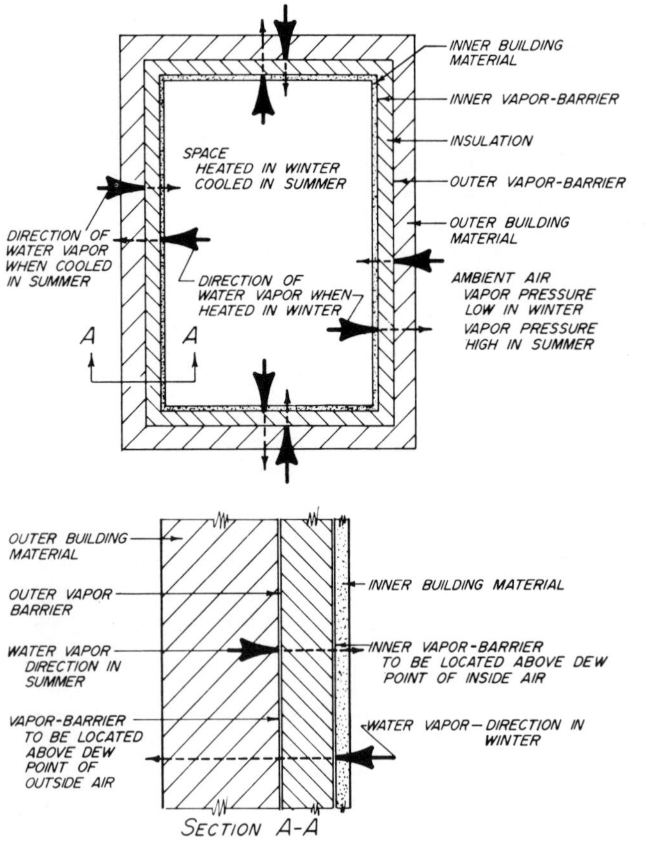

Figure 50 Schematic diagram of insulated wall heated and cooled building.

Three basic methods which have proved successful in insulating buildings which are both heated and cooled are:

1. Non-absorbent, but not vapor-resistant, insulations are faced on each side with excellent vapor-barriers such as metal foil.
2. The cavity is insulated with reflective sheets of aluminum or aluminum foil.
3. The insulation is a very vapor-resistant one, such as cellular inorganic or organic materials, which require no vapor-barriers but must be sealed to itself.

In all cases the thermal resistance of the insulation should be sufficient so that its outer surface temperature, when calculated in conjunction with the other building materials, is above the dew-point temperatures of both the inside and outside air for a major part of the time during the day.

To explain further—when the relative humidity gets to 100% outside, a building, the inside of which is cooled below the ambient air temperature, can in no way be designed so that the outer vapor-barrier will be above the dew-point temperature. Under these conditions condensate will form on the outer vapor-barrier. If the ambient conditions are such that most of the time the relative humidity is below 100%, then the vapor-barrier in the wall should be above the dew-point temperature of the 80% relative humidity conditions. Therefore, if condensate occurs at excessively high humidity conditions, it will have a chance to dissipate during more average conditions.

As may be noted from the above discussion about the uses of insulation, in most every case control of moisture was mentioned. In most instances the moisture was that which was in the vapor phase which could be condensed into a liquid where it would damage the insulation. This does not imply that liquid moisture is not important, but only that the importance of the control of *vapor* is not as well understood by many as that of liquid moisture. In most industrial and building applications of insulation the knowledge, design, and application of weather and vapor-barriers are the most important factors for the successful installation of insulation. More complete discussion of these factors will follow in another chapter.

Insulation Used as Fire Protection

It is impossible to present here more than the bare fundamentals of insulation used for fire protection, as this is quite a specialized field in itself.

Where insulation is used over liquid-filled vessels and pipes which have considerable weight and high specific heat, these form a large heat sink. If the insulation has a low conductivity and will resist the maximum fire temperature, it will perform well as a fire-protective medium. As long as the insulation stays in place and retards the flow of heat inward, the temperature rise of the vessel or pipe and its contents will be slow.

The temperature of gas or vapor-filled vessels and pipes will rise very rapidly with small amounts of heat input, as they have small heat storage capacity. This is also true of their supporting steel such as beams, columns, tank legs, and skirts. Low conductivity, light-weight materials are of little value for fire protection of these. Regardless of the low conductivity, the temperature rise can be quite rapid, as the light-weight materials have a high diffusivity.

As explained, relatively small amounts of heat will elevate the temperature of light masses rapidly. Conductivity is a measure of *amount of heat flow.* Diffusivity is the measure of *rate of temperature flow* and is explained later in this chapter.

Protection of structural steel is a problem of variable heat flow, and the important property of the material in relation to its ability to retard an increase in temperature is not thermal conductivity but temperature conductivity, referred to as thermal diffusivity. Materials used as fire protection where there is no interior heat sink or cooling medium must have low thermal diffusivities. In addition, they must also have the ability to physically withstand the fire for a long time with only slow breakdown of their physical properties.

Note that the term fireproofing has not been used! There is no fireproof material. Exposed to sufficiently high fire temperature and time, all materials will eventually fail. Therefore fire protection must always be related to time and temperature.

Other factors enter into the ability of a material to act as a fire-protection material. One of these is the amount of moisture it contains. Gypsum contains a great amount of water of crystallization, and when heated by fire roughly 1000 Btu/lb (2.326×10^6 J/kg) of water contained will be required to vaporize this water. For this reason, when heated in a fire the temperature will rise to 212° F (100° C) rapidly. It will stay at this temperature for a relatively long period of time until the water is driven out. When it is dried the temperature will again rise rapidly, and the material will begin to disintegrate at approximately 1000° F (538° C). However, this long drying-out period makes gypsum a valuable material in fire protection.

Of course, water is the most common material used for fire protection because it does absorb so much heat in its vaporization and because a water-wetted surface, at atmospheric pressure, cannot rise above 212° F (100° C). When used in connection with insulated pipes and vessels, water-sprayed surfaces of the insulation also cannot be greater than 212° F (100° C). The combination makes an excellent method for fire protection of industrial piping and equipment.

In locations where water is unavailable and the entire burden of obtaining time lag must be borne by fire-protective materials, then high strength, low thermal diffusivity materials must be used. For this reason an explanation of thermal diffusivity is in order.

Thermal diffusivity

An elevation of temperature at one surface of a material progresses from that layer to the next, necessitating the storage of heat in the first layer. An amount of heat, Q, will cause an increase in temperature in a layer inversely proportional to the specific heat of the material. The amount of temperature displacement (rise) is the thermal diffusivity, A_d. Thus, if S_{vol} is the specific heat of that volume under temperature change and C is a proportionality factor, then in

English Units

$$A_d = \frac{C}{S_{vol}} \tag{63}$$

The rate of the temperature progression is greater the faster the temperature of the adjacent layer increases, and the temperature, in turn, is proportional to the amount of heat flowing into the material. For a given temperature difference, this amount is proportional to the thermal conductivity K.

The *temperature* conductivity must be $A_d = \frac{C'K}{S_{vol}}$ where C′ is a new proportionality factor. C′ = 1 if the following units are used:

K = conductivity in Btu in./ft², hr, °F
S_{vol} = specific heat in Btu/cu ft, °F
s = specific heat per unit weight in Btu/lb, °F
p = density in lb/cu ft

The specific heat of unit volume is:

$$S_{vol} = sp \text{ Btu/cu ft, ft} \tag{64}$$

The expression for thermal diffusivity is therefore:

$$A_d = \frac{K}{sp} \text{ ft}^2\text{/hr} \tag{65}$$

(Note K is given in Btu/sq ft, *ft*, hr, °F

Metric Units

The temperature conductivity A_{md} is:

$$A_{md} = \frac{C' K_m}{S_{mvol}} \tag{63a}$$

C′ = 1
k_m = conductivity in W/mK
S_{mvol} = specific heat in W/m³K
s_{mvol} = specific heat per unit weight in W/kgK
p_m = density in kg/m³

The specific heat per unit of volume:

$$S_{mvol} = s_m p_m \text{ W/m}^3 \text{ K} \tag{64a}$$

The expression for thermal diffusivity is therefore:

$$A_{dm} = \frac{k}{s_m p_m} \text{ m}^2\text{/hr} \tag{65a}$$

From inspection of the equation it is apparent that temperature progression will be slowest in a material which combines low conductivity, high specific heat, and heavy density. This is the type of material which will retard the rise of temperature of steel which it is protecting from accidental fire for the longest period of time. Conversely, once hot, it takes a long time to cool.

It is evident that fire protection of pipes, vessels, structural steel, or buildings is a complex subject. Where insulation is used for this purpose the conditions and requirements must be studied with care to engineer and install an effective system.

General

Individual functions of insulation have been described, such as the insulation of surfaces to conserve energy, control surface temperatures, and protect from fire. In many instances the insulation is required to serve more than one of these functions for a given installation. In such cases the best for one function may not be the best for the other function to be served. The engineering design then becomes one of compromise to obtain an installation which will serve both requirements reasonably well.

4 Properties of Thermal Insulation

Basic Types of Thermal Insulation

The characteristics of a thermal insulation are mainly determined by its composition. For that reason thermal insulations are divided into five major types, although in many instances particular insulations are hybrids of these types.

The five major types are:

1. Flake
2. Fibrous
3. Granular
4. Cellular
5. Reflective

Flake insulation is composed of small particles or flakes which finely divide the air space. These flakes may or may not be bonded together. Vermiculite, or expanded mica, is commonly used for flake insulation.

Fibrous insulation is composed of small diameter fibers which finely divide the air space. These fibers may be organic or inorganic and may or may not be bonded together. Organic fibers may be of hair, wood or cane, or synthetic. The inorganic fibers may be of glass, rock wool, slag wool, alumina silica, asbestos, or carbon.

Granular insulation is composed of small nodules which contain voids or hollow spaces. It is not considered a true cellular material since gas can be transferred between the individual spaces. The material may be magnesia, calcium silicate, diatomaceous earth, and vegetable cork.

Cellular insulation is composed of small individual cells sealed from each other. It is produced of glass, rubber, and plastics.

Reflective insulation is composed of parallel thin sheets, or foil, of high thermal reflectance and spaced to reflect radiant heat back toward its source. The spacing also is designed to provide restricted air (or gas) spaces. The restricted air space reduces heat transfer which would be caused by convection and conduction. In most instances the thin reflective sheets are made of aluminum or stainless steel. Reflective insulation obtains its results from its design and construction; it is a system rather than a single homogeneous material. For this reason its properties will be discussed separately from those of the mass insulations.

Forms of Mass Insulation

Many mass insulations are produced using two or more of the types listed above. For example, fibers are added to granular material to increase the tensile strength of the product.

To obtain desired properties, frequently one type of insulation is used in combination with others, such as systems using lightweight fibers to separate reflective sheets. Another method used to reduce heat transfer is to install various types of insulation in a system by which the amount of air is reduced to small quantities. These *systems* of insulation will be discussed further on in the Manual.

The numerous mass insulations produced have varied characteristics and are intended for many uses. For this reason insulations are produced in many forms. Some of these are:

1. Rigid boards, blocks, sheets, and preformed shapes such as pipe covering, curved segments, etc.
2. Semi-rigid boards, sheets, and preformed shapes
3. Flexible boards, sheets and preformed shapes
4. Blankets
5. Batts
6. Felts
7. Tapes
8. Ropes
9. Cements
10. Loose
11. Fill
12. Mastics

Eventually each of these forms of insulation is installed and becomes a part of an insulation *system*. Its properties then become important to ensure that it will function properly in the system. Thus the insulation's properties become of utmost importance to its economic installation, service life, and thermal efficiency.

The large number of types of insulation and forms in which it is produced almost ensures that their comparative properties will vary widely. This is to be expected and desired, as the uses of the material also vary widely.

Properties of Mass Insulations

Knowledge of the properties of insulations is necessary for selection of the proper insulation for specific uses. Many times, in a given installation, a certain property may be of little importance, whereas that same property may be of utmost importance in a different application. One of the problems in the insulation industry is that some of the manufacturers resist providing information as to their material's properties. They feel that this information may limit the use of their material; this is an error. Without this information their materials may be misused, resulting in failures and a loss of money to the manufacturer, the contractor, and the user.

Many of the following listed properties are not freely available from the manufacturers or in their published literature; therefore they have not been included in the Tables of Properties 00-00, which follow later in this chapter. Upon occasion, however, some of the omitted properties may assume extreme importance. In this case effort should be made to obtain them on an individual basis from the appropriate manufacturers.

Properties (and Their Significance) of Rigid and Semi-Rigid Insulations

The basic properties of mass insulations and the significance of these properties in choosing an insulation are as follows:

Abrasion resistance. The property of a material measuring its ability to withstand abrasion without wearing away. Abrasion is caused by the rubbing together of two abutting objects or by an external force or object rubbing against the surface of one of the objects.

Alkalinity (pH) or acidity. The tendency of a material to have a basic (alkaline) or acidic reaction. It is measured on the pH scale, with all readings above 7.0 alkaline and below 7.0 acidic. A reading of 7.0 indicates neutral. This indicates to which metals the material may be applied without causing corrosion.

Breaking load. The property of a material which indicates its strength in flexure when loaded as a simply supported beam with a constantly increasing concentrated load at its center.

This property is of significance in installations where the material must "bridge" over a discontinuity in its support.

Capillarity. The property of a material which enables it to suck liquid up into or through itself.

Should a water leak occur in installed insulation, high capillarity will spread the damage.

Chemical reaction. The property of a material which measures its tendency to chemically combine (or react) with other materials which may come into contact with or be absorbed by it.

This information is very important as such reaction may cause a fire hazard. In addition to holding the combustible liquids, the intermixing may change the combustibility of the chemicals by lowering their flash, fire, and self-ignition points.

Coefficient of expansion (Contraction). The property of a material which measures its dimensional change corresponding to temperature change.

This is necessary information in solving problems related to spacing and design of expansion and contraction joints in the insulation system.

Combustibility. The property of a material which measures (on an arbitrary scale referred to some common material) its tendency to burn.

This information is vital in determining any possible fire hazard. If a material is combustible its potential contribution to a fire hazard is determined by the combination of the following:

1. Flash point temperature
2. Flame point temperature
3. Self-ignition point temperature (if any)
4. Burning rate
5. Rate of flame travel
6. Heat contribution
7. Explosion index
8. Auto-internal heating
9. Toxicity of products of combustion
10. Smoke density
11. Melting point temperature

Because of the emphasis on energy conservation, more insulation is being used in plants, buildings and homes, including large quantities of combustible organic foams and cellulose fibers. Where these are used their fire hazard should be evaluated. Unfortunately, the properties of combustibility as listed above are difficult to obtain for individual products. It should be noted that such classifications as flame spread index, smoke index determined from a single tunnel test method are of little value in the determination of the fire hazard as related to a particular installation. Such indexes only relate as a comparison of combustibles, and can be very misleading. It should be noted, tests where such indexes are determined, such as ASTM E84 and E286, contain the following statement in their scope:

> "This standard should be used to measure and describe the properties of materials, products or systems in response to heat and flame under controlled laboratory conditions and should not be used for the description or appraisal of the fire hazard of materials, products, or systems under actual fire conditions."

Compressive strength. The property of a material which measures its ability to resist a load tending to squeeze or shorten it.

It is one of the properties of insulation which determines its performance in service. The desirable compressive strength may be high or low. Where the insulation should support a load without crushing, high compression strength is needed. Conversely when insulation is used to take up dimensional change, a low compressive strength may be needed. In the latter case the percentage of recovery to original size upon the relief of stress becomes important.

Corrosion to substrates. The property of a material which indicates its chemical effect on various metals upon which it may be installed.

This is of upmost importance. If an insulation may cause corrosion to a substrate metal, either another insulation should be selected or the substrate metal should be protected by a proper coating. The fact that certain combinations of insulations and metals may cause a corrosion of the metal makes it necessary to determine the corrosive effect of each insulation on, for example, carbon steel, stainless steel, monel, copper, and aluminum. Of particular importance is the stress-corrosion effect on austenitic stainless steels.

Density. The weight of a unit volume of insulation, such as lb cu ft (kg/m^3).

It is necessary to know the density to calculate loadings and the heating rate when mass is one of the functions.

Dimensional stability. The property of a material which indicates its ability to retain its size or shape after aging, cutting, or being subjected to temperature or moisture.

As every insulation must be installed to given dimensions, this property affects the ease of installation and also its service life.

Emittance. The ability of a body to emit radiant energy as a consequence of temperature only to the corresponding emittance of a perfect emitter or black body at same temperature.

Flexural strength. The property of a material which measures its ability to resist bending (flexing) without breaking.

It is one of the mechanical properties which determines the suitability of an insulation during application and in service.

Hardness. The property of a material which measures its ability to resist penetration.

In some applications a hard surface is needed, while in others a soft surface is desired. It is, again, one of the mechanical properties to be considered in the selection of a material for a given application.

Hygroscopicity. The property of a material which measures its ability to absorb and retain water, in either the liquid or vapor state from the ambient air, with said retained moisture being in a state of equilibrium with the moisture in the surrounding air.

This property points up the fact that water in insulation, however it got there, must be allowed to escape without damage to the outer insulation protection. In other words, the higher the hygroscopicity, the more water the insulation will initially absorb and eventually have to reject when subjected to heat. This emphasizes the need for near-perfect moisture protection in the first place. If moisture does get in, however, then the outer barrier must be of the "breathing" type to let it back out when heated.

Incidence of cracking. The property of a material which indicates its tendency to crack when applied to a hot surface at its recommended maximum temperature.

This cracking affects the strength of the insulation and its thermal conductivity.

Reflectance. The ratio of radiant flux reflected by a body to that incident upon it.

Resistance to acids. The property of a material which indicates its ability to resist decomposition by various acids to which it may be subjected.

As insulation is used in all kinds of atmospheres and conditions of chemical contamination, its resistance to the acids to which it may be subjected affects its service life.

Resistance to caustics. The definition and significance of resistance to caustics is the same as that given for "resistance to acids," but it is paraphrased to read "caustics" instead of "acids."

Resistance to solvents. The definition and significance of resistance to solvents is the same as that given for "resistance to acids," but it is paraphrased to read "solvents" instead of "acids."

Shear strength. The property of a material which indicates its ability to resist cleavage.

It is another of the mechanical properties which indicates the suitability of an insulation during application and in service.

Shrinkage. The property of a material which indicates its proportionate loss in dimensions or volume when its temperature is changed.

It is a major cause of cracks in high-temperature applications. It will occur both as linear (length and width) and volumetric loss. Insulation's shrinkage in relation to temperature should be determined in order to properly design insulation expansion-contraction joints for vessels and piping.

Specific gravity—apparent. The ratio of the specific weight of a material, including all its voids, to the weight of an equal volume of water. This can be obtained by the simple device of dividing the weight/cu ft of the material (density) by the weight of a cubic foot of water.

Specific gravity—real. The ratio of the specific weight of the mass of solids making up a material to the weight of an equal volume of water.

To illustrate the difference between these two specific gravities, glass fiber insulation will have an approximate *apparent specific gravity of 0.05* and a *real specific gravity of 2.5.*

Specific heat. The ratio of the amount of heat required to raise a unit mass of material 1 ° F to that required to raise a unit mass of water 1 ° F at some specified temperature. It is essential to know this property for solution of problems related to heat storage and temperature-time lag. In Metric System specific enthropy is in Joules per kilogram degree Kelvin.

Temperature limits. The maximum temperatures, as determined by its properties, up to and down to which a material will experience no essential change in its properties.

Within its limits a material should not fail mechanically, chemically, or thermally. These limits are determined by the effect of the temperature on the properties listed in this chapter. Time exposure to temperature and the number of changes in temperature have an effect on some materials. For this reason, in most instances, it is necessary to determine the following:

Max. temp.—continuous operation
Max. temp.—intermittent operation
Max. temp.—cycle operation
Min. temp.—cycle operation
Min. temp.—continuous cycle operation

Temperature rise—self-internal heating. A property of some materials indicating that when heated above a certain temperature an internal reaction will occur, causing an internal temperature rise far in excess of the temperatures to which they are subjected.

This internal temperature not only causes damage to the material itself, but can be very hazardous, as serious fires can result.

Tensile strength. The property of a material which measures its ability to resist a stress tending to pull it apart.

It is another of the mechanical properties which evaluate the service performance of the insulation.

Thermal conductivity. The property of a homogeneous body measured by the ratio of steady state heat flux (time rate of heat flow per unit area) to the temperature gradient (temperature difference per unit length of heat flow path) in the direction perpendicular to the area.

As the property varies with mean temperature it is essential to know the conductivity curve of an insulation to calculate heat transfer. However, the prime value of insulation is its *resistance* to heat transfer. The "thermal resistivity" of a material is the reciprocal of thermal conductivity and as it is more meaningful should be the fundamental property listed for insulations. For example: An insulation with a conductivity of 0.25 as compared to another of 0.45 appears to be not too much different. Whereas if the thermal resistivity was stated it would be 4.00 for the first material and 2.22 for the second.

In Metric Units the conductivity of one insulation given in English Units as 0.25 Btu, in./ft^2, hr °F is 0.03605 W/mk and the one given as 0.45 Btu, in./ft^2, hr °F is 0.06489 W/mk. However, the first still has 1.8 more thermal resistivity than the second. The use of thermal resistance values is now being done extensively and as such includes the thickness of the insulation. For example, an insulation of conductivity of 0.25 4″ thick is being advertised as an insulation of 16.0. Thermal resistance (4/0.25 = 16.00)—English Units. In Metric Units the thermal resistance would be 0.0254 × 4/0.03605 = 0.1016/0.03605 = 2.818. For this reason it is essential that "thermal resistance" where used must be identified, as no terms have yet been assigned to differentiate between thermal resistance in English and Metric Units.

Thermal diffusivity. The property of a material which measures the time rate of temperature movement through it. It is *not* a measure of amount of heat or heat transfer.

In many cases this is an important factor. For example, in cycle operations where rapid dissipation of temperature is desired a high rate of thermal diffusivity is important. Conversely, when insulation is used as fire protection a slow rate of thermal diffusivity is most important. This little understood factor is of utmost importance in engineering these types of applications.

Thermal resistance. The property of a body by the ratio of the difference between the average temperatures of two surfaces to the steady heat flux.

Thermal shock resistance. The property of a material which indicates its ability to be subjected to rapid temperature changes without physical failure.

This property is important when insulations are used on cycle operations, fast heat ups, and for fire protection.

Vapor (water) migration. The property of a material which measures the rate at which water vapor will penetrate it due to vapor pressure differences between its surfaces.

The water vapor which enters insulation has serious effects on its thermal conductivity and on some of its physical and chemical properties. It is well known that water vapor entering low temperature insulation will at some point condense into liquid or freeze into ice and will eventually destroy the thermal efficiency of the insulation. However, vapor migration into insulation on ambient and hot service may cause swelling of the insulation and contribute to rusting of steel or stress corrosion of stainless steel.

Vapor (water) permeability. The water vapor transmission of a homogeneous body under vapor pressure difference between two specific surfaces.

Vibration resistance. The property of a material which indicates its ability to resist mechanical vibration, without wearing away, settling, or dusting off.

Almost any insulation used in an industrial application will be subjected to some vibration, such as compressor vibration, fan pulsations, or vibrations caused by the fluids or gases passing through the lines or vessels.

Warpage. The change in dimension of one surface of insulation as compared to that of another surface due to difference in temperature of the two surfaces.

Warpage will cause cracks in the finished insulation and damage to the weather and vapor barriers.

Water absorption. The property of a material which measures the amount of water it will soak up when submerged in water.

It is a measure of the amount of water which may be taken into an insulation due to water leaks in weather barrier or during construction.

It is also important as an indicator of the amount of combustible or toxic liquids which an insulation could absorb in case of spillage. However, a material may be water repellent and still be quite absorptive when subjected to hydrocarbons, solvents, or other penetrating liquids. In special cases like these, the absorptivity of insulations must be tested with the chemical to which it may be subjected.

Properties of Insulating Cements, Foams, and Insulating Mastics

After they are applied and dried, insulating cements, insulating mastics, foamed in place insulations, sprayed foam, and sprayed inorganic insulations all become rigid or semi-rigid and their properties are the same as those listed for the rigid or semi-rigid material. However, from the wet state to dried (or cured) state additional properties must be considered, as follows:

Adhesion—wet. The property of a material which indicates its ability to stick to the surface to which it has been applied without sliding or falling off.

As the temperature at the time of application and the temperature of the surface to which the material is applied affect the adhesion, either the limits of these two temperatures must be established or the adhesion for various temperatures determined. This property has direct bearing on the capability and cost of an installation.

Adhesion—dry. The property of a material which indicates its ability to bond to the surface to which it is applied and remain in place in service.

Temperature may also affect this surface adhesion so limits must be set, or adhesion must be determined at various temperatures.

Shrinkage—wet to dry. The property of a material which measures the difference in volumetric and linear change which occurs in the drying of insulating cements and mastics.

This change affects not only the quantities of material required, but it can also affect the finished, dried material if cracks or breaks develop because of excessive shrinkage.

Expansion—wet state to cured state. The property of a material which measures the difference in volumetric change in poured or sprayed, foamed-in-place organics.

This information is necessary in the calculation of quantities of material necessary to obtain a given amount of mass insulation.

Coverage—wet. The property of a material which measures the amount of material necessary to cover a given area to obtain a specific dried or cured thickness.

Of course this information is necessary for calculation of quantities necessary for a particular application.

Properties of Blanket and Batt Insulation

Since blankets and batt insulations are essentially non-rigid, such physical properties as flexural strength and shear, listed as properties of rigid materials, are not applicable. However, other properties given for rigid materials are also properties of the non-rigid materials. These include:

- Alkalinity (pH)
- Capillarity
- Combustibility (or non-combustibility)
- Chemical reaction
- Compressive strength (although generally of relatively low value as compared to rigid material)
- Corrosion to substrates
- Density
- Hygroscopicity
- Resistance to acids

Resistance to caustics
Resistance to solvents
Specific gravity—apparent
Specific gravity—real
Specific heat
Tensile strength
Thermal conductivity
Thermal diffusivity
Temperature limits
Temperature rise (self internal heating)
Vapor migration
Vibration resistance
Water absorption

In addition, properties other than those listed for rigid materials do become important in this form of insulation, namely:

Compaction or settling. The property of the blankets or batts which measures their change in density and thickness resulting from loading or vibration thus producing change of thermal efficiency.

Some materials tend to fluff up when subjected to vibration whereas others tend to compact. This factor has a definite relation to suitability of the insulation for certain installations.

Recovery of thickness after compression of blankets or batts after having been compressed is a vital factor in the use of material as cushion blankets, or in expansion joints.

Resistance to air movement. The property which indicates the ability of a blanket-type material to resist erosion by air currents over its surface.

The importance of a blanket being able to resist air velocities across its surface is demonstrated most frequently when blankets are used to line the inside of ducts.

Properties of Fill Insulations

Although the particular set of properties which is important to fill material is slightly different from the forms of insulation previously discussed, each of the individual properties have already been mentioned. The important properties of fill insulations follows:

Adsorptivity
Alkalinity
Capillarity
Combustibility (or non-combustibility)
Chemical reaction
Compaction or settling
Corrosion to substrates
Hygroscopicity
Resistance to acids
Resistance to caustics
Resistance to solvents
Specific gravity—apparent
Specific gravity—real
Specific heat
Thermal conductivity
Thermal diffusivity
Temperature limits
Temperature rise—self internal heating
Vapor migration

Measurement of Properties of Thermal Insulation

Determining what insulation properties are important only indicates the information needed and provides the basis of investigation. The next question then becomes, how can these properties be measured and under what conditions should they be measured? Another question which follows is whether or not measurements made by laboratory test are truly indicative of the properties of the material in service. Does aging affect the properties? If so, how much and at what rate? What effect does temperature have on test results? What effect does moisture content have on the properties?

Unfortunately, testing methods and studies of properties of material are not yet able to answer all these questions. Nor is there a laboratory available which is accepted as *the* authority on testing of properties of insulation materials. This statement does not imply that there are no standard test methods and laboratories to perform these standard test methods, but it does mean that the insulation industry lacks even the minimum requirements to provide reliable information. In many instances the manufacturers of insulation have developed their own test methods and can furnish some information regarding their materials. However, the test method used by one manufacturer is not always identical with that of another, so that the values of one cannot be compared with the results obtained by another. This makes direct comparison of material properties quite difficult. Another problem is that most of the mechanical properties are measured at atmospheric conditions and not at the temperatures to which the material is subjected in service. Inadequate information leads to poor design, the use of improper materials, and the installation failures to which the insulation industry has become accustomed.

In recent years progress has been made in correcting this confusion and presently there exists a number of American Society of Testing and Materials Standard Test Methods. However, even with standard test methods, sometimes it is necessary, because of the wide range in characteristics, to perform more than one test for a single property, as one test may be incapable of measuring all the characteristics for the wide variety of insulation materials. Thus, duplication of test methods for a single property of insulations does exist and the proper selection must be made for the insulation being tested. Following is a list of American Society of Testing and Materials Test Methods for various properties:

ASTM Test Methods No.	For
	ADHESION OF THERMAL INSULATING CEMENTS
C-353	Adhesion of Dried Thermal Insulation or Finishing Cement

ASTM Test Methods No.	For
C-383	Adhesion, Wet, of Thermal Insulating Cements to Metal
	BREAKING LOAD
C-203	Test for Breaking Load and Calculated Flexural Strength of Block Type Thermal Insulation
C-446	Test for Breaking Load and Calculated Modulus of Rupture of Preformed Insulation for Pipes
	COMBUSTIBILITY
D-92	Flash and Fire Points by Cleveland Open Cup Tester
D-93	Flash Point by Pensky-Martens Closed Tester
D-568	Test for Flammability of Flexible Plastics
D-635	Test for Flammability of Self-Supporting Plastics
D-1525	Vicat Softening Temperature of Plastics
D-2015	Gross Calorific Value of Solid Fuel by the Adiabatic Bomb Calorimeter Tester
D-2582	Test for Heat of Combustion of Hydrocarbon Fuels by Bomb Calorimeter
D-2843	Measure of Density of Smoke From Burning or Decomposition of Plastics
D-2863	Test for Flammability of Plastics Using Oxygen Index Method
D-3014	Flammability of Rigid Cellular Plastics (Vertical Position)
D-3211	Test for Relative Density of Black Smoke (Ringlemann Method)
E-84	Test for Surface Burning Characteristics of Building Materials
E-136	Test for Noncombustibility of Elementary Materials
E-162	Test for Surface Flammability of Materials Using Radiant Heat Energy Source
E-286	Test for Surface Flammability of Building Materials Using 8′ (244 m) Tunnel Furnace
	COMPRESSIVE STRENGTH
C-165	Recommended Practice for Measuring Compressive Properties of Thermal Insulation
C-354	Tests for Compression Strength of Thermal Insulating or Finishing Cements
C-495	Test for Compressive Strength of Lightweight Insulating Concrete
D-1621	Test for Compressive Properties of Rigid Cellular Plastics
	CORROSION
C-464	Corrosion Effect of Thermal Insulating Cements on Base Metal
C-692	Stress Corrosion Effect of Wicking-Type Thermal Insulation on Stainless Steel
	COVERING CAPACITY
C-166	Covering Capacity and Volume Change upon Drying of Thermal Insulating Cements
	DENSITY
C-519	Density of Fibrous Loose Fill Building Insulations
C-520	Density of Granular Loose Fill Insulations
C-303	Density of Preformed Block-Type Insulation
C-302	Density of Preformed Pipe-Covering Type Thermal Insulation
D-1622	Apparent Density of Rigid Cellular Plastics
	DIMENSIONAL STABILITY
C-548	Dimensional Stability of Low Temperature Thermal Block and Pipe Insulation
	DROPPING (RESISTANCE TO)
C-487	Resistance to Dropping of Preformed Block Type Thermal Insulation

ASTM Test Methods No.	For
	EMITTANCE
C-445	Normal Total Emittance of Surfaces of Materials 0.01″ or Less in Thickness at Approx. Room Temperature
C-835	Test for Total Hemispherical Emittances of Surfaces from 20 to 1400° C (68 to 2550° F)
	HARDNESS
C-569	Test for Indention Hardness of Preformed Thermal Insulation
	HEAT FLUX
C-745	Heat Flux Through Evacuated Insulations Using a Flat Plate Boiloff Calorimeter
	HOT SURFACE PERFORMANCE
C-411	Hot-surface Performance of High-Temperature Thermal Insulation
	MAXIMUM USE TEMPERATURE
C-447	Recommended Practice for Estimating Maximum Use Temperature of Preformed Homogeneous Thermal Insulation
	MECHANICAL STABILITY
C-421	Mechanical Stability of Preformed Thermal Insulation, Tumbling Test
	RESISTANCE TO EXTERNAL LOADS
C-854	Resistance to External Loads on Metal Reflective Insulation
	SHRINKAGE DUE TO HEAT
C-356	Shrinkage, Linear of Preformed High-Temperature Thermal Insulation Subjected to Soaking Heat
	SPECIFIC HEAT
C-351	Mean Specific Heat of Thermal Insulation
	THERMAL HEAT TRANSFER
C-177	Steady State Thermal Conductivity Properties by Means of the Guarded Hot Plate
C-518	Steady State Thermal Conductivity Properties by Means of the Heat Flow Meter
C-236	Thermal Conductance and Transmittance of Built Up Sections by Means of the Guarded Hot Plate
C-335	Thermal Conductivity of Pipe Insulation
C-691	Thermal Transference of Nonhomogeneous Pipe Insulation Above Ambient Temperatures
	WATER ABSORPTION
C-209	Insulating Board (part 13)
	WATER VAPOR TRANSMISSION
E-96	Water Vapor Transmission of Materials in Sheet Form
C-355	Water Vapor Transmission of Thick Materials

As can be seen from the above, standard test methods do not exist for all the properties of insulation listed previously. One reason is that some of these properties are difficult to determine. In other cases test equipment is quite expensive, few manufacturers or laboratories have the required equipment, and an industry standard has never been devised.

It should be remembered that any laboratory test is a measure of a small sample. For certain characteristics, a small sample will provide an accurate appraisal, whereas for other materials the answer may only *indicate* the characteristics of that insulation. Other tests may attempt to predict the service life of insulation. Such service life tests are quite difficult to devise and evaluate. Yet this type of test is essential, and as the demand for it increases, eventually such tests will be devised by the industry.

Service records also assist in evaluating materials. Unfortunately, an early failure is costly and a successful installation requires many years to prove its success. There is no one method of engineering all insulation installations, and the really difficult applications still depend upon engineering judgement. An engineer with an extensive knowledge of the properties of *many* insulations is in a position to make the best decision, *based on available knowledge at the time of the decision.*

Reflective Insulation

Reflective insulation, which consists of metallic sheets spaced to direct the heat back to its source, must be constructed to provide space between each of the reflective sheets. As its efficiency depends upon the arrangement of the sheets and their means of support, a *system* of construction is necessary. Its effectiveness and properties are those of the system and not the properties of any single material used in its construction.

Reflective insulation systems have been employed in the side walls and ceilings of building, and when proper spacing of layers and methods of construction are used they have proven to be extremely efficient.

Another use of reflective insulations has been in the construction of cold storage rooms. This type of construction is similar to that used for insulation of buildings, the only difference being that tight vapor seals must be constructed on the warm side for cold service.

Reflective insulation used to insulate pipes and vessels, is a factory made product, and more closely approaches a form which can be evaluated by a list of properties. This type of reflective insulation is formed into sectional pipe insulation, fitting covers, and shaped segments to fit equipment. These shapes consist of reflective sheets of aluminum, or stainless steel, so formed as to have little direct contact between the sheets, or with spacers of a low conductive metal or other material to form isolated air chambers between the sheets and end closures.

As a finished product these formed reflective insulation structures can be evaluated to have certain characteristics as follows:

- Aklalinity (pH)
- Capillarity
- Coefficient of expansion
- Combustibility (or non-combustibility)
- Corrosion to substrates
- Cracking (in service)
- Hygroscopicity
- Leachable material content
- Resistance to acids
- Resistance to caustics
- Resistance to solvents
- Resistance to nuclear radiation effects
  1. Damage to material
  2. Loss in thermal efficiency
  3. "Half life"
- Strength
  1. Crushing
     a. Perpendicular to longitudinal axis
     b. Parallel to longitudinal axis
  2. Flexural—beam across supports
  3. Impact resistance
- Specific gravity—apparent
- Specific gravity—real
- Temperature limits
  1. Maximum—continuous
  2. Maximum—cyclic
  3. Maximum—intermittent
  4. Minimum—continuous
  5. Minimum—cyclic
- Thermal emittance
  1. Inner surfaces
  2. Outer surfaces
- Thermal conductance
  1. Across total thickness
  2. Per inch displacement
- Thermal shock resistance
- Vibration resistance
- Water vapor transmission

The reflective insulation systems are fabricated insulation structures and do not lend themselves as easily to testing as do homogeneous materials. Thus there does not exist a series of standard test methods for their evaluation, as exists for other materials. Since these are designed systems, the physical properties and strength required for any given installation can be engineered into the finished product, whereas with homogeneous materials a product naturally having the properties must be selected to fulfill the installation requirements.

Although each of the individual materials used in the construction may be tested by standard test methods, there presently exist few ASTM test methods which are suitable for the testing of reflective insulation *systems.* However as these and other cryogenic insulation systems increase in importance tests will be developed.

Cryogenic Insulation

When the term *cryogenic insulation* is used it does not designate a specific type or system of insulation, but refers to the use of insulation in controlling heat gains to cryogenic liquids. There is not established temperature below which insulation is termed cryogenic. However, it is generally understood to be in the range of −150° F down to absolute zero (−459.6° F).

Although homogeneous mass insulations are used for applications at the upper part of the cryogenic temperature range, to obtain the efficiency necessary for proper control of very low temperatures, a system of insulation using more than one material is generally employed. Maximum efficiency to retard the flow of heat by the combined action of all heat transfer mechanisms (radiation, conduction and convection) is the aim of the cryogenic insulation *systems.*

Combining the principles of the control of heat transfer by mass, by reflective substances, and by vacuum, systems have been designed to obtain thermal resistances unobtainable with ordinary mass insulations. Development of these efficient insulations essentially started with the discovery of the Dewar Flask by James Dewar in 1898. Further contributions to the development of these systems were made by I. A. Black, A. A. Fowle and P. E. Glaser of Arthur D. Little Company; and by P. E. Riede, D. Wang, L. C. Matsch, C. R. Lindquist and L. R. Niendorf of the Linde Company.

The systems of cryogenic insulations made possible by these developments are basically of two types—one type is composed of powders in a partial vacuum and the other multilayered, spaced reflective sheets in a partial vacuum.

Powders such as perlite and "Santocel" used in vacuum have a conductivity of approximately 0.012 Btu/sq ft, hr, in., °F at temperatures from 70° F to −300° F. Powders used with reflective and absorptive opacifiers of the proper formula can improve the performance. Such a formula, developed by Linde, using "Santocel" and copper powder in a vacuum has a conductivity of 0.0026 Btu/sq ft, hr, in., °F in a vacuum of 0.05 to 5.0, microns of mercury, absolute pressure.

Multilayered reflective sheets consist of layers of reflective sheets such as aluminum foil or metal-coated plastics film, separated by micro glass-fiber mats in a vacuum space, can attain even lower thermal conductivity than opacified powders. However, their effectiveness is much more dependent upon the almost complete absence of pressure than the powders. Depending upon the layers per inch, conductivities of 0.00108 to 0.00012 Btu/sq ft, hr, in., °F can be atained when in spaces under a vacuum of 0.5, or less, microns of mercury, absolute pressure.

Of course, the problem is to provide the vacuumized space for these insulations to function. This is done by utilizing double-shell vessels, with the insulation installed between the two shells and the vacuum generated in the insulated space. "Permanent vacuum" vessels of such extremely low leak rate that they do not require any re-evacuation for many years have been developed.

As with reflective insulations, the cryogenic vacuum insulated vessels are designed and constructed to withstand any mechanical and chemical abuses to which they may be subjected. Thus the properties are those of a design rather than that of a material. A chart of thermal conductivities of the cryogenics will be given later in this chapter.

Underground Insulation Systems

Pipe or equipment located underground which must be insulated, in most instances uses the more conventional mass insulations in some system which not only insulates, but also protects the insulation from water.

As these systems use many combinations of conduit and thermal insulations, they are not a "class" of insulation in themselves. The properties of the insulation used in their construction can be obtained from the properties of that insulation as listed. These systems will be discussed in a later chapter.

Property Tables—Not Including Conductivity

The properties in Tables 39 through 48 are not all of the relevant properties of the materials listed previously in this chapter. The tables do list, however, the most important properties and those most commonly needed to make an intelligent selection of an insulation for a specific purpose.

Note that many of the materials have only a few properties listed and that none of them are complete. The properties were compiled from information obtained from many sources, but most were secured directly from the manufacturers of the insulations, or from their published literature. As can easily be understood, further testing and checking by the author of each individual value obtained, involved a task much beyond the scope of this book. However, based upon the author's knowledge of insulation materials, the values presented were reviewed to establish that each one was apparently reasonable for that particular insulation and that particular property.

All properties listed are not significant for all materials. For example, the water vapor transmission rate of an insulation used for high temperature service is unimportant because the high temperature pipe or vessel causes a high vapor pressure which drives moisture outward and the insulation remains dry and effective. For this reason, the vapor transmission rate on high temperature materials is seldom measured.

For certain properties, such as alkalinity, there has been no test method agreed upon by the insulatin industry. Some manufacturers provide the information in terms of pH and others in respect to percentage of Na_2O. Again because of these limitations, despite the fact that this may cause confusion, the only recourse was to list the values obtainable.

Another factor which should be remembered in the use of these tables is that a particular generic material may be made by several manufacturers. As there are differences in the properties of a generic material produced by different manufacturers, the properties are a mean of the values listed for the products of the various manufacturers. Thus, for those products made by several manufacturers, no single manufacturer's product will have exactly the values as listed. It follows then, that these tables are presented to serve as a *guide* for selection of the correct *generic material* for an installation. However, the final choice of the particular manufactured product must be made by the individual by comparing competitive products.

The property "Maximum Temperature Limits" is given first as this is the first requirement that must be fulfilled. "Density" is second, as it frequently identifies or classifies a particular material as being in a given generic class, and within a generic class it affects the thermal conductivity.

All other properties are listed in alphabetical order.

Property Tables—Conductivity

The thermal conductivities given in Tables 00 through 42 have been gathered from various sources and while deemed accurate, are not the result of tests by the author. In most cases, they are the results of tests performed upon laboratory-dried samples by the manufacturers. They are based upon representative values for generic material within their ranges of densities and are neither the highest nor the lowest of a particular listing of competitive materials. They represent mean averages which we expect will provide relatively accurate information for use in heat transfer calculations.

As all values listed are taken as representative of the means of generic materials available and are provided for engineering calculation only, they are not deemed suitable for use as material specification limits. Any one particular manufacturer's product is not likely to have the mean values of all the materials manufactured in that particular generic classification.

For a particular installation, the material specification must be written by deciding what property values are essential and, with knowledge of materials available, the specification must be written to obtain those essential properties within the ranges of the materials available.

Hopefully, this book will indicate the need for more complete information on properties of materials and that any missing information will be made available by the manufacturers. When insulation consumers understand the need for designing insulation systems on a scientific basis the manufacturers will quickly respond by providing the necessary technical information.

The abbreviations in the Tables have the following meanings:

amts.	amounts
Exp. Co.	Coefficient of expansion
Comb.	Combustible
hex.	hexagon
Insul.	Insulation
Neg.	Negligible
NC	Noncombustible
PC	Pipe Covering
RH	Relative Humidity
Temp.	Temperature
thk.	thickness
SS	Stainless steel

Listing of Thermal Insulation and Their Properties

An identification number is given in the listing of individual generic materials. This same number is used to identify the properties of that material in Tables 39 through 48. That same number is used to identify the material in the Conductivity Tables 49 through 54.

Wherever possible the physical properties listed were properties established by standard ASTM Test Methods. Unfortunately, standard test methods have not yet been established for a number of properties needed for good engineering design. It should also be pointed out that all the strength properties listed were obtained at atmospheric temperatures and not at the temperature at which the material may be expected to operate. Many of the important properties for effective insulation design were not obtainable from the manufacturers of the insulation materials. In some instances, to guide the user, descriptive wording was entered on the property sheets where exact test results were not obtainable. In other instances where information was not obtainable the space for entry of that property was left blank. Where such information is essential the consumer may be able to obtain the information from the manufacturer, or he may have the tests for these properties done by others.

It should also be noted that the heat transfer test results in accordance with the listed ASTM Test Methods are based on laboratory dried samples. For absorbent fibrous, cellular, or granular insulation the heat transfer will be greater than that obtained from these test methods. The amount of conductivity raise depends upon degree of absorbence, and adsorbence, of the insulation in question and also the particular service conditions involved.

Properties related to fire hazards are difficult to obtain. Few manufacturers of combustible insulations list or publish the basic combustible properties (as listed under combustibility) of their materials. Instead they list flame spread index, smoke index, etc., which do not provide a guide for good engineering design. Where combustible insulations are used in buildings, spray systems and other fire protection measures should be provided. Likewise combustible insulations should be used with caution in industrial installations.

It must be remembered that the following listing of properties and conductivities was the most accurate that could be collected at the time of the preparation of this book and that manufacturers change their products. Thus where a property is critical to design it is well to check with the manufacturer or manufacturers of that product as to characteristics at that time.

Rigid and Semi-Rigid Insulations

Material identification number	Insulation material, generic type	Description and general characteristics
1	Alumina-silica fibers, combined with binders.	Block insulation for exceptionally high temperature service, resists thermal shock.
2	Alumina-silica fibers, combined with binders.	Block insulation—stronger than (1) above. For exceptionally high temperature services. Resists thermal shock.
3	Refractory-fiber molded, predominantly alumina and silica.	High temperature lightweight board.
4	Alumina-silica ceramic fibers, combined with binders.	High temperature board, resists thermal shock.
5	Calcium-silicate, hydrous, reinforced with small amounts of mineral fiber. Type 1.	Molded into rigid block and pipe covering. Strong, for its weight. Good cutting characteristics. Water absorbent, but retains most of its strength when wet.
6	Calcium, silicate, hydrous, reinforced with small amounts of mineral fiber, extra high temperatures. Type 2.	Molded into rigid block and pipe covering. Strong, for its weight. Good cutting characteristics. Water absorbent.
7	Diatomaceous silica, blended and bonded. Class I.	Formed into rigid block and pipe covering. Good for high temperatures. Excellent thermal shock resistance.

Rigid and Semi-Rigid Insulations

Material identification number	Insulation material, generic type	Description and general characteristics
8	Diatomaceous silica, blended and bonded. Extra high temperature. Class II.	Formed into rigid block and pipe covering. Good for higher temperatures than material (7). Higher density. Excellent thermal shock resistance.
9	Cork, vegetable, compressed. Board.	Molded into rigid board or pipe covering. Strong, for its weight. Good cutting characteristics. Will resist liquid water, but has relatively high rate of vapor transmission.
10	Cork, vegetable, compressed. Pipe covering.	Molded into rigid board or pipe covering. Strong for its weight. Good cutting characteristics. Will resist liquid water, but has relatively high rate of vapor transmission.
11	Glass, cellular	Rigid block and pipe covering. Good strength characteristics. Poor abrasion resistance. Can be formed and fabricated rapidly. Water resistant. Has almost perfect resistance to the migration of moisture vapor.
12	Glass, cellular, between layers of felt	Same a material (11) except it is fabricated into sheets between layers of felt for roof insulation.
13	Diatomaceous earth, Type 1	Light density rigid block and pipe covering.
14	Diatomaceous earth, Type 2	Moderate density rigid block and pipe covering.
15 16 17	Glass, fiber, with organic binder	Semi-rigid formed board or sheets of various densities. The higher the density the stronger the material. The higher densities also provide a lesser conductivity. Available with various types of factory attached outer facings.
18	Glass, fiber, with organic binder, faced one side with vapor resistant facing	Formed into rigid sheets for constructing air conditioning and heating ducts.
19	Glass, fiber, with organic binder	Semi-rigid formed pipe covering. Light density pipe insulation. Available with various types of factory attached jacketing.
20	Glass, fiber, with organic binder	Formed into rigid pipe covering. Heavy density pipe insulation. Available with various types of factory attached jacketing.
21	Mineral fiber, Class 1, lower temperature limit	Semi-rigid formed block or board, moderate density. Available with various types of factory attached jacketing.

Rigid and Semi-Rigid Insulations

Material identification number	Insulation material, generic type	Description and general characteristics
22	Mineral fiber, Class 2, inorganic binder	Formed semi-rigid board and block.
23	Mineral fiber, Class 3, with inorganic binders	Formed into semi-rigid block and board.
24	Mineral fiber, Class 4, with inorganic binder	Semi-rigid formed block and board.
25	Mineral fiber, with organic binder, Class 5	Formed into semi-rigid board and block.
26	Mineral fiber, with inorganic binders, Class 1	Formed into pipe covering.
27 29	Mineral fiber, with inorganic binders, Class 2	Formed into pipe covering.
28 30	Mineral fiber, with inorganic binders, Class 3	Formed into pipe covering.
31	Perlite, expanded, blended and bonded with binders	Roof board. Low shrinkage at high temperatures. Water repellent.
32	Phenolic foam	Formed into board and pipe covering.
33	Polystyrene, expanded beads, molded to shape	Formed into rigid block, board and pipe coverings. Light weight, excellent cutting characteristics.
34	Polystyrene, cellular foam, molded	Molded rigid board and pipe covering. Light weight.
35	Polystyrene, cellular foam	Rigid board, sheets and pipe covering cut from slab stock. Light weight. Good cutting characteristics. Combustible.
36	Polystyrene, cellular foam	Rigid boards. Light weight. Good cutting characteristics.
37	Polystyrene, cellular	Molded to shape for roof boards.
38	Polyurethane, cellular foam	Formed into rigid boards, sheets, and pipe covering. Light weight. Good cutting characteristics.
39	Polyurethane, cellular foam	Formed into rigid boards, sheets, and pipe covering. Light weight. Good cutting characteristics.
40	Polyurethane, cellular foam	Formed into block, board and pipe covering.
41	Rubber resin, cellular foam	Rigid board. Good cutting characteristics.
42	Expanded silica and binders	4' x 8' (approx.) semi-rigid panels for insulation of large diameter and flat surfaces.
43	Wood fiber and binder	Formed into rigid panels for roof deck insulation.
44	Borosilicate closed cell inorganic foam	Formed into rigid board.
45	Elastomeric sheet	Seamless flexible pipe covering.
46 47	Glass, fiber and inorganic binder, available with various	Flexible rolls or sheets. All are of similar material

Rigid and Semi-Rigid Insulations

Material identification number	Insulation material, generic type	Description and general characteristics
48	facings	except for density. As the density affects both strength and flexibility and also conductivity, divisions in density were provided to establish the relationships.
49	Kaolin cermaic fiber in S.S. metal mesh	Flexible rolls or sheets.
50	Kaolin ceramic fiber in S.S. metal mesh	Flexible rolls or sheets.
51	Mineral and spun glass fibers-metal mesh	Formed into blankets.
52	Mineral fibers-metal mesh facing, Class 1	Formed into blankets with metal mesh facing.
53	Mineral fibers-metal mesh facing, Class 2	Formed into blankets with metal mesh facing.
54	Urethane cellular foam	Formed into block and pipe covering.
55	Urethane cellular foam	Formed into block and pipe covering.

Blanket, Felt and Batt Insulations

Material identification number	Insulation material, generic type	Description and general characteristics
56 57 58	Alumina-silica fiber	Formed into blanket and strip for high temperature service.
59	Alumina-silica fiber	Formed into rope, cord and yarn for high temperature service.
60	Glass, fine fiber, (no binder)	Formed into blanket and matt.
61	Glass, fine fiber with binder	Formed into blanket.
62	Glass fiber, batts with various facings, with binder	Formed into batts and encased in various facings. Mainly used for building ceiling and wall cavity insulation.
63	Glass fiber, bonded with thermosetting organic binders. Available in various facings	Formed into blankets. Numerous densities available. General purpose blanket.
64	Mineral fiber	Formed into blankets and batts.
65	Mineral fiber with binders, coated one side	Blanket specially coated on one side to resist air erosion. For lining heating or air conditioning ducts.
66	Silica fibers	Formed into woven blankets and felts for high temperature installations.

Insulating and Finishing Cements

Material identification number	Insulation material, generic type	Description and general characteristics
67	Alumina-silica, semi-refractory hydraulic setting cement	Semi-refractory and insulating cement. Water mix—hydraulic setting.
68	Alumina-silica, ceramic fibers, air setting mold cement	High temperature—water mix—cement.
69	Alumina-silica ceramic spray mix, cement	Water mix, insulating and finishing cement.
70	Diatomaceous silica and binders, air setting cement	Water mix, finishing cement.
71	Kaowool ceramic fiber and inorganic binders-spray cement	Water mix, hard surface finishing cement
72	Kaowool and mineral fiber blend-spray cement	Water mix, finishing cement.
73	Mineral fiber insulating cement	Water mix, insulating and finishing cement.
74	Mineral fiber, binders, hydraulic setting cement	Water mix—hydraulic setting cement.
75	Vermiculite-expanded with binders, water mix-air setting	Water mix insulating cement. High shrinkage.

Formed or Foamed Sprayed-in-Place Insulations

Material identification number	Insulation material, generic type	Description and general characteristics
76	Urethane, two part liquid mix, rigid foam. Self-extinguishing	Liquids mixed by spray gun. Liquid on surface foams to form rigid insulation. Formulas obtainable for insulation will not continue burning after fire is removed.
77	Urea-formaldehyde foamed in place	Formulated to be used as fire protection and high temperature insulation.
78	Polyvinyl acetate cork filled mastic	Premixed mastic. Can be applied by spray, trowel or palm. Used to control condensation.

Fill and Loose Insulation

Material identification number	Insulation material, generic type	Description and general characteristics
79	Alumina-silica fibers, bulk	For fill applications such as packing furnace joints.
80	Alumina-silica, chopped and milled	Same as (79)
81 83	Alumina-silica, long fibers, bulk cellulose fiber—loose fill	Same as (79)

Fill and Loose Insulation

Material identification number	Insulation material, generic type	Description and general characteristics
82	Cork, granulated	Available in various grades and particle sizes.
84	Diatomaceous silica, fine powder	Milled to as fine powder, for filling cavities.
85	Diatomaceous silica, calcined	Calcined to withstand higher temperatures. Used as fill, or with concrete to form insulating concrete.
86	Diatomaceous silica, coarse powder	Milled to a coarse powder, for filling cavities.
87	Gilsonite granules	Available in various grades for various service temperatures. Used as fill around underground hot piping.
88	Glass fibers, loose or shredded	Various grades available.
89	Glass fibers, loose and blowing wool	For pouring or blowing into cavity spaces.
90	Gypsum pellets—bulk	For pouring into cavity spaces.
91	Mineral fibers, with small amount of oil lubricant	Fibers in loose form for filling cavity spaces.
92	Mineral fibers, oil free, loose	Loose fiber for filling cavity spaces.
93	Mineral fiber, granulated	For pouring or blowing in cavity spaces.
94	Perlite, expanded, loose	Loose for filling cavity spaces. Available in various size spheres.
95	Quartz fibers-loose	For filling cavity spaces.
96	Silica fibers	Soft fibers for high temperature service.
97	Silica-aerogel spheres	Small spheres. Very low conductivity. Will flow through a very small opening.
98	Vermiculite flakes	Expanded mica. Available in various sizes. For filling cavity spaces.

20 types

Reflective Insulation/Foils and Sheets

Material identification number	Insulation material, generic type	Description and general characteristics
99	Aluminum Sheet	Polished sheet, thickness as required for strength. Installed to enclose and separate air spaces.
100	Aluminum Foil	Thickness as required for strength. Installed to enclose or separate air spaces.
101	Aluminum Foil on membrane paper, or other reinforcements such as glass fiber	Foil with various reinforcements. Installed to enclose or separate air spaces.

Reflective Insulation/Foils and Sheets

Material identification number	Insulation material, generic type	Description and general characteristics
102	Preformed aluminum with kraft paper, accordian formed. Wall and ceiling insulation	Preformed into strips or rolls, then when opened and installed, layers of foil will form enclosed spaces in cavity installation.

Reflective Insulation/Preformed Pipe and Equipment Insulation

Material identification number	Insulation material, generic type	Description and general characteristics
103	Preformed aluminum case with aluminum reflective sheets	Custom formed to fit piping or equipment. Non absorbent. Can be removed and replaced rapidly. Light weight.
104	Preformed stainless steel case with aluminum reflective sheets	Same as (103) above. Slightly more efficient, but limited to lesser service temperatures.
105	Preformed stainless steel case with stainless steel reflective sheets	Custom formed to fit piping or equipment. Non absorbent. Can be removed and replaced rapidly. Light weight.

Cryogenic Insulation-Reflective, Partial Vacuum-Powder in Partial Vacuum

Material identification number	Insulation material, generic type	Description and general characteristics
106	Aluminum foil, spaced with glass mat—in vacuum	Multilayers of aluminum reflective foil in vacuum cavity. Very low thermal conductivity.
107	Aluminum foil, spaced with glass paper. In vacuum.	Same as above.
108	Carbon-silica in vacuum	Powder in vacuum cavity.
109	Perlite in vacuum	Same as above.
110	"Santocel" in vacuum	Same as above.
111	"Santocel" and aluminum powder in vacuum	"Santocel" opacified with aluminum powder in vacuum cavity.
112	"Santocel" and copper powder in vacuum	"Santocel" opacified with copper powder in vacuum cavity.

Cryogenic Insulation-Mass-Atmospheric Pressure

Material identification number	Insulation material, generic type	Description and general characteristics
113	Cellular Glass	Rigid block and pipe covering.

Table 39 Rigid and Semi-Rigid Insulation

Material Identification Number				1	2	3
Composition				Alumina-Silica fiber & binder-Block	Alumina-silica fiber & binder-Block	Alumina-silica cermaic fiber & binders-Board
Abrasion Resistance % of Weight Loss			1st Run	30		
			2nd Run	60		
Alkalinity pH						
Capillarity				Will Wick	Will Wick	Will Wick
Coefficient of Expansion			English	Will Shrink	Will Shrink	Will Shrink
			Metric	Will Shrink	Will Shrink	Will Shrink
Combustibility				Non-combustible	Non-combustible	Non-combustible
If Combustible	Flash Point Temp.		°F	"	"	"
			°C	"	"	"
	Flame Point Temp.		°F	"	"	"
			°C	"	"	"
	Heat Release		Btu/lb	"	"	"
			J/kg	"	"	"
	Self Internal Heating			"	"	"
	Smoldering – Afterglow			"	"	"
Corrosion – Rusting						
Corrosion – Stress						
Density			lb/ft^3	15 to 17	18 to 22	10 to 16
			kg/m^2	240 to 272	288 to 352	160 to 256
Hardness – mm penetration						
Hygroscopicity % Volume				High	High	High
Resistance to Acids				Good, except HF & H_3PO_4	Good, except HF & H_3 PO_4	Good, except HF & H_3PO_4
Resistance to Caustics				Fair to Poor	Fair to Poor	Fair to Poor
Resistance to Solvents				Excellent	Excellent	Excellent
Shrinkage %			Linear	2.5 at 1800°F (982°C)	2 to 5 at 2300°F (1260°C)	3.0 at 1800°F (982°C)
			Volumetric			
Specific Gravity			Real	180 to 200	190 to 210	195 to 210
			Apparent	0.25	0.32	0.24
Specific Heat			Btu/lb	0.27	0.27	0.27
			w/kg	0.15	0.15	0.15
Strength	Compressive		psi			
			kPa			
	Flexural		psi/m			
			kPa			
	Modulus of Rupture		psi		125 after 24 hrs 2300°F	31 after 24hrs. 1500°F
			kPa		858 after 24hrs 1260°C	213 after 24hrs. 816°C
	Tensile		psi			
			kPa			
Temperature Limits, Service	Short Time		°F	2300	2300	2300
			°C	1260	1260	1260
	Continuous		°F	2300	2300	2300
			°C	1260	1260	1260
Melting Temperature			°F	3200	3200	3200
			°C	1980	1980	1980
Thermal Diffusivity			ft^2/hr			
			m^2/hr			
Thermal Shock Resistance				Excellent	Excellent	Excellent
Vibration Resistance				Good	Good	Good
Water Absorption			% by Volume	91	90	90 to 92
Water Vapor Transmission			perm-inch	Very High	Very High	Very High
			perm-cm	Very High	Very High	Very High

Table 39 Rigid and Semi-Rigid Insulation

Material Identification Number				4	5	6
Composition				Alumina-silica cermaic fiber & binder-Board	Calcium Silicate-Type 1: bulk & Pipe Covering	Calcium Silicate-Type 2: Block & Pipe Cover
Abrasion Resistance % of Weight Loss			1st Run		(max.) 30	(max.) 20
			2nd Run		(Max.) 50	(max.) 45
Alkalinity pH					8 to 10.5	8 to 10.5
Capillarity				Will Wick	Will Wick	Will Wick
Coefficient of Expansion			English	Will Shrink	0	0
			Metric	Will Shrink	0	0
Combustibility				Non-combustible	Non-combustible	Non-combustible
If Combustible	Flash Point Temp.		°F	"	"	"
			°C	"	"	"
	Flame Point Temp.		°F	"	"	"
			°C	"	"	"
	Heat Release		Btu/lb	"	"	"
			J/kg	"	"	"
	Self Internal Heating			"	"	"
	Smoldering – Afterglow			"	"	"
Corrosion – Rusting					Will not contribute	Will not contribute
Corrosion – Stress					Will contribute	Will contribute
Density			lb/ft^3	21 to 90	(max.) 14.5	(max.) 15
			kg/m^2	336 to 640	(max.) 230	(max.) 240
Hardness – mm penetration					40	60
Hygroscopicity % Volume				High	10 to 15	10 to 14
Resistance to Acids				Good, except HF & H_8 Po_4		
Resistance to Caustics				Fair to Poor		
Resistance to Solvents				Excellent		
Shrinkage %			Linear		2.0 at 1200° F (650° C)	2.5 at 1500° F (982° F)
			Volumetric			
Specific Gravity			Real	200 to 210	4.4	4.16
			Apparent	0.48	0.22	0.24
Specific Heat			Btu/lb	0.27	0.28	0.28
			w/kg	0.15	0.16	0.16
Strength	Compressive		psi		60 to 100 at 5% defor	60 to 250 at 5% defor
			kPa		414 to 600 at 5% defor	412 to 1725 at 5% defor
	Flexural		psi/m		35 to 50	38
			kPa		241 to 300	262
	Modulus of Rupture		psi	125 after 24hrs. 2300° F		
			kPa	858 after 24hrs. 1260° C		
	Tensile		psi			
			kPa			
Temperature Limits, Service	Short Time		°F	2300	1200	1500
			°C	1260	649	816
	Continuous		°F	2300	1200	1500
			°C	1260	649	816
Melting Temperature			°F	3200		
			°C	1980		
Thermal Diffusivity			ft^2/hr		0.016	0.016
			m^2/hr		4.12×10^{-7}	4.12×10^{-7}
Thermal Shock Resistance				Excellent	Excellent	Excellent
Vibration Resistance				Good	Fair	Fair
Water Absorption			% by Volume	84	90	90
Water Vapor Transmission			perm-inch	Very High	Very High	Very High
			perm-cm	Very High	Very High	Very High

Table 39 Rigid and Semi-Rigid Insulation

Material Identification Number			7	8	9
Composition			Diatomaceous-Silica Type 1: Block & Pipe Covering	Diatomaceous-Silica Type 2: Block & Pipe Covering	Cork Vegetable Board
Abrasion Resistance % of Weight Loss		1st Run	(max.) 55	(max.) 55	
		2nd Run	(max.) 80	(max.) 80	
Alkalinity pH			7 to 9	7 to 9	7
Capillarity			Will Wick	Will Wick	0
Coefficient of Expansion		English	0	0	
		Metric	0	0	
Combustibility			Non-combustible	Non-combustible	Combustible
If Combustible	Flash Point Temp.	°F	"	"	575
		°C	"	"	303
	Flame Point Temp.	°F	"	"	575
		°C	"	"	303
	Heat Release	Btu/lb	"	"	8000 to 9000
		J/kg	"	"	1.8 to 2.0 × 10^7
	Self Internal Heating		"	"	
	Smoldering – Afterglow		"	"	
Corrosion – Rusting			Will not contribute	Will not contribute	Steel should be protected
Corrosion – Stress					Will not contribute
Density		lb/ft^3	(max.) 24	(max.) 26	6 to 8
		kg/m^2	(max.) 384	(max.) 416	96 to 128
Hardness – mm penetration			50 to 60	50 to 60	
Hygroscopicity % Volume			High	High	1
Resistance to Acids					Poor
Resistance to Caustics					Good
Resistance to Solvents					Good
Shrinkage %		Linear	2.5 at 1600°F (871°C)	2.5 at 1900°F (1038°C)	
		Volumetric	5 to 7 at 1600°F (871°C)	5 to 7 at 1900°C (1038°C)	
Specific Gravity		Real	2.2 to 2.3	2.2 to 2.3	
		Apparent	0.34 to 0.40	0.34 to 0.40	0.096 to 0.128
Specific Heat		Btu/lb	0.22 to 0.28	0.22 to 0.28	.43
		w/kg	0.12 to 0.16	0.12 to 0.16	.25
Strength	Compressive	psi	50 at 5% defor	65 at 5% defor	5 at 5% defor
		kPa	345 at 5% defor	448 at 5% defor	34 at 5% defor
	Flexural	psi/m	60	65	15
		kPa	414	448	103
	Modulus of Rupture	psi			
		kPa			
	Tensile	psi			
		kPa			
Temperature Limits, Service	Short Time	°F	1600	1900	200
		°C	871	1038	93
	Continuous	°F	1600	1900	180
		°C	871	1038	80
Melting Temperature		°F			Combustible
		°C			"
Thermal Diffusivity		ft^2/hr	0.008 to 0.01 at 500°F	0.008 to 0.01 at 500°F	0.007 at 70°F
		m^2/hr	2.06 to 2.58 × 10 at 260°C	2.06 × 2.58 × 10 at 260°C	1.08 × 10^{-7} at 21°C
Thermal Shock Resistance			Excellent	Excellent	
Vibration Resistance			Fair	Fair	Good
Water Absorption		% by Volume	85	83	5% (surface only)
Water Vapor Transmission		perm-inch	Very High	Very High	3 to 7
		perm-cm	Very High	Very High	5.4 to 12.6

Table 39 Rigid and Semi-Rigid Insulation

Material Identification Number			10	11	12
Composition			Cork Vegetable Pipe Covering	Cellular glass Block & Pipe Covering	Cellular glass between layers of felt-Board
Abrasion Resistance % of Weight Loss		1st Run		Poor	Poor
		2nd Run		Poor	Poor
Alkalinity pH			7	7 to 8	7 to 8
Capillarity			0	0	0
Coefficient of Expansion		English		46 X 10^{-7}	46 X 10^{-7}
		Metric		83 X 10^{-7}	83 X 10^{-7}
Combustibility			Combustible	Non-combustible	Felt cov. is combustible
If Combustible	Flash Point Temp.	°F	575	"	Felt-550°F Ins. non-combustible
		°C	303	"	Felt-288°C Ins. non-combustible
	Flame Point Temp.	°F	575	"	Felt-575°F Ins. non-combustible
		°C	303	"	Felt-302°C Ins. non-combustible
	Heat Release	Btu/lb	8000 to 9000	"	Ins. non-combustible
		J/kg	1.8 to 2.0 X 10^7	"	Ins. non-combustible
	Self Internal Heating			"	
	Smoldering – Afterglow			"	
Corrosion – Rusting			Steel should be protected	Will not contribute	Will not contribute
Corrosion – Stress			Will not contribute	Will not contribute	Will not contribute
Density		lb/ft³	7 to 14	7.5 to 9.5	7.5 to 9.5
		kg/m²	112 to 224	120 to 152	120 to 152
Hardness – mm penetration					
Hygroscopicity % Volume			1	0	0
Resistance to Acids			Poor	Excellent except HF & H_3PO_4	Excellent except HF & H_3PO_4
Resistance to Caustics			Good	Excellent	Excellent
Resistance to Solvents			Good	Excellent	Excellent
Shrinkage %		Linear		0 at 800°F (427°C)	0 at 800°F (427°C)
		Volumetric		0 at 800°F (427°C)	0 at 800°F (427°C)
Specific Gravity		Real		2.5	2.5
		Apparent	0.096 to 0.128	0.012 to 0.152	0.012 to 0.152
Specific Heat		Btu/lb	.43	0.18	0.18
		w/kg	.25	0.10	0.10
Strength	Compressive	psi	5 at 5% defor.	75 to 100 (max.)	75 to 100 (max.)
		kPa	34 at 5% defor.	517 to 690 (max.)	517 to 690 (max.)
	Flexural	psi/m	18 to 21	60 to 80	60 to 80
		kPa	123 to 144	414 to 552	414 to 552
	Modulus of Rupture	psi			
		kPa			
	Tensile	psi		84	84
		kPa		580	580
Temperature Limits, Service	Short Time	°F	200	940	350
		°C	93	504	427
	Continuous	°F	180	800	350
		°C	80	427	427
Melting Temperature		°F	Combustible	1500	(insul.) 1500
		°C	"	816	816
Thermal Diffusivity		ft²/hr	0.007 at 70°F	0.018 at 70°F	0.018 at 70°F
		m²/hr	1.08 X 10^{-7} at 21°C	1.08 X 10^{-7} at 21°C	1.08 X 10^{-7} at 21°C
Thermal Shock Resistance				Poor	Surface protected by felt
Vibration Resistance			Good	Poor	Surface protected by felt
Water Absorption		% by Volume	3% (surface only)	5% (surface only)	2% (surface only)
Water Vapor Transmission		perm-inch	3 to 7	0.002 at 80°F	0.002 at 80°F
		perm-cm	5.4 to 12.6	0.0036 at 27°C	0.0036 at 27°C

Table 39 Rigid and Semi-Rigid Insulation

Material Identification Number			13	14	15
Composition			Diatomaceous earth-Type 1-Block & pipe Cov.	Diatomaceous earth-Type 2-Block & pipe Cov.	Glass fiber with organic binder-Boards
Abrasion Resistance % of Weight Loss		1st Run	30 to 37		
		2nd Run			
Alkalinity pH			7 to 9	7 to 9	8 to 10
Capillarity			Will Wick	Will Wick	Negligable
Coefficient of Expansion		English	12×10^{-8} on reheat	28×10^{-7} on reheat	
		Metric			
Combustibility			Non-combustible	Non-combustible	Binder is combustible
If Combustible	Flash Point Temp.	°F	"	"	(Binder) 750°F
		°C	"	"	(Binder) 399°C
	Flame Point Temp.	°F	"	"	Smolders
		°C	"	"	"
	Heat Release	Btu/lb	"	"	Fibers Non-combustible
		J/kg	"	"	"
	Self Internal Heating		"	"	
	Smoldering – Afterglow		"	"	Binder will smolder
Corrosion – Rusting			Will not contribute	Will not contribute	Will not contribute
Corrosion – Stress					Will not contribute
Density		lb/ft³	17 to 24	26	1.4 to 1.6
		kg/m²	217 to 384	416	22 to 26
Hardness – mm penetration			0.6 to 0.8	0.5 to 0.6	
Hygroscopicity % Volume			2 to 8	2 to 8	2
Resistance to Acids			Good	Good	Good, except HF & H_3PO_4
Resistance to Caustics			Good	Good	Good from pH 7 to 10
Resistance to Solvents			Resistant	Resistant	Good
Shrinkage %		Linear	2% at 1600°F (870°C)	2% at 1900°F (1040°C)	0 up to 450°F (232°C)
		Volumetric	6 to 18% at 1600°F (870°C)	5 to 7% at 1900°F (1040°C)	0 up to 450°F (232°C)
Specific Gravity		Real	2.2 to 2.3	2.2 to 2.3	2.5
		Apparent	0.38	0.41	0.02
Specific Heat		Btu/lb	.22 to .28	.22 to .28	0.20
		w/kg	.12 to .16	.12 to .16	0.11
Strength	Compressive	psi	50 at 5% deform.	65 at 5% deform.	0.03 at 10% deform.
		kPa	344 at 5% deform.	448 at 5% deform.	0.21 at 10% deform.
	Flexural	psi/m	60	65	
		kPa	413	448	
	Modulus of Rupture	psi			
		kPa			
	Tensile	psi	10 to 20	15 to 20	
		kPa	69 to 137	103 to 137	
Temperature Limits, Service	Short Time	°F	1600	1900	max. 450
		°C	870	1040	max. 232
	Continuous	°F	1600	1900	max. 450
		°C	870	1040	max. 232
Melting Temperature		°F			1200
		°C			649
Thermal Diffusivity		ft²/hr	0.008 to 0.01	0.009 to 0.012	0.04 at 75°F
		m²/hr	2.06×10^{-7} to 2.58×10^{-7}	2.32×10^{-7} to 3.1×10^{-7}	1.03×10^{-6} at 24°C
Thermal Shock Resistance			Excellent	Excellent	Excellent
Vibration Resistance					
Water Absorption		% by Volume	High	High	93
Water Vapor Transmission		perm-inch	150 to 250	High	Very High
		perm-cm	270 to 450		Very High

Table 39 Rigid and Semi-Rigid Insulation

Material Identification Number		16	17	18
Composition		Glass fiber with organic binder-Boards	Glass fiber with organic binder-Boards	Glass fiber w/organic binder-Duct board foil reinforced
Abrasion Resistance % of Weight Loss	1st Run		40	
	2nd Run			
Alkalinity pH		8 to 10	8 to 10	8 to 10
Capillarity		Negligable	Negligable	Negligable
Coefficient of Expansion	English			
	Metric			
Combustibility		Binder is combustible	Binder is combustible	Binder is combustible
If Combustible: Flash Point Temp.	°F	(Binder) 750°F	(Binder) 750°F	(Binder) 750°F
	°C	(Binder) 399°C	(Binder) 399°C	(Binder) 399°C
If Combustible: Flame Point Temp.	°F	Smolders	Smolders	Smolders
	°C	"	"	"
If Combustible: Heat Release	Btu/lb	Fibers Non-combustible	Fibers Non-combustible	Fibers Non-combustible
	J/kg	"	"	"
If Combustible: Self Internal Heating				
If Combustible: Smoldering – Afterglow		Binder will smolder	Binder will smolder	Binder will smolder
Corrosion – Rusting		Will not contribute	Will not contribute	Will not contribute
Corrosion – Stress		Will not contribute	Will not contribute	Will not contribute
Density	lb/ft^3	2.5 to 3.5	5 to 7	2.5 to 3.5
	kg/m^2	40 to 56	80 to 112	40 to 56
Hardness – mm penetration				
Hygroscopicity % Volume		2	2	2
Resistance to Acids		Good, except HF & H_3PO_4	Good, except HF & H_3PO_4	Good, except HF & H_3PO_4
Resistance to Caustics		Good from pH 7 to 10	Good from pH 7 to 10	Good from pH 7 to 10
Resistance to Solvents		Good	Good	Good
Shrinkage %	Linear	0 up to 450°F (232°C)	0 up to 450°F (232°C)	0 up to 250°F (121°C)
	Volumetric	0 up to 450°F (232°C)	0 up to 450°F (232°C)	0 up to 250°F (121°C)
Specific Gravity	Real	2.5	2.5	2.5
	Apparent	o.5	1.0	0.5
Specific Heat	Btu/lb	0.20	0.20	0.20
	w/kg	0.11	0.11	0.11
Strength – Compressive	psi	0.7 at 10% defor.	2.4 at 10% defor.	
	kPa	4.8 at 10% defor.	16.6 at 10% defor.	
Strength – Flexural	psi/m			
	kPa			
Strength – Modulus of Rupture	psi			
	kPa			
Strength – Tensile	psi			
	kPa			
Temperature Limits, Service – Short Time	°F	max. 450	max. 450	max. 250
	°C	max. 232	max. 232	max. 121
Temperature Limits, Service – Continuous	°F	max. 450	max. 450	max. 250
	°C	max. 232	max. 232	max. 121
Melting Temperature	°F	1200	1200	1200
	°C	649	649	649
Thermal Diffusivity	ft^2/hr	0.02 at 75°F	0.015 at 75°F	0.015 at 75°F
	m^2/hr	5.16×10^{-7} at 24°C	3.8×10^{-7} at 24°C	3.87×10^{-7} at 24°C
Thermal Shock Resistance		Excellent	Excellent	Excellent
Vibration Resistance				
Water Absorption	% by Volume	92	93	93
Water Vapor Transmission	perm-inch	Very High	Very High	Depends on covering
	perm-cm	Very High	Very High	Depends on covering

Table 39 Rigid and Semi-Rigid Insulation

Material Identification Number			19	20	21
Composition			Glass fiber-organic binder semi-rigid light weight pipe cov.	Glass fiber with organic binder-heaver density pipe cov.	Mineral fiber (rock slab or glass) Class I block or Board
Abrasion Resistance % of Weight Loss		1st Run	50	40	
		2nd Run	95	90	
Alkalinity pH			8 to 10	8 to 10	7 to 9
Capillarity			Will Wick	Negligable	Will Wick
Coefficient of Expansion		English	0 up to 450°F	0 up to 450°F	Will Shrink
		Metric	0 up to 232°C	0 up to 232°C	Will Shrink
Combustibility			Binder is combustible	Binder is combustible	Binder is combustible
If Combustible	Flash Point Temp.	°F	(Binders) 750°F	(Binders) 750°F	Fibers non-combustible
		°C	(Binders) 399°C	(Binders) 399°C	"
	Flame Point Temp.	°F	Smolders	Smolders	"
		°C	"	"	"
	Heat Release	Btu/lb	Fibers non-combustible	Fibers non-combustible	
		J/kg	"	"	
	Self Internal Heating		Can self heat	Can self heat	Can self heat
	Smoldering – Afterglow		Will smolder	Will smolder	Will smolder
Corrosion – Rusting			Will not contribute	Will not contribute	Will not contribute
Corrosion – Stress			Depends on binder	Depends on binder	Depends on binder
Density		lb/ft^3	3 to 5	5 to 10	9 to 11
		kg/m^2	54 to 80	80 to 160	144 to 176
Hardness – mm penetration					
Hygroscopicity % Volume			2	2	2.5
Resistance to Acids			Good, except HF & H_3PO_4	Good, except HF & H_3PO_4	Good, except HF & H_3PO_4
Resistance to Caustics			Good, pH 7 to 10	Good, pH 7 to 10	Good, pH 7 to 10
Resistance to Solvents			Good	Good	Good
Shrinkage %		Linear	0 up to 450°F (232°C)	0 up to 450°F (232°C)	(max.) 2 up to 400°F (204°C)
		Volumetric	0 up to 450°F (232°C)	0 up to 450°F (232°C)	
Specific Gravity		Real	2.5	2.5	2.5
		Apparent	0.08 (max.)	0.16 (max.)	0.14 (max.)
Specific Heat		Btu/lb	0.20	0.20	0.20
		w/kg	0.12	0.12	0.12
Strength	Compressive	psi			
		kPa			
	Flexural	psi/m			
		kPa			
	Modulus of Rupture	psi			
		kPa			
	Tensile	psi			
		kPa			
Temperature Limits, Service	Short Time	°F	(max.) 500	(max.) 500 to 650	(max.) 400 to 500
		°C	(max.) 260	(max.) 260 to 343	(max.) 204 to 260
	Continuous	°F	(max.) 450 to 500	(max.) 450 to 500	(max.) 400 to 500
		°C	(max.) 232 to 260	(max.) 232 to 260	(max.) 204 to 200
Melting Temperature		°F	1200	1200	1200
		°C	649	649	649
Thermal Diffusivity		ft^2/hr	0.018 at 70°F	0.020 at 70°F	0.030 at 70°F
		m^2/hr	4.6×10^{-7} at 21°C	5.16×10^{-7} at 21°C	7.7×10^{-7} at 21°C
Thermal Shock Resistance			Good	Good	Good
Vibration Resistance			Fair	Fair	Fair
Water Absorption		% by Volume		0.2	
Water Vapor Transmission		perm-inch	Depends on Jacket	Depends on Jacket	Very High
		perm-cm	Depends on Jacket	Depends on Jacket	Very High

Table 39 Rigid and Semi-Rigid Insulation

Material Identification Number		22	23	24
Composition		Mineral fiber (rock slab or glass) Class 2 Block or Board	Mineral fiber Class 3 Block or Board	Mineral fiber Class 4 Block or Board
Abrasion Resistance % of Weight Loss	1st Run			
	2nd Run			
Alkalinity pH		7 to 9	7 to 9	7 to 9
Capillarity		Will Wick	Will Wick	Will Wick
Coefficient of Expansion	English	Will Shrink	Will Shrink	Will Shrink
	Metric	Will Shrink	Will Shrink	Will Shrink
Combustibility		Binder is combustible	Some binders will burn	Some binders oxidize
If Combustible: Flash Point Temp.	°F	Fibers non-combustible	Fibers non-combustible	Fibers non-combustible
	°C	"	"	"
Flame Point Temp.	°F	"	"	"
	°C	"	"	"
Heat Release	Btu/lb			
	J/kg			
Self Internal Heating		Can self heat	Can self heat	Can self heat
Smoldering – Afterglow		Will smolder	Can smolder	Can smolder
Corrosion – Rusting		Will not contribute	Will not contribute	Will not contribute
Corrosion – Stress		Depends on binder	Depends on binder	Will not contribute
Density	lb/ft^3	11 to 13	7 to 12	8 to 13
	kg/m^2	176 to 192	112 to 192	128 to 208
Hardness – mm penetration				
Hygroscopicity % Volume		2.5	2.5	2.5
Resistance to Acids		Good, except HF & H_3PO_4	Good, except HF & H_3PO_4	Good, except, HF & H_3PO_4
Resistance to Caustics		Good pH 7 to 10	Good	Good
Resistance to Solvents		Good	Good	Good
Shrinkage %	Linear	(max.) 2 up to 400°F (204°C)	2.5% at 350°F (434°C)	2% at 1000°F (649°C)
	Volumetric			
Specific Gravity	Real	2.5	2.5	2.5
	Apparent	0.20 (max.)	0.11 to 0.19	0.12 to 0.20
Specific Heat	Btu/lb	0.20	0.22	0.22
	w/kg	0.12	0.13	0.13
Strength: Compressive	psi	25	0.15 to 0.25 at 10% defor.	1.0 to 2.0 at 10% defor.
	kPa	172	2.4 to 4.4 at 10% defor.	33.5 to 67 at 10% defor.
Flexural	psi/m			25
	kPa			172
Modulus of Rupture	psi			
	kPa			
Tensile	psi			
	kPa			
Temperature Limits, Service: Short Time	°F	(max.) 400 to 500	850 to 1000	1000 to 1200
	°C	(max.) 204 to 260	454 to 540	540 to 649
Continuous	°F	(max.) 400 to 500	850 to 1000	1000 to 649
	°C	(max.) 204 to 260	454 to 590	540 to 649
Melting Temperature	°F	1200		
	°C	649		
Thermal Diffusivity	ft^2/hr	0.035 at 70°F	0.01 to 0.07 at 70°F	0.005 to 0.01 at 70°F
	m^2/hr	9.03×10^{-7} at 21°C	2.5 to 5.1×10^{-7} at 21°C	1.25 to 2.6×10^{-7} at 21°C
Thermal Shock Resistance		Good	Good	Good
Vibration Resistance		Fair		
Water Absorption	% by Volume		85 to 93	85 to 93
Water Vapor Transmission	perm-inch	Very High	Very High	Very High
	perm-cm	Very High	Very High	Very High

Table 39 Rigid and Semi-Rigid Insulation

Material Identification Number				25	26	27
Composition				Mineral fiber Class 5 Board & Block	Mineral fiber Class 1 Pipe Cov.	Mineral fiber Class 2 Pipe Cov.
Abrasion Resistance % of Weight Loss			1st Run			
			2nd Run			
Alkalinity pH				7 to 9	7 to 9	7 to 9
Capillarity				Will Wick	Will Wick	Will Wick
Coefficient of Expansion			English	Will Shrink	Will Shrink	Will Shrink
			Metric	Will Shrink	Will Shrink	Will Shrink
Combustibility				Non-combustible	Some binders will burn	Some binders oxidize
If Combustible	Flash Point Temp.		°F	"	Fibers non-combustible	Fibers non-combustible
			°C	"	"	"
	Flame Point Temp.		°F	"	"	"
			°C	"	"	"
	Heat Release		Btu/lb	"	"	"
			J/kg	"	"	"
	Self Internal Heating			"	Can self heat	Can self heat
	Smoldering – Afterglow			"	Can smolder	Can smolder
Corrosion – Rusting				Will not contribute	Will not contribute	Will not contribute
Corrosion – Stress				Will not contribute	Depends on binder	Depends on binder
Density			lb/ft^3	12 to 25	9 to 11	11 to 14
			kg/m^2	192 to 400	144 to 176	176 to 224
Hardness – mm penetration						
Hygroscopicity % Volume				2	2	2
Resistance to Acids				Good, except HF & H_3PO_4	Good, except HF & H_3PO_4	Good, except HF & H_3PO_4
Resistance to Caustics				Good	Good	Good
Resistance to Solvents				Good	Good	Good
Shrinkage %			Linear	5% at 1900°F (1038°C)	2% at 450°F (232°C)	2% at 650°F (345°C)
			Volumetric			
Specific Gravity			Real	2.5	2.5	2.5
			Apparent	0.19 to 0.40	0.11 to 0.17	0.17 to 0.22
Specific Heat			Btu/lb	0.22	0.22	0.22
			w/kg	0.13	0.13	0.13
Strength	Compressive		psi	10 to 15 at 10% defor.		
			kPa	69 to 104 at 10% defor.		
	Flexural		psi/m	40		
			kPa	276		
	Modulus of Rupture		psi			
			kPa			
	Tensile		psi			
			kPa			
Temperature Limits, Service	Short Time		°F	1800 to 1900	450	650
			°C	982 to 1038	232	345
	Continuous		°F	1800 to 1900	450	650
			°C	982 to 1038	232	345
Melting Temperature			°F			
			°C			
Thermal Diffusivity			ft^2/hr	0.004 to 0.005 at 70°F	0.005 to 0.01 at 70°F	0.004 to 0.008 at 70°F
			m^2/hr	$1.03 \text{ to } 2.06 \times 10^{-7}$ at 21°C	$1.25 \text{ to } 2.6 \times 10^{-7}$ at 21°C	$1.03 \text{ to } 2.06 \times 10^{-7}$ at 21°C
Thermal Shock Resistance				Good	Good	Good
Vibration Resistance						Good
Water Absorption			% by Volume	85 to 90	85 to 90	90
Water Vapor Transmission			perm-inch	Very High	Very High	Very High
			perm-cm	Very High	Very High	Very High

Table 39 Rigid and Semi-Rigid Insulation

Material Identification Number			28	29	30
Composition			Mineral Fiber Class 3 Pipe Cov.	Mineral Fiber Class 2 Pipe Cov.	Mineral Fiber Class 3 Pipe Cov.
Abrasion Resistance % of Weight Loss		1st Run		35 to 40	30 to 40
		2nd Run		60 to 70	60 to 70
Alkalinity pH			7 to 9	7 to 8	7 to 8
Capillarity			Will Wick	Negligible	Will Wick
Coefficient of Expansion		English	Will Shrink	Will Shrink	Will Shrink
		Metric	Will Shrink	Will Shrink	Will Shrink
Combustibility			Non-combustible	Non-combustible	Non-combustible
If Combustible	Flash Point Temp.	°F	"	"	"
		°C	"	"	"
	Flame Point Temp.	°F	"	"	"
		°C	"	"	"
	Heat Release	Btu/lb	"	"	"
		J/kg	"	"	"
	Self Internal Heating		"	"	"
	Smoldering – Afterglow		"	"	"
Corrosion – Rusting			Will not contribute	Will not contribute	Will not contribute
Corrosion – Stress			Will not contribute	Will not contribute	Will not contribute
Density		lb/ft^3	16 to 20	13 to 15	12 to 15
		kg/m^2	256 to 320	208 to 240	240
Hardness – mm penetration					
Hygroscopicity % Volume			2	4	16
Resistance to Acids			Good, except HF & M_3PO_4	Excellent	Excellent
Resistance to Caustics			Good	Excellent	Excellent
Resistance to Solvents			Good	Good	Good
Shrinkage %		Linear	2% at 1200°F (650°C)	2% at 1500°F (816°C)	2% at 1500°F (816°C)
		Volumetric			
Specific Gravity		Real	2.5	1.8	1.8
		Apparent	0.25 to 0.32	0.14 to 0.16	0.14 to 0.16
Specific Heat		Btu/lb	0.22	0.216	0.216
		w/kg	0.13	0.13	0.13
Strength	Compressive	psi		60 at 5% deform.	55 at 5% deform.
		kPa		415 at 5% deform.	379 at 5% deform.
	Flexural	psi/m		35	
		kPa		241	
	Modulus of Rupture	psi			
		kPa			
	Tensile	psi		32	
		kPa		221	
Temperature Limits, Service	Short Time	°F	1200	1500	1500
		°C	650	816	816
	Continuous	°F	1200	1500	1500
		°C	650	816	816
Melting Temperature		°F			
		°C			
Thermal Diffusivity		ft^2/hr	0.007 to 0.008 at70°F	0.014 at 70°F	0.014 at 70°F
		m^2/hr	1.8 to 2.06 × 10^{-7} at 21°C	3.6 × 10^{-7} at 21°C	3.6 × 10^{-7} at 21°C
Thermal Shock Resistance			Good	Good	Good
Vibration Resistance			Good	Fair	Fair
Water Absorption		% by Volume	90	9	180
Water Vapor Transmission		perm-inch	Very High	18	40
		perm-cm	Very High	32	72

Table 39 Rigid and Semi-Rigid Insulation

Material Identification Number			31	32	33
Composition			Perlite expanded with binders and fibers Roof board	Phenolic foam Board and pipe Cov.	Polystyrene expanded beads molded to shape block, board pipe Cov.
Abrasion Resistance % of Weight Loss		1st Run	30 to 40		
		2nd Run	65 to 70		
Alkalinity pH			7 to 8	6.5 to 7.5	6.5 to 7.5
Capillarity			Negligable	Negligable	0.5
Coefficient of Expansion		English	0.2 to 0.9% at 50% RH	Shrinks	35×10^{-6}
		Metric	0.2 to 0.9% at 50% RH	Shrinks	63×10^{-6}
Combustibility			Resists combustion	Slow burning	Combustible
If Combustible	Flash Point Temp.	°F		780	645
		°C		416	345
	Flame Point Temp.	°F		850	756
		°C		454	357
	Heat Release	Btu/lb			16,000
		J/kg			3.71×10^{7}
	Self Internal Heating				Melts and burns
	Smoldering – Afterglow				Melst and burns
Corrosion – Rusting			Will not contribute	Will not contribute	Does not contribute
Corrosion – Stress			Will not contribute	Will not contribute	Does not contribute
Density		lb/ft^3	10	2 to 3	1.0 to 1.5
		kg/m^2	160	32 to 48	16.01 to 24.01
Hardness – mm penetration			(Brinnel) 60 (X 1000)		
Hygroscopicity % Volume					0.5 to 1.5
Resistance to Acids			Good	Good	Good to most acids
Resistance to Caustics			Good	Good	Good to most caustics
Resistance to Solvents			Good	Fair	Poor to many solvents
Shrinkage %		Linear			Expands to temp. limits
		Volumetric			"
Specific Gravity		Real	2.2	2.3	
		Apparent	.16	0.3 to 0.5	0.016 to 0.024
Specific Heat		Btu/lb	.25		0.27 at 40°F
		w/kg	.14		0.15 at 4°C
Strength	Compressive	psi	35 at 10% defor.	13 to 22 at 10% defor.	12 at 10% defor.
		kPa	240 at 10% defor.	70 to 151 at 10% defor.	82 at 10% defor.
	Flexural	psi/m	40		
		kPa	276×10^{3}		
	Modulus of Rupture	psi			
		kPa			
	Tensile	psi	4		
		kPa	28		
Temperature Limits, Service	Short Time	°F	200	250	180
		°C	93	121	82
	Continuous	°F	200	250	180
		°C	93	121	82
Melting Temperature		°F			200 to 230
		°C			93 to 110
Thermal Diffusivity		ft^2/hr			0.063 at 40°F
		m^2/hr			1.6×10^{-6} at 4°C
Thermal Shock Resistance					
Vibration Resistance					Excellent
Water Absorption		% by Volume	1.5 max.	0.4 to 0.8	3.8 to 4.0
Water Vapor Transmission		perm-inch	25	High	2.0
		perm-cm	45	High	3.6

Table 39 Rigid and Semi-Rigid Insulation

Material Identification Number			34	35	36
Composition			Polystyrene expanded foam-cut Shapes Board & Pipe Cov.	Polystyrene expanded foam-molded in shape Board & Pipe Cov.	Polystyrene expanded foam-molded in shape board
Abrasion Resistance % of Weight Loss		1st Run			
		2nd Run			
Alkalinity pH			6.5 to 7.5	6.5 to 7.5	6.5 to 7.5
Capillarity			0	0	0
Coefficient of Expansion		English	35×10^{-6}	35×10^{-6}	35×10^{-6}
		Metric	63×10^{-6}	63×10^{-6}	63×10^{-6}
Combustibility			Combustible	Combustible	Combustible
If Combustible	Flash Point Temp.	°F	730	730	730
		°C	388	388	388
	Flame Point Temp.	°F	750	750	750
		°C	399	399	399
	Heat Release	Btu/lb	16,000	16,000	16,000
		J/kg	3.71×10^{7}	3.71×10^{7}	3.71×10^{7}
	Self Internal Heating		Melts and burns	Melts and burns	Melts and burns
	Smoldering – Afterglow		Melts and burns	Melts and burns	Melts and burns
Corrosion – Rusting			Does not contribute	Does not contribute	Does not contribute
Corrosion – Stress			Does not contribute	Does not contribute	Does not contribute
Density		lb/ft^3	1.75 to 2.0	2.0 to 2.2	2.25 at 2.5
		kg/m^2	28.02 to 32.2	32.02 to 35.22	36.02 to 40.03
Hardness – mm penetration					
Hygroscopicity % Volume			0.5 to 1.0	0.6 to 1.1	0.6 to 1.0
Resistance to Acids			Good to most acids	Good to most acids	Good to most acids
Resistance to Caustics			Good to most caustics	Good to most caustics	Good to most caustics
Resistance to Solvents			Poor to many solvents	Poor to many solvents	Poor to many solvents
Shrinkage %		Linear	Expands to temp. limit	Expands to temp. limit	Expands to temp. limit
		Volumetric	"	"	"
Specific Gravity		Real			
		Apparent	0.028 to 0.032	0.032 to 0.035	0.036 to 0.04
Specific Heat		Btu/lb	0.27 at 40°F	0.27 at 40°F	0.27 at 40°F
		w/kg	0.15 at 4°C	0.15 at 4°C	0.15 at 4°C
Strength	Compressive	psi	40 at 10% deform.	40 at 10% deform.	45 at 10% deform.
		kPa	275 at 10% deform.	275 at 10% deform.	310 at 10% deform.
	Flexural	psi/m	40	60	100
		kPa	275	413	689
	Modulus of Rupture	psi			
		kPa			
	Tensile	psi	70	70	85
		kPa	482	482	586
Temperature Limits, Service	Short Time	°F	165	165	165
		°C	74	74	74
	Continuous	°F	165	165	165
		°C	74	74	74
Melting Temperature		°F	200 to 230	200 to 230	200 to 230
		°C	93 to 110	93 to 110	93 to 110
Thermal Diffusivity		ft^2/hr	0.017 at 40°F	0.017 at 40°F	0.017 at 40°F
		m^2/hr	4.4×10^{-7} at 4°C	4.4×10^{-7} at 4°C	4.4×10^{-7} at 4°C
Thermal Shock Resistance					
Vibration Resistance			Excellent	Excellent	Excellent
Water Absorption		% by Volume	0.5 to 0.7	0.6 to 0.8	0.5 to 0.7
Water Vapor Transmission		perm-inch	1.0 to 1.4	0.4 to 1.0	0.4 to 0.8
		perm-cm	1.8 to 2.5	0.7 to 1.8	0.7 to 1.4

Table 39 Rigid and Semi-Rigid Insulation

Material Identification Number			37	38	39
Composition			Polystrene expanded foam molded to shape roof boards	Polyurethane expanded foam Block, Board & Pipe Cov.	Polyurethane expanded foam Block, Board & Pipe Cov.
Abrasion Resistance % of Weight Loss		1st Run		6 to 20	6 to 20
		2nd Run			
Alkalinity pH			6.5 to 7.5	6.5 to 7.5	6.5 to 7.5
Capillarity			0	0	0
Coefficient of Expansion		English	35×10^{-6}	5×10^{-5}	5×10^{-5}
		Metric	63×10^{-6}	9×10^{-5}	9×10^{-5}
Combustibility			Combustible	Combustible	Combustible
If Combustible	Flash Point Temp.	°F	730	590 to 750	590 to 750
		°C	388	310 to 399	310 to 399
	Flame Point Temp.	°F	750	600 to 750	600 to 750
		°C	399	316 to 399	316 to 399
	Heat Release	Btu/lb	16,000	16,000 to 17,000	16,000 to 17,000
		J/kg	3.71×10^{7}	3.71×10^{7} to 3.9×10^{7}	3.71×10^{7} to 3.9×10^{7}
	Self Internal Heating		Melts and burns	Will react with some chemicals	Will react with some chemicals
	Smoldering – Afterglow		Melts and burns	Chars	Chars
Corrosion – Rusting			Will not contribute	Does not contribute	Will contribute
Corrosion – Stress			Does not contribute	Does not contribute	Does not contribute
Density		lb/ft³	2.5 to 2.7	Up to 1.7	1.7 to 2.5
		kg/m²	40.0 to 43.2	Up to 27.2	27.2 to 40.0
Hardness – mm penetration					
Hygroscopicity % Volume			0.6 to 0.7	1.0 to 1.8	0.9 to 1.6
Resistance to Acids			Good to most acids	Resistant, to most dilute acids	Resistant to dilute acids
Resistance to Caustics			Good to most caustics	Resistant to most dilute caustics	Resistant to dilute caustics
Resistance to Solvents			Poor to many solvents	Resistant to most solvents	Resistant to most solvents
Shrinkage %		Linear	Expands to temp. limit	Expands to temp. limit	Expands to temp. limit
		Volumetric	Exp Expands to temp. limit	Expands to temp. limit	Expands to temp. limit
Specific Gravity		Real			
		Apparent	0.04 to 0.043	0.027	0.027 to 0.04
Specific Heat		Btu/lb	0.27 at 40°F	0.4 at 40°F	0.4 at 40°F
		w/kg	0.15 at 4°C	0.22 at 4°C	0.22 at 4°C
Strength	Compressive	psi	60 at 10% deform.	15 at 10% deform.	17 at 10% deform.
		kPa	413 at 10% deform.	103 at 10% deform.	117 at 10% deform.
	Flexural	psi/m	125	15	20
		kPa	707	105	140
	Modulus of Rupture	psi			
		kPa			
	Tensile	psi	200	10 to 30	15 to 50
		kPa	1378	69 to 205	130 to 344
Temperature Limits, Service	Short Time	°F	165	230	220
		°C	74	110	104
	Continuous	°F	165	200	200
		°C	74	94	94
Melting Temperature		°F	200 to 230	Chars	Chars
		°C	93 to 110	Chars	Chars
Thermal Diffusivity		ft²/hr	0.017 at 40°F	0.011 to 0.15 at 40°F	0.015 to 0.02 at 40°F
		m²/hr	4.4×10^{-7} at 4°C	2.8×10^{-7} to 3.6×10^{-7} at 4°C	3.6×10^{-7} to 5.1×10^{-7} at 4°C
Thermal Shock Resistance					
Vibration Resistance			Excellent	Excellent	Excellent
Water Absorption		% by Volume	0.5 to 0.7	1.5	1.3
Water Vapor Transmission		perm-inch	0.5 to 0.8	4 to 6	3 to 5
		perm-cm	0.9 to 1.4	6.7 to 10	5 to 8.4

Table 39 Rigid and Semi-Rigid Insulation

Material Identification Number			40	41	42
Composition			Polyurethane expanded foam Block, Board & Pipe Cov.	Expanded rubber Rigid Board	Expanded silica with binders-Block & Pipe Cov.
Abrasion Resistance % of Weight Loss		1st Run	6 to 20		35 to 40
		2nd Run			65 to 70
Alkalinity pH			6.5 to 7.5		8 to 9.5
Capillarity			0	10%	
Coefficient of Expansion		English	5×10^{-5}	Shrinks	Shrinks
		Metric	9×10^{-5}	Shrinks	Shrinks
Combustibility			Combustible	Combustible	Non-combustible
If Combustible	Flash Point Temp.	°F	590 to 750	560	"
		°C	310 to 399	293	"
	Flame Point Temp.	°F	600 to 750	590	"
		°C	316 to 399	310	"
	Heat Release	Btu/lb	16,000 to 17,000	15,000	"
		J/kg	3.71×10^7 to 3.9×10^7	3.486×10^7	"
	Self Internal Heating		Will react with some chemicals		"
	Smoldering – Afterglow		Chars		"
Corrosion – Rusting			Will contribute	Does not contribute	Does not contribute
Corrosion – Stress			Does not contribute	Does not contribute	Does not contribute
Density		lb/ft³	2.5 to 5.0	4 to 6	13
		kg/m²	40 to 80	64 to 96	208
Hardness – mm penetration					
Hygroscopicity % Volume			0.8 to 1.4		0.5
Resistance to Acids			Resistant to dilute acid	Good	Excellent
Resistance to Caustics			Resistant to dilute caustic	Good	Excellent
Resistance to Solvents			Resistant to most solvents	Poor	Good
Shrinkage %		Linear	Expands to temp. limit	10% at 200°F	1% at 1500°F
		Volumetric	Expands to temp. limit		
Specific Gravity		Real			2.2
		Apparent	0.04 to 0.08	0.07	0.22
Specific Heat		Btu/lb	0.4 at 40°F	0.20	0.14 to 0.18
		w/kg	0.22 at 4°C		0.08 to 0.10
Strength	Compressive	psi	30 at 10% deform.	40 at 10% deform.	82 at 5% deform.
		kPa	207 at 10% deform.	275 at 10% deform.	565 at 5% deform.
	Flexural	psi/m	40		46
		kPa	275		179
	Modulus of Rupture	psi			
		kPa			
	Tensile	psi	20 to 108	60	
		kPa	137 to 670	413	
Temperature Limits, Service	Short Time	°F	220	200	1500
		°C	104	93	815
	Continuous	°F	200	200	1500
		°C	94	93	815
Melting Temperature		°F	Chars		
		°C	Chars		
Thermal Diffusivity		ft²/hr	0.02 to 0.03 at 40°F		0.015
		m²/hr	5.1×10^7 to 7.7×10^7 at 4°C		3.8×10^7
Thermal Shock Resistance					Excellent
Vibration Resistance			Excellent		Good
Water Absorption		% by Volume	1.1	1	
Water Vapor Transmission		perm-inch	2.5 to 5	0.1	High
		perm-cm	4.2 to 8.4	0.2	High

Table 39 Rigid and Semi-Rigid Insulation

Material Identification Number			43	44
Composition			Wood fibers with binders-Board	Borosilica closed cell inorganic foam-Block
Abrasion Resistance % of Weight Loss		1st Run		Poor abrasion
		2nd Run		Resistance
Alkalinity pH			4.5 to 6	6 to 7
Capillarity				None
Coefficient of Expansion		English		1.6×10^{-6} / °F
		Metric		2.8×10^{-8} / °C
Combustibility			Combustible	Non-combustible
If Combustible	Flash Point Temp.	°F	500	"
		°C	260	"
	Flame Point Temp.	°F	500	"
		°C	260	"
	Heat Release	Btu/lb	8,000	"
		J/kg	1.859×10^7	"
	Self Internal Heating			"
	Smoldering – Afterglow			"
Corrosion – Rusting				Does not contribute
Corrosion – Stress				Does not contribute
Density		lb/ft³	15 to 18	12
		kg/m²	240 to 288	190
Hardness – mm penetration				
Hygroscopicity % Volume			14	0
Resistance to Acids				Impervious
Resistance to Caustics				Impervious
Resistance to Solvents				Impervious
Shrinkage %		Linear	5.4% at 200° F	None
		Volumetric		None
Specific Gravity		Real	0.95	2.3
		Apparent	0.24	0.19
Specific Heat		Btu/lb		0.2
		w/kg		0.11
Strength	Compressive	psi		200 to 210
		kPa		1400 to 1970
	Flexural	psi/m		90 to 120
		kPa		580 to 770
	Modulus of Rupture	psi	50	
		kPa	344	
	Tensile	psi		
		kPa		
Temperature Limits, Service	Short Time	°F	200	960 (unloaded)
		°C	93	517 (unloaded)
	Continuous	°F	200	800 (with applied load)
		°C	93	425 (with applied load)
Melting Temperature		°F		3100
		°C		1704
Thermal Diffusivity		ft²/hr		
		m²/hr		
Thermal Shock Resistance				Excellent
Vibration Resistance			Fair	Fair
Water Absorption		% by Volume	10	2% (surface only)
Water Vapor Transmission		perm-inch	5	0.00000+
		perm-cm	9	0.00000+

Table 40 Flexible and Faced Insulation

Material Identification Number			45	46
Composition			Elastomeric sheet and pipe covering	Glass fiber, flexible with various facings
Alkalinity pH				7 to 8
Capillarity			Negligable	Will Wick
Combustibility			Combustible	Binder is combustible
If Combustible	Flash Point Temp.	°F	560	(Binders) 750
		°C	293	(Binders) 399
	Flame Point Temp.	°F	590	Smolders
		°C	310	Smolders
	Heat Release	Btu/lb	15000	Fibers non-combustible
		J/kg	3.486×10^7	Fibers non-combustible
	Self Internal Heating			
	Smoldering or Afterglow			Can Smolder
Corrosion – Rusting			Does not contribute	Does not contribute
Corrosion – Stress			Does not contribute	Does not contribute
Density of Insulation (only)		lb/ft^3	4.5 to 8.5	0.6 to 3.0
		kg/m^3	36 to 72	10.4 to 48.0
Hygroscopicity % Weight				
Maximum Temperature Limits	Short	°F	220	400
		°C	104	204
	Continuous	°F	220	400
		°C	104	204
Resistance to Acids			Resistant to dilute acids	Good from pH 5 to 7 except HF & H_3PO_4
Resistance to Caustics			Resistant to dilute acids	Good from pH 7 to pH 10
Specific Heat		Btu/lb		.20
		w/kg		.11
Strength Compression		psi		902 to 0.09 at 10% deform.
		Pa		
Thermal Diffusivity		ft^2/hr		0.13 to 0.62 at 10% deform.
		m^2/hr		0.06 at 70°F
Water Absorption % Weight			4.6 to 8.4	9.126×10^{-6} at 21°C
Water Vapor Transmission		perm-inches	0.15 to 0.30	Depends on facing
		perm-cm	0.27 to 0.54	Depends on facing

Material Identification Number			47	48
Composition			Glass fiber flexible with various facings**	Glass fiber, flexible semi-rigid w/various facings**
Alkalinity pH			7 to 8	7 to 8
Capillarity			Will Wick	Will Wick
Combustibility			Binder is combustible	Binder is combustible
If Combustible	Flash Point Temp.	°F	(Binder) 750°F	(Binder) 750°F
		°C	(Binder) 399	(Binder)
	Flame Point Temp.	°F	Smolders	Smolders
		°C	Smolders	Smolders
	Heat Release	Btu/lb	Fibers non-combustible	Fibers non-combustible
		J/kg	Fibers non-combustible	Fibers non-combustible
	Self Internal Heating			
	Smoldering or Afterglow		Can Smolder	Can Smolder
Corrosion – Rusting			Does not contribute	Does not contribute
Corrosion – Stress			Depends on binder	Depends on binder
Density of Insulation (only)		lb/ft^3	3 to 4.5	4.5 (min.)
		kg/m^3	48 to 72	72 (min.)
Hygroscopicity % Weight				
Maximum Temperature Limits	Short	°F	400	400
		°C	204	204
	Continuous	°F	400	400
		°C	204	204
Resistance to Acids			Good from pH 5 to 7 except HF & H_3PO_4	Good from pH 5 to 7 except HF & H_3PO_4
Resistance to Caustics			Good from pH 7 to pH 10	Good from pH 7 to pH 10
Specific Heat		Btu/lb	.20	.20
		w/kg	.11	.11
Strength Compression		psi	0.5 at 10% deform.	0.5 to 1.5 at 10% deform.
		Pa	3.45 at 10% deform.	3.45 to 10.35 at 10% deform.
Thermal Diffusivity		ft^2/hr	0.16 at 70°F	0.05 to 0.15 at 70°F
		m^2/hr	4.126×10^{-6} at 21°C	1.25 to 3.75×10^{-6} at 21°C
Water Absorption % Weight				
Water Vapor Transmission		perm-inches	Depends on facing	Depends on facing
		perm-cm	Depends on facing	Depends on facing

**Facings may be woven wire mesh, expanded metal lath, (copper bearing) on one or both sides.

Table 40 Flexible and Faced Insulation

Material Identification Number			49	50
Composition			Kaolin ceramic fiber in ss* metal mesh	Kaolin ceramic fiber in ss* metal mesh
Alkalinity pH				
Capillarity			Will Wick	Will Wick
Combustibility			Non-combustible	non-combustible
If Combustible	Flash Point Temp.	°F	"	"
		°C	"	"
	Flame Point Temp.	°F	"	"
		°C	"	"
	Heat Release	Btu/lb	"	"
		J/kg	"	"
	Self Internal Heating		"	"
	Smoldering or Afterglow		"	"
Corrosion – Rusting			Does not contribute	Does not contribute
Corrosion – Stress			Does not contribute	Does not contribute
Density of Insulation (only)		lb/ft³	3 to 4	6 to 8
		kg/m³	48 to 64	96 to 128
Hygroscopicity % Weight				
Maximum Temperature Limits	Short	°F	3000	3000
		°C	1649	1649
	Continuous	°F	2300	2300
		°C	1260	1260
Resistance to Acids			Fair except HF & H_3PO_4	Fair except HF & H_3PO_4
Resistance to Caustics			Fair to Poor	Fair to Poor
Specific Heat		Btu/lb	0.285 at 1800°F	0.285 at 1800°F
		w/kg	0.158 at 982°C	0.158 at 982°C
Strength Compression		psi		
		Pa		
Thermal Diffusivity		ft²/hr		
		m²/hr		
Water Absorption % Weight				
Water Vapor Transmission		perm-inches	High	High
		perm-cm	High	High

Material Identification Number			53	54
Composition			Mineral fibers w/metal mesh facing** Class II	Urethane foam, block, pipe cov.
Alkalinity pH			7 to 9	6.5 to 7.5
Capillarity			Will Wick	Negligable
Combustibility			Non-combustible	Combustible
If Combustible	Flash Point Temp.	°F	"	590 to 750
		°C	"	310 to 399
	Flame Point Temp.	°F	"	600 to 750
		°C	"	316 to 300
	Heat Release	Btu/lb	"	15,000
		J/kg	"	3.71×10^7
	Self Internal Heating		"	Will react with various liquids and gases
	Smoldering or Afterglow		"	Chars
Corrosion – Rusting			Will not contribute	Will not contribute
Corrosion – Stress			Will not contribute	Will not contribute
Density of Insulation (only)		lb/ft³	12 (ave. max.)	Up to 1.5
		kg/m³	192 (ave. max.)	Up to 24
Hygroscopicity % Weight			1.5	
Maximum Temperature Limits	Short	°F	1200	210
		°C	650	99
	Continuous	°F	1200	185
		°C	650	85
Resistance to Acids			Poor	Resistant to dilute acids
Resistance to Caustics			Poor	Resistant to dilute caustics
Specific Heat		Btu/lb	.22	.25
		w/kg	.12	.14
Strength Compression		psi	25 at 10% deform.	8 at 5% deform.
		Pa	172 at 10% deform.	5.5 at 5% deform.
Thermal Diffusivity		ft²/hr	0.22 at 70°F	0.011 to 0.027 at 70°F
		m²/hr	5.67×10^{-6} at 21°C	2.83×10^{-7} to 6.93×10^{-7} at 21°C
Water Absorption % Weight			96	5
Water Vapor Transmission		perm-inches	High	5
		perm-cm	High	8.3

* ss = Stainless steel metal mesh.
** Facings may be woven wire mesh, expanded metal lath (copper bearing) on one, or both sides.

Table 40 Flexible and Faced Insulation

Material Identification Number			51	52
Composition			Mineral fibers-spun glass fibers metal mesh**	Mineral fibers w/metal mesh facing-** Class I
Alkalinity pH			7 to 8	7 to 9
Capillarity			Will Wick	Will Wick
Combustibility			Non-combustible	Non-combustible
If Combustible	Flash Point Temp.	°F	"	"
		°C	"	"
	Flame Point Temp.	°F	"	"
		°C	"	"
	Heat Release	Btu/lb	"	"
		J/kg	"	"
	Self Internal Heating		"	"
	Smoldering or Afterglow		"	"
Corrosion – Rusting			Does not contribute	Does not contribute
Corrosion – Stress			Does not contribute	Does not contribute
Density of Insulation (only)		lb/ft^3	10 to 14	6 (ave. max.)
		kg/m^3	160 to 224	96 (ave. max.)
Hygroscopicity % Weight			1.5	2.0
Maximum Temperature Limits	Short	°F	1200	1000
		°C	649	540°C
	Continuous	°F	1200	1000
		°C	649	540°C
Resistance to Acids			Fair to Poor	Fair to Poor
Resistance to Caustics			Fair to Poor	Fair to Poor
Specific Heat		Btu/lb	.20	.22
		w/kg	.11	.12
Strength Compression		psi		10 at 10% deform.
		Pa		69 at 10% deform.
Thermal Diffusivity		ft^2/hr	0.02 to 0.04 at 70°F	0.02 to 0.04 at 70°F
		m^2/hr	5.16×10^{-7} to 1.03×10^{-6} at 21°C	5.16×10^{-7} to 1.03×10^{-7} at 21°C
Water Absorption % Weight				96
Water Vapor Transmission		perm-inches	High	High
		perm-cm	High	High

Material Identification Number			55	56
Composition			Urethane foam, block, pipe cov.	Alumina-silica fiber, blanket and strip
Alkalinity pH			6.5 to 7.5	
Capillarity			Negligable	Will Wick
Combustibility			Combustible	Non-combustible
If Combustible	Flash Point Temp.	°F	590 to 750	"
		°C	310 to 750	"
	Flame Point Temp.	°F	600 to 750	"
		°C	316 to 399	"
	Heat Release	Btu/lb	16,000	"
		J/kg	3.71×10^7	"
	Self Internal Heating		Will react with various liquids and gases	"
	Smoldering or Afterglow		Chars	"
Corrosion – Rusting			Will contribute	Does not contribute
Corrosion – Stress			Will contribute	Does not contribute
Density of Insulation (only)		lb/ft^3	Over 1.5	3 to 4
		kg/m^3	Over 24	48 to 64
Hygroscopicity % Weight			210	3
Maximum Temperature Limits	Short	°F	99	2300
		°C	185	1260
	Continuous	°F	85	2300
		°C	.24	1260
Resistance to Acids			Resistant to dilute acids	Good, except HF & H_3PO_4
Resistance to Caustics			Resistant to dilute caustics	Good to poor
Specific Heat		Btu/lb	.25	0.25
		w/kg	.14	0.14
Strength Compression		psi	30 at 5% deform.	
		Pa	205 at 5% deform.	
Thermal Diffusivity		ft^2/hr	0.011 to 0.027	
		m^2/hr	2.83×10^{-7} to 6.99×10^{-7}	
Water Absorption % Weight			2	Saturates
Water Vapor Transmission		perm-inches	1.5	High
		perm-cm	2.5	High

* ss = Stainless steel metal mesh.
** Facings may be woven wire mesh, expanded metal lath (copper bearing) on one, or both sides.

Table 40 Flexible and Faced Insulation

Material Identification Number			57	58
Composition			Alumina-silica fiber, blanket and strip	Alumina silica fiber, blanket and strip
Alkalinity pH				
Capillarity			Will Wick	Will Wick
Combustibility			Non-combustible	Non-combustible
If Combustible	Flash Point Temp.	°F	"	"
		°C	"	"
	Flame Point Temp.	°F	"	"
		°C	"	"
	Heat Release	Btu/lb	"	"
		J/kg	"	"
	Self Internal Heating		"	"
	Smoldering or Afterglow		"	"
Corrosion — Rusting			Does not contribute	Does not contribute
Corrosion — Stress			Does not contribute	Does not contribute
Density of Insulation (only)		lb/ft^3	6 to 8	12
		kg/m^3	96 to 128	192
Hygroscopicity % Weight			3	3
Maximum Temperature Limits	Short	°F	2300	2300
		°C	1260	1260
	Continuous	°F	2300	2300
		°C	1260	1093
Resistance to Acids			Good, except HF & H_3PO_4	Good, except HF & H_3PO_4
Resistance to Caustics			Fair to poor	Fair to poor
Specific Heat		Btu/lb	0.25	0.25
		w/kg	0.14	0.14
Strength Compression		psi		
		Pa		
Thermal Diffusivity		ft^2/hr		
		m^2/hr		
Water Absorption % Weight			Saturates	Saturates
Water Vapor Transmission		perm-inches	High	High
		perm-cm	High	High

Material Identification Number			61	62
Composition			Glass fine-fiber with binder, Blanket	Glass fiber, with binder
Alkalinity pH			6 to 10	7 to 9
Capillarity			None	Negligable
Combustibility			Non-combustible	Binder-Combustible
If Combustible	Flash Point Temp.	°F	650 to 700	650 to 700
		°C	343 to 371	343 to 371
	Flame Point Temp.	°F	1100	1100
		°C	593	593
	Heat Release	Btu/lb	Only binder burns	Only binder burns
		J/kg	Only binder burns	Only binder burns
	Self Internal Heating			
	Smoldering or Afterglow		Smolders- (Binder only)	Smolders- (Binder only)
Corrosion — Rusting			Will not contribute	Will not contribute
Corrosion — Stress			s.s. should be protected	s.s. should be protected
Density of Insulation (only)		lb/ft^3	1/2 to 2	1 to 2
		kg/m^3	8 to 32	16 to 32
Hygroscopicity % Weight			0.2	0.2
Maximum Temperature Limits	Short	°F	370	400
		°C	188	204
	Continuous	°F	370	400
		°C	188	204
Resistance to Acids			Good, except HF & H_3PO_4	Good, except HF & H_3PO_4
Resistance to Caustics			Good from pH 7 to pH 10	Good from pH 7 to pH 10
Specific Heat		Btu/lb	0.20	0.20
		w/kg	0.11	0.11
Strength Compression		psi	*0.02 at 10% deform.	4.5 at 10% deform.
		Pa	**0.13 at 10% deform.	31 at 10% deform.
Thermal Diffusivity		ft^2/hr	0.10 at 100°F (1 lb)	6.10 at 100°F
		m^2/hr	2.58×10^{-6} at 38°C	2.58×10^{-6} at 38°C
Water Absorption % Weight			98	98
Water Vapor Transmission		perm-inches	High	High
		perm-cm	High	High

* 2 pound per cubic foot density.
** 32 kg/m^3 density.

Table 40 Flexible and Faced Insulation

Material Identification Number			59	60
Composition			Alumina silica fiber, rope, cord, yarn	Glass, fine fiber (no binder) Blanket and Matt
Alkalinity pH				6 to 8
Capillarity			Will Wick	None
Combustibility			Non-combustible	Non-combustible
If Combustible	Flash Point Temp.	°F	"	"
		°C	"	"
	Flame Point Temp.	°F	"	"
		°C	"	"
	Heat Release	Btu/lb	"	"
		J/kg	"	"
	Self Internal Heating		"	"
	Smoldering or Afterglow		"	"
Corrosion — Rusting			Does not contribute	Will not contribute
Corrosion — Stress			Does not contribute	Will not contribute
Density of Insulation (only)		lb/ft^3	25	11 to 12
		kg/m^3	400	176 to 192
Hygroscopicity % Weight			3	5
Maximum Temperature Limits	Short	°F	2300	1200
		°C	1260	649
	Continuous	°F	2000	1200
		°C	1093	649
Resistance to Acids			Good except HF & H_3PO_4	Excellent except, HF & H_3PO_4
Resistance to Caustics			Fair to Poor	Good from pH 7 to pH 10
Specific Heat		Btu/lb	0.25	0.20
		w/kg	0.14	0.11
Strength Compression		psi		20 at 63% deform.
		Pa		137 at 63% deform.
Thermal Diffusivity		ft^2/hr		0.08 at 150°F
		m^2/hr		2.06×10^{-6} at 66°C
Water Absorption % Weight			Saturates	93
Water Vapor Transmission		perm-inches	High	High
		perm-cm	High	High

Material Identification Number			63	64
Composition			Glass fiber with binder	Mineral fiber, **Blanket & Batt**
Alkalinity pH			7 to 9	7 to 9
Capillarity			Negligable	Will Wick
Combustibility			Binder-Combustible	*Non-combustible
If Combustible	Flash Point Temp.	°F	650 to 700	Fibers-non-combustible
		°C	343 to 371	"
	Flame Point Temp.	°F	1100	"
		°C	593	"
	Heat Release	Btu/lb	Only binder burns	"
		J/kg	Only binder burns	"
	Self Internal Heating			*Depends on manufacturer
	Smoldering or Afterglow		Smolders- (Binder only)	*Depends on manufacturer
Corrosion — Rusting			Will not contribute	Will not contribute
Corrosion — Stress			s.s. should be protected	Will not contribute
Density of Insulation (only)		lb/ft^3	2.5 to 3.5	8 to 12
		kg/m^3	40 to 56	128 to 192
Hygroscopicity % Weight			0.2	0.5 to 1.0
Maximum Temperature Limits	Short	°F	400	1200
		°C	204	649
	Continuous	°F	400	1200
		°C	204	649
Resistance to Acids			Good, except HF & H_3PO_4	Poor to fair
Resistance to Caustics			Good from pH 7 to pH 10	Poor to good
Specific Heat		Btu/lb	0.20	0.20 to 0.22
		w/kg	0.11	0.11 to 0.12
Strength Compression		psi	100 at 10% deform.	4 to 6 at 10% deform.
		Pa	689 at 10% deform.	25 to 41 at 10% deform.
Thermal Diffusivity		ft^2/hr	0.35 at 100°F	0.012 at 200°F
		m^2/hr	9.03×10^{-6} at 38°C	3.09×10^{-7} at 93°C
Water Absorption % Weight			96	90
Water Vapor Transmission		perm-inches	High	No resistance
		perm-cm	High	No resistance

* Some oil is used to lubricate fibers. Incorrectly used oxidation can cause self internal heating and smoldering.

Table 40 Flexible and Faced Insulation

Material Identification Number			**65**	**66**
Composition			Mineral fibers with binders, **Blanket**	Silica fibers, Blankets & Felts
Alkalinity pH			7 to 9	5 to 7
Capillarity			Will Wick	Will Wick
Combustibility			*Non-combustible	Non-combustible
If Combustible	Flash Point Temp.	°F	Fibers- non combustible	"
		°C	"	"
	Flame Point Temp.	°F	"	"
		°C	"	"
	Heat Release	Btu/lb	"	"
		J/kg	"	"
	Self Internal Heating		*Depends on binders	"
	Smoldering or Afterglow		*Depends on binders	"
Corrosion – Rusting			Will not contribute	
Corrosion – Stress			Will not contribute	
Density of Insulation (only)		lb/ft³	7 to 9	6 to 9.5
		kg/m³	112 to 144	96 to 152
Hygroscopicity % Weight			2	10
Maximum Temperature Limits	Short	°F	1000	1800
		°C	540	982
	Continuous	°F	1000	1800
		°C	540	982
Resistance to Acids			Poor to Fair	Good except HF & H_3PO_4
Resistance to Caustics			Poor to Good	Poor
Specific Heat		Btu/lb	0.24	0.19
		w/kg	0.13	0.10
Strength Compression		psi	120 at 10% deform.	10 at 50% deform.
		Pa	703 at 10% deform.	69 at 50% deform.
Thermal Diffusivity		ft^2/hr	0.01 to 0.026 at 200°F	0.039 to 0.054 at 1000°F
		m^2/hr	2.6×10^{-7} to 6.7×10^{-7} at 93°C	1.0×10^{-6} to 1.39×10^{-6} at 540°C
Water Absorption % Weight			93	High
Water Vapor Transmission		perm-inches	No resistance	No resistance
		perm-cm	No resistance	No resistance

* Some oil is used to lubricate fibers. Incorrectly used oxidation can cause self internal heating and smoldering.

Table 41 Insulating and Finishing Cements

Material Identification Number			67	68	69
Composition			Alumina-silica semi-refractory hydraulic setting	Alumina-silica ceramic fiber, air setting mold mix cement	Alumina-silica ceramic-spray mix cement
Abrasion Resistance % of Weight Loss		1st Run			
		2nd Run			
Alkalinity pH			9 to 10	9.7	9 to 10
Capillarity			Will Wick	Will Wick	Will Wick
Coefficient of Expansion		English	Shrinks 2% at 1800°F	Shrinks 2.1% at 1800°F	Shrinks 1.4% at 1800°F
		Metric	Shrinks 2% at 987°C	Shrinks 2.1% at 987°C	Shrinks 1.4% at 982°C
Combustibility			Non-combustible	Non-combustible	Non-combustible
If Combustible	Flash Point Temp.	°F	"	"	"
		°C	"	"	"
	Flame Point Temp.	°F	"	"	"
		°C	"	"	"
	Heat Release	Btu/lb	"	"	"
		w/kg	"	"	"
	Self Internal Heating		"	"	"
	Smoldering — Afterglow		"	"	"
Corrosion — Rusting					
Corrosion — Stress					
Coverage Dried or Set		board ft/100 lbs	15 to 25		7 to 8
		m^2 cm/100 kgs	7.8 to 13		6.2 to 6.8
Density Dried or Set		lbs/ft^3	48 to 70	75	12 to 13
		kgs/m^3	768 to 1120	1200	192 to 208
Hardness — mm penetration					
Hygroscopicity % Weight					
Maximum Temperature Limits	Short	°F	2000	2300	2300
		°C	1093	1257	1257
	Continuous	°F	2000	2300	1800
		°C	1093	1257	982
Shrinkage Wet to Dry		Volumetric			
Specific Gravity		Real			
		Apparent	.77 to 1.12	1.2	0.21
Specific Heat		Btu/lb	.27	.27	.27
		w/kg	.15	.15	.15
Strength	Compressive	psi		6500	0.6 at 5% deform.
		kPa		451,500	4.1 at 5% deform.
	Shear	psi			
		kPa			
	Tensile	psi			
		kPa			
Thermal Diffusivity		ft^2/hr			
		m^2/hr			
Thermal Shock Resistance			Excellent	Excellent	Excellent
Vibration Resistance					
Water Absorption % Weight			High	High	High
Water Vapor Transmission		perm-inch	High	High	High
		perm-cm	High	High	High

Table 41 Insulating and Finishing Cements

Material Identification Number			70	71	72
Composition			Diatomateous silica & binders air setting cement	Kaowool ceramic fiber & inorganic binders-spray cement	Kaowool and mineral fiber blend-spray cement
Abrasion Resistance % of Weight Loss		1st Run	10		
		2nd Run	20		
Alkalinity pH			9		
Capillarity			Will Wick		
Coefficient of Expansion		English	Shrinks 5% at 1900°F	Shrinks 1.5% at 2000°F	
		Metric	Shrinks 5% at 1040°C	Shrinks 1.5% at 1093°C	
Combustibility			Non-combustible	Non-combustible	Non-combustible
If Combustible	Flash Point Temp.	°F	"	"	"
		°C	"	"	"
	Flame Point Temp.	°F	"	"	"
		°C	"	"	"
	Heat Release	Btu/lb	"	"	"
		w/kg	"	"	"
	Self Internal Heating		"	"	"
	Smoldering – Afterglow		"	"	"
Corrosion – Rusting					
Corrosion – Stress			s.s. should be protected		
Coverage Dried or Set		board ft/100 lbs	35	100	170 to 180
		m^2 cm/100 kgs	18		
Density Dried or Set		lbs/ft^3	23 to 32	12 to 13	7 to 15
		kgs/m^3	368 to 512	192 to 206	112 to 288
Hardness – mm penetration			0.95 to 1.20		
Hygroscopicity % Weight					
Maximum Temperature Limits	Short	°F	1900	2000	800 to 1600
		°C	1040	1093	427 to 871
	Continuous	°F	1900	1000	800 to 1600
		°C	1040	982	427 to 871
Shrinkage Wet to Dry		Volumetric	30%		
Specific Gravity		Real	2.3		
		Apparent	0.37 to 51		
Specific Heat		Btu/lb	0.22 to 0.28		
		w/kg	0.12 to 0.16		
Strength	Compressive	psi	25	0.6 at 5% deform.	55 at 5% deform.
		kPa	173		2630 at 5% deform.
	Shear	psi			
		kPa			
	Tensile	psi			
		kPa			
Thermal Diffusivity		ft^2/hr	0.01		
		m^2/hr	2.58×10^{-7}		
Thermal Shock Resistance			Excellent	Excellent	Excellent
Vibration Resistance			Poor		
Water Absorption % Weight			40 to 50		
Water Vapor Transmission		perm-inch	High	High	High
		perm-cm	High	High	High

Table 41 Insulating and Finishing Cements

Material Identification Number			73	74	75
Composition			Mineral fiber water mix insulating cement	Mineral fiber binders hydraulic setting cement	Vermiculte expanded with binders water mix-air setting
Abrasion Resistance % of Weight Loss		1st Run	25 to 45	15 to 35	35
		2nd Run	50 to 75	25 to 65	60
Alkalinity pH			8.5 to 11	8½ to 12½	6.9
Capillarity			High	High	High
Coefficient of Expansion		English	Shrinks 3% to 1800°F	Shrinks 2% to 1200°F	Shrinks 3% at 1800°F
		Metric	Shrinks 3% to 980°C	Shrinks 2% to 689°C	Shrinks 3% at 982°C
Combustibility			Non-combustible	Non-combustible	Non-combustible
If Combustible	Flash Point Temp.	°F	"	"	"
		°C	"	"	"
	Flame Point Temp.	°F	"	"	"
		°C	"	"	"
	Heat Release	Btu/lb	"	"	"
		w/kg	"	"	"
	Self Internal Heating		"	"	"
	Smoldering — Afterglow		"	"	"
Corrosion — Rusting			Holds Water	Holds Water	Holds Water
Corrosion — Stress			s.s. must be protected	s.s. must be protected	
Coverage Dried or Set		board ft/100 lbs	40	30	65
		m^2 cm/100 kgs	21	15.3	34
Density Dried or Set		lbs/ft^3	22 to 30	27 to 44	18 to 19
		kgs/m^3	352 to 512	432 to 704	288 to 304
Hardness — mm penetration			1 to 2.5		
Hygroscopicity % Weight			2.5 to 5	0.8 to 2.0	
Maximum Temperature Limits	Short	°F	1800	1200	1800
		°C	980	650	980
	Continuous	°F	1800	1200	1800
		°C	980	650	980
Shrinkage Wet to Dry		Volumetric	25%	10%	20%
Specific Gravity		Real	2.3	2.5	2.4
		Apparent	0.35 to .48	0.43 to 0.70	0.28 to 0.30
Specific Heat		Btu/lb	0.22 to 0.23	0.22 to 0.23	0.22
		w/kg	0.12 to 0.13	0.12 to 0.13	0.12
Strength	Compressive	psi	10 at 5% deform.	100 at 5% deform.	15 at 5% deform.
		kPa	69 at 5% deform.	689 at 5% deform.	103 at 5% deform.
	Shear	psi			
		kPa			
	Tensile	psi			
		kPa			
Thermal Diffusivity		ft^2/hr	0.006 to 0.01 at 200°F	0.007 at 200°F	0.015 at 200°F
		m^2/hr	1.54×10^{-7} to 2.58×10^{-7} at 93°C	1.8×10^{-7} at 93°C	3.87×10^{-7} at 93°C
Thermal Shock Resistance			Good	Good	Good
Vibration Resistance					
Water Absorption % Weight			High	High	High
Water Vapor Transmission		perm-inch	High	High	High
		perm-cm	High	High	High

Table 42 Foamed or Sprayed Insulations

Material Identification Number			76	77	78
Composition			Polyurethune – 2 part mix foamed air-sprayed in place	Urea-formaldehyde foamed in place	Poly Vinyl-acetate cork filled mastic sprayed or trowelled
Abrasion Resistance % of Weight Loss		1st Run			
		2nd Run			
Alkalinity pH			6.5 to 7.5		7 to 7.5
Capillarity					
Coefficient of Expansion		English	Expands	1.8% to 3% at 70°F	
		Metric	Expands	1.8% to 3% at 21°C	
Combustibility			Combustible	Slow burning	Slow burning
If Combustible	Flash Point Temp.	°F	590		
		°C	310		
	Flame Point Temp.	°F	630	1280°F	
		°C	332	654	
	Heat Release	Btu/lb	16,000		
		w/kg	3.71×10^7		
	Self Internal Heating		Will react with some chemicals		Will not self internal heat
	Smoldering – Afterglow		Will smolder	Will char	Will char
Corrosion – Rusting			Will contribute	Will contribute	Pre coat steel
Corrosion – Stress			Must be protected		Must be pre-coated
Coverage Dried or Set		board ft/100 lbs	20 to 30		¼" thk. 100 ft^2 – 24 gal.
		m^2 cm/100 kgs	11 to 16		6.35 mm thk. 10 m^2 – 100 litre
Density Dried or Set		lbs/ft^3	1.6 to 3.0	0.6 to 1.0	53
		kgs/m^3	25 to 48	9.6 to 16	848
Hardness – mm penetration					
Hygroscopicity % Weight			0.8 to 1.4	2	
Maximum Temperature Limits	Short	°F	220	212	180
		°C	104	100	82
	Continuous	°F	200	180	180
		°C	94	82	82
Shrinkage Wet to Dry		Volumetric		3	30
Specific Gravity		Real			
		Apparent		0.009 to 0.016	25 to 35
Specific Heat		Btu/lb	.4 at 40°F		
		w/kg	0.22 at 4°C		
Strength	Compressive	psi	15 at yield		
		kPa	103 at yield		
	Shear	psi			
		kPa			
	Tensile	psi			
		kPa			
Thermal Diffusivity		ft^2/hr	0.02 to 0.03 at 40°F		
		m^2/hr	5.1×10^{-7} to 7.7×10^{-7} at 4°C		
Thermal Shock Resistance					Excellent
Vibration Resistance			Excellent		Excellent
Water Absorption % Weight					
Water Vapor Transmission		perm-inch	2.5 to 5.0	15 to 38	
		perm-cm	4.2 to 8.4	24 to 61	

Table 43 Loose and Fill Insulations

Material Identification Number		79	80	81	82
Composition		Alumina-silica fibers-Bulk	Alumina-silica fibers-Chopped and milled	Alumina-silica fibers-long fibers-bulk	Cork, vegetable Granulated
Alkalinity pH					6.5 to 7.5
Combustibility		Non-combustible	Non-combustible	Non-combustible	Combustible
If Combustible: Flash Point Temp.	°F	"	"	"	570
	°C	"	"	"	299
If Combustible: Flame Point Temp.	°F	"	"	"	620
	°C	"	"	"	327
If Combustible: Heat Release	Btu/lb	"	"	"	8000
	w/kg	"	"	"	1.85×10^7
If Combustible: Self Internal Heating		"	"	"	
If Combustible: Smoldering or Afterglow		"	"	"	
Density As Received	lb/ft^3	3 to 4	7.5	3	8 to 12
	kg/m^3	48 to 64	120	48	128 to 192
Packed	lb/ft^3	6	7.5	6	12 to 14
	kg/m^3	96	120	96	192 to 224
Hydroscopicity % Weight					
Maximum Temperature Limits: Short Time	°F	2300	2300	2300	200
	°C	1256	1256	1256	93
Maximum Temperature Limits: Continuous	°F	2300	2300	2300	200
	°C	1256	1256	1256	93
Specific Heat	Btu/lb	0.20	0.20	0.20	0.27
	w/kg	0.11	0.11	0.11	0.15
Thermal Diffusivity	ft^2/hr				
	m^2/hr				
Water Absorption % Weight		High	High	High	Low

Material Identification Number		83	84	85	86
Composition		Cellulose fiber Loose Fill	Diatomaceous silica fine powder	Diatomaceous silica-Caloined powder	Diatomaceous silica-Coarse powder
Alkalinity pH		4.5 to 6.0	5 to 7	6 to 7	5 to 7
Combustibility		Combustible	Non-combustible	Non-combustible	Non-combustible
If Combustible: Flash Point Temp.	°F	500	"	"	"
	°C	260	"	"	"
If Combustible: Flame Point Temp.	°F	500	"	"	"
	°C	260	"	"	"
If Combustible: Heat Release	Btu/lb	8000	"	"	"
	w/kg	1.859×10^7	"	"	"
If Combustible: Self Internal Heating			"	"	"
If Combustible: Smoldering or Afterglow		Can smolder	"	"	"
Density As Received	lb/ft^3		10 to 12	22	25 to 27
	kg/m^3		160 to 192	352	400 to 432
Packed	lb/ft^3	10 to 15	13 to 17	32	25 to 31
	kg/m^3	160 to 240	208 to 272	496	400 to 496
Hydroscopicity % Weight					
Maximum Temperature Limits: Short Time	°F	200	1600	2000	1600
	°C	93	871	1093	871
Maximum Temperature Limits: Continuous	°F	200	1600	2000	1600
	°C	93	871	1093	871
Specific Heat	Btu/lb		0.25	0.25	0.25
	w/kg		0.14	0.14	0.14
Thermal Diffusivity	ft^2/hr		0.01	0.01	0.009
	m^2/hr		2.58×10^{-7}	2.58×10^{-7}	2.32×10^{-7}
Water Absorption % Weight		High	High	High	High

Table 43 Loose and Fill Insulations

Material Identification Number			87	88	89	90
Composition			Gilsonite granules, for packing around underground pipe	Glass fiber, unbonded no binders loose fibers	Glass fibers Loose & Blowing Wool	Gypsum pellets Bulk
Alkalinity pH			7	8 to 10	8 to 10	
Combustibility			Combustible	Non-combustible	Fiber coating or lubi. maybe combustible	Non-combustible
If Combustible	Flash Point Temp.	°F	600	"	Fibers non-combustible	"
		°C	316	"	"	"
	Flame Point Temp.	°F	600	"	"	"
		°C	316	"	"	"
	Heat Release	Btu/lb	17000	"		"
		w/kg	3.95×10^7	"		"
	Self Internal Heating			"		"
	Smoldering or Afterglow		Will smolder	"	May smolder	"
Density As Received		lb/ft^3	40	2 to 12		12 to 20
		kg/m^3	640	32 to 192		
Packed		lb/ft^3	44 to 50	2 to 12	2 to 6	20 to 30
		kg/m^3	704 to 800	32 to 192	32 to 96	320 to 480
Hydroscopicity % Weight			Very Low	0	0	
Maximum Temperature Limits	Short Time	°F	300 to 460	1000	350	800
		°C	149 to 238	538	177	427
	Continuous	°F	300 to 460	1000	350	800
		°C	149 to 238	538	177	427
Specific Heat		Btu/lb	0.5 at 200°F	0.2 at 200°F	0.2 at 100°F	
		w/kg	0.27 at 93°C	0.11 at 93°C	0.11 at 38°C	
Thermal Diffusivity		ft^2/hr	0.0023 at 200°F	0.0167 at 100°F		
		m^2/hr	5.93×10^{-8} at 93°C	4.3×10^{-7} at 38°C		
Water Absorption % Weight			Low	800	1000	High

Material Identification Number			91	92	93	94
Composition			Mineral fiber with small amounts of oil, Loose	Mineral fiber, oil free, Loose	Mineral fiber granulated	Perlite, expanded Loose powder
Alkalinity pH			7 to 9	7 to 9	7 to 9	
Combustibility			Lub. oil is combustible	Non-combustible	Non-combustible	Non-combustible
If Combustible	Flash Point Temp.	°F	Fibers are non-combus.	"	"	"
		°C	"	"	"	"
	Flame Point Temp.	°F	"	"	"	"
		°C	"	"	"	"
	Heat Release	Btu/lb	"	"	"	"
		w/kg	"	"	"	"
	Self Internal Heating		Can develop self internal heating	"	"	"
	Smoldering or Afterglow		Will smolder	"	"	"
Density As Received		lb/ft^3	4 to 11	4 to 11	7 to 10	
		kg/m^3	64 to 176	64 to 176	112 to 160	
Packed		lb/ft^3	4 to 12	10 to 15	6 to 12	5 to 8
		kg/m^3	64 to 192	160 to 240	96 to 192	80 to 128
Hydroscopicity % Weight						
Maximum Temperature Limits	Short Time	°F	1000	1200	1000 to 1200	1800
		°C	538	649	538 to 649	982
	Continuous	°F	1000	1200	1000 to 1200	1800
		°C	538	649	.22	982
Specific Heat		Btu/lb	.22	.22	.12	
		w/kg	.12	.12		
Thermal Diffusivity		ft^2/hr				
		m^2/hr				
Water Absorption % Weight			High	High	High	High
Capillarity			Negligable	Negligable	Negligable	Negligable

Table 43 Loose and Fill Insulations

Material Identification Number			95	96	97	98
Composition			Quartz fibers Loose	Silica fibers Loose	Silica aerogel granuals	Vermiculite Flakes
Alkalinity pH			5 to 7	5 to 7	3.5 to 4.0	6 to 10
Combustibility			Non-combustible	Non-combustible	Non-combustible	Non-combustible
If Combustible	Flash Point Temp.	°F	"	"	"	"
		°C	"	"	"	"
	Flame Point Temp.	°F	"	"	"	"
		°C	"	"	"	"
	Heat Release	Btu/lb	"	"	"	"
		w/kg	"	"	"	"
	Self Internal Heating		"	"	"	"
	Smoldering or Afterglow		"	"	"	"
Density As Received		lb/ft³		6 to 12	3.5 to 5.5	4 to 10
		kg/m³		96 to 192	56 to 88	64 to 160
Packed		lb/ft³	6 to 12		4 to 5.5	4 to 10
		kg/m³	96 to 192		64 to 88	64 to 160
Hydroscopicity % Weight				2 to 10	10 to 15	2
Maximum Temperature Limits	Short Time	°F	2500	1800	1300	1400
		°C	1371	982	704	760
	Continuous	°F	2500	1800	1300	1400
		°C	1371	982	704	760
Specific Heat		Btu/lb		0.19 at 300°F	0.2 at 150°F	0.24 at 300°F
		w/kg		0.11 at 149°F	0.11 at 66°C	0.13 at 149°C
Thermal Diffusivity		ft²/hr		0.054 at 1000°F*	0.02 at 150°F	0.025 at 150°F
		m²/hr		1.39×10^{-6} 538°C**	5.16×10^{-7} at 66°C	6.45×10^{-7} at 66°C
Water Absorption % Weight			High	High	100	300
Capillarity			Will Wick	Will Wick	Will Wick	Will Wick

*At 6 lb/ft³ density
**At 96 kg/m³ density

Table 44 Reflective Insulation — foils and sheets

Material Identification Number			99	100
Composition			Aluminum sheet- Sheet	Aluminum foil, various thickness
Form			Any thickness sheet	Rolls
Capilarity			Negligable	Negligable
Combustibility			Non-combustible	Non-combustible
If Combustible	Flash Point Temp.	°F	"	"
		°C	"	"
	Flame Point Temp.	°F	"	"
		°C	"	"
	Heat Release	Btu/lb	"	"
		w/kg	"	"
	Self Internal Heating		"	"
Melting Temp.		°F	1200	1200
		°C	650	650
Corrosion — Rusting			Will not contribute to	Will not contribute to
Corrosion — Stress			Will not contribute to	Will not contribute to
Emissivity — Heat			0.05	0.05
Maximum Temperature Limits	Short Time	°F	1000	600
		°C	538	316
	Continuous	°F	1000	600
		°C	538	316
Resistance to Acids			Resistant to acids	Resistant to Acids
Resistance to Caustics			Not resistant	Not resistant to Caustics
Resistance to Solvents			Resistant	Resistant
Reflectance — Heat			0.95	0.95
Thermal Shock Resistance			Excellent	Excellent
Vibration Resistance			Excellent	Excellent
Water Absorption % Weight			0	0
Construction			Installed to form air spaces which are enclosed and/or separated by sheets of aluminum or aluminum surfaced paper	

Material Identification Number			101	102
Composition			Aluminum foil or membrane paper or other reinforcement	Aluminum-kraft paper—Accordian formed
Form			Foil attached to kraft paper or scrim	Reflective sheet separated to form spaces
Capilarity			Negligable	Negligable
Combustibility			Paper-combustible	Paper is combustible
If Combustible	Flash Point Temp.	°F	500	500
		°C	260	260
	Flame Point Temp.	°F	500	500
		°C	260	260
	Heat Release	Btu/lb	Paper — 8000	Paper — 8000
		w/kg	Paper — 1.859×10^7	Paper — 1.854×10
	Self Internal Heating		Not self heating	Not self heating
Melting Temp.		°F	Aluminum — 1200	Aluminum — 1200
		°C	Aluminum — 650	Aluminum — 650
Corrosion — Rusting			Will not contribute to	Will not contribute to
Corrosion — Stress			Will not contribute to	Will not contribute to
Emissivity — Heat			0.05	0.05
Maximum Temperature Limits	Short Time	°F	200	200
		°C	93	93
	Continuous	°F	200	200
		°C	93	93
Resistance to Acids			Paper-not resistant	Paper-not resistant
Resistance to Caustics			Paper-not resistant	Paper-not resistant
Resistance to Solvents			Resistant to	Resistant to
Reflectance — Heat			0.95	0.95
Thermal Shock Resistance			Excellent	Excellent
Vibration Resistance			Excellent	Excellent
Water Absorption % Weight			0	0
Construction			Installed to form air spaces which are enclosed and/or separated by sheets of aluminum or aluminum surfaced paper	To be installed into cavities to form spaces separated by sheets

Table 45 Reflective Insulation – Preformed to Fit Pipe or Equipment

Material Identification Number			103	104	105
Composition			Aluminum casing & aluminum reflective sheets	Stainless steel casing & aluminum reflective sheets	Stainless steel casing & ss reflective sheets
Form			Factory fabricated to fit equipment and pipe	Factory fabricated to fit equipment and pipe	Factory fabricated to fit equipment and pipe
Capillarity % by Weight			Negligable	Negligable	Negligable
Coefficient of Expansion		English	0.142×10^4	Casing 0.961×10^5 Sheets 0.142×10^4	0.961×10^5
		Metric	0.257×10^4	Casing 1.72×10^5 Sheets 0.257×10^4	1.72×10^5
Combustibility			Non-combustible	Non-combustible	Non-combustible
If Combustible	Flash Point Temp.	°F	"	"	
		°C	"	"	
	Flame Point Temp.	°F	"	"	
		°C	"	"	
	Heat Release	Btu/lb	"	"	
		w/kg	"	"	
	Self Internal Heating		"	"	
Melting Temp.		°F	1200	Casing 2600 Sheets 1200	2600
		°C	649	Casing 1423 Sheets 649	1423
Corrosion – Rusting			Will not contribute	Will not contribute	Will not contribute to
Corrosion – Stress			Will not contribute	Will not contribute	Will not contribute to
Density – Apparent			5 to 7 lbs/ft^3, 80 to 112 kgs/m^3	6 to 8 lbs/ft^3, 96 to 123 kgs/m^3	8 to 10 lbs/ft^3, 128 to 160 kgs/m^3
Hygroscopicity % Weight			0	0	0
Maximum Temperature Limits	Short Time	°F	1000	1000	1500
		°C	538	538	816
	Continuous	°F	1000	1000	1500
		°C	538	538	816
Resistance to Acids			Excellent	Excellent	Excellent
Resistance to Caustics			Poor	Fair	Excellent
Resistance to Solvents			Excellent	Excellent	Excellent
Specific Gravity		Real	2.7	3.0 to 5.2	5.76
		Apparent	.17	.16	.12
Shrinkage			No shrinkage	No shrinkage	No shrinkage
Thermal Shock Resistance			Excellent	Excellent	Excellent
Vibration Resistance			Excellent	Excellent	Excellent
Water Absorption % Weight			0	0	0

Table 46 Reflective Cryogenic Insulation – Partial Vacuum

Material Idenficication Number		106			107	
Composition		Aluminum foil, glass matt			Aluminum foil, glass paper	
Construction		Fluffy glass matts separating aluminum foil sheets			Glass paper separating aluminum foil sheets	
Number of Layers		10 to 20	15 to 30	35 to 70	50 to 100	75 to 150
Density	lbs/ft^3	2.5	3.0	4.7	5.5	7.5
	kgs/m^3	40	48	75	88	120
Recommended Vacuum – Absolute Pressure in Microns of Hg		0.1	0.1	0.1	0.1	0.1

Table 47 Cryogenic Insulation, Powder, in Partial Vacuum

Material Identification Number		108	109	110	111	112
Composition		Carbon-silica powder mixture	Perlite powder	Santocel® powder	Santocel® and aluminum powder-opacified	Santocel® and copper powder-opacified
Density	lbs/ft^3	8.0	8.0	6.0	10.0	11.0
	kgs/m^3	128	128	96	160	176
Recommended Vacuum – Absolute Pressure in Microns of Hg		2.0	2.0	2.0	2.0	2.0

Table 48 Mass Cryogenic Insulation – Atmosphere Pressure

Material Identification Number	113
Rigid Bulk and Pipe Covering	Cellular Glass
Propertied as listed under identification number II.	

Table 49 Thermal Conductivities of Insulations – *English Units* Btu, in./ft., hr., °F

Mean Temp.		Identification No. 1	Identification No. 2	Identification No. 3	Identification No. 4	Identification No. 5	Identification No. 6	Identification No. 7
		Generic Type: Alumina-silica fiber and binder Form: Block	Generic Type: Alumina-silica fiber and binder Form: Block	Generic Type: Alumina-silica fiber and binder Form: Board	Generic Type: Alumina-silica fiber and binder Form: Board	Generic Type: Calcium-Silicate Type 1 Form: Block and Pipe Cov.	Generic Type: Calcium-Silicate Type 2 Form: Block and Pipe Cov.	Generic Type: Diatomaceous-silica Type 4 Form: Block and Pipe Cov.
°R	°F	Density lbs/ft.^{3} 15 to 17	Density lbs/ft.^{3} 18 to 22	Density lbs/ft.^{3} 10 to 16	Density lbs/ft.^{3} 21 to 40	Density lbs/ft.^{3} 14 (max.)	Density lbs/ft.^{3} 15 (max.)	Density lbs/ft.^{3} 24 (max.)
260	−200							
360	−100							
460	0							
560	100					0.35		
660	200					0.43		
760	300					0.50		
860	400	0.45	0.55	0.40	0.65	0.55	0.66	
960	500	0.50	0.60	0.46	0.70	0.60	0.70	0.70
1060	600	0.56	0.65	0.52	0.75	0.66	0.76	0.72
1160	700	0.62	0.70	0.58	0.80	0.72	0.82	0.75
1260	800	0.68	0.75	0.64	0.85		0.94	0.77
1360	900	0.74	0.80	0.70	0.90		0.01	0.80
1460	1000	0.81	0.85	0.76	0.95		1.08	0.83
1560	1100	0.88	0.91	0.83	1.00		1.16	0.87
1660	1200	0.96	0.97	0.90	1.05		1.25	0.91

Metric Units w/mK

°K	°C	Density kgs/m 240 to 272	Density kgs/m 288 to 352	Density kgs/m^3 160 to 256	Density kgs/m^3 336 to 640	Density kgs/m^3 224 (max.)	Density kgs/m^3 240 (max.)	Density kgs/m^3 384 (max.)
173	−100							
223	−50							
273	0							
323	50							
373	100					0.066		
423	150					0.072		
273	200	0.065	0.079	0.058		0.079	0.095	
523	250	0.073	0.085	0.066		0.085	0.100	0.100
573	300	0.081	0.091	0.074	0.106	0.093	0.109	0.103
623	350	0.089	0.098	0.082	0.114	0.100	0.118	0.106
673	400	0.097	0.105	0.090	0.120		0.128	0.109
723	450	0.107	0.112	0.098	0.126		0.136	0.112
773	500	0.116	0.119	0.106	0.132		0.144	0.115
823	550	0.126	0.126	0.114	0.138		0.152	0.118
873	600	0.137	0.133	0.122	0.144		0.161	0.122
923	650	0.138	0.140	0.130	0.151		0.170	0.126

Table 49 Thermal Conductivities of Insulations — *English Units* Btu, in./ft., hr., °F

Mean Temp.		Identification No. 8	Identification No. 9	Identification No. 10	Identification No. 11	Identification No. 12	Identification No. 13	Identification No. 14
		Generic Type: Diatomaceous-silica Type 2 Form: Block and Pipe Cov.	Generic Type: Cork, vegetable Form: Board	Generic Type: Cork, vegetable Form: Pipe Cov.	Generic Type Cellular Glass Form:	Generic Type: Cellular Glass between layers of felt Roof Form: Board	Generic Type: Diatomaceous-silica Type 1 Class I Form: Block and Pipe Cov.	Generic Type: Diatomaceous-silica Type 2 Class II Form: Block and Pipe Cov.
°R	°F	Density lbs/ft.^{3} 26 (max.)	Density lbs/ft.^{3} 6 to 8	Density lbs/ft.^{3} 7 to 14	Density lbs/ft.^{3} 7.5 to 9.5	Density lbs/ft.^{3} 7.5 to 9.5	Density lbs/ft.^{3} 17 to 24	Density lbs/ft.^{3} 26
260	−200				0.22			
360	−100				0.28			
460	0		0.31	0.35	0.35	0.35		
560	100		0.35	0.39	0.43	0.43		
660	200				0.51	0.51		
760	300				0.58			
860	400				0.69			
960	500	0.78			0.81		0.72	0.80
1060	600	0.80					0.74	0.82
1160	700	0.82					0.76	0.84
1260	800	0.84					0.78	0.87
1360	900	0.87					0.80	0.90
1460	1000	0.90					0.83	0.93
1560	1100	0.93					0.86	0.96
1660	1200	0.96					0.90	0.99

Metric Units w/mK

°K	°C	Density kgs/m^3 416 (max.)	Density kgs/m^3 96 to 128	Density kgs/m^3 112 to 224	Density kgs/m^3 120 to 152	Density kgs/m^3 120 to 152	Density kgs/m^3 272 to 584	Density kgs/m^3 416
173	−100				0.031			
223	−50				0.039	0.039		
273	0		0.046	0.051	0.047	0.047		
323	50		0.051	0.056	0.055	0.055		
373	100				0.063	0.063		
423	150				0.073			
273	200				0.085			
523	250	0.112			0.098		0.103	0.115
573	300	0.115					0.105	0.117
623	350	0.118					0.108	0.119
673	400	0.121					0.111	0.123
723	450	0.124					0.114	0.127
773	500	0.127					0.116	0.131
823	550	0.130					0.121	0.135
873	600	0.133					0.125	0.140
923	650	0.136					0.130	0.143

Table 49 Thermal Conductivities of Insulations — *English Units* Btu, in./ft., hr., °F

		Identification No. 15	Identification No. 16	Identification No. 17	Identification No. 18	Identification No. 19	Identification No. 20	Identification No. 21
Mean Temp.		Generic Type: Glass fiber with organic binder Form: Board	Generic Type: Glass fiber with organic binder Form: Board	Generic Type: Glass fiber with organic binder Form: Board	Generic Type: Glass fiber with organic binder foil reinforced Form: Duct Board	Generic Type: Glass fiber with organic binder Form: Pipe Cov.	Generic Type: Glass fiber with organic binder Form: Pipe Cov.	Generic Type: Mineral fiber (rock, slag, or glass) Form: Block or Board
°R	°F	Density lbs/ft.^{3} 1.4 to 1.6	Density lbs/ft.^{3} 2.5 to 3.5	Density lbs/ft.^{3} 5 to 7	Density lbs/ft.^{3} 2.5 to 3.5	Density lbs/ft.^{3} 3 to 5	Density lbs/ft.^{3} 5 to 10	Density lbs/ft.^{3} 9 to 11
260	−200							
360	−100							
460	0	0.22						
560	100	0.29	0.28	0.27	0.30	0.28	0.28	0.28
660	200	0.36	0.32	0.30	0.34	0.32	0.32	0.35
760	300	0.47	0.39	0.35		0.39	0.42	0.43
860	400	0.62	0.50	0.46		0.48	0.50	0.54
960	500							
1060	600							
1160	700							
1260	800							
1360	900							
1460	1000							
1560	1100							
1660	1200							

Metric Units w/mK

°K	°C	Density kgs/m^3 22 to 26	Density kgs/m^3 40 to 56	Density kgs/m^3 80 to 112	Density kgs/m^3 40 to 56	Density kgs/m^3 54 to 80	Density kgs/m^3 80 to 160	Density kgs/m^3 144 to 176
173	−100							
223	−50							
273	0	0.033			0.030			
323	50	0.040	0.039	0.038	0.043	0.040	0.040	0.040
373	100	0.053	0.047	0.045		0.047	0.047	0.052
423	150	0.068	0.056	0.051		0.056	0.059	0.062
273	200	0.089	0.072	0.066		0.069	0.071	0.078
523	250							
573	300							
623	350							
673	400							
723	450							
773	500							
823	550							
873	600							
923	650							

Table 49 Thermal Conductivities of Insulations — *English Units* Btu, in./ft., hr., °F

Mean Temp.		Identification No. 22	Identification No. 23	Identification No. 24	Identification No. 25	Identification No. 26	Identification No. 27	Identification No. 28
		Generic Type: Mineral fiber (rock, slag or glass) Form: Block or Board	Generic Type: Mineral fiber Class 3 Form: Block and Board	Generic Type: Mineral fiber Class 4 Form: Block and Board	Generic Type: Mineral fiber Class 5 Form: Block and Board	Generic Type: Mineral fiber Class 1 Form: Pipe Cov.	Generic Type: Mineral fiber Class 2 Form: Pipe Cov.	Generic Type: Mineral fiber Class 3 Form: Pipe Cov.
°R	°F	Density lbs/ft.^{3} 11 to 13	Density lbs/ft.^{3} 7 to 12	Density lbs/ft.^{3} 8 to 13	Density lbs/ft.^{3} 12 to 25	Density lbs/ft.^{3} 9 to 11	Density lbs/ft.^{3} 11 to 14	Density lbs/ft.^{3} 16 to 20
260	−200							
360	−100							
460	0					0.24		
560	100	0.28	0.30	0.36		0.27	0.33	0.36
660	200	0.35	0.36	0.42	0.40	0.33	0.39	0.42
760	300	0.43	0.42	0.50	0.46	0.42	0.46	0.49
860	400	0.54	0.49	0.57	0.51		0.53	0.56
960	500			0.64	0.57			0.64
1060	600				0.62			
1160	700				0.68			
1260	800				0.75			
1360	900				0.82			
1460	1000				0.90			
1560	1100				0.98			
1660	1200				1.07			

Metric Units w/mK

°K	°C	Density kgs/m^3 176 to 208	Density kgs/m^3 112 to 192	Density kgs/m^3 128 to 208	Density kgs/m^3 192 to 400	Density kgs/m^3 144 to 176	Density kgs/m^3 176 to 224	Density kgs/m^3 256 to 320
173	−100							
223	−50							
273	0							
323	50	0.040	0.046	0.054		0.036	0.046	
373	100	0.052	0.053	0.062		0.049	0.056	0.061
423	150	0.062	0.060	0.072	0.066	0.060	0.066	0.071
273	200	0.078	0.072	0.082	0.073		0.076	0.081
523	250			0.091	0.082			0.099
573	300				0.087			
623	350				0.095			
673	400				0.105			
723	450				0.111			
773	500				0.121			
823	550				0.132			
873	600				0.140			
923	650				0.154			

Table 49 Thermal Conductivities of Insulations — *English Units* Btu, in./ft., hr., °F

Mean Temp.		Identification No. 29	Identification No. 30	Identification No. 31	Identification No. 32	Identification No. 33	Identification No. 34	Identification No. 35
		Generic Type: Perlite expanded and water repellent binder Form: Block and Pipe Cov.	Generic Type: Perlite expanded with binders Form: Block and Pipe Cov.	Generic Type: Perlite expanded with binder and fibers Form: Roof Board	Generic Type: Expanded Phenolic foam Form: Block and Pipe Cov.	Generic Type: Polystrene expanded beads, molded to shape Form: Block, Board Pipe Cov.	Generic Type: Polystyrone expanded foam-cut shapes Form: Board and Pipe Cov.	Generic Type: Polystyrone expanded foam-molded shapes Form: Board and Pipe Cov.
°R	°F	Density lbs/ft.³ 13 to 15	Density lbs/ft.³ 13 to 15	Density lbs/ft.³ 10	Density lbs/ft.³ 2 to 3	Density lbs/ft.³ 1.0 to 1.5	Density lbs/ft.³ 1.75 to 2.0	Density lbs/ft.³ 2.0 to 2.2
260	−200							
360	−100							
460	0			0.36	0.22	0.24	0.20	0.20
560	100	0.40	0.40	0.40	0.26	0.28	0.24	0.24
660	200	0.45	0.45					
760	300	0.50	0.50					
860	400	0.55	0.55					
960	500	0.60	0.60					
1060	600	0.65	0.65					
1160	700	0.71	0.71					
1260	800	0.77	0.77					
1360	900	0.83	0.83					
1460	1000	0.90	0.90					
1560	1100							
1660	1200							

Metric Units w/mK

°K	°C	Density kgs/m³ 208 to 240	Density kgs/m³ 208 to 240	Density kgs/m³ 160	Density kgs/m³ 32 to 48	Density kgs/m³ 16.01 to 24.01	Density kgs/m³ 28.01 to 32.01	Density kgs/m³ 32.02 to 35.22
173	−100							
223	−50							
273	0			0.053	0.032	0.035	0.026	0.026
323	50			0.057	0.058	0.041	0.032	0.032
373	100	0.068	0.068					
423	150	0.072	0.072					
273	200	0.078	0.078					
523	250	0.085	0.085					
573	300	0.091	0.091					
623	350	0.098	0.098					
673	400	0.105	0.105					
723	450	0.112	0.112					
773	500							
823	550							
873	600							
923	650							

Table 49 Thermal Conductivities of Insulations — *English Units* Btu, in./ft., hr., °F

Mean Temp.		Identification No. 36	Identification No. 37	Identification No. 38	Identification No. 39	Identification No. 40	Identification No. 41	Identification No. 42
		Generic Type: Polystyrone expanded foam-molded into shape Form: Board	Generic Type: Polystyrone foam molded into shapes Form: Roof Board	Generic Type: Polyurethane expanded foam Form: Block, Board Pipe Cov.	Generic Type: Polyurethane expanded foam Form: Block, Board Pipe Cov.	Generic Type: Polyurethane expanded foam Form: Block, Board Pipe Cov.	Generic Type: Expanded rubber Form: Rigid Board	Generic Type: Expanded silica and binders Form: Block and Pipe Cov.
°R	°F	Density lbs/ft.^{3} 2.25 to 2.5	Density lbs/ft.^{3} 2.5 to 2.7	Density lbs/ft.^{3} Up to 1.7	Density lbs/ft.^{3} 1.7 to 2.5	Density lbs/ft.^{3} 2.5 to 5.0	Density lbs/ft.^{3} 4 to 6	Density lbs/ft.^{3} 13
260	−200							
360	−100							
460	0	0.20	0.20	0.23	0.22	0.20	.26	
560	100	0.24	0.24	0.27	0.26	0.24	.28	.40
660	200							.42
760	300							.45
860	400							.50
960	500							.55
1060	600							.60
1160	700							.65
1260	800							.70
1360	900							.76
1460	1000							.82
1560	1100							
1660	1200							

Metric Units w/mK

°K	°C	Density kgs/m^3 36.02 to 40.03	Density kgs/m^3 90.03 to 43.22	Density kgs/m^3 27.3	Density kgs/m^3 27.2 to 40.0	Density kgs/m^3 40 to 80	Density kgs/m^3 64 to 96	Density kgs/m^3 208
173	−100							
223	−50							
273	0	0.026	0.026	0.033	0.032	0.028	0.037	
323	50	0.032	0.032	0.039	0.038	0.034	0.041	0.057
373	100							0.061
423	150							0.065
273	200							0.072
523	250							0.078
573	300							0.085
623	350							0.091
673	400							0.098
723	450							0.105
773	500							0.113
823	550							
873	600							
923	650							

Table 49 Thermal Conductivities of Insulations — *English Units* Btu, in./ft., hr., °F

Mean Temp.		Identification No. 43	Identification No. 44	Identification No. 45	Identification No. 46	Identification No. 47	Identification No. 48	Identification No. 49
		Generic Type: Wood fibers and binders Form: Board	Generic Type: Borosilica closed cell inorganic foam Form: Block	Generic Type: Elastomeric Flexible Form: Sheet and Pipe Cov.	Generic Type: Glass fiber, flexible with various facings Form: Roll or Sheets	Generic Type: Glass fiber, flexible with various facings** Form: Sheets	Generic Type: Glass fiber, semi-rigid with various facings** Form: Sheets	Generic Type: Kaolin ceramic fiber in stainless steel metal mesh Form: Strips & Shaped
°R	°F	Density lbs/ft.^{3} 15 to 18	Density lbs/ft.^{3} 12	Density lbs/ft.^{3} 9.5 to 8.5	Density lbs/ft.^{3} 0.6 to 3.0	Density lbs/ft.^{3} 3 to 4.5	Density lbs/ft.^{3} 4.5 (min.)	Density lbs/ft.^{3} 3 to 4
260	−200							
360	−100							
460	0	0.40		0.29	0.30	0.28	0.27	
560	100	0.42	0.58	0.32	0.40	0.36	0.33	
660	200		0.64		0.54	0.46	0.40	
760	300		0.69		0.68	0.60	0.50	
860	400		0.75					0.45
960	500		0.82					0.57
1060	600							0.70
1160	700							0.82
1260	800							0.96
1360	900							1.11
1460	1000							1.30
1560	1100							1.47
1660	1200							1.65

Metric Units w/mK

°K	°C	Density kgs/m^3 240 to 288	Density kgs/m^3 190	Density kgs/m^3 72	Density kgs/m^3 10.4 to 8	Density kgs/m^3 98 to 72	Density kgs/m^3 72 (min.)	Density kgs/m^3 48 to 64
173	−100							
223	−50							
273	0			0.0425	0.044	0.042	0.040	
323	50	0.059	0.0865	0.0476	0.059	0.053	0.049	
373	100	0.062	0.0937		0.079	0.067	0.059	
423	150		0.0995		0.101	0.082	0.071	
273	200		0.1082					0.065
523	250		0.1168					0.084
573	300							0.099
623	350							0.114
673	400							0.128
723	450							0.146
773	500							0.170
823	550							0.194
873	600							0.211
923	650							0.237

**Facing may be woven wire mesh, expanded metal lath (copper bearing) on one or both sides.

Table 49 Thermal Conductivities of Insulations — *English Units* Btu, in./ft., hr., °F

Mean Temp.		Identification No. 50	Identification No. 51	Identification No. 52	Identification No. 53	Identification No. 54	Identification No. 55	Identification No. 56
		Generic Type: Kaolin ceramic fiber in stainless steel metal mesh Form: Strips & Shaped	Generic Type: Mineral wool-spun glass-stainless steel metal mesh Form: Various Shapes	Generic Type: Mineral fibers with metal mesh facing** Class I Form: Sheets	Generic Type: Mineral fibers with metal mesh facing** Class II Form: Sheets	Generic Type: Urethane foam Form: Block, Pipe Cov.	Generic Type: Urethane foam Form: Block, Pipe Cov.	Generic Type Alumina fibers Form: Blanket and Felts
°R	°F	Density lbs/ft.^{3} 6 to 8	Density lbs/ft.^{3} 10 to 14	Density lbs/ft.^{3} 6 (max. ave.)	Density lbs/ft.^{3} 12 (ave. max.)	Density lbs/ft.^{3} Up to 1.5	Density lbs/ft.^{3} Over 1.5	Density lbs/ft.^{3} 3 to 4
260	−200							
360	−100							
460	0		0.26			0.29	0.27	
560	100		0.30		0.31	0.31	0.29	
660	200		0.34	0.29	0.37			
760	300		0.40	0.36	0.44			
860	400	0.35	0.47	0.46	0.51			.40
960	500	0.40	0.55	0.56	0.59			.49
1060	600	0.47	0.62	0.67				.60
1160	700	0.56	0.76	0.67				.74
1260	800	0.68	0.87					.90
1360	900	0.81	0.98					1.05
1460	1000	0.95	1.10					1.25
1560	1100	1.07						1.42
1660	1200	1.20						1.65

Metric Units w/mK

°K	°C	Density kgs/m^3 96 to 128	Density kgs/m^3 160 to 224	Density kgs/m^3 96 (max. ave.)	Density kgs/m^3 96 (ave. max.)	Density kgs/m^3 Up to 24	Density kgs/m^3 Over 24	Density kgs/m^3 40 to 64
173	−100							
223	−50							
273	0		0.039			0.043	0.040	
323	50		0.044	0.036	0.039	0.046	0.043	
373	100		0.050	0.044	0.046			
423	150		0.057	0.057	0.054			
273	200	0.051	0.067	0.069	0.063			0.058
523	250	0.059	0.078	0.082	0.072			0.069
573	300	0.066	0.088					0.084
623	350	0.077	0.103					0.097
673	400	0.089	0.118					0.118
723	450	0.106	0.132					0.140
773	500	0.125	0.150					0.163
823	550	0.142	0.162					0.187
873	600	0.157						0.209
923	650	0.173						0.238

**Facing may be woven wire mesh, expanded metal lath (copper bearing) on one or both sides.

Table 49 Thermal Conductivities of Insulations — *English Units* Btu, in./ft., hr., °F

Mean Temp.		Identification No. 57	Identification No. 58	Identification No. 59	Identification No. 60	Identification No. 61	Identification No. 62	Identification No. 63
		Generic Type: Alumina fibers Form: Blanket and Felts	Generic Type: Alumina fibers Form: Blanket and Felts	Generic Type: Alumina fibers Form: Rope, cord, yarn	Generic Type: Glass fiber, fine fiber (no binder) Form: Blanket and Batt	Generic Type: Glass fiber, fine fiber with binder Form: Blanket	Generic Type: Glass fiber with binder Form: Blanket	Generic Type: Glass fiber with binder Form: Blanket
°R	°F	Density lbs/ft.³ 6 to 8	Density lbs/ft.³ 12	Density lbs/ft.³ 25	Density lbs/ft.³ 11 to 12	Density lbs/ft.³ ½ to 2	Density lbs/ft.³ 1 to 2	Density lbs/ft.³ 2.5 to 3.5
260	−200							
360	−100							
460	0				0.26	0.23		
560	100				0.30	0.24	0.38	0.35
660	200				0.34	0.30	0.54	0.48
760	300				0.40	0.38		
860	400	0.33	0.30	0.40	0.47			
960	500	0.38	0.36	0.49	0.55			
1060	600	0.44	0.42	0.60	0.62			
1160	700	0.53	0.49	0.74	0.76			
1260	800	0.62	0.57	0.90	0.87			
1360	900	0.72	0.65	1.05	0.98			
1460	1000	0.83	0.72	1.25	1.10			
1560	1100	0.94	0.82	1.42				
1660	1200	1.06	0.92	1.65				

Metric Units w/mK

°K	°C	Density kgs/m³ 96 to 128	Density kgs/m³ 192	Density kgs/m³ 400	Density kgs/m³ 176 to 192	Density kgs/m³ 8 to 32	Density kgs/m³ 16 to 32	Density kgs/m³ 40 to 56
173	−100							
223	−50							
273	0				0.039	0.033		
323	50				0.044	0.037	0.060	0.055
373	100				0.050	0.044	0.079	0.071
423	150				0.057	0.054		
273	200	0.047	0.430	0.058	0.067			
523	250	0.053	0.050	0.069	0.078			
573	300	0.061	0.058	0.084	0.088			
623	350	0.072	0.066	0.097	0.103			
673	400	0.083	0.077	0.118	0.118			
723	450	0.095	0.086	0.140	0.132			
773	500	0.104	0.099	0.163	0.150			
823	550	0.123	0.104	0.187	0.162			
873	600	0.140	0.121	0.209				
923	650	0.153	0.133	0.238				

Table 49 Thermal Conductivities of Insulations — *English Units* Btu, in./ft., hr., °F

Mean Temp.		Identification No. 64	Identification No. 65	Identification No. 66	Identification No. 67	Identification No. 68	Identification No. 69	Identification No. 70
		Generic Type: Mineral Fibers Form: Blanket and Batt	Generic Type: Mineral Fibers with binders Form: Blankets	Generic Type: Silica fibers Form: Blanket, Felt	Generic Type: Alumina-silica, semi refractory hydraulic setting Form: Cement	Generic Type: Alumina silica-ceramic fiber air setting cement Form: Ready-mix	Generic Type: Alumina silica-ceramic Form: Spray mix cement	Generic Type: Diatomaceous silica and binder Form: Water mix cement
°R	°F	Density lbs/ft.^{3} 8 to 12	Density lbs/ft.^{3} 7 to 9	Density lbs/ft.^{3} 6 to 9.5	Density lbs/ft.^{3} 48 to 70	Density lbs/ft.^{3} 115 to 122	Density lbs/ft.^{3} 12 to 13	Density lbs/ft.^{3} 23 to 32
260	−200							
360	−100							
460	0							
560	100		0.30					
660	200		0.34					0.75
760	300	0.49	0.38					0.80
860	400	0.57	0.44				0.40	0.85
960	500	0.68	0.52	0.47			0.46	0.90
1060	600	0.82		0.50	1.42		0.52	0.95
1160	700			0.56	1.47		0.58	1.00
1260	800			0.67	1.52		0.65	1.06
1360	900			0.77	1.58		0.71	1.12
1460	1000			0.90	1.64	3.5	0.78	1.19
1560	1100			1.02	1.70	4.2	0.86	
1660	1200			1.20	1.76	5.0	0.94	

Metric Units w/mK

°K	°C	Density kgs/m^3 128 to 192	Density kgs/m^3 112 to 140	Density kgs/m^3 96 to 152	Density kgs/m^3 768 to 1120	Density kgs/m^3 1840 to 1952	Density kgs/m^3 192 to 208	Density kgs/m^3 368 to 512
173	−100							
223	−50							
273	0							
323	50		0.045					
373	100		0.051					0.109
423	150	0.071	0.055					0.115
273	200	0.082	0.063				0.058	0.123
523	250	0.097	0.076	0.067			0.004	0.129
573	300	0.121		0.071	0.203		0.073	0.135
623	350			0.078	0.210		0.081	0.140
673	400			0.086	0.216		0.088	0.146
723	450			0.102	0.223		0.097	0.153
773	500			0.120	0.231		0.107	0.162
823	550			0.133	0.239	0.505	0.115	0.174
873	600			0.152	0.248	0.610	0.125	
923	650			0.173	0.254	0.721	0.136	

Table 49 Thermal Conductivities of Insulations — *English Units* Btu, in./ft., hr., °F

Mean Temp.		Identification No. 71	Identification No. 72	Identification No. 73	Identification No. 74	Identification No. 75	Identification No. 76	Identification No. 77
		Generic Type: Kaowool ceramic fiber and inorganic binders Form: Spray cement	Generic Type: Kaowool fiber and mineral fiber blend Form: Spray cement	Generic Type: Mineral fiber, binder and hydraulic setting cement Form: Cement	Generic Type: Mineral fiber Form: Cement	Generic Type: Vermiculite expands, with binders Form: Water mix-air setting	Generic Type: Polyurethane, 2 part mix Form: Foamed or sprayed	Generic Type: Ureaformaldehyde Form: Foamed in place
°R	°F	Density lbs/ft.^{3} 12 to 13	Density lbs/ft.^{3} 7 to 15	Density lbs/ft.^{3} 22 to 30	Density lbs/ft.^{3} 27 to 44	Density lbs/ft.^{3} 18 to 19	Density lbs/ft.^{3} 1.6 to 3.0	Density lbs/ft.^{3} 0.6 to 1.0
260	−200							
360	−100							
460	0						0.22	0.25
560	100						0.26	0.29
660	200		0.48	0.93	0.73	0.98		
760	300		0.52	0.98	0.78	1.03		
860	400		0.57	1.03	0.84	1.08		
960	500		0.65	1.08	0.90	1.15		
1060	600	0.61	0.70	1.10	0.96	1.22		
1160	700	0.67	0.76		1.02	1.29		
1260	800	0.73	0.83					
1360	900	0.80	0.91					
1460	1000	0.86						
1560	1100	0.93						
1660	1200	0.99						

Metric Units w/mK

°K	°C	Density kgs/m^3 192 to 208	Density kgs/m^3 112 to 240	Density kgs/m^3 352 to 512	Density kgs/m^3 482 to 704	Density kgs/m^3 288 to 304	Density kgs/m^3 25 to 48	Density kgs/m^3 9.6 to 16.0
173	−100							
223	−50							
273	0						0.032	0.035
323	50						0.038	0.042
373	100		0.071	0.134	0.103	0.140		
423	150		0.075	0.139	0.109	0.147		
273	200		0.082	0.144	0.115	0.154		
523	250		0.092	0.148	0.123	0.162		
573	300	0.086	0.099	0.153	0.129	0.170		
623	350	0.094	0.108		0.136	0.182		
673	400	0.101	0.115					
723	450	0.109	0.124					
773	500	0.118	0.134					
823	550	0.125						
873	600	0.136						
923	650	0.143						

Table 49 Thermal Conductivities of Insulations — *English Units* Btu, in./ft., hr., °F

Mean Temp.		Identification No. 78	Identification No. 79	Identification No. 80	Identification No. 81	Identification No. 82	Identification No. 83	Identification No. 84
		Generic Type: Poly-vinyl cork filled mastic Form: Sprayed or trowelled	Generic Type: Alumina-silica fibers Form: Bulk	Generic Type: Alumina silica fiber Form: Chopped and milled	Generic Type: Alumina silica fiber-long fiber Form: Bulk	Generic Type: Cork, vegetable granulated Form: Suitable for pouring	Generic Type: Cellulose fiber Form: Loose-Fill	Generic Type: Diatomaceous silica Form: Fine powder
°R	°F	Density lbs/ft.^{3} 53	Density lbs/ft.^{3} 6	Density lbs/ft.^{3} 7.5	Density lbs/ft.^{3} 6	Density lbs/ft.^{3} 12 to 14	Density lbs/ft.^{3} 10 to 15	Density lbs/ft.^{3} 13 to 17
260	−200							
360	−100							
460	0					0.31	0.42	
560	100	.96				0.34	0.46	
660	200							0.42
760	300							0.45
860	400							0.48
960	500				0.50			0.51
1060	600				0.61			0.55
1160	700		0.85	0.55	0.75			0.58
1260	800		0.97	0.65	0.90			0.62
1360	900		1.09	0.75	1.05			0.65
1460	1000		1.22	0.85	1.20			0.69
1560	1100		1.35	0.95	1.35			0.74
1660	1200		1.55	1.05	1.52			0.80

Metric Units w/mK

°K	°C	Density kgs/m^3 848	Density kgs/m^3 96	Density kgs/m^3 120	Density kgs/m^3 96	Density kgs/m^3 124 to 224	Density kgs/m^3 160 to 240	Density kgs/m^3 208 to 272
173	−100							
223	−50							
273	0					0.046		
323	50	0.138				0.047	0.059	
373	100							0.062
423	150							0.064
273	200							0.069
523	250				0.071			0.073
573	300				0.085			0.078
623	350		0.121	0.078	0.107			0.082
673	400		0.133	0.088	0.120			0.087
723	450		0.147	0.099	0.139			0.091
773	500		0.166	0.114	0.161			0.097
823	550		0.180	0.125	0.177			0.101
873	600		0.200	0.139	0.200			0.108
923	650		0.224	0.151	0.219			0.115

Table 49 Thermal Conductivities of Insulations — *English Units* Btu, in./ft., hr., °F

Mean Temp.		Identification No. 85	Identification No. 86	Identification No. 87	Identification No. 88	Identification No. 89	Identification No. 90	Identification No. 91
		Generic Type: Diatomaceous silica calcined Form: Powder	Generic Type: Diatomaceous silica Form: Course Powder	Generic Type: Gilsonite granules-for packing around underground pipe Form:	Generic Type: Glass fiber, unbonded Form: Loose fibers	Generic Type: Glass fibers Form: Loose and Blowing Wool	Generic Type: Gypsum pellits Form: Bulk	Generic Type: Mineral fiber, with small amounts of oil Form: Loose
°R	°F	Density lbs/ft.³ 32 (max.)	Density lbs/ft.³ 25 to 31	Density lbs/ft.³ 44 to 50	Density lbs/ft.³ 6 (ave.)	Density lbs/ft.³ 2 to 3	Density lbs/ft.³ 20 to 30	Density lbs/ft.³ 4 to 12
260	−200							
360	−100							
460	0				0.25			
560	100			0.75	0.26	0.35	0.45 to 0.95	0.30
660	200	0.92	0.63	0.85	0.31			0.35
760	300	0.97	0.68	1.00	0.36			0.44
860	400	1.03	0.77		0.45			0.54
960	500	1.08	0.84					0.65
1060	600	1.13	0.92					0.75
1160	700	1.18	0.99					0.86
1260	800	1.25	1.06					0.97
1360	900	1.31	1.14					
1460	1000	1.37	1.22					
1560	1100	1.43	1.32					
1660	1200	1.49	1.42					

Metric Units w/mK

°K	°C	Density kgs/m³ 512 (max.)	Density kgs/m³ 400 to 496	Density kgs/m³ 704 to 800	Density kgs/m³ 86 (ave.)	Density kgs/m³ 32 to 43	Density kgs/m³ 320 to 480	Density kgs/m³ 64 to 192
173	−100							
223	−50							
273	0							
323	50			0.111	0.038	0.052	0.067 to 0.143	0.045
373	100	0.134	0.092	0.124	0.046			0.052
423	150	0.140	0.098	0.144	0.052			0.063
273	200	0.148	0.111		0.064			0.078
523	250	0.155	0.120					0.092
573	300	0.162	0.131					0.105
623	350	0.169	0.141					0.120
673	400	0.177	0.150					0.134
723	450	0.184	0.158					
773	500	0.192	0.168					
823	550	0.200	0.183					
873	600	0.207	0.193					
923	650	0.215	0.204					

Table 49 Thermal Conductivities of Insulations — *English Units* Btu, in./ft., hr., °F

Mean Temp.		Identification No. 92	Identification No. 93	Identification No. 94	Identification No. 95	Identification No. 96	Identification No. 97	Identification No. 98
		Generic Type: Mineral fiber, oil free Form: Loose	Generic Type: Mineral fiber, granulated Form: Loose	Generic Type: Perlite expanded Form: Loose powder	Generic Type: Quartz fibers Form: Loose	Generic Type: Silica fibers Form: Loose	Generic Type: Silica aerogel granuals Form: Loose	Generic Type: Vermiculite flakes Form: Loose
°R	°F	Density lbs/ft.^{3} 10 to 15	Density lbs/ft.^{3} 6 to 12	Density lbs/ft.^{3} 5 to 8	Density lbs/ft.^{3} 6 to 12	Density lbs/ft.^{3} 9	Density lbs/ft.^{3} 4 to 5.5	Density lbs/ft.^{3} 8 to 10
260	−200							
360	−100							
460	0							
560	100		0.30	0.29			0.21	0.47
660	200	0.53	0.35	0.35	0.35		0.23	0.57
760	300	0.58	0.44		0.41		0.24	0.63
860	400	0.64	0.54		0.47			0.74
960	500	0.70	0.65		0.53			0.84
1060	600	0.76	0.75		0.59	0.45		0.94
1160	700	0.83	0.86		0.66	0.52		1.05
1260	800	0.91	0.97		0.73	0.60		1.15
1360	900				0.80	0.71		1.25
1460	1000				0.87	0.81		1.36
1560	1100				0.95	0.92		
1660	1200				1.04	1.02		

Metric Units w/mK

°K	°C	Density kgs/m^3 160 to 240	Density kgs/m^3 96 to 192	Density kgs/m^3 80 to 120	Density kgs/m^3 96 to 192	Density kgs/m^3 144	Density kgs/m^3 64 to 88	Density kgs/m^3 128 to 160
173	−100							
223	−50							
273	0							
323	50		0.045	0.042			0.032	0.069
373	100	0.078	0.052	0.049	0.049		0.034	0.079
423	150	0.084	0.063		0.059		0.036	0.091
273	200	0.092	0.078		0.067			0.106
523	250	0.099	0.092		0.073			0.119
573	300	0.108	0.105		0.083	0.063		0.133
623	350	0.117	0.120		0.093	0.072		0.146
673	400	0.131	0.134		0.103	0.082		0.160
723	450				0.113	0.094		0.173
773	500				0.123	0.106		0.186
823	550				0.133	0.119		0.200
873	600				0.142	0.133		
923	650				0.150	0.148		

Table 50 Thermal Conductances of Reflective Insulations —

Material Identification number	Generic type and class or description of insulating material	Form
99	Aluminum Sheet	Various Thicknessess
100	Aluminum Foil	Various Thicknessess
101	Aluminum foil, both sides, or membrane paper or other reinforcement	Laminate
102	Aluminum foil accordian formed roll insulation-Opened in place	Expandable roll for cavities

English Units

Position	Direction of Heat Flow	Conductance in Btu/ft² hr °F, 75° Mean Temperature, Air spaces 3/4" $\epsilon = 0.05$ (Aluminum), Air spaces enclosed* and separated by reflective shield — Number of spaces: 1	2	3	4	5	6	7	8	9	10
Vertical	Across	0.35	0.18	0.12	0.09	0.07	0.06	0.05	0.045	0.04	0.035
Horizontal	Upward	0.48	0.25	0.17	0.12	0.10	0.08	0.07	0.06	0.055	0.05
Horizontal	Downward	0.35	0.18	0.12	0.09	0.08	0.07	0.055	0.055	0.04	0.035

Metric Units

Position	Direction of Heat Flow	Conductance in w/m²k, 24°C Mean Temperature, Air spaces 19mm $\epsilon = 0.05$ (Aluminum), Air spaces enclosed* and separated by reflective shield — Number of spaces: 1	2	3	4	5	6	7	8	9	10
Vertical	Across	1.987	1.022	0.681	0.511	0.397	0.341	0.284	0.256	0.227	0.159
Horizontal	Upward	2.725	1.419	0.965	0.681	0.568	0.454	0.397	0.341	0.312	0.284
Horizontal	Downward	1.997	1.022	0.681	0.511	0.454	0.397	0.312	0.256	0.277	0.189

*All spaces constructed to prevent air movement from one space to another.
Note: Heat transfer is given in conductance not conductivity.

Table 51 Thermal Conductivities of Preformed Reflective Insulations

English Units Btu, in/ft, hr, °F

Mean Temp.		Identification Number 103	Identification Number 104	Identification Number 105
		Generic Type: Aluminum casing & aluminum reflective sheets Form: Factory formed for equipment & pipe	Generic Type: Stainless steel casing & aluminum reflective sheets Form: Factory formed for equipment & pipe	Generic Type: Stainless steel casing & reflective sheets Form: Factory formed for equipment & pipe
°R	°F	Density Lbs/ft³ 5 to 7	Density Lbs/ft³ 6 to 8	Density Lbs/ft³ 8 to 10
260	−200			
360	−100			
460	0			
560	100	0.250	0.265	Varied to suit
660	200	0.307	0.328	need
760	300	0.378	0.400	
860	400	0.444	0.466	
960	500	0.507	0.537	
1060	600	0.570	0.600	
1160	700			
1260	800			
1360	900			
1460	1000			
1560	1100			
1660	1200			

Metric Units w/mK

°K	°C	Density kgs/m³	Density kgs/m³	Density kgs/m³
173	−100			
223	−50			
273	0			
323	50	0.037	0.040	
373	100	0.046	0.049	Varied to suit
423	150	0.055	0.058	need
273	200	0.064	0.067	
523	250	0.073	0.076	
573	300	0.082	0.085	
623	350	0.091	0.094	
673	400			
723	450			
773	500			
823	550			
873	600			
923	650			

Based on spacing of one reflective sheet per 1/2" or 1.27 cm enclosed space.

Table 52 Thermal Conductance of Reflective Insulation in Partial Vacuum – Cryogenic Temperatures

Identification Number	106			107	
Composition	Aluminum foil glass matts	Aluminum foil and glass matts	Aluminum foil and glass matts	Aluminum foil and glass paper	Aluminum foil and glass paper
Number of Layers	10 to 12	15 to 30	35 to 70	50 to 100	75 to 150
Conductance *Btu/ft^2, hr °F	0.00108	0.00078	0.00024	0.00022	0.00012
Conductance **w/m^{2}K	6.13 X 10^{-3}	4.429 X 10^{-3}	1.363 X 10^{-3}	1.249 X 10^{-3}	6.814 X 10^{-4}

*Based on – 200°F **Based on – 129°C, 144 K mean temperature.
The thermal conductance noted is for the number of reflective layers per one foot thickness under 1 micron pressure.

Table 53 Thermal Conductivities of Powder – in Partial vacuum – in Cryogenic Temperatures

Identification Number		108	109	110	111	112
Composition		Carbon-Silica Powder	Perlite Powder	Santocel® Powder	Santocel® & Aluminum Powder	Santocel® & Aluminum Powder
Density	lbs/ft^3	8.0	8.0	6.0	10.0	11.0
	kgs/m^3	128	128	96	160	176
Conductivity *Btu, in/ft^2, hr, °F		0.0048	0.0087	0.0115	0.0027	0.0028
Conductivity **w/mK		69.21 X 10^{-4}	1.254 X 10^{-3}	1.658 X 10^{-3}	3.893 X 10^{-4}	4.038 X 10^{-4}

*Based on mean temperature of – 200°F, under partical vacuum.
**Based on mean temperature of – 129°C, 144 K, under partical vacuum (50 microns or less).

Table 54 Thermal Conductivity of Rigid Mass Cryogenic Insulation

Identification Number	11 and 113						
Composition	Cellular Glass						
Conductivity	Temperature °F						
English Units	0	–40	–50	–120	–160	–200	–240
Btu, in/ft^2, hr, °F	0.35	0.32	0.29	0.27	0.25	0.22	0.19
Metric Units	Temperature °C						
	–17.8	–40	–62.2	–84.4	–106.7	–128.9	–151.1
	Temperature °K						
	255	233.2	211	188.8	166.5	144.3	122.1
w/mK	0.050	0.046	0.042	0.039	0.036	0.032	0.027

5 Installation Requirements

"Installation Requirements" per se may be misunderstood to mean only the requirements during the application or service life of an "installation." What is intended, however, is a discussion of the requirements that an insulation must fulfill from the time it is manufactured until it is in service. After insulation is produced, it passes through several phases; i.e., it is shipped, stored, possibly fabricated, reshipped to the job, installed, and finally used in the service for which it was originally specified.

For lack of a more descriptive term, these various operations that the insulation must pass through from the time of its production until its ultimate discard will be called "phases." They may differ, depending upon the method of manufacture, type of material, and use, but in general they may be listed as follows:

Shipping
Storage
Fabrication
Application
Service

Each of these phases imposes a set of requirements upon the insulating material. It must meet these requirements in order to fulfill successfully its economic function. Evaluation of these requirements and the selection of the insulating material which will best perform the specified function is an engineering design problem. This problem is complicated, not only by the large number of requirements, but also by the lack of reliable data on the requirements themselves. Because of this lack of data, it is impossible to provide both an exact list of the requirements and of the properties whereby a precise selection of material can be determined. Therefore, this discussion must be kept general, for the final selection will be determined by engineering judgment. This chapter will be helpful in determining the factors affecting that judgment. It must be understood that an error in judgment here may make the difference between a successful or an unsuccessful installation—a profitable or unprofitable operation.

Other than the purely technical aspects, there are reasons of business and procurement (such as availability) which influence the selection of materials. As this is a technical manual, however, these factors will not be discussed here.

Each "phase" includes one or more of the following:

Mechanical requirements
Chemical requirements

Thermal requirements
Moisture requirements
Safety requirements
Economic requirements

Evaluation of the Requirements

Lack of factual information makes the evaluation of these "phase" requirements most difficult. The past tendency of the industry has been to hide problems rather than to examine them and arrive at a solution. Also, a requirement of one phase may be exactly opposite to the requirement of another phase. Thus, a compromise must sometimes be made in the desired properties of a material in order to partially fulfill the requirements of each phase. To better understand these total requirements, each of the phases will be examined.

Phases

Shipping

To be usable, an insulation produced in one location and used in another must be capable of being transported.

Rigid insulation must have a combination of compressive, tensile, and shear strengths to permit shipment without excessive breakage, or excessively costly protective packaging.

Non-rigid materials, such as blankets, should have sufficient strength in length, width, and depth to resist delamination. Their binders should withstand vibration without dusting or excessive fiber release. Fibrous materials must resist permanent compaction caused by either compressive forces or vibration.

Fill materials must also withstand excessive compression or compaction, even though packaged in bags.

Practically all insulating materials must be packaged in cartons, bags, rolls, or some other type of acceptable shipping container. Proper selection of container and packaging influences the cost of handling and space requirements. The use of mechanical handling makes pallet shipping economical. This can be used, economically, not only for large bulk handling, but also for storage. However, such handling may require materials (or packaging) of higher compressive strength.

During loading, shipping, and unloading, materials may encounter water damage caused by rain or water vapor due to high humidity. The more highly absorbent or adsorbent the insulation is, the greater the hazard of excessive moisture pick-up. Thus, the greater the absorbency and adsorbency, the greater the need for care in handling and weather-tight packaging.

Although few insulations are subject to self internal heating, those which have this property become a fire hazard during shipment unless they are shipped in quantities that are sufficiently small to be below their critical mass. Of course, combustible insulations are a hazard because they will support combustion if ignited. Even those which burn slowly may produce very toxic smoke and gases.

Storage

Insulation is seldom used immediately upon receipt from the manufacturer. In most instances it is stored first by a distributor, and later on the job. During these periods of storage it often receives more physical abuse than it will receive in service. Thus, rigid materials must have sufficient strength to withstand such handling. Flexible materials must withstand the same type of handling without compaction, or pulling apart. Of course, all insulations should be dimensionally stable and not warp or shrink from moisture or age.

Few inorganic insulating materials are subject to chemical change during storage. However, some of the organic foam-spray chemicals may be affected by shelf life or temperature. These must be stored and used within their limits, or they may become unusable. The same applies to some accessory materials such as adhesives, sealers and weather barriers.

The problem of keeping absorbent insulations dry during this storage period is most important. An insulation which contains a large percentage of water may lose its strength, or chemically deteriorate, and will always lose its thermal efficiency. Improper warehousing and handling of absorbent insulation can be costly. Not only the insulation itself, but its packaging may be adversely affected by water. Water damage to cartons, with resultant loss of identification and damage to the contents, may be very expensive. Needless to say, the more absorbent an insulation is, the greater the need for its protection from water in any form.

During storage, safety to property and personnel must be considered. Any combustible insulation should be stored with care. Fire will spread rapidly in such light density material. Sealers, weather-vapor-barriers, or other liquid combustibles, are always a potential fire hazard. Many insulations which do not support combustion by themselves will burn in the presence of flame. Any organic insulation, weather-barrier, sealer, adhesive, cement, or jacketing will produce toxic fumes when burning. Safe storage is, therefore, essential for these materials.

Fabrication

Advanced techniques have made it practical and economical to prefabricate, preform, or premold insulation shapes prior to their installation in the field.

A rigid material requires sufficient mechanical strength to permit handling without breakage. It should have good cutting characteristics. Its cut surfaces should be sufficiently smooth and free of dust in order than pieces can be cemented together into strong, integral finished shapes.*

For ease of fabrication, the original blocks of insulation should be dimensionally true within acceptable tolerances. Out of squarre, non-parallel, and untrue planes in the original blocks will cause excessive waste of both material and labor. Pipe covering used in fabrication should have true, concentric cylindrical surfaces. Any material not dimensionally stable is difficult to fabricate

*Dimensions for fabrication are available in ASTM Recommended Dimensional Standards for "Prefabrication and Field Fabrication of Thermal Insulation Fitting Covers" Part of ASTM C-450.

into usable fittings or shapes. Materials which delaminate when cut also cause excessive labor and waste. In most instances, materials that have good tensile strength in all directions are more suitable for cutting into compatible parts, and can then be bonded together into the finished shape.

Light density, fibrous materials generally do not lend themselves to this cutting and cementing type of fabrication. Fittings composed of these materials are formed most frequently by molding and heat-curing uncured batts or blankets into the desired shapes. Inorganic insulations are treated in a similar manner. Organic insulations may be molded into desired shapes. Although the application of these products is steadily increasing due to their usefulness in reducing field labor and installation time, the molding techniques will not be discussed further as specialty molding is beyond the scope of this Manual.

During fabrication, or molding, safety is an important consideration. Dust, besides being annoying, may be a serious health hazard. The dust count and type of dust should be carefully controlled to prevent serious disability claims. The same is true of toxic fumes from molding. Combustible fumes from solvents, or the solvents themselves used in adhesives or coatings, may constitute a fire hazard and should be treated with care.

Application

Since application requirements depend upon the individual installation, this also, must be considered. To illustrate:

Building insulation. Insulation in walls needs little strength after the wall is completed, but must be able to withstand the necessary handling up to the point of installation. The dimensions should meet standard building construction. Wherever possible, vapor barriers (if required) should be a part of the manufactured product, or so designed that they can be installed effectively with a minimum of effort.

Insulations placed on the underside of ceilings or roofs should be of relatively light weight for ease in overhead handling. However, insulations light enough to be blown away by a moderate breeze are difficult to install on the top of a roof.

Insulations applied to duct work by adhesives must resist delamination. The surfaces should be suitable for adhesives, so that good, fast, and permanent bonds are obtained. If installed with pins and clips, the surface of the insulation should be soft enough to allow clips to be embedded below the surface, but strong enough to prevent the insulation from being easily pulled off.

Adhesives and sealers used in the installation should not be toxic or extremely flammable as their use within the confined spaces of a building could create a definite hazard.

Insulations whose surfaces are excessively dustty may cause a health hazard, or at the very least an excessive cleanup bill.

Cold storage insulation applications require a similar procedure. In addition they must be able to be installed on an existing vapor-barrier, or on insulation with its own integral preformed vapor-barrier, in such a manner that efficient seals can be constructed between vapor-barrier sections. When walls of insulation must be free-standing, the insulation must have such properties that the workman can cut, fit, and install it to form such walls. Movement will occur as soon as the room is put into service, therefore, expansion and contraction joints must be suitably designed regarding both materials and ease of installation. Any supporting structures, such as a suspended ceiling system, or supports for doors and inside paneling, must be designed so that insulation can be installed without heat short circuits. A consideration of the items which must pass through any insulation—beams, electrical conduits, refrigeration pipes, etc.—must be made to insure proper design and specification of the material and accessories, as well as efficiency of application. In many installations, lack of design of these details improperly places the burden of how to accomplish a completed system upon the field insulator. Later, when failure occurs, the workman, lacking knowledge, time, and proper material for the construction of these junctions, is frequently blamed for doing a poor job.

Sprayed, or foamed in place insulations present another set of problems. When they are sprayed or foamed from the interior, the vapor-barrier must be installed first. It must be sufficiently strong to resist the expansion forces of the foam. Also, the foam must have excellent adhesion to the barrier. Proper formulation, guns, and mixing are necessary to obtain even foaming. As the temperatures of the surface and the ambient air affect the reaction, a proper formulation is necessary to allow for the conditions under which the application will be made. During application, most of these materials are flammable and toxic so provisions for protection of the workman and the property are required.

Industrial applications impose additional requirements on the insulation materials which are important to a successful installation. The materials must have the necessary strength to resist an excessive amount of handling. Storage space at most of these installations is limited, and the materials may be moved many times before they are finally installed. They may have to be hoisted to elevated scaffolds and be used in positions where the workmen are unable to hold a piece other than by one edge.

When secured by wires materials must resist the tendency to crack into pieces due to cleavage along the wire, yet be of such texture that twisted wire ends can be embedded into their surface. Vessel block insulation and rigid pipe covering must withstand considerable force when pulled up to a tight fit by straps. The strapping tool will exert a 600-800 lb tensile pull with an insulation strap to draw the joints tight.

The dustier a material is the more difficult it becomes for field personnel to work with and install it. Dust, besides being a health hazard, irritates the eyes, makes scaffold boards slick, and causes a cleanup problem. In addition, dusty surfaces resist good bonding with insulation and finishing cements, and weather-barrier mastics.

The tendency of some insulating materials to absorb moisture adds to the cost of installation. Such installations must be protected from water, rain, and snow, both before and during their installation. The need for weather protection before, during, and after installation until the weather-barrier is installed adds considerably to costs.

Trueness of dimensions of block pipe covering is essential to efficient field installation. If the ends of pipe covering are not square and true, the gaps must be plugged with insulating cement, or the end recut to fit. Either of these operations is estimated to cost between $1.00 and $2.50 (depending upon the size and conditions) in labor time. Poorly manufactured insulations are a luxury that few can afford, and one that no one should have thrust upon him.

Generally, the insulation industry has done an excellent job in producing efficient materials at a reasonable price within their own plants, but have overlooked the various phases through which they must pass from manufacture to final service. Especially overlooked are the problems of the insulation mechanic who attempts to effectively install material which is difficult to handle, cut, and fit. Considerable concern has been expressed about the excessive cost of installation labor, but a good percentage of this cost stems from the fact that the materials themselves were not properly designed or produced for ease of application.

The requirements just discussed are those imposed upon an insulation before it has to perform as such. Service requirements, although more tangible, are many, and must be presented in more detail and with more technical classification.

Service

To perform its intended service, an insulation must remain where it was installed. Its properties must fulfill all the requirements imposed upon it during its service life. It is impossible to present all the various installation conditions and their corresponding requirements. Typical installations are presented to assist and guide the insulation designer.

As the service phase is complex, the requirements are divided into various technical divisions. The order in which these are given may seem incorrect to some, for one would expect the first division discussed to be thermal requirements. However, in the evaluation of an installation, the physical requirements most frequently dictate the type of insulation neccessary to serve the needs. Then, the insulation of this type which best serves the balance of the requirements—Thermal, Safety and Economic—is selected. Thus, the Mechanical Requirements are considered first.

Mechanical Requirements

Building installations

Walls. Insulations installed in the cavities of walls, protected and supported on each side by structural building materials, require little physical strength. In such installations, fill insulations, blanket insulations, batt insulations, reflective foils, and light density rigid and semi-rigid insulations are quite useful, as they are not required to resist external physical forces. However, walls do move and vibrate, and these insulations must have the ability to expand and contract with the movements of the walls and must not compact or settle.

Panel sections, such as organic foams bonded to internal and external surfaces, are subjected to internal stresses caused by the independent movement of the panel surfaces. Such applications require that the insulation be sufficiently rigid to hold the sandwich construction together, but sufficiently flexible that does not shear, fracture, or lose its adhesion to the panels.

Where insulation used on the surface of a wall is exposed to mechanical abuse, it must possess the properties of both an insulation and of a structural material. It must have the rigidity and dimensional stability necessary to withstand the mechanical forces which may be imposed upon it.

Floors. Direct compression, vibration, and lateral movement are forces imposed upon insulations used under slab floors placed on the ground. Such insulations need a particular set of properties. They should be sufficiently resilient to withstand uneven loading. They must not compact, or break down under vibration, and must have sufficient compressive strength to support both the weight of the slab and all external loads imposed on it.

The required physical characteristics of insulation installed below floors, but above grade, depends upon the type of construction of the floors, and their location. Construction where insulating board is secured to wood joists below a floor requires insulation to be sufficiently rigid so it will not sag excessively. However, if a structural board is used under the joists, the space may be filled with blanket type insulation which will be supported by the board. These considerations are the same for ceiling installations.

Roofs. Roof deck insulations are subjected to many stresses. A building roof constantly changes its dimensions, due to temperature changes inside the building. These dimension changes tend to be transmitted to the insulation by the adhesives (generally asphalt) holding the deck and its insulation together. On the other side, the roofing also attempts to change dimensions because of atmospheric temperature fluctuations. The insulation is sandwiched between these two differential and often directly opposed expansion changes. The insulation itself must either be flexible enought to conform to these dimensional changes, or it must be so constructed (with expansion and contraction joints) to allow its parts to move relative to one another without damage as their dimensions change. On large area roofs where dimensional changes can be extremely large, almost all insulations must be installed with such expansion joints. An insulation having excessive shrinkage or expansion characteristics greater than those of the deck materials will prove unsatisfactory.

The air which is contained in an insulation and is free to escape from its inner and outer surfaces can cause excessive pressure under the roofing materials in the partial vacuum caused by winds blowing across the roof. Unless the insulations are properly adhered to the deck and have sufficient tensile strength normal to the surface of the roof, the partial vacuum will either pull the roofing or the insulation off, or destroy the insulation itself. This force, capable of pulling insulation off a deck, is frequently overlooked in roof design.

Concentrated compressive loads are imposed upon deck insulation by people walking on the roof and by items resting on the roofing. The insulation must resist these compressive forces without permanent deformation. Point loading, resulting from persons walking on the crushed rock surfacing, tends to puncture roofing, especially if the insulation is soft. The insulation should have sufficient hardness of surface and compressive strength so that point loading becomes distributed over a sufficient area to prevent puncture under common roofing traffic.

Cold Storage. Insulation most frequently is installed inside, or as part of, a building. It is either used with, or becomes part of the building structure. Thus, normal forces acting on, and contained within building structures must be considered in the design of the insulation, or its supporting structure. Besides these normal forces, additional ones, caused by differential expansion of the building in relation to that of the storage room, are also present. Such differential movements, when incorrectly calculated and provided for in the design, are responsible for many of the major failures in cold storage rooms.

Industrial insulation

Industrial insulation, like building insulation, covers a wide range of various requirements, depending upon the particulars of the design and service.

Cavity insulation. Small forces are generally exerted upon insulation used in cavities, as most insulations are quite compressible in relation to the differential movement between the walls of the cavity. However, furnaces constructed with insulated cavity walls, double shelled vessels, and totally enclosed units filled with insulation do have their own sets of physical requirements. These applications require insulation that can be poured, blown, or tamped into place. They require an insulation that is easily handled and resilient; one that resists compaction and excessive dusting. Many such installations also require that the insulation be such that it can be suction pumped out of an enclosure for equipment repairs.

Another type of cavity application employs fill insulation around buried, hot pipes. In addition to the requirements mentioned above, this insulation must be free of cracks and crevices, as it consolidates around the pipe. It must have the strength to withstand the backfill pressure, yet be sufficiently flexible in service to resist the formation of cracks or cleavage lines when the dimension of the pipe changes with temperature.

Insulation on ducts and flat surfaces. The stresses imposed upon duct insulation depend upon the size, location, and operating temperature of the ducts. Small size ducts may have insulation secured to them by straps, bands, wire, or an outer jacketing completely encircling the duct. Insulation so installed must be capable of withstanding the compressive force of the wires or bands without being crushed or excessively deformed.

Large ducts or flat surfaces make it impractical to attempt to secure insulation to the outside by the tension of straps or wire. Thus, the insulation must, in some way, be secured to the surface. Frequently, the insulation is bonded to the surface if the temperature of the duct is within the temperature range of adhesives. Or, pins may be attached to the duct surface, upon which the insulation is impaled, and then secured in place by clips forced over the pins. In other instances, the insulation is secured to the surface by wires attached to blank nuts or eyes welded to the duct surface. In all cases, expansive and contractive movement of the duct (or flat surface) is transmitted directly to the insulation by its securements. If the insulation is in large panels, they must have the flexibility to withstand this movement. When insulation is composed of, and installed in blocks, they will move independently of each other, and the joints will open and close with temperature change. In such instances, one inch hexagonal wire mesh is frequently secured over the entire surface, then plastered with insulating cement, thus providing a flexible insulation layer over the block. Insulation is unable to resist the forces of expanding metal, so it must either flex with the movement, or be provided with expansion joints to allow for the needed movement.

Often when large hot ducts are insulated steel wire, road mesh, or similar materials are placed over the duct ribs to which the insulation is secured by pins or wires. This provides an air space between the duct and the insulation (except at the ribs). Originally, many thought such an application would eliminate some of the expansion-contraction problems. Unfortunately, the road mesh changes dimensions along with the duct, therefore the same dimensional changes are transmitted to the insulation. In addition, such an application produces some more problems: (1) the ingassing and outgassing of air through the insulation due to the air pressure on the space between the duct and insulation attempting to maintain equal pressure with ambient air, and (2) the movement of enclosed air due to the convection currents. If the insulation does not have the physical property to withstand this movement of air from face to face and across its inside face, it will be eroded away.

Large, outside ducts are subject to similar lifting and tearing forces of wind as described for insulation installed on roof decks. The ''free'' air in insulation plus the weather-barrier can exert a large force when wind creates a partial vacuum over any of the outer surfaces.

The internal pressure from a leak in a pressure duct can build up under the insulation and weather-barrier and cause rips in the protective cover, or force off the insulation. Insulation contractors should advise engineers and general contractors that ordinary insulation is not designed to contain pressure.

Insulation on the tops of large ducts must be sufficiently strong so that the weight of snow or rain does not cause the center to depress further than the sides. Such a depression will create a water pocket which eventually will make or find a crack and leak into the insulation.

Insulation on Equipment and Pipes

Mechanical requirements

Insulation on equipment located in industrial plants is subject to external and internal physical forces. The external forces may be bumps, persons walking upon the insulation, vibration, wind, or even explosions. Although this last may seem an extraordinary force to consider, many insulations are expected to protect pipe and equipment from fire. Many fires are started by an explosion, and if the insulation is blown off of surrounding items and piping, it cannot protect against fire.

The insulation should be sufficiently strong to resist ordinary usage. This may even entail a workman walking or standing upon insulated vessels and pipes. The compressive strengths of most insulations are less than the several hundred pounds per square inch pressure occurring when a man places his heel down to take a step. Therefore, the amount of depression in the insulation, related to the ability of the weather-barrier to spread without puncture over the permanent depression in the insulation, is most critical.

Vibration is frequently overlooked in the selection of insulation. This force causes abrasion between the insulation and the surface to which it is secured, and between fibers in fibrous materials. The insulation must either be sufficiently vibration-resistant, or special, anti-abrasive coatings must be made part of the system to prevent it from dusting or wearing away.

The force of wind has already been discussed. The fact has been brought out that both wind pressures and partial vacuums, respectively, tend to compress, or pull the insulation apart.

The forces most frequently damaging to insulation applications are those built into the system by poor design and application, which ignore the movement of vessels and pipes caused by thermal expansion (or contraction) of the metal. In considering the requirements of an application made under these conditions, it must be realized that the bad effects apply not only to the insulation, but also to the weather-barrier (and vapor-barriers on low temperature installations). For this reason, an explanation of the action of weather-barriers must be included.

Expansion and contraction of vessels and pipes can cause serious damage to thermal insulation and weather-barrier coverings. In addition to the growth of high temperature vessels and pipes, the problem is amplified by the shrinkage of most high temperature insulations as the temperature rises. The fact that metal vessels and pipes expand with temperature increase and insulation shrinks, indicates some method must be provided to allow for these dimensional changes. If not, the insulation (if rigid) and the weather-barrier will be fractured.

From a practical standpoint, the *size* of the crack is insignificant. The additional heat loss from heated vessels and pipes, due to cracks, would be quite small, if all other factors remained the same. The big difficulty is water entering the cracks. Wet insulation has a thermal conductivity approximately fifteen times that of dry insulation. Thus, in addition to heat loss resulting from the crack, wet insulation becomes inefficient. Each pound of water which enters a crack will absorb approximately 1070 Btu's (313.5 Watts) as it is vaporized by the heat of the vessel or pipe. These losses are significant.

Cracks in low temperature insulation, and its weather-vapor-barrier, can cause even greater losses than in high temperature insulation. Entry of water causes ice formation, rapidly and completely ruining such insulation. Iced-up low temperature insulation will transmit approximately forty-five times as much heat as dry conventional mass insulation. So, efficiency demands that either hot or cold insulation be kept free of cracks.

The amount of growth of a vessel or pipe depends upon its coefficient of expansion and its temperature rise from ambient temperature. Of course, low temperature operation is the reverse.

Most coefficients of expansion are presented at a given mean temperature, such as steel at 70° F (21.1° C) (mean) being 6.5 × 10^{-6}. These figures can be confusing. For this reason the following table is presented.

Table 55 Thermal expansion of metals (inches/100 ft—based on 70°F ambient)

Material	Operating temperature °F						
	200	300	400	600	800	1000	1400
Steel	0.99	1.82	2.70	4.60	6.70	8.89	13.34
Stainless Steel	1.46	2.61	3.80	6.24	8.80	11.48	16.92
Aluminum	2.00	3.66	5.39	9.03			
Copper	1.51	2.69	3.89	6.40			

Table 55a Thermal expansion of metals (mm/10 metres, based on 20°C ambient)

Metric Units

Material	Operating Temperature °C					
	100	150	200	300	400	500
Steel	8.51	14.33	16.94	33.87	48.60	63.94
Stainless Steel	12.54	20.56	23.83	45.94	63.83	82.60
Aluminum	17.18	28.63	33.80	66.48		
Copper	12.98	21.19	24.39	47.12		

As previously stated, most high temperature insulations shrink as they are heated. The following table gives the typical shrinkage of some commonly used high temperature insulations.

Table 56 Heat shrinkage of insulation (inches 100/ft—based on 70°F ambient)

Material	Temperature °F				
	700	1000	1200	1500	1900
Asbestos fibers (rigid)	0.12	0.30	0.60		
Calcium silicate	5.70	10.80	18.00		
Calcium glass (has no shrinkage, but a coefficient of expansion)					
Diatomaceous earth (rigid)	*	*	*	*	20.00
Expanded silica	3.00	4.50	5.00	15.00	24.00
Mineral wood block (rigid)	7.00	9.80	12.00	15.50	

*No test data available

Table 56a Heat shrinkage of insulations (mm/10 metres, based on 20°C ambient)

Metric Units

Material	Temperature °C				
	400	500	600	800	1000
Calcium Silicate	47.91	77.68	128.75		
Cellular Glass (Has no shrinkage—but a coefficient of expansion.)					
Diatomaceous Earth (Rigid)	*	*	*	*	149.0
Expanded Silica	25.21	32.37	35.76	114.0	179.0
Mineral Wool (Rigid)	58.83	70.49	85.83	117.0	

With the exception of cellular glass, the materials listed will shrink, when heated to their service temperature.

The magnitude of the problem can be determined from the above tables. For example: if a steel column, operating at 750° F was 100 feet high, the change from an ambient temperature of 70° F to its operating temperature would increase its height 6.2″. If it were insulated with calcium silicate, the mean shrinkage (half way through its thickness) would be 3.85″. Thus, the total dimensional difference would be 10.05″. If the vessel were 10 feet in diameter, the change would be 0.62″ (or 1.95″ in circumference) and the linear mean difference in the insulation circumference would be 0.9 inches.

In Metric Units the steel column would be operating at 399° C and is 30.48 metres in height, the change from ambient temperature of 21 ° C to its operating temperature would increase its height 0.1575 metres. If it were insulated with calcium silicate the mean shrinkage would be 0.0978 metres. Thus the total dimensional difference would be 0.2553. If the vessel were 3.05 metres in diameter the change would be 0.01575 metres (or 0.0495 in circumference) and the linear mean difference would be 0.0229 metres.

From the preceding, it is easily understood that it would be impractical to attach each piece of insulation directly to the shell of this vessel. It shows the need for allowing for differential movement between the vessel and the insulation, and the necessity for providing expansion joints in the insulation.

On low temperature vessel and pipe applications, a similar problem exists. The amount of shrinkage of a vessel or pipe when reduced from a 70° F (21.1° C) ambient temperature to various operating temperatures is given in the following table.

Table 57 Thermal contraction of metals (inches/100 ft–based on 70°F ambient)

Material	Operating temperature °F					
	32	0	−50	−100	−150	−200
Steel	0.29	0.50	0.83	1.12	1.38	1.62
Stainless steel	0.45	0.77	1.30	1.76	2.16	2.51
Aluminum	0.58	0.99	1.66	2.30	2.91	3.46
Copper	0.42	0.73	1.24	1.70	2.10	2.44
Nickel	0.30	0.53	0.89	1.26	1.56	1.85
Monel	0.34	0.55	0.96	1.31	1.66	1.87

Table 57a Thermal contraction of metals (mm/10 metres, based on 20° ambient)

Metric Units

Material	Operating Temperature °C				
	0	−20	−40	−80	−100
Steel	2.24	3.98	5.78	9.18	12.31
Stainless Steel	3.48	6.13	9.06	14.43	19.07
Aluminum	4.49	7.88	11.57	18.85	26.30
Copper	3.25	5.81	8.64	13.94	18.55
Nickel	2.32	4.22	6.20	10.33	14.06
Monel	2.63	4.37	6.69	10.74	14.21

Most cellular insulations used for low temperature service do have a definite coefficient of expansion (contraction). Thus, the problem becomes one of differential contraction.

The following tables gives the thermal contraction of various cellular, insulation materials.

Table 58 Thermal contraction of cellular insulations (inches/100 ft–based on 70°F ambient)

Material	Operating temperature °F					
	32	0	−50	−100	−150	−200
Cellular glass	0.21	0.37	0.61	0.94	1.21	1.48
Cellular styrene	1.6	2.94	5.03	7.13	9.23	11.34
Cellular urethane	2.3	4.2	7.20	10.18	13.18	16.17

Table 58a Thermal contraction of cellular insulation (mm/10 metres, based on 20°C ambient)

Metric Units

Material	Operating Temperature °C				
	0	−20	−40	−80	−100
Cellular Glass	1.63	2.94	4.72	7.71	9.19
Cellular Styrene	12.38	23.41	38.94	58.45	70.16
Cellular Urethane	17.80	33.44	55.74	83.45	100.18

From these tables, the difference in contraction between the metal and insulation, when reduced from ambient temperature (70° F, 21.1° C) to operating temperature, can be determined.

Metal contracts slightly more than cellular glass, whereas organic foams contract much more than metals, thus the design of low temperature installations using cellular glass is fundamentally different from installations of cellular organic foam. In the case of cellular glass installations on cylindrical metal pipe or equipment the outside diameter of the metal contracts slightly more than the inside diameter of the cellular glass. Cellular organics contract much more than the cylindrical surfaces to which they are applied and for this reason overlapping slip contraction joints are necessary to prevent breaks and openings in the cellular organic insulations.

Past practices of insulation design have ignored this differential expansion and contraction between the insulation and the metal to which it is applied. This lack of proper design has probably contributed to more insulation failures, poor operating service, poor service life, and greater heat losses (or gains) than any other single factor.

Chemical requirements

The chemical requirements which determine the choice of insulation may be one or many. As there is always a possibility of spillage or leaks, insulations should be used which do not react to the chemical contained in the vessel or piping to which they are attached. Such a reaction may cause a lowering of the ignition temperature of the chemical in contact with the insulation, or, over a longer period of time, spontaneous combustion may occur. One such reaction occurs when oxygen comes into contact with various materials. Many organic materials, when in contact with oxygen, can be detonated by simple, physical shock.

Another factor to be considered is that the insulation should be non-absorptive when used on toxic processes. This is a frequently overlooked factor. However, if leakage occurs, a highly absorptive insulation may soak up a poisonous material. Even if it is then removed safely, disposal still remains a major problem. These same problems must be considered when insulation is used in nuclear work. Radioactive insulation, or its dust, is highly dangerous.

A selected insulation should not be chemically corrosive to the metal to which it is applied. Basically, insulation installed on steel should be neutral, or slightly alkaline. That installed on aluminum should be neutral, or slightly acidic. However, these are only very broad guide lines, because chemical attack may be caused if salts or other components leach out of the insulation.

A major factor of concern in the past few years has been the failure of stainless steel process equipment and piping caused by external stress corrosion. The stress corrosion cracking of stainless steel results from the presence of a concentration of halogen ions (most frequently chloride ions) on the surface of a tensile stressed, austenitic, stainless steel alloy at warm to hot temperatures. However, the mere presence of chloride ions on the surface is not enough to cause stress corrosion. A certain concentration of the ions is necessary. When the stainless is held above ambient, and a material traps the moisture containing the ions in place on the surface, concentration of ions occurs as the moisture evaporates. When insulation, or components used in the insulation system, contain chlorides they can cause stress corrosion of austtenitic stainless steel.

However, stress corrosion can occur under completely chloride-free insulation, if the installation is located in a salt-laden atmosphere, chloride contaminated atmosphere, or where spillage or fumes contain organic or inorganic chlorides. In such cases, an insulation holds these contaminates to the surface where the concentration can occur. If such contaminated conditions exist, the best protection for the austenitic stainless steel is to paint it with a suitable chloride free and chloride resistant coating prior to the application of the insulation.*

Thermal requirements

The thermal requirements of an installation are related to a number of properties of materials. It may be easier to understand if these properties are explained before listing the requirements. These thermal properties of materials are as follows:

Temperature limits
Thermal Shock Resistance
Thermal Diffusivity
Thermal Specific Heat
Thermal Conductivity

The temperature requirements of an installation are the maximum and minimum temperatures to which the insulation will be subjected. This appears simple. However, there are degrees of subjection. For example, many materials may be satisfactory if heated on one surface, but may fail if subjected to a soaking heat. Frequently a material will withstand a high temperature for a short period of time, but will begin to disintegrate if subjected to that same temperature over a long period of time. Another factor is the time-rate of raising or lowering the temperature. For this reason, when an installation's temperature requirement is evaluated, not only the high and low service temperature limits are necessary. The following questions must also be answered: is it on only one side, is it continuous, intermittent, cyclic, or rapidly changed?

The effect the temperature has on the materials must also be considered. With the exception of combustible materials which will completely burn when subjected to excessive heat (or fire), the limit of usefulness of an insulation most frequently occurs at the temperature at which some physical property begins to break down. This imposes a caution on the design engineer in that the physical properties listed for most insulations are values determined at room temperature (70° F), not at the maximum temperature for which the insulation is recommended.

In many high temperature installations, the maximum temperature is set above that point at which excessively high shrinkage will occur. However, before this point is reached, the compressive, tensile, and flexural strengths may be only small percentages of that which the material possessed when it was tested at room temperature. For this reason, it must be remembered that the practice of the industry is to test the physical properties of insulation at conditions when the material is *not* in service.

In installations where rapid heat-up and cool-off is a factor, an insulation of high thermal diffusivity, low specific heat, and low density is desirable. Conversely, where high heat retention and slow temperature change is desired, low thermal diffusivity, high specific heat, and high density are desirable. To illustrate, let us consider an example of the two extremes.

1. A process changing from hot to cold every few minutes requires an insulation that has the ability to change temperature quickly and has very small mass to retain heat.
2. Insulation used to resist fire (fireproofing) must have a low rate of heat diffusivity, very high mass, and high specific heat in order to slow down the temperature attempting to penetrate it and reach the metal it is attempting to protect.

An installation which has sudden temperature changes can cause thermal shock to the insulation. Sudden thermal changes in insulation can be caused by rain or snow. Where insulation is used as fireproofing, it, naturally, must be able to withstand thermal shock. This is due not only to the rapid rise of temperature caused

*For more detailed information on this subject, see "External Stress Corrosion Cracking of Stainless Steel" by W. G. Ashbaugh, Materials Protection, May 1965.

by the fire, but also to its sudden reduction when it is hit by the water from a hose.

Emittance is the thermal factor which effects the surface temperature of an insulation exposed to hot air or hot gas. When the insulation is directly exposed, its emittance is an important factor. The emittance of reflective insulation, where metal composes both the inner and outer surfaces is definitely a deciding factor in its function as an insulation. Most mass insulations are jacketed or coated on the exposed surface so that the emittance of the jacket becomes the deciding factor. An installation that has a critical surface temperature, either high or low, must have the selection of the insulation (or its jacketing) made with care. Typical installations where emittance is vital occur where insulation is used for personnel protection, or where excessive surface temperatures might cause ignition of fumes or gases. On low temperature installations, surface temperatures should be above dew point to prevent condensation and drip. A general rule to remember: the lower the emittance, the closer the outer surface temperature will be to the temperature of the surface under the insulation. Conversely, the higher the emittance, the closer the outer surface temperature will be to the temperature of the ambient air.

Moisture requirements

Insulation, to be efficient, must be kept dry. Moisture in two forms, the liquid or the vapor state, can saturate the insulation. It can enter by various means, water pressure, vapor pressure, ingassing, and simple leaks. The protection required will be determined by the installation conditions.

When insulated piping or equipment is submerged in water, a completely water-tight outer shell is required around the insulation. Such systems must be encased in metal, fabricated by welded sections, and/or flanged with jacketed flanges, so that the outer jacket is sufficiently tight and strong to withstand the water pressure.

Underground installations where insulation systems may not be directly subjected to definite water pressure must still be protected to prevent the entry of water. Such protection must not only prevent the said entry, but it must also resist corrosion to remain water-tight. The corrosion may be ordinary rust, chemical corrosion, or galvanic corrosion.

If the function of the two previously mentioned examples is the insulation of hot surfaces, vapor migration will be of no concern. However, if the installation is for low temperature service, water vapor migration will be an additional problem. Many conduit systems may be water-tight and *not* be vapor tight. Practically, it is relatively easy to construct water-tight systems, but *almost impossible* to construct vapor-tight ones. For this reason, it is suggested that, wherever possible, the cold equipment and piping be installed above grade.

The moisture protection necessary on equipment and piping located above grade is determined by whether the location is indoors or outdoors, and the operating temperature above or below ambient.

High temperature insulated equipment and piping located indoors and not subjected to rain, snow, or sleet does not require water or moisture protection. A word of caution—the insulation must be adequately protected from moisture due to possible spillage or the washing down of vessels.

Low temperature insulated equipment and piping located indoors must be protected against moisture vapor. The necessity for such protection was explained in Chapter 3, and illustrated in Figure 45.

High temperature insulated equipment and piping located outdoors must be protected from liquid moisture in the form of rain, sleet, or snow.

Low temperature insulated equipment and piping located outdoors must be protected both from liquid water, and also moisture in the vapor phase.

These moisture requirements are summarized below:

Service		Moisture Protection Required		
Temperature	Location	Water pressure	Water	Vapor
High temp.	Underwater	Required	Required	Not required
Cyclic temp.	Underwater	Required	Required	Required
Low temp.	Underwater	Required	Required	Required
High temp.	Below grade	Depends	Required	Not required
Cyclic temp.	Below grade	Depends	Required	Required
Low temp.	Below grade	Depends	Required	Required
High temp.	Above grade indoors	Not required	Depends	Not required
Cyclic temp.	Above grade indoors	Not required	Depends	Required
Low temp.	Above grade indoors	Not required	Depends	Required
High temp.	Above grade outdoors	Not required	Required	Not required
Cyclic temp.	Above grade outdoors	Not required	Required	Required
Low temp.	Above grade outdoors	Not required	Required	Required

Safety requirements

The major safety requirements may be separated into the following divisions:

Safe surface temperature for personnel protection or ignition.
Safety from radioactivity or chemical reaction.
Safety from toxic conditions.
Fire protection.

When hot equipment or piping is located where its insulation may be touched by personnel, the insulation should be so designed that its thermal resistance, surface emittance, and jacket conductivity together create a condition which will not produce skin burns when its outer surface is accidentally touched. In general, metallic surfaces at a temperature over 145° F (00.0° C) can cause skin burns. A specific figure is not possible as this depends upon the individual, the roughness of the surface, the cleanness of the surface, and the conductivity and mass of the metal. The higher the conductivity of the metal, the smoother the surface, the cleaner the surface, and the greater the mass of metal, the more likely a person is to be burned at a particular temperature.

Where combustible materials are present and chances of

leakage or spillage prevail, a suitable low surface temperature of insulated vessels and piping may be very important in preventing the ignition of these materials. When this condition prevails, it becomes necessary that all hot surfaces, such as valves, flanges, and metal projections, be insulated so as to maintain all exposed surfaces below the maximum allowable surface temperature. Some areas which cannot be insulated, such as valve bonnets and stuffing boxes, may require shielding to prevent contact with the chemical.

As previously mentioned, non-combustible but absorptive insulation may be a fire hazard in two ways. First, in the case of combustible material leaks, the insulation's tremendous internal surface area can assist in causing rapid oxidation which will result in spontaneous combustion. Second, also in the case of leaks, any absorbent can hold large quantities of combustibles and, if an accidental fire occurs, the saturated insulation will feed and spread it. To illustrate the potential hazard—a cubic foot of saturated insulation can hold approximately six gallons of combustible material. In case of ignition, the insulation then acts like the wick of a lantern. Even though it does not itself burn, it conveys the vaporizing combustible to the surface where it can burn. For these reasons, it is evident that the process, the materials being processed, the potential of leaks, and the degree of hazard of the location, all must be considered in the selection and design of an insulation system.

The potential fire hazard of an insulation, as related to an installation, has been shown. However, properly used insulation can be used as fire protection. An installation on equipment and piping must be considered from two basically different viewpoints. First, the combustibility of the insulation must be considered, including the speed at which the insulation will spread fire, or the amount of heat and smoke it will contribute. Second, the time rate at which temperature will pass through the insulation from the fire side to the vessel or pipe must be thought of.

Where fire *spread* is hazardous, the system should be of non-combustible insulations which do not contribute fuel to the fire. In many instances light weight insulations and thin metallic jackets will meet these requirements.

If vessels and piping are to be protected for a period of time from accidental fire, then the time required for the high temperature of the fire to pass through the insulation to the pipe or vessel is of utmost importance. Where this is a requirement, light weight materials of high diffusivity are of little advantage even though they are incombustible. To illustrate: a piece of wood will provide much greater time-temperature protection to a vessel than any insulation of 2 to 3 lbs. density, even while burning on its outer surface. Of course, the use of wood is not recommended for fire protection as it can add considerable fuel to a fire. It was mentioned only to show that, for temperature retardation, the rate of diffusivity and mass are the deciding factors, not conductivity (by itself) or combustibility.

Economic requirements

The economic value of the insulation depends upon the installation requirements, and, as this has been previously discussed, a simple listing of the factors to consider is given.

1. Heat transmission allowable as determined by the process.
2. Savings in boil-off or vaporization of the product.
3. Temperature which must be maintained in the process.
4. Fire protection required by the process. Insulation as related to savings in additional fire protection such as spray heads, larger safety valves, etc.
5. Savings in investment in heat or refrigeration equipment.
6. Savings in heat or refrigeration energy.

Summary of requirements

The preceding was presented in very general form for a number of reasons. Little work has been done in the evaluation of insulation requirements by any educational institution or independent body to obtain sufficient *facts* that could be transformed into any numerical values. Some of this type of investigation has been done by private industry. However, the facts obtained, for the most part, have been for an individual process and for individual sets of costs and design conditions.

Proper evaluation of installation requirements as related to selection of insulation materials and design of systems is a difficult task. For this reason, many insulation installations are incorrectly designed. Hopefully, this discussion will guide engineers in the evaluation of installation requirements.

Miscellaneous

In addition to the installation requirements important enough to justify individual headings, many occur only in certain segments of the industry, and in this context merit mention.

Marine work

This branch of the industry, while coming under the general heading "Industrial", is still a distinct field and has some special facets which should be considered.

1. The moisture laden air characteristic of this service is especially conducive to fungus growth. Care should be taken to select insulations, weather- and vapor-barriers, and protective coatings that are resistant or immune to fungus attack.
2. The legendary attraction of ships for rats is no legend. It is hard fact. And another hard fact is that rats, when hungry, will develop an appetite for many strange things—insulation and weather- and vapor-barriers included. Some means of combatting this must be devised.
3. *Clearances.* Due to the crowded quarters inherent in marine construction, the working clearances available, in which to install the insulation, are much less than in normal "Industrial" work. This fact should be kept in mind in the design of such a job.

Food industry

Another branch of the "Industrial" classification, this industry also has its own special problems.

1. *Contamination.* It takes no special knowledge to realize that our food must be kept free from any contaminate—including insulation and its accessories. Insulation used for this service should be strong enough to prevent slivers, as nearly dust-free as possible, and non-toxic, in case of accidental contamination. The same precautions apply also to the weather- and vapor-barriers and the protective coatings.
2. *Rat-Proofing.* The same comments as given under "Marine Work" also apply here even more stringently in the contamination and cleanliness phases.

Soap industry

This branch of the "Industrial" tree, is generally about the same in its requirements as the balance, but it does have a couple of special requirements.

1. *Contamination* which would affect the color or purity of their formulations must be avoided at all costs. Again, no slivers or dust, to prevent, as nearly as possible, accidental contamination.
2. *Glass.* This highly regarded material is used, in the soap industry, only on storage tanks and pipes removed from the actual formulation area, due to the possibility of severe skin lacerations in the event of accidental contamination.

6 Weather-Barriers, Vapor-Retarders, Indoor Coverings, and Finishes

Weather-barriers, vapor-retarders, indoor coverings, and finishes all have one basic function—to protect insulation. Each provides protection in a different manner. Thus, the designation of each is a specific name for a *protective function.* These names are not names of materials.

A weather-barrier. A material, or materials, which, when installed on the outer surface of thermal insulation, protects the insulation from the weather, such as rain, snow, sleet, wind, solar radiation, or atmospheric contamination, and mechanical damage.

A vapor-retarder. A material or materials, which, when installed on the high vapor pressure side, retards the passage of the moisture vapor to the lower vapor pressure side.

A weather-barrier, vapor-retarder. A material, or materials, which serves both as a weather-barrier and a vapor-retarder.

A condensate-barrier. A material, normally used as an inner lining for the metal jacket weather-barrier of an insulation installation, which will bar the alkaline condensate, which normally tends to form on the inner surface of the metal jacket, from contact with the metal jacket.

An indoor covering. A material which protects the insulation from mechanical damage and wear and tear on indoor applications.

Note: There is no commonly accepted name for this function of materials and, in order to establish a title for it, this name was chosen.

An insulation finish is a material, or materials, applied to the insulation to provide the final contour, a smooth, even finish, and to strengthen the outer surface.

Note: This term may be applied to this function, since it has been well established by usage, i.e., insulating "finishing" cement, asbestos "finishing" cement.

An appearance covering. A material, or materials, used over insulation, the weather-barrier or indoor covering to provide the desired color or texture, for decorative purposes.

These are the names of functions, not the names of materials. However, frequently the names of these functions are combined with generic names of materials to designate the material and its function, i.e., weather-barrier mastic, or weather-barrier jacket, or even weather-barrier vapor-retarder mastic.

One name commonly used in the insulation industry which should be explained is "lagging adhesive." Such material is used to adhere canvas or other reinforcing cloth to the surface of insulation, and in that sense it is an adhesive. However, the same material is used as a coating over the cloth, the two of which form an *indoor covering* and *appearance covering.*

Service Requirements

In the performance of the functions listed previously, the materials used as weather-barriers vapor-retarders, indoor coverings, or appearance coverings are subjected to many of the forces of the installation, as mentioned in the previous chapter, in addition to the elements they are resisting for the benefit of the insulation. These service requirements, classified technically, are the same as those listed for insulation in the previous chapter, and include:

Mechanical
Chemical
Thermal
Moisture
Safety
Economic
Application

It might be assumed that weather requirements were forgotten in the above listing. However, the requirements imposed by weather, from a technical point of view, may be broken down into the same categories as listed above.

Of course, not all requirements are applicable to every installation, nor all for each of the functions. Also, the importance of each for a given installation must be evaluated. For this reason, these requirements will be given generally and noted specifically only when applicable to a particular function.

It must be stressed that the purpose of the barriers and coverings is to *protect* the insulation. Poorly designed barriers and coverings which fail to protect the insulation result in failure of the entire system. In the past few years, due to lack of understanding of this principle, there has been a tendency to cheapen the protective systems to the point where they are ineffective. Protective systems which allow insulation to become wet are protective in name only, and economically unsound.

Mechanical requirements—Internal forces

The weather-barriers and coverings are subjected to the movements of the equipment and piping caused by their thermal expansion and contraction. This motion is, in turn, transmitted by the insulation to its outer surface. In addition to the forces of expansion and contraction, *differential* movement is caused by the shrinkage of high temperature insulation, which tends to concentrate all these dimensional differences in lines, rather than spreading them over the entire area. On hot surfaces, the lateral movement transmitted to the outer barrier, or covering, is tension. This is illustrated in Figure 51. The circumferential expansion problem is illustrated in Figure 52.

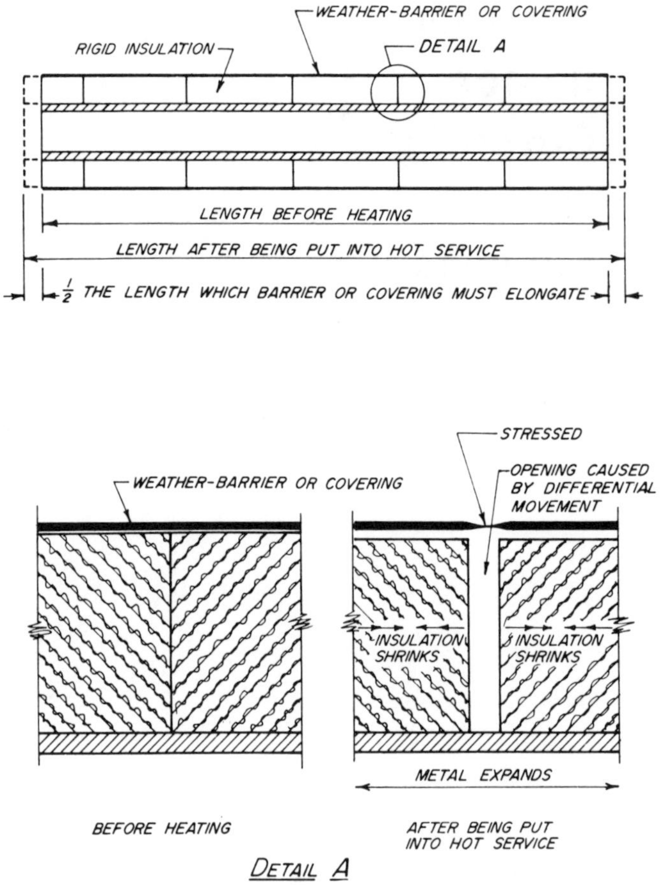

Figure 51 Effect of differential movement on weather-barrier or covering.

On low temperature applications where metal contracts rather than expands, joint openings will occur when the coefficient of expansion (contraction) of the insulation is greater than that of the metal pipe or vessel to which it is applied. In such instances, the effects are similar to those shown in Figures 39 and 40. When the coefficient of expansion of the insulation is the same, or less, than that of the metal, the metal tends to shrink away from the insulation circumferentially, and the metal tries to compress the insulation longitudinally. Unless this compression causes fractures in the insulation itself, it may cause slight wrinkles in mastics or felts, and wrinkles in metal jackets if contraction joints are not provided.

Vibration of pipes and equipment will produce internal forces that can damage weather-barriers and coverings. Such vibration can cause insulation blocks to work loose from their securement and move in an up and down direction with a resultant shearing force between their abutting surfaces. This small movement on the otuer surface will flex weather-barrier mastics or coverings at this moving joint, eventually causing mechanical fatigue and failure.

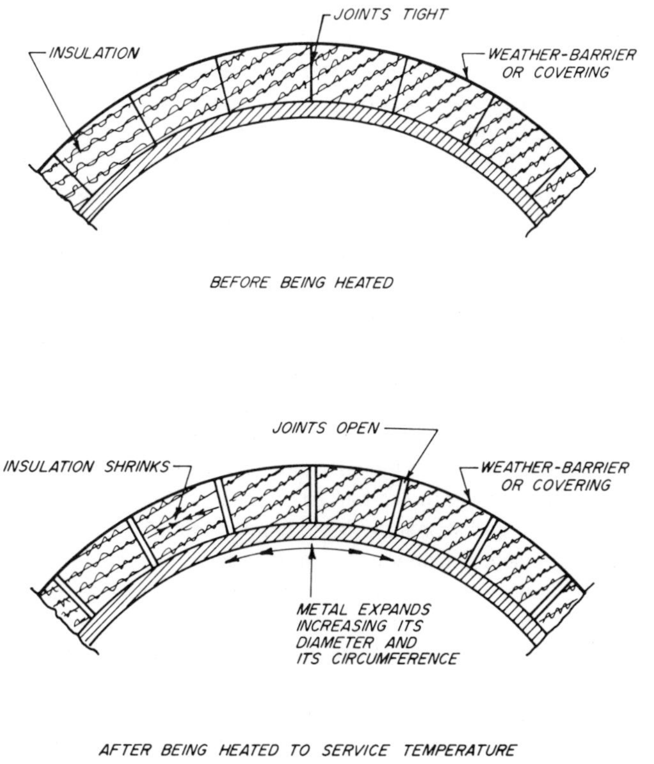

Figure 52 Effect of expansion movement on weather-barrier or covering.

Mechanical requirements—External forces

The external mechanical forces to which weather-barriers and coverings are subjected, and must resist, are more apparent. These are cutting or shearing, abrasion, impact, and compression.

Cutting or shearing forces occur when sharp objects are dropped or driven against the surface.

Abrasion results from objects rubbing against the surface, expansion and contraction, or from vibration causing the insulated surface to move against a stationary object.

Impact can be caused by the surface being bumped, being walked upon, by having objects dropped upon it, or even by being in the path of the shock wave of an explosion.

Compressive force may result from objects resting on the insulation's surface, or the insulation surface resting on a stationary object.

Wind can cause these forces to act on the insulation surface. High velocity winds can drive sharp objects into the surface. Sand and grit blown by wind cause abrasion. Wind creates pressure on one side and a partial vacuum on the other. When a weather-barrier is on an insulation and has little resistance to air movement, the trapped air attempts to move to the partial vacuum. In so doing, it attempts to push the weather-barrier from the pipe or equipment surface, with resultant tensile or tearing action. These are illustrated in Figure 53.

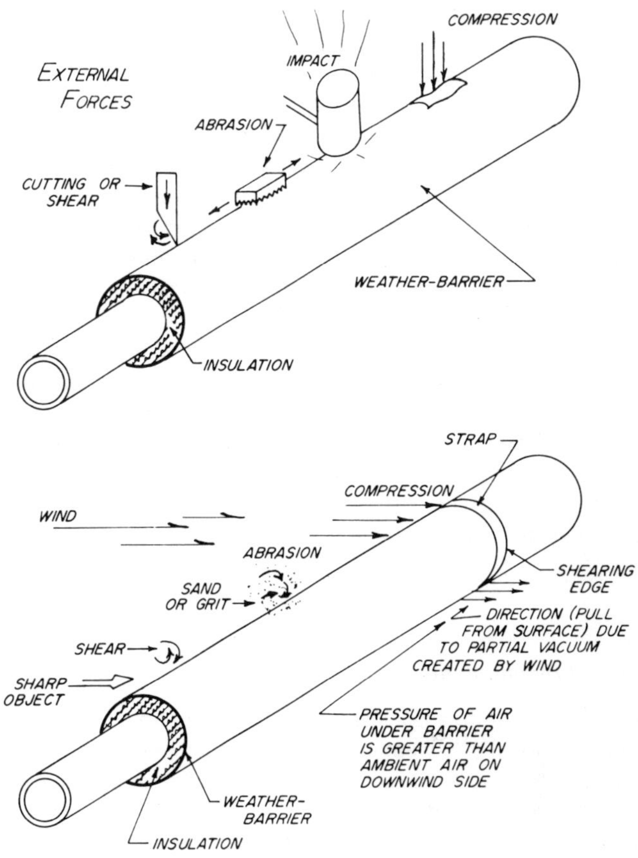

Figure 53 External forces to which weather-barrier are subjected.

Chemical requirements

One of the first chemical requirements is that the barrier, or covering, be compatible with the insulation to which it is applied. If either causes a chemical reaction, or exerts a solvent action on the other, a different selection of materials is indicated.

Primarily, the weather-barriers, or covers, should not contribute to rusting or corrosion of the insulated metal surfaces. Weather-barriers or coverings, adhesives or sealers which contain halogen salts soluble in water (especially chlorides) are just as capable of causing stress corrosion of stainless steel as the basic insulation containing such ions.

The atmospheric environment must be considered in the choice of materials. Most industrial environments contain some contamination which may cause reaction, or corrosion, of barriers or coverings. Plants located near coast lines will definitely be surrounded by salt-laden air, which, of course, will subject the exterior surfaces to corrosion because of the collected salt. In addition to the atmosphere, attach may be caused by accidental spillage or blow-off from vent valves. Such direct chemical attach may ruin barriers or coverings in a few minutes.

The effects of the ultra-violet and infra-red rays in the atmosphere and the oxidizing effects of the air are some of the biggest causes of deterioration of barriers and coverings.

Another factor to be considered is galvanic corrosion. This phenomenon occurs when two pieces of metal of different electrical potential are either in contact, or electrically connected by an electrolyte. This will become a factor when a metallic jacket is a different metal than the pipe and the insulation becomes soaked, thus providing the electrolyte. It will also occur between the metal jacket of underground metal jacketed, insulated pipe lines and other metallic buried pipes, and the salts-laden moisture in the earth acting as the electrolyte. In this particular instance the other buried pipes do not need to be close, as galvanic currents have been known to travel long distances to do their nefarious work.

All these factors—compatibility, chemical resistance, atmospheric resistance, and galvanic corrosion—must be considered in making the proper choice of materials to be used as barriers or coatings over insulation. Frequently, for a given difficult installation, the advice of the Corrosion or Paint Engineer may be helpful in obtaining a satisfactory solution.

Thermal requirements

On low temperature applications, the temperature on the surface of the insulation (the weather-vapor-barrier temperature) will be less than the surrounding air. For this reason, the temperature at which these materials must function satisfactorily is somewhat less than the expected ambient temperature. To illustrate, if the ambient air temperature is expected to be as low as 0 °F (−17.8 °C), the insulation surface temperature of a vessel or pipe operating at −100° F (−73.8° C) may be expected to be as low as −10° F (−23.3° C). Excessive brittleness of a mastic weather-barrier and vapor-barrier resulting in cracking, may cause an early failure of the insulation installation.

As shown in Chapter 1, ("Determination of Insulation Thickness Required to Prevent Condensation") the emittance of the outer surface of the weather-barrier-vapor-retarder on insulation makes considerable difference in the thickness of insulation required to prevent condensation, due to its effect on surface temperature. One of the most common errors in insulation design is looking up the insulation thickness to prevent condensation in a table based on an emittance value of 0.9, then selecting a weather-vapor-barrier of much lower emittance, especially low-emittance metallic jackets. When such error is made, the insulation thickness is most frequently only 35 to 50% of the required thickness. Even if the proper high emittance barrier is selected, the entire design can be upset if the color is changed. To illustrate: if an insulation design is based on a weather-barrier-vapor-retarder having a surface emittance of 0.9 and the installation is made in accordance with these factors, and if, after installation, it is decided to "dress up" the application with a coat of nice, glossy, white paint, with an emittance of 0.4, the surface will then sweat at temperatures well below design conditions.

On high temperature applications, the temperature of the surface of the insulation (the weather-barrier or covering) will be greater than the temperature of the surrounding air. Depending upon the operating temperature and insulation thickness, this surface temperature can be from 10° to 75° F (5.5° C to 42° C) higher than the surrounding air—not considering the solar radiation effect. Surfaces directly exposed to the sun can increase an additional 75° F (42° C). Because of these two effects, the temperature of the weather-barrier or insulation covering can be as much as 150° F (83° C) above ambient air temperature. Thus, barriers and coverings used on high temperature must be suitable for constant operation at 200° F (93.3° C) and intermittent high temperatures up to 250° F (121° C).

As in the case of low temperature applications, the surface emittance of the barrier or covering affects the outer insulation (barrier or covering) surface temperature. On high temperature installations, the lower the surface emittance the higher the surface temperature. An extremely low surface emittance can increase surface temperature 50° to 75° F (28° to 42° C). This is due to heat passing out through the insulation and not being transferred on to the ambient air by the barrier or covering. The low surface emittance is an additional barrier to heat transfer. Thus, the temperature gradient is higher at the surface than if the heat, which passed through the insulation, was more rapidly dissipated. Unless the insulation thickness is increased to compensate for this surface resistance, highly reflective metallic jacket surfaces on hot insulation may be a personnel hazard, for accidental contact may produce severe skin burns.

Thermal conductivity of the weather-barrier, or covering, is seldom thought to be of importance. However, there are several places where this property affects their function. In the preceding paragraph, personnel burns are considered. Here, one of the deciding factors determining the safety of a surface is its conductivity. When a high temperature surface is in direct contact with a lower temperature body (in this case, skin) the greater the conductivity of the high temperature surface, the greater the amount of heat that will be transmitted. Thus, a high conductive metal will cause burns at a much lower temperature than low conductive materials. To illustrate: insulation materials which have been dried out in an oven at 225° F (107° C) can be taken from the oven and readily handled bare handed (if careful not to crush) with little danger of skin burns. However, metal being handled at that temperature would blister the skin immediately. Low conductivity barriers and coverings have an advantage when they are used over insulation which may be damaged by thermal shock or temperature deterioration. In instances of accidental fire, this restriction of rapid temperature change and heat input is sometimes the difference between rapid failure or slow disintegration.

If openings in the insulation, caused by pipe or vessel expansion and insulation shrinkage, occur, then hot spots can develop. This can cause rapid disintegration of the barriers, and dangerous open spots in the insulation. If combustible materials are used, such breaks in the insulation may produce combustion in the barriers. This is one major reason for the use of the multiple layer, broken joint method of installation on extremely high temperature surfaces.

As both high and low temperatures affect the physical and chemical properties of barriers and coatings, much of the data given regarding these materials has little significance unless related to the extremes of temperature to which they will be subjected. This is another area where improved test methods and material evaluation is needed to provide sound technical information for the design engineer.

Moisture requirements

The function of a weather-barrier is to prevent the entry of water or other liquids into the insulation system. Liquid protection, as it refers to weather-barriers, is restricted to rain, sleet, snow, hosed water, or spillage. The weather-barrier is not designed to resist water and pressure when submerged. Weather-barriers, properly installed with flashings, rain drips, overlapping downward joints, and the outer lap on outside, can prevent entry of liquids above grade, but would be completely unsatisfactory if submerged or if used underground.

The weather-barrier keeps insulation protected from liquids above grade in either hot or cold installations. They may be required indoors as well, if equipment or piping is washed down with water, or subjected to an excessively concentrated moisture laden atmosphere.

Rain and snow are generally accompanied by wind acting as a driving force to push the liquid into every crack and crevice. Water and liquids are the prime enemy of an effective insulation system. Unfortunately, the weather-barrier hides its own short-comings from view. Thus, a completely ineffective weather-barrier may look excellent from the outside, but may be wasting heat. Because this waste and inefficiency is hidden, the tendency in the past few years has been to use inefficient, poorly designed, and poorly installed weather-barriers. *Ineffective weather-barriers are today's single greatest cause of insulation inefficiencies and failures in service.* More about this subject will be discussed under economics.

When used on high temperature insulation, weather-barriers should allow moisture *vapor* to pass outward while still keeping *liquid* moisture from coming inward. By the time they are installed, almost all high temperature insulations contain a fair percentage of moisture. Then, when the weather-barrier is installed, this moisture is trapped. As soon as heat is applied to equipment, or pipe, this moisture becomes vapor (or steam) at a relatively high pressure in relation to the ambient vapor pressure. If the heat is applied too rapidly, or the weather-barrier mastic has an exceedingly low vapor migration rate, blisters will form, and the vapor will condense in these blisters. For this reason, weather-barrier mastics should *not* have a low vapor permability. If the weather-barrier is constructed of metal, suitable vapor vents are required.

The function of a vapor-retarder is to prevent as nearly as possible, the passage of moisture *vapor.* It is used to prevent moisture in the atmosphere, in the form of vapor, from entering insulation installed on surfaces operating at temperatures lower than ambient. The following table of vapor pressures can be used to determine the pressure that can be exerted on the vapor barrier when the ambient air is in a saturated state. To illustrate: If a pipe is operating at 0 °F (−17.8 °C) in an atmospheric temperature of 70 °F (21.1 °C), the vapor pressure difference in lbs/sq ft will be:

In English Units

Vapor pressure at 70° F saturated =	52.28	lbs sq ft
Vapor pressure at 0° F saturated =	2.67	lbs sq ft
Vapor pressure difference =	49.61	lbs sq ft—Ans.

In Metric Units the pipe would be operating at −17.8° C in an atmospheric temperature of 21.1° C and the vapor pressure difference in pascals (Pa) would be:

In Metric Units

Vapor pressure at 21.1° C saturated =	2503.2	Pa
Vapor pressure at −17.8° C saturated =	127.7	Pa
Vapor pressure difference =	2375.5	Pa—Ans.

If the atmospheric temperature is 90° F, and the pipe temperature remains at 0° F, then:

In English Units

Vapor pressure at 90° F =	99.58	lbs sq ft
Vapor pressure at 0° F =	2.67	lbs sq ft
Vapor pressure difference =	96.91	lbs sq ft—Ans.

If the atmospheric temperature is 32.2° C and the pipe temperature remains at −17.8° C then:

In Metric Units

Vapor pressure at 32.2° C saturated =	4815.4	Pa
Vapor pressure at −17.8° C saturated =	127.7	Pa
Vapor pressure difference =	4687.7	Pa—Ans.

The vapor pressure difference at which a vapor-barrier must operate is significant. From this it becomes apparent that, to resist these pressures, the retarder must be continuous and free of pin holes or cracks, otherwise it is ineffective. The idea held by many that a few pinhole leaks in vapor-barriers is unimportant is as senseless as feeling that a pinhole in an automobile tire is not important, for it represents only about one millionth of the total area of the tire. Pinholes or cracks allow rapid migration of vapor. It has been estimated by one manufacturer of drying equipment that moisture vapor will migrate through small orifices at a speed of 200 miles per hour (321.8 kilometres/hr). In other words, by positive pressure of dry gas, moisture can be prevented from entering a hole if the dry gas is forced out of the opening at a rate faster than 200 miles per hour (321.8 kilometres/hr).

Table 59 Vapor pressures in air at 100% relative humidity

English and Metric Units

Above ice at temperature indicated				
Temperature		Vapor Pressure		
°F	°C	In. of Hg	Lbs/sq ft	Pa
−40	−40	.0039	.2758	13.2
−35	−37.2	.0052	.3678	17.6
−30	−34.4	.0070	.4951	23.7
−25	−31.7	.0094	.6648	31.8
−20	−28.9	.0126	.8912	42.7
−15	−26.1	.0167	1.181	56.6
−10	−23.3	.0220	1.556	74.5
− 5	−20.6	.0289	2.044	97.8
0	−17.8	.0377	2.667	127.7
1	−17.2	.0397	2.808	134.4
2	−16.7	.0419	2.964	141.8
3	−16.1	.0441	3.119	149.3
4	−15.6	.0464	3.282	157.1
5	−15	.0488	3.452	165.3
6	−14.4	.0514	3.636	174.1
7	−13.9	.0542	3.869	183.5
8	−13.3	.0570	4.032	193.0
9	−12.8	.0599	4.237	202.8
10	−12.2	.0629	4.449	213.0
11	−11.7	.0661	4.675	223.8
12	−11.1	.0695	4.916	235.4
13	−10.6	.0730	5.163	247.2
14	−10.0	.0767	5.425	259.7
15	− 9.4	.0806	5.701	272.9
16	− 8.9	.0847	5.990	286.8
17	− 8.3	.0889	6.288	301.0
18	− 7.6	.0933	6.599	316.9
19	− 7.2	.0979	6.924	331.5
20	− 6.7	.1028	7.271	348.1
21	− 6.1	.1078	7.625	365.1
22	− 5.6	.1131	7.999	383.0
23	− 5.0	.1186	8.388	401.6
24	− 4.4	.1243	8.792	420.9
25	− 3.9	.1303	9.216	441.2
26	− 3.3	.1366	9.662	462.6
27	− 2.8	.1432	10.128	484.9
28	− 2.2	.1500	10.609	507.9
29	− 1.7	.1571	11.112	532.0
30	− 1.1	.1645	11.635	557.0
31	− 0.6	.1723	12.186	583.5
32	− 0.0	.1803	12.753	610.6
				N/m^2

Table 60 Vapor pressures in air at 100% relative humidity

English and Metric Units

Above water at temperature indicated				
Temperature		Vapor Pressure		
°F	°C	In. of Hg	Lbs/sq ft	Pa
32	0.0	.1803	12.752	610.6
33	0.6	.1878	13.283	636.0
34	1.1	.1955	13.827	662.0
35	1.7	.2035	14.393	689.1
36	2.2	.2118	14.981	717.2
37	2.8	.2203	15.582	746.0
38	3.3	.2292	16.211	776.2
39	3.9	.2383	16.855	807.0
40	4.4	.2478	17.527	839.1
41	5.0	.2576	18.220	872.3
42	5.6	.2677	18.934	906.5
43	6.1	.2782	19.677	942.1
44	6.7	.2891	20.448	979.0
45	7.2	.3004	21.247	1017.3
46	7.8	.3120	22.068	1056.6
47	8.3	.3240	22.916	1097.2
48	8.9	.3364	23.794	1139.2
49	9.4	.3493	24.706	1182.9
50	10.0	3626	25.646	1227.9
51	10.6	.3764	26.623	1274.6
52	11.1	.3906	27.627	1322.7
53	11.7	.4052	28.659	1372.2
54	12.2	.4203	29.728	1423.3
55	12.8	.4359	30.831	1476.1
56	13.3	.4520	31.970	1530.6
57	13.9	.4686	33.144	1588.9
58	14.4	.4858	34.360	1645.1
59	15.0	.5035	35.612	1705.0
60	15.6	.5218	36.906	1767.0
61	16.1	.5407	38.244	1831.0
62	16.7	.5601	39.616	1896.7
63	17.2	.5802	41.037	1964.8
64	17.8	.6009	42.501	2034.9
65	18.3	.6222	44.008	2107.0
66	18.9	.6442	45.564	2181.5
67	19.4	.6669	47.169	2258.4
68	20.0	.6903	48.825	2337.6
69	20.6	.7144	50.529	2419.2
70	21.1	.7392	52.283	2503.2
71	21.7	.7648	54.044	2589.9
72	22.2	.7912	55.961	2679.3
73	22.8	.8183	57.878	2771.1
74	23.3	.8462	57.311	2865.6
75	23.9	.8750	61.889	2963.1
76	24.4	.9046	63.982	3063.3
				N/m^2

Table 60 Vapor pressures in air at 100% relative humidity

English and Metric Units

Above water at temperature indicated				
Temperature		Vapor Pressure		
°F	°C	In. of Hg	Lbs/sq ft	Pa
77	25.0	.9352	66.147	3166.9
78	25.6	.9666	68.367	3273.3
79	26.1	.9989	70.652	3382.7
80	26.7	1.032	72.993	3494.7
81	27.2	1.066	75.398	3609.9
82	27.8	1.102	77.944	3731.8
83	28.3	1.138	80.490	3853.7
84	28.9	1.175	83.108	3979.0
85	29.4	1.213	85.795	4107.7
86	30.0	1.253	88.625	4243.1
87	30.6	1.293	91.454	4378.6
88	31.1	1.335	92.425	4520.8
89	31.7	1.378	97.466	4666.4
90	32.2	1.422	99.578	4815.4
91	32.8	1.467	103.76	4967.8
92	33.3	1.513	107.01	5123.5
93	33.9	1.561	110.41	5286.1
94	34.4	1.610	113.87	5452.1
95	35.0	1.660	117.41	5621.4
96	35.6	1.712	121.09	5797.5
97	36.1	1.765	124.84	5977.0
98	36.7	1.819	128.65	6159.8
99	37.2	1.875	132.62	6349.0
100	37.8	1.933	136.72	6545.0
101	38.3	1.992	140.89	6746.0
102	38.9	2.052	145.14	6949.0
103	39.4	2.114	149.52	7158.0
104	40.0	2.178	154.05	7375.0
105	40.6	2.24	158.43	7585.0
106	41.1	2.31	163.39	7823.0
107	41.7	2.38	168.34	8060.0
108	42.2	2.45	173.29	8297.0
109	42.8	2.52	178.23	8533.0
110	43.3	2.60	183.89	8805.0
112	44.4	2.75	194.51	9313.0
114	45.6	2.91	205.82	9854.0
116	46.7	3.08	217.85	10430.0
118	47.8	3.26	230.58	11039.0
120	48.9	3.45	244.02	11683.0
122	50.0	3.64	257.45	12326.0
124	51.1	3.85	272.31	13038.0
126	52.2	4.06	287.16	18749.0
128	53.3	4.29	303.43	14528.0
130	54.4	4.53	320.41	15340.0
132	55.6	4.77	337.38	16153.0
				N/m^2

Table 60 Vapor pressures in air at 100% relative humidity

English and Metric Units

Above water at temperature indicated				
Temperature		Vapor Pressure		
°F	°C	In. of Hg	Lbs/sq ft	Pa
134	56.7	5.03	355.77	17033.0
136	57.8	5.30	374.87	17948.0
138	58.9	5.59	395.38	18930.0
140	60.0	5.88	415.89	19912.0
142	61.1	6.19	437.82	20962.0
144	62.2	6.51	460.45	22045.0
146	63.3	6.85	484.50	23197.0
148	64.4	7.20	509.25	24381.0
150	65.6	7.57	535.42	25634.0
152	66.7	7.95	562.20	26922.0
154	67.8	8.35	590.59	28279.0
156	68.9	8.77	620.30	29669.0
158	70.0	9.20	650.71	31155.0
160	71.1	9.65	682.54	32679.0
162	72.2	10.12	715.78	34270.0
164	73.3	10.61	750.44	35929.0
166	74.4	11.12	786.52	37657.0
168	75.6	11.65	824.00	39451.0
170	76.7	12.20	862.90	41314.0
172	77.8	12.77	903.22	43244.0
174	78.9	13.37	945.66	45276.0
176	80.0	13.98	949.41	47341.0
178	81.1	14.63	1034.78	49542.0
180	82.2	15.29	1081.46	51778.0
182	83.3	15.98	1130.26	54114.0
184	84.4	16.70	1181.19	56553.0
186	85.6	17.44	1233.53	59058.0
188	86.7	18.21	1287.99	61666.0
190	87.8	19.01	1344.58	64375.0
192	88.9	19.84	1403.28	67186.0
194	90.0	20.70	1464.11	70098.0
196	91.1	21.59	1527.06	73113.0
198	92.2	22.52	1592.84	76261.0
200	93.3	23.47	1660.03	79478.0
202	94.4	24.46	1730.06	82831.0
204	95.6	25.48	1802.20	86285.0
206	96.7	26.53	1876.47	89840.0
208	97.8	27.63	1954.27	93566.0
210	98.9	28.76	2034.19	97392.0
212	100.0	29.92	2116.24	101320.0
214	101.1	31.13	2201.82	105418.0
216	102.2	32.38	2290.24	109651.0
218	103.3	33.66	2380.77	113986.0
220	104.4	34.90	2214.48	118185.0
				N/m^2

Safety requirements

One safety requirement has already been mentioned under the thermal property of emissivity—that is, the importance of having surfaces below a temperature able to cause burns to personnel. Also, a surface must be provided which is below the ignition point of materials which may come into contact with the surface.

Fire should be a major concern in all installations. The external barriers or coverings are the first part of an insulation system which may be subjected to fire. The characteristics of materials as related to fire are many and varied. A list of these properties and their definitions follow:

Noncombustible. A material which will not contribute fuel or heat to a fire to which it is exposed.

Combustible. A material which will contribute fuel or heat to a fire to which it is exposed.

Combustion (general). A chemical process of oxidation that occurs at a rate fast enough to produce both heat and light either as glow or flames.

Note: Some oxidation, such as that of hydrogen, emits radiation outside the visible spectrum.

Fire resistance. That property which enables a material to resist decomposition or deterioration when exposed to a fire.

Fire retardance.* That property of a material which delays the spread of fire, either through or over it.

Flame (general). A hot luminous zone of gas, or matter, or both, in gaseous suspension, that is undergoing combustion.

Flame resistance. That property which enables a material to resist decomposition or deterioration when exposed to flame.

Flame retardance. That property of a material which delays the spread of flame, either through or over it.

Flammable. The property of a material which permits it to oxidize rapidly and release heat of combustion when exposed to flame or fire, and allows continuous burning after the external ignition source is removed.

Note: Many people do not take into account the marked difference of properties of materials when exposed to fire as compared to flame. A material may be nonflammable when exposed to a flame only, but completely flammable when exposed to fire. To illustrate, aluminum with a melting point of 1200° F can be exposed to a small flame continuously with no harmful deterioration, but when exposed to fire it melts in 45-50 seconds.

Nonflammable. That property of a material which prevents it from oxidizing rapidly and releasing heat of combustion when exposed to flame or fire.

Self-extinguishing.* That property of a material which enables it to stop its own ignition after external ignition sources are removed.

There are various degrees of flammability of combustibility. The properties which indicate degree of flammability are:

Flash-ignition temperature (flash point). The lowest temperature of a material at which i t gives off vapor, which, when combined with air near its surface, forms an ignitable mixture.

Fire point temperature. The lowest temperature of a material at which it gives off vapor, which, when combined with air near its surface, forms an ignitable mixture at a rate sufficient to support combustion continuously after the external ignition source is removed.

Ignition. The initiation of combustion. Combustion may be evident by glow, flame, detonation or explosion.

Auto ignition temperature, piloted ignition temperature. The temperature of a substance of which combustion occurs without an external ignition source. Such self-ignition temperature may be obtained either by surrounding ambient temperature or self-heating within the material itself.

Surface flame spread. The rate, expressed in distance-time at which a material will propagate flame across its surface. For a single material this time-rate will vary in respect to position, location, direction and temperature. Flame-spread index only provides a comparison of surface flame spread of one combustible with that of another, and as such is not a measure of any property of combustion. For this reason the use of flame spread index for engineering safety of insulation systems is not recommended. Organic films or foams that have melting temperatures, or gasification temperatures, which are lower than flash ignition temperature cannot be evaluated by Surface Burning Test ASTM E 84 as most of the combustibles have dripped away or been blown out the vent before combustion occurs.

Diffusivity. The property of a material which determines the time-temperature rate of change.

Flammability index is a measure of flame spread and heat released by a flammable material. This term has no significance.

Fuel contribution. The quantity of heat released by a unit of material by combustion of that material under specified conditions. (Fuel contribution factor (or index) is only a comparison of one combustible material to another, thus it provides no design information.)

*This term refers only to the flame or fire spread due to its own combustion—not the retardance of flame or fire from another source.

*This term, as used by industry, does not indicate the material will extinguish or tend to extinguish flame or fire from another source, other than its own flammability.

Heat release rate. The quantity of heat released per unit of weight (or volume) in a unit of time.

Heat retardant material. A material which restricts the flow of heat from one of its surfaces to another, i.e., thermal insulation.

Melting temperature. The minimum temperature at which a solid changes to liquid state.

Oxygen depletion rate. The quantity of oxygen removed from air per unit quantity of material consumed by combustion per unit of time.

Smoke. The airborn solid and liquid particulates and gases, evolved when a material undergoes pyrolysis or combustion. (Chemical smokes are excluded from this definition.)

Smoke density, optical. The opacity of the smoke per unit of length of light path.

Smoke emission. The volume of smoke produced by combustion per unit weight of material.

Smoke emission rate. Volume of smoke produced per unit weight of material burned in flaming or non-flaming conditions in respect to time.

Smoke index (classification). Comparison of the opaqueness of smoke of one combustible to that from another combustible under a set of specified conditions. (As a comparison, it provides no technical data.)

Smoke toxicity. The degree of health hazard of smoke.

Smoldering. Combustion of a solid material without flames.

Temperature retardant material. A material which delays change of the temperature of one of its surfaces as related to any temperature change of the temperature of its outer surface. (Materials effective in delaying temperature change must have low thermal diffusivity.)

Thermal diffusivity. The property of a material which determines the time-temperature rate change. (See equations 57 and 57a.)

Heat resistance. That property of a material which enables it to withstand extremely high temperatures without deterioration or failure.

Thermal shock resistance. The property of a material which enables it to withstand rapid changes in temperature without deterioration, cracking, spalling or other failure.

These safety properties listed in most instances relate to temperature resistances or combustibility. Following is a list of American Society for Testing and Materials Test Methods for various properties.

ASTM Test Methods No.	Property Title of Test Method
D-56	Flash Point by Tag Closed Tester
D-87	Melting Point of Petroleum Wax
D-92	Flash and Fire Point by Cleveland Open Cup
D-93	Flash Point by Pensky-Martens Closed Tester
D-127	Drop Melting Point of Petroleum Wax
D-568	Flammability of Flexible Plastics
D-635	Flammability of Self-Supporting Plastics
D-777	Flammability of Treated Paper and Cardboard
D-781	Test for Puncture and Stiffness of Paperboard, Corrugated and Solid Fiberboard
D-828	Test for Tensile Breaking Strength of Paper and Paperboard
D-1310	Flash Point of Liquids by Tag Open Cup Apparatus
D-1525	Vicat Softening Temperature of Plastics
D-1929	Ignition Properties of Plastics
D-2015	Grass Calorific Value of Solid Fuel by Adiabatic Bomb Calorimeter
D-2117	Melting Point of Semicrystalline Polymers
D-2026	Test for Mildew Resistance of Paper and Paperboard
D-2382	Heat of Combustion of Hydrocarbon Fuels by Bomb Calorimeter
D-2843	Density of Smoke from Burning or Decomposition of Plastics
D-2863	Flammability of Plastics Using Oxygen Index Method
D-3014	Flammability of Rigid Cellular Plastics (Vertical Position)
D-3211	Relative Density of Black Smoke (Ringelmann Method)
E-69	Combustion Properties of Treated Wood by Fire-Tube Apparatus
E-84	Surface Burning Characteristics of Building Materials (Comparison with dry wood with fire forced in horizontal travel by tunnel)
E-136	Noncombustibility of Elementary Materials
E-162	Surface Flammability of Materials Using a Radiant Heat Energy Source
E-286	Surface Flammability of Building Materials Using an 8 ft (2.44 m) Tunnel Furnace (Comparison with dry wood)
E-324	Relative Initial and Final Melting Points and the Melting Range of Organic Chemicals

It may be noticed that most of the tests are for mass materials not just sheets or coatings, and should be used, where applicable, to establish the burning characteristics and properties of combustible insulations. However, as fires—in most instances—start and impinge on the outside surfaces of most pipe or equipment insulation this presentation on combustibility was located in this chapter.

An installation must be evaluated as to its potential fire hazard. If an accidental fire can cause no great loss nor imperil life, then the fire resistant or fire retardant properties of weather- or vapor-barriers may be of little significance. However, in areas where fire might be of major concern, these properties will have great significance.

Depending upon hazard conditions, the desired functions which the barriers or jackets must fulfill may differ widely. In the case where fire is of little concern, combustible barriers or jackets may be quite satisfactory. Should there be some concern as to the magnitude of the fire, then the amount of allowable fuel contribution by the barrier may be the limiting factor. If the hazard must be confined to a given area, then the rate of fire spread must be considered.

Where hazards dictate that consideration be given to the fire properties of the insulation system, it is necessary to evaluate the application to determine what must be accomplished by the barriers or coverings. If the insulation is fire and heat retardant, it may be that the major function of the barriers would be to prevent the water from a fire hose from washing the insulation from the pipe or vessel. In this case, the barriers need to be fire resistant, but would not be required to be fire or heat retardant.

In another situation, the insulation may disintegrate at fire temperature. Then the weather-barrier or coating should be fire and heat retardant. In *most instances,* barriers and coatings which are fire and heat *retardant* are *not* fire *resistant,* despite the common misuse and intermingling of the terms by the industry. They undergo many changes when subjected to fire. They release moisture, become cellulated, or intumesce, thereby slowing the temperature rise of the surface to which they are applied. However, their properties are changed by the fire and consequently, they cannot be called fire resistant. In other words, they are designed to be sacrificed to the fire to protect the substrate.

To illustrate: a pipe insulated with 2 inches of combustible foam and jacketed with stainless steel and subjected to fire will slow the temperature rise of the pipe. Insulated with such a system, the pipe will reach 1000° F (538° C) in from 9-10 minutes. A similar insulation system with a weather-barrier of polyvinyl acetate, heavy-build mastic, overcoated with an intumescent coating will extend the time of the temperature rise of the pipe to 1000° F (538° C) to 29 minutes. However, in the first system, except for discoloration, the stainless steel jacket will be undamaged; whereas, in the second system, the polyvinyl acetate-intumescent barrier will be puffed, and deteriorated to the point where it is valueless as a weather-barrier. In both cases the foam insulation would be burned completely, with only char remaining.

Another factor in fire protection is the ability of the insulation to resist thermal shock. Fire itself can cause considerable shock, but during fire fighting, when the surfaces are quenched with water from a fire hose or sprinkler heads, many materials will shatter. If the insulation itself is unable to resist thermal shock, then it is quite important that the barrier or covering be sufficiently fire and shock resistant to stay in place and contain the insulation during the fire and fire fighting operation. This will prevent the protected metal surface from being directly exposed to the fire.

Conversely, if an insulation is combustible and can burn and propagate fire under a weather-barrier or vapor-barrier, the most hazardous arrangement that can be made is to provide a strong fire-resistant outer protective covering. If fire starts between the vessel, or pipe, and the outer covering, and the outer covering resists efforts by the water stream from the fire hose to knock it off, then the fire can spread, as it is protected from the water stream. This is similar to the problem of fighting a fire that is in the interior of a cavity wall.

The melting point of a barrier and/or coating can also be an extremely important factor in the spread of fire. A material which melts and produces combustible drippings can spread fire to lower levels. Such melting and combustible drip has been responsible for the spread of many fires, particularly roof fires.

Metals of low melting point can contribute to fire spread, even though their hot drippings are not combustible. However, molten metal at 1200° F (649° C) will act as a secondary ignition source wherever it drops on any combustible material.

Chemicals may penetrate the weather-barriers and/or coverings by accidental spillage or atmospheric contamination and change their fire characteristic from noncombustible to combustible. For this reason, it is frequently necessary to evaluate these barrier materials after subjecting them to chemicals they may come into contact with in service.

Another danger stems from weather-barriers or vapor retarders which contain solvents of low flash point. During application, and for a short time afterward, until these low flash point solvents are evaporated, the solvent vapor-air mixture forms a highly hazardous, flammable combination and will readily ignite from a stray spark or flame. Vapors, other than water vapors, are hazardous to health, and precautions, such as the use of adequate ventilation and/or fresh air masks, should be observed.

For these reasons, a careful evaluation of service requirements in regard to fire exposure, and other hazards is necessary to determine what weather-barriers or vapor-retarder coatings are suitable for the insulation system.

Economic requirements

The first economic requirement of weather-barriers, vapor-barriers, and indoor coverings is to protect the insulation from mechanical damage and moisture entry. Although this is so fundamental that it seems unnecessary to even mention it, it is the most disregarded fact in the insulation industry. If barriers allow the insulation to become wet, due to either liquids or vapors, the system definitely will fail.

On hot service, moisture which enters the insulation may be revaporized by the heat and driven out again. This simple fact leads many persons to believe that a few leaks are unimportant. However, each pint of water that enters the insulation will require a minimum of 1000 Btu to revaporize it. This revaporization is necessary to force it back out of the insulation after the source of the water (rain, etc.) has been removed. Logically, each pint of water vaporized from insulation per hour will require at least one lb per hour in steam generating capacity. Depending upon capital cost of steam capacity, this represents a capital investment of from $20.00 to $75.00. If insulation becomes completely wet, its thermal conductivity increases and its efficiency decreases. For the most absorbent high temperature insulations, completely wet insulations are approximately 10% as efficient as dry insulations. In other words, the heat loss per hour can be multiplied by ten. These two factors make leaky weather-barriers a very expensive luxury.

On low temperature applications, moisture, in the form of vapor, which enters the insulation system and condenses into water or ice is trapped and will continue to build up until the insulation system is ruined and must be replaced. The effectiveness of vapor-barriers and vapor seals determine the efficient service life of insulation installed on low temperature vessels and pipes. Insulation which is iced up has only 3-7% of the efficiency of dry insulation. Thus, low temperature vapor-barriers, or vapor seals, are all-important to efficient use of low temperature insulation.

In today's economy, so much emphasis has been placed on initial cost that we have overlooked the true economics.

First and foremost, the weather and vapor systems must protect the insulation as planned. If they do not, they are worthless. We must remember that the weather-barrier and vapor-barrier which keeps the *insulation effective* for the *lowest cost per year* is the proper system to use.

Application requirements

The location, shape, and size of the insulated surfaces to be protected by weather-barriers or coverings exert a very definite influence on the choice of such materials. Clean cylindrical surfaces can be easily encased with jackets, but when jackets must be fitted to various contours, or around fittings, valves, and odd shaped equipment, the labor of tailoring to fit, and sealing joints becomes relatively difficult as compared to the application of mastics.

It is frequently difficult to install jackets properly on large flat surfaces because it is almost impossible to pull them sufficiently taut so that they will not sag on the underside, or be pulled loose on the sides and/or top by wind. Such applications generally require weather-barriers that cling to the surface by adhesion.

Size makes a difference in the selection of mastic weather- or vapor-barriers. Large surfaces lend themselves to economic spray application, whereas smaller and complex surfaces are more efficiently applied by trowel or palm.

As these characteristics will be discussed at greater length under the properties of materials, a detailed discussion will not be presented here.

Types of Weather-Barriers

Weather-barriers may be jackets or mastics.

Jackets may be felts, films, laminates or metal.

Felt weather-barriers may be of rag felt or asbestos felt. These are generally saturated with asphalt or tar. However, some are made with the asbestos exposed on one side. As would be expected, saturated rag felts are more flammable than the asbestos felts.

Most of these felts were developed for the roofing industry, and the insulation industry, naturally, started using them as a weather-barrier over outdoor piping. As a group, these materials provide very satisfactory service life and were almost universally used as weather-barriers until industry became more concerned about labor cost, fire resistance, and appearance.

In an installation, these felts must be cut to fit, and all laps sealed in order to become an effective barrier. As regards labor, the most time consuming of installations is the water-sealing of all joints at the overlaps with an asphalt or tar base cement to obtain rain-tight joints. Holding the joints in position, especially longitudinal joints, until the cement has set is difficult and expensive.

To reduce the application expense, the industry began to simply wire the felts in place and left the joints unsealed. By this maneuver, they succeeded in installing felts in such a manner that they ceased to be a weather-barrier, but only hid the water soaked insulation. However, where fire resistance is not required and appearance not a factor, a good grade asphalt-asbestos felt—properly applied—is still one of the best weather-barriers available.

Film weather-barriers. With the development of the plastics industry, a number of plastic films were introduced as weather-barriers. These all had one problem—cold flow. After a period of time, they became loose and sagged. This was overcome by the addition of reinforcing strands. However, these films, like the felts, required that the joints be sealed to be effective.

Laminated weather-barriers. In the past few years, a number of excellent laminates have been developed for weather-barriers. These laminates consist of films laminated to felts or metal foil, with or without reinforcing strands. Some manufacturers have supplied these materials with contact adhesives for their joint areas, or have produced contact adhesive tapes to seal the joints.

Metal jackets may be galvanized steel, stainless steel, galvanized and prepainted steel, or aluminum.

Galvanized steel jackets were the first of the metal jackets used in the insulation industry. Although the joint problem, true of all jackets, was given relatively little consideration, its other properties seemed to favor it as a strong, fire resistant, long lasting weather-barrier. Its disappointing service stemmed from the fact that on hot lines and equipment the vapor pressure outward carried highly corrosive alkaline condensate to its inner surface where the galvanizing was attacked. Then the jacket rusted from *the inside outward.* Regardless of the maintenance painting on the outside, in a few years the galvanized steel jacket corroded and ceased to function. This is illustrated in Figure 54.

Stainless steel jackets are the most chemically resistant, weather resistant and physically strongest of any insulation jacketing. These jackets are generally furnished in 0.010 inch (0.25 mm) thickness metal. In very chemically corrosive atmospheres, they provide long lasting insulation protection. Where insulation may be subjected to extreme mechanical abuse, they prove to be a good investment. As they resist corrosion they require practically no maintenance to keep their very pleasing appearance. The melting temperature of stainless steel is approximately 2650° F (1454° C), thus making it very fire resistant. Stainless, as furnished, has an emittance of approximately 0.4.

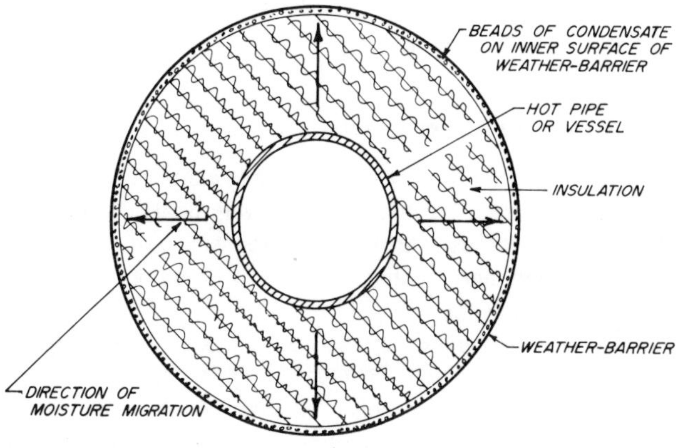

Figure 54 Insulation installed on hot pipe or equipment-direction of moisture migration.

Treated steel jackets are of galvanized steel 0.01 in. (0.25 mm) thick, phosphorized, then prime and finish coated on both sides to protect the galvanizing. The interior surface is provided with an additional protective "vapor-barrier" of impregnated craft paper. These jackets are physically strong, but not as corrosion resistant as stainless steel. After some years of outdoor service, the outer surface may require repainting. The melting point of steel is approximately 2700° F (1482° C) so it is also quite fire resistant. The surface emittance of these jackets is approximately 0.8, and for this reason they will not cause excessively high surface temperature when used in hot services.

Aluminum jackets are made of various aluminum alloys and various thicknesses of metal. Most often thicknesses range between 0.006-0.021 (0.152-0.533 mm). The light gauges have little impact or tear resistance, and should be used only when located where they are immune to physical damage, or sheltered from high winds. In almost all cases, these jackets are also furnished with a "vapor-barrier" on the inner surface, as are the treated steel jackets. This so called, but misnamed, "vapor-barrier" is supposed to protect the metal from condensate that forms on the inner surface due to heat driving moisture from the insulation, as illustrated in Figure 54. A more nearly correct term for this so called "vapor-barrier" would be "condensate-barrier." As this condensate is alkaline in nature, it will cause corrosion of unprotected steel or aluminum. Due to its light weight and ductility, aluminum is the easiest of all metals used for jackets to cut and fit. Conversely, it is the most easily damaged. Unless of a very heavy gauge, it should not be used where physical abuses are expected, as punctured jackets cease to be weather-barriers. Aluminum has a surface emittance of 0.05-0.10 and, consequently, provides considerable resistance to heat dissipation from its surface. The result is that its surface temperature, with the same operating temperature and the same thickness of insulation, will be much higher than other jackets (or mastics). On low temperature service, this same low surface emittance completely changes *the insulation thickness required to prevent condensation.* Quite frequently, in the published tabulations of such thicknesses, they are based on an emittance of 0.9. When aluminum jacketing is used as the outer surface of the insulation, the thickness necessary to prevent condensation, as ordinarily published, should be multiplied 2 to 2½ times to provide dry surfaces.

Joints of weather-barrier jackets

All jackets, no matter of what type, are not weather-barriers until they are installed over the insulation with watertight or water-shed joints. These joints must withstand the expansion and contraction movements, wind, and mechanical forces without opening up to allow the entry of water.

In the past few years, the insulation industry has tended to overlook the basic principles that were originally borrowed from, and are well known in the roofing business. First, laps must be sealed, and sealed carefully, to prevent water leakage. Second, wind blown rain will run into any gaps. Third, overlapped, unsealed metal never provides a water tight joint. Fourth, water-shed joints overlapped, and pointing downward, must be overlapped a minimum of 3 inches to prevent wind from driving the rain upward and over the inner layer. This is illustrated in Figure 55.

Most felts, films, or laminates depend upon an adhesive seal between the overlaps to adhere the surfaces. As most of these materials are capable of some degree of elongation, they must not exceed this elongation and pull the joints apart or tear the weather-barrier.

Metal, stronger and requiring much higher tensile forces to obtain even a slight degree of elongation, will generally break any adhesive used to seal its overlaps. Felt and film jackets will usually elongate to accommodate this tensile force.

To obtain rigidity for light weight metals, it became the practice to corrugate with small pitch corrugations. This material has been misused since its introduction. With complete disregard for basic principles, this type of jacket has been used by overlapping the joints without sealing them. It seems that no consideration was given to its fulfilling the function it was to serve. Simple, overlapped, corrugated metal on horizontal lines or equipment *is not a weather-barrier.*

Any metal installed in a vertical position, with simple, non-watertight overlaps is *not a weather-barrier.*

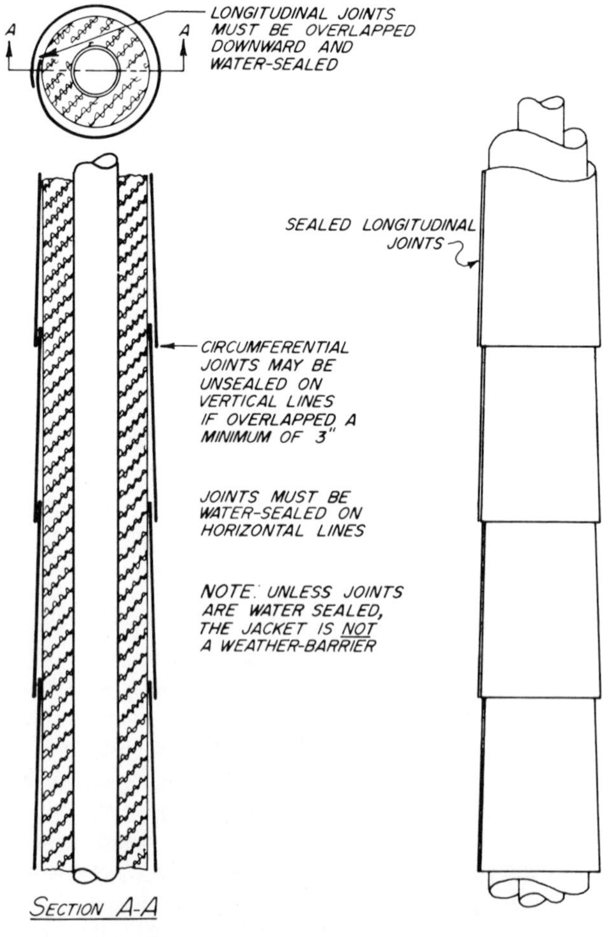

Figure 55 Proper sealing of felt, film, or laminated jacket weather-barrier.

These two conditions are illustrated in Figure 56.

From the preceding, it is evident that the major problem occuring in using metal jackets as weather-barriers is in the joints. The longitudinal joint on cylindrical pipe insulation is the most easily solved. The use of a well-formed, modified Pittsburgh point provides a good water-repellent joint. Illustration of this joint is difficult, so the illustration shown in Figure 57 is incorrect, as the various layers of metal can not be clearly shown in contact.

Several manufacturers make closure bands for circumferential joints, similar to that shown in Figure 57. The sealer at the edge of this band must be of a type which does not harden and become brittle with age. This is necessary, for the independent movements of the adjacent metal jackets must be compensated for at this junction, while a water seal is still maintained. In the vertical position, it is imperative that the top of these closures be sealed. Otherwise, in addition to leaking, the closure ring becomes a collector ring funneling water into the covering.

Of course, other designs for water repellent closures are available, and the one illustrated was selected only to show the movements which must be considered in a design with good water repellency.

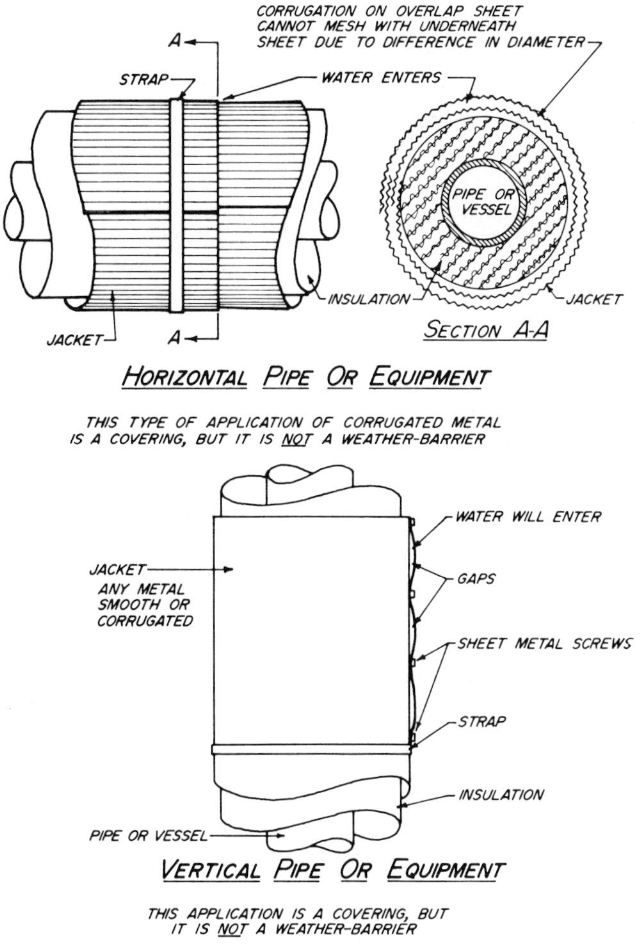

Figure 56 Misuse of corrugated metal as weather-barrier.

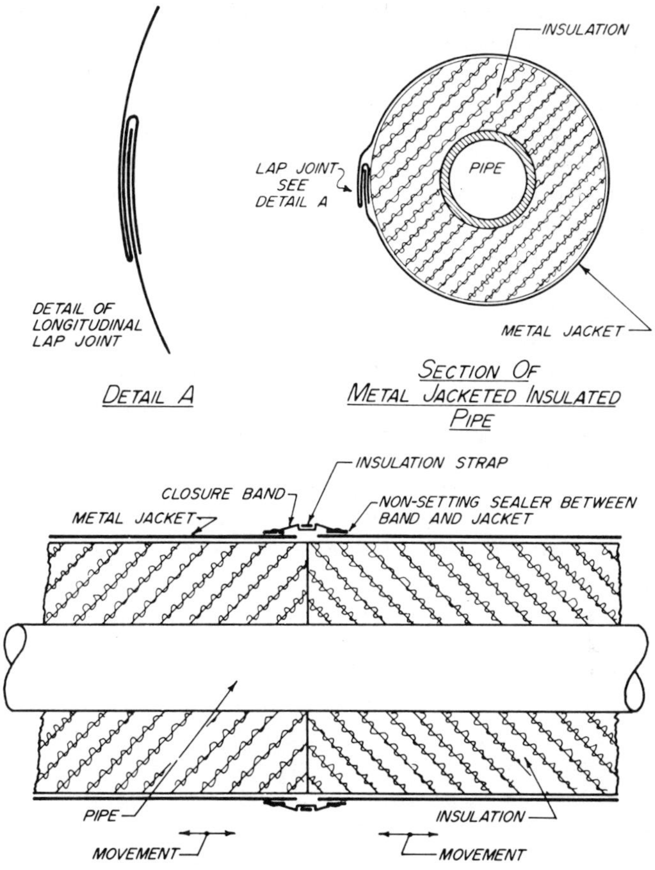

Figure 57 Circumferential closure band.

The inside and outside corners, where the various planes of insulation butt together, are also difficult to make water tight. Inside corners can be sealed by caulking or by the use of inside metal strips with non-setting sealers. However, outside corners present very difficult metal jacket sealing problems, particularly on vertical cylindrical ends. Methods of sealing inside and outside corners are shown in Figure 46. In spite of the necessity of keeping insulation dry, there will be many who think that they cannot afford the cost of sealing joints properly. However, the only purpose of the jacket is to keep the insulation dry, and if it does not accomplish this, the entire expenditure is useless.

Like in the roof of a house, a small leak can cause damage and monetary loss. If anyone said that a few leaks in a roof were unimportant because the major part did not leak, he would be considered to be mentally deficient. Remember, what the roof is to a house, the weather-barrier is to insulation.

Mastics. Weather-barriers may be made of many materials and combinations of materials. The two most basic classifications are bituminous or resin based.

Bituminous weather-barriers are further classified into emulsion and cut-back (solvent) types.

Emulsion bituminous weather-barriers have been used for many years. They are a thick, black, heavy build material, with a high percentage of inorganic fillers. They are most frequently applied by trowel. Naturally, they are applied thickly, and when dry are about ¼ in. (6.35 mm) thick. Shrinkage is high, and drying frequently causes shrinkage cracks. These cracks must then be filled and sealed. The ¼ in. (6.35 mm) thickness is necessary for long-life service since these barriers are very brittle when dry. Unless specially treated, they burn readily, contribute to fire, and have a high rate of flame spread.

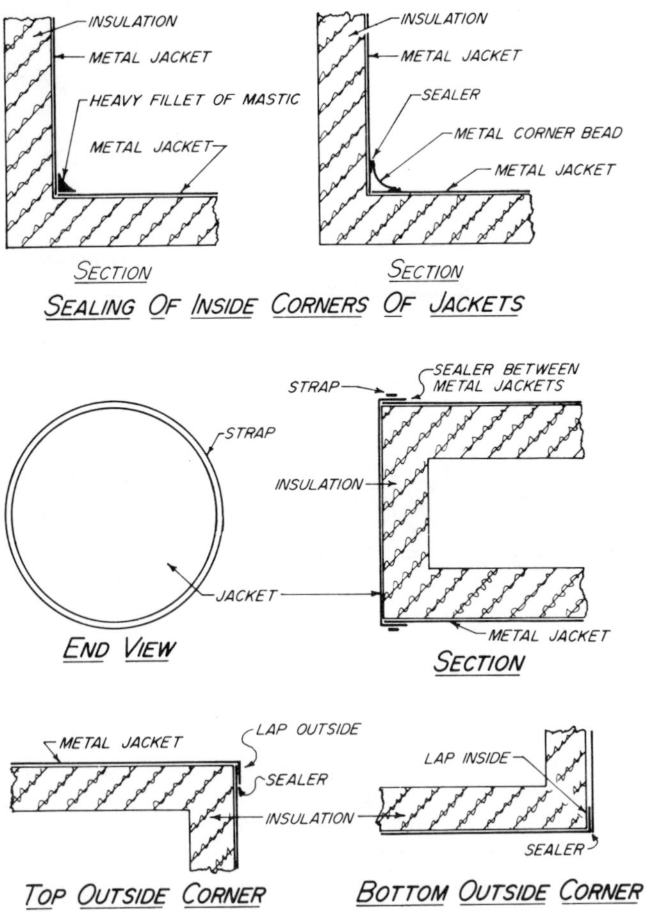

Figure 58 Sealing of inside and outside corners of jacketed insulation.

Cut-back bituminous weather-barriers are also thick, black, heavy build mastic. This material, when dry, forms a stronger, tougher film than that produced by the emulsion. These barriers can be applied by trowel, spray, or, by using thinner consistency grades, by brush. Because they produce tougher films than the emulsion types, the manufacturers most frequently recommend a 1/8 in. (3.17 mm) thick dried film. Some of the cut-back bituminous mastics are formulated with chemicals to reduce rate of fire spread. However, all are combustible in the wet state, during application, and until the material has dried enough to expel the major portion of the solvents. Although less brittle than the emulsions at atmospheric temperatures above 45° F (7.2° C) these materials are temperature sensitive, and will become very brittle in temperatures below freezing. Care must be taken that the insulation be dry when applying them on high temperature installations. Otherwise, when the vessel or piping is heated, the moisture pushing outward may cause severe vapor blistering. Although the shrinkage (from wet to dry states) of these materials is approximately 50% due to the stronger film, the shrinkage occurs in thickness rather than in length, and for this reason dry cracks are less troublesome than with the emulsions.

Emulsion resin weather-barrier mastics potentially can be made from any one of the numerous resins produced in emulsion form. The practice currently most prevalent, however, uses either polyvinyl acetate, acrylic, or a combination of these two resins. These materials may be made to a consistency and grade to be applied by the palm, trowel, spray, or brush. Shrinkage of the heavier consistency material is approximately 40%, and of the lighter, approximately 50%. All dry to a tough resilient film and, when properly formulated, dry without shrinkage cracks. As a group, the polyvinyl acetates become stiff, then brittle, at temperatures less than 35° F (1.7° C). The acrylics are less temperature sensitive and can be formulated to be flexible at temperatures below 20° F (−6.7° C).

This low temperature flexibility makes them much more resistant to cracking and deterioration in climates where atmospheric temperatures go below the freezing point. As these films are tough, the thickness necessary for good service life is less than that required for the other heavy build mastics. Recommended dried thickness is 1/16 in. (1.59 mm). These resin emulsions can be produced in almost any color, thus eliminating the need for painting to match existing color schemes. Being water-mix materials, they are not a fire hazard during application. After drying, they have a low rate of flame spread and do not melt in a fire, but slowly disintegrate. Thus, they are considered fire retardant, and can be safely used in hazardous installations. However, although they are resistant to many chemicals, some chemicals and solvents do saturate the film. Where the film becomes saturated with a combustible material, it looses its fire retardancy and low flame spread because of the presence of a particular combustible.

Vapor Retarders

By definition, the function of a vapor-retarder is to resist the flow of moisture vapor. Thus, the major property related to vapor-barriers is vapor transmission. A table giving poor pressures for the determination of vapor pressure differences was given previously in this chapter. This table gives vapor pressure in two units—inches of mercury and lbs per square foot. A confusing factor is that vapor transmission through materials is presented in various units determined by various test methods. The following indicates some of the units in which vapor transmission is given:

WATER VAPOR TRANSMISSION UNITS

WATER VAPOR TRANSMISSION

$$\text{Water vapor transmission} = \frac{\text{Weight Change in Grains}}{\text{Time} \times \text{Area}}$$

IN ENGLISH UNITS
In Unit Time

$$\text{Water Vapor Transmission, per hour} = \frac{\text{Weight Change in Grains per Hour}}{\text{Area in Square Feet}}$$

In Unit Time, Unit Area
WVT = Rate of water vapor transmission in number of grains per hour per square foot, or
= Grains per hour square foot

$$= \frac{G}{\text{hr, sq ft}}$$

When G = Grains

IN METRIC UNITS
Water vapor transmission is expressed in unit time of 24 hours.
Thus:
WVT = Rate of vapor transmission in number of grams per hour per square meter
= Grams per 24 hr sq metre

$$= \frac{g}{\text{24 hrs, sq m}}$$

When g = Grams

CONVERSION BETWEEN THE TWO SYSTEMS
Grams per 24 hrs, sq meter × 0.0598 = Grains/hr, sq ft
Grains per hr, sq ft × 16.7 = Grams/24 hr, sq metre

PERMEANCE OF A MATERIAL
Permeance =

$$\frac{\text{Water Vapor Transmission}}{\text{Vapor Pressure Difference between the Two Sides of the Material}}$$

$$= \frac{WVT}{\Delta P}$$

When ΔP = Vapor pressure difference between the two sides of the material

IN ENGLISH UNITS

$$\text{Permeance (Perms)} = \frac{G}{\Delta P\text{, hr, sq ft}}$$

IN METRIC UNITS

$$\text{Permeance (Metric Perms)} = \frac{g}{\Delta p\text{, hr, sq ft}}$$

When Δp = Vapor pressure difference between the two sides of the material

CONVERSION BETWEEN THE TWO SYSTEMS
Metric Perms × 1.52 = Perms
Perms × 0.659 = Metric Perms

PERMEABILITY OF A MATERIAL
For thick materials permeability is given in perm-inches for *English Units* and in Metric Perm-Centimeters for *Metric Units.*

CONVERSION BETWEEN THE TWO SYSTEMS
Metric Perm-Centimeters × 0.598 = Perm-inches
Perm-inches × 1.67 = Metric Perm Centimeters

Vapor-retarders, like weather-barriers, are produced in jacket and mastic forms. Metal vapor-barriers can be used, but a major difficulty, here, is vapor sealing of their joints. Tapes and non-setting vapor-sealers have been tried as sealers for metal joints, with varying degrees of success. Over a period of time, the adhesive of a tape tends to dry out and become brittle, and non-setting seals tend to harden. Movements due to expansion or contraction then crack the seal, causing a leak. Vapor will then enter through this crack, bypassing the metal, in its attempt to equalize the vapor pressure difference. As *vapor* molecules are small, they easily penetrate seams and small cracks that are completely *water* tight.

Heavier outer metal casings completely hermetically sealed by welding, as used to produce cryogenic insulation, are completely satisfactory as outer weather-barriers. Such hermetic sealing, by welding, is not an easy task. The welding techniques necessary are more difficult than those for welding pipes and vessels for high pressure service. Because of these difficulties satisfactory installations are most frequently done in manufacturing plants having the proper welding, handling, and test equipment available to achieve the degree of perfection required.

Films are produced which are non-reinforced, or laminated with metal foils and have very low vapor transmission. Their main problem, as regards efficient installation and satisfactory service life is their joints. Adhesives and solvents are used to bond and seal the joints, as well as tapes of various descriptions, but, again, the perfection necessary to obtain vapor proof joints is time consuming and costly in field applications. Systems have been developed for pipe covering and cylindrical surfaces which do provide economic service life, but the treatment of inside corners, outside corners, and double curved surfaces, necessary for vapor sealing of pipe valves and fittings, are their weak link. Vapor leakages in these areas are the limiting factor affecting the service life of the installation. For this reason, many people are misled into an improper choice of materials, as the selection is based upon laboratory tests of the film alone. This test value is of little significance unless vapor migration rates are also provided for the "sealed" joints at overlaps, inside, and outside corners. In practical application, the vapor migration through these joints will be many times that of the migration through the vapor-retarder material itself.

Mastic vapor-retarders

Mastic vapor-retarders, like weather-barriers, can be divided generically into bituminous, materials, resins, and combinations of the two.

Bituminous mastic vapor-retarders also black, and formulated containing few wicking fibers and fillers, do produce films of low vapor transmission. Almost all of these are cut-back type compounds and are combustible during and after drying. They are temperature sensitive and become brittle at freezing temperatures. Although some can be troweled, the most usual application is by spray.

Resin mastic vapor-retarders are made of various types of resins. Similar to bituminous vapor-barriers, they have less fillers and fibers than the high-build weather-barrier mastics, and, to obtain low vapor transmission, are of the solvent type. Where combustible solvents are used, the barriers are combustible in the wet state, but most frequently are formulated to be fire resistive in the dry state. Solid content is only about 40%, and, thus, they require considerable material to obtain the dry film thickness desired for low permeability. Application is by brush or spray. Some of the more vapor resistant materials have poor weather resistance and require an overcoating when used outdoors.

All vapor-retarders, be they metals, films, or mastics, are no better than their continuity. Joints, cracks, pinholes, or voids, allow the vapor to bypass the vapor resistant area and enter around it. Effective vapor-retarders on low temperature permeable insulation require the utmost effort to obtain perfection. In an auto tire, the internal pressure attempts to escape to the atmosphere. Everyone knows that, regardless of how good the tire is over most of its surface, the presence of a small leak, will allow the air to escape! Consequently, regardless of how good a vapor-retarder is, if it has holes, cracks, or is unsealed at any spot or area, the moisture vapor will leak in. This is illustrated in Figure 59.

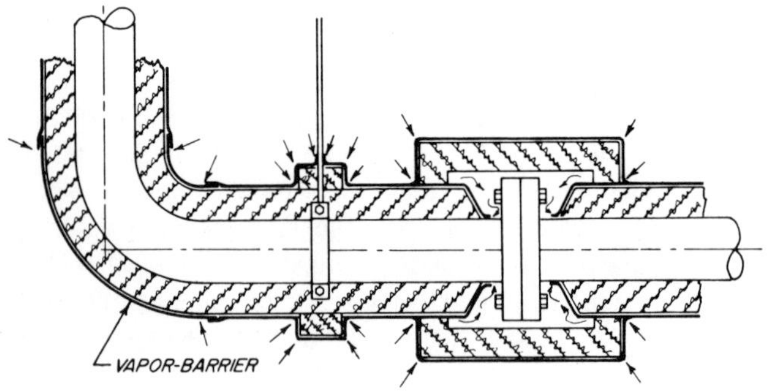

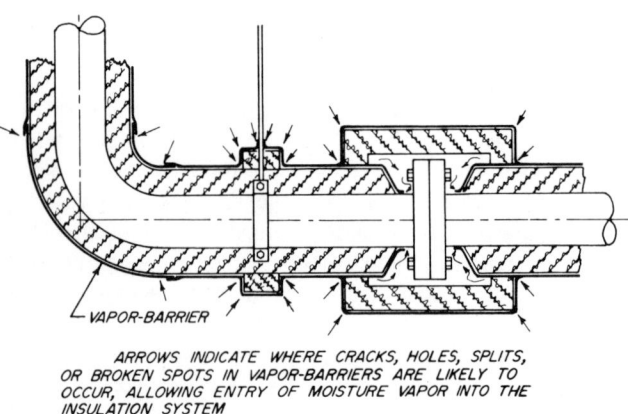

Figure 59 Potential vapor leaks.

Weather-barrier, vapor-retarder mastics are either bituminous or resin mastics formulated to serve a dual function as both weather-barrier and vapor-barrier. As such, their physical application propperties are closer to those described for vapor-retarders than for weather-barriers. Properly used, these materials serve to eliminate one extra material where the service is for low temperature insulations located out of doors.

In some instances these materials have been used on insulation for high temperatures, probably with the mistaken idea that the low vapor transmission would also be *advantageous* for high temperature installations if it was *necessary* for cold installations. Where these materials are installed on bone dry high temperature insulation they will serve satisfactorily. Where the insulation contains water and the line or vessel is heated, the moisture will probably be pushed to the insulation surface causing water blisters to appear. This will also happen with ordinary weather-barriers when insulation is wet and the low vapor transmission of the weather-barrier retards the outward vapor migration to the point where the vapor pressure exceeds the surface adhesion.

Under certain weather conditions, even with the best of systems, blisters can occur. They occur most often when an application must be done in wet weather, or when a sudden rain storm wets freshly applied mastic and retards its curing. When this is followed by bright sun, the rapid vaporization of moisture directly under the surface-dried film will almost certainly result in vapor blisters. In most instances, if the size of the blisters is small, a few days of warmth will dry out the moisture and they will disappear.

Reinforcement for Weather-Barriers, Vapor-Retarders

The emulsion bituminous weather-barriers require internal reinforcement to have sufficient flexibility and tensile strength to remain in a continuous film over the insulation surface. One inch hexagonal wire netting (chicken wire) is most frequently used for this reinforcement. This netting is fastened to the insulation securements with wire. Therefore, the bituminous emulsion could lose its adhesion to the insulation and still remain in place because of the embedded secured wire netting. Until about 15 years ago, this netting was always made of galvanized steel wire, as it was thought that the heat from the insulated surface would keep the system dry and prevent rusting. This was not true. The alkaline condensate attacked the galvanized steel wire and corroded through in about ten years, at which time the emulsion bituminous weather-barrier coating fell of the insulation. To correct this difficulty, one inch hexagonal wire mesh is now available in either stainless steel or Monel. To obtain long lasting netting reinforcement for emulsion mastics, or for insulating cements, the use of one or the other of these two metals is recommended.

In the latter part of the 1940's, cut-back bituminous coatings were developed, which produced stronger films than the emulsion types. Considerable quantities of these coatings were applied without any reinforcement. After a period of time, these coatings began to crack at the line where the insulation pieces butted together. To correct this, installation practice was revised as follows. A coat of mastic was applied, and then a reinforcing glass fiber fabric was embedded into the fresh mastic. After this coat dried, a second coat of cut-back bituminous mastic was applied, completely covering and filling the voids in the cloth. The glass fiber cloth served several purposes. First, it bridged voids and cracks. Second, it provided a reliable builtin thickness gage to indicate whether sufficient mastic had been applied. Third, it added tensile strength to the weather-barrier system. Even so, the reinforcement did not correct winter cracking of the weather-barrier system—still a major problem.

However, low temperature flexibility and elongation can be formulated into resin systems which will reduce winter cracking. Application of these resin weather-barriers follows the same procedure as that used with the cut-back asphalts. Glass fiber cloth was used for reinforcement in the weather-barrier system and, oddly enough, even with these low temperature flexibility mastics, the winter cracking failures were not corrected.

Investigation revealed that the resin mastic had excellent elongation and flexibility properties. However, glass fabric will elongate very little before it tears. It was evident that the elongation properties of the reinforcement did not match the elongation properties of the mastic. This mismatch was the cause of excessive weather-barrier cracking at the butt joints of the insulation where the movements exceeded the elongation limits of the system. These functions are shown in Figure 60.

Tests were initiated to establish the facts related to obtaining satisfactory weather-barrier systems. The first conclusion was that, where expansion and contraction forces are involved, no coating or reinforcement possessed sufficient tensile strength to restrain these forces. Thus, an excessively high tensile strength coating system is relatively unimportant. However, it is important that the system possess the ability to elongate without rupture. It is also important that the properties of the barrier and those of its reinforcement complement each other to obtain the best results.

At one point, a change from woven to knitted glass fabric was tried. It was thought that the knitted construction would provide the desired flexibility. Unfortunately, when embedded in mastic, the ability of the knitted cloth to elongate was no better than that of the woven variety. This indicated that, as soon as the individual thread and fibers were bonded together, the property of elongation was that of the fiber, rather than that of the type of weave. Because of this, the properties of the needed fiber had to be as follows: (1) Good tensile strength, (2) good elongation over a broad temperature range (from 250° F to −50° F) (121° C to −45.6° C) and (3) inability to support combustion.

Tensile-elongation tests of various mastics reinforced with glass cloth and with monocrylic fiber established the properties of each system. The results of these tests, run at 70° F (21.1° C), are shown in Figure 61. A reinforcing fabric must be selected which will allow the weather-barrier system to elongate sufficiently to accommodate, without failure, the full expansion movement imposed upon it.

As the properties of some mastics are quite sensitive to temperature, it should be noted that the tests shown are valid only if the temperature of the weather-barrier does not drop below 70° F (21.1° C). In winter climates, similar curves should be plotted for the mastics under consideration—based on the temperature at which they are expected to function.

The above exposition of the importance of proper selection of weather-barrier mastic and reinforcing fabric to obtain optimum results, is equally true for reinforced vapor-barriers, or weather-vapor-barriers. A polyvinyl acetate emulsion weather-barrier reinforced with glass fiber cloth, installed on a pipe is shown in Figure 62. The expansion broke the weather-barrier.

A vessel also weather protected with polyvinyl acetate mastic with modacrylic fiber cloth is shown in Figure 61. This weather-barrier did not break due to expansion.

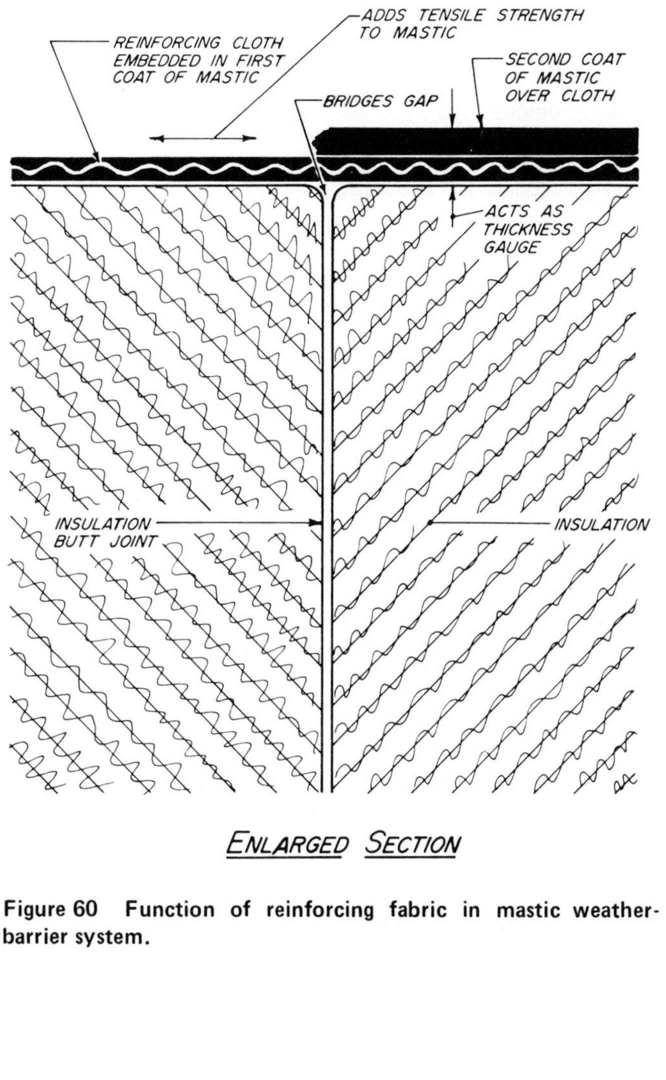

Figure 60 Function of reinforcing fabric in mastic weather-barrier system.

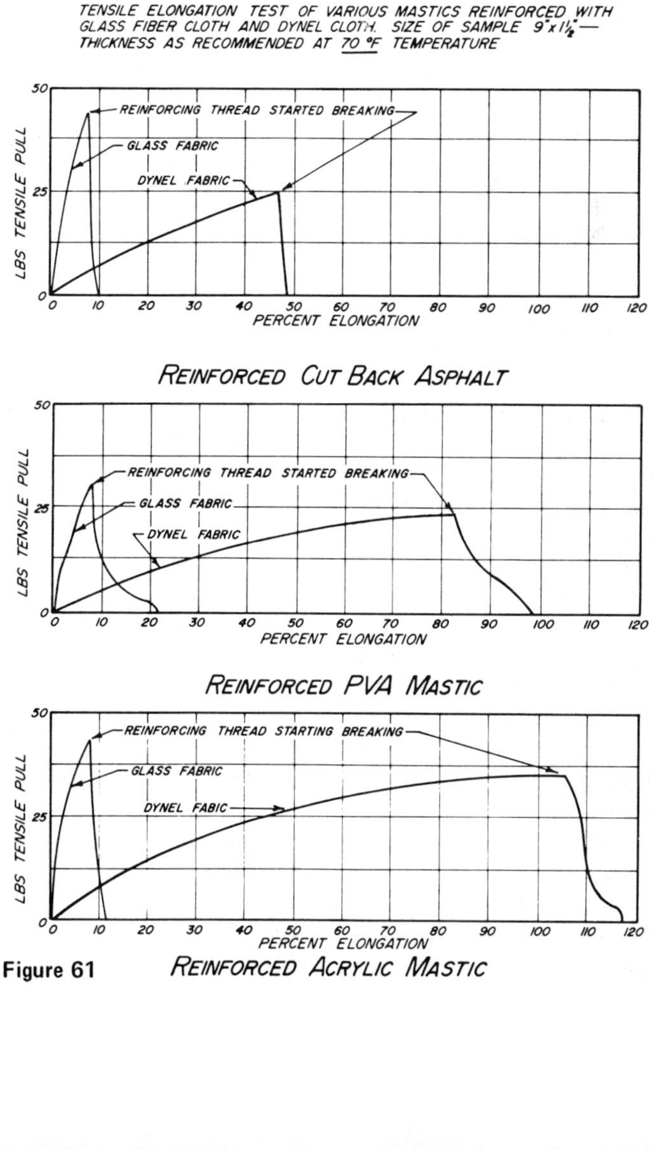

Figure 61

Types of Indoor Coverings

Indoor coverings may be jackets or mastics.

Jacket indoor coverings may be felts, films, laminates, or metals.

The various types of jackets described as weather-barriers may also be used as indoor coverings. In most indoor applications, the insulation is usually not subjected to as much mechanical abuse as it receives in outdoor applications. Thus, materials of less mechanical strength are frequently satisfactory for indoor applications. Unless indoor equipment and piping is washed down with water, sealing each joint against water penetration is not necessary.

Combustibility and fire spread may be much more critical indoors than in some outdoor applications. This is due to the fact that the heat of a fire indoors is contained and is not as readily dissipated as it is outdoors. If a building has human occupancy, any interior coverings should meet building and insurance codes and regulations.

Appearance may be more important indoors than outdoors. Where insulation is exposed to view indoors it, naturally, receives closer scrutiny than when located outdoors in an industrial complex. For this reason, even color may be very important in the final selection of the material. Conversely, where warm or hot piping is enclosed in pipe chases, or otherwise hidden from view and protected from mechanical damage, no covering may be required at all.

Mastic indoor coverings may be made of various based materials, the types being the same as described for weather-barriers. Due to the need for more fire resistive materials used indoors and the need for interior coatings of better appearance, the trend has been toward resin mastic coverings rather than bituminous coverings. However, fire resistive coatings are now available.

As the mastic coverings used indoors do not have to withstand outdoor weathering, they frequently are not as high build as the same basic materials used outdoors. Again, in normal indoor uses, the need for water tight covering is not as necessary as it is four outdoor applications. A crack in the indoor coverings is not as important as one outdoors. When located indoors, the insulation will not become wet because of rain.

Reinforcing fabrics for indoor coverings

Although cracks are unsightly, an expansion crack indoors can be tolerated, for extensive damage to the high temperature insulation will not occur. Thus, the elongation ability of the reinforcement indoors is less important than it would be outdoors. Also, the reinforcement is not subject to weathering, so materials which would rot outdoors may be used indoors. As appearance may be important, more closely woven fabrics are sometimes also desirable. For this reason, brattice cloth, canvas, and closely woven glass cloth are also used as reinforcing fabrics. At present a closely woven, extensible cloth is not available.

On moderate and high temperature applications, the reinforcement is often first adhered to the insulation with the mastic indoor covering, and then, when this has dried, the reinforcement is given a second coat of the same mastic. Most often these mastic indoor coverings are of a consistency suitable to easy application by brush or spray.

Applications on equipment and piping operating at low temperature are subject to moisture damage either from washing down or from the atmosphere. For this reason, cracks, pinholes, and voids are as critical, here, as they are in outdoor installations. The only major difference is that the barriers are not subjected to weathering. But they require the same care in application indoors, as outdoors.

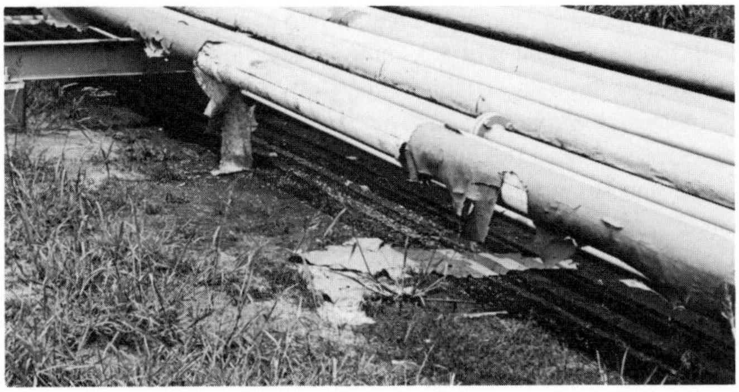

Figure 62 PVA mastic reinforced with glass fiber cloth on pipe operating at 380° F (193° C) temperature.

Appearance coatings

Appearance coatings are used either to obtain the desired aesthetic appearance or for color coding.

Weather-barriers, and vapor-retarders may not be available in the required color, texture, or gloss. Thus, they may have to be overcoated to obtain the desired appearance.

When these appearance coatings are applied over asphaltic felts or bituminous coatings, either a "seal prime coat" to act as a primer, or a coat to prevent "bleeding" of the asphalts into the finish appearance coating, or both, may be required.

In coating weather-barriers, vapor-retarders, or indoor coverings, it should first be determined whether the solvents in the appearance coating will be detrimental to the system. These coatings may be acrylic, alkyds, polyvinyl-acetates, polyvinyl chlorides, or urethanes. When used on water based mastics, care must be taken that the mastic is completely dry before application. High temperature insulation must also be dry before the application of appearance coatings, or blistering will occur.

Special overcoatings

Where required in hazardous areas, a fire resistive, or fire and heat retardant, overcoating may be used to control fire spread, or to obtain greater length-of-time protection for the insulation and the surface to which it is applied. These overcoatings are of two different types.

The fire resistive overcoating is one which does not disintegrate rapidly in fire. Even though it stays intact and retards flame spread, it has little value in retarding temperature rise of the substrate to which it is applied. For this reason, it cannot be considered fire retardant nor heat retardant.

Fire and heat retardant coatings are not fire resistive, as the fire (temperature) changes their form radically and rapidly. These fire and heat retardant paints are designed to intumesce when heated above about 350° F (177° C). At this temperature, they swell and cellulate to form a layer of thermal insulation of excellent low conductivity. The insulation formed is relatively fragile, with just sufficient strength to resist the high velocity air currents occurring in a fire. After the fire is extinguished, the insulation, called "puff," may be washed off with water, as it is water soluble.

Used outdoors, intumescent paint requires a top coat to protect it from weather and from solar radiation. Without such protection, radiation reacts with the intumescent materials within the paint. After this occurs, it will not produce the insulation puff when heated.

Intumescent paints are used frequently on surfaces other than insulation to provide fire protection. It is typically used on wooden doors, or on other surfaces which are combustible, to make these materials fire resistive. In case of a fire, the finish itself is sacrificed to protect the more vital material.

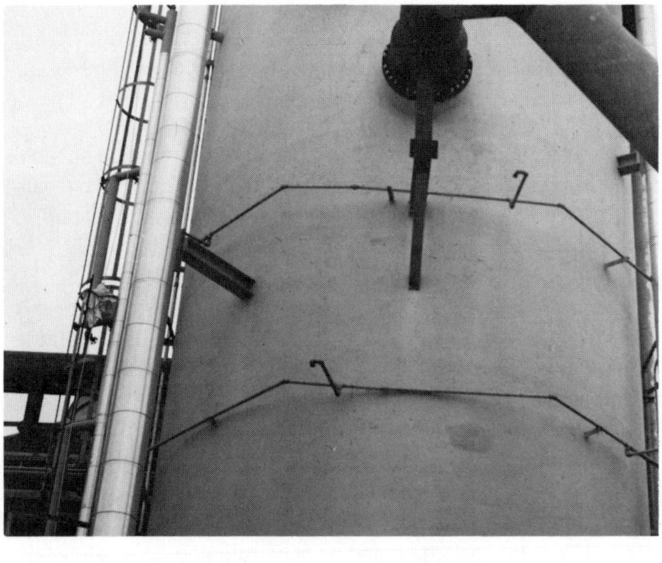

Figure 63 PVA mastic reinforced with modacrylic fiber cloth on vessel operating at 410° F (210° C) temperature.

the temperature range within which they are expected to function, and the materials between which they are to provide a seal. In many instances, they serve both as a sealer and an adhesive.

The need for sealing the joints of weather-barrier metal jackets, felts, films, and laminates has already been mentioned. Where the sealers are used between two surfaces, they should have good adhesion to each surface. If movement exists between these two surfaces, the sealer must be sufficiently flexible to withstand this movement within itself without cracking, or losing its adhesion to the boundary surfaces. To be effective in these services, the sealers should have little shrinkage and must remain flexible from the lowest to the highest temperature to which it is subjected. Action of expansion and movement is shown in Figure 64.

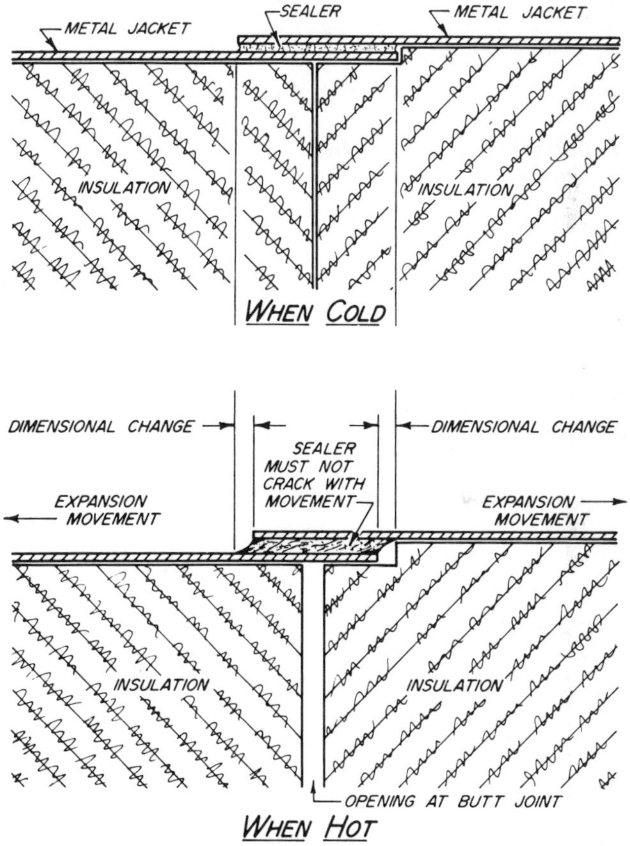

Figure 64 Movement of sealed metal jackets due to expansion.

Sealers, Anti-Abrasive Coatings, Caulkings, Adhesives

In addition to the fundamental weather-barriers, vapor-barriers, and coverings, there are other mastic and plastic materials which are used independently, or in conjunction with the barriers.

Sealers

Sealers, in insulation practice, are used as water seals and vapor seals. They may be made of various materials, depending upon

If the sealer is used as a water seal, it may have a high rate of vapor transmission. This is especially desirable on high temperature applications. Of course, when used on low temperature service, an extremely low rate of vapor transmission is essential.

Where metal passes through high temperature insulations, special high temperature sealers are frequently required to prevent water from entering through the discontinuity where the metal passes through the insulation. Where the temperatures of these metal projections exceed the allowable temperatures of available high temperature sealers, it becomes necessary to insulate the projections to such a distance that the heat dissipation through the projecting insulation is sufficient to lower the temperature of the metal surface at its point of exit to a level low enough for it to be sealed. This is illustrated in Figure 65.

Another large use of vapor sealers, and non-setting vapor sealers, is to seal abutting parts of a vapor resistant material to themselves to form a vapor-barrier. Cellular glass is more vapor resistant than all vapor-barriers, other than hermetically welded steel, or heavy metallic outer shells. For this reason, it may be sealed to itself to form an effective vapor-barrier. Sealers used for this service must be relatively solvent free, or be of such a nature that contained solvents do not adversely affect the function of the sealer. Shrinkage of the sealer should be small to prevent cracks through which the vapor can pass. Solvent sealers used in these butt joints are not effective as an adhesive, because the trapped solvents can not pass through the cellular glass and can only dry thorugh the edges from the outside in. If the sealer is an excellent one, this may take years. This is shown in Figure 66.

Anti-abrasive coatings and bedding compounds

Anti-abrasive coatings are applied to the surfaces of insulations which may be worn away by abrasion, due to movement or vibration of the substrates to which they are applied. Insulations which most frequently require this surface protection are organic and inorganic foams.

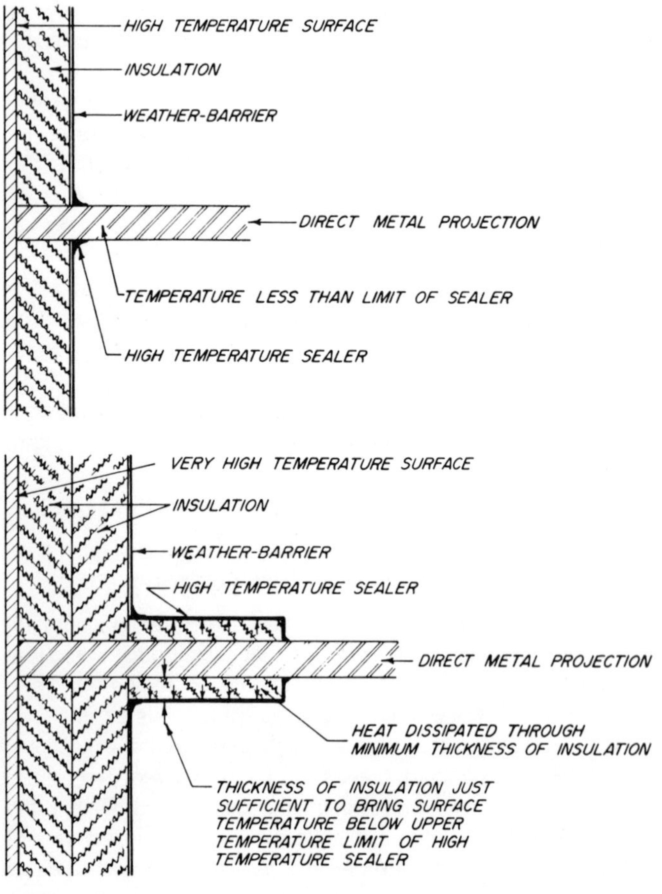

Figure 65 Sealing of high temperature projections.

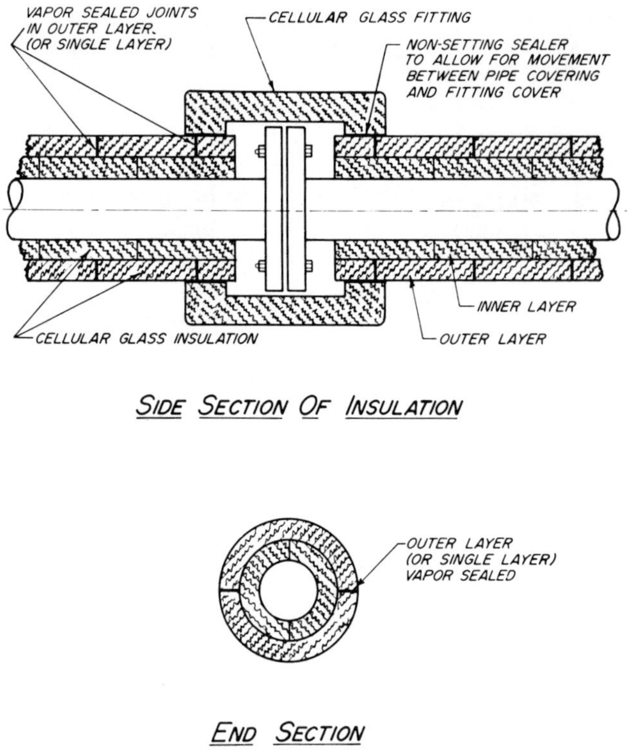

Figure 66 Vapor sealing of cellular glass.

The particular anti-abrasive coating selected must be suitable for the temperature to which it will be subjected, compatible with the insulation to which it is applied, and, of course, resistant to abrasion.

Three general types of anti-abrasion coatings are available: (1) the solvent resin type, (2) the catalyst type, and (3) the water mix type. The solvent resin type is most frequently used for the temperature range of −200° F to +200° F (−178° C to +93° C), the catalyst type between −100° F to +300° F (−73° C to +149° C), and the water mix type between 70° F and 750° F (399° C).

Frequently, this type of coating is misused by applying it just before the insulation is installed on the equipment or piping, as if it was an adhesive. When applied in this manner, it bonds to the surfaces of both the insulation and its substrate and completely fails to function as designed. It must be applied to the proper surface and allowed to dry prior to erection.

Another method of reducing abrasion is the use of a bedding compound between the foam and the surface. This is a heavy bodied material which is designed to adhere to both the surface of the insulation and to its substrate. Its function is to provide a soft cushion mastic between the two surfaces. To be effective bedding compounds should never harden or become brittle.

Caulking compounds

Caulking compounds are made of many basic materials and mixtures. They are primarily designed to fill voids and provide a watertight seal between two pieces of material. Their function is to patch small cracks and crevices in the insulation before the weather-barrier is installed. They also round out inside corners of

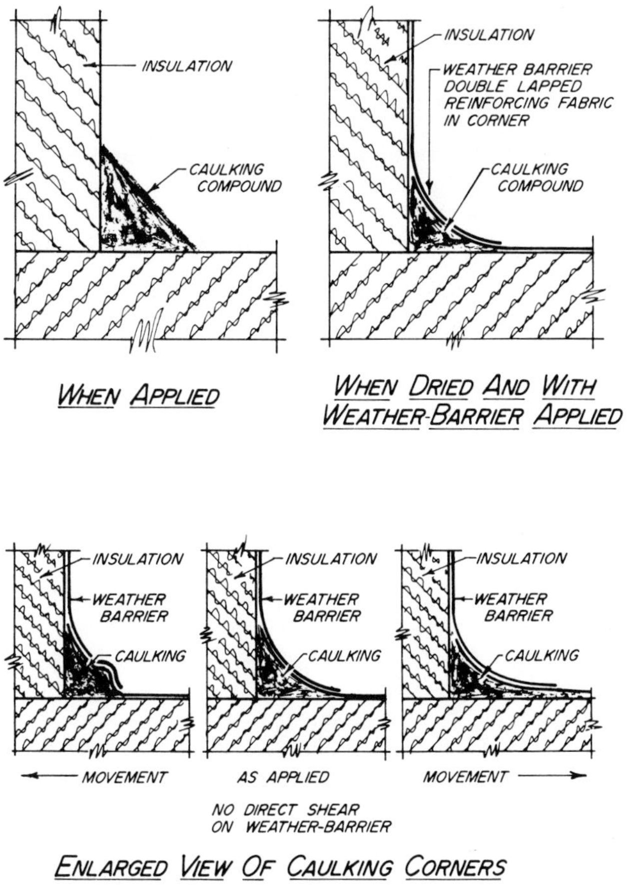

Figure 67 Effect of caulking inside corners prior to application of weather-barrier or vapor-retarder.

insulation so that movement does not create direct shearing forces in the weather or vapor-retarder. An inside corner that is rounded out with a caulking compound has a much better chance of remaining water tight than one which allows the barrier to intersect at an inside sharp angle. This action is illustrated in Figure 67, and illustrates the necessity of strong adhesion to the insulation surface, a low percentage of shrinkage, and excellent flexibility after drying. Without these properties it will not perform its function. Heavy build resin mastics with finely ground cork have proved to be an excellent compound for this service.

Adhesives

Adhesives which are used to cement weather-barriers, vapor-barriers, and indoor coverings to the insulation are part of the insulation protective system. Those used to fasten pieces of insulation together in fabricating, or to bond insulation to the surface to which it is applied serve a different purpose, and will be discussed in their area of function.

One of the oldest known adhesives used to attach brattice cloth or canvas to insulation is wheat paste. Like any material of this kind, it provided food for rats, mice, and bugs, and caused many warehouse and field storage problems. It was always resoluble in water, and, if the fabric became wet, would dissolve, lose its adhesion, and allow the indoor covering to come off, no matter how long after application. It had little to recommend it other than an inexpensive price.

The development of the so called "lagging-adhesives," made most frequently of polyvinyl acetate based resins, made the use of wheat paste obsolete. The "lagging adhesive" was not only used as an adhesive, but also served as a paint over the cloth. As they are produced in colors, they eliminated the need for sizing and overcoating with paint. They also have a low rate of fire spread, and with these advantages are in wide use.

Asphaltic adhesives are used to seal the overlapping joints of asphalt, rag, and asbestos felts where these materials are used as weather-barriers. These asphaltics dry and set relatively slowly. Therefore, to obtain a good seal, it is necessary to temporarily hold the surfaces to be joined with tie wires, or other securement, until the adhesive has sufficient strength to hold the felts together by its own efforts. After the asphaltic adhesive has bonded, the holding wires are removed. Because this practice is time consuming, many of the so called "reasonable cost" installations do not call for the sealing operation. Without proper sealing, some money may be "saved" on the original installation, but the heat loss each year due to water soaked insulation will amount to many times that of the "saving."

Special types of pressure sensitive tapes and adhesives are produced to seal laminates and coverings for factory jacketed pipe, preinsulated ducts, and the like. In most instances, the manufacturer supplies the proper adhesive for these products.

Vapor sealing of vapor-retarders composed of various films and laminates is sometimes done with adhesives and at other times by the use of solvents which soften the surface of the film so that it adheres both to itself or to other surfaces.

When adhesives are used to bond and vapor seal films and laminates together, this bonding must be done with extreme care. Vapor tight joints using adhesives have, in most applications, proven to be very unsuccessful. As the adhesive dries, its shrinkage generally causes channeling—lines and areas—between the two surfaces which are not bonded together. Then air spaces exist that provide easy ingress of the moisture vapor. Regardless of the vapor resistance of any film laminate, as a vapor-retarder it is *no more effective than the vapor resistance of the joint that can be produced on the job site.*

To sum up, any pinhole, crack, or break in a weather- or vapor-barrier where moisture can enter is a hole, crack, or break where *money* pours out.

Service Requirements Affecting the Selection of Weather-Barriers, Jackets, and Mastic Weather-Barriers

The *mechanical* abuses, external or internal, that the jackets are expected to withstand dictate the following properties of jackets:

- Abrasion resistance
- Hardness
- Impact strength
- Puncture resistance
- Shear strength
- Tear strength
- Tensile strength
- Percent of Elongation before rupture

Any expected exposure to chemical atmosphere or spillage determines the following properties which the barrier must have:

Acid resistance
Caustic resistance
Solvent resistance
Flammability when contaminated

The temperatures to which the barriers are subjected affect the mechanical properties and chemical resistance. For this reason, the properties listed above must be such that they are suitable at the *extremes* of temperature to which the barriers will be exposed in service. The temperature involved may be due to atmospheric temperature, the temperature of the substrate, or a combination of both. Besides the effect of temperature on the mechanical and chemical properties, there are some other basic temperature properties which must also be considered in the selection, such as:

Heat stability
Coefficient of expansion
Low temperature embrittlement
Melting point
Softening point
Thermal shock resistance

Moisture requirements pertain to the ability of the material to resist penetration by liquid and vapor moisture. However, it should be remembered that some very vapor resistant materials can be troublesome when used as weather-barriers on high temperature applications. To remain moisture resistant, the materials must be weather resistant. The physical, chemical, and thermal properties, as well as moisture resistance, change with weathered material. Thus, weather resistance is not a property unto itself, but is the effect of weathering on other properties.

Under moisture resistance, then, we have:

Resistance to liquid water
Resistance to moisture in the vapor phase
Vapor transmission
Resistance to weather (as it effects the other properties)

Safety requirements in service pertain to properties of the materials after application. The following apply to metal jackets:

Sharpness of corners and cut edges
Emittance of surface (as it affects surface temperature)
Conductivity—thermal
Reflectance—thermal

Fire hazard is an area of both personnel and property safety. The properties of materials which must be considered are:

Flash point
Fire point
Fire resistance
Fire reardance
Fire—self ignition point
Fire spread
Flammability
Melting point
Softening point
Thermal shock resistance

Economic considerations are part of the total evaluation as regards its ability to protect the insulation and its length of service life. After the materials have been selected which will provide the service required, then consideration must be given to the following:

Service life needed vs. service life obtainable
Cost of material
Cost of application
Cost of maintenance
Hazard risk—(fire)
Value of fire protection
Reliability
Cost of possible shutdown
Effectiveness in protecting insulation

The last is most important, because it has been observed many times that consideration has been given to almost every aspect of weather-barriers, metal vs. mastic, cost of application, cost of mastics, etc., except its prime function of insulation protection. The fact that some leak water, or do not provide the desired protection to the insulation causing loss of efficiency of the primary material, seems never to be considered or questioned. *If a weather-barrier does not keep insulation dry, it is not a weather-barrier.*

One unsealed joint in a weather-barrier can easily leak a pound of water into the insulation in a rainstorm. When the insulation is wet, it will transmit *15 times* more heat than when dry. If this saturates 10 square feet, it can waste fuel worth from $0.45-$1.25 for every 24 hours of rain. In addition, it will require approximately 1070 Btu (1.6×10^6 joules) to vaporize each lb of water from the insulation, or increase the load on the boiler by one pound of steam per hour. (This represents an investment of $25.00-$75.00.) Conservatively, each crack or unsealed weather-barrier joint probably costs $15.00-$30.00 per year in steam cost plus interest on capital investment.

If heat were visible, industrial management would make certain that this costly item was not so carelessly wasted. As it is invisible, they do not realize the extent of their monetary loss. However, engineers, knowing the value of energy, should not ignore this expensive waste.

Application Requirements

Application requirements are last in this list, because they differ as to whether the barriers are jackets or mastics. In general, most of the other requirements apply to either jackets or mastics.

Jacket application requires that the jackets themselves have a number of properties for their successful application. This might be termed the applicability of a material. No single property can be the measure of applicability, for it is a combination of:

Cutting characteristics
Ductility

Elongation
Elasticity
Flexural strength
Tear strength
Shear strength
Tensile strength
Hardness
Surface compatibility with adhesives and sealers
Puncture resistance

Felts and films are most always tailored to fit in the field and the joints sealed with mastics or adhesives.

Many laminates are factory applied to pipe covering and sheet insulation, with laps coated with pressure sensitive adhesive. On these installations, curved surfaces are weather protected with tape of various types. Most of these are designed for use where abuses are relatively moderate.

Metal jackets vary from paper-thin, soft metal 0.006″ (0.1524 mm) thick, to strong steel and stainless steel jacketing sometimes greater than 0.036″ (0.9144 mm) thick. The thick, heavy gage materials require the services of the sheet metal craft for their application. There is no doubt that soft, thin, metal jacketing costs less and can be applied with less labor than stronger jacketing material. However, if these light weight, soft materials are subjected to any mechanical abuse, chemical attack, or strong winds, they will not protect the insulation, and thus, are valueless.

The need for heavier, stronger metal jackets has promoted the need and use of insulation panels with factory attached preformed pipe jackets, fitting jackets, and vessel jackets. Several areas of improvement are needed to make these systems better. First, few of the joints of the fitting covers are water tight after installation, and expansion movement occurs. Second, is where field cutting and fitting is required. Special tools, not always available, are required to make these field fitted joints water tight. In other words, most of the metal jacketing systems still have not been completely designed for practical and efficient field installation, nor are these incomplete systems a complete weather-barrier.

Hardly any users have had sufficient curiosity to remove a few sections of jacketing after a rain storm to determine the degree of wetness of their insulation. It is recommended that such investigations be made to discover if a weather-barrier really exists, or only a cover which hides water-embezzled profits.

Mastic application requires almost a completely different set of material properties than jackets, although when dried and cured, the reinforced mastic becomes a field made, reinforced film, and the properties needed in service become identical. Even the method of application has a bearing on the properties needed. Mastics may be applied by brush, spray, palm, or trowel. Brush and spray mastics are of a somewhat lower consistency than those applied by palm or trowel. Application of mastics requires that the following properties be evaluated for the installation:

Consistency
Coverage
Build
Flammability (during application)
(a) Flash point
(b) Fire point
Toxicity
Temperature and humidity range (during application)
Surface wetting and wet adhesion
Gap filling
Bridging
Sealing
Corrosion to substrate
Solvent attack on substrate
Drying time
Curing time
Shrinkage
Thixotropicity
Odor

After the application, during the drying and curing stages, and finally proceeding to service conditions, the character of the mastic changes. During this change, faults may occur. The following should not occur:

Alligatoring
Cracking
Blistering
Checking
Delamination
Discoloration
Sagging

After application, jackets and mastics should provide the necessary appearance for the installation.

Appearance properties to be considered, depending upon job requirements, are:

Color
Sheen
Smoothness
Gloss
Reflection factor—light

Due to the large number of requirements and properties of weather-barrier-vapor-retarders that should be understood, evaluated, and matched, the obtaining of the most efficient and economic system is quite complex. Many of the items mentioned have not been studied sufficiently to develop standard methods of evaluation. Thus, much of the evaluation becomes a matter of observation and judgment. Unfortunately, this is a very difficult subject on which to find a nice clean-cut set of values for engineering decisions.

Tests of the Properties of Weather-Barrier-Vapor-Retarders

The evaluation of weather-barrier-vapor-retarders is complex, not only because of the number of properties involved, but also because the effect of the interrelationship of these properties is also hard to establish. Even if all the individual properties of each material were known, they would still be only indicative of the actual service life and effectiveness of the barrier. Thus, for final answers, field service is the only true measure of the life of a barrier system. The big problem, then, is that, by the time a barrier has proved its excellence, a period of at least ten years has passed. This is a long time to wait to make a decision. In the meantime, new materials which might be better have already been introduced to the market. Presently, there is no acceptable solution to this dilemma.

Test methods to obtain values for various properties must, therefore, be used to assist in the evaluation of these materials. Unfortunately, the number of test methods available for evaluation is still very small. In addition, some methods suitable for testing one type of material may not be suitable for another, although both are used in the same service. Where several test methods exist for the testing of one property, the values of one test may not be interchangeable with the values of the other. It is, therefore, important to know not only the test result, but also the method by which that result was obtained.

The available ASTM test methods which might be used to evaluate various properties are as follows:

ASTM Test Methods No.	Property Title of Test Method
D-658	Abrasion Resistance of Coatings of Paint, Varnish, Lacquer and Related Products with the Air Blast Abrasion Tester
D-968	Abrasion Resistance of Coatings of Paint, Varnish, Lacquer and Related Products by the Falling Sand Method
C-590	Action on Substrates by Coatings, Adhesives and Joint Sealants Used on or with Thermal Insulation
D-903	Peel or Stripping Strength of Adhesive Bonds
C-419	Test Specimens of Mastic Thermal Insulation Coatings—Making and Curing
D-543	Resistance of Plastics to Chemical Reagents
D-1729	Visual Evaluation of Color Differences of Opaque Materials
C-461	Bituminous Mastic Coatings Used in Conjunction with Thermal Insulation, Testing of
D-217	Cone Penetration of Lubricating Grease
D-1823	Apparent Viscosity of Plastisols and Organosols at High Shear Rates by Castor-Severs Viscometer
D-1824	Apparent Viscosity of Plastisols and Organosols at Low Shear Rates by Brookfield Viscometer
D-1654	Evaluation of Painted or Coated Specimens Subjected to Corrosive Environments
C-692	Stress Corrosion Effect of Wicking-Type Thermal Insulations on Stainless Steel
D-1640	Drying, Curing or Film Formation of Organic Coatings at Room Temperature
D-638	Tensile Properties of Plastics
C-488	Exterior Exposure Tests of Finishes for Thermal Insulation
D-747	Stiffness of Plastics by Means of Cantilever Beam
D-790	Flexural Properties of Plastics
G-20	Chemical Resistance of Pipeline Coatings
G-21	Resistance of Synthetic Polymeric Materials to Fungi, Determining for
G-22	Resistance of Plastics to Bacteria, Determining of
D-1354	Concentration of Odorous Vapors—Adsorption Method
D-781	Puncture and Stiffness of Paperboard, Corrugated and Solid Fiberboard
D-70	Specific Gravity of Semi Solid Bituminous Materials
D-71	Specific Gravity of Solid Pitch and Asphalt—Displacement Method
D-828	Tensile Breaking Strength of Paper and Paperboard
C-355	Water Vapor Transmission of Thick Materials
E-96	Water Vapor Transmission of Materials in Sheet Form

A number of the test methods listed were developed for other materials than weather-barriers, vapor-retarders or coverings for insulation. However, they can be used to determine properties of materials where they are applicable. Likewise not all test methods listed are suitable for determining a particular property for all materials. In all cases tests must be chosen which are meaningful for the material and its application requirements.

Glossary of Terms Related to Weather-Vapor-Barriers and Coverings

A number of terms used in connection with the weather-vapor-barriers and coverings have been defined where they have occurred in the text. As there are a number of terms used in the field which may bear different meanings than the *generally understood* meaning of the same word, a glossary of these terms is included. It was prepared by the Benjamin Foster Company. Many of the definitions used in this Manual have been taken, by permission, from this glossary.

Definitions of the following terms will be found in the Glossary in the Appendix A.

Adherend
Adhesive
Alligatoring
Ambient temperature
Application temperature limits
Asphalt
Asphalt emulsion
Bitumen
Bleeding
Blister
Bond
Bond age
Bond strength
Bonding time
Brattice cloth
Bubble
Canvas
Cast film
Caulking compound
Cement
Chalking
Checking
Chemical resistance
Coal tar
Coating
Color
Combustible
Conditioning
Consistency
Contact adhesive
Coverage
Creep
Cure
Curing agent
Cut-back products
Crater
Delamination
Dew point
Deterioration
Di-electric strength
Dilatancy
Discoloration
Dispersion
Drying time (adhesives)
Drying time (finishes)
Durability
Duck
Efflorescence (bloom)
Elasticity
Elastomer
Elongation
Emulsion
Exposure
Fading
False body
Filler
Fillet
Film (wet)
Fire point
Fire-resistance
Fire-resistive
Flame-spread
Flammability, wet
Flammable
Flammable (or explosive) limits
Flammable (or explosive) range
Flash point
Flow
Gel
Gloss
Holiday
Humidity
Humidity, absolute
Humidity, relative
Ignition
Ignition temperature
Impact resistance
Intumescence
Joint
Joint, lap
Laminate
Lap adhesive (lap cement)
Mastic
Membrane reinforcement

Mil
Mildew
Mold
Mud cracking
Muslin
Non-combustible
Non-flammable
Non-volatile content
Opacity
Open time maximum (adhesives)
Open time minimum (adhesives)
Open time optimum (adhesives)
Open time range (adhesives)
Orange-peel
Penetration
Penetrometer
Permanence
Permeability
Permeance
Pigment
Pinhole
Pit
Plastic
Plasticizer
Plasticizer migration
Pot life
Primer
Right angle test
Rust blush
Sag
Sealer
Self-ignition temperature
Service temperature limits
Set
Shade
Shear strength
Sheen
Shelf life
Sizing
Skinning
Slip (trowel application)
Solids content
Solvent
Specimen
Storage life
Strength, dry
Strength, wet
Stress-crack
Substrate
Suction
Tack
Tensile strength
Thermoplastic
Thermoset
Thinner
Thixotropy
Tint
Total solids
Translucent
Ultimate strength
Viscometer (viscosimeter)
Viscosity
Water absorption
Water-resistant
Water-vapor-barrier
Water-vapor permeability
Water-vapor permeance
Water-vapor pressure
Water-vapor transmission (WVT)
Weather-barrier
Weathering
Wet
Working life (pot life)

List of Materials and Properties

The type of weather-barrier, vapor-retarder, covering or finish depends upon the installation and application requirements. Weather barriers may be of felts, plastic sheets, metals or mastics. Likewise, sealers may be tapes or adhesives. Because of the very wide variety of materials used for these purposes, the basic division of materials and their properties will be divided into their basic function. It must be pointed out that any one material may serve more than one basic function; e.g., an outdoor weather-barrier may be used as an indoor finish over insulation. It must also be noted that properties listed are only "typical" properties. Each manufacturer formulates coatings and sealers differently, thus the properties of a type of coating, adhesive or sealer will differ with the various manufacturer. Where one particular property is critical, it is advisable to obtain specific information from the producer.

Some sealers or coatings may be used to protect substrate metal to which the insulation is installed. The properties as listed are not directed toward that function. In all cases, substrate metal should be corrosion protected, as necessary, prior to the installation of the insulation system.

Listing of weather-barriers, (mastic lagging adhesive, coatings, metal jackets and felts) vapor-retarders, bonding and insulation adhesives and accessories follow:

LAGGING ADHESIVE AND COATING

POLYVINYL ACETATE—HEAVY BUILD—EMULSION, LAGGING ADHESIVE AND COATING

A combination lagging adhesive, sizing for cloth and coating. Applied by brush, dipping or spray. Non pigmented or white available.

Properties—Wet

Flammability: Nonflammable
Freeze-Thaw Stability: Keep from freezing
pH: 6.5 to 7.5
Storage Stability: Two years at temperatures above freezing
Total Solids Content: 53 to 57% by wt., 41 to 45% by volume
Viscosity: Heavy brushing
Weight: 9.3 to 10.6 lbs/gal (1.1 to 1.3 kg/liter)

Properties—Application

Adhesion to Substrate: Excellent
Coverage: 1.0 to 3 gal/100 ft^2 (0.41 to 1.2 liter/m^2)
Drying Time: To touch—½ to 2 hrs, through 8 to 12 hrs
Solvent (while still wet): Water
Temperature Limits—Installation: Above 40°F (4° C)

Properties—After Drying

Abrasion Resistance: Excellent
Adhesion to Substrate: Excellent
Flammability: Will carbonize but will not support flame spread
Flexibility: Excellent
Resistance to Mild Acids: Excellent
Mild Alkalis: Excellent
Salts: Excellent
Solvents: Good (most solvents)
Temperature Limits, Max: 200° F (93° C)
Water Vapor Permeance: 1.0 English perm (1.6 Metric perm) as applied

POLYVINYL ACETATE EMULSION LAGGING ADHESIVE AND COATING

A combination coating and adhesive for adhering canvas and other cloths to insulation and to overcoat the cloth to present a pleasing finished surface. White and colors available.

Properties—Wet

Flammability: Nonflammable
Freeze-Thaw Stability: Must be protected from freezing
pH: 6.5 to 7.5
Storage Stability: Two years at temperatures above freezing
Total Solids Content: 50 to 55% by wt., 36 to 40% by volume
Viscosity: Brushing 100 ± 10 KU
Weight: 10 to 11 lbs/gal (1.2 to 1.3 kg/liter)

Properties—Application

Adhesion to Substrate: Excellent
Build: 0.02" (.58 mm) wet
Coverage: As adhesive, 1.4 to 3 gal/100 ft^2 (0.51 to 1.2 liter/m^2)
Drying Time: To touch—2 hrs, through 12 hrs at 50% R.H.
Solvent (while still wet): Water
Temperature Limits—Installation: 40° F (4° C) to 100° F (38° C)

Properties—After Drying

Abrasion Resistance: Excellent
Adhesion to Substrate: Excellent
Flammability: Will carbonize but will not support flame spread
Resistance to Acids: Good
Alkalis: Good
Salts: Good
Solvents: Good
Temperature Limit, Max: 200° F (93° C)
Water Vapor Permeance: 1.0 English perm (1.6 Metric perm) as applied

LAGGING, DUCT AND LAP RUBBER BASE ADHESIVE AND COATINGS

RUBBER BASE, SOLVENT TYPE, DUCT AND LAP ADHESIVES

An adhesive to bond low density fibrous glass to metal surfaces, or to itself. Long "open time" for shop and field fabrication—to be applied by brush or spray. Colors are amber, red or clear.

Properties—Wet

Flammability: Flash point approximately 10° F (−12° C)
Odor: Strong
Storage Stability: 2 Years
Total Solids Content: 27 to 33% by wt., 19 to 21% by volume
Toxicity: Requires good ventilation
Weight: 6.2 to 6.9 lbs/gal (0.73 to 0.79 kg/liter)

Properties—Application

Adhesion to Surfaces: Excellent
Coverage: 1 to 2 gal 100 ft^2 (0.41 to 0.81 liters/m^2)
Drying Time: 2 to 15 minutes
Solvent: Hexane

Properties—After Drying

Adhesion: Excellent
Flammability: Flame spread index less than 25
Flexibility: Excellent
Resistance to Acids: Good
Alkalis: Good
Salts: Good
Temperature Limits: Min. 5° F (−16° C) to 200° F (93° C)

LAGGING, DUCT AND LAP RUBBER BASE ADHESIVE (WHITE COLOR) COATING

RUBBER BASE, SOLVENT TYPE LAGGING DUCT AND LAP ADHESIVE

An adhesive to bond lagging, films and laminates to insulation. Is rapid setting, suitable for application by brush or spray. The color is white.

Properties—Wet

Flammability: Flash point approx. 10° F (−12° C)
Odor: Strong
Storage Stability: 2 years
Total Solids Content: 32 to 36% by wt., 20 to 23% by volume
Toxicity: Requires good ventilation
Weight: 6.6 to 7 lbs/gal (0.78 to 0.84 kg/liter)

Properties—Application

Adhesion to Surfaces: Excellent
Coverage, Attaching Insulation: 1.5 to 2 gal/100 ft^2 (0.61 to 0.81 liters/m^2)
Coverage, Sealing Laps: 1.2 to 1.7 gal/100 ft^2 (0.49 to 0.69 liters/m^2)
Drying Time: 15 minutes
Solvent: Hexane

Properties—After Drying

Adhesion to Surfaces: Excellent
Flammability: Flame spread index 25
Flexibility: Excellent
Resistance to Acids: Good
Alkalis: Good
Salts: Good
Solvents: Fair
Temperature Limits: Min. 5° F (−16° C) Max. 200° F (93° C)

INSULATION ADHESIVE AND COATING—HIGH TEMPERATURE

FILLED SILICATE, EMULSION ADHESIVE AND COATING

A combination coating and adhesive used to bond calcium-silicate, expanded silica, fiberglass and similar insulations to themselves and other surfaces. Applied by brush or trowel. Color is white.

Properties—Wet

Flammability: Nonflammable
Freeze-Thaw Stability: Excellent
pH: 10.5 to 11.5
Storage Stability: Over 2 years
Total Solids: 62% by wt., 35% by volume
Weight: 10.5 to 11.5 lbs/gal (1.25 to 1.35 kg/liter)

Properties—Application

Build: 0.125″ (3.175 mm)
Coverage: 8 gal/100 ft^2 (3.2 liter/m^2)
Curing Time: To touch—3 to 4 hours, through 36-48 hours at 50% R.H.
Shrinkage—Wet to Dry: 60% by volume

Properties—After Drying

Abrasion Resistance: Excellent
Coverage Dry: .125″ (3.175) per 12.5 ft^2 (1.16m^2) per gal (3.785 liters)
Flammability: Nonflammable
Impact Resistance: Poor
Resistance to Acids: Excellent
Solvents: Excellent
Temperature Limits: 0° F (−18° C) to 450° F (232° C)

MASTIC INSULATION INDOOR COATING

POLYVINYL-ACETATE, HEAVY BUILD WATER EMULSION COATING

A heavy build coating of paste consistency. Applied by glove, brush or spray. Can be used with reinforcing fabric or cloth. White and colors available.

Properties—Wet

Flammability: Will not support combustion
Freeze-Thaw Stability: Must be protected from freezing
pH: 7.1 to 7.5
Total Solids: 57 to 61% by wt., 40 to 44% by volume
Viscosity: Soft paste, 130 ± 10 KU
Weight: 11.4 to 11.8 lbs/gal (1.36 to 1.4 kg/liter)
Storage Stability: Over 2 years at temperatures above freezing

Properties—Application

Adhesion to Substrate: Excellent
Built, Wet: 1/32″ (0.8 mm) wet on vertical surface will not slip, slide or sag.
Coverage: 2 gal/100 ft^2 (.81 liter/m^2) for 1/64″ (0.4 m) of dry thickness
Curing Time: To touch—1 hour, through 6 hours
Solvent (while still wet): Water
Temperature Limits—installation: 40° F (4° C) to 100° F (38° C)

Properties—After Drying

Abrasion Resistance: Excellent
Adhesion to Substrate: Excellent
Flammability: Flame spread index 10
Resistance to Acids: Good
Alkalis: Good
Salts: Good
Solvents: Good
Temperature Limit, Max: 180° F
Water Vapor Permeance: 1.5 English perm (2.4 Metric perm)

MASTIC WEATHER BARRIER COATINGS

ACRYLIC POLYMER EMULSION WEATHER BARRIER MASTIC

A heavy build water emulsion mastic. To be applied by trowel, brush, glove or spray. Applied in two coats, a tack coat into which reinforcing fabric is installed into wet coat and then after drying, a second coat shall be applied. White and colors available.

Properties—Wet

Flammability: Will not support combustion

Freeze-Thaw Stability: Will withstand temp. down to 30° F for short periods.
pH: 6.3 ± 0.3
Total Solids: 64 to 68% by wt., 56 to 60% by volume
Weight: 9.8 to 10.5 lbs/gal (1.17 to 1.25 kg/liter)

Properties—Application
Adhesion to Substrate: Excellent
Build: 1/8" (3.175 mm) will not slip or slide on vertical surface
Coverage: 6 to 8 gal/100 ft^2 (2.44 to 3.25 liters/m^2)
Shrinkage—Wet to Dry: 1/8" (3.175 mm/wet—1/16" [1.58 mm] dry)
Curing Time: To touch—2 hours, Firm—24 hours
Solvent (while still wet): Water

Properties—After Drying
Abrasion Resistance: Excellent
Elongation: 70% at 70° F (21° C) without breaking
Flammability: Flame spread index less than 15
Flexibility: Bend 180° over 1/2" (12.5 mm) mandrel in 10 sec. no cracking
Resistance to Acids: Good for mild acids
Alkalis: Weak, excellent; strong, poor
Salts: Excellent
Temperature Limit, Max: 200° F (93° C) continuous
Tensile Strength: 115 psi (790 kPa) at 70° F (21° C)
Water Vapor Permeance: 1 English perm (1.6 Metric perm)
Weather Resistance: Excellent

ASPHALT, EMULSION TYPE, WEATHER BARRIER MASTIC

A water emulsion asphaltic mastic coating. To be applied by trowel, brush, spray or glove. To be applied in two coats; a tack coat into which reinforcing fabric is applied to wet coat and then after drying, overcoated with second coat.

Properties—Wet
Flammability: Nonflammable in liquid state
Total Solids: 70% by volume
Weight: 8 lbs/gal (0.95 kg/liter)

Properties—Application
Build: (Two coats) total 3/16" (4.76 mm)
Coverage: 2 to 3 gal/100 sq ft (.81 to 1.22 liter/m^2) each coat
Drying Time: Touch—1 hour, Firm—24 hours
Temperature Limits: 60° F (16° C) to 100° F (38° C)
Solvent (while still wet): Water
Solvent (clean up): Mineral spirits, kerosene

Properties—After Drying
Adhesion to Substrate: Good
Color: Black
Dried Film Thickness: Approx. 1/16" (1.58 mm)
Resistance to Acids: Excellent
Caustics: Excellent
Solvents: Not resistance
Flash Point: 100° F (37.8° C)
Service Temperature: −20° F (−29° C) to 200° F (93° C)
Water Vapor Permeance: 0.032 English perms (0.051 Metric perms)

ASPHALT, SOLVENT TYPE, WEATHER BARRIER AND VAPOR RETARDANT MASTIC

A solvent type asphaltic mastic coating. To be applied by spray, brush, trowel or glove. To be applied in two coats; a tack coat into which reinforcing fabric is applied to wet coat, then overcoated with second coat.

Properties—Wet
Flammability: Flash point 100° F (37.8° C)
Total Solids: 60 to 65% by volume
Weight: 9.3 to 10.3 lbs/gal (1.11 to 1.2 kg/m^3)

Properties—Application
Build: (Two coats) total 1/8" (3.175 mm)
Coverage: 2 to 3 gal/100 ft^2 (.81 to 1.22 liter/m^2) each coat
Drying Time: Touch—2 hours, Firm—24 hours through 14 days
Temperature Limits: 50 to 120° F (10 to 50° C)
Solvent (while still wet): Mineral spirits, kerosene

Properties—After Drying
Adhesion to Substrate: Excellent
Color: Black
Dried Film: Approx. 1/16" (1.58 mm)
Resistance to Acids: Excellent
Alkalis: Excellent
Salts: Excellent
Solvents: Not resistant
Flame Spread Index: (ASTM-84) 40
Service Temperature: 90° F (−40° C) to 200° F (93° C)
Water Permeance: 0.032 English perms (0.051 Metric perms)

ELASTOMERIC WEATHER BARRIER AND VAPOR RETARDANT COATING

A vapor retarder coating for installation over cold or cycling temperature service made of elastomers. To be applied over insulation with brush, glove or spray. Furnished in white, gray or special colors.

Properties—Wet
Flammability: Solvent type, combustible in wet state
Chlorinate solvent type, non combustible
Freeze-Thaw Stability: Excellent
Storage Stability: 2 years
Total Solids: 45 to 58 by volume
Weight: Solvent type, 10.5 to 10.9 lbs/gal (1.25 to 1.3 kg/liter)
Chlorinated solvent type, 12.8 to 13.8 lbs/gal (1.53 to 1.65 kg/liter)

Properties—Application
Build: 0.0625" (1.58 mm) for 1 gal (3.785 liters) per 25 ft^2 (2.32 m^2)
Coverage: 4 gal per 100 ft^2 (1.62 liters/m^2)
Curing Time: To touch—2 hours (at 50% R.H.) through 24 hours (at 50% R.H.)
Shrinkage: Wet to dry 67% (fabrics or matts not in measurement)

Properties—After Drying
Abrasion Resistance: Excellent
Coverage Dry: 0.02" (0.508 mm) per 1 gal (3.785 liters) per 12.5 ft^2 (1.16 m^2)
Emissivity, surface (e): Varies with color
Resistance to Acids: Excellent
Alkalis: Excellent
Salts: Excellent
Temperature Limits: 40° F (−40° C) to 190° F (88° C)
Water Vapor Permeance: At 0.02" (0.508 mm) thick 0.02 English perms (0.032 Metric perms)
Weather Resistance: Excellent

HYPALON RESIN, SOLVENT TYPE WEATHER-BARRIER, VAPOR RETARDER MASTIC

A heavy build solvent type mastic coating. To be applied by trowel, rubber glove or spray. White, gray and black colors available.

Properties—Wet
Flammability: Flash point 100° F (38° C)
Freeze-Thaw Stability: Stable (does not freeze)
Storage Stability: 6 months
Total Solids: 41% by volume
Weight: 9.3 lbs/gal (1.12 kg/liter)

Properties—Application
Coverage: 6 gal/100 ft^2 (2.4 liters/m^2)
Curing Time: To touch—3 hours, through 48 hours at 50% R.H.

MASTIC WEATHER BARRIER COATINGS (Cont.)

Shrinkage: Wet to dry thickness 70%
Solvent: Organic solvent

Properties—After Drying
Adhesion to Substrate: 75 lbs/in.^{2} (5300 kg/m^2)
Flammability: Flame spread index 20
Flexibility: Bend 180° over 1/2″ mandrel in 1 sec. at (−4° C) 5° F without cracking
Resistance to Acids: Very good
Alkalis: Excellent
Salts: Excellent
Solvent: Depends on particular solvent
Tensile Strength: 900 psi (6200 kPa)
Water Vapor Permeance: 0.08 English perm (0.128 Metric perms)

PETROLEUM RESIN (NON ASPHALTIC) WEATHER BARRIERS, VAPOR RETARDER MASTIC

A heavy build solvent petroleum resin mastic coating. To be applied by trowel, brush or spray.

Properties—Wet
Flammability: Combustible
Storage Stability: Very good
Total Solids: 48% by volume
Weight: 10.2 lbs/gal (1.22 kg/liter)

Properties—Application
Coverage: 8 to 10 gal/100 ft^2 (3.2 to 4.8 liter/m^2)
Curing Time: 6 hours to touch at 50% R.H., 36 hours through at 50% R.H.
Solvent Release: Formulated with blend of flammable and nonflammable solvents to fire safety in application
Shrinkage: Wet to dry thickness is 57%

Properties—Dry
Adhesion to Substrate: Excellent
Combustibility: Flame spread index = 30
Flexibility: Bend 180° over 1/2″ mandrel in 1 sec. at 25° F (−4° C) without cracking
Resistance to Acids: Good
Alkalis: Very good
Salts: Excellent
Solvents: Depends on particular solvent
Tensile Strength: 300 psi (2100 kPa)
Temperature Limits: Max, 200° F (93° C)—continuous
Water Vapor Permeance: 0.03 English perms (.048 Metric perms)
Weather Resistance: Excellent

POLYVINYL-ACETATE EMULSION WEATHER BARRIER MASTIC

A heavy build water emulsion mastic. To be applied by trowel, brush, glove or spray. Applied in two coats, a tack coat into which reinforcing fabric is installed into wet 1st coat, then after drying a second coat is applied. White and colors available.

Properties—Wet
Flammability: Will not support combustion
Freeze-Thaw Stability: Will withstand temperatures down to 15° F) (−9° C)
pH: 6.5 to 7.5
Total Solids: 60 to 66% by wt., 51 to 60% by volume
Weight: 10 to 10.8 lbs/gal (1.19 to 1.28 kg/liter)

Properties—Application
Adhesion to Substrate: Excellent
Build: 1/8″ (3.17 mm) wet thick on vertical surface will not slip, slide or sag
Coverage: 6 to 8 gal/100 ft^2 (2.44 to 3.25 liters/m^2)
Curing Time: To touch—3 hours through 24 hours at 50% R.H.
Shrinkage: Wet to dry thickness—38 to 48%
Solvent (while still wet): Water

Properties—After Drying
Abrasion Resistance: Excellent
Elongation: 70% at 70° F (21° C) without breaking
Flammability: Radiant Panel index less than 25 (special grade available with index less than 5)
Flexibility: Bend 180° over 1/2″ mandrel in 10 sec at 25° F (−4° C) without breaking
Resistance to Acids: Good for mild acids
Alkalis: Weak-excellent; strong-poor
Salts: Excellent
Temperature Limits: Max. 200° F (93° C) continuous
Tensile Strength: 115 psi (790 kPa) at 70° F (21° C)
Water Vapor Permeance: Over 1.0 English perm (1.6 Metric perm)
Weather Resistance: Excellent

VAPOR RETARDER INSULATION COATING

RESIN, HEAVY BUILD MASTIC FOR COATING OVER THERMAL INSULATION

A heavy build organic solvent coating to be installed on surfaces to vapor-seal semi permeable thermal insulations installed on substrates which has temperatures which are below ambient air temperature. Applied by brush or spray. Color is white.

Properties—Wet
Flammability: Combustible
Storage Stability: 3 years
Total Solids: 65 to 70% by wt., 92 to 46% by volume
Toxicity: TLV 500 ppm; poison. Requires good ventilation
Weight: 10.5 to 11.0 lbs/gal (1.25 to 1.32 kg/liter)

Properties—Application
Adhesion to Substrate: Excellent
Build: 1/16″ (1.58 mm) on vertical surface will not slip, slide or sag.
Coverage: 4 gal/100 ft^2 (1.92 liters/m^2) to obtain 1/32″ (0.79 mm) dried thickness
Curing Time: To touch—4 hours through 24 hours
Solvent: Organic solvents

Properties—After Drying
Adhesion to Surfaces: 50 psi (35000 kg/m^2)
Flammability: Flammable
Flexibility: 180° over 1/2″ mandrel in 8 sec at 25° F without cracking
Resistance to Acids: Fair
Alkalis: Fair
Salts: Good
Solvents: Poor
Temperature Limits: Max. 150° F (66° C)
Water Vapor Permeance: At 0.055″ (1.39 mm) thick 0.05 English perms (0.08 Metric perms)

FIRE PROTECTIVE COATINGS

FIRE-RETARDANT (INTUMESCENT) PAINT OR COATING

A protective paint or coating which when exposed to heat foams to multi-celled insulation that thermally insulates the substrate from the fire. The paint is generally used on surfaces requiring less protection (time) than surfaces requiring longer fire protection, on which the coating is used. Applied by brush or spray. Available in white and other colors.

Properties—Wet
Flammability: Flash point 90° F (32° C)
Storage Stability: 1 year
Total Solids, Paint: 66 to 68% by wt.
Coating: 68 to 70% by wt
Viscosity: Paint, 80 to 90 KU; Coating, 95 to 100 KU
Weight per Gallon: 10.9 to 11.2 lbs/gal (1.3 to 1.4 kg/liter)

Properties—Application

Adhesion to Substrate: Excellent
Coverage, Paint: 0.5 gal/100 ft^2 (0.2 liters/m^2) per coat
Coating: 1.25 to 1.5 gal/100 ft^2 (0.5 to 0.61 liter/m^2)
Drying Time: To touch—1 to 4 hours, through 24 to 48 hours
Solvent: Aromatic Solvent

Properties—

Abrasion Resistance: Good
Adhesion: Excellent
Flammability: Surface of intumescent foam, less than 25 flame spread index.*
Temperature Limit of Substrate: Max. 340° F (171° C)

JACKETING AND SHEATHING FABRICS

GLASS FIBER FABRIC FORMED WITH FIRE RETARDANT ADHESIVE BOND, COATED WITH HYPOLON

Cloths to be applied over pipe insulation, fittings or equipment. Film or aluminized barriers available on one or both sides—White on one side, aluminum or black color on other side, depending upon installation requirements.

Properties—

Thickness: 0.010 to 0.103″ (0.25 to 0.33 mm)
Weight: 11 to 15 oz/sq yd (0.26 kg to 0.35 kg/m^2)
Puncture: ASTM-D781 300 min ave
Water Vapor Permeability: 0.05 to 0.2 perms

Properties—In Service

Corrosion Resistance: Excellent
Temperature Limits: Min −50° F (−46° C); Max, 200° F (93° C)
Weather Resistance: Excellent

GLASS FIBER AND DILICONE FABRIC

Cloth for insulation pad covers, flange and valve covers. Color, silver (aluminum), satin weave.

Properties—

Thickness: 0.037″ ± 10%
Weight: 30 to 36 oz/sq yd (0.71 kg to 0.82 kg/m^2)
Flame Resistance: Flame out 1 second
(small flame) Afterglow 1 second
Char length 1″ (25.4 mm) max

Properties—In Service

Temperature Limits: Max 500° F (260° C) continuous
700° F (371° C) intermittent

WEATHER BARRIERS
Metal Jackets

ALUMINUM SHEETS

Surface Treatment: Inner, Polykraft moisture retarder
Gage: 0.016″ (0.4 mm) to 0.04″ (1.01 mm) thickness
Configurations: Deep corrugated, 1¼ × ¼″ (24.1 mm × 6.3 mm) or 2½ × 5/8″ (63.5 mm × 15.8 mm)
Smooth
Cross crimped
Stucco embossed
Sheet Sizes: 6′ 0″ × 33″ (1.8 m × 0.84 m) to 12′ 0″ × 33 (3.6 m × 0.84 m)
Type: 3003 or 5005
Metal Properties:
Conductivity, Thermal 1130 to 1340 Btu in./ft^2 hr, °F (162 to 193 w/m K)
Expansion, Thermal, 1.3 × 10^{-5} per °F at 68° F to 212° F or (2.34 × 10^{-5} per °C (°K) or 20° C to 100° C)
Density, 169 lbs/ft^3 (2700 kg/m^3)
Melting Range, 1180 to 1215° F (720 to 755° C)

*After exposure to fire expanded foam must be removed and replaced as it has little mechanical strength.

ALUMINUM ROLL JACKETING

Surface Treatment: Inner, Polykraft, or plastic moisture retarder
Gage: 0.006″ (0.1524 mm) to 0.05″ (1.27 mm)
Configurations: 3/16″ (4.76 mm) corrugations (longitudinal)
Smooth
Stucco embossed
Type: 3003 or 5005
Metal Properties:
Conductivity, Thermal 1130 to 1340 Btu, in./ft^2, hr, °F (162 to 193 W/m K)
Expansion, Thermal 1.3 × 10^{-5} per °F at 68° F to 212° F (2.34 × 10^{-5} per °C (or K) at 20° C to 100° C)
Density, 169 lbs/ft^3, 2700 kg/m^3
Melting Range, 1180 to 1215° F (or 720 to 755° C)

ALUMINUM CIRCUMFERENTIALLY CORRUGATED JACKETS

Surface Treatment: Inner, Polykraft moisture retarder
Gage: 0.010″ (0.254 mm) or 0.016″ (0.406 mm)
Configuration: Circumferentially corrugated in 36″ wide sections to fit nominal size insulation or in 36″ (0.914 m) × 100′0″ (30.48 m) roll jacketing.
Type: 3003 or 5005
Metal Properties:
Conductivity, Thermal 1130 to 1340 Btu in./ft^2, hr, °F (162 to 193 W/mK)
Expansion, Thermal 1.3 × 10^{-5} per °F at 68° F to 212° F (2.34 × 10^{-5} per °C at 20° C to 100° C)
Density, 169 lbs/ft^3 (2700 kg/m^3)
Melting Range, 1180 to 1215° F (720 to 755° C)

ALUMINUM COATED ROLL JACKETING

Surface Treatment: Inner, Polykraft moisture retarder or epoxy coating outer epoxy, acrylic or vinyl coating. Colors available—many.
Gage of Metal: 0.016″ (0.406 mm) to 0.040″ (1.016 mm)
Configuration: Smooth or corrugated in 36″ (0.914 m) × 100′ 0″ (30.48 m) rolls
Metal Properties: Same as listed for corrugated jackets
Emittance: (e) depends on color and surface
Coating Resistance to Acids: Good to Excellent
Alkalis: Good to Excellent
Salts: Good to Excellent
Combustibility of Coatings: Depends upon coating used on outer surface.

STAINLESS STEEL ROLL JACKETING

Gage: 0.010″ (0.254 mm) to 0.019″ (0.483 mm)
Configuration: Smooth or 3/16″ (4.76 mm) corrugated in 36″ (0.914 mm) × 50′0″ (15.24 m) roll jacketing
Metal Properties:
Conductivity 212 Btu in./ft^2, hr, °F or 30.6 W/mK
Density, 501 lbs/ft^3 or 8021 kg/m^3
Emittance (e) = 0.08 (new and clean)
Hardness, 150 Brinell No.
Melting Range 2600° F (1426° C)
Expansion 9.61 × 10^{-7} °F at 54° F to 212° F (1.53 × 10^{-8} °C at 12° C to 100° C)
Resistance to Acids: Excellent
Alkalis: Excellent
Salts: Excellent (except Halogens)
Max Temperature Limits, 2000° F (1093° C)

WEATHER BARRIERS—Metal Jackets (Cont.)

STAINLESS STEEL DEEP CORRUGATED SHEETS
Gage: 0.010″ (0.254 mm) to 0.019″ (0.484 mm)
Configurations: 1-1/9″ × 1/4″ (31.75 mm × 6.35 mm) corrugation
2-1/2″ × 5/8″ (53.8 mm × 15.8 mm) corrugation
Stucco embossed
Cross crimped
Smooth
Inner Surface: Available with Polykraft moisture barrier
Metal Properties: Same as shown for stainless steel roll jacketing

BONDERIZED, ELECTROGALVANIZED COLD ROLLED STEEL ROLLS
Surface Treatment: Inner surface electrogalvanized, outer surface coated with vinyl organosol
Gage of Metal: 0.010″ (0.254 mm) to 0.015″ (0.376 mm)
Configurations: Smooth or 3/16″ (4.76 mm) corrugated
Color of Outer Surface: White and colors available
Rolls: 36″ (0.914 m) wide × 50′ 0″ (15.2 m) or 75′ (22.8 m) long
Metal Properties:
Conductivity: 620 Btu in./ft^2, hr, °F (89.4 W/mK)
Density: 484 lb/ft^3 (7749 kg/m^3)
Melting Range: 2800° F (1537° C)
Max Temperature Range: 1150° F (621° C) (metal)
Max Temperature Range: 300° F (149° C) (organosol)
Hardness: 35-60 Rockwell B
Corrosion Resistance to Acids: Fair to Excellent depending on acid and concentration
Alkalis: Fair to excellent depending on alkali and concentration

PREFORMED METAL JACKETING
Fabricated to fit standard nominal insulation sizes, 36″ long sections (0.914 m) with modified Pittsburgh "Z" lock on longitudinal seam. Butt strap with high temperature sealant formed to fit over circumferential joint. (As illustrated in Figure 45.)

Metal as Specified: Aluminum, stainless steel or electrogalvanized cold rolled steel
Properties of Metals: As given previously for each under jacketing

WEATHER BARRIERS
Metal Lagging

RIBBED, BOX BEAM, AND CORRUGATED
Gage of Metal: 0.015″ (0.376 mm) to 0.05″ (1.25 mm)
Configurations: 7″ Ribbed siding, 8″ ribbed siding
5.33″ box beam siding, 2.67 corrugated siding,
and 4″ box rib siding
Metals Available: Aluminum, aluminized steel, steel, or stainless steel
Properties of Metals: Same as listed for metal under jacketing

Felts and Laminated Jackets

LAMINATE KRAFT, REINFORCEMENT, ADHESIVE ATTACHED ALUMINUM FOIL
Fiberglass reinforcement foil available 0.0007 to 0.001 thick

Weather Barrier and Vapor Retarder—In rolls

Physical Properties
Tensile Strength: (ASTM-D-828) 30 to 50 MD—30 to 50 × D
Puncture Resistance: (ASTM-D-781) 20 to 50
Weight: 25 to 35 lbs/100 sq ft (1.22 to 1.71 kg/m^2)
Flame Spread Index: 25 (ASTM E-84)
Flame Resistance: Less than 2″ (ASTM D-777)
Vapor Migration: 0.009 to 0.027 grains/ft^2 hr (5.4×10^{-6} to 1.62×10^{-5} kg/m^2 hr)

LAMINATED TEDLAR/NEOPRENE IMPREGNATED FELT

Weather Barrier and Vapor Retarder—In rolls 35½″ (0.91 m) wide × 166.7′ (50.8 m) long

Physical Properties—
Tensile Strength: (ASTM-D-828) 42MD
Tear Strength: (ASTM-D-689) 448 CMG gms
Weight: 12 lbs/100 sq ft (0.586 kg/m^2)
Max Temperature Limits: 225° F (107° C)
Resistance to Acids: Excellent
Alkalis: Excellent
Salts: Excellent
Vapor Migration: 0.05 grains/ft^2 hr (3.01×10^{-5} kg/m^2 hr)

LAMINATED TEDLAR/NEOPRENE IMPREGNATED FELT/SARAN

Weather Barrier and Vapor Retarder—In rolls 35½″ (0.91 m) wide × 166.7 (50.8 m) long

Physical Properties—
Tensile Strength: (ASTM-D-828) 42MD
Tear Strength: (ASTTM-D-689) 448 CMG gms
Weight: 12 lbs/100 sq ft (0.586 kg/m^2)
Max Temperature Limits: 225° F (107° C)
Resistance to Acids: Excellent
Alkalis: Excellent
Salts: Excellent
Vapor Migration: 0.05 grains/ft^2 hr, (3.0×10^{-5} kg/m^2 hr)

PLASTIC WEATHER BARRIER PIPE AND FITTING COVERS

Weather Barrier—In rolls 42″ (1.06 m) wide by 114′ (34.8 m) long. Polyvinyl chloride plastic sheet.

Fittings molded of polyvinyl chloride plastic

Physical Properties—
1/8″ thickness—Tensile Strength: 4900 psi (1.39 kg/m^2)
Hardness: (ASTM-D-785) 89 Rockwell R
Flash Point: 870° F (466° C)
Max Temperature Limit: −150° F (65° C)
Vapor Migration: (0.028″ thick) 130 grains/ft^2, hr, (7.826×10^{-3} kg/m^2 hr)

Used outdoors or indoors as covering. Installed with strap, special adhesives or tape.

PLASTIC INDOOR JACKET AND FITTING COVERS
Jacketing in rolls or preformed to fit pipe and fitting insulation.

Physical Properties—
1/8″ thickness—Tensile Strength: 4900 psi (1.39 kg/m^2)
Hardness: (ASTM-D-785) 89 Rockwell R
Flash Point: 870° F (65° C)
Max Temperature Limit: 160° F (71° C)
Vapor Migration: (.028″ thick) 130 grains/ft^2 hr, (7.826×10^{-5} kg/m^2 hr)

Used indoors as covering. Installed with strap, special adhesives or tape.

ACCESSORIES
Lap and Insulation Adhesive and Sealer

RUBBER OR NEOPRENE BASE ADHESIVE WITH NON-FLAMMABLE SOLVENTS
An adhesive to bond laps of vapor barrier sheets and to adhere fibrous insulation to metal ducts. Has quick tack time. Application by brush or spray. Color is off-white.

Properties—Wet
Flammability: Nonflammable
Odor: Strong

Storage Stability: 2 years
Total Solids Content: 20 to 30% by wt., 19 to 25% by volume
Toxicity: Requires good ventilation
Viscosity: 75 ± 5 KU
Weight: 10.6 to 11.2 lbs/gal (1.26 to 1.34 liters/liter)

Properties—Application
0.6 to 1.5 gal/100 ft^2 (0.25 to 6.1 liters/m^2)
Drying Time: Tack free 2 hours, through 8 hours
Solvent: Chlorinated solvents

Properties—After Drying
Adhesion to Surfaces: Excellent
Flammability: Flame spread index less than 10
Flexibility: Excellent
Resistance to Acids: Good
Alkalis: Good
Salts: Good
Solvents: Fair
Temperature Limits: Min −10° F (−23° C), Max 225° F (107° C)

VINYL-ASPHALT SEALER AND ADHESIVE FOR CELLULAR INSULATION
A vinyl-asphalt sealer adhesive for sealing and bonding of surfaces of cellular glass and polyurethane to themselves to produce a vapor resistant joint.

Properties—Wet
Flammability: Combustible, Flash point 60° F (17° C)
Freeze-Thaw Stability: Will not freeze down to 0° F (−17° C)
Odor: Strong
Storage Stability: 2 years
Total Solids Content: 48 to 50% by wt., 36 to 40% by volume
Toxicity: Solvent fumes are toxic. Requires good ventilation
Weight: 9 to 9.3 lbs/gal (1.07 to 1.11 kg/liter)

Properties—Application
Adhesion to Cellular Glass and Polyurethane: Excellent
Build, Wet: 1/16″ (1.6 mm) will not slip or sag
Coverage: 6½ gal/100 ft^2 (3.12 liters/m^2) to obtain 1/8″ thick wet
Solvent: Ketone

Properties—After Drying
Abrasion Resistance: Excellent
Adhesion to Cellular Glass and Polyurethane: Excellent
Flammability: Slow burning
Resistance to Acids: Excellent
Alkalis: Excellent
Solvents: Hydrocarbons: poor
Temperature Limits: Min −50° F (−46° C), Max 160° F (71° C)
Toxicity: Not toxic when dry
Water Vapor Permeance: 0.6 English perms (0.96 Metric perms)
Weather Resistance: Excellent

SEALANT, OIL BASED
A non-hardening moisture vapor sealant for joints and caulking of insulation. Color is gray..

Properties—
Flammability: Combustible, Flash point 105° F (41° C)
Weight: 11 to 13.4 lbs/gal (1.31 to 1.6 kg/liter)
Coverage: 900 to 1200 lin ft for 1/8″ × 2″ wide joint per gal (72 to 100 lin m for 3 mm × 50.8 mm wide joint per liter)
Drying Time: Skins over in 24 hours—non drying
Temperature, Application: 40 to 100° F (4 to 38° C)
Service: −40 to 180° F (−40° C to 60° C)
Water Vapor Permeance: 1/16″ (1.6 mm) thickness is 0.15 English perms (0.25 Metric perms)

SEALANT, BUTYL BASE
A non-hardening moisture vapor sealant for joints and caulking. Colors available are black, white and gray.

Properties—
Flammability: Combustible, Flash point 105° F (41° C)
Weight: 10.2 to 12.6 lbs/gal (1.21 to 1.5 kg/liter)
Coverage: 1/8″ film, 13 ft^2/gal (3.17 mm film, 1.02 m^2/liter)
Drying Time: Skins over in 3 hours—non hardening
Temperature, Application: 40 to 100° F (4 to 38° C)
Service: −60 to 180° F (−51° C to 60° C)
Water Vapor Permeance: 1/16″ (1.6 mm) thickness is 0.10 English perms (0.16 Metric perms)

VINYL-CHLORIDE ANTI-ABRASIVE COATING FOR CELLULAR GLASS INSULATION
A coating for coating the inner surfaces of cellular glass insulation to minimize fracturing and powdering of the glass due to vibration and movement. Applied by brush or spray.

Properties—Wet
Flammability: Flash point 40° F (4° C)
Storage Stability: 2 to 3 years
Weight: 8.0 to 8.4 lbs/gal (0.95 to 1.00 kg/liter)

Properties—Application
Coverage: 1/2 to 1 gal/100 ft^2 (0.2 to 0.41 liters/m^2)
Drying Time: 10 minutes to touch
Solvent: Keytone

Properties—After Drying
Abrasion Resistance: Excellent
Adhesion to Cellular Glass: Excellent
Flammability: Will not spread flame when dry
Temperature Limits: Min −150° F (−101° C), Max 300° F (149° C)

SYNTHETIC POLYMER ANTI-ABRASIVE COATING FOR CELLULAR GLASS INSULATION
A coating for coating the inner surfaces of cellular glass insulation to minimize fracturing and powdering of the glass due to vibration and movement. Apply by brush or spray.

Properties—Wet
Flammability: Nonflammable
Storage Stability: Excellent
Total Solids: 68% by weight
Toxicity: TLV = 350 pg m
Weight: 15.2 lbs/gal (1.82 kg/liter)

Properties—Application
Coverage: 1 gal/100 ft^2 (0.4 liter/m^2)
Toxicity: Requires good ventilation
Solvent: Chlorinated organic solvent

Properties—After Drying
Flammability: Radiant Panel ASTM E-162 Flame Spread Index = 5
Corrosion: Not recommended to be used with stainless steel substrate
Temperature Limits: Max 250° F (121° C)

GYPSUM-GLASS FIBER REINFORCED BONDING ADHESIVE FOR CELLULAR GLASS INSULATION
Inorganic-hydraulic setting cement gypsum reinforced with glass fibers for coating the inner surfaces of cellular glass insulation to minimize fractures and powdering of glass due to vibration and movement. Also used to bond surfaces of cellular glass to themselves. For use at temperatures above ambient air temperature.

Properties—Application
Powder to mix with water: Brush or trowel consistency
Setting Time: Full strength 24 hours, hydraulic setting

ACCESSORIES (Cont.)

Properties—After Drying
Adhesion to Cellular Glass: Excellent
Flammability: Nonflammable
Temperature Limits: To be used above ambient air temperature
Max 500° F (200° C)
Weather Resistance: Poor

SODIUM SILICATE AND FILLERS HIGH TEMPERATURE ADHESIVE AND BONDING CEMENT

A cement-adhesive to bond calcium-silicate, expanded silica insulation block and board to themselves and to nonporous surfaces. Of particular use to bond cut high temperature insulation pieces together to form fabricated insulation fittings and valve covers.

Properties—Wet
Flammability: Nonflammable
Freeze-Thaw Stability: Shall not be allowed to freeze
Total Solids: 50 to 57% by weight
Viscosity: Brush grade—soft paste; Trowel grade—thick paste
Weight: 9.4 to 11.3 lbs/gal (1.12 to 1.34 kg/liter)
Storage Stability: 2 years

Properties—Application
Adhesion to Porous Insulation: Excellent
Bonding Time: 10 to 20 minutes
Coverage: 1.25 to 4 gal/100 ft^2 (0.51 to 1.6 liter/m^2)
Drying Time in Joints: Approximately 2 hours
Solvent: Water

Properties—After Drying
Flexibility: Brittle
Adhesion: Excellent
Flammability: Noncombustible
Temperature Limit: Max, 1000° F (538° C)
Toxicity: Nontoxic

TWO COMPONENT ADHESIVE AND BONDING CEMENT FOR CELLULAR GLASS

A two part adhesive for bonding cellular glass surfaces to each other and to other materials of different thermal expansion. Cures to form a flexible bond that absorbs mechanical and thermal shock. Component No. 1 is stirred then Component No. 2 is added using electric or air operated 5″ mixer. Application by notched trowel. Color is black.

Properties—Unmixed
Flammability of Liquid Content: Flash point 100° F (38° C)
Solids: Component No. 1 95%; Component No. 2 100%
Weight: 10 lbs/gal (1.193 kg/liter)

Properties—Application
Coverage: 2.5 to 5 gal/100 ft^2 (1.0 to 2 liters/m^2)
Pot Life: 1 hour
Shrinkage: Max 5%
Temperature of Material: 70° F to 95° F (21° C to 35° C)
of Surface: 40° F to 95° F (4.4° C to 35° C)

Properties—After Drying
Adhesion to Cellular Glass: Excellent
Temperature, Service: −70° F to 180° F (−57° C to 82° C)
Water Vapor Transmission: 0.0048 perm-inch (0.0064 perm cm)
Weather Resistance: Must be protected from sunlight

POLYVINYL-ACETATE, CORK-FILLED, CAULKING AND FILLET COMPOUND

A compound used as fillet to round out inside corners where two planes of insulation meet. Also used to caulk joints in moderate and low temperature insulations. Applied by trowel, color is white.

Properties—Wet
Flammability: Nonflammable
Freeze-thaw Stability: Shall not be allowed to freeze
Solids: 76 to 80% by weight, 70 to 74% by volume
Storage Stability: 2 years
Viscosity: Troweling
Weight: 10.7 to 11 lbs/gal. (1.27 to 1.31 kg/liter)

Properties—Application
Adhesion to Insulation Surfaces: Good
Build: 3/8″ (9.5 mm) thickness will not slip or sag on vertical surface
Drying Time: To touch—4 to 8 hrs; through—48 to 96 hrs
Solvent (while still wet): Water
Temperature: Min. surface and ambient air, 35° F (2° C)

Properties—After Drying
Adhesion to Insulation: Excellent
Flammability: Slow burning
Flexibility: Excellent
Impact Resistance: Excellent
Resistance to Acids: Good
Alkalis: Good
Salts: Good
Temperature Limit, Max.: 200° F (93° C)
Weather Resistance: Excellent

EXTENSIBLE REINFORCING CLOTH

A modacrylic fiber cloth for reinforcing mastics and coatings to insure even thickness and to bridge gaps.

Description
A modacrylic fiber cloth, treated with 8% solution of polyvinyl acetate, weighing 2.65 ounces per sq yd (0.0627 kg/m^2), 8 threads per inch (25.4 mm)

Properties
Elongation, Ultimate: At 70° F (21° C) 37%
At 0° F (−18° C) 35%
At −60° F (−51° C) 30%
Chemical Resistance to Acids: Excellent
Alkalis: Excellent

GLASS FIBER REINFORCING CLOTH

A glass fiber cloth for reinforcing mastics and coatings.

Description
White resin binder 10 × 10 weave for outdoor applications. White resin binder 20 × 20 weave for indoor applications with light viscosity coatings.

Properties
Elongation, Ultimate: At 70° F (21° C) 8%
At 0° F (−18° C) 7%

POLYESTER (40%) AND NYLON® (60%) REINFORCING CLOTH

A resin fibers cloth for reinforcing heavy build mastics.

Description
A resin fibers cloth, weighing 3.08 ounces per sq yd (0.0728 kg/m^2) 6.5 × 6.0 threads per inch.

Properties
Elongation, Ultimate: At 70° F (21° C) 37%

Tapes

ALUMINUM FOIL PRESSURE SENSITIVE TAPES

A tape for sealing foil faced insulation; 2 mil thickness aluminum available in 2″ (50 mm), 2½″ (62 mm) and 3″ (76 mm) widths.

Properties—Physical
Tensile Break: 26 to 28 lbs (117 to 126 kg's)
Adhesion: 53 oz/in. (1.49 kg/25 mm) to steel
47 oz/in. of width (1.33 kg/25 mm) to own backing
Temperature Limits: Application, above 40° F (4.4° C)
Service, max. 180° F (82° C)

ALUMINUM FOIL QUICK STICK ADHESIVE TAPE

A tape for sealing foil faced insulation, 2 mil thickness aluminum with synthetic rubber adhesive for quick stick.

Properties—Physical

Tensile Break: 25 to 30 lbs (113 to 136 kg's)
Adhesion to Steel: 60 to 90 oz/in. of width (1.7 to 2.35 kg per mm width)
Flammability: Flame spread index 25
Temperature Limits: Application, above 40° F (4.4° C)
Service, max. 180° F (82° C)

ALUMINUM FOIL, GLASS FIBER SCRIM REINFORCING, KRAFT PAPER AND HIGH-TACK ADHESIVE TAPE

A tape for use on hot, cold or dual temperature services. Available in widths 2½″ (63 mm), 3″ (76 mm), 4″ (102 mm) and 5″ (127 mm)

Properties

Tensile Break: 47 to 50 lbs (213 to 226 kg's)
Adhesion to Steel: 121 oz per in. of width (3.43 kg per mm width)
Temperature Limits: Application, above 40° F (4.4° C)
Service, min. 0° F (−18° C) to 180° F (82° C)

POLYVINYL CHLORIDE FILM PRESSURE SENSITIVE SEALING TAPE

A film which will stretch so that it can be used to seal elbows and circumferential joints of molded or preshaped pipe insulation coverings. Available in widths of 1″ (25 mm), 1½″ (37 mm) and 2″ (50 mm). Colors, white or black. 5.5 mil and 10 mil thickness.

Properties

Elongation: 250%
Tensile, Warp: 5.5″ mils thick 15 lbs (break), 10 mils thick 30 lbs (break)
Adhesion to Own Back: 21 oz per in. width
Temperature Limits, Service: Above 40° F (4° C) to 125° F (52° C)

TEDLAR® FILM REINFORCED WITH POLYESTER MAT AND INCOMBUSTIBLE PRESSURE SENSITIVE ADHESIVE

A pressure sensitive tape for sealing jacketing on pipes and ducts. Film thickness 1½ mils. Total thickness: min. 0.0065″ (0.16 mm), max. 0.008″ (0.20 mm). Colors, white or gray.

Properties

Corrosion Resistance: No corrosion after 30 days at 120° F (49° C) at 96 RH
Flame Resistance: Flame out 5 seconds
Afterglow 2 seconds
Char Length 5″ (127 mm)
Oil Resistance: No delamination after soaking for 24 hours
Temperature Limits: 40° F (4° C) to 110° F (43° C)
Water Permeance: No delamination after soaking for 24 hours

TEDLAR® FILM, WHITE COATED WITH INCOMBUSTIBLE PRESSURE SENSITIVE ADHESIVE

A tape for sealing jackets on pipe and duct insulation. Supplied on release paper. Color, white, thickness 1½ mil.

Properties

Corrosion Resistance: No corrosion after 30 days at 120° F (49° C) at 96 RH
Flame Resistance: Flame out 5 seconds
Afterglow 2 seconds
Char length 5 inches (127 mm)
Oil Resistance: No delamination after soaking for 24 hours
Temperature Limits: 40° F (4° C) to 110° F (43° C)
Thickness: Min, 0.0045″ (0.114 mm), max 0.006″ (0.152 mm)
Water Resistance: No delamination after soaking for 24 hours

TEDLAR® FILM, GLASS FABRIC AND HYPALON RUBBER COATING AND INCOMBUSTIBLE PRESSURE SENSITIVE ADHESIVE

A tape for sealing and securing of jackets on pipe, equipment and duct insulation. Color is white.

Properties

Corrosion Resistance: No corrosion after 30 days at 120° F (49° C) at 96 RH
Flexibility after: 250° F (121 ° C) slight hardening after 120 hours
Oil Resistance: No delamination after soaking 24 hours
Temperature Limits: 40° F (40° C) to 110° F (43° C)
Water Resistance: No delamination after 24 hours in boiling water
Weather Resistance: Excellent
Water Vapor Permeance: 0.2 English perms (0.36 Metric perms)

7 Accessories

Other than the primary insulations, barriers, and coverings there are a number of other items necessary for a complete insulation application. These items are determined by the installation. The major functions of these various accessories are:

Supports for insulation
Securements for insulation
Attachment adhesives for insulation
Fasteners for insulation
Reinforcement for insulation
Flashings for weather-barriers
Expansion and contraction slip joints
Cradles for insulated pipe
Insulating pipe slide supports

In addition to some of the structural insulating boards manufactured for the building industry, most insulations are not structural materials. They must be supported, secured, or fastened to other materials to maintain their position in service.

Insulation Supports

An insulation support is any device which, by attachment to a more rigid, or stronger, material, transmits the weight of, and the external load on, the insulation to the stronger and more rigid material.

The insulation may rest on supports, or hang from them by various types of securements.

Building and cold storage insulation supports

Insulation supports in building construction may be clips or pins attached to walls, ceilings, or ducts. Cold storage rooms may use similar types on walls, and may use "tee" sections, or other supporting members, to support the ceiling insulation. A major part of cold storage design is to provide insulation supports which do not cause thermal "short circuits" and which do not cause cracks in the insulation as a result of expansion and contraction. Because such systems are so much a part of the building design, an engineering discussion would involve architecture as well as civil and mechanical engineering. Thus, it is beyond the scope of this manual.

Duct and flat surface insulation supports

Supports for insulation on duct work in buildings are (1) pins, either welded or cemented to the duct surface, (2) blank nuts, slotted rectangular studs, split studs, lagging studs welded to the surface, and (3) grip nails clinched to the metal surface. In all cases these items transmit the weight of the insulation to the surface being insulated.

Insulation supports on vessels

Vertical vessels, insulated with block insulation, require a support at the bottom, and, if the vessel is over a certain height, intermediate supports to prevent the insulation from sliding downward. These supports may be angles, plates, rods, or other projections welded or bolted into position around the vessel.

The distance these angles or plates project from the vessel and the distance between supports depends upon the thickness of the insulation, the height of the insulation being supported, and its compressive strength. The supporting area under the first layer of insulation must be such that the weight of material supported, in lbs per square inch, is well within the allowable compressive unit strength of the insulation. Insulations with little compressive strength must have supports spaced more closely than insulation of higher compressive strength. The distance the support projects from the surface should always be less than the insulation thickness, to reduce the thermal "short circuit" effect. Remember: except for vessels operating at very moderate temperatures, the insulation should be sufficiently friction-free of the vessel surface so that expansion and contraction movement of the vessel is independent of the insulation. Stated another way, the cylindrical vessel should be free to slide up and down within the cylinder of insulation. To obtain this freedom of movement, insulation should be installed slightly oversize. A cushion blanket is recommended where vessels operate at very high or very low temperatures.

Insulation supports should be installed on vertical vessels on 12 ft-18 ft (3.65 to 5.5 metres) centers. Compressive strength, and change in length caused by expansion or contraction of the vessel compared to the insulation change in length are two factors controlling the support centers. In some instances, the support distances are set by vessel flanges. In all cases, a support should be installed above a vessel flange to prevent the insulation from sliding down and resting on the flange studs. The support should be a sufficient height above the flange to permit easy removal of the studs. On low temperature vessels, the differential movements which occur at the contraction joint should not exceed the dimensional flexibility of the caulking-weather-vapor seal at these joints.

Insulation on long runs of vertical pipe should be supported in the same manner as that described for vertical equipment.

Factory attached, metal jacketed insulation panels are supported by a vessel in several ways. The first panel side rests on the tank base, and successive panels are supported on "S" clips attached to the top of the installed panel. In addition, the metal jackets are supported by pins welded to the side of the vessel. The jackets in turn support the insulation cemented to it. The panels are secured by clips or straps.

Large horizontal vessels (over 6 ft or 1.8 metre in diameter) require supports to which securement straps, used to pull the bottom insulation into position, can be attached. Such supports are necessary, for, if straps were simply brought around the vessel insulation on top, the pull required to draw the bottom insulation into position would exceed the compressive limits of the insulation over which it passes. The supports are so located that bands can be drawn to secure the bottom third sector of insulation into position. Smaller horizontal vessels are their own support for the cylindrical insulation. The straps drawn around the insulation transmit the load of the lower insulation to that on top, which in turn is supported by the vessel. This is also true of all insulation installed on horizontal pipes.

The head insulation on horizontal vessels must also be supported. In most instances, the support for the head insulation is the same as that for the support of the cylindrical insulation, with slots or holes punched for the attachment of straps. The straps then pass from the cylindrical support on one side, over the insulation to the support on the other side, and when drawn tight secure the insulation in position.

Depending upon the method of installation various types of insulation supports are required. In summary, insulation supports are:

- Structural angles
- Formed strapping
- Metal plates and rectangular bars
- Metal rods
- Road mesh
- Pins—metal and plastic (welded or adhered to surface)
- Blank nuts
- Rectangular studs (punched, split, etc.)
- Grooved studs
- "S" clips
- Grip nails
- Accessories for the above are: bolts, nuts, rivets, attachment by welding, or adhesives.

All these are used to transfer the weight of the insulation to the vessels, pipe, equipment, walls, or structural steel. They are structural members, regardless of whether they support the load in compression or tension.

Securements for Insulation

An insulation securement is any device, which, by itself, or attached to a support, holds the insulation in position.

For clarity in discussing supports, it was necessary to mention securements.

Duct work and flat surfaces

When pins are attached to ducts or flat surfaces and extend through the insulation, the securement is obtained through the use of a clip that slides over the pin, or a stud cap that is placed on the stud.

Where punched studs or blank nuts are welded to the surface, the securement may be wire which is threaded through the stud slot or nut, brought through the insulation, then laced with other wires to hold the insulation in place.

Insulation securements on vessels

The insulation on vertical cylindrical vessels insulated with block insulation is secured to the vessel with straps, which are tightened and their ends fastened together with strap clips (sometimes called seals). The metal or plastic of the strap depends upon chemical atmospheric conditions, the metal of the vessel and the vessel location (indoors or outdoors). The strength of the strap must be sufficient to hold the tensile load imposed upon it. This load depends upon the weight of the insulation, size of vessel, and the load which may be exerted on it by expansion of the vessel and the tension induced by the insulator with his strapping tool. The limiting strength factor in service is the strap clip rather than the strap.

In many instances, expander straps or expansion springs are used with straps to allow for expansion as a high temperature vessel increases in circumference. Where such expansion pushes the insulation to a larger circle, cracks between insulation blocks are opened. Thus, the use of expansion straps or springs in such applications indicates a compensation for poor design. The insulation should be sufficiently oversize, as installed, to prevent this motion from being transmitted to its outer surface. This cavity between the vessel wall and its insulation can be obtained in several ways, by insulation cut and fitted to a larger diameter than the vessel, by soft asbestos rope wrapped around the vessel before application of the block, or by the application of a soft cushion blanket of insulation prior to application of the block.

On horizontal, cylindrical vessels, the same procedure is used as for vertical vessels with the exception of securements at the bottom quadrant. In this case, the securement for the bottom insulation is fastened to supports which are directly attached to the vessel. As the vessel will expand and contract, regardless of any space between it and the insulation, this movement is directly imposed upon the straps. A vessel which operates through any great temperature range (high or low) requires that this securement strap be expander strap or strap and expander springs, otherwise the insulation may be crushed (high temperature) or allowed to sag excessively (low temperature).

Another use of flexible strap is in low temperature applications where both the vessel and the insulation contract more than the contraction of the strap on the outer surface at approximately atmospheric temperature. In this case, nylon strap can be used. It will stretch under tension, and, as the subsequent contraction occurs, the "elastic memory" of the strap will cause it to constrict, thus keeping the insulation snug and tight regardless of the reduced circumferential dimension.

Insulation securements for piping

Wire is used as a securement for most sectional insulation on pipes up to approximately 12″ (305 mm) OD of the insulation. Above this diameter, strap is used. Except on long vertical runs of pipe, the pipe itself is the insulation support. The weight of the insulation is supported by the top half of the pipe. The upper section of the insulation rests on the pipe, and the weight of the lower half of the insulation is transferred by the strap or wire to this top half of the insulation, thus imposing the full load on the pipe. For this reason, wire, strap, and other materials used in this manner are classified as securements rather than supports.

Wire, when pulled tight and the ends twisted together, will provide 40-80 lbs (181 to 362 kg) tensile pull to secure the insulation in place. As it is pulled taut it will slip over the insulation surface and draw the edges together.

Because of its high tensile pull and small size, wire is not suitable for use on friable insulations, such as foams, as it will cut the material as it attempts to draw the circumferential joints together.

On foam insulation used in low temperature service, glass filament pressure sensitive tape is recommended for securement of all inner layers and outer layers up to 7 in. (178 mm) diameter. Glass filament pressure sensitive tape cannot be "pulled up" as can wire and strap. For this reason, the two sections must be held tightly together while the tape is rapped in position to secure them. The 7 in. diameter is about the maximum that an applicator can hold together and install the tape and still obtain a tight fit. When used on inner layers above this diameter, slight looseness is less important and a tight vapor seal is obtained by the outer layer which is drawn tight by strap.

Location of wire, strap, tape

Some pipe insulations are furnished with jackets attached at the factory, with circumferential and longitudinal laps which are to be cemented together with fast setting cements. Thus, the securement of the insulation is its jacket and adhesive joints.

Where pipe insulation is installed in pipe chases, or in places where it receives no mechanical abuse, and is in moderate temperature service, it may be installed with straight butt circumferential joints. Where it may be subjected to mechanical abuse, and where the temperature differences (high or low) cause changes in pipe dimensions, then the pipe insulation should be installed with broken joint construction as shown in Figure 68. Wires, straps, or tape should be installed to secure each half section to the preceding half section, so as to obtain a continuous, tied-together cylinder in which the pipe can slide without disturbing the individual half sections in their relationship to each other. Of course, at proper intervals, this cylinder must be interrupted by an expansion or contraction joint to allow for differential movement between the pipe and the insulation. Wire, strap, or tape should be located half way from each of the staggered butt ends. This is also shown in Figure 68.

The metal of the wire or strap should be chosen to resist atmospheric conditions, chemical contamination, and spillage. Plain steel and galvanized steel wire and straps are frequently specified for outdoor applications of high temperature insulation, based on the assumption that the weather-barrier protects them. It does not. Due to vapor being forced outward by the difference in vapor pressures, the surface of the insulation under the barrier is always damp. Steel and galvanized steel under these conditions will eventually rust. Compared to the insulation weather-barrier and labor cost, the relatively small additional cost involved in using stainless steel wire or strap is low cost assurance of a long, satisfactory service life.

Installation of insulation blocks to a vessel should be staggered joint construction similar to that described for pipe, as shown in Figure 68. When a second layer of insulation is required,

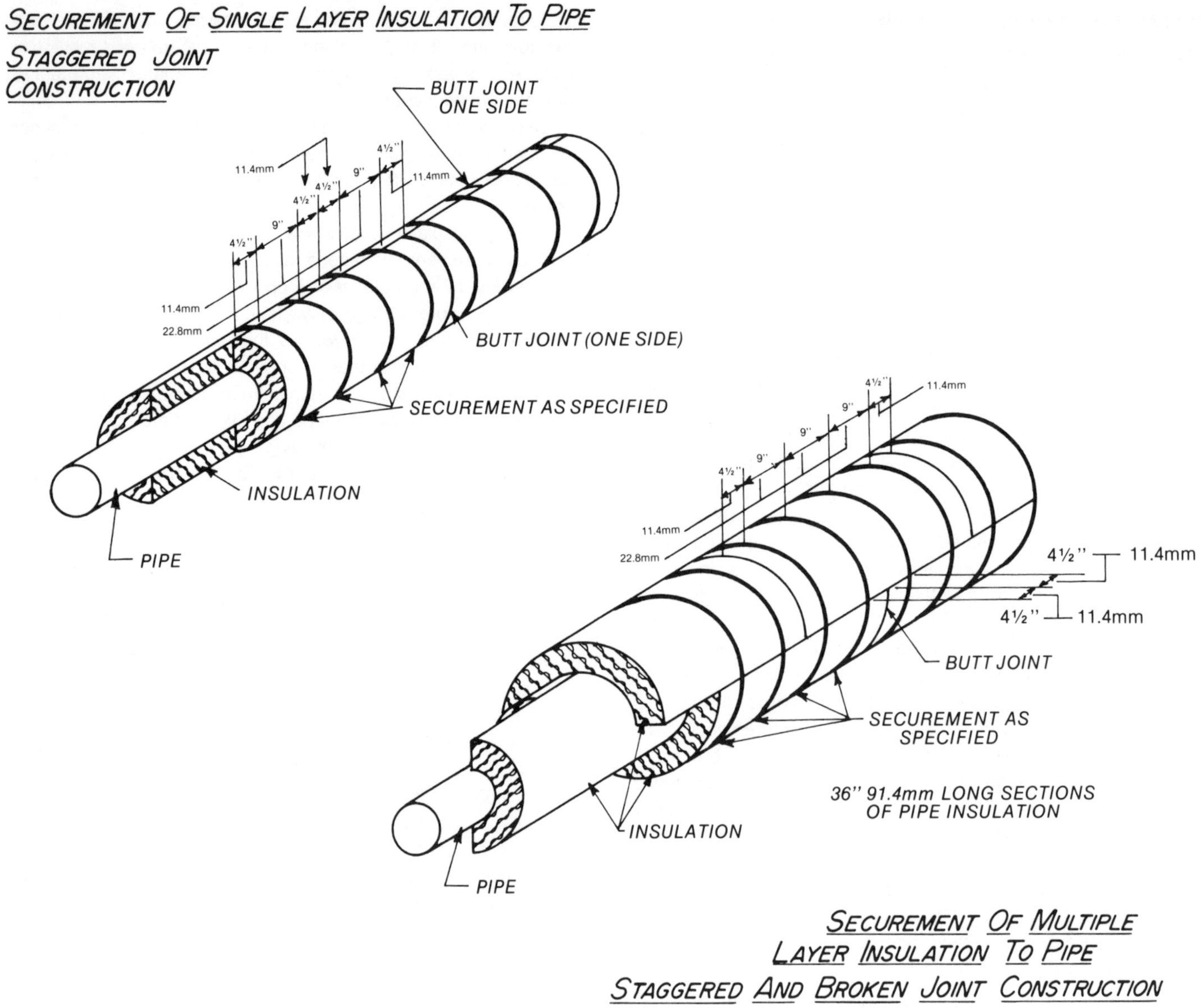

Figure 68

the multiple layers should be applied so that the butt joints of one layer do not coincide with those of the other layer.

The commonly used insulation securements for the various types of installations have been shown to be:

- Pin clips
- Steel clips
- Wire
- Metal strap and strap clips
- Metal expansion straps
- Expansion springs
- Nylon straps (metal strap clips)
- Polyester corded strapping
- Glass filament tape and strapping
- Outer jacket and adhesive

Attachment Adhesives

Attachment adhesives serve a dual function, acting both as a support and a securement.

Although the technology of adhesives has made big strides in the last few years, their use in the insulation field has been relatively limited. This is due mainly to the fact that most adhesives fail at the extreme temperatures where insulations are required.

A few, relatively small installations of block high temperature insulation have been installed on equipment by the use of refractory adhesive. Most of these installations have been made quite recently. Thus, it is too soon to tell whether they will have sufficient service life to be economically feasible.

In the moderate temperature range, the use of adhesives to support and secure glass fiber insulation to the inside and outside of heating and air-conditioning ducts has proved to be satisfactory. These adhesives are of an air drying type, and are suitable for service temperatures of 20° F to 180° F (−6.7° C to 82.2° C), although the temperature range for an individual product may be different from that given.

Where cellular glass is to be installed on flat surfaces such as roofs, ducts, etc., asphalt is frequently used as an attachment adhesive. As the asphalt is thermoplastic, it is heated to the liquid state and immediately applied to the surface of the cellular glass. In a short time, the temperature of the asphalt lowers and it returns to the solid state and forms an attachment between the cellular glass and the surface to be insulated. Of course, this application is limited to surfaces that will never in the future reach a temperature at which asphalt melts.

Cellular glass, being vapor resistant, effectively retards solvent adhesives from drying. For this reason, when it is necessary to use adhesives which have better temperature stability or better flexural properties than asphalt, it is necessary to use a catalytically cured adhesive. Where this type of adhesive is required for attachment of celluar glass to the surface, the temperature range of the surface may be from −100° F to +200° F (−73° C to 93° C). Thus, differential movement between the surface and the cellular glass must be absorbed by the elastic properties of the adhesive. Because of this, a satisfactory adhesive must be heavy bodied and must retain its elasticity over the full service temperature range.

Contact adhesives are used to attach organic foams to themselves and to other surfaces. Split, elastic, organic foam pipe insulation is applied to pipe, the split edges coated with contact adhesives, and after slight drying the edges are placed together, bonding the split cylinder back together. In such applications, the contact adhesive replaces tape and strap as the insulation securement. Where necessary to attach the insulation to the substrate surface, the contact adhesive is applied both to the substrate and to the insulation surface. When the insulation is put into position, the contact adhesive bonds the organic foam to the surface, serving as the insulation support. Edges are bonded together in the same manner as described for pipe insulation.

Rigid, organic foams may be bonded together in the same manner as described for elastic foams. However, being rigid, the problem of expansion-contraction movement makes it necessary to consider the differential movement between the substrate surface and the insulation. This movement may cause stresses in the insulation which may exceed its strength and/or the strength of the contact adhesive. Thus, glass filament tape, nylon bands, or other securements may be required.

When any adhesive is used to attach insulation to itself, or to another surface, it must be compatible with each. Also, the insulation must have sufficient strength in shear so that the attached surface will not *peel* from the insulation itself.

When adhesives are used to bond insulation to a substrate surface, the type, condition, and finish on that substrate is of utmost importance. If the substrate is dirty, rusty, or oily, a bond cannot be obtained. If the surface rusts or corrodes after application, the bond will be lost. One common error is to use an adhesive to bond insulation to a primed or painted metal surface. A few weeks later, the insulation falls off because the paint had insufficient adhesion to the metal, so that the paint adhered to the insulation, but left nothing on the metal surface. Adhesives will be used more frequently to install insulation, but their use is tricky. An application must be studied to make certain that the substrate surface is suitable for the adhesive, the adhesive is suitable for the temperature range, the adhesive will have good service life, the insulation will not spall from the adhesive, the movements do not exceed the elasticity of the system, and that all are compatible.

Attachment adhesives for various services are:

Refractory type adhesives
Water base adhesives
Solvent base adhesives
Thermoplastic adhesives
Catalytic cured adhesives
Contact adhesives

Insulation Fasteners

Insulation fasteners are mechanical devices for holding two pieces of insulation together.

Mechanical fasteners are not as extensively used for insulation as nails and screws are used for holding wood together. Insulation is of light density, fairly soft, and does not lend itself to such fasteners. However, sometimes it is necessary to temporarily hold blocks together until cements set, or securements can be placed. In other cases, fasteners are used to help hold insulation in position. An example of this is skewers or pins used to hold flange insulation in place on cylindrical vessel, high temperature insulation.

Wooden skewers are used to hold low temperature block insulation in cold storage and low temperature applications. As these skewers have a much higher conductivity than insulation the number used should be kept to a minimum.

Insulation fasteners are:

Stainless steel pins (4d stainless steel nails)
Stainless steel skewers (16d stainless steel nails)
Wooden skewers (meat skewers)

Insulation Reinforcements

Insulation reinforcements serve to add mechanical strength to thermal insulation with which they are installed.

Typical of an application which requires insulation reinforcement is that in which block insulation is used on very high temperature flat surfaces of considerable area. On such installations, the block must be attached to the flat surface by pins and clips, slotted studs and tie wires, or other such means of support and securement. Thus, each block moves with the surface to which it is attached. As the metal expands, slight gaps open up between the edges of the insulation blocks. Even if multiple layer, broken joint construction is used, this dimensional change is conveyed to the outer surface. For this reason, it is advisable to provide an expansional layer of insulation on the outside surface. This is provided by a layer of 1/2″ in. (12.7 mm) thick insulation cement. When it is dry and the expansion movement has spread evenly, it will give sufficiently to prevent leaks. Unfortunately, the movement can cause separation between the block and the insulating cement. To spread the movement evenly and reinforce and support the insulating cement where separation might occur, a reinforcement is needed. This reinforcing is accomplished by securing a wire netting, or expanded metal lath, to the insulation securement, then applying the insulating cement through the netting, so that when it dries it has the hexagonal wire as a reinforcement. Due to the hexagonal pattern of the wire mesh, this reinforcement can elongate in both directions more than the dimensional expansion for the same distances.

Used indoors, the galvanized wire mesh will give long lasting service. However, again due to vapor pressure outward, when installed under a weather-barrier, this galvanized wire will have limited service life. Therefore, stainless steel or Monel hexagonal wire netting should be used for long service life.

Fastening adhesives

Fastening adhesives are adhesives or cements used to bond one piece of insulation to another.

A fastening adhesive must bond to each of the surfaces which are to be attached. It must be suitable for the temperature to which it is to be subjected without failure, or loss of bond to the surfaces. When used as an adhesive for low temperature insulation, the adhesive must be vapor resistant. The drying or curing characteristics of the adhesive are dependent upon the insulations being attached. Because of these varied requirements, there are many types of fastening adhesives.

These adhesives are used in the field erection of insulation, but their major use is in the shop or plant where insulation is fabricated into its various shapes.

In most instances, the predominate requirement in the selection of an adhesive is the temperature to which the adhesive and the insulation will be subjected. The temperature not only affects the adhesive, it also indicates the general type of insulation which is suitable. Thus, the types of adhesive and insulation tend to group together.

Low temperature fastening adhesives

Thermoplastic fastening adhesives. Asphalt is used extensively as a thermoplastic adhesive for fastening cellular glass to itself. It is heated to a liquid state and deposited on the insulation surface either by rollers, or by dipping the insulation into the hot liquid. While still hot, the surfaces to be adhered are pressed together. The asphalt hardens to a bond when the temperature drops. Asphalt has very high vapor resistance, thus provides an almost vapor tight joint. The use of asphalt for such bonding of cellular glass used in *high* temperature service is limited to the melting point of the asphalt. This same adhesive can be used on organic foams having sufficient temperature resistance to withstand the temperature of the hot asphalt without deterioration. The service temperature range in which asphalt adhesive is used is approximately −300° F to +212° F (−184.4° C to +100° C).

Catalyst type (vapor resistant) fastening adhesives. Catalyst adhesives consist of one part powdered resin and one part liquid. After the two parts are mixed, the resultant mix will set up within a given time—called "pot life." For this reason, only the amount of material that can be used in less time than the pot life should be mixed. This type of adhesive is used to fasten cellular glass and organic foams. It is most frequently applied to the insulation surfaces by brush. (When the adhesive sets in the brush, cleaning is very expensive and time consuming, so inexpensive brushes are used and discarded.) There are quite a number of catalyst cements available, and the one chosen must be suitable for the insulation and the service temperature at which it will be used. Depending upon the individual adhesive selected, the temperature range can be from −300° F to +400° F (−184.4° C to +204° C).

Contact fastening adhesive. Contact adhesive is a single compound adhesive which is applied in a thin coat to each surface to be joined. The adhesive is allowed to dry until non-tacky to the touch before joining the surfaces. The surfaces are joined under slight pressure and bonding is immediate. Contact adhesives are used to bond organic foams, both rigid and flexible. The service temperature range of these adhesives is −100° F to +200° F (−73.3° C to +93.3° C).

Resin fastening adhesives. Resin adhesives may be of the water emulsion or solvent types. Basically, either the water emulsion or the solvent type resin adhesive is to be used on insulation which will allow the passage of vapor. They are used as received from the manufacturer and have unlimited pot life. Depending upon the particular product to which they are applied, application is by brush or trowel. Temperature range of these materials is 50° F to 250° F (10° C to 121° C).

Hydraulic-setting cement adhesive. This adhesive is water-mix, hydraulic-setting Keene's cement. The cement powder is mixed with water. Being hydraulic-setting, the mixed cement must be used before it begins to harden. The cement can be applied by brush or trowel. It is used to bond cellular glass insulation to itself, and is suitable for service temperatures up to 750° F (399° C).

Refractory adhesives. These adhesives are mixed silicates made of fire clays, binders, and, sometimes, asbestos fibers. They depend upon dissipation of moisture for their setting. They are best applied to the insulation surface by trowel, and are generally used on calcium silicate, asbestos fiber, or similar, absorbent, high temperature insulations. Depending upon the manufacturer, they are suitable to be used on services from 1200° F to 2000° F (649° C to 1093° C).

Refractory adhesive (for non-absorbent high temperature insulation). This adhesive is made of fire clays, binders, and special solvents. It depends upon dissipation of moisture for setting, and is applied to the insulation by trowel. It is suitable for use on service temperatures up to 1500° F (816° C).

Sodium silicate adhesive. This consists of sodium silicate or sodium silicate and sodium silicate flour. It is used for bonding the absorbent type of high temperature insulations. As this material will shrink and crack when dried, it is not suitable as a permanet adhesive for high temperature insulations. This adhesive is frequently used to temporarily bond insulation pieces together, which, after they are installed, require no future bond.

Flashings for Insulation

Flashings for insulation are materials which either direct the flow of water to prevent it from getting on the inside of the insulation, or direct the flow of water from the inside to the outside.

Flashings to prevent water from entering the insulation from the outside are commonly used and understood. The flashings used around roof insulations are a basic part of roofing practice. Not so well known are flashings used to direct water outward from the inside of insulation. Typically, such flashings (or water stops) are required on large vessels, especially spheres, which operate below ambient temperature and above 32° F (0° C). Vapor will enter and condense on the vessel and channel downward as liquid water. If allowed to continue this path, it collects and is trapped at the bottom, and continues to build up a water head until sufficient pressure is exerted to break through, or force off, the insulation. Water stop flashings, built in to control the build up of this water head, prevent this from occurring. This type of flashing is illustrated in Figure 69.

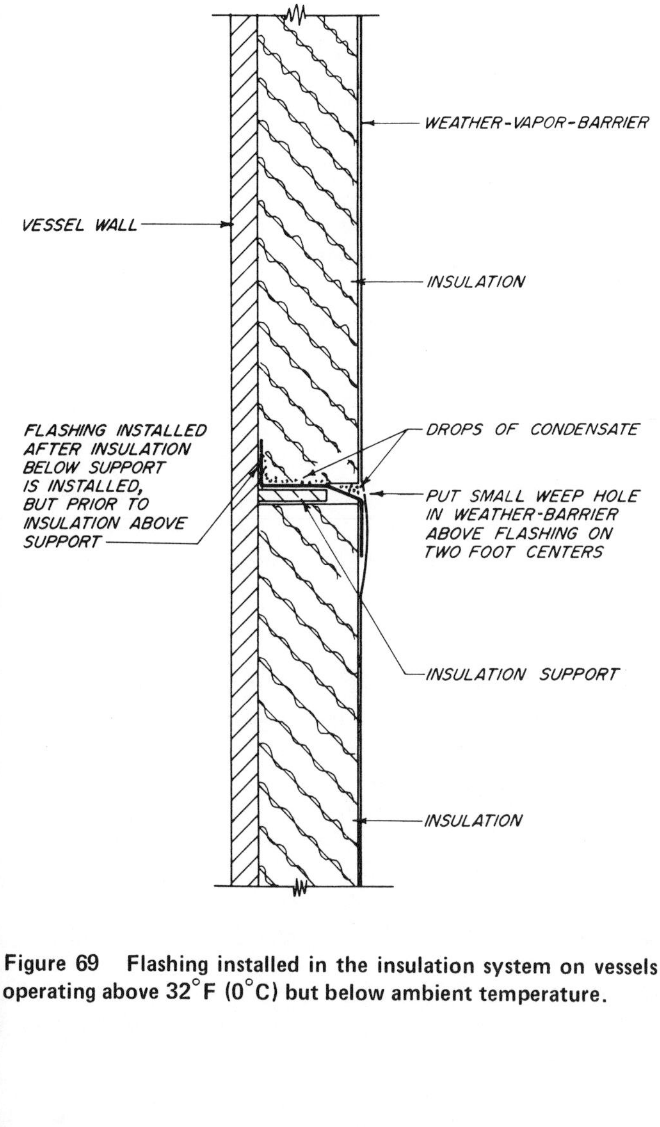

Figure 69 Flashing installed in the insulation system on vessels operating above 32°F (0°C) but below ambient temperature.

Flashings may be of metal, plastic sheet material, or they may be built up of mastic and reinforcing cloth, depending upon the installation requirements—service temperature, necessary flexibility, and corrosion resistance.

Metal flashings may be of:
- Stainless steel
- Treated steel (galvanized and prepainted)
- Galvanized steel
- Aluminum
- Copper

Plastic flashings may be of:
- Polyvinyl chloride sheets
- Rubber sheets
- Other plastics and plastic laminates

Built up flashings may be of mastics and reinforcing cloth.

Insulation Expansion-Contraction Joints

An insulation expansion-contraction joint should be placed where one area of insulation can move in relation to an adjacent area of insulation, and the covering joining these two areas must accommodate this movement. The term "contraction joint" is used for joints in which the metal substrate contracts more than the insulation. "Expansion joint" is used when the metal substrate expands more than the insulation.

Joints for low temperature installations must be treated differently than joints for high temperature installations. This is due to the fact that the low temperature contraction joint must be installed to allow the insulation to come together when the pipe contracts and, also, it must be installed so as to be vapor resistant.

When the contraction movement is within the limits of the compressibility of low temperature flexible insulation, a flexible vapor-barrier can be installed over the entire system. Then, the movement is absorbed by the insulation assembly as a whole. If metal is used on the outer surface of the low temperature insulation, each joint should be sealed with non-setting sealer, to attempt to allow for this movement. Such sealing is very difficult and expensive. With rigid insulation, there is no possibility that the movement can be taken up in the assembly. The movement must be taken up either by allowing movement beneath the insulation, by overlapping slip joints in the insulation, or by forming vapor resistant contraction joints between two areas of insulation. This is illustrated in Figure 70.

Although expansion joints may be required in low temperature service if the insulation has a much higher coefficient of expansion than that of the metal, *expansion* joints will be required most frequently on *high temperature installations.* In these cases, the pipe or vessels expand, whereas most high temperature insulations shrink. The circumferential movement may be compensated for by using cushion blanket on vessels, or slightly oversize insulation on pipes, but the difference between the metal expansion and the insulation shrinkage in axial dimension must be compensated for in some manner. If properly engineered expansion joints are not provided, the expanding metal will cause openings in the insulation at the weakest points. Openings cause excessive heat loss, and possible elevated temperatures on the surface, which

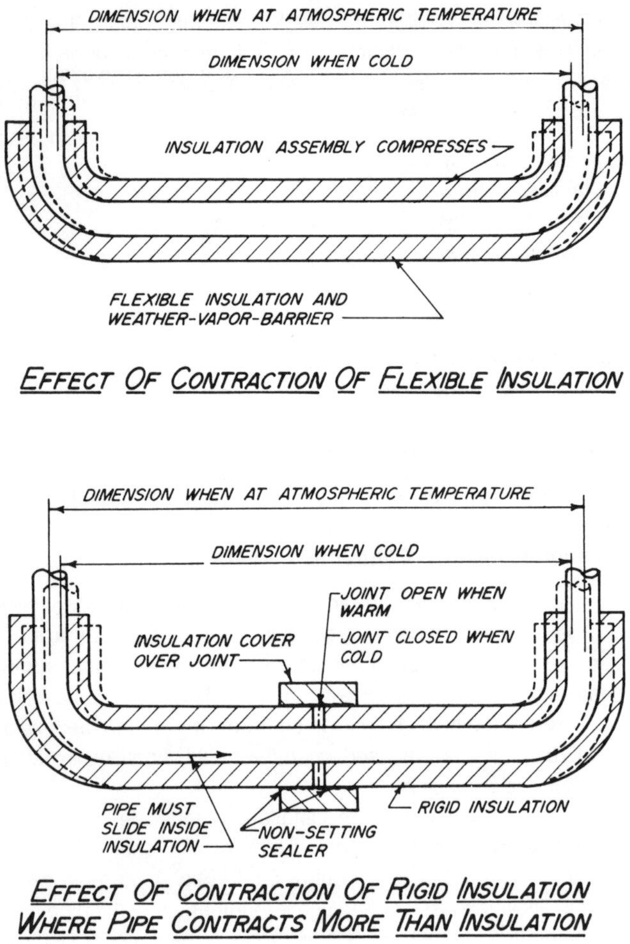

Figure 70 Low temperature contraction.

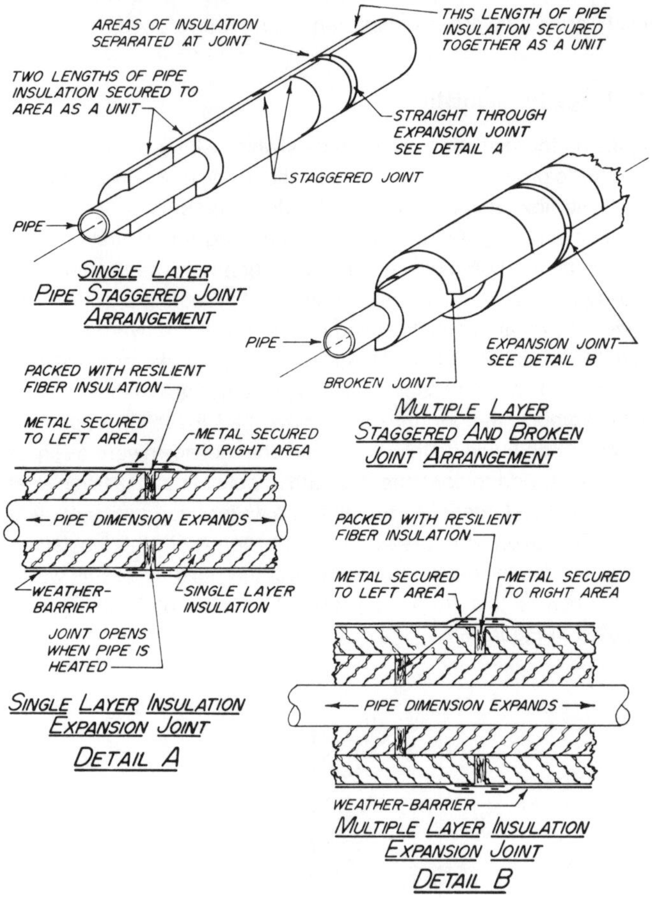

Figure 71 High temperature fiber expansion joints.

rapidly deteriorate mastic and felt weather-barriers, or hot spots on metal. Properly designed expansion joints serve to cut costly waste and extend service life.

In effect, the particular area, or stretch of insulation to be held together as a unit, should be installed in staggered joint construction. Then, a complete break, or sliding joint, should be installed between it and the adjacent area or stretch of insulation. This joint should be left open and *over packed* with resilient fiber insulation so that, as the joint becomes larger, the packed insulation will expand as the pressure on it is lessened. To contain this packed insulation, two pieces of overlapped metal, one attached to each of the adjacent sections, are installed as a slip joint. Various types of slip joints are illustrated in Figure 71.

The distance between the points reached by the expansion and contraction dimensional movements of the contraction joint must be calculated, based on the expansion of the substrate metal and the characteristics of the insulation, as presented in Chapter 5.

On vertical installations, the overlapped metal must be pointed down, to act as a rain shed. On horizontal applications, such slip joints will not be completely rain tight and some water leakage will occur. Unfortunately, the author does not know of a water tight expansion joint being produced for this service.

As these joints are field constructed, expansion-contraction joints are not listed as a specific accessory item.

Insulation Cradles, Pipe Supports and Hangers

All pipe must be held in position by hangers or supports. Although this function of engineering is the responsibility of the piping engineer, the interrelationship between supports and insulation should be established to obtain efficient operation of the supports and the insulation. Because of the insulation requirements, high and low temperature piping require different treatment of the supporting members.

High temperature pipe supports and hangers

Pipes held up by hangers have flexibility—due to bending or the hanger rod—which, in most cases, is sufficient to allow for the movement of expansion. Pipes which are supported on fixed members must have supports which provide for the expansion and contraction caused change in length, which occurs between certain anchor points. So, to support piping on fixed members requires anchors, guides, and slide supports. Anchors secure the pipe in a fixed position to a fixed member, guides allow the pipe to move along one axis, and slides allow the pipe to move in two directions.

The anchors, guides, and slides should elevate the pipe above the fixed members to allow sufficient space for the full thickness of the insulation.

Anchors, guides, and slides are designed with various configurations of metal, such as "T" bars, curved plates with extended legs, and channels. For the insulation requirements, the various metals will not be discussed, and only the simple "T" section will be used for illustration. Shown in Figure 72 is a "T" section, which by modification at the fixed member, may be used as an anchor, guide, or slide. To prevent damage to the insulation, the minimum distance which the "T" elevates the insulation above any supporting surface or projections above it should be equal to the insulation thickness plus 1/2″ (12.7 mm).

Where the top of the "T" support is attached to the pipe, its temperature, when in service, will be the same as the pipe. The metal, being a high conductor of heat, will transfer the heat outward and, where it projects through the insulation, will be at a relatively high temperature. At this location, it is necessary to use heat resistant sealers to seal the juncture of the steel and insulation. The support metal beyond this point should be painted with

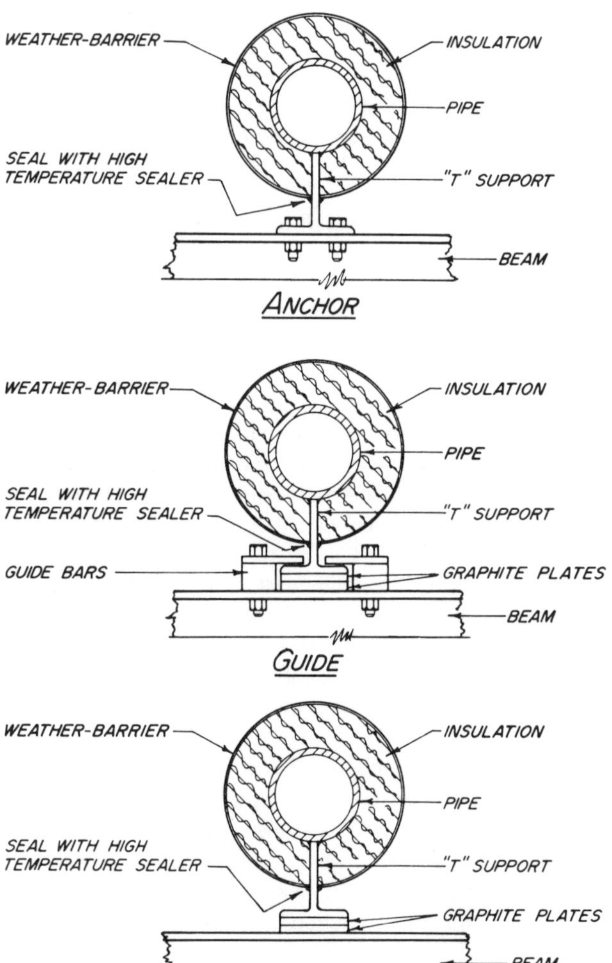

Figure 72 Anchors, guides and slides for high temperature insulation pipe.

heat resisting paint, such as inorganic zinc-silicate or graphite-silicone paint, to resist rust and corrosion.

An insulating slide-support, made of graphite, is available for the support of hot insulated lines. The graphite is of a low thermal conductive variety, so it provides a moderate amount of thermal resistance. The graphite block is strapped to the pipe and slides on a graphite plate which is cemented to the fixed member. Graphite requires no paint or other protection. However, the juncture of the insulation with the graphite support should be sealed with high temperature sealer to prevent rain from entering. This type of support is illustrated in Figure 73.

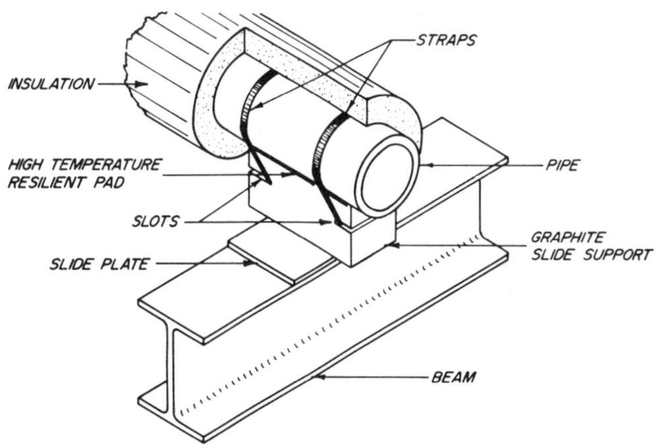

Figure 73 Insulating type of pipe slide support.

The insulation itself can be used as a slide support for hot lines if a metal cradle is used under the insulation to distribute the load. Due to the difficulty of guiding lines using such supports, and the strength limitations of the insulation, this system is usually restricted to pipes of 4 NPS (114.3 mm) and smaller, and service temperatures not exceeding 300° F (149° C). As the expansion movement shifts the position of the cradle in relation to the fixed member, graphite slide shoes and plates should be used between the cradle and the fixed member. If not, when rust bonds the metal cradle to the fixed member, any movement will try to occur between the cradle and the insulation, resulting in tearing the weather-barrier, and abrading and wearing away the insulation. Specific consideration in the design of cradles will be presented in the discussion of cradles for low temperature piping.

Piping is also installed by the use of hangers, rods, and clevis hangers with cradles or saddles. In the case where hangers are attached directly to the pipe and the rod extends directly through the insulation, special care must be taken to seal around the rod with heat resistant sealer reinforced with fabric. This will seal it against water and provide a flexible seal to compensate for the movement that will occur at this point. Typical of this type of pipe support is that shown in Figure 74.

One major thing to avoid is the practice of installing pipes directly on fixed members (such as pipe racks) and then installing the insulation up to and around the fixed member, as shown in Figure 75. Although apparently inexpensive, this is the most costly way to install hot piping. Based on $30.00 per lb per hour of capital investment in steam producing equipment, depending upon the

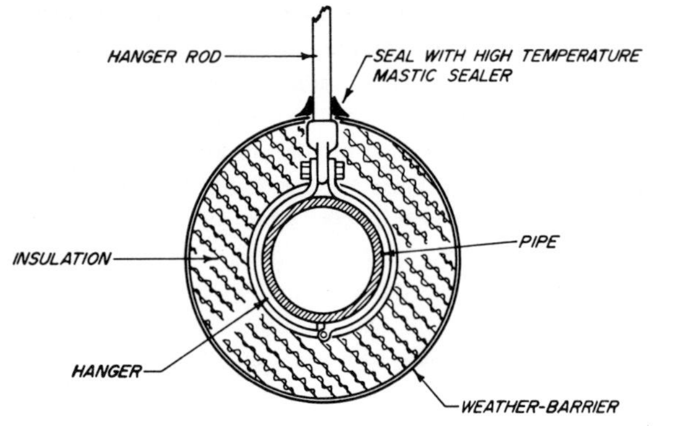

Figure 74 Pipe hanger installed directly to high temperature pipe.

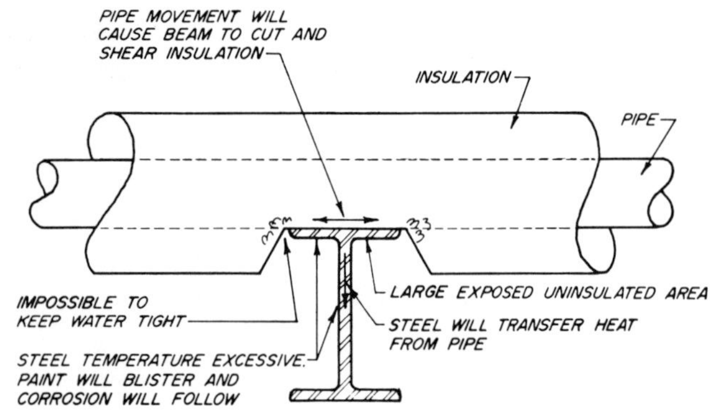

Figure 75 Incorrect way to support insulated pipe.

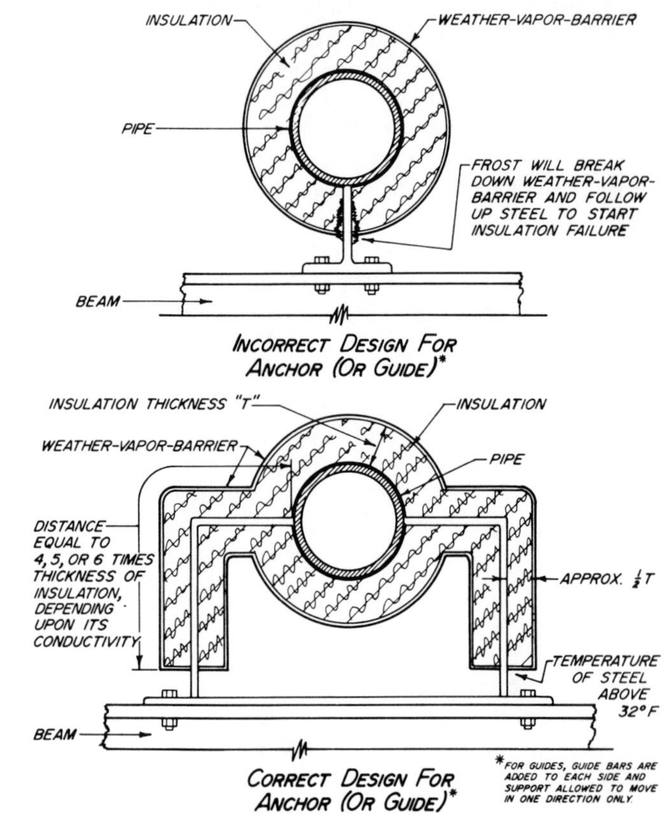

Figure 76 Anchors (or guides) for low temperature piping.

size of pipe and operating temperature, the capital investment necessary to supply the heat loss through this exposed steel will be $9.00 to $36.00 and the production cost of this lost steam will be $6.00 to $21.00 per year.

Other than the principal types of pipe supports mentioned, there are many special types for special purposes, such as spring hangers, adjustable rollers, etc. However, the selection and use of these are more a piping than an insulation problem.

Low temperature pipe supports

Wherever possible, metal directly projecting through low temperature insulation should be avoided. However, in the case of pipe anchors and pipe guides, only metal has proved to have sufficient strength to withstand the forces. Because of its high conductivity, the steel acts as a "thermal short circuit" and will frost where it projects through the covering. This frost will build up along the metal, eventually ruining the vapor seals, and causing early failure of the insulation system. The solution is to design the anchor so that its temperature, at the point at which it projects through the insulation, will be above the frost point. If the conductivity of the insulation is .3-.4, this distance should be 4 times the insulation thickness, if .2, the distance should be 5 times the insulation thickness, and 6 times the thickness if the insulation conductivity is less than .2. The insulation on these projections should be approximately one-half the thickness of the insulation installed on the pipe. A correctly and an incorrectly designed anchor are shown in Figure 76.

A low temperature insulated guide support should be designed in the same manner as that described for anchors, but must allow for movement in only one direction. Guide bars installed at the base, and graphite slide plates between the bottom of the support and the top of the beam, produce a thermally and mechanically efficient pipe guide.

The most efficient low temperature pipe slides consist of cradles which support the insulation, which in turn supports the pipe. The weather-vapor-barrier should be installed on the pipe insulation prior to setting the assembly into the cradle. The cradle under the insulation should be a metal of the correct length and thickness to support the pipe load without crushing the insulation. It should be just sufficiently strong to resist indentation at the center, and have enough flexibility in its length so that it will not cut into the weather-vapor-barrier and the insulation at the ends. For low temperature insulations which do not have excessive deformation at a loading of 20 lbs per sq in., the following equations can be used to determine the required length and thickness.

When: *In English Units*

T_c = thickness of cradle in inches
L = length of cradle in inches
D_2 = OD of insulation in inches

S = span between cradled supports in feet
W = unit weight per foot of pipe, liquid (or gas), and insulation +20% safety factor

$$\text{Then } T_c = 0.0125\, D_2 \tag{66}$$

$$\text{Then } L = 0.15\, \frac{WS}{D_2} \tag{67}$$

When: *In Metric Units*

T_{cm} = thickness of cradle in mm
L_m = length of cradle in mm
D_{2m} = OD of insulation in mm
S_m = span between cradled supports in feet
W_m = kilograms weight per metre of pipe, liquid (or gas) and insulation +20% safety factor

$$\text{Then } T_{cm} = 0.0125\, D_{2m} \tag{66a}$$

$$\text{Then } L_m = 700\, \frac{W_m\, S_m}{D_{2m}} \tag{67a}$$

A graphite slide shoe and a graphite slide plate should be installed between the cradle and beam to prevent erosion due to abrasion, and to provide a low coefficient of friction. The insulation-cradle assembly is illustrated in Figure 77.

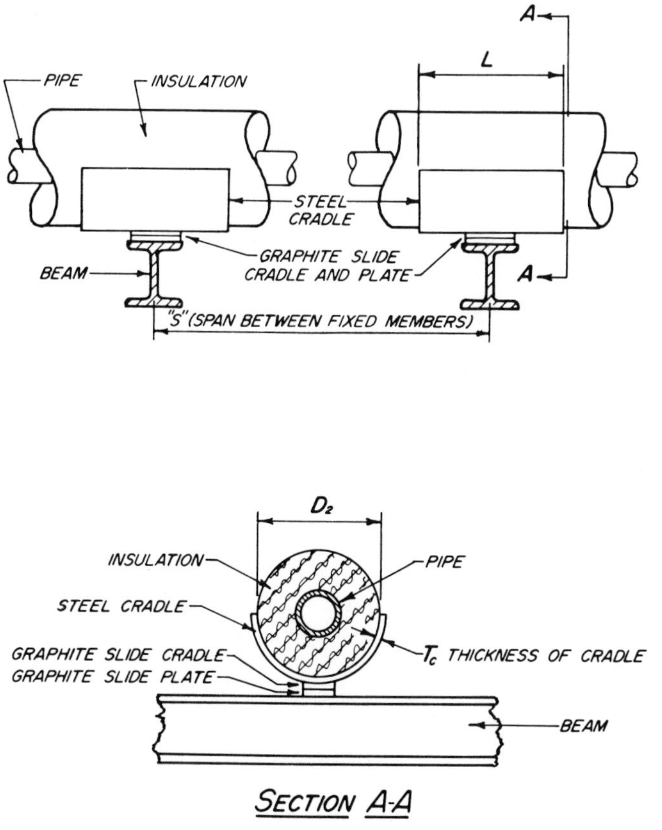

Figure 77 Cradles for insulation functioning as pipe slide supports.

A number of manufacturers are producing such preinsulated cradles for use with clevis hangers, or to be directly installed on fixed members. The insulation provided by some of these is organic foam, and by others, heavy density glass fibers. Those made of glass fibers have a vapor-barrier totally enclosing the insulation with overlaps for sealing to the vapor-barrier of the adjacent insulation. Within their limitations these can save field installation labor.

If a low temperature cradled line supports a high riser, or some other load which causes the weight to exceed that which can be supported by the insulation, then a wooden block the shape of the insulation can be substituted for the insulation to obtain the needed compressive strength. Yellow pine treated with "Penta" preservative is recommended for this service. Although the wood block has three to four times the heat conductivity of the insulation, the high conductivity of the metal and its contact with the beam most frequently keeps the cradle surface temperature above dew point, even when the wood is the same thickness as the insulation it replaced. Should heat gain be the controlling factor, then the thickness of the wood between the pipe and the cradle would have to be increased in the proportional ratio of its conductivity to that of the insulation. The wood block must be protected with a vapor-barrier identical to permeable low temperature insulation. The use of this wood block in cradles is shown in Figure 78.

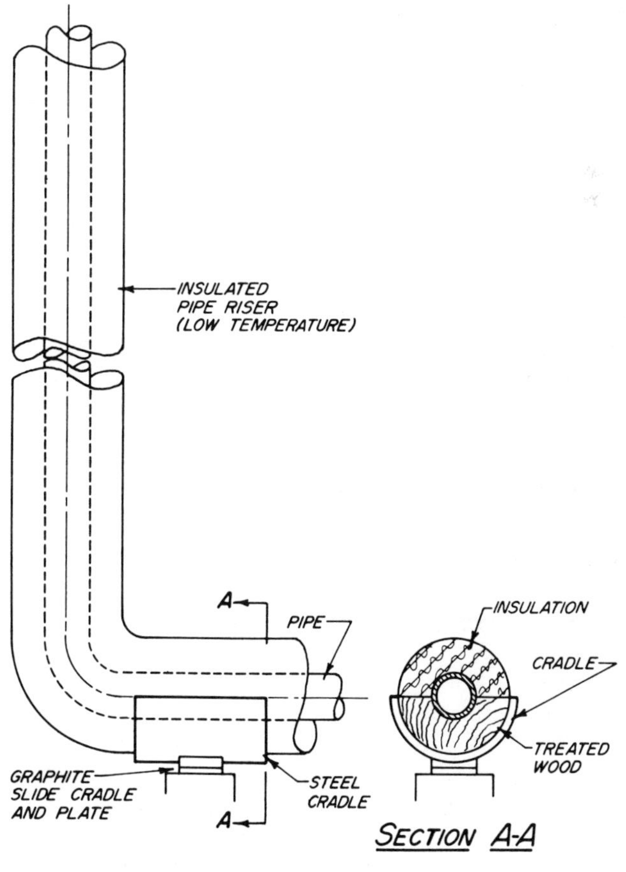

Figure 78 Use of treated wood to support heavy loads low temperature insulated pipe.

Again, due to loading, it may be necessary to attach hangers directly to low temperature pipe. When this occurs, then the rod should be insulated upward with insulation approximately one half the thickness as that on the pipe. The distance it should be carried up the rod is the same as given previously for other metal projections.

Materials used in the supporting and hanging of insulated pipe are:

Pipe anchors
Pipe guides
Pipe slides
Pile saddles
Insulation cradles (sometimes called shields)
Graphite pipe cradles and plates
Hangers
Clevis hangers
Low temperature insulations
Weather and vapor-barrier mastics and films
Adhesives and sealers
Treated wood blocks

8 Relation of System Requirements to the Design of Insulation Systems and to the Properties of the Materials Used

For any specific set of installation requirements, the properties of a material determine its suitability. If there were only one, or a limited number of sets of installation requirements, selection of material would be simple, and the need for all the vaious types of insulations and weather-barriers reduced. However, this is not the case. Each installation must be considered, and its requirements evaluated to allow the selection of the best suited material (or materials) for the individual installation under consideration.

Not only do the installation requirements change with the individual case, but the relative importance of the requirements also vary. Some of the possible variations of the properties required of individual materials, as the installation requirements change, will be treated in the short discussion following.

In the transportation phase, weight is always a factor. Although light density material is desirable, light weight insulation is of greater importance in air craft than in ships, trucks, or railroad cars. Conversely, when insulation is used in chemical and petroleum processing plants, where fire protection is required, a higher density insulation is essential to obtain the low diffusivity necessary for fire protection.

Where insulation may be contaminated by toxic or highly combustible chemicals, the most important single requirement is that it is completely non-absorbent.

When used in nuclear applications, two predominant requirements decide the materials selected. First, to prevent a problem of radioactive dust, the material must not dust. Second, the material must have a very short radioactive half-life.

In a process where cyclic application is required, the single most important requirement is that the insulation be such that it can be quickly and economically removed and reapplied.

Other cyclic operations may require that the insulation have low mass and specific heat so that the operating temperature can be changed without an excessive amount of heat and time being lost in "heating up" the insulation.

Just the opposite property is desirable on *cold storage* and storage of materials which must be held at relatively constant temperatures. In these installations, in case of power failure, the insulation's retention of the existing temperature is of utmost importance.

Even on the same installation, the specific end use of an insulation may dictate its properties. A high temperature pipe may require very rigid, strong insulation to resist the mechanical stresses imposed upon it, but the insulation used in the expansion joint should be soft, fluffy, and resilient to cushion the movements of the adjacent rigid insulation.

On low temperature installations, the single most important property is resistance to the passage of water vapor, or ability to construct the insulation system to be as near vapor-proof as possible.

Economics always enters into the selection process. If a process is critical, the most important single consideration may be reliability. If conservation of heat or power is the deciding factor, then the savings per year as compared to the installed cost is the most important factor. Again, and almost opposite, when insulation is used for a temporary function such as holding the heat in while a lining is being heat cured, then the lowest possible installed cost would be decisive.

Thus, because of conflicting requirements, there can be no "all purpose" insulation. Nor is there a "perfect" insulation for each set of requirements. It is essential that the engineer evaluate the installation and determine which requirements must be fulfilled, and which are of lesser importance. At the present state of technology, it is almost impossible to obtain every fact relating to an insulation installation requirement. Therefore, some requirements must be left to engineering judgment. Also, conflicting requirements may exist even in the same installation. Because of these difficulties, too many insulation jobs have been poorly engineered. Also, which is true of any complex subject, those who know the least about the problem always seem to have the most positive opinions as to the correct answer. For this reason, it is hoped that the engineers responsible for insulation design will develop, not only their own methods of evaluating requirements for future installations, but will also analyze past installations to help determine where improvements are possible. Only through better engineering will the insulation industry change from a "craft" to a "science."

Although the installation requirements and properties of materials have been presented in previous chapters, a list of installation requirements as related to properties of materials is presented in outline form to assist the engineer in his evaluation.

RELATION OF SYSTEMS REQUIREMENTS TO PROPERTIES OF MATERIALS

Transportation and storage phase

REQUIREMENT	INSULATION PROPERTIES	ACCESSORY PROPERTIES
Shipping, Movement, and Storage	Containers Required:	Containers Required:
To be able to move, be handled with ease without damage, and be transported without excessive care in protecting the materials from rain or moisture.	Size, shape, strength, weight, water resistance, mold and mildew resistance, cost	Size, shape, strength, weight, water resistance, cost
To be able to be moved without special handling equipment	**All insulations** **Strength properties:** Compressive, tensile, shear, flexure, impact resistance, elasticity, abrasion resistance	
To be able to be stacked for conservation of space	**Moisture properties** Absorptivity, adsorptivity, capillarity, mold and mildew resistance	**Weather, vapor barriers sealers, etc:**
To be able to be handled and stored without excessive danger of fire, or toxic effects.	**Fire properties:** Flash point, fire point, self ignition point, flame spread, fuel contribution, smoke density, smoke toxicity	**Fire properties** Flash point, fire point, self ingition point, flame spread, fuel contribution, smoke density, smoke toxicity.
To be able to be stored without loss due to aging or deterioration	**Shelf life** (Water mixed insulation)	**Shelf life** Shelf life, mold and mildew resistance, freeze-thaw resistance

Fabrication phase

REQUIREMENT	INSULATION PROPERTIES	ADHESIVE OR BONDING CEMENT PROPERTIES
Fabrication Operation To be able to be cut, or ground to desired shape. To be able to maintain true surfaces and surfaces suitable for application and retention together with adhesives or cements	**Rigid Insulations** **Dimensional properties** Sizes and shapes, straightness and smoothness of surfaces, squareness and trueness	**Pre-Use Properties** Shelf life, freeze-thaw resistance, mixing or storing time
Pieces able to be handled without excessive breakage, dusting, or wear	**Cutting properties** Hardness, delamination characteristics, ease of cutting, smoothness and trueness of cut, dusting characteristics, grinding characteristics, compatibility of cut surface with adhesive or cement	**Use properties** Troweling or brushing characteristics, surface wetting and bonding characteristics, wet adhesive strength, compatibility with surface, drying time, curing time, wet to dry shrinkage, toxicity (solvent, fumes, etc.) pot life
To be able to be assembled and bonded together rapidly, accurately, and tightly		
To be able to be handled in a short time after bonding assembly	**Handling properties:** Tensile strength, flexural strength, compressive strength	**Dried properties** Adhesive strength, shear strength, temperature resistance (maximum and minimum), resistance to moisture (vapor and liquid), resistance to solvents, acids, or caustics, flexibility and elasticity
To have relatively short setting, drying, or curing time so that fabricated and assembled pieces can be used in a short time	**Assembly properties** Surface suitability to spreading or brushing of cement or adhesive. Surface compatibility to adhesive or cement, resistance to surface shear on drying of cement or adhesive, original tack of surface to bond to surface, absorbency	
	Flexible materials Cutting characteristics, tensile strength, bonding characteristics, surface or layer delamination	

Application Phase

REQUIREMENT	INSULATION PROPERTIES	ACCESSORY PROPERTIES
Cutting and fitting To be able to field cut, fit, and place into position efficiently and accurately	**Rigid insulations** **Dimensional properties** Size and shape, straightness and smoothness of surfaces, squareness and trueness, within dimensional tolerances **Cutting properties:** Ease of cutting, dusting, resistance to abrasion and cracking **Handling properties:** Tensile strength, resistance to breakage due to load or impace, flexural strength **Surface properties:** Smoothness, trueness, compatibility with cements or sealers which may be required **Non-rigid insulation** **Dimensional properties** Trueness in width, length, and thickness **Cutting properties:** Cutting characteristics, dusting **Strength properties** Tear strength, resiliency, flexibility	**Adhesive mastics, bonding cements, or sealers (where required)** Brushing characteristics, troweling characteristics, wet adhesion, surface wetting, gap filling, bridging, sizing and sealing, drying time, curing time, shrinkage, compatibility with insulation surface, solvent toxicity
Securement To be able to be secured in position by wires, bands, pins, and clips, or adhesives, as required by the installation	**Rigid insulations** Compressive strength, hardness, shear resistance, tensile strength	**Fastener properties** **Wire and metal strap** Hardness, ductility, tensile strength, flexibility **Tape** Tensile strength, flexibility, elasticity, peel back, adhesive strength, shear strength **Pins and clips** Tensile strength, pull resistance, clamping strength **Adhesives** Wet and dry adhesive strength, shear strength
Trowel applications To be able to mix quickly and efficiently trowel into position insulating cements or mastics To obtain good attachment to surface and obtain even thickness and smoothness of surface required by installation	**Water mix cements** **Mixing properties** Mixing time, consistency **Application properties** Wet adhesion, build, wet coverage, dry coverage, shrinkage wet to dry, trowelability, corrosiveness to substrate metal, drying time **Dried properties** Compressive strength, tensile strength, flexibility *Note:* These same properties are of importance when the troweled cements are applied over other insulations. **Insulating mastics** **Mixing properties** Consistency in can **Application properties** Trowelability, wet adhesion, build, wet coverage, shrinkage wet to dry, corrosiveness to substrate metal, drying time, curing time **Dried properties** Compressive strength, tensile strength, extensibility	

REQUIREMENT	INSULATION PROPERTIES	ACCESSORY PROPERTIES
Sprayed applications To be able to be applied to insulation on a surface by spray equipment in a fast, efficient manner, with good adhesion to surface To be able to obtain a reasonably smooth, even surface without sags, hills, and valleys, suitable for weather-barrier or coverings To be able to obtain density and related thermal efficiency To be able to obtain desired strengths for permanent installation To be able to make spray application with least amount of cost involved in clean up or cleaning of spatter or overspray.	**Sprayed mass insulations during spraying** Wet adhesion, build rebound, overspray, compaction, corrosion to substrate metal, drying time, shrinkage wet to dry, toxic dust or fibers **Temperature limits application** Ambient air **Dried properties** Compressive strength, tensile strength, flexibility, dimensional tolerance, surface evenness and smoothness **Sprayed foam insulations during spraying** Wet adhesion, spray ratio, build, reaction time, curing time, expansion ratio, density, sag, toxic fumes, corrosiveness to substrate metal **Limits of application** Condition of surface, temperature of surface, temperature of air, humidity of air, wind, flash point, fire point **After application** Dry adhesion, compressive strength, density, tensile strength, dimensional stability, surface evenness and smoothness, bridging, elongation, flexibility	
Poured applications To be efficient, pour or ram fibrous, granular, or powder insulation into cavities, so that all voids are completely filled. *Note:* As poured insulations require some receptacle, requirements in reference to weather-vapor-barrier or coverings will not apply to this type of insulation system.	Compaction, compressibility, resiliency, dusting, density, toxicity of dust	
Application of weather-barrier, vapor-barrier, coverings **Jackets** To be able to cut, fit, form, seal, and apply jackets over flat surface insulation, vessel insulation, pipe and fitting insulation		**Jackets** Cutting characteristics, tear resistance, forming characteristics, handleability **Jacket accessories** **Lap adhesives** Adhesion, shear strength, shrinkage, flexibility **Circumferential closures** Type and means of attachment, sealing compound, flexibility **Longitudinal laps** Method of securing and sealing
Mastic weather barriers To be able to brush, trowel, palm, or spray smooth, even water and weather-barrier mastic or interior coating to obtain a smooth dried, film of even texture & thickness over entire surface Reinforcing fabrics which can be cut, fitted, and formed to fit contours, to reinforce gage and bridge gaps so as to be able to obtain maximum protection and life of the mastic coating	**All insulations** Surface compatibility Surface suitability Smoothness Dryness Dust free Clean	**Mastics or Coatings during application** Mixing or stirring time, consistency, brushing characteristics, troweling characteristics, palming characteristics, wet adhesion, wet covering capacity, surface wetting, gap filling, bridging, sizing and sealing, shrinkage wet to dry, drying time, curing time, solvent toxicity **Limits of application** Material temperature, air temperature, humidity of air **Reinforcing fabric for mastics** Compatibility with mastic, tensile strength, elongation, tear strength, formability, stiffness, permittance of mastic penetration

REQUIREMENT	INSULATION PROPERTIES	ACCESSORY PROPERTIES
Safety during application To be able to apply insulation system in safe manner, in accordance with safe scaffolding practice and in accordance with safe practice in handling dusty and flammable materials	**All Insulation** **Combustibility** Flash point, fire point, self ignition point, flame spread, smoke density, smoke toxicity, fuel contribution **Toxicity** Dust, fumes	**Metal jackets and securements** Sharp points, sharp edges **Mastics, adhesives, jacket films, sealers and coverings** **Combustibility** Flash point, fire point, self ignition point, flame spread, smoke density, smoke toxicity, fuel contribution **Toxicity** Fumes

Service phase

REQUIREMENT	INSULATION PROPERTIES	ACCESSORY PROPERTIES
Physical To withstand dead loads, wear, impact and mechanical damage, forces of expansion and contraction, and vibration	**All insulations** Compressive strength, flexural strength, shear strength, tensile strength, flexibility, shrinkage, coefficiency of expansion, compaction resistance, abrasion resistance, vibration resistance	**Weather, vapor-barrier, coverings** Impact resistance, indentation resistance, tear resistance, abrasion resistance, flexure, elongation, tensile strength, adhesion, elasticity, shear strength
Chemical Compatibility with metal to which it is applied, resistance to atmospheric and spillage contamination	Alkalinity, acidity, inhibitors to corrosion, acid resistance, caustic resistance, solvent resistance	Acid resistance, caustic resistance, solvent resistance
Moisture Resistance to moisture in liquid or vapor form or both	Absorptivity, adsorptivity, hygroscopicity, capillarity, vapor permeability	Absorptivity, adsorptivity, hygroscopicity, capillarity, vapor permeability (These properties relate to a system, rather than to a material, and must include the joints and seals of jackets or films.)
Electrical To resist flow of galvanic currents	Dielectric constant (when wet)	Dissimilar metals in jacket and substrate, producing a potential difference which causes galvanic current to flow
Weather resistance To resist solar radiation, rain, sleet, snow, wind, and atmospheric contamination, maximum and minimum temperatures	*Note:* If properly weather-protected this function is one of the weather-barrier.	Water resistance, solar radiation resistance, temperature stability, impact resistance, contamination resistance, wind resistance, mold resistance, freeze-thaw resistance *Note:* Deterioration of a weather-barrier system caused by weather is a deterioration of the properties of the system.
Safety To maintain standards of fire safety; personnel protection from burns (As controlled by insulation thickness vs. surface emittance)	**Hazard properties** Absorptivity (of combustible or toxic liquids), adsorptivity (of combustible or toxic vapors) **Combustibility** Flash point, flame spread, fire point, self ignition point, fuel contribution, smoke density, smoke toxicity **Melting point**	**Absorptivity** (same) **Adsorptivity** (same) **Combustibility** Flash point, flame spread, fire point, self ignition point, fuel contribution, smoke density, smoke toxicity **Melting point emittance** (surface temperature)
To protect from fire exposure	**Protection properties** Noncombustible conductivity, diffusivity, density,, specific heat Fire resistivity	Resistance to flame

REQUIREMENT	INSULATION PROPERTIES	ACCESSORY PROPERTIES
Thermal properties To resist service temperature.	**Maximum service temperature** Coefficient of expansion, shrinkage	**Maximum surface temperature** Coefficient of expansion (metals)
Control of heat flow, control of temperature flow, heat storage.		Embrittlement
Control of surface temperature	**Minimum service temperature** Coefficient of expansion Warpage Conductivity Diffusivity Specific heat Density Conductivity and thickness	**Minimum surface temperature** Reduction in elongation Emittance
Miscellaneous To resist rodents, insects, termites, etc. (particularly in building and cold storage insulation)	Vermin resistance Mold resistance	Vermin resistance Mold resistance
To be odor free	Odor	Odor
Aesthetics		Color, texture, smoothness
Maintenance Phase To resist damage in service	Physical strength Dimensional stability Temperature strength Moisture resistance (Above are general terms for individual properties listed in each of above.)	Weather resistance Abuse resistance Toughness Temperature stability (same rate)
To remove and replace	Removability Replaceability Reuseability	Removability Replaceability Reuseability
Ease of maintenance	Ease of patching Permanence of patching	East of patching Blending of patch

Economics phase/Investment cost

ITEM	INSULATION COST	ACCESSORY COST
Purchase price	Cost of all insulation, including pipe covering, block, blanket and all accessories	Cost of weather-vapor-barrier or covering, including sealers, caulking and adhesives
Shipping and handling cost	Cost of shipping, handling, and storage	Cost of shipping, handling and storage
Fabrication cost	Cost of cutting and assembly of prefabricated pieces, or molded preformed pieces	
Application cost	Cost of installation	Cost of installation
Overhead cost	Cost of supervision, contingencies, etc.	Cost of supervision, contingencies, etc.
Maintenance cost	Yearly cost of maintenance	Yearly cost of maintenance
Service life or depreciation period		

From these costs the capital investment and cost of insulation per year can be determined.

Operational Savings Phase

	SAVINGS BY INSULATION SYSTEM	
ITEM	CAPITAL SAVINGS	YEARLY SAVINGS
Dollar savings in heat producing equipment, or refrigeration equipment	Savings in capital investment due to small size	Savings per year in maintenance and investment. Yearly cost for difference.
Savings in dollars Heat or refrigeration (energy)	Reduced capital investment required to make process work	Savings per year in dollars
Operational savings		Value of additional product (if any) per year
Fire and personnel protection		Value of savings per year—reduction in overhead and insurance cost.

From these savings, the capital savings and savings per year can be determined. Comparison with cost will provide capital savings, or return on investment, if insulation capital cost is more than amount saved.

Design of Insulation Systems

The design of an insulation system may be divided into the following parts:

1. Setting of the criteria and establishing the basic functions that the insulation system must perform.
2. Determination of the properties of the insulation and accessories necessary to fulfill the requirements of the system.
3. Selection of the insulation and accessories which will fulfill the requirements of the system.
4. Determination of the thicknesses of insulation required.
5. The physical design of the insulation materials and accessories into an insulation system (as presented in Chapter 9).

Chapter 5 presented "Installation Requirements," and from these the properties of insulation and accessories to fulfill these requirements must be determined. The outline previously presented in this chapter will assist in converting requirements into needed material properties.

1. Setting of the Criteria and Establishing the Functions that the Insulation must Perform

Insulations are used to perform many functions. In designing an insulation system, it first must be established what specific function, or functions, the insulation must perform by virtue of its ability to retard heat flow. A number of these functions are:

Conserve heat energy for economics
Conserve energy, either heat or refrigeration, for process reasons
Maintain desired temperature in a space, vessel or pipe
Retard any change in temperature in a vessel, pipe, or object
Prevent condensation of vapors on either inner or outer surfaces
Limit temperatures of exposed surfaces of hot equipment or pipe to safe values for personnel protection, or to prevent flash fires
Provide fire protection

In most instances, more than one function is served by an insulation system. Typical of this may be the insulation system on a hot processing vessel which serves to conserve energy, maintain the required temperature in the vessel, protect personnel from burns, and provide fire protection. Careful consideration of these functions may make possible the design of an insulation system which will serve more than one purpose for little or no additional cost over the single purpose system.

It must be further established under what specific conditions the insulation will operate. Discussion and expansion of this follows.

Location of the insulation system must be established.

In building insulation or cold storage: is it on walls, floors, or ceiling?
Does it have inside or outside exposure, or both?
The indoor or outdoor location of insulated ducts, equipment, vessels, or pipes, must be established.

If located indoors, the maximum and minimum temperatures of the surroundings should be established, and, if condensation is important, then the extremes of relative humidity should be determined. If high level radiation sources, such as furnaces, boilers, or hot equipment, are in the area, they should be identified and located in their relationship to the system being insulated.

These same factors must also be known for outdoor installations, plus the information regarding wind conditions and solar exposure.

As the most severe conditions of heat transfer on hot insulation occur under exposure to rain, snow, and sleet, these must also be considered if peak loads are important.

If underground systems are to be used, the ground temperatures at various times of the year are the determining factors in consideration of surrounding temperature.

If insulation is expected to serve as protection from accidental fire, then the expected fire temperature of the possible flammable fuel must be determined.

The next thing to be established is the temperature of the space, equipment, vessel, or pipe that the insulation is to serve.

In many instances, this is not a single temperature. Thus, there may be:

maximum temperature
average temperature
minimum temperature

From these three, it may be necessary to establish *design* temperature.

If the operation is not continuous, then the amount of down time expected when the temperature of the insulated item will fall (or rise) into balance with the surrounding temperature must be determined.

If operation is cyclic, then the time at high and low level temperature, and time to reach one from the other must be established.

With these ambient and operating temperatures established, the *temperature differences* and *mean temperatures* may be determined to allow the solution of the heat transfer equations.

Also required for heat transfer calculations is information as to size, and shape of equipment or vessel, and sizes of pipe to be insulated.

Finally, for heat calculations, information as to conductivity of insulation and the surface emittance of exposed surfaces is necessary. However, the latter may be unknown until the insulation weather-barrier, or interior covering is selected.

If the function of the insulation is to obtain the best economics in heat conservation, information as to the capital investment necessary to produce that heat, the cost of producing the heat, the cost of insulation, its maintenance cost per year, and years of amortization, will all be required.

When insulation is to control either the heat loss or heat gain of a process system, then the maximum heat gain or loss must be established by the process requirements.

If an insulation system is to control the temperature of a process, then all the factors which affect the temperature of the system are needed. These include the weight of vessels and pipes, specific heat of materials composing the vessels and pipes, the flow of material in lbs per hour, the specific heat and weight of the material. With the input and output temperature of the system and the ambient temperatures, it is then possible to set up the heat transfer equation to obtain the heat balance of the system for the required operating temperature.

Retardation of temperature change requires the time element to be part of the heat transfer equation.

Thus, depending upon the function of the insulation, the designer must obtain specific information on each item of the following which effects his calculations.

Operating temperature
  Maximum, minimum, average, design
Surrounding (or ambient) temperature
  Maximum, minimum, average, design
Rate of change of operating temperature
Rate of change of ambient temperature
Location of insulation system
Thermal characteristics of process materials
  Specific heat, density, conductivity
Weather conditions
  Solar, wind, rain, snow
Moisture conditions
  Relative humidity in relation to temperature
  Underground systems
    Water table
    Soil moisture content, with resultant conductivity of soil
Capital cost of heat, or refrigeration
Product cost of heat, or refrigeration
Cost of insulation
Maintenance cost of insulation
Time of plant amortization
Time of insulation amortization
Thermal characteristics of insulation
  Conductivity
  Diffusivity
  Specific heat
  Density—apparent
  Surface emittance of weather-barrier
Thermal characteristics of process system
  Specific heat
  Density
Thermal characteristics of process materials
  Specific heat, density, conductivity
Heat input from process sources
Heat input from coils, tracers, etc.
Size and shape of equipment, or vessel
Size and length of pipe
Projections and supports through insulation
Hazards
  Fire—temperature
  Minimum fire protection requirements
  Allowable surface temperatures
    Flash point and self ignition point of process materials
    Personnel protection
    Condensation

All of the thermal factors either will be used directly in the heat transfer calculation, or will affect the design temperatures, while the cost factors directly affect the monetary efficiency of the system.

In Chapter 5, the phases of operations an insulation must fulfill were given as: (1) shipping, (2) storage, (3) fabrication, (4) application, and (5) service. Now it becomes necessary to evaluate these phases to determine the properties of materials which will fulfill these phase requirements. The basic divisions of the properties of materials are:

Thermal	Moisture
Physical	Fire
Chemical	Toxicity

The order of listing bears no relation to the order of importance, as the installation itself establishes which properties are of the most importance. This order will change from one installation to another.

2. Determination of the Required Properties

Thermal properties of insulation materials as related to installation requirements

The thermal properties of materials are interrelated to the other properties listed. For example, the temperature to which a material is subjected affects its strength, chemical properties, moisture content, and fire properties. There is no clear line of demarcation between some properties. For example: is expansion and contraction a thermal or physical property? Therefore, the divisions which will be listed are only for convenience in correlating the installation requirements and properties of materials. For the thermal properties, the major divisions have been taken to be:

Temperature limitation
Thermal shock resistance
Thermal diffusivity
Specific heat
Thermal emittance (conversely, thermal reflectance)
Thermal conductivity (or thermal transmittance)

Although low thermal conductivity (or thermal transmittance) is the essential property in classifying a material as an insulation, it was purposely placed last in the listing because, frequently, the insulation selection is based upon other properties, and conductivity should be the basic consideration in the determination of insulation *thickness*. (See "The Physical Design of Insulation System," p. 363.)

Temperature limitations of materials should never be exceeded if lasting service is expected. Each insulation is suitable only within a given temperature range. The maximum temperature limit of some insulations may be two different values depending upon the period of exposure. However, exposure of any material to temperatures above its maximum continuous temperature value should be avoided, as exposure to such temperatures for even short periods of time can cause gradual breakdown of the material. From this fact comes the determination that the *max-*

imum temperature limit of the insulation must be above the maximum operating temperature.

When an insulation is used to prevent heat from entering into a space, pipe, or equipment from the surrounding air, or objects, then the minimum operating temperature is important for two reasons: (1) the ability of the insulation to function at this low service temperature, and (2) the tendency of the vapor from the ambient air to migrate to the cooler surface. *The minimum temperature limitation of the insulation must be below* the minimum operating temperature. Vapor migration to the cooler surface will be discussed under moisture requirements.

If operating temperature is cyclic then three additional thermal considerations are necessary. The first of these is *thermal shock resistance.* The rate of operating temperature change may cause some materials to crack or spall. For this reason, if the change of temperature is rapid, the thermal insulation should have sufficient *thermal shock resistance* to resist this sudden change without physical deterioration.

Another item of interest in cyclic operation is the rate of time that the insulation takes to adjust to temperature change. Where fast adjustment to temperature change is desired, a high rate of *thermal diffusivity* is necessary. If a slow rate of temperature change is desired, such as to retard the freezing of pipes, then a low rate of thermal diffusivity is essential.

Specific heat, the second of the three additional considerations, is the measure of the quantity of heat required to raise the temperature of a body. As such, it enters into calculations of an unsteady temperature state. In cyclic operations, it determines the amount of heat gain or loss necessary to raise or lower the temperature of the insulation.

Surface emittance, the third additional consideration, is the amount of heat which a surface will radiate to surrounding bodies at a lower temperature. On hot surface applications, when the surface emittance is low, the surface temperature of the insulation (or surface) will be higher than when the surface emittance is high. For this reason, if the surface temperature must be below a certain temperature, the surface emittance is very important. On low temperature applications, the lower the surface emittance, the lower the surface temperature. If dew point is a vital consideration, then this surface emittance is of utmost importance. In almost all instances, the surface emittance is a property of the outer weather-barrier, or jacket, which is applied over insulation.

Thermal conductivity as a thermal property will be discussed in "Determination of the Thickness of Insulation Required," page 338.

From the evaluation of operating temperature, (maximum and minimum, cyclic time, ambient conditions and the requirements of surface temperature) the following needed properties of insulation can be established:

The minimum and maximum temperature range of the insulation
The minimum requirement of thermal shock resistance of the insulation
The approximate thermal diffusivity of the insulation
The range of specific heat of the insulation
The desired surface emittance of the insulation coating or jacketing

Physical properties of insulation materials as related to installation requirements

If an insulation application is to be made above grade, the first consideration which must be decided is the basic form of material which would best lend itself to economic application and fulfill the installation requirements. The forms to be considered are:

Rigid
Semi-rigid and flexible
Blankets or felts
Cements or sprayed-on fibrous
Sprayed or poured-in-place foams
Poured or loose granular or fibrous
Reflective
Mastics

In general, these forms of materials are used for the following types of installation.

Rigid insulation used where structural strength is needed and where it is exposed to mechanical abuses: e.g., cold storage walls and ceilings, hot or cold equipment and vessels, hot or cold piping, or exposed ducts, or where the insulation must be at least partly self supporting.

Semi-rigid, or flexible, insulation used in walls, ceilings, and other areas where it obtains support from other members, where mechanical abuses are light, or where its flexibility is desired to conform to curvatures.

Blanket and felt insulation used where it is supported by other members; such as in cavity walls, wrapping or lining of ducts, and where compressibility and expansibility are necessary. It is frequently used in combination with rigid insulation to provide a thermal and mechanical cushion.

Insulating cements used to insulate small fittings on contoured equipment, and in combination with rigid insulations to fill voids, and sometimes as a leveling coat over block or curved segments. Sprayed-on fibrous insulation is most advantageously used on large irregular surfaces.

Sprayed-on foam insulation used advantageously on large moderate temperature areas. Foamed-in-place is used to fill cavities, but the walls of the cavity must be sufficiently strong (or reinforced) to withstand forming pressure, which is 1½-5 psi (4.4^{-4} to 1.3^{-3} kg m^2).

Poured and loose insulation used for filling of cavities or tight enclosures.

From these general guide lines the form of insulation may be selected, but there are cases where there may be a choice of forms. For example, either a rigid or sprayed-on insulation may be used on a storage vessel. The other installation requirements must then be evaluated to determine which form is most suitable. The final choice between the two materials may even be dependent upon the time of year and expected weather conditions at the time of installation. Where more than one form of insulation meets all the physical, thermal, chemical, and hazard requirements, the choice is determined by efficiency and economics.

After the form (or forms) of insulation is established, it is necessary to choose the type of material which will be most nearly in accord with the requirements of the installation. This selection must be based upon the properties of the materials under consideration, and how nearly those properties meet the requirements.

RIGID MATERIALS

The following properties are those of rigid insulating materials. Insulating cements, sprayed-on fibrous insulations, and sprayed on foamed-in place foam insulations are, essentially, rigid insulation after being installed.

The following properties all relate, directly or indirectly, to the physical strength of the materials:

Breaking load
Deflection at breaking load
Flexural strength
Impact strength
Compressive strength
Shear strength
Tensile strength
Transverse strength
Indentation hardness
Resistance to dropping
Resistance to vibration
Resistance to tumbling

Although all the strength factors of a material effect its ability to be shipped, stored, handled, and applied, the *breaking load, impact strength, resistance to dropping* and *resistance to tumbling* are major properties. As could be expected, except for the rare case of a bad batch of material, all materials presently marketed are produced to withstand shipping and handling without excessive breakage.

Materials of high *compressive strength, tensile strength, impact strength* and *shear strength,* should be used on installations where there may be equipment and piping, and in locations such as cold storage walls and ceilings where the insulation is depended upon for part of their structural strength.

On piping and vessels, the *compressive strength* of the insulation must be sufficient to carry its own weight when supported from the bottom. To illustrate: suppose insulation 3″ (7.6 mm) thick is installed on the sides of a vessel, with supports every 18′ 0″ (45.2 mm). The weight of every sq ft of insulation would be about 5 lbs (allowing some safety factor) and the stacked load would be 18 × 5 = 90 lbs per linear foot (248 × 69 = 1,712 kg/lin m) of the support. If this was supported on a 1″ (2.54 mm) wide insulation support, the load per square inch would be 90 divided by 12, or 7½ lbs per sq in. (5273.2 kg/m^2). Therefore, the insulation should have a compressive strength of 7½ lbs per sq in. (5273.2 kgs/m^2) with no appreciable deformation. Where insulated horizontal pipe is supported by a cradle, the insulation's compressive strength, in relation to the cradle supporting area, must be sufficient to support the pipe, the insulation, and the contents of the pipe. The power the compressive strength of the insulation, the larger the necessary supporting area of the cradle required to carry the load without excessive deformation of the insulation. The method for calculating cradle size is given in Chapter 7.

The compressive strength of insulation provides the bearing on which the weather-barrier and vapor-barrier depend for support. If a concentrated load causes excessive deflection, the barrier may be sheared, dented, or otherwise damaged. This is illustrated in Figure 79.

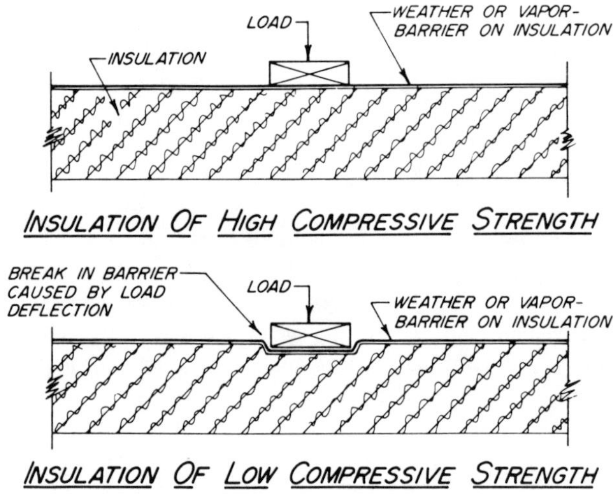

Figure 79 Loads on insulation.

Installations which require the insulation to bridge gaps in the substrate whould be of materials of high strength, particularly as regards *breaking load, flexural strength, impact strength, shear strength, tensile strength* and *transverse strength.* Typical of this type of installation are enclosures around pipe heat exchangers, heat traced pipe, and walls and ceilings of insulated areas where the insulation is used as a structural board or sheet. As shown in Figure 80 load or impact on the insulation bridging a void tends to deflect and eventually break it, if it has insufficient strength. It must have sufficient compressive, flexural, shear, tensile, and transverse strength to resist the load or impact and transmit it to the pipe or structural member to which it is attached, without fracture, breaking, or excessive permanent deformation.

Installations indoors, located where there is less chance of mechanical damage, such as in pipe chases, or near the ceiling, do not need the insulations which have high physical strength. In this type of application, the strengths required for shipping, handling, and application are more than sufficient to fulfill the service requirements.

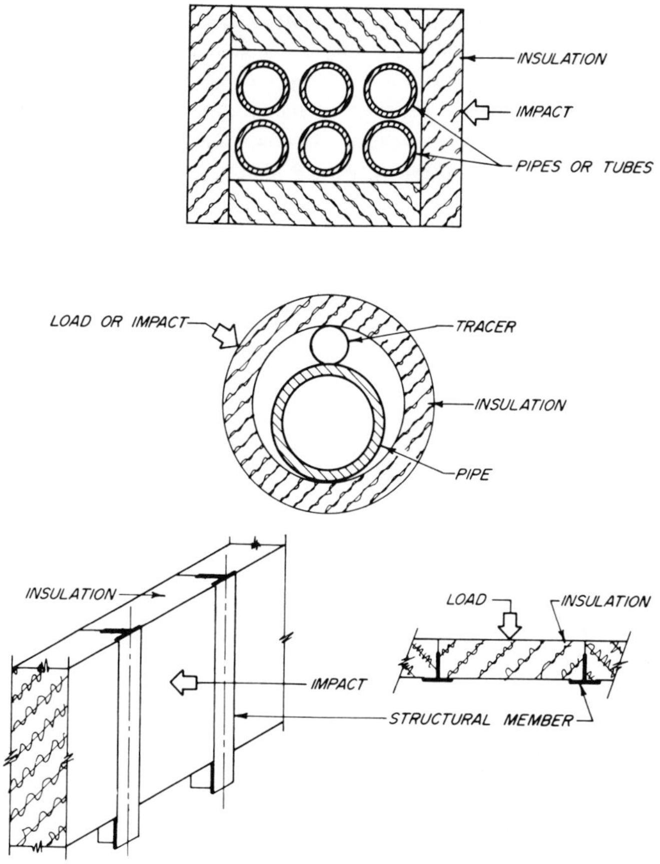

Figure 80 **Loads and impact to which insulation is subjected.**

Tensile strength is important for heavy duty applications because, when pipe and vessels expand and contract, the cylindrical insulation covering must remain together as a unit. Their tensile strength is what enables the individual pieces of the insulation to hold together as a unit and resist the friction caused by the vessel or pipe which tends to pull the pieces apart. The movement should be taken up at an expansion joint in the insulation, rather than having the individual pieces separate as illustrated in Figure 81.

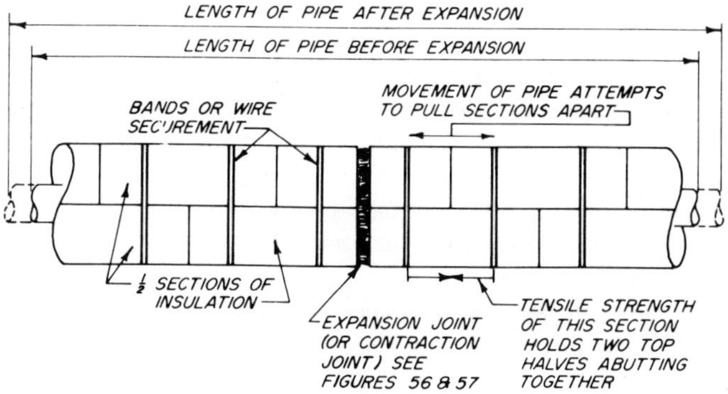

Figure 81 **Need for tensile strength in insulation materials.**

Vibration resistance is necessary to prevent the insulation from destroying itself when it is installed on equipment, vessel, or pipe which vibrates. Compressors, piping to and from receivers, steam lines, and lines of high velocity flow all have some degree of vibration. Where necessary, the insulation may have to be treated with an anti-abrasive coating to resist this vibration.

Insulations which are cut, fitted, and the pieces cemented together to form prefabricated shapes, require most of the strength properties listed for installation applications where insulation is subjected to mechanical abuse, and where it bridges gaps. They also require additional properties such as *indentation hardness.* The degree of indentation hardness depends upon the generic type of material, but the fabricated shapes require a material which, when joined together, is not so soft that it parts on either side of a joint when subjected to a fair amount of movement or handling. Other properties required for this service, in addition to the common ones, are *ease of cutting, smoothness of cut surfaces, lack of dustiness on cut surfaces, compatibility of surfaces with adhesives and cements, resistance to delamination and dimensional stability.* Of course, *dimensional trueness, dimensional squareness,* and *dimensional tolerances,* although not true material properties, are important as they affect the efficiency of the fabrication operation.

In addition to strength, other physical factors are important in the selection of insulation. These are:

Coefficient of expansion
Dimensional stability after soaking heat or humidity
Density—real, and specific gravity—real
Density—apparent, and specific gravity—apparent

Note: *English Units*

$$\text{Specific gravity} = \frac{\text{Density in lbs cu ft}}{62.43}$$

(either real or apparent)

Metric Units

$$\text{Specific gravity} = \frac{\text{Density in kgs w metric}}{2204}$$

Coefficient of expansion and shrinkage under soaking heat are necessary to be able to calculate the differential movement between the insulation and the vessel, pipe, or materials of construction, with which it is used. This has previously been discussed in Chapter 7.

Dimensional stability under soaking heat is a measure of factors limiting the use of a material, and, essentially, establishes the temperature above which a material should not be used. Although it may appear to be a physical property of material, it, basically, is a measure of temperature limitation.

Density and specific gravity each measure weight—the first in terms of lbs per cu ft and the second in terms of the ratio of its weight to the weight of water. The *density-real* and *specific* *gravity-real* indicate the weight of the solids of an insulation in lbs per cu ft, or ratio in respect to water, with all the entrapped air or gas excluded. The *density-apparent,* or *specific gravity-apparent,* is the weight per cu ft, or ratio in respect to water, including the gas or air making up its volume. The *density-apparent* is the density most frequently given for an insulation. To illustrate: fibrous materials of a density (apparent) of 2-20 lbs per cu ft (32 to 320 kgs/m^3) means that 2 to 20 lbs of fibers occupy one cubic foot space or 32 to 320 kgs of fiber occupy one cubic metre. As the *density-real* of the fibers is 120 to 150 lbs per cu ft (1920 to 2400 kgs/m^3); if these materials are placed in water, as the water replaces the trapped air, the material will sink. If the air or gas is hermetically sealed by closed cells, as in foam insulation, then the material cannot sink because the water is unable to replace the gas or air. As a physical property, the *density-apparent* (or specific gravity-apparent) is important, as weight affects design loads of insulated vessels and pipe. Low density materials have many advantages as insulation for many forms of transportation, ships, air planes, trucks, or railroad equipment, as the light weight means additional pay load.

Relatively high unit weight insulation should be used on top of large domed, coned, or flat roof vessels where wind blowing over the top can cause lifting. On this type of application, partial vacuums can occur. Differential pressures of over 5 psi (3515.5 kgs/m^2) are possible when winds of over 25 mph (40.2 km/h) occur.

Additional physical properties of materials which are in liquid, mastic or plastic form before application. This type of material regardless of whether it is insulating cement, sprayed-on-fibrous insulation, insulating mastic, or spray-on organic foam, has physical properties to be considered in addition to the properties in their final form. These properties are:

Adhesion—dry
Adhesion—wet
Build
Expansion ratio or shrinkage

Most of these materials rely upon *dry adhesion* to the substrate for their support in service. Sometimes they are reinforced with netting, or other securements, but, basically, for long lasting application, the bond between them and the substrate must remain unbroken. This means that this adhesion must be of sufficient strength to resist the forces imposed upon it by expansion and contraction, vibration and load, and the materials should have sufficient ability to *elongate* without shearing this bond.

The property of *wet adhesion* is a measure of the ability of the material to stay in place as it is applied. Unless the insulation is completely saturated with water during its operation, this property is not a factor in service requirements. The *build* is also a property of application.

Shrinkage of insulating cements or *expansion ratios* are properties affecting application and required quantity of materials. Although they effect the density and other properties of the materials, they, as properties, are not significant in the finished form of the material.

Physical properties particularly related to blankets, felts, and flexible insulation. In the basic classification of materials, blankets and felts were treated as flexible insulations, but, in service, the essential properties of them and rigid materials are the same. The properties of these materials, which must be considered, in addition to those of rigid (and semi-rigid) insulations are:

Elongation (without breaking)
Flexibility
Resiliency

In rigid insulations, differential movement between the metal and insulation, caused by expansion and contraction, is compensated for by slippage, or movement between the two. By contrast, in most applications of blankets and flexible insulations, the expansion and contraction movement is expected to be transmitted to these flexible materials that can *elongate* and recover without breaking. As these insulations are expected to expand and contract with the vessel, pipe, or equipment, they are attached directly to the surfaces by clips, adhesives, or by wrapping in place and securing edges and end joints. For this reason, the ability of the insulation to *elongate* and recover must exceed the expected movement of the substrate caused by expansion and contraction. Typical applications using flexible insulation are ducts, vessels, tubing, and flexible connections.

Flexibility is the property of a material enabling it to be bent or deformed from its original shape to another without cracking, breaking, wrinkling, or buckling. This property is essential where installation conditions require that the insulation be wrapped around various contours. Another type of installation where this property is essential is where duct metal is preinsulated before it is formed and assembled into its final shape.

Recovery is the ability of a material to return to its original size or form after it has been forced into a shape other than its original one. In installations where it is installed to a permanent shape other than its original, excessive ability to recover is not an asset, and may be even detrimental. However, where the substrate changes position, size, or shape, recovery may be a most important property.

Because of these particular properties, blanket and flexible insulations are frequently used with rigid insulations to act as mechanical, and sometimes thermal "cushions." Figure 82 shows the method of using a flexible insulation (most frequently fibrous blanket) to compensate for expansion of a vessel so that the joints in the rigid insulation remain tight. In the case of hot service, the flexible insulation should not be compressed during application. On low temperatures it should be compressed sufficiently so that the insulation's recovery faculty will expand it as the vessel contracts. Such applications are not necessary at moderate temperature differences and on small diameter equipment, but are essential where differentials from atmospheric temperature and equipment diameters are large. The larger the diameter, the smaller the temperature difference at which this type of construction becomes necessary. Care should be taken not to have expansion

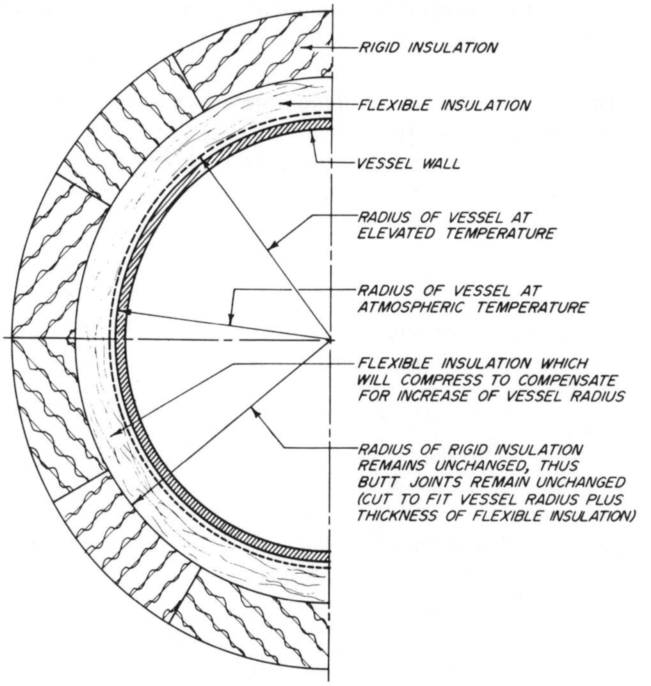

Figure 82 Use of flexible insulation in combination with rigid insulation to compensate for vessel expansion.

forces exceed the strength of the band securement. The stress which will be exerted on the band is:

> Pressure per unit area to compress insulation to required deflection, times the spacing in unit distance, times the circumferential length.

■ EXAMPLE 44

English Units

If an expected expansion of 1/4 inch needs 10 lbs per sq ft to compress the insulation the required 1/4 inch, the band spacing is 18 inches, and the circumferential length of the bands and clips is 60 ft, then the stress on the bands is

$$10 \times \frac{18}{12} \times 60 = 10 \times 1.5 \times 60 = 900 \text{ lbs}$$

As the ordinary 1/2 inch double pronged strap clip will open at approximately 700 lbs, a more flexible (or thicker) cushion blanket is required. Allowing some safety factor for clip opening (500 lbs), the compressive strength required for 1/4 inch deflection would be

$$500 = 1.5 \times 60 \times (\text{lb/sq ft max})$$

$$\text{lb/sq ft max} = \frac{500}{90}$$

= 5.57 lb/sq ft max to compress flexible insulation 1/4 inch ■

■ EXAMPLE 44a

Metric Units

If an expected expansion of 6.35 mm needs 48.82 kgs/m^2 to compress the insulation the required 6.35 mm, and the band spacing is 0.457 metres and the circumferential length of the bands and clips is 18.29 metres, then the stress on the band is:

$$48.82 \times 0.457 \times 18.29 = 408.2 \text{ kgs}$$

As the ordinary 12.7 mm double pronged clip will open at approximately 310 kgs a more flexible (or thicker) cushion blanket is required. Allowing some safety factor for clip opening (225 kgs) the compressive strength required for 6.35 mm deflection should be less than:

$$225 = 0.457 \times 18.29 \times (\text{kgs/m}^2 \text{ max})$$

$$\text{kgs/m}^2 \text{ max} = \frac{225}{8.35}$$

= 26.92 kgs/m^2 to compress flexible insulation 6.35 mm. ■

On linear expansion joints, the situation is reversed from cylindrical expansion or contraction. The flexible insulation used in linear butt joints is compressed as temperatures drop below and

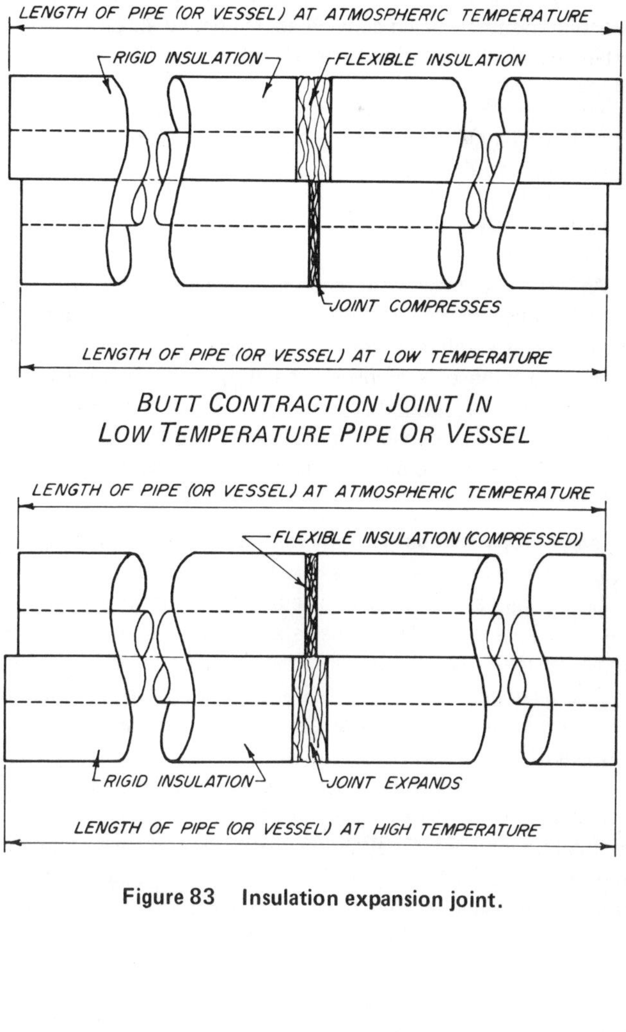

Figure 83 Insulation expansion joint.

expand as they rise above atmospheric. This is illustrated in Figure 83.

Physical properties related to fill and poured materials. Other than *compressive strength* and *recovery,* these materials have no physical strength properties. They are used where they can be placed in an enclosure, cavity, or space and must be contained by surrounding materials. Other than the two properties mentioned above, only one other must be considered. This is the *resistance to settling or compaction.*

These materials are fibrous, granular, or spherical, and various ones do have different characteristics so far as application goes, such as the ability to pump, blow, pour, or ram them into the spaces to be filled. The type of installation determines the allowable *aggregate size, fiber size,* and *fiber flexibility.*

Within its confined space, the insulation may be required to support a load which is determined by its *compressive strength* at its *packed density.* The property of the insulation enabling it to resist compaction or settling is its ability to preserve its voids and maintain its integrity after a period of service. The property of resisting vibration without physical breakdown is one factor in resisting compaction.

Physical properties particularly related to preformed reflective insulation. Preformed, industrial, reflective insulation for pipe and vessels is not a homogeneous material, but is a fabricated assembly. For that reason its physical properties depend upon the particular assembly being considered. In general, its physical properties are its abilities to withstand

Concentrated load
Deflection
Resistance to impact
Resistance to vibration
Edge compression loading

As these assemblies are made from various gages of metals and supporting members, they can be designed and manufactured to the strength requirements of the insulation. The most unique feature of these assemblies is that they are a system of insulation which can be installed and removed as units. However, other than for straight pipe insulation, the units must be custom fabricated to very accurate dimensions for the individual installation.

Physical properties particularly related to insulation mastics. Insulating mastics are a cross-breed between insulations and coatings. They are particularly useful where relatively little thermal resistance can fulfill the insulation requirements. A good example is their use to prevent condensation of moisture on moderate temperature indoor lines and equipment to prevent dripping. They serve a dual purpose as a coating to retard rusting and corrosion.

As insulation mastics are applied directly to the substrate surface, the *adhesion to this surface* is of utmost importance, as it is the total attachment. To attain the necessary adhesion, the surface must be properly prepared and clean. Being attached to the surface, the insulation must have the ability to *elongate* and *compress* with the expansion and contraction of the surface. Also important are its properties to *resist abrasion, impact,* and tearing.

Chemical properties of insulation materials as related to installation requirements. The chemical properties of insulation materials which must be considered in relation to the effect on their installation are common to most forms of insulation. They are:

Corrosion effect on substrate metal to which it is applied
Resistance to acids
Resistance to caustics
Resistance to solvents
Reaction with various chemicals
Odor emission

The *corrosion effect of insulation* on metals may be caused by several actions. *Chemical corrosion* may be caused by insulation, if its components chemically attack the metal to which it is attached. For this reason, an insulation material should be selected which will not chemically attack the substrate metal. As this phenomenon varies with various metals and insulations, the combination selected must be compatible. For example, an insulation suitable for steel may not be suitable for aluminum. Thus, insulation suitable for use on steel must be used on steel, and those suitable for aluminum on aluminum.

Other than corrosion caused by chemical attack, rusting of steel may be caused by the insulation absorbing water and holding it against the steel surface. If the atmosphere is contaminated with rust or corrosive chemicals, this same absorption may result in rust or corrosion of metals to which the insulation is applied. The absorptive characteristics will be discussed in more detail later in this chapter. *Stress corrosion* of austenitic stainless steel is another problem. Insulations which can contribute the chloride ion necessary to this type of corrosion should never be used on stainless steel.

The *chemical resistance* required of insulation is completely dependent upon the chemicals to which the insulation may be exposed. If it is exposed to an acid, caustic, or solvent it must be resistant to that specific acid, caustic, or solvent. The exposure may be from direct liquid or gas contact, accidental spillage, or contamination of the atmosphere. It will be found that, because of the thousands of potential exposure combinations, manufacturers are unlikely to have data about the results of exposure of their insulation to a particular chemical. Thus, it frequently becomes necessary for the user of insulation to test a number of insulations for his particular exposure to determine one or more insulations suitable for his conditions.

In addition to being able to resist chemical attack, insulations which might become saturated with chemicals should not react with them in a manner which might cause combustion, or a lowering of the flash point or self-ignition point to the hazardous range. As previously mentioned, the only safe way to determine which insulations are safe is to test them in contact with the chemicals to which they may be exposed.

Odor emission is seldom a problem in the use of insulation in an industrial installation, but it may be a most important property when insulation is used on the interior of office buildings, restaurants and dairies, and in the food handling and cosmetic industries.

Moisture properties of insulation materials as related to installation requirements

The moisture properties of insulation which must be considered in relation to the installation are:

Absorption
Hygroscopicity
Capillarity
Water vapor transmission

As explained previously, mass insulation depends upon trapped air or gas to be thermally efficient. If this trapped air or gas is replaced by liquid (or ice), the insulation is no longer efficient. For this reason the amount of the liquid that an insulation will absorb if exposed to it is most important.

If an insulation is located indoors, on hot lines or equipment, with no chance of being wetted by liquid, then the insulation may be *highly absorbent* without damage to its effectiveness. When located where it may become saturated with water (or other liquids) it must be *non-absorbent,* or *protected* by a water-barrier. In the language of the industry, a water-barrier is known as a weather-barrier, as it most frequently serves to protect the insulation, not only from water, but also from the other elements of weather.

Non-absorbent insulation should be used on installations which are exposed to weather, and where the insulation's temperature is less than 212° F (100° C), or where heat energy is limited. This is because a relatively high temperature and a great amount of heat are required to dry out insulation once it becomes wet. This is particularly true of electrically traced installations where heat input is set by the electrical resistance of the heating cable.

To illustrate the importance of non-absorbency of insulation in this low temperature range, consider the following. A highly absorbent insulation of 10 lbs per cu ft apparent-density can soak up approximately 52 lbs of water (6.3 gallons or 23.8 liters). To dry out, such soaked insulation would require a minimum of 52,000 Btu's (54,863,120 Joules) of heat even if no recondensation occurred in the process.

The absorbency of insulation in regard to water has little relation to its absorbence of acids, caustics, or solvents. Insulations which are water repellent may freely absorb other liquids. If the insulation may be subjected to liquids other than water, then the absorbence for those liquids must be determined by testing.

In cases where absorbed liquids or adsorbed gases in insulation are combustible, the entrapped combustible materials may constitute a severe fire hazard. The materials contained would be vaporized and expelled from the insulation by the heat from a fire, and add considerable fuel to the fire, even though the insulation itself was incombustible. Aside from adding fuel to the fire, insulation saturated with these combustibles would spread any fire very rapidly. Even steel jacketing over insulation will not provide an effective barrier to prevent the fuel contribution, as the fire will cause enough internal pressure within the jacket to cause joint rupture. This will release the combustible vapors, and, in addition to contributing fuel to the exterior fire, may combine with the insulation (if it is a low melting point material) to start a fire under the jacketing. Such a fire is almost impossible to extinguish, as the jacketing then becomes a barrier between the fire and the hose stream or spray head water.

Hygroscopicity of an insulation is the amount of moisture a completely dry sample will absorb from the ambient air at any given air temperature and relative humidity. Highly absorbent insulations do not give up this moisture readily. Therefore, insulations of high hygroscopicity and high absorbency should not be used in highly humid surroundings unless they are on hot (212° F) surfaces where sufficient heat is available to maintain the vapor pressure outward.

Capillarity is the amount of water that an insulation will draw upward into itself from water with which its base is in contact. Weather-barriers shed water, but seldom are watertight. For this reason, wherever insulation extends down into areas which might be flooded with water by rain, or other water sources, the insulation can "wick" up water into itself by capillarity. Where these conditions exist, or where the possibility exists of cracks or joints allowing water entry, the use of insulation which has extremely low capillarity should be used.

Water vapor transmission is the amount of moisture in vapor phase which can pass through the insulation due to a vapor pressure difference existing between its two sides. When air in an enclosed space is at a lower temperature than the surrounding outside air, or the temperature of a surface to which the insulation is applied is also lower than the surrounding outside air, the moisture travels either into the space, or to the cold surface. Where the vapor is prevented from travelling further by this cold surface, it becomes trapped and condenses in the insulation. For this reason, when insulation is used on refrigerated spaces or on cold vessels and pipe, the vapor transmission of the insulation must be very low, or it must be protected with an absolutely tight vapor-barrier which has sufficiently high vapor resistance to retard gradual filling up with moisture over a long enough time period to provide economic service. Thus, the rate of vapor transmission into an insulation installed on a low temperature metal is a direct measure of the time that the insulation will remain efficient as a thermal barrier.

Combustibility properties of insulations as related to installation requirements

Without question, the safest of all insulations, as regards fire hazards, are those which are noncombustible. Unfortunately, there is no exact definition as to what way an insulation must perform when exposed to fire to justify its classification as noncombustible. For this reason, and for purposes of clarity in this manual, the following conditions are presented as the author's intent when speaking of a noncombustible material. When exposed to fire, the material should

Contribute no fuel to the fires,
Emit very little smoke,
Emit no toxic gases, and
Have a melting point above 1600° F (870° C)

If the insulation is also to be considered a fire *protection* material, in addition to the above characteristics it should have low *thermal diffusivity, low shrinkage,* the *ability to resist fire without excessive cracking, spalling,* or *deformation,* and be *thermal shock resistant,* and have sufficient mass to provide a heat sink for a considerable amount of heat. To illustrate: steel is a noncombustible material, but will fail rapidly as the temperature of a fire impinging upon its surface is transmitted throughout the steel exceedingly fast. In a short time, the steel temperature rises above 1000° F (538° C) and loses its structural strength. A fire protection material must have thermal properties which *retard the rise of temperature,* and physical properties which *resist the fire* for an extended period of time.

When insulation is used in areas, or buildings where combustible materials may be used, the problem shifts as to the degree of combustibility which is acceptable. Most materials will emit quantities of smoke when subjected to fire, so that another factor is their *smoke density.* Means to measure the toxic effects of smoke need considerable development, and relatively little data is available in this field for insulation materials. One reason for this lack of information is that combustion does produce toxic gases, and the danger from fumes and smoke is so great that degree of difference is unimportant. Most deaths caused in building fires are due to suffocation and the effects of toxic fumes, rather than by heat from the fire.*

Outdoor installations, and installations of lesser hazard can tolerate more combustible materials. The amount of combustibility acceptable must be determined by the allowable tolerance of risk. The properties which measure the degree of combustibility of material are *flame spread, fuel contribution, melting point, flash point, fire point,* and *auto-ignition point.**

Under "Absorbency" it was mentioned that a noncombustible insulation, when saturated with combustibles, is a fire hazard. Such a saturataed insulation must be evaluated by the same properties as those given for a combustible insulation.

3. Selection of the Insulation and Accessories which Fulfill the Requirements of the System

Insulation of buildings and cold storage and refrigeration rooms will not be discussed in this manual for two reasons. First, the insulation in these structures is used with, and is a part of, their architectural and engineering design. Second, there has been considerable design information already presented on this subject in greater depth than could possibly be given in this manual. The "ASHRAE Guide" presents the conductances of many combinations of materials, and "Thermal Design of Buildings" by T. S. Rogers presents an excellent guide to economical building insulation design. Consequently, this manual will concentrate on insulation design for industrial equipment and piping.

One of the first points to be clarified is that, in the insulations available for industrial application, there are systems which are custom made in a manufacturing plant for installation at a given job, and those in which the components are assembled into a system at the job site. Usually, those insulation systems designed for a particular functional use where it would be almost impossible to form and fabricate the needed system at the job site are those which fall under the first category. The areas most frequently using the factory formed systems are:

Cryogenic—double shell vacuum insulation
Underground conduit systems
Above grade conduit systems
Preformed reflective insulation systems

Where these systems are used, their designs are the responsibility of the manufacturers, and cannot be dictated by others. Where these are most suitable, they will be so recommended, but details of design will not be discussed. Insulation systems which require the selection and specification of materials and their application are of prime interest to this manual.

It must be understood that the suggestions can be only of a very general nature, as any specific installation may have one single factor which is of utmost importance, but which is not covered in a general recommendation. Another factor which cannot be covered is that the given recommendations are based on obtaining a service life generally expected in a permanent installation, while the use of less expensive and shorter lasting installations may be warranted for the temporary type of installation.

The following examples are given only to illustrate the *method* of selecting a material and determining its proper thickness. It must be pointed out that the material selected, in most instances, is *not the only possible choice,* and that others may serve equally well. Thus, in the examples, the term "*selected*" means "selected to illustrate the example," and is not to be taken to mean "recommended choice."

All the installation requirements which could effect the choice of an insulation and weather-barrier are not given. To have done so would have resulted in a complex, lengthy, list with so many different selections that the examples would be too numerous to include in a manual. In order to illustrate by examples, it was necessary to condense, and present only the *major* requirements which effect choice.

4. Determination of the Thickness of Insulation Required

The method of calculating heat transfer and examples of such calculations are presented in Chapter 1. This further presentation, with the following examples, is made to tie in the theory with practical problems involving the insulation function. These examples will illustrate the method of selecting insulation and its weather-barrier (an insulation system) based on fulfilling the installation requirements through the properties of the materials used. Accessories, such as adhesives, sealers, or securements, are not listed in the examples, but each of these must also fulfill the requirements as do the basic materials.

5. Safe Surface Temperature

If insulation is used where its surface temperature must be safe for personnel that may touch it then surface temperature of metal jacketed insulation must be below 140° F (60° C) and mastic coated below 150° F (67° C). Precalculated tables are given in Appendix B-54, B-55, and B-56.

*In Volume 2, October 1978, of "Journal of Thermal Insulation," Dr. Carlos Hilado, Editor, published by Technomic Publishing Co., Inc. An article has been presented regarding recent work as to "The Toxicity of Pyrolyis Gases from Thermal Insulation Materials."

■ EXAMPLE 45

English Units

Installation conditions

1. Operating temperature—300° F
2. Vessel 10′ 0″ dia, 30 ft long, horizontal
3. Located outdoors
4. Located where it receives solar radiation
5. Ambient temp maximum 105° F, average 63° F

Installation requirements

1. Shall provide very low heat transmission
2. Shall have exceptionally low vapor permeability
3. Shall be fire resistant
4. Shall not react with liquefied gas to cause ignition

Insulation function

1. To conserve costly liquefied gas by reducing heat input that would cause vaporization

Insulation selection

1. Multi-layered reflective, or opacified powder installed between double shells in a vacuum. Cryogenic insulation system installed by manufacturer

Insulation properties

1. Thermal conductance 0. __ __ to 0. __ __, depending upon system
2. Vapor transmission—hermetically sealed in vacuum space with moisture migration of less than __ __ __ in __ __ __ years
3. Inert, will not burn
4. Noncombustible

Thermal calculations

In accordance with determination of insulation thickness required to prevent condensation as presented in Chapter 1.

■

■ EXAMPLE 45a

Metric Units

Installation conditions

1. Operating temperature—184.4° C (88.8° K)
2. Vessel 3.048 m dia, 9.144 metres long, horizontal
3. Located outdoors
4. Located where it receives solar radiation
5. Ambient temperature maximum 40.6° C, average 17.2° C

Installation requirements

1. Shall provide very low heat transmission
2. Shall have exceptionally low vapor permeability
3. Shall be fire resistant
4. Shall not react with liquified gas to cause ignition

Insulation function

1. To conserve costly liquified gas by reducing heat input that would cause vaporization

Insulation selection

1. Multilayered reflective or opacified powder installed between double shells, in a vacuum. Cryogenic insulation system installed by manufactuer

Installation properties

1. Thermal conductance 0.0 __ __ to 0.0 __ __ W/m^2, depending upon system
2. Vapor transmission—hermetically sealed in vacuum space with moisture migration of less than __ __ __ in __ __ __ years
3. Inert, will not burn
4. Noncombustible

Thermal calculations

1. In accordance with determination of insulation thickness required to prevent condensate as presented in Chapter 1

■

■ EXAMPLE 46

English Units

Installation conditions

1. Operating temperature—150° F
2. Located outdoors
3. Ambient temperature max 110° F, average 60° F
4. Design temperature 90° F
5. Design relative humidity 80% at 90° F
6. Pipe 12″ NPS horizontal

Installation requirements

1. Shall provide minimum effective life of 12 years
2. Shall provide strong mechanically resistant installation
3. Shall be fire resistant
4. Shall be non-absorbent with organics
5. Shall not contribute fuel to an accidental fire
6. Shall have minimum fire spread
7. Shall be prefabricated before installation for fast installation
8. Shall not react with chemicals with which it comes into contact

Insulation function

1. To prevent condensation on its surface at design ambient temperature and relative humidity

Insulation selection

1. Cellular glass

Insulation properties

1. Low vapor transmission to retard vapor pickup for an extremely long period of time. Vapor trans = 0.00000 + perm inches
2. High compressive strength—ultimate is 100 psi
3. Noncombustible—melting point over 1600° F
4. Non-absorbent
5. Flame spread classification less than 15.
6. Has physical properties which permit it to be accurately cut or ground into shape
7. Inert inorganic glass

Vapor-barrier selection

1. As the insulation selected is very vapor-resistant, it need only be sealed to itself to provide a continuous vapor-barrier. The joint sealer should have low vapor-transmission and remain resilient under the temperatures to which it is exposed.

Weather-barrier selection

1. Resilient, fire-resistant weather-barrier mastic, reinforced with open weave resilient fiber cloth. Mastic should be of color

which has high surface emittance, otherwise excessive insulation thickness is required to prevent condensation.

Thickness calculations

The calculations necessary to determine the thickness of insulation to prevent condensation are presented in Chapter 1.

Dry bulb temp 90° F
Relative humidity 80%
Dew point 83° F (from Table 2)
Design surface temp 84° F
Dry bulb temp °F (to absolute zero)
radiation = 159.2 Btu/sq ft, hr (Table 3)

Surface temp °F (to absolute zero)
radiation = 152.4 Btu/sq ft, hr (Table 3)
Net radiation = 6.8 Btu/sq ft, hr

Absorptance (emittance) of surface

$$\text{Total radiation} = \frac{0.9}{6.12}\ \text{Btu/sq ft, hr}$$

Dry bulb temp − surface temp = 90° F − 84° F = 6° F
Heat gain by convection = 1.67 Btu/sq ft, hr (Table 5)
Convection factor 0
Total convection heat gain 1.67 Btu/sq ft, hr
Total radiation and convection heat gain is 6.1 + 1.7 = 7.8 = q_a
Operating temperature −150° F
$\Delta t_2 = t_2 - t_1 = 84 - (-150) = 234°$ F

$$R = \frac{\Delta t_2}{q_a} = \frac{234}{7.8} = 30$$

Insulation conductivity at

$$\frac{84 + (-150)}{2} = \frac{-66}{2} = -33°\text{ F mean temp is } 0.34$$

Flat thickness (equivalent) L, L = Rk = 30 × .34 = 10.2 inches
Nominal Pipe Insulation Thickness of 12″ pipe
6″ thickness = 8.33″ equivalent thickness L
6½″ thickness = 9.17″ equivalent thickness L
7″ thickness = 10.03″ equivalent thickness L
Thus, answer is 7″ thickness required.

■

■ EXAMPLE 46a

Metric Units

Installation Conditions

1. Operating temperature = −101.1° C (172.1° K)
2. Located outdoors
3. Ambient temperature max 43.3° C (316.5° K)
4. Design temperature 32.2° C (305.4° K)
5. Design relative humidity 80% at 32.2° C (305.4° K)
6. Pipe 304.3 mm (12″ OD Tube)

Installation requirements

1. Shall provide minimum effective life of 12 years
2. Shall provide strong mechanically resistant insulation
3. Shall be fire resistant
4. Shall be non-absorbant with organics
5. Shall not contribute fuel to accidental fire
6. Shall have minimum fire spread
7. Shall be prefabricated for fast installation
8. Shall not react with chemical with which it may come into contact

Insulation function

1. To prevent condensation on its surface at design ambient temperature and relative humidity.

Insulation selection

1. Cellular glass

Insulation properties

1. Low vapor transmission to retard vapor pick up for an extremely long period of time. Vapor transmission—0.00000 + perm—cm's
2. High compressive strength—ultimate 70310 kgs/m^2
3. Noncombustible—melting point over 871° C (1144° K)
4. Nonabsorbent
5. Flame spread classification less than 15
6. Has physical properties which permit it to be accurately cut or ground to shape
7. Is composed of cellulated inorganic glass

Vapor-retardant selection

1. As the cellular glass insulation is more vapor retardant than any mastic or sheet material it need only to be sealed to itself to provide a continuous vapor-retarder. The joint sealer should have low vapor transmission and remain resilient under the temperature to which it is exposed.

Weather-barrier selection

1. Resilient, fire-resistant weather-barrier mastic reinforced with open weave resilient fiber cloth. Mastic should be of color which has high surface emittance, otherwise excessive insulation is required to prevent surface condensation.

Thickness calculations

In this case due to the fact that pipe and pipe insulation standard dimensions have not yet been established in Metric Units it is easier to determine the thickness of insulation, to prevent condensation, using English Units. This is shown in Chapter 1 and illustrated by Example 34.

Dry bulb temperature 32.2° C (90° F)
Relative humidity 80%
Dew point 83° F (from Table 2) = 28.3° C
Design surface temperature 29° C (84° F)
Dry bulb temperature (to absolute zero, Table 3) using 90° F
radiation = 159.2 Btu/ft^2, hr × 3.15248 = 501.87 W/m^2

Surface temperature (to absolute zero, Table 3) using 84° F
radiation = 152.4 Btu/ft^2, hr × 3.15248 = 480.43 W/m^2
Net radiation = 21.44 W/m^2

Absorption (emittance) of surface—0.9
Total radiation = 0.9 × 21.44 = 19.3 W/m^2

Dry bulb temp − surface temp 32.2° C − 28.3° C = 3.9° C = 6° F

Heat gain by convection using 6° F—Tables = 1.67 Btu/ft^2, hr
1.67 × 3.15248 = 5.26 W/m^2
Total radiation and convection gain (q_{ma}) is
5.26 + 19.3 = 24.56 W/m q_a = or 7.8 Btu/ft^2, hr
Operating temperature = −101.1 ° C (−150 ° F)
$\Delta t_{m2} - t_{m1} - t_{m1}$ = 29 − (−101.1 ° C) = 130.1 ° C (234 ° F)

$$R \text{ (English Units)} = \frac{\Delta t}{q_a} = \frac{234}{7.8} = 30$$

Insulation conductivity at mean temp

$$\frac{29 + (-101)}{2} = -36° \text{ C } (-33° \text{ F})$$

has conductivity of 0.049 W/mK or 0.34 Btu in./ft^2, hr °F

Flat thickness (equivalent) L is L = Rk *English Units*
L = 30 × 0.34
= 10.2 inches (254 mm)

Nominal Pipe Insulation Thickness of 12″ NPS pipe
6″ thickness = 8.33 equivalent thickness L (in.)
6½″ thickness = 9.17 equivalent thickness L (in.) Table B-6a
7″ thickness = 10.03 equivalent thickness L (in.)

The answer is 7″ thickness is required = 177.8 mm insulation thickness.

■

■ EXAMPLE 47

English Units

Installation conditions

1. Installed inside duct
2. Subjected to 600 fps, min air velocity
3. Insulation inside of 24″ by 36″ rectangular galvanized steel duct
4. Position horizontal
5. Air temperature in duct is 40° F
6. Ambient air temperature surrounding duct is 75° F
7. Relative humidity of ambient air is 60%
8. Surface of duct painted light green—emittance 0.5

Installation requirements—major

1. Shall have flame spread index of less than 25
2. Shall have smoke density index of less than 50
3. Shall have fuel contribution index of less that 50
4. Shall have air friction coefficient of resistance of less than 0.025.

Insulation function

1. To retard heat gain to air inside of duct
2. To prevent condensation of moisture on outside of duct

Insulation selection

1. Glass fiber insulation board

Properties of the insulation

1. Can be cut, fitted, and installed in ducts
2. Will resist air velocities well above that given without abrading
3. Has flame spread index of less than 25
4. Has smoke density index of less than 50
5. Has fuel contribution index of less than 50
6. Has air friction coefficient of resistance of approximately 0.021
7. Conductivity of insulation at 50° F mean temp approximated 0.24 Btu/sq ft, hr, in. °F
8. Also serves as sound absorbent insulation

Vapor-barrier selection

1. None needed, as metal duct acts as vapor-barrier and covering over insulation

Thickness calculations

Although most ducts insulated to prevent condensation use one constant thickness of insulation for the bottom, sides, and top, the minimum thickness required for each is different. The following calculations will determine the minimum for each, even though the practical application is a single thickness. The calculations to prevent condensation are presented in Chapter 1.

Ambient air temperature is 75 dry bulb
Relative humidity is 60%
Dew point is 60° F
Design surface temperature 61° F
Dry bulb temp °F (to absolute zero)
radiation = 142.6 Btu/sq ft, hr (Table 3)

Surface temp °F (to absolute zero)
radiation = 128.2 Btu/sq ft, hr (Table 3)
Net radiation = 14.4

Absorption (emittance factor of surface) ε = 0.5

Thus, total radiation = 14.4 × 0.5 = 7.2 Btu/ft^2, hr
Dry bulb temp—surface temp = 75 − 61 = 14° F
Heat gain by convection (from Table 5)
Bottom surface 8.57 Btu/sq ft, hr plus radiation
7.2 = 15.77 Btu/sq ft, hr = q_a
Vertical surface 6.68 Btu/sq ft, hr plus radiation
7.2 = 13.88 Btu/sq ft, hr = q_a
Top surface 4.26 Btu/sq ft, hr plus radiation
7.2 = 11.46 Btu/sq ft, hr = q_a
Operating temp 40° F
t_2 = 61° F − 40° F = 21° F

$$\text{Resistance is } R = \frac{\Delta t_2}{q_a}$$

$$\text{For bottom } R = \frac{21}{15.77} = 1.34$$

$$\text{For sides } R = \frac{21}{13.88} = 1.52$$

$$\text{For top } R = \frac{21}{11.46} = 1.83$$

Thickness required
For bottom = 1.34 × .24 = 32″
sides = 1.52 × .24 = 36″
top = 1.83 × .24 = 44″
Thus, this application would use 3/4″ thick insulation.

■ EXAMPLE 47a

Metric Units

Installation conditions

1. Installed inside duct
2. Subjected to 182.68 metres per minute air velocity
3. Insulation inside 609.5 mm × 914.4 mm rectangular galvanized steel duct
4. Position horizontal
5. Air temperature in duct is 4.4° C (277.6° K)
6. Ambient air temperature surrounding duct is 23.9° C (297.1° K)
7. Relative humidity of ambient air is 60%
8. Surface of duct painted light green—emittance ε is 0.5

Installation requirements

1. Shall not cause fire hazard
2. Shall have very little air friction to flow of air in duct
3. Shall provide sound absorption and provide a sound reduction barrier for fan and duct noise
4. Shall have air friction coefficient of resistance less than 0.025

Insulation function

1. To retard heat gain to air inside of duct
2. To prevent condensation of moisture on outside of duct

Insulation selection

1. Glass fiber insulation board

Properties of the insulation

1. Can be cut, fitted and installed in ducts
2. Will resist air velocities well above that given without abrading
3. Has flame spread classification by ASTM E-84 of less than 25
4. Has fuel contribution by ASTM E-84 of not over 50
5. Has smoke contribution by ASTM E-84 of not over 50
6. Has air friction coefficient of resistance of approximately 0.021
7. Conductivity of insulation at 10° C (283.2° K) of approximately 0.0346 W/mK
8. Also serves as sound absorbent insulation

Vapor-retarder selection

1. None needed, as metal duct acts as vapor-retarder and covering over insulating

Thickness calculations

Although most ducts insulated to prevent condensation use one constant thickness of insulation for the bottom, sides, and top, the minimum thickness required for each is different. The following calculations will determine the minimum for each, even though the practical application is a single thickness. The equations for calculations to prevent condensation are presented in Chapter 1.

Ambient air temperature is 23.9° C (297.1° K) dry bulb
Relative humidity is 60%
Dew point is 15.6° C (288.8° K) (Table 2a)
Design surface temperature 16.1° C (289.3° K)
Dry bulb temperature (to absolute zero) radiation =
At 23.9° C (297.1° K) = 488 W/m^2 (Table 3a)
Surface temperature (to absolute zero)
At 16.1° C (289.3° K) = 442 W/m^2 (Table 3a)
Net radiation = 488 − 442 = 46 W/m^2
Absorptance (emittance factor of surface) = 0.5

Thus, total radiation = 46 × 0.5 = 23 W/m^2
Dry bulb temperature—surface temperature = 23.9 − 16.1 = 7.8 °C or °K
Heat gain by convection (Table 5a)
Bottom surface 27 W/m^2 plus radiation 23 = 50 W/m^2 = q_{ma}
Vertical surface 21 W/m^2 plus radiation of 23 W/m^2 = 44 W/m^2 = q_{ma}
Top surface 13 W/m^2 plus radiation of 23 W/m^2 = 36 W/m^2 = q_{ma}
Operating temperature is 4.4° C (277.6° K) Δt_m = (16.1 − 4.4) = 11.7° C or °K

$$\text{Resistance is } R = \frac{\Delta t_{m2}}{q_{ma}}$$

$$\text{For bottom } R = \frac{11.7}{50} = 0.234$$

$$\text{For side } R = \frac{11.7}{44} = 0.266$$

$$\text{For top } R = \frac{11.7}{36} = 0.325$$

Thickness required
For bottom = 0.234 × 0.0346 = 0.0081 metres (8.1 mm)
sides = 0.266 × 0.0346 = 0.0092 metres (9.2 mm)
top = 0.325 × 0.0346 = 0.0112 metres (11.2 mm)

Check:
8.1 mm × 25.4 = 0.318″
9.2 mm × 25.4 = 0.362″
11.2 mm × 25.4 = 0.44″

Thus this application would use 19.05 mm or 3/4″ thick insulation.

■ EXAMPLE 48

English Units

Insulation conditions

1. 3″ NPS pipe
2. Operating temperature is atmospheric
3. Pipe electric traced to maintain contents of pipe above 40° F
4. Ambient air temperature—low, minimum temperature is −20°F, wind 20 mph

Installation requirements

1. Insulation must remain thermally efficient at atmospheric conditions
2. Insulation must not become saturated if exposed to water, due to cracks in weather-barrier
3. Insulation must have sufficient strength to withstand mechanical abuses to which it may be subjected in exposed industrial installation
4. Insulation shall be prefabricated before installation
5. Although installed on a line containing noncombustible liquid, it must not spread fire

Insulation function

1. To maintain temperature of line above 40° F when atmospheric temperature is −20° F

Insulation selected

1. Cellular glass

Insulation properties

1. Has low hygroscopicity
2. Is non-absorbent to water
3. Has compressive strength of approximately 50 lbs per sq in. (which is low for abuse resistance in industrial applications)
4. Can be cut and cemented together to form prefabricated shapes, or can be molded to desired shapes
5. Is noncombustible
6. Its conductivity at 10° F mean is 0.35 Btu in./ft^2, hr °F

Weather-barrier selection

1. Treated steel with closure bands

Weather-barrier properties

1. The strength of the steel weather-barrier will protect the soft insulation, and will spread any point loading or impact. The steel jacketing may dent, but it resists puncture.

Thermal calculations

In this instance, the function of the installation is to prevent freezing of line contents during winter. Conservation of heat is not of prime importance, but its loss must be kept below that which the electrical tracer can supply in order to keep the contents from freezing.

Because of the tracer system, a 4″ NPS pipe insulation is used around the 3″ NPS pipe and tracer. This leaves some voids and to obtain sufficient strength, 1½″ thickness of insulation is chosen.

$$q = \frac{t_1 - t_a}{\dfrac{r_2\left(\log_e \dfrac{r_2}{r_1}\right)}{k} + \dfrac{1}{f}} = \frac{40 - (-20)}{\dfrac{r_2\left(\log_e \dfrac{r_2}{r_1}\right)}{0.35} + \dfrac{1}{f}}$$

$$= \frac{60}{\dfrac{2.01}{0.35} + \dfrac{1}{f}} = \frac{60}{5.73 + \dfrac{1}{f}}$$

From Table B-6 for 4″ NPS with 1½″ nominal thick

$$r_2\left(\log_e \frac{r_2}{r_1}\right) = 2.01$$

From Table B-13 at $\varepsilon = 0.4$ and operating temp of 40° F value of $\frac{1}{f}$ by extrapolation = 0.38. Entered in the equation:

$$q = \frac{60}{5.73 + 0.38} = \frac{60}{6.11}$$

$$= 9.81 \text{ Btu/ft}^2\text{, hr of outer surface}$$

Area of 4″ 1½″ thick insulation which is 7.62″ outside diameter is $(7.62 \div 12) \times \pi = 1.99$ ft^2/lin, ft.

Thus loss per linear foot of pipe = 1.99 × 9.81 = 19.54 Btu/lin, ft, hr.

However the tracer system must have some additional capacity for heating of pipe, insulation, and itself so as to obtain 40° F. The loss as calculated only indicates minimum loss and the true loss can not be determined without complete information on the electric tracer system. ■

■ EXAMPLE 48a

Metric Units

Installation conditions

1. 88.9 mm outside diameter pipe (3″ NPS)
2. Operating temperature is atmospheric
3. Pipe electrically traced to maintain contents of pipe above 4.4°C (277.6° K)
4. Ambient air conditions, temperature −28.9° C (244.3° K), wind at 32.19 km/hr

Installation requirements

1. Installation must remain efficient at atmospheric conditions
2. Insulation must not become saturated if exposed to water, due to cracks in weather-barrier
3. Insulation must have sufficient strength to withstand mechanical abuses to which it may be subjected in exposed industrial installation
4. Insulation shall be prefabricated before installation
5. Although installed on a line containing noncombustible liquid it must not spread fire

Insulation function

1. To maintain temperature of line above 4.4° C (277.6° K) when atmospheric temperature is −28.9° C (244.3° K)

Insulation properties

1. Has low hygroscopicity
2. Is non absorbant to water
3. Has compressive strength of not less than 517 k Pa
4. Can be cut and cemented together to form prefabricated shapes, or can be molded to desired shape
5. Is noncombustible
6. Its conductivity at −12.2° C (261° K) mean temperature is not over 0.05047 W/mK

Weather-barrier selection

1. Treated steel with closure bands

Weather-barrier properties

1. The strength of the steel weather-barrier will protect the insulation and will spread any point loading or impact. The steel jacketing may dent, but it resists puncture.

Thermal calculations

In this instance, the function of the installation is to prevent freezing of line during winter. Conservation of heat is not of prime importance, but its loss must be kept below that which the electric tracer can supply in order to keep contents from freezing.

The heat loss q_m can be determined from Equation 17a, Chapter 1. Because of the tracer system a 104.8 mm (4″ NPS) pipe size insulation is used around the 88.9 mm (3″ NPS) pipe and tracer. This leaves some voids and to obtain sufficient strength 38 mm (1½″) nominal thickness of insulation is chosen.

Outside diameter of insulation = 193.7 mm
r_2 = 193.7 ÷ 2 = 96.85 mm (.09685 metres) r_1 = 114.3 ÷ 2 = 57.15 = .05714 metres

$$q_m = \frac{t_{m1} - t_{ma}}{\dfrac{r_{2m} \log_e\left(\dfrac{r_{2m}}{r_{1m}}\right)}{k_m}}$$

$$= \frac{4.4 - (-28.9)}{\dfrac{0.09685\left(\log_e \dfrac{0.09685}{.05715}\right)}{0.05047} + \dfrac{1}{f_m}}$$

$$= \frac{-33.3}{\dfrac{0.09685 \times 0.527}{0.05047} + \dfrac{1}{f_m}}$$

From Table B-13 for ε = 0.4 and operating temperature of 4.4° C (40° F) value of $\frac{1}{f}$ = 0.38 (in English Units) this is converted to metric thermal resistance by multiplying by conversion factor 0.1761 = 0.067 = $\frac{1}{f_m}$

$$q_m = \frac{-33.3}{\dfrac{0.05101}{0.05047} + (0.176 \times 0.38)}$$

$$= \frac{-33.3}{1.011 + 0.067} = \frac{-33.3}{1.078} = 30.89 \text{ W/m}^2$$

Area of the 104.8 mm (inside diameter) by 193.5 mm (outside diameter), in metres/lin metre is .608 m²/lin m.
Thus loss per lin metre of pipe = 30.89 × 0.608 = 18.78 W/lin m.

Check:

19.54 Btu/lin, ft, hr × 0.9609 = 18.77 W/lin m. ■

■ EXAMPLE 49

English Units

Installation conditions

1. 3″ NPS pipe
2. Operating temperature is atmospheric
3. Pipe is electric traced to maintain contents of pipe above 40° F
4. Ambient air temperature—low, minimum temperature −20° F, wind 12 mph

Installation requirements

1. Must remain thermally efficient at atmospheric conditions
2. Shall not become saturated with water due to cracks in weather-barrier
3. Shall have sufficient strength to withstand mechanical abuses to which it will be subjected in exposed industrial application
4. Shall be prefabricated or preformed before installation
5. Pipe contents is highly combustible, and insulation must provide fire protection for a minimum of 1/2 hr, during which pipe temperature shall not exceed 300° F

Note: These requirements, with the exception of 5, are identical to those of Example 48.

Insulation function

1. To maintain temperature of contents of line above 40° F when atmospheric temperature is −20° F
2. To provide fire protection of pipe and contents

Insulation selected

1. Water repellent expanded silia 4″ NPS by 1½″ thick

Insulation properties

1. Has low hygroscopicity
2. Is moisture repellent
3. Has minimum compressive strength of 75 psi
4. Can be cut and cemented together to form prefabricated shapes, or can be molded into desired shapes
5. Is noncombustible and has low diffusivity. Will retard temperature rise within limits as stated, when tested in accordance with standard fire curve.
6. The conductivity at 10° F mean temperature is 0.31 Btu/sq ft, hr, in., °F and at 1000° F mean temperature is 0.75 Btu/sq ft, hr, in., °F

Weather-barrier

1. Treated steel, or stainless steel, with closure bands

Weather-barrier properties

1. Strong and resists mechanical abuse
2. Noncombustible
3. Will stay in place under direct flame impingement

Thermal calculations

The minimum Btu/sq ft, hr, heat loss, which is incorrect as it ignores the higher temperature of the tracer, is

$$q = \frac{t_1 - t_a}{r_2\left(\log_e \dfrac{r_2}{r_1}\right) + \dfrac{1}{f}}$$

$\frac{1}{f}$ = 0.4 (12 mph) (ε of stainless steel = 0.4) (from interpolation Table 57)

$$q = \frac{60}{\dfrac{1.91}{.31} + 0.4} = \frac{60}{6.16 + 0.4} = \frac{60}{6.54}$$

$$= 9.15 \text{ Btu/sq ft, hr}$$

and heat loss per lin ft of pipe per hour = 9.15 × 2.0 = 18.3 Btu/sq ft, hr. ■

■ EXAMPLE 49a

Metric Units

Installation conditions

1. 88.9 mm (3″ NPS) pipe
2. Operating temperature is atmospheric
3. Pipe is electric traced to maintain contents of pipe above 4.4 °C (277.6° K)
4. Ambient air temperature, low minimum temperature is −28.9 °C (244.3° K) with wind of 19.31 km per hour

Installation requirements

1. Must remain thermally efficient at atmospheric conditions
2. Shall not become saturated with water due to cracks in weather-barrier
3. Shall have sufficient strength to withstand mechanical abuses to which it will be subjected in exposed industrial application
4. Shall be prefabricated or preformed before installation
5. Pipe contents is highly combustible and insulation must provide fire protection for a minimum of 1/2 hour during which pipe temperature shall not exceed 149° C (422° K)

Insulation function

1. To maintain temperature of contents of line above 4.4° C (277.6° K) when atmospheric temperature is −28.9° C (244.3° K) with 19.31 kgs/hr, wind
2. To provide fire protection of pipe and contents

Insulation selected

1. Water repellent expanded silica 114.3 mm (4″ NPS), 38 mm (1½″) thick pipe insulation

Insulation properties

1. Has a low hygroscopicity
2. Is moisture repellent
3. Has a minimum compressive strength of 344.7 k Pa
4. Can be cut and cemented together to form prefabricated shapes, or can be molded into desired shapes
5. Is noncombustible and has low thermal diffusivity. Wil retard temperature rise within the limits stated when tested in accordance with standard fire curve
6. The conductivity at −12.2° C (261° K) is 0.041790 W/mK and at 538° C (811° K) is 0.10815 W/mK

Weather-barrier

1. Treated steel or stainless steel with closure bands

Weather-barrier properties

1. Strong and resists mechanical abuses
2. Noncombustible
3. Will stay in place under direct flame impingement

Thermal calculations

The calculated loss is loss based on pipe temperature just being 4.4° C (277.6° K). Tracer temperature must be higher than 4.4° C to prevent freezing of product.

$$q_m = \frac{t_{1m} - t_{am}}{\dfrac{r_{2m}\left(\log_e \dfrac{r_{2m}}{r_{1m}}\right)}{k_m} + \dfrac{1}{f_m}}$$

$$= \frac{4.4 - (-28.9)}{\dfrac{0.09685\left(\log_e \dfrac{0.09685}{0.05715}\right)}{0.044702} + \dfrac{1}{f_m}}$$

From Table (using English Units) $\dfrac{1}{f} = 0.4$

$$0.1762 \times 0.4 = \frac{1}{f_m} = 0.07048$$

$$q_m = \frac{33.3}{\dfrac{0.09685 \times 0.527}{0.044702} + 0.07048}$$

$$= \frac{33.3}{\dfrac{0.05101}{0.044702} + 0.07048}$$

$$= \frac{33.3}{1.1412 + 0.17048}$$

$$= \frac{33.3}{1.2117} = 27.48 \text{ W/mK}$$

and heat loss per lin metre = 27.48 × 0.608 = 16.70 W/lin m

Check:

17.35 Btu/lin, ft, hr × 0.9609 = 16.67 W/lin m ■

■ EXAMPLE 50

English Units

Installation conditions

1. Storage vessel 10′-0″ in diameter and 20′-0″ high, with two dished heads
2. Storage vessel located in nonhazardous area. Contents of vessel is noncombustible
3. Operating temperature 150° F
4. Ambient—low, minimum temperature −10° F, 12 mph wind
5. Interior of vessel lined to prevent corrosion by contents, outer surface of vessel corrosion protected by coating suitable as base, with good adhesion for sprayed on insulation

Installation requirements

1. Shall be of such formulation that it can be sprayed on outer surface of vessel
2. Shall have flexibility to expand and contract with vessel without cracking or breaking
3. Shall be moisture repellent, as little heat is available to keep it dry
4. Shall be slow burning

Insulation functions

1. To conserve heat
2. To reduce size of heating coil necessary to keep material at 150° F

Insulation selection

1. Foamed-in-place urethane—slow burning type, minimum thickness 1″

Properties of material

1. Can be sprayed as liquid onto vessel where it expands into rigid foam insulation
2. Has fair degree of flexibility
3. Has low hygroscopicity and high water resistance
4. Has flame spread index less than 50
5. Has conductivity of less than 0.23 at 75° F mean temperature and Δt of 50° F

Weather-barrier selection

1. Sprayable low flame spread mastic weather-barrier white color $\varepsilon = 0.4$

Weather-barrier properties

1. Can be sprayed on the irregular surfaces of the foamed-in-place urethane
2. Has sufficient elongation that it can withstand expansion without cracking. Its minimum elongation at 20° F is approximately 15%
3. Will not drip or run in fire
4. Has flame spread index of less than 35

Thermal calculations

Areas of nozzles and manholes left uncovered by insulation is 65 sq ft

Balance of area of vessel is insulated

Diameter of insulated vessel = 10′-0″ + 1″ + 1″ = 10′-2″

Area of vessel

cylindrical section 10′-2″ × π × 20.0 = 639 sq ft

area of two dished 10′-2″ heads = 174 sq ft

total insulated area = 639 + 174 = 813 sq ft

Heat loss insulated area

$$q = \frac{t_1 - t_a}{\frac{1}{0.23} + \frac{1}{f}} = \frac{150 - (-10)}{4.34 + \frac{1}{f}}$$

$$\frac{1}{f} = 0.38$$ (at 12 mph, operating temp 150° F from Table 61)

$$\frac{160}{4.34 + 0.38} = \frac{160}{4.72} = 33.89 \text{ Btu/ft}^2\text{, hr}$$

Btu loss insulated area

$$A \times q = 813 \times 33.89 = 27.552 \text{ Btu/hr}$$

Heat loss uninsulated area

By radiation

Radiation level surface temperature at 150° F (to absolute zero) = 244 Btu/sq ft, hr

Radiation level ambient air at −10° F (to absolute zero) = 72 Btu/sq ft, hr

Gross radiation = 172 Btu/sq ft, hr

Emittance of black painted steel = .9

Net radiation = 155 Btu/sq ft, hr

Convection at 150° F to −10° F at 12 mph = 620 Btu/sq ft, hr

Total bare loss = 775 Btu/sq ft, hr

Area is 65 sq ft Total loss 50,400 Btu/hr

Total loss then is 50,400 + 20,325 = 70,725 Btu loss per hour.

Note: The 65 sq ft of bare surface loss as compared to the 813 sq ft of insulated surface loss. ■

■ EXAMPLE 50a

Metric Units

Installation conditions

1. Storage vessel 3.048 metres in diameter and 6.096 metres high with two dished heads
2. Storage vessel located in non hazardous area. Contents of vessel is noncombustible.
3. Operating temperature is 65.6° C (338.8° K)
4. Ambient—low, minimum temperature is −23.3° C (249.9° K) with 22.2 km per hour wind
5. Interior of vessel lined to prevent corrosion by contents. Outer surface of vessel must be corrosion protected prior to application of insulation. This coating must serve as suitable base, with good adhesion, for sprayed-on insulation.

Installation requirements

1. Shall be of such formulation that i t can be sprayed to vessel wall and top without additional securement.
2. Shall have flexibility to expand and contract with vesssel without cracking or breaking
3. Shall be moisture repellent, as little heat is available to keep it dry.

Insulation functions

1. To conserve heat
2. To reduce size of heating coil necessary to keep material at 65.6° C (338.8° K).

Insulation selection

1. Sprayed on polyurethane insulation, low flame-spread type, minimum installed and set thickness is one inch

Properties of material

1. Can be sprayed as liquid onto vessel where it expands into rigid foam insulation
2. Has fair degree of flexibility
3. Has low hygroscopicity and high water resistance
4. Has flame spread classification of less than 50

5. Has a thermal conductivity, when tested at 23.9° C (297.1° K), and 63.3° C or °K temperature differential of not more than 0.0332 W/mK

Weather-barrier selection

1. Sprayable, low flame spread classification mastic weather-barrier white color $\varepsilon = 0.4$

Weather-barrier properties

1. Can be sprayed on irregular surfaces of the foamed-in-place polyurethane insulation
2. Has sufficient elongation so that it can withstand expansion without cracking. Its minimum elongation at −6.7° C (266.5° K) is approximately 15%
3. Will not run or drip in fire
4. Has flame spread classification, with ASTM E-84, of less than 35.

Thermal calculations

Areas of nozzles and man holes left uncovered is 6.039 m^2

Balance of vessel is insulated

Diameter of insulated vessel = 3.048 m + 0.0254 m + 0.0254 m = 3.0966 m

Area of vessel

cylindrical section = 3.0966 × π × 6.096 = 59.35 m^2

area of two dished heads = 3.048 m diameter = 16.16 m^2

total insulated area = 59.35 + 16.16 = 75.51 m^2

Heat loss from insulated area (Equation 11a) when ℓm is in metres

$$q_m = \frac{t_{1m} - t_{am}}{\frac{\ell_m}{k_m} + \frac{1}{f_m}} = \frac{65.6 - (-23.3)}{\frac{.0254}{0.0332} + \frac{1}{f_m}}$$

$$= \frac{88.9}{.765 + \frac{1}{f_m}}$$

From Table B-13 the surface resistance $\frac{1}{f}$, at 65.6° C (150° F) with air velocity of 22.2 km per hour (12 mph), is 0.38 (English Units). To convert to Metric Units multiply 0.38 by 0.1761 thus

$$\frac{1}{f_m} = 0.38 \times 0.1761 = 0.067$$

$$q_m = \frac{88.9}{0.765 + 0.067} = \frac{88.9}{0.832} = 106.86 \text{ W/m}^2$$

Check:

33.89 Btu/ft^2, hr × 3.152591 = 106.85 W/m^2

106.85 × 75.51 = 8068.24 W = heat loss of insulated area

Check:

8068.24 × 3.412141 = 27530 Btu/hr

Heat loss uninsulated area

Radiation level, surface temperature at 65.6° C (338.7° K)
(to absolute zero) = 769.2 W/m^2

Radiation level, ambient air at −23.3° C (249.9° K)
(to absolute zero) = 227.0 W/m^2

Gross radiation from uninsulated surface
= 769.2 − 227.0
= 542.2 W/m^2 hr

Emittance of black painted steel $\varepsilon = 0.9$
542.2 × 0.9 = 488 W/m^2

Convection at 65.6° C (150° F) to − 23.3° C (−10° F) at 22.2 km/hr (12 mph) is 1954 W/m^2

Total bare loss per hour per m^2 = 1954 + 488 = 2442

Area of uninsulated surface is 6.039 m^2 = 2442 × 6.039 = 14747 W

Total loss then is 14747 + 8069 = 22816 W

Check:

22816 × 3.412141 = 77851 Btu/hr ■

■ EXAMPLE 51

English Units

Installation conditions

1. Installation is in air-conditioned building
2. Hot water tank, steam heated
3. Size of tank 4′-0″ dia by 10′-0″ long
4. Temperature of tanks 150° F
5. Ambient air 75° F

Installation requirements

1. Insulation shall be slow burning

Function of insulation

1. To conserve heat in the tank
2. To reduce heat load in air-conditioned space

Insulation selection

1. Foamed styrene—slow burning type

Properties of insulation

1. Self-extinguishing when flame is removed
2. Low conductivity, clean, easy to fabricate

Selection of indoor covering

1. Low flame spread classification (not over 35) installed with extensible reinforcing cloth. Color to harmonize with surroundings.

Thermal calculations

$$\text{Mean temperature} = \frac{150 + 75}{2} = \frac{225}{2} = 112$$

Conductivity of insulation at this mean temp is 0.31 Btu/sq ft, hr, in. °F

Depreciation period of building is 20 years

Cost of 1″ × 1″ insulation is $3.20/lin ft

Capital cost of steam per lb is $40.00

Cost of steam per 1000 lbs is $2.00

From "How To Determine Economic Thickness of Insulation," the economic thickness of insulation necessary to conserve the steam is 1/2". However, in this particular set of conditions, any heat leaked to the ambient air by the hot water tanks tends to overload the air-conditioning system. Then, the answer becomes one of practicality. The labor cost to apply either 1/2" or 1" thick insulation is practically the same, and so is the cost of the indoor covering. Thus, the cost of 1/2" more thickness of insulation is insignificant, and the extra 1/2" would effectively reduce the heat loss from the vessel to the conditioned air.

Heat loss through 1/2" insulation per sq ft

$$Q = \frac{150 - 75}{\frac{.5}{.31} + \frac{1}{f}}$$

$$= \frac{75}{1.6 + .65} = \frac{75}{2.25} = 33.3 \text{ Btu/sq ft, hr}$$

$\frac{1}{f} = 0.65$ in still air conditions (Table B-13)

Heat loss through 1" insulation per sq ft

$$Q = \frac{75}{\frac{1}{.31} + \frac{1}{f}}$$

$$= \frac{75}{3.2 + .65} = \frac{75}{3.85} = 19.4 \text{ Btu/sq ft, hr}$$

Thus, for the cost of only 1/2" extra thickness of insulation, the heat loss to the air-conditioned space was reduced 13.9 Btu per sq ft, hr. Consequently, the answer should be 1.0" thickness. ■

■ EXAMPLE 51a

Metric Units

Installation conditions

1. Installation is in air-conditioned building
2. Hot water tank, steam heated
3. Size of tank 1.22 metres diameter × 3.05 metres long
4. Temperature of tank is 65.6° C (338.8° K)
5. Ambient air temperature is 23.9° C (297° K)

Installation requirements

1. Insulation shall be slow burning organic foam

Function of insulation

1. To conserve heat in tank
2. To reduce heat load in air-conditioned space

Insulation selection

1. Foamed styrene—low flame-spread type

Properties of insulation

1. Insulation shall have a flame spread classification of less than 50 when tested in accordance with ASTM E-84
2. Low conductivity, clean, easy to fabricate

Selection of indoor covering

1. Low flame-spread mastic, classification (not over 35) installed with extensible cloth. Color to harmonize with surroundings

Thermal calculations

$$\text{Mean temperature} = \frac{65.6 + 23.9}{2} = \frac{89.5}{2}$$

$$= 44.7° \text{ C } (317.7° \text{ K})$$

Conductivity of insulation at this mean temperature is 0.0447 W/mK

Depreciation period of building is 20 years

Cost of 33.4 mm pipe (1" NPS) insulation 25.4 mm thick installed per metre of pipe is $5.25.

Capital cost investment to produce steam is $54.00 per 1000 watt-hours

Cost of steam production is $5.44 per million watts

From "Economic Design of Thermal Insulation" the economic thickness of insulation necessary to conserve steam would be 12.7 mm. However in this particular set of conditions, any heat loss to ambient air by the hot water tanks tends to overload the air-conditioning system. Then, the answer becomes one of practicality. The labor cost to apply 12.7 mm or 25.4 mm thick insulation is practically the same, and so is the cost of indoor covering. Thus, the cost of 12.7 mm more thickness of insulation is insignificant, and the extra 12.7 mm would effectively reduce the heat loss from the vessel to the conditioned air.

Heat loss through 12.7 mm insulation per square metre is:

$$Q_m = \frac{t_{m1} - t_{ma}}{\frac{\ell_m}{k_m} + \frac{1}{f_m}} = \frac{65.6 - 23.9}{\frac{.0127}{.0447} + \frac{1}{f_m}}$$

$\frac{1}{f_m}$ is determined by multiplying the English Unit $\frac{1}{f}$ from Table B-13 0.65 × 0.1761 = 0.1145

$$Q_m = \frac{41.7}{0.2841 + 0.1145} = \frac{41.7}{.3986} = 104.6 \text{ W/m}^2$$

Check:

$$33.3 \text{ Btu/ft}^2\text{, hr} \times 3.152 = 104.9 \text{ W/m}^2$$

Heat loss through 25.4 mm thickness of insulation per square metre

$$Q_m = \frac{41.7}{\frac{0.0254}{0.0447} + 0.1145} = \frac{41.7}{0.5682 + 0.1145}$$

$$= \frac{41.7}{0.6827} = 61.1 \text{ W/m}^2$$

Thus for the cost of only 12.7 mm extra thickness of insulation the heat loss to the air-conditioned space was reduced 43.5 W/m². Consequently the answer should be 25.4 mm thickness of insulation. ■

■ **EXAMPLE 52**

English Units

Installation conditions

1. Service, cycle water, temperature 50° F to 120° F in 1/2″, 3/4″, 1″ OD copper tubes
2. Indoors, in pipe chases, and outside of air-conditioning cabinets, hidden from view
3. Ambient air temperature 75° F, relative humidity 70%

Installation requirements

1. Insulation shall be flexible, to allow for bending of tubing without breaking of the insulation
2. As the service is above and below ambient, the insulation shall be vapor resistant, and water repellent
3. The insulation shall be elastic so that, as tubing is installed and the insulation is compressed out of shape or dented, it will recover its shape
4. Shall be slow burning

Insulation function

1. To prevent condensation of moisture on its surface
2. To control temperature of water in supply tubing

Insulation selection

1. Flexible cellular foam insulation

Insulation properties

1. Is flexible, elastic, and can be bent
2. Is formed into cylinders to fit OD tubing
3. Is vapor-resistant—vapor transmission is 0.1 perm inches
4. Self-extinguishing
5. Water resistant
6. Thermal conductivity at 75° F mean temperature = 0.26 Btu/sq ft, hr, in. °F

Vapor-barrier selection

1. Vinyl chloride coating (emittance = 0.7)

Thermal calculations

Dry bulb temperature 75° F
Relative humidity 70%
Dew point temperature 64° F
Design surface temperature 65° F
Dry bulb temperature (to absolute zero) radiation 142.6 Btu/sq ft, hr (Table 3)

Surface temperature (to absolute zero) radiation

	132.2	Btu/sq ft, hr (Table 3)
Gross radiation	10.4	Btu/sq ft, hr
Surface emittance is 0.7	.7	Btu/sq ft, hr
Net radiation	7.3	

Ambient air temperature − Surface temperature = 75° F − 65° F = 10° F
Convection gain 3.18 Btu/sq ft, hr (Table 5)
Convection factor 1.64
Total convection gain 3.18 × 1.64 = 5.2 Btu/sq ft, hr
Radiation and convection total gain 5.2 + 7.3 = 12.5 Btu/sq ft, hr = q_a
Operating temperature 50° F

Surface temperature − operating temperature = 65° F − 50° F = 15° F

$$\text{Thermal resistance} = R = \frac{t_2 - t_1}{q_a} = \frac{15}{12.5} = 1.2$$

Equivalent thickness = R × k = 1.2 + .26 = 0.31″
Thus, 1/2″ thick insulation will prevent condensation on insulation surface. ■

■ **EXAMPLE 52a**

Metric Units

Installation conditions

1. Service, cyclce water, temperature 10° C (323.2° K) to 48.9° C (322.1° K) in 12.7 mm, 19.05 mm and 25.4 mm OD
2. Indoors in pipe chases and inside air-conditioning cabinets, hidden from view
3. Ambient air temperature 23.9° C (297.1° K), relative humidity is 70%

Installation requirements

1. Insulation shall be flexible to allow for bending of tubing without breaking of the insulation
2. As the service is above and below ambient air temperature, the insulation shall be vapor resistant and water resistant
3. The insulation shall be elastic so that, as tubing is installed and the insulation is compressed out of shape, or dented, it will recover its shape
4. Insulation shall have flame-spread classification of less than 50

Insulation function

1. To prevent condensation of moisture on its surface
2. To control temperature of water in supply tubing

Insulation selection

1. Flexible cellular foam insulation

Insulation properties

1. is flexible, elastic and can be bent without breaking
2. Is formed into cylinders to fit OD tubing
3. Is vapor resistant—its vapor transmission shall be less than 0.1 perm-inches
4. Shall be slow burning
5. Shall be water resistant
6. Thermal conductivity at 23.9° C (297.1° K) mean temperature shall not be greater than 0.0375 W/mK

Vapor-retarder selection

1. Shall be vinyl coated of color having emittance ε not greater than 0.7

Thermal calculations

1. Dry bulb temperature of ambient air is 23.9° C (297.1° K)
2. Relative humidity 70%
3. Dew point temperature is 17.8° C (291° K)
4. Design surface temperature 18.3° C (291.5° K)

From Table 3a

Dry bulb temperature (to absolute zero) radiation is 449.6 W/m^2
Surface temperature (to absolute zero) radiation is 416.6 W/m^2
Gross radiation is 449.6 − 416.8 = 32.8 W/m^2
As surface emittance is $\varepsilon = 0.7$, net radiation is 32.8 × 0.7 = 23 W/m^2
Ambient air temperature − surface temperature is 23.9 − 18.3 = 5.6 °C

From Table 5a
Convection gain is 17.65 W/m^2
Convection factor is 1.64
17.65 × 1.64 = 28.95

Radiation and convection total gain is 23.0 + 28.95 = 51.95 W/m^2 $= q_{am}$

At minimum operating temperature of 10° C temperature difference of tube to surface is 18.3 − 10 = 8.3° C.

The thermal resistance required (min.) is

$$R_m = \frac{18.3 - 10}{q_{am}} = \frac{8.3}{51.95}$$

$$R_m = 0.1597$$

The equivalent thickness of insulation must be greater than $R_m \times k_m$ or 0.1597 × 0.0375 = 0.005988.

From Table B-6a Tube in English Units the equivalent thickness of 1″ (25.4 mm) with 1/2″ (12.7 mm) thickness of insulation is .69 or 0.015, thus 12.7 mm (1/2″) of insulation will prevent condensation on its surface under the stated conditions.

■

■ EXAMPLE 53

English Units

Installation conditions

1. Storage vessel 60′ dia × 40′ high cone head, flat bottom on foundation
2. Temperature of stored liquid 350° F
3. Ambient temperature—minimum 0° F, average 60° F, with wind-design for heat transfer −10 mph
4. Ground temperature—average 50° F
5. Vessel not heated when empty
6. Set on foundation, bottom of vessel 6″ above grade
7. Vessel is protected by water spray system

Installation requirements

1. Insulation shall be of such shape that it can be rapidly installed on large storage vessel
2. Exposed weather-barrier over insulation shall be noncombustible or have flame spread classification of less than 25
3. Exposed insulation shall not support combustion
4. Insulation on head shall resist a wind of 70 mph minimum
5. Insulation bottom shall be water resistant
6. Insulation on side wall of vessel shall have sufficient flexibility to expand and contract with changes of dimensions of vessel

Insulation function

1. To conserve heat
2. To reduce the size of coils necessary to heat vessel
3. Economics

Insulation selection

1. Under bottom—cellular glass, with cushion of dry sand between insulation and tank bottom. The cellular glass shall have joints vapor-water sealed with hot steep asphalt
2. Wall—bottom 18″ shall be insulated with cellular glass with inner 1/2″ of spun glass fiber cushion blanket. Above 18″ walls shall be insulated with mineral fiber insulation factor attached to deep corrugated aluminum panels
3. Top—expanded silica

Properties of insulation

1. Bottom—cellular glass
   a. Is water and vapor-resistant to resist ground water and vapors which may be generated by contact of any water leakage with hot vessel
   b. Has good compressive strength, ultimate of 100 psi. (The maximum load, composed of the liquid and bottom steel, which must be supported by this material, is approximately 50 psi.)
   c. Conductivity at 200° F mean temperature shall not be greater than 0.5 Btu, in./ft^2, hr, °F
2. Side—mineral fiber insulation, factory attached to aluminum panels
   a. Panels can be so installed so as to provide expansion and contraction joints at sides and top of each panel
   b. Fast erection
   c. Will not support combustion
   d. Conductivity at 200° F mean temperature shall not be greater than 0.33 Btu, in./ft^2, hr, °F
3. Top—expanded silica block
   a. Insulation block does not allow free passage of air as does a fibrous material
      Note: On ordinary high wind, pulsating air pressures will pull top insulation off at a much lower wind velocity than it will the side walls.
   b. Has higher density to resist air lift
      Note: Block shall be securely fastened to steel top by mechanical fasteners
   c. Insulation is water repellent
      Note: On a top of this size the expansion of the steel is almost certain to cause some cracks in the weather-barrier, or wind-blown rain will penetrate under flashings. Water repellent insulation is necessary for continued high thermal efficiency.
   d. Conductivity of insulation at 200° F mean temperature shall not be greater than 0.37 Btu, in./ft^2, hr, ° F

Weather-barrier selection

1. Bottom—water sealing
   Plastic sheet under cellular glass block on soil—1/2″ dry sand cushion over sheet. Edges of blocks buttered with steep asphalt and 1/2 in dry sand cushion to spread load evenly on insulation.

2. Side walls
   Factory applied insulation to deep corrugated aluminum panels
3. Top
   Polyvinyl acetate mastic, reinforced with modacrylic fiber-fabric pressed into first coat, then overcoated with second coat—total thickness dry, not less than 1/16″

Properties of weather-barriers

1. Wall
   a. Flexible joint construction allows for expansion and contraction. However, if bands are used, the bands must be of the expander type, or have springs at given intervals.
   b. Will melt at 1250° F, but will not burn
2. Top
   a. May be bonded to surface of block to prevent being lifted by the wind
   b. Has flexibility (should have not less than 20% elongation at 20° F)
   c. Fire-resistant—has flame spread classification less than 35

Thermal calculations

Economics

1. For simplicity, the cost of all the systems is averaged at $3.00 per sq ft
2. Cost of capital investment in steam is taken as $20.00
3. Cost of steam production—$1.00 per M lbs
4. Depreciation period—15 years
5. Temperature difference—average 350° F − 50° F = 300° F

From "How To Determine Economic Thickness of Insulation," published by NIMA, these factors determine that the economic thickness is 2½″ if the vessel is used almost all the time during a year.

Heat loss—based on minimum atmospheric temperature

Bottom

$$q_b = \frac{350 - 50}{\frac{2.5}{.5}} = \frac{300}{5} = 60 \text{ Btu/sq ft, hr}$$

Area of bottom

$30^2 \times \pi = 900 \times \pi = 2830$

Heat loss through bottom = 2830 × 60 = 169.800 Btu/hr

Sides

$$q_s = \frac{350 - 0}{\frac{2.5}{.33} + \frac{1}{f}}$$

$$\frac{1}{f} = .43 \text{ at 10 mph.}$$

Emittance of surface = 0.9

$$q_s = \frac{350 - 0}{7.5 + 0.43} = \frac{350}{7.93} = 44 \text{ Btu/sq ft, hr}$$

Area of sides

$60 \times \pi \times 40 = 7550$ sq ft

Heat loss sides = 7550 × 44 = 332,200 Btu/sq ft, hr

Top

$$q_T = \frac{350 - 0}{\frac{2.5}{.37} + \frac{1}{f}} = \frac{350}{6.75 + .35} = \frac{350}{7.10}$$

$$= 49.3 \text{ Btu/sq ft, hr}$$

Area of cone head = 3100 sq ft

Heat loss of head

3100 × 49.3 = 153,100 Btu/hr

Total loss of insulated area

Bottom	169,800
Side walls	332,200
Top head	153,100
	654,100 Btu/hr

To this should be added the loss of any bare manholes, flanges or projections in a similar manner to that shown in a previous example. ■

■ EXAMPLE 53a

Metric Units

Installation conditions

1. Storage vessel 18.28 metres diameter × 12.19 metres high, cone head, flat bottom on foundation
2. Temperature of stored liquid is 177° C (450° K).
3. Ambient air temperature, the minimum is −17.8° C (255.4° K), the average temperature is 15.6° C (288.8° K) with 16.1 km/hr wind, this is basis for thermal and economic design
4. Ground temperature is average of 10.0° C (283.2° K)
5. Vessel not heated when empty
6. Set on foundation, bottom of vessel is 0.1524 metres above grade

Installation requirements

1. Insulation shall be of such shape that it can be rapidly installed on large storage vessel
2. Exposed weather-barrier shall be noncombustible or low flame-spread classification
3. Insulation shall be noncombustible
4. Installation of insulation shall be such as to resist a 112 km/hr wind
5. Bottom insulation shall be water and vapor resistant
6. Installation of insulation on side wall of vessel shall have sufficient flexibility to withstand changes of dimensions of vessel

Insulation function

1. To conserve heat
2. To reduce the size of coils necessary to heat vessel
3. To maintain the temperature of the contents of the vessel

Insulation selection

1. Under bottom, cellular glass insulation shall be installed over flat ground surface with joiints vapor-water sealed and the top surface coated with hot-steep asphalt, over which a cushion of dry sand is placed between insulation and tank bottom

2. Wall-bottom .45 m shall be insulated with cellular glass with inner 12.7 mm of spun glass wool blanket. Above the 0.45 m level the wall shall be insulated with mineral fiber insulation factory attached to deep corrugated aluminum panels
3. Top, insulation to be water repellent expanded silica

Properties of insulation

1. Bottom—cellular glass insulation
   a. Insulation must be water and vapor-resistant to resist ground water and vapors which may be generated by contact of any water with hot vessel. The insulation hygroscopicity is 0 and its vapor permeability is 0.00000+ perm cm
   b. Has good compressive strength, ultimate of 70300 kgs/m^2. (The maximum load, composed of the liquid and vessel which must be suppported by the insulation is approximately 35,150 kgs/m^2)
   c. The conductivity of the insulation at 93.3° C (366.5° K) mean temperature shall not be greater than 0.072 W/mK
2. Side—mineral fiber, factory attached to aluminum panels
   a. Panels can be installed so as to provide expansion and contraction joints at sides and top of each panel
   b. Fast erection
   c. Will not support combustion
   d. Conductivity at 93.3° C (366.5° K) mean temperature shall be no greater than 0.0476 W/mK
3. Top—Insulation water repellent expanded silica block
   a. Insulation block does not allow free passage of air as does a fibrous material
      Note: An ordinary high wind may cause pulsating air pressures which will pull top insulation off at much lower wind velocity than it will the side walls.
   b. Has higher density to resist air lift
      Note: Block shall be securely fastened to steel top by mechanical fasteners.
   c. Insulation is water repellent
      Note: On a top of this size, the expansion of steel is almost certain to cause some cracks in the weather-barrier, or wind blown rain will penetrate under flashings. Water repellent insulation is necessary for continued high thermal efficiency.
   d. Conductivity of insulation at 93.3° C (366.5° K) mean temperature shall be no greater than 0.0534 W/mK

Weather-barrier selection

1. Bottom—water sealing
   Plastic sheet under cellular glass block on soil—13 mm dry sand cushion over sheet. Edges of blocks buttered with heat resistant sealer, surfaced with steep asphalt, at 13 mm dry sand cushion to spread load evenly on insulation,
2. Side walls
   Factory applied insulation to deep corrugated heavy gage aluminum panels
3. Top
   Polyvinyl acetate mastic to be installed with extensible fabric pressed into first coat, then overcoated with second coat—the total thickness—dry—not less than 1.6 mm

Properties of weather-barriers

1. Wall
   a. Flexible joint construction allows for expansion and contraction. However, if bands are used, the bands must be of the expander type or have springs at given intervals.
   b. Will melt at 677° C (950° K), but will not burn
2. Top
   a. May be bonded to surface of block to prevent being lifted by wind
   b. Has flexibility (should have not less than 20% elongation) at −6.7° C (266.5° K)
   c. Fire resistant—has flame spread classification less than 35

Thermal calculations

Economics

1. For simplicity of determination of economic thickness, the cost is averaged at $32.30 per square metre
2. The cost of capital investment of steam is $27.00 per 1000 watt hours
3. Cost of steam produced is $1.35 per million watts
4. Depreciation period is 15 years
5. Temperature difference average 177° C − 10° C = 167° C or °K

From "Economics of Thermal Insulation Design," these factors determine that the economic thickness is 63.5 mm if the vessel is used almost all the time during the year.

Heat loss—based on minimum atmospheric temperature

Bottom

$$q_{bm} = \frac{t_{1m} - t_{gm}}{\dfrac{\ell_{mcg}}{k_{mcg}}}$$

When: t_{1m} = temperature of tank, t_{gm} = ground temp ℓ_{mcg} = thick of insulation in metres and k_{mcg} = conductivity of insulation in W/mK

$$q_{bm} = \frac{177 - 10}{\dfrac{0.0635}{0.072}} = \frac{167}{0.8819} = 189.35 \text{ W/m}^2$$

Area of bottom

$r_{m2} \times \pi = (9.14)^2 \times \pi = 262.45 \text{ m}^2$

Heat loss from bottom of tank to ground

= 262.45 × 189.35 = 49694 W

Sides

$$q_{3m} = \frac{177 - (-17.8)}{\dfrac{0.0635}{0.0476} + \dfrac{1}{f_m}}$$

when air temperature = −17.8° C from Table B-13 the value of $\frac{1}{f}$ = .43 English Units. To convert to $\frac{1}{f_m}$ multiply by 0.1761.

$$= \frac{194.8}{1.410} = 138.16 \text{ W/m}^2$$

$$\frac{1}{f_m} = 0.076$$

Area of sides = diameter × π × height
= 18.28 × π × 12.19 = 700 m^2
Heat loss of sides = 700 × 138.16 = 96719 W

Top

$$q_{tm} = \frac{177 - (-17.8)}{\frac{0.0635}{0.0534} + \frac{1}{f_m}}$$

$\frac{1}{f}$ in English Units—heat flow up is 0.35. Multiplied by 0.1761

$$= .0616 = \frac{1}{f_m}$$

$$q_{tm} = \frac{194.8}{1.1891 + 0.0616}$$

$$= \frac{194.8}{1.2507} = 155.75 \text{ W/m}^2$$

Area of cone head = 288 m^2
Heat loss of head = 288 × 155.75 = 44854 W
Total loss of insulated area

Bottom	49,694
Side walls	96,719
Top head	44,854
	191,267 W

Check:

191,267 × 3.412 = 652,600 Btu/hr

To this should be added the loss of any bare manholes, flalnges or projections as calculated in a similar manner as shown in previous example. ■

■ EXAMPLE 54

English Units

Installation conditions

1. Pipe is 3″ NPS stainless steel
2. Operating temperature is 325° F (intermittent)
3. Located outdoors
4. Ambient air temperature, minimum −5° F, average 55° F, maximum 105° F
5. Heat gain to pipe should not exceed 90 Btu/lin. ft hr

Installation requirements

1. Insulation is installed on line containing very toxic materials, thus, in case of a leak the insulation must be non-absorbent and non-adsorbent. (Disposal of contaminated insulation would be very difficult.)
2. Insulation shall not support combustion
3. In case of fire, the insulation shall protect the line

Insulation selection

1. Cellular glass

Insulation properties

1. Non-absorbent and non-adsorbent
2. Vapor-resistant
3. Inorganic glass with sealed cells which resists contamination
4. Will not contribute to stress corrosion of stainless steel
5. Noncombustible
6. Conductivity at 160° F mean temperature is 0.45 Btu/sq ft, hr, in, °F

Selection of weather-barrier

1. Stainless steel jacket

Weather-barrier properties

1. Non-absorbent
2. Fire-resistant
3. Flame spread index of 0

Thermal calculations

Heat loss for 1½″ of insulation—disregarding film factor—at minimum ambient temperature

$$q = \frac{t_1 - t_a}{r_2\left(\log_e \frac{r_2}{r_1}\right)} = \frac{325 - (-5)}{\frac{2.08}{0.45}} = \frac{330}{4.62}$$

$$= 71.5 \text{ Btu/sq ft, hr}$$

Sq ft/lin ft of 3″ × 1½″ insulation = 1.73 sq ft/lin ft
1.73 × 71.5 = 123.4 Btu/lin ft, hr
Heat loss for 2½″ of insulation—disregarding film factor—at minimum ambient temperature

$$q = \frac{330}{\frac{3.85}{0.45}} = \frac{330}{} = 38.6 \text{ Btu/sq ft, hr}$$

Sq ft/lin ft of 3″ × 2½″ insulation = 2.26
2.26 × 38.6 = 87.4 Btu/lin ft
Thus, 2½″ of insulation thickness is required. ■

■ EXAMPLE 54a

Metric Units

Installation conditions

1. Pipe is 88.9 mm (3″ NPS) stainless steel
2. Operating temperature is 163° C (436° K) intermittent operation
3. Located outdoors
4. Ambient air temperature—minimum −20.6° C (252.6° K), average 12.8° C (286° K), maximum 40.6° C (313.8° K)
5. Heat gain to pipe shall not exceed 86.5 Watts/lin metre

Installation requirements

1. Insulation is installed on line containing very toxic material, thus in case of leak the insulation must be non-absorbent and non-adsorbent. (Disposal of contaminated insulation would be very difficult.)
2. Insulation shall be non-combustible
3. In case of fire, the insulation shall protect the line

Insulation selection

1. Cellular glass

Insulation properties

1. Non-absorbent and non-adsorbent
2. Vapor resistance – permeability = 0.00000+ perm inches
3. Inorganic glass sealed cells which resist contamination
4. Will not contribute to stress corrosion of stainless steel
5. Conductivity at 71° C (344° K) mean temperature to be not greater than 0.0649 W/mK
6. Noncombustible

Selection of weather-barrier

1. Stainless steel jacket

Weather-barrier properties

1. Non-absorbent
2. Fire-resistant
3. Noncombustible

Thermal calculations

Heat loss for 38.1 mm thick insulation—disregarding film factor—at minimum temperature −20.6° C (252.6° K).

$$q_m = \frac{t_{1m} - t_{am}}{\dfrac{r_{2m}\left(\log_e \dfrac{r_{2m}}{r_{1m}}\right)}{k_m}}$$

$$q_m = \frac{163 - (-20.6)}{\dfrac{0.083\left(\log_e \dfrac{0.083}{0.0449}\right)}{0.0649}}$$

$$= \frac{183.6}{\dfrac{0.083 \times 0.614}{0.0649}} = \frac{183.6}{\dfrac{.05099}{0.0649}}$$

$$= \frac{183.6}{.78575}$$

$$q_m = 233.66 \text{ W/m}^2$$

$r_1 = 44.9$ mm = 0.0449 metres
$r_2 = 83.0$ mm = 0.083 metres

Square metres per lin metre of 88.9 mm diameter pipe × 38.1 mm thick

Insulation is 165.1 mm outside diameter (0.165 metres)

$0.165 \times \pi = .518$ m/lin m, $.518 \times 233.66 = 121.19$ W/lin m

This is too great a heat gain per lin metre—exceeds the 86.5 limit

Check:

$$121.19 \times 1.04068 = 126 \text{ Btu/lin, ft}$$

Heat loss for 63.5 mm thick insulation—disregarding film factor—at minimum temperature −20.6° C r_{2m} of insulation = 108 mm (.0108 metres).

$$q_m = \frac{183.6}{\dfrac{0.108\left(\log_e \dfrac{0.108}{0.0449}\right)}{0.0649}} = \frac{183.6}{\dfrac{0.108 \times 0.878}{0.0649}}$$

$$= \frac{183.6}{\dfrac{0.0958}{0.0649}} = 125.7 \text{ W/m}^2$$

Square metres per lin metre of 88.9 mm diameter pipe × 63.5 mm thick insulation is 88.9 + 127 = 215.9 mm (0.2159 m)

$0.2159 \times \pi = 0.678$ m/lin metre, $0.678 \times 125.7 = 85$ W/lin metre

This is within the maximum limit of 86.5

Thus 63.5 mm thickness of insulation is required. ■

■ EXAMPLE 55

English and Metric Units

Installation conditions

1. Outdoors—horizontal line, 2″ NPS (60.3 mm) stainless steel pipe
2. Minimum ambient temperature −20° F (−23.9° C)
3. Process line to be steam traced with 200 lb (1378 k Pu) steam in 5/8″ OD (15.9 mm) tubing
4. Oversized installation supported only at top and bottom
5. Combustible liquid in line
6. Corrosive atmosphere
7. Process line temperature to be not less than 200° F (93.3° C) under static conditions. No heat gain or loss from process liquid.

Installation requirements

1. Requires fire protection for minimum of 30 minutes
2. Insulation system must be fire-resistant
3. Must not cause stress corrosion of stainless steel

Insulation selected

1. Water repellent expanded silica

Insulation properties

1. Has good compressive strength
2. Will resist accidental fire for the period of time required
3. Will not support combustion
4. Has low thermal diffusivity
5. Will inhibit stress corrosion

Weather-barrier selection

1. Stainless steel jacketing with closure bands

Weather-barrier properties

1. Resistant to corrosive atmospheres
2. Fire-resistant

3. Strong, and has high tensile strength

Thermal calculations

According to graph in Chapter 4, a thermally bonded tracer of this size will be able to maintain this pipe above 200° F (93.3° C) with 1½" (38.1 mm) thickness of insulation.

■

■ EXAMPLE 56

English Units

Installation conditions

1. Pipe 2", 300' long, 22 90° welded elbows, 16 flanges
2. Contents of pipe entering at 500° F temp
3. Temperature of contents at leaving pipe not to be less than 490° F, at minimum ambient conditions
4. Contents flow at 300 lbs per minute
5. Specific heat of contents is 0.6
6. Contents of pipe are combustible
7. Pipe outdoors—ambient temperature 0° F minimum, 20 mph wind
8. Ambient temperature to be used to design safe surface temperature is 90° F
9. Conductivity of insulation at 250° F mean temperature is .42

Installation requirements

1. Industrial installation where insulation is subjected to considerable mechanical abuse
2. In hazardous area—insulation should serve to protect pipe from fire
3. Pipe located where it may be touched by personnel, so surface temperature should be below 150° F for safety
4. Line has many elbows and fittings, so the insulation for these must be prefabricated to ensure good economic application

Insulation function

1. Control temperature in pipe
2. Provide safe surface temperature

Insulation selection

1. Calcium Silicate

Insulation properties

1. High compressive strength
2. Good impact resistance
3. Will withstand fire for long period of time
4. Noncombustible
5. Can be cut and fabricated

Weather-barrier selection

1. Heavy build poly-vinyl acetate emulsion mastic which shall have first coat applied and while still wet, an extensible modacrylic fiber reinforcing cloth embedded in it. After dry, a second coat of mastic is applied.

Weather-barrier properties

1. Flame spread less than 35
2. Will not melt or drip when exposed to fire
3. Gray (or other medium dark color) with emittance of 8
4. Will elongate and remain flexible at low temperatures

Thermal calculations

Allowable heat loss from entry to exit of line = wt of material per hour flowing into line, times specific heat, times allowable temperature drop (300 × 60) × 0.6 × 10 = 18,000 × 6 = 108,000 Btu/hr.

1 in thick insulation

$$q = \frac{t_1 - t_a}{\dfrac{r_2\left(\log_e \dfrac{r_2}{r_1}\right)}{k} + \dfrac{1}{f}} = \frac{500 - 0}{\dfrac{1.41}{.42} + \dfrac{1}{f}}$$

$\frac{1}{f}$ at 0 amb 12 mph is approx .27

$\varepsilon = .8$ (from Table B-13)

$$q = \frac{500}{3.35 + .27} = \frac{500}{3.62} = 138 \text{ Btu/sq ft, hr}$$

Sq ft per lin ft of 2 × 1" insulation = 1.18
Thus, total Btu loss for 300 ft of pipe
138 × 1.18 × 300 = 48,852 Btu/hr

Loss of elbows
sq ft 2" 90° welded—1" insulation = .76 sq ft each 22 × .76 × 138 = 2300 Btu/hr
Btu/loss per hour from flanges
sq ft per pair of 2" flanges—1" insulation = 1.9 sq ft pr 16 × 1.9 × 138 = 4200 Btu/hr
Total loss

pipe Btu/hr =	48,852
elbows Btu/hr	2,300
flanges Btu/hr	4,200
	55,350 Btu/hr

which is less than allowable.

Thus, 1" insulation will maintain temperature.

However, the other requirement is that the surface temperature must not exceed 150° F at 90° F ambient.

Calculate maximum allowable Btu/hr heat transfer based on 150° F surface temperature

Radiation
150° F surface temperature—ambient air (90° F)
From Table B-8 = 65 at .8 emittance (192 − 127 = 65)

Convection
From Table 5
Horizontal pipe 30.7 and convection factor for 4½" insulation diameter is 1.4 or 30.7 × 1.4 = 43 Btu/hr
Radiation and convection = 65 + 43 = 108 Btu/sq ft, hr

But 1" insulation gives a heat transfer of 138 Btu/sq ft, hr which is too great to satisfy surface temperature requirements.

Try 1½″ insulation

$$q = \frac{t_1 - t_a}{\dfrac{r_2\left(\log_e \dfrac{r_2}{r_1}\right)}{k}} = \frac{500 - 0}{\dfrac{2.33}{.42} + .27}$$

$$= \frac{500}{5.55 + .27} = \frac{500}{5.82} = 86 \text{ Btu/sq ft, hr}$$

which is below the allowable limit of 108 Btu/sq ft, hr.

Thus, 1½″ thickness of calcium silicate is required. ■

■ EXAMPLE 56a

Metric Units

Installation conditions

1. Pipe 60.3 mm (2″ NPS), 91.44 metres long, 22 90° welded elbows and 16 flanges
2. Contents of pipe entering at 260° C (533° K) temperature
3. Temperature of contents at leaving pipe not to be less than 254° C (527° K) at minimum ambient air temperature
4. Contents flow at 136.1° Kg's per minute
5. Specific heat of contents is 0.698
6. Contents of pipe are combustible
7. Pipe located outdoors—minimum ambient air temperature is −17.8° C (255.4° K) with 32.2 kms/hr wind
8. Ambient temperature to be used to design safe surface temperature of jacket is 32.2° C (305.4° K)
9. Conductivity of insulation at 121° C (394° K) is 0.0606 W/mK

Installation requirements

1. Industrial installation where insulation is subjected to considerable mechanical abuse
2. In hazardous area—insulation should serve to protect pipe in fire
3. Pipe is located where it may be touched by personnel, so surface temperature should be lower than 65.6° C (338.8° K) for safety
4. Line has many elbows and fittings, so insulation for these must be prefabricated to ensure good economic application

Insulation function

1. Control temperature in pipe
2. Provide safe surface temperature

Insulation selection

1. Calcium silicate

Insulation properties

1. High compressive strength
2. Good impact resistance
3. Will withstand fire for long period of time
4. Noncombustible
5. Can be cut and prefabricated

Weather-barrier selection

1. Heavy build polyvinyl-acetate emulsion mastic which shall have first coat applied and while still wet, an extensible modacrylic fiber reinforcing cloth shall be embedded in it. After drying a second coat of mastic shall be applied.

Weather-barrier properties

1. Flame-spread classification shall be less than 35
2. Will not melt or drip when exposed to fire
3. Gray (or other medium dark color) with emittance $\varepsilon = 0.8$
4. Will elongate and remain flexible at low temperature

Thermal calculations

Allowable heat loss calculated from entry to exit of line = weight of material per hour flowing into line, times specific heat, times allowable temperature drop.

136.1 × 60 × 0.698 × 5.55 = 31,634 W

25.4 mm thick insulation

$$q_m = \frac{t_{m1} - t_{ma}}{\dfrac{r_{2m}\left(\log_e \dfrac{r_{2m}}{r_{1m}}\right)}{k} + \dfrac{1}{f_m}}$$

$$r_{m1} = 30.15 \text{ mm}, \ r_{m2} = 57.15$$

$\frac{1}{f}$ English Units from Table B-13 = 0.27 × 0.1761 = $\frac{1}{f_m}$ = 0.04755

$$q_m = \frac{260 - (-17.8)}{\dfrac{0.05715\left(\log_e \dfrac{0.05715}{0.03015}\right)}{0.0606} + 0.04755}$$

$$= \frac{278.8}{\dfrac{0.05715 \times 0.6395}{0.0606} + 0.04755}$$

$$q_m = \frac{278.8}{\dfrac{0.03655}{0.0606} + 0.04755}$$

$$= \frac{278.8}{0.6031 + 0.04755} = \frac{278.8}{0.65063}$$

$$= 428.5 \text{ W/m}^2$$

Square metre per linear metre of 60.3 mm pipe with 25.4 mm of insulation is 0.114 metre (diameter) × π = 0.359 m²/lin metre = 153 W/lin m.

91.44 metres in length × 153 = 14066 W

Check:

14066 W × 3.412 = 47996 Btu/hr

Loss from elbows (*Note:* Obtain sq ft per fitting from Table B-28)
= .76 sq ft insulation surface × 0.0929 = 0.0706 sq metres/elbow
= each .0706 × 928.5 × 22 = 665 W

Loss from flanges (1.9 sq ft from Table B-28)
= 1.9 × 0.0929 = 0.1765 sq metres per each flange cover
0.1765 × 428.5 × 16 = 1210 W

Loss from pipe W =	14066
elbows W =	665
flanges W =	1210
	15941 W

Which is less than the maximum calculated allowable loss of 31,634 W. Thus, 25.4 mm of insulation thickness will maintain temperature of contents.

However, the other requirement is that the surface temperature must not exceed 65.6° C (338.8° K) at 32.2° C (305.4° K) ambient air temperature.

Calculate maximum allowable W/m, based on 65.6° C (338.8° K) surface temperature.

Radiation 65.6° C surface temperature − 32.2° C = 33.4° C Δt_m

From Table B-8 Radiation in English Units
At 65.6° C (150° F) = 192 Btu/ft², hr at $\varepsilon = 0.8$
At 32.2° C (90° F) = 127 Btu/ft², hr at $\varepsilon = 0.8$
65 Btu/ft², hr at $\varepsilon = 0.8$

Converting to metric units 65 Btu/ft², hr × 3.15259 = 204 W/m²

Convection
From Table 5a at 33.4 Δt_m heat loss in W/m² = 97 W/m²
Radiation and convection = 97 + 204 = 301 W/m²

But the heat transferred from surface of 25.4 mm thick insulation was previously calculated to be 428.5 W/m which exceeds the allowable maximum. Thus, 1″ has a heat transfer which is too great to satisfy surface temperature requirements.

Try 38.1 mm insulation

when t_{am} = 32.2° C (90° F) d_m when insulation thickness is 38.1 = 141.2 mm r_{m2} = 70.6

$$q_m = \frac{t_{1m} - t_{am}}{\dfrac{r_{m2}\left(\log_e \dfrac{r_{m2}}{r_{m1}}\right)}{k_m} + \dfrac{1}{f_m}}$$

$$q_m = \frac{260 - 32.2}{\dfrac{0.0706\left(\log_e \dfrac{0.0706}{0.03015}\right)}{0.0605} + \dfrac{1}{f_m}}$$

$$= \frac{227.8}{\dfrac{0.0706 \times 0.8508}{0.0606} + 0.04755}$$

$$= \frac{227.8}{\dfrac{0.06}{0.0606} + 0.04755} = \frac{227.8}{0.9912 + 0.04755}$$

$$= \frac{227.8}{1.03875} = 219 \text{ W/m}^2$$

which is below the allowable 301 W/m² radiation and convection less to maintain 65.6° C surface temperature. Thus, 38.1 mm thickness of calcium silicate insulation is required.

■

■ EXAMPLE 57

English and Metric Units

Installation requirements

1. Installed in reactor in "hot area" where insulation is subject to nuclear radiation
2. Indoors
3. External temperature of reactor 550° F (288° C)
4. Ambient temperature 100° F (37.8° C)

Installation requirements

1. Shall not dust off under vibration
2. Shall be able to install and take insulation off with remote control equipment
3. Shall be a custom fit
4. Shall be noncombustible
5. Shall not cause stress corrosion in stainless steel

Insulation function

1. Conserve energy
2. Control temperature in reactor and retard heat loss into confined space

Insulation selection

1. S.S. Reflective insulation system

Properties of insulation

1. 100% stainless steel
2. Factory prefabricated shapes
3. Will not cause stress corrosion of stainless steel
4. Vibration resistant—no fibrous materials to dust
5. Thermal shock resistant
6. Units removable and reusable
7. Low thermal transmittance for thickness and mass

Outer covering selection

1. None required—outer casing supplied as part of manufactured unit

Thermal calculations

1. As units are custom manufactured, and the spacing of reflective shields adjusted to fulfill the thermal requirements, the thickness and heat transfer calculations are a part of the manufacturer's design

■

■ EXAMPLE 58

English Units

Installation conditions

1. Ducts 6′-0″ × 10′-0″
2. Indoors and outdoors

3. Operating temperature +600° F
4. Ambient temperature—maximum 110° F, average 50° F, minimum −0° F

Installation requirements

1. Ducts of such size that expansion is a major problem and shall be allowed for
2. Shall resist vibration
3. Shall have good compressive strength on top so that weight of rain water does not cause sagging in center, with resultant water pools
4. Shall be suitable for pin impalement application

Insulation function

1. Control temperature inside duct
2. Retard excessive heat transmission to surrounding air for personal comfort
3. Provide safe surface temperature

Insulation selection

1. High strength mineral wool block, over which is installed 1/2″ thick insulation cement reinforced with Monel (or stainless steel) hexagonal wire mesh

Insulation properties

1. Insulation block can be impaled on pins welded to duct, or pins can be forced through insulation and welded to duct by welding gun. Block should be individually secured to substrate.
2. Fair resistance to vibration
3. Compressive strength of 20 to 30 psi at 5% deformation
4. Sufficient weight (15 to 22 lbs per cu ft) that wind will not tend to lift it from surface
5. Resistance to free passage of air. Thus, retards "puffing" and subsequent retraction of the outer weather-barrier, which is caused by wind flowing over and around duct.
6. The reinforced insulating cement has flexibility to bridge cracks caused by expansion
7. Conductivity of mineral wool block insulation is 0.42 Btu/sq ft, hr, in °F at 350° F mean temperature

Weather-barrier selection

1. Polyvinyl-acetate heavy build mastic reinforced with modacrylic fiber cloth

Properties of weather-barrier

1. To resist the "puffing" due to air movements over large flat surfaces the weather-barrier should be adhered to the surface of the insulation
2. Good elongation over wide temperature range
3. Fire resistant
4. Low flame spread
5. Emittance of 0.8

Thermal calculations

The minimum thickness of insulation is sufficient to control the temperature in the duct. In most instances, the steel must be kept above 275° F to prevent acid fumes in the flue gas from condensing and corroding the duct.

For *safe* surface temperature for personnel, and one which will be within the range to prevent the rapid deterioration of the weather-barrier, it should be not over 160° F.

To determine minimum thickness

Heat transfer from surface by radiation and convection at 160° F, 0.8 emissivity, to ambient air at 110° F, 0 mph wind

From Tables B-8 and B-10 98.2 Btu/sq ft, hr = q

From Table B-8

the radiation at 160° F to absolute zero when $\varepsilon = 0.8 =$ 205.7

the radiation at 110° F to absolute zero when $\varepsilon = 0.8 =$ 146.9

58.8 Btu/ft², hr

Convection from Table B-10

At 50° F Δt still air is 39.4 Btu/ft², hr.

Total allowable maximum heat transfer through insulation is 39.4 + 58.8 = 48.2 Btu/ft², hr

$$q = \frac{t_1 - t_s}{\frac{\ell}{k}} = 98.2 = \frac{600 - 160}{\frac{\ell}{.42}}$$

$$\frac{\ell}{k} = \frac{440}{98.2} = 4.48$$

$$\ell = 4.48 \times k$$

$$\ell = 4.48 \times .42 = 1.88''$$

The thickness required for this insulation is 1½″ mineral wool block, and 1/2″ insulating cement. Thus, 1½″ thick mineral wool block and 1/2″ insulating cement, reinforced with Monel hexagonal mesh, is the correct thickness of insulation.

■

■ EXAMPLE 58a

Metric Units

Installation conditions

1. Ducts 1.83 metres × 3.05 metres
2. Indoors and outdoors
3. Operating temperature 316° C (589° K)
4. Ambient temperature, maximum 43.3° C (316.5° K), average 10° C (282.2° K), minimum −17.8° C (255.4° K)

Installation requirements

1. Ducts of such size that expansion is a major problem and shall be allowed for
2. Shall resist vibration
3. Shall have good compressive strength on top so that weight of rain water does not cause sagging in center, with resultant water pools
4. Shall be suitable for pin impalement application

Insulation function

1. Control temperature inside duct
2. Retard excessive heat transmission to surrounding air, for personnel comfort
3. Provide safe surface temperature

Insulation selection

1. High strength mineral wool block over which is installed 12.7 mm thick insulation cement reinforced with Monel (or stainless steel) hexagonal wire mesh

Insulation properties

1. Insulation block can be impaled on pins welded to duct, or pins can be forced through insulation and welded to duct by welding gun. Block should be individually secured to substrate.
2. Fair resistance to vibration
3. Compressive strength of 140 to 205 k Pa at 5% deformation
4. Sufficient weight (240 to 350 kgs/m^3) that wind will not tend to lift it from surface
5. Resistance to free passage of air. This retards "puffing" and subsequent retraction of outer weather-barrier, which is caused by wind flowing over and around duct.
6. The reinforced insulation cement has flexibility to bridge cracks caused by expansion
7. Conductivity of the mineral wool block is 0.0606 W/mK at 177° C (450° K) mean temperature

Weather-barrier selection

1. Heavy build polyvinyl-acetate emulsion mastic which shall have first coat applied, then while still wet an extensible mod-acrylic fiber reinforcing cloth shall be embedded in it. After drying a second coat of mastic shall be applied.

Properties of weather-barrier

1. To resist the "puffing" due to air movements over large flat surfaces the weather-barrier should be adhered to the surface of the insulation
2. Good elongation over wide temperature range
3. Flame-spread in accordance with ASTM E-84 of less than 35
4. Emittance $\varepsilon = 0.8$

Thermal calculations

The minimum thickness of insulation is sufficient to control the temperature of the duct and its contents. In most cases the steel must be kept above 135° C (408° K) to prevent acid fumes in flue gas from condensing and corroding the duct.

For *safe* insulation surface temperature for personnel and one which will be within the range to prevent rapid deterioration of weather-barrier, it should not be over 71.1° C (344.3° K).

To determine minimum thickness

Heat transferred by radiation and convection. Surface temperature maximum 71.1° C with emittance $\varepsilon = 0.8$. By convection the heat transfer is to air at 43.3° C.

From Table B-8 the surface radiation in English Units is 205.7 Btu/ft^2, hr

at $\varepsilon = 0.8$ 205.7 × 3.152591 = 648.5 W/m^2

the air radiation in English Units is 146.9 Btu/ft^2, hr

at $\varepsilon = 0.8$ 146.9 × 3.152591 = 463.1 W/m^2

at $\varepsilon = 0.8$ radiation surface to ambient = 648.5 − 463.1 = 185.4 W/m^2

Convection

from Table B-10 at Δt 27.7° C (50.0° F) at 0 kgs/hr wind

In English Units is 39.4 Btu/ft^2, hr

In Metric Units is 39.4 × 3.15291 = 124.2 W/m^2

Maximum total heat loss by radiation and convection is:

$$185.4 + 124.2 = 309.6 \text{ W/m}^2$$

$$q_m = \frac{t_{m1} - t_{m2}}{\frac{\ell_m}{k_m}} = 309.6 = \frac{316 - 71.1}{\frac{\ell_m}{k_m}}$$

$$\frac{\ell_m}{k_m} = \frac{244.9}{309.6} = 0.791$$

$$\ell_m = 0.791 \times k_m$$

$$k_m = 0.0606$$

$$\ell_m = 0.791 \times 0.0606 = 0.0479 \text{ metres or } 47.9 \text{ mm}$$

The thickness required for this installation is 38.1 mm of mineral wool block and 12.7 mm insulation cement. Thus, 38.1 mm thick mineral wool block and 12.7 mm of insulation cement, reinforced with Monel hexagonal mesh will satisfy the requirements. ■

■ EXAMPLE 59

English and Metric Units

Installation conditions

1. Operating temperature, −150° F (−101.1° C)
2. Located outdoors
3. Ambient temperature maximum 110° F (43.3° C), average 60° F (15.6° C)
4. Design temperature 90° F (32.2° C)
5. Design relative humidity 80% at 90° F (32.2° C)
6. Pipe 12 NPS (323.8 mm dia)—horizontal
7. A cold line containing a hazardous material and is located in a hazardous area
8. It may be subjected to mechanical abuses
9. It may be subjected to accidental fire
10. If fire occurs, it will be subject to water streams from fire hose

Installation requirements

1. Shall provide minimum effective life of 12 years
2. Shall provide strong mechanically resistant installation
3. Shall be fire resistant
4. Shall be non-absorbent with organics
5. Shall not contribute fuel to an accidental fire
6. Shall have minimum fire spread
7. Shall be prefabricated before installation for fast installation
8. Shall not react with chemicals with which it comes into contact

Insulation function

1. To prevent condensation on its surface at design ambient temperature and relative humidity
2. To protect line from fire for a period of time until the fire fighters can wet down the line by means of hose streams

Insulation selection

1. Inner layer—cellular glass
2. Outer layer—water repellent silica

Insulation properties

1. Cellular glass
   a. Low vapor transmission which retards vapor pick up for an extremely long period of time
   b. High compressive strength (ultimate is 100 psi or 690 k Pa)
   c. Noncombustible
   d. Nonabsorbent
2. Expanded silica
   a. Has high temperature—1200° F (649° C)
   b. Is water-repellent
   c. Will not swell when saturated with moisture
   d. Noncombustible

Weather-barrier selection

1. Fire-resistant, heavy build mastic, reinforced with Dynel reinforcing cloth or stainless steel jacketing

Weather-barrier properties

1. Fire-resistant

Thermal calculations

The thickness calculations for the cellular glass part should be the same as shown in Example 46. No credit should be taken for the thickness of the expanded silica on the outside of the cellular glass. The moisture vapor pressure inward will cause the outer layer of expanded silica to adsorb a considerable amount of moisture, with corresponding loss in thermal efficiency.

The expanded silica should have a minimum thickness of 1½″ (38.1 mm). This will provide approximately one hour protection to the cellular glass. The moisture which the expanded silica may contain will assist in this fire protection.

■

■ EXAMPLE 60

English and Metric Units

Installation conditions

1. Vessels and pipe operating at ambient temperature to less than 212° F (100° C)
2. Indoors and outdoors
3. Fire may be caused by various combustible fluids or gases

Installation requirements

1. Require fire protection
2. Are subjected to ordinary plant abuses
3. Conservation of heat is not important

Insulation selection

1. Expanded silica

Insulation properties

1. Water repellent, will not become water soaked and mushy if slight leaks occur in weather-barrier
2. Will not swell due to vapor migration in and out, due to change of atmospheric conditions

Weather-barrier selection

1. Fire resistant mastic on heads and irregular surfaces and steel jackets (treated or stainless) for regular and cylindrical surfaces

Properties of weather-barriers

1. Mastic
   a. Will not melt
   b. Fire-resistant
   c. Low flame spread index
2. Steel
   a. Noncombustible
   b. High melting point

Thermal calculations

As various chemicals burn at different temperatures, the amount of time obtained by the fire protective insulation is a function of the temperature of the fire and the thickness of the insulation. Only by tests can the expected time of protection be obtained.

If the time of protection needed is short, then a single layer of insulation may be satisfactory. For long time exposure, double layer, broken-joint construction is recommended, so that shrinkage of individual pieces of insulation will not provide a straight-through open joint.

Note: Where fire protection is necessary on vessels and pipes operating above 212° F (100° C) which provide sufficient heat to induce a vapor pressure outward to keep the insulation dry, more absorbent, high temperature insulation can be used satisfactorily.

■

■ EXAMPLE 61

English and Metric Units

Installation conditions

1. Underground steam line to be insulated
2. Installation to pass through sandy soil with excellent drainage
3. Pipe can be placed well below frost line
4. Distances and elevations are such that steam trapping can be installed in building and not in underground line
5. Average soil temperature 60° F (15.6° C), minimum 40° F (4.4° C)
6. Steam line is in continuous operation
7. Excellent soil conditions, requiring no galvanic protection
8. Temperature of steam 325° F (163° C)
9. Pipe size 4″ NPS (114.3 mm dia.)

Installation requirements

1. Requires continuous operation to supply heat to process
2. Not required that insulation system work below water table, as level of water table is well below level of pipe

Insulation selection

1. Granular bituminous fill insulation

Insulation properties

1. Granular material can be poured in trench around pipe
2. Can be packed and, after curing, will form a protective finish around pipe, which will remain in place if not subjected to excessive changes in dimension and temperature
3. Water repellent

Thermal calculations

Due to the nature of the material, the thickness required is based on its curing, non-consolidated, and loose distance. Thus, thickness is a function of pipe size. In this case of 4″ NPS (114.3 mm dia.), the minimum thickness recommended by the manufacturers is 4 inches.

■

■ EXAMPLE 62

English Units

Installation conditions

1. Underground hot water line
2. Average temperature of soil 50° F, minimum temperature 20° F in places where line is above frost line
3. Temperature of hot water 180° F
4. Length of line 1000 ft
5. Pipe size 2″
6. Flow of water through line 400 lbs per minute
7. Intermittent operation
8. Wet soil
9. Surface water may cause very soaked ground conditions

Installation requirements

1. To hold temperature drop in pipe from inlet to outlet to below 5° F

Insulation function

1. To control heat loss in line so that temperature drop of water is less than 5° F

Insulation selection

1. Urethane foam preinsulated conduit system, with heavy plastic conduit and pipe, including factory preinsulated fittings and compression couplings

Insulation properties

1. Efficient thermal insulation, with conductivity of 0.17 at 100° F mean temperature
2. Water-resistant
3. Conduit is effective in preventing water from entering the system
4. Conduit and insulation effectively protect pipe from rusting

Thermal calculations

Allowable heat loss is weight of water per hour, times specific heat, times allowable temperature drop in °F, or 400 × 60 × 1 × 5 = 120,000 Btu/hr

Heat loss per sq ft outer surface if 1″ of insulation is used (at minimum ground temperature of 20° F)

$$q = \frac{t_1 - t_a}{\dfrac{r_2 \log_e\left(\dfrac{r_2}{r_1}\right)}{k}} = \frac{180 - 20}{\dfrac{1.41}{.17}} = \frac{160}{8.3}$$

= 19.3 Btu/sq ft, hr

Outside diameter of conduit for 1″ insulation on 1″ pipe is 5 in.

$$\text{Sq ft per lin ft} = \frac{5}{12} \times \pi = 1.3$$

Total loss through conduit equals
Btu/sq ft × sq ft/lin ft × length
19.3 × 1.3 × 1000 = 25,100 Btu/hr

which is sufficiently low to fulfill requirements ■

■ EXAMPLE 62a

Metric Units

Installation conditions

1. Underground hot water line
2. Average temperature of soil is 10° C (283.2° K) minimum temperature is −6.7° C (267.1° K) where line is above frost line
3. Temperature of hot water is 82.2° C (355.4° K)
4. Length of line is 304.8 metres
5. Pipe size is 60.3 mm (2″ NPS)
6. Flow of water through line is 181.4 kg per minute
7. Intermittent operation
8. Wet soil
9. Surface water may cause very soaked ground conditions

Installation requirements

1. To hold temperature drop in pipe from inlet to outlet at less than 2.77° C (or °K)

Insulation function

1. To control heat loss in line so that temperture drop of water is less than 2.77° C (or °K)

Insulation selection

1. Urethane foam preinsulated conduit system, with heavy plastic conduit and pipe, including factory preinsulated fittings and compression couplings

Insulation properties

1. Thermal insulation with conductivity of 0.0245 W/mK at 37.8° C (311.0° K)
2. Water resistant
3. Conduit is effective in preventing water from entering the system
4. Conduit and insulation to protect pipe from rusting

Thermal calculations

Allowable heat loss is weight of water per hour, times specific heat, times allowable temperature drop.

Specific heat of water in English Units = 1.0
$1 \times 1.163 = Sh_m = 1.163$
$181.4 \times 1.163 \times 2.777 = 585.9$ W

$$q_m = \frac{t_{m1} - t_{ma}}{\dfrac{r_{2m} \log_e\left(\dfrac{r_{2m}}{r_{1m}}\right)}{k_m}}$$

If 25.4 mm thickness is used and ground temperature is taken to be −6.7° C (267.1° K)

$$r_{m1} = 30.15 \text{ mm}$$
$$r_{m2} = 55.85 \text{ mm}$$

$$q_m = \frac{82.2 - (-6.7)}{\dfrac{0.05585 \log_e \dfrac{0.05585}{0.03015}}{0.0245}}$$

$$= \frac{88.9}{\dfrac{0.05585 \times 0.61648}{0.0245}} = \frac{88.9}{\dfrac{0.03443}{0.0245}}$$

$$= \frac{88.9}{1.4053} = 63.25 \text{ W/m}^2$$

Outside diameter = 111.7 mm or 0.1117 metre × π = 0.35 m/lin m.

Total loss through conduit equals

$W/m^2 \times m^2$/lin m × length (in metres)

63.25 × 0.35 × 304.8 = 6745.5 W

which is sufficiently low to fulfill the requirements. ■

■ EXAMPLE 63

English and Metric Units

Installation conditions

1. Underground steam line to be insulated
2. Installation up and down grade
3. Ground conditions are clay, where trenching can cause water collection after installation
4. Poor drainage
5. Some levels of installation are above frost line
6. Average soil temperature 45° F (7.2° C) minimum 10° F (−12.2° C)
7. Steam line in continuous operation
8. Temperature of steam 425° F (219° C)

Installation requirements

1. To conserve steam at most economic cost of installation versus savings

Insulation selection

1. Factory prefabricated treated steel, or cast iron, conduit system
2. Insulation used may be asbestos fibers, heavy density glass fiber, or calcium silicate
3. Expansion loops and fittings to be factory preinsulated

Properties of system

1. Shall be completely water tight
2. Shall be designed for movement of steam piping within the conduit
3. Shall be corrosion resistant

Thermal calculations

As this is a factory system, the choice of insulation and its thickness affects the entire system. For this reason, insulation thickness cannot be calculated as a single factor. ■

9 The Physical Design of Industrial Insulation Systems

The properties of insulation materials, weather-barriers, and coverings have been described in previous chapters, as have the effects of expansion and contraction. In addition, the various functions of insulation have been established. The means of calculating the insulation thickness required for a given function have also been discussed. What remains is to bring all these together to form industrial insulation systems.

The factory manufactured systems, such as vacuum cryogenic, insulated conduit, and reflective systems will not be presented, as these are designed by their manufacturers, and detailed information can be obtained from that source.

The systems used by industry can be broken down into divisions which are basically determined by the form of the insulation. Because these types of installations are so numerous, those which are presented represent only a few of the typical possible applications. These systems are:

Poured-in-place loose fill insulation
Blanket insulations
Sprayed-in-place insulations
Rigid insulations

Poured-in-place Insulation Systems

These are difficult to present independently because they depend upon materials other than the insulation to form a cavity, box, or some type of enclosure.

To illustrate a few: Poured insulation, frequently perlite, mineral wool, and other loose insulations are used to fill a sealed metal box containing cold equipment and piping. This cold box is often equipped with a dry nitrogen purge, serving two purposes. First, to remove moisture that enters the box, and second, as a fire snuffing agent. Such boxes are commonly used in the liquefied gas industry.

In hot application of poured insulation, high temperature insulation granules are poured into the cavity between the refractory and outside masonry of kiln, or furnace walls. Diatomaceous earth granules is one insulation suitable for these applications.

These two applications are shown in Figure 84.

Loose insulation is also used for the insulation of hot underground lines. The loose fill is packed around the pipe as shown in

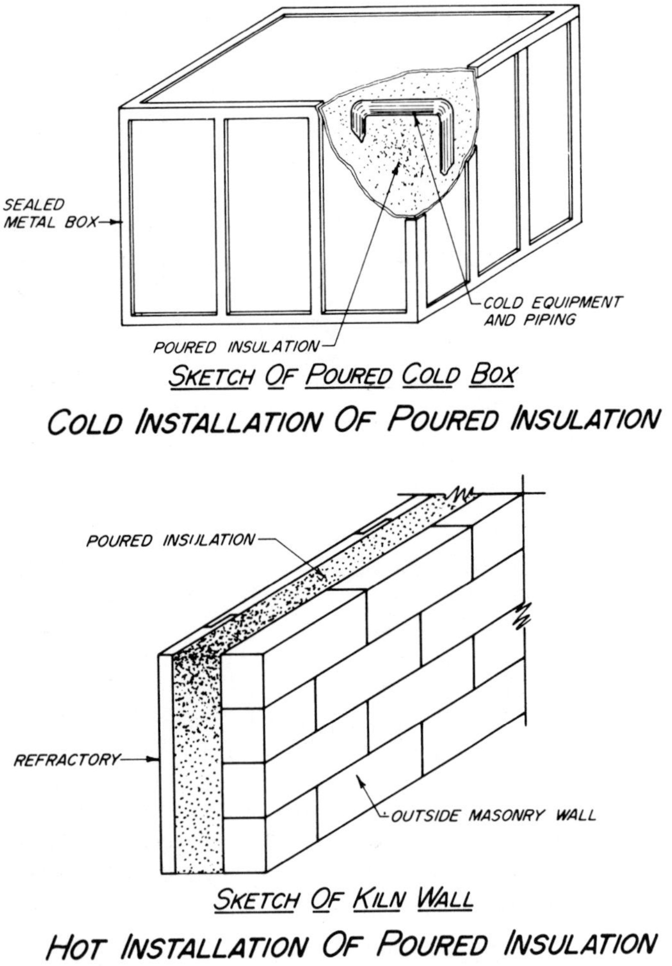

Figure 84 Poured in place insulation systems.

Figure 85. It is of the utmost importance that, when this loose fill bituminous insulation is installed, the material be packed tightly on the underside of the pipe. As the pipe is heated, the material next to it solidifies, providing a corrosion resistant layer. Any voids underneath the pipe leave unprotected pipe surfaces where moisture seepage can cause rusting. After packing, the line must be heated for a period of time to "cure" the insulation. After this the trench can be backfilled. Although the bituminous loose fill insulation tends to protect steel or iron pipe it cannot be absolutely crack free. Thus, all underground lines should be properly galvanic protected.

Expansion and movement in expansion loops must be taken up by using greater masses of insulation on the side towards which movement will occur. A typical pipe loop so insulated is also shown in Figure 85.

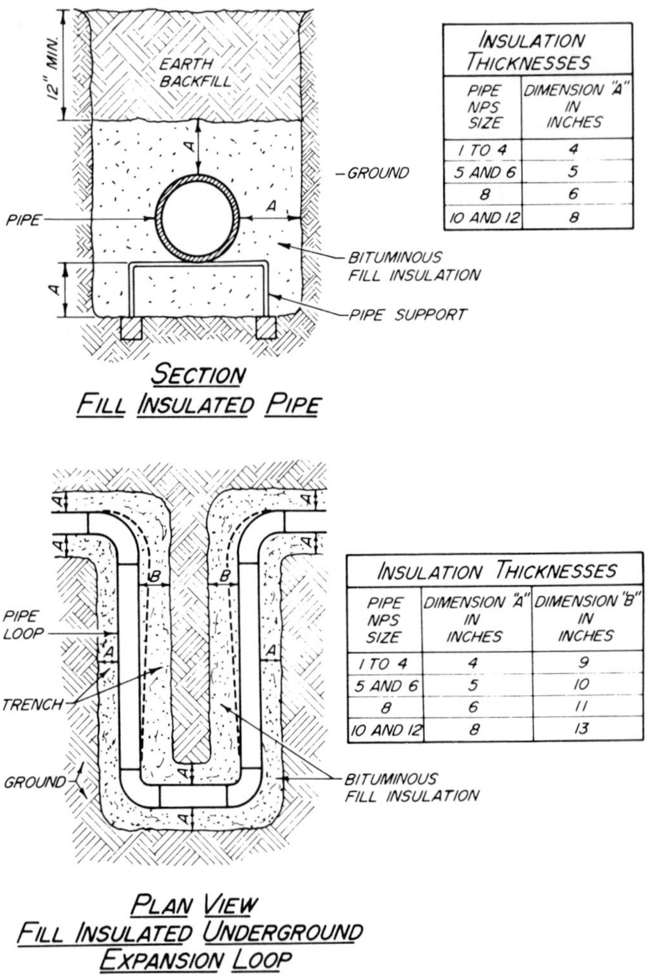

INSULATION THICKNESSES	
PIPE NPS SIZE	DIMENSION "A" IN INCHES
1 TO 4	4
5 AND 6	5
8	6
10 AND 12	8

INSULATION THICKNESSES		
PIPE NPS SIZE	DIMENSION "A" IN INCHES	DIMENSION "B" IN INCHES
1 TO 4	4	9
5 AND 6	5	10
8	6	11
10 AND 12	8	13

Figure 85 Loose fill bituminous underground pipe insulation system.

Blanket Insulation

Fibrous blanket insulations in industry are used to wrap and line ducts. Where used as an external wrap, the blankets are most often supplied with a factory-applied facing. If the operating temperature of these ducts is occasionally lower than ambient air temperature for a period of time, then the facing must be a vapor-barrier. A typical installation is shown in Figure 86.

Flexible plastic foam sheets can be used on ducts and large spheres by means of contact adhesives. The adhesives are coated onto the substrate metal and also on the insulation sheet. The sheet is so positioned that it requires approximately 1/4 in. (6.35 mm) lateral compression to bring it into line with the adjacent sheet edge. These two edges should also have been coated with the adhesive. Such an application is useful in insulating moderate temperature spheres not exceeding 180° F (82.2° C) in temperature. As this is in the corrosion range of steel, and, as the installation depends upon adhesion to the surface, the steel substrate must be sandblasted and treated with an inorganic zinc coating prior to the application of the foam sheet. A heavy build fire resistant mastic weather-barrier, reinforced with Elastafab® reinforcing cloth, provides an outer barrier with sufficient elasticity to expand and contract with the movement of the spheres without over-compressing the insulation.

® Trade Mark—Vimasco Coatings Co., Nitro, W. Va.

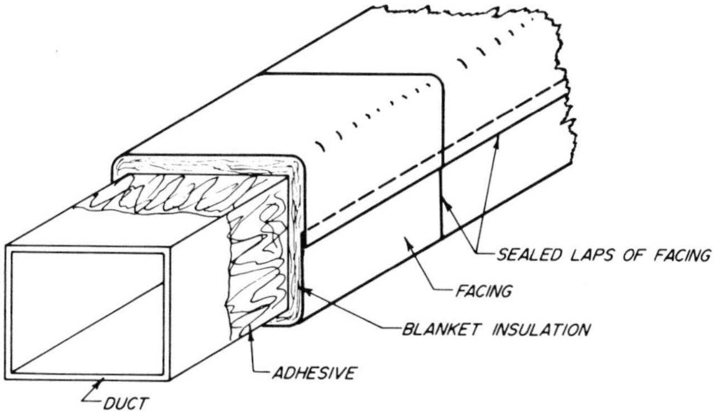

FIBROUS BLANKET INSULATION ON EXTERIOR OF DUCT

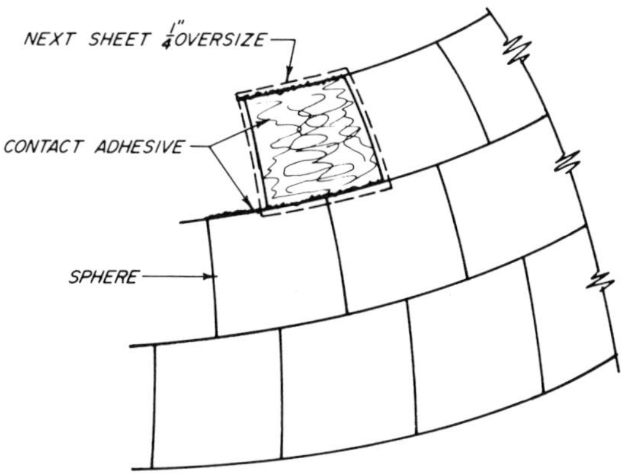

FLEXIBLE PLASTIC FOAM SHEETS ON SPHERE

Figure 86 Flexible insulation systems.

Spray Insulations

There are two major types of spray insulations. One is the inorganic fiber and binder type, and the other is sprayed organic urethane, which foams after its application.

The inorganic type is composed of mineral wool fibers and inorganic binders. It is installed by a specially designed machine which pumps dry material and water to the spray gun. The water mixes with the fiber and binder before it impinges upon the surface to be insulated. This plastic mass of insulation is then pressed by hand to conform to the surface and to compress it to the desired density. When dry, it forms a semi-rigid insulation suitable for service temperatures up to 700° F to 1350° F (371° C to 816° C) depending upon the fibers used in the mix. As it can be pressed to relatively heavy densities and has a low rate of diffusivity, it is used as fire protection where fire exposure may be of limited duration. On bottom and sides of ducts, or bottom of vessels, welded pins and spud nuts, or studs over which wire mesh is drawn tightly, are required to reinforce the insulation and hold it in place. Almost any of the conventional weather-barriers can be used as the protective outer covering. An application of this insulation is shown in Figure 87.

The second type of sprayed insulation is urethane organic foam. This is a two-package system, suitable for spray application. The two separate liquids are mixed at the spray gun in proper ratio. Mixing is done in a chamber- or internal-mix gun. Each of the liquids is supplied to the gun by a metering pump. These pumps supply the liquids to the gun in the ratio as recommended by the manufacturer of the liquids. The resultant insulation is generally recommended for operating temperatures in the range of 50° F to 180° F (10° C to 82° C).

As these insulations form on the vessel or pipe surface and depend upon the adhesion to the surface for securement, the type of surface is important. If the surface is masonry, it must be clean and free of dust. Steel surfaces must be clean and free of mill scale, dirt, oil, or other contaminates. If steel is painted, the paint must be completely dry, clean and be of a type which resists the solvent action of the urethane mix. In case of doubt, the foam should be sprayed on a test patch to see if the paint is affected and that the paint adhesion to the metal is sufficiently tight to secure the foam in position.

In addition to preparation of the surface to make it suitable for application of the foam, it is essential that surrounding equipment, piping, instruments, and surfaces not to be insulated, be covered, wrapped, or screened to protect them from overspray or drift, as would be done for ordinary spray painting.

For best results, the application should be done in dry weather, at relatively moderate humidity conditions, with ambient air temperature above 50° F (10° C) and very low wind velocity (less than 5 mph). The temperature of the substrate should be between 70° F to 120° (21.1° C to 98.9° C). In cases where the ambient temperature is less than 50° F (10° C), the feed tanks and hose to gun should be heated. When substrate temperature is below 50° F (10° C), a thin layer of "froth" should be applied to the surface prior to the main application. This "froth" can be obtained from the gun by adding a propellant into the liquid stream just before the mixing chamber or heating section, depending upon the spray equipment.

The foam liquid should be spray applied over the surface, with even passes over the area, to provide an even thickness approximately 1/30 of the desired insulation thickness. To obtain a nearly uniform final expanded thickness of foam requires a skilled gun operator as variations in thickness of the applied liquid are multiplied approximately 30 times by the expansion of the liquid into a rigid foam. Application is illustrated in Figure 87.

Because any variations in application of the liquid will show up as multiplied variations on the surface of the finished, cured rigid foam, variations on the outer surface can be expected. Should the installation demand a smooth surface, the high spots can be ground off even with the low spots with a disk sander.

For outdoor installations the outer weather-barrier can be spray applied heavy build mastic. On installations where variations

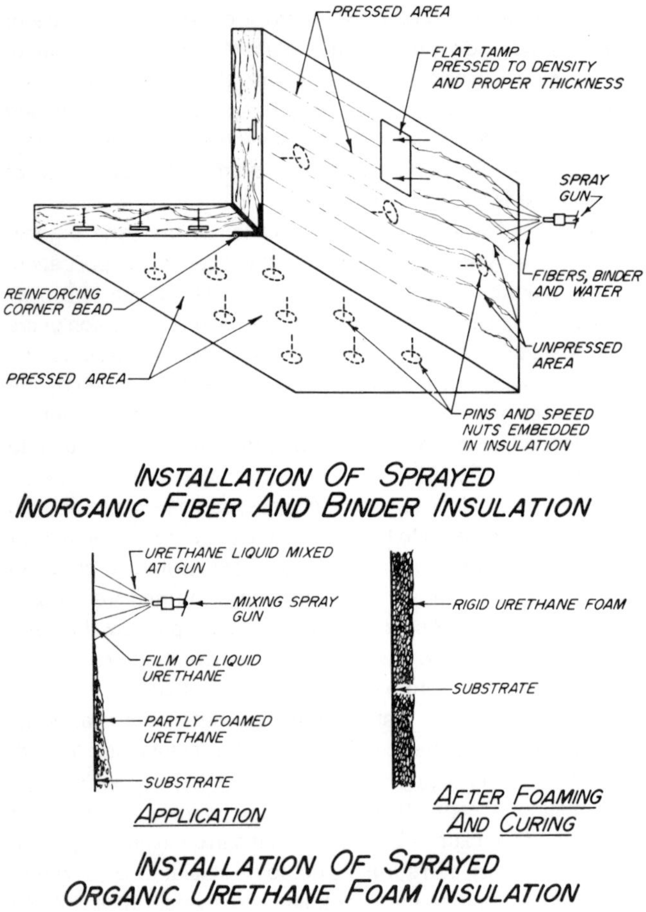

Figure 87 Insulations applied by spray-mixing guns.

of surface are allowable, an outer "crust" on the foam develops. This "crust" adds considerable surface strength and so it is practical to spray the mastic directly over the foam without using cloth fabric reinforcing. As a matter of fact, installing cloth on such a surface would be difficult and unsatisfactory. However, due to surface nonuniformity, the coverage of the mastic will be 30%-50% less, for the same thickness, as on a smooth surface.

Urethane foam surfaces which have been ground smooth may be given the same coverings, or weather-barriers, as any other insulation.

As the urethane foam is a combustible material its use is only recommended on vessels containing noncombustible materials and located in a non-hazardous area. The use of sprayed urethanes on the interior of metal buildings is also not recommended. Many feel that such installations, when fire protected with gypsum board or similar fire resistant materials would constitute a safe installation. However, a small fire on the outside of the building can raise the temperature of the steel to be above ignition temperature of the organic foam, thus starting a fire inside the building.

Likewise, metal jacketing over polyurethane or ppolystyrene insulation installed on vessels or pipes does not provide a fire barrier over the organic foam insulation. Once ignited, urethane insulation will burn between the metal substrate and the metal jacketing. This is because in the enclosed space the solid material is only 4% to 5% of the enclosed volume. The balance is made up of combustible gases and air.

Over styrene foam, aluminum jacketing can become a secondary ignition source as it melts and drips from its point of installation downward as molten aluminum and styrene. In addition, the inner barrier protecting aluminum from corrosion is most frequently of combustible material and when used over combustible insulation it adds to the fire hazard.

Corrugated metals increase the fire hazard potential in two ways. First, as the corrugations can never be water or air tight, they allow for entry of air and the exit of burning gases. Second, if the area is equipped with sprinkler systems for fire protection, the corrugations prevent an even spread of water over the surface. Where corrugated metal is used over insulation where it is necessary to protect with water spray systems, then the spray heads must be located much closer together to obtain good water dispersal.

Spray or trowel insulating mastics

These mastics are made to serve a dual purpose, as both coating and insulation. Their major use is on vessels and piping, such as cold water tanks or pipes, which have an operating temperature low enough to condense moisture from the ambient air.

Besides causing rusting of the vessel and pipe, the condensate will collect and drip, with resultant damage to objects below. As there is no money to be saved by keeping the vessels or pipe cold, all that is required of the insulation is that it prevent condensation. For this purpose high thermal resistance is not necessary. Thus, mastics which contain cork granules or mica generally can provide the small thermal resistance required.

The cork or mica filled mastics may be solvent or emulsion types, and may be applied by spray or trowel to the desired thickness. The thickness required depends upon the operating and ambient temperatures and relative humidity. In most instances, if a thickness greater than 3/8" (9.55 mm) is required, it becomes more practical to use conventional insulations.

The surfaces to which these mastics are applied should be clean and dry. This last is especially important. It often happens that when sweating pipe or vessels do become a problem, an attempt is made to apply these insulating mastics to correct it. Naturally, if the temperature of the pipe or vessel is sufficiently low to condense moisture out of the air when the mastic is applied, the condensate continues to keep it wet so that it never dries. Therefore, the vessels or pipes must be taken out of service, cleaned and dried before the mastic is applied, and must be kept out of service until the mastic has dried.

Most of these mastics dry to an irregular pebble type surface rather than a smooth and even one. This is the correct approach, as the irregular surface has a high outer surface area and high heat transfer by convection. Thus, it has a low surface resistance and is effective in retarding condensation. If such a surface is smoothed, or painted a reflective color such as white or aluminum, condensation is like to reoccur.

Insulating and finishing cements

Insulating cements are mixed with water, then applied as a plastic material to the surfaces to be insulated. They have two major uses, first, as insulation on irregular and curved surfaces, and second, as a finishing coat over block, or field applied segments of rigid insulation. On items located where appearance is essential, finishing cements are applied over rigid insulation or insulating cements, to obtain a smooth, even-textured surface. Some cements are designed to be both an insulating and a finishing cement.

These types make excellent maintenance materials when insulations have been broken, or torn off, as they can be applied to them to fill voids or exposed areas up to existing adjacent surfaces.

Illustration of the use of insulating cement on a flat surface insulated with block is illustrated later in this chapter in Figure 88.

In this particular use, the insulating cement is reinforced with hexagonal netting which is tied to pins welded to the metal surface. The cement, being more flexible than the block, provides a semi-rigid surface over the block when reinforced with hexagonal netting. This tends to prevent straight-through voids which occur when expansion of the metal separates each block from the adjacent one. In this service, it provides an insulating bridging layer over the block, which acts as a base for the outer finish coating or weather-barrier. The stretch of the hexagonal netting and the flexibility of the cement converts the movement from lines between blocks to an area expansion, thus giving the weather-barrier an opportunity to expand and contract both longitudinally and transversely instead of concentrating all its movement in a straight line.

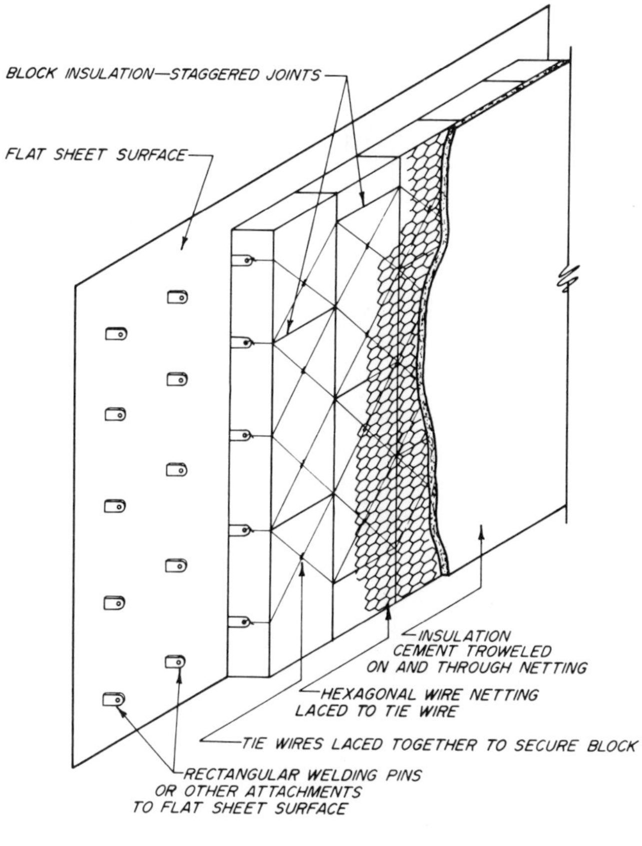

Figure 88 Method of installing insulation cement to block on flat sheet surface.

Rigid Insulations

The fact that rigid insulations *are* rigid makes it necessary that they be molded, fabricated, or otherwise shaped at the point of manufacture, fabrication shop, or in the field, to dimensions and shapes to fit the equipment, or pipe upon which it is to be installed. As soon as dimensions are involved, standardization of sizes, tolerances, and clearances become necessary for efficient assembly.

Although some of the semi-rigid and flexible insulations are not manufactured in accordance with dimensional standards, rigid pipe insulations are made in accordance with Recommended Practice of Inside and Outside Dimensions ASTM C-585. These dimensions are such that the outer diameter of any pipe insulation is the same as that of some specific NPS pipe. Thus, an insulation for th eparticular size of pipe which matches that outside diameter will "nest" over a preceding insulation. A listing of these sizes is shown in Tables B-18, B-19, B-20, and B-21.

Contrary to a popularly held misconception, insulation for neither hot nor cold application should fit absolutely snugly on a pipe. Some clearance should be provided to allow for slippage between the pipe and its insulation to accommodate differential movements.

The method of handling these differential movements of pipe and equipment must be designed into the insulation. Otherwise, cracks in the insulation and outer barriers will develop. This was previously mentioned in Chapter 7 in regard to insulation supports and accessories, but it must be extended to the physical design of the entire insulation system. As bringing consideration of contraction and expansion into the design is the key to long lasting insulation systems, the following discussion will be broken down into cold, medium, and high temperature installations. This is because of the effect the service temperature has on this problem.

Cold equipment insulated with cellular glass will be considered first. **Although the intent of this manual is not to discuss specific materials and their application, cellular glass has properties so different from others that it must be considered in a separate generic classification.**

Cellular glass, which has a vapor permeability of 0.000000+, is more vapor resistant than most vapor-barriers. By sealing cellular glass to itself with vapor sealers to form a complete enclosure around the vessel or equipment, the insulation forms its own vapor-barrier, and none is needed on its outer surface. However, it is important that it be installed in such a manner that these joints remain unbroken. The cellular glass insulation system should be so designed that movement, due to contraction of the vessel or equipment, is not transmitted to the insulation to cause opening of the sealed joints. The vapor sealer chosen to seal cellular glass must be selected with care. Because of its vapor resistance, sealers containing high solvent content will not dry when placed between butt ends of the insulation. For this reason, the sealer should be of high solid content and it must remain elastic and flexi-

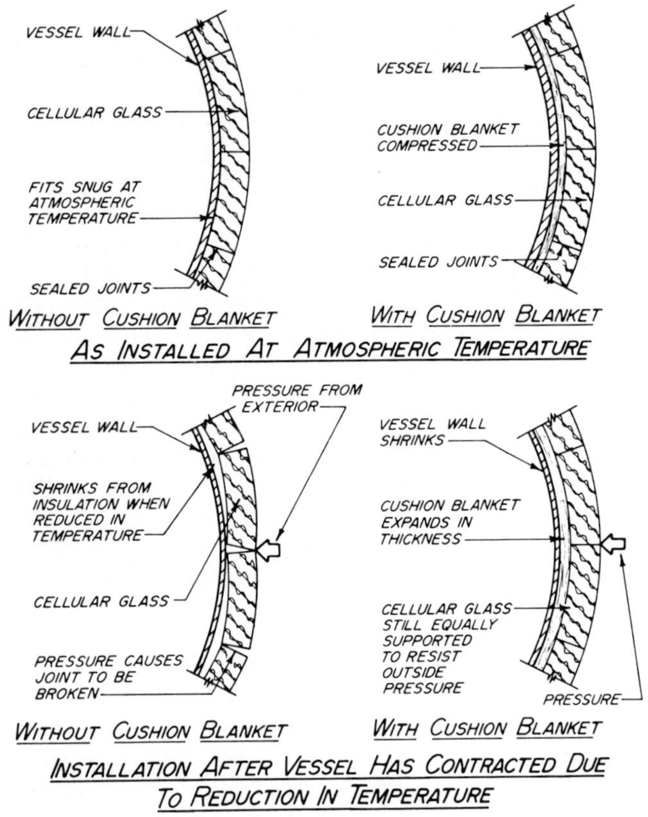

Figure 89 Effect of temperature reductions in the diameter of vessel wall and cellular glass insulation.

ble to the lowest temperature to which it is to be subjected. Sealers containing water, or which become brittle and shrink at low temperatures are not suitable.

The coefficient of expansion (or contraction) of cellular glass is 0.0000046 per °F or 0.0000074° C and that of steel, which has the lowest coefficient of the commonly used construction metal, is 0.0000065 per °F or 0.0000104 per °C. Thus, when cellular glass is installed on a vessel at atmospheric temperature, and the vessel's temperature lowers, the circumferential dimension of the vessel reduces more than the circumferential dimension of the cellular glass for two reasons. First, the higher rate of contraction of the vessel, and second, the service temperature of the vessel is reduced almost twice the reduction of the mean temperature of the insulation. Thus, the vessel "pulls away" from the insulation, leaving a void. This void causes no harm to the insulation itself, but it does leave it with no "backing." Therefore, any pressure on the outer surface may break the vapor sealed joints of the installed insulation. These factors are illustrated in Figure 89.

The insulation must be also designed to accommodate any axial contraction of the vessel. On the cylindrical surface, this change in dimension must be taken up by insulation contraction joints. Two such contraction joints are shown in Figure 90. The spacing of these joints is based upon the movement of the vessel and the weight of insulation to be supported.

On vessels operating below −100 °F (−73.3° C), these joints should be spaced on approximately 12 foot (3.66 metres) centers. On vessels which operate above this temperature, the spacing is generally determined by the weight of the insulation being supported by itself at its bottom. The maximum spacing of supports should be 18 feet on centers when based on weight alone.

A contraction joint is always required just below the top on vertical vessels, as the contracting vessel will pull away from the top, leaving a void below if a joint is not located at this position. This is illustrated in Figure 91.

Also illustrated in this figure, is the installation of cellular glass in broken joint construction. On all installations of very low temperature insulations, broken joint construction is used for several reasons. First, it reinforces the joint by providing a back-up under it, and second, if the service temperature is below −50° F (−45.6° C), many sealers will freeze, harden, and cause spalling of the cellular glass. Frequently, one sealer possessing extremely low temperature properties is used on the inner layer, and another to seal the butt joints of the outer layer.

Cold equipment and piping insulated with organic foam has contraction problems just the reverse of that of equipment insulated with the cellular glass just described. This is because its coefficient of expansion (and contraction) is 0.00004 to 0.00005 per °F (0.000064 to 0.00006 per °C). In this case, the insulation contracts more than the vessel.

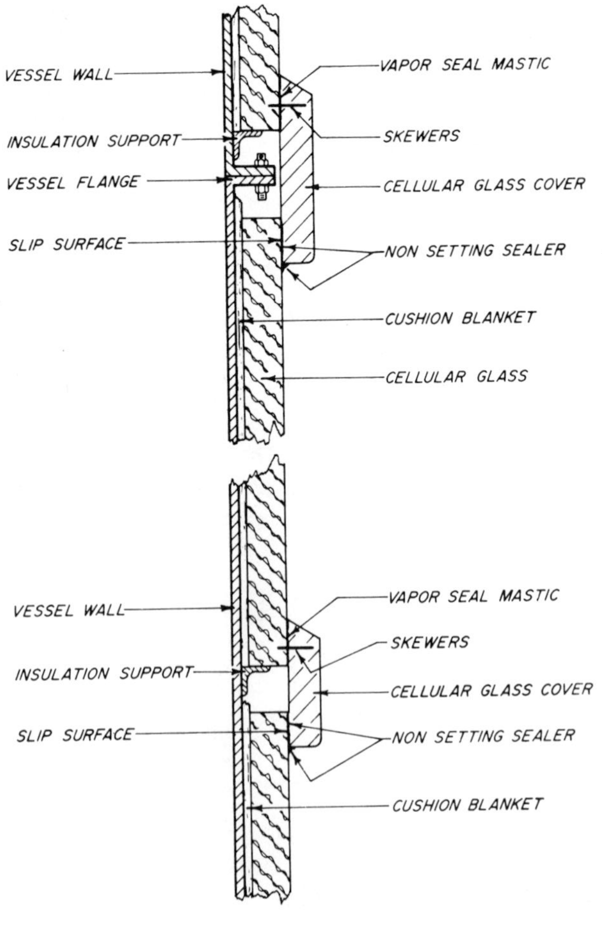

Figure 90 Contraction joints to allow for axial movement of cylindrical vessel.

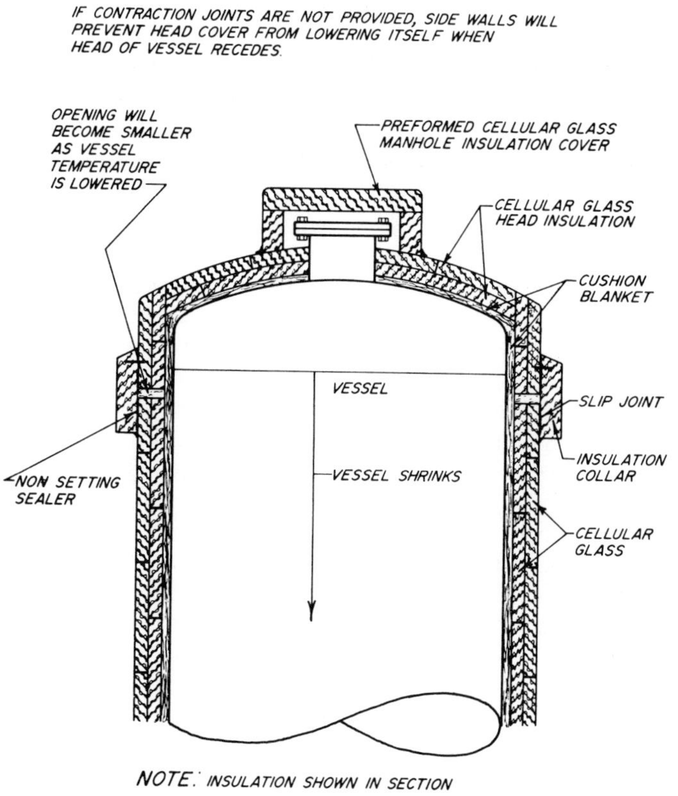

Figure 91 Contraction joints below head cover to allow head insulation to "float" with vessel.

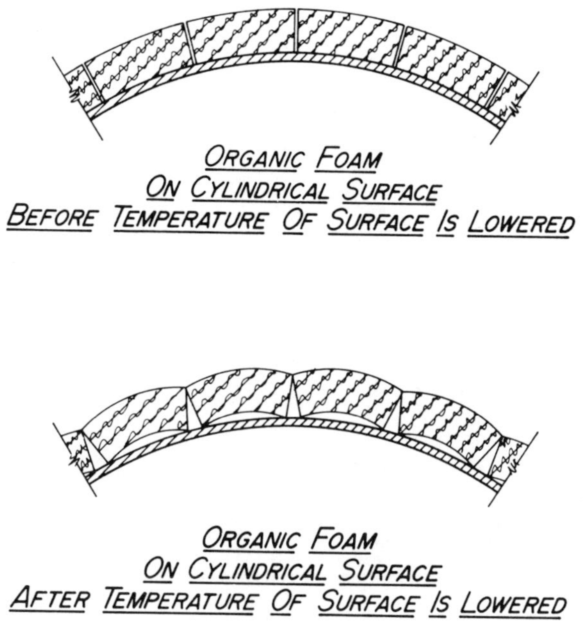

ORGANIC FOAM
ON CYLINDRICAL SURFACE
BEFORE TEMPERATURE OF SURFACE IS LOWERED
ORGANIC FOAM
ON CYLINDRICAL SURFACE
AFTER TEMPERATURE OF SURFACE IS LOWERED

Figure 92 Effect of lowering the temperature of organic foam having a high coefficient of expansion.

In most instances, these materials are more resilient than cellular glass, but they have a much large coefficient of expansion than the metal itself. This creates another set of problems. The outside of the insulation surface at ambient temperature is about stable in dimension, whereas its inner dimension shrinks. Around a circular surface this creates a condition which leaves "V" shaped cracks. This is illustrated in Figure 92. On small and moderate pipe sizes, when insulation can be foamed oversize and the bands tend to draw and compress it to a smaller circle, this action can be controlled relatively easily. On vessels, the problem must be dealt with as a basic problem of contraction. The way this is done is to provide a thickness of cushion blanket which can be compressed down to the circumference of the foamed insulation at the temperature of the vessel. The insulation must be pulled into this position by use of a stretched expansion or nylon band. This is shown in Figure 93.

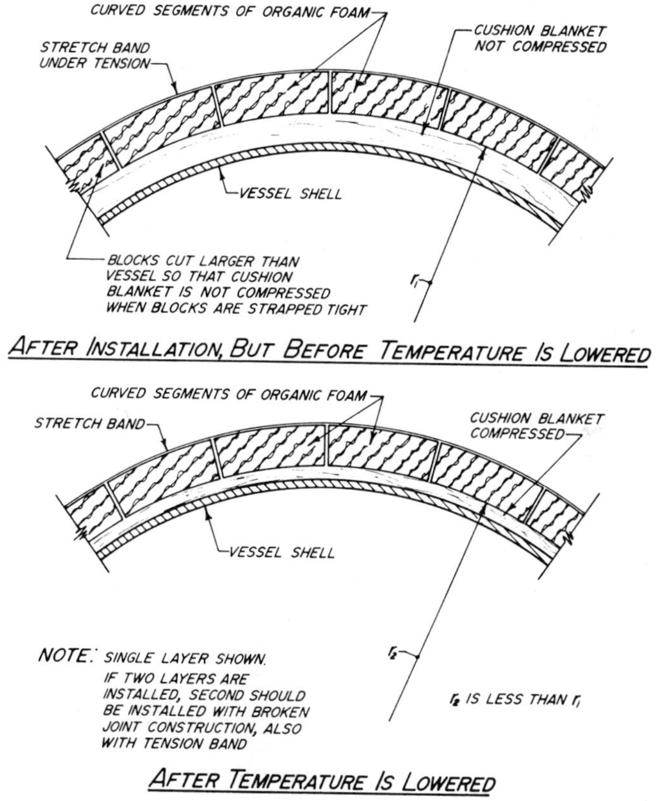

Figure 93 Method of installing organic foam on low temperature vessel to provide for contraction.

To compensate for these factors on vessels operating below 32° F (0° C), the recommended construction would be to have a thick layer of non-compressed cushion blanket, double layer, sealed joint construction with the best sealed vapor-barrier possible. If located outdoors, a weather-barrier over the vapor-barrier is recommended. All projections, nozzle insulation covers, inside and outside corners, must be carefully vapor sealed to prevent the vapor from by-passing the areas covered with the vapor-barrier. An illustration of this construction on a vessel is shown in Figure 94.

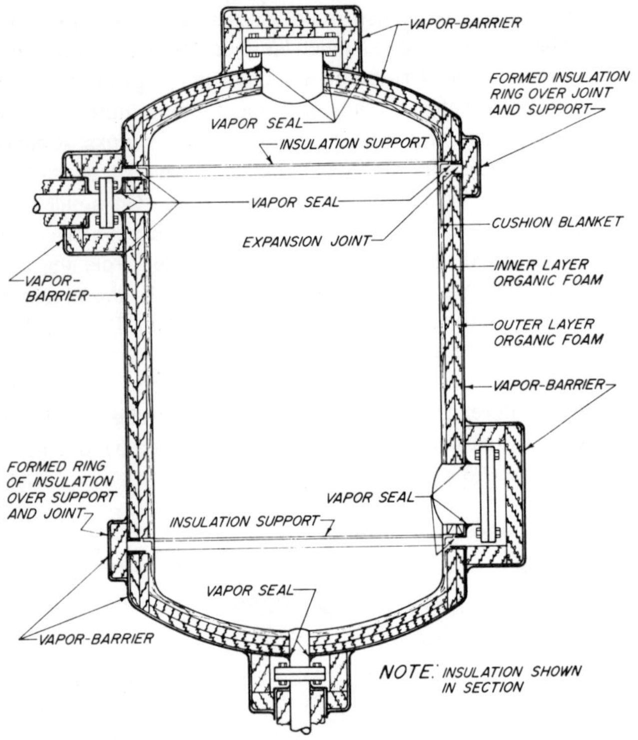

Figure 94 Method of installing vapor-barriers at nozzles, projections and supports.

Moderate temperature equipment (above ambient temperatures to 212° F [100° C]) because of its smaller problem of expansion or contraction, has less supporting and securing problems than either low or high temperature service equipment. If the insulation is dimensionally stable and has even a moderate compressibility, the blocks or curved segments can be applied directly to the surface and secured in position without space allowance for vessel movement. However, because of the limited heat available for water evaporating purposes, if located outdoors, or where the equipment can become wet, it is vital that either the insulation be moisture repellent, or that the weather-barrier keep the insulation dry. Once the insulation is water soaked, the temperature is not high enough to provide a vapor-driving force that will dry out the insulation in a reasonable length of time.

High temperature equipment (above 212° F [100° C]) gets back into the range where expansion movement must be considered. Again, because of the number of variables, there is no clear cut guide to follow. The effects of circumferential expansion on the tightness of the insulation joint is a function of the size of the vessel, its temperature, the shrinkage (or expansion) of the insulation, and its compressibility.

Such commonly used high temperature materials as calcium silicate, or expanded silica, should be installed in wedge-shaped or curved segments. Kerfing and breaking blocks around cylindrical vessels cannot give as efficient an insulation application as properly cut and fitted segments. The resulting voids from such kerfing will reduce the thermal efficiency of the installation. If it is attempted to fill the voids *completely* with insulating cement, the resulting labor cost would be much greater than the proper cutting of the insulation. For vessels up to 10 ft (3.66 metres) in diameter and 400° F (204° C) in temperature, if the circumferential block is cut and fitted to be 1/2 in. greater in circumference than the vessel, and secured so that the tension of bands produces the compression on the butt edges rather than on the surface towards the vessel, this little space and the compressiveness of the blocks will most often suffice to take care of the expansion of the vessel. On vessels of greater diameter, definite provisions for expansion should be made.

Of course, the metal of the vessel is the controlling factor in the expansion of that vessel. To illustrate the dimensional changes of a vessel, Tables 61 and 61a follow. This table gives the dimensional change in diameter and circumference of various metals at various temperatures and diameters of vessels. As can be seen from this table, the changes in these dimensions are significant, and the installation design must make certain that the increases in dimensions do not open joints and ruin the installation. Of course,

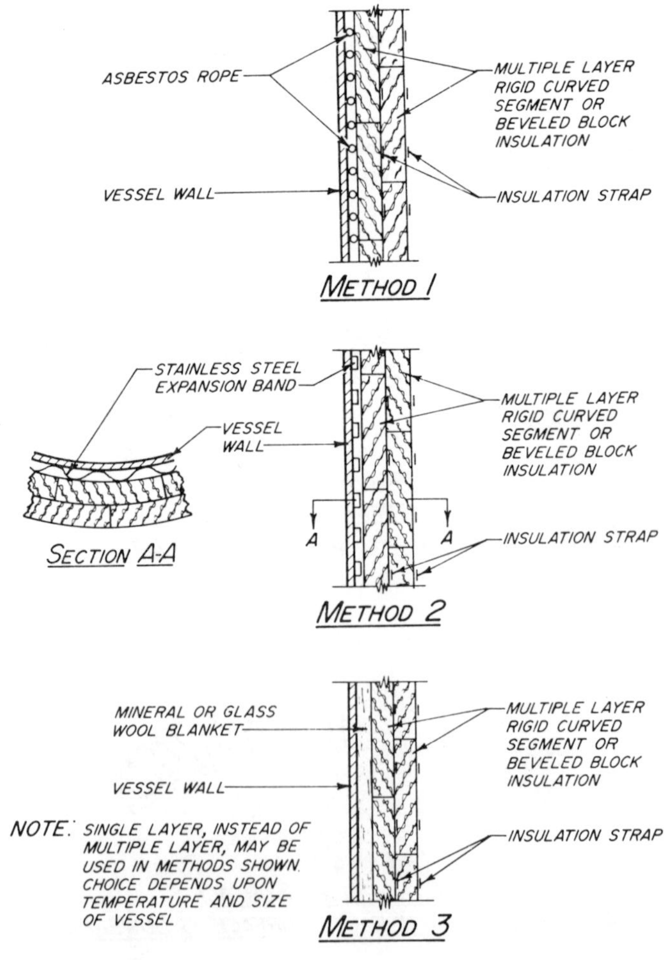

Figure 95 Three methods used to provide for high temperature vessel cylindrical expansion.

to these *expansion* figures must be added any *shrinkage* of insulation, caused by temperature.

Numerous methods are used to provide an expansion area around the shell of a vessel. First, is to wrap asbestos rope around the vessel, second, to place stainless steel, spring type bands around the vessel, and third, to install compressible blanket around the vessel, prior to the application of the insulation. The use of multiple layer, broken joint construction rigid insulation also helps, as there is bound to be an added degree of flexibility with multiple layers as compared to a single layer. In addition, any joint opening is restricted to only one layer, and there is always at least one layer left to act as a heat barrier. This reduces hot spots and heat disintegration of the weather-barrier. These methods are illustrated in Figure 95. Of course, the block must be cut and fitted

Table 61 Expansion of cylindrical vessel based on 70° F ambient *English Units*

Metal of vessel	Operating temp. of vessel °F	Cylindrical Dimensional increase in inches due to coefficient of expansion									
		Circumferential expansion, inches					Diametral expansion, inches				
		5′ dia.	10′ dia.	15′ dia.	20′ dia.	30′ dia.	5′ dia.	10′ dia.	15′ dia.	20′ dia.	30′ dia.
Steel	200	0.15	0.28	0.44	0.57	0.85	0.05	0.09	0.14	0.18	0.27
	300	0.28	0.57	0.85	1.14	1.70	0.09	0.18	0.27	0.36	0.54
	400	0.44	0.85	1.29	1.70	2.54	0.14	0.27	0.41	0.54	0.81
	600	0.72	1.44	2.17	2.79	4.33	0.23	0.46	0.69	0.92	1.38
	800	1.07	2.10	3.17	4.21	5.98	0.34	0.67	1.01	1.34	1.91
	1000	1.42	2.79	4.18	5.57	8.38	0.45	0.89	1.33	1.77	2.67
Stainless Steel	200	0.22	0.44	0.66	0.88	1.34	0.07	0.14	0.21	0.28	0.42
	300	0.41	0.82	1.22	1.63	2.45	0.13	0.26	0.39	0.52	0.78
	400	0.60	1.19	1.63	2.39	3.58	0.19	0.38	0.52	0.76	1.14
	600	1.01	1.95	2.92	3.90	5.83	0.32	0.62	0.93	1.24	1.86
	800	1.38	2.76	4.14	5.53	8.29	0.44	0.88	1.32	1.76	2.64
	1000	1.82	3.51	5.43	7.22	10.83	0.58	1.15	1.73	2.30	3.45
Aluminum	200	0.31	0.63	0.94	1.26	1.88	0.10	0.20	0.30	0.40	0.60
	300	0.60	1.16	1.60	2.32	3.49	0.19	0.37	0.51	0.74	1.11
	400	0.85	1.69	2.54	3.39	5.08	0.27	0.54	0.81	1.08	1.62
	600	1.41	2.83	4.24	5.65	8.48	0.45	0.90	1.35	1.80	2.70
Copper	200	0.25	0.47	0.72	0.94	1.41	0.08	0.15	0.23	0.30	0.45
	300	0.44	0.85	1.29	1.69	2.54	0.14	0.27	0.41	0.54	0.81
	400	0.63	1.22	1.85	2.45	4.63	0.20	0.39	0.59	0.78	1.47
	600	1.01	2.01	3.01	4.02	6.03	0.32	0.64	0.96	1.28	1.92

Table 61a Expansion of cylindrical vessel based on 20° C ambient *Metric Units*

Metals of Vessel	Operating Temperature of Vessel °C	Coefficient of Expan. per °C X 10[6]	Cylindrical Vessel Dimensional increase in mm due to coefficient of expansion									
			Circumferential Expansion mm					Diametral Expansion mm				
			2m dia.	4m dia.	6m dia.	8m dia.	10m dia.	2m dia.	4m dia.	6m dia.	8m dia.	10m dia.
Steel	100	9.9	4.98	9.96	14.94	19.92	24.9	1.58	3.17	4.75	6.34	7.92
	200	10.8	8.82	17.64	26.47	44.11	44.11	2.81	5.62	8.42	11.23	14.04
	300	11.5	20.23	40.45	60.69	101.15	101.15	6.44	12.88	19.32	25.76	32.20
	400	11.8	28.17	56.34	84.52	140.86	140.86	8.97	17.94	26.90	35.82	44.84
	500	12.1	36.49	72.98	109.48	182.46	182.46	11.62	23.23	34.85	46.46	58.08
Stainless Steel Ave. for Various Alloys	100	14.9	7.49	14.98	22.47	29.96	37.45	2.38	4.76	7.14	9.52	11.9
	200	15.2	17.19	34.38	51.57	68.76	85.95	5.47	10.94	16.42	21.89	27.36
	300	15.5	27.27	54.54	81.81	109.08	136.35	8.68	17.36	26.04	34.72	43.40
	400	15.8	37.72	75.45	113.17	150.90	188.62	12.01	24.02	36.03	48.04	60.04
	500	16.1	48.56	97.11	145.67	194.22	242.78	15.46	30.91	46.37	61.82	77.28
Aluminum Ave. for Various Alloys	50	20.1	3.79	7.58	11.36	15.15	18.94	1.21	2.41	3.62	4.82	6.03
	100	20.9	10.51	21.01	31.52	42.02	52.53	3.34	6.69	10.63	13.38	16.72
	200	21.8	24.66	49.31	73.97	98.62	123.28	7.85	15.70	23.54	31.39	39.24
	300	22.5	39.58	79.17	118.75	158.34	197.92	12.60	25.20	37.80	50.40	63.00
Copper and Bronze Ave.	50	15.2	2.86	5.72	8.59	11.46	14.32	0.91	1.82	2.73	3.64	4.56
	100	15.4	7.74	15.48	23.22	30.96	38.70	2.46	4.93	7.39	9.86	12.32
	200	15.7	17.76	35.51	53.27	71.02	88.78	5.65	11.30	16.96	22.61	28.26
	300	16.0	28.15	56.30	84.44	112.59	140.74	8.96	17.92	26.88	35.84	44.30

to the outside radius of the vessel *plus the thickness of the spacer used.* The insulation, when installed, should not compress the rope, spring band, or the blanket, otherwise they will not serve their purpose in allowing for the expansion of the vessel.

Where circumferential expansion control is necessary on the side walls, it is also necessary on the heads of the equipment. Use of compressible blanket under the rigid insulation on heads is easier to install than rope or expansion bands, and for this reason is more frequently used than the other two methods.

Longitudinal expansion of vessels also requires expansion joints in the insulation. The installation of joints in vertical vessels is somewhat easier than in horizontal ones, because the weight of the insulation will cause all movement to take place where desired, and joints can be designed to shed water more positively than those on vessels in a horizontal position. A section showing joints in a vertical vessel and methods of insulating the heads is presented in Figure 96. The vessel shown in the Figure is not typical, as the various parts are drawn to indicate conditions which might occur on a number of various types of vessels. Also shown in the installation of insulation around nozzels and supports.

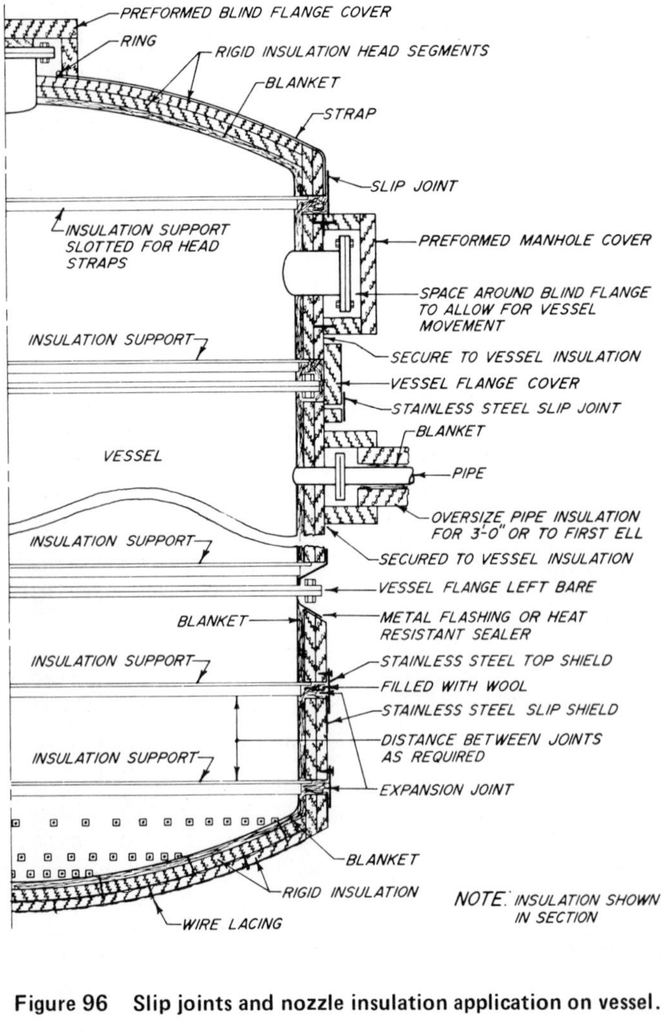

Figure 96 Slip joints and nozzle insulation application on vessel.

The installation of expansion joints in horizontal vessels, where the sliding surfaces are in a horizontal plane, depends upon non-setting sealers or flexible mastics to ensure water tightness of the joint. Even the best of these will deteriorate with time. Thus, these joints do require maintenance as leaks occur. Such joints are shown in Figure 97.

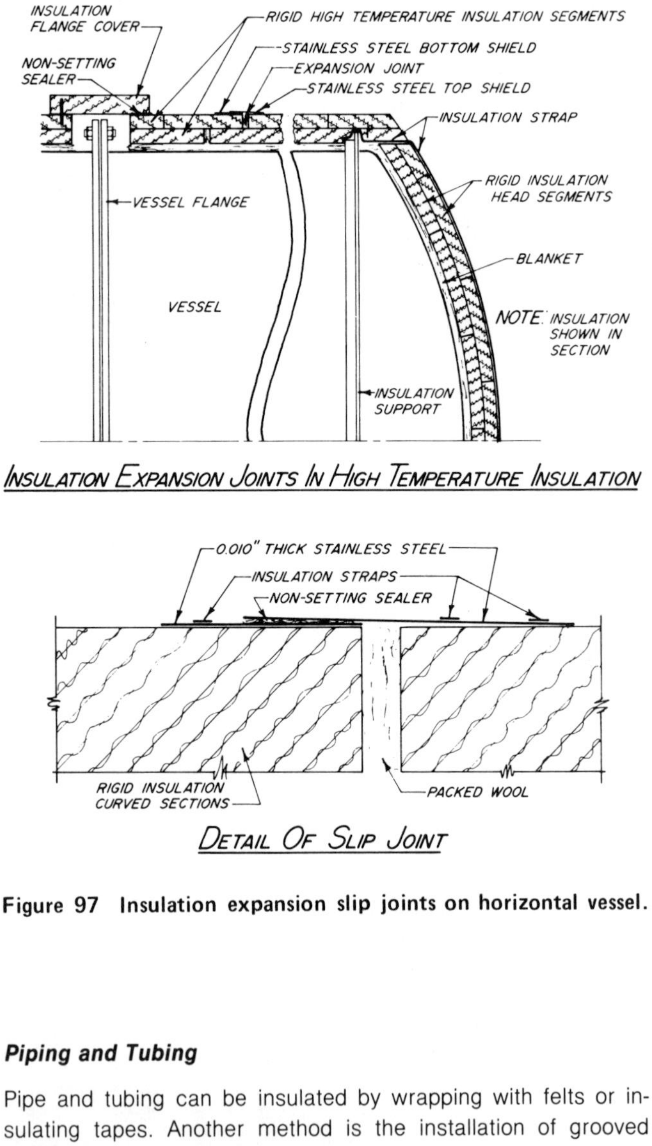

Figure 97 Insulation expansion slip joints on horizontal vessel.

Piping and Tubing

Pipe and tubing can be insulated by wrapping with felts or insulating tapes. Another method is the installation of grooved blocks cut to fit around the particular size pipe (or vessel). Such a system can be used very effectively on long straight runs, however where fittings occur, it is quite difficult to form an insulation fitting cover that is effective.

The most common form of pipe and tube insulation is rigid insulation premolded, or precut sections to fit the pipe or tubing. Unfortunately, sizes and thicknesses produced for many years were in accordance with what a particular manufacturer desired to produce. Even today many of the manufacturers of organic foam insulations produce their products with thicknesses of 1/2", 3/4" or 1" increments which fit no standard. For this reason it makes these pipe coverings, which need fire protection, have outside diameters on which high temperature insulations will not fit.

In January 1941, one of the writers (W. C. Turner) conceived the idea of standardization of insulation dimensions so that the outer diameter always was approximately the same as the standard outside diameter of a Nominal Pipe Size pipe. Thus, a second layer of insulation will fit (or nest) over the first layer, providing its inner diameter is sized for the pipe size of the outer diameter of the first layer. This system also makes it possible to prefabricate jackets as the outside diameters are fixed to relatively few sizes. In the 1960's ASTM committee C-16 on Thermal Insulation adopted this system as standard. This is ASTM C-585 Standard Recommended Practice of "Inner and Outer Diameters of Rigid Thermal Insulation for Nominal Sizes of Pipe and Tubing."

The insulation dimensions for NPS pipe sizes are given in Table B-18 and B-19 and tube dimensions are given in Table B-20 and B-21. These are in English Units. Metric dimensions are given in Tables B-19a and B-21a.

The use of standard size insulation made possible the more efficient installation of insulation and jackets, reduced stock and provided a basis for rational calculation of economic insulation thicknesses—based on standardized thicknesses. The next needed standardization was fitting covers. Turner, with the assistance of Mr. Robert Estep, designed insulation fitting covers for high and low temperature applications for NPS pipe and tubing. This was adopted by ASTM as Tentative Recommended Practice C-450 in 1960, the title being "ASTM Recommended Dimensional Standards for Prefabrication and Field Fabrication of Thermal Insulation Fitting Covers." The basic principle for prefabrication is shown in Figure 98 and Figure 99.

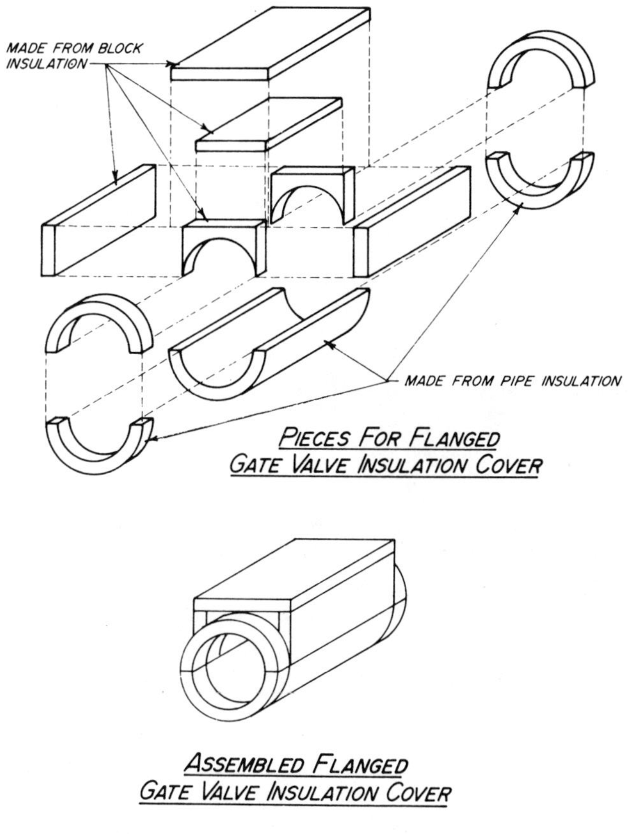

Figure 98 Fitting prefabrication system.

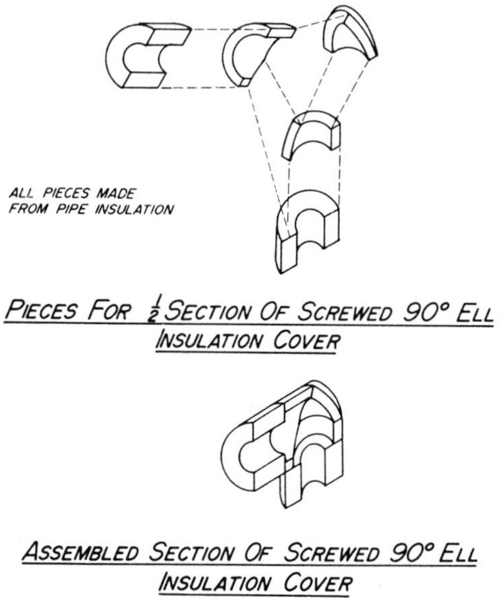

Figure 99 Fitting prefabrication system.

During the 1950's and 1960's considerable progress was made in the use of tracing by use of tubing or electric cable installed on pipes to keep them at desired temperature. To satisfy the need for fittings which would encase the line and tracer, Turner and Mr. K. B. Lanham revised ASTM C-450 to include dimensions for the traced systems, also included dimensions for fabricating vessel lagging and dished head segments. This information is now published in the ASTM manual, "ASTM Recommended Practice C-450," the title being "ASTM Recommended Dimensional Standards for Prefabrication and Field Fabrication of Thermal Insulation Fitting Covers for NPS Piping, Vessel Lagging and Dished Head Segments."

Because of the differences such as design for contraction on low temperature piping and design for expansion of high temperature lines, the dimensions and configuration of low temperature fittings differ from high temperature fittings. The correctly designed fittings for each of these services are contained in the ASTM C-450 manual of sizes. Also included are the dimensions for fittings used on traced piping. The system is based on using pipe insulation produced to dimensions in ASTM Standard C-585 and flat block to produce the fitting covers. Where materials are molded they can be formed to comply with these basic dimensions. The ASTM Recommended Dimensions C-450 provides the information in English Units for dimensions for:

PIPE COVERING, SINGLE, DOUBLE AND TRIPLE LAYERS

Low Temperature Fittings

Pressures	*Fittings*
Std. 150	Screwed 45° and 90° Ells, Tees, Valves
Std. 150	Flanged 45° and 90° Ells, Tees, Valves
150-1500	Welded Ells and Tees
150, 300, 400, 600, 900 & 1500	Flanged Gate Valves, Flanged Globe Valves
150, 300, 400, 600, 900 & 1500	Line Flanges

High Temperature Fittings

Pressures	*Fittings*
Std. 150	Screwed 45° and 90° Ells, Tees, Valves
Std. 150	Flanged 45° and 90° Ells, Tees, Valves
150-1500	Welded Ells and Tees
150, 300, 400, 600, 900 & 1500	Line Flanges

Traced NPS Pipe (with space for up to 3/4″ OD tracer)

Pressures	*Fittings*
Std. 150	Screwed 45° and 90° Ells, Tees, Valves
Std. 150	Flanged 45° and 90° Ells, Tees, Valves
150-1500	Welded Ells and Tees
150, 300, 400, 600, 900 & 1500	Flanged Gate Valves, Flanged Globe Valves, Line Flanges

Also given are the dimensions for the same fitting covers used for traced systems, allowing a space of 3/4 in. for tracer tubing or electrical conduit.

The system using pipe insulation and block insulation for valves and elbows are shown in Figures 98 and 99. The equipment for fabricating curved side wall segments pipe covering and fittings is shown in Figures 100 and 101.

The use of the standardized dimensions and fabrication equipment which is available makes the insulation of fitting more efficient both in respect to installed cost and thermal efficiency. In addition, such fittings can be reinstalled. The assembled pieces of calcium silicate to insulate a flanged globe valve and an elbow pipe are shown in Figure 102. The manner in which the valve insulation fits over a valve is shown in Figure 103. The valve insulation cover coated with reinforced PVA emulsion type weather-barrier mastic is shown in Figure 104.

These same type saws, with proper blades can also be used to fabricate cellular glass fittings. Such fittings are shown in Figure 105.

Use of these dimensions for shop or field fabrication, depending upon the economics of where fabrication is to be done, will result in standardized shapes, giving excellent job appearance. In addition, where these type fittings are used they can be removed for pipe repairs then be reinstalled.

The use of the dimensions provided in ASTM 450 is not restricted to cutting of fittings at the job site or job shop, they can also be used to produce factory made fittings. These fittings can be fabricated by cutting, grinding or molding: typical of these fittings is illustrated in Figures 102 and 105.

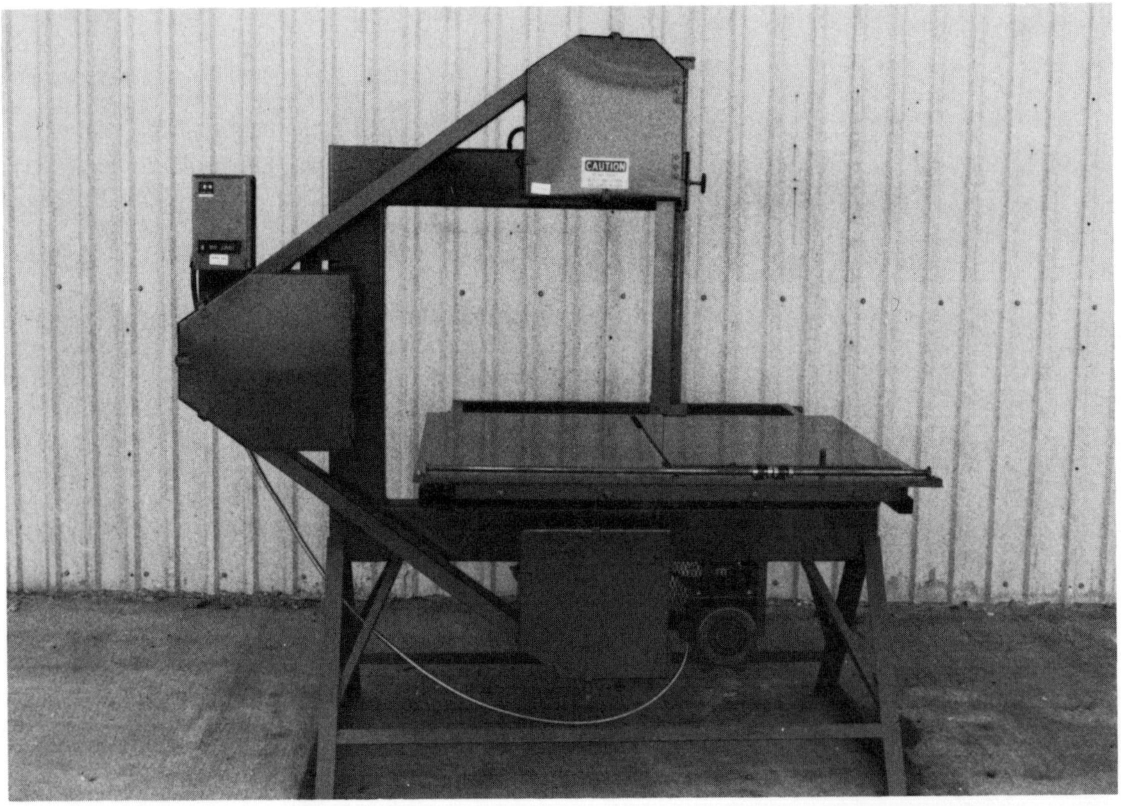

Reproduced by permission of Howard C. Forrest Co., Houston, Texas

Figure 100 A fitting saw. The rolling table is calibrated to cut ells, tees, valves, etc.

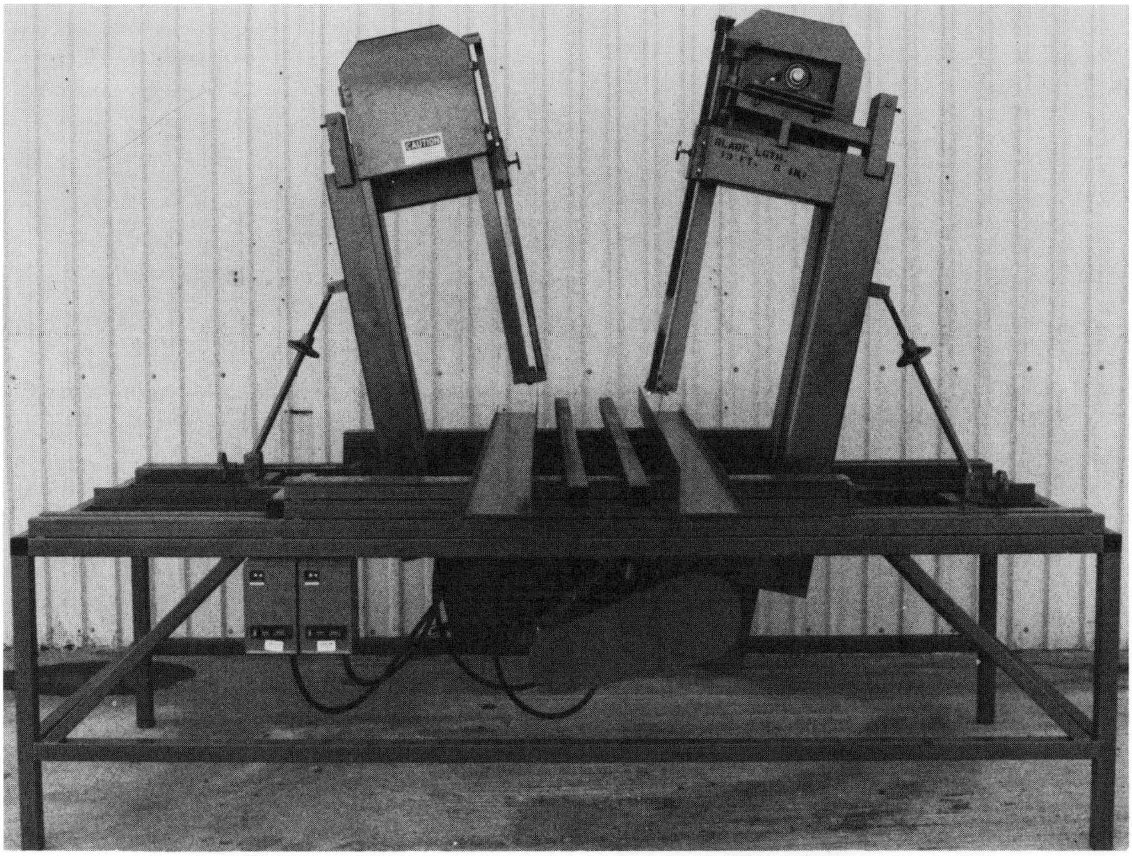

Reproduced by permission of Howard C. Forrest Co., Houston, Texas

Figure 101 Beveling and trim saw. Two band saws with tilting blades and adjustable bases. This is a versatile combination, usable for many cutting operations such as curved sidewall segment beveling, cant strip, pipe covering, trimming, block sizing, etc.

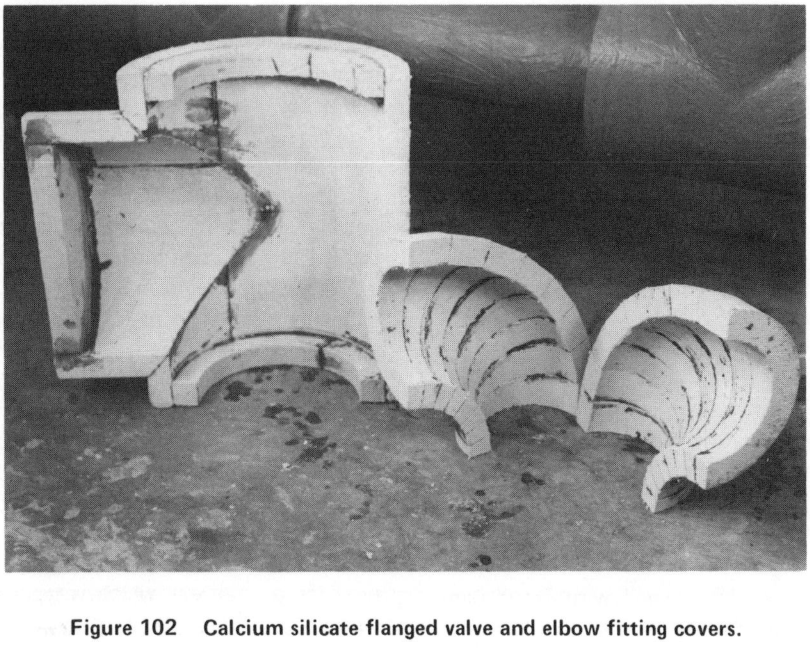

Figure 102 Calcium silicate flanged valve and elbow fitting covers.

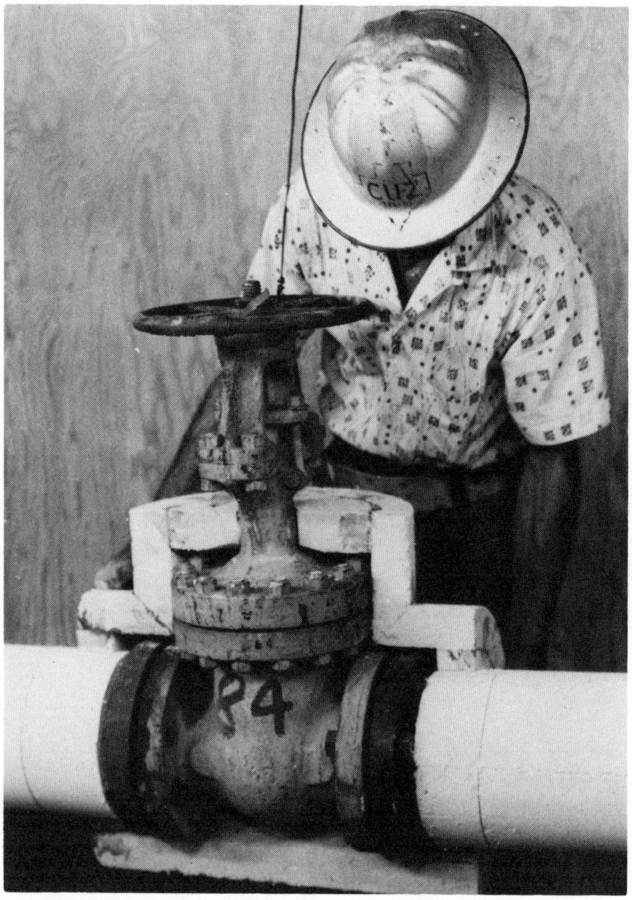

Figure 103 **Calcium silicate prefabricated flanged valve cover being placed in position on valve.**

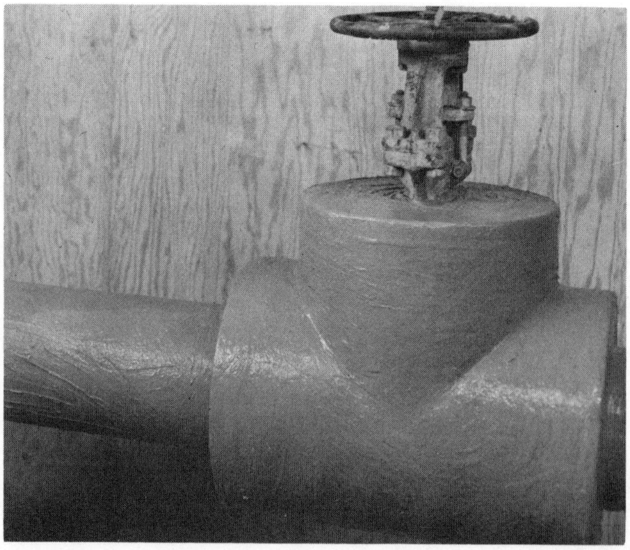

Figure 104 **Prefabricated valve cover coated with fabric reinforced PVA emulsion type weather-barrier mastic.**

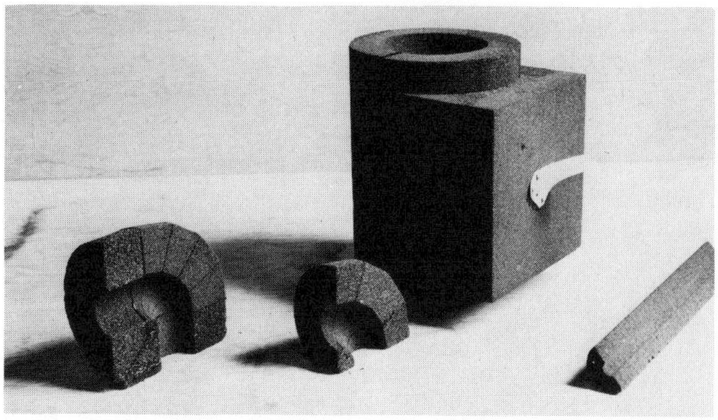

ELBOWS, PIPE AND FLANGED VALVE FITTING COVERS

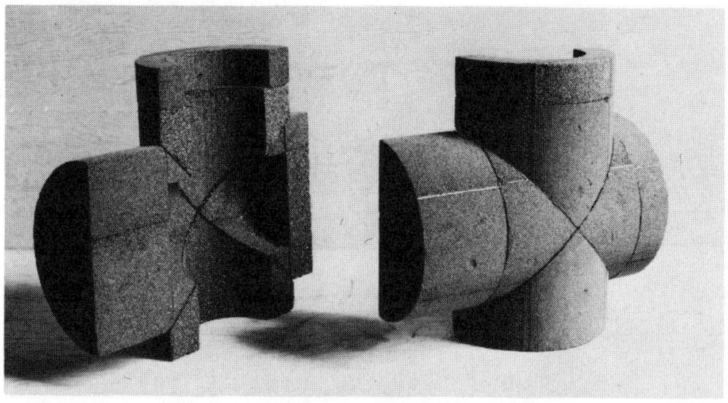

VALVE FITTING COVER

Figure 105 **Cellular glass prefabricated fitting insulation.**

Low Temperature Piping Insulation Systems

Low temperature piping installations, like equipment, must be divided into two types of applications, one giving the application of cellular glass, and another presenting the application of insulations that require a vapor-barrier for their protection against vapor migration.

Low temperature pipe insulated with cellular glass follows the same principles as stated for low temperature equipment. The insulation should be sealed to itself to form its own vapor-barrier, and installed with proper contraction joints so that pipe movement does not break open the sealed joints.

Pipe fitting covers should be fabricated in accordance with the dimensions given in ASTM Recommended Practice C-450. Cementing of pieces together in shop fabricated covers is most frequently done with hot melt (steep) asphalt. As asphalt is a thermoplastic material, it contains no solvents to dry out, and is applied in the hot liquid state to surfaces to be joined. When the pieces which have been pressed together cool down, a tight bond is obtained. However, surfaces must fit smoothly together, and the joint itself should be very thin. Excessive thickness of asphalt should be avoided, as asphalt is not an insulation, and a thick joint provides a heat leak which will cause cold spots on the outer surface.

In the field, the application of hot asphalt on pipe is difficult and dangerous to use (because of possible spillage from scaffolds and the fire needed to keep it hot) and other types of joint vapor-sealers are used. In places where no movement is desired, vapor-sealers which set should be used, but where movement is necessary between two surfaces non-setting vapor-sealers must be used.

The same problems of contraction stated for vessel insulation are also true for pipe insulation. In most cases, cylindrical contraction can be ignored as cellular glass sectional pipe insulation is strong enough in its encirclement of the pipe that it does not require support like equipment insulation, which is made of many segments sealed together. However, if the pipe has considerable vibration, or is heated rapidly, then a glass wool cushion layer should be installed prior to the application of the cellular glass.

Longitudinal contraction of the piping systems must be provided for in the design and application of cellular glass insulation. The total contraction can be determined from the tables on contraction in Chapter 5. With this information, the proper contraction joints and their location can be made a part of the insulation system. Flanged fitting covers can be used as contraction joints if they are installed with non-setting sealers so that contraction movement of the pipe will cause slipping at the surfaces between the fitting and the pipe covering.

The decision for the use of single or multiple layers depends upon a number of factors. In general, large size pipes are recommended for the use of multiple layer, broken joint construction on very low temperature, as each layer acts as a backup and strengthens the joint of the other layer. Another facctor is that some of the less expensive sealers have a limit of only −50° F (−95.6° C) before they stiffen and become brittle. On the inner

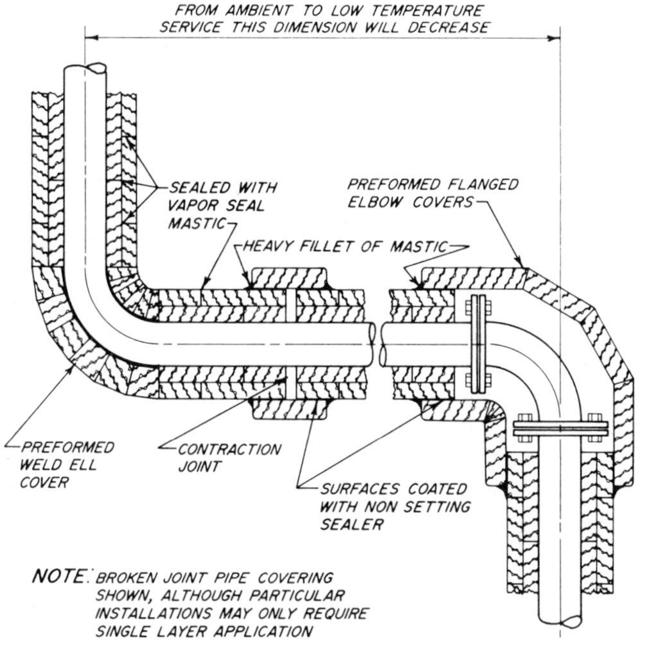

Figure 106 Cellular glass insulated piping system illustrating elbows and contraction joint.

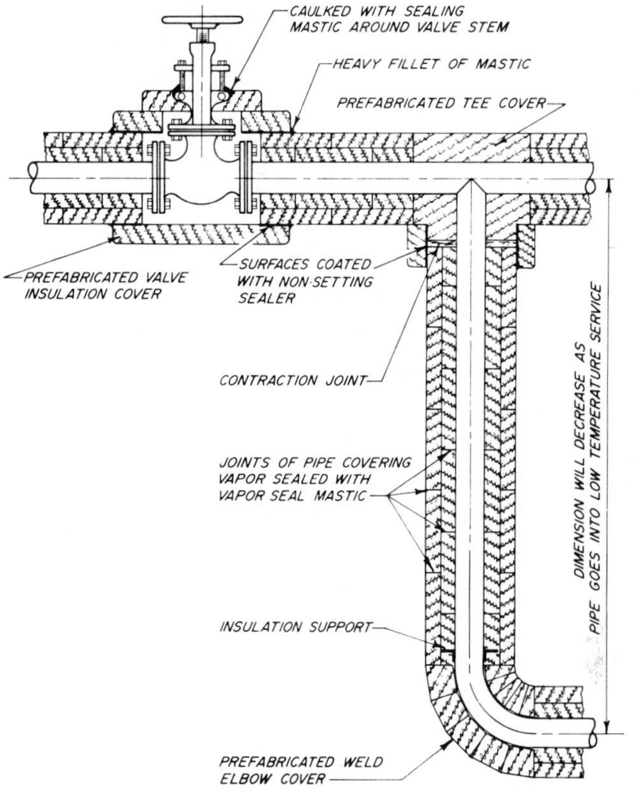

Figure 107 Cellular glass insulated piping system illustrating valve, tee, elbow, and contraction joint.

layer, which is below −50° F (−45.6° C) more expensive sealers must be used which remain flexible and elastic at extremely low temperatures. Again, broken joints do prevent direct heat leaks and tend to prevent cold spots on the outer surface.

In most instances, unless thickness exceeds 4½ in. (1.14 mm) fitting covers are made in a single layer. One reason for this is that joints formed in the shop can be made very tight and thin, thus they will not cause excessive heat losses as will field fitted joints.

Typical installations on piping systems are shown in Figures 106 and 107.

Because organic foam insulation depends upon a vapor-barrier for efficient service on low temperature pipe, the location and effectiveness of the vapor-barrier is most important in the installation. The vapor-barrier may be made of film, or a laminate, installed in sheet form with the joints sealed tightly together, or of vapor-resistant mastics. There exists considerable difference of opinioon as to which is better. Sheets and laminates can be obtained which have lower vapor transmission than mastics, but the sealing of the joints of these materials to make them *vapor tight* is most difficult. Most adhesives commonly used for these purposes are quite deceptive in that, even when a strong bond is obtained, the joint itself is very often not vapor tight. Mastics can be applied as continuous cover to eliminate joints, but their dried film does not have the vapor resistance of the sheets and laminates.

As the service life does depend upon the vapor-barrier, and the vapor-barrier must be located on the outer surface, it is sub-

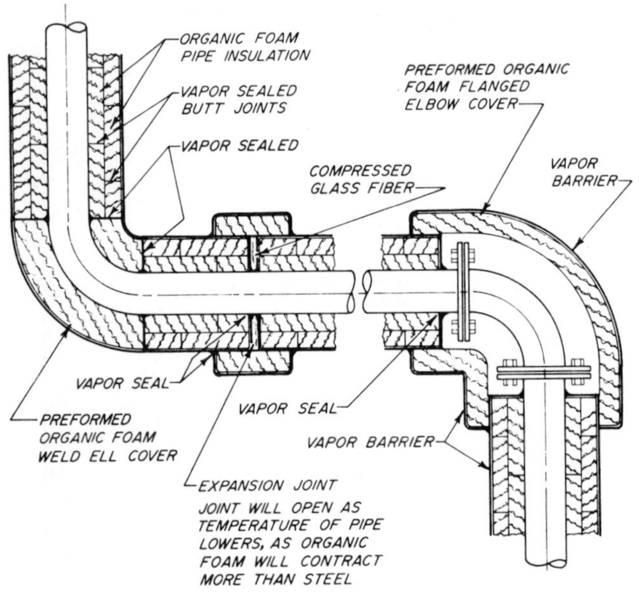

Figure 108 Organic foam insulation system for pipe illustrating elbows and expansion joints.

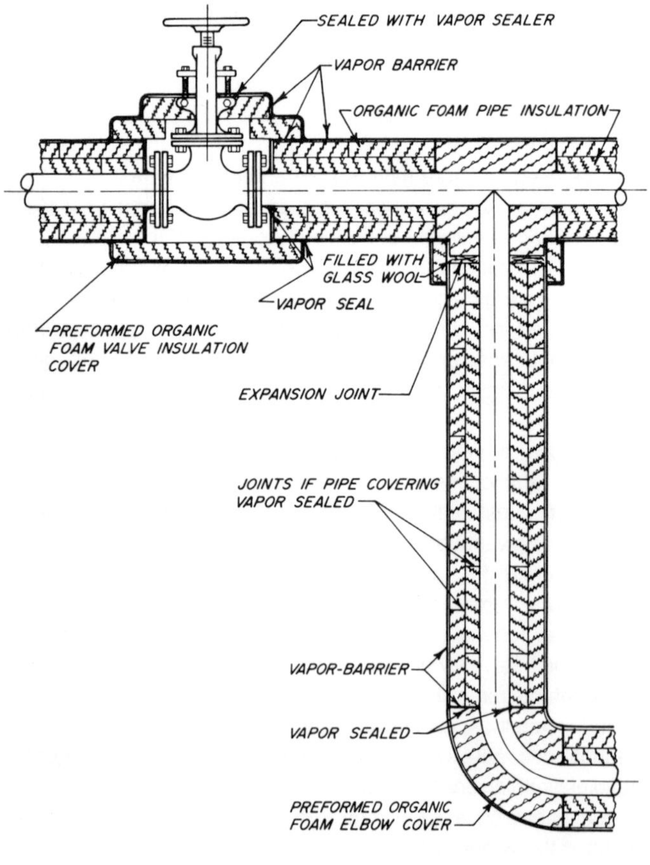

Figure 109 Organic foam insulation system for pipe illustrating valve, tee, elbow and expansion joint.

jected to the normal abuses of an external surface and may become torn, cracked, or punctured. When this occurs, the vapor will migrate inward and can spread in all directions. Thus, these applications should have internal vapor-barriers to sectionalize the installation so that one break will not ruin the entire system. This method of dividing a piping insulation system is shown in Figures 108 and 109. At each fitting and expansion joint, the vapor-barrier is brought down over the butt ends of the fitting or pipe covering and sealed to the pipe. Because these seals do go clear to the pipe, they must be capable of staying resilient and effective over the complete range of temperature to which the pipe may be subjected.

Although weather-barriers have not been discussed in this chapter, a word of caution must be injected here. Metal jacket weather-barriers should not be used over vapor-barriers, for, during their application, sharp edges or screws are likely to puncture the vapor-barrier. In service, differential movement is also likely to cause the metal to cut or shear the vapor-barrier.

Moderate temperature piping. The application of rigid insulation on moderate temperature piping is the least critical of the three temperature ranges. Because it *is* moderate temperature, the insulation thicknesses almost never requires multi-layer construction except for fire protection.

In most instances, moderate temperature insulation should be installed in the same manner as will be described for high temperature installations. The need for insulation expansion joints is much less, as the temperature will not cause excessive increase in metal dimensions, nor is it sufficient to cause much shrinkage of insulation.

One common practice, which should be changed for two reasons, is the excessive use of water mixed insulating cement on these installations. First, because of the lack of sufficient heat to completely dry the cement to its desired state, and second, because this is the temperature range at which the rusting of steel is most severe. Insulation which is wet or damp accelerates steel rusting.

Although not an insulation problem, steel located outdoors which is to be insulated, should be protected with a suitable paint system before it is insulated if its service temperature is less than 175° F (79.9° C).

Preformed, flexible plastic foam pipe insulation used on moderate temperature pipe and tubing, because of its unique propperties, is installed differently than rigid insulation. This particular insulation is used on moderate temperature tubing which may be bent to fit into position. It can be obtained in 6 ft (1.96 mm) lengths, and longer, in tubular form. In addition, it may be pushed over the tubing, or pipe, *before it is erected.* As it can be compressed, it can be pushed back from the tube (or pipe) joints to allow for their connections. After the connections are made, the insulation can be released to expand and cover the exposed metal.

Split sections of this material can be installed around erected pipe and the joint sealed by contact adhesive which was previously applied to the surface of the split. Correctly installed, additional securements are not required. Fittings can be made of the material using the dimensions of ASTM C-450 for cutting and bonding the pieces together with contact adhesive.

High temperature piping. Rigid insulation applied to high temperature piping may be single or double layer, depending upon the service conditions and temperature. When there are long runs of piping, and where the piping is large, the problem of expansion occurs again. To make the problem more troublesome, most high temperature insulations shrink as they are subjected to higher temperatures. This has been previously explained, and the following discussion shows how this is physically handled in the installation.

The circumferential expansion of pipe within the covering does not present a problem as the covering is manufactured with an inside diameter greater than the OD of the pipe. High temperature pipe covering should not fit snugly on the pipe at atmospheric temperature, but should have sufficient clearance to allow for expansion of the pipe. Properly installed, the insulation should be free to slip on the pipe without binding, even after the line is in service.

Single layer pipe insulation should be installed with staggered joints, and multiple layer insulation should be installed with staggered broken joints as shown in Figure 68 (Chapter 7). This secured cylinder of insulation should be free to slip as the pipe expands and the insulation itself shrinks. Of course, this requires that, at the correct distances, one insulation cylinder be separated from the adjacent cylinder by an expansion joint. This is to prevent the frictional pull exerted by expansion of the pipe from pulling the

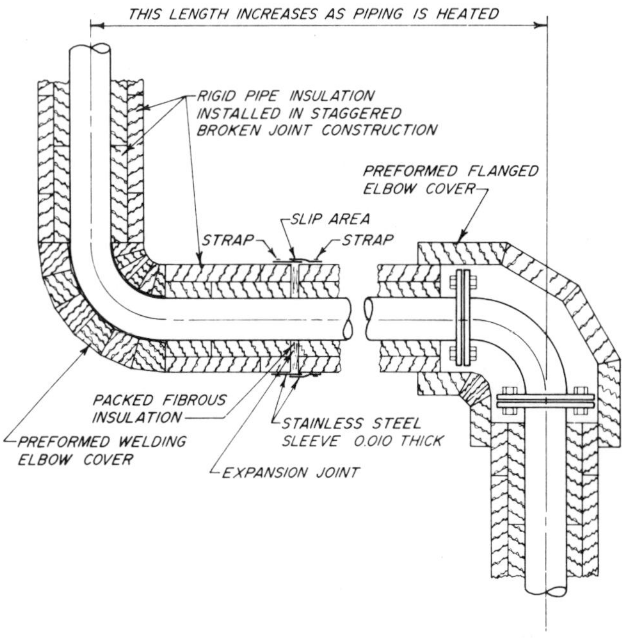

Figure 111 Rigid insulation system for high temperature piping illustrating elbows and expansion joints.

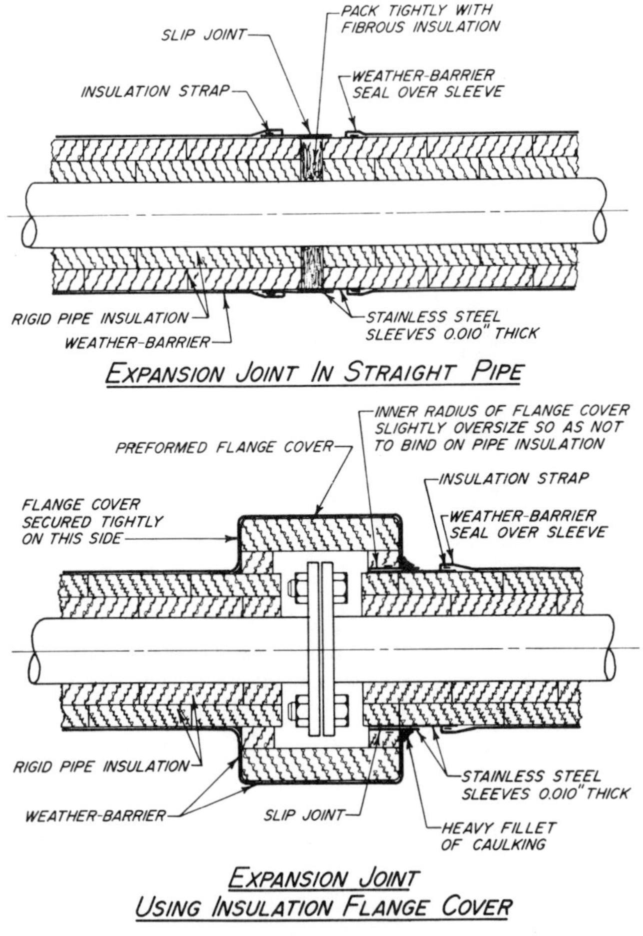

Figure 110 Typical expansion joints.

insulation assembly apart. This can be accomplished by an expansion joint, or by using a flanged insulation cover as an expansion joint.

Not every straight run of pipe needs an expansion joint. The need must be determined by the length of the pipe, its temperature, metal of pipe and shrinkage of the insulation which can be determined from the information presented in Chapter 5.

A typical expansion joint, and an expansion joint using a flanged insulation cover are shown in Figure 110.

Flanged fitting coverrs, preformed either in the field or in the shop in accordance with dimensions given in ASTM C-450, lend themselves to use as expansion joints as shown. In addition, the use of these preformed fittings is helpful in that they may be removed for maintenance, and they also provide for uniformity in appearance.

Securement of rigid pipe insulation and fittings is by wire and strap. As the straps are on the outside, they will be subjected to water from vapor condensing under the vapor-barrier and are, thus, subject to rusting. Galvanized steel will give long service only on indoor applications which are not subjected to moisture.

Although the ability of the material to resist cutting does make a difference, wire can be used to secure most high temperature insulations up to 12 in. (305 mm) OD. Above this diameter, the tension needed to pull the edges tight will begin to cause the wire to dig into the insulation. With the exception of the inner layers, which should always be wired into position, on insulation above 12 in. (305 mm) OD, strap should be used to secure it in position.

Wired inner layers are used even on large diameters because, even though originally the joints are not completely tight, they will be drawn up tight by the outer layer when it is secured by the strap.

Typical high temperature piping systems insulated with rigid

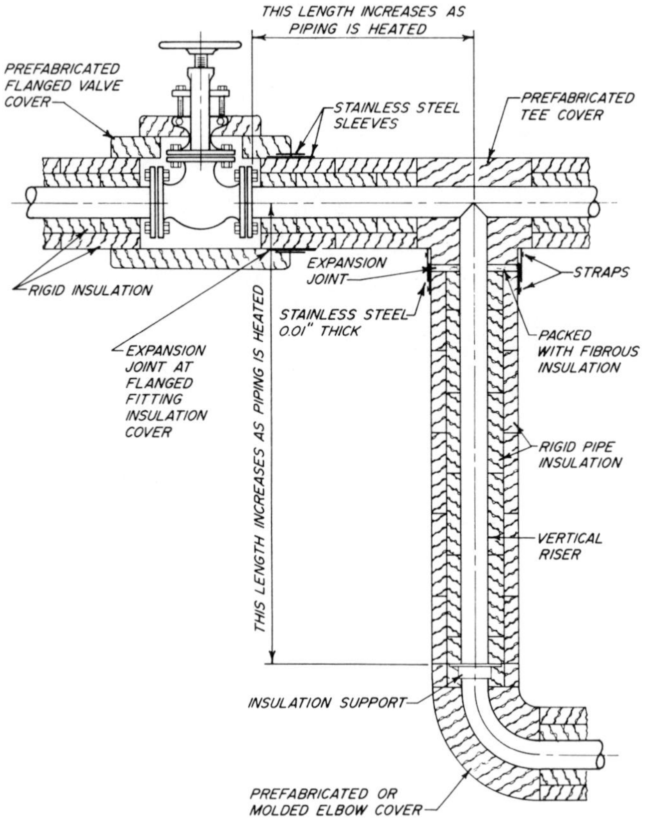

Figure 112 Rigid insulation system for high temperature piping illustrating valve, tee, elbow and expansion joints.

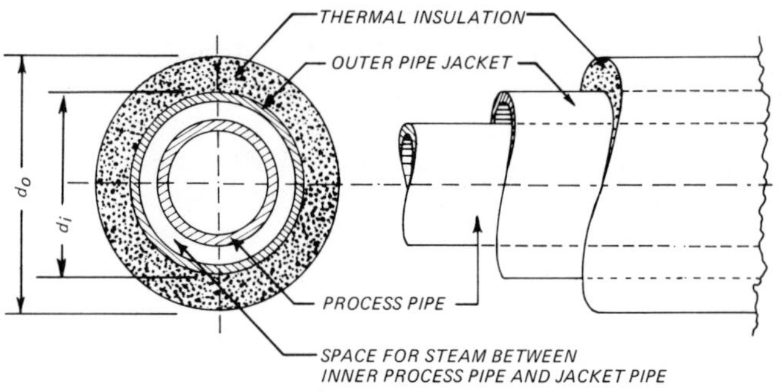

Figure 113 Jacketed process pipe.

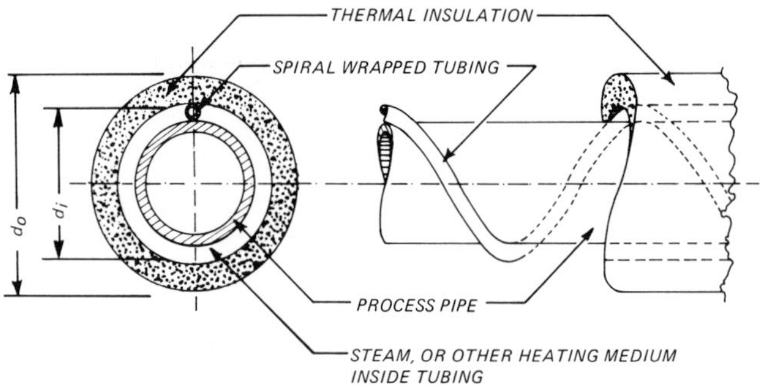

Figure 114 Spiral wrapped tracing.

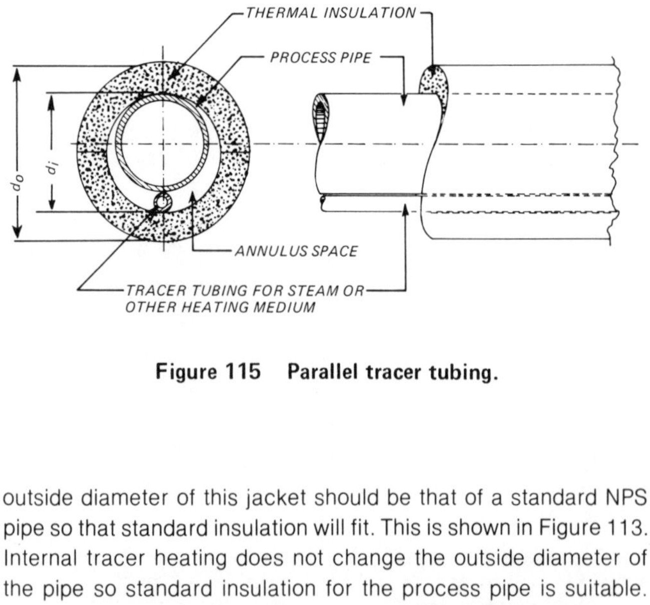

Figure 115 Parallel tracer tubing.

insulation and preformed fitting covers are shown in Figures 111 and 112.

The system illustrated is multiple layer, staggered broken joint construction. However, this should not be taken to indicate that all high temperature applications require this multiple layer construction. In the majority of installations, single layer construction is satisfactory, and the multiple layer is used only where quite thick insulation is required, or where no joints completely penetrating the insulation can be tolerated.

Only slight reference to weather-barriers has been made in this chapter, and this only where it was necessary as it tied in with the physical application of the insulation. The selection of weather-barriers over the installed system should be in accordance with the principles established in Chapter 6.

Presentation in this chapter was to establish the general principles of the physical design of insulation systems. Various installation details will be presented in the next chapter.

Insulation on Externally Heated or Cooled Piping

Where heating (or cooling) of piping is necessary, it may be accomplished in a number of ways as illustrated in Figures 113, 114 and 115. Where the piping is provided with an external pipe jacket the inside diameter (d_i) of the insulation is determined by the external diameter (d_o) of the jacket. Wherever possible, the outside diameter of this jacket should be that of a standard NPS pipe so that standard insulation will fit. This is shown in Figure 113. Internal tracer heating does not change the outside diameter of the pipe so standard insulation for the process pipe is suitable. Likewise, electric strip tape heat, or electrical induction heating will not change the size of insulation to be installed on the process pipe.

Where heating (or cooling) of pipe is accomplished by the addition of external tubing or electrical conduit, this additional attached appendage must be considered in the sizing and installation of pipe insulation. Illustrated in Figure 114 is spiral wrapped tubing as the heating medium (this could also be electrical con-

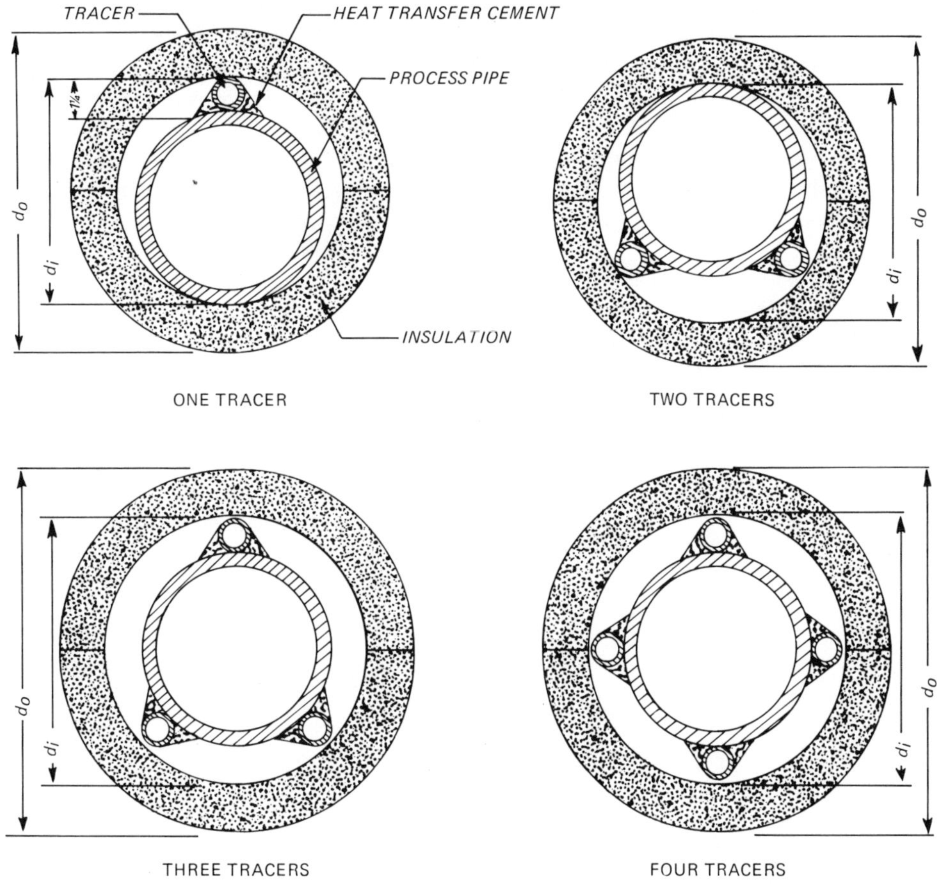

Figure 116 Suggested location of tracers on process pipe.

duit). The inside diameter of the insulation (d_i) must be pipe diameter + twice the tube diameter.

Where tubing or electric cable is installed in parallel position as illustrated in Figures 115 and 116, and depend on air convection, or be thermally bonded to the pipe. On very moderate requirements such as to prevent water from freezing in a pipe, thermal bonding may not be required, and the heater may be placed below the process pipe. Where heating requirements are more severe or when electric heating cables are used for maintaining relatively high temperatures, thermal bonding is most frequently required. In the cases of parallel tracers, bonded or unbonded, if the extension of the project is 1¼" or less then the dimensions provided in ASTM C-450 are suitable for the fabrication of the fitting covers. The selection of pipe insulation sizes are shown in Table 8 on page 58 (1st Edition).

Where more than one tracer is required as shown in Figure 116, then the correct NPS pipe insulation must be determined from d_i as illustrated. Where these exceed those shown in Table 8, the dimensions for prefabrication of the insulation fitting covers must be determined for the individual case, as those in ASTM C-450 will not be correct.

Where insulation is installed on traced piping it does not fit snugly to process pipe. This is illustrated in Figures 113, 114 and 115. Due to this it bridges over from line contact to line contact. The insulation should have sufficient physical strength to fulfill these requirements. The only exception to using oversize insulation on traced piping is when insulation is grooved out to make space for the tracer. However, as the insulation is thinned at the hottest point the formulas previously presented for calculation of heat losses will not give correct answers.

In all cases heat loss (or heat gain) is critical for the function of external traced lines and for this reason the insulation must be kept dry. Where insulation is used on tracing systems operating below 212° F (100° C), water repellent expanded silica, cellular glass insulation, or similar moisture resistant insulation, should be used. Likewise correctly designed weather-barrier systems are of utmost importance. Mastic systems should be such that they can expand and contract without cracking or breaking. Metal jackets, where selected, should be provided with corrected installed and sealed water proof joints and closure bands. Corrugated metal jacketing should never be used on horizontal piping as overlapping joints cannot be kept water tight under service conditions.

Prefabricated Underground Insulated Pipe Systems

The fill type installation of fill insulation as shown in Figure 85 is limited to temperature above soil in which it is buried and the upper temperature limit of the bituminous insulation. Where the piping system operates below soil temperature the system must be insulated with cellular foam insulation and the insulation itself must be protected with an outer jacket. This is also true of high temperature installations. To be effective, the jacket must be made of metal or rigid plastics that can withstand the mechanical forces and the moisture (water and vapor) imposed upon them. As all fittings or multiple piping in the same confinement must be protected in similar manner, prefabricated systems are very practical.

The prefabricated underground systems have all the insulation requirements as does insulation above grade. The systems must fulfill the thermal, economic and physical requirements previously described. Likewise, these requirements make it necessary to have a number of different types of underground systems. Other than where reinforced mastic jackets are installed after the insulation is installed, these systems are designed for water-tight joint closure of the insulation and jacket in the field with special jacket fitting and sealing techniques. It is necessary that such systems be designed to provide suitable expansion-contraction joints or loops for the process pipe and the jackets.

Illustrated in Figure 117a is an underground system using prefabricated cellular glass insulation. In this particular system, the pipe insulation, fitting insulation and expansion loop (or joints) are installed after the pipe is in position in the trench. It is held up by temporary supports which are removed as the insulation is installed. The inner surface of the insulation must be protected from abrasion by anti-abrasion coating or glass matt to prevent any movement of pipe in relation to insulation from wearing away the insulation. Special loop insulation, to take care of change in configuration of pipe must be provided. The insulation surface should be protected with a sealed laminate of glass fabric, aluminum bonded with bituminous, high molecular weight-polymers. Care must be taken that this protective jacket is water-tight before earth is back filled over the insulated pipe.

Heat transfer through this system of insulation may be calculated by the equation for heat transfer through one thickness of cylindrical insulation (Equations 8 and 8a) with temperature t_2 being equal to the soil temperature.

If the line contains gases or liquids lower than soil temperature, then the heat loss must be kept very low to prevent excessive

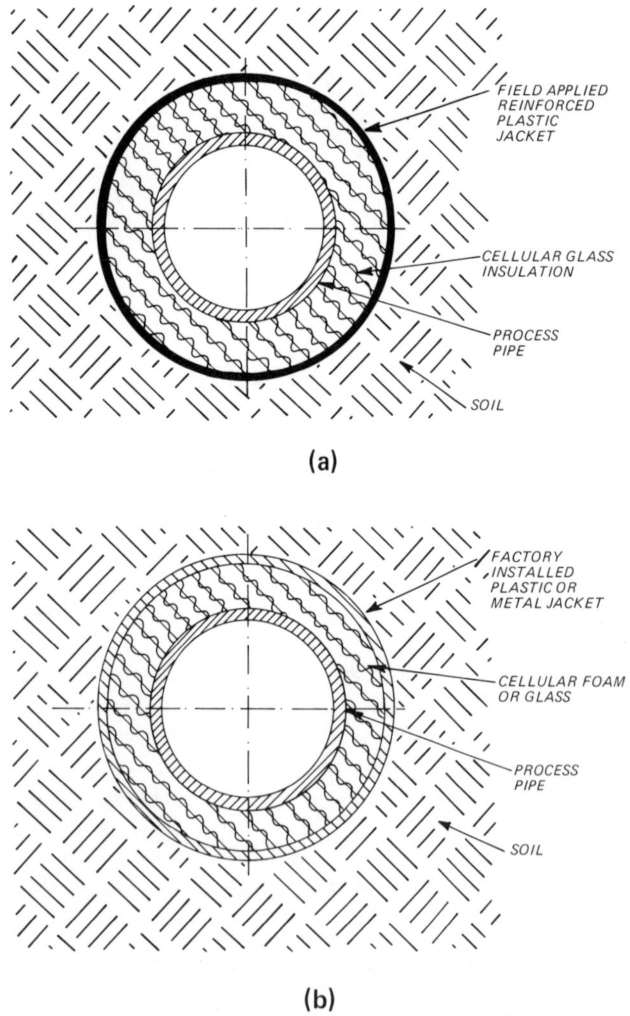

Figure 117 Underground insulated pipe.

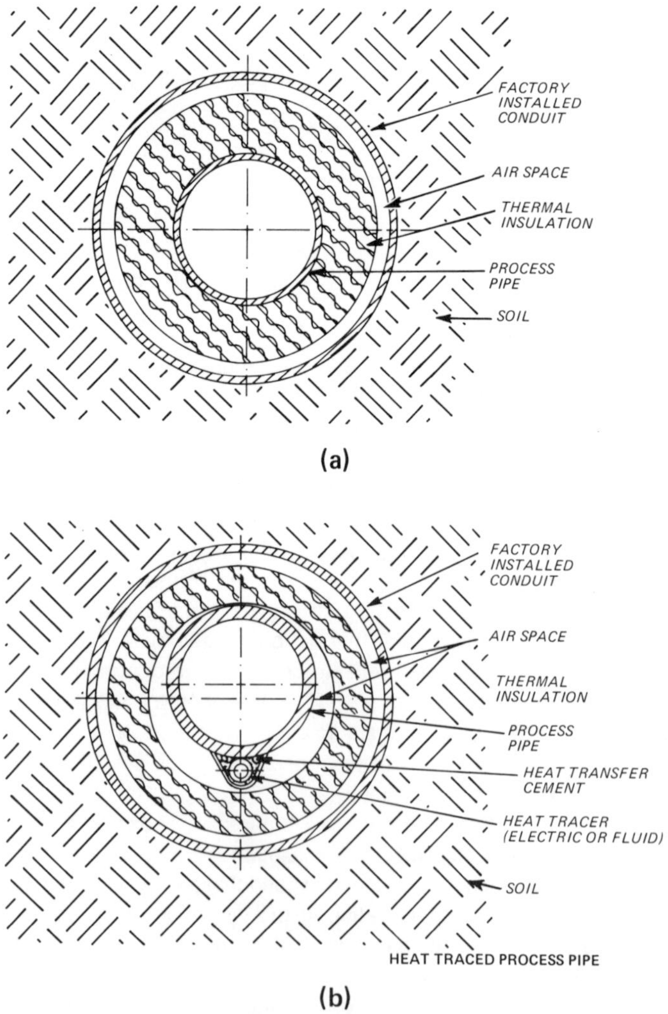

Figure 118 Underground insulated pipe.

condensation or ice forming around the insulation. It might also be necessary to provide restraints to prevent flotation where water level may be above the bottom of the insulated process pipe.

Where pipe contains gas, steam, or liquid where energy loss must be supplied by heaters etc., then the thickness of the insulation should be determined in accordance with recommendations for economic thickness of insulation as given in Chapter 2.

Figure 117b illustrates a factory prefabricated insulated pipe system where the insulation fills all the void between the pipe and outer jacket or conduit. In most instances, the insulation used in these systems is cellular organic foams or glass. Because of the problem of differential expansion between the process pipe and outer plastic jacket (or metal conduit), the temperature range is limited. Fittings and connections are preformed for the particular system and must be designed to fulfill the piping requirements.

The heat transfer from pipe, through the insulation, to the soil is calculated by heat transfer Equations 8 and 8a.

Being a preformed system, the cost of installed insulation by itself is not readily available. It is suggested that economic thicknesses be determined by taking the cost of installed jacketed underground pipe from the cost of installed insulated jacketed pipe so as to determine insulation cost for use in economic thickness calculations.

Figure 118a illustrates a conduit system where the process pipe is supported by cradles or shoes secured to the conduit and the insulation is installed to provide an air space between it and the conduit. Such arrangement makes possible the choice of insulation which most nearly serves all the physical and thermal requirements of the system. The insulation may be cellular glass, glass fiber, mineral fiber, calcium-silicate, expanded silicate, or diotomaceous earth rigid type insulations.

The heat transfer equations for determination of heat loss are Equations 11 and 11a, where t_a is taken as the temperature of outer conduit being that of soil with which it is in contact.

The thickness of insulation should be determined in accordance with the principles as stated in Chapter 2, "Economics of Thermal Insulation."

Figure 118b shows a pipe in a conduit system where heat is supplied by an external electric or fluid tracer. The thickness and type of insulation should be that which fulfills the economic requirements and the tracer requirements as given in Chapter 1, "Heating and Cooling of Piping, Equipment, and Vessels."

Prefabricated underground piping systems for one or more pipes, of one or more metals, insulated or not insulated, are available from a number of manufacturers. In all cases, the major problem is the design of the system to fulfill the pipiing and installation conditions, and as such goes beyond the scope of this book.

10 Installation Systems Details

In addition to the physical design fundamentals discussed in Chapter 9, the success of an insulation system depends upon individual consideration of design and insulation for the many details of the system. These details are part of the physical design. However, they have been separated from the previous chapter so as not to confuse the more basic considerations.

In all insulation design, thermal "short circuits" should be avoided or minimized. These thermal short circuits are most frequently the result of voids or projections of metal or other high conducive materials through the insulation. They are illustrated in Figure 119. The illustration shows the excessive heat transfer from a hot surface to a lower ambient temperature. If the conditions were reversed, with the temperature of the surface lower than ambient, the same excessive heat transfer would exist, but the heat flow, as shown, would be reversed.

Voids in Insulation

Some voids in insulation are the result of insulation design and specifications. Typical of voids in the insulation system, which are a result of design, are those specifications which state that flanges and valves are not to be insulated. As demonstrated in Chapter 2, unless the temperature difference is relatively small, or the heat is of little value, these specifications are in economic error.

Voids which are not deliberately *designed* in, but *develop* in an insulation system are most often the result of expansion of the substrate to which the insulation is applied, shrinkage of the insulation, or poor design or workmanship. The first two of these have been discussed in Chapter 2 under "Mechanical Requirements." Suggested solutions for preventing voids caused by expansion and shrinkage of insulation, from forming in pipe covering on hot piping are shown in Figure 68 (Chapter 7), Figure 83 (Chapter 8), and Figure 111 (Chapter 9). Suggested solutions for preventing voids from being formed in insulation installed on hot equipment are shown in Figures 95, 96, and 97 (Chapter 9. Voids left in insulation, such as those occuring when flat blocks are used around curved surfaces, are the result of poor design specifications. Proper understanding of the importance of obtaining accurately fitted insulation should result in better design. Voids due to poor workmanship can only be corrected by better application, training, and supervision.

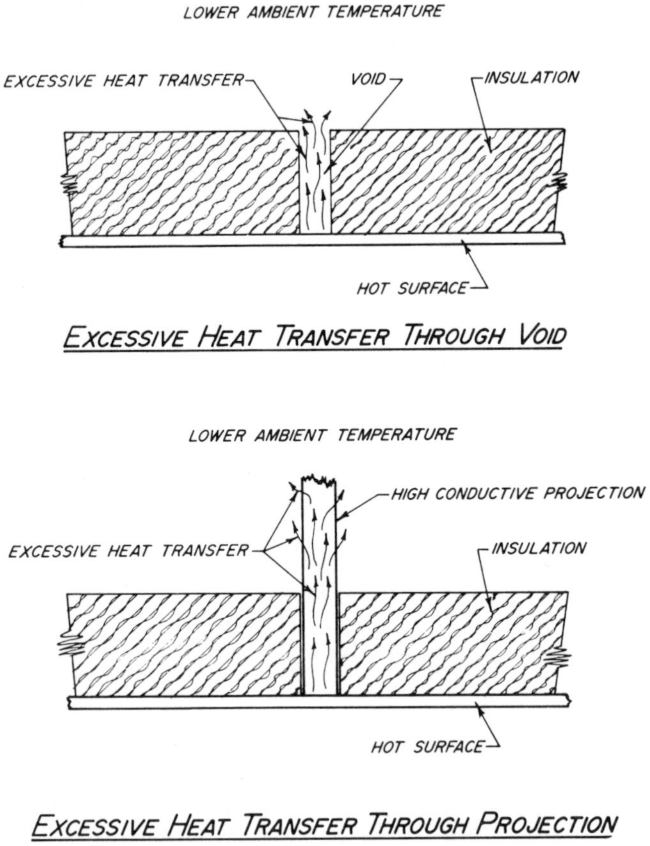

Figure 119 Thermal "short circuits".

Where rigid insulation is installed on low temperature surfaces, space must be provided for the rigid insulation to move into when the substrate contracts. This space must be filled with compressible insulation, such as fibrous or flexible foam. Just as vapor-barriers must be provided to prevent entry of moisture, these contraction joints must be properly designed and installed to prevent them from constituting a void, rather than being an insulation contraction joint. Design of equipment and piping contraction joints has been discussed, and suggested methods are shown in Figure 51 (Chapter 6), Figure 71 (Chapter 7), and Figures 95, 96 and 97 (Chapter 9). Design of contraction joints on flat surfaces, such as walls and ceilings in refrigerated spaces, are shown in Figures 124 and 125 of this chapter. It should be remembered that voids in insulation on low temperature surfaces in which moisture can condense and freeze, rapidly change from a void to a high conductive projection. As ice fills the voids, its conductivity, which is approximately 15.5 Btu per sq ft per hr, in., °F (2.235 W/mK), makes it a high conductive means to transfer heat.

Projections through insulation

Metal projections through insulation should be avoided, if possible. Metals of two different temperatures should be isolated by insulation where it is practical. Figure 77 (Chapter 7) illustrates a means by which pipe insulation is supported by a cradle, and the hot, or cold, pipe is completely isolated by insulation from metal contact with the beam.

Such isolation is not always possible, due to mechanical requirements, and projections through the insulation, such as equipment supports, tank legs, vessel skirts, pipe anchors, and pipe guides are necessary, due to imposed loads and stresses beyond the capability of the insulating materials.

Where it is necessary to have high conductive materials, such as metal, project through insulation, the heat transfer can be minimized by several methods. First, keep the cross section of the projection to a minimum. Second, use a material of the lowest possible thermal conductivity. Third, provide a long flow path through insulation. Fourth, provide a thermal barrier in some phase of the projection connections or bearing.

Besides heat loss, or gain, in the consideration of projections through insulation, the temperature at which the projection makes its exit from the surface of the insulation may be of vital concern. On extremely hot installations, the projection may be sufficiently hot at the point where it leaves the insulation to be a hazard for personnel, or a potential ignition source for gases or chemicals of low ignition points. Excessively hot projections also deteriorate the weather-barriers or seals necessary to keep them water tight. On low temperature applications, projections may cause excessive sweating, with associated rusting and corrosion, or ice formation.

Where a choice exists between holding an insulated item in place by supports in compression, or tension, the use of the latter is preferable as the load can be supported with less metal cross section.

The use of stainless steel, which, besides having a lower conductivity than carbon steel, also has higher strength, can reduce any thermal transfer to one-fourth that of carbon steel.

The use of pads of highly compressed, woven asbestos or low conductive graphite to break direct metal-to-metal contact between a metal of one temperature and one at another temperature can be very effective in reducing heat loss or gain.

Methods of supporting piping and equipment usually are not the responsibility of the insulation engineer or designer. However, cooperation with the structural engineer to help devise *thermally efficient* equipment and piping supports will result in more efficient and trouble free process systems.

The length of flow path of a projection through insulation can be extended by insulating the projection beyond the basic insulation installed on a pipe or vessel. On hot service, besides the savings in heat, it is frequently necessary to extend the insulation on a projection so that the temperature of the projection where it leaves the insulation is low enough to be below the maximum allowable temperature of the coating or weather-barrier. In such instances, the projection should be insulated just sufficiently to conserve heat, but also to allow sufficient heat loss that the temperature will drop down to a required exit temperature. The insulation of hot projections are illustrated in Figure 65 (Chapter 6).

Projections from low temperature pipe and vessels are insulated out their length to control condensation and frosting where they leave the insulation. Typical of the insulation of a projection which must pass through insulation is that shown in Figure 76 (Chapter 7). This also illustrates that some projections must be lengthened to provide a sufficiently long heat path through the in-

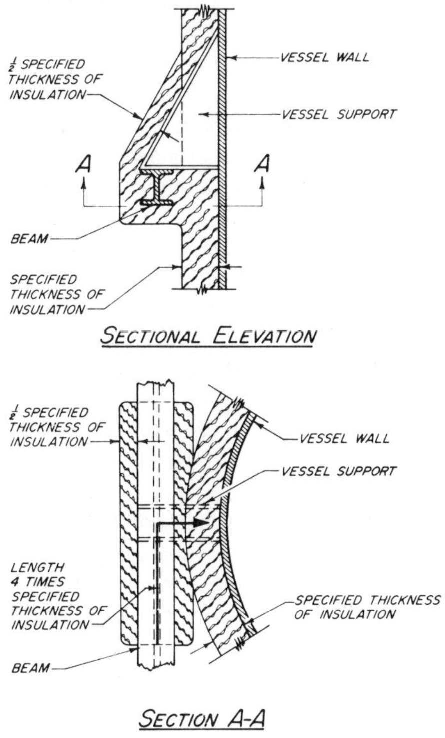

Figure 120 Application of insulation to projections installation to equipment lug and beam.

sulation. An application to an equipment lug and beam is shown in Figure 120. The length of covering of the projection, shown as four times the insulation thickness, is generally sufficient to prevent frosting at the end of the projected insulation. The thickness is reduced to 1/2 the specified thickness so that the metal can retain sufficient heat to bring its temperature up to above frost point at its exit from the insulation. These same proportions are frequently used for insulation on hot surfaces, but they may need to be altered for very high temperature services.

In some instances, combinations of these methods of controlling thermal short circuits are required. Partial insulation, using materials (to separate metals) which are not specifically insulation, but which have lower conductivity than metals, in combination with insulation extended out the metal supports may be the only way in which both thermal and mechanical requirements may be satisfied. One such problem is illustrated in Figure 121. This figure shows a horizontal vessel, *at low temperature*, which is supported by metal cradles which are too short to obtain sufficient heat gain from the atmosphere to prevent frost formation on them, and, thus, would frost up beyond the ends of the anchor and slide plates. Wood blocks, treated to prevent rotting, moisture absorption, and termites, are used as partial insulation isolators between the metal and the grout plates bearing on the footers. The wood should be carefully vapor protected on all sides and ends with a thick coat of barrier coating before being secured in position. Antifriction slide plates should be installed between the grout plate and steel on bottom of wood supports to prevent binding of slide assembly due to rust or corrosion. Insulation should be installed down around edges and sides of block and carefully sealed to retard vapor and liquid moisture leaks.

While it may appear that great emphasis has been placed on voids and projections through insulation on low temperature service, the reason for this is that these voids and projections will cause more rapid insulation failure on such service than on high temperature service. With the greater temperature differences that are generally present in high temperature service, the heat transfer through voids and projections on this service is most likely to be much greater than on low temperature service. However, the great heat losses due to voids and projections in the high temperature range is not nearly as visually apparent (as ice formations) but they may be just as costly.

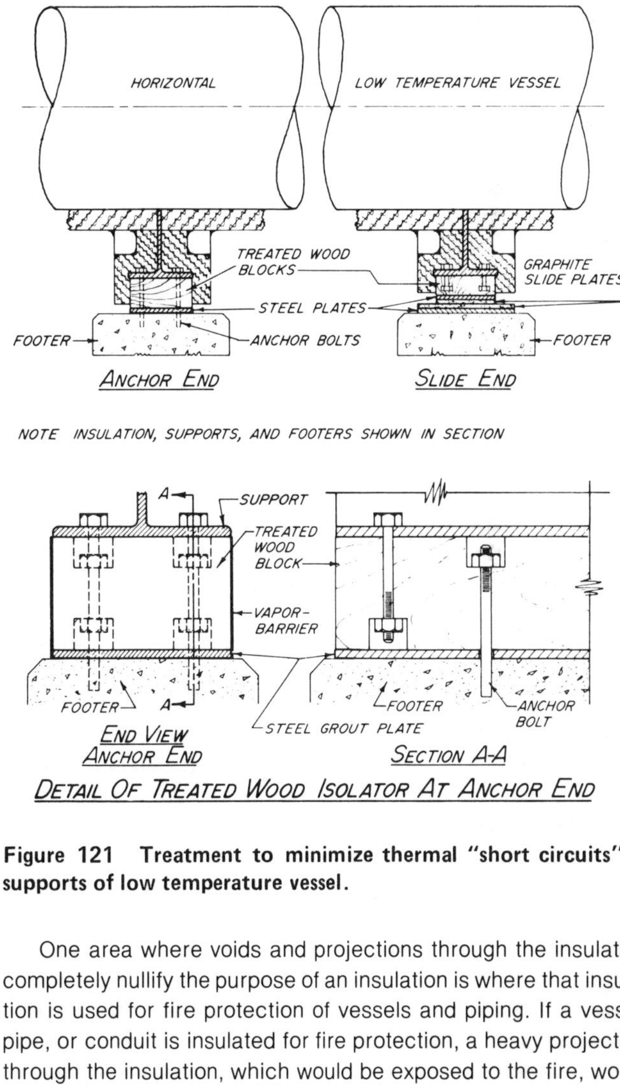

Figure 121 Treatment to minimize thermal "short circuits" at supports of low temperature vessel.

One area where voids and projections through the insulation completely nullify the purpose of an insulation is where that insulation is used for fire protection of vessels and piping. If a vessel, pipe, or conduit is insulated for fire protection, a heavy projection through the insulation, which would be exposed to the fire, would transfer the fire heat through the insulation at that point and fire protection would not be achieved. This is also true of voids in insulation.

Voids in the insulation used for fire protection can be caused by the fire itself. Some insulations used for this purpose are calcium-silicate, diatomaceous earth, expanded silica, and asbestos fibers. For ordinary insulation services, these insulations

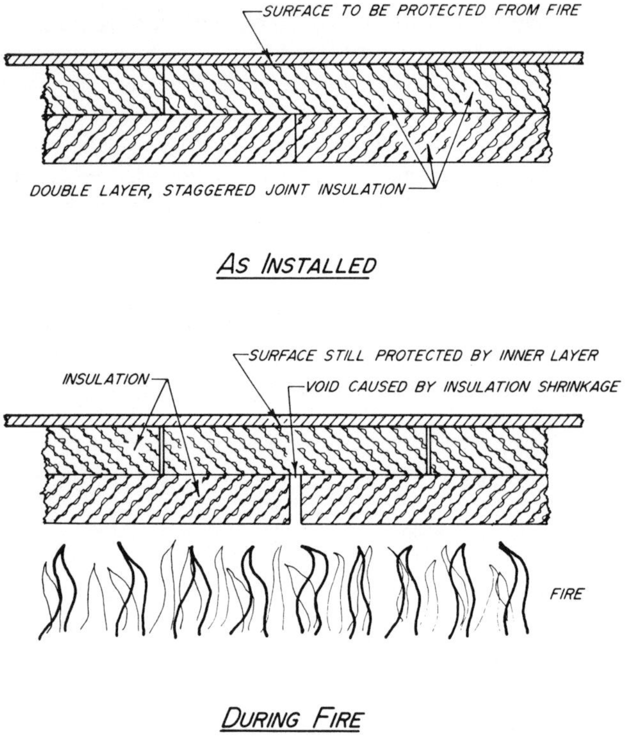

Figure 122 Installation of insulation for fire protection.

are suitable for service temperatures of 1000° F to 1800° F (538° C to 982° C). All will resist a fire of 2300° F (1260° C) for a considerable period of time, (over one hour) thus making them suitable as a fire protection material. All of them, however, will shrink when subjected to fire temperatures. The shrinkage will cause voids at the butt joints. For this reason, when fire protection is a critical function of the insulation, double layer, broken joint construction should be used. Thus, as the outer layer gets hot and shrinkage occurs, the resultant void will exist only half way through the total insulation thickness. This is illustrated in Figure 122. Single layer application should be restricted to fire protection of pipes or conduits which require only short time protection (15-20 minutes), as this is the amount of time required to shrink the insulation.

Fire protection of cold lines and equipment requires careful consideration. It must be remembered that non-burning and flame spread properties are not a measure of the length of time which a material will resist fire. When completely enveloped by flame, organic foam insulation will completely disintegrate within a few minutes. Cellular glass, when subjected to fire, will stress crack into pieces 3″-6″ (76.2 to 152.4 mm) in size. However, if the cracked cellular glass can be held in position, these pieces will resist fire for a considerable length of time. This property provides the means for three practical systems for obtaining fire protection of cold lines and equipment. The choice of which system depends upon the length of time of protection required.

The thickness of the cellular glass should be not less than that required for the refrigeration load. It should be installed with vapor sealed joints to resist the vapor pressure inward, and the outer layer should be secured in position with stainless steel strap. This method of application should be used for all three systems, as the difference in the systems is in the outer treatment after the cellular glass is installed.

System 1 consists of covering the cellular glass with either a fire resistant weather-barrier mastic reinforced with fire resistant cloth, or a weather-barrier jacket of stainless steel. The mastic should be not less than 1/16″ (1.59 mm) thick after it has dried. This type of mastic will deteriorate in a fire, but will stay in position and hold the thermally cracked cellular glass in place.

System 2 is installed in an identical manner as described above, with a coat of intumescent paint applied over the weather-barrier mastic, after the weather-barrier has completely dried. The fire retardant intumescent paint cellulates at approximately 350°-400° F (177° to 204° C) and forms a light density carbonized insulation which will resist fire for a period of time. This installation will provide approximately 15-20 minutes more fire protection than System 1.

System 3 provides for the installation of 1½″ (38 mm) thick expanded-silica insulation over the cellular glass, then weather protecting the expanded-silica insulation with fire resistant and fire retardant mastic reinforced with fire resistant fabric. As the vapor pressure is inward, the moisture vapor will eventually saturate the expanded-silica insulation. However, this insulation will not swell

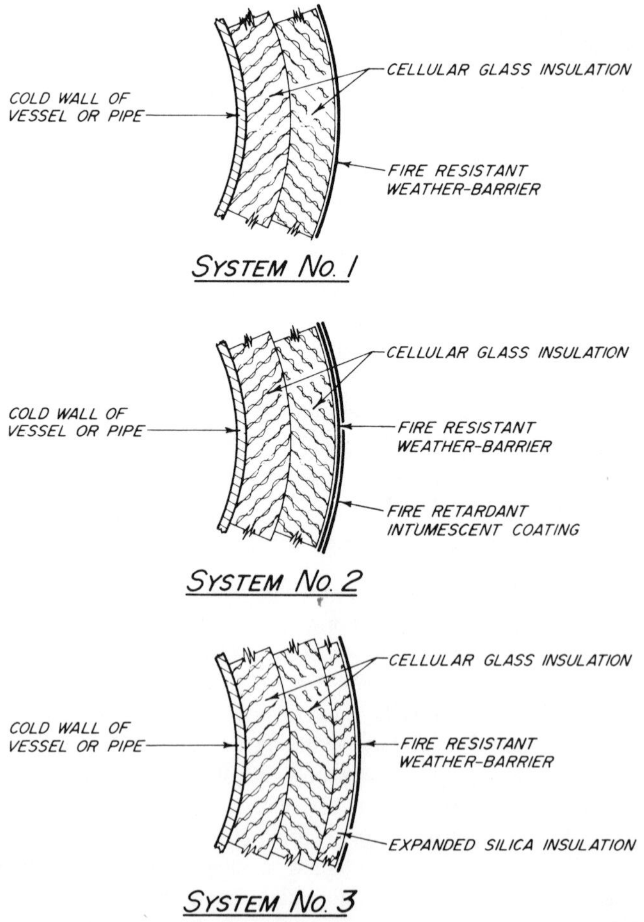

Figure 123 Methods for providing fire protection for insulated low temperature pipes and vessels.

when moisture saturated, and, if the cellular glass is properly vapor sealed at every joint, the moisture is restricted to being trapped in the outer layer of the expanded-silica.

These systems are illustrated in Figure 123.

There exists no method of fire *proofing.* All materials will be destroyed in time when exposed to a fire of sufficient temperature. The fire *protection* which can be obtained is a matter of the time which insulation materials, or other fire protection materials, will prevent an excessive temperature rise, or failure of the substrate they are to protect.

Fire protection insulation for pipe, both high and low temperature, should be supported by cradles *outside* the insulation wherever possible. Wherever projections, such as anchors for pipe, or support legs, or skirts for equipment must extend through the insulation, then these must also be fire protected in the same manner as specified for the pipe or equipment.

Voids and Projections—Cold and Refrigeration Storage Rooms

The problems related to voids and projections through the insulation also exist in insulation used to construct cold storage and refrigerated spaces. Because the insulation is built in as a part of the structure, it is impossible to make complete and definite recommendations, as the combinations with the structural design provides an extremely large number of solutions. For this reason, only the most fundamental recommendations can be given in this manual.

To reduce the number of structural members which pass through the insulation, the design should be such that the insulation is either completely inside the structure, or completely outside it as nearly as possible. Due to the temperature differential involved, differential expansion will occur between the structural members and the insulation itself, and expansion-contraction joints in the walls, ceilings, or roofs, are essential.

A typical roof insulation with the structural steel on the interior of the refrigerated space is shown in Figure 124. In this figure, if the insulation will transmit vapor, then a vapor-barrier must be provided on the hot side. Cellular glass requires no vapor-barrier, but must be vapor sealed at its butt joints. The same is true for the figures presented for corners, and walls, shown in Figure 125.

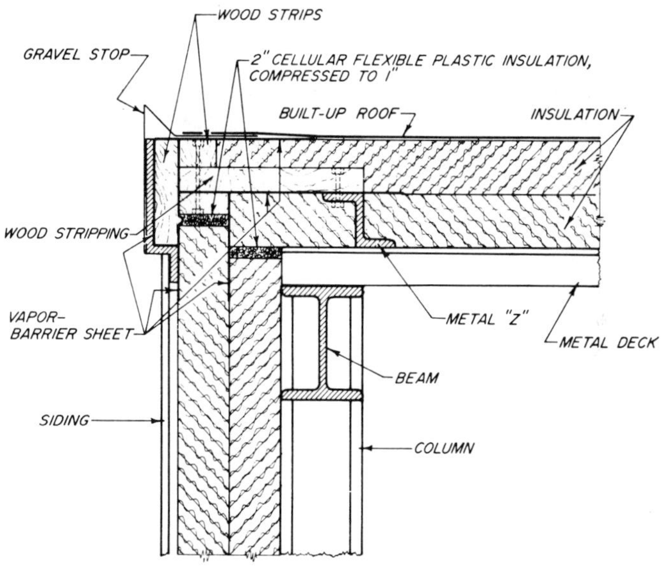

Figure 125 Expansion-contraction joint of low temperature space in corner of wall and roof insulation.

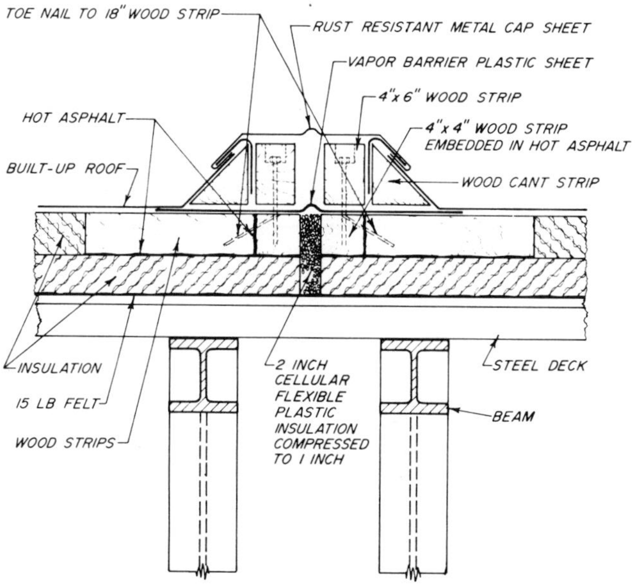

Figure 124 Expansion-contraction joint low temperature space roof insulation.

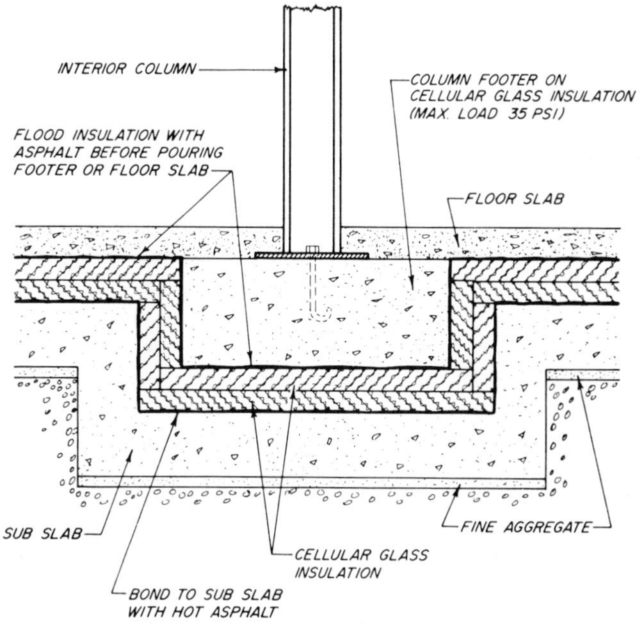

Figure 126 Isolation of interior column and footer with cellular glass insulation.

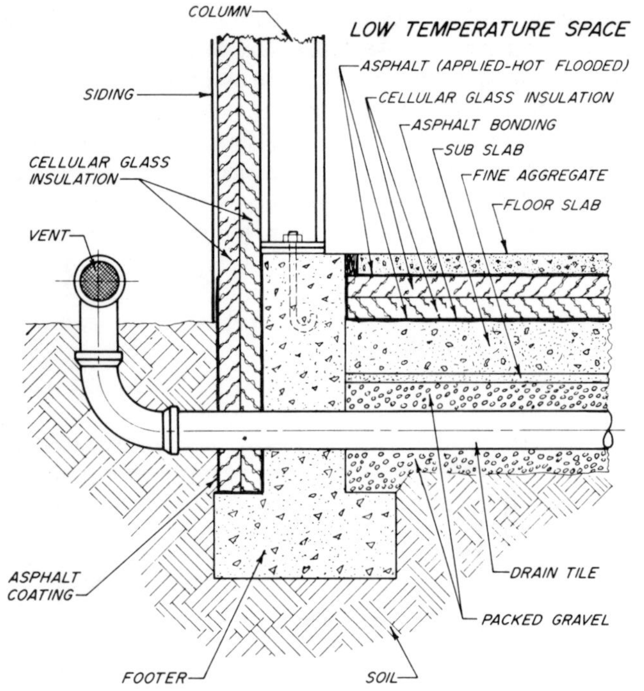

Figure 127 Venting under floor and wall footer detail low temperature space insulation.

Interior columns in low temperature insulated spaces which go to ground, can cause considerable difficulty. As the columns provide a large cross sectional area for thermal short circuit, they become quite cold and can cause freezing of the soil below them. This, in turn, will cause ice heaving, and push the column out of its original position. Thus, it becomes quite important that the column and its footer be isolated by insulation from the ground on which it rests. The method of doing this is illustrated in Figure 126. Cellular glass insulation, being inorganic, is termite proof, rot resistant, and has sufficient compressive strength to support loads, if the load is properly distributed over sufficient area.

It is essential to keep the ground under refrigerated spaces dry and above freezing. Where soil conditions are wet, it becomes necessary to vent the moisture from below the slab-insulation assembly. A suggested method for such venting, and a detail of insulation around the wall footers, is shown in Figure 127.

In areas where climatic and soil conditions, the size of the low temperature space over the ground, and its temperature, would cause freezing of the soil beneath the slab, it is necessary to add heat to the ground to prevent freezing and ice heaving. Two ways of adding this heat are, one, to bury hot water pipes, and two, electric heating cables in the ground beneath the slab.

The hot liquid is generally composed of a solution of antifreeze and water to prevent it from inadvertently freezing. It is distributed through a piping system so designed as to provide even distribution of the heat, by the use of a pump or pumps.

Recently, electric heated systems have proved very effective. One reason being that the heat output is equal for each linear foot, regardless of distance from the power supply.

Both systems require careful heat transfer calculations to obtain the correct heat balance. Also, both systems should be thermostatically controlled to compensate for variations in atmospheric conditions.

Doors in low temperature space enclosures provide closable voids through the insulation walls. There are so many types of doors available to serve individual needs that even a general recommendation is impossible. There are sliding doors, swinging doors, flush fitting doors, and surface fitting doors. The method of their installation, the amount of framing, and structural reinforcement required depends upon the weight, size, temperature differential, usage, and other factors. It is essential that the doors be selected to fulfill the operating requirements, and that the structural requirements be followed so that proper thermal and structural design can be accomplished.

On a very low temperature space, it may be most economical to install a vestibule type of entrance, using a door at each end, so as to minimize the excessive heat gain which would occur if the space was opened directly to the ambient air.

Under Slab Insulation for Heated and Air-Conditioned Building

Considerable information is available regarding numerous systems for the insulation of walls and roofs of buildings, however, almost no consideration is given the correct method of insulating under concrete slab floors. This is due to a number of incorrect assumptions which are: (1) Most consider concrete slab good insulation, which it is not; (2) Heat loss to ground is small and; (3) Insulation will get wet and be ineffective. All these assumptions are incorrect.

1. The thermal conductivity of concrete is very high, 12.0 Btu in./ft^2, hr, °F (1.73 W/mK).
2. The heat loss is high and with the higher energy cost, insulation is justified by economics.
3. Cellular glass has sufficient strength to support most slab loadings and its low vapor migration rate will keep it effective as an insulation for a very long period of time.

As the heat from the building tends to heat the soil upon which it rests, the temperature of soil is higher towards the center of the building and colder towards the perimeter. For this reason, the heat loss per square unit of area changes in regards to location. For these reasons a heat transfer chart and surface temperature of tile surface has been prepared for a number of soil temperatures under the slab.

The heat loss and interior surface temperature of the floor tile of uninsulated slab, as shown in Figure 128, is given in the Table below. The controlled temperature of the space is 72° F (22.2° C).

UNINSULATED SLAB

English Units			*Metric Units*		
Soil Temp °F	**Heat Loss Btu/ft^2 hr**	**Surface Temp-Tile °F**	**Soil Temp °C**	**Heat Loss W/m^2 hr**	**Surface Temp-Tile °C**
40	25	49	4.4	78.8	9.4
45	21	53	7.2	66.2	11.7
50	17	56	10	53.6	13.3
55	13	60	12.8	41.0	15.6

The heat loss and interior surface temperature of the floor tile of insulated slab, as shown in Figure 129, is given in the Table below.

INSULATED SLAB (2″ CELLULAR GLASS)

English Units			***Metric Units***		
Soil Temp °F	**Heat Loss Btu/ft^2 hr**	**Surface Temp-Tile °F**	**Soil Temp °C**	**Heat Loss W/m^2 hr**	**Surface Temp-Tile °C**
40	4.8	67	4.4	15.1	19.4
45	4.0	68	7.2	12.6	20.0
50	3.3	69	10.0	10.4	20.6
55	2.5	70	12.8	7.9	21.1

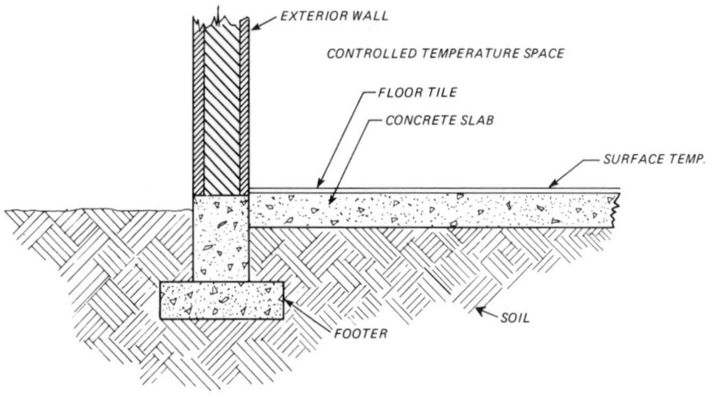

Figure 128 No insulation under floor slab.

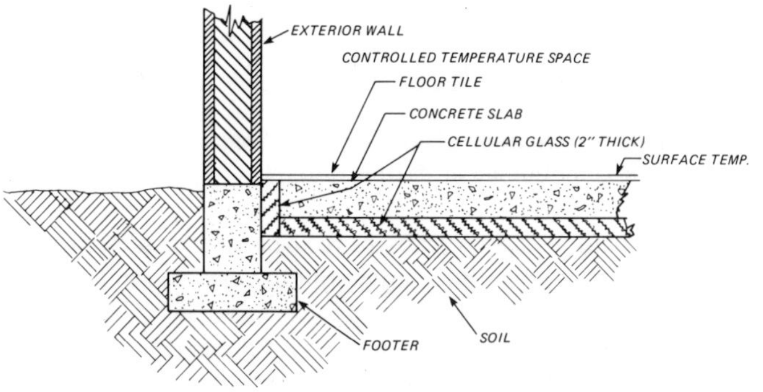

Figure 129 Perimeter and under slab insulation of 2″ thickness of cellular glass.

Voids and Projections through Insulation Enclosing Hot Temperature Spaces

The considerations regarding voids and projections through insulation enclosing high temperature spaces, such as drying rooms, ovens, etc. are almost identical to those of insulated low temperature spaces. Instead of contraction joints, expansion joints must be provided to prevent direct-through voids. Projections through the insulation cause an excessive heat loss, however, rather than an excessive heat gain. One major difference does exist, and that is that no vapor-barrier should be put on the outside. However, in the case of drying ovens which produce vapor at very high pressure, it becomes necessary to provide vapor-barriers on the *inside* to prevent the excessive moisture from passing into and condensing within the high temperature insulation. This is a factor which is frequently overlooked.

Internal Insulation of Air-Conditioning and Heating Ducts

This particular use of insulation is presented because it brings up two factors not often associated with thermal insulation. First is the ability of the insulation to resist erosion due to velocity of air moving over its surface, and second is its ability to absorb and attenuate sound.

Selection of insulation, over and above its thermal properties, depends upon its characteristics as a sound absorber and attenuator, and its ability to resist the velocity of air moving through the duct. Both rigid and flexible insulations are used in this service. Various densities of these types are available. Although this manual is not primarily on sound, a single fundamental regarding sound attenuation, the ability of materials to reduce the sound level in its passing through the material, should be presented. *That* is that high density materials are necessary to restrict sound level in its passage through the material.

To explain: If the sound level inside a duct is high, and the duct passes through a quiet area where the noise is objectionable, then a very heavy density insulation is rrequired to contain the sound

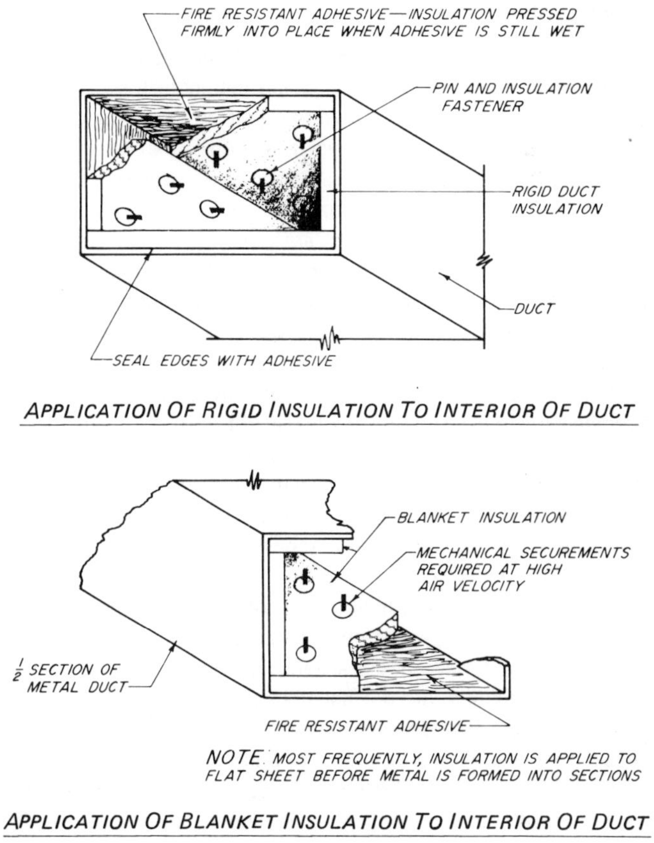

Figure 130 Installation of thermal and sound insulation to interior of air conditioning or heating duct.

within the duct. If the duct leaves air-conditioning or heating equipment, travels some distance before reaching an outlet, (through areas where noise is of no importance) and it is desired that the sound be absorbed during its travel through the duct, then lighter density materials are suitable.

Fastening of thermal and accoustical insulation inside of duct work is usually done with adhesives, or adhesives and mechanical fasteners. Fastenings are dependent upon the size of the duct and the velocity of the air passing through it. For small ducts with air velocities up to 3000 fpm (914 m/min), adhesives, if correctly chosen and applied, are sufficient to secure the insulation inside the ducts. Large ducts and higher velocities require mechanical fasteners such as pins and clips or grip nails to assure that the insulation does not pull loose. At the higher velocities, it is also essential that the edges of the insulation be carefully adhered to the duct and sealed so that insulation cannot be pulled apart by the air stream. Typical examples of such installations are shown in Figure 130.

Because such installations are completely hidden and difficult to inspect, there has been some tendency to use very lightweight insulations with insufficient binder to withstand the air velocity and also to skimp on the adhesives and fasteners. However, if the insulation becomes loose and blocks the air passage, correction of the failure can be extremely expensive.

External Insulation on Air-Conditioning and Heating Ducts

Most ducts inside a building which warrant insulation on their exterior are air-conditioning ducts, which distribute the cooled air which might cause condensation to occur on the surface of the metal. As the insulation is installed on a surface whose temperature is lower than the ambient air, it requires a vapor-barrier to resist the vapor pressure inward. For this reason, the insulations, both rigid and flexible, are most often supplied with a factory applied vapor-barrier covering. As with all insulation involving the retardation of vapor passage, the most important and difficult part of the installation is the vapor sealing of the vapor-barrier.

With the development of better adhesives, the practice has become almost universal to apply blanket (flexible) insulation to ducts, using fire resistant insulation adhesive as the securement of the blankets to the duct without resorting to mechanical fasteners. Laps of the vapor-barrier are sealed with the same adhesive. A typical installation is shown in Figure 131.

For large ducts, and where more mechanical strength is required, rigid insulation is held in position by welded or adhesive fastener pins. These pins are then impaled through the insulation as it is forced on them, and a speed clip is forced on each pin to fasten the insulation into position. Pins should be cut to be flush with the surface. Points where the pins penetrate the vapor-barrier should be sealed with joint sealer, or a patch of the vapor-barrier sealed to the vapor-barrier by joint sealer. All joints should be sealed with joint sealer. A typical application is shown in Figure 131.

Exposed ducts may be coated with suitable mastics or coverings to obtain additional mechanical protection for the insulation or for decorative purposes.

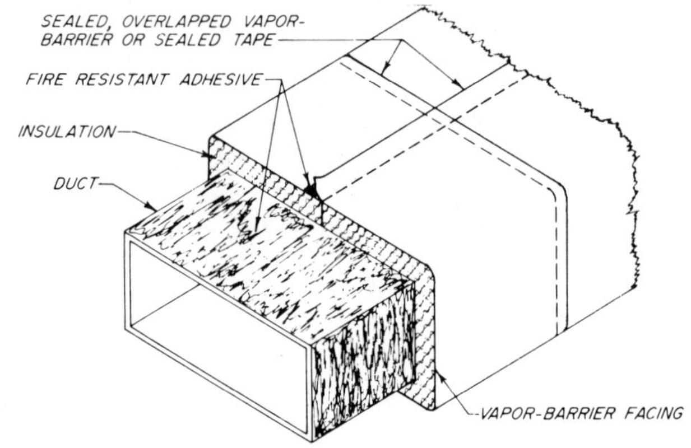

INSTALLATION OF BLANKET INSULATION ON DUCT

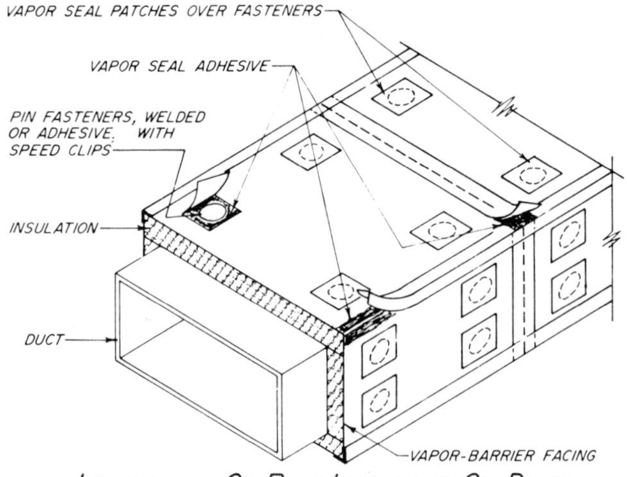

INSTALLATION OF RIGID INSULATION ON DUCT

Figure 131 Installation of insulation and vapor-retarder on exterior of air conditioning or heating duct.

Rigid board insulation made of resin-bonded glass fiber and faced with fire resistant foil Scrim-Kraft can be used to form air handling ducts. These boards are available in 1″ (25 mm) thickness, 48″ (1219 mm) wide and 96″ (2438 mm) or 120″ (3048 mm) in length. Standard tools are available to cut shiplap and grooves in the sheets so as to form duct systems. These are shown in Figure 133. Boards can be cut to form rectangular ducts. At the junction of where two pieces are being joined the shiplap joint is recommended. The same type construction can be used to form branches, elbows, tees or transitions.

Where two pieces join, the longitudinal junction should be provided with stapling falp of foil Scrim-Kraft facing material. After edges are trued-up and grooves properly seated, the stapling flap is folded around the corner and stapled to the mating side on 2″ (50.8 mm) centers. The corner must then be sealed to prevent air leakage. This can be accomplished by heat sealing thermal sensitive tape with hot iron or by the use of pressure sensitive tape. These ducts should be reinforced by channels and tee bars in accordance with the recommendations of the manufacturer of the insulation board. A typical straight duct construction is shown in Figure 133. Scored block installed around duct or large pipe is shown in Figure 134.

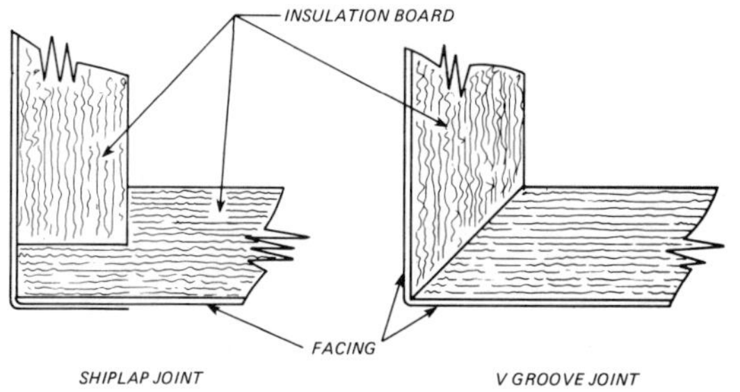

Figure 132 Methods for fabricating rigid insulation board into ducts (corner joint).

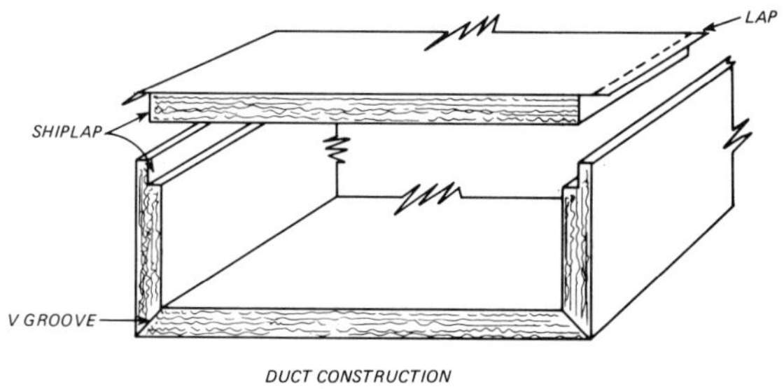

Figure 133 Typical duct construction of glass fiber insulation board.

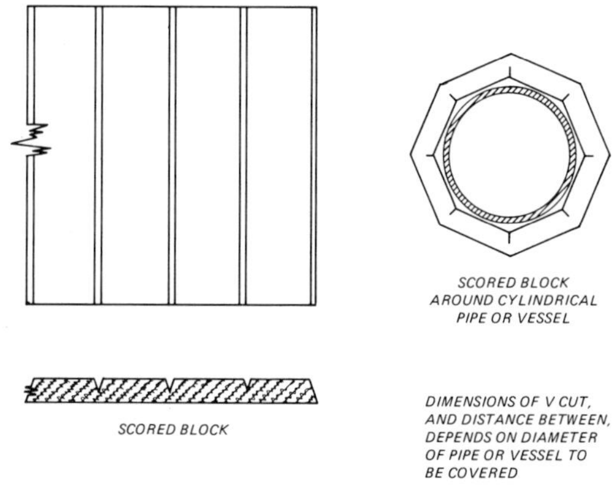

Figure 134 Prescored block to fit around pipe or vessel.

Traced Piping and Vessels

Where it is necessary to add or remove heat from piping or equipment, the external container tubing, pipe, or electric cables are called tracers. Pipe and tubing tracers containing hot water, hot vapor, or steam, is a medium to supply heat to the piping or equipment. In addition, heat may be supplied by electric cables. To remove heat the tracers may use refrigerants, cold water or brine.

If the additional temperature to be supplied by the tracers is small then they may simply be fastened to the pipe or vessel and the air space between the insulation and metal substrate will distribute the heat over the pipe or vessel surface. However, where high temperatures or close tolerance of temperature is required, then the tracers should be thermally bonded to the pipe or vessel, as presented in Chapter 1.

For efficient operation of tracing systems they must be installed correctly. Tracer tubing and all its fittings shall be suitable for the pressure of the steam vapor, gas or liquid which they contain. Should the operating temperatures of the process line exceed the temperature of the steam (or liquids) used in the tracers, the tracer shall be protected with relief valves to prevent bursting due to excessive pressure as a result of the higher temperatures.

The fluid or steam tracers shall be of a metal which resists corrosion. To illustrate: Steel is not recommended for tracers of any extended shut down period or when the exterior of the tracers are exposed to a moist temperature below 175° F (79.4° C). On cooling tracers, if brine is used and the brine contains chloride or fluoride ions, stainless steel is unsuitable.

Where fluid and steam tracers are required to keep the process pipe at temperatures greater than what can be obtained by a single unbonded tracer, then the tracer must be thermally bonded by heat transfer cement, or premolded strip, to the process pipe or equipment.

Where heat transfer cement is used it must be suitable for the process pipe and tracer temperatures to which it is to be subjected. The heat transfer cement should not cause corrosion or rusting of the tracers or piping and care should be taken to heat up water mix cements within the period of time recommended by the manufacturer so as to minimize rusting of steel or wetting of the insulation. The heat transfer cement, depending upon conditions, may or not be contained in channels.

Tracers may also be bonded to process pipe and equipment

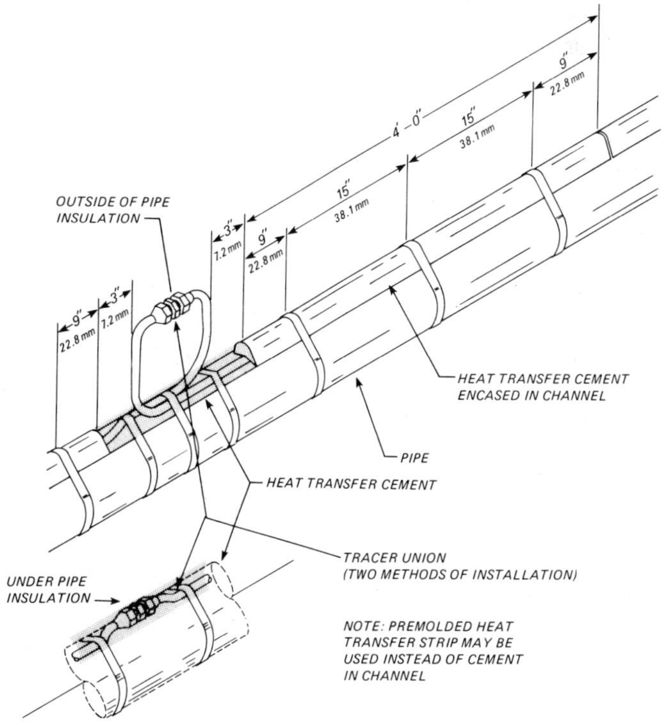

Figure 135 Installation of tubular tracer, thermally bonded, to pipe.

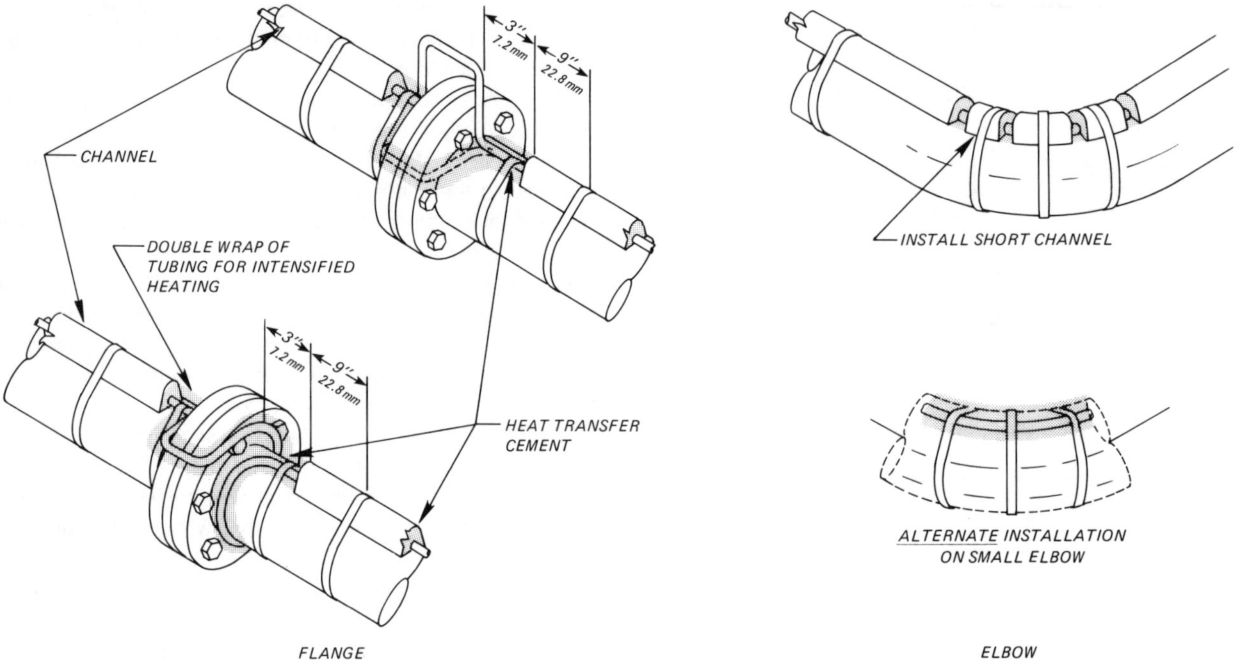

Figure 136 **Installation of tubular tracer, thermally bonded, to pipe and fittings.**

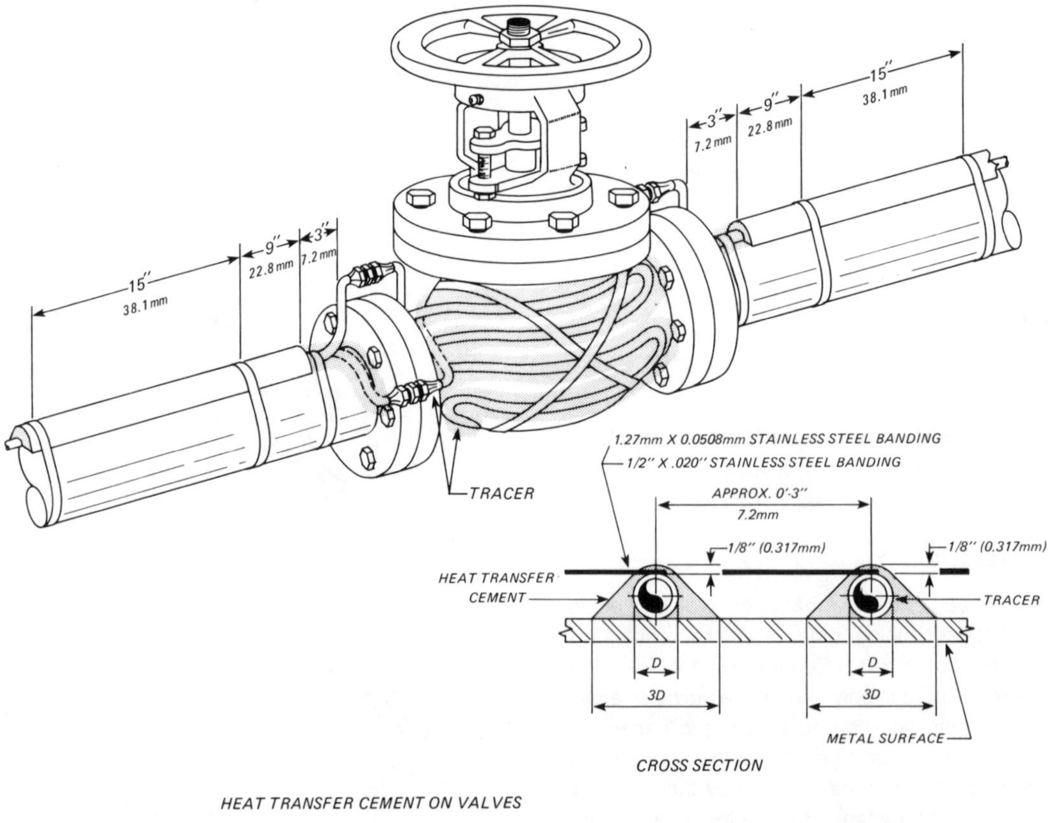

Figure 137 **Installation of tubular tracer, thermally bonded, to flanged valve.**

with premolded heat transfer strip of dimensions to fit the tracer and the channel.

It is quite important that tracers be bonded to clean pipe and fitting and be securely strapped in place. Figure 135 illustrates the method of installing tracers to pipe and the methods used for installing unions in the tracer systems. Figure 136 shows proper installation of tracers to elbows and flanges and Figure 137 shows how tracers are looped to obtain sufficient heat transfer at valve.

The selection of insulation for traced lines depends upon temperatures to be maintained in tracer system. The insulation shall be strong and rigid, of a type that can be obtained in sectional pipe insulation and which can be prefabricated into required shapes for ells, tees, flange and valve covers. The dimensions for prefabrication of these fittings for single traced piping systems are available in the ASTM Recommended Practice C-450 "ASTM Recommended Dimensional Standards for Prefabrication and Field Fabrication of Thermal Insulating Fitting Covers for NPS Piping, Vessel Lagging, and Dished Head Segments." The manner in which the tracing, pipe covering and fitting covers are installed is shown in Figure 138.

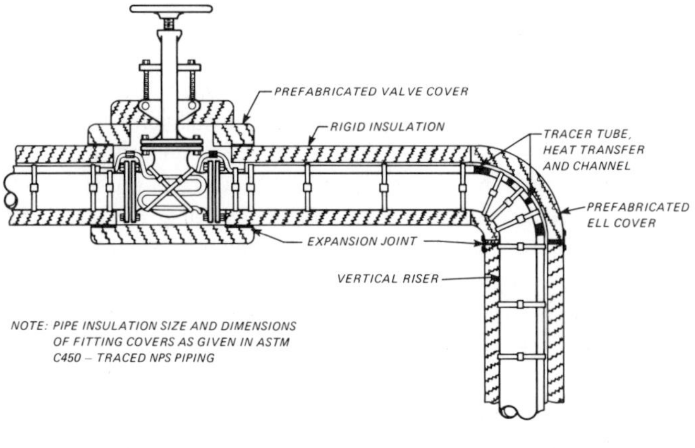

Figure 138 Rigid insulation system for traced NPS piping.

Summary

The installation requirements have been studied and evaluated, and the material chosen to fulfill the requirements, thus determining the properties of the material for basis of design.

From the thermal conductivity and insulation function and thermal demands, the insulation thicknesses can be determined and the mechanical design considered.

With the knowledge of expansion-contraction, material shrinkage, and physical limitations considered, a suitable system can be devised.

Protection from water or water vapor as a system requirement and the function of those water- and vapor-barriers must also be incorporated into the system.

With this combined information, including fastening and securements, and including the detail considerations presented in this chapter, the engineering of an installation is possible. The possible number of combinations and solutions for various individual problems is so large that specific recommendations for all individual possibilities are beyond the scope of any insulation manual. However, with this information, an engineer can design his system.

When the system is designed to the engineer's satisfaction, it then must be written in the form of specifications so that its installation can be carried out by the field erectors. Preparation of specifications are presented in the next chapter.

Weather-Vapor-Barrier Details

Vapor-barriers

The details of sealing a vapor-resistant insulation such as cellular glass have been discussed in detail. The sealing of other insulations which do transfer vapor has been mentioned, but the details of their application were not stressed. This is not because it was unimportant, but it was given only general discussion, because the details of such an application are complex, and many materials and systems may be used.

As mentioned in Chapter 6, a vapor-barrier is the protective barrier over a permeable insulation, and even a small leak will allow the entry of vapor into the entire system. This is illustrated in Figure 139. Because vapor-barriers on the outer surface of insulation are subject to all the ravages of time, mechanical abuse, and stresses imposed upon the system, cracks, breaks, punctures, or tears are likely to occur. This being true, the problem is to provide a system which is compartmentalized so that a break or puncture does not ruin the entire system. By the use of such a system a break in a vapor-barrier at one point will limit the saturation of the insulation with vapor to one compartment. Single and double layer pipe insulation and one flange cover are illustrated in Figure 140. Such methods of installing vapor-barriers were common over 25 years ago, but the significance was forgotten when cellular glass became an almost universal material for low temperature insulations on pipes and vessels, and eliminated the need for such a system. However, with cellular organics now being used in these services, the practices developed for cork and mineral cork again become important.

Many people believe that vapor cannot be transferred through cellular plastics, due to their cellular nature. Unfortunately this is not the case. The vapor transmission, measured by dry cup methods, establishes most foamed plastics as having a vapor transmission of 0.3-1.5 perm (0.54 to 0.81 per cm) compared to vegetable cork, which has a perm inch rating of 1.6 (0.86 per cm). Thus, the foamed plastics differ only slightly in vapor transmission from vegetable cork. The principles of devising vapor-barriers to extend the effective service life of vegetable cork also apply to the foamed organics. For this reason, the practice of compartmentalizing, as shown in Figure 140, is essential for efficient long service of permeable organic foam insulations.

The vapor-barrier may consist of mastic, plastic film, or laminates with vapor sealed joints, or a combination. Selection of proper vapor-barriers was discussed in Chapter 6. However, it must be remembered that the vapor-barriers must be suitable for the temperature which is imposed on them, and they must be suffi-

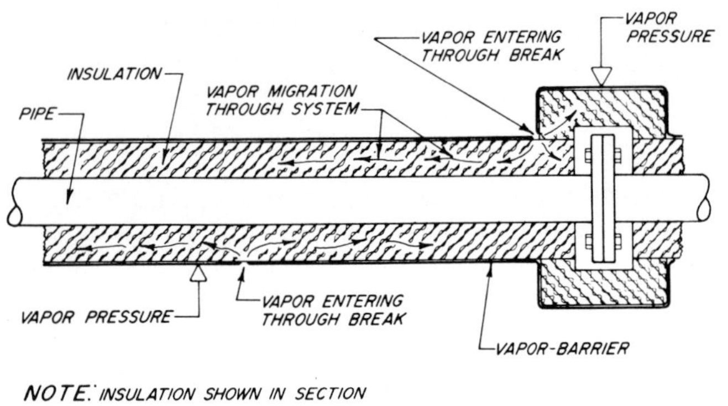

Figure 139 Vapor migration through permeable insulation after entry through break in vapor-barrier.

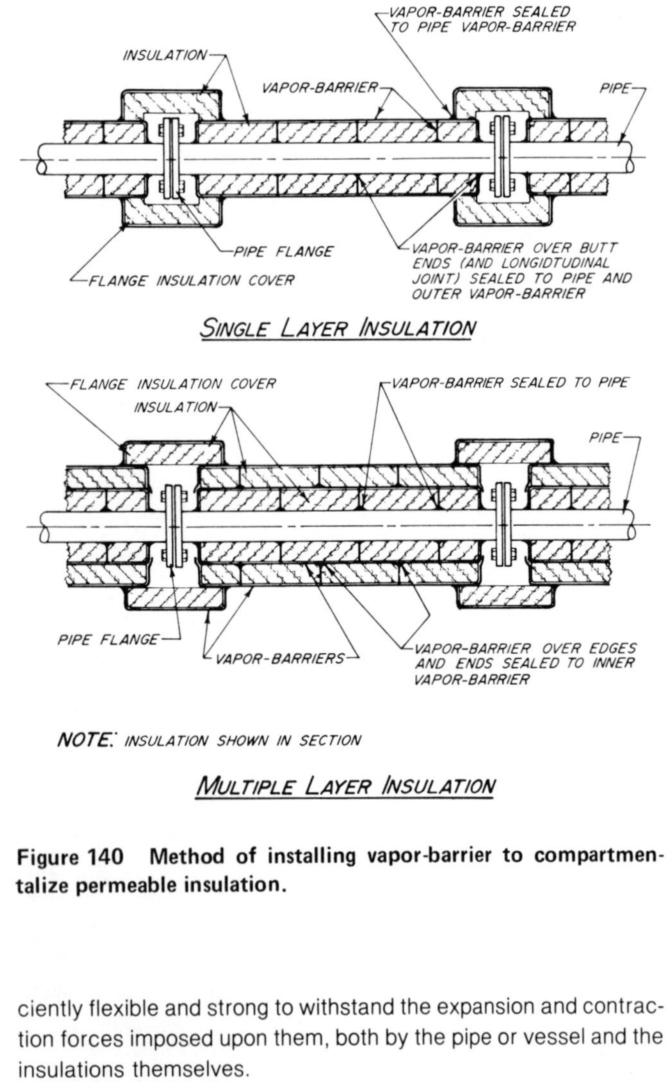

Figure 140 Method of installing vapor-barrier to compartmentalize permeable insulation.

ciently flexible and strong to withstand the expansion and contraction forces imposed upon them, both by the pipe or vessel and the insulations themselves.

The major property a vapor-barrier must have is its ability to retard the flow of vapor. This is its most important property, but its other properties, such as its ability to withstand mechanical abuse or weathering, may not be sufficient to fulfill the installation requirements. Where this is true, it may be necessary to provide a weather-barrier over the vapor-barrier. Although there are mastic materials which are formulated to serve both as vapor- and weather-barriers, in many low temperature applications it is desirable to install a vapor-barrier and protect it with a weather-barrier.

The vapor sealing of cellular glass is important, even though it has greater resistance to vapor transmission than most materials used as vapor-barriers. In Figure 66 the cellular glass is shown with only the outer layer vapor sealed, whereas in Figures 106 and 107 both inner and outer layers are shown sealed. This apparent discrepancy stems from the fact that no operating temperature was given, nor was the vapor sealer identified. Many vapor sealers become hard and shrink at a specific low temperature. Where the operating temperature of the pipe or vessel is lower than this critical temperature of the vapor-barrier mastic selected, it is better not to use it on the inner layer of cellular glass, as by its embrittlement it may cause spalling of the cellular glass. In such instances, it is advisable to precoat the cellular glass with a suitable antiabrasive compound to prevent wear caused by expansion or contraction movement. This is true on both vessel and pipe insulation.

Where the operating temperature is higher than the critical temperature of the vapor seal mastic being used, sealing of the inner layer is advisable as it provides additional resistance to the flow of vapor.

The use of non-setting sealers to provide vapor resistance for cellular glass joints at vessel flanges and insulation supports is shown in Figures 90 and 91, and the flanged fitting of pipe and contraction joints in Figures 106 and 107. Care in obtaining full coverage over the insulation slip area, and filling the complete void between the two surfaces of the insulation is essential to obtain high vapor resistance.

With proper joint sealing of cellular glass, and proper design to prevent its cracking or shearing due to expansion and contraction movement, there is no need to apply vapor-barrier material over its exterior surface. However, it does require a weather-barrier outdoors, and a finish indoors to protect it from ordinary weathering and mechanical abuse.

When insulating for temperatures in the cryogenic range, the control of the passage of vapor into the insulation system is vital. Because these systems are critical in regard to vapor control and thermal heat gain, most of the insulation systems used in this range must be carefully engineered and manufactured. Thus, most of these systems are intended for very specific purposes and are produced in a manufacturing plant, rather than being assembled in the field. As a result, this is a specialized field of insulation practice not falling within the scope for which this manual was written.

A recent book published by the National Aeronautics and Space Administration, titled "Thermal Insulation Systems" by Peter A. Glaser, Igor A. Black, Richard S. Lindstrom, Frank E. Ruccia, and Alfred E. Wichsler, presents very complete information on these systems. This book can be obtained from the Superintendent of Documents, U.S. Government Printing Office, Washington, D.C. 20402. In addition to the cryogenic systems discussed, the book also presents very interesting information concerning high-temperature protection systems for spacecraft structures.

Weather-barriers

Insulation in today's specialized applications must be protected by physical, chemical, and weather barriers. Under the terminology of the insulation industry, such protection is called the weather-barrier. The prime function of the insulation weather-barrier is to provide conditions in which the insulation can function to retard heat flow, in spite of the fact that the environmental conditions may be totally alien and even hostile to those required by the insulation.

The weather-barrier covering, jacket, or finish forms the necessary protection around the insulation. So, the design engineer is obliged to determine existing conditions, and design and specify a weather-barrier system which is adequate, economical, and functional.

In response to specialized conditions and a need for materials which will provide several types of protection simultaneously, an enormous variety of products has been developed. Many of these application requirements are listed and discussed in Chapter 6. Almost each weather-barrier material has its place where it is more suitable than any other. Conversely, each has its own limitations. A knowledge of these limitations is as important to the successful installation of an insulation weather-barrier or finish as is a thorough knowledge of the relative advantages of one material as opposed to another for a particular application.

For a successful installation, the details of application are of utmost importance. Regardless of the weather-barrier used, it is ineffective if it is installed in a manner which will allow water to leak into the insulation.

A proper method of installing metal jackets is shown in Figures 55, 56, and 57. Other methods of making tight joints are possible, such as the production of corrugated jackets with the ends flattened out and with circumferential ribs formed into the metal. These, if properly installed and used with non-setting sealers, can produce water tight joints.

Fitting and valve covers present a difficult problem in obtaining water tightness. Too frequently metal covers are installed by slipping the halves together without sealing the overlapping metal with mastic sealers. This type of construction will not provide a water tight installation, and only serves to hide the water as it soaks into the insulation with every rain.

Another common incorrect installation practice is the use of mastics as the weather-barrier on valves and fittings while the straight pipe is jacketed with metal which projects over the extended mastic of the fittings, but with no sealing mastic used to prevent water from entering the overlap of the two. Nonsetting sealers must be used in these spots, as the expansion and contraction movement will break any sealer that sets and hardens. This is illustrated in Figure 141.

Another installation detail which requires considerably more attention than it has been given is the sealing off of projections through insulation. In most instances, a bead of weather-barrier is run around the projection to seal it. This heavy thickness of mastic shrinks as it dries, and when the equipment or pipe is put into service, a large crack develops. Then, all the rain running down or over the projection follows it right into the insulation system, where it causes excessive heat transfer and damage to the insulation itself.

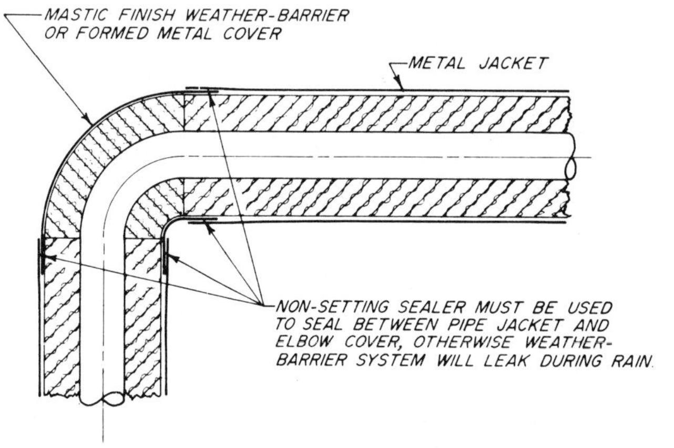

Figure 141 Method of sealing weather-barrier at fitting covers and jackets.

When projections are so hot where they emerge from the insulation that suitable mastics cannot be found to seal this joint, then the projection should be insulated a sufficient distance outward to a location where its exit point through the insulation can be sealed. This is illustrated in Figure 65 (Chapter 6). Such an extension of the insulation is also required on cold insulation to get the projection above frost temperature.

Regardless of the type of weather-barrier used, the inside corner of these projections should be caulked with a caulking compound of high solids content to eliminate the sharp inside corner.

This same procedure should be used at *all* projections through the insulation, such as where the pipe itself projects through the insulation, and where vessel nozzles, manholes, legs, or skirts project through the insulation.

After the projections are properly insulated and caulked, the inside corners should be weather protected with a flexible mastic coating reinforced with extensible fiber cloth. If the weather-barrier is mastic reinforced with cloth, the weather-barrier system should extend a minimum of 4 in. (10.8 mm) out over the projection. If metal jacketing, felts, or plastic films are used as the basic weather-barrier system then the flexible mastic with the extensible fiber cloth should extend a minimum of 4 in. (10.8 mm) in each direction from the inside corner. This is illustrated in Figure 142. Where projections are quite hot, the mastic used may be required to be high temperature sealer instead of ordinary weather-barrier mastic.

Understandably, there will be some people who disagree with these recommendations, for two basic reasons. First, they will state that such sealing, especially where metal jackets are used, will not have the clean, neat appearance they desire. However, the main purpose of the weather-barrier is to protect the insulation—appearance must be secondary. Second, others will state that this construction adds to the expense of the installation—and it will. But *this leakage through and around weather-barriers is a major cause of insulation failure* and the resultant excessive heat transfer. This additional expense necessary to obtain an installation which will give good service, as compared with one which will fail, is small.

There are other ways to seal inside joints, such as metal inside bands with non-setting sealers which are useful with metal

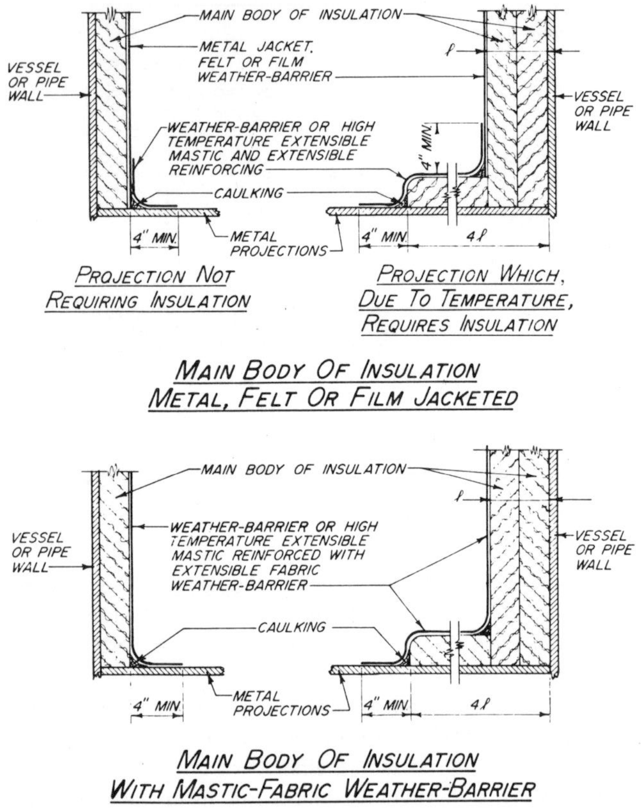

Figure 142 Method of weather-sealing projections.

jacketing systems. The general method presented in Figure 142 does not imply that the metal band system and other systems are not good and suitable, as there are many excellent methods of sealing inside joints. The major point is that inside joints must be sealed and stay sealed in service.

Whatever method is used to seal the inside corners of insulation and projections through insulation, must be designed and specified correctly for the system of insulation and weather-barriers used in the installation. *These details must be specified by the design engineer.* If they are not so specified, they will not be done by the insulation contractor, for the simple reason that under competitive bidding the contractor will not add the cost of doing the sealing corectly, unless it is specified, so that all the contractors are bidding on an equal basis.

Large flat surfaces, such as flue gas ducts, present a difficult problem as the flat surface is difficult to separate into areas where expansion can be easily and correctly handled, as can be done with cylindrical vessels and pipe.

The greater the expanse of flat surface, the more difficult it becomes to install insulation and weather-barrier systems which will not be pulled apart or cracked due to the thermal expansion of the metal duct. For this reason, it is desirable to use the reinforcing ribs of the duct system to "work in" expansion joints for the insulation and weather-barrier system. Where metal jacketing is used, overlapping slip joints should be installed at these ribs. The details of the securement and installation depend completely upon the design and temperature of the duct being insulated.

When mastic is used as a weather-barrier, a slightly different practice, used in the insulation of cylindrical vessels, can be used to advantage. This is to use reinforcing netting on the outside of the insulation. This netting must be secured to punched pins attached to the duct itself. These pins must be spaced on relatively close centers on the duct to spread the expansion evenly to the insulation system.

Insulation blocks or boards should be secured to the pins by stainless steel wire. Over the blocks a coating of insulation cement should be installed. Or, if sprayed high temperature fibrous insulation is used, it should be sprayed and compacted flush with the top of the pins, which should be equal in length to the specified thickness of the insulation. Rectangular or half cylindrical shaped insulators should be installed over the duct ribs. Metal wire mesh with 1 inch hexagonal openings shall be secured to the pins by stainless steel wire pulled taut over the flat surface and formed over the rib sections.

If the area has an at all corrosive atmosphere, the wire netting should be Monel, as all condensate of vapors will be on the inside of the weather-barrier.

The taut metal mesh should then be covered with weather-barrier mastic in sufficient thickness to completely fill all voids and cover the wire netting. After the mastic dries, the weave pattern of the wire mesh will be apparent regardless of the thickness of the mastic. It is important that the proper weather-barrier mastic be selected which will withstand the operating conditions and have excellent elongation properties even at low temperature, to which it may be subjected during winter months.

Summary

Because each installation has many individual requirements and must be designed to fulfill them, the number of systems possible is quite large. The details necessary for each system are also quite numerous, so the potential number of details which might be discussed is staggering. For this reason, only the details common to many systems and conditions could be presented in this manual.

The intent of this manual is to present the basic theory, an insight into problems related to installations, properties of materials which would serve to fulfill installation requirements, some of the details of installation which are most troublesome, and to direct an approach to working out the details of a system.

All too frequently, insulation systems have been specified in the most general terms without regard to the fact that, if the details of installation are not properly worked out and specified, failure is almost bound to occur. It should be remembered by the design engineer that, if he does not solve his design problems, including the details of installation, he should not be surprised that they remain unsolved and he ends up with a failed insulation system on his hands.

11 Specifications

The first thing to understand about a specification for the insulation of a structure, a piece of equipment, a pipe line, a building, etc., is that its preparation is not a simple operation. It is a tremendously complicated process, involving the consideration of many variables and the making of many engineering oriented choices. Most specifications must be individually tailored to the specific job and are not universally applicable.

It must be realized tthat a well written specification will convey to the Insulation Contractor, clearly and without equivocation, the INTENT of the writer—what he intends to be accomplished by the installation about which he is writing. A well written specification will convey to the Contractor the information he needs as to the results required, but *not* how to do it. That is his prerogative. Another pit into which many Contractors fall is looking to the Specifications to tell him what the writer does *not* want done. The listt of the things not wanted is obviously much too extensive to include in a specification, and the specification writer should avoid any tendency in this direction. The well written specification lists only the things you wish the Contractor to accomplish, and leaves the manner in which he does it up to him.

For the above reasons, this manual will not attempt to present Insulation Specifications as such, but will present a method for constructing a specification, combined with a check list from which the items for the individual specifications may be selected. It is manifestly impossible to cover the multitude of details inherent in a complex installation, and there should be no attempt to cover other than the major items applying.

An insulation specification may be broken down into four major divisions, which are:

General Conditions
Scope
Material Specifications—covering Type and Quality of Insulations, Weather-Barriers, and Accessories
Application Specifications—Covering Type and Quality of the Insulation System

The *General Conditions* state what job conditions are to be expected, such as storage, warehousing, and responsibilities. It further acts to tie the specifications to the contract agreement between the Contractor and his customer. As such, the General Conditions section of a specification is a legal document or agreement, not a technical matter, and is out of place in an engineering manual.

The *Scope* of the Specifications sets the boundaries within which the specifications are to be used. This section should be relatively short, but sufficiently directive to indicate to the reader what is the general purpose of the detail specifications which will follow, a general classification of the material to be used, the boundary limits of facilities to be served, and their location. To illustrate, if the specifications are to be used to insulate the roof of a building, say so in the Scope. Also state the conditions under which the insulation is to serve. Example:

Scope: This specification describes the material and application of rigid roof deck insulation for a flat roof building. Insulation shall be suitable to economically insulate an air-conditioned building during extreme summer and winter conditions from −20° F to +110° F (−28.9° C to 43.3° C) ambient air temperature.

An example of the scope of industrial insulation might be for hot piping and equipment, and as such might read: This specification describes the essential requirements of the materials and application of calcium silicate insulation on outdoor equipment and piping normally operating at temperatures up to 1000° F (538° C).

From these two examples, it may be seen that the overall boundaries of the installation have been established for each, and that the material and application specifications to follow have been directed toward the proper installation.

The *Material Specifications* may specify materials by three different methods.

1. A material may be specified by what it is made of and the manner in which the material was manufactured.
2. A material may be specified by its manufacturer and trade name or number.
3. A material may be specified by its generic classification and its properties, to fulfill given conditions.

Wherever of sufficient importance, or money is involved, number 3 is recommended, as it allows for competition and improvements. Judgment must be exercised in its use. Were it attempted for every item required, pages might be written to obtain small amounts of material which could not warrant such careful definition. In most good insulation specifications all three methods are used.

A typical example of each of the three methods follows:

Stainless Steel Wire: 0.0475″ diameter (0.1207 mm) (18 gage), dead soft annealed, Type 304 stainless steel.

Such a specification gave the type of stainless steel of which the wire was to be made, its diameter, and the fact that it was to be soft annealed. In this case, the strength of the wire was not given, as it was known from past experience that wire of that description is sufficiently strong to fulfill the required function. Also no statement was made to the effect that the wire had to resist weather corrosion, because the corrosion resistance of 304 stainless steel is well established.

An example of specification by manufacturer and trade number is:

Contact Adhesive: No. __________
manufactured by______________________________ Co.

In this case, a single source of material was specified and might be justified in that the quantity of material to be purchased was small, or possibly the manufacturer of the insulation would not be responsible for the securement of the insulation if any other contact adhesive were used.

It could well have been that when 85% magnesia was the most used insulation, the accepted manner, typical of that time, for specifying its purchase was: insulation consisting of not less than 85% magnesium carbonate, 10% asbestos fibers, and 5% inorganic binders. With such a specification, was any improvement in material possible? If a change in the components was made to improve the material, the resultant material would not pass the specifications. An example of how the specifications by classification and properties should have been written follows.

Insulation consisting mainly of magnesium carbonate, and having the following properties:

1. Shall be noncombustible
2. Shall have a compressive strength of not less than 60 psi (41.369 Pa) at 5% deformation
3. Shall have a density of not less than 9 lbs (144 kg) nor more than 14 lbs (224 kg) per cu ft
4. Shall have a flexural strength of not less than 40 psi (27579 Pa)
5. Shall have less than 1% shrinkage when subjected to a soaking heat of 600° F (316° C)
6. Shall have a specific heat of 0.25 to 0.30 Btu per lb (0.033 to 0.0399 W/kg)
7. Shall have a tensile strength of not less than 15 psi (10,362 Pa)
8. Shall be suitable for continuous soaking heat up to 600° F (316° C)
9. Shall have a thermal conductivity not greater than: 0.35 at 100° F mean temperature (0.05 W/mK at 37.0° C mean) 0.46 at 400° F mean temperature (0.066 W/mK at 204° C mean)

With such a specification, the manufacturer could improve the properties and not be precluded by the specifications, and each manufacturer would have an incentive to improve his product to obtain the "edge" on his competition.

Of course, the engineer must know enough about the material he is specifying that he does not write a specification which is impossible to meet. However, it is the engineer's job to know the properties of materials in order that he may properly specify and use them.

The properties listed must be based on some method of testing. For this reason, methods of testing properties must be established so that both the manufacturer and the consumer will be using the same criteria. Otherwise, differences between them can arise. Following this reasoning, where such are available, American Society for Testing and Material test methods have been given in this Manual, as universally acceptable test methods for evaluating properties.

It should be noted here that, for specific installations, ASTM Insulation Material Specifications may not be suitable. This is due to the fact that these specifications must be written with such a broad scope that possibly the most significant property for an individual installation may not be covered. In a test method for an individual property, this single property is the only result sought, thus, test methods of properties can be very explicit. In most instances, a test method does not establish acceptable or nonacceptable levels, but an engineer can establish these levels based on the installation requirements and materials available. With this knowledge available, the Engineer can write complete and good material specifications.

The *Application Specification,* which is the identification most often used, is slightly misnamed. Basically, this part of the specification is not written to tell the workman *how* to apply materials, but attempts to state that, after certain provisions are followed, the assembly of insulation, accessories, and weather-barriers, etc., will *result* in an installation as specified.

In many specifications, other than those for insulation, the proper function of a finished piece of equipment or system is also specified. Unfortunately, at this stage of development in the insulation industry, it is so difficult and time consuming to prove the final results in an insulation system that the minimum performance of such a system is rarely made a part of the specifications. In time this will become standard practice, but at present it has not become a part of construction specifications.

In general, Insulation Application Specifications should follow the pattern shown below:

List of Material Required
Preparation
Insulation Application
Weather-Barrier Application

Where there are a number of individual specifications which share a number of common practices this might be modified to have a "General Application Specifications" section to which the individual specifications refer, rather than repeat the same items in each individual specification. This "General Application Specification" may have parts of each, thus, a set of industrial specifications may have the following format.

General Specification
- List of Materials Required
- Preparation
- Insulation Application
- Weather-Barrier or Coverings Application

Individual Specification
- Preparation
- Insulation Application
- Weather-Barrier, or Covering Application

Returning to the consideration of one specification, rather than a set of specifications, the *Preparation Section* should state what must have been done before the insulation is applied. In almost every instance, these are operations which must have been performed by other crafts, and possibly other contractors, prior to the application of the insulation. The specifications must state what these operations are and place the responsibility for them correctly, so as to prevent conflict and delay. For example, if a vessel requires insulation supports, are these supports to be furnished and installed by the vendor, the steel contractor, or the insulation contractor? This must be established. However, regardless of who will furnish and install them it must be the responsibility of the insulation contractor to make certain that the specified supports are installed before the insulation is applied. Likewise, if vessels are to be painted prior to insulation, the preparation section should state that the insulation will not be applied until after the proper painting is finished and dry.

The *Insulation Application Section* shall spell out the intent as to the quality of the installation, and what is expected from the finished product of assembled insulation and accessories. Of course, no specification can detail each and every piece of the installation, but sufficient typical details must be provided that the applicators understand what is expected. For this reason, it is extremely helpful to include drawings in specifications to illustrate typical assemblies. The old saying that a picture equals a thousand words is worth remembering when writing specifications. Most workmen, when shown a drawing, can get the idea of what is desired much more quickly than by reading a detailed description.

The *Weather-Barrier or Covering Application Section* may be made a part of the Insulation Application Section, as the barrier is a part of the insulation system, but because these are installed last, it does lend itself to being a separate section, especially as a different set of materials is being used. Another advantage in making such a separate section is that, under various conditions, insulation installed as called for under the Insulation Application Section may require one type or system of Weather-Barriers in one location, and another in a different location. Conversely, one set of Weather-Barrier Application Specifications may be suitable for use on many various types and forms of insulation.

This particular part of specifying has not been given the careful consideration it deserves. Poorly designed and specified weather-barrier systems have been one of the major causes of poor insulation systems. Contributing to this is the fact that it is often overlooked that pipes, vessels, or any other substrate surfaces to which the insulation is applied move by expansion and contraction. This movement must be compensated for in some manner in the weather-barrier or covering system. If it is not, cracks and resultant leaks will develop.

Also, the material and application specifications must be written definitely and exactly to obtain the proper quality of materials and assembled system.

From this, then, we can establish a general format:

1. General Conditions
2. Scope
3. Material Specifications, as to Type and Quality of Insulations, Weather-Barriers, and Accessories
   a. Insulation Material and Accessories
   b. Weather-Barrier Materials and Accessories
4. Application Specifications, to obtain type and quality of Insulation System
   a. List of materials required for individual system
   b. Preparation
   c. Insulation Application
   d. Weather-Barrier, or Covering Application

Now that the format is established, it will be expanded by a number of individual items which may appear in an insulation in their proper location. Not all these items will be present in any one individual specification. The following is prepared for convenience, to be used as a check list. As a specification is being prepared, it may assist as a reminder to include each individual item needed.

SPECIFICATION OUTLINE

1. General Conditions

1.1 Material Storage
1.2 Material Handling (on job site)
1.3 Fabrication Facilities (on job site)
1.4 Installation Hazards
1.5 Safety Requirements
1.6 Scaffolding (who furnishes, reuse and multiple use)
1.7 Services Required—Heat, Light, Water, Air
1.8 Special Installation Requirements
1.9 Insulation Design Schedule

2. Scope

2.1 Use
2.1.1 Where
2.1.2 For What Service
2.2 Temperature Limits
2.2.1 Operating Temperatures
2.2.2 Surrounding Temperatures
2.3 Location
2.3.1 As Part of Building
2.3.2 Indoors
2.3.3 Outdoors
2.3.4 Underground
2.3.5 In Trenches

3. Materials

3.1 Insulation Materials
3.1.1 Generic Identification
3.1.2 Properties
3.1.2.1 Thermal
3.1.2.2 Physical
3.1.2.3 Chemical
3.2 Accessory Materials
3.2.1 Supports
3.2.2 Securements
3.2.3 Cements
3.2.4 Sealers
3.2.5 Adhesives
3.2.6 Finishes
3.2.7 Vapor-Retarders
3.2.8 Weather-Barriers
3.3 Composite Materials (such as preinsulated conduit)

4. Application

4.1 Building
4.1.1 List of Materials Required (as specified in Material Specification)
4.1.1.1 Insulations
4.1.1.2 Accessories
4.1.2 Application
4.1.2.1 Floors
4.1.2.2 Walls
4.1.2.2.1 Interior
4.1.2.2.2 Exterior
4.1.2.3 Ceilings
4.1.2.4 Roofs
4.1.2.4.1 Above
4.1.2.4.2 Below
4.2 Cold Storage and Refrigeration (Same last as above from 4.2.1 through 4.2.2.4.2)
4.3 Industrial Equipment and Piping
4.3.1 Preparation
4.3.1.1 Insulation Supports

SPECIFICATION OUTLINE

4.3.1.2 Surfaces—Condition of
4.3.1.3 Heat Tracing
4.3.2 List of Materials Required (as specified in Material Specifications)
4.3.3 Equipment
4.3.3.1 Cushion Blanket (if required)
4.3.3.1.1 Vertical Surfaces
4.3.3.1.2 Horizontal Surfaces
4.3.3.2 Curved Surfaces
4.3.3.2.1 Shop Fabrication
4.3.3.2.1.1 Size, Shape, Tolerances, Clearances
4.3.3.2.2 Field Fabrication (Kerfing, etc.)
4.3.3.2.3 Securement
4.3.3.2.3.1 Welding Pins and Clips
4.3.3.2.3.2 Strap
4.3.3.2.3.3 Wire
4.3.3.2.3.4 Adhesive
4.3.3.3 Heads
4.3.3.3.1 Shop Fabrication
4.3.3.3.2 Field Fabrication
4.3.3.3.3 Securement (same list as 4.3.3.2.3)
4.3.3.4 Vessel Flanges
4.3.3.4.1 Insulation Form and Fitting
4.3.3.4.2 Securement (same as 4.3.3.2.3)
4.3.3.5 Insulation Expansion Joints
4.3.3.6 Nozzles
4.3.3.6.1 Form and Fitting
4.3.3.6.2 Securement (same as 4.3.3.2.3)
4.3.3.7 Legs
4.3.3.7.1 Insulation Form and Fitting
4.3.3.7.2 Securement (same as 4.3.3.2.3)
4.3.3.8 Skirts
4.3.3.8.1 Insulation Form and Fitting
4.3.3.8.2 Securement (same as 4.3.3.2.3)
4.3.3.9 Flat Surfaces
4.3.3.9.1 Insulation Form and Fitting
4.3.3.9.2 Securement (same as 4.3.3.2.3)
4.3.4 Piping
4.3.4.1 Cushion-Blanket or Premolded (if required)
4.3.4.2 Straight Pipe
4.3.4.2.1 Insulation Form
4.3.4.2.2 Securement (same as 4.3.3.2.3)
4.3.4.2.2.1 Strap
4.3.4.2.2.2 Wire
4.3.4.2.2.3 Tape
4.3.4.2.2.4 Adhesive
4.3.4.3.1 Flanged Valves and Fittings
4.3.4.3.1.1 Insulation Form and Shape
4.3.4.3.1.2 Securement (same as 4.3.3.2.3)
4.3.4.4.1 Welded Fittings
4.3.4.4.1.1 Insulation Form
4.3.4.4.1.2 Securement (same as 4.3.3.2.3)
4.3.4.5.1 Screwed Fittings
4.3.4.5.1.1 Insulation Form and Shape
4.3.4.6.1.2 Securement (same as 4.3.3.2.3)
4.3.4.6 Insulation Expansion
4.3.4.7 Projections
4,3.4.8 Pipe Supports
4.3.4.9 Cradles
4.3.5 Weather-Barriers
4.3.5.1 Jackets
4.3.5.1.1 Erection
4.3.5.1.1.1 Joint Sealing and Caulking
4.3.5.1.1.2 Securement
4.3.5.2 Mastics
4.3.5.2.1 Reinforcing Cloth
4.3.5.2.2 Application

Many may feel that the care necessary to prepare a good tight specification is unwarranted. Also, there are those who feel that tight specifications may cause high job cost because of contractors' "fear" of tight specifications. However, it has been observed that the best and most responsible contractors like good tight specifications because then they are assured that all competition is bidding on the same quality of installation.

It is well known that it requires effort on the part of the engineer and the contractor to obtain the quality of materials and workmanship called for by a complete and definite specification. It is practically impossible to obtain good quality and workmanship where poorly prepared specifications are used. An insulation installation will only be as good as the specifications from which it is built. A representative insulation specification following through on the ideas presented in the above discussion follows:

Insulation Specification

The following specification is presented only to illustrate the *form* of an insulation specification. *It is not a recommendation of material, or a recommendation that the materials mentioned be used for any given installation.* As a matter of fact, the combination of materials were selected to illustrate a variety of methods of specifying. Thus, the specification and insulation thicknesses, as presented, are unlikely to be the proper selection for any particular installation.

Other than the Insulation Design Schedule, which is given only to establish a basis for reference in the balance of specification, no presentation will be given for "General Conditions," as this particular part must be written for the special set of conditions for each construction installation. Many of the items in this section relate to business arrangements between the consumer and supplier, and, although necessary in the individual company's specification, it is not strictly an engineering decision.

For this particular example, the insulation thickness will be based upon economic thickness of insulation to conserve heat. The specific set of conditions will be the same as used in Chapter 2 to illustrate the economics of insulation. For this problem, the operating temperature of the vessels and pipe will be taken to be 800° F (427° C). From Table 13, the thicknesses of insulation for the vessels and pipe can be determined. Again, these thicknesses are *not* a thickness recommendation, but are valid only if the economic conditions happen to be exactly as given for the example in Chapter 2. With these restrictions understood, an example of the preparation of a Thermal Insulation Material and Application Specification follows.

THERMAL INSULATION MATERIAL AND APPLICATION SPECIFICATION

1. General Conditions

1.1 through 1.8. General job conditions stating responsibilities and business agreements between the consumer and the contractor. This should include the proper references to and in the legal contract making this section and the balance of the Specifications a part of the contract agreement between the consumer and contractor.

1.9 Insulation Design Schedule

		Temp °F	Temp °C	Insulation thickness nominal—inches	mm	Insulation specification designation	Weather-barrier specification designation	
Equipment and vessels up to 36" dia (914.4 mm) to be insulated the same thickness as pipe of same diameter		800	427			A	II	
Over 36" (914.4 mm)		800	427	5"	127.0	A		
							Weather-barrier designation	
Pipe size NPS	Pipe size diameter mm						Straight pipe	Fittings
½" and ¾"	21.3 and 26.7 mm	800	427	1½"	38.1	A	I	II
1" through 2"	33.4 through 60.3 mm	800	427	2"	50.8	A	I	II
2½" through 5"	73.0 through 141.3 mm	800	427	2½"	63.5	A	I	II
6" through 8"	168.3 through 209.1 mm	800	427	3"	76.2	A	I	II
10" through 12"	273.0 through 323.8 mm	800	427	3½"	88.9	A	I	II
14" through 20"	355.6 through 508 mm	800	427	4"	101.6	A	I	II
24" through 30"	609.6 through 762 mm	800	427	4½"	114.3	A	I	II
36" and over	914.4 and over	800	427	5"	127.0	A	I	II

2. Scope—Calcium Silicate—Designation A

This material and application specification describes the material required and the application of insulation of Individual Specification Designation A and Weather-Barrier Specification ________ and ________, for use on industrial equipment and piping located outdoors, normally up to and including 1200° F (649° C).

3. Individual Specifications—Designation A

3.1 Thermal Insulations

3.1.1 *Calcium Silicate* (standard)

Shall be composed of hydrous calcium silicate, blended with asbestos fibers in suitable proportions, with the following properties:

3.1.1.1 Thermal Properties

3.1.1.1.1 Shall be suitable for use up to 1200° F (649° C) without mechanical, chemical or thermal failure

3.1.1.1.2 Shall have a specific heat of not less than 0.21 nor more than 0.29 at 400° F (204° C) mean temperature in accordance with ASTM Test Method C-350, latest revision

3.1.1.1.3 Shall have a thermal conductivity not exceeding the values listed below, when tested in accordance with ASTM Test Methods C-177 or C-335, latest revision

Mean Temperature °F	°C	Conductivity, Btu/sq ft in., hr, °F	W/mK
100	37.8	0.33	0.048
200	93.3	0.37	0.053
300	149.0	0.41	0.059
400	204.0	0.46	0.066
600	316.0	0.47	0.068

3.1.1.2 Physical Properties

3.1.1.2.1 Percent of volume of water absorbed after 24 hours submersion shall not exceed 85%

3.1.1.2.2 Shall be noncombustible

3.1.1.2.3 Shall have a compressive strength of not less than 100 psi (689 Pa) when tested in accordance with ASTM Test Method C-165 latest revision

3.1.1.2.4 Shall have a density not less than 10 nor more that 13 when tested in accordance with ASTM Test Methods C-302 or C-303, latest revisions

3.1.1.2.5 Shall have hardness such that a 1/8" (3 mm) ball under a 1 kg load does not penetrate more than 0.5 mm when tested in accordance with ASTM Test Method C-569, latest revision

3.1.1.2.6 Shall not lose more than 20% of its weight after 10 minutes nor more than 40% of its weight after 20 minutes in the tumbling test, ASTM Test Method C-421, latest revision

3.1.1.2.7 Shall not have more than 1-6% linear shrinkage when tested in accordance with ASTM Test Method C-356, latest revision

3.1.1.3 Chemical Properties

3.1.1.3.1 Shall not contribute to the corrosion of stainless steel when tested in accordance with A.W. Dana method, ASTM Bulletin, October 1957

3.1.2 Cushion Blanket—Fibrous Glass, metal encased Wool Blanket with the following properties

3.1.2.1 Thermal Properties

3.1.2.1.1 Shall be suitable for use up to 1200° F (649° C) without mechanical, chemical or thermal failure

3.1.2.1.2 Shall have a thermal conductivity not exceeding the values listed below when tested in accordance with ASTM Test Method C-177, latest revision

Mean Temperature °F	°C	Conductivity, Btu/sq ft in., hr, °F	W/mK
200	93.3	0.33	0.048
300	149.0	0.44	0.063
400	204.0	0.49	0.071
500	260.0	0.55	0.079

3.1.2.2 Physical Properties

3.1.2.2.1 Shall be reinforced with 1" (2.54 mm) hexagonal wire netting on one side

3.1.2.2.2 Shall be noncombustible

3.1.2.2.3 Shall have a density of not less than 8 nor more than 10 lbs per cu ft (128.1 nor more than 160.2 kg per m^3) when tested in accordance with ASTM Test Method C-167, latest revision

3.1.2.3 Chemical Properties

3.31.2.3.1 Shall not contribute to stress corrosion of stainless steel when tested in accordance with A. W. Dana Method, ASTM Bulletin, October 1957

3.1.3 Insulating Cement, Mineral Wool and Binders, Hydraulic Setting, having the following properties

3.1.3.1 Thermal Properties

3.1.3.1.1 Shall be noncombustible

3.1.3.1.2 Shall be suitable for temperature up to 1800° F (982° C) without mechanical, chemical, or thermal failure

3.1.3.1.3 Shall have a thermal conductivity not exceeding the values listed below when tested in accordance with ASTM Test Method C-177, latest revision

Mean Temperature °F	°C	Conductivity, Btu/sq ft in., hr, °F	W/mK
200	93.3	0.70	0.101
600	316.0	0.82	0.118
800	427.0	0.94	0.136

3.1.3.2 Physical Properties

3.1.3.2.1 Shall have dry coverage capacity of 16 board feet for 50 lbs (6.72 m^2 per 100 kg) of dry material when tested in accordance with ASTM Test Method C-166, latest revision

3.1.3.2.2 Shall have dried density of not less than 30 nor more than 40 lbs per cu ft (not less than 480 nor more than 641/m^3) when tested in accordance with ASTM Test Method C-167, latest revision

3.2 Accessories

3.2.1 Insulation Strap, 1/2″ by 0.020″ (13.4 mm by 0.5 mm), Type 304, dead soft, stainless steel strap

3.2.2 Insulation Clips, 1/2″ by 7/8″ (13.4 mm by 5 mm), double pronged, Type 304 stainless steel

3.2.3 Stainless steel wire, 0.0475″ (1.2 mm) diameter (18 gage), dead soft annealed, Type 304

3.2.4 Wire netting shall be one inch hexagonal mesh, 20 gage, Type 304 stainless steel or Monel

3.2.5 Stainless steel pins shal be 4d stainless steel nails

3.2.6 Stainless steel skewers shall be 6d stainless steel nails

3.2.7 Welding pins shall be 1/8″ by 1″ by 1-3/16″ (3 mm by 25.4 mm by 33 mm) rectangular mild steel welding pins with 5/8″ hole punched at projecting end

3.2.8 Fabrication cement—high temperature, shall be semi-refractory water dispersed No. ________ as manufactured by ________________

3.3 Weather-barrier and weather-barrier accessory material specifications shall be given with the individual weather-barrier specifications

4.1 Application Specifications—Equipment and Piping

4.1.1 Preparation

4.1.1.1 Equipment insulation supports, of same metal as vessel, projecting to a point 1″ (25.4 mm) less than the thickness of the insulation, shall be welded or bolted to the vessels to support vertical insulation. They should be located at base of insulation, at top of vessel, and at 15′ (457.2 mm) centers above the base support. A support shall be located above each vessel flange, a sufficient distance above the flange bolts to allow for their easy removal. The bottom and top suppports shall be slotted with 1″ by 1/4″ (25.4 mm by 7.6 mm) slots for attachment of straps or wire

4.1.1.2 Pipe above 3″ (88.9 mm dia) NPS shall be supplied with insulation supports welded or bolted just above lower elbow of risers. Projection of this support shall be a minimum of 1½″. On long vertical runs of pipe, additional supports shall be installed on 15′ (457.2 mm) centers.

4.1.1.3 Flat surfaces and large bottom heads shall be equipped with rectangular punched welding pins to support the insulation. These pins shall be welded to surface in an approved manner. On the top surfaces the pins shall be located on 24″ (0.61 m) centers—diamond pattern, on sides 18″ (0.457 m) centers—diamond pattern, and 12″ (0.305 m) centers—diamond pattern on bottom surfaces.

4.1.1.4 All surfaces shall be clean and dry.

4.1.1.5 Where metal surfaces are to be painted to prevent corrosion, the paint shall be completely dry prior to the installation of the insulation.

4.1.1.6 Where heat tracing is required it shall be installed prior to application of the insulation.

4.1.1.7 Where heat tracing requires the installation of heat transfer cement, this cement shall be installed as called for in "Specifications for Materials and Application of Heat Transfer Cement."

4.2.1 Equipment

4.2.1.1 Application to Curved Surfaces

4.2.1.1.1 On equipment operating at over 500° F (260° C) and 6′ (1.83 m) and larger in diameter, a one inch cushion blanket shall be applied over the entire surface before application of the calcium silicate curved segmental insulation. The blanket shall not be compressed, and shall be secured with a minimum amount of stainless steel wire.

4.2.1.1.2 On all equipment, calcium silicate shall be molded, or cut, into curved segments to fit the vessel surface, with one inch space allowed for cushioning blanket, where used. Curved block shall be applied to vessel sides in staggered position with all joints tightly butted. Insulation shall be secured with straps on 9″ (0.229 m) centers. Multiple layers shall be applied so that the butt joints of one layer do not coincide with those of the other. This is illustrated in Figure 68 (Chapter 7).

4.2.1.1.3 All layers of insulation shall be secured with straps where the contour of the vessel permits firm attachment. Insulation applied to irregular surfaces, where use of straps is impractical, shall be secured with wire.

4.2.1.2 Application to Flat Surfaces

4.2.1.2.1 Block insulation shall be applied to flat surfaces in staggered position with all joints tightly butted. Blocks shall be secured with wire attached to welding pins welded to surface. Multiple layers shall be applied so that the butt joints of one layer do not coincide with those of any other layer.

4.2.1.2.2 After the specified thickness of insulation has been applied, wire netting shall be stretched tightly over block and secured with tie wires.

4.2.1.3 Application to Vessel Heads

4.2.1.3.1 Vessel heads shall be insulated with preformed block (and cushion blanket where required) in the same manner as described in 4.2.1.1.

4.2.1.3.2 Insulation on top heads up to 12′ (3.658 m) diameter shall be secured with straps. Top heads larger than 12′ (3.658 m) diameter, all bottom heads, and vertical heads shall have insulation secured with wire attached to rectangular pins located as called for in 4.1.1.3.

4.2.1.4 Application to Vessel Flanges

4.2.1.4.1 All vessel flanges shall be insulated unless specifically excepted. The flange cover shall be a preformed—slip type flange cover of the same thickness as the specified insulation thickness. Flange cover shall be of step construction, so designed that it functions as an insulation expansion joint. The upper and lower sections of the cover shall be secured to the curved side wall insulation with straps and skewers. The mid section shall be secured at its bottom so that it may slide in respect to the upper section. This flange cover is illustrated in Figure 96 (Chapter 9).

4.2.1.5 Application of Expansion Joints

Insulation expansion joints shall be provided on 15′ (4.572 m) centers. The insulation support shall be one in. above the termination of the insulation below. This void shall be packed tightly with fibrous glass wool. Slip sleeves of stainless steel sheet shall cover this opening and protect it from the weather. This is illustrated in Figure 00 (Chapter 9). Where a vessel flange occurs on a vessel, expansion shall be considered as taking place, and be compensated for by the flange cover, at that location.

4.2.1.6 Application to Manholes and Nozzles

4.2.1.6.1 All manholes, blind nozzles, and connecting piping flanges shall be insulated (unless specifically excepted) with precut oversize covers. Thickness shall be the same as specified for the vessel insulation. These covers shall be *attached to the vessel insulation* by wire, straps, and skewers in such a manner that movement of the vessel does not cause them to break loose. This application is illustrated in Figure 96 (Chapter 9).

4.2.1.7 Application to Vessel Legs

4.2.1.7.1 Channel legs for vessels shall be insulated with blocks which shall extend over channel flanges and down to fireproofing or footer. The thickness of insulation shall be a minimum of 1/2 that specified for the vessel, but in no lase less than 2″ (50.8 mm).

4.2.1.7.2 Vessel insulation shall be butted firmly against the tank leg flange. Reentrant space between the channel flanges and over the channel flange between body insulation shall be filled with block.

4.2.1.7.3 Leg insulation shall be secured with straps and wire.

4.2.1.8 Application to Vessel Skirts

4.2.1.8.1 Skirts of vertical vessels shall be insulated down a minimum of 4 times the insulation thickness specified for the vessel from their junction points with the vessels. Thickness of insulation shall be the same as specified for the vessel. Where required for fire protection, the outside and inside of skirts shall be insulated entirely, with a minimum of 2″ (50.8 mm) of block insulation precut to fit skirt curvature. No cushion blanket shall be used under this block insulation.

4.2.1.8.2 Outer insulation on skirt shall be secured with straps. Inner insulation on skirt shall be secured with wire attached to rectangular pins welded to inside of the skirt on 6″ centers.

4.2.1.9 Surface Finish

4.2.1.9.1 Regular cylindrical surfaces, limited area flat surfaces, and preformed covers do not require any insulation or finishing cement surfacing. Any slight voids may be pointed up with insulating cement troweled flush with adjacent surfaces.

4.2.1.9.2 Large flat surfaces and irregular surfaces over which wire netting is installed shall be given a 1/2″ (12.7 mm) thick, smooth and uniformly applied trowel coat of hydraulic setting insulating cement.

4.2.1.10 Weather barrier shall be as designated in insulation schedule. Materials and application shall be as called for under that designation.

4.2.2 Piping

4.2.2.1 Application to Straight Pipe

4.2.2.1.1 Vertical pipe over 3″ (76.2 mm) NPS shall have insulation supported by an insulation support, welded or bolted to pipe directly above the lowest pipe fitting. Additional insulation supports shall be located 15′ on centers above the bottom support. An insulation support shall also be installed above each valve or pair of line flanges located in the vertical run of the pipe.

4.2.2.1.2 Insulation shall be sectional up to 12″ NPS (323.8 mm dia), and may be sectional or curved segments above this diameter. Insulation shall be applied in staggered joint construction. Multiple layers shall be installed so that the butt joints of one layer do not coincide with that of another.

4.2.2.1.3 Securement of insulation shall be by wire up to 12″ (304.8 mm) OD. Above this diameter the insulation shall be secured with straps, except that all inner layers shall be secured with wire. If metal jacketing is specified for the weather-barrier and it is secured by straps, then all layers of insulation, regardless of size shall be secured with wire. Arrangement of insulation and location of securements are illustrated in Figure 68 (Chapter 7).

4.2.2.2 Expansion Joints

4.2.2.2.1 Expansion joints in the insulation shall be installed every 15′ (4.572 mm) of uninterrupted straight pipe in both the horizontal and vertical. The insulation of single and each of multiple layers shall be terminated in a straight cut. A space of 1″ (25.4 mm) shall be left between the insulation terminations. This void shall be packed *tightly* with glass wool blanket. The expansion joint shall be protected by stainless steel sleeves. This is illustrated in Figure 71 (Chapter 7).

4.2.2.3 Application to Flanged Fittings

4.2.2.3.1 All flanged valves and fittings, with the exception of ball and plug valves, shall be insulated with preformed covers in accordance with the dimensions given in ASTM Recommended Practice C-450, latest revision. Ball and plug valve covers shall be field fabricated of the proper sectional pipe insulation. These covers shall be secured in position with straps. These are illustrated in Figures 98 and 111 (Chapter 9).

4.2.2.4 Application of Welded Fitting Covers

4.2.2.4.1 All welded fittings over 3″ (88.9 mm dia) NPS shall be insulated with preformed covers in accordance with ASTM Recommended Practice C-450, latest revision. These covers shall be secured in position by straps or wires, depending upon pipe diameter. These are illustrated in Figures 99 and 111 (Chapter 9).

4.2.2.5 Application to Small Welded or Screwed Fittings

4.2.2.5.1 Welded fittings under 3″ (88.9 mm dia) NPS and screwed fitting covers may be preformed or field fabricated and secured in position with wire. Fittings less than 1½″ (48.3 mm dia) NPS may be insulated with insulating cement installed to the specified thickness.

4.2.2.6 Finish

No insulating cement or finishing cement shall be used to cover any preformed pipe or fitting covers. Slight voids shall be pointed up with cement to bring flush with adjacent surfaces.

4.2.2.7 Application to Heat Traced Pipe (Where Required)

4.2.2.7.1 Piping requiring tracing by tubing or electric conduit up to 5/8″ (16 mm dia) OD shall be insulated with oversize insulation as given in Table 8. The size of preformed fittings for this traced piping is given in the "Traced" section of ASTM Recommended Practice C-450, latest revision. Illustrated in Table 8 (Chapter 1).

4.2.2.7.2 After tracing is installed and is thermally connected by heat transfer cement, the pipe and fittings shall be insulated as previously specified.

Note: Although the insulation schedule presented did not include heat tracing, heat tracing was included in the specification as it is commonly used, and most frequently is a part of all industrial insulation specifications.

5.1 **Weather-Barriers**

5.1.1 Metal Jacket Pipe Weather-Barrier—Designation I

5.1.1.1 Materials Required

5.1.1.1.1 Metal Jacket shall be preformed of smooth austenitic stainless steel, one quarter hard, not less than 0.008″ (0.02 mm) thick, preformed to fit designated outside diameter of insulation, with preformed interlocking "Z" longitudinal joint.

5.1.1.1.2 Closure bands for circumferential butt joints shall be austenitic stainless steel, not less than 0.008″ (0.2 mm) thick, precut and preformed to fit the outside diameter of the insulation jacket of designated size. Closure bands shall be furnished with factory installed, high temperature non-setting sealing compound on each exterior lip of its inside surface to provide a water seal when installed on the metal jacket.

5.1.1.1.3 Insulation straps shall be 1/2″ by 0.020″ (12.7 mm by 0.5 mm). Type 304 dead soft, stainless steel.

5.1.1.1.4 Insulation clips shall be 1/2″ by 7/8″ (13.4 mm by 5 mm) double pronged. Type 304 stainless steel.

5.1.1.1.5 Heat Resistant sealer shall be silicone base sealing compound No. ________ as manufactured by ________

5.1.1.2 Application of Metal Jacket to Insulated Straight Pipe

5.1.1.2.1 Jacket shall be installed by placing it around the pipe insulation and engaging the "Z" joint. The "Z" joint on horizontal piping shall be on the side of the insulation, with the open edge of the "Z" joint pointed down.

5.1.1.2.2 The butt joints between adjacent jackets shall be sealed with a closure band. Closure band sealing compound shall be used to seal voids across interior of closure bands where they lap over "Z" joint. Closure bands shall be secured in place with insulation strap.

5.1.1.2.3 On pipe insulation less than 12″ (0.3048 m) outside diameter an insulation strap shall be installed at the half way point from each end to secure the entire assembly. On pipe insulation above 12″ (0.3048 m) outside dimater it shall be secured in position by two straps. Straps shall be spaced 12″ (0.3048 m) from each end. Illustration of proper assembly of jacket on straight pipe is shown in Figure 71 (Chapter 7).

5.1.1.2.4 Weather-Barrier for fittings and irregular shapes of piping insulation is given in Weather-Barrier Designation II which follows.

5.2 **Weather-Barrier, Reinforced Mastic—Designation II**

5.2.1 Material

5.2.1.1 Weather-Barrier Mastic shall be acrylic latex based heavy build mastic having the following properties.

5.2.1.1.1 It shall be suitable for palming or troweling.

5.2.1.1.2 It shall be suitable for application at atmospheric and substrate temperatures from 32° F to 150° F (0°C to 65.6° C).

5.2.1.1.3 It shall be nonflammable when wet.

5.2.1.1.4 It shall be fire retardant when dry. Dried film flame spread index, when tested in accordance with ASTM Test Method E-162 shall not exceed 35.

5.2.1.1.5 Solids content, as received, shall not be less than 50% by volume.

5.2.1.1.6 After being dried and completely cured it shall have a minimum elongation of 15% when tested at 20° F (−6.7° C).

5.2.1.1.7 Material shall be suitable for application on vertical surfaces, without sliding, slipping or sagging, to a 1/8″ (3.2 mm) wet thickness in one application that, after drying, will achieve a minimum thickness of 1/16″ (1.6 mm).

5.2.1.1.8 Mastic shall have a minimum shelf life of six months after it is received by the purchaser.

5.2.1.2 Reinforcing Cloth shall be a 100% resin fiber which is extensible at temperatures from −100° F to +250° F (−73.3° C to +121° C). Cloth shall be 44″ wide. Leno weave, weighing approximately 2.65 ounces per sq yd, with count of 16 × 8, employing 18/2 in the warp and 9/2 in the filling, and fabric wet-out with 8% solution of PVA, then dried at 250° F (121° C).

5.2.1.3 Mastic, caulking; shall be cork filled PVA mastic No. ______________ as manufactered by ______________

5.2.1.4 Heat-Resistant Sealer shall be silicone based sealer of paste consistency, No. ____ as manufactured by ____

5.2.1.5 Expansion Sleeve shall be Type 304 stainless steel strip, smooth, 9″ (228.6 mm) wide, a minimum of 0.008″ (0.2 mm) thick.

5.2.2 Application

5.2.2.1 Surface of insulation shall be smooth, even, and free of voids, and in a relatively dry state.

5.2.2.2 On outside installations, insulation shall be sloped for water drainage.

5.2.2.3 Sharp outside corners of insulation shall be rounded off to not less than a 1″ (25.4 mm) radius.

5.2.2.4 Inside corners shall be caulked with caulking mastic to obtain a minimum 1″ (25.4 mm) radius inside corner as illustrated in Figure 67 (Chapter 6).

5.2.2.5 A heavy fillet of heat resistant sealer shall be installed around all metal which projects through the insulation. The sealer shall extend 6″ over the insulation and 6″ over the projection. Application shall be as shown in Figure 65 (Chapter 6).

5.2.2.6 Insulation shall be protected from weather as soon as possible. However, the barrier shall not be applied when atmospheric temperature is below 32° F (0° C), or when the temperature is expected to be as low as 25° (−3.9° C) within the ensuing 24 hours.

5.2.2.7 Reinforcing cloth shall be bonded, taut and smooth, to the insulation surface with the weather-barrier mastic. All joints shall be overlapped a minimum of 2″ (5.08 mm). All inside and outside corners shall be rounded and overlapped with two layers of cloth. Cloth shall extend a minimum of 4″ (10.16 mm) out on all projections through the insulation.

5.2.2.8 Mastic shall be troweled or palmed over the entire surface, pressing it through the mesh of the cloth to obtain bond with the insulation. The weather-barrier mastic shall be carried out 6″ onto metal, beyond termination of the insulation on supports, skirts, or other metal projections. Care must be taken that mastic completely seals the openings in the cloth. After the weather-barrier has partly set, it shall be water brushed to a smooth even surface. The combined thickness of the weather-barrier coating and the reinforcing cloth shall not be less than 1/16" (1.6 mm) when dry.

5.2.2.9 Drip, on floors, concrete, or splatter on gages, valve stems, instruments or other items must be immediately washed clean with water, then dried.

5.2.2.10 Expansion joints in weather-barriers over insulation expansion joints shall be constructed with expansion sleeves as illustrated in Figure 71 (Chapter 7).

It will be noticed that the decimal divisions of this sample specification do not concur with the decimal divisions in the outline. This is because the outline was prepared for general scope and contains more elements than would occur in any one specification. It would never be possible to directly use each of the numbers in the outline for an identical element in an individual specification, as, by necessity, there would be decimals omitted and additional ones required.

Due to the nature of this presentation, reference illustrations were located in various Chapters in the manual. When used in a specification, illustrations would be either inserted in their proper location in the specification, or assembled at the back of the specification, arranged and numbered for easy reference. The use of drawings to illustrate specifications is highly recommended as being a ver effective means of communicating information.

In the preparation of specifications, various means may be used to reduce the space and amount of written matter. For example, the properties of material may be presented in tabular form rather than by referring to each property and test method in written form. These methods of presentation are quite efficient and desirable. This is particularly true when a number of individual specifications are involved. The presentation in this manual, necessarily, had to be made in depth, thus making the specification longer than it would be if short cuts were used.

Here it should be reemphasized that the specification, as presented, was a sample of the items to be covered by a specification and a suggested format by which all these items can be presented to communicate the desires of the designer-engineer to the contractor and his personnel. *The presentation given is not a recommendation of materials or their application.* Only the engineer who knows his installation requirements, the properties of materials, and how they should be applied is in a position to prepare material and application specifications for his installation.

12 Contracts

General Considerations

As an engineering manual, this book will make no attempt to state how contracts should be written, or the language to be used. That is a matter for the legal profession. It will discuss, however, the matters which should be considered in the preparation of a contract between an Insulation Contractor and a Consumer.

Like insulation installations where no single material and application specification can serve all installations, no one standard contract can be written which will serve the needs of all installations. A contract can range from simple verbal instructions from a consumer to a contractor to install a few feet of pipe insulation on a piece of pipe, to a complex legal contract of many pages, a book of specifications and hundreds of drawings. In the first instance, the cost may be less than $20.00, whereas in the latter the cost may be $1,000,000.00 or more. In the first case, a formal written and signed contract would cost more money than the entire cost of the material and work on the job, so it could not be justified. In the second case, if complete contract agreements, detailed specifications, and plans are not used, the installation will likely end in utter confusion and possible failure.

Thus, the need for contract agreements between the Insulation Contractor and the Consumer ranges from zero to very complex. The correct contract can only be one which provides a fair agreement between the parties involved, based on the requirements of the installation.

The more complex an installation, the greater the probability that some points of difference will arise that have not been cleared up in the Contract. These differences, if not resolved, may lead to bad feelings, or law suits between the parties involved. In most instances of serious disagreement, both parties are losers in the end. For this reason, a good clear contract with provisions defining the responsibilities of both parties in all particulars is an advantage to both.

All the items mentioned in this discussion will never be included in any single contract. No contract would ever be written to contain *all* the various methods which could be used to determine the amount to be paid by the consumer to the Contractor. Nor would any one installation be likely to have all the possible individual items discussed within its particular scope. However, these many points will be included, so that they may serve as a reference guide, to see if such items are of significance in the contract being considered.

The ethics involved in requesting bids, submitting bids, and awarding of contracts will be discussed only so far as to state that good ethics are good business. If a contractor attempts to make excessive profits, he will soon realize that our competitive system works, and that other contractors will be found who will work on a reasonable profit basis. On the other hand, a consumer who attempts to force contractors to a price level below their cost will eventually find that he has forced the contractor into short cuts which result in a poor installation. The resultant losses in heat, and the increased maintenance cost will rapidly wipe out any supposed savings.

Another item frequently forgotten by consumers who require excessive number of bids is that their preparation costs money. All the money contractors spend in preparing unsuccessful bids must eventually be paid by the consumer. In other words, all the Contractor's cost, plus a profit, must be paid by the consumer, or the Contractor will be bankrupt. Thus, the consumer saves himself money if he saves money for the Contractor. Such consideration by the consumer may not be apparent in a single installation, but will be reflected in future bids.

In addition to these general considerations, one other item frequently overlooked by both the consumer and contractor is the question of who is responsible for the engineering design. If a contractor is requested to recommend and install an insulation system, then he is responsible for the successful functioning of that system. If a contractor is furnished material and application specifications by the consumer, he is then responsible only for furnishing the quality of materials specified, and to install them as called for. When he fulfills this obligation, his responsibility ceases. Too frequently, a contractor recommends the substitution of materials and methods of application without realizing that, if his recommendations are accepted, he then becomes responsible for the proper functioning of the installed system. In many instances, this turns out to be a costly error in judgement, as in attempting to make a small additional profit he assumes responsibility for the proper functioning of an entire costly installation.

The capability of insulation contractors to do insulation engineering design varies considerably, as would be expected. Thus, if several contractors are requested to bid self-designed systems, the quality of the design will likely vary from poor to excellent. If a contractor's engineering is paid for only if he obtains the job, it must be assumed that his design is based upon materials and methods of application which are at his disposal, regardless of whether or not these may be the ones which will most closely fulfill the installation requirements. For this reason, engineers other than contractors should be responsible for the design of major investment insulation installations if at all possible.

Where it is impractical to obtain engineering from other than the insulation contractor, it becomes advisable that the most competent person available be chosen to do the engineering design, and he be paid for that service. He must also be instructed to supply a list of all competitive materials for the installation so that all can bid on the installation on as nearly an equal basis as possible. It should also be noted that many contractors do not have an engineering staff. However, such a lack need not affect their competence as suppliers and installation contractors.

Many of these considerations must be decided by the customer before the consideration of a contract, and may thus never appear in the Contract itself. However, they must be analyzed, and a course of action decided upon before the Contract is written.

The provisions which must be included are divided into a number of sections, having many subsections and items, any one or all of which, may be important to the installation under consideration. Each of these sections will be considered individually. The Sections are as follows:

- A Scope
- B Types of Contracts
- C Specifications, Drawings, Insulation Lists, Work Orders, Changes
- D Responsibilities for Workmanship and Materials
- E Job Site Trucking
- F Scaffolding
- G Hoisting
- H Preparation for Application
- I Starting and Completion
- J Application
- K Duration Time
- L Work Week
- M Safety
- N Liabilities
- O Taxes
- P Escalation
- Q Progress Billing
- R Additional Work
- S Extra Work
- T Authorization for Additional and Extra Work
- U Changes
- V Billing of Extras and Additional Work
- W Payment
- X Accounting Breakdown
- Y Unused Materials
- Z Authorized Personnel of Owner or General Contractor
- AA Bond
- BB Guarantee
- CC Definition of Terms

With the number of details covered in each of these sections, a consumer could question whether or not this section was written for the benefit of the Contractor. Actually, the clearer the Contract Documents (Contract, Specifications, List of Materials, and Drawings) are, the smaller the amount of money the Contractor must add to his bid for contingencies to protect himself from possible loss. More important, if the Contractor knows exactly what is expected of him he can better accomplish the work quickly and economically and prevent job delay.

A SCOPE

The Scope of the Contract should contain the general facts concerning the installation. These are as follows:

1. *Contracting Parties.* It should be established definitely who are the parties to the Contract. Does the Contract specify more than one person or organization to whom the Contractor is responsible? Multiple responsibility should be eliminated if possible. Is the Contractor responsible to a general contractor? If the Contractor is responsible to someone other than the Owner, who is responsible for payments and approval of materials and workmanship? As a good basic principle, the Contractor's work should be subject to approval only by the party who is responsible for the payment for the work.

   There is another item which should be investigated by a subcontractor. It is the condition occurring when under a contract with a general contractor, he completes his work satisfactorily, and other parts of the work done by the general contractor are not satisfactory. This condition will often cause the Owner to hold up payments to the general contractor. Procedure should be established as to how the subcontractor can collect directly from the Owner.

   Another item which should be cleared in advance occurs when such a contract exists between a subcontractor and the general contractor, and the subcontractor is given conflicting instructions by the Owner and the general contractor. Whose instructions should he follow?
2. *Site:* The site of the work, name of plant unit area, and location should be stated. Permission should be granted to all potential contractors to look over the site prior to bidding to determine access roads and rail facilities. The general area alloted to the contractor's base of operations should be located.
3. *Time Schedule.* Statement should be made as to the date through which the prices stated in the Contract will remain firm. Also, state whether this clause is unconditional, or whether it hinges upon the receipt of the order or letter of intent within a specified period of days.

   Or, state the number of days during which the quotation will remain open, and is subject to acceptance during that period.

   Or, state the date through which the prices stated in the Contract will remain firm, subject to the General Contractors completion date (which should be stated). Also state whether the firm price date is unconditional, or whether it is predicated upon receipt of an order from the General Contractor within a specified period of days. State whether work performed, or materials delivered after a specified date will be subject to a surcharge.

   State the percentage surcharge for the first month and any additional percentage for the following months until completion.
4. *Contract Documents.* Documents which are part of the contract should be named and listed in the Contract so that all parties are assured that each has the same information. These documents may be (1) Detailed Installation Information, (2) Material Specifications, (3) General and Detailed Application Specifications, (4) Insulation Lists, and (5) Drawings. It is important that proper identification and number of each document be given.
5. *Subcontracts to this Contract.* If there are areas of the work that must be subcontracted to others by the Contractor, the rules and restrictions governing such subcontracts should be given.

B TYPES OF CONTRACTS

There are many systems by which the amount the Contractor is entitled to can be established. The correct choice of system depends upon many conditions, including type of work, time urgency, working and economic conditions. The bid prices are influenced by all the factors of the Contract, type of work, time allotment, job conditions, cost of labor, and material. Thus, there is no simple way to establish cost. Some systems used are:

1. Lump Sum

   This type of contract states that the Contractor will furnish, deliver, and install in the Owner's plant or building all the insulation, complete with all accessories, in accordance with the Contract Documents, for a stipulated sum of money. When it is possible to furnish complete and final information, this is a very clear cut method for obtaining an installation. Unfortunately, there are almost always some job changes. Thus, in lump sum contracts it is almost always essential that a basis for potential charges for alterations and additions be stated. These additions may be charged on a unit price basis, materials plus a unit price of installation, unit price of material plus labor cost and percentage, or by cost plus a percentage or fee. The same basis may, or may not be used, for any deletions. Each of these methods will be discussed separately.

   a. Unit Prices

      Contracts based on unit prices, meaning that the Contractor will furnish and install the insulation for certain prices for definite units of piping, equipment, etc., appear to be quite clear cut, but actually are quite complex.

First, it is necessary to define the units. One of the problems here is that there has never been a truly uniform unit basis established, especially for pipe insulation. For example, taken from one suggested method of measurement is the following statement: ''Measurement of straight pipe shall be taken through all insulated fittings. Insulation on valve body, including bonnet, shall be counted as fitting. The bonnet flange shall be counted as one flange. Each line flange on flanged valve shall be counted as one pair of flanges.'' Thus, the unit price of a flanged valve is the combined total of the length of pipe covering the length of the valve, plus one fitting, plus three pairs of flanges. For some unaccountable reason, the industry has held on to this ancient and potentially inaccurate type of unit pricing. It then wonders why its customers become agitated. The unit price for insulating a flanged valve should be the price the Contractor feels is fair for insulating that valve.
Because of this confusion, it is suggested that the following unit price system be considered, as being fair to both the customers and contractors.

b. Measurement and Units

(1) Piping—Fittings
Insulated fittings should be quoted and counted as individual fittings. Flanged valves and fittings should be counted as a complete unit only. Line flanges, welded and screwed valves, and fittings should be counted separately, each as a separate and complete unit. To prevent confusion, valves should be listed separately from fittings, and each be given separate unit prices.

(2) Tees—line nozzles present the problem of ascertaining when a branch line should be considered a tee, or only an insulated pipe joining another insulated pipe. It is suggested that the following be used to govern this question. For any initial pipe size, as listed in Column A, when the branch line is equal to, or larger than, the corresponding size listed in Column B, the unit shall be considered a tee. If the connecting branch in Column B is smaller than the corresponding size branch in Column A, the unit shall be considered an insulated pipe joining another insulated pipe. The smaller branch size should be used for determining nozzle size for unit pricing.

Column				Column			
A		B		A		B	
English	*Metric*	*English*	*Metric*	*English*	*Metric*	*English*	*Metric*
3/4"	19.1 mm	3/8"	9.53 mm	5"	127 mm	2"	50.8 mm
1"	25.4 "	3/8"	9.53 "	6"	152.4 "	2 1/2"	63.6 "
1 1/4"	31.8 "	1/2 "	12.7 "	8"	203.2 "	3"	76.2 "
1 1/2"	38.1 "	1/2"	12.7 "	10"	254.0 "	4"	101.6 "
2"	50.8 "	3/4"	19.1 "	12"	304.8 "	5"	127.0 "
2 1/2"	63.6 "	1"	25.4 "	14"	355.6 "	6"	152.4 "
3"	76.2 "	1"	25.4 "	16"	406.4 "	6"	152.4 "
3 1/2"	88.9 "	1 1/2"	38.1 "	18"	457.2 "	8"	203.2 "
4"	101.6 "	1 1/2"	38.1 "	20"	508.0 "	8"	203.2 "
				24"	609.6 "	10"	254.0 "

At time of contract, it should be decided whether or not steam traps and strainers are to be counted as fittings.

(3) Pipe Flanges, Supports and Cradles
Pipe hangers on pipes operating at temperatures below ambient air temperature, where it is necessary that they be insulated up the hanger rod to prevent condensation or frosting, should be given their proper unit price. Hangers requiring no insulation should be ignored in pricing. The same should be true for pipe supports. If insulation is required, they should be so priced, but should be ignored in pricing if no insulation is required. On the latter, the small amount of cutting of the insulation to fit around these supports should be included in the cost of installing the straight pipe insulation. Where pipe is supported by cradles on the underside of the insulation, additional work is required by the Insulation Contractor, even if the cradles are supplied by others. It is suggested that unit prices be furnished for this work to avoid later confusion if an attempt is made to obtain additional compensation for this, based on classifying it as ''extra'' work. If cradle assemblies are to be furnished and installed as part of the insulation contract, they should be considered as individual units.

(4) Bent Piping
Bent piping, where the application of mitered insulation is required, should be calculated on the basis of feet of insulation measured on the outside radius of the bend. Units should be expressed in linear feet.

(5) Straight Piping

Pipe insulation should be calculated and measured between fittings along the center line of the pipe. No flanges, valves, ells, or other fittings should be included in this footage. Units should be expressed in linear feet.

Diagrams illustrating this suggested measurement system are shown in Figures 143 and 144. A suggested form for listing these quantities is shown in Figure 145.

(6) Equipment

(a) Vessel Bodies

Vessel area should be calculated and priced based on *outside* area of the insulation.

(b) Vessel Heads

Area of the heads, like the body, should be based upon the outside area of the insulation. Where the head is inside a skirt, individual prices should be included to indicate the additional cost for this slower installation.

(c) Manholes, Handholes, Blind Flanges, Piping Flanges

Insulated manholes, handholes, blind flanges, and flanges of piping connected to equipment should be priced in accordance with their size and insulation thickness. The last line flange on a line of insulated piping, which connects directly to the flange on the equipment, shall be counted with the vessel, not with the piping.

There should be no charge for the cutting of insulation to fit around any flanged manholes or vessel flanges.

(d) Insulated Vessel Flanges and Insulation Supports

Where additional insulation is required at the vessel flanges and insulation supports, it shall be provided under separate unit price figures. Where insulation expansion joints are supplied, these also should be separately priced.

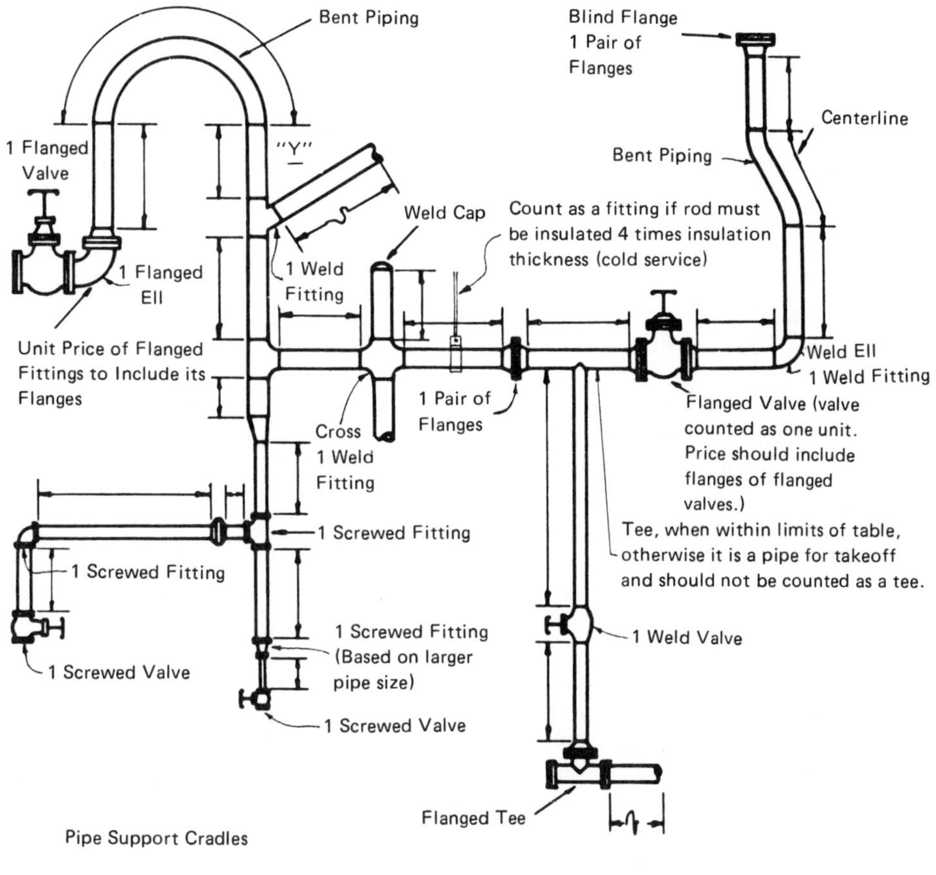

Figure 143 Piping—method of measuring.

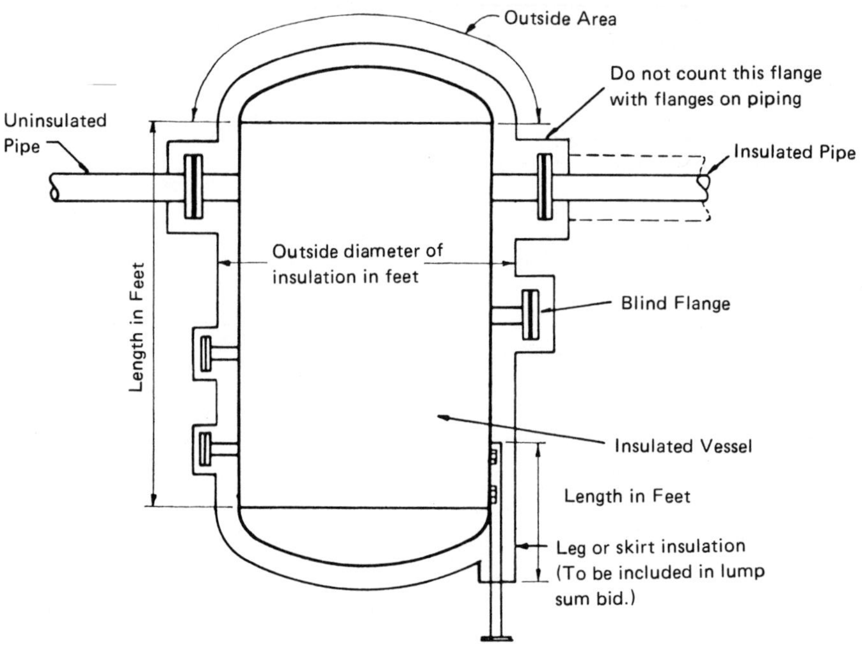

Figure 144 Equipment—method of measuring.

(e) Insulation Supports

In many installations the insulation supports on the vessels and equipment are supplied by the equipment vendor. However, in other instances the Insulation Contractor may be expected to furnish, or have installed by others, these necessary insulation supports. This item should be clarified in the Contract.

(f) Thermocouple Wells and Instrument Connections

Where these items project through the insulation, a charge should be made for the necessary cutting, fitting and sealing of the insulation around them. If it is necessary to remove the insulation to allow the installation of them, such removal and reinsulation warrants an extra charge.

(7) Factors Affecting Unit Prices

(a) Scaffolding

Having the scaffolding furnished by others will affect the unit prices. For this reason, the question of who is to furnish the scaffolding must be decided prior to submission of unit prices. When the scaffolding is to be furnished by the Contractor and must be dismantled at the Owner's request, then re-erected to complete the installation, such extra cost warrants additional charges.

(b) Height Above Grade or Floor

The higher the work is above grade or floor, the greater the cost of scaffolding, and the slower the installation. For this reason, it may be necessary to provide correction factors for unit prices due to height.

(c) Time Duration

The duration of time from the start to the finish of the installation may be a vital factor for both parties. The consumer may require the installation within a given length of time, thus requiring a large labor force. This would markedly influence the cost of the installation. On the other hand, delays which extend the period of the installation may also prove costly. Any stipulation by either party as to time should be clearly expressed in the Contract, with provision for adjusting the prices accordingly.

(d) Time of Year

The time of year, especially in Northern climates, affects productivity, and the unit prices for a summer installation may not be suitable if the application is in the outdoors during the winter. Correction Factors for these differences could be made a part of the unit prices, or otherwise cared for.

Unit Price List – Piping

Pipe Size	Insulation thickness		Straight pipe		Bent pipe		★
	inches	mm	$/lin ft	$/lin m	$lin. ft	$ lin m	

★All divisions from here on to end of table are same as originally printed.

Unit Price List – Equipment

Insulation thickness		Curved side walls		Flat surface		Heads contoured		Heads inside skirts		Skirts inside surface		Vessel flanges		Cont'd below
inches	mm	$/sq ft	$/sq m	$/sq ft	$/sq m	$/sq ft	$/sq m	$/sq ft	$/sq m	$/sq ft	$/sq m	$/sq ft	$/sq m	

Cont'd from above	Insulation support insulation		Insulation support slip joint		Vessel legs		Nozzles by size	Manholes by size	Vessel supports	
	$/sq ft	$/sq m	$/lin ft	$/lin m	$/sq ft	$/sq m	$/each	$/each	$/sq ft	$/sq m

Figure 145 Typical unit price lists.

(e) Location
Indoor installation, where the work is protected from the rain, naturally costs less than outdoor applications. However, if an installation is in tight crawl spaces, over ceilings, inside confined spaces, or in trenches a special effort should be exerted to see that separate unit prices are provided to take care of these difficult installations.

2. Unit Prices
   a. Unit Prices For Material, Plus Labor Cost
   Another type of contract is one in which the material costs are based on unit prices, and the labor is based on the actual labor cost plus a percentage or fixed fee. The material may, or may not, include preformed fittings. Where the Contract does require preformed fittings, it is recommended that the same system of units suggested for unit prices of installed insulations be used to quote the *prices of the materials only*.
   When such units are used to quote materials, special care must be exercised to be sure that the material being quoted is stated. If the units are only for the basic insulation, such as block, blanket, pipe covering, and preformed fittings, then this should be stated. If the unit prices include all the materials needed for an installation of these items, including the jacketing or mastic, the securements, and all accessories, then this should be definitely stated.
   The labor costts may be priced on direct labor, plus necessary overhead, accounting, and other expenses, plus a fixed percentage for profit, or a fixed fee. To correctly determine these labor expenses, proper accounting procedures agreeable to both the Contractor and the Owner must be a part of the Contract.
   b. Unit Prices For Labor Only—Materials On Lump Sum Cost, or Furnished by Owner
   In some instances the Owner either owns, produces, or has arrangements with a Distribution-Contractor for the purchase of materials, and when some individual installation is required he requests unit prices for *application only*.
   Where such an arrangement is desired, the Contractor may be requested to furnish unit prices for installation only. It is suggested that the system of unit prices listed in this chapter be used as the basis for bidding. Of course, the units would be for application only. Again, these prices must be clarified as to what they include: items such as scaffolding, spray equipment, warehousing, delivery of materials, etc.
3. Cost Plus
   a. Contracts to furnish labor and materials on the basis of actual cost, plus either a percentage or fee, would seem to be quite direct and simple. The major problem in such contracts is the setting up of suitable accounting procedures, and control of both labor and material costs.
   Direct labor of application is relatively easy to check and determine. The associated labor costs of materials, stocking, handling, and delivery to points of application is much more difficult to determine. These associated costs, including that of the accounting of costs, is a fairly high percentage of the total. The method of determining, accounting, and charging of this cost should be established in the Contract.
   Also, the basis of payment for the use of equipment such as sprays, trucks, building fabrication equipment, and scaffolding must be determined and provided for in the Contract.

C SPECIFICATIONS, DRAWINGS, LISTS, WORK ORDERS, CHANGES

The Specifications, Drawings, Insulation Lists, and Work Orders are the "working" documents. These must be referred to in the Contract so that they are recognized as part of it. If certain of these are submitted as part of the request for bids, they should be listed in the Contract so that the bidder can check to see that he has the entire package from which to prepare his prices. Where additional specifications, drawings, work orders, and lists are to be supplied at a later time, the manner of transmittal and the fact that these are acknowledged as being additional must be included in the Contract.

1. Specifications

Specifications for material and application are the basis for calculating bids. Several facts should be remembered. Specifications provide the *intent* of the customer regarding what he requires as to quality of materials and workmanship in the resultant assembly. The Specifications should not attempt to state *how* this is to be accomplished, only *what* is to be accomplished. This must be very specific. A contractor should not be expected to guess what the consumer requires.

Specifications cannot be expected to state everything which should *not be done,* but only what *is to be done.* In a like manner, the Contractor can only be expected to perform the services which are specifically stated to be done. Statements in the Specifications to the effect that the Contractor will be held responsible for the insulation of any steam lines, or other items, which have been omitted on the Drawings shows complete irresponsibility on the part of those who expect anyone to bid on contracts whose specifications contain such statements.

2. Additional or Extra Work

With any contract, it is almost impossible to foresee, or have available at the time of awarding contracts, the details of every section of piping or equipment. Thus, "additional" work provisions should be made a part of the Contract. In most instances, even with firm bid contracts, this additional work can be priced, based on unit prices which were submitted as part of the Contract.

"Extra" work, such as removal or replacement of insulation at the Owner's request, is different in that it is work performed on insulation that was in the original contract. It is suggested that the difference between "additional" work and "extra" work be kept clear and distinct. Many times an installation gets blamed for "extras" which in reality were "additional" work. The preceding discussion shows the necessity for having the Drawings and Lists recorded in the Contract. Otherwise, differences will arise as to which drawings and lists were furnished at the time of bidding, and which were added later.

3. Drawings, Lists

The Drawings, Sketches, and Lists should specify the equipment and lines to be insulated, and should be definite as to what vessel flanges, nozzles, legs, etc. are to be insulated. Likewise, line lists and drawings should be definite as to the fittings, flanges, and valves to be insulated. Any interferences, floors, or walls should be located in reference to the insulated items. Pipe supports, cradles, or hangers should be identified and noted in the Specifications, Lists, or Drawings if they are to be insulated. It should be remembered that it takes much longer to question, run down and answer these uncertainties than it does to properly note them originally.

D RESPONSIBILITIES FOR MATERIALS AND WORKMANSHIP

1. Responsibilities

The first and foremost thing to be established in any contract is the determination of the responsibilities of the contracting parties. If the Insulation Contractor's contract is with the General Contractor, who has the responsibilities of procuring and furnishing information and making acceptance and payments? If a situation exists in which there is a General Contractor, or other contractors, on the job in addition to the Insulation Contractor who has the Contract directly with the Owner, what are the relative responsibilities of each, and their relationship with one another? In the case of differences between them, who is responsible for the final decision? Although these questions appear elementary, a surprising number of contractors sign contracts without these responsibilities being determined in advance. This results, at times, in the Contractor finding himself in receipt of contradictory orders.

Beside the general responsibilities to be established, there are a number of specific things which should be determined between the contracting parties before the signing of a contract. Some of these are:

a. Job Site Facilities

(1) Storage

It should be determined if the Owner, or General Contractor, is to provide storage space for materials. Are the spaces heated? If not, does the Contractor need to erect a storage shed, and "change-house" at the job site? If so, what provision is made for this cost, especially in Unit Price Bids? Are unloading facilities, railroad spurs, and roads available? Who pays for these facilities? Will trucks for material be furnished, or will the Insulation Contractor be required to furnish his own?

Does the Owner or General Contractor furnish receiving and storage personnel? Does the Insulation Contractor furnish this personnel?

If the Insulation Contractor needs to set up a fabrication shop on the job site, is space available? Are there restrictions regarding dust or fumes? Are there areas considered hazardous where no open flame is permitted? Are there any building restrictions?

(2) Utilities

What utilities will be furnished by the Owner to the Sub-Contractor? These may be Electricity/Steam/Gas/Compressed Air/Telephone.

At who's cost are these facilities piped or wired up to the Insulation Contractor's facilities?

(3) Change of Location of Storage Facilities

In commercial installations, as a building is being constructed it may be necessary to change the storage location of the materials as the job progresses. Where this could be arranged to work in with the insulation work, extra charges would be unnecessary. However, if numerous changes werre required either to accommodate other crafts, or by the General Contractor's directive, such costs should be borne by others. The same is true on industrial work. Where it is necessary to change location of storage or fabrication shop to satisfy the requirements of others, this cost should be borne by others unless it was stated in the original contract to be the responsibility of the Insulation Contractor.

(4) Responsibility for Equipment

Another item which must be considered is, who is responsible for materials, tools and equipment on the job site. In case of loss or damage, who pays for replacement or repairs? Is plant protection sufficient, or does the Contractor need to furnish watchmen outside of working hours?

b. Materials

(1) Quality

Responsibility for quantity and quality depends upon who furnishes the materials. If the Owner or General Contractor furnishes the materials to the Insulation Contractor, then they are responsible. This is an infrequent arrangement.

If the Insulation Contractor furnishes the material, then he is responsible for the quantity and for the quality being that specified. Any poor quality or broken materials received are his responsibility. He should make proper claims and secure replacement from either the carrier or the manufacturer. Should he desire to substitute materials different from those specified, he is responsible for obtaining approval from the General Contractor, or Owner, in accordance with the terms of the Contract.

(2) Broken or Rejected

If the quality of any material is suspected, it is the Insulation Contractor's responsibility to bring such to the attention of the Owner or General Contractor, with whom he has his contract, to obtain either approval or rejection. If rejected, the manufacturer should be notified so as to immediately obtain acceptable replacement material. Where the Owner or General Contractor has furnished approved material, and the material fails in service, the Insulation Contractor should not be held responsible for repair or replacement.

(3) Left-over Material

On fixed price bids, all materials left over at completion of the installation belong to the Insulation Contractor. Materials purchased from the Insulation Contractor, either by Unit Prices, or other purchase arrangement with the Owner, belong to the Owner at the completion of the installation. It is the Contractor's responsibility to inform his employees as to ownership of materials so that they do not inadvertently remove the Onwer's material to another job.

(4) Cancellation

On a fixed fee, or other type of contract for which the Insulation Contractor has purchased material for an installation which is later cancelled, he is entitled to compensation for his cost of purchasing, handling, and stocking of material. The amount of such allowed compensation should be a part of the original contract.

(5) Substitution

Material shall be substituted for that specified only when the substitution is approved in writing by the Owner's Authorized Representative.

E JOB SITE TRUCKING

The furnishing of trucks for material handling on large jobs falls into the same category as does hoisting. The most economical means of distributing materials on the job site should be investigated prior to final quotation on the installation.

F SCAFFOLDING

The Insulation Contractor is responsible for the furnishing of proper scaffolding for his work, unless specifically stated otherwise in the bid request. However, there must be an understanding between the contracting parties as to *use* of the scaffolding. In most instances, it is assumed that the Contractor may install all piping insulation and insulation covers for fittings, flanges, and valves connecting to the vessel and adjacent to vessel scaffolding, while it is still in place and needed for the vessel insulation. If, by the Owner's or the General Contractor's request, any scaffolding is removed and then has to be reerected, such additional cost should be reimbursed to the Contractor.

In the event that other crafts or contractors use the scaffolding they should be allowed to do so as long as they do not interfere with the Insulation Contractor's work and progress. Others using the scaffolding should be required to give the Insulation Contractor a "save-harmless" waiver, guaranteeing the Insulation Contractor against liability for any injuries which might occur during the others' use of the scaffolding.

If the General Contractor desires to use the scaffold for a period of time after the insulation is completed, he should expect to pay a reasonable "rental" rate during the period of his use and until he releases the scaffolding back to the Insulation Contractor who will dismantle it.

Where individual drawings of equipment are furnished that do not locate it in relation to its height from the grade, the Contractor should be advised of the height involved so that he can properly calculate his scaffolding and application cost. Failure to do this should be considered an omission of information needed for bidding and the Contractor should be compensated for all extra costs.

G HOISTING

On high rise buildings, or tall industrial columns, materials must be transported to the level of installation. It is good judgement to determine in advance which of the contracting parties will furnish hoisting equipment, and the charges for such service if it must be leased or rented. Frequently, the General Contractor can arrange to furnish such equipment for all sub-contractors at a more economical rate than if each were to furnish his own, or lift the materials by hand methods.

H PREPARATION FOR APPLICATION

1. Surface Preparation

In most insulation specifications, it is stated that the surface to be insulated should be clean and dry before insulation is applied. What constitutes "clean" metal can lead to considerable disagreement. It can be argued that only metal which is sandblasted to clear it of all mill scale, rust and other contamination is truly clean. In the insulation industry the Insulation Contractors expect only to wipe metal with waste or rags to remove surface dirt or moisture. When more cleaning than this is expected it should be so stated in the Contract.

When sandblasting, painting, or coating of surfaces is to be done prior to the installation of insulation, it is suggested that a separate contract be awarded for this work, even though the Insulation Contractor may also be a painting contractor. Among other advantages, such separate contracts tend to remove craft conflict which may delay an application.

Any painting of surfaces should be scheduled to be completed in sufficient time to allow the paint to dry before the insulation is applied, and to reduce confusion and conflict in the use of scaffolds.

2. Insulation Supports

Insulation supports on vessels and equipment are most frequently installed by the equipment vendors, or installed on the vessels and equipment prior to starting the insulation application. For this reason, it is seldom expected that the Insulation Contractor will furnish or install these supports. If, for some reason, it is expected that the Insulation Contractor will furnish such supports, this fact should be definitely called for in the Contract, pointing out that this additional function is expected of him.

3. Pins, Clips, etc.

Likewise, the furnishing of angles, rings, clips, nuts, wood blocks, steel saddles, and cradles for the support of insulation and/or pipe is not generally part of an insulation contract, as these are fabricated and installed by crafts other than asbestos workers. However, pins, welded or adhered to a surface, and speed clips to secure insulation in place are accepted as part of the insulation application.

I STARTING AND COMPLETION

1. Time of Start

A definite statement shall be made in the Contract that the work will be started in a specified number of days after the receipt of the order, and manned to keep pace with the overall construction progress.

2. Notice of Readiness

A statement in the Contract shall specify that the Owner, or General Contractor, shall notify the Insulation Contractor a specified number of days in advance when the premises will be ready for the installation of his material, and that he will at all times be given free and unobstructed access to the place where the work is to be done. It shall also be stated that any losses incurred by the Insulation Contractor due to the premises not being ready, after the Insulation Contractor has been notified to begin work, shall be borne by the Purchaser.

3. Delays

A statement shall also be made in the Contract to the effect that the Insulation Contractor shall not be responsible for any delays occasioned by transportation difficulties, priorities of any kind, acts of God, strikes, fires, floods, storms, walk-outs, boycotts, acts of war, rebellion, or any items of similar nature, nor shall the Insulation Contractor be liable for any consequential damages. (See ''Duration Time'')

J APPLICATION

1. Intent of Specification

It is the Insulation Contractor's responsibility to furnish capable workmen, and the necessary supervision, to apply the material in a manner that accomplishes the intended results as specified. Two things should be remembered: (1) An application specification cannot cover each and every *detail* of construction, thus leaving the Contractor responsible for details being completed in accordance with the *intent* of the specification. (2) An application specification should not attempt to direct the Contractor as to *how* he should do his work. Regardless of the work methods, if the Contractor provides the quality of assembly specified he has completed his responsibility.

2. Unusual Working Conditions

When any unusual working conditions are expected, it is only fair that the Contractor be advised of such. If excessive fumes, heat, dust, or other factors are involved which require special care, safety equipment required should be pointed out by the customer in the invitation to bid. When it is necessary to insulate high or low temperature equipment or piping in service, the care required by the workmen will, by necessity, slow down the application.

One exception to this is the application of insulation to hot flanged valves and line flanges. As it is standard practice that these valves and flanges must be inspected after being placed in service, it follows that the insulation must be installed after the inspection. Scaffolding must remain in place until after the inspection and the application of insulation, and this longer use of scaffolding, and the return to apply insulation after the piping is completed should be expected on all industrial installations, and so this extra cost should be included in his original bid by the Insulation Contractor.

3. Inspection

When inspection reveals that workmanship is not up to specified quality, or that improper material was used, the Contractor, or his representative, should be notified as soon as practical to prevent any additional work from being done in the same manner. Failure to notify the Contractor to this effect without delay is poor inspection practice. Poor material or workmanship must be replaced at Contractor's expense.

4. Replacement

Any removal of material for inspection, at the customer's request, which is then found to be of proper quality and workmanship should be replaced at the customer's expense. If it is found to be faulty, it must be replaced at the Contractor's expense.

5. Clean-up

Clean-up after application is a responsibility of the Contractor. However, if there are unusual conditions, such as in processing plants or occupied offices, where excessive dust is present and immediate clean-up is required, this should be pointed out in the original scope of the installation in the Contract.

K. DURATION TIME

1. Penalties

If a contract contains provisions for penalties against the Insulation Contractor, it should be stated that penalties will not be charged against him for delays over which he has no control. For example, if the insulation application was held up at the start of the job, or by delays during the job because of pipes or vessels not being released for work, these delays should be added to the required completion time without penalty. As insulation is one of the last crafts to work on a major installation, an understanding of this is very urgent, otherwise all job delay caused by other crafts, or late deliveries of pipe or equipment may cause total job delay to a point that the Insulation Contractor has no possibility of finishing the work at the time desired by the General Contractor, or Owner. In other words, the Insulation Contractor should take care not to be penalized for the faults of others.

2. Delayed Starting Time

It is also only fair that the Insulation Contractor have some protection, both as to time of job start-up and its duration. Many installations planned to start in the summer end up starting in late fall or winter. On outside work, application is much slower in cold northern climates. If a contractor figures his cost based on summer installation, he is justified in insisting that his charges be increased a fair amount if the work is done after a certain date. Also, if a Contractor was advised that all work would be available for execution within a certain period of time and, due to delays by others, this does not transpire, he should be paid for the additional time required for his supervisory personnel and equipment. (See "Starting and Completion.")

L. WORK WEEK

1. "Spot" Overtime

Basic bids are usually based on a normal forty-hour work week. When overtime is required it is usually true that the General Contractor or Owner expects to pay for the premium time plus other charges such as taxes and overhead. This is a fair arrangement for moderate amounts of "spot" overtime.

2. Scheduled Overtime

Many times, due either to urgency or to speed completion, the General Contractor or Owner will require the Contractor to schedule overtime for a period of time. However, the greater the number of hours worked per day by an individual the less he can produce per hour. When contracts are based on firm bids or unit prices, this overtime places an unfair burden on the Contractor. For this reason, increased compensation should be paid to the Contractor, based on an increasing percentage in direct relation to the number of hours worked per day per man.

M. SAFETY

1. Regulations

The Insulation Contractor is responsible to see that his employees abide by all Governmental Agency safety codes which apply to his work and its location. He should also require that the Owner or General Contractor provide him with all local or plant rules and regulations with which he must abide. These should all be established prior to his submission of a bid.

2. Smoking

If there are plant smoking rules the Contractor should determine what they are and when and where his men can take their smoking break. If this requires more than ordinary time away from the work, it should be included in the original contract. Requests for additional compensation because of time lost in smoking periods will most generally be rejected by owners.

3. Safety Clothes and Equipment

The Contractor should provide his workmen with adequate protection from accidents and for their health. The Contractor should furnish such safety equipment as safety helmets, safety goggles, gloves, dust masks, and dust filters where they are needed.

4. Hazardous Areas

If there are areas in a plant where the General Contractor or Owner knows that, due to fumes, chemicals, or heat, special rubber suits will be required and/or limits of human endurance are involved, it should be his responsibility to call this to the attention of the Insulation Contractor in advance of the bidding. Otherwise, extra charges by the Insulation Contractor are legitimate.

N. LIABILITIES

1. Accidents

Any statement in this manual as to who would be responsible for damages because of accidental injury, death, or loss of health would be meaningless. These responsibilities can only be determined by due legal process, and each case has its own individual aspect.

2. Insurance

The General Contractor and/or the Owner has the right to specify accident and public liability insurance regarding coverage and amount which he requires the Contractor to carry to minimize any chance of loss. This should be clearly stated in the invitation to bids, since this liability insurance represents a cost, and is eventually reflected in the job cost.

The procurement of Workman's Compensation and liability and property damage insurance on all vehicles, owned or leased, which are used by the Contractor on the Owner's property is the Contractor's responsibility. These are a part of the Contractor's overhead cost.

Should the Owner or General Contractor require additional insurance for any purpose, he must state this requirement prior to the bidding so that the cost for such can be included in the job cost.

3. Strikes
The Insulation Contractor should not be held responsible for any labor strikes, or loss of time, or the unavoidable delay in completion of work due to strikes.

O. TAXES
In almost all contracts it is made clear that the Owner or General Contractor, depending upon who enters into the contract with the Insulation Contractor, will be responsible for any *additional* Local, State, or Federal taxes, levied by any governmental agency upon the merchandise, labor, manufacture, use, sale, or delivery, which would be added to existing taxes during the course of the Contract.

P. ESCALATION
1. Labor
Should the life of the Contract extend beyond a period during which labor negotiations are expected to occur, the Contract should contain provisions to escalate prices in accordance with any increased labor cost. An Owner or General Contractor should insist that all bidding be in accordance with *current* labor rates, and that escalation clauses be included, especially when it is known that labor negotiations are expected before completion of the Contract. Otherwise, the Contractor has no alternative but to figure his bid on what he believes to be the maximum rate increase possible.
2. Materials
On most installations, unless they extend over a very long period of time (at least over one year), the escalation of job cost because of increased material prices is unjustified for the original scope of the installation. Such increase is justified on additional work, which might be added at a time months after the signing of the original contract.

Q. PROGRESS BILLING
1. Materials and Labor
Proper procedure should be set forth in the Contract for the Contractor to follow in his billing for materials and labor. It is frequently arranged that the Contractor be paid 90% of the cost of any material at the time of its delivery to the jobsite. It is also customary that he be paid 90% of the cost of all work completed on a unit price basis at the time of its completion, or 90% of the amount of work completed on a firm bid at the time of its completion. The manner by which the Contractor and Owner's Representative reach agreement as to the amount of work completed is difficult to set any guides for, because in the midst of a large installation opinions as to the percentage completed may differ markedly. These percentage payments are usually made on a monthly basis.
2. Retention
In almost all large contracts, the final 10% of payment for materials and labor is withheld until completion of the work covered in the scope of the original Contract. Additional or extra work beyond the original scope should not be a basis for withholding payment of the final 10% of the completed major contract.

R. ADDITIONAL WORK
1. Unit Prices, Time and Material, and Firm Bid Additional work may be added in terms of the unit prices contained in the original contract, or based on the cost of time and material plus a percentage, or it may be added on the basis of a firm bid for that portion of the work. Where unit prices are clearly defined, compensation for additional work, such as some pipe and fittings in accordance with the unit prices, is in most instances preferable to both parties. Where additional work requires the tearing out of existing insulation, or large scale patching of existing work, one of the other methods may be desirable or necessary.

S. EXTRA WORK
Notice again that there is a distinction between "Additional" Work and "Extra" Work. This distinction is necessary because, frequently, Owner management is appalled at the "extras" on a construction job. In many instances, the "extras" were truly "additional" work and not "extras" as related to the original scope of the work. "Extras" should be restricted to that work which is made necessary because of removal of insulation and/or its replacement within the original scope of the installation. In almost all cases, because of its nature, the only way for fair charges is by cost of time and material, plus a percentage.

T. AUTHORIZATION FOR ADDITIONAL AND EXTRA WORK
In most contracts it is stated that no consideration will be given to claims for extra or additional work unless approved in writing before the work is started. For additional work, such a requirement is reasonable. Fortunately for

the General Contractor and/or Owner, most Insulation Contractors will follow verbal orders from a designated responsible representative of the Owner or General Contractor. Many "extras" are emergencies caused by such things as leaky flanges or removal of insulation for other crafts so that their work can be continued without interruption. Should the Insulation Contractor insist that he be given written orders before doing this service work, job delays would develop causing excessive cost and delay in total job completion. It is reasonable to state that for such extras he obtain written orders within a reasonable period, say within three days.

U. CHANGES

Any changes in the materials or method of application made by the Owner or General Contractor should be made in writing before the Insulation Contractor is required to perform the work covered by the change. Physical changes made by the Owner or General Contractor in the field, such as field run pipe, shall be handled as "Additional" or "Extra" Work depending upon which previously given definition fits the situation.

V. BILLING OF EXTRAS AND ADDITIONAL WORK

All extra and additional work should be invoiced at the end of each month in which the work was done. It is suggested that a proper form for such billing, including space stating by whom the work was authorized and reference to the field work order number, be decided upon by the contracting parties prior to start of installation.

W. PAYMENT

Payment of progress billing, and invoices for additional and extra work should be made within ten days of receipt of the billing and invoices.

X. ACCOUNTING BREAKDOWN

When the General Contractor or Owner requires a breakdown of costs in respect to work orders, item numbers, individual pipes, etc., such requirement shall be stated in the original contract so that this extra cost of accounting may be figured into the job cost. The proper forms and procedures for such breakdowns shall be furnished to the Contractor.

Y. UNUSED MATERIALS

1. Due to Cancellation or Changes

   Materials unused by the Contractor due to cancellations of, or changes in the work after completion of the job shall be charged to the General Contractor or the Owner at current market value, provided sufficient prior notice was not given the Insulation Contractor to enable him to cancel his purchase of these materials. Should the Owner or General Contractor desire to return unused materials to the Insulation Contractor, these should be accepted at a fair market value less a handling and restocking charge of ___________ %.
2. Due to Overage on Firm Bid

   Unused materials which are still on hand at completion of the job which were not the result of cancellations or changes and which had not been ordered at the request of the Owner shall be returned for full credit to the Insulation Contractor, if these had been billed to and paid for by the Owner.

Z. AUTHORIZED PERSONNEL OF OWNER OR GENERAL CONTRACTOR

In almost all contracts it states that the Insulation Contractor shall obtain changes, orders, extra work orders, etc., from an Authorized Representative. This statement is of no value until the Contractor is provided with a statement in writing as to who are the Authorized Representatives. Such personnel should not be named in the Contract, but, before the start of each project, the Contractor should obtain from the General Contractor, or Owner, a letter giving him the names and titles of the persons authorized to make decisions on the various phases of the work. On large installations and large companies, the scope of each Representative may be limited to his function within the organization. Thus, the Insulation Contractor should have a letter stating the authorized person or persons from whom he may request decisions in engineering for technical questions, and the field personnel authorized to request changes or extra work, and the job superintendent or Resident Engineer in charge of decisions on scheduling, and the proper individuals in contract and material procurement.

AA. BOND

If a performance bond is required by the General Contractor or Owner, this should be stated in the original request for bids, as the cost of such bond must be added into the cost of the job.

BB. GUARANTEE

1. Time Guarantee

Most commonly, the Insulation Contractor guarantees to use materials called for in the Specifications, and guarantees these materials and his workmanship for a period of one year after completion of the Contract. He agrees to replace, or repair, defects in material and workmanship which appear within this one year period. However, it is expected that the Owner will remove and replace all fixtures attached to the surface to be repaired without cost to the Contractor. The Contractor does not expect to be held liable for any losses, damages, or delays caused by such defects.

2. Performance Guarantee

If performance guarantees, such as maximum heat loss, or heat gain, or maximum or minimum surface temperatures, are required, the Owner and the Contractor should make prior agreement as to the test method or procedure which is to be used to determine these performances. As such measurements are quite difficult to make because of all the variables involved, tests of this nature, if accurate, are quite expensive, time consuming, and troublesome. All too frequently, the design specification, over which the Contractor has no control, is the major factor affecting the performance of the installation. Because of these factors, most performance field tests leave much in doubt and should be avoided, both by the consumer and the Contractor.

CC. DEFINITION OF TERMS

In all specifications and contracts there are terms which are clear to the writer, but may not convey a definite meaning to the Contractor. It is advisable that those terms common to the industry be given an industry definition. Until all such definitions are available, it is suggested that the Contractor request clarification in writing by the General Contractor as to the definite meaning of these words which appear in the Contracts and Material and Application Specifications which are not clear to him. A few samples of such words follow:

"surface-clean and dry"—generally means that surface dirt is wiped off. It does not mean that the surface is sandblasted, solvent cleaned, or wire brushed.

"condensate lines"—Condensate is a liquid. They are not steam lines.

"exposed to view"—etc.

"heated rooms"—etc.

"crawl spaces"—etc.

Summary

This presentation has been written in general, non-legal language. It was done purposely. This is a technical manual, not a legal one. If some of the phrasing appears to be in legal terms, the tendency might be for one to use the wording in a contract, and the user might find himself in legal difficulties.

The presentation is made to familiarize the consumer and the Contractor with some of the various questions that the author has experienced in the installation of a number of large insulation applications. To some consumers it might appear that a contractor had written this chapter to cover all possible areas of the work for his protection. However, experience has shown the opposite to be true. Too many contractors gladly accept wide open and vague contracts, with the result that many of the items mentioned in this chapter come up as questions later and result in time lost both by the consumer and the Contractor.

Of course, all the points raised will not apply to each installation, but it is well to make certain that the points that do apply are decided upon in advance, not after the job is in progress.

An outline of the items covered in this chapter is included. It may be used as a check list for the preparation of contracts. After such items are selected which are to be part of the contract, proper legal advice should be obtained to put the contract into correct legal form and language.

OUTLINE AND CHECK LIST

I. General Considerations
II. Contract Provisions
  A. Scope
    1. Contracting Parties
    2. Site
    3. Time Schedule
    4. Contract Documents
    5. Subcontracts
  B. Types of Contracts
    1. Lump Sum
      a. Unit Prices
      b. Measurement and Units
        (1) Piping—Fittings
        (2) Tees
        (3) Pipe Hangers, Supports and Cradles
        (4) Bent Piping
        (5) Straight Piping
        (6) Equipment
          (a) Vessel Bodies
          (b) Vessel Heads
          (c) Manholes, Blind Flanges, Piping Flanges
          (d) Insulated Vessel Flanges and Insulation Supports

(e) Insulation Supports
(f) Thermocouple Wells and Instrument Connections
(7) Factors Affecting Unit Prices
(a) Scaffolding
(b) Height Above Grade or Floor
(c) Time Duration
(d) Time of Year
(e) Location
2. Unit Prices
a. Unit Prices for Material, plus Labor Cost
b. Unit Prices for Labor only, Materials on a Lump Sum Basis, or furnished by Owner
3. Cost Plus
a. Cost Plus a Percentage, or Plus a Fixed Fee

C. Specifications, Drawings, Insulation Lists, Work Orders, Changes
1. Specifications
2. Additional or Extra Work
3. Drawings, Lists

D. Responsibilities for Workmanship and Materials
1. Responsibilities
a. Job Site Facilities
(1) Storage
(2) Utilities
(3) Change of Location of Storage Facilities
(4) Responsibility for Equipment
b. Materials
(1) Quality
(2) Broken or Rejected
(3) Left over Material
(4) Cancellation
(5) Substitution

E. Job Site Trucking

F. Scaffolding

G. Hoisting

H. Preparation for Application
1. Surface Preparation
2. Insulation Supports
3. Pins, Clips, etc.

I. Starting and Completion
1. Time of Start
2. Notice of Readiness
3. Delays

J. Application
1. Intent of Specification
2. Unusual Working Conditions
3. Inspection
4. Replacement
5. Clean-up

K. Duration Time
1. Penalties
2. Delayed Starting Time

L. Work Week
1. "Spot" Overtime
2. Scheduled Overtime

M. Safety
1. Regulations
2. Smoking
3. Safety Clothes and Equipment
4. Hazardous Areas

N. Liabilities
1. Accidents
2. Insurance
3. Strikes

O. Taxes

P. Escalation
1. Labor
2. Materials

Q. Progress Billing
1. Materials and Labor
2. Retention

R. Additional Work
1. Unit Prices, Time and Material, and Firm Bid

S. Extra Work

T. Authorization for Additional and Extra Work

U. Changes

V. Billing of Extras and Additional Work

W. Payment

X. Accounting Breakdown

Y. Unused Materials
1. Due to Cancellation or Changes
2. Due to Overage on Firm Bid

Z. Authorized Personnel of Owner or General Contractor

AA. Bond

BB. Guarantee
1. Time Guarantee
2. Performance Guarantee

CC. Definition of Terms

13 Inspection and Maintenance of Insulation

Insulation

Whether a company buys and installs its own insulation, or contracts for the materials and their installation with an insulation contractor, proper and intelligent inspection can be advantageous. Contrary to the common conception, careful inspection by the purchaser can be advantageous to the contractor.

The inspection procedures described here are directed to industrial installations rather than building or cold storage installations, which are generally handled by the overall building inspector.

A consumer's inspector or field representative should be much more knowledgeable regarding his company's specifications, drawings, process, safety requirements, and time schedules than the contractor. By keeping the contractor properly informed about these, and assisting him with problems that arise, the inspector can help to prevent mistakes, avoid loss of time, and to make the best utilization of labor. Of course, he is expected to reject materials and applications which do not fulfill specifications, but his major function is to try to prevent such rejections from occurring.

Although the customer's representative is commonly called an inspector, the term does not completely describe his function. The customer's inspector, or representative, in cooperation with others, should act as coordinator, advisor, interpreter, and inspector. His functions during an installation may be broken down as follows:

Advice and assistance in obtaining needed information
Advice and assistance on the storage of materials
Advice and assistance on the equipment required
Initial inspection
Pre-application inspection
Planning of work
Application period
Authorization of extra work
Final inspection
Measurements

Advice and assistance in obtaining needed information

Any Insulation Materials and Application Specification provides, (1) the listing of acceptable materials and (2) the *intent* as to what is desired in the finished application. No specification can provide absolutely detailed information on each and every piece to be installed.

Drawings are marked to show insulation in various ways. The manner in which the marked drawings refer back to the specification differs with various companies.

Insulation line and equipment lists are frequently furnished to direct the contractor or field applicators to work to be done. Again, like drawings, there is no standard practice in industry. Thus, the inspector serves a real purpose in assisting the Contractor in correct interpretation of these guides to the work to be done.

Another vital function that the inspector serves is to make certain that any *design changes* come to the contractor's attention so that immediate material and labor requirement changes may be made to prevent future delay of the installation.

Information regarding the detail dimensions for fittings, flanges, valves, vessel and vessel heads is needed for the ordering of material and its fabrication. Standard dimensions for NPS pipe fitting, flange, and valve insulation, including those used on traced piping, are availabe from the American Society for Testing and Materials publication entitled "ASTM Recommended Practice for Prefabrication and Field Fabrication of Thermal Insulating Fitting Covers for NPS Piping, Vessel Lagging, and Dished Head Segments," Recommended Practice C-450-65T. This information should be available to the fabrication shop workmen and to workmen fabricating fittings in the field.

The inspector should be able to help field personnel in using this information to obtain a uniform, clean installation.

Advice and assistance on the storage of materials on job site

Provisions for storage and necessary services should have been agreed upon in the contract. In keeping with this agreement, the inspector should assist the contractor in making arrangements for a suitable area for adequate storage, and facilities for fabrication where it is done on the job site.

The area should be accessible to trucks, and, if possible, to the railroad. It should be located in a non-hazardous area where open flame heat is permitted.

All materials adversely affected by moisture must be protected from the elements. Any such materials which become wet should not be used. They should be rejected by the inspector. Shipping cartons which are marked to denote contents, size, and thickness should also be protected.

Materials which are subject to freezing must be stored in a heated area or building where ambient temperature is above freezing. Any material that has frozen must not be used until it is determined that the freezing did not damage its quality.

If required for the job, electric power, gas, air, and steam shall be available in accordance with the contract agreement. The cost of installation of these services and arrangement for the payment of the periodic charges for them should also be in accordance with the contract agreement. The consumer's engineer or inspector shall assist in obtaining these services and seeing that they are extended to a specified point, to avoid any delays in fabrication or installation.

Advice and assistance on the equipment required

Hand tools common to the trade shall either be furnished by the applicator or the insulation contractor. It is the contractor's responsibility to see that such tools are available to facilitate efficient and quality workmanship.

Power tools necessary for shop fabrication, spray insulating, or coating in accordance with the specification requirements are to be furnished by the contractor.

The inspector or customer's field representative shall offer his assistance in the planning of requirements for tools and equipment for the fabrication shop. He should also assist in planning a shop layout which will provide good working conditions and be arranged so that the fabrication can be accomplished with minimum lost motion and material handling. These recommendations should be confined to the minimum requirements necessary to conform to the specifications. The inspector must be sure that the equipment is in accordance with safety rules prescribed by the Company, National Safety Council, the construction industry, and State law safety regulations and codes.

Initial inspection

Inspection should determine that the materials received are those which will fulfill the material specification. Where necessary, the inspector should check the acceptability of various brands or trade names purchased by the contractor, with the Engineering Department responsible for the specifications, if such brands and trade names are not provided in the specifications.

The condition of lots of materials should be spot-checked when received to determine whether it is as ordered and was not damaged in shipment. Any damaged or unsuitable material should be rejected and immediately removed from the job site.

Determination to see that materials comply with the specifications depends to a large extent upon the Inspector's experience. Many of the physical properties can be evaluated from this experience. If the inspector should suspect and cannot evaluate any of the properties he should then request that samples be submitted to a qualified testing laboratory, or to the engineering department for evaluation.

When weather conditions are such that emulsion coatings may have been frozen in transit, they should be spot checked as soon as possible after delivery. These coatings should be checked for proper viscosity, adhesion, and workability. No thinning of material is to be permitted other than as recommended by the manufacturer.

All accessories should be checked to ensure that they are of the type and quality specified and of sufficient quantity for the installation. This is important, because lack of proper accessories is one of the most common causes of job delay.

All factory or field prefabricated insulation fitting covers, valve covers, and equipment curved segments and heads shall be inspected to make certain that:

1. All insulation, fabrication cements, and sealers used are in accordance with the material specifications

2. The insulation shapes are thoroughly dry
3. Shapes are in accordance with "ASTM Recommended Practice" C-450 latest revision.
4. Details of the fabrication, such as end dutchmen and inserts, are workmanlike.
5. Broken or miscut insulation is rejected
6. Proper fabrication cement is used to correspond with the insulation and temperature requirements
7. The joints are thin, tight, and strong
8. Any low temperature fittings, when held to a strong light, are rejected if pinholes can be observed
9. The outer surfaces are smooth and clean, suitable for the application of the weather-barrier finish.

Pre-application inspection

Inspectors should determine when equipment or pipe is ready for insulation, and make certain that:

Equipment is in its proper location, mill-wrighted and grouted. Its surface, if it requires no corrosion protection, is dry, clean, and free of rust. Surfaces requiring corrosion protection are properly coated, and the coating is dry. Insulation supports are of the correct size, and properly located in accordance with the specifications.

Piping above grade, is properly located. All supports and anchors are installed, and all field welds have been tested. Like equipment, surfaces requiring no corrosion protection are dry, clean and free of rust. Surfaces requiring corrosion protection are coated in accordance with coating specifications.

Long vertical runs are equipped at specified intervals with insulation supports, to prevent insulation at the bottom from being crushed by excessive weight.

Piping has sufficient clearance to provide enough space for application of the insulation.

Low temperature piping supports, anchors, guides, or hangers are free from obstructions, in order to provide a clear length (minimum of four times the insulation thickness) for application of insulation.

Heat tracing on equipment and piping. The inspector shall make certain that:

Heat tracers, steam or electrical, are installed and attached as specified.

Where tracers are to be thermally bonded to surfaces with heat transfer cement, the surfaces are clean of rust, paint, oil, and other foreign matter, prior to installation of the tracers. On aluminum surfaces, they should also be treated with primer as recommended by the manufacturer of the transfer cement.

A specified type of heat transfer cement is applied to the installed tracer. This shall be properly cured prior to the installation of insulation. If the heat transfer cement is water-soluble, it must be protected from weather and moisture until the insulation and its weather-barrier are installed.

Planning of work

The inspector shall assist in determining the time sequence of insulating the equipment and piping.

The use and location of scaffolding should be determined to obtain its maximum use and minimum interference with other crafts. The scaffolding and planking must be erected in accordance with the safety codes previously mentioned.

Process requirements often make it necessary that certain lines and equipment be insulated before others. The inspector should advise the contractor of these conditions, and assist him in arranging to install as much as possible within a given area to effect the most efficient operation.

Hazardous areas require special attention, and it is the responsibility of the contractor to see that they get it. However, it is the responsibility of the inspector to make sure that all necessary precautions are taken for safe application. When Hazardous Work Permits are needed, the inspector shall assist in obtaining these permits and seeing that their provisions are carried out.

Application Period

Insulation must be inspected during its various phases of installation, because after the weather-barrier is installed the details of application are hidden.

The inspector shall see that the following requirements are met.

On applications of cellular glass for low temperatures, the most important factor is sealing the individual pieces to their adjacent pieces to form a completely sealed enclosure. All sealed joints are to be as thin as possible while still obtaining a complete seal. Sealers have poor insulating value, thus the cellular glass should be cut and fitted to be free of voids. Plugging of holes and gaps with sealer is not be be allowed.

Proper location of contraction joints is essential to prevent equipment shrinkage from crushing the insulation. The installation of contraction joints in accordance with the specifications is frequently ignored by the applicators, as many do not understand the importance of these joints.

On all low temperature insulations, other than cellular glass, their economic life depends completely upon the o uter vapor-barrier. On insulations using a vapor-barrier jacket or film the inspector shall make certain that:

1. After insulation securement, the circumferential and longitudinal joints of the vapor-barrier are thoroughly sealed.
2. All fittings are vapor sealed with vapor-barrier jacket or vapor-barrier tape.
3. Any vapor-barrier which becomes punctured must be replaced and resealed.
4. The vapor-barriers of all pipe, fittings, and equipment are sealed to form a continuous, tight envelope.

On sprayed-on installations for moderate and high temperature installations, the inspector must determine that the proper amount and thickness of insulation is installed on all surfaces. The outer surface should be sufficiently smooth and even to provide a good substrate for the weather-barrier. If necessary, samples of the material should be taken to check its properties with those listed in the material specification.

Rigid insulations should be inspected as they are installed to make sure they are in accordance with the specifications; that broken, badly chipped insulation is not used, that joints fit tightly and snugly. Excessive packing and filling of voids with insulating cements shall not be permitted. Insulation expansion joints must be installed in accordance with the specifications, both circumferentially by use of blanket, and longitudinally by filling butt ends with tightly packed mineral fibers.

Securement of all insulation systems shall be done with the type and style of securements specified. The wire, bands, and clips are to be inspected to assure that they are of the proper type, quality, and metal. Securements shall be installed in a neat workmanlike manner with no sharp projections which could cut through the weather-barrier, or present a personnel hazard.

The inspector shall also see that the following requirements are met.

After insulation is secured in place, it is imperative that any absorbent insulations be kept dry until the weather-barrier is installed. For this reason, any insulation that is not weather protected directly after application must be protected with temporary sheeting or canvas until the weather-barrier is installed. Proper installation of the weather-barrier is essential to efficient operation of the insulation system, as any water leakage causes excessive losses of heat in high temperature insulation and will rapidly ruin low temperature insulation.

Where metal jacketing is used as a weather-barrier, the inspector shall make certain that all circumferential joints in horizontal pipe and equipment are *sealed* with non-setting sealer, or closure bands, in such a manner that driving rain cannot penetrate the insulation through the joint. On vertical piping, the circumferential joint must have the adjacent top jacket overlapping the lower jacket to provide a water shed, or be sealed with closure bands. When closure bands are used, they must be drawn very tightly at their top side and completely sealed so that water running downward from above does not collect in a cavity left between the band and the jacket.

Longitudinal joints should be either of "Z" joint construction or overlapped. On horizontal lines, either the "Z" joint, or overlapped joints, should be on the side of the pipe or equipment, so installed as to shed water. Overlapped joints in a vertical position must be sealed with non-setting sealer to remain tight. The lap must lay flat and be free of gaps. Unless very heavy gage metal is used to obtain tight, flat longitudinal joints, it is necessary to have at least one fold back on the outer leg of the jacket to give it sufficient strength to resist bending as the jacket is drawn tight and secured around the insulation.

All inside corners should be caulked, and at outside corners where they join the elbow's joint, the straight piping should be overlapped, or otherwise caulked, and sealed so as to remain tight after movement occurs at these junctions.

Jacketing materials other than metal, such as felts, or laminates, shall be installed in a similar manner as described for metal jacketing, making certain that all joints are installed facing downward and are overlapped to shed water. Other joints where wind driven rain can penetrate are sealed water tight with suitable non-setting sealers of adhesives.

Where *mastics* are used as weather-barriers they shall be applied by brush, trowel, palm, or spray, whichever is specified. The inspector shall be certain that the insulation surface is suitable for the application, in that it is clean, dry and smooth. The containers of the mastics shall be kept sealed until ready for use. If, during use, a film is found over the mastic surface in a can or drum, this film must be removed and not reworked into the mass of the material. Any loss of solvent which causes material to become too heavy may be replaced, but the addition of solvent must be kept to a minimum and shall be added only as recommended by the manufacturer. All inside corners shall be caulked and flashed prior to application of mastic, and the caulking allowed to dry sufficiently to take its initial set and shrinkage prior to mastic application.

When mastic is to be sprayed, the adjacent surfaces of equipment, piping, and instruments must be protected from overspray or splatter.

Mastic must only be applied when ambient and substrate temperatures are within the limits set by the specifications of the manufacturer.

Where it is specified that reinforcing cloth is to be embedded in the first coat of mastic, the mastic must be applied in an even smooth coat of specified thickness, and the cloth embedded into the mastic with edges overlapped the specified minimum distance. A double layer of cloth should be installed over the inside and outside corners.

Installation must be in a workmanlike manner to provide a smooth, even, wrinkle-free surface for application of the second coat. The second coat shall be applied smoothly and evenly, free of runs or pot marks, sufficiently thick to allow the insulation to attain its specified, dried thickness.

If the mastic is to be applied by palm or trowel in one coat over the reinforcing cloth, the cloth must first be installed by attaching it to the surface with spot applications of mastic. The cloth must be installed to be smooth, even, wrinkle free, and properly overlapped. The mastic shall be applied over the cloth with sufficient pressure to force the mastic through the weave of the cloth to obtain good adhesion to the insulation surface. The amount applied should be sufficient to produce the specified dried thickness.

Finished application must be protected from heavy rains long enough for them to take their initial set.

Any overspray, spillage, or drip of mastics must be cleaned off immediately with solvents (water for emulsions), before they can dry and set.

Interior coverings. The inspection of interior coverings shall be made in a similar manner as for outside weather-barriers. However, on jackets (metal or felts), it is not necessary that all joints be rain tight unless equipment and piping is washed down with water streams. In that case, the installation must be treated as an outdoor one, and all precautions to prevent water entry must be taken.

Underground fill insulation shall be inspected to determine that it is of the type, grade, and quality called for on the drawings and specifications. The inspector shall make sure that the pipes are installed in accordance with the drawings and specifications, that the trenches and forms are such to provide the proper installing space for the specified thickness of insulation, and that they are free of standing water. After insulation is poured and packed into position leaving no voids under the pipe, the material must be heat cured in accordance with the specifications. After curing, the inspector should have the insulation checked at random points to determine that particles around pipe have "zoned" properly and adhered to the pipe. Only then may the trench be backfilled.

Authorization of extra and additional work

Although the terms "extra" and "additional" have almost the same meaning, their usage in field construction practice takes on slightly different meanings. If, in the construction of a plant, due to testing, process requirements, or other reasons it becomes necessary to remove insulation already installed, with subsequent replacement—or when insulation is damaged by other than the insulation contractor, thus calling for repairs, this type of work is frequently referred to as extra work.

When pipe or equipment is to be insulated beyond the confines of the original contract, this is most often called additional work. Most of the differences between the contractor and consumer occurs in the area of "extra" work.

When extra work is required, the inspector should obtain all the information necessary to describe the scope of the work requested and the specifications to be used, and to designate whether or not the work shall be done on time and material basis. He should assist the Contractor in the proper accounting of such work and, when completed, make sure that it is properly invoiced for payment. The invoicing of such work should be done as quickly as possible, for, on a large installation, the reasons for these extras are soon forgotten in the rush for completion.

Additional work most frequently flows through the regular work channels, and provisions for payment are within the terms of the contract.

Final inspection

Final inspection shall determine that all pipe and equipment called for to be insulated was both insulated and coated as specified in the contract.

The weather-barrier, or other covering, shall be examined to make sure that coating was installed to the specified thickness, and presents a smooth even surface. Special care must be taken to assure that the caulking and coating of all inside and outside corners are completely crack free, and that flashing around uninsulated flanges and projections are also watertight.

All construction damage to insulation, vapor-barriers, or weather-barriers must have been repaired. The inspector shall make certain that damage was repaired as part of his contract, and repairs of damage caused by others is covered by an extra field work order.

In addition, the inspector shall make certain that all mastic, adhesive, sealers, or cements dropped, spilled, or spattered on equipment, instruments, valves, operating handles, or building structures be removed, and that the area is clean.

The inspector must make sure that all unused, broken insulation, left over mastics, cement, or other materials are removed from the job.

If the contractor has no additional installations, the inspector must see that all tools and equipment, including storage buildings, sheds, fabrication shop, and change rooms belonging to the contractor are removed from the job.

Unused material and equipment belonging to the customer should be sent to the stores department, or disposed of in a proper manner.

Any materials or equipment which the contractor wishes to sell to the consumer must be appraised as to its quality, quantity, and condition. Such information shall be given to the Purchasing Department for their use in arriving at an agreement with the contractor.

Measurement of work by unit prices. Where drawings are available, the measurements for equipment and piping insulation installed at unit prices can be estimated from these instead of field measurements. Where changes make such estimates impractical, or where pipe is field run, it is necessary for the inspector and a representative of the contractor to make these measurements for the estimate. These must conform to the method of measurement for unit prices as established in the contract.

Inspection by infrared camera

The infrared camera "sees" infrared radiation much as a normal camera "sees" and records visible light. The infrared camera detector then converts the invisible infrared energy electronically into light rays visible on a "thermograph." Thus, it provides a rapid means of determining and recording surface temperatures of pipe and equipment in industrial plants.

It should be noted that the infrared camera "sees" the indication of temperature producing energy, but not the temperature itself. However, knowing the surface emittance, air movement, position and location, these temperature "indicating" pictures can be used to determine heat transfer. As energy conservation becomes more important, the use of these instruments to indicate areas which are losing heat will increase.

At present, infrared measuring systems can be put to use by thermal insulation technology in four basic ways to assist in conserving energy. These are:

1. Temperature indicating pictures of existing equipment and piping to indicate where thermal insulation, or additional thickness of it, can be used to conserve energy and to save money.
2. Temperature indicating pictures of completed insulation systems to determine faulty installation, or hot spots (such as projections) where insulation, or more of it, is needed.
3. Temperature indicating pictures, used as a tool to determine where repair, replacement or additional thickness of insulation is required.
4. Temperature indicating pictures which might indicate faulty process operation.

The camera is such that it is sensitive to the radiant energy producing the temperature of the surface at which the camera is pointed. The impulses of temperature producing energy picked up by the camera are transformed, electronically amplified, and displayed on a screen as light waves. The brightness of the light rays then indicates the temperature of the surfaces. This is shown in Figure 147.

The brightness of the light rays on the screen can then be compared to a scale of known temperatures, thus establishing the temperatures indicated in the picture. A diagram illustrating this relationship is shown in Figure 145. The picture shown on the screen can be recorded as a single photograph or be recorded on movie film. This method of inspection can determine in a short time the temperatures of various components in a system which would require a tremendous number of individual surface temperature measurements. It also provides a means whereby temperature measurements can be studied in an office, where engineering of the proper insulation systems to conserve heat is done. An ordinary light ray photograph of a fuse box is shown in Figure 148. The same fuse box, photographed by its radiant energy, is shown as a Thermogram in Figure 149.

To determine the heat loss of surfaces photographed by infrared radiation, it is necessary to know the surface emittances, position of the surfaces, ambient air temperatures, and air movement. Knowing these (which should be recorded at time of taking the infrared pictures), it is then possible to calculate the heat

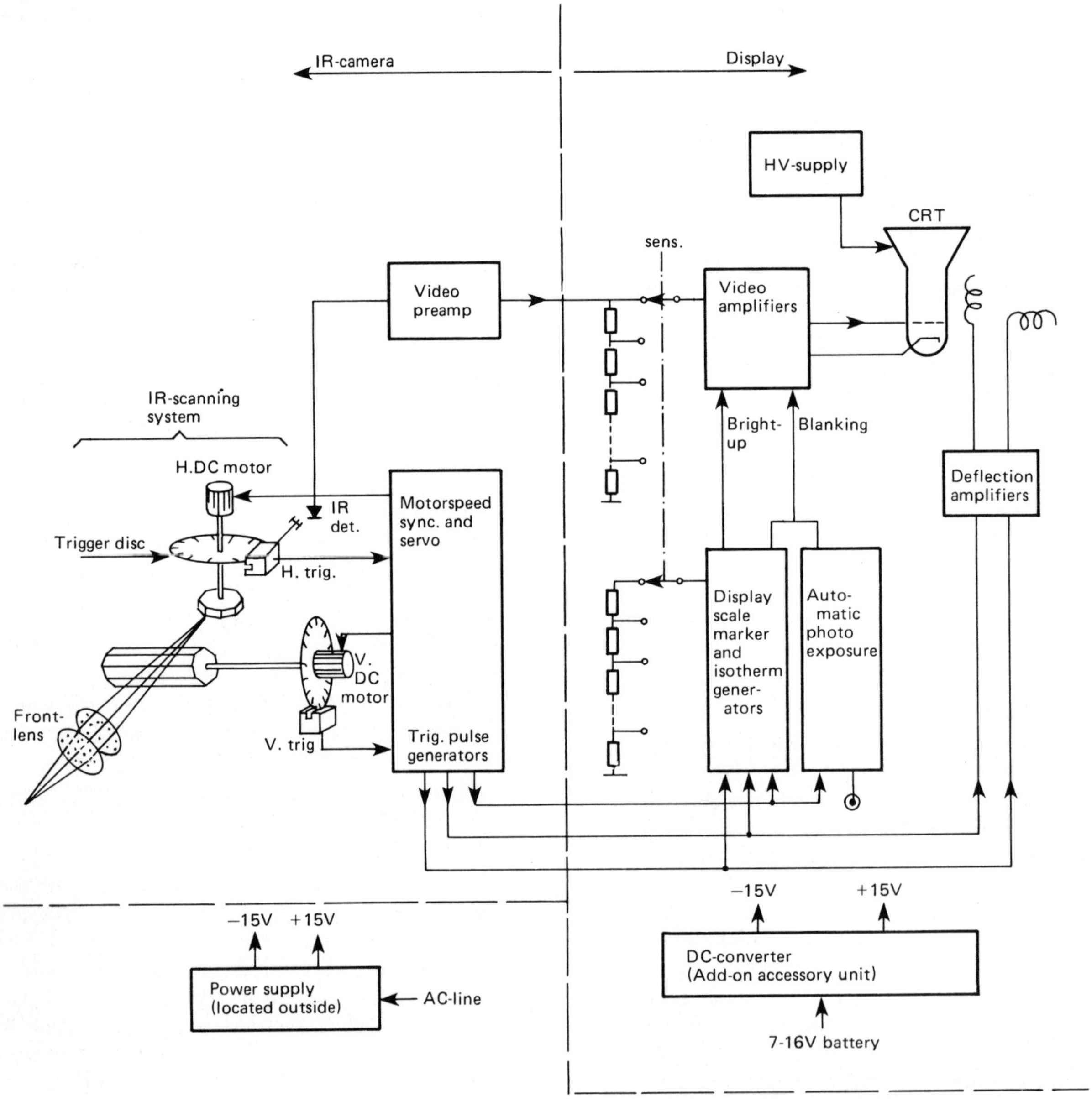

Figure 146 Electrical and optical diagram of infrared camera detecting system.

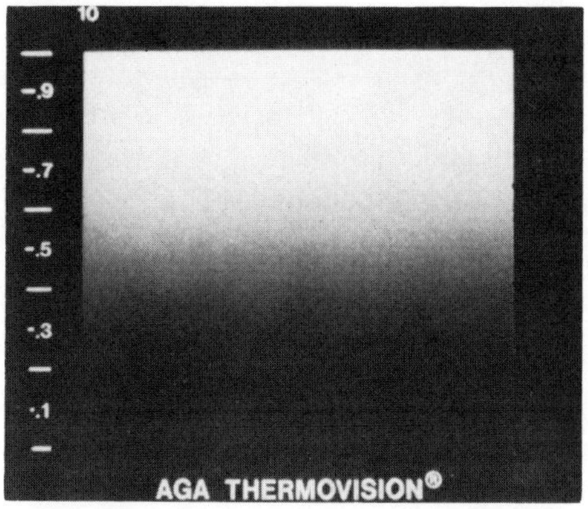

Figure 147 Relative brightness of rising temperatures on a thermovision.

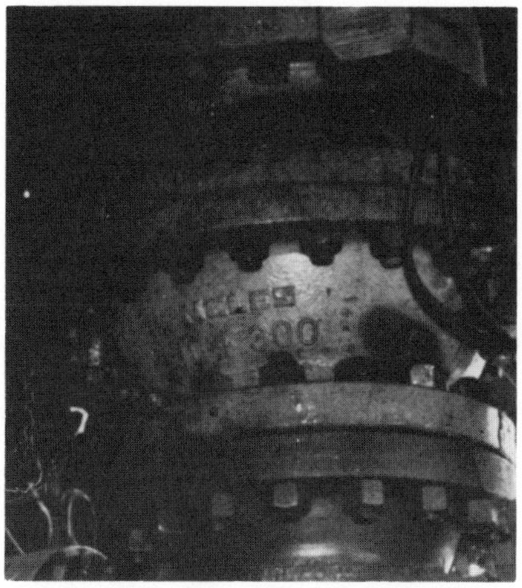

Figure 148 Normal photograph of a flange.

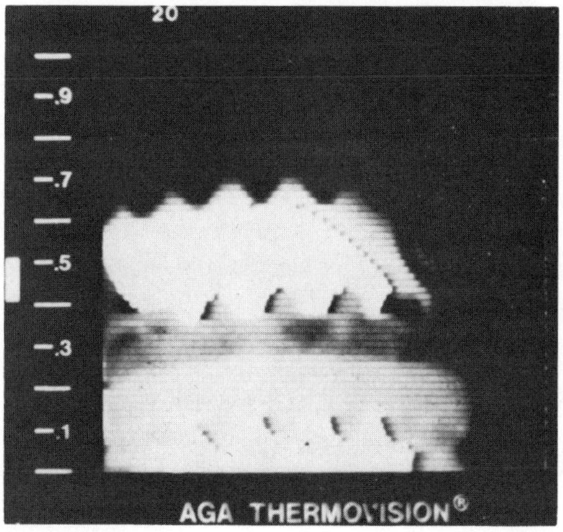

Figure 149 Photograph of the same flange by infrared on a thermovision.

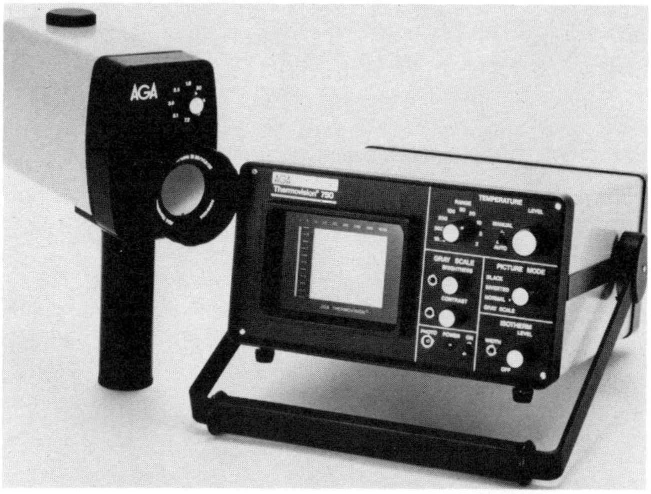

Figure 150 Photo of infrared camera scanning device.

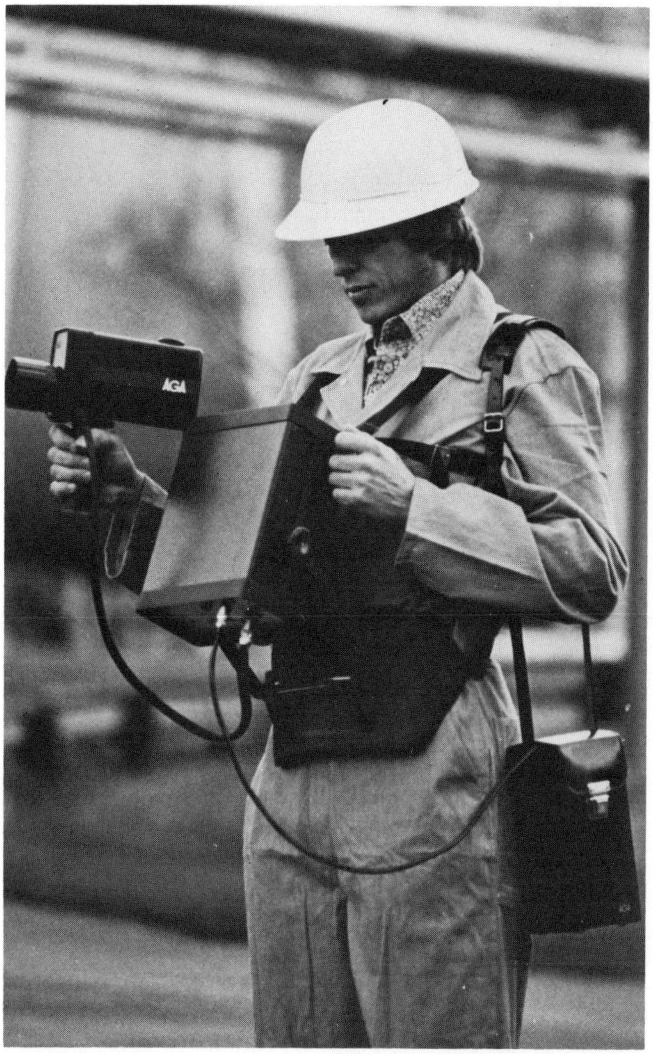

Figure 151 Infrared camera scanning device in use.

transferred from, or to, the surface. The equation for such calculations is given by Equations 25 or 26 for English units, and Equations 25a or 26a for Metric units.

It must be remembered that correct use of these infrared cameras depends upon the surface being absolutely dry. Where thermal insulation is involved, the insulation must also be dry, thus inspections using these cameras on outdoor equipment should be made only after a minimum of 24 hours of dry weather. The reason is that if moisture is present energy is used to convert liquid to vapor with no increase in temperature. All energy used to change the state of a liquid to a vapor is subtracted from the energy impinging on a surface causing a rise in temperature of that surface.

A typical type of this equipment which is available is shown in Figure 150 and a picture of the equipment in use is shown in Figure 151.

(Pictures and diagrams courtesy of AGA Corporation, Secaucus, N.J.)

Maintenance

There are two reasons why insulation may require maintenance. First, the repair or replacement of damaged, or water soaked insulation caused by weathering, abuse, or vapor migration. Second, the removal and replacement of insulation because the equipment, or valves, on which it is installed require repairs. Where piping, or process changes, require the insulation to be replaced or changed, this is, strictly speaking, not maintenance, but it is frequently done by these personnel.

Throughout this manual, the importance of keeping all hot surfaces insulated, to prevent heat loss from the product, and cold surfaces insulated, to prevent heat gain to the product, has been stressed. Likewise, it has been stressed that insulation must be kept dry to remain effective. The prompt replacement of insulation whhich for some reason has been removed or broken off, and the proper maintenance of moisture barriers to keep insulation dry, result in significant savings in plant production and operation costs.

To illustrate: If the insulation is removed from a 6″ NPS (168.3 mm dia) line flange cover, with steam at 400° F (20.4° C) and not replaced for three months, the cost for allowing it to remain bare for this period will be $10.64 in steam at $1.00 per M pounds. Also, during that period of time, it will require 4.8 lb of steam per hour more than if it had been insulated.

Wet insulation will transfer approximately 12 times more heat than if it is dry. To dry insulation after it has become wet, heat is required to vaporize and drive the moisture outward. This requires a minimum of 1150 Btu to vaporize each pound of water contained in the insulation. Thus, when insulation becomes wet because of leaks in the weather-barrier, the loss in heat energy is twofold; first, the efficiency of the insulation is greatly reduced until it is dried, and second, considerable heat is used to dry the insulation.

So, the major concern in the maintenance of high temperature systems is to make sure that any insulation damaged or removed is replaced promptly, and that all hot surfaces requiring insulation are so insulated and kept dry.

Insulation on low temperature pipes and equipment poses a similar problem. When low temperature insulation gets wet, either due to water or vapor migration, there is no practical or economical method of drying it. For this reason, wet or iced-up low temperature insulation must be promptly removed and replaced with new insulation.

The length of service of low temperature insulation can be prolonged if the vapor-barrier and weather-barrier protecting it are kept in good condition, and punctures and cracks are sealed very soon after they occur.

Maintenance of high temperature insulation

One of the major common faults in maintenance is tha, when it is necessary to remove insulation to repair a valve or piece of equipment, the insulation is simply broken up and pulled off without concern for its salvage value, nor is proper protection given the adjacent insulation until the repair is completed.

When valves and fittings are insulated with preformed or prefabricated covers, these can be taken off by careful removal of the weather-barrier at the joints and at the securement straps or wire. The securement can then be cut and the insulation removed by separating its two half sections. After repairs to the valve are completed, this same insulation cover can be reinstalled, resecured, and the weather-barrier patched or replaced.

When insulation is removed, the adjacent insulation should be protected from rain, water, or contamination, by a temporary weather-barrier until the removed insulation is replaced and the weather-barrier restored to its original condition.

Where insulation is removed, the replacement installation most frequently is done in almost the same manner as the original application, unless service has proved it to be faulty. Where this occurs, the reapplication should be done according to improved material and application specifications.

Insulation that has been damaged may be repaired with insulating cement. Care must be taken that it is not applied to water saturated insulation, and that, after application, the cement is protected until it is dry, and the weather-barrier is reapplied.

There are many places where damage commonly occurs: Where (1) foot traffic breaks the weather-barrier and crushes the insulation, (2) cracks are caused by expansion, (3) spillage ruins the weather-barriers, and soaks into the insulation, (4) gaps or unsealed joints occur in the weather-barrier jacketing, (5) hot projections through the insulation cause disintegration of the weather-barrier and result in peel-back, and (6) weather-barriers shrink and crack at inside corners, and are broken on outside corners.

Where foot traffic causes fast breakdown of the weather-barrier, the insulation and weather-barrier damage should be repaired, and a sheet of metal installed over the existing jacket. Heavy gage metal is not required, since the metal is supported by the insulation and its purpose is only to spread the load. If galvanized steel is used, 26 gage metal strapped into position is sufficient for the purpose. If conditions will cause corrosion of galvanized steel, then stainless steel 0.01 in. in thickness is sufficient. Aluminum is not recommended as it is too soft to spread the load over a sufficient area unless it is quite thick.

Where cracks are caused by expansion of the pipe, simple patching will not correct the problem for any length of time. The very existence of the crack shows that the insulation needs an ex-

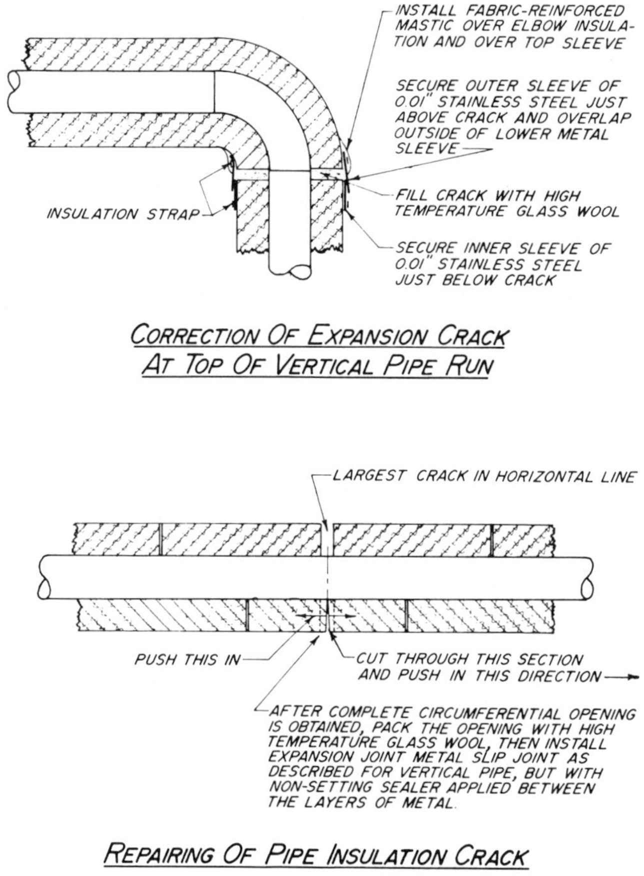

Figure 152 Repairing of pipe insulation crack.

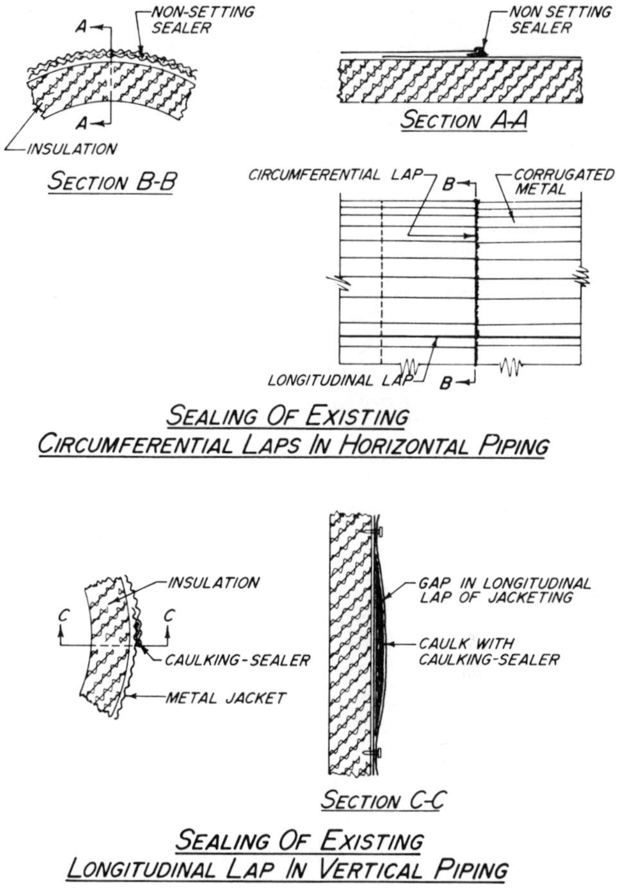

Figure 153 Jacket lap sealing.

pansion joint at that location. Therefore, the solution is to construct expansion joints into the system where they are needed. The method of constructing these joints is shown in Figure 152.

In any process, places will develop where spillage and leakage occur. Depending upon the material spilled, these may cause damage to the weather-barriers and eventually soak into the insulation. Where they do occur, corrosive resisttant metal should be installed to catch the leak, drip, or vapor and be guttered to provide run-off without damage to the insulation and its weather-barrier. Besides ruining the insulation, leaks of a combustible material constitute a great fire hazard, as combustible-soaked insulation will provide a considerable amount of fuel to a fire. In some cases, the soaked insulation will take fire by spontaneous combustion.

As mentioned in the section on weather-barriers, circumferential overlapped metals do not provide a rain tight joint when located in the horizontal position. Logitudinal, simple, overlapped metal in a vertical position will not keep out rain, as it gaps open between bands or screws. Although, in most instances these are faults of the original installation, they should be corrected. Correction of these ills, however, does fall to maintenance personnel, if good plant efficiency is to be obtained.

As there is expansion and contraction movement between sections of a metal jacket, any unsealed circumferential joints in horizontal pipe should be caulked and sealed with a non-setting sealer. Forcing sealer between two sheets of metal and sealing lapped corrugated metal is difficult. If these laps are not sealed, the barrier does not serve the function for which it was intended.

The longitudinal laps in vertical runs must also be water tight. Simple overlapped metal jacketing in vertical position should be caulked and sealed with a high solids type of caulking-sealer. These sealing methods are shown in Figure 153.

Where hot projections, such as supports or uninsulated nozzles, penetrate the insulation, the water seal at these junctures requires relatively frequent repair, as the heat will cause disintegration of the weather-barrier and sealer. Movement causes tensile stress to break the bond. In some instances, the best solution is to install metal flashing around circular projections, whereas, with irregular surfaces, only high temperature sealers can be used to seal the cracks. These are illustrated in Figure 154.

Inside and outside corners are another source of weather-barrier failures. The inside corner fails, because the expansion stresses pull the weather-barriers away from that corner. To make the problem more acute, the mastics and caulkings used on inside corners shrink and tend to crack at this juncture. Where inside corner weather-barrier cracks appear, they should be patched by first sealing the crack with mastic, then installing a flexible fabric reinforcing cloth to extend at least 3" (76.2 mm) on each side of the crack, and be embedded in the mastic. When this is dry, the

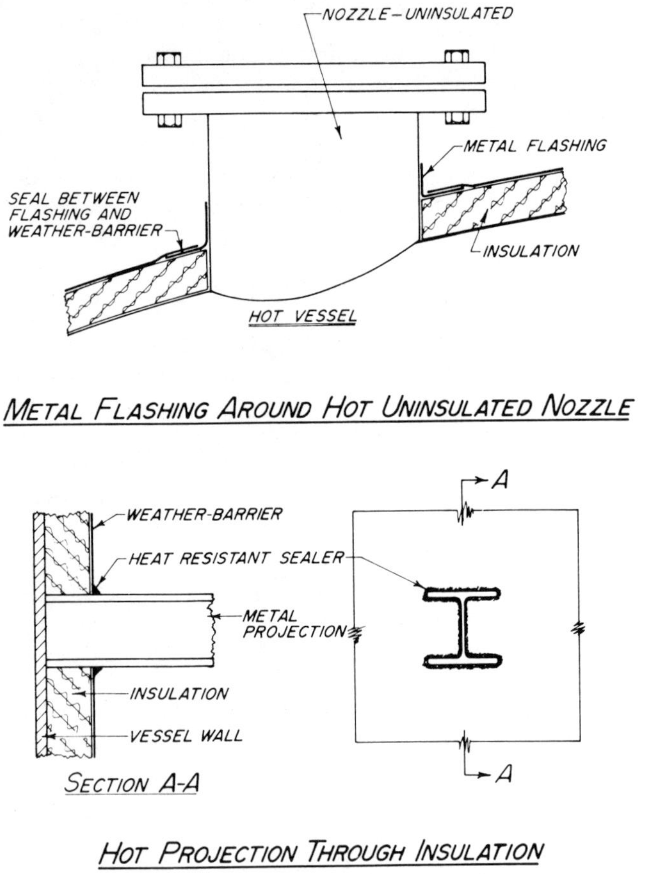

Figure 154 Projection sealing.

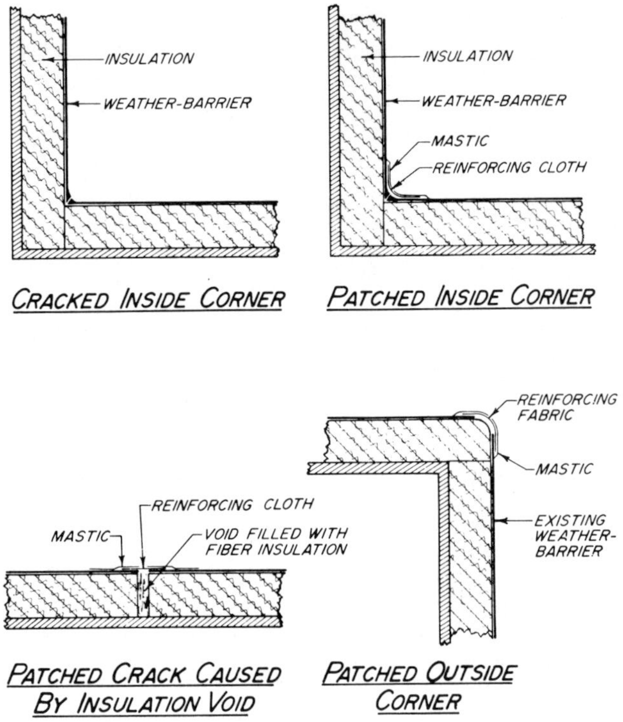

Figure 155 Repairing of cracks in weather-barrier.

cloth should be coated with the mastic, and the mastic extended at least 1″ (25.4 mm) beyond the edges to seal the existing weather-barrier. Simple coating of the crack with mastic will not make a permanent seal, for, when the mastic dries and hardens with age, the crack will reopen at the same spot. Only by reinforcing it and providing excellent bond over sufficient area to the adjacent weather-barrier, will a permanent patch be obtained.

The weather-barriers on outside corners frequently require repair because they are located where they get the most mechanical abuse. When these corners are broken, their repair should be done in a similar manner as described for inside corners. They need reinforcing to withstand the abuses.

Small weather-barrier cracks, due to shrinkage or movement of the insulation which causes opening below the weather-barrier, should be repaired in the following manner: the insulation void should be packed with high temperature glass, mineral wool, or asbestos fibers. After the insulation void is filled, the weather-barrier should be patched with reinforced mastic as stated for inside corners.

The method of patching the cracks in weather-barriers is illustrated in Figure 154.

Repairs on moderate temperature equipment and piping require the same techniques as presented for high temperatures. The major difference between high and medium temperature is that insulation on medium temperature service is much more difficult to dry if it becomes wet. For this reason, it is very important to keep the weather-barriers and water seals in excellent condition.

The maintaining of insulation on low temperature service is different from that on high temperature service, in that, if moisture vapor or *water does get into the insulation and saturates it, the only remedy is to strip the old wet insulation off and reinsulate with new materials.*

Just as the installation of cellular glass is different, its maintenance is also different from that of other low temperature insulations having a higher rate of vapor transmission. Unless the weather-barrier has been damaged by external abuse, in which case it can be easily patched, any cracks indicate that the insulation has pulled apart at a joint. This joint should be sealed with vapor sealer, after which the weather-barrier can be patched. Because of its high degree of vapor resistance, cellular glass resists water vapor as well as liquid water, and for this reason, its failures do not necessarily originate at openings in the weather-barrier on its outer surface. Where joints become unsealed, moisture penetration causes condensation and ice formation between the edges and ends of the insulation. During temperature changes, both in the service and the ambient air, the freeze-thaw cycle of moisture causes a breakdown of individual cells. Continuation of this process will eventually cause the ruin of cellular glass insulation. If the joints are kept sealed by proper maintenance, then almost unlimited service life of cellular glass can be achieved.

Low temperature insulations having a significant rate of vapor transmission must be maintained with much greater care than cellular glass. Their service life depends almost completely upon the maintenance of a completely unbroken and unpunctured highly effective vapor-barrier over the entire exposed surface, from beginning to end. Pin holes, small cracks, and partly sealed joints too small to allow the entry of liquid water will allow the entry of water vapor. The gradual pickup of moisture increases the conductivity of the insulation. Maintaining the vapor-barrier, to slow down this moisture pickup, is essential to satisfactory service life of these materials.

On outdoor applications, the maintenance of the vapor-barrier may be additionally complicated, as it may be covered by a weather-barrier. Where this is true, the weather-barrier around any crack, hole, split, or other fault in the vapor-retarder, must be removed sufficiently to allow the vapor-barrier to be patched with a vapor-barrier material. After this patch has set, dried, or cured, the weather-barrier can be patched.

The reason for this procedure is that weather-barriers as a group are not vapor resistant, although *some* materials are both weather- and vapor-barriers. Of course, should the outer covering be a weather- and vapor-retarder material, the statement in the preceding paragraph would not apply. However, if a vapor-barrier patch is made over the top of an ordinary weather-barrier, it will transmit vapor to the crack or opening, by-passing the vapor-retarder patch. This is illustrated in Figure 156 along with the correct method of patching vapor-barriers.

From the outside, low temperature insulation gives little indication of its loss of efficiency until it has lost a high percentage of its thermal resistance. When the outer surface begins to show beads of condensate at a dew point temperature lower than that for which it was designed, it indicates that the insulation is collecting moisture. When frost appears in spots on the outside surface, the insulation has become completely ruined. When this occurs, that portion of the insulation which is water- and ice-soaked must be removed and replaced with new material; as there is no way to repair or salvage it.

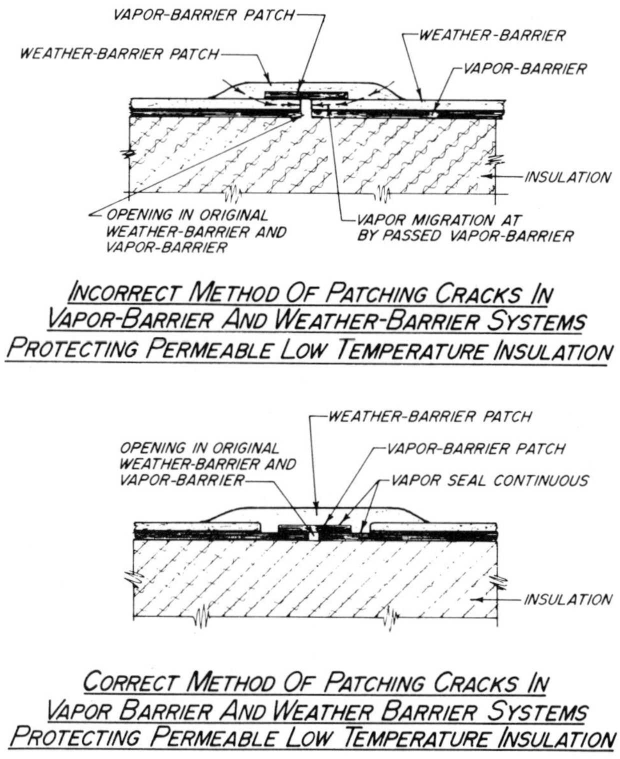

Figure 156 Repairing cracks in vapor retarder.

Economics of Maintenance

Properly maintained insulation saves money. Unfortunately, in many instances management may know what it costs to maintain insulation, but they do not know what savings this maintenance returns to them. This is quite understandable, as it is almost impossible to determine how much heat or refrigeration energy is being properly used, and how much is being wasted. It should be pointed out here that, in the calculation of economic thickness of hot insulation, it was assumed that, if insulation could be *properly maintained at a cost per year of 10% of its capital investment, it would return no less than 20% per year on that investment.* Reversing the statement, if insulation can be maintained at a cost of 10% per year or less of its capital investment, return on investment of from 25% to several hundred percent will be assured. Actual savings would have to be determined for each individual set of insulation conditions and application.

In the determination of the cost of insulation maintenance, in the case where insulation must be removed and replaced to allow process changes, or to repair valves, pumps, or malfunctioning equipment, these costs should *not* be considered as *cost to maintain* the insulation. These costts should be assigned to process change, or added to the repair cost of the malfunctioning piece of equipment.

Reliable records regarding the actual *cost to maintain* insulation can be very valuable, since such costs can indicate where improvements are necessary in the design of insulation systems.

As a fitting end to this presentation:

Insulation is an excellent investment, and good maintenance protects that investment.

Appendix A — Glossary and Definitions

The following definitions sometimes are not as strict as those which would be applied by a physicist, but are phrased in the terms familiar to and commonly used by engineers and the people of the construction industry. Users of this Glossary should not that, in many cases, dictionary definitions are not closely followed, but, instead, words are used in the connotations familiar to the industry, and the definitions are restricted to this limited usage.

Abrasion Force: A force caused by the rubbing together of two abutting objects, or by an external object rubbing against the surface of one of them.

Abrasion Resistance: The ability of a material to withstand abrasion without wearing away.

Absolute Humidity: The mass of water vapor present in a unit volume of atmospheric air.

Absolute Zero: The point at which all molecular motion ceases, with the resultant complete absence of heat. This point is −459.6° Fahrenheit, and −273.2° Celsius.

Absorbency: That property of a material which measures, in unit terms, its capacity to take up and assimilate liquids (from either the liquid or vapor form).

Absorptance: That property of a material which measures its total capacity to take up and assimilate liquids (from either the liquid or vapor form).

Absorption: That property of a material which enables it to take up liquids (in either the liquid or vapor form), especially by suction, and to assimilate them.

Acrylic Resin: Resins made by the polymerization of acrylic monomers, such as ethyl acrylate and methyl methacrylate.

Adherend: A body which is held to another body by an adhesive.

Adhesion—Dry: The property of a material which indicates its ability to bond to the surface to which it is applied and remain in place in service.

Adhesion—Wet: The property of a material which indicates its ability to stick to the surface to which it has been applied without sliding or falling off.

Adhesive: A substance capable of holding materials together by surface attachment.

Adsorption: That property of a material which enables it to retain liquids (in either liquid or vapor form) upon its surfaces—both internal and external.

Afterglow: The incandescence in a material after removal of an external flame or fire, or after an integral flaming has been extinguished.

Aggregate Size: The size of the coarsest particles in a cement, concrete, loose fill insulation, or similar product. The size is usually given in a percentage range of those particles which will or will not (with a maximum) pass through the mesh of a given size screen.

Alkalinity: The tendency of a material to have a basic (alkaline) reaction. The tendency is measured on the pH scale, with all readings above 7.0 alkaline, and below 7.0 acidic.

Alkyd Resins: Resins composed principally of polymeric esters (polyesters) in which the recurring ester groups are an integral part of the main polymer chain, and in which ester groups occur in most cross-links that may be present between chains.

Alligatoring: A term describing the action of a coating or mastic when it cracks into large segments. When the action is fine and incomplete it is usually referred to as "checking."

Ambient: Surrounding.

Ambient Temperature: The temperature of the medium, usually air, surrounding the object under consideration.

Anti-Abrasive Coating: A coating used on both of the mating surfaces of an insulation and its substrate, to prevent or retard the wearing away of the surface.

Appearance Covering: A material, or materials, used over insulation, the weather-barrier, or indoor covering, to provide the desired color or texture, for decorative purposes.

Application Temperature Limits: Temperatures between which it is usually safe to apply finishes.

Asbestos (Asbestos Fiber): A group of fibrous minerals which occur as small veins in the massive body of natural hydrous silicates of serpentine or amphobole, and have heat-, fire-, and solvent-resistant properties. Used as a reinforcement in the manufacture of mastics.

Asperigillus niger: One of the most common mold growths found on vegetable tanning vats and on leather, usually greenish or blackish in color.

Asphalt: A dark brown to black cementitious material, solid or semisolid in consistency, in which the predominating constituents are bitumens which occur in nature as such, or are obtained as residue in refining petroleum. The principal ingredient in asphalt mastics.

Asphalt Emulsion: A colloidal dispersion of petroleum asphalt in water. The emulsifying agent may be a colloidal clay or a chemical soap.

Batt: A piece of insulation, of the flexible type, cut into easily handleable sizes, square or rectangular in shape, usually 24″ or 48″ long and with a vapor-barrier on one side, and with, or without, a container sheet on the other side.

Bedding Compound: A plastic material, composed of various ingredients, spread on the substrate and used as a medium in which to embed the insulation layer. The compound acts as a cushion, anti-abrasive, and adhesive.

Binder: The cementing material used to bond fibers, flakes, granular materials together.

Bitumen: Hydrocarbon material of natural or pyrogenous origin, or combinations of both, which may be liquid, semisolid, or solid, and which is completely soluble in carbon disulfide.

Blackbody: An ideal, perfect emitter and absorber of thermal radiated energy.

Blanket: Insulation, of the flexible type, formed into sheets or rolls, usually with a vapor-barrier on one side and with or without a container sheet on the other side.

Blanket Insulation: A flat flexible type of insulation formed into sheets or rolls. These blankets may be faced or coated on one or both sides.

Blanket Insulation—Metal Mesh: Blanket insulation covered with flexible metal mesh on one or both sides. Metal mesh secured to blanket with tie wires or other temperature resistant securements.

Bleeding: The diffusion of coloring matter through a coating from the substrate. (Such as bleeding of asphalt mastic through a topcoat of paint.)

Blister: Undesirable rounded elevation of the surface of a mastic whose boundaries may be either more or less sharply defined, somewhat resembling, in shape, a blister on the human skin. A blister may burst and become flattened.

Block: Rigid or semi-rigid insulation formed into sections, rectangular both in plan and cross section, usually 36″-48″ long, 6″-24″ wide, and 1″-6″ thick.

Board: Rigid or semi-rigid insulation formed into sections, rectangular both in plan and cross section, usually more than 48″ long, 24″-30″ wide and up to 4″ thick.

Boardy: Adjective applied to stiff inflexible mastic or coating resembling a board.

Body: The degree of consistency and internal cohesion. An increase in "body" indicates an increase in consistency, internal cohesion or both.

Bond: The union of materials by adhesives.

Bond Age: Time period elapsed since bonding specimens prior to testing.

Bond Strength: The unit load applied in tension, compression, peel, impact, cleavage, or shear required to break an adhesive assembly with failure occurring in or near the plane of the bond.

Bonding Time: Time period after application of adhesive, during which the adherends may be combined.

Brattice Cloth: A coarse plain weave heavy fabric of cotton or jute fiber used as an air curtain to control mine ventilation.

Breaking Load: That load, concentrated in the middle of the span, which will just break a measured sample of insulation under test, according to ASTM C-203 or C-446.

Bridging Ability: The ability of an insulation to span a gap in the substrate to which it is applied, or the ability of a weather-barrier, or vapor-barrier, to span a gap in the insulation.

British Thermal Unit: Originally the amount of heat necessary to raise one pound of water one degree Fahrenheit at standard atmospheric pressure. Now by international agreement, the Btu has been established as 778.26 ft lbs.

Bubble: An internal void or a trapped globule of air or other gas in a mastic application.

Build—Dry: The dry thickness attained by a coating or cement by the application of a given number of coats.

Build—Wet: The wet thickness attained by a coating or cement by the application of a given number of coats, combined with its ability to be applied to that wet thickness on a vertical surface without slipping, sliding, or sagging.

Canvas: A light, plain weave, coarse, cotton cloth with hard twisted yarns, usually not more than 8 oz per square yard. See "Duck."

Capillarity: That property of a material which will enable it to suck a liquid up into, or through itself, with the driving force of the liquid being its surface tension.

Cast Film: A film made by depositing a layer of a coating or adhesive onto a surface, stabilizing this form, and removing the film from the surface.

Caulking Compound: A soft, plastic material, consisting of pigment and vehicle, used for sealing joints in buildings, and other structures, where normal structural movement may occur.

Celsius: The temperature measuring scale (formerly Centigrade) in which the ice point of water is taken at 0° and the steam point at 100°. The absolute zero on this scale is −273.2°.

Cellular Plastic: A plastic whose apparent density is decreased substantially by the presence of numerous cells disposed throughout its mass.

Cement: See Adhesive.

Cement—Insulating: See Insulating Cement.

Cement—Fabrication: See Fabrication Cement.

Cement—Finishing: See Finishing Cement.

Cement—Finishing and Insulation: See Finishing and Insulation Cement.

Centigrade: See Celsius.

Chaetomium globosum: A type of mold growth found chiefly on cellulosic materials, and particularly on paper, usually black, and characterized by long stiff hairs variously straight, branched or curled.

Chalking: Dry, chalk-like appearance or deposit on the surface of a weathered finish.

Checking: A defect in a coated surface characterized by the appearance of fine fissures in all directions. Designated as "surface checking" if superficial, or "through checking" if extending deeply into, or through to an adjoining surface.

Chemical Reaction: The property of a material which measures its tendency to chemically combine (or react) with other materials which may come into contact with, or be absorbed by it.

Chemical Resistance: Capability of withstanding limited exposure to designated acids, alkalies, and salts and their solutions.

Chemically Foamed Plastic: A cellular plastic produced by gases generated from chemical interaction of constituents.

Chlorinated Solvent: An organic chemical liquid characterized by a high chlorine content and used in coating products to impart non-flammability.

Closed-Cell Foamed Plastic: A cellular plastic in which there is a predominance of non-interconnecting cells.

Coal Tar: Tar produced by the destructive distillation of bituminous coal.

Coating: A liquid, or semi-liquid, protective finish capable of application to thermal insulation or other surfaces, usually by brush or spray, in moderate thickness, less than 30 mils (0.030″).

Cobwebbing: A phenomenon observed during spray application, characterized by the formation of web-like threads in addition to the usual droplets leaving the nozzle of the spray gun.

Coefficient of Expansion (Contraction): The increase (decrease) in length of a material, one unit long, due to the increase (decrease) of its temperature one degree. In the English System the unit is usually one foot, and the temperature Fahrenheit.

Cold: The absence of heat.

Color: The aspect of appearance dependent upon the spectral composition of the incident light, the spectral reflectance or transmittance of the object, and the spectral response of the observer.

Hue: The attribute by which a perceived color is distinguished, as red, yellow, green, blue, purple, or a combination of these. (White, gray, and black colors possess no hue.)

Lightness: The attribute by which a perceived color is judged to be equivalent to a member of the continuous series of grays ranging from black to white.

Saturation: The attribute by which a perceived color is judged to depart from gray of equal lightness toward a pure hue.

Combustible: Capable of uniting with air or oxygen in a reaction initiated by heating, accompanied by the subsequent evolution of heat and light. Capable of burning.

Combustibility: That property of a material which measures its tendency to burn. It is normally expressed in the arbitrary terms of "Flame Spread Index" and "Smoke Density Index," according to ASTM Test E-84.

Combustion: A chemical process, usually involving oxygen, which produces light and heat, either as glow or flames.

Compaction or Settling: The property of the blankets, or batts, which measures their change in density and thickness resulting from loading, or vibration, with a resultant change of thermal efficiency.

Compaction Resistance: That property of a fibrous or loose fill material to resist compaction under load or vibratory conditions.

Compressive Force: A force tending to cause the fibers of the material to be pressed or squeezed into more intimate contact, as opposed to a tensile force which tends to pull them apart.

Compressive Strength: That property of a material which enables it to resist any change in dimensions when acted upon by a force tending to squeeze or shorten it.

Concentrated Load: A load applied at a point, or over a very small percentage of the possible load bearing area.

Condensate-Barrier: A material, normally used as an inner lining for the metal jacket weather-barrier of an insulation installation, which will bar the alkaline condensate, which normally tends to form on the inner surface of the metal jacket, from contact with it.

Condensation: The act of water vapor turning into liquid water upon contact with a surface at a lower temperature than the dew point of the vapor.

Conditioning: The exposure of a material to the influence of a prescribed atmosphere for a stipulated period of time, or until a stipulated relation is reached between material and atmosphere.

Conductance: See Thermal Conductance.

Conduction: The transfer of energy within a body, or between two bodies in physical contact, from a higher temperature region to a lower temperature region by tangible contact.

Conductivity: See Thermal Conductivity.

Consistency: The resistance of a non-Newtonian material to deformation. Note: Consistency is not a fundamental property, but is comprised of viscosity, plasticity, and other phenomena.

Contact Adhesive: An adhesive which is apparently dry to the touch, and which will adhere to itself instantaneously upon contact; also called contact bond adhesive, or dry bond adhesive.

Convection—Forced: See Forced Convection.

Convection—Natural: See Natural Convection.

Copolymer: See Polymer.

Corrosion Effect: The wearing away, or destruction, of a substrate caused by acid or alkaline reactions between materials contained in the insulation and the substrate.

Covering—Appearance: See Appearance Covering.

Covering Capacity—Dry: The volume occupied in square area, expressed as thickness times the area of the material after being dried.

Covering Capacity—Wet: For materials which are mixed with water before application—the volume occupied in square area, exprexxed as thickness, for a unit of the dry material before being mixed with water. For materials which are pre-mixed (in cans)—the volume occupied in square area expressed as thickness of unit volume as received.

Crater: Small, shallow, crater-like surface imperfection in finishes.

Crawl Space: The space, usually 2-4′ between the original undisturbed grade level and the bottom of the floor construction above, or the space between the top of the ceiling joists and the bottom of the roof rafters above. Neither space is ever high enough for a man to stand upright, hence the name "crawl" space.

Creep: The dimensional change with time of a material under load apart from, and following, the initial instantaneous elastic or rapid deformation.

Cryogenic: Pertaining to the extremely low temperatures, such as the liquefaction points of gaseous elements, usually below −150° F (−101° C) on down to absolute zero.

Cure: To change the properties of a plastic or resin by chemical reaction, which for example may be condensation, polymerization, or addition; usually accomplished by the action of either heat, catalyst, or both, and with or without pressure.

Curing Agent: An additive incorporated in a coating or adhesive resulting in increased chemical activity between the components, with an increase or decrease in the rate of cure.

Curing Time: The length of time necessary to affect a cure of a plastic or resin by chemical reaction. See Cure.

Curved Segmental Block: A piece of rigid insulation, rectangular in plan, and the sector of a tube, in cross section, molded, or cut from block of the proper thickness.

Cut-Back Products: Petroleum or tar residuums which have been blended with distillate solvents.

Cutting Force (Shearing Force): A force tending to cleave the object to which it is applied.

Dauber: A spatula-like instrument (such as the Foster Gooper) for smear application of adhesives.

Deflection: The distance by which an object deviates from its original axis or shape after the application of a load.

Delamination: The separation of the layers of material in a laminate.

Density—Apparent: The weight of a unit volume of a material in its manufactured state, including all voids. Usually expressed in pounds per cubic foot.

Density—Real: The weight of a unit volume of a material, excluding all voids. Usually expressed in pounds per cubic foot.

Deterioration: A permanent change in physical properties evidenced by impairment of these properties.

Dew Point: The temperature at which the quantity of water vapor in a material would cause saturation, with resultant condensation of the vapor into liquid water by any further reduction of temperature.

Di-electric Strength: The rate of electric stress (volts per cm) required to puncture a film of coating.

Diffusivity: The time rate of temperature change within a body, or between two of its surfaces.

Dilatancy: The property of some highly filled coating materials of increasing in viscosity when the system is subjected to a distortional stress.

Dimensional Stability: That property of a material which enables it to hold its original size, shape, and dimensions when subjected to aging, load, heat, cold, moisture, or cutting.

Dimensional Trueness: That property of a material which enables it to be manufactured or fabricated, without internal stresses causing it to depart from size or shape with the passage of time.

Discoloration: Any change from the initial color. A lack of uniformity in color, where it should be uniform over the whole area.

Dispersion: A heterogeneous system in which a finely divided material is distributed in another material. Note: A dispersion is usually the distribution of a finely divided solid in a liquid or a solid; for example, pigments or fillers in coatings. A dispersion of a solid in a liquid only is a suspension.

Doctor Blade: A device consisting of a fixed metal blade or blades with which coatings or adhesives may be applied with a scraping action.

Dropping Resistance: That property of a material which enables it to withstand being dropped without fracture, breaking, or crumbling.

Dry: To change the physical state by the loss of solvent constituents by evaporation, absorption, oxidation, or a combination of these factors.

Drying Oil: An oil which possesses, to a marked degree, the property of readily taking up oxygen from the air and changing to a relatively hard, tough, elastic substance when exposed to the airr in a thin film.

Drying Time (Adhesives): Time elapsed since bonding at the optimum time when no further increase in bond strength is realized.

Drying Time (Finishes): Time elapsed after which no further significant changes take place in appearance or performance properties, due to drying.

Duck: A compact, firm, heavy, plain weave cotton fabric.

Ductility: That property of a material which enables it to undergo large deformations without rupture.

Durability: As applied to finishes, it is the lasting quality or permeance in service with particular reference to deterioration. May be related directly to an exposure condition.

Efflorescence (Bloom): A white powdery substance occurring on the surface of coated insulation products, caused by the migration of soluble salts from the insulation, followed by precipitation and carbonation.

Elasticity: The tendency of a material to recover its original size and shape after deformation.

Elastomer: A material which at room temperature can be stretched repeatedly to at least twice its original length and, immediately upon release of the stress, will return with force to its approximate original length.

Elongation (Extensibility): The extension between bench marks produced by a tension force applied to a specimen. It is expressed by a percentage of the original distance between the marks on the unstretched specimen (also known as Stretch).

Emissivity: The total heat lost per unit of time through a unit area of the surface of a body.

Emittance: The ratio of the total heat lost per unit of time through a unit area of the surface of a body to the total heat which would be lost in the same unit of time through the same unit area of a perfect black body.

Emittance—Directional: The ratio of the total heat transferred per unit of time and unit of surface area in a particular direction to that from a unit of blackbody surface of same area, temperature and conditions.

Emittance—Hemispherical: The ratio of the total heat radiant flux density from a body to that of a blackbody under same temperature conditions.

Emittance—Spectral: The emittance ratio based on energy emitted per unit wavelength interval.

Emulsion: Strictly stated, a colloidal suspension of one liquid in another. Often loosely used for "dispersion."

Energy: The measure of the amount of work a body (or system of bodies) can do, by virtue of its motion, or position, against forces applied to it. It is also a measure of the work it can do by virtue of its chemical composition, or as a result of having been heated. See Mechanical Potential Energy, Mechanical Kinetic Energy, and Internal Energy.

Epoxy Resins: Resins made by the reaction of epoxides or oxiranes with other materials such as amines, alcohols, phenols, carboxylic acids, acid anhydrides, and unsaturated compounds. Epoxy compounds may be cured at ambient temperature to form finishes which are highly resistant to solvents and chemicals.

Evaporation: The loss, from a liquid, through the transformation of a portion of it to its vapor state, caused by the application of heat.

Expandable Plastic: A plastic which can be made cellular by thermal, chemical, or mechanical means.

Expansion Ratio: The ratio of the Coefficients of Expansion of any two abutting materials.

Expansion—Wet State to Cured State: The property of a material which measures the difference in volumetric change in poured, sprayed, or foamed in place organics.

Exposure: The action by which a protective finish is exposed to the weather elements.

Facing: A thin layer, usually factory applied, on the surface of an insulating panel, variously acting as a vapor-barrier, weather-barrier, protector from damage, and a decorative coating.

Fading: Any lightening of an initial color.

Fahrenheit: The temperature scale of the English system of units in which the ice point of water is assigned the value of 32° and the steam point the value of 212°, with 180 even divisions between, and corresponding divisions above and below. Absolute zero on this scale is −459.6°.

False Body: Thixotropic flow property of a suspension or dispersion coating. When a compound "thins down" on stirring, or "builds up" on standing, it is said to exhibit false body.

Fatigue Resistance: That property of a material which enables it to be flexed back and forth, with reversal of stress each time, without rupture. Usually expressed as the number of cycles without rupture.

Felt: An insulation material composed of fibers of one or more kinds, in which they are interlocked, and have been compacted under pressure.

Fiber (Fibrated): See Asbestos.

Fill (in Fabrics): Yarn running from selvage to selvage at right angles to the warp yarn. See Pick Count.

Fill Insulation: An insulating material consisting of loose granules, fibers, beads, flakes, etc., which must be contained, and is usually placed in cavities of some description.

Filler: A relatively inert material added to a mastic or coating to modify its strength, permeance, working properties, or other qualities.

Fillet: That portion of an adhesive, mastic coating, or sealant which fills the corner, or angle, where two adherends or surfaces are joined.

Film: An optional term for sheeting having nominal thickness not greater than 0.010″.

Film—Wet: The freshly applied layer of mastic, coating, or adhesive before curing or drying has occurred.

Finishing Cement: A mixture of fibrous or powdery materials, or both, with suitable binders, that, when mixed with suitable proportion of water will develop a plastic consistency and can be used on the surface of insulations to provide a medium-hard to hard, even finish.

Finishing and Insulating Cement: A mixture of fibers and binders, water-mixed to a plastic mass on the job, and used as a finishing cement, and for an insulation for situations where only a small insulating effect is desired.

Fire Endurance: That property of a material which measures the elapsed time during which it continues to exhibit resistance to fire, under specified conditions of test and performance.

Fire Point Temperature: The lowest temperature of a material at which it gives off vapor, which, when combined with air near its surface, forms an ignitable mixture at a rate sufficient to support combustion continuously after the external ignition source is removed.

Fire Resistance: That property of a material which enables it to resist decomposition or deterioration when exposed to a fire.

Fire Resistive: Having fire resistance.

Fire Retardance: That property of a material which delays the spread of fire, either through or over itself.

Fire Retardant: Having fire retardance.

Fish-Mouth: A transverse gap between layers of sheet materials caused by warping or bunching of one or both layers.

Flame Spread: The rate, expressed in distance-time, at which a material will propagate flame on its *surface*. As this is a difficult property to measure in time and distance, the measure is now by *flame spread index* to enable the comparison of materials by test methods.

Flame Spread Requirements G.S.A.: United States General Services Administration, Public Buildings Service, Guide Specification, Section 301-1,

"For all concealed horizontal and vertical ductwork, and for all concealed piping, covering materials and accessories shall have a fire hazard rating not to exceed 25 for flame spread and 50 for fuel contributed and smoke developed. Ratings shall be determined by Underwriters' Laboratories, Inc. Method of Test of Surface Burning Characteristics of Building Materials or by the method in Interim Federal Standard No. 00136, Flame-Spread Properties of Materials."

Flame Resistance: That property of a material which enables it to resist decomposition or deterioration when exposed to flame.

Flame Retardance: That property of a material which delays the spread of flame either through or over itself.

Flammable: That property of a material which permits it to oxidize rapidly and release heat of combustion when exposed to flame or fire, and allows continuous burning after the external ignition source is removed.

Flammable (or Explosive) Limits: In the case of solvent vapors which form flammable mixtures with air or oxygen, there is a minimum concentration of vapor in air or oxygen below which propagation of flame does not occur on contact with a source of ignition. There is also a maximum proportion of vapor or gas in air, above which propagation of flame does not occur. These boundary-line mixtures of vapor with air, which, if ignited, will just propagate flame, are known as the "lower and upper flammable," usually expressed in terms of percentage by volume of vapor in air.

Flammable (or Explosive) Range: The range of combustible vapor and air mixtures between the upper and lower flammable limits is known as the "flammable range," sometimes referred to as the "explosive range."

Flammability Index: The comparison of the flammability of a material with that of an arbitrarily chosen standard material. Expressed as an *index*.

Flash-Ignition Temperature (Flash Point): The lowest temperature of a material at which it gives off vapor, which, when combined with air near its surface, forms an ignitable mixture.

Flashing: A thin strip of material, usually, but not always, metal, inserted at the junction of two materials, or two parts of a material, to divert liquid water to a specific direction.

Flexibility: That property of a material which allows it to be bent (flexed) without loss of strength.

Flexural Resistance: That property of a material which enables it to resist bending (flexure).

Flexural Strength: That property of a material which measures its resistance to bending (flexing). Usually expressed in pounds per square inch.

Flow: Movement of an adhesive, coating, or sealant during the application process before set has occurred. See Sag.

Fly Ash: Extremely fine inorganic dust from combustion of solid fuel in large power boilers.

Force: The potential for movement, or relative movement, of physical mass or temperature.

Forced Convection: The movement of a body, with its associated energy, from one location to another, with the rate of this movement increased by some outside influence such as wind, or a fan.

Freezing Point: The temperature at which a liquid will change from the liquid to the solid stage, with a simultaneous loss of energy.

Freeze-Thaw Resistance: The property of a material which permits it to be alternately frozen and thawed—through many cycles—without damage from rupture or cracking.

Friction: The resistance to relative motion between two bodies in contact.

Fuel Contribution: Flammable by-products of fire generated by, and emitted from, a burning object.

Gallon Weight: Weight of the standard U.S. volumetric unit of 231 cubic inches.

Galvanic Corrosion: Pitting or eating away of one of the metals when two metals of different electric potential are in direct contact, or electrically connected by an electrolyte.

Gel: A semisolid systtem consisting of a network of solid aggregates in which liquid is held. Note: Gels have very low strength and do not flow like a liquid. They are soft, flexible and will rupture under their own weight unless supported externally.

Geodesic: Of, or pertaining to the shape of the earth; therefore, cut or made in the shape of a sphere.

Gloss: A term used to express the shine, sheen, or luster of a dried film.

"Gooper": A special disposable type of dauber designed for easy smear application of adhesives. Registered trade mark, Benjamin Foster Co.

Graybody: A body having the same spectral emittance (less than unity) at all wave lengths.

Grit: Hard, relatively large inclusions in a coating composition.

Hanger (Insulation): A device to carry the weight of insulation in which the load carrier is positioned *above* the insulation. See Support.

Hardness: See Indentation Hardness.

Hazard—Fire: The susceptibility of a material to ignition and consequent potential for spread of flame and release of toxic gases and smoke.

Heat: The result of molecular motion and interacting forces. Energy in transient form.

Heat Capacity: The amount of heat required to raise a unit mass of a material 1 degree in temperature.

Heat—Latent: See Latent Heat.

Heat Conduction: See Conduction.

Heat Flux: The time rate of heat flow per unit area, in a direction perpendicular to the isothermal shield and spacer surfaces.

Heat Convection: See Convection.

Heat Radiation: See Radiation.

Heat Retardance: That property of a material which enables it to delay the flow of heat from a hot surface to a cold surface of a body.

Heat Transfer Cement: A soft, plastic material, which under use quickly solidifies to a rock-like hardness, having a high coefficient of heat transfer, which is used to bond tubes, or other heat-conveying devices, to the pipe or equipment to which it is desired to transfer the heat.

Heat Resistance: That property of a material which enables it to withstand heat without deterioration or failure.

Hexagonal Wire Mesh: Generic term for poultry netting, chicken wire, etc., usually made from pregalvanized wire woven in 1" mesh size. Also available in post-galvanized and rustless metal alloys.

Holiday: In a coating application a place not covered by coating compound.

Humidity: The condition of the atmosphere in respect to water vapor. See also Humidity—Absolute, Humidity—Relative.

Humidity—Absolute: See Absolute Humidity.

Humidity—Relative: See Relative Humidity.

Hydrocarbon Resins: Resins composed of carbon and hydrogen alone.

Hygroscopicity: That property of a material which enables it to readily absorb and retain water in either its liquid or vapor state.

Ignition: The initiation of combustion as evidenced by glow, flame, or explosion.

Ignition Temperature: The minimum temperature to which a solid, liquid, or gas must be heated in order to initiate or cause self-sustained combustion independently of the heating element.

Impact: The single instantaneous stroke of a body in motion against another either in motion or at rest.

Impact Force: The force resulting from an impact of one body upon another.

Impact Resistance: Capability of a finish to withstand mechanical or physical abuse under severe service conditions. Resistance to blows, bumps, and shocks incident to plant operation.

Indentation Hardness: That property of a body which enables it to resist, to a measurable degree, the tendency of a moving body, or a force, to dent its surface.

Indoor Covering: A material which serves to protect insulation from mechanical damage and wear and tear on indoor applications.

Insulating Mastic: A premixed soft, plastic material of various consistencies, applied by spray, trowel, brush or palm, which possesses some insulating value in addition to its other vapor- or weather-barrier characteristics.

Insulation Finish: A material, or materials, applied to the insulation to provide the final contour and a smooth even finish, and to strengthen the outer surface.

Insulation—Thermal: See Thermal Insulation.

Insulating Cement: A mixture of various fibers and binders, to be mixed with water to form a soft, plastic mass, and used to insulate small irregular surfaces and fill the cracks and crevices between the units insulating larger surfaces. (See Chapter 4 for types and properties.)

Insulation Cover: The cover for a flange, pipe fitting, or valve, composed of the specified thickness insulating material, and preformed into its proper shape before application.

Insulation—Fill: See Fill Insulation.

Insulation—Loose: See Loose Insulation.

Internal Energy: Energy stored within a body such as a gas, liquid, or solid, or within any material from which it can be released by chemical reaction.

Intumescence: The process of swelling or expanding on fire exposure to form a cellular charred layer which insulates and retards flaming.

Isocyanate Resins: Resins made by the condensation of organic isocyanates with other compounds. See Urethane.

Jacket: A covering placed around an insulation to protect it from mechanical damage, and, insofar as it is intrinsically able, from weather, water, ultra violet light, etc.

Joint: The location at which two mating surfaces are in juxtaposition.

"k" Factor: See Thermal Conductivity.

Kelvin: A temperature scale on which the absolute zero point is taken as 0°, the ice point of water 273.2°, and the steam point 373.2°, with corresponding divisions on up the scale. The degree divisions of Kelvin are the same as those of Celsius.

Lag: A long, narrow piece of rigid insulation, rectangular in plan, trapezoidal in cross section, molded, or cut from block of the proper thickness.

Lagging n.: An insulation layer, on a cylindrical surface, composed of lags.

Lagging v.: Action of covering something, as a boiler with insulation, now generally used to mean covering installed pipe insulation with canvas, using a combination adhesive-coating. This is a common, but incorrect usage of the word.

Laminate: A product made by bonding together two or more layers of material or materials.

Lap Adhesive (Lap Cement): The adhesive material used to seal the side and end laps of insulation jackets.

Latent Heat: The energy which must be added to a liquid at its boiling point to change its state from liquid to gas at a constant pressure. Conversely, it is the energy which must be removed from a gas at the boiling point of a liquid to change its state to a liquid, at a constant pressure. The latent heat to or from a liquid or gas causes a change in state without change in temperature.

Leno Weave: See Weave.

Linear Expansion: Increase in length in an axial or lengthwise direction, due to increase in temperature of the material.

Load: The force exerted on a material, or a support, by other material placed upon it, or from some external object.

Load—Breaking: See Breaking Load.

Load—Superimposed: See Superimposed Load.

Loose Insulation: Insulation in the form of loose granules, fibers, beads, flakes, etc., which must be contained and is usually placed in cavities of some description.

Low Temperature Bending: That property of a material which allows it to be bent (flexed) without rupture at low temperatures, with complete recovery to line and shape upon removal of the force causing the bending.

m.a.c. (Maximum Allowable Concentration): See Threshold Limit Value.

Mastic: A relatively thick consistency protective finish capable of application to thermal insulation or other surfaces, usually by spray or trowel, in thick coats, greater than 30 mils (0.030″).

Mat: A piece of insulation, of the semi-flexible type, cut into easily handled sizes, usually square or rectangular in shape, composed of fibers of one or more kinds, in which the fibers are in random arrangement and are compacted and held together by an adhesive. This material is used both as a reinforcement and an insulation, thus, the word is often used indiscriminately. Care should be taken to use the proper qualifying adjective, i.e., "reinforcing" or "insulating," when using the term.

Mean: The arithmetical average of a set of numbers.

Mechanical Abuse: See Impact Resistance.

Mechanically Foamed Plastic: A cellular plastic in which the cells are formed by the physical incorporation of gases.

Mechanical Kinetic Energy: Energy possessed by a body by virtue of the relative motion between it and other parts of a system.

Mechanical Potential Energy: Energy possessed by a body by virtue of its vertical distance above a horizontal plane.

Median: If the numerical values for a given property are arranged in ascending order, the median is (1) the middle value of the series if the number of values is odd, or (2) the mean of the two middle values if the number of values is even.

Melting Point: The temperature at which a material will change from the solid to the liquid state by the application of additional heat.

Membrane Reinforcement: Woven or non-woven fabrics used for saturation and embedment in mastic and coating applications to provide strength, continuity, and impact resistance.

Mil: A unit used in measuring thickness, being 0.001″. (British equivalent: thou.)

Mildew: Any discoloration caused by parasitic fungi on vegetable matter or other substances. See mold.

Mineral Spirits (Petroleum Spirits): A refined petroleum distillate with volatility, flash point, and other properties making it suitable as a thinner and solvent in coatings, mastics, and similar products.

Moisture Vapor: See Water Vapor.

Mold: A growth produced by fungi on various forms of organic matter, especially when damp or decaying. See Mildew.

Mold and Mildew Resistance: That property of a material which enables it to resist the formation of fungus growths under unfavorable conditions of temperature and humidity.

Monomer: A relatively simple compound which can react to form a polymer.

Mud Cracking: A form of alligatoring, or stress cracking, which may occur during drying in thick applications of water-base mastics or coatings, usually caused by shrinkage from excessive volatile content.

Muslin: Any of various plain weave coarse cotton fabrics having a weight per square yard usually not more than 2 oz.

Natural Convection: The movement of a body, with its associated energy, from one location to another.

Naphtha (Petroleum): A generic term applied to refined, partly refined, or unrefined petroleum products and liquid products of natural gas, not less than 10 percent of which distills below 347° F (175° C) and not less than 95 percent of which distills below 464° F (240° C), when subjected to distillation in accordance with ASTM Method D86, Test for Distillation of Petroleum Products. The "naphthas" used for specific purposes, such as manufacture of rubber cements, paints and varnishes, etc., are made to conform to specifications which may require products of considerably greater volatility than set by the limits of this generic definition.

Netting: Interwoven wires of some metal, usually either galvanized steel or Monel, woven into either a rectangular, square, or hexagonal pattern, used as a reinforcement in the application of insulation on large surfaces.

Newtonian (Simple) Liquid: A liquid in which the rate of sheer is proportional to the shearing stress.

Non-Newtonian (Complex) Liquid: A liquid in which the rate of shear is not proportional to the shearing stress.

Noncombustible: A material which will not contribute fuel or heat to a fire to which it is exposed.

Noncombustibility: That property of a material which prevents it from contributing fuel or heat to a fire to which it is exposed.

Nonflammable: That property of a material which prevents it from oxidizing rapidly and releasing heat of combustion when exposed to fire or flame.

Non-Volatile Content: That portion of a material which does not evaporate at ordinary temperatures.

Nylon Resins: Resins composed principally of a long-chain synthetic polymeric amide which has recurring amide groups as an integral part of the main polymer chain.

Odor Emission: That property of a material which indicates its relative scent, smell, or fragrance.

Opacity: The degree of obstruction to the transmission of visible light.

Open-Cell Foamed Plastic: A cellular plastic in which there is a predomi nance of interconnected cells.

Open Time Maximum (Adhesives): That open time which corresponds to 90 percent of the optimum strength after the maximum value has been reached.

Open Time Minimum (Adhesives): That open time which corresponds to 90 percent of the optimum strength prior to reaching the maximum value.

Open Time Optimum (Adhesives): That open time which gives the optimum strength at a bond age of 24 hours.

Open Time Range (Adhesives): Time spread between minimum open time and maximum open time.

Orange-Peel: Uneven surface of a spray-applied coating, somewhat resembling an orange peel.

Panel: An insulation, prefabricated at the factory into a rectangular shape, of relatively thin material, usually with a facing of some material on one of its rectangular surfaces, erected and secured as one piece. Panels are made in almost any erectable size, and in thicknesses up to about 2".

Peel Resistance: That property of a material which imparts to it the maximum bond to its substrate, to enable it to resist any external forces tending to peel them apart.

Penetration: The consistency of a mastic material, expressed as the distance that a standard cone vertically penetrates a sample of the material under known conditions of loading, time, and temperature. The units of penetration indicate hundredths of a centimeter.

Penetrometer: The instrument used for determinating penetration values of consistency.

Perm: The accepted unit of Water Vapor Permeance. Is expressed as 1 grain per square foot, hr, inch of mercury.

Perm-Inch: The accepted unit of Water Vapor Permeability. Is expressed as 1 grain per square foot, hour, inch of mercury, inch of thickness.

Permanence: The property of a finish describing its resistance to appreciable changes in characteristics with time and environment.

Permeability: See Water Vapor Permeability.

Permeance: See Water Vapor Permeance.

pH: The negative logarithm of the hydrogen ion concentration. A solution at pH 7 is neutral; lower numbers indicate increasing acidity, higher numbers, increasing alkalinity.

Phenolic Resins: Resins made by the condensation of phenols, such as phenol and cresol, with aldehydes.

Phoma pigmentovora: A form of mold which attacks freshly applied coating surfaces, characterized by pink to purple spots up to 1" or 2" in diameter.

Pick Count (Fabrics): The number of fill yars per inch of fabric.

Pigment: The fine solid color particles used in the preparation of colored coatings and substantially insoluble in the vehicle.

Pinhole: Very small hole through a mastic or coating.

Pipe: A circular conduit for the convenience of liquids or semi-solids. Nominal Pipe Sizes (NPS) are expressed as the *nominal inside* diameter of the conduit through 12" NPS, and the actual outside diameter and Schedule from 14" on up. See Tube.

Pipe Insulation: Rigid or semi-rigid preformed thermal insulation to fit over NPS pipe and tubing. Thickness and dimensions for pipe insulation should be in accordance with ASTM Standard C-585.

Pit: Small regular or irregular crater in the surface of a plastic, usually with its width approximately of the same order of magnitude as its depth.

Plastic: A material that contains, as an essential ingredient, an organic substance of large molecular weight, is solid in its finished state, and, at some state in its manufacture or in its processing into finished articles, can be shaped by flow.

Plastic Foam: See Cellular Plastic.

Plasticizer: A material incorporated in a plastic, or coating, to increase its flexibility or distensibility.

Plasticizer Migration: The transfer of a constituent from a plastic body to other contracting solids.

Polyamid Resins: See Nylon Resins.

Polyester Resins: Synonymous with alkyd resins.

Polymer: A compound formed by the reaction of simple molecules having functional groups that permit their combination to proceed to high molecular weights under suitable conditions. Polymers may be formed by polymerization (addition polymer) or polycondensation (condensation polymer). When two or more monomers are involved, the product is called a copolymer.

Polymer Emulsion: Colloidal dispersion of a polymer in a water base, with or without other ingredients.

Polystyrene: A resin made by polymerization of styrene as the sole monomer.

Polyurethane: See Urethane Resins.

Polyvinyl Acetate Resins: Resins made by the polymerization of vinyl acetate or copolymerization of vinyl acetate with minor amounts (not over 50 percent) of other unsaturated compounds.

Pot Life: See Working Life.

Power: The time rate of doing work.

Primer: The first of two or more coats of a finish system.

Pullularia pullulans: A fairly common mold occurring particularly on damp cellulosic materials, such as cotton or paper, varying in color from dirty white to greenish and eventually becoming black and leathery all over.

Puncture Resistance: That property of a material which enables it to resist a tendency to puncture or perforate under blows or pressure from sharp objects.

Punking: The incandescence, or glow, which lingers in some materials after any flame, or other evidence of fire, has departed.

Radiant Flux Density: The rate of radiant energy emissions from a unit area of a source in all the radial directions of the overspreading hemisphere.

Radiation: The transfer of energy from a higher temperature body, through space, to another body, or bodies, some distance away at a lower temperature, without raising the temperature of the medium through which the energy passes.

Rankine: A temperature scale on which the absolute zero point is assigned the value of 0° the ice point of water 491.6, and the steam point 671.6°, with corresponding divisions on up the scale. The degree divisions of Rankine are the same as those of Fahrenheit.

Recovery of Thickness After Compression: This is a vital factor of blankets or botts in the use of these materials as cushion blankets, or in expansion joints.

Reflectance: The ration of the radiant energy reflected by a body to that incident upon it.

Reflective Insulation: Insulation, composed of closely spaced sheets of either aluminum or stainless steel, which obtains its insulating value from the ability of the sheets to reflect a large part of the radiant energy incident on them.

Reinforcing Cloth or Fabric: A loosely woven cloth or fabric of glass or resilient fibers, placed approximately in the center of the vapor- or weather-barrier to act as reinforcing to the mastic of the barrier.

Relative Humidity: The ratio of the actual pressure of existing water vapor in the atmosphere at the same temperature, expressed as a percentage. (See dew point)

Resin: A solid, semisolid, or pseudo-solid organic material which has an indefinite and often high molecular weight, exhibits a tendency to flow when subjected to stress, usually has a softening or melting range, and fractures conchoidally.

Resistance (Thermal): See Thermal Resistance.

Resistance to Abrasion: Ability to withstand scuffing, scratching, rubbing, or wind-driven particles without loss of mechanical protection properties.

Resistance to Air Flow: The quotient of the air pressure difference across the surface divided by the volume velocity of air flow across the surface.

Resistance to Air Movement: The property which indicates the ability of a blanket type material to resist erosion by air currents over its surface.

Resistance to Freeze-Thaw: The resistance to change in application, thermal and/or mechanical properties from exposure to alternate cycles of freezing and thawing.

Resistance to Impact: The ability to withstand mechanical blows without loss of physical integrity and protective properties.

Resistance to Mold and Mildew: The ability to resist deterioration by fungi.

Resistance to Plastic Flow: That property of a material which opposes flow (deformation) beyond the elastic range of the material.

Right Angle Test: A method designed by DuPont Engineering to determine cracking tendencies of emulsion mastics when applied over an interior right angle, formed by fastening together two sections of insulation board.

Rigidity: That property of a material which opposes any tendency for it to bend (flex) under load.

Room Temperature: A temperature in the range of 20°C-30°C (68°F-85°F).

Rubber Cement: A natural or synthetic elastomeric material suitably compounded to form an effective adhesive for specified uses.

Rust Blush: The earliest stage of rusting characterized by the orange or red color ferric hydroxide (common rust). Occurs frequently on freshly sandblasted steel if allowed to stand too long before coating.

Sag: Excessive flow in material after application to a surface, resulting in "curtaining" or running.

Self-Ignition Temperature (Autogenous Ignition): The lowest temperature of a material which will cause it to ignite without other ignition source.

Self-Extinguishing: That property of a material which enables it to stop its own ignition after external ignition sources are removed.

Sealer: A putty-like substance, composed of various materials, used as a barrier to the passage of water vaport or liquid water into the joint formed by the mating surfaces of jackets and water- and vapor-barriers over insulation. A good sealer will possess relatively little shrinkage. There are several types of sealers, such as nonsetting, setting, and heat resisting.

Securements—Insulation: Any device, wire, strap or adhesive used to fasten insulation into its service position and hold it there.

Self-Heating: The process whereby, due to exothermic reactions, heat is liberated within a material at a rate sufficient to raise its temperature.

Self-Ignition: Ignition resulting from self-heating.

Service Temperature Limits: The limiting temperatures at a coated surface, within which limits the applied coating will have satisfactory service performance.

Set: To convert into a fixed or hardened state by chemical or physical action.

Shade: A term descriptive of a lightness difference between surface colors, the other attributes of color being essentially constant. A lighter shade of a color is one that has higher lightness, but approximately the same hue and saturation. A darker shade is one that has a lower lightness.

Shear Strength: The maximum stress tending to cleave a material which it is capable of sustaining without destruction. Shear strength is calculated from the maximum load during a shear or torsion test, and is based on the original dimensions of the cross-section of the specimen.

Sheen: Shiny or lustrous appearance at, or near, grazing incidence of a surface that appears to have no gloss for near perpendicular incidence; also, the 85-degree specular gloss of such a surface.

Sheet: A piece of material which is very thin in relation to its length and breadth.

Shelf Life: See Storage Life.

Shrinkage: The property of a material which causes its dimensions to become smaller. The material may become smaller due to thermal contraction due to reduction in temperature or it may become smaller due to reduction of moisture content.

Shrinkage—Wet to Dry: The property of a material which measures the difference in volumetric and linear change which occurs in the drying of insulating cements and mastics.

Silicone Resins: Resins in which the main polymer chain consists of alternating silicon and oxygen atoms, with carbon-containing side groups.

Sizing: Any of various glutinous materials, used to fill the pores in the surface of a paper, fiber, or cloth.

Skinning: The formation of a relatively dense film on the surface of a mastic or coating material while stored in containers.

Smoke Density: The Smoke Density Factor is the amount of smoke given off by the burning material compared to the amount of smoke given off by the burning of a standard material.

Smoke Toxicity: The degree of hazard to health of the smoke.

Smoldering: The combustion of solid materials without the accompaniment of flame.

Soaking Heat Stability: That property of a material which enables it to endure a soaking heat over an appreciable length of time, without materially changing its dimensions or properties.

Softening Point: That temperature at which a material will change its property from firm or rigid to soft or malleable.

Solar Resistance: 1. The resistance of a material to decomposition by the ultra-violet rays from the sun. 2. The resistance of an insulation to the passage of radiant heat from the sun.

Solids Content: The percentage of the non-volatile matter. Note: The determined value of non-volatile matter in any adhesive, coating, or sealant will vary somewhat, according to the analytical procedure used. A standard test method must be used to obtain consistent results.

Solvent: Any substance, usually a liquid, which dissolves other substances. In the coatings industry, normally a liquid organic compound used to make a coating work more freely. See Thinner.

Specific Gravity—Apparent: The ratio of the weight of a unit volume (usually a cubic foot) of the material as manufactured—including all voids—to the weight of a unit volume (usually a cubic foot) of water.

Specific Gravity—Real: The ratio of the weight of a unit volume (usually a cubic foot) of the actual material—excluding all voids—to the weight of a unit volume (usually a cubic foot) of water.

Specific Heat (at Constant Pressure): The ratio of the amount of heat required to raise a unit mass of a material 1 degree, to that required to raise a unit mass of water 1 degree at some specified temperature. In the English system the unit of measurement is Btu/lb degree F; in the Metric system the unit of measurement is Cal/gg degree C. The numerical value for specific heat in these units of measurement is the same.

Specimen: A portion of a uanit taken for a single measurement of a given property or characteristic.

Sprayed-On Insulation: Insulation of the fibrous, or foam, type which is applied to the substrate by means of any one of a large number of powered spray devices, and which is secured to its substrate by its own properties of adhesion.

Stability Under Edge Compression: That property of a material which enables it to retain its shape and dimensions during the application of a compressive load on its edge, or edges. This is sually applicable only to rigid reflective insulation.

Standard Deviation: The square root of the mean square of the deviations of a set of values from their mean. It is a measure of the dispersion of the data.

Standard Time—Temperature Curve: A curve depicting the allowable temperature rise in a material, as related to time. The Standard Time—Temperature Curve is published by ASTM.

Static: Motionless. At rest or in equilibrium.

Static Load Test: The application of a constant load at rest, in testing.

Storage Life: The period of time during which a packaged adhesive, coating, or sealant can be stored under specified temperature conditions and remain suitable for use. Sometimes called shelf life.

Storage Stability: The ability of a material to retain its shape, dimensions, and properties while in storage.

Strain: A measure of the change, due to force, in the size or shape of a body referred to its original size or shape. Strain is a nondimensional quantity, but it is frequently expressed in inches per inch, etc.

Strength—Dry: The strength of an adhesive joint determined immediately after drying under specified conditions, or after a period of conditioning in a standard atmosphere.

Strength—Wet: The strength of an adhesive joint determined immediately after bonding adherends under specified conditions.

Strength—Compressive: See Compressive Strength.

Strength—Flexural: See Flexural Strength.

Strength—Tensile: See Tensile Strength.

Strength—Transverse: See Transverse Strength.

Strength—Tear: See Tear Strength.

Stress: The intensity, at a point in a body, of the internal forces or components of force that act on a given plane through the point. Stress is expressed in force per unit of area (pounds per square inch, kilograms per square millimeter, etc.)

Stress Corrosion: Intergranular corrosion and cracking in a metallic material, caused by the combination of a minimum temperature, tensile stress, and a specific corrodent (the chloride ion in the case of Austenitic stainless steel).

Stress-Crack: External or internal cracks, caused by tensile stresses less than those concerned in the short-time mechanical strength. Note: The development of such cracks is frequently accelerated by the environment. The stresses which cause cracking may be present internally or externally or may be combinations of these stresses. The appearance of a network of fine cracks is called crazing.

Structural Insulation: Insulation which is used as a part of the loadcarrying frame of a structure such as the walls of a cold room, and which has the necessary physical properties to perform its structural function.

Substrate: A material upon the surface of which an adhesive or coating is spread. A broader term than Adherend.

Suction: The absorptive effect exerted on coating materials by a highly porous substrate.

Superimposed Load: A load on, and carried by, an insulation, over and above the dead weight of the insulation itself.

Support—Insulation: A device to carry the weight of insulation in which the load carrier is positioned *under* the insulation. See Hanger.

Surface Wetting: The property of a material applied to a substrate which enables it to thoroughly wet the substrate to produce a good bond.

Tack: The property of an adhesive that enables it to form a bond of measurable strength immediately after adhesive and adherend are brought into contact under low pressure.

Tape: A narrow strip or band of any flexible material.

Tar: Brown or black bituminous material, liquid or semi-solid in consistency, in which the predominating constituents are bitumens obtained as condensates in the destructive distillation of coal, petroleum, oil-shale, wood, or other organic materials, and which yields substantial quantities of pitch when distilled.

Tear: To divide, disrupt, or pull apart by the action of opposing forces.

Tear Strength: That property of a material which enables it to resist being pulled apart by opposing forces.

Temperature: The level of the thermal state as indicated on a designated scale. (For the English system—Fahrenheit.)

Temperature Limits: The upper and lower temperatures at which a material will experience no essential change in its properties.

Temperature Retardance: That property of a material which delays any change of the temperature of one of its surfaces as related to any change in the temperature of its other surface.

Temperature Rise—Internal: The rise in the internal temperature of a material, due to application of heat, to fire, or to contamination.

Tensile Strength: The force per unit of the original cross sectional area (of an unstretched specimen) which is applied at the time of rupture of the specimen. It is calculated by dividing the breaking force, in pounds by the cross-section of the unstretched specimen, in square inches.

Test Area: The designated location from which specimens for physical and chemical testing shall be taken.

Test Result: A single numerical quantity determined by measuring a test specimen for a given property. In the case of a physical characteristic, the test result shall be considered representative of the test unit from which the specimen was taken. In the case of a chemical property, the test result, being obtained from a composite sample, is not associated with any particular test unit.

Thermal Conductance: The amount of heat transferred through a unit area of a material in a unit time, through its *total* thickness, with a unit of temperature difference between the surfaces of the two opposite sides.

Thermal Conductivity: The amount of heat transferred through a unit area of a material in a unit time, through a *unit* thickness, with a unit of temperature difference between the surfaces of the two opposite sides.

Thermal Diffusivity: See Diffusivity.

Thermal Insulation: Material having air- or gas-filled pockets, void spaces, or heat-reflective surfaces, which, when properly applied, will retard the transfer of heat with reasonable effectiveness under ordinary conditions.

Thermal Insulation—Forms of:

Blanket Thermal Insulation: A flat, flexible thermal insulation which can be placed over flat, curved, or irregular surfaces. May or may not be faced, coated, or reinforced on one or both sides.

Block Thermal Insulation: Rigid thermal insulation preformed into rectangular units.

Board Thermal Insulation: Rigid and semi-rigid thermal insulation preformed into relatively large area rectangular units.

Cement Thermal Insulation: A prepared composition, in dry form, composed of granular, flaky, fibrous or powdery materials, which when mixed with correct proportion of water develops into a mix of plastic consistance. After application and becoming dry in place it forms a coherent semi-rigid insulation.

Loose Fill Thermal Insulation: Material in granular, nodular, fibrous, or powder form suitable for installation, dry, by pouring, blowing or hand placement into confined spaces or areas.

Pipe Thermal Insulation: Rigid or semi-rigid insulation preformed for application to pipe or tubing.

Reflective Thermal Insulation: Sheets or preformed shapes to provide spaces, and reflective sheets which obtain their insulating resistance from the ability of the sheets to reflect a large part of the radiant energy incident on them.

Foamed Thermal Insulation: Liquids sprayed or mixed together which foam and set into cellular or semicellular organic rigid or semi-rigid insulation.

Thermal Resistance: That property of a material which enables it to withstand the passage of heat through it, due to a temperature difference between its two opposite surfaces.

Thermal Shock Resistance: That property of a material which enables it to retain its shape and not distort, crack, or shatter, due to a sudden change in its temperature.

Thermal Transference: The steady-state heat flow from (or to) a body through applied thermal insulation and to (or from) the external surroundings by conduction, convection, and radiation. It is expressed as the time rate per unit area of the body surface per unit temperature difference between body surface and the external surroundings.

Thermal Transmission, Heat: The quantity of heat flowing due to all modes of heat transfer under the prevailing conditions.

Thermal Transmittance: The ratio of the steady flow of heat energy from ambient air on one side of a body, through the body, to the external surroundings on the opposite side of the body, due to the temperature difference between the two surroundings.

Thermally Foamed Plastic: A cellular plastic, produced by applying heat to effect gaseous decomposition or volatilization of a constituent.

Thermoplastic: Capable of being repeatedly softened by increase of temperature, and hardened by decrease of temperature. Note: Thermoplastic applies to those materials whose change upon heating is substantially physical.

Thermoset: A plastic or coating which, when cured by application of heat or chemical means, changes into a substantially infusible and insoluble product.

Thinner: Volatile organic liquid used to adjust consistency, or to modify other properties of mastics and coatings, which volatilizes during the drying process.

Thixotropy: The property of decreasing in consistency upon being sheared or worked, followed by a gradual recovery of consistency when the shearing stress is removed.

Threshold Limit Value: The maximum allowable concentration (m.a.c.) of vapor to which nearly all industrial workers may be repeatedly exposed, day after day, without adverse physiological effect.

Tint: A color produced by a mixture of white pigment or coating in predominating amount with a colored pigment or coating, not white. The tint of a color is, therefore, much lighter and much less saturated than the color itself.

Tolerance: The allowable variation from given dimensions in the manufacturing or fabrication of an object.

Total Solids: See Solids Content.

Toxicity: The degree of hazard to health.

Traced: The supplying of auxiliary heat (or refrigeration) to a line, or piece of equipment, by means of a comparison line containing a hot (or cold) liquid or gas, thermally bonded to the line or equipment, with the resultant assembly completely encased by the insulation.

Translucent: Allowing the passage of some light, but not a clear view of any object.

Transverse Strength: That property of a material which enables it to resist loads or impact, normal to its primary axis.

Tube: A circular conduit for the conveyance of liquids or semi-solids. Tube sizes are expressed as the *outside* diameter and wall thickness. See Pipe.

Ultimate Strength: That load in pounds per square inch (in the English system) at which a material will rupture.

Ultimate Strength (as applied to Adhesives): Bond strength after the established drying time has been determined.

Uniformity: The state of being unvarying in form or composition.

Urethane Resins: Resins made by the condensation of organic isocyanates with compounds or resins that contain hydroxol groups. Note: Urethanes are a type of isocyanate resins.

Vapor-Barrier: A material, or materials, which, when installed on the high vapor pressure side, retards the passage of the moisture vapor to the lower vapor pressure side.

Vapor Density: The relative density of a vapor or gas (with no air present) as compared with air. A figure less than 1 indicates that a vapor is lighter than air, and a figure greater than 1 that a vapor is heavier than air.

Vapor Migration: That property of a material which measures the rate at which water vapor will penetrate it, due to vapor pressure differences between its surfaces.

Vapor Pressure: The gas pressure exerted by the water vapor present in the air.

Vehicle: The liquid portion of a mastic or coating. Anything that is dissolved in the liquid portion is a part of the vehicle.

Vibration Resistance: That property of a material which enables ti to stay whole a nd not disintegrate when subject to vibration.

Vinyl Resins: Resins made from vinyl monomers, except those specifically covered by other classifications such as acrylic and styrene resins. Typical vinyl resins are polyvinyl chloride, polyvinyl alcohol, and polyvinyl butyral as well as copolymers, or vinyl monomers, with unsaturated compounds.

Vinyl Chloride Resins: Resins made by the polymerization of vinyl chloride or copolymerization of vinyl chloride with minor amounts (not over 50 percent) of other unsaturated compounds.

Viscometer (viscosimeter): An instrument for measuring viscosity or consistency.

Viscosity: The property of resistance to flow exhibited within the body of a material. Note: This property can be expressed in terms of the relationship between applied shearing stress and resulting rate of strain in shear. Viscosity is usually taken to mean "Newtonian Viscosity," in which case the ratio of shearing stress to the rate of shearing strain is constant. In non-Newtonian behavior, which is the usual case with adhesives, coatings, and sealants, the ratio varies with the shearing stress. Such ratios are often called the "apparent viscosities" at the corresponding shearing stresses. See Consistency.

Volatile Content: The proportional part (usually expressed as a percentage) of a material comprised of volatile (readily vaporizable) compounds.

Volatile Loss: Weight loss by vaporization.

Warp: The yarn running lengthwise in a woven fabric.

Warpage: The change in dimension of one surface of insulation as compared to that of another surface, due to differences in temperature of the two surfaces.

Water Absorption: The increase in weight of a test specimen, expressed as a percentage of its dry weight after immersion in water for a specified time.

Waterproof: Impervious to prolonged exposure to water.

Water-Repellency: That property of a material which prevents it from adhering to, or mixing with, water.

Water-Repellent: Having the property of water repellency.

Water-Resistant: Capable of withstanding limited exposure to water, or wet conditions, without failure.

Water Vapor: Water in a gaseous state.

Water-Vapor-Barrier: A material, or materials, which serves both as a weather-barrier and a vapor-barrier.

Water Vapor Permeability: The water vapor permeability of a homogeneous material is a property of the substance. This property may vary with conditions of exposure. The average permeability of a specimen is the product of its permeance and thickness. An accepted unit of permeability is a perm inch, or 1 grain per square foot, hour, inch of mercury per inch of thickness. The test conditions must be stated.

Water Vapor Permeance: The water vapor permeance of a body between two specified parallel surfaces is the ratio of its WVT to the vapor pressure difference between the two surfaces. An accepted unit of permeance is a perm, or 1 grain per square foot, hour, inch of mercury. The test conditions must be stated.

Water Vapor-Retarder (Barrier): The correct term is "Retarder," as these materials or systems retard the transmission of water vapor. It should be noted that no material or system can completely stop water vapor transmission when there is a vapor pressure difference between each of its surfaces.

Water Vapor Diffusion: The process by which water vapor spreads or moves through materials caused by a difference in water vapor pressure.

Water Vapor Transmission (WVT): The rate of water vapor transmission of a body between two specified parallel surfaces is the time rate of water vapor flow normal to the surfaces under steady condition through unit area, under the conditions of test. An accepted unit of WVT is 1 grain per square foot, hour (with test conditions stated).

Weather-Barrier (Weathercoat): A material or materials, which, when installed on the outer surface of thermal insulation, protects the insulation from the ravages of weather, such as rain, snow, sleet, wind, solar radiation, atmospheric contamination, and mechanical damage.

Weathering: The exposure of mastics or coatings outdoors.

Weather-Ometer:* A machine device used to determine the life of coatings by subjection to ultraviolet light, water, and sometimes other conditions.

Weather-Vapor-Barrier: A material which combines the properties of a weather-barrier and a vapor-barrier.

Weave—Leno: A fabric pattern in which warp yarns are arranged in pairs, twisting around one another between picks of filling yarns.

Weave—Plain: A fabric pattern in which each yarn of the filling passes alternately over and under a yarn of warp, and each yarn of the warp passes alternately over and under a yarn of the filling.

Weight: The force (pounds in the English system) with which a body is attracted toward the earth.

Wet: In the freshly applied state before drying.

Work: Energy in transient form. Expressed as the force applied to an object multiplied by the distance through which the object moves. $W = F \times D$.

Working Life (Pot Life): The period of time during which an adhesive or coating, after mixing with catalyst, solvent, or other compounding ingredients remains suitable for use.

*T.M. Reg. Atlas Electric Devices Co.

Appendix B — Tables

Relations between the temperature scales

Temp. (t)	Fahrenheit (°F)	Rankine (°R)	Celsius (°C)	Kelvin (°K)
°F =	1	t°R − 459.6	1.8 t°C + 32	1.8(t°K − 273.16) + 32
°R =	t°F + 459.69	1	1.8 t°C + 491.69	1.8 t°K
°C =	(t°F − 32) × 5/9	5/9(t°R) − 273.16	1	t°K − 273.16
°K =	(t°F − 32) × 5/9 + 273.16	5/9(t°R)	t°C + 273.16	1

Table B-1 Energy, work and heat conversion factors

Multiply number of ⟶ Btu

By / To obtain ↓	Btu	Centi-meter-grams	Ergs or centi-meter dynes	Foot-pounds	Horse-power-hours	Joules or watt-seconds	Kilogram-calories	Kilowatt hours	Meter-kilograms	Watt-hours
BTU	1	9.297×10^{-8}	9.480×10^{-11}	1.285×10^{-3}	2545	9.480×10^{-4}	3.969	3413	9.297×10^{-3}	3.413
Centimeter-grams	1.076×10^{7}	1	1.020×10^{-3}	1.383×10^{4}	2.737×10^{10}	1.020×10^{4}	4.269×10^{7}	3.671×10^{10}	10^{5}	3.671×10^{7}
Ergs or centimeter-dynes	1.055×10^{10}	980.7	1	1.356×10^{7}	2.684×10^{13}	10^{7}	4.186×10^{10}	3.6×10^{13}	9.807×10^{7}	3.6×10^{10}
Foot-pounds	778.3	7.233×10^{-3}	7.367×10^{-8}	1	1.98×10^{6}	0.7376	3087	2.655×10^{6}	7.233	2655
Horsepower-hours	3.929×10^{-4}	3.654×10^{-11}	3.722×10^{-14}	5.050×10^{-7}	1	3.722×10^{-7}	1.559×10^{-3}	1.341	3.653×10^{-6}	1.341×10^{-3}
Joules or watt seconds	1054.8	9.807×10^{-6}	10^{-7}	1.356	2.684×10^{6}	1	4186	3.6×10^{6}	9.807	3600
Kilogram-calories	0.2520	2.343×10^{-8}	2.389×10^{-11}	3.239×10^{-4}	641.3	2.389×10^{-4}	1	860.0	2.343×10^{-3}	0.8600
Kilowatt hours	2.930×10^{-4}	2.724×10^{-11}	2.778×10^{-14}	3.766×10^{-7}	0.7457	2.778×10^{-7}	1.163×10^{-3}	1	2.724×10^{-6}	0.001
Meter-kilograms	107.6	10^{-5}	1.020×10^{-8}	0.1383	2.737×10^{5}	0.1020	426.9	3.671×10^{5}	1	367.1
Watt-hours	0.2930	2.724×10^{-8}	2.778×10^{-11}	3.766×10^{-4}	745.7	2.778×10^{-4}	1.163	1000	2.724×10^{-3}	1

Table B-2 Temperature conversion

FAHRENHEIT SCALE LISTED IN EVEN NUMBERS

KELVIN K	CELSIUS C	FAHREN-HEIT F	RANKINE R	KELVIN K	CELSIUS C	FAHREN-HEIT F	RANKINE R	KELVIN K	CELSIUS C	FAHREN-HEIT F	RANKINE R
0	−273.2	−459.7	0	38.6	−234.4	−390	69.7	77.6	−195.6	−320	139.7
0.4	−272.8	−459	0.7	39.3	−233.9	−389	70.7	78.2	−195.0	−319	140.7
1.0	−272.2	−458	1.7	39.9	−233.3	−388	71.7	78.8	−194.4	−318	141.7
1.5	−271.7	−457	2.7	40.4	−232.8	−387	72.7	79.3	−193.9	−317	142.7
2.1	−271.1	−456	3.7	41.0	−232.2	−386	73.7	79.9	−193.3	−316	143.7
2.6	−270.6	−455	4.7	41.5	−231.7	−385	74.7	80.4	−192.8	−315	144.7
3.2	−270.0	−454	5.7	42.1	−231.1	−384	75.7	81.0	−192.2	−314	145.7
3.8	−269.4	−453	6.7	42.6	−230.6	−383	76.7	81.5	−191.7	−313	146.7
4.3	−268.9	−452	7.7	43.2	−230.0	−382	77.7	82.1	−191.1	−312	147.7
4.9	−268.3	−451	8.7	43.8	−229.4	−381	78.7	82.6	−190.6	−311	148.7
5.4	−267.8	−450	9.7	44.3	−228.9	−380	79.7	83.2	−190.0	−310	149.7
6.0	−267.2	−449	10.7	44.9	−228.3	−379	80.7	83.8	−189.4	−309	150.7
6.5	−266.7	−448	11.7	45.4	−227.8	−378	81.7	84.3	−188.9	−308	151.7
7.1	−266.1	−447	12.7	46.0	−227.2	−377	82.7	84.9	−188.3	−307	152.7
7.6	−265.6	−446	13.7	46.5	−226.7	−376	83.7	85.4	−187.8	−306	153.7
8.2	−265.0	−445	14.7	47.1	−226.1	−375	84.7	86.0	−187.2	−305	154.7
8.8	−264.4	−444	15.7	47.6	−225.6	−374	85.7	86.5	−186.7	−304	155.7
9.3	−263.9	−443	16.7	48.2	−225.0	−373	86.7	87.1	−186.1	−303	156.7
9.9	−263.3	−442	17.7	48.8	−224.4	−372	87.7	87.6	−185.6	−302	157.7
10.4	−262.8	−441	18.7	49.3	−223.9	−371	88.7	88.2	−185.0	−301	158.7
11.0	−262.2	−440	19.7	49.9	−223.3	−370	89.7	88.8	−184.4	−300	159.7
11.5	−261.7	−439	20.7	50.4	−222.8	−369	90.7	89.3	−183.9	−299	160.7
12.1	−261.1	−438	21.7	51.0	−222.2	−368	91.7	89.9	−183.3	−298	161.7
12.6	−260.6	−437	22.7	51.5	−221.7	−367	92.7	90.4	−182.8	−297	162.7
13.2	−260.0	−436,	23.7	52.1	−221.1	−366	93.7	91.0	−182.2	−296	163.7
13.8	−259.4	−435	24.7	52.6	−220.6	−365	94.7	91.5	−181.7	−295	164.7
14.3	−258.9	−434	25.7	53.2	−220.0	−364	95.7	92.1	−181.1	−294	165.7
14.9	−258.3	−433	26.7	53.8	−219.4	−363	96.7	92.6	−180.6	−293	166.7
15.4	−257.8	−432	27.7	54.3	−218.9	−362	97.7	93.2	−180.0	−292	167.7
16.0	−257.2	−431	28.7	54.9	−218.3	−361	98.7	93.8	−179.4	−291	168.7
16.5	−256.7	−430	29.7	55.4	−217.8	−360	99.7	94.3	−178.9	−290	169.7
17.1	−256.1	−429	30.7	56.0	−217.2	−359	100.7	94.9	−178.3	−289	170.7
17.6	−255.6	−428	31.7	56.5	−216.7	−358	101.7	95.4	−177.8	−288	171.7
18.2	−255.0	−427	32.7	57.1	−216.1	−357	102.7	96.0	−177.2	−287	172.7
18.8	−254.4	−426	33.7	57.6	−215.6	−356	103.7	96.5	−176.7	−286	173.7
19.3	−253.9	−425	34.7	58.2	−215.0	−355	104.7	97.1	−176.1	−285	174.7
19.9	−253.3	−424	35.7	58.8	−214.4	−354	105.7	97.6	−175.6	−284	175.7
20.4	−252.8	−423	36.7	59.3	−213.9	−353	106.7	98.2	−175.0	−283	176.7
21.0	−252.2	−422	37.7	59.9	−213.3	−352	107.7	98.8	−174.4	−282	177.7
21.5	−251.7	−421	38.7	60.4	−212.8	−351	108.7	99.3	−173.9	−281	178.7
22.1	−251.1	−420	39.7	61.0	−212.2	−350	109.7	99.9	−173.3	−280	179.7
22.6	−250.6	−419	40.7	61.5	−211.7	−349	110.7	100.4	−172.8	−279	180.7
23.2	−250.0	−418	41.7	62.1	−211.1	−348	111.7	101.0	−172.2	−278	181.7
23.8	−249.4	−417	42.7	62.6	−210.6	−347	112.7	101.5	−171.7	−277	182.7
24.3	−248.9	−416	43.7	63.2	−210.0	−346	113.7	102.1	−171.1	−276	183.7
24.9	−248.3	−415	44.7	63.8	−209.4	−345	114.7	102.6	−170.6	−275	184.7
25.4	−247.8	−414	45.7	64.3	−208.9	−344	115.7	103.2	−170.0	−274	185.7
26.0	−247.2	−413	46.7	64.9	−208.3	−343	116.7	103.8	−169.4	−273	186.7
26.5	−246.7	−412	47.7	65.4	−207.8	−342	117.7	104.3	−168.9	−272	187.7
27.1	−246.1	−411	48.7	66.0	−207.2	−341	118.7	104.9	−168.3	−271	188.7
27.6	−245.6	−410	49.7	66.5	−206.4	−340	119.7	105.4	−167.8	−270	189.7
28.2	−245.0	−409	50.7	67.1	−206.1	−339	120.7	106.0	−167.2	−269	190.7
28.8	−244.4	−408	51.7	67.6	−205.6	−338	121.7	106.5	−166.7	−268	191.7
29.3	−243.9	−407	52.7	68.2	−205.0	−337	122.7	107.1	−166.1	−267	192.7
29.9	−243.3	−406	53.7	68.8	−204.4	−336	123.7	107.6	−165.6	−266	193.7
30.4	−242.8	−405	54.7	69.3	−203.9	−335	124.7	108.2	−165.0	−265	194.7
31.0	−242.2	−404	55.7	69.9	−203.3	−334	125.7	108.8	−164.4	−264	195.7
31.5	−241.7	−403	56.7	70.4	−202.8	−333	126.7	109.3	−163.9	−263	196.7
32.1	−241.1	−402	57.7	71.0	−202.2	−332	127.7	109.9	−163.3	−262	197.7
32.6	−240.6	−401	58.7	71.5	−201.7	−331	128.7	110.4	−162.8	−261	198.7
33.2	−240.0	−400	59.7	72.1	−201.1	−330	129.7	111.0	−162.2	−260	199.7
33.8	−239.4	−399	60.7	72.6	−200.6	−329	130.7	111.5	−161.7	−259	200.7
34.3	−238.9	−398	61.7	73.2	−200.0	−328	131.7	112.1	−161.1	−258	201.7
34.9	−238.3	−397	62.7	73.8	−199.4	−327	132.7	112.6	−160.6	−257	202.7
35.4	−237.8	−396	63.7	74.3	−198.9	−326	133.7	113.2	−160.0	−256	203.7
36.0	−237.2	−395	64.7	74.9	−198.3	−325	134.7	113.8	−159.4	−255	204.7
36.5	−236.7	−394	65.7	75.4	−197.8	−324	135.7	114.3	−158.9	−254	205.7
37.1	−236.1	−393	66.7	76.0	−197.2	−323	136.7	114.9	−158.3	−253	206.7
37.6	−235.6	−392	67.7	76.5	−196.7	−322	137.7	115.4	−157.8	−252	207.7
38.2	−235.0	−391	68.7	77.1	−196.1	−321	138.7	116.0	−157.2	−251	208.7

Table B-2 (continued)

KELVIN K	CELSIUS C	FAHRENHEIT F	RANKINE R	KELVIN K	CELSIUS C	FAHRENHEIT F	RANKINE R	KELVIN K	CELSIUS C	FAHRENHEIT F	RANKINE R
116.5	−156.7	−250	209.7	156.5	−116.7	−178	281.7	196.5	−76.7	−106	353.7
117.1	−156.1	−249	210.7	157.1	−116.1	−177	282.7	197.1	−76.1	−105	354.7
117.6	−155.6	−248	211.7	157.6	−115.6	−176	283.7	197.6	−75.6	−104	355.7
118.2	−155.0	−247	212.7	158.2	−115.0	−175	284.7	198.2	−75.0	−103	356.7
118.8	−154.4	−246	213.7	158.8	−114.4	−174	285.7	198.8	−74.4	−102	357.7
119.3	−153.9	−245	214.7	159.3	−113.9	−173	286.7	199.3	−73.9	−101	358.7
119.9	−153.3	−244	215.7	159.9	−113.3	−172	287.7	199.9	−73.3	−100	359.7
120.4	−152.8	−243	216.7	160.4	−112.8	−171	288.7	200.4	−72.8	−99	360.7
121.0	−152.2	−242	217.7	161.0	−112.2	−170	289.7	201.0	−72.2	−98	361.7
121.5	−151.7	−241	218.7	161.5	−111.7	−169	290.7	201.5	−71.7	−97	362.7
122.1	−151.1	−240	219.7	162.1	−111.1	−168	291.7	202.1	−71.1	−96	363.7
122.6	−150.6	−239	220.7	162.6	−110.6	−167	292.7	202.6	−70.6	−95	364.7
123.2	−150.0	−238	221.7	163.2	−110.0	−166	293.7	203.2	−70.0	−94	365.7
123.8	−149.4	−237	222.7	163.8	−109.4	−165	294.7	203.8	−69.4	−93	366.7
124.3	−148.9	−236	223.7	164.3	−108.9	−164	295.7	204.3	−68.9	−92	367.7
124.9	−148.3	−235	224.7	164.9	−108.2	−163	296.7	204.9	−68.3	−91	368.7
125.4	−147.8	−234	225.7	165.4	−107.8	−162	297.7	205.4	−67.8	−90	369.7
126.0	−147.2	−233	226.7	166.0	−107.2	−161	298.7	206.0	−67.2	−89	370.7
126.5	−146.7	−232	227.7	166.5	−106.7	−160	299.7	206.5	−66.7	−88	371.7
127.1	−146.1	−231	228.7	167.1	−106.1	−159	300.7	207.1	−66.1	−87	372.7
127.6	−145.6	−230	229.7	167.6	−105.6	−158	301.7	207.6	−65.6	−86	373.7
128.2	−145.0	−229	230.7	168.2	−105.0	−157	302.7	208.2	−65.0	−85	374.7
128.8	−144.4	−228	231.7	168.8	−104.4	−156	303.7	208.8	−64.4	−84	375.7
129.3	−143.9	−227	232.7	169.3	−103.9	−155	304.7	209.3	−63.9	−83	376.7
129.9	−143.3	−226	233.7	169.9	−103.3	−154	305.7	209.9	−63.3	−82	377.7
130.4	−142.8	−225	234.7	170.4	−102.8	−153	306.7	210.4	−62.8	−81	378.7
131.0	−142.2	−224	235.7	171.0	−102.2	−152	307.7	211.0	−62.2	−80	379.7
131.5	−141.7	−223	236.7	171.5	−101.7	−151	308.7	211.5	−61.7	−79	380.7
132.1	−141.1	−222	237.7	172.1	−101.1	−150	309.7	212.1	−61.1	−78	381.7
132.6	−140.6	−221	238.7	172,6	−100.6	−149	310.7	212.6	−60.6	−77	382.7
133.2	−140.0	−220	239.7	173.2	−100.0	−148	311.7	213.2	−60.0	−76	383.7
133.8	−139.4	−219	240.7	173.8	−99.4	−147	312.7	213.8	−59.4	−75	384.7
134.3	−138.9	−218	241.7	174.3	−98.9	−146	313.7	214.3	−58.9	−74	385.7
134.9	−138.3	−217	242.7	174.9	−98.3	−145	314.7	214.9	−58.3	−73	386.7
135.4	−137.8	−216	243.7	175.4	−97.8	−144	315.7	215.4	−57.8	−72	387.7
136.0	−137.2	−215	244.7	176.0	−97.2	−143	316.7	216.0	−57.2	−71	388.7
136.5	−136.7	−214	245.7	176.5	−96.7	−142	317.7	216.5	−56.7	−70	389.7
137.1	−136.1	−213	246.7	177.1	−96.1	−141	318.7	217.1	−56.1	−69	390.7
137.6	−135.6	−212	247.7	177.6	−95.6	−140	319.7	217.6	−55.6	−68	391.7
138.2	−135.0	−211	248.7	178.2	−95.0	−139	320.7	218.2	−55.0	−67	392.7
138.8	−134.4	−210	249.7	178.8	−94.4	−138	321.7	218.8	−54.4	−66	393.7
139.3	−133.9	−209	250.7	179.3	−93.9	−137	322.7	219.3	−53.9	−65	394.7
139.9	−133.3	−208	251.7	179.9	−93.3	−136	323.7	219.9	−53.3	−64	395.7
140.4	−132.8	−207	252.7	180.4	−92.8	−135	324.7	220.4	−52.8	−63	396.7
141.0	−132.2	−206	253.7	181.0	−92.2	−134	325.7	221.0	−52.2	−62	397.7
141.5	−131.7	−205	254.7	181.5	−91.7	−133	326.7	221.5	−51.7	−61	398.7
142.1	−131.1	−204	255.7	182.1	−91.1	−132	327.7	222.1	−51.1	−60	399.7
142.6	−130.6	−203	256.7	182.6	−90.6	−131	328.7	222.6	−50.6	−59	400.7
143.2	−130.0	−202	257.7	183.2	−90.0	−130	329.7	223.2	−50.0	−58	401.7
143.8	−129.4	−201	258.7	183.8	−89.4	−129	330.7	223.8	−49.4	−57	402.7
144.3	−128.9	−200	259.7	184.3	−88.9	−128	331.7	224.3	−48.9	−56	403.7
144.9	−128.3	−199	260.7	184.9	−88.3	−127	332.7	224.9	−48.3	−55	404.7
145.4	−127.8	−198	261.7	185.4	−87.8	−126	333.7	225.4	−47.8	−54	405.7
146.0	−127.2	−197	262.7	186.0	−87.2	−125	334.7	226.0	−47.2	−53	406.7
146.5	−126.7	−196	263.7	186.5	−86.7	−124	335.7	226.5	−46.7	−52	407.7
147.1	−126.1	−195	264.7	187.1	−86.1	−123	336.7	227.1	−46.1	−51	408.7
147.6	−125.6	−194	265.7	187.6	−85.6	−122	337.7	227.6	−45.6	−50	409.7
148.2	−125.0	−193	266.7	188.2	−85.0	−121	338.7	228.2	−45.0	−49	410.7
148.8	−124.4	−192	267.7	188.8	−84.4	−120	339.7	228.8	−44.4	−48	411.7
149.3	−123.9	−191	268.7	189.3	−83.9	−119	340.7	229.3	−43.9	−47	412.7
149.9	−123.3	−190	269.7	189.9	−83.3	−118	341.7	229.9	−43.3	−46	413.7
150.4	−122.8	−189	270.7	190.4	−82.8	−117	342.7	230.4	−42.8	−45	414.7
151.0	−122.2	−188	271.7	191.0	−82.2	−116	343.7	231.0	−42.2	−44	415.7
151.5	−121.7	−187	272.7	191.5	−81.7	−115	344.7	231.5	−41.7	−43	416.7
152.1	−121.1	−186	273.7	192.1	−81.1	−114	345.7	232.1	−41.1	−42	417.7
152.6	−120.6	−185	274.7	192.6	−80.6	−113	346.7	232.6	−40.6	−41	418.7
153.2	−120.0	−184	275.7	193.2	−80.0	−112	347.7	233.2	−40.0	−40	419.7
153.8	−119.4	−183	276.7	193.8	−79.4	−111	348.7	233.8	−39.4	−39	420.7
154.3	−118.9	−182	277.7	194.3	−78.9	−110	349.7	234.3	−38.9	−38	421.7
154.9	−118.3	−181	278.7	194.9	−78.3	−109	350.7	234.9	−38.3	−37	422.7
155.4	−117.8	−180	279.7	195.4	−77.8	−108	351.7	235.4	−37.8	−36	423.7
156.0	−117.2	−179	280.7	196.0	−77.2	−107	352.7	236.0	−37.2	−35	424.7

Table B-2 Temperature conversion (continued)

KELVIN K	CELSIUS C	FAHRENHEIT F	RANKINE R	KELVIN K	CELSIUS C	FAHRENHEIT F	RANKINE R	KELVIN K	CELSIUS C	FAHRENHEIT F	RANKINE R
236.5	-36.7	-34	425.7	276.5	3.3	38	497.7	316.5	43.3	110	569.7
237.1	-36.1	-33	426.7	277.1	3.9	39	498.7	317.1	43.9	111	570.7
237.6	-35.6	-32	427.7	277.6	4.4	40	499.7	317.6	44.4	112	571.7
238.2	-35.0	-31	428.7	278.2	5.0	41	500.7	318.2	45.0	113	572.7
238.8	-34.4	-30	429.7	278.8	5.6	42	501.7	318.8	45.6	114	573.7
239.3	-33.9	-29	430.7	279.3	6.1	43	502.7	319.3	46.1	115	574.7
239.9	-33.3	-28	431.7	279.9	6.7	44	503.7	319.9	46.7	116	575.7
240.4	-32.8	-27	432.7	280.4	7.2	45	504.7	320.4	47.2	117	576.7
241.0	-32.2	-26	433.7	281.0	7.8	46	505.7	321.0	47.8	118	577.7
241.5	-31.7	-25	434.7	281.5	8.3	47	506.7	321.5	48.3	119	578.7
242.1	-31.1	-24	435.7	282.1	8.9	48	507.7	322.1	48.9	120	579.7
242.6	-30.6	-23	436.7	282.6	9.4	49	508.7	322.6	49.4	121	580.7
243.2	-30.0	-22	437.7	283.2	10.0	50	509.7	323.2	50.0	122	581.7
243.8	-29.4	-21	438.7	283.8	10.6	51	510.7	323.8	50.6	123	582.7
244.3	-28.9	-20	439.7	284.3	11.1	52	511.7	324.3	51.1	124	583.7
244.9	-28.3	-19	440.7	284.9	11.7	53	512.7	324.9	51.7	125	584.7
245.4	-27.8	-18	441.7	285.4	12.2	54	513.7	325.4	52.2	126	585.7
246.0	-27.2	-17	442.7	286.0	12.8	55	514.7	326.0	52.8	127	586.7
246 5	-26.7	-16	443.7	286.5	13.3	56	515.7	326.5	53.3	128	587.7
247.1	-26.1	-15	444.7	287.1	13.9	57	516.7	327.1	53.9	129	588.7
247.6	-25.6	-14	445.7	287.6	14.4	58	517.7	327.6	54.4	130	589.7
248.2	-25.0	-13	446.7	288.2	15.0	59	518.7	328.2	55.0	131	590.7
248.8	-24.4	-12	447.7	288.8	15.6	60	519.7	328.8	55.6	132	591.7
249.3	-23.9	-11	448.7	289.3	16.1	61	520.7	329.3	56.1	133	592.7
249.9	-23.3	-10	449.7	289.9	16.7	62	521.7	329.9	56.7	134	593.7
250.4	-22.8	-9	450.7	290.4	17.2	63	522.7	330.4	57.2	135	594.7
251.0	-22.2	-8	451.7	291.0	17.8	64	523.7	331.0	57.8	136	595.7
251.5	-21.7	-7	452.7	291.5	18.3	65	524.7	331.5	58.3	137	596.7
252.1	-21.1	-6	453.7	292.1	18.9	66	525.7	332.1	58.9	138	597.7
252.6	-20.6	-5	454.7	292.6	19.4	67	526.7	332.6	59.4	139	598.7
253.2	-20.0	-4	455.7	293.2	20.0	68	527.7	333.2	60.0	140	599.7
253.8	-19.4	-3	456.7	293.8	20.6	69	528.7	333.8	60.6	141	600.7
254.3	-18.9	-2	457.7	294.3	21.1	70	529.7	334.3	61.1	142	601.7
254.9	-18.3	-1	458.7	294.9	21.7	71	530.7	334.9	61.7	143	602.7
255.4	-17.8	0	459.7	295.4	22.2	72	531.7	335.4	62.2	144	603.7
256.0	-17.2	1	460.7	296.0	22.8	73	532.7	336.0	62.8	145	604.7
256.5	-16.7	2	461.7	296.5	23.3	74	533.7	336.5	63.3	146	605.7
257.1	-16.1	3	462.7	297.1	23.9	75	534.7	337.1	63.9	147	606.7
257.6	-15.6	4	463.7	297.6	24.4	76	535.7	337.6	64.4	148	607.7
258.2	-15.0	5	464.7	298.2	25.0	77	536.7	338.2	65.0	149	608.7
258.8	-14.4	6	465.7	298.8	25.6	78	537.7	338.8	65.6	150	609.7
259.3	-13.9	7	466.7	299.3	26.1	79	538.7	339.3	66.1	151	610.7
259.9	-13.3	8	467.7	299.9	26.7	80	539.7	339.9	66.7	152	611.7
260.4	-12.8	9	468.7	300.4	27.2	81	540.7	340.4	67.2	153	612.7
261.0	-12.2	10	469.7	301.0	27.8	82	541.7	341.0	67.8	154	613.7
261.5	-11.7	11	470.7	301.5	28.3	83	542.7	341.5	68.3	155	614.7
262.1	-11.1	12	471.7	302.1	28.9	84	543.7	342.1	68.9	156	615.7
262.6	-10.6	13	472.7	302.6	29.4	85	544.7	342.6	69.4	157	616.7
263.2	-10.0	14	473.7	303.2	30.0	86	545.7	343.2	70.0	158	617.7
263.8	-9.4	15	474.7	303.8	30.6	87	546.7	343.8	70.6	159	618.7
264.3	-8.9	16	475.7	304.3	31.1	88	547.7	344.3	71.1	160	619,7
264.9	-8.3	17	476.7	304.9	31.7	89	548.7	344.9	71.7	161	620.7
265.4	-7.8	18	477.7	305.4	32.2	90	549.7	345.4	72.2	162	621.7
266.0	-7.2	19	478.7	306.0	32.8	91	550.7	346.0	72.8	163	622.7
266.5	-6.7	20	479.7	306.5	33.3	92	551.7	346.5	73.3	164	623.7
267.1	-6.1	21	480.7	307.1	33.9	93	552.7	347.1	73.9	165	624.7
267.6	-5.6	22	481.7	307.6	34.4	94	553.7	347.6	74.4	166	625.7
268.2	-5.0	23	482.7	308.2	35.0	95	554.7	348.2	75.0	167	626.7
268.8	-4.4	24	483.7	308.8	35.6	96	555.7	348.8	75.6	168	627.7
269.3	-3.9	25	484.7	309.3	36.1	97	556.7	349.3	76.1	169	628.7
269.9	-3.3	26	485.7	309.9	36.7	98	557.7	349.9	76.7	170	629.7
270.4	-2.8	27	486.7	310.4	37.2	99	558.7	350.4	77.2	171	630.7
271.0	-2.2	28	487.7	311.0	37.8	100	559.7	351.0	77.8	172	631.7
271.5	-1.7	29	488.7	311.5	38.3	101	560.7	351.5	78.3	173	632.7
272.1	-1.1	30	489.7	312.1	38.9	102	561.7	352.1	78.9	174	633.7
272.6	-0.6	31	490.7	312.6	39.4	103	562.7	352.6	79.4	175	634.7
273.2	0	32	491.7	313.2	40.0	104	563.7	353.2	80.0	176	635.7
273.8	0.6	33	492.7	313.8	40.6	105	564.7	353.8	80.6	177	636.7
274.3	1.1	34	493.7	314.3	41.1	106	565.7	354.3	81.1	178	637.7
274.9	1.7	35	494.7	314.9	41.7	107	566.7	354.9	81.7	179	638.7
275.4	2.2	36	495.7	315.4	42.2	108	567.7	355.4	82.2	180	639.7
276.0	2.8	37	496.7	316.0	42.8	109	568.7	356.0	82.8	181	640.7

Table B-2 (continued)

KELVIN K	CELSIUS C	FAHRENHEIT F	RANKINE R	KELVIN K	CELSIUS C	FAHRENHEIT F	RANKINE R	KELVIN K	CELSIUS C	FAHRENHEIT F	RANKINE R
356.5	83.3	182	641.7	594	321	610	1070	983	710	1310	1770
357.1	83.9	183	642.7	600	327	620	1080	989	716	1320	1780
357.6	84.4	184	643.7	605	332	630	1090	994	721	1330	1790
358.2	85.0	185	644.7	611	338	640	1100	1000	727	1340	1800
358.8	85.6	186	645.7	616	343	650	1110	1005	732	1350	1810
359.3	86.1	187	646.7	622	349	660	1120	1011	738	1360	1820
359.9	86.7	188	647.7	627	354	670	1130	1016	743	1370	1830
360.4	87.2	189	648.7	633	360	680	1140	1022	749	1380	1840
361.0	87.8	190	649.7	639	366	690	1150	1027	754	1390	1850
361.5	88.3	191	650.7	644	371	700	1160	1033	760	1400	1860
362.1	88.9	192	651.7	650	377	710	1170	1039	766	1410	1870
362.6	89.4	193	652.7	655	382	720	1180	1044	771	1420	1880
363.2	90.0	194	653.7	661	388	730	1190	1050	777	1430	1890
363.8	90.6	195	654.7	666	393	740	1200	1055	782	1440	1900
364.3	91.1	196	655.7	672	399	750	1210	1061	788	1450	1910
364.9	91.7	197	656.7	677	404	760	1220	1066	793	1460	1920
365.4	92.2	198	657.7	683	410	770	1230	1072	799	1470	1930
366.0	92.8	199	658.7	689	416	780	1240	1077	804	1480	1940
366.5	93.3	200	659.7	694	421	790	1250	1083	810	1490	1950
367.1	93.9	201	660.7	700	427	800	1260	1089	816	1500	1960
367.6	94.4	202	661.7	705	432	810	1270	1094	821	1510	1970
368.2	95.0	203	662.7	711	438	820	1280	1100	827	1520	1980
368.8	95.6	204	663.7	716	443	830	1290	1105	832	1530	1990
369.3	96.1	205	664.7	722	449	840	1300	1111	838	1540	2000
369.9	96.7	206	665.7	727	454	850	1310	1116	843	1550	2010
370.4	97.2	207	666.7	733	460	860	1320	1122	849	1560	2020
371.0	97.8	208	667.7	739	466	870	1330	1127	854	1570	2030
371.5	98.3	209	668.7	744	471	880	1340	1133	860	1580	2040
372.1	98.9	210	669.7	750	477	890	1350	1139	866	1590	2050
372.6	99.4	211	670.7	755	482	900	1360	1144	871	1600	2060
373.2	100.0	212	671.7	761	488	910	1370	1150	877	1610	2070
377	104	220	680	766	493	920	1380	1155	882	1620	2080
383	110	230	690	772	499	930	1390	1161	888	1630	2090
389	116	240	700	777	504	940	1400	1166	893	1640	2100
394	121	250	710	783	510	950	1410	1172	899	1650	2110
400	127	260	720	789	516	960	1420	1177	904	1660	2120
405	132	270	730	794	521	970	1430	1183	910	1670	2130
411	138	280	740	800	527	980	1440	1189	916	1680	2140
416	143	290	750	805	532	990	1450	1194	921	1690	2150
422	149	300	760	811	538	1000	1460	1200	927	1700	2160
427	154	310	770	816	543	1010	1470	1205	932	1710	2170
433	160	320	780	822	549	1020	1480	1211	938	1720	2180
439	166	330	790	827	554	1030	1490	1216	943	1730	2190
444	171	340	800	833	560	1040	1500	1222	949	1740	2200
450	177	350	810	839	566	1050	1510	1227	954	1750	2210
455	182	360	820	844	571	1060	1520	1233	960	1760	2220
461	188	370	830	850	577	1070	1530	1239	966	1770	2230
466	193	380	840	855	582	1080	1540	1244	971	1780	2240
472	199	390	850	861	588	1090	1550	1250	977	1790	2250
477	204	400	860	866	593	1100	1560	1255	982	1800	2260
483	210	410	870	872	599	1110	1570	1261	988	1810	2270
489	216	420	880	877	604	1120	1580	1266	993	1820	2280
494	221	430	890	883	610	1130	1590	1272	999	1830	2290
500	227	440	900	889	616	1140	1600	1277	1004	1840	2300
505	232	450	910	894	621	1150	1610	1283	1010	1850	2310
511	238	460	920	900	627	1160	1620	1289	1016	1860	2320
516	243	470	930	905	632	1170	1630	1294	1021	1870	2330
522	249	480	940	911	638	1180	1640	1300	1027	1880	2340
527	254	490	950	916	643	1190	1650	1305	1032	1890	2350
533	260	500	960	922	649	1200	1660	1311	1038	1900	2360
539	266	510	970	927	654	1210	1670	1316	1043	1910	2370
544	271	520	980	933	660	1220	1680	1322	1049	1920	2380
550	277	530	990	939	666	1230	1690	1327	1054	1930	2390
555	282	540	1000	944	671	1240	1700	1333	1060	1940	2400
561	288	550	1010	950	677	1250	1710	1339	1066	1950	2410
566	293	560	1020	955	682	1260	1720	1344	1071	1960	2420
572	299	570	1030	961	688	1270	1730	1350	1077	1970	2430
577	304	580	1040	966	693	1280	1740	1355	1082	1980	2440
583	310	590	1050	972	699	1290	1750	1361	1088	1990	2450
589	316	600	1060	977	704	1300	1760	1366	1093	2000	2460

Table B-2a Temperature conversion

SCALE LISTED IN EVEN NUMBERS

KELVIN K	CELSIUS C	FAHRENHEIT F	RANKINE R	KELVIN K	CELSIUS C	FAHRENHEIT F	RANKINE R
0	−273.16	−459.60	0	77.16	−196	−320.8	138.9
1.16	−272	−457.6	2.1	78.16	−195	−319.0	140.7
2.16	−271	−455.8	3.9	79.16	−194	−317.2	142.5
3.16	−270	−454.0	5.7	80.16	−193	−315.4	144.3
4.16	−269	−452.2	7.5	81.16	−192	−313.6	146.1
5.16	−268	−450.4	9.3	82.16	−191	−311.8	147.9
6.16	−267	−448.6	11.1	83.16	−190	−310.0	149.7
7.16	−266	−446.8	12.9	84.16	−189	−308.2	151.5
8.16	−265	−445.0	14.7	85.16	−188	−306.4	153.3
9.16	−264	−443.2	16.5	86.16	−187	−304.6	155.1
10.16	−263	−441.4	18.3	87.16	−186	−302.8	156.9
11.16	−262	−439.6	20.1	88.16	−185	−301.0	158.7
12.16	−261	−437.8	21.9	88.16	−185	−301.0	158.7
13.16	−260	−436.0	23.7	89.16	−184	−299.2	160.5
14.16	−259	−434.2	25.5	90.16	−183	−297.4	162.3
15.16	−258	−432.4	27.3	91.16	−182	−295.6	164.1
16.16	−257	−430.6	29.1	92.16	−181	−293.8	165.9
17.16	−256	−428.8	30.9	93.16	−180	−292.0	167.7
18.16	−255	−427.0	32.7	94.16	−179	−290.2	169.5
19.16	−254	−425.2	34.5	95.16	−178	−288.4	171.3
20.16	−253	−423.4	36.3	96.16	−177	−286.6	173.1
21.16	−252	−421.6	38.1	97.16	−176	−284.8	174.9
22.16	−251	−419.8	39.9	98.16	−175	−283.0	176.7
23.16	−250	−418.0	41.7	99.16	−174	−281.2	178.5
24.16	−249	−416.2	43.5	100.16	−173	−279.4	180.3
25.16	−248	−414.4	45.3	101.16	−172	−277.6	182.1
26.16	−247	−412.6	47.1	102.16	−171	−275.8	183.9
27.16	−246	−410.8	48.9	103.16	−170	−274.0	185.7
28.16	−245	−409.0	50.7	104.16	−169	−272.2	187.5
29.16	−244	−407.2	52.5	105.16	−168	−270.4	189.3
30.16	−243	−405.4	54.3	106.16	−167	−268.6	191.1
31.16	−242	−403.6	56.1	107.16	−166	−266.8	192.9
32.16	−241	−401.8	57.9	108.16	−165	−265.0	194.7
33.16	−240	−400.0	59.7	109.16	−164	−263.2	196.5
34.16	−239	−398.2	61.5	110.16	−163	−261.4	198.3
35.16	−238	−396.4	63.3	111.16	−162	−259.6	200.1
36.16	−237	−394.6	65.1	112.16	−161	−257.8	201.9
37.16	−236	−392.8	66.9	113.16	−160	−256.0	203.7
38.16	−235	−391.0	68.7	114.16	−159	−254.2	205.5
39.16	−234	−389.2	70.5	115.16	−158	−252.4	207.3
40.16	−233	−387.4	72.3	116.16	−157	−250.6	209.1
41.16	−232	−385.6	74.1	117.16	−156	−248.8	210.9
42.16	−231	−383.8	75.9	118.16	−155	−247.0	212.7
43.16	−230	−382.0	77.7	119.16	−154	−245.2	214.5
44.16	−229	−380.2	79.5	120.16	−153	−243.4	216.3
45.16	−228	−378.4	81.3	121.16	−152	−241.6	218.1
46.16	−227	−376.6	83.1	122.16	−151	−239.8	219.9
47.16	−226	−374.8	84.9	123.16	−150	−238.0	221.7
48.16	−225	−373.0	86.7	124.16	−149	−236.2	223.5
49.16	−224	−371.2	88.5	125.16	−148	−234.4	225.3
50.16	−223	−369.4	90.3	126.16	−147	−232.6	227.1
51.16	−222	−367.6	92.1	127.16	−146	−230.8	228.9
52.16	−221	−365.8	93.9	128.16	−145	−229.0	230.7
53.16	−220	−364.0	95.7	129.16	−144	−227.2	232.5
54.16	−219	−362.2	97.5	130.16	−143	−225.4	243.3
55.16	−218	−360.4	99.3	131.16	−142	−223.6	236.1
56.16	−217	−358.6	101.1	132.16	−141	−221.8	237.9
57.16	−216	−356.8	102.9	133.16	−140	−220.0	239.7
58.16	−215	−355.0	104.7	134.16	−139	−218.2	241.5
59.16	−214	−353.2	106.5	135.16	−138	−216.4	243.3
60.16	−213	−351.4	108.3	136.16	−137	−214.6	245.1
61.16	−212	−349.6	110.1	137.16	−136	−212.8	246.9
62.16	−211	−347.8	111.9	138.16	−135	−211.0	248.7
63.16	−210	−346.0	113.7	139.16	−134	−209.2	250.5
64.16	−209	−344.2	115.5	140.16	−133	−207.4	252.3
65.16	−208	−342.4	117.3	141.16	−132	−205.6	254.1
66.16	−207	−340.6	119.1	142.16	−131	−203.8	255.9
67.16	−206	−338.8	120.9	143.16	−130	−202.0	257.7
68.16	−205	−337.0	122.7	144.16	−129	−200.2	259.5
69.16	−204	−335.2	124.5	145.16	−128	−198.4	261.3
70.16	−203	−333.4	126.3	146.16	−127	−196.6	263.1
71.16	−202	−331.6	128.1	147.16	−126	−194.8	264.9
72.16	−201	−329.8	129.9	148.16	−125	−193.0	266.7
73.16	−200	−328.0	131.7	149.16	−124	−191.2	268.5
74.16	−199	−326.2	133.5	150.16	−123	−189.4	270.3
75.16	−198	−324.4	135.3	151.16	−122	−187.6	272.1
76.16	−197	−322.6	137.1	152.16	−121	−185.8	273.9

Table B-2a (continued)

KELVIN K	CELSIUS C	FAHRENHEIT F	RANKINE R	KELVIN K	CELSIUS C	FAHRENHEIT F	RANKINE R
153.16	−120	−184.0	275.7	231.16	−42	−43.6	416.1
154.16	−119	−182.2	277.5	232.16	−41	−41.8	417.9
155.16	−118	−180.4	279.3	233.16	−40	−40.0	419.7
156.16	−117	−178.6	281.1	234.16	−39	−38.2	421.5
157.16	−116	−176.8	282.9	235.16	−38	−36.4	423.3
158.16	−115	−175.0	284.7	236.16	−37	−34.6	425.1
159.16	−114	−173.2	286.5	237.16	−36	−32.8	426.9
160.16	−113	−171.4	288.3	238.16	−35	−31.0	428.7
161.16	−112	−169.6	290.1	239.16	−34	−29.2	430.5
162.16	−111	−167.8	291.9	240.16	−33	−27.4	432.3
163.16	−110	−166.0	293.7	241.16	−32	−25.6	434.1
164.16	−109	−164.2	295.5	242.16	−31	−23.8	435.9
165.16	−108	−162.4	297.3	243.16	−30	−22.0	437.7
166.16	−107	−160.6	299.1	244.16	−29	−20.2	439.5
167.16	−106	−158.8	300.9	245.16	−28	−18.4	441.3
168.16	−105	−157.0	302.7	246.16	−27	−16.6	443.1
169.16	−104	−155.2	304.5	247.16	−26	−14.8	444.9
170.16	−103	−153.4	306.3	248.16	−25	−13.0	446.7
171.16	−102	−151.6	308.1	249.16	−24	−11.2	448.5
172.16	−101	−149.8	309.9	250.16	−23	−9.4	450.3
173.16	−100	−148.0	311.7	251.16	−22	−7.6	452.1
173.16	−100	−148.0	311.7	252.16	−21	−5.8	453.9
174.16	−99	−146.2	313.5	253.16	−20	−4.0	455.7
175.16	−98	−144.4	315.3	254.16	−19	−2.2	457.5
176.16	−97	−142.6	317.1	255.16	−18	−0.4	459.3
177.16	−96	−140.8	318.9	256.16	−17	1.4	461.1
178.16	−95	−139.0	320.7	257.16	−16	3.2	462.9
179.16	−94	−137.2	322.5	258.16	−15	5.0	464.7
180.16	−93	−135.4	324.3	258.16	−15	5.0	464.7
181.16	−92	−133.6	326.1	259.16	−14	6.8	466.5
182.16	−91	−131.8	327.9	260.16	−13	8.6	468.3
183.16	−90	−130.0	329.7	261.16	−12	10.4	470.1
184.16	−89	−128.2	331.5	262.16	−11	12.2	471.9
185.16	−88	−126.4	333.3	263.16	−10	14.0	473.7
186.16	−87	−124.6	335.1	264.16	−9	15.8	475.5
187.16	−86	−122.8	336.9	265.16	−8	17.6	477.3
188.16	−85	−121.0	338.7	266.16	−7	19.4	479.1
189.16	−84	−119.2	340.5	267.16	−6	21.2	480.9
190.16	−83	−117.4	342.3	268.16	−5	23.0	482.7
191.16	−82	−115.6	344.1	269.16	−4	24.8	484.5
192.16	−81	−113.8	345.9	270.16	−3	26.6	486.3
193.16	−80	−112.0	347.7	271.16	−2	28.4	488.1
194.16	−79	−110.2	349.5	272.16	−1	30.2	489.9
195.16	−78	−108.4	351.3	273.16	0	32.0	491.7
196.16	−77	−106.6	353.1	274.16	1	33.8	493.5
197.16	−76	−104.8	354.9	275.16	2	35.6	495.3
198.16	−75	−103.0	356.7	276.16	3	37.4	497.1
199.16	−74	−101.2	358.5	277.16	4	39.2	498.9
200.16	−73	−99.4	360.3	278.16	5	41.0	500.7
201.16	−72	−97.6	362.1	279.16	6	42.8	502.5
202.16	−71	−95.8	363.9	280.16	7	44.6	504.3
203.16	−70	−94.0	365.7	281.16	8	46.4	506.1
204.16	−69	−92.2	367.5	282.16	9	48.2	507.9
205.16	−68	−90.4	369.3	283.16	10	50.0	509.7
206.16	−67	−88.6	371.1	284.16	11	51.8	511.5
207.16	−66	−86.8	372.9	285.16	12	53.6	513.3
208.16	−65	−85.0	374.7	286.16	13	55.4	515.1
209.16	−64	−83.2	376.5	287.16	14	57.2	516.9
210.16	−63	−81.4	378.3	288.16	15	59.0	518.7
211.16	−62	−79.6	380.1	289.16	16	60.8	520.5
212.16	−61	−77.8	381.9	290.16	17	62.6	522.3
213.16	−60	−76.0	383.7	291.16	18	64.4	524.1
214.16	−59	−74.2	385.5	292.16	19	66.2	525.9
215.16	−58	−72.4	387.3	293.16	20	68.0	527.7
216.16	−57	−70.6	389.1	294.16	21	69.8	529.5
217.16	−56	−68.8	390.9	295.16	22	71.6	531.3
218.16	−55	−67.0	392.7	296.16	23	73.4	533.1
219.16	−54	−65.2	394.5	297.16	24	75.2	534.9
220.16	−53	−63.4	396.3	298.16	25	77.0	536.7
221.16	−52	−61.6	398.1	299.16	26	78.8	538.5
222.16	−51	−59.8	399.9	300.16	27	80.6	540.3
223.16	−50	−58.0	401.7	301.16	28	82.4	542.1
224.16	−49	−56.2	403.5	302.16	29	84.2	543.9
225.16	−48	−54.4	405.3	303.16	30	86.0	545.7
226.16	−47	−52.6	407.1	304.16	31	87.8	547.5
227.16	−46	−50.8	408.9	305.16	32	89.6	549.3
228.16	−45	−49.0	410.7	306.16	33	91.4	551.1
229.16	−44	−47.2	412.5	307.16	34	93.2	552.9
230.16	−43	−45.4	414.3	308.16	35	95.0	554.7

Table B-2a Temperature conversion (continued)

CELSIUS				CELSIUS			
KELVIN K	C	FARENHEIT F	RANKINE R	KELVIN K	C	FARENHEIT F	RANKINE R
309.16	36	96.8	556.5	513	240	464	923
310.16	37	98.6	558.3	523	250	482	941
311.16	38	100.4	560.1	533	260	500	959
312.16	39	102.2	561.9	543	270	518	977
313.16	40	104.0	563.7	553	280	536	995
314.16	41	105.8	565.5	563	290	554	1013
315.16	42	107.6	567.3	573	300	572	1031
316.16	43	109.4	569.1	583	310	590	1049
317.16	44	111.2	570.9	593	320	608	1067
318.16	45	113.0	572.7	603	330	626	1085
319.16	46	114.8	574.5	613	340	644	1103
320.16	47	116.6	576.3	623	350	662	1121
321.16	48	118.4	578.1	633	360	680	1139
322.16	49	120.2	579.9	643	370	698	1157
323.16	50	122.0	581.7	653	380	716	1175
324.16	51	123.8	583.5	663	390	734	1193
325.16	52	125.6	585.3	673	400	752	1211
326.16	53	127.4	587.1	683	410	770	1229
327.16	54	129.2	588.9	693	420	788	1247
328.16	55	131.0	590.7	703	430	806	1265
329.16	56	132.8	592.5	713	440	824	1283
330.16	57	134.6	594.3	723	450	842	1301
331.16	58	136.4	596.1	733	460	860	1319
332.16	59	138.2	597.9	743	470	878	1337
333.16	60	140.0	599.7	753	480	896	1355
334.16	61	141.8	601.5	763	490	914	1373
335.16	62	143.6	603.3	773	500	932	1391
336.16	63	145.4	605.1	783	510	950	1409
337.16	64	147.2	606.9	793	520	968	1427
338.16	65	149.0	608.7	803	530	986	1445
339.16	66	150.8	610.5	813	540	1004	1463
340.16	67	152.6	612.3	823	550	1022	1481
341.16	68	154.4	614.1	833	560	1040	1499
342.16	69	156.2	615.9	843	570	1058	1517
343.16	70	158.0	617.7	853	580	1076	1535
343.16	70	158.0	617.7	863	590	1094	1553
344.16	71	159.8	619.5	873	600	1112	1571
345.16	72	161.6	621.3	883	610	1130	1589
346.16	73	163.4	623.1	893	620	1148	1607
347.16	74	165.2	624.9	903	630	1166	1625
348.16	75	167.0	626.7	913	640	1184	1643
349.16	76	168.8	628.5	923	650	1202	1661
350.16	77	170.6	630.3	923	650	1202	1661
351.16	78	172.4	632.1	933	660	1220	1679
352.16	79	174.2	633.9	943	670	1238	1697
353.16	80	176.0	635.7	953	680	1256	1715
354.16	81	177.8	637.5	963	690	1274	1733
355.16	82	179.6	639.3	973	700	1292	1751
356.16	83	181.4	641.1	983	710	1310	1769
357.16	84	183.2	642.9	993	720	1328	1787
358.16	85	185.0	644.7	1003	730	1346	1805
359.16	86	186.8	646.5	1013	740	1364	1823
360.16	87	188.6	648.3	1023	750	1382	1841
361.16	88	190.4	650.1	1033	760	1400	1859
362.16	89	192.2	651.9	1043	770	1418	1877
363.16	90	194.0	653.7	1053	780	1436	1895
364.16	91	195.8	655.5	1063	790	1454	1913
365.16	92	197.6	657.3	1073	800	1472	1931
366.16	93	199.4	659.1	1083	810	1490	1949
367.16	94	201.2	660.9	1093	820	1508	1967
368.16	95	203.0	662.7	1103	830	1526	1985
369.16	96	204.8	664.5	1113	840	1544	2003
370.16	97	206.6	666.3	1123	850	1562	2021
371.16	98	208.4	668.1	1133	860	1580	2039
372.16	99	210.2	669.9	1143	870	1598	2057
373.16	100	212.0	671.7	1153	880	1616	2075
383	110	230	689	1163	890	1634	2093
393	120	248	707	1173	900	1652	2111
403	130	266	725	1183	910	1670	2129
413	140	284	743	1193	920	1688	2147
423	150	302	761	1203	930	1706	2165
433	160	320	779	1213	940	1724	2183
443	170	338	797	1223	950	1742	2201
453	180	356	815	1233	960	1760	2219
463	190	374	833	1243	970	1778	2237
473	200	392	851	1253	980	1796	2255
483	210	410	869	1263	990	1814	2273
493	220	428	887	1273	1000	1832	2291
503	230	446	905	1283	1010	1850	2309

Table B-2a (continued)

CELSIUS				CELSIUS			
KELVIN K	C	FARENHEIT F	RANKINE R	KELVIN K	C	FARENHEIT F	RANKINE R
1293	1020	1868	2327	1543	1270	2318	2777
1303	1030	1886	2345	1553	1280	2336	2795
1313	1040	1904	2363	1563	1290	2354	2813
1323	1050	1922	2381	1573	1300	2372	2831
1333	1060	1940	2399	1583	1310	2390	2849
1343	1070	1958	2417	1593	1320	2408	2867
1353	1080	1976	2435	1603	1330	2426	2885
1363	1090	1994	2453	1613	1340	2444	2903
1373	1100	2012	2471	1623	1350	2462	2921
1383	1110	2030	2489	1633	1360	2480	2939
1393	1120	2048	2507	1643	1370	2498	2957
1403	1130	2066	2525	1653	1380	2516	2975
1413	1140	2084	2543	1663	1390	2534	2993
1423	1150	2102	2561	1673	1400	2552	3011
1433	1160	2120	2579	1683	1410	2570	3029
1443	1170	2138	2597	1693	1420	2588	3047
1453	1180	2156	2615	1703	1430	2606	3065
1463	1190	2174	2633	1713	1440	2624	3083
1473	1200	2192	2651	1723	1450	2642	3101
1483	1210	2210	2669	1733	1460	2660	3119
1493	1220	2228	2687	1743	1470	2678	3137
1503	1230	2246	2705	1753	1480	2696	3155
1513	1240	2264	2723	1763	1490	2714	3173
1523	1250	2282	2741	1773	1500	2732	3191
1533	1260	2300	2759				

Table B-2b Relationship of temperature scales

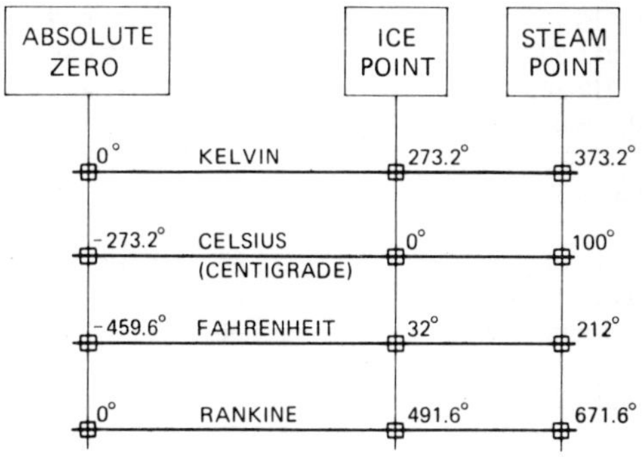

Table B-3 **Fourth Power of Numbers**

n	n^4	n	n^4	n	n^4	n	n^4
0.01	0.000 000 01	0.56	0.098 344 96	1.11	1.518 070 41	1.66	7.593 331 36
0.02	0.000 000 16	0.57	0.105 560 01	1.12	1.573 519 36	1.67	7.777 963 21
0.03	0.000 000 81	0.58	0.113 164 96	1.13	1.630 473 61	1.68	7.965 941 76
0.04	0.000 002 56	0.59	0.121 173 61	1.14	1.688 960 16	1.69	8.157 307 21
0.05	0.000 006 25	0.60	0.129 600 00	1.15	1.749 006 25	1.70	8.352 100 00
0.06	0.000 012 96	0.61	0.138 458 41	1.16	1.810 639 36	1.71	8.550 360 81
0.07	0.000 024 01	0.62	0.147 763 36	1.17	1.873 887 21	1.72	8.752 130 56
0.08	0.000 040 96	0.63	0.157 529 61	1.18	1.938 777 76	1.73	8.957 450 41
0.09	0.000 065 61	0.64	0.167 772 16	1.19	2.005 339 21	1.74	9.166 361 76
0.10	0.000 100 00	0.65	0.178 506 25	1.20	2.073 600 00	1.75	9.378 906 25
0.11	0.000 146 41	0.66	0.189 747 36	1.21	2.143 588 81	1.76	9.595 125 76
0.12	0.000 207 36	0.67	0.201 511 21	1.22	2.215 334 56	1.77	9.815 062 41
0.13	0.000 285 61	0.68	0.213 813 76	1.23	2.288 866 41	1.78	10.038 758 56
0.14	0.000 384 16	0.69	0.226 671 21	1.24	2.364 213 76	1.79	10.266 256 81
0.15	0.000 506 25	0.70	0.240 100 00	1.25	2.441 406 25	1.80	10.497 600 00
0.16	0.000 655 36	0.71	0.254 116 81	1.26	2.520 473 76	1.81	10.732 831 21
0.17	0.000 835 21	0.72	0.268 738 56	1.27	2.601 446 41	1.82	10.971 993 76
0.18	0.001 049 76	0.73	0.283 982 41	1.28	2.684 354 56	1.83	11.215 131 21
0.19	0.001 303 21	0.74	0.299 865 76	1.29	2.769 228 81	1.84	11.462 287 36
0.20	0.001 600 00	0.75	0.315 406 25	1.30	2.856 100 00	1.85	11.713 506 25
0.21	0.001 944 81	0.76	0.333 621 76	1.31	2.944 999 21	1.86	11.968 832 16
0.22	0.002 342 56	0.77	0.351 530 41	1.32	3.035 957 76	1.87	12.228 309 61
0.23	0.002 798 41	0.78	0.370 150 56	1.33	3.129 007 21	1.88	12.491 983 36
0.24	0.003 317 76	0.79	0.389 500 81	1.34	3.224 179 36	1.89	12.759 898 41
0.25	0.003 906 25	0.80	0.409 600 00	1.35	3.321 506 25	1.90	13.032 100 00
0.26	0.004 569 76	0.81	0.430 467 21	1.36	3.421 020 16	1.91	13.308 633 61
0.27	0.005 314 41	0.82	0.452 121 76	1.37	3.522 753 61	1.92	13.589 544 96
0.28	0.006 146 56	0.83	0.474 583 21	1.38	3.626 739 36	1.93	13.874 880 01
0.29	0.007 072 81	0.84	0.497 871 36	1.39	3.733 010 41	1.94	14.164 684 96
0.30	0.008 100 00	0.85	0.522 006 25	1.40	3.841 600 00	1.95	14.459 006 25
0.31	0.009 235 21	0.86	0.547 008 16	1.41	3.952 541 61	1.96	14.757 890 56
0.32	0.010 485 76	0.87	0.572 897 61	1.42	4.065 868 96	1.97	15.061 384 81
0.33	0.011 859 21	0.88	0.599 695 36	1.43	4.181 616 01	1.98	15.369 536 16
0.34	0.013 363 36	0.89	0.627 422 41	1.44	4.299 816 96	1.99	15.682 392 01
0.35	0.015 006 25	0.90	0.656 100 00	1.45	4.420 506 25	2.00	16.000 000 00
0.36	0.016 796 16	0.91	0.685 749 61	1.46	4.543 718 56	2.01	16.322 408 01
0.37	0.018 741 61	0.92	0.716 392 96	1.47	4.669 488 81	2.02	16.649 664 16
0.38	0.020 831 36	0.93	0.748 052 01	1.48	4.797 852 16	2.03	16.981 816 81
0.39	0.023 134 41	0.94	0.780 748 96	1.49	4.928 844 01	2.04	17.318 914 56
0.40	0.025 600 00	0.95	0.814 506 25	1.50	5.062 500 00	2.05	17.661 006 25
0.41	0.028 257 61	0.96	0.849 346 56	1.51	5.198 856 01	2.06	18.008 140 96
0.42	0.031 116 95	0.97	0.885 292 81	1.52	5.337 948 16	2.07	18.360 368 01
0.43	0.034 188 01	0.98	0.922 368 16	1.53	5.479 812 81	2.08	18.717 736 96
0.44	0.037 480 96	0.99	0.960 596 01	1.54	5.625 486 56	2.09	19.080 297 61
0.45	0.041 006 25	1.00	1.000 000 000	1.55	5.772 006 25	2.10	19.448 100 00
0.46	0.044 774 56	1.01	1.040 604 01	1.56	5.922 408 96	2.11	19.821 194 41
0.47	0.048 796 81	1.02	1.082 432 16	1.57	6.075 732 01	2.12	20.199 631 36
0.48	0.053 084 16	1.03	1.125 508 81	1.58	6.232 012 96	2.13	20.583 461 61
0.49	0.057 648 01	1.04	1.169 858 56	1.59	6.391 289 61	2.14	20.972 736 16
0.50	0.062 500 00	1.05	1.215 506 25	1.60	6.553 600 00	2.15	21.367 506 25
0.51	0.067 652 01	1.06	1.262 476 96	1.61	6.718 982 41	2.16	21.767 823 36
0.52	0.073 116 16	1.07	1.310 796 01	1.62	6.887 475 36	2.17	22.173 739 21
0.53	0.078 904 81	1.08	1.360 488 95	1.63	7.059 117 61	2.18	22.585 305 76
0.54	0.085 030 56	1.09	1.411 581 61	1.64	7.233 948 16	2.19	23.002 575 21
0.55	0.091 506 25	1.10	1.464 100 00	1.65	7.412 006 25	2.20	23.425 600 00

Table B-3 (Continued) **Fourth Power of Numbers**

n	n^4	n	n^4	n	n^4	n	n^4
2.21	23.854 432 81	2.76	58.027 829 76	3.31	120.036 127 21	3.86	221.998 080 16
2.22	24.289 126 56	2.77	58.873 394 41	3.32	121.493 301 76	3.87	224.307 533 61
2.23	24.729 734 41	2.78	59.728 166 56	3.33	122.963 703 21	3.88	226.634 959 36
2.24	25.176 309 76	2.79	60.592 212 81	3.34	124.447 411 36	3.89	228.980 450 41
2.25	25.628 906 25	2.80	61.465 600 00	3.35	125.944 506 25	3.90	231.344 100 00
2.26	26.087 577 76	2.81	62.398 395 21	3.36	127.455 068 16	3.91	233.726 001 61
2.27	26.552 378 41	2.82	63.240 665 16	3.37	128.979 171 61	3.92	236.126 248 96
2.28	27.023 362 56	2.83	64.142 479 21	3.38	130.516 915 36	3.93	238.544 936 01
2.29	27.500 584 81	2.84	65.053 903 36	3.39	132.068 362 41	3.94	240.982 156 96
2.30	27.984 100 00	2.85	65.975 006 25	3.40	133.633 600 00	3.95	243.438 006 25
2.31	28.473 963 21	2.86	66.905 856 16	3.41	135.212 709 61	3.96	245.912 578 56
2.32	28.970 229 76	2.87	67.846 521 61	3.42	136.805 772 96	3.97	248.405 968 81
2.33	29.472 955 21	2.88	68.797 071 36	3.43	138.412 872 01	3.98	250.918 272 16
2.34	29.982 195 36	2.89	69.757 574 41	3.44	140.034 088 96	3.99	253.449 589 01
2.35	30.498 006 25	2.90	70.728 400 00	3.45	141.669 506 25	4.00	256.000 000 00
2.36	31.020 444 16	2.91	71.708 717 61	3.46	143.319 206 56	4.01	258.569 616 01
2.37	31.549 565 61	2.92	72.699 496 96	3.47	144.983 272 81	4.02	261.158 528 16
2.38	32.085 427 36	2.93	73.700 508 01	3.48	146.661 788 16	4.03	263.766 832 81
2.39	32.628 086 41	2.94	74.711 820 96	3.49	148.359 836 01	4.04	266.394 626 56
2.40	33.177 600 00	2.95	75.733 506 25	3.50	150.062 500 00	4.05	269.042 006 25
2.41	33.734 025 61	2.96	76.765 634 56	3.51	151.784 864 01	4.06	271.709 065 96
2.42	34.297 420 96	2.97	77.808 276 81	3.52	153.522 012 16	4.07	274.395 912 01
2.43	34.867 844 01	2.98	78.861 504 16	3.53	155.274 028 81	4.08	277.102 632 96
2.44	35.445 352 96	2.99	79.925 388 01	3.54	157.040 998 56	4.09	279.829 324 61
2.45	36.030 006 25	3.00	81.000 000 00	3.55	158.823 006 25	4.10	282.576 100 00
2.46	36.621 862 56	3.01	82.085 412 01	3.56	160.620 136 96	4.11	285.343 042 41
2.47	37.220 980 81	3.02	83.181 696 16	3.57	162.432 476 01	4.12	288.130 255 36
2.48	37.827 420 16	3.03	84.288 924 81	3.58	164.260 108 96	4.13	290.937 837 61
2.49	38.441 240 01	3.04	85.407 170 56	3.59	166.103 121 61	4.14	293.765 888 16
2.50	39.062 500 00	3.05	86.536 506 25	3.60	167.961 600 00	4.15	296.614 506 25
2.51	39.691 260 01	3.06	87.677 004 96	3.61	169.835 630 41	4.16	299.483 791 36
2.52	40.327 580 16	3.07	88.828 740 01	3.62	171.725 299 36	4.17	302.373 843 21
2.53	40.971 520 81	3.08	89.991 784 96	3.63	173.630 693 61	4.18	305.284 761 76
2.54	41.623 142 56	3.09	91.166 213 61	3.64	175.551 900 16	4.19	308.216 647 21
2.55	42.282 506 25	3.10	92.352 100 00	3.65	177.489 006 25	4.20	311.169 600 00
2.56	42.949 672 96	3.11	93.549 518 41	3.66	179.442 009 36	4.21	314.143 720 81
2.57	43.624 704 01	3.12	94.758 543 36	3.67	181.411 267 21	4.22	317.139 110 56
2.58	44.307 660 96	3.13	95.979 249 61	3.68	183.396 597 96	4.23	320.155 870 41
2.59	44.998 605 61	3.14	97.211 712 16	3.69	185.398 179 21	4.24	323.194 101 76
2.60	45.697 600 00	3.15	98.456 006 25	3.70	187.416 100 00	4.25	326.253 906 25
2.61	46.404 706 41	3.16	99.712 207 36	3.71	189.450 448 81	4.26	329.335 385 76
2.62	47.119 987 36	3.17	100.980 391 21	3.72	191.501 314 56	4.27	332.438 642 41
2.63	47.843 505 61	3.18	102.260 633 76	3.73	193.568 786 41	4.28	335.563 778 56
2.64	48.575 320 16	3.19	103.553 011 21	3.74	195.652 953 76	4.29	338.710 896 81
2.65	49.315 506 25	3.20	104.857 600 00	3.75	197.753 906 25	4.30	341.880 100 00
2.66	50.064 115 36	3.21	106.174 476 81	3.76	199.871 733 76	4.31	345.071 491 21
2.67	50.821 215 21	3.22	107.503 718 56	3.77	202.006 525 91	4.32	348.285 173 76
2.68	51.586 869 76	3.23	108.845 402 91	3.78	304.158 376 56	4.33	351.521 251 21
2.69	52.361 143 21	3.24	110.199 605 76	3.79	206.327 368 81	4.34	354.779 827 36
2.70	53.144 100 00	3.25	111.566 406 25	3.80	208.513 600 00	4.35	358.061 006 25
2.71	53.935 804 81	3.26	112.945 881 16	3.81	210.717 159 21	4.36	361.364 982 16
2.72	54.736 322 56	3.27	114.338 110 41	3.82	212.938 137 76	4.37	364.691 589 61
2.73	55.545 718 41	3.28	115.743 170 56	3.83	215.176 621 2	4.38	368.041 203 36
2.74	56.364 057 16	3.29	117.161 140 81	3.84	217.432 719 36	4.39	371.413 838 41
2.75	57.191 406 25	3.30	118.592 100 00	3.85	219.706 506 25	4.40	374.809 600 00

Table B-3 (Continued)

Fourth Power of Numbers

n	n^4	n	n^4	n	n^4	n	n^4
4.41	378.228 593 61	4.96	605.238 722 56	5.51	921.735 672 01	6.06	1348.622 796 96
4.42	381.670 924 96	4.97	610.134 460 81	5.52	923.445 276 16	6.07	1357.546 656 01
4.43	385.136 700 01	4.98	615.059 840 16	5.53	935.191 444 81	6.08	1366.514 728 96
4.44	388.626 026 96	4.99	620.014 980 01	5.54	941.974 310 56	6.09	1375.527 161 61
4.45	392.139 002 5	5.00	625.000 000 00	5.55	948.794 006 25	6.10	1384.584 100 00
4.46	395.675 750 56	5.01	630.015 020 01	5.56	955.650 664 96	6.11	1393.685 690 41
4.47	399.236 364 81	5.02	635.060 160 16	5.57	962.544 420 01	6.12	1402.833 079 36
4.48	402.820 956 16	5.03	640.135 540 81	5.58	969.475 404 96	6.13	1412.023 413 61
4.49	406.429 632 01	5.04	645.241 282 56	5.59	976.443 753 61	6.14	1421.259 840 16
4.50	410.062 500 00	5.05	650.377 506 25	5.60	983.449 600 00	6.15	1430.541 506 25
4.51	413.719 663 01	5.06	655.544 332 96	5.61	990.493 078 41	6.16	1439.868 559 36
4.52	417.401 244 16	5.07	660.741 884 01	5.62	997.574 323 36	6.17	1449.241 147 21
4.53	421.107 336 81	5.08	665.970 280 96	5.63	1006.693 469 61	6.18	1458.659 417 76
4.54	424.838 054 56	5.09	671.229 645 61	5.64	1011.850 632 16	6.19	1468.123 519 21
4.55	428.593 506 25	5.10	676.520 100 00	5.65	1019.046 006 25	6.20	1477.633 600 00
4.56	432.373 800 96	5.11	681.841 766 41	5.66	1026.279 667 36	6.21	1487.189 808 81
4.57	436.179 048 01	5.12	687.194 767 36	5.67	1033.551 771 21	6.22	1496.792 294 56
4.58	440.009 356 96	5.13	692.579 225 61	5.68	1040.862 453 76	6.23	1506.441 206 41
4.59	443.864 837 61	5.14	697.995 264 16	5.69	1048.211 851 21	6.24	1516.136 693 76
4.60	447.745 600 00	5.15	703.443 006 25	5.70	1055.600 100 00	6.25	1525.878 906 25
4.61	451.651 754 41	5.16	708.922 575 36	5.71	1063.027 336 81	6.25	1535.667 993 76
4.62	455.583 411 36	5.17	713.323 095 21	5.72	1070.493 698 56	6.27	1545.504 106 41
4.63	459.540 681 61	5.18	719.977 689 76	5.73	1077.999 322 41	6.28	1555.337 394 56
4.64	463.523 676 16	5.19	725.553 483 21	5.74	1035.544 345 76	6.29	1565.318 008 81
4.65	467.532 506 25	5.20	731.161 600 00	5.75	1093.128 906 25	6.30	1575.296 100 00
4.66	471.567 283 36	5.21	736.802 164 81	5.76	1100.753 141 76	6.31	1585.321 819 21
4.67	475.628 119 21	5.22	742.475 302 56	5.77	1108.417 190 41	6.32	1595.395 317 76
4.68	479.715 125 76	5.23	748.181 138 91	5.78	1116.121 190 56	6.33	1605.516 747 21
4.69	483.828 415 2	5.24	753.919 797 76	5.79	1123.865 280 81	6.34	1615.686 259 36
4.70	487.968 100 00	5.25	759.691 406 25	5.80	1131.649 600 00	6.35	1625.904 006 25
4.71	492.134 292 81	5.26	765.496 089 76	5.81	1139.474 287 21	6.36	1626.170 140 16
4.72	496.327 106 56	5.27	771.333 974 41	5.82	1147.339 481 76	6.37	1646.484 813 61
4.73	500.546 654 41	5.28	777.205 186 56	5.83	1155.245 323 21	6.38	1656 848 179 36
4.74	504.793 049 76	5.29	783.109 852 81	5.84	1163.191 951 36	6.39	1667.260 390 41
4.75	509.066 406 25	5.30	789.048 100 00	5.85	1171.179 506 25	6.40	1677.721 600 00
4.76	513.366 837 76	5.31	795.020 055 21	5.86	1179.208 128 16	6.41	1688.231 961 61
4.77	517.694 458 41	5.32	801.025 845 76	5.87	1187.277 957 61	6.42	1698.791 623 96
4.78	522.049 382 56	5.33	807.065 599 2	5.88	1195.389 135 36	6.43	1709.400 756 01
4.79	526.431 724 81	5.34	813.139 443 36	5.89	1203.541 802 41	6.44	1720.059 496 96
4.80	530.841 600 00	5.35	819.247 506 25	5.90	1211.736 100 00	6.45	1730.768 006 25
4.81	535.279 123 21	5.36	825.389 916 16	5.91	1219.972 169 61	6.46	1741.526 438 56
4.82	539.744 409 76	5.37	831.566 801 61	5.92	1228.250 152 96	6.47	1752.334 998 81
4.83	544.237 575 21	5.38	837.778 291 36	5.93	1236.570 192 01	6.48	1763.193 692 16
4.84	548.758 735 36	5.39	844.024 514 41	5.94	1244.932 428 96	6.49	1774.102 824 01
4.85	553.308 006 25	5.40	850.305 600 00	5.95	1253.337 006 25	6.50	1785.062 500 0
4.86	557.885 504 16	5.41	856.621 677 61	5.96	1261.784 066 54	6.51	1796.072 876 01
4.87	562.491 345 61	5.42	862.972 786 96	5.97	1270.273 752 81	6.52	1807.134 108 16
4.88	567.125 647 36	5.43	869.359 328 01	5.98	1278.806 208 16	6.53	1818.246 352 81
4.89	571.788 526 41	5.44	875.781 160 96	5.99	1287.381 576 01	6.54	1829.409 766 56
4.90	576.480 100 00	5.45	882.238 506 25	6.00	1296.000 000 00	6.55	1840.624 506 25
4.91	581.200 485 61	5.46	888.731 494 56	6.01	1304.661 624 01	6.56	1851.890 728 96
4.92	585.949 800 96	5.47	895.260 256 81	6.02	1313.366 592 26	6.57	1863.208 592 01
4.93	590.728 164 01	5.48	901.824 924 16	6.03	1322.115 048 81	6.58	1874.578 252 96
4.94	595.535 692 96	5.49	908.425 628 01	6.04	1330.907 138 56	6.59	1885.990 869 61
4.95	600.372 506 25	5.50	915.062 500 00	6.05	1339.743 006 25	6.60	1897.473 600 00

Table B-3 (Continued)

Fourth Power of Numbers

n	n^4	n	n^4	n	n^4	n	n^4
6.61	1908.999 602 41	7.16	2628.161 743 36	7.71	3533.601 024 81	8.26	4655.005 401 76
6.62	1920.578 035 36	7.17	2642.874 999 21	7.72	3551.969 282 56	8.27	4677.588 770 41
6.63	1932.209 057 61	7.18	2657.649 945 76	7.73	3570.409 058 41	8.28	4700.254 210 56
6.64	1943.892 828 16	7.19	2672.486 735 21	7.74	3588.920 537 76	8.29	4723.001 920 81
6.65	1955.629 506 25	7.20	2687.385 600 00	7.75	3607.503 906 25	8.30	4745.832 100 00
6.66	1967.419 251 36	7.21	2702.346 652 81	7.76	3626.159 349 76	8.31	4768.744 947 21
6.67	1979.262 223 21	7.22	2717.370 086 56	7.77	3644.887 034 41	8.32	4791.740 661 76
6.68	1991.158 581 76	7.23	2732.456 074 41	7.78	3663.687 206 56	8.33	4814.819 443.21
6.69	2003.108 487 21	7.24	2747.604 789 76	7.79	3682.559 992 81	8.34	4837.981 491 36
6.70	2015.112 100 00	7.25	2762.816 406 25	7.80	3701.505 600 00	8.35	4861.227 006 25
6.71	2027.169 580 81	7.26	2778.091 096 76	7.81	3720.524 215 21	8.36	4884.556 188 16
6.72	2039.281 090 56	7.27	2793.429 038 41	7.82	3739.616 025 76	8.37	4907.969 237 61
6.73	2051.446 790 41	7.28	2808.830 402 56	7.83	3758.781 219 21	8.38	4931.466 358 36
6.74	2063.666 841 76	7.29	2824.295 364 81	7.84	3778.019 903 36	8.39	4955.047 742 41
6.75	2075.941 406 25	7.30	2839.824 100 00	7.85	3797.332 506 25	8.40	4978.713 600 00
6.76	2088.270 645 76	7.31	2855.416 783 21	7.86	3816.718 976 16	8.41	5002.464 129 61
6.77	2100.654 722 41	7.32	2871.073 589 76	7.87	3836.179 581 61	8.42	5026.299 532 96
6.78	2113.993 798 56	7.33	2886.794 695 21	7.88	3955.714 511 36	8.43	5050.220 012 01
6.79	2125.588 035 81	7.34	2902.580 275 36	7.89	3875.323 954 41	8.44	5074.225 768 96
6,80	2133.137 600 00	7.35	2918.430 506 25	7.90	3895.008 100 00	8.45	5098.317 006 25
6.81	7150.742 651 21	7.36	2934.345 564 16	7.91	3914.767 137 61	8.46	5122.493 926 56
6.82	2163.403 353 76	7.37	2950.325 625 61	7.92	3934.601 256.96	8.47	5146.756 732 81
6.83	2175.119 871 21	7.38	2966.370 867 36	7.93	3954.510 648 01	8.48	5171.105 623 16
6.84	2188.892 367 36	7.39	2982.481 466 91	7.94	3974.495 500 96	8.49	5195.540 816 01
6.85	2201.721 006 25	7.40	2998.657 600 00	7.95	3994.556 006 25	8.50	5220.062 500 0
6.86	2214.605 952 16	7.41	3014.899 445 61	7.96	4014.692 354 56	8.51	5244.670 884 01
6.87	2227.547 369 61	7.42	3031.207 180 96	7.97	4034.904 736 81	8.52	5269.366 172 16
6.88	2240.545 423 36	7.43	3047.580 984 01	7.98	4055.193 344 16	8.53	5294.148 568 81
6.89	2253.600 278 41	7.44	3064.021 032 96	7.99	4075.558 368 01	8.54	5319.018 278 56
6.90	2266.712 100 00	7.45	3080.527 506 25	8.00	4095.000 000 00	8.55	5343.975 506 25
6.91	2279.881 053 61	7.46	3097.100 582 56	8.01	4116.518 432 01	8.56	5369.020 456 96
6.92	2293.107 304 96	7.47	3113.740 440 81	8.02	4137.113 855 16	8.57	5394.153 336 01
6.93	2306.391 020 01	7.48	3130.447 260 16	8.03	4157.786 464 81	8.58	5419.374 348 96
6.94	2319.732 364 96	7.49	3147.221 220 01	8.04	4178.536 450 56	8.59	5444.683 701 61
6.95	2333.131 506 25	7.50	3164.062 500 00	8.05	4199.364 006 25	8.60	5470.081 600 00
6.96	2345.588 610 56	7.51	3180.971 280 01	8.06	4220.269 325 96	8.61	5495.568 250 41
6.97	2360.103 844 81	7.52	3197.947 740 16	8.07	4241.252 600 01	8.62	5521.143 859 36
6.98	2373.677 376 16	7.53	3214.992 060 81	8.08	4262.314 024 96	8.63	5546.808 633 61
6.99	2387.309 372 01	7.54	3232.104 422 56	8.09	4283.453 793 51	8.64	5572.562 780 16
7.00	2401.000 000 00	7.55	3249.285 006 25	8.10	4304.672 100 00	8.65	5598.406 506 25
7.01	2414.749 428 01	7.56	3266.533 992 96	8.11	4325.969 138 41	8.66	5624.340 019 36
7.02	2428.557 824 16	7.57	3283.851 564 01	8.12	4347.345 103 36	8.67	5650.363 527 21
7.03	2442.425 356 81	7.58	3301.237 900 96	8.13	4368.800 189 61	8.68	5676.477 237 76
7.04	2456.352 194 56	7.59	3318.693 185 61	8.14	4390.334 592 16	8.69	5702.681 359 21
7.05	2470.338 506 25	7.60	3336.217 600 00	8.15	4411.948 506 25	8.70	5728.976 100 00
7.06	2484.384 460 96	7.61	3353.811 326 41	8.16	4433.642 127 36	8.71	5755.361 668 81
7.07	2498.490 228 01	7.62	3371 474 547 36	8.17	4455.415 641 21	8.72	5781.838 274 56
7.08	2512.655 976 96	7.63	3389.207 445 61	8.18	4477.269 273 76	8.73	5808.406 126 41
7.09	2526.881 877 61	7.64	3407.010 204 16	8.19	4499.203 191 21	8.74	5835.065 433 76
7.10	2541.168 100 00	7.65	3424.883 006 25	8.20	4521.217 600 00	8.75	5861.816 406 25
7.11	2555.514 814 41	7.66	3442.826 035 36	8.21	4543.312 696 81	8.76	5888.659 253 76
7.12	2569.922 191 36	7.67	3460.839 4752	8.22	4565.488 678 56	8.77	5915.594 184 41
7.13	2584.390 401 61	7.68	3478.923 509 76	8.23	4587.745 742 41	8.78	5942.621 414 56
7.14	2598.919 616 16	7.69	3497.078 323 21	8.24	4610.084 085 76	8.79	5969.741 148 81
7.15	2613.510 006 25	7.70	3515.304 100 00	8.25	4632.503 906 2	8.80	5996.953 600 00

Table B-3 (Continued) **Fourth Power of Numbers**

n	n^4	n	n^4	n	n^4	n	n^4
8.81	6024.258 979 21	9.36	7675.442 012 16	9.91	9644.830 905 61	14.6	45437.1856
8.82	6051.657 497 76	9.37	7708.295 649 61	9.92	9683.819 560 96	14.7	46694.8881
8.83	6079.149 367 21	9.38	7741.254 643 36	9.93	9722.926 304 01	14.8	47978.5216
8.84	6106.734 799 36	9.39	7774.319 218 41	9.94	9762.151 392 96	14.9	49288.4401
8.85	6134.414 006 25	9.40	7807.489 600 00	9.95	9801.495 006 25	15.0	50625.0000
8.86	6162.187 200 16	9.41	7840.766 013 61	9.96	9840.957 442 56	15.1	51988.5601
8.87	6190.054 593 61	9.42	7874.148 654 96	9.97	9880.538 920 81	15.2	53379.4816
8.88	6218.016 399 36	9.43	7907.637 840 01	9.98	9920.239 680 16	15.3	54798.1281
8.89	6246.072 830 41	9.44	7941.233 704 96	9.99	9960.059 960 01	15.4	56244.8656
8.90	6274.224 100 00	9.45	7974.936 506 25	10.00	10000.000 000 00	15.5	57720.0625
8.91	6302.470 421 61	9.46	8008.746 470 56	10.1	10406.0401	15.6	59224.0896
8.92	6330.812 008 96	9.47	8042.663 824 31	10.2	10824.3216	15.7	60757.3203
8.93	6359.249 076 01	9.48	8076.588 796 16	10.3	11255.0881	15.8	62320.1296
8.94	6387.781 836 96	9.49	8110.821 612 01	10.4	11698.5856	15.9	63912.8961
8.95	6416.410 506 25	9.50	8145.062 500 00	10.5	12155.0625	16.0	65536.0000
8.96	6445.135 298 56	9.51	8179.411 688 01	10.6	12624.7696	16.1	67189.8241
8.97	6473.956 428 81	9.52	8213.869 404 16	10.7	13107.9601	16.2	68875.7530
8.98	6502.874 112 16	9.53	8248.435 876 81	10.8	13604.8896	16.3	70591.1761
8.99	6531.888 564 01	9.54	8283.111 334 56	10.9	14115.8161	16.4	72339.4816
9.00	6561.000 000 00	9.55	8317.896 006 25	11.0	14641.0000	16.5	74120.0625
9.01	6590.208 636 01	9.56	8352.790 120 96	11.1	15180.7041	16.6	75933.3136
9.02	6619.514 688 16	9.57	8387.793 908 01	11.2	15735.1936	16.7	77779.6321
9.03	6648.918 372 81	9.58	8422.907 596 96	11.3	16304.7361	16.8	79659.4176
9.04	6678.419 906 56	9.59	8458.131 417 61	11.4	16889.6016	16.9	81573.0721
9.05	6708.019 506 25	9.60	8498.465 600 00	11.5	17490.0625	17.0	83521.0000
9.06	6737.717 388 96	9.61	8528.910 374 41	11.6	18106.3936	17.1	85503.6081
9.07	6767.513 772 01	9.62	8564.465 971 36	11.7	18738.8721	17.2	87521.3050
9.08	6797.408 872 96	9.63	8600.132 621 61	11.8	19387.7776	17.3	89574.5041
9.09	6827.402 909 61	9.64	8635.910 556 16	11.9	20053.3921	17.4	91063.6176
9.10	6857.496 100 00	9.65	8671.800 006 25	12.0	20736.0000	17.5	93789.0625
9.11	6887.688 662 41	9.66	8707.801 203 36	12.1	21435.8881	17.6	95951.2576
9.12	6917.980 815 36	9.67	8743.914 379 21	12.2	22153.3456	17.7	98150.6241
9.13	6948.372 777 61	9.68	8780.139 765 76	12.3	22868.6641	17.8	100387.5856
9.14	6978.864 768 16	9.69	8816.477 595 21	12.4	23642.1376	17.9	102662.5681
9.15	7009.457 006 25	9.70	8852.928 100 00	12.5	24414.0625	18.0	104976.0000
9.16	7040.149 711 36	9.71	8889.491 512 81	12.6	25204.7376	18.1	107328.3121
9.17	7070.943 103 21	9.72	8926.168 066 56	12.7	26014.4641	18.2	109719.9376
9.18	7101.837 401 76	9.73	8962.957 994 41	12.8	26843.5456	18.3	112515.3127
9.19	7132.832 827 21	9.74	8999.861 529 76	12.9	27692.2881	18.4	114622.8736
9.20	7163.929 600 00	9.75	9036.878 906 25	13.0	28561.0000	18.5	117130.0625
9.21	7195.127 940 81	9.76	9074.010 356 76	13.1	29449.9921	18.6	119688.3216
9.22	7226.428 070 56	9.77	9111.256 118 41	13.2	30359.5776	18.7	122283.0961
9.23	7257.830 210 41	9.78	9148.616 422 56	13.3	31290.0721	18.8	124919.8336
9.24	7289.334 581 76	9.79	9186.091 504 81	13.4	32241.7936	18.9	127598.9841
9.24	7320.941 406 25	9.80	9223.681 600 00	13.5	33215.0625	19.0	130321.0000
9.26	7352.650 905 76	9.81	9261.386 943 21	13.6	34210.2016	19.1	133086.3361
9.27	7384.463 301 41	9.82	9299.207 769 76	13.7	35227.5361	19.2	135895.4496
9.28	7416.378 818 56	9.83	9337.144 315 21	13.8	36267.3936	19.3	138748.8001
9.29	7448.397 676 81	9.84	9375.196 845 36	13.9	37330.1041	19.4	141646.8496
9.30	7480.520 100 00	9.85	9413.365 506 25	14.0	38416.0000	19.5	144590.0625
9.31	7512.746 311 21	9.86	9451.650 624 16	14.1	39525.4161	19.6	147578.9056
9.32	7545.076 533 76	9.87	9490.052 405 61	14.2	40658.6895	19.7	150613.8401
9.33	7577.510 991 21	9.88	9588.571 087 36	14.3	41816.1601	19.8	153695.3616
9.34	7610.049 907 36	9.89	9567.206 906 41	14.4	42998.1696	19.9	156823.9201
9.35	7642.693 506 25	9.90	9605.960 100 00	14.5	44205 0625	20.0	160000.0000

Table B-3 (Continued)

Fourth Power of Numbers

n	n^4
20.1	163224.0801
20.2	166496.6416
20.3	169818.1681
20.4	173189.1456
20.5	176610.0625
20.6	180081.4096
20.7	183603.6801
20.8	187177.3696
20.9	190802.9761
21.0	194481.0000
21.1	198211.9447
21.2	201996.3136
21,3	205834.6167
21.4	209727.3616
21.5	213675.0625
21.6	217678.2836
21.7	221737.3921
21.8	225853.0576
21.9	230025.7527
22.0	234256.0000
22.1	238544.3281
22.2	242891.2656
22.3	247297.3447
22.4	251763.0976
22.5	256289.0625
22.6	260875.7776
22.7	265523.7841
22.8	270233.6256
22.9	275005.8487
23.0	279841.0000
23.1	284739.6321
23.2	289702.2976
23.3	294729.5527
23.4	299821.9536
23.5	304980.0625
23.6	310204.4416
23.7	315495.6561
23.8	320854.2786
23.9	326280.8647
24.0	331776.0000
24.1	337340.2561
24.2	342974.2096
24.3	348678.4401
24.4	354453.5296
24.5	360300.0625
24.6	366218.6256
24,7	372209.8081
24.8	378274.2016
24.9	384412.4001
25.0	390625.0000
25.1	396912.6001
25.2	403275.8016
25.3	409715.2081
25.4	416231.4256
25.5	422825.0625

n	n^4
25.6	429996.7296
25.7	436247.0401
25.8	443076.6096
25.9	449986.0561
26.0	456976.0000
26.1	464047.0641
26.2	471199.8736
26.3	478435.0561
26.4	485753.2416
26.5	493155.0625
26.6	500641.1536
26.7	508212.1521
26.8	515868.6976
26.9	523611.4321
27.0	531441.0000
27.1	539358.0481
27.2	547363.2256
27.3	555457.1841
27.4	563640.5776
27.5	571914.0625
27.6	580278.2976
27.7	588733.9441
27.8	597281.6656
27.9	605922.1281
28.0	614656.0000
28.1	623488.9521
28.2	632406.6596
28.3	641424.7921
28.4	650539.0336
28.5	659750.0625
28.6	669058.5616
28.7	678465.2167
28.8	687970.7136
28.9	697575.7441
29.0	707281.0000
29.1	717087.1761
29.2	726994.9606
29.3	737005.0801
29.4	747118.2096
29.5	757335.0625
29.6	767656.3456
29.7	778082.7681
29.8	788615.0416
29.9	799253.8801
30.0	810000.0000
30.1	820854.1201
30.2	831816.9616
30.3	842889.2481
30.4	854071.7056
30.5	865365.0625
30.6	876 770.0496
30.7	888 287.4001
30.8	899 917.8496
30.9	911 662.1361
31.0	923 521.0000

n	n^4
31.1	935 495.1841
31.2	947 585.4336
31.3	959 792.4961
31.4	972 117.1216
31.5	984 560.0625
31.6	997 122.0736
31.7	1009 803.9121
31.8	1022 606.3376
31.9	1035 530.1121
32.0	1048 576.0000
32.1	1061 744.7687
32.2	1075 037.1856
32.3	1088 454.0241
32.4	1101 996.0576
32.5	1115 664.0625
32.6	1129 458.8176
32.7	1143 381.1041
32.8	1157 431.7056
32.9	1171 611.4081
33.0	1185 921.0000
33.1	1206 361.2721
33.2	1214 933.0176
33.3	1229 637.0321
33.4	1244 474.1136
33.5	1259 445.0625
33.6	1274 550.6816
33.7	1289 791.7761
33.8	1305 169.1536
33.9	1320 683.6241
34.0	1336 336.0000
34.1	1352 127.0961
34.2	1368 057.7296
34.3	1384 128.7201
34.4	1400 340.8896
34.5	1416 695.0625
34.6	1433 192.0056
34.7	1449 832.7281
34.8	1466 617.8016
34.9	1483 548.3601
35.0	1500 625.0000
35.1	1517 848.6401
35.2	1535 220.1216
35.3	1552 740.2881
35.4	1570 409.9856
35.5	1588 230 0625
35.6	1606 201.3696
35.7	1624 324.7601
35.8	1642 601.0896
35.9	1661 031.2161
36.0	1679 616.0000
36.1	1698 356.3041
36.2	1717 252.9936
36.3	1736 306.9361
36.4	1755 519.0016
36.5	1774 890.0625

n	n^4
36.6	1794 420.9936
36.7	1814 112.6721
36.8	1833 965.9776
36.9	1853 981.7921
37.0	1874 161.0000
37.1	1894 504.4881
37.2	1915 913.1456
37.3	1935 687.8641
37.4	1956 529.5376
37.5	1977 539.0625
37.6	1998 717.3
37.7	2020 065.2641
37.8	2041 583.7456
37.9	2063 273.6881
38.0	2085 136.0000
38.1	2107 171.5921
38.2	2129 381.3776
38.3	2151 766.2721
38.4	2174 327.1936
38.5	2179 065.0625
38.6	2219 980.8016
38.7	2243 075.3361
38.8	2266 349.5936
38.9	2289 804.5041
39.0	2313 441.0000
39.1	2337 260.0161
39.2	2361 262.4896
39.3	2385 449.3601
39.4	2409 821.5696
39.5	2434 380.0625
39.6	2459 125.7856
39.7	2484 059.6881
39.8	2509 182.7216
39.9	2534 495.8401
40.0	2560 000.0000
40.1	2585 696.1601
40.2	2611 585.2816
40.3	2637 668.3201
40.4	2663 946.2656
40.5	2690 420.0625
40.6	2717 090.6896
40.7	2743 959.1201
40.8	2771 026.3296
40.9	2798 293.2961
41.0	2825 761.0000
41.1	2853 430.4241
41.2	2881 302.5536
41.3	2909 378.3761
41.4	2937 658.8816
41.5	2966 145.0625
41.6	2994 837.9136
41.7	3023 738.4321
41.8	3052 847.6176
41.9	3082 166.4721
42.0	3111 696.0000

Table B-3 (Continued) **Fourth Power of Numbers**

n	n^4
42.1	3141 437.2081
42.2	3171 391.1050
42.3	3201 558.7041
42.4	3231 941.0176
42.5	3262 539.0625
42.6	3293 353.8576
42.7	3324 386.4241
42.8	3355 637.7856
42.9	3387 108.9681
43.0	3418 801.0000
43.1	3450 714.9121
43.2	3482 851.7376
43.3	3515 212.5121
43.4	3947 798.2736
43.5	3584 610.0625
43.6	3610 648.9216
43.7	3646 915.8961
43.8	3680 412.0336
43.9	3714 138.3841
44.0	3718 096.0000
44.1	3782 285.9361
44.2	3816 709.2486
44.3	3851 367.0001
44.4	3886 260.2496
44.5	3921 390.0625
44.6	3956 757.5056
44.7	3992 363.6481
44.8	4028 209.5016
44.9	4064 296.3201
45.0	4100 625.0000
45.1	4137 196.6801
45.2	4174 012.4416
45.3	4211 073.3681
45.4	4248 380.5456
45.5	4285 935.0625
45.6	4328 738.0696
45.7	4367 790.4801
45.8	4400 093.5696
45.9	4438 648.3761
46.0	4477 456.0000
46.1	4516 517.5441
46.2	4555 834.1136
46.3	4595 406.8161
46.4	4635 236.7616
46.5	4675 325.0625
46.6	4715 672.8336
46.7	4756 281.1921
46.8	4797 151.2576
46.9	4838 284.1521
47.0	4879 681.0000
47.1	4921 342.9281
47.2	4963 271.0656
47.3	5005 466.5441
47.4	5047 930.4976
47.5	5090 664.0625
47.6	5133 668.3976
47.7	5176 944.5841
47.8	8220 493.8256
47.9	5264 317.2481
48.0	5308 416.0000
48.1	5352 791.2321
48.2	5397 444.0976
48.3	5442 375.7521
48.4	5487 587.3536
48.5	5533 080.0625
48.6	5578 855.0416
48.7	5624 913.4561
48.8	5671 256.4736
48.9	5717 885.2641
49.0	5764 801.0000
49.1	5812 004.8561
49.2	5859 498.0096
49.3	5907 281.6401
49.4	5955 356.9296
49.5	6003 725.0625
49.6	6052 387.2256
49.7	6101 344.6081
49.8	6150 598.4016
49.9	6200 149.8001
50.0	6250 000.0000
50.1	6300 150.2001
50.2	6350 601.6016
50.3	6401 355.4081
50.4	6452 412.8256
50.5	6503 775.0625
50.6	6555 443.3296
50.7	6607 418.8401
50.8	6659 702.8096
50.9	6712 296.4551
51.0	6765 201.0000
51.1	6818 417.6641
51.2	6871 947.6736
51.3	6925 792.2561
51.4	6979 952.6416
51.5	7034 430.0625
51.6	7089 225.7536
51.7	7144 340.9521
51.8	7199 776.8976
51.9	7255 534.8321
52.0	7811 616.0000
52.1	7368 021.6481
52.2	7424 753.0256
52.3	7481 811.3841
52.4	7539 197.9776
52.5	7596 914.0625
52.6	7654 960.8976
52.7	7713. 339.7441
52.8	7772 051.8656
52.9	7831 098.5281
53.0	7895 481.0000
53.1	7950 200.5521
53.2	8010 258.4576
53.3	8070 655.9921
53.4	8131 394.4836
53.5	8192 475.0625
53.6	8253 899.1616
53.7	8315 668.0161
53.8	8377 782.9136
53.9	8440 245.1441
54.0	8503 056.0000
54.1	8566 216.7761
54.2	8629 728.7696
54.3	8693 593.2801
54.4	8757 811.6036
54.5	8822 385.0625
54.6	8887 314.9456
54.7	8952 602.5481
54.8	9018 249.2416
54.9	9084 256.2801
55.0	9150 625.0000
55.1	9217 356.7201
55.2	9284 452.7616
55.3	9351 914.4481
55.4	9419 743.1056
55.5	9487 940.0625
55.6	9556 506.6496
55.7	9825 444.2001
55.8	9694 754.0496
55.9	9764 437.5361
56.0	9834 496.0000
56.1	9904 930.7841
56.2	9975 743.2336
56.3	10046 934.6961
56.4	10118 506.4216
56.5	10190 460.0625
56.6	10262 796.6736
56.7	10335 517.7121
56.8	10408 624.5376
56.9	10482 118.5121
57.0	10556 001.0000
57.1	10630 273.3681
57.2	10704 936.9856
57.3	10779 993.2241
57.4	10885 443.4576
57.5	10931 289.0625
57.6	11007 531.4176
57.7	11084 171.9041
57.8	11161 211.9056
57.9	11238 652.8081
58.0	11316 496.0000
58.1	11394 742.8721
58.2	11473 394.8176
58.3	11552 453.2321
58.4	11631 919.5136
58.5	11711 795.0625
58.6	11792 081.2816
58.7	11872 779.5761
58.8	11953 891.3536
58.9	12035 418.0241
59.0	12117 361.0000
59.1	12199 721.6961
59.2	12282 501.5296
59.3	12365 701.9201
59.4	12449 324.2896
59.5	12533 370.0625
59.6	12617 840.6656
59.7	12702 737.5281
59.8	12788 062.0816
59.9	12873 815.7601
60.0	12960 000.0000
60.1	13046 616.2401
60.2	13133 665.9216
60.3	13221 150.4881
60.4	13309 071.3856
60.5	13397 430.0625
60.6	13486 227.9696
60.7	13575 466.5601
60.8	13665 147.2896
60.9	13755 271.6161
61.0	13845 841.0000
61.1	13936 856.9041
61.2	14028 320.7936
61.3	14120 234.1361
61.4	14212 598.4016
61.5	14305 415.0625
61.6	14398 685.5936
61.7	14492 411.4721
61.8	14586 594.1776
61.9	14681 235.1921
62.0	14776 336.0000
62.1	14871 898.0081
62.2	14967 922.9456
62.3	15064 412.0641
62.4	15161 366.9376
62.5	15258 789.0625
62.6	15356 679.9376
62.7	15455 041.0641
62.8	15553 873.9456
62.9	15653 180.0881
63.0	15752 961.0000
63.1	15853 218.1921
63.2	15953 953.1776
63.3	16055 167.4721
63.4	16156 862.5936
63.5	16259 040.0625
63.6	16361 701.4016
63.7	16464 848.1361
63.8	16568 401.7936
63.9	16672 603.9041
64.0	16777 216.0000

Table B-3 (Continued) **Fourth Power of Numbers**

n	n^4	n	n^4	n	n^4	n	n^4
64.1	16882 319.6161	69.6	23465 886.1056	75.1	31809 712.8001	80.6	42202 693.2496
64.2	16987 916.2896	69.7	23601 038.4481	75.2	31979 477.4016	80.7	42412 526.0001
64.3	17094 007.5001	69.8	23736 773.7616	75.3	32149 920.6081	80.8	42623 140.2496
64.4	17200 594.9606	69.9	23873 093.7201	75.4	32321 044.2256	80.9	42834 537.9361
64.5	17307 680.0625	70.0	24010 000.0000	75.5	32492 850.0625	81.0	43046 721.0000
64.6	17415 264.3856	70.1	24147 494.2801	75.6	32665 339.9296	81.1	43259 691.3841
64.7	17523 349.4881	70.2	24285 478.2916	75.7	32838 515.6401	81.2	43473 451.0336
64.8	17631 936.9216	70.3	24424 253.6081	75.8	33012 379.0096	81.3	43688 001.8961
64.9	17741 028.2401	70.4	24563 521.9456	75.9	33186 931.8561	81.4	43903 345.9216
65.0	17850 625.0000	70.5	24703 385.0625	76.0	33621 176.0000	81.5	44119 485.0625
65.1	17960 728.7601	70.6	24843 844.6096	76.1	33538 113.2641	81.6	44336 421.2736
65.2	18071 341.0816	70.7	24984 902.2801	76.2	33714 745.4136	81.7	44554 156.5121
65.3	18182 463.5281	70.8	25126 559.7006	76.3	33892 074.4561	81.8	44772 692.7376
65.4	18294 097.6656	70.9	25268 818.7761	76.4	34070 102.0416	81.9	44992 031.9121
65.5	18406 245.0625	71.0	25411 681.0000	76.5	34243 830.0625	82.0	45212 176.0000
65.6	18518 907.2896	71.1	25555 148.1441	76.6	34428 260.3536	82.1	45433 126.9681
65.7	18632 085.9201	71.2	25699 221.9136	76.7	34605 394.7521	82.2	45654 886.7856
65.8	18745.782.9296	71.3	25843 904.0161	76.8	34789 235.0976	82.3	45877 457.4241
65.9	18859 998.6961	71.4	25989 196.1616	76.9	34970 783.2321	82.4	46100 840.8576
66.0	18974 736.0000	71.5	26135 100.0625	77.0	35153 041.0000	82.5	46325 039.0625
66.1	19089 996.0241	71.6	26281 617.4336	77.1	35336 010.3481	82.6	46550 054.0176
66.2	19205 780.3536	71.7	26428 749.9921	77.2	35519 692.8256	82.7	46775 887.7041
66.3	19322 090.5761	71.8	26576 499.4576	77.3	35704 090.5841	82.8	47002 542.1056
66.4	19438 928.2816	71.9	26724 867.5521	77.4	35889 205.3776	82.9	47230 019.2081
66.5	19556 295.0625	72.0	26873 856.0000	77.5	36075 039.0625	83.0	47458 321.0000
66.6	19674 192.5136	72.1	27023 466.5281	77.6	35261 593.4976	83.1	47687 449.4721
66.7	19792 622.2321	72.2	27173 700.8656	77.7	36448 870.5441	83.2	47917 406.6176
66.8	19911 585.8176	72.3	27324 560.7441	77.8	36636 872.0656	83.3	48148 194.4321
66.9	20031 084.8721	72.4	27476 047.8976	77.9	36825 599.9681	83.4	48379 814.9136
67.0	20151 121.0000	72.5	27628 164.0625	78.0	37015 056.0000	83.5	48612 270.0625
67.1	20271 695.8081	72.6	27780 910.9776	78.1	37205 242.1521	83.6	48845 561.8816
67.2	20392 810.9056	72.7	27934 290.3841	78.2	37396 160.2576	83.7	49079 692.3761
67.3	20514 467.9041	72.8	28088 304.0256	78.3	37587 812.1921	83.8	49314 663.5536
67.4	20636 668.4176	72.9	28242 953.6481	78.4	37780 199.8336	83.9	49550 477.4241
67.5	20759 414.0625	73.0	28398 241.0000	78.5	37973 325.0625	84.0	49787 136.0000
67.6	20882 706.4576	73.1	28554 167.8321	78.6	38167 189.7616	84.1	50024 641.2961
67.7	21006 547.2241	73.2	28710 735.8976	78.7	38361 795.8161	84.2	50262 995.3296
67.8	21130 937.9856	73.3	28867 946.9521	78.8	38557 145.1136	84.3	50502 200.1201
67.9	21255 880.3681	73.4	29025 802.1531	78.9	38753 239.5441	84.4	50742 257.6898
68.0	21381 376.00	73.5	29184 305.0625	79.0	38950 081.0000	84.5	50983 170.0625
68.1	21507 426.5121	73.6	29343 455.6416	79.1	39147 671.3761	84.6	51224 939.2656
68.2	21634 033.5376	73.7	29503 256.2561	79.2	39346 012.5696	84.7	51467 567.3281
68.3	21761 198.7121	73.8	29663 708.6136	79.3	39545 106.4801	84.8	51711 056.2816
68.4	21888 923.6736	73.9	29824 814.6641	79.4	39744 955.8096	84.9	51955 408.1601
68.5	22017 210.0625	74.0	29986 576.0000	79.5	39945 560.0625	85.0	52200 625.0000
68.6	22146 059.5216	74.1	30148 994.4561	79.6	40146 923.5456	85.1	52446 708.8401
68.7	22275 473.6961	74.2	30312 071.8096	79.7	40349 047.3681	85.2	52693 661.7216
68.8	22405 454.2336	74.3	30475 809.8401	79.8	40551 933.4416	85.3	52941 485.6881
68.9	22536 002.7841	74.4	30640 210.3296	79.9	40755 583.6801	85.4	53190 182.7856
69.0	22667 121.0000	74.5	30805 276.0625	80.0	40960 000.0000	85.5	53439 756.0625
69.1	22798 810.5361	74.6	30971 005.8256	80.1	41165 184.3201	85.6	53690 204.5696
69.2	22931 073.0496	74.7	31137 404.4081	80.2	41371 138.5616	85.7	53941 533.3601
69.3	23063 910.2001	74.8	31304 472.6016	80.3	41577 864.6481	85.8	54193 743.4896
69.4	23197 323.6496	74.9	31472 212.2001	80.4	41785 369.5056	85.9	54446 837.0161
69.5	23331 315.0625	75.0	31640 625.0000	80.5	41993 640.0625	86.0	54700 816.0000

Table B-3 (Continued) **Fourth Power of Numbers**

n	n^4
86.1	54955 682.5041
86.2	55211 438.5936
86.3	55468 086.3361
86.4	55725 627.8016
86.5	55984 065.0625
86.6	56243 400.1936
86.7	56503 635.2721
86.8	56764 772.3776
86.9	57026 813.5921
87.0	57289 761.0000
87.1	57553 616.6881
87.2	57818 382.7456
87.3	58084 061.2641
87.4	58350 654.3376
87.5	58618 164.0625
87.6	58886 592.5376
87.7	59155 941.8641
87.8	59426 214.1456
87.9	59697 411.4881
88.0	59969 536.0000
88.1	60242 598.7921
88.2	60516 574.9776
88.3	60791 493.6721
88.4	61067 347.9936
88.5	61344 140.0625
88.6	61621 872.0061
88.7	61900 545.9361
88.8	62180 163.9936
88.9	62460 728.3041
89.0	62742 241.0000
89.1	63024 704.2161
89.2	63308 120.0896
89.3	63592 490.7601
89.4	63877 818.3696
89.5	64164 105.0625
89.6	64451 352.9856
89.7	64739 564.2881
89.8	65028 741.1216
89.9	65318 885.6401
90.0	65610 000.0000
90.1	65902 086.3601
90.2	66195 146.8816
90.3	66489 183.7281
90.4	66784 199.0636
90.5	67080 195.0625
90.6	67377 173.8896
90.7	67675 137.7201
90.8	67974 088.7296
90.9	68274 029.0961
91.0	68574 961.0000
91.1	68876 886.6241
91.2	69179 808.1536
91.3	69483 727.7761
91.4	69788 647.6816
91.5	70094 570.0635

n	n^4
91.6	70401 497.1136
91.7	70709 431.0321
91.8	71018 374.0176
91.9	71328 328.2721
92.0	71639 296.0000
92.1	71951 279.4081
92.2	72264 280.7056
92.3	72578 302.1041
92.4	72893 345.8176
92.5	73209 414.8625
92.6	73526 509.0516
92.7	73844 633.0241
92.8	74163 788.1856
92.9	74483 976.7681
93.0	74805 201.0000
93.1	75127 463.1121
93.2	75450 765.3376
93.3	75775 109.9121
93.4	76100 499.0736
93.5	76426 935.0625
93.6	76754 420.1216
93.7	77082 956.4961
93.8	77412 546.4336
93.9	77743 192.1841
94.0	78714 896.0000
94.1	78401 660.1361
94.2	78741 486.8496
94.3	79076 378.4001
94.4	79412 337.0496
94.5	79749 365.0625
94.6	80087 464.7056
94.7	80426 638.2481
94.8	80766 887.1916
94.9	81108 216.1201
95.0	81450 625.0000
95.1	81794 116.8881
95.2	82138 694.0416
95.3	82484 358.7681
95.4	82831 113.3456
95.5	83178 960.0625
95.6	83527 901.2096
95.7	83877 939.0001
95.8	84229 075.9696
95.9	84581 314.1761
96.0	84934 656.0000
96.1	85289 103.7441
96.2	85644 659.7136
96.3	86001 326.2161
96.4	86359 105.5616
96.5	85718 000.0625
96.6	87078 012.0336
96.7	87439 143.7921
96.8	87801 397.6576
96.9	88164 775.9521
97.0	88529 281.0000

n	n^4
97.1	88894 915.1281
97.2	89261 580.6656
97.3	89629 579.9441
97.4	89998 615.2976
97.5	90368 789.0615
97.6	90740 103.5776
97.7	91112 561.1841
97.8	91486 164.2256
97.9	91860 915.0481
98.0	92236 816.0000
98.1	92613 869.4321
98.2	92992 077.6976
98.3	93371 443.1521
98.4	93751 968.1536
98.5	94133 655.0625
98.6	94516 506.2416
98.7	94900 524.0561
98.8	95285 710.8736
98.9	95672 069.0641
99.0	96059 601.0000
99.1	96448 309.0561
99.2	96838 195.6096
99.3	97229 263.0401
99.4	97621 513.7296
99.5	98014 950.0615
99.6	98409 574.4256
99.7	98805 389.2081
99.8	99202 396.8016
99.9	99600 599.6001
100.0	100000 000.0000
101	104 060 401
102	108 243 216
103	112 550 881
104	116 985 856
105	121 550 625
106	126 247 696
107	131 079 601
108	136 048 896
109	141 158 161
110	146 410 000
111	151 807 041
112	157 351 936
113	163 047 361
114	168 896 016
115	174 900 625
116	181 063 936
117	187 388 721
118	193 877 776
119	200 533 921
120	207 360 000
121	214 358 881
122	221 533 456
123	228 886 641
124	236 421 376
125	244 140 625

n	n^4
126	252 047 376
127	260 144 641
128	268 435 456
129	276 922 881
130	235 610 000
131	294 499 921
132	303 595 716
133	312 900 721
134	322 417 936
135	332 150 625
136	342 102 016
137	352 275 361
138	362 673 936
139	373 301 041
140	384 160 000
141	395 254 161
142	406 586 896
143	418 161 601
144	429 981 696
145	442 050 625
146	454 371 856
147	466.948 881
148	479 785 216
149	492 884 401
150	506 250 000
151	519 885 601
152	533 794 816
153	541 981 281
154	562 448 656
155	577 200 625
156	592 240 896
157	607 573 201
158	623 201 296
159	639 128 961
160	655 360 000
161	671 898 241
162	688 747 536
163	705 911 761
164	723 394 816
165	741 200 625
166	759 333 136
167	777.796 321
168	796 594 176
169	815 730 721
170	835 210 000
171	855 036 081
172	875 213 056
173	895 745 041
174	916 636 176
175	937 890 625
176	959 512 576
177	981 506 241
178	100 387 5856
179	102 662 5681
180	104 976 0000

Table B-3 (Continued) **Fourth Power of Numbers**

n	n^4	n	n^4	n	n^4	n	n^4
181	107 328 3121	236	3102 044 416	291	7170 871 761	346	14331 920 656
182	109 719 9376	237	3154 956 561	292	7269 949 696	347	14498 327 281
183	112 151 3121	238	3208 542 730	293	7370 050 801	348	14666 178 816
184	114 622 8736	239	3262 808 641	294	7471 182 096	349	14835 483 601
185	117 135 0625	240	3317 760 000	295	7573 350 625	350	15006 250 000
186	119 688 3216	241	3373 402 561	296	7676 563 456	351	15178 486 401
187	122 283 0961	242	3429 742 096	297	7780 827 681	352	15352 201 216
188	124 919 8336	243	3486 784 401	298	7886 150 416	353	15527 402 881
189	127 598 9841	244	3544 535 296	299	7992 538 801	354	15704 099 856
190	130 321 0000	245	3603 000 625	300	8100 000 000	355	15882 300 625
191	133 086 3361	246	3662 186 256	301	8208 541 201	356	16062 013 696
192	135 895 4496	247	3722 098 081	302	8318 169 616	357	16243 247 601
193	138 748 8001	248	3782 742 016	303	8428 892 481	358	16426 010 896
194	141 646 8496	249	3844 124 001	304	8540 717 056	359	16610 312 161
195	144 590 0625	250	3906 250 000	305	8657 650 625	360	16796 160 000
196	147 578 9056	251	3969 126 001	306	8767 700 496	361	16983 563 041
197	150 613 8481	252	4032 758 016	307	8882 874 001	362	17172 529 936
198	153 695 3616	253	4097 152 081	308	8999 178 496	363	17363 069 361
199	156 823 9201	254	4162 314 256	309	9116 621 361	364	17555 190 016
200	160 000 0000	255	4228 250 625	310	9235 210 000	365	17748 900 625
201	1632 240 801	256	4294 967 296	311	9354 951 841	366	17944 209 936
202	1664 966 416	257	4362 470 401	312	9475 854 336	367	18141 126 721
203	1698 181 681	258	4430 766 096	313	9597 924 961	368	18339 659 776
204	1731 891 456	259	4499 600 561	314	9721 171 216	369	18539 817 921
205	1766 100 625	260	4569 760 000	315	9845 600 625	370	18741 610 000
206	1800 814 096	261	4640 470 641	316	9971 220 736	371	18945 044 881
207	1836 036 801	262	4711 998 736	317	10098 039 121	372	19150 131 456
208	1871 773 696	263	4784 350 561	318	10226 063 376	373	19356 878 641
209	1908 029 761	264	4857 532 416	319	10355 301 121	374	19565 295 376
210	1944 810 000	265	4931 550 625	320	10485 760 000	375	19775 390 625
211	1982 119 441	266	5006 411 536	321	10617 447 681	376	19987 173 376
212	2019 963 136	267	5082 121 521	322	10750 371 856	377	20200 652 641
213	2058 346 161	268	5158 686 976	323	10884 540 241	378	20415 837 456
214	2097 273 616	269	5236 114 321	324	11019 960 576	379	20632 736 881
215	2136 750 625	270	5314 410 000	325	11156 640 625	380	20851 360 000
216	2176 782 386	271	5393 580 481	326	11294 588 176	381	21071 715 921
217	2217 373 921	272	5473 632 256	327	11433 811 041	382	21293 813 776
218	2258 530 576	273	5554 571 841	328	11574 317 056	383	21517 662 721
219	2300 257 521	274	5636 405 776	329	11716 114 081	384	31743 271 936
220	2342 560 000	275	5719 140 625	330	11859 210 000	385	21970 650 625
221	2385 443 281	276	5802 782 976	331	12003 612 721	386	22199 808 016
222	2428 912 656	277	5887 339 441	332	12149 330 176	387	22430 753 361
223	2472 973 441	278	5972 816 656	333	12296 370 321	388	22663 495 936
224	2517 630 976	279	6059 221 281	334	12444 741 136	389	22898 045 041
225	2562 890 625	280	6146 560 000	335	12594 450 625	390	23134 410 000
226	2608 757 776	281	6234 839 521	336	12745 506 816	391	23372 600 161
227	2655 237 841	282	6324 066 576	337	12897 917 761	392	23612 624 896
228	2702 336 256	283	6414 247 921	338	13051 691 536	393	23854 493 601
229	2750 058 481	284	6505 390 336	339	13206 836 241	394	24098 215 696
230	2798 410 000	285	6597 500 625	340	13363 360 000	395	24343 800 625
231	2847 396 321	286	6690 585 616	341	13521 270 961	396	24591 257 856
232	2897 022 976	287	6784 652 161	342	13680 577 296	397	24840 996 881
233	2947 295 521	288	6879 707 136	343	13841 287 201	398	25091 827 216
234	2998 219 536	289	6975 757 441	344	14003 408 896	399	25344 958 401
235	3049 800 625	290	7072 810 000	345	14166 950 625	400	25600 000 000

Table B-3 (Continued)

Fourth Power of Numbers

n	n^4
401	25856 961 601
402	26115 852 816
403	26376 683 281
404	26639 462 656
405	26904 200 625
406	27170 906 896
407	27439 591 201
408	27110 263 296
409	27982 932 961
410	28257 610 000
411	28534 304 241
412	28813 025 536
413	29093 783 761
414	29376 588 816
415	29661 450 625
416	29948 379 136
417	30237 384 321
418	30528 476 176
419	30821 664 721
420	31116 960 000
421	31414 372 081
422	31713 911 056
423	32015 587 041
424	32319 410 116
425	32625 390 625
426	32933 538 576
427	33243 864 241
428	33556 377 856
429	33871 089 681
430	34188 010 000
431	34507 149 121
432	34828 517 376
433	35152 125 121
434	35477 982 776
435	35800 100 625
436	36136 489 216
437	36469 158 961
438	36804 120 336
439	37141 383 841
440	37480 960 000
441	37822 859 361
442	38167 092 496
443	38513 670 001
444	38862 602 496
445	39213 900 625
446	39567 575 056
447	39923 636 481
448	40282 095 616
449	40642 963 201
450	51006 250 000
451	41371 966 801
452	41740 124 416
453	42110 733 681
454	42483 805 456
455	42859 350 625
456	43237 380 096
457	43617 904 801
458	44000 935 696
459	44386 483 761
460	44774 560 000
461	45165 175 441
462	45558 341 136
463	45954 068 161
464	46352 367 616
465	46753 250 625
466	47156 728 336
467	47562 811 921
468	47971 512 516
469	48382 841 521
470	48796 810 000
471	49213 429 281
472	49632 710 656
473	50054 665 441
474	50479 304 976
475	50906 640 625
476	51336 683 116
477	51769 445 841
478	52204 938 256
479	52643 172 481
480	53084 160 000
481	53527 912 321
482	53974 440 976
483	54423 757 521
484	54875 873 536
485	55330 800 625
486	55788 550 416
487	56249 134 561
488	56712 564 736
489	57178 852 641
490	57648 010 000
491	58120 048 561
492	58594 980 096
493	59072816 401
494	59553 569 296
495	60037 250 625
496	60523 872 256
497	61013 446 081
498	61505 984 016
499	62001 498 001
500	62500 000 000
501	63001 502 001
502	63506 016 016
503	64013 554 081
504	64524 128 256
505	65037 750 625
506	65554 433 296
507	66074 188 401
508	66597 028 096
509	67122 964 561
510	67652 010 000
511	68 184 176 641
512	68 719 476 736
513	69 257 922 561
514	69 799 526 416
515	70 344 300 625
516	70 892 257 536
517	71 443 409 521
518	71 997 768 976
519	72 555 348 321
520	73 116 160 000
521	73 680 216 481
522	74 241 530 256
523	74 818 113 341
524	75 391 979 776
525	75 969 140 625
526	76 549 608 976
527	77 133 397 441
528	77 720 518 656
529	78 310 985 281
530	78 904 810 000
531	79 502 005 521
532	80 102 584 576
533	80 706 559 921
534	81 313 944 336
535	81 924 750 625
536	82 538 991 616
537	83 156 680 161
538	83 777 829 136
539	84 402 451 441
540	85 030 560 000
541	85 662 167 761
542	86 297 287 696
543	86 935 932 801
544	87 578 116 096
545	88 223 850 625
546	88 873 149 456
547	89 526 025 681
548	90 182 492 416
549	90 842 562 801
550	91 506 250 000
551	92 173 567 201
552	92 844 527 616
553	93 519 144 481
554	94 197 431 056
555	94 879 400 625
556	95 565 066 996
557	96 254 442 001
558	96 947 540 496
559	97 644375 361
560	98 344 960 000
561	99 049 307 341
562	99 757 432 336
563	100 469 346 961
564	101 185 065 216
565	101 904 600 625
566	102 627 966 736
567	103 355 111 121
568	104 086 245 376
569	104 821 185 121
570	105 560 010 000
571	106 302 733 681
572	107 049 369 856
573	107 799 932 241
574	108 554 434 577
575	109 312 890 625
516	110 075 314 116
577	110 841 719 041
578	111 612 119 056
579	112 386 528 081
580	113 164 950 000
581	113 947 428 721
582	114 744 948 116
583	115 524 532 321
584	116 319 195 136
585	117 117 950 625
586	117 920 812 816
587	118 727 795 161
588	119 538 913 536
589	120 354 180 241
590	121 173 610 000
591	121 997 216 961
592	122 825 015 296
593	123 657 019 201
594	124 493 242 896
595	125 333 700 625
596	126 178 406 656
597	127 027 375 281
598	127 880 620 816
599	128 738 157 601
600	129 600 000 000
601	130 466 162 401
602	131 336 659 216
603	132 211 504 881
604	133 090 713 856
605	133 974 300 625
606	134 862 279 698
607	135 754 665 601
608	136 651 472 896
609	137 552 716 161
610	138 458 410 000
611	139 368 569 041
612	140 283 207 936
613	141 202 341 361
614	142 125 984 016
615	143 054 150 625
616	143 986 855 936
617	144 924 114 721
618	145 865 941 776
619	146 812 351 921
620	147 763 360 000

Table B-3 (Continued) **Fourth Power of Numbers**

n	n^4	n	n^4	n	n^4	n	n^4
621	148 718 980 881	676	208 827 064 516	731	285 541 678 321	786	381 671 897 616
622	149 679 229 456	677	210 065 472 241	732	287 107 358 976	787	383 617 958 161
623	150 644 120 641	678	211 309 379 856	733	288 679 469 561	788	385 571 451 136
624	151 613 669 376	679	212 558 803 681	734	290 258 027 536	789	387 532 395 441
625	152 587 890 625	680	213 813 760 000	735	291 843 050 625	790	389 500 810 000
626	153 566 799 376	681	215 074 265 121	736	293 434 556 416	791	391 476 713 781
627	154 550 410 641	682	216 340 335 376	737	295 032 562 561	792	393 460 125 696
628	155 538 739 456	683	217 611 987 121	738	296 637 086 736	793	395 451 064 801
629	156 531 800 881	684	218 889 236 736	739	298 248 146 641	794	397 449 550 096
630	157 529 610 000	685	220 172 100 625	740	299 865 760 000	795	399 455 600 625
631	158 532 181 921	686	221 460 595 216	741	301 589 944 561	796	401 469 235 456
632	159 539 531 776	687	222 754 736 961	742	303 120 718 096	797	403 490 473 681
633	160 551 674 721	688	224 054 542 336	743	304 758 098 401	798	405 519 334 416
634	161 568 625 936	689	225 360 027 841	744	306 402 103 296	799	407 555 836 801
635	162 590 400 625	690	226 671 210 000	745	308 052 750 625	800	409 600 000 000
636	163 617 014 016	691	227 988 105 361	746	309 710 058 256	801	411 651 843 201
637	164 648 481 361	692	229 310 730 496	747	311 374 044 081	802	413 711 385 616
638	165 684 811 936	693	230 639 102 001	748	313 044 726 016	803	415 778 646 481
639	166 726 039 041	694	231 973 236 496	749	314 722 122 001	804	417 853 645 056
640	167 772 160 000	695	233 313 150 625	750	316 406 250 000	805	419 936 400 625
641	168 823 196 161	696	234 658 861 056	751	318 097 128 001	806	422 026 932 496
642	169 879 162 896	697	236 010 384 481	752	319 794 774 016	807	424 125 260 001
643	170 940 075 601	698	237 367 737 616	753	321 409 206 081	808	426 231 402 496
644	172 005 949 696	699	238 730 937 201	754	323 210 442 256	809	428 345 379 361
645	173 076 800 625	700	240 100 000 000	755	324 928 500 625	510	430 467 210 000
646	174 152 643 856	701	241 474 942 801	756	326 653 399 296	811	432 596 913 841
647	175 233 494 881	702	242 855 782 416	757	328 385 156 401	812	434 734 510 336
648	176 319 369 216	703	244 242 593 681	758	330 123 790 096	813	436 880 018 961
649	177 410 282 401	704	245 635 219 456	759	331 869 318 561	814	439 033 459 216
650	178 506 250 000	705	247 033 850 625	760	333 621 760 000	815	441 194 850 625
651	179 607 281 601	706	258 438 446 096	761	335 381 132 641	816	443 364 212 736
652	180 713 410 816	707	249 849 022 801	762	337 147 454 136	817	445 541 565 121
653	181 824 635 281	708	251 265 591 696	763	338 920 744 561	818	447 726 927 376
654	182 940 976 656	709	252 688 187 761	764	340 701 020 416	819	449 920 319 121
655	184 062 450 625	710	254 116 810 000	765	342 488 300 625	820	452 121 760 000
656	185 189 072 896	711	255 551 481 441	766	344 282 603 536	821	454 331 269 687
657	186 320 859 201	712	256 992 219 136	767	346 083 947 521	822	456 548 867 836
658	187 457 825 296	713	258 439 040 161	768	347 892 350 976	823	458 774 574 241
659	188 599 986 961	714	259 891 961 616	769	349 707 832 321	824	461 008 408 576
660	189 747 360 000	715	261 351 000 625	770	351 530 410 000	825	463 250 390 625
661	190 899 960 241	716	262 816 174 336	771	353 360 102 481	826	465 500 540 116
662	192 057 803 536	717	264 287 499 921	772	355 196 928 256	827	467 758 871 041
663	193 220 905 161	718	265 764 994 576	773	357 040 905 841	828	470 025 421 056
664	194 389 282 816	719	267 248 675 521	774	358 892 053 776	829	472 300 192 081
665	195 562 950 625	720	268 738 560 000	775	360 750 390 625	830	474 583 210 000
666	196 741 925 136	721	270 234 665 281	776	362 615 934 916	831	476 874 494 727
667	197 926 222 321	722	271 737 008 656	777	364 488 705 441	832	479 174 066 116
668	199 115 858 116	723	273 245 601 441	778	366 368 720 656	833	481 481 944 321
669	200 310 838 721	724	274 760 478 976	779	386 255 999 281	834	483 798 149 136
670	201 511 210 000	725	276 281 640 625	780	370 150 560 000	835	486 122 700 625
671	202 716 958 081	726	277 809 109 116	781	372 052 421 521	836	488 455 618 816
672	203 928 109 050	727	279 342 903 841	782	373 961 602 576	837	490 796 923 761
673	205 144 679 041	728	280 883 040 256	783	375 878 121 921	838	493 146 635 536
674	206 366 684 176	729	282 429 536 481	784	377 801 998 336	839	495 504 774 241
675	207 594 140 625	730	283 982 410 000	785	379 733 250 625	840	497 871 360 000

Table B-3 (Continued) **Fourth Power of Numbers**

n	n^4	n	n^4	n	n^4
841	500 246 412 961	896	644 513 529 856	951	817 941 168 801
842	502 629 953 296	897	647 395 642 881	952	821 386 940 416
843	505 022 001 201	898	650 287 411 216	953	824 843 581 681
844	507 422 576 396	899	653 188 856 401	954	828 311 133 456
845	509 831 700 625	900	656 100 000 000	955	831 789 600 625
846	512 249 392 656	901	659 020 863 601	956	835 279 012 096
847	514 675 673 281	902	661 951 468 816	957	838 779 390 801
848	517 110 562 816	903	664 891 837 281	958	842 290 759 696
849	519 554 081 601	904	667 841 990 656	959	845 813 141 761
850	522 006 250 000	905	670 801 950 625	960	849 346 560 000
851	524 467 088 401	906	673 771 738 896	961	852 891 031 441
852	526 936 611 216	907	676 751 377 601	962	856 446 597 136
853	529 414 856 881	908	679 740 887 296	963	860 013 262 161
854	531 901 821 856	909	682 740 290 961	964	863 591 055 616
855	534 397 550 625	910	685 749 610 000	965	867 180 000 625
856	536 902 045 696	911	688 768 866 241	966	870 780 120 336
857	539 415 333 601	912	691 798 081 536	967	874 391 437 921
858	541 937 434 896	913	694 837 277 761	968	878 013 976 576
859	544 468 370 161	914	697 886 476 816	969	881 647 759 521
860	547 008 160 000	915	700 945 700 625	970	885 292 810 000
861	549 556 825 041	916	704 014 971 136	971	888 949 151 281
862	552 114 385 936	917	707 094 310 321	972	892 616 806 656
863	554 680 863 361	918	710 183 740 176	973	896 295 799 441
864	557 256 278 016	919	713 283 282 721	974	899 986 156 916
865	559 840 650 625	920	716 392 960 000	975	903 687 890 625
866	562 434 001 936	921	719 512 794 081	976	907 401 035 776
867	565 036 352 121	922	722 642 801 056	977	911 125 611 841
868	567 647 723 776	923	725 783 021 041	978	914 861 642 256
869	570 268 135 921	924	728 933 458 176	979	918 609 150 481
870	572 897 610 000	925	732 094 140 625	980	922 368 160 000
871	575 536 166 881	926	735 265 090 576	981	926 138 694 321
872	578 183 827 456	927	738 446 330 241	982	929 920 776 976
873	580 840 612 641	928	741 637 881 856	983	933 714 431 521
874	583 506 543 376	929	744 839 767 681	984	937 519 681 536
875	586 181 640 625	930	748 052 010 000	985	941 336 550 625
876	588 865 925 376	931	751 274 631 121	986	945 165 062 416
877	591 559 418 641	932	754 507 653 376	987	949 005 240 561
878	594 262 141 456	933	757 151 099 121	988	952 857 100 736
879	596 794 114 881	934	761 004 990 136	989	956 720 690 641
880	599 695 360 000	935	764 269 350 025	990	960 596 010 000
881	602 425 897 921	936	767 544 201 216	991	964 483 090 561
882	605 165 749 775	937	770 829 564 961	992	968 381 956 096
883	607 914 936 721	938	774 125 464 336	993	972 292 630 401
884	610 673 479 936	939	777 431 921 041	994	976 215 137 296
885	613 441 400 625	940	780 748 960 000	995	980 149 500 625
886	616 218 720 016	941	784 076 601 361	996	984 095 744 256
887	619 005 459 361	942	787 414 868 486	997	988 053 892 081
888	621 801 639 936	943	790 763 784 001	998	992 023 968 016
889	624 607 283 041	944	794 123 370 496	999	996 005 996 001
890	627 422 410 000	945	797 493 650 625	1000	1000 000 000 000
891	630 247 042 161	946	800 874 641 056		
892	633 081 200 896	947	804 266 382 481		
893	635 924 907 601	948	807 668 879 616		
894	638 778 183 696	949	811 082 161 201		
895	641 641 050 625	950	814 506 250 000		

Table B-4 Five-fourths power of numbers

N	0	1	2	3	4	5	6	7	8	9
0	0.000	1.000	2.3784	3.9482	5.6569	7.4767	9.3905	11.386	13.454	15.588
10	17.783	20.033	22.335	24.685	27.081	29.520	32.000	34.519	37.076	39.668
20	42.295	44.955	47.646	50.369	53.121	55.910	58.711	61.547	64.409	67.297
30	70.210	73.148	76.109	79.094	82.101	85.130	88.182	91.254	94.347	97.461
40	100.59	103.75	106.92	110.11	113.32	116.55	119.80	123.06	126.34	129.64
50	132.96	136.29	139.64	143.00	146.38	149.78	153.19	156.62	160.06	163.52
60	166.99	170.48	173.98	177.49	181.02	184.56	188.12	191.69	195.27	198.87
70	202.48	206.10	209.73	213.38	217.04	220.71	224.40	228.09	231.80	235.52
80	239.26	243.00	246.76	250.52	254.30	258.09	261.89	265.70	269.53	273.73
90	277.21	281.06	284.93	288.80	292.69	269.59	300.50	304.41	308.34	312.28
100	316.23	320.19	324.15	328.13	332.12	336.11	340.12	344.14	348.16	352.19
110	356.24	360.29	364.35	368.42	372.50	376.59	380.69	384.80	388.91	393.04
120	397.17	401.31	405.46	409.62	413.79	417.96	422.15	426.34	430.54	434.75
130	438.96	443.19	447.42	451.66	455.91	460.17	464.43	468.71	472.99	477.27
140	481.57	485.87	490.19	494.50	498.83	503.16	507.51	511.85	516.21	520.57
150	524.95	529.32	533.71	538.10	542.50	546.91	551.32	555.74	560.17	564.61
160	569.05	573.50	577.95	582.42	586.89	591.36	595.85	600.34	604.83	609.34
170	613.85	618.37	622.89	627.42	631.96	636.50	641.05	645.60	650.17	654.74
180	659.31	663.89	668.48	673.08	677.68	682.28	686.90	691.52	696.14	700.77
190	705.41	710.05	714.70	719.36	724.02	728.69	733.37	738.04	742.73	747.42
200	752.12	756.82	761.53	766.25	770.97	775.70	780.43	785.17	789.91	794.66
210	799.42	804.18	808.95	813.72	818.50	823.28	828.07	832.87	837.67	842.47
220	847.28	852.10	856.92	861.75	866.58	871.42	876.27	881.11	885.97	890.83
230	895.69	900.56	905.44	910.32	915.21	920.21	925.00	929.90	934.81	939.72
240	944.63	949.56	954.48	959.42	964.36	969.30	974.25	979.20	984.16	989.12
250	994.09	999.06	1004.0	1009.0	1014.0	1019.0	1024.0	1029.0	1034.0	1039.0
260	1044.0	1049.1	1054.1	1059.1	1064.2	1069.2	1074.2	1079.3	1084.3	1089.4
270	1094.5	1099.5	1104.5	1109.7	1114.8	1119.9	1125.0	1130.1	1135.2	1140.3
280	1145.4	1150.5	1155.6	1160.7	1165.9	1171.0	1176.1	1181.3	1186.4	1191.6
290	1196.7	1201.9	1207.1	1212.2	1217.4	1222.6	1227.8	1232.9	1238.1	1243.3
300	1248.5	1253.7	1259.0	1264.2	1269.4	1274.6	1279.8	1285.1	1290.3	1295.5
310	1300.8	1306.0	1311.3	1316.5	1321.8	1327.1	1332.3	1337.6	1342.9	1348.2
320	1353.4	1358.7	1364.0	1369.3	1374.6	1379.9	1385.2	1390.5	1395.9	1401.2
330	1406.5	1411.8	1417.2	1422.5	1427.9	1433.2	1438.5	1443.9	1449.3	1454.6
340	1460.0	1465.4	1470.7	1476.1	1481.5	1486.9	1492.3	1497.7	1503.1	1508.5
350	1513.9	1519.3	1524.7	1530.1	1535.5	1540.9	1546.4	1551.8	1557.2	1562.7
360	1568.1	1573.6	1579.0	1584.5	1589.9	1595.4	1600.9	1606.3	1611.8	1617.3
370	1622.8	1628.2	1633.7	1639.2	1644.7	1650.2	1655.7	1661.2	1666.7	1672.2
380	1677.8	1683.3	1688.8	1694.3	1699.9	1705.4	1710.9	1716.5	1722.0	1727.6
390	1733.1	1738.7	1744.1	1749.8	1755.4	1760.9	1766.5	1772.1	1777.7	1783.3
400	1788.9	1794.4	1800.0	1805.6	1811.2	1816.8	1822.5	1828.1	1833.7	1839.3
410	1844.9	1850.6	1856.2	1861.8	1867.5	1873.1	1878.7	1884.4	1890.0	1895.7
420	1901.3	1907.0	1912.7	1918.3	1924.0	1929.7	1935.4	1941.0	1946.7	1952.4
430	1958.1	1963.8	1969.5	1975.2	1980.9	1986.6	1992.3	1998.0	2003.7	2009.5
440	2015.2	2020.9	2026.6	2032.4	2038.1	2043.9	2049.6	2055.3	2061.1	2066.8
450	2072.6	2078.4	2084.1	2089.9	2095.7	2101.4	2107.2	2113.0	2118.8	2124.5
460	2130.3	2136.1	2141.9	2147.7	2153.5	2159.3	2165.1	2170.9	2176.7	2182.6
470	2188.4	2194.2	2200.0	2205.9	2211.7	2217.5	2223.4	2229.2	2235.0	2240.9
480	2246.7	2252.6	2258.4	2264.3	2270.2	2276.0	2281.9	2287.8	2293.6	2299.5
490	2305.4	2311.3	2317.2	2323.0	2328.9	2334.8	2340.7	2346.6	2352.5	2358.4

Table B-4 Five-fourths power of numbers (continued)

N	0	1	2	3	4	5	6	7	8	9
500	2364.4	2370.3	2376.2	2382.1	2388.0	2393.9	2399.9	2405.8	2411.7	2417.7
510	2423.6	2429.5	2335.5	2441.4	2447.4	2453.3	2459.3	2465.3	2471.2	2477.2
520	2483.2	2489.1	2495.1	2501.1	2507.1	2513.0	2519.0	2525.0	2531.0	2537.0
530	2543.0	2549.0	2555.0	2561.0	2567.0	2573.0	2579.0	2585.0	2591.1	2597.1
540	2603.1	2609.1	2615.2	2621.2	2627.2	2633.3	2639.3	2645.4	2651.4	2657.5
550	2663.5	2669.6	2675.6	2681.7	2687.7	2693.8	2699.9	2705.9	2712.1	2718.1
560	2724.2	2730.3	2736.3	2742.4	2748.5	2754.6	2760.7	2766.8	2772.9	2779.0
570	2785.1	2791.2	2797.3	2803.5	2809.6	2815.7	2821.8	2827.9	2834.1	2840.2
580	2846.3	2852.5	2858.6	2864.7	2870.9	2877.0	2883.2	2889.3	2895.5	2901.6
590	2907.8	2914.0	2920.1	2926.3	2932.5	2938.6	2944.8	2951.0	2957.2	2963.4
600	2969.5	2975.7	2981.9	2988.1	2994.3	3000.5	3006.7	3012.9	3019.1	3025.3
610	3031.5	3037.7	3044.0	3050.2	3056.4	3062.6	3068.9	3075.1	3084.3	3087.5
620	3093.8	3100.0	3106.3	3112.5	3118.8	3125.0	3131.3	3137.5	3143.8	3150.0
630	3156.3	3162.5	3168.8	3175.1	3181.4	3187.6	3193.9	3200.2	3206.5	3212.7
640	3219.0	3225.3	3231.6	3237.9	3244.2	3250.5	3256.8	3263.1	3269.4	3275.7
650	3282.0	3288.3	3294.7	3301.0	3307.3	3313.6	3319.9	3326.3	3332.6	3338.9
660	3345.3	3351.6	3357.9	3364.3	3370.6	3377.0	3383.3	3389.7	3396.0	3402.4
670	3408.7	3415.1	3421.5	3427.8	3434.2	3440.6	3446.9	3453.3	3459.7	3466.1
680	3472.5	3478.8	3485.2	3491.6	3498.0	3504.4	3510.8	3517.2	3523.6	3530.0
690	3536.4	3542.8	3549.2	3555.6	3562.0	3568.5	3574.9	3581.3	3587.7	3594.2
700	3600.6	3607.0	3613.4	3919.9	3626.3	3632.8	3639.2	3645.6	3652.1	3658.5
710	3665.0	3671.4	3677.9	3684.4	3690.8	3697.3	3703.7	3710.2	3716.7	3723.2
720	3729.6	3736.1	3742.6	3749.1	3755.5	3762.0	3768.5	3775.0	3781.5	3788.0
730	3794.5	3801.0	3807.5	3814.0	3820.5	3827.0	3833.5	3840.0	3846.5	3853.1
740	3859.6	3866.1	3872.6	3879.1	3885.7	3892.2	3898.7	3905.3	3911.8	3918.3
750	3924.9	3914.4	3939.0	3944.5	3951.1	3957.6	3964.2	3970.7	3977.3	3983.8
760	3990.4	3997.0	4003.5	4010.1	4016.7	4023.2	4029.8	4036.4	4043.0	4049.6
770	4056.1	4062.7	4069.3	4075.9	4082.5	4089.1	4095.7	4102.3	4108.9	4115.5
780	4122.1	4128.7	4135.3	4141.9	4148.5	4155.2	4161.8	4168.4	4175.0	4181.6
790	4188.3	4194.9	4201.5	4208.1	4214.8	4221.4	4228.1	4234.7	4241.3	4248.0
800	4254.6	4261.3	4267.9	4274.6	4281.2	4287.9	4294.6	4301.2	4307.9	4314.5
810	4321.2	4327.9	4334.6	4341.2	4347.9	4354.6	4361.3	4367.9	4374.6	4381.3
820	4388.0	4394.7	4401.4	4408.1	4414.8	4421.5	4428.2	4434.9	4441.6	4448.3
830	4455.0	4461.7	4468.4	4475.1	4481.8	4488.6	4495.3	4502.0	4508.7	4515.5
840	4522.2	4528.9	4535.7	4542.4	4549.1	4555.9	4562.6	4569.3	4576.1	4582.8
850	4589.6	4596.3	4603.1	4609.8	4616.6	4623.4	4630.1	4636.9	4643.6	4650.4
860	4657.2	4663.9	4670.7	4677.5	4684.3	4691.0	4697.8	4704.6	4711.4	4718.2
870	4725.0	4731.8	4738.6	4745.3	4752.1	4758.9	4765.7	4772.5	4779.3	4786.1
880	4793.0	4799.8	4806.6	4813.4	4820.2	4827.0	4833.8	4840.7	4847.5	4854.3
890	4861.1	4868.0	4874.8	4881.6	4888.5	4895.3	4902.1	4909.0	4915.8	4922.7
900	4929.5	4936.4	4943.2	4950.1	4956.9	4963.8	4970.6	4977.5	4984.3	4991.2
910	4998.1	5004.9	5011.8	5018.7	5025.5	5032.4	5039.3	5046.2	5053.0	5059.9
920	5066.8	5073.7	5080.6	5087.5	5094.4	5101.3	5108.2	5115.0	5121.9	5128.0
930	5135.7	5142.7	5149.6	5156.5	5163.4	5170.3	5177.2	5184.1	5191.0	5198.0
940	5204.9	5211.8	5218.7	5225.6	5232.6	5239.5	5246.4	5253.4	5260.3	5267.2
950	5274.2	5281.1	5288.1	5295.0	5301.9	5308.9	5315.8	5322.8	5329.8	5336.7
960	5343.7	5350.6	5357.6	5364.5	5371.5	5378.5	5385.4	5392.4	5399.4	5406.4
970	5413.3	5420.3	5427.3	5434.3	5441.3	5448.2	5455.2	5462.2	5469.2	5476.2
980	5483.2	5490.2	5497.2	5504.2	5511.2	5518.2	5525.2	5532.2	5539.2	5546.2
990	5553.2	5560.2	5567.3	5574.2	5581.3	5588.3	5595.3	5602.3	5609.4	5616.4
1000	5623.4									

Table B-5 Hyperbolic logarithms

These two pages give the natural (hyperbolic, or Napierian) logarithms ($\log_e$) of numbers between 1 and 10, correct to four places. Moving the decimal point n places to the right [or left] in the number is equivalent to adding n times 2.3026 [or n times 3.6974] to the logarithm. Base e = 2.71828+

n	n(2.3026)	n(0.6974–3)
1	2.3026	0.6974–3
2	4.6052	0.3948–5
3	6.9078	0.0922–7
4	9.2103	0.7897–10
5	11.5129	0.4871–12
6	13.8155	0.1845–14
7	16.1181	0.8819–17
8	18.4207	0.5793–19
9	20.7233	0.2767–21

Number.	0	1	2	3	4	5	6	7	8	9	Avg. diff.
1.0	0.0000	0100	0198	0296	0392	0488	0583	0677	0770	0862	95
1.1	0953	1044	1133	1222	1310	1398	1484	1570	1655	1740	87
1.2	1823	1906	1989	2070	2151	2231	2311	2390	2469	2546	80
1.3	2624	2700	2776	2852	2927	3001	3075	3148	3221	3293	74
1.4	3365	3436	3507	3577	3646	3716	3784	3853	3920	3988	69
1.5	0.4055	4121	4187	4253	4318	4383	4447	4511	4574	4637	65
1.6	4700	4762	4824	4886	4947	5008	5068	5128	5188	5247	61
1.7	5306	5365	5423	5481	5539	5596	5653	5710	5766	5822	57
1.8	5878	5933	5988	6043	6098	6152	6206	6259	6313	6366	54
1.9	6419	6471	6523	6575	6627	6678	6729	6780	6831	6881	51
2.0	0.6931	6981	7031	7080	7129	7178	7227	7275	7324	7372	49
2.1	7419	7467	7514	7561	7608	7655	7701	7747	7793	7839	47
2.2	7885	7930	7975	8020	8065	8109	8154	8198	8242	8286	44
2.3	8329	8372	8416	8459	8502	8544	8587	8629	8671	8713	43
2.4	8755	8796	8838	8879	8920	8961	9002	9042	9083	9123	41
2.5	0.9163	9203	9243	9282	9322	9361	9400	9439	9478	9517	39
2.6	9555	9594	9632	9670	9708	9746	9783	9821	9858	9895	38
2.7	0.9933	9969	*0006	*0043	*0080	*0116	*0152	*0188	*0225	*0260	36
2.8	1.0296	0332	0367	0403	0438	0473	0508	0543	0578	0613	35
2.9	0647	0682	0716	0750	0784	0818	0852	0886	0919	0953	34
3.0	1.0986	1019	1053	1086	1119	1151	1184	1217	1249	1282	33
3.1	1314	1346	1378	1410	1442	1474	1506	1537	1569	1600	32
3.2	1632	1663	1694	1725	1756	1787	1817	1848	1878	1909	31
3.3	1939	1969	2000	2030	2060	2090	2119	2149	2179	2208	30
3.4	2238	2267	2296	2326	2355	2384	2413	2442	2470	2499	29
3.5	1.2528	2556	2585	2613	2641	2669	2698	2726	2754	2782	28
3.6	2809	2837	2865	2892	2920	2947	2975	3002	3029	3056	27
3.7	3083	3110	3137	3164	3191	3218	3244	3271	3297	3324	27
3.8	3350	3376	3403	3429	3455	3481	3507	3533	3558	3584	26
3.9	3610	3635	3661	3686	3712	3737	3762	3788	3813	3838	25
4.0	1.3863	3888	3913	3938	3962	3987	4012	4036	4061	4085	25
4.1	4110	4134	4159	4183	4207	4231	4255	4279	4303	4327	24
4.2	4351	4375	4398	4422	4446	4469	4493	4516	4540	4563	23
4.3	4586	4609	4633	4656	4679	4702	4725	4748	4770	4793	23
4.4	4816	4839	4861	4884	4907	4929	4951	4974	4996	5019	22
4.5	1.5041	5063	5085	5107	5129	5151	5173	5195	5217	5239	22
4.6	5261	5282	5304	5326	5347	5369	5390	5412	5433	5454	21
4.7	5476	5497	5518	5539	5560	5581	5602	5623	5644	5665	21
4.8	5686	5707	5728	5748	5769	5790	5810	5831	5851	5872	20
4.9	5892	5913	5933	5953	5974	5994	6014	6034	6054	6074	20
5.0	1.6094	6114	6134	6154	6174	6194	6214	6233	6253	6273	20
5.1	6292	6312	6332	6351	6371	6390	6409	6429	6448	6467	19
5.2	6487	6506	6525	6544	6563	6582	6601	6620	6639	6658	19
5.3	6677	6696	6715	6734	6752	6771	6790	6808	6827	6845	18
5.4	6864	6882	6901	6919	6938	6956	6974	6993	7011	7029	18

Table B-5 Hyperbolic logarithms (continued)

Number	0	1	2	3	4	5	6	7	8	9	Avg. diff.
5.5	1.7047	7066	7084	7102	7120	7138	7156	7174	7192	7210	18
5.6	7228	7246	7263	7281	7299	7317	7334	7352	7370	7387	18
5.7	7405	7422	7440	7457	7475	7492	7509	7527	7544	7561	17
5.8	7579	7596	7613	7630	7647	7664	7681	7699	7716	7733	17
5.9	7750	7766	7783	7800	7817	7834	7851	7867	7884	7901	17
6.0	1.7918	7934	7951	7967	7984	8001	8017	8034	8050	8066	16
6.1	8083	8099	8116	8132	8148	8165	8181	8197	8213	8229	16
6.2	8245	8262	8278	8294	8310	8326	8342	8358	8374	8390	16
6.3	8405	8421	8437	8453	8469	8485	8500	8516	8532	8547	16
6.4	8563	8579	8594	8610	8625	8641	8656	8672	8687	8703	15
6.5	1.8718	8733	8749	8764	8779	8795	8810	8825	8840	8856	15
6.6	8871	8886	8901	8916	8931	8946	8961	8976	8991	9006	15
6.7	9021	9036	9051	9066	9081	9095	9110	9125	9140	9155	15
6.8	9169	9184	9199	9213	9228	9242	9257	9272	9286	9301	15
6.9	9315	9330	9344	9359	9373	9387	9402	9416	9430	9445	14
7.0	1.9459	9473	9488	9502	9516	9530	9544	9559	9573	9587	14
7.1	9601	9615	9629	9643	9657	9671	9685	9699	9713	9727	14
7.2	9741	9755	9769	9782	9796	9810	9824	9838	9851	9865	14
7.3	1.9879	9892	9906	9920	9933	9947	9961	9974	9988	*0001	13
7.4	2.0015	0028	0042	0055	0069	0082	0096	0109	0122	0136	13
7.5	2.0149	0162	0176	0189	0202	0215	0229	0242	0255	0268	13
7.6	0281	0295	0308	0321	0334	0347	0360	0373	0386	0399	13
7.7	0412	0425	0438	0451	0464	0477	0490	0503	0516	0528	13
7.8	0541	0554	0567	0580	0592	0605	0618	0631	0643	0656	13
7.9	0669	0681	0694	0707	0719	0732	0744	0757	0769	0782	12
8.0	2.0794	0807	0819	0832	0844	0857	0869	0882	0894	0906	12
8.1	0919	0931	0943	0956	0968	0980	0992	1005	1017	1029	12
8.2	1041	1054	1066	1078	1090	1102	1114	1126	1138	1150	12
8.3	1163	1175	1187	1199	1211	1223	1235	1247	1258	1270	12
8.4	1282	1294	1306	1318	1330	1342	1353	1365	1377	1389	12
8.5	2.1401	1412	1424	1436	1448	1459	1471	1483	1494	1506	12
8.6	1518	1529	1541	1552	1564	1576	1587	1599	1610	1622	12
8.7	1633	1645	1656	1668	1679	1691	1702	1713	1725	1736	11
8.8	1748	1759	1770	1782	1793	1804	1815	1827	1838	1849	11
8.9	1861	1872	1883	1894	1905	1917	1928	1939	1950	1961	11
9.0	2.1972	1983	1994	2006	2017	2028	2039	2050	2061	2072	11
9.1	2083	2094	2105	2116	2127	2138	2148	2159	2170	2181	11
9.2	2192	2203	2214	2225	2235	2246	2257	2268	2279	2289	11
9.3	2300	2311	2322	2332	2343	2354	2364	2375	2386	2396	11
9.4	2407	2418	2428	2439	2450	2460	2471	2481	2492	2502	11
9.5	2.2513	2523	2534	2544	2555	2565	2576	2586	2597	2607	10
9.6	2618	2628	2638	2649	2659	2670	2680	2690	2701	2711	10
9.7	2721	2732	2742	2752	2762	2773	2783	2793	2803	2814	10
9.8	2824	2834	2844	2854	2865	2875	2885	2895	2905	2915	10
9.9	2925	2935	2946	2956	2966	2976	2986	2996	3006	3016	10
10.0	2.3026										

$\log_e x = (2.3026) \log_{10} x \qquad \log_{10} x = (0.4343) \log_e x$

where $2.3026 = \log_e 10$ and $0.4343 = \log_{10} e$

Moving the decimal point n places to the right [or left] in the number requires adding n times 2.3026 [or n times (0.6974−3)] in the body of the table. See auxiliary table of multiples on top of the preceding page.

Table B-6 NPS Pipe

EQUIVALENT THICKNESS – INCHES – ENGLISH

PIPE		NOMINAL INSULATION THICKNESS – INCHES																			
NPS	Diam."	½	1	1½	2	2½	3	3½	4	4½	5	5½	6	6½	7	7½	8	8½	9	9½	10
¼	0.540	0.81	1.98	3.32	4.84	6.45	8.13	9.05	11.70	13.70	15.55	17.66	19.69	21.80	23.92	26.11	28.28	30.52	32.82	35.07	37.38
⅜	0.675	0.76	1.84	3.12	4.53	6.04	7.64	9.32	11.06	12.87	14.74	16.64	18.57	20.65	22.60	24.69	26.76	28.90	31.00	33.15	33.33
½	0.840	0.72	1.73	2.92	4.23	5.66	7.18	8.74	10.39	12.10	13.82	15.69	17.53	19.38	21.30	23.28	25.26	27.29	29.29	31.35	33.02
¾	1.030	0.69	1.63	2.73	3.96	5.29	6.73	8.21	9.73	11.36	12.98	14.64	16.44	18.19	20.02	21.91	23.78	25.63	27.62	29.57	31.57
1	1.315	0.66	1.52	2.57	3.72	4.96	6.29	7.65	9.12	10.62	12.16	13.79	15.44	17.11	18.76	20.56	22.33	24.08	25.98	27.83	29.73
1¼	1.660	0.63	1.45	2.40	3.45	4.63	5.86	7.14	8.50	9.91	11.37	12.85	14.41	15.90	17.54	19.24	20.93	22.58	24.28	26.03	27.83
1½	1.900	0.62	1.40	2.30	3.33	4.45	5.65	6.85	8.17	9.54	10.89	12.38	13.83	15.35	16.85	18.42	20.05	21.74	23.38	25.08	26.83
2	2.375	0.59	1.33	2.21	3.16	4.17	5.28	6.47	7.62	8.88	10.21	11.50	12.93	14.39	15.80	17.24	18.83	20.34	21.90	23.51	25.03
2½	2.875	0.58	1.27	2.09	2.99	3.95	4.98	6.07	7.23	8.42	9.66	10.96	12.20	13.57	14.93	16.35	17.74	19.18	20.66	22.20	23.16
3	3.500	0.56	1.24	2.02	2.85	3.78	4.75	5.78	6.84	7.94	9.11	10.30	11.54	12.79	14.09	15.36	16.77	18.14	19.46	20.93	22.33
3½	4.000	0.53	1.22	1.96	2.76	3.64	4.60	5.56	6.60	7.67	8.75	9.90	11.12	12.32	13.50	14.82	16.10	17.43	18.70	20.15	21.48
4	4.500	0.53	1.19	1.91	2.71	3.56	4.46	5.41	6.38	7.42	8.47	9.61	10.73	11.90	13.04	14.53	15.58	16.77	18.11	19.39	20.70
4½	5.000	0.54	1.18	1.88	2.66	3.45	4.35	5.28	6.24	7.21	8.25	9.28	10.37	11.52	12.73	13.90	15.12	16.28	17.60	18.84	20.13
5	5.563	0.54	1.17	1.84	2.58	3.38	4.28	5.15	6.03	6.99	8.01	9.03	10.10	11.14	12.32	13.47	14.55	15.79	16.96	18.30	19.43
6	6.625	0.53	1.13	1.78	2.50	3.25	4.10	4.91	5.78	6.72	7.65	8.64	9.59	10.70	11.65	12.76	13.91	15.00	16.12	17.30	18.50
7	7.625	0.53	1.12	1.75	2.44	3.22	3.95	4.75	5.62	6.48	7.40	8.29	9.22	10.31	11.24	12.33	13.35	14.41	15.50	16.64	17.82
8	8.625	0.53	1.11	1.74	2.40	3.13	3.88	4.61	5.49	6.26	7.17	8.05	8.97	9.95	10.86	11.93	12.93	13.97	15.04	16.02	17.17
9	9.625	0.53	1.10	1.70	2.38	3.07	3.75	4.57	5.29	6.15	6.97	7.84	8.76	9.62	10.63	11.66	12.56	13.58	14.50	15.60	16.59
10	10.750	0.52	1.09	1.69	2.34	3.05	3.66	4.50	5.22	6.09	6.81	7.69	8.61	9.22	10.35	11.28	12.25	13.13	14.18	15.12	16.24
12	12.750	0.52	1.08	1.68	2.33	3.00	3.63	4.32	5.06	5.85	6.57	7.50	8.26	8.98	9.99	10.78	11.75	12.60	13.49	14.56	15.50
14	14.000	0.52	1.07	1.65	2.26	2.90	3.56	4.26	4.97	5.71	6.47	7.25	8.05	8.86	9.70	10.30	11.43	12.25	13.22	14.19	15.07
16	16.000	0.52	1.06	1.63	2.23	2.85	3.50	4.17	4.86	5.57	6.31	7.06	7.83	8.62	9.42	10.23	11.09	12.05	12.81	13.65	14.59
18	18.000	0.51	1.05	1.62	2.21	2.82	3.45	4.10	4.78	5.47	6.19	6.91	7.66	8.42	9.21	9.90	10.81	11.55	12.48	13.32	14.19
20	20.000	0.51	1.05	1.61	2.19	2.80	3.41	4.05	4.71	5.44	6.08	6.80	7.52	8.16	9.02	9.80	10.58	11.47	12.20	13.06	13.86
24	24.000	0.51	1.04	1.59	2.16	2.74	3.35	3.96	4.60	5.25	5.92	6.58	7.30	7.99	8.73	9.56	10.22	11.07	11.74	12.47	13.29
28	28.000	0.51	1.03	1.58	2.14	2.71	3.30	3.91	4.52	5.15	5.80	6.46	7.12	7.82	8.51	9.25	9.94	10.80	11.32	12.22	12.93
32	32.000	0.51	1.03	1.57	2.12	2.68	3.27	3.85	4.46	5.07	5.71	6.34	7.01	7.66	8.35	9.17	9.73	10.54	11.15	11.98	12.62
36	36.000	0.51	1.03	1.56	2.11	2.66	3.24	3.81	4.41	5.02	5.63	6.25	6.90	7.53	8.11	8.92	9.56	10.34	10.94	11.83	12.37
40	40.000	0.51	1.02	1.55	2.10	2.65	3.21	3.79	4.37	4.97	5.58	6.20	6.82	7.46	8.10	8.80	9.42	10.26	10.78	11.80	12.16
48	48.000	0.49	1.02	1.55	2.08	2.62	3.18	3.74	4.31	4.89	5.48	6.08	6.69	7.30	7.92	8.51	9.20	9.75	10.54	11.06	11.84

Note: Tabular values are those of the Equivalent Thickness $L = r_2 \log_e \frac{r_2}{r_1}$ for the Nominal Insulation Thicknesses given in inches.

r_1 = Inside Radius of Insulation in inches r_2 = Outside Radius of Insulation in inches $r_2 - r_1$ = Actual Thickness (ℓ) in inches
L = Equivalent Thickness in inches

Table B-6a Tube sizes

EQUIVALENT THICKNESS – INCHES – ENGLISH

TUBE SIZE		NOMINAL INSULATION THICKNESS – INCHES																			
Nom."	Diam.	½	1	1½	2	2½	3	3½	4	4½	5	5½	6	6½	7	7½	8	8½	9	9½	10
¼	0.250	1.04	2.47	4.17	6.29	7.99	10.04	13.21	14.82	16.70	18.78	21.41	23.84	25.88	28.81	31.34	33.91	35.49	39.15	40.76	44.49
½	0.500	0.82	2.01	3.40	4.93	6.58	8.33	10.15	12.04	13.98	15.98	18.03	20.11	22.25	24.44	26.58	28.86	31.15	33.39	35.68	38.07
1	1.000	0.69	1.65	2.77	4.02	5.34	6.81	8.32	9.89	11.51	13.18	14.90	16.67	18.67	20.31	22.16	24.08	26.01	27.97	30.00	31.97
2	2.000	0.61	1.39	2.29	3.30	4.38	5.55	6.77	8.05	9.38	10.75	12.17	13.62	15.21	16.63	18.19	19.77	21.38	23.03	24.68	26.38
3	3.000	0.57	1.28	2.08	2.97	3.92	4.94	6.02	7.15	8.32	9.53	10.78	12.07	13.34	14.73	16.21	17.48	19.00	20.43	21.89	23.32
4	4.000	0.56	1.22	1.96	2.77	3.64	4.58	5.56	6.59	7.71	8.76	9.91	11.09	12.30	13.54	14.82	16.10	17.33	18.75	20.13	21.50
5	5.000	0.55	1.18	1.88	2.65	3.47	4.34	5.25	6.21	7.21	8.24	9.31	10.40	11.53	12.67	13.90	15.07	16.28	17.55	18.84	20.12
6	6.000	0.54	1.15	1.82	2.55	3.32	4.16	5.03	5.93	6.87	7.85	8.84	9.88	10.95	12.04	13.13	14.29	15.41	16.64	17.88	19.06
7	7.000	0.53	1.13	1.78	2.49	3.23	4.02	4.85	5.72	6.61	7.54	8.50	9.49	10.50	11.54	12.54	13.68	14.76	15.91	17.03	18.22
8	8.000	0.53	1.12	1.75	2.43	3.16	3.92	4.73	5.55	6.41	7.30	8.22	9.16	10.13	11.13	12.08	13.18	14.25	15.32	16.47	17.54
9	9.000	0.53	1.10	1.72	2.39	3.10	3.83	4.60	5.40	6.24	7.09	7.99	8.90	9.83	10.79	11.76	12.77	13.78	14.83	15.82	16.97
10	10.000	0.52	1.09	1.70	2.35	3.04	3.76	4.51	5.29	6.10	6.93	7.79	8.67	9.58	10.51	11.50	12.42	13.37	14.41	15.37	16.47
12	12.000	0.52	1.08	1.67	2.30	2.95	3.65	4.36	5.11	5.98	6.65	7.48	8.32	9.17	10.05	10.93	11.86	12.76	13.74	14.73	16.18
14	14.000	0.52	1.07	1.65	2.26	2.90	3.56	4.26	4.97	5.71	6.47	7.25	8.05	8.86	9.70	10.55	11.43	12.25	13.22	14.19	15.07
16	16.000	0.52	1.06	1.63	2.23	2.85	3.50	4.17	4.86	5.57	6.31	7.06	7.83	8.62	9.42	10.23	11.09	12.05	12.81	13.65	14.59
18	18.000	0.51	1.05	1.62	2.21	2.82	3.45	4.10	4.78	5.47	6.19	6.91	7.66	8.42	9.21	9.90	10.81	11.65	12.48	13.32	14.19
20	20.000	0.51	1.05	1.61	2.19	2.80	3.41	4.05	4.71	5.44	6.08	6.80	7.52	8.16	9.02	9.80	10.58	11.47	12.20	13.06	13.86
24	24.000	0.51	1.04	1.59	2.16	2.74	3.35	3.96	4.60	5.25	5.92	6.59	7.30	7.99	8.73	9.56	10.22	11.07	11.75	12.47	13.29
28	28.000	0.51	1.03	1.58	2.14	2.71	3.30	3.91	4.52	5.15	5.80	6.46	7.12	7.82	8.51	9.25	9.94	10.80	11.32	12.22	12.93
32	32.000	0.51	1.03	1.57	2.12	2.68	3.27	3.85	4.46	5.07	5.71	6.34	7.01	7.66	8.35	9.17	9.73	10.54	11.15	11.98	12.62
36	36.000	0.51	1.03	1.56	2.11	2.66	3.24	3.81	4.41	5.02	5.63	6.25	6.90	7.53	8.21	8.92	9.56	10.34	10.94	11.83	12.37
40	40.000	0.51	1.02	1.55	2.10	2.65	3.21	3.79	4.37	4.97	5.58	6.20	6.82	7.46	8.10	8.80	9.42	10.26	10.78	11.80	12.16
48	48.000	0.50	1.02	1.55	2.08	2.62	3.18	3.74	4.31	4.90	5.48	6.08	6.69	7.30	7.92	8.51	9.20	9.75	10.51	11.06	11.84
56	56.000	0.50	1.02	1.54	2.06	2.60	3.15	3.71	4.27	4.82	5.42	5.99	6.60	7.18	7.81	8.42	9.05	9.67	10.31	10.88	11.60
64	64.000	0.50	1.02	1.54	2.04	2.59	3.14	3.67	4.24	4.78	5.36	5.95	6.51	7.11	7.72	8.30	8.92	9.57	10.15	10.79	11.40
72	72.000	0.50	1.02	1.53	2.03	2.56	3.11	3.65	4.21	4.76	5.33	5.91	6.48	7.07	7.62	8.22	8.82	9.43	10.04	10.65	11.28

Note: Tabular values are those of the Equivalent Thickness $L = r_2 \log_e \frac{r_2}{r_1}$ for the Nominal Insulation Thicknesses given in inches.

r_1 = Inside Radius of Insulation in inches r_2 = Outside Radius of Insulation in inches $r_2 - r_1$ = Actual Thickness (ℓ) in inches

L = Equivalent Thickness in inches

Table B-6b NPS Pipe

Equivalent Thickness — Metres (Metric)

Pipe Size		NOMINAL INSULATION THICKNESS — mm																			
NPS inches	Dia. mm	13	25	38	51	64	76	89	102	114	127	140	152	165	178	190	203	216	229	241	250
1/4	13.7	0.021	0.050	0.084	0.123	0.164	0.206	0.229	0.297	0.348	0.395	0.448	0.500	0.553	0.607	0.663	0.718	0.775	0.833	0.891	0.949
3/8	17.1	0.019	0.047	0.079	0.115	0.153	0.194	0.236	0.281	0.327	0.374	0.422	0.471	0.524	0.574	0.627	0.680	0.734	0.787	0.842	0.846
1/2	21.3	0.018	0.045	0.074	0.107	0.144	0.182	0.221	0.264	0.307	0.351	0.398	0.447	0.492	0.541	0.591	0.641	0.693	0.744	0.796	0.839
3/4	26.7	0.017	0.041	0.069	0.101	0.134	0.171	0.208	0.247	0.288	0.329	0.372	0.417	0.462	0.508	0.556	0.604	0.651	0.701	0.751	0.802
1	33.4	0.017	0.039	0.065	0.094	0.126	0.160	0.194	0.232	0.270	0.309	0.350	0.392	0.434	0.476	0.522	0.567	0.611	0.660	0.707	0.753
1 1/4	42.2	0.016	0.037	0.061	0.088	0.117	0.149	0.181	0.216	0.251	0.288	0.326	0.366	0.403	0.445	0.488	0.531	0.573	0.617	0.661	0.707
1 1/2	48.3	0.016	0.036	0.058	0.085	0.113	0.143	0.174	0.231	0.242	0.276	0.314	0.351	0.390	0.428	0.468	0.509	0.552	0.594	0.637	0.681
2	60.3	0.015	0.034	0.056	0.080	0.106	0.134	0.164	0.215	0.225	0.259	0.292	0.328	0.365	0.401	0.438	0.478	0.516	0.556	0.597	0.636
2 1/2	73.0	0.015	0.032	0.053	0.076	0.100	0.126	0.154	0.183	0.214	0.245	0.278	0.309	0.344	0.379	0.415	0.450	0.487	0.524	0.564	0.588
3	88.9	0.014	0.031	0.051	0.073	0.096	0.121	0.147	0.173	0.201	0.231	0.261	0.293	0.324	0.357	0.390	0.426	0.460	0.494	0.531	0.567
3 1/2	101.6	0.013	0.031	0.050	0.070	0.092	0.117	0.141	0.167	0.195	0.222	0.251	0.282	0.312	0.342	0.376	0.408	0.443	0.474	0.512	0.545
4	114.3	0.013	0.030	0.049	0.069	0.090	0.113	0.137	0.162	0.188	0.215	0.244	0.272	0.302	0.331	0.369	0.396	0.426	0.460	0.492	0.526
4 1/2	127.0	0.014	0.030	0.048	0.068	0.088	0.110	0.134	0.158	0.183	0.209	0.235	0.263	0.292	0.323	0.353	0.384	0.413	0.447	0.478	0.511
5	141.3	0.014	0.030	0.048	0.065	0.086	0.108	0.131	0.153	0.177	0.203	0.229	0.256	0.282	0.312	0.342	0.369	0.401	0.431	0.464	0.493
6	168.3	0.013	0.029	0.045	0.063	0.082	0.104	0.124	0.146	0.170	0.194	0.219	0.242	0.271	0.295	0.324	0.353	0.381	0.409	0.439	0.470
7	193.7	0.013	0.028	0.044	0.062	0.082	0.100	0.120	0.142	0.164	0.188	0.210	0.234	0.261	0.285	0.313	0.339	0.366	0.394	0.422	0.452
8	219.1	0.013	0.028	0.044	0.061	0.079	0.098	0.117	0.139	0.159	0.182	0.204	0.228	0.252	0.276	0.303	0.328	0.355	0.382	0.407	0.436
9	244.5	0.013	0.028	0.043	0.060	0.078	0.095	0.116	0.134	0.156	0.177	0.199	0.222	0.244	0.270	0.296	0.319	0.345	0.368	0.396	0.421
10	273.5	0.013	0.028	0.043	0.059	0.077	0.093	0.114	0.132	0.154	0.172	0.195	0.218	0.234	0.263	0.286	0.311	0.333	0.360	0.384	0.412
12	323.8	0.013	0.027	0.043	0.059	0.076	0.092	0.109	0.128	0.148	0.167	0.190	0.210	0.228	0.253	0.274	0.298	0.320	0.342	0.370	0.394
14	355.6	0.013	0.027	0.042	0.057	0.074	0.090	0.108	0.126	0.145	0.164	0.184	0.204	0.225	0.246	0.261	0.290	0.311	0.336	0.360	0.382
16	406.4	0.013	0.027	0.041	0.056	0.072	0.089	0.106	0.123	0.141	0.160	0.179	0.199	0.218	0.239	0.260	0.281	0.306	0.325	0.347	0.370
18	457.2	0.013	0.027	0.041	0.056	0.071	0.087	0.104	0.121	0.139	0.157	0.175	0.194	0.214	0.234	0.251	0.274	0.293	0.317	0.338	0.360
20	508.0	0.013	0.027	0.041	0.055	0.071	0.086	0.100	0.119	0.138	0.154	0.172	0.191	0.207	0.229	0.248	0.269	0.291	0.310	0.332	0.352
24	609.6	0.013	0.026	0.040	0.054	0.069	0.085	0.100	0.116	0.133	0.150	0.167	0.185	0.203	0.221	0.242	0.259	0.281	0.298	0.317	0.337
28	711.2	0.013	0.026	0.040	0.054	0.068	0.084	0.099	0.114	0.131	0.147	0.164	0.180	0.198	0.216	0.235	0.252	0.274	0.287	0.310	0.328
32	812.8	0.013	0.026	0.040	0.054	0.068	0.083	0.097	0.113	0.128	0.145	0.161	0.178	0.194	0.212	0.233	0.247	0.268	0.283	0.304	0.320
36	914.4	0.013	0.026	0.039	0.053	0.067	0.082	0.096	0.112	0.127	0.143	0.158	0.175	0.191	0.206	0.226	0.242	0.262	0.278	0.300	0.314
40	1016.0	0.013	0.026	0.039	0.053	0.067	0.081	0.096	0.111	0.126	0.141	0.157	0.173	0.189	0.205	0.223	0.239	0.260	0.274	0.299	0.309
48	1219.2	0.013	0.026	0.039	0.053	0.066	0.080	0.095	0.110	0.124	0.139	0.154	0.170	0.185	0.201	0.216	0.234	0.248	0.268	0.280	0.301

Note: Tabular values are those of the Equivalent Thickness $L_m = r_{2m} \log_e \frac{r_{2m}}{r_{1m}}$ for the Nominal Insulation Thickness given.

r_{1m} = Inside radius of insulation in metres

$r_{1m} - r_{2m}$ = Actual thickness (ℓ_m) in metres.

r_{2m} = Outside radius of insulation in metres.

L_m = Equivalent thickness in metres.

Table B-6c Tube Sizes

Equivalent Thickness — Metres (Metric)

Tube Size		NOMINAL INSULATION THICKNESS – mm																			
Dia. inches	Dia. mm	13	25	38	51	64	76	89	102	114	127	140	152	165	178	190	203	216	229	241	250
1/4	6.4	0.026	0.063	0.106	0.160	0.202	0.255	0.335	0.376	0.424	0.477	0.544	0.605	0.657	0.731	0.796	0.861	0.901	0.994	1.035	1.130
1/2	15.9	0.021	0.051	0.086	0.125	0.167	0.211	0.258	0.306	0.355	0.406	0.458	0.511	0.565	0.620	0.675	0.733	0.791	0.848	0.906	0.967
1	22.2	0.017	0.042	0.070	0.102	0.135	0.172	0.211	0.251	0.292	0.334	0.378	0.423	0.474	0.516	0.562	0.612	0.661	0.710	0.762	0.812
2	59.0	0.015	0.035	0.058	0.084	0.111	0.141	0.172	0.204	0.238	0.273	0.309	0.346	0.386	0.422	0.462	0.502	0.543	0.585	0.627	0.670
3	79.4	0.014	0.032	0.053	0.075	0.099	0.125	0.153	0.182	0.211	0.242	0.274	0.306	0.339	0.374	0.412	0.444	0.482	0.519	0.556	0.592
4	104.8	0.014	0.031	0.050	0.070	0.092	0.116	0.141	0.167	0.195	0.222	0.252	0.282	0.312	0.344	0.376	0.409	0.440	0.476	0.511	0.546
5	130.2	0.014	0.030	0.048	0.067	0.088	0.110	0.133	0.158	0.183	0.209	0.236	0.264	0.293	0.322	0.353	0.382	0.413	0.446	0.478	0.511
6	155.7	0.014	0.029	0.046	0.065	0.084	0.105	0.128	0.150	0.174	0.199	0.224	0.251	0.278	0.306	0.333	0.363	0.391	0.422	0.454	0.484
7	177.8	0.013	0.029	0.045	0.063	0.082	0.102	0.123	0.145	0.168	0.191	0.216	0.241	0.267	0.293	0.318	0.347	0.375	0.404	0.432	0.463
8	203.2	0.013	0.028	0.044	0.062	0.080	0.099	0.120	0.141	0.163	0.185	0.209	0.233	0.257	0.282	0.307	0.335	0.362	0.389	0.418	0.445
9	228.6	0.013	0.028	0.044	0.061	0.079	0.097	0.117	0.137	0.158	0.180	0.203	0.226	0.250	0.274	0.299	0.324	0.350	0.377	0.401	0.431
10	254.0	0.013	0.028	0.043	0.060	0.077	0.095	0.115	0.134	0.155	0.176	0.198	0.220	0.243	0.266	0.292	0.315	0.340	0.366	0.390	0.418
12	304.8	0.013	0.027	0.042	0.058	0.075	0.093	0.111	0.130	0.152	0.169	0.190	0.211	0.233	0.255	0.277	0.290	0.324	0.349	0.374	0.411
14	355.6	0.013	0.027	0.042	0.057	0.079	0.090	0.108	0.126	0.145	0.164	0.184	0.204	0.225	0.246	0.261	0.290	0.311	0.336	0.360	0.382
16	406.4	0.013	0.027	0.041	0.056	0.072	0.089	0.106	0.123	0.141	0.160	0.179	0.199	0.218	0.239	0.260	0.281	0.306	0.325	0.347	0.370
18	437.2	0.013	0.027	0.041	0.056	0.071	0.087	0.104	0.121	0.139	0.157	0.175	0.194	0.214	0.234	0.251	0.274	0.293	0.317	0.338	0.360
20	508.0	0.013	0.027	0.041	0.055	0.071	0.086	0.100	0.119	0.138	0.159	0.172	0.191	0.207	0.229	0.248	0.269	0.291	0.310	0.332	0.352
24	609.6	0.013	0.026	0.040	0.054	0.069	0.085	0.100	0.116	0.133	0.150	0.167	0.185	0.203	0.221	0.242	0.259	0.281	0.298	0.317	0.337
28	711.2	0.013	0.026	0.040	0.054	0.068	0.084	0.099	0.118	0.131	0.197	0.164	0.180	0.198	0.216	0.235	0.252	0.279	0.297	0.310	0.328
32	812.8	0.013	0.026	0.040	0.054	0.068	0.083	0.097	0.113	0.128	0.145	0.161	0.178	0.194	0.212	0.233	0.247	0.268	0.283	0.304	0.320
36	914.4	0.013	0.026	0.039	0.053	0.067	0.082	0.096	0.112	0.127	0.143	0.158	0.175	0.191	0.206	0.226	0.242	0.262	0.278	0.300	0.314
40	1016.0	0.013	0.026	0.039	0.053	0.067	0.081	0.096	0.111	0.126	0.141	0.157	0.173	0.189	0.205	0.223	0.239	0.260	0.274	0.299	0.309
48	1219.2	0.012	0.026	0.039	0.053	0.066	0.080	0.095	0.110	0.124	0.139	0.154	0.170	0.185	0.201	0.216	0.234	0.248	0.268	0.280	0.301
56	1422.4	0.012	0.026	0.039	0.052	0.066	0.080	0.094	0.108	0.122	0.138	0.152	0.168	0.182	0.198	0.214	0.230	0.245	0.262	0.276	0.295
64	1625.6	0.012	0.026	0.039	0.052	0.066	0.080	0.093	0.108	0.121	0.136	0.151	0.165	0.180	0.196	0.210	0.226	0.243	0.259	0.274	0.289
72	1828.8	0.012	0.026	0.039	0.052	0.065	0.079	0.093	0.107	0.121	0.135	0.150	0.164	0.179	0.194	0.209	0.224	0.239	0.255	0.270	0.286

Note: Tabular values are those of the Equivalent Thickness $L_m = r_{2m} \log_e \frac{r_{2m}}{r_{1m}}$ for the Nominal Insulation Thickness given.

r_{1m} = Inside radius of insulation in metres

$r_{1m} - r_{2m}$ = Actual thickness (ℓ_m) in metres.

r_{2m} = Outside radius of insulation in metres.

L_m = Equivalent thickness in metres.

Table B-7 Data for use in heat transfer formulae on combination (two materials) pipe insulation — English Units

Note: Values for $r_3 \log_e \frac{r_2}{r_1}$ and $r_3 \log_e \frac{r_3}{r_2}$ are based on actual thicknesses of Pabco Pipe Insulation.

Pipe size	Sq ft area of pipe per lin ft	Nominal thickness of two layers	Equivalent Thickness $r_3 \log_e \frac{r_2}{r_1}$	Equivalent Thickness $r_3 \log_e \frac{r_3}{r_2}$
2"	.622	1" x 1"	2.114	1.280
2"	"	1" x 1½"	2.433	2.009
2"	"	1" x 2"	2.752	2.804
2"	"	1" x 2½"	3.071	3.659
2"	"	1" x 3"	3.410	4.622
2"	"	1½" x 1"	3.241	1.203
2"	"	1½" x 1½"	3.666	1.890
2"	"	1½" x 2"	4.091	2.638
2"	"	2" x 1"	4.422	1.134
2"	"	2" x 1½"	4.934	1.794
2"	"	2" x 2"	5.479	2.554
2"	"	2½" x 1½"	6.230	1.801
2"	"	2½" x 2"	6.813	2.492
2"	"	2½" x 2½"	7.396	3.226
2½"	.753	1" x 1"	2.089	1.563
2½"	"	1" x 1½"	2.383	2.351
2½"	"	1½" x 1½"	4.015	1.794
2½"	"	1½" x 2"	4.458	2.554
2½"	"	2" x 1½"	5.165	1.847
2½"	"	2" x 2"	5.618	2.498
2½"	"	2½" x 1½"	6.384	1.731
2½"	"	2½" x 2"	6.967	2.444
2½"	"	2½" x 2½"	7.653	3.340
3"	.917	1" x 1"	1.765	1.200
3"	"	1" x 1½"	1.996	1.889
3"	"	1½" x 1½"	3.068	1.794
3"	"	1½" x 2"	3.406	2.554
3"	"	2" x 1½"	4.159	1.801
3"	"	2" x 2"	4.548	2.492
3"	"	2½" x 1½"	5.269	1.775
3"	"	2½" x 2"	5.720	2.448
3"	"	2½" x 2½"	6.283	3.340
3½"	1.047	1" x 1"	2.117	1.136
3½"	"	1" x 1½"	2.365	1.799
3½"	"	1½" x 1½"	3.446	1.801
3½"	"	1½" x 2"	3.768	2.492
3½"	"	2" x 1½"	4.445	1.813
3½"	"	2" x 2"	4.802	2.448
3½"	"	2½" x 1½"	5.499	1.752
3½"	"	2½" x 2½"	6.044	2.577
4"	1.179	1" x 1"	1.667	1.134
4"	"	1" x 1½"	1.860	1.794
4"	"	1½" x 1½"	2.816	1.801
4"	"	1½" x 2"	3.079	2.492
4"	"	1½" x 2½"	3.343	3.226
4"	"	2" x 1½"	3.799	1.775
4"	"	2" x 2"	4.124	2.448
4"	"	2" x 2½"	4.551	3.389
4"	"	2½" x 1½"	4.819	1.751
4"	"	2½" x 2"	5.319	2.620
4"	"	2½" x 2½"	5.651	3.235
4"	"	3" x 1½"	6.052	1.890
4"	"	3" x 2"	6.429	2.454
4"	"	3" x 2½"	6.862	3.138
4"	"	3½" x 1½"	7.094	1.789
4"	"	3½" x 2"	7.571	2.428
4½"	1.310	1" x 1½"	2.196	1.802
4½"	"	1½" x 1½"	3.182	1.775
4½"	"	1½" x 2"	3.454	2.448
4½"	"	2" x 1½"	4.150	1.751
4½"	"	2" x 2"	4.559	2.579
4½"	"	2½" x 2½"	5.959	3.143
5"	1.458	1" x 1½"	1.682	1.801
5"	"	1½" x 1½"	2.561	1.771
5"	"	1½" x 2"	2.780	2.444
5"	"	1½" x 2½"	3.067	3.389
5"	"	2" x 1½"	3.477	1.751
5"	"	2" x 2"	3.819	2.574
5"	"	2" x 2½"	4.076	3.235
5"	"	2½" x 1½"	4.546	1.849
5"	"	2½" x 2"	4.852	2.454
5"	"	2½" x 2½"	5.178	3.138
5"	"	3" x 1½"	5.518	1.789
5"	"	3" x 2"	5.889	2.429
5"	"	3" x 2½"	6.283	3.140
5"	"	3½" x 1½"	6.542	1.777
6"	1.736	1" x 1½"	1.537	1.771
6"	"	1½" x 1½"	2.365	1.751
6"	"	1½" x 2"	2.598	2.574
6"	"	1½" x 2½"	2.773	3.235
6"	"	2" x 1½"	3.317	1.809
6"	"	2" x 2"	3.556	2.454
6"	"	2" x 2½"	3.795	3.138
6"	"	2½" x 1½"	4.217	1.789
6"	"	2½" x 2"	4.500	2.429
6"	"	2½" x 2½"	4.784	3.094
6"	"	3" x 1½"	5.153	1.777
6"	"	3" x 2"	5.477	2.406
6"	"	3" x 2½"	5.822	3.100
6"	"	3½" x 1½"	6.272	1.608
8"	2.262	1½" x 1½"	2.254	1.789
8"	"	1½" x 2"	2.405	2.429
8"	"	1½" x 2½"	2.557	3.094
8"	"	2" x 1½"	3.058	1.777
8"	"	2" x 2"	3.250	2.406
8"	"	2" x 2½"	3.455	3.100
8"	"	2½" x 1½"	4.086	1.573
8"	"	2½" x 2"	4.358	2.256
8"	"	2½" x 2½"	4.600	2.900
8"	"	3" x 1½"	4.901	1.715
8"	"	3" x 2"	5.174	2.322
8"	"	3" x 2½"	5.446	2.956
8"	"	3½" x 1½"	5.792	1.700

Metric Units: When radii r_1, r_2, r_3 are r_{1m}, r_{2m}, r_{3m} given in metres, the equivalent thickness values given above can be converted to meters by multiplying by 0.0254.

Table B-7 Data for use in heat transfer formulae on combination (two materials) pipe insulation (continued) – English Units

Pipe size	Sq ft area of pipe per lin ft	Nominal thickness of two layers	Equivalent Thickness $r_3 \log_e \frac{r_2}{r_1}$	Equivalent Thickness $r_3 \log_e \frac{r_3}{r_2}$
10″	2.817	1½″ x 1½″	2.195	1.649
10″	″	1½″ x 2″	2.325	2.262
10″	″	1½″ x 2½″	2.454	2.897
10″	″	2″ x 1½″	2.946	1.635
10″	″	2″ x 2″	3.110	2.240
10″	″	2″ x 2½″	3.275	2.872
10″	″	2½″ x 1½″	3.724	1.631
10″	″	2½″ x 2″	3.921	2.232
10″	″	2½″ x 2½″	4.117	2.850
10″	″	3″ x 1½″	4.528	1.624
10″	″	3″ x 2″	4.755	2.217
10″	″	3″ x 2½″	4.982	2.835
10″	″	3½″ x 1½″	5.355	1.616
12″	3.344	1½″ x 1½″	2.105	1.631
12″	″	1½″ x 2″	2.216	2.232
12″	″	1½″ x 2½″	2.327	2.850
12″	″	2″ x 1½″	2.827	1.624
12″	″	2″ x 2″	2.969	2.217
12″	″	2″ x 2½″	3.120	2.877
12″	″	2½″ x 1½″	3.615	1.612
12″	″	2½″ x 2″	3.787	2.205
12″	″	2½″ x 2½″	3.959	2.811
12″	″	3″ x 1½″	4.349	1.642
12″	″	3″ x 2″	4.547	2.230
12″	″	3½″ x 1½″	5.135	1.637
12″	″	3½″ x 2″	5.344	2.182
12″	″	3½″ x 2½″	5.568	2.782
12″	″	4″ x 1½″	5.936	1.601
14″	3.665	1½″ x 1½″	1.939	1.621
14″	″	1½″ x 2″	2.036	2.216
14″	″	1½″ x 2½″	2.133	2.834
14″	″	2″ x 1½″	2.632	1.612
14″	″	2″ x 2″	2.758	2.205
14″	″	2″ x 2½″	2.883	2.811
14″	″	2½″ x 1½″	3.357	1.604
14″	″	2½″ x 2″	3.510	2.192
14″	″	3″ x 1½″	4.097	1.607
14″	″	3″ x 2″	4.275	2.187
14″	″	3½″ x 1½″	4.865	1.593
14″	″	3½″ x 2″	5.068	2.174
14″	″	3½″ x 2½″	5.271	2.775
14″	″	4″ x 1½″	5.646	1.594
14″	″	4½″ x 1½″	6.455	2.102
16″	4.189	1½″ x 1½″	1.885	1.604
16″	″	1½″ x 2″	1.971	2.192
16″	″	1½″ x 2½″	2.056	2.801
16″	″	2″ x 1½″	2.566	1.607
16″	″	2″ x 2″	2.677	2.187
16″	″	2″ x 2½″	2.789	2.789
16″	″	2½″ x 1½″	3.258	1.593
16″	″	2½″ x 2″	3.394	2.174
16″	″	3″ x 1½″	3.980	1.593
16″	″	3″ x 2″	4.139	2.162
16″	″	3½″ x 1½″	4.713	1.588
16″	″	3½″ x 2″	4.894	2.153
16″	″	3½″ x 2½″	5.075	2.749
16″	″	4″ x 1½″	5.474	1.590
18″	4.713	1½″ x 1½″	1.842	1.593
18″	″	1½″ x 2″	1.919	2.174
18″	″	1½″ x 2½″	1.996	2.775
18″	″	2″ x 1½″	2.506	1.593
18″	″	2″ x 2″	2.606	2.162
18″	″	2″ x 2½″	2.706	2.761
18″	″	2½″ x 1½″	3.178	1.588
18″	″	2½″ x 2″	3.300	2.153
18″	″	3″ x 1½″	3.880	1.589
18″	″	3″ x 2″	4.023	2.149
18″	″	3½″ x 1½″	4.589	1.586
18″	″	3½″ x 2″	4.753	2.152
18″	″	3½″ x 2½″	4.917	2.734
18″	″	4″ x 1½″	5.327	1.578
20″	5.238	1½″ x 1½″	1.816	1.588
20″	″	1½″ x 2″	1.886	2.153
20″	″	1½″ x 2½″	1.956	2.749
20″	″	2″ x 1½″	2.461	1.589
20″	″	2″ x 2″	2.552	2.149
20″	″	2″ x 2½″	2.643	2.739
20″	″	2½″ x 1½″	3.124	1.586
20″	″	2½″ x 2″	3.235	2.152
20″	″	3″ x 1½″	3.804	1.578
20″	″	3″ x 2″	3.935	2.135
20″	″	3½″ x 1½″	4.501	1.578
20″	″	3½″ x 2″	4.651	2.139
20″	″	3½″ x 2½″	4.801	2.715
20″	″	4″ x 1½″	5.215	1.575
24″	6.286	1½″ x 1½″	1.766	1.578
24″	″	1½″ x 2″	1.825	2.139
24″	″	1½″ x 2½″	1.884	2.715
24″	″	2″ x 1½″	2.380	1.575
24″	″	2″ x 2″	2.457	2.124
24″	″	2″ x 2½″	2.533	2.702
24″	″	2½″ x 1½″	3.023	1.568
24″	″	2½″ x 2″	3.117	2.118
24″	″	3″ x 1½″	3.681	1.572
24″	″	3″ x 2″	3.793	2.122
24″	″	3½″ x 1½″	4.342	1.558
24″	″	3½″ x 2″	4.469	2.123
24″	″	3½″ x 2½″	4.597	2.686
24″	″	4″ x 1½″	5.030	1.572
30″	7.854	1½″ x 1½″	1.716	1.566
30″	″	1½″ x 2″	1.763	2.112
30″	″	1½″ x 2½″	1.810	2.671
30″	″	2″ x 1½″	2.309	1.563
30″	″	2″ x 2″	2.372	2.102
30″	″	2″ x 2½″	2.434	2.674
30″	″	2½″ x 1½″	2.917	1.549
30″	″	2½″ x 2″	2.994	2.104
30″	″	3″ x 1½″	3.555	1.554
30″	″	3″ x 2″	3.646	2.105
30″	″	3½″ x 1½″	4.188	1.557
30″	″	3½″ x 2″	4.294	2.105
30″	″	3½″ x 2½″	4.398	2.659
30″	″	4″ x 1½″	4.851	1.559

Metric Units: When radii r_1, r_2, r_3 are r_{1m}, r_{2m}, r_{3m} given in metres, the equivalent thickness values given above can be converted to meters by multiplying by 0.0254.

Table B-8 Radiation heat transfer table Btu/sq ft, hr to absolute zero — English Units — Radiation in Btu/ft², hr

Temperature		Emittance										Temperature	
°F	°R	1.0	0.9	0.8	0.7	0.6	0.5	0.4	0.3	0.2	0.1	°C	°K
−210	250	6.80	6.12	5.44	4.76	4.08	3.40	2.72	2.04	1.36	0.68	−134.4	138.8
−205	255	7.36	6.62	5.89	5.15	4.41	3.68	2.94	2.21	1.47	0.74	−131.7	141.0
−200	260	7.95	7.16	6.36	5.57	4.77	3.98	3.18	2.39	1.59	0.80	−129.9	144.3
−195	265	8.58	7.72	6.86	6.01	5.15	4.29	3.43	2.57	1.72	0.86	−126.1	147.1
−190	270	9.25	8.32	7.40	6.47	5.55	4.62	3.70	2.77	1.85	0.92	−123.3	149.9
−185	275	9.95	8.96	7.96	6.97	5.97	4.98	3.98	2.99	1.99	1.00	−120.6	152.6
−180	280	10.70	9.63	8.56	7.49	6.42	5.35	4.28	3.21	2.14	1.07	−117.8	156.0
−175	285	11.48	10.33	9.18	8.04	6.89	5.74	4.59	3.44	2.30	1.15	−113.0	158.2
−170	290	12.31	11.08	9.85	8.61	7.38	6.15	4.92	3.69	2.46	1.23	−112.2	161.0
−165	295	13.18	11.86	10.54	9.22	7.91	6.59	5.27	3.95	2.64	1.32	−109.4	163.8
−160	300	14.09	12.68	11.28	9.87	8.46	7.05	5.64	4.23	2.82	1.41	−106.7	166.5
−155	305	15.06	13.55	12.05	10.54	9.03	7.53	6.02	4.52	3.01	1.51	−103.9	169.3
−150	310	16.07	14.46	12.86	11.25	9.64	8.03	6.43	4.82	3.21	1.61	−101.1	172.1
−145	315	17.13	15.42	13.71	11.99	10.28	8.57	6.85	5.14	3.43	1.71	−98.3	174.9
−140	320	18.25	16.42	14.60	12.77	10.95	9.12	7.30	5.47	3.65	1.82	−95.6	171.6
−135	325	19.41	17.47	15.53	13.59	11.65	9.71	7.77	5.82	3.88	1.94	−92.8	180.4
−130	330	20.64	18.57	16.51	14.44	12.38	10.32	8.25	6.19	4.13	2.06	−90.0	183.2
−125	335	21.91	19.72	17.53	15.34	13.15	10.96	8.77	6.57	4.38	2.10	−87.2	186.0
−120	340	23.25	20.93	18.60	16.28	13.95	11.63	9.30	6.98	4.65	2.33	−84.4	188.8
−115	345	24.65	22.19	19.71	17.26	14.79	12.33	9.86	7.40	4.93	2.47	−81.7	191.5
−110	350	26.11	23.50	20.89	18.28	15.67	13.05	10.44	7.83	5.22	2.61	−78.9	194.3
−105	355	27.64	24.87	22.11	19.34	16.58	13.82	11.05	8.29	5.53	2.76	−76.1	197.1
−100	360	29.23	26.30	23.38	20.46	17.54	14.61	11.69	8.77	5.85	2.92	−73.3	199.9
−95	365	30.88	27.79	24.71	21.62	18.53	15.44	12.35	9.26	6.18	3.09	−70.6	202.6
−90	370	32.61	29.35	26.09	22.83	19.57	16.31	13.04	9.78	6.52	3.26	−67.8	206.0
−85	375	34.41	30.97	27.53	24.09	20.65	17.20	13.76	10.32	6.88	3.44	−65.0	208.2
−80	380	36.28	32.65	29.03	25.40	21.77	18.14	14.51	10.88	7.26	3.63	−62.8	211.0
−75	385	38.23	34.41	30.58	26.76	22.94	19.11	15.29	11.47	7.65	3.82	−59.4	213.8
−70	390	40.25	36.23	32.20	28.18	24.15	20.13	16.10	12.08	8.05	4.03	−56.7	216.5
−65	395	42.36	38.12	33.89	29.65	25.41	21.18	16.94	12.71	8.47	4.24	−53.9	219.3
−60	400	44.54	40.09	35.64	31.18	26.73	22.27	17.82	13.36	8.91	4.45	−51.1	222.1
−55	405	46.81	42.13	37.45	32.77	28.09	23.41	18.73	14.04	9.36	4.68	−48.3	224.9
−50	410	49.17	44.25	39.33	34.42	29.50	24.58	19.67	14.75	9.83	4.92	−45.6	227.6
−45	415	51.61	46.45	41.29	36.13	30.97	25.81	20.64	15.48	10.32	5.16	−42.8	230.4
−40	420	54.14	48.73	43.31	37.90	32.49	27.07	21.66	16.24	10.83	5.41	−40.0	233.2
−35	425	56.77	51.09	45.41	39.74	34.06	28.38	22.71	17.03	11.35	5.68	−37.2	236.0
−30	430	59.49	53.54	47.59	41.64	35.69	29.74	23.79	17.85	11.90	5.95	−34.4	238.8
−25	435	62.30	56.07	49.84	43.61	37.38	31.15	24.92	18.69	12.46	6.23	−31.7	242.1
−20	440	65.22	58.70	52.17	45.65	39.13	32.61	26.09	19.57	13.04	6.52	−28.9	244.3
−15	445	68.23	61.41	54.59	47.76	40.94	34.12	27.29	20.47	13.65	6.82	−26.1	247.1
−10	450	71.35	64.22	57.08	49.95	42.81	35.68	28.54	21.41	14.27	7.14	−23.2	249.9
−5	455	74.58	67.12	59.66	52.20	44.75	37.29	29.83	22.37	14.92	7.46	−20.6	252.6
0	460	77.91	70.12	62.33	54.54	46.74	38.95	31.16	23.37	15.58	7.79	−17.8	255.4
5	465	81.35	73.22	65.08	56.95	48.81	40.68	32.54	24.41	16.27	8.14	−15.0	258.2
10	470	84.91	76.42	67.93	59.43	50.94	42.45	33.96	25.47	16.98	8.49	−12.2	261.0
15	475	88.58	79.72	70.86	62.00	53.15	44.29	35.43	26.57	17.72	8.86	−9.4	263.8
20	480	92.37	83.13	73.89	64.66	55.42	46.18	39.95	27.71	18.47	9.24	−6.7	266.5
25	485	96.28	86.65	77.02	67.39	57.77	48.14	38.51	28.88	19.26	9.63	−3.9	269.3
30	490	100.31	90.28	80.25	70.22	60.18	50.15	40.12	30.09	20.06	10.03	−1.1	272.1
35	495	104.46	94.02	83.57	73.13	62.68	52.23	41.79	31.34	20.89	10.45	−1.7	274.9
40	500	108.75	97.88	87.00	76.13	65.25	54.38	43.50	32.63	21.75	10.88	4.4	277.6
45	505	113.17	101.85	90.53	79.22	67.90	56.58	45.27	33.95	22.63	11.32	7.2	280.4
50	510	117.71	105.94	94.17	82.40	70.63	58.86	47.09	35.31	23.54	11.77	10.0	283.2
55	515	122.40	110.16	97.92	85.68	73.44	61.20	48.96	36.72	24.48	12.24	12.8	286.0
60	520	127.22	114.50	101.78	89.06	76.33	63.61	50.89	38.17	25.44	12.72	15.6	288.8
65	525	132.19	118.97	105.75	92.53	79.31	66.09	52.87	39.66	26.44	13.22	18.3	291.5
70	530	137.29	123.56	109.84	96.11	82.38	68.65	54.92	41.19	27.46	13.73	21.1	294.3
75	535	142.55	128.29	114.04	99.78	85.53	71.27	57.02	42.76	28.51	14.25	23.9	297.1
80	540	147.95	133.16	118.36	103.57	88.77	73.98	59.18	44.39	29.59	14.80	26.7	299.9
85	545	153.51	138.16	122.81	107.46	92.11	76.75	61.40	46.05	30.70	15.35	29.4	302.6
90	550	159.22	143.30	127.38	111.45	95.53	79.61	63.69	47.77	31.84	15.92	32.2	305.4
95	555	165.09	148.58	132.07	115.56	99.05	82.55	66.04	49.53	33.02	16.51	35.0	308.2
100	560	171.12	154.01	136.90	119.78	102.67	85.56	68.45	51.34	34.22	17.11	37.8	311.0
105	565	177.31	159.58	141.85	124.12	106.39	88.66	70.93	53.19	35.46	17.73	40.6	313.8
110	570	183.67	165.31	146.94	128.57	110.20	91.84	73.47	55.10	36.73	18.37	43.3	316.5
115	575	190.20	171.18	152.16	133.14	114.12	95.10	76.08	57.06	38.04	19.02	46.1	319.3
120	580	196.91	177.82	157.53	137.83	118.14	98.45	78.76	59.07	39.38	19.69	48.9	322.1
125	585	203.79	183.41	163.03	142.65	122.27	101.89	81.51	61.14	40.76	20.38	51.7	324.9
130	590	210.84	189.76	168.67	147.59	126.51	105.42	84.34	63.25	42.17	21.08	54.4	327.6
135	595	218.08	196.27	174.46	152.66	130.85	109.04	87.23	65.42	43.62	21.81	57.2	330.4
140	600	225.50	202.95	180.40	157.85	135.30	112.75	90.20	67.65	45.10	22.55	60.0	333.2
145	605	233.12	209.80	186.49	163.18	139.87	116.56	93.25	69.93	46.62	23.31	62.8	336.0
150	610	240.92	216.83	192.73	168.64	144.55	120.46	96.37	72.28	48.18	24.09	65.5	338.8
155	615	248.91	224.02	199.13	174.24	149.35	124.46	99.57	74.67	49.78	24.89	68.3	341.5
160	620	257.11	231.40	231.40	205.69	179.98	154.26	128.55	77.13	51.42	25.71	71.1	344.3
165	625	265.50	238.95	212.40	185.85	159.30	132.75	106.20	79.65	53.10	26.55	73.9	347.1
170	630	274.10	246.69	219.28	191.87	164.46	137.05	109.64	82.23	54.82	27.41	76.7	349.1
175	635	282.91	254.62	226.33	198.04	169.74	141.45	113.16	84.87	56.58	28.29	79.4	352.6
180	640	291.92	262.73	233.54	204.35	175.15	145.96	116.77	87.58	58.38	29.19	82.2	355.4
185	645	301.15	271.04	240.92	210.81	180.69	150.58	120.46	90.35	60.23	30.12	85.0	358.2
190	650	310.60	279.54	248.48	217.42	186.36	155.30	124.24	93.18	62.12	31.06	87.8	361.0
195	655	320.27	288.24	256.21	224.19	192.16	160.13	128.11	96.08	64.05	32.03	90.6	363.8

Note: To convert to Metric Units, multiply listed Btu/ft², hr by 3.152591 to obtain W/m².

Note: Radiation values were calculated from the following formula: (English Units)

$$q_{ra} = \epsilon \times .174 \times 10^{-8} \times (460 + t)^4$$

Table B-8 Radiation heat transfer table Btu/sq ft, hr to absolute zero (continued) – English Units – Radiation in Btu/ft², hr

Temperature		Emittance										Temperature	
°F	°R	1.0	0.9	0.8	0.7	0.6	0.5	0.4	0.3	0.2	0.1	°C	°K
200	660	330.16	297.14	264.13	231.11	198.10	165.08	132.06	99.05	66.03	33.02	93.3	366.5
210	670	350.6	315.6	280.5	245.4	210.4	175.3	140.3	105.2	70.1	35.1	98.9	372.1
220	680	372.0	334.8	297.6	260.4	223.2	186.0	148.8	111.6	74.4	37.2	104.5	377.7
230	690	394.4	355.0	315.5	276.1	236.6	197.2	157.8	118.3	78.9	39.4	110.1	383.3
240	700	417.8	376.0	334.2	292.4	250.7	208.9	167.1	125.3	83.6	41.8	115.7	388.9
250	710	442.2	397.9	353.7	309.5	265.3	221.1	176.9	132.6	88.4	44.2	121.3	394.5
260	720	467.6	420.8	374.1	327.3	280.6	233.8	187.0	140.3	93.5	46.8	127	400
270	730	494.1	444.7	395.3	345.9	296.5	247.1	197.7	148.2	98.8	49.4	132	405
280	740	521.8	469.6	417.4	365.2	313.1	260.9	208.7	156.5	104.4	52.2	138	411
290	750	550.5	495.5	440.4	385.4	330.3	275.3	220.2	165.2	110.1	55.1	143	416
300	760	580.5	522.5	464.4	406.4	348.3	290.3	232.2	174.2	112.1	58.1	149	422
310	770	611.7	550.5	489.3	428.2	367.0	305.8	244.7	183.5	122.3	61.2	154	427
320	780	644.1	579.7	515.2	450.8	386.4	322.0	257.6	193.2	128.8	64.4	160	433
330	790	677.7	610.0	542.2	474.4	406.6	338.9	271.1	203.3	135.5	67.8	166	439
340	800	712.7	641.4	570.2	498.9	427.6	356.4	285.1	213.8	142.5	71.3	171	444
350	810	749.0	674.1	599.2	524.3	449.4	374.5	299.6	224.7	149.8	74.9	177	450
360	820	786.7	708.0	629.4	550.7	472.0	393.3	314.7	236.0	157.3	78.7	182	455
370	830	825.8	743.2	660.6	578.0	495.5	412.9	330.3	247.7	165.2	82.6	188	461
380	840	866.3	779.7	693.0	606.4	519.8	433.1	346.5	259.9	173.3	86.6	193	466
390	850	908.3	817.5	726.6	635.8	545.0	454.1	363.3	272.5	181.7	90.8	199	472
400	860	951.8	856.6	761.4	666.3	571.1	475.9	380.7	285.5	190.4	95.2	204	477
410	870	996.8	897.2	797.5	697.8	598.1	498.4	398.7	299.1	199.4	99.7	210	483
420	880	1043.5	939.1	834.8	730.4	626.1	521.7	417.4	313.0	208.7	104.3	216	489
430	890	1091.7	982.5	873.4	764.2	655.0	545.9	436.7	327.5	218.3	109.2	221	494
440	900	1141.6	1027.5	913.3	799.1	685.0	570.8	456.6	342.5	228.2	114.2	227	500
450	910	1193.2	1073.9	954.6	835.2	715.9	596.6	477.3	358.0	238.6	119.3	232	505
460	920	1246.5	1121.9	997.2	872.6	747.9	623.3	498.6	374.0	249.3	124.7	238	511
470	930	1301.6	1171.4	1041.3	911.1	781.0	650.8	520.6	390.5	260.3	130.2	243	516
480	940	1358.5	1222.7	1086.8	951.0	815.1	679.3	543.4	407.6	271.7	135.9	249	522
490	950	1417.2	1275.5	1133.8	992.1	850.3	708.6	566.9	425.2	283.4	141.7	254	527
500	960	1477.9	1330.1	1182.3	1034.5	886.7	738.9	591.1	443.4	295.6	147.8	260	533
550	1010	1811.	1630.	1449.	1267.	1086.	905.	543.	543.	362.	181.	288	561
600	1060	2197.	1977.	1757.	1538.	1318.	1098.	659.	659.	439.	220.	316	589
650	1110	2641.	2377.	2113.	1849.	1585.	1321.	792.	792.	528.	264.	343	616
700	1160	3151.	2835.	2520.	2205.	1890.	1575.	945.	945.	630.	315.	371	644
750	1210	3730.	3357.	2984.	2611.	2238.	1865.	1119.	1119.	746.	373.	399	672
800	1260	4386.	3947.	3508.	3070.	2631.	2193.	1316.	1316.	877.	439.	427	700
850	1310	5124.	4612.	4099.	3587.	3075.	2562.	1537.	1537.	1025.	512.	454	727
900	1360	5953.	5357.	4762.	4167.	3572.	2976.	1786.	1786.	1191.	595.	482	755
950	1410	6877.	6190.	5502.	4814.	4126.	3439.	2063.	2063.	1375.	688.	510	783
1000	1460	7906.	7115.	6325.	5534.	4744.	3953.	2372.	2372.	1581.	791.	538	811
1050	1510	9046.	8141.	7237.	6332.	5428.	4523.	2714.	2714.	1809.	905.	566	839
1100	1560	10305.	9274.	8244.	7213.	6183.	5152.	3091.	3091.	2061.	1030.	593	866
1150	1610	11691.	10522.	9353.	8184.	7015.	5846.	3507.	3507.	2338.	1169.	621	894
1200	1660	13212.	11891.	10570.	9249.	7929.	6606.	3964.	3964.	2642.	1321.	649	922
1250	1710	14878.	13390.	11902.	10414.	8927.	7439.	4463.	4463.	2976.	1488.	677	950
1300	1760	16696.	15026.	13356.	11687.	10017.	8348.	5009.	5009.	3339.	1679.	704	977
1350	1810	18675.	16808.	14940.	13073.	11205.	9338.	5603.	5603.	3735.	1868.	732	1005
1400	1860	20826.	18743.	16661.	14578.	12495.	10413.	6248.	6248.	4165.	2083.	760	1033
1450	1910	23157.	20841.	18526.	16210.	13894.	11579.	6947.	6947.	4631.	2316.	788	1061
1500	1960	25679.	23111.	20543.	17975.	15407.	12839.	7704.	7704.	5136.	3568.	816	1089
1550	2010	28401.	25561.	22721.	19881.	17041.	14200.	8520.	8520.	5689.	2840.	843	1116
1600	2060	31334.	28201.	25067.	21934.	18800.	15667.	9400.	9400.	6267.	3133.	871	1144
1650	2110	34489.	31040.	27591.	24142.	20693.	17244.	10347.	10347.	6898.	3449.	899	1172
1700	2160	37876.	34088.	30301.	26513.	22726.	18933.	11363.	11363.	7575.	3788.	927	1200
1750	2210	41507.	37356.	33205.	29055.	24904.	20753.	12452.	12452.	8301.	4151.	954	1227
1800	2260	45392.	40853.	36314.	31775.	27235.	22696.	13618.	13618.	9078.	4539.	982	1235
1850	2310	49545.	44590.	39636.	34681.	29727.	14863.	14863.	14863.	9909.	4954.	1010	1283
1900	2360	53976.	48578.	43180.	37783.	32385.	26988.	16193.	16193.	10795.	5398.	1038	1311
1950	2410	58697.	52827.	46958.	41088.	35218.	29349.	17609.	17609.	11739.	5870.	1066	1339
2000	2460	63722.	57350.	50978.	44605.	38233.	31861.	19117.	19117.	12744.	6372.	1093	1366
2050	2510	69063.	62157.	55250.	48344.	41438.	34531.	20719.	20719.	13813.	6906.	1121	1394
2100	2560	74732.	67259.	59786.	52313.	44839.	37366.	22420.	22420.	14946.	7473.	1149	1422
2150	2610	80744.	72670.	64595.	56521.	48447.	40372.	24223.	24223.	16149.	8074.	1176	1450
2200	2660	87112.	78400.	69689.	60978.	52267.	43556.	26133.	26133.	17422.	8711.	1204	1477
2250	2710	93848.	84463.	75079.	65694.	56309.	46924.	28154.	28154.	18770.	9385.	1232	1505
2300	2760	100968.	90872.	80775.	70678.	60581.	50484.	30291.	30291.	20194.	10097.	1260	1533
2350	2810	108486.	97638.	86789.	75940.	65092.	54243.	32546.	32546.	21697.	10849.	1288	1561
2400	2860	116416.	104775.	93133.	81491.	69850.	58208.	34925.	34925.	23283.	11642.	1316	1588
2450	2910	124773.	112296.	99819.	87341.	74864.	62387.	37432.	37432.	24955.	12477.	1344	1616
2500	2960	133572.	120215.	106858.	93501.	80143.	66786.	40072.	40072.	26714.	13357.	1371	1644
2550	3010	142829.	128546.	114262.	99980.	85697.	71414.	42849.	42849.	28566.	14283.	1399	1672
2600	3060	152558.	137302.	122046.	106791.	91535.	76279.	45767.	45767.	30512.	15256.	1427	1700
2650	3110	162776.	146499.	130221.	113943.	97666.	81388.	48833.	48833.	32555.	16278.	1455	1728
2700	3160	173499.	156149.	138799.	121449.	104100.	86750.	52050.	52050.	34700.	17350.	1482	1756
2750	3210	184744.	166269.	147795.	129320.	110846.	92372.	55423.	55423.	36949.	18474.	1510	1783
2800	3260	196526.	176873.	157221.	137568.	117915.	98263.	58958.	58958.	39305.	19653.	1538	1810
2850	3310	208863.	187977.	167090.	146204.	125318.	104431.	62659.	62659.	41773.	20886.	1566	1839
2900	3360	221772.	199595.	177417.	155240.	133063.	110886.	66532.	66532.	44354.	22177.	1594	1867
2950	3410	235270.	211743.	188216.	164689.	141162.	117635.	70581.	70581.	47054.	23527.	1621	1894
3000	3460	249375.	224438.	199500.	174563.	149625.	124688.	74813.	74813.	49875.	24938.	1649	1922

Note: To convert to Metric Units, multiply listed Btu/ft², hr by 3.152591 to obtain W/m².

Note: Radiation values were calculated from the following formula: (English Units)

$q_{ra} = \in \times .174 \times 10^{-8} \times (460 + t)^4$

Table B-9 Radiation heat transfer insulation to air for surface emittances $\epsilon = 1.0$ to 0.1 in Btu/ft, hr

English Units

Temp. difference surface to air °F or °R	Emittance of outer weather barrier or jacket in reference to thermal black body										Temp. difference surface to air °C or °K
	1.0	0.9	0.8	0.7	0.6	0.5	0.4	0.3	0.2	0.1	
2	1.8	1.6	1.4	1.3	1.1	0.9	0.7	0.5	0.4	0.2	1.1
4	3.5	3.2	2.8	2.5	2.1	1.8	1.4	1.1	0.7	0.4	2.2
6	5.6	5.0	4.5	3.9	3.4	2.8	2.2	1.7	1.1	0.6	3.3
8	7.5	6.8	6.0	5.2	4.5	3.8	3.0	2.3	1.5	0.8	4.4
10	9.5	8.6	7.6	6.6	5.7	4.8	3.8	2.9	1.9	1.0	5.6
15	14.4	13.0	11.5	10.1	8.6	7.2	5.8	4.3	2.9	1.4	8.3
20	19.6	17.6	15.7	13.7	11.8	9.8	7.8	5.9	3.9	2.0	11.1
25	24.7	22.2	19.8	17.3	14.8	12.4	9.9	7.4	4.9	2.5	13.3
30	30.2	27.2	24.2	21.1	18.1	15.1	12.1	9.1	6.1	3.0	16.7
35	35.8	32.2	28.6	25.1	21.5	17.9	14.4	10.8	7.2	3.6	19.4
40	41.5	37.4	33.2	28.0	24.9	20.8	16.6	12.4	8.3	4.2	22.2
45	47.4	42.6	37.9	31.5	28.4	23.7	19.0	14.2	9.5	4.7	25.0
50	53.1	47.8	42.4	37.2	31.9	26.6	21.2	15.9	10.6	5.3	27.7
60	65.9	59.4	52.7	46.1	39.5	33.0	26.4	19.8	13.2	6.6	33.3
70	79.2	71.3	63.3	55.5	47.5	39.6	31.6	23.8	15.8	7.9	33.9
80	93.1	83.9	74.4	65.2	55.8	46.6	37.2	27.9	18.6	9.3	44.4
90	107.8	97.2	86.3	75.5	64.7	53.9	43.2	32.2	21.6	10.8	50.0
100	123.2	111.1	95.5	87.2	73.9	61.6	49.3	37.0	24.6	12.3	55.5
110	139.4	125.6	111.5	95.5	83.6	69.7	55.8	41.8	27.9	13.9	61.1
120	156.4	141.0	125.1	109.0	93.2	78.2	62.6	46.9	31.3	15.6	66.6
130	174.2	157.1	139.3	122.0	104.5	87.1	69.8	52.4	34.8	17.4	72.2
140	192.9	173.7	154.0	135.0	115.8	96.5	77.2	57.8	38.6	19.3	77.7
150	212.4	191.3	169.9	148.7	127.7	106.2	85.1	63.8	42.6	21.2	88.3
160	232.9	209.0	185.5	163.0	139.8	116.5	93.3	70.0	46.7	23.3	88.8
170	254.3	229.1	203.5	178.0	152.6	127.2	101.9	76.4	50.8	25.4	94.4
180	276.7	249.3	221.1	192.9	166.0	138.4	110.8	83.1	55.4	27.7	100.0
190	300.1	270.0	240.0	210.0	180.0	150.0	120.0	90.0	60.0	30.0	105.5
200	324.5	292.5	259.5	227.5	194.8	162.3	130.0	97.5	64.8	32.4	111.1

Note: To convert to Metric Units, multiply listed Btu/ft^2, hr by 3.152591 to obtain W/m^2.

Table B-10 Evaluation of Langmuir's formula for convection

English Units

$Q_{cv} = 0.296\,(t_s - t_a)^{5/4} \left(\frac{V + 68.9}{68.9}\right)^{1/2}$ Btu/ft², hr

Slow convection currents

$t_s - t_a$	$.296\,(t_s - t_a)^{5/4}$											
Velocity metre/sec ft/min mi/hr		.0508 10 .1136	.1016 20 .2272	.1524 30 .3408	.2032 40 .4544	.254 50 .5680	.3048 60 .6816	.3556 70 .7952	.4064 80 .9088	.4572 90 1.0224	.508 100 1.1360	m/s ft/min mph
$\left(\frac{V + 68.9}{68.9}\right)^{1/2}$		1.070	1.136	1.198	1.257	1.313	1.367	1.419	1.470	1.518	1.565	$t_{sm} - t_{am}$
2° F or R	.703	.752	.798	.842	.884	.922	.961	.996	1.033	1.066	1.100	1.1 °C or °K
4	1.672	1.790	1.900	2.001	2.101	2.195	2.285	2.370	2.457	2.537	2.617	2.2
5	2.210	2.365	2.511	2.614	2.777	2.901	3.020	3.132	3.247	3.353	3.459	2.7
6	2.777	2.971	3.157	3.327	3.490	3.648	3.798	3.926	4.081	4.217	4.350	3.3
8	3.986	4.262	4.530	4.775	5.010	5.233	5.450	5.655	5.860	6.050	6.240	4.4
10	5.26	5.63	5.98	6.31	6.62	6.90	7.19	7.47	7.74	7.99	8.23	5.6
15	8.74	9.35	9.94	10.47	10.98	11.45	11.94	12.38	12.83	13.25	13.67	8.3
20	12.71	13.61	14.44	15.23	16.00	16.70	17.40	18.04	18.70	19.20	19.90	11.7
25	16.53	17.68	18.78	19.80	20.77	21.70	22.60	23.43	24.28	25.08	25.85	13.9
30	21.78	22.20	23.61	24.90	26.12	27.29	28.42	29.50	30.54	31.55	32.52	16.7
35	25.18	26.95	28.62	30.16	31.65	33.05	33.67	35.70	37.00	38.20	29.40	19.4
40	29.80	31.90	33.85	35.72	37.50	39.13	40.76	42.30	43.80	45.24	46.61	22.2
45	34.51	36.93	39.22	41.33	43.40	45.30	47.20	48.92	50.70	52.40	54.02	25.0
50	39.35	42.10	44.70	47.19	49.48	51.68	53.80	55.82	57.83	59.76	61.58	27.7
60	49.41	52.90	56.18	59.23	62.21	64.93	67.60	70.18	72.70	75.06	77.37	33.3
70	59.90	64.10	68.05	71.80	75.35	78.70	81.95	85.00	88.10	91.00	93.70	38.9
80	70.85	75.90	80.50	84.95	90.40	93.10	96.90	100.4	104.0	107.4	110.7	44.4
90	82.10	87.95	93.30	95.84	103.1	107.7	112.2	116.4	120.6	124.5	128.4	50.0
100	93.55	100.2	106.2	112.0	117.5	122.8	127.8	132.6	137.5	141.9	146.3	55.5
110	105.4	112.7	119.7	126.3	132.5	137.1	144.1	149.5	154.9	160.0	164.8	61.1
120	117.5	125.7	133.5	140.8	147.7	154.4	160.6	166.7	172.8	178.4	183.8	66.6
130	130.0	139.1	147.7	155.8	163.5	170.7	177.7	184.5	191.1	197.4	203.3	72.2
140	142.5	152.5	161.9	170.8	179.2	187.0	194.8	202.1	209.5	216.3	223.0	77.7
150	155.4	166.3	176.5	186.2	195.3	204.0	212.3	220.5	228.2	236.0	243.0	83.3
160	168.4	180.2	191.3	201.9	211.8	221.0	230.1	239.0	247.5	255.7	263.3	88.8
170	181.6	194.3	206.3	217.5	228.3	238.4	248.2	257.5	267.0	275.8	284.2	94.4
180	195.2	208.9	221.8	234.0	245.5	256.3	266.9	277.0	286.9	296.4	305.3	100.0
190	208.8	222.3	237.1	250.1	262.3	273.9	285.2	296.2	306.9	317.0	326.7	105.5
200	222.7	238.2	253.0	266.9	280.0	292.2	304.2	315.9	327.2	338.0	348.2	111.1

Tabular values are those of Q_{cv}.

Note: To convert to Metric Units, multiply listed Btu/ft², hr by 3.152591 to obtain W/m².

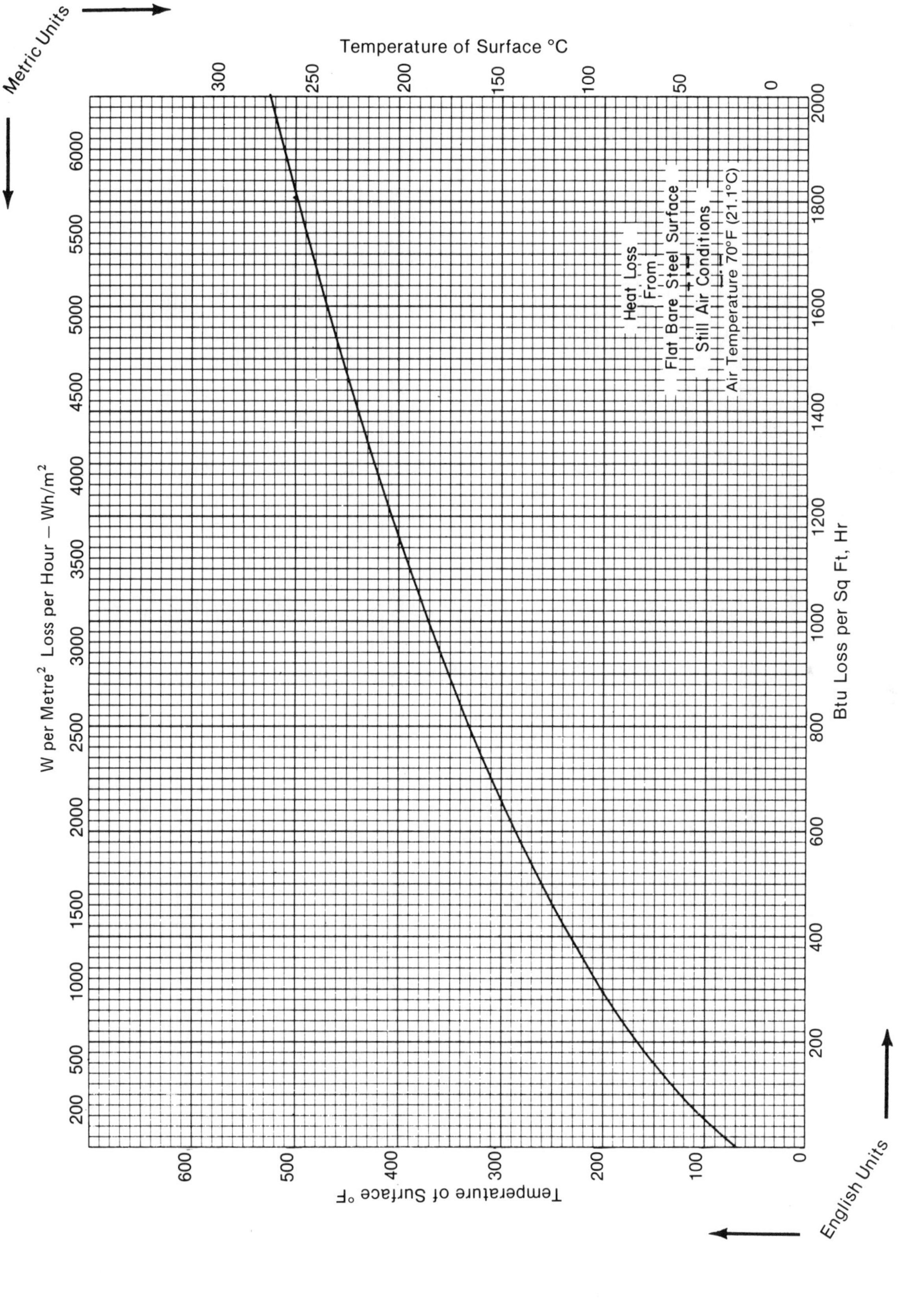
Metric Units
Temperature of Surface °C
300
250
200
150
100
50
0
W per Metre² Loss per Hour — Wh/m²
200
500
1000
1500
2000
2500
3000
3500
4000
4500
5000
5500
6000
Heat Loss
From
Flat Bare Steel Surface
Still Air Conditions
Air Temperature 70°F (21.1°C)
Btu Loss per Sq Ft, Hr
200
400
600
800
1000
1200
1400
1600
1800
2000
Temperature of Surface °F
0
100
200
300
400
500
600
English Units

Chart B-1

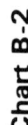

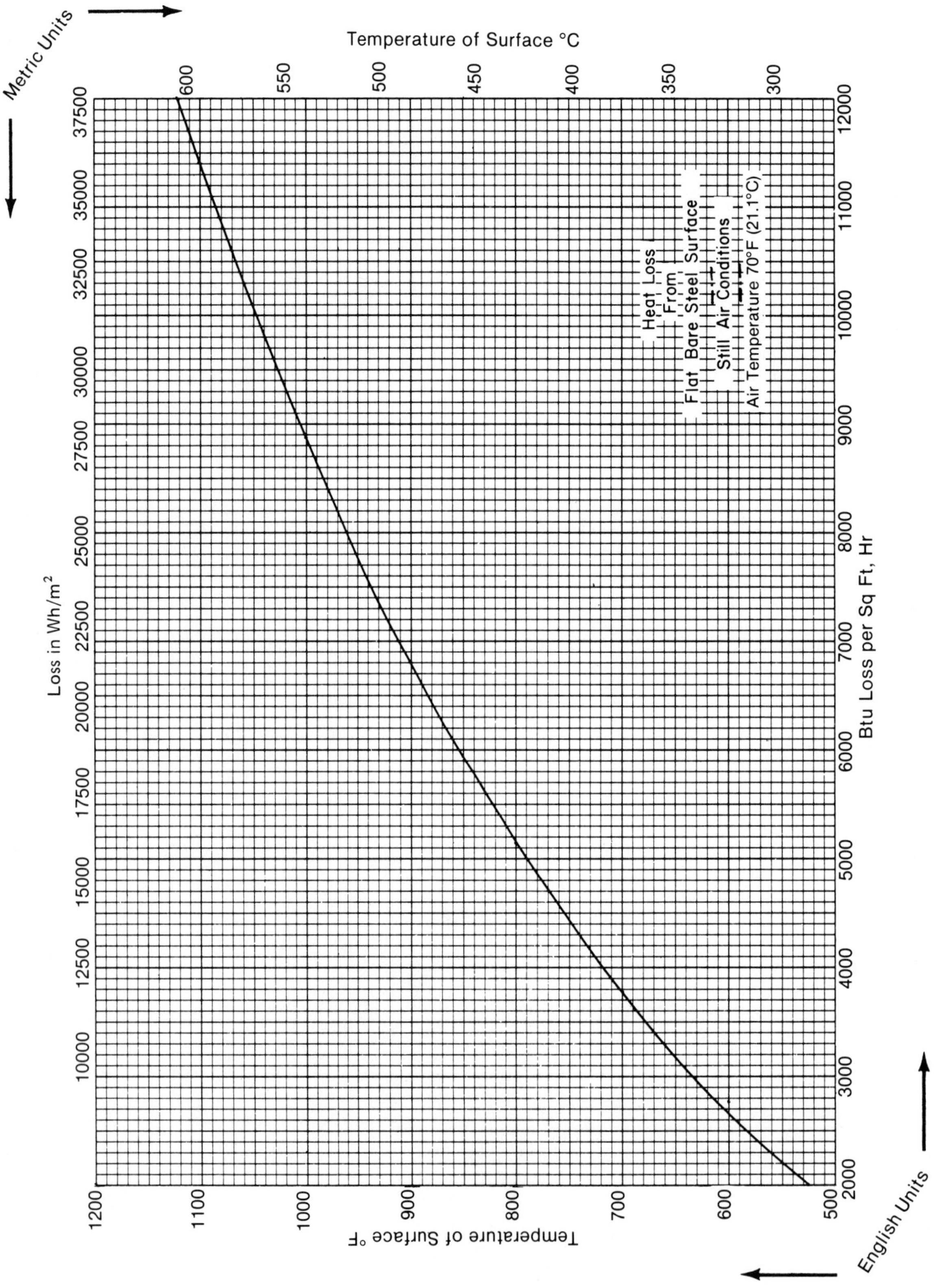
Metric Units
Temperature of Surface °C
600
550
500
450
400
350
300
Loss in Wh/m²
37500
35000
32500
30000
27500
25000
22500
20000
17500
15000
12500
10000
Heat Loss
From
Flat Bare Steel Surface
Still Air Conditions
Air Temperature 70°F (21.1°C)
Btu Loss per Sq Ft, Hr
12000
11000
10000
9000
8000
7000
6000
5000
4000
3000
2000
Temperature of Surface °F
1200
1100
1000
900
800
700
600
500
English Units

Chart B-2

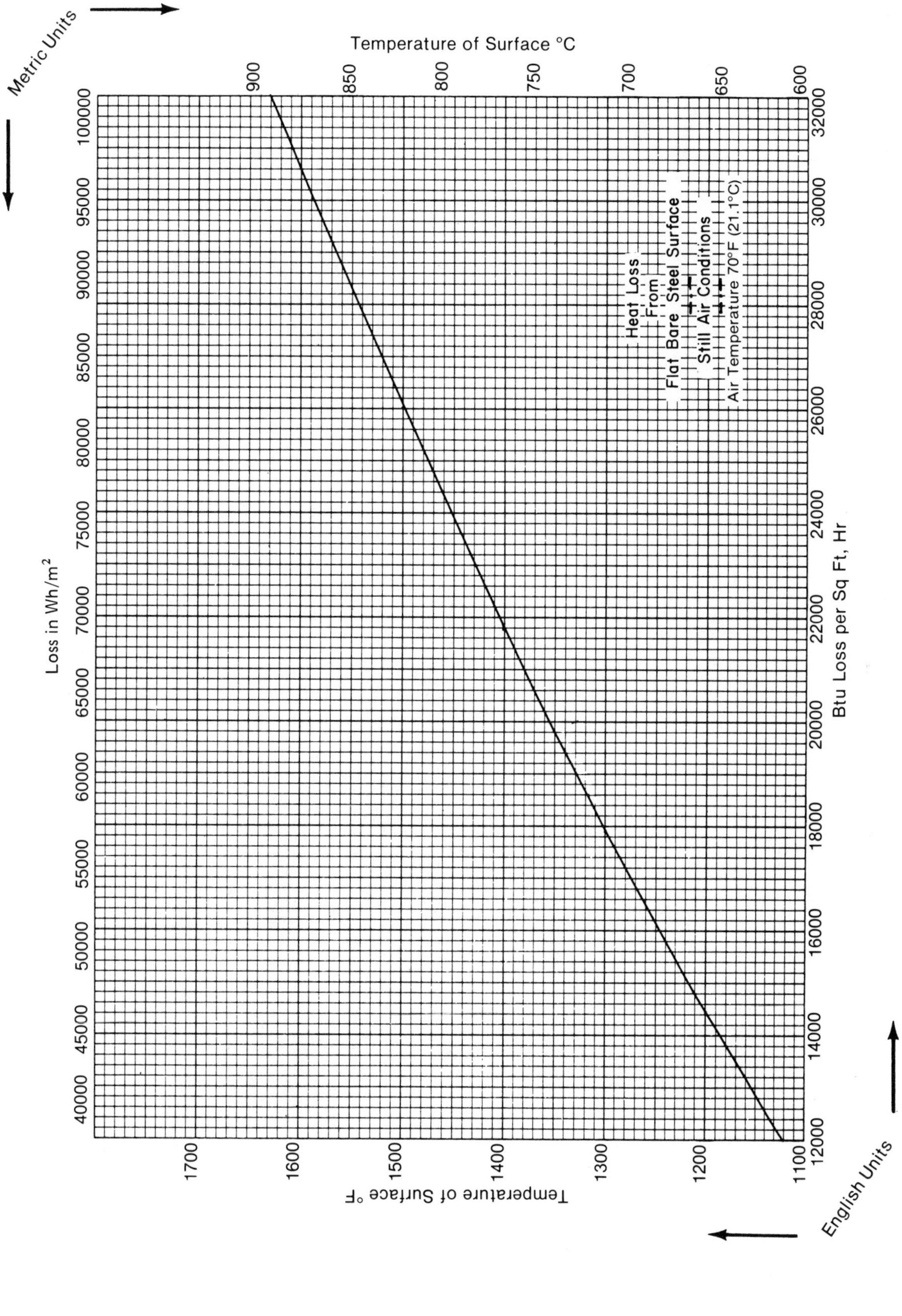
Metric Units
Temperature of Surface °C
900
850
800
750
700
650
600
Loss in Wh/m²
100000
95000
90000
85000
80000
75000
70000
65000
60000
55000
50000
45000
40000
Heat Loss
From
Flat Bare Steel Surface
Still Air Conditions
Air Temperature 70°F (21.1°C)
Btu Loss per Sq Ft, Hr
32000
30000
28000
26000
24000
22000
20000
18000
16000
14000
12000
Temperature of Surface °F
1700
1600
1500
1400
1300
1200
1100
English Units

Chart B-3

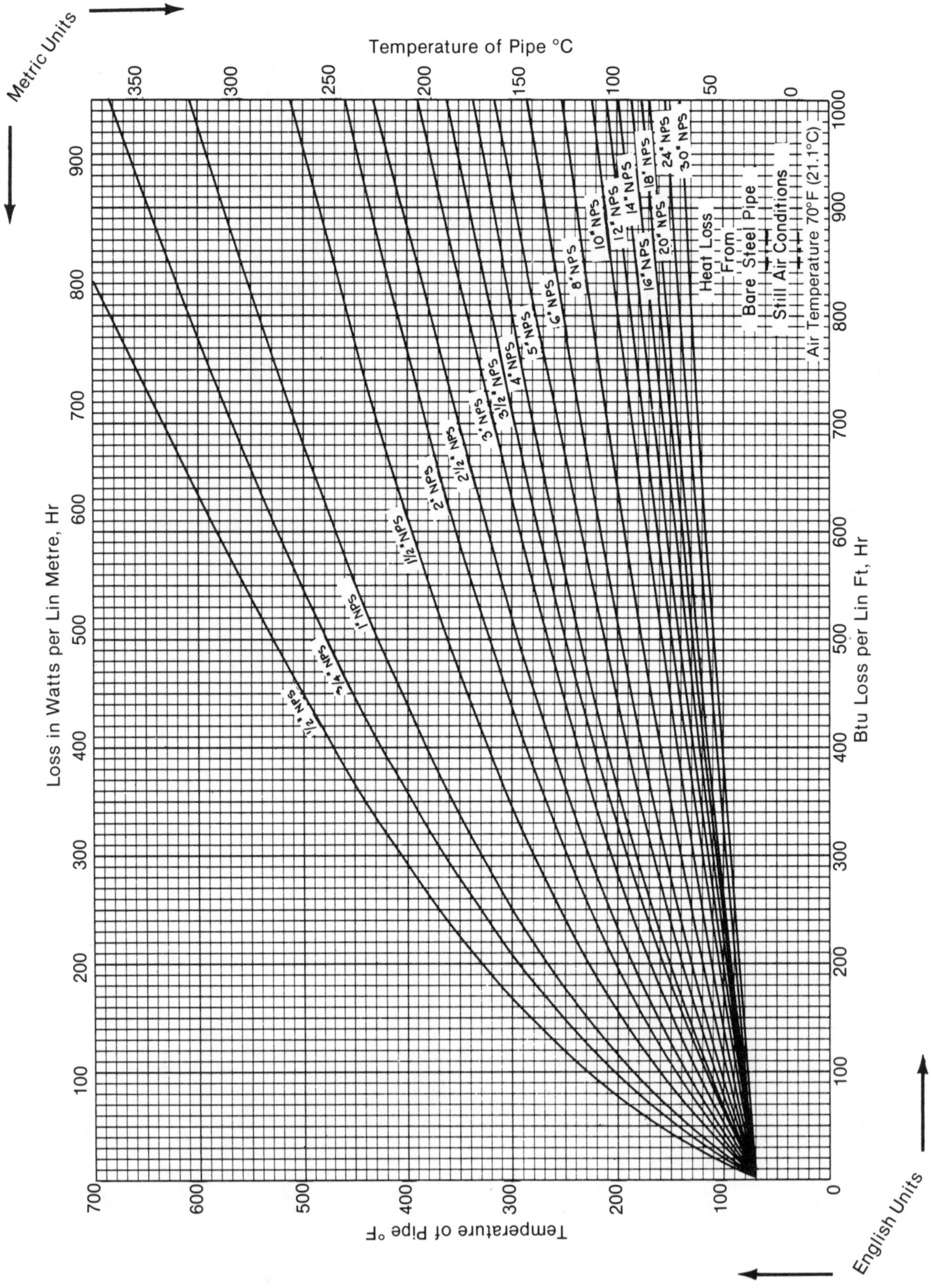
Metric Units
Temperature of Pipe °C
Loss in Watts per Lin Metre, Hr
English Units
Temperature of Pipe °F
Btu Loss per Lin Ft, Hr
Heat Loss From Bare Steel Pipe
Still Air Conditions
Air Temperature 70°F (21.1°C)
½" NPS
¾" NPS
1" NPS
1½" NPS
2" NPS
2½" NPS
3" NPS
3½" NPS
4" NPS
5" NPS
6" NPS
8" NPS
10" NPS
12" NPS
14" NPS
16" NPS
18" NPS
20" NPS
24" NPS
30" NPS

Chart B-4

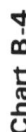

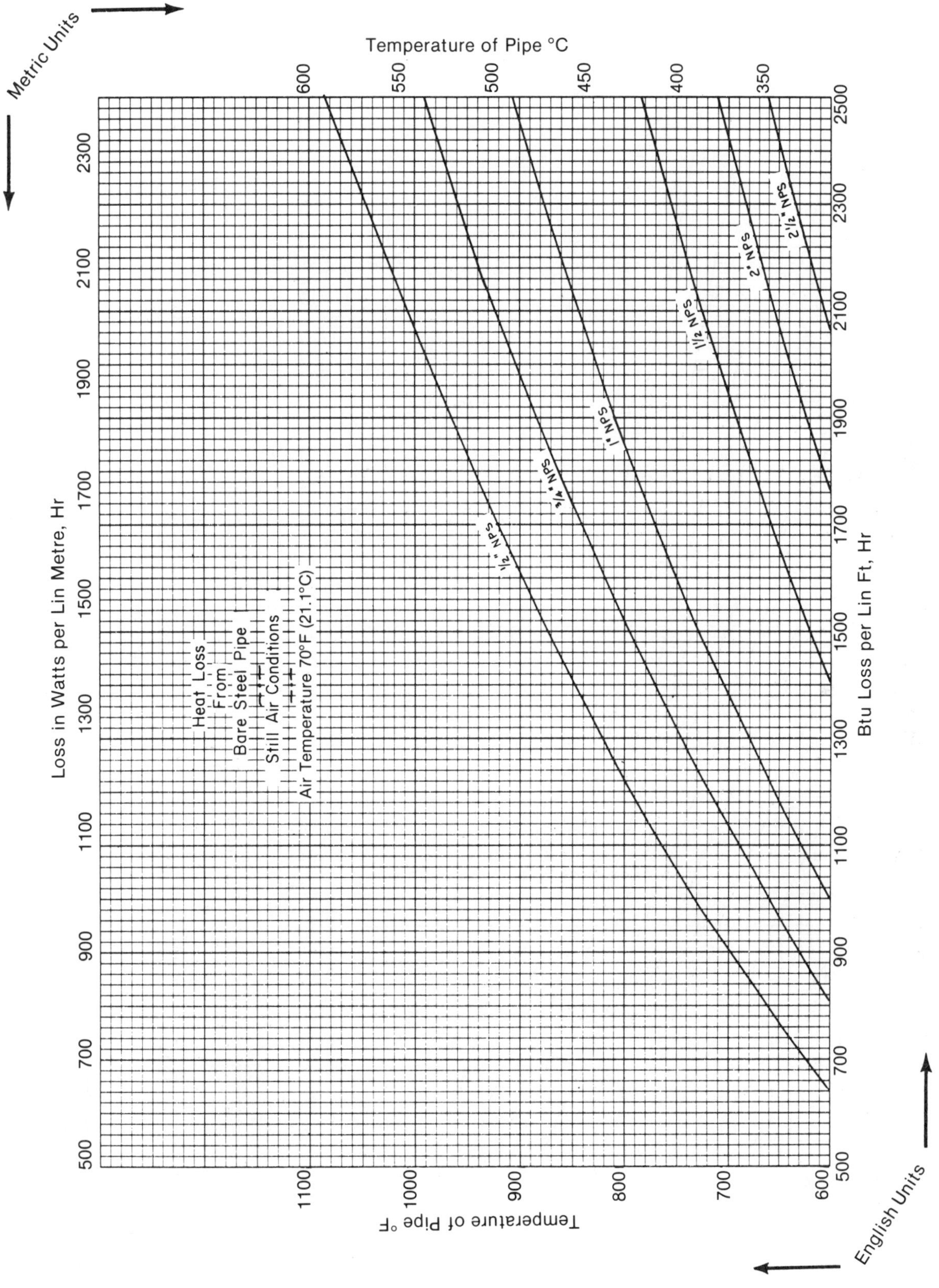
Metric Units
Temperature of Pipe °C
600
550
500
450
400
350
Loss in Watts per Lin Metre, Hr
500
700
900
1100
1300
1500
1700
1900
2100
2300
Heat Loss
From
Bare Steel Pipe
Still Air Conditions
Air Temperature 70°F (21.1°C)
1/2" NPS
3/4" NPS
1" NPS
1 1/2 NPS
2" NPS
2 1/2" NPS
Btu Loss per Lin Ft, Hr
500
700
900
1100
1300
1500
1700
1900
2100
2300
2500
Temperature of Pipe °F
1100
1000
900
800
700
600
English Units

Chart B-5

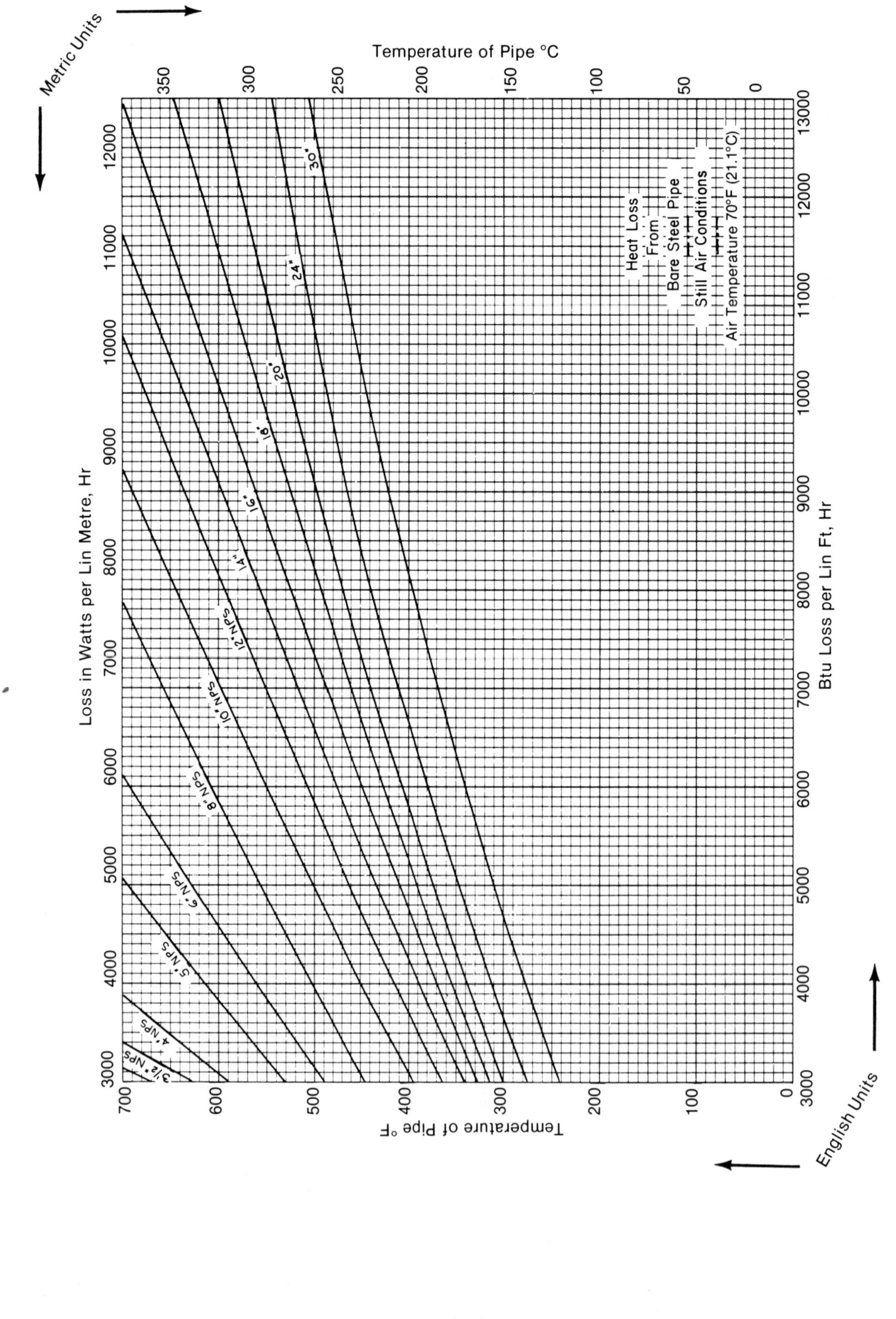
Metric Units
Temperature of Pipe °C
Loss in Watts per Lin Metre, Hr
English Units
Temperature of Pipe °F
Btu Loss per Lin Ft, Hr
Heat Loss
From
Bare Steel Pipe
Still Air Conditions
Air Temperature 70°F (21.1°C)
3½" NPS
4" NPS
5" NPS
6" NPS
8" NPS
10" NPS
12" NPS
14"
16"
18"
20"
24"
30"

Chart B-6

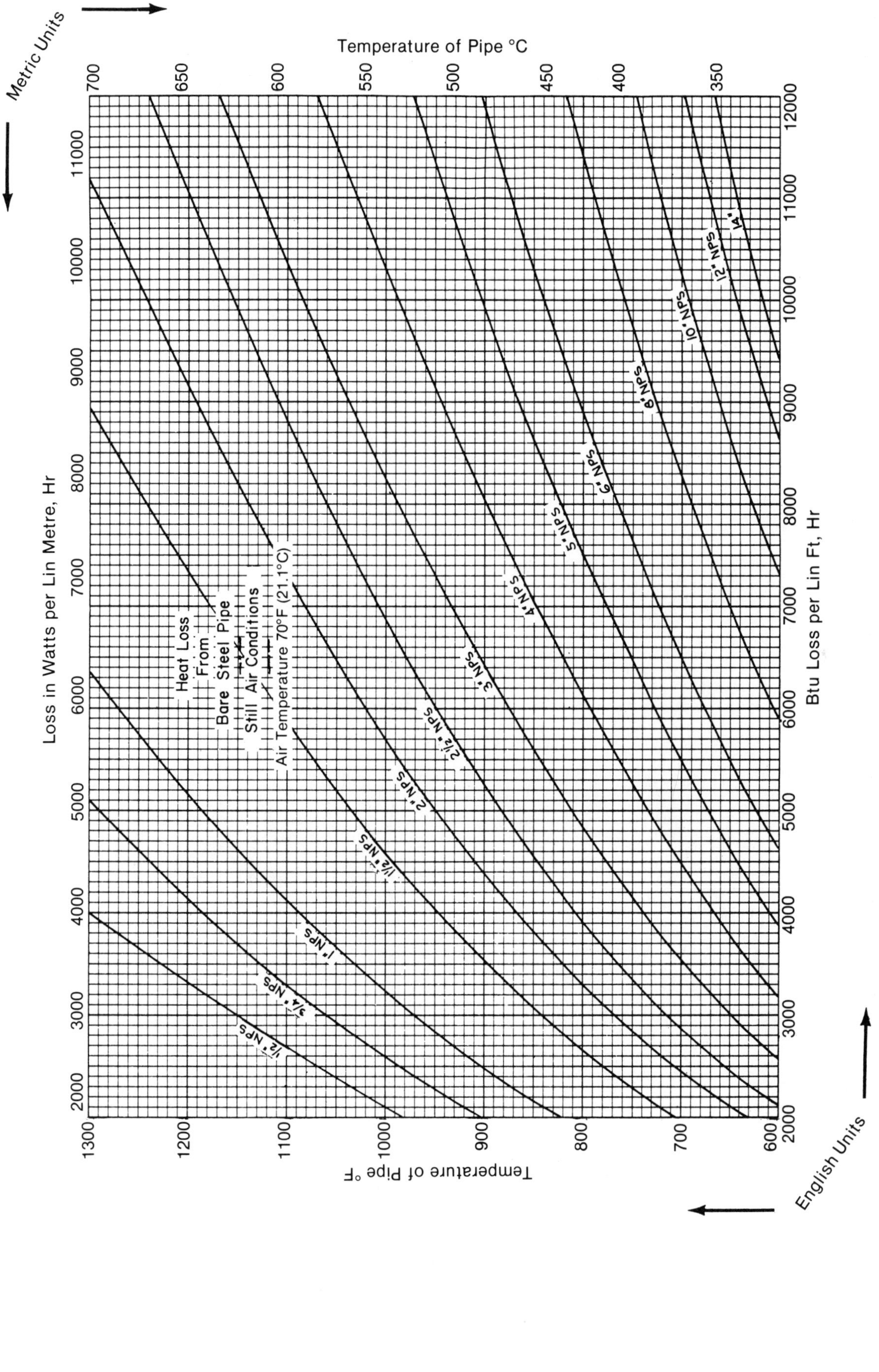
Metric Units
Temperature of Pipe °C
700
650
600
550
500
450
400
350
Loss in Watts per Lin Metre, Hr
2000
3000
4000
5000
6000
7000
8000
9000
10000
11000
Heat Loss
From
Bare Steel Pipe
Still Air Conditions
Air Temperature 70°F (21.1°C)
1/2" NPS
3/4" NPS
1" NPS
1 1/2" NPS
2" NPS
2 1/2" NPS
3" NPS
4" NPS
5" NPS
6" NPS
8" NPS
10" NPS
12" NPS
14"
Btu Loss per Lin Ft, Hr
12000
Temperature of Pipe °F
1300
1200
1100
1000
900
800
700
600
English Units

Chart B-7

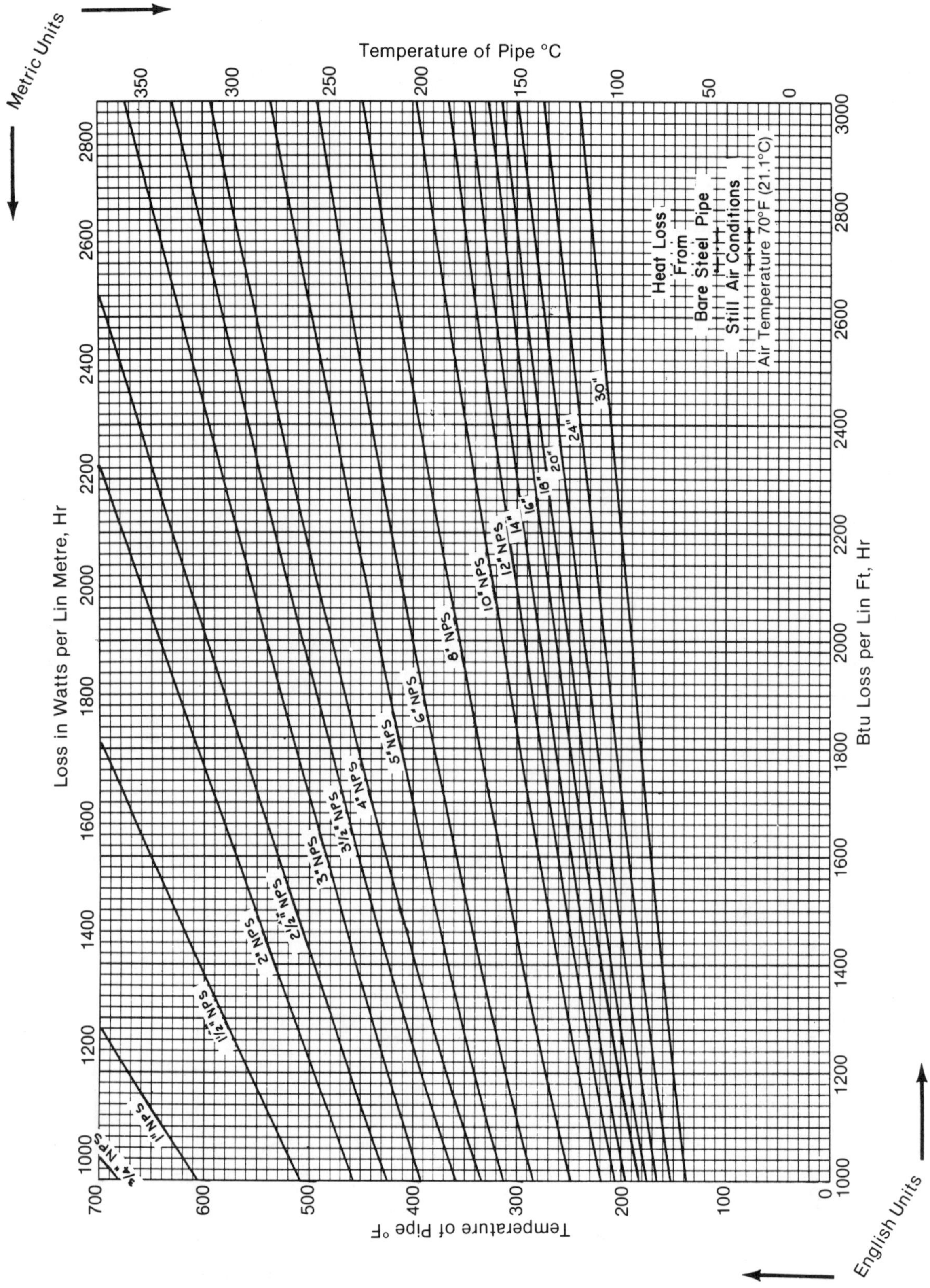
Metric Units
English Units
Temperature of Pipe °C
Temperature of Pipe °F
Loss in Watts per Lin Metre, Hr
Btu Loss per Lin Ft, Hr
Heat Loss
From
Bare Steel Pipe
Still Air Conditions
Air Temperature 70°F (21.1°C)
1" NPS
1½" NPS
2" NPS
2½" NPS
3" NPS
3½" NPS
4" NPS
5" NPS
6" NPS
8" NPS
10" NPS
12" NPS
14"
16"
18"
20"
24"
30"
350
300
250
200
150
100
50
0
700
600
500
400
300
200
100
0
1000
1200
1400
1600
1800
2000
2200
2400
2600
2800
3000

Chart B-8

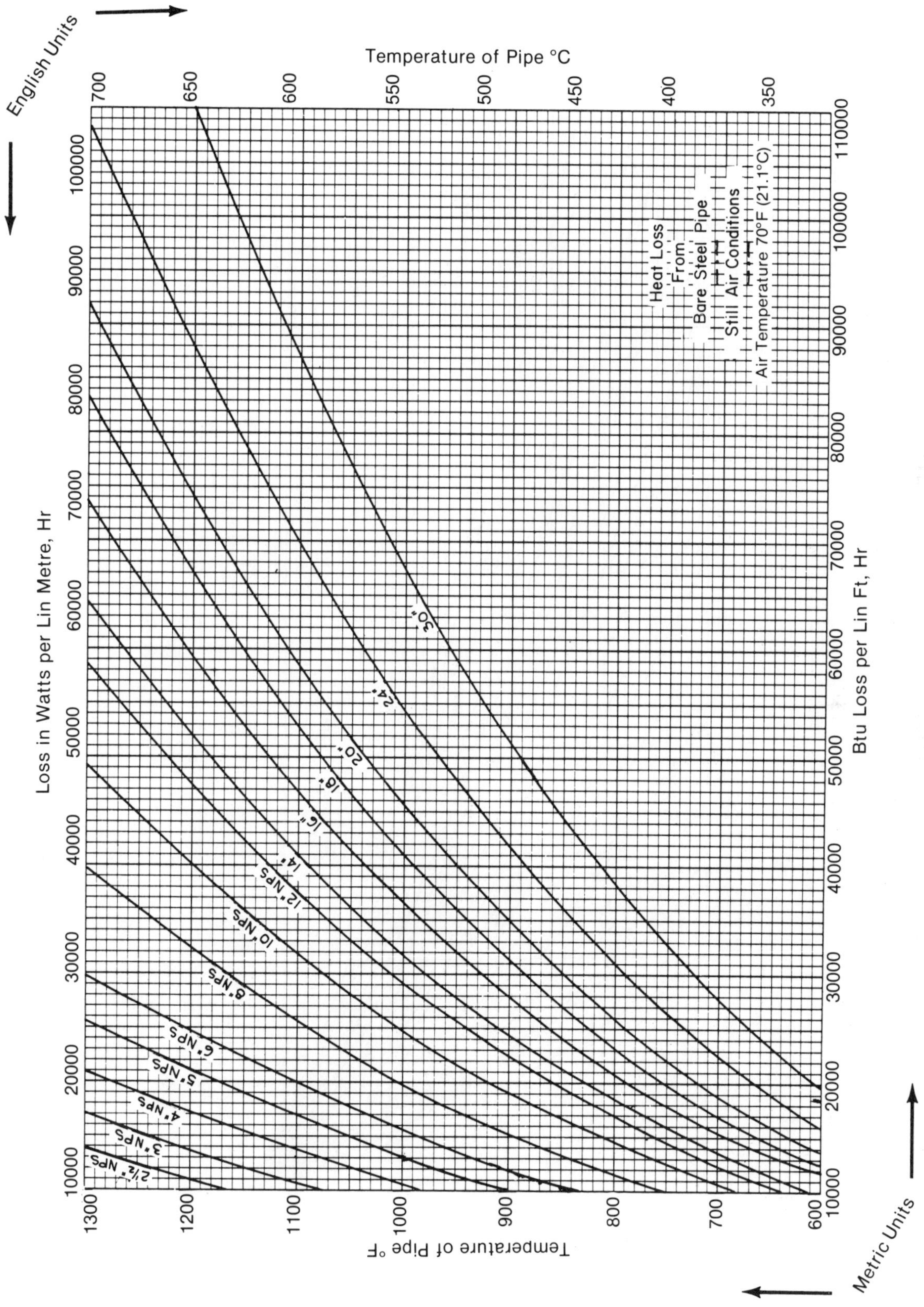
English Units
Metric Units
Temperature of Pipe °C
700
650
600
550
500
450
400
350
Loss in Watts per Lin Metre, Hr
10000
20000
30000
40000
50000
60000
70000
80000
90000
100000
Btu Loss per Lin Ft, Hr
10000
20000
30000
40000
50000
60000
70000
80000
90000
100000
110000
Temperature of Pipe °F
1300
1200
1100
1000
900
800
700
600
Heat Loss
From
Bare Steel Pipe
Still Air Conditions
Air Temperature 70°F (21.1°C)
2½" NPS
3" NPS
4" NPS
5" NPS
6" NPS
8" NPS
10" NPS
12" NPS
14"
16"
18"
20"
24"
30"

Chart B-9

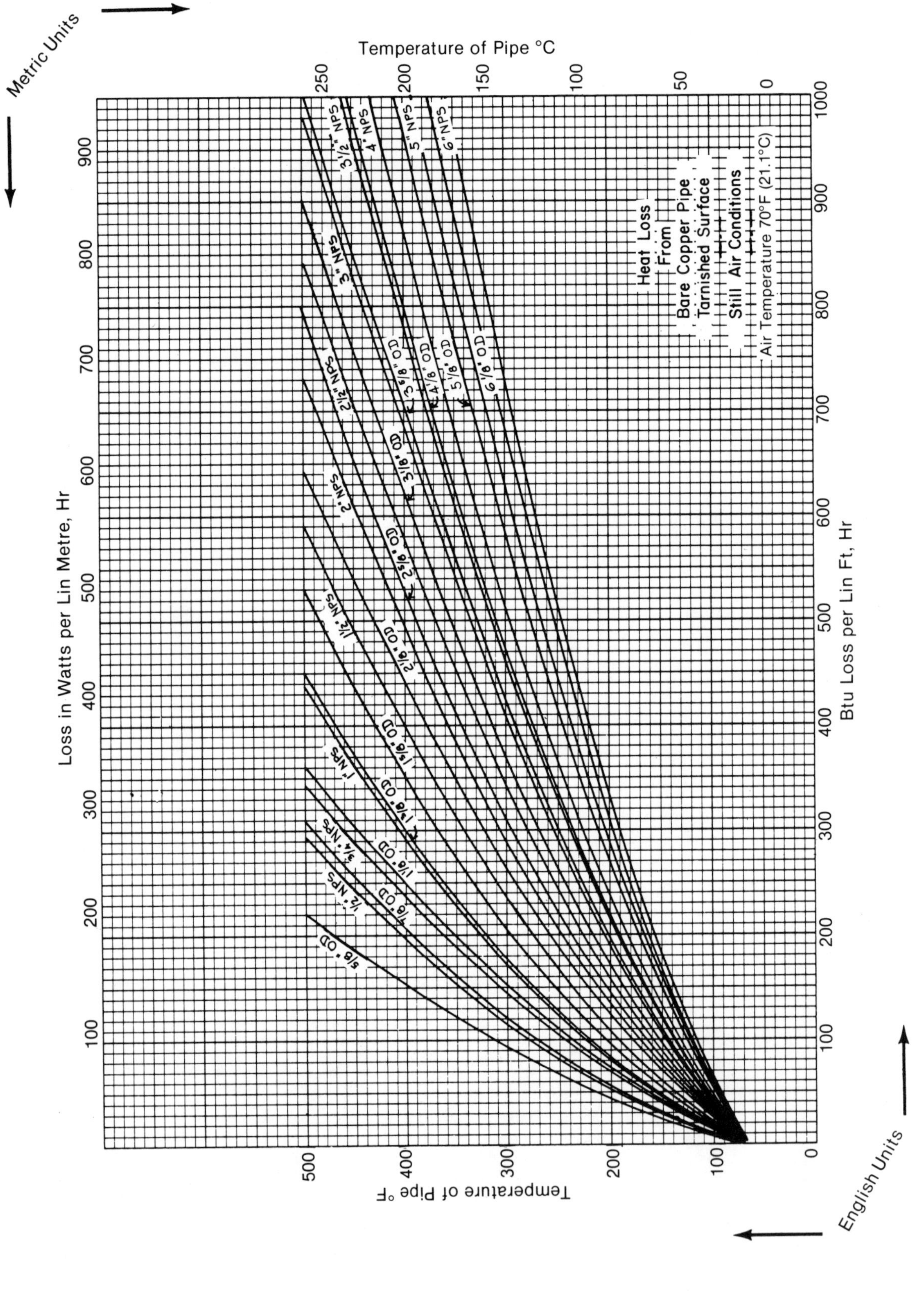
Metric Units
Temperature of Pipe °C
Loss in Watts per Lin Metre, Hr
Btu Loss per Lin Ft, Hr
Temperature of Pipe °F
English Units
Heat Loss
From
Bare Copper Pipe
Tarnished Surface
Still Air Conditions
Air Temperature 70°F (21.1°C)
6" NPS
5" NPS
4" NPS
3½" NPS
3" NPS
2½" NPS
2" NPS
1½" NPS
1" NPS
¾" NPS
½" NPS
6⅛" OD
5⅛" OD
4⅛" OD
3⅝" OD
3⅛" OD
2⅝" OD
2⅛" OD
1⅝" OD
1⅜" OD
1⅛" OD
⅞" OD
⅝" OD

Chart B-10

Table B-11 Heat losses from bare flat surfaces

Still air conditions—air temperature 70° F, (21.1° C) black surface $\epsilon = 1.0$

Losses in Btu per hour, per square foot, or watts per hour per square metre, of surface for *total* temperature difference between hot surface and air

Hot Surface Temp. °F	Temperature Diff. °F	Btu/ft², hr Loss	Hot Surface Temp °C	Temperature Diff. °C	W/m² Loss
100	30	50	38	17	157
200	130	295	93	72	930
300	230	654	149	128	2061
400	330	1145	204	183	3609
500	430	1806	260	239	5693
600	530	2666	316	295	8404
700	630	3765	371	350	11869
800	730	5126	427	406	16160
900	830	6799	482	461	21434
1000	930	8885	538	517	28010
1100	1030	11400	593	572	35939
1200	1130	14394	649	628	45378
1300	1230	17896	704	683	56418
1400	1330	21758	760	739	68594
1500	1430	25983	816	795	81913

Table B-12 Heat losses from bare NPS pipe in Btu per hr per linear ft and W per hr per linear metre

Still Air Conditions—Air Temperature 70° F, 21.1° C, Black Surface on pipe

English and Metric Units

Pipe size	Pipe temp °F	Btu/lin ft hr loss	Pipe temp °C	W/lin m loss
1/2'' NPS				
	100	13	38	12
	200	75	93	72
	300	165	149	158
	400	287	204	275
outside	500	444	260	426
dia	600	649	316	624
0.084''	700	901	371	866
21.3 mm	800	1218	427	1170
	900	1602	482	1539
	1000	2075	538	1993
	1100	2644	593	2340
	1200	3317	649	3187
	1300	3990	704	3834
	1350	4326	732	4157
3/4'' NPS				
	100	15	38	14
	200	92	93	88
	300	204	149	196
	400	353	204	339
outside	500	547	260	525
dia	600	801	316	765
1.050''	700	1113	371	1069
26.7 mm	800	1508	427	1449
	900	1984	482	1906
	1000	2576	538	2475
	1100	3282	593	3153
	1200	4122	649	3961
	1300	4962	704	4768
	1350	5382	732	5171
1'' NPS				
	100	19	38	18
	200	113	93	108
	300	250	149	240
	400	433	204	416
outside	500	674	260	647
dia	600	989	316	950
1.315''	700	1379	371	1325
33.4 mm	800	1865	427	1792
	900	2462	482	2365
	1000	3194	538	3069
	1100	4080	593	3920
	1200	5123	649	4922
	1300	6166	704	5924
	1350	6687	732	6425

Pipe size	Pipe temp °F	Btu/lin ft hr loss	Pipe temp °C	W/lin m loss
1 1/4'' NPS				
	100	24	38	23
	200	141	93	135
	300	312	149	299
	400	540	204	518
outside	500	843	260	810
dia	600	1237	316	1188
1.66''	700	1728	371	1660
42.2 mm	800	2342	427	2250
	900	3091	482	2970
	1000	4010	538	3853
	1100	5123	593	4922
	1200	6438	649	6186
	1300	7800	704	7495
	1350	8410	732	8081
1 1/2'' NPS				
	100	27	38	26
	200	159	93	152
	300	352	149	338
	400	612	204	588
outside	500	955	260	914
dia	600	1403	316	1348
1.90''	700	1960	371	1883
48.3 mm	800	2661	427	2557
	900	3514	482	3376
	1000	4568	538	4389
	1100	5841	593	5612
	1200	7345	649	7057
	1300	8849	704	8503
	1350	9601	732	9226
2'' NPS				
	100	33	38	32
	200	195	93	187
	300	431	149	414
	400	753	204	723
outside	500	1176	260	1130
dia	600	1732	316	1664
2.375''	700	2423	378	2328
60.3 mm	800	3295	427	3166
	900	7201	482	4184
	1000	5665	538	5443
	1100	7251	593	6967
	1200	9133	649	8775
	1300	11015	704	10584
	1350	11956	732	11489

Pipe size	Pipe temp °F	Btu/lin ft hr loss	Pipe temp °C	W/lin m loss
2 1/2'' NPS				
	100	39	38	37
	200	232	93	222
	300	515	149	494
	400	899	204	863
outside	500	1408	260	1352
dia	600	2077	316	1995
2.875''	700	2907	371	2793
73.0 mm	800	3956	427	3801
	900	5235	482	5030
	1000	6817	538	6550
	1100	8732	593	8390
	1200	11005	649	10575
	1300	13278	704	12759
	1350	14414	732	13850
3'' NPS				
	100	48	38	46
	200	280	93	269
	300	619	149	594
	400	1083	204	1040
outside	500	1695	260	1629
dia	600	2505	316	2407
3.50''	700	3511	371	3373
88.9 mm	800	4784	427	4597
	900	6337	482	6089
	1000	8259	538	7936
	1100	10582	593	10168
	1200	13344	649	12822
	1300	16106	704	15476
	1350	17487	732	16803
3 1/2'' NPS				
	100	54	38	51
	200	317	93	304
	300	700	149	672
	400	1226	204	1178
outside	500	1922	260	1847
dia	600	2841	316	2730
4.0''	700	3987	371	3831
101.6 mm	800	5434	427	5221
	900	7201	482	6919
	1000	9390	538	9022
	1100	12039	593	11568
	1200	15189	649	14595
	1300	18339	704	17622
	1350	19914	732	19135

Table B-12 Heat losses from bare NPS pipe (continued)

English and Metric Units

Pipe size	Pipe temp °F	Btu/lin ft hr loss	Pipe temp °C	W/lin m loss
4″ NPS	100	60	38	57
	200	354	93	340
	300	783	149	752
	400	1370	204	1316
outside	500	2149	260	2064
dia	600	3179	316	3054
4.5″	700	4467	371	4292
114.3 mm	800	6094	427	5856
	900	8079	482	7763
	1000	10545	538	10132
	1100	13513	593	12985
	1200	17049	649	16382
	1300	20585	704	19780
	1350	22353	732	21479
4 1/2″ NPS	100	66	38	63
	200	388	93	372
	300	861	149	827
	400	1513	204	1453
outside	500	2377	260	2284
dia	600	3513	316	3376
5.0″	700	4939	371	4746
127.0 mm	800	6742	427	6478
	900	8944	482	8594
	1000	11676	538	11219
	1100	14982	593	14396
	1200	18892	649	18153
	1300	22802	704	21911
	1350	24757	732	23789
5″ NPS	100	73	38	70
	200	428	93	411
	300	952	149	194
	400	1674	204	1608
outside	500	2633	260	2530
dia	600	3893	316	3741
5.563″	700	5473	371	5259
141.3 mm	800	7482	427	7189
	900	9925	482	9537
	1000	12960	538	12453
	1100	16639	593	15988
	1200	20983	649	20162
	1300	25327	704	24337
	1350	27499	732	26424
6″ NPS	100	86	38	82
	200	503	93	493
	300	1122	149	1078
	400	1976	204	1898
outside	500	3105	260	2583
dia	600	4604	316	4424
6.625″	700	6479	371	6225
168.3 mm	800	8858	427	8512
	900	11769	482	11308
	1000	15366	538	14765
	1100	19740	593	18968
	1200	24909	649	23935
	1300	30078	704	28902
	1350	32662	732	31385
7″ NPS	100	99	38	95
	200	574	93	551
	300	1279	149	1229
	400	2251	204	2163
outside	500	3552	260	3413
dia	600	5179	316	4976
7.625″	700	7303	371	7017
193.7 mm	800	9995	427	9604
	900	13300	482	12780
	1000	17395	538	16715
	1100	22343	593	21469
	1200	28227	649	27123
	1300	34111	704	32777
	1350	37053	732	35604
8″ NPS	100	111	38	106
	200	643	93	617
	300	1436	149	1379
	400	2530	204	2431
outside	500	3988	260	3832
dia	600	5927	316	5695
8.625″	700	8356	371	8029
219.1 mm	800	11433	427	10986
	900	15211	482	14616
	1000	19895	538	19117
	1100	25568	593	24568
	1200	32293	649	31030
	1300	39018	704	37492
	1350	42380	732	40723

Pipe size	Pipe temp °F	Btu/lin ft hr loss	Pipe temp °C	W/lin m loss
9″ NPS	100	122	38	117
	200	712	93	684
	300	1597	149	1534
	400	2808	204	2698
outside	500	4429	260	4256
dia	600	6464	316	6211
9.625″	700	9134	371	8777
244.5 mm	800	12515	427	12025
	900	16648	482	15997
	1000	21820	538	20967
	1100	28051	593	26594
	1200	35471	649	34084
	1300	42891	704	41214
	1350	46601	732	44779
10″ NPS	100	136	38	130
	200	790	93	759
	300	1769	149	1700
	400	3114	204	2992
outside	500	4918	260	4725
dia	600	7312	316	7026
10.75″	700	10325	371	9921
273.0 mm	800	14147	427	13594
	900	18812	482	18077
	1000	24621	538	23658
	1100	31679	593	30440
	1200	40051	649	38485
	1300	48423	704	46530
	1350	52609	732	50552
11″ NPS	100	147	38	141
	200	859	93	825
	300	1917	149	1842
	400	3380	204	3248
outside	500	5357	260	5147
dia	600	7969	316	7657
11.75″	700	11259	371	10819
298.4 mm	800	15434	427	14831
	900	20530	482	19277
	1000	26877	538	25826
	1100	34590	593	33238
	1200	43740	649	42030
	1300	52890	704	50822
	1350	57465	732	55219
12″ NPS	100	159	38	152
	200	930	93	894
	300	2076	149	1994
	400	3664	204	3520
outside	500	5809	260	5582
dia	600	8627	316	8290
12.75″	700	12194	371	11717
323.8 mm	800	16721	427	16067
	900	22248	482	21378
	1000	29133	538	27994
	1100	37501	593	36035
	1200	47429	649	45575
	1300	57357	704	55115
	1350	62321	732	59885
14″ NPS	100	173	38	166
	200	1009	93	970
	300	2258	149	2170
	400	3989	204	3833
outside	500	6316	260	6069
dia	600	9404	316	9036
14.0″	700	13301	371	12781
355.6 mm	800	18236	427	17523
	900	24279	482	23330
	1000	31844	538	30599
	1100	40965	593	39363
	1200	51856	649	49829
	1300	62477	704	60034
	1350	67787	732	65137
16″ NPS	100	197	38	189
	200	1142	93	1097
	300	2560	149	2460
	400	4529	204	4352
outside	500	7160	260	6880
dia	600	10697	316	10278
16.0″	700	15124	371	14532
406.4 mm	800	20769	427	19957
	900	27663	482	26582
	1000	36257	538	34840
	1100	46689	593	44863
	1200	59089	649	56779
	1300	71489	704	68694
	1350	77689	732	74652

Pipe size	Pipe temp °F	Btu/lin ft hr loss	Pipe temp °C	W/lin m loss
18″ NPS	100	220	38	211
	200	1282	93	1231
	300	2875	149	2763
	400	5074	204	4876
outside	500	8032	260	7718
dia	600	12010	316	11540
18.0″	700	16992	371	16328
457.2 mm	800	23346	427	22433
	900	31109	482	29893
	1000	40788	538	39194
	1100	52539	593	50485
	1200	66456	649	73858
	1300	80373	704	77231
	1350	87331	732	83917
20″ NPS	100	243	38	233
	200	1416	93	1360
	300	3168	149	3044
	400	5618	204	5398
outside	500	8897	260	8549
dia	600	13270	316	12751
20.0″	700	18777	371	18043
508.0 mm	800	25810	427	24801
	900	34432	482	33086
	1000	45147	538	43387
	1100	58159	593	55885
	1200	73691	649	70810
	1300	89223	704	85735
	1350	96989	732	93197
24″ NPS	100	288	38	276
	200	1683	93	1617
	300	3772	149	3624
	400	6679	204	6418
outside	500	10595	260	10180
dia	600	15825	316	15206
24.0″	700	22375	371	21500
609.6 mm	800	30836	427	29630
	900	41112	482	39504
	1000	53958	538	51849
	1100	69536	593	66817
	1200	88221	649	84772
	1300	106906	704	102727
	1350	116248	732	111703
30″ NPS	100	351	38	337
	200	2042	93	1962
	300	4642	149	4460
	400	8242	204	7920
outside	500	13103	260	12590
dia	600	19606	316	18840
30.0″	700	27758	371	26673
762.0 mm	800	38299	427	36802
	900	51107	482	49109
	1000	67126	538	64502
	1100	86558	593	83170
	1200	109872	649	105577
	1300	133186	704	127980
	1350	144843	732	139181

Table B-13 Hot Service—Approximate Air Film Surface Resistance—Values of $\frac{1}{f}$ or (R_s) — English Units

∈ Emittance	Wind Velocity MPH	Insulation Resistance Min.	Insulation Resistance Max.	Operating Temperature (t_1) Degree F 200	250	300	350	400	500	600	700	800	900	1000	1100	1200	1300	1400	1500	1600	Wind Velocity km/hr
1.0	0	1.0	2.5	0.61	0.58	0.55	0.54	0.51	0.48	0.46	0.44	0.42									0
	Natural convec-tion	2.5	5.0	0.65	0.63	0.60	0.58	0.55	0.54	0.52	0.49	0.47	0.46	0.45							Natural convec-tion
		5.0	10.0	0.69	0.65	0.63	0.61	0.59	0.58	0.56	0.54	0.52	0.51	0.50	0.50	0.49					
		10.0	15.0	0.72	0.69	0.67	0.65	0.63	0.61	0.59	0.58	0.57	0.56	0.55	0.54	0.53	0.52	0.51			
		15.0	20.0	0.77	0.72	0.69	0.67	0.65	0.63	0.62	0.61	0.60	0.59	0.58	0.58	0.57	0.56	0.55	0.55	0.54	
		20.0	25.0	0.80	0.75	0.72	0.69	0.66	0.65	0.64	0.63	0.61	0.60	0.59	0.59	0.59	0.58	0.58	0.57	0.56	
	2	1.0	2.5	0.48	0.46	0.44	0.43	0.41	0.39	0.37	0.36	0.35									3.2
		2.5	5.0	0.50	0.48	0.46	0.45	0.42	0.41	0.40	0.39	0.38	0.37	0.37							
		5.0	10.0	0.56	0.52	0.50	0.49	0.47	0.45	0.44	0.43	0.42	0.41	0.40	0.40	0.39					
		10.0	15.0	0.60	0.56	0.53	0.52	0.50	0.48	0.47	0.46	0.45	0.44	0.43	0.42	0.42	0.41	0.40			
		15.0	20.0	0.62	0.60	0.57	0.56	0.54	0.52	0.50	0.48	0.47	0.46	0.45	0.45	0.44	0.44	0.43	0.43	0.43	
		20.0	25.0	0.64	0.62	0.60	0.58	0.56	0.54	0.52	0.50	0.48	0.48	0.47	0.47	0.46	0.46	0.45	0.45	0.45	
	4	1.0	2.5	0.41	0.38	0.37	0.36	0.34	0.32	0.31	0.30	0.29									6.4
		2.5	5.0	0.44	0.42	0.41	0.39	0.37	0.36	0.35	0.33	0.32	0.32	0.31							
		5.0	10.0	0.48	0.46	0.44	0.43	0.41	0.40	0.38	0.37	0.36	0.35	0.35	0.34	0.34					
		10.0	15.0	0.53	0.51	0.48	0.47	0.45	0.43	0.42	0.41	0.40	0.39	0.38	0.37	0.37	0.36	0.35			
		15.0	20.0	0.55	0.53	0.51	0.49	0.47	0.45	0.43	0.42	0.42	0.41	0.40	0.39	0.38	0.38	0.38	0.37	0.37	
		20.0	25.0	0.56	0.54	0.53	0.51	0.48	0.46	0.44	0.43	0.42	0.42	0.42	0.41	0.40	0.40	0.39	0.39	0.38	
	8	1.0	2.5	0.34	0.33	0.31	0.30	0.28	0.27	0.26	0.25	0.25									12.9
		2.5	5.0	0.38	0.36	0.34	0.33	0.31	0.30	0.29	0.28	0.27	0.27	0.26							
		5.0	10.0	0.43	0.40	0.38	0.36	0.34	0.33	0.32	0.31	0.30	0.30	0.29	0.29	0.29					
		10.0	15.0	0.45	0.43	0.41	0.39	0.37	0.36	0.35	0.34	0.33	0.33	0.32	0.31	0.31	0.30	0.30			
		15.0	20.0	0.46	0.44	0.43	0.41	0.39	0.38	0.37	0.36	0.35	0.34	0.33	0.33	0.33	0.32	0.32	0.31	0.31	
		20.0	25.0	0.48	0.46	0.45	0.43	0.41	0.39	0.38	0.37	0.36	0.36	0.35	0.34	0.34	0.33	0.33	0.33	0.32	
	12	1.0	2.5	0.30	0.29	0.28	0.27	0.25	0.24	0.24	0.23	0.23									19.3
		2.5	5.0	0.33	0.32	0.30	0.29	0.28	0.26	0.26	0.25	0.24	0.24	0.23							
		5.0	10.0	0.38	0.36	0.33	0.32	0.31	0.29	0.29	0.29	0.28	0.28	0.27	0.27	0.27					
		10.0	15.0	0.41	0.39	0.37	0.36	0.33	0.32	0.31	0.30	0.30	0.29	0.29	0.29	0.28	0.28	0.28			
		15.0	20.0	0.42	0.40	0.39	0.38	0.36	0.34	0.32	0.31	0.31	0.30	0.30	0.30	0.29	0.29	0.29	0.29	0.28	
		20.0	25.0	0.43	0.41	0.40	0.39	0.37	0.35	0.34	0.32	0.32	0.32	0.31	0.31	0.30	0.30	0.30	0.30	0.29	
	20	1.0	2.5	0.26	0.25	0.24	0.23	0.22	0.21	0.20	0.20	0.19									32.2
		2.5	5.0	0.29	0.27	0.26	0.25	0.24	0.23	0.22	0.21	0.21	0.20	0.20							
		5.0	10.0	0.33	0.31	0.29	0.28	0.26	0.25	0.25	0.24	0.24	0.23	0.23	0.23	0.23					
		10.0	15.0	0.36	0.34	0.32	0.31	0.29	0.27	0.27	0.26	0.25	0.25	0.24	0.24	0.24	0.24	0.23			
		15.0	20.0	0.38	0.36	0.34	0.33	0.31	0.29	0.28	0.27	0.26	0.26	0.25	0.25	0.25	0.24	0.24	0.24	0.24	
		20.0	25.0	0.40	0.38	0.36	0.34	0.32	0.30	0.29	0.28	0.27	0.27	0.27	0.26	0.26	0.26	0.25	0.25	0.25	
0.9	0	1.0	2.5	0.65	0.61	0.58	0.56	0.53	0.50	0.48	0.47	0.46									0
	Natural convec-tion	2.5	5.0	0.68	0.63	0.61	0.59	0.56	0.54	0.52	0.51	0.50	0.49	0.48							Natural convec-tion
		5.0	10.0	0.72	0.69	0.68	0.63	0.62	0.60	0.58	0.57	0.56	0.55	0.54	0.53	0.52					
		10.0	15.0	0.77	0.75	0.72	0.69	0.67	0.66	0.64	0.63	0.61	0.59	0.58	0.57	0.57	0.56	0.55			
		15.0	20.0	0.80	0.77	0.74	0.72	0.70	0.68	0.67	0.66	0.65	0.64	0.63	0.62	0.61	0.60	0.59	0.58	0.57	
		20.0	25.0	0.83	0.79	0.76	0.74	0.72	0.70	0.69	0.68	0.67	0.66	0.65	0.64	0.63	0.62	0.61	0.60	0.59	
	2	1.0	2.5	0.50	0.47	0.45	0.43	0.41	0.39	0.37	0.36	0.36									3.2
		2.5	5.0	0.52	0.50	0.49	0.47	0.44	0.43	0.41	0.40	0.39	0.38	0.38							
		5.0	10.0	0.58	0.54	0.52	0.51	0.49	0.47	0.46	0.45	0.44	0.43	0.42	0.41	0.40					
		10.0	15.0	0.63	0.59	0.56	0.54	0.52	0.51	0.49	0.48	0.47	0.46	0.45	0.44	0.43	0.42	0.41			
		15.0	20.0	0.63	0.61	0.58	0.56	0.54	0.53	0.51	0.50	0.49	0.48	0.47	0.46	0.45	0.45	0.44	0.44	0.44	
		20.0	25.0	0.67	0.64	0.61	0.59	0.56	0.55	0.53	0.52	0.51	0.50	0.49	0.48	0.47	0.47	0.46	0.46	0.46	
	4	1.0	2.5	0.42	0.40	0.38	0.36	0.34	0.32	0.31	0.31	0.30									6.4
		2.5	5.0	0.45	0.43	0.41	0.40	0.38	0.37	0.35	0.34	0.33	0.32	0.32							
		5.0	10.0	0.50	0.48	0.46	0.44	0.42	0.41	0.40	0.39	0.38	0.37	0.26	0.35	0.35					
		10.0	15.0	0.56	0.53	0.49	0.46	0.44	0.43	0.42	0.41	0.41	0.40	0.39	0.38	0.38	0.37	0.36			
		15.0	20.0	0.60	0.56	0.52	0.50	0.47	0.46	0.45	0.44	0.43	0.42	0.41	0.40	0.40	0.39	0.39	0.38	0.38	
		20.0	25.0	0.63	0.59	0.55	0.53	0.49	0.47	0.46	0.45	0.44	0.43	0.42	0.41	0.41	0.41	0.40	0.40	0.40	
0.9	8	1.0	2.5	0.35	0.34	0.32	0.30	0.28	0.27	0.26	0.26	0.26									12.9
		2.5	5.0	0.39	0.37	0.35	0.34	0.32	0.31	0.30	0.29	0.28	0.27	0.27							
		5.0	10.0	0.44	0.41	0.39	0.37	0.35	0.34	0.33	0.32	0.31	0.31	0.30	0.30	0.29					
		10.0	15.0	0.46	0.44	0.42	0.41	0.38	0.37	0.36	0.35	0.34	0.33	0.33	0.32	0.32	0.31	0.31			
		15.0	20.0	0.48	0.46	0.45	0.43	0.41	0.39	0.38	0.37	0.36	0.35	0.35	0.34	0.33	0.33	0.32	0.32	0.32	
		20.0	25.0	0.50	0.48	0.47	0.45	0.42	0.40	0.39	0.38	0.37	0.36	0.36	0.35	0.35	0.34	0.34	0.34	0.33	
				93	121	149	177	204	260	316	371	427	482	538	593	649	704	760	816	871	
				Degree C Operating Temperature (t_{m_1})																	

Note: Above resistances are in English Units °F, ft^2, hr, Btu.
Multiply by 0.1761 to obtain Metric Units K, m^2, W = R_{ms}.

Table B-13 Hot Service—Approximate Air Film Surface Resistance—Values of $\frac{1}{f}$ or (R_s) – English Units

∈ Emittance	Wind Velocity MPH	Insulation Resistance Min.	Insulation Resistance Max.	Operating Temperature (t_1) Degree F 200	250	300	350	400	500	600	700	800	900	1000	1100	1200	1300	1400	1500	1600	Wind Velocity km/hr
	12	1.0	2.5	0.31	0.29	0.28	0.27	0.25	0.24	0.24	0.24	0.23									19.3
		2.5	5.0	0.35	0.33	0.31	0.30	0.28	0.27	0.26	0.25	0.25	0.24	0.24							
		5.0	10.0	0.39	0.37	0.35	0.34	0.32	0.31	0.30	0.29	0.28	0.28	0.27	0.27	0.27					
		10.0	15.0	0.42	0.40	0.38	0.37	0.35	0.34	0.33	0.32	0.31	0.30	0.30	0.29	0.29	0.28	0.28			
		15.0	20.0	0.44	0.42	0.40	0.39	0.37	0.35	0.34	0.33	0.32	0.32	0.31	0.31	0.30	0.30	0.29	0.29	0.29	
		20.0	25.0	0.45	0.43	0.42	0.40	0.38	0.36	0.35	0.34	0.34	0.33	0.32	0.32	0.31	0.31	0.30	0.30	0.30	
	20	1.0	2.5	0.27	0.26	0.24	0.23	0.22	0.21	0.20	0.20	0.20									32.2
		2.5	5.0	0.30	0.28	0.27	0.26	0.25	0.24	0.23	0.22	0.22	0.21	0.21							
		5.0	10.0	0.34	0.32	0.30	0.29	0.27	0.26	0.25	0.24	0.24	0.24	0.23	0.23	0.23					
		10.0	15.0	0.37	0.35	0.33	0.32	0.30	0.29	0.28	0.27	0.26	0.25	0.25	0.24	0.24	0.24	0.23			
		15.0	20.0	0.39	0.37	0.35	0.33	0.32	0.30	0.29	0.28	0.27	0.27	0.26	0.26	0.25	0.25	0.25	0.24	0.24	
		20.0	25.0	0.41	0.38	0.36	0.35	0.33	0.32	0.30	0.29	0.28	0.28	0.28	0.27	0.27	0.26	0.26	0.26	0.26	
0.8	0 Natural convection	1.0	2.5	0.69	0.65	0.62	0.60	0.55	0.52	0.51	0.50	0.49									0 Natural convection
		2.5	5.0	0.73	0.69	0.67	0.65	0.62	0.60	0.58	0.56	0.54	0.53	0.52							
		5.0	10.0	0.78	0.75	0.73	0.71	0.69	0.66	0.64	0.62	0.60	0.59	0.58	0.57	0.56					
		10.0	15.0	0.82	0.79	0.77	0.75	0.73	0.70	0.68	0.67	0.65	0.64	0.63	0.62	0.61	0.60	0.58			
		15.0	20.0	0.87	0.83	0.79	0.78	0.77	0.74	0.72	0.70	0.68	0.67	0.65	0.65	0.64	0.63	0.62	0.61	0.61	
		20.0	25.0	0.89	0.85	0.81	0.80	0.79	0.76	0.74	0.72	0.70	0.69	0.68	0.67	0.66	0.65	0.64	0.63	0.63	
	2	1.0	2.5	0.51	0.49	0.46	0.45	0.43	0.40	0.38	0.37	0.37									3.2
		2.5	5.0	0.54	0.52	0.51	0.49	0.46	0.45	0.43	0.41	0.40	0.39	0.39							
		5.0	10.0	0.60	0.57	0.54	0.53	0.51	0.49	0.48	0.46	0.45	0.44	0.43	0.44	0.44					
		10.0	15.0	0.65	0.61	0.57	0.55	0.53	0.51	0.50	0.48	0.47	0.46	0.46	0.45	0.45	0.45	0.44			
		15.0	20.0	0.68	0.65	0.62	0.60	0.58	0.56	0.54	0.52	0.51	0.50	0.49	0.48	0.47	0.47	0.46	0.46	0.46	
		20.0	25.0	0.70	0.68	0.65	0.62	0.60	0.58	0.56	0.54	0.53	0.52	0.51	0.50	0.49	0.49	0.48	0.47	0.47	
	4	1.0	2.5	0.43	0.41	0.40	0.38	0.36	0.34	0.32	0.32	0.31									6.4
		2.5	5.0	0.47	0.45	0.43	0.41	0.40	0.38	0.36	0.35	0.34	0.34	0.33							
		5.0	10.0	0.52	0.50	0.48	0.46	0.44	0.42	0.41	0.40	0.39	0.38	0.37	0.37	0.36					
		10.0	15.0	0.57	0.54	0.52	0.50	0.48	0.46	0.44	0.43	0.42	0.41	0.40	0.40	0.39	0.39	0.39			
		15.0	20.0	0.61	0.57	0.54	0.52	0.50	0.48	0.46	0.45	0.44	0.43	0.42	0.42	0.41	0.41	0.40	0.40	0.39	
		20.0	25.0	0.64	0.59	0.56	0.54	0.52	0.50	0.48	0.47	0.46	0.45	0.44	0.43	0.42	0.42	0.41	0.41	0.41	
	8	1.0	2.5	0.36	0.35	0.33	0.31	0.29	0.28	0.27	0.27	0.27									12.9
		2.5	5.0	0.40	0.38	0.36	0.35	0.33	0.32	0.31	0.30	0.29	0.28	0.28							
		5.0	10.0	0.45	0.42	0.40	0.38	0.37	0.36	0.35	0.34	0.33	0.32	0.31	0.31	0.30					
		10.0	15.0	0.48	0.46	0.44	0.42	0.40	0.38	0.37	0.36	0.35	0.34	0.34	0.34	0.33	0.32	0.32			
		15.0	20.0	0.50	0.48	0.46	0.44	0.42	0.40	0.39	0.38	0.37	0.36	0.36	0.35	0.35	0.34	0.34	0.33	0.33	
		20.0	25.0	0.52	0.50	0.48	0.46	0.44	0.42	0.40	0.39	0.38	0.37	0.37	0.36	0.36	0.36	0.35	0.35	0.34	
	12	1.0	2.5	0.32	0.30	0.29	0.28	0.26	0.25	0.24	0.24	0.24									19.3
		2.5	5.0	0.36	0.34	0.31	0.31	0.29	0.28	0.27	0.26	0.25	0.25	0.25							
		5.0	10.0	0.40	0.38	0.36	0.35	0.33	0.32	0.31	0.30	0.29	0.29	0.28	0.28	0.28					
		10.0	15.0	0.44	0.41	0.39	0.38	0.36	0.35	0.34	0.32	0.31	0.30	0.30	0.29	0.29	0.29	0.29			
		15.0	20.0	0.46	0.44	0.42	0.40	0.38	0.36	0.35	0.34	0.33	0.32	0.32	0.31	0.30	0.30	0.30	0.30	0.30	
		20.0	25.0	0.48	0.46	0.44	0.42	0.40	0.38	0.36	0.35	0.34	0.33	0.33	0.32	0.32	0.31	0.31	0.31	0.31	
	20	1.0	2.5	0.28	0.26	0.25	0.24	0.22	0.21	0.20	0.20	0.20									32.2
		2.5	5.0	0.30	0.29	0.28	0.27	0.25	0.24	0.23	0.22	0.22	0.22	0.22							
		5.0	10.0	0.34	0.33	0.31	0.30	0.28	0.27	0.26	0.25	0.25	0.24	0.24	0.24	0.24					
		10.0	15.0	0.38	0.36	0.34	0.32	0.30	0.29	0.28	0.28	0.27	0.26	0.26	0.25	0.25	0.24	0.24			
		15.0	20.0	0.40	0.38	0.36	0.34	0.32	0.31	0.30	0.29	0.28	0.28	0.27	0.27	0.26	0.26	0.25	0.25	0.25	
		20.0	25.0	0.42	0.40	0.38	0.36	0.34	0.32	0.31	0.30	0.29	0.29	0.28	0.28	0.28	0.27	0.27	0.27	0.26	
0.7	0 Natural convection	1.0	2.5	0.73	0.69	0.66	0.64	0.59	0.56	0.53	0.52	0.51									0 Natural convection
		2.5	5.0	0.78	0.74	0.71	0.68	0.66	0.63	0.60	0.58	0.56	0.55	0.54							
		5.0	10.0	0.83	0.81	0.78	0.76	0.73	0.70	0.68	0.65	0.63	0.62	0.61	0.60	0.59					
		10.0	15.0	0.88	0.85	0.82	0.80	0.77	0.74	0.72	0.70	0.69	0.68	0.67	0.65	0.64	0.63	0.62			
		15.0	20.0	0.93	0.90	0.86	0.84	0.81	0.78	0.76	0.74	0.73	0.71	0.70	0.69	0.68	0.67	0.66	0.65	0.64	
		20.0	25.0	0.95	0.92	0.88	0.86	0.83	0.80	0.78	0.76	0.75	0.73	0.72	0.71	0.70	0.69	0.68	0.67	0.66	
	2	1.0	2.5	0.53	0.50	0.48	0.47	0.43	0.41	0.39	0.38	0.38									3.2
		2.5	5.0	0.57	0.54	0.53	0.51	0.49	0.47	0.45	0.43	0.42	0.41	0.40							
		5.0	10.0	0.63	0.61	0.58	0.56	0.54	0.52	0.50	0.49	0.48	0.47	0.46	0.46	0.45					
		10.0	15.0	0.69	0.66	0.63	0.61	0.57	0.55	0.54	0.52	0.51	0.50	0.49	0.49	0.48	0.47	0.46			
		15.0	20.0	0.71	0.69	0.65	0.63	0.60	0.58	0.57	0.56	0.54	0.53	0.52	0.51	0.50	0.49	0.49	0.49	0.48	
		20.0	25.0	0.74	0.71	0.69	0.67	0.63	0.60	0.59	0.57	0.56	0.55	0.54	0.53	0.52	0.51	0.50	0.50	0.50	
			Degree C Operating Temperature (t_{m_1})	93	121	149	177	204	260	316	371	427	482	538	593	649	704	760	816	871	

Note: Above resistances are in English Units °F, ft², hr, Btu.
Multiply by 0.1761 to obtain Metric Units K, m², W = R_{ms}.

Table B-13 Hot Service—Approximate Air Film Surface Resistance—Values of $\frac{1}{f}$ or (R_s) – English Units

∈ Emittance	Wind Velocity MPH	Insulation Resistance Min.	Insulation Resistance Max.	Operating Temperature (t_1) Degree F 200	250	300	350	400	500	600	700	800	900	1000	1100	1200	1300	1400	1500	1600	Wind Velocity km/hr
	4	1.0	2.5	0.44	0.42	0.41	0.39	0.37	0.35	0.33	0.32	0.32									6.4
		2.5	5.0	0.50	0.47	0.45	0.44	0.41	0.39	0.38	0.37	0.36	0.35	0.34							
		5.0	10.0	0.54	0.51	0.49	0.47	0.45	0.44	0.42	0.41	0.40	0.39	0.38	0.38	0.37					
		10.0	15.0	0.61	0.56	0.54	0.52	0.49	0.47	0.46	0.45	0.44	0.43	0.42	0.41	0.40	0.40	0.40			
		15.0	20.0	0.63	0.59	0.57	0.55	0.52	0.50	0.48	0.47	0.46	0.43	0.44	0.43	0.42	0.42	0.41	0.40	0.40	
		20.0	25.0	0.66	0.64	0.61	0.57	0.55	0.52	0.50	0.48	0.47	0.46	0.45	0.45	0.44	0.44	0.43	0.43	0.42	
	8	1.0	2.5	0.37	0.36	0.34	0.42	0.30	0.29	0.28	0.28	0.28									12.9
		2.5	5.0	0.41	0.39	0.37	0.36	0.34	0.33	0.32	0.31	0.30	0.29	0.29							
		5.0	10.0	0.46	0.43	0.41	0.39	0.38	0.37	0.36	0.35	0.34	0.33	0.32	0.32	0.31					
		10.0	15.0	0.49	0.47	0.45	0.43	0.41	0.39	0.38	0.37	0.36	0.35	0.34	0.34	0.33	0.33	0.33			
		15.0	20.0	0.52	0.50	0.48	0.46	0.44	0.42	0.40	0.39	0.38	0.37	0.37	0.36	0.36	0.35	0.34	0.34	0.33	
		20.0	25.0	0.54	0.52	0.50	0.48	0.46	0.44	0.42	0.40	0.39	0.39	0.38	0.37	0.37	0.36	0.36	0.35	0.35	
	12	1.0	2.5	0.33	0.31	0.30	0.29	0.27	0.26	0.25	0.25	0.25									19.3
		2.5	5.0	0.37	0.35	0.33	0.32	0.30	0.29	0.28	0.27	0.26	0.26	0.26							
		5.0	10.0	0.41	0.39	0.37	0.36	0.34	0.33	0.32	0.31	0.30	0.29	0.29	0.29	0.29					
		10.0	15.0	0.45	0.42	0.41	0.39	0.36	0.35	0.34	0.33	0.32	0.31	0.31	0.30	0.30	0.30	0.30			
		15.0	20.0	0.47	0.45	0.43	0.41	0.39	0.37	0.36	0.35	0.34	0.33	0.32	0.32	0.31	0.31	0.31	0.30	0.30	
		20.0	25.0	0.49	0.47	0.45	0.43	0.41	0.39	0.37	0.36	0.35	0.34	0.33	0.33	0.33	0.32	0.32	0.31	0.31	
	20	1.0	2.5	0.28	0.26	0.25	0.25	0.23	0.22	0.21	0.21	0.21									32.2
		2.5	5.0	0.31	0.30	0.29	0.27	0.26	0.25	0.23	0.23	0.22	0.22	0.22							
		5.0	10.0	0.35	0.33	0.31	0.30	0.29	0.27	0.27	0.26	0.26	0.25	0.24	0.24	0.24					
		10.0	15.0	0.38	0.36	0.35	0.33	0.31	0.30	0.29	0.28	0.27	0.27	0.26	0.26	0.26	0.25	0.25			
		15.0	20.0	0.40	0.38	0.37	0.35	0.33	0.31	0.31	0.30	0.29	0.28	0.27	0.27	0.27	0.26	0.26	0.26	0.26	
		20.0	25.0	0.42	0.40	0.38	0.37	0.35	0.33	0.32	0.31	0.30	0.30	0.29	0.29	0.28	0.28	0.27	0.27	0.27	
0.6	0 Natural convection	1.0	2.5	0.78	0.73	0.69	0.66	0.63	0.60	0.57	0.56	0.55									0 Natural convection
		2.5	5.0	0.83	0.80	0.76	0.75	0.70	0.68	0.75	0.63	0.61	0.60	0.59							
		5.0	10.0	0.91	0.87	0.84	0.82	0.77	0.75	0.72	0.70	0.68	0.67	0.65	0.64	0.63					
		10.0	15.0	0.95	0.91	0.89	0.87	0.83	0.80	0.77	0.75	0.73	0.71	0.70	0.69	0.68	0.67	0.66			
		15.0	20.0	0.99	0.95	0.92	0.90	0.87	0.84	0.82	0.80	0.78	0.76	0.74	0.73	0.72	0.71	0.70	0.69	0.68	
		20.0	25.0	1.04	0.98	0.95	0.93	0.90	0.87	0.85	0.83	0.81	0.79	0.77	0.76	0.75	0.74	0.73	0.72	0.71	
	2	1.0	2.5	0.55	0.52	0.50	0.48	0.45	0.42	0.41	0.40	0.40									3.2
		2.5	5.0	0.60	0.58	0.55	0.53	0.50	0.48	0.47	0.45	0.44	0.43	0.42							
		5.0	10.0	0.66	0.63	0.61	0.59	0.56	0.53	0.51	0.50	0.49	0.48	0.47	0.47	0.46	0.46				
		10.0	15.0	0.71	0.68	0.66	0.63	0.60	0.58	0.56	0.54	0.53	0.52	0.51	0.50	0.49	0.48	0.47			
		15.0	20.0	0.76	0.71	0.68	0.67	0.63	0.61	0.59	0.57	0.56	0.55	0.54	0.53	0.52	0.51	0.50	0.50	0.40	
		20.0	25.0	0.79	0.73	0.72	0.69	0.66	0.63	0.61	0.59	0.58	0.57	0.56	0.55	0.54	0.53	0.52	0.51	0.51	
	4	1.0	2.5	0.45	0.43	0.42	0.40	0.38	0.36	0.35	0.34	0.34									6.4
		2.5	5.0	0.52	0.49	0.47	0.45	0.42	0.41	0.39	0.38	0.37	0.36	0.35							
		5.0	10.0	0.57	0.53	0.51	0.50	0.47	0.45	0.44	0.43	0.42	0.41	0.40	0.39	0.38					
		10.0	15.0	0.64	0.58	0.56	0.53	0.51	0.49	0.48	0.46	0.45	0.44	0.43	0.42	0.42	0.41	0.41			
		15.0	20.0	0.68	0.63	0.58	0.56	0.53	0.51	0.50	0.49	0.48	0.47	0.46	0.45	0.44	0.43	0.43	0.42	0.42	
		20.0	25.0	0.71	0.68	0.64	0.60	0.57	0.54	0.52	0.50	0.50	0.49	0.48	0.47	0.46	0.35	0.45	0.44	0.43	
0.6	8	1.0	2.5	0.38	0.36	0.34	0.33	0.31	0.30	0.29	0.29	0.29									12.9
		2.5	5.0	0.42	0.40	0.38	0.37	0.35	0.34	0.33	0.32	0.31	0.30	0.30							
		5.0	10.0	0.47	0.44	0.42	0.40	0.39	0.38	0.37	0.36	0.35	0.34	0.33	0.32	0.32					
		10.0	15.0	0.50	0.48	0.46	0.44	0.42	0.40	0.39	0.38	0.37	0.36	0.35	0.35	0.34	0.34	0.34			
		15.0	20.0	0.54	0.52	0.49	0.47	0.45	0.43	0.41	0.40	0.39	0.38	0.38	0.37	0.37	0.36	0.36	0.35	0.34	
		20.0	25.0	0.56	0.54	0.52	0.50	0.47	0.45	0.43	0.41	0.40	0.40	0.39	0.39	0.38	0.37	0.37	0.26	0.36	
	12	1.0	2.5	0.34	0.32	0.31	0.30	0.28	0.26	0.25	0.25	0.25									19.3
		2.5	5.0	0.38	0.36	0.34	0.33	0.31	0.30	0.29	0.28	0.27	0.27	0.27							
		5.0	10.0	0.42	0.40	0.38	0.37	0.35	0.33	0.32	0.31	0.31	0.30	0.30	0.30	0.29					
		10.0	15.0	0.46	0.44	0.42	0.40	0.38	0.36	0.35	0.34	0.33	0.32	0.32	0.31	0.31	0.30	0.30			
		15.0	20.0	0.48	0.46	0.44	0.42	0.40	0.38	0.37	0.36	0.35	0.34	0.33	0.33	0.32	0.32	0.32	0.31	0.31	
		20.0	25.0	0.50	0.48	0.46	0.44	0.42	0.40	0.38	0.37	0.36	0.35	0.34	0.34	0.34	0.33	0.33	0.32	0.32	
	20	1.0	2.5	0.28	0.27	0.25	0.25	0.23	0.22	0.21	0.21	0.21									32.2
		2.5	5.0	0.32	0.30	0.29	0.28	0.26	0.25	0.24	0.23	0.22	0.22	0.22							
		5.0	10.0	0.35	0.34	0.33	0.32	0.30	0.28	0.27	0.26	0.26	0.25	0.24	0.24	0.24					
		10.0	15.0	0.39	0.37	0.35	0.34	0.32	0.30	0.29	0.29	0.28	0.28	0.26	0.26	0.26	0.25	0.25			
		15.0	20.0	0.41	0.39	0.37	0.36	0.34	0.33	0.32	0.31	0.30	0.39	0.28	0.28	0.27	0.27	0.26	0.26	0.26	
		20.0	25.0	0.43	0.41	0.39	0.37	0.35	0.34	0.33	0.32	0.31	0.30	0.39	0.39	0.28	0.28	0.28	0.28	0.27	
			Degree C, Operating Temperature (t_{m_1})	93	121	149	177	204	260	316	371	427	482	538	593	649	704	760	816	871	

Note: Above resistances are in English Units °F, ft², hr, Btu.
Multiply by 0.1761 to obtain Metric Units K, m², W = R_{ms}.

Table B-13 Hot Service—Approximate Air Film Surface Resistance—Values of $\frac{1}{f}$ or (R_s) – English Units

∈ Emittance	Wind Velocity MPH	Insulation Resistance Min.	Insulation Resistance Max.	Operating Temperature (t_1) Degree F 200	250	300	350	400	500	600	700	800	900	1000	1100	1200	1300	1400	1500	1600	Wind Velocity km/hr
0.5	0	1.0	2.5	0.82	0.77	0.74	0.71	0.67	0.64	0.61	0.60	0.59									0
	Natural convection	2.5	5.0	0.88	0.84	0.81	0.79	0.76	0.71	0.68	0.66	0.64	0.65	0.64							Natural convection
		5.0	10.0	0.98	0.95	0.90	0.87	0.83	0.79	0.77	0.75	0.73	0.70	0.68	0.67	0.66					
		10.0	15.0	1.03	1.00	0.97	0.94	0.89	0.85	0.82	0.80	0.78	0.77	0.76	0.75	0.74	0.73	0.72			
		15.0	20.0	1.07	1.03	1.00	0.97	0.92	0.87	0.85	0.83	0.81	0.80	0.79	0.78	0.77	0.76	0.75	0.72	0.73	
		20.0	25.0	1.11	1.06	1.03	1.00	0.95	0.90	0.88	0.87	0.84	0.83	0.82	0.81	0.80	0.79	0.78	0.77	0.76	
	2	1.0	2.5	0.57	0.54	0.51	0.48	0.46	0.44	0.42	0.41	0.41									3.2
		2.5	5.0	0.64	0.60	0.57	0.55	0.52	0.50	0.48	0.46	0.45	0.44	0.43							
		5.0	10.0	0.70	0.66	0.64	0.62	0.58	0.55	0.54	0.52	0.51	0.50	0.49	0.48	0.48					
		10.0	15.0	0.75	0.71	0.68	0.66	0.63	0.60	0.57	0.56	0.55	0.54	0.53	0.52	0.51	0.51	0.50			
		15.0	20.0	0.80	0.75	0.72	0.69	0.66	0.64	0.62	0.60	0.58	0.57	0.56	0.55	0.54	0.54	0.53	0.52	0.51	
		20.0	25.0	0.83	0.78	0.75	0.72	0.69	0.66	0.64	0.62	0.60	0.59	0.58	0.57	0.56	0.55	0.55	0.54	0.54	
	4	1.0	2.5	0.48	0.45	0.43	0.31	0.39	0.37	0.36	0.35	0.35									6.4
		2.5	5.0	0.54	0.51	0.49	0.47	0.44	0.42	0.40	0.39	0.38	0.37	0.36							
		5.0	10.0	0.60	0.56	0.54	0.52	0.49	0.47	0.46	0.44	0.43	0.42	0.41	0.40	0.40					
		10.0	15.0	0.67	0.61	0.59	0.56	0.53	0.51	0.49	0.48	0.47	0.46	0.45	0.44	0.43	0.42	0.42			
		15.0	20.0	0.72	0.67	0.62	0.59	0.56	0.54	0.52	0.50	0.49	0.48	0.47	0.47	0.46	0.46	0.45	0.45	0.44	
		20.0	25.0	0.75	0.72	0.67	0.62	0.59	0.56	0.54	0.52	0.51	0.50	0.49	0.48	0.47	0.47	0.47	0.46	0.35	
	8	1.0	2.5	0.39	0.37	0.35	0.34	0.32	0.30	0.29	0.29	0.29									12.9
		2.5	5.0	0.42	0.40	0.38	0.37	0.35	0.34	0.33	0.32	0.31	0.31	0.31							
		5.0	10.0	0.49	0.47	0.45	0.43	0.41	0.39	0.37	0.36	0.35	0.35	0.34	0.34	0.33					
		10.0	15.0	0.54	0.52	0.49	0.47	0.44	0.42	0.41	0.40	0.39	0.38	0.37	0.36	0.35	0.35	0.35			
		15.0	20.0	0.57	0.54	0.52	0.50	0.47	0.45	0.43	0.42	0.40	0.40	0.39	0.39	0.38	0.38	0.37	0.36	0.35	
		20.0	25.0	0.59	0.56	0.54	0.52	0.49	0.47	0.45	0.43	0.42	0.41	0.40	0.40	0.39	0.39	0.39	0.38	0.37	
	12	1.0	2.5	0.35	0.34	0.32	0.31	0.28	0.27	0.26	0.26	0.25									19.3
		2.5	5.0	0.38	0.36	0.35	0.34	0.32	0.31	0.29	0.28	0.28	0.28	0.27							
		5.0	10.0	0.44	0.41	0.39	0.38	0.35	0.34	0.34	0.33	0.33	0.31	0.30	0.30	0.30					
		10.0	15.0	0.48	0.45	0.43	0.41	0.38	0.37	0.36	0.35	0.34	0.34	0.33	0.32	0.32	0.31	0.31			
		15.0	20.0	0.52	0.49	0.46	0.45	0.42	0.39	0.38	0.37	0.36	0.34	0.34	0.34	0.34	0.34	0.33	0.32	0.32	
		20.0	25.0	0.54	0.51	0.48	0.46	0.43	0.41	0.39	0.38	0.37	0.36	0.35	0.35	0.35	0.34	0.34	0.34	0.33	
	20	1.0	2.5	0.29	0.27	0.26	0.25	0.23	0.22	0.21	0.21	0.21									32.2
		2.5	5.0	0.33	0.31	0.29	0.28	0.27	0.26	0.25	0.24	0.23	0.23	0.22							
		5.0	10.0	0.37	0.35	0.33	0.32	0.30	0.29	0.28	0.27	0.26	0.26	0.25	0.25	0.25					
		10.0	15.0	0.40	0.38	0.36	0.35	0.33	0.31	0.30	0.29	0.29	0.28	0.27	0.27	0.26	0.26	0.26			
		15.0	20.0	0.42	0.40	0.38	0.37	0.35	0.33	0.32	0.31	0.30	0.29	0.29	0.29	0.28	0.28	0.27	0.27	0.27	
		20.0	25.0	0.45	0.43	0.41	0.39	0.37	0.35	0.34	0.32	0.31	0.31	0.30	0.30	0.29	0.29	0.29	0.28	0.28	
0.4	0	1.0	25.	0.87	0.83	0.79	0.75	0.71	0.68	0.65	0.64	0.63									0
	Natural convection	2.5	5.0	0.96	0.90	0.87	0.84	0.79	0.76	0.74	0.71	0.69	0.68	0.67							Natural convection
		5.0	10.0	1.07	1.01	0.98	0.95	0.91	0.85	0.83	0.81	0.79	0.77	0.75	0.74	0.73					
		10.0	15.0	1.13	1.08	1.04	1.01	0.96	0.90	0.88	0.86	0.84	0.82	0.80	0.79	0.78	0.77	0.76			
		15.0	20.0	1.18	1.13	1.08	1.05	1.01	0.97	0.94	0.91	0.88	0.86	0.85	0.84	0.83	0.82	0.81	0.80	0.79	
		20.0	25.0	1.24	1.17	1.12	1.09	1.04	1.00	0.97	0.94	0.91	0.89	0.88	0.87	0.86	0.85	0.84	0.83	0.82	
	2	1.0	2.5	0.60	0.57	0.54	0.51	0.48	0.46	0.45	0.44	0.43									3.2
		2.5	5.0	0.68	0.64	0.61	0.59	0.55	0.52	0.50	0.48	0.47	0.46	0.45							
		5.0	10.0	0.74	0.71	0.68	0.65	0.61	0.59	0.57	0.55	0.53	0.52	0.51	0.50	0.50					
		10.5	15.0	0.81	0.75	0.73	0.71	0.68	0.64	0.62	0.60	0.58	0.57	0.56	0.55	0.54	0.53	0.52			
		15.0	20.0	0.85	0.80	0.75	0.73	0.70	0.68	0.66	0.64	0.62	0.60	0.59	0.58	0.57	0.56	0.55	0.54	0.53	
		20.0	25.0	0.90	0.84	0.81	0.77	0.73	0.70	0.68	0.66	0.64	0.62	0.61	0.60	0.59	0.59	0.58	0.57	0.57	
	4	1.0	2.5	0.50	0.47	0.45	0.43	0.40	0.39	0.38	0.37	0.36									6.4
		2.5	5.0	0.57	0.53	0.50	0.48	0.46	0.44	0.43	0.41	0.39	0.38	0.37							
		5.0	10.0	0.63	0.60	0.58	0.56	0.53	0.49	0.47	0.46	0.45	0.44	0.43	0.42	0.42					
		10.0	15.0	0.69	0.63	0.61	0.59	0.56	0.53	0.51	0.50	0.49	0.48	0.47	0.46	0.45	0.44	0.43			
		15.0	20.0	0.73	0.68	0.65	0.62	0.59	0.57	0.54	0.52	0.51	0.50	0.49	0.48	0.47	0.47	0.46	0.46	0.45	
		20.0	25.0	0.77	0.72	0.69	0.65	0.61	0.59	0.57	0.55	0.53	0.52	0.51	0.50	0.49	0.49	0.48	0.47	0.47	
	8	1.0	2.5	0.40	0.38	0.36	0.35	0.33	0.31	0.30	0.30	0.29									12.9
		2.5	5.0	0.44	0.42	0.40	0.38	0.35	0.34	0.33	0.32	0.32	0.31	0.31							
		5.0	10.0	0.51	0.48	0.46	0.44	0.42	0.40	0.38	0.37	0.36	0.35	0.35	0.34	0.34					
		10.0	15.0	0.56	0.53	0.50	0.48	0.45	0.43	0.42	0.41	0.40	0.39	0.38	0.38	0.37	0.36	0.36			
		15.0	20.0	0.60	0.57	0.53	0.51	0.48	0.46	0.44	0.43	0.41	0.41	0.40	0.40	0.39	0.39	0.38	0.37	0.36	
		20.0	25.0	0.63	0.59	0.56	0.54	0.51	0.49	0.47	0.45	0.44	0.43	0.42	0.41	0.40	0.40	0.39	0.38	0.38	
			Degree C Operating Temperature (t_{m_1})	93	121	149	177	204	260	316	371	427	482	538	593	649	704	760	816	871	

Note: Above resistances are in English Units °F, ft^2, hr, Btu.
Multiply by 0.1761 to obtain Metric Units K, m^2, W = R_{ms}.

Table B-13 Hot Service—Approximate Air Film Surface Resistance—Values of $\frac{1}{f}$ or (R_s) – English Units

∈ Emittance	Wind Velocity MPH	Insulation Resistance Min.	Insulation Resistance Max.	Operating Temperature (t_1) Degree F 200	250	300	350	400	500	600	700	800	900	1000	1100	1200	1300	1400	1500	1600	Wind Velocity km/hr
	12	1.0	2.5	0.36	0.34	0.33	0.31	0.29	0.28	0.27	0.26	0.25									19.3
		2.5	5.0	0.40	0.38	0.37	0.35	0.33	0.31	0.30	0.29	0.28	0.28	0.27							
		5.0	10.0	0.46	0.43	0.40	0.39	0.37	0.35	0.34	0.33	0.32	0.32	0.31	0.31	0.31					
		10.0	15.0	0.50	0.47	0.45	0.42	0.40	0.38	0.37	0.36	0.35	0.34	0.33	0.33	0.33	0.32	0.32			
		15.0	20.0	0.54	0.51	0.48	0.46	0.43	0.40	0.39	0.38	0.37	0.37	0.36	0.35	0.34	0.34	0.33	0.33	0.32	
		20.0	25.0	0.56	0.53	0.50	0.48	0.45	0.42	0.40	0.40	0.39	0.38	0.37	0.37	0.36	0.35	0.34	0.34	0.34	
	20	1.0	2.5	0.30	0.28	0.26	0.26	0.25	0.24	0.23	0.22	0.22									32.2
		2.5	5.0	0.34	0.32	0.30	0.29	0.28	0.26	0.25	0.24	0.24	0.24	0.23							
		5.0	10.0	0.38	0.36	0.34	0.32	0.31	0.30	0.29	0.28	0.27	0.26	0.25	0.25	0.25					
		10.0	15.0	0.42	0.40	0.38	0.36	0.34	0.32	0.31	0.30	0.29	0.29	0.29	0.28	0.27	0.26	0.26			
		15.0	20.0	0.44	0.42	0.40	0.38	0.35	0.34	0.33	0.32	0.31	0.31	0.30	0.30	0.29	0.28	0.28	0.27	0.27	
		20.0	25.0	0.47	0.45	0.43	0.40	0.38	0.35	0.34	0.33	0.32	0.32	0.31	0.31	0.30	0.30	0.29	0.29	0.29	
0.3	0	1.0	2.5	0.95	0.89	0.86	0.81	0.76	0.72	0.69	0.67	0.66									0
	Natural convection	2.5	5.0	1.08	1.00	0.96	0.92	0.88	0.83	0.80	0.77	0.75	0.74	0.73							Natural convection
		5.0	10.0	1.17	1.12	1.05	1.02	0.97	0.92	0.88	0.86	0.84	0.82	0.80	0.79	0.78					
		10.0	15.0	1.25	1.19	1.15	1.11	1.05	1.00	0.96	0.94	0.91	0.89	0.87	0.85	0.84	0.83	0.82			
		15.0	20.0	1.31	1.25	1.21	1.16	1.07	1.04	1.02	0.99	0.96	0.94	0.92	0.91	0.89	0.88	0.87	0.86	0.85	
		20.0	25.0	1.34	1.28	1.25	1.20	1.15	1.10	1.07	1.04	1.01	0.98	0.96	0.94	0.92	0.91	0.90	0.89	0.88	
	2	1.0	2.5	0.64	0.60	0.57	0.54	0.51	0.48	0.47	0.46	0.45									3.2
		2.5	5.0	0.71	0.67	0.64	0.61	0.58	0.55	0.53	0.51	0.50	0.49	0.48							
		5.0	10.0	0.78	0.75	0.72	0.68	0.66	0.62	0.60	0.58	0.56	0.54	0.53	0.52	0.52					
		10.0	15.0	0.85	0.81	0.77	0.75	0.71	0.67	0.66	0.63	0.61	0.60	0.59	0.58	0.56	0.55	0.54			
		15.0	20.0	0.90	0.85	0.82	0.78	0.75	0.72	0.69	0.67	0.66	0.64	0.62	0.60	0.69	0.59	0.58	0.57	0.57	
		20.0	25.0	0.95	0.88	0.85	0.82	0.77	0.75	0.73	0.70	0.68	0.67	0.66	0.64	0.61	0.60	0.60	0.60	0.59	
	4	1.0	2.5	0.53	0.50	0.48	0.45	0.42	0.40	0.39	0.38	0.37									6.4
		2.5	5.0	0.59	0.56	0.53	0.51	0.48	0.45	0.43	0.42	0.40	0.39	0.38							
		5.0	10.0	0.65	0.62	0.59	0.57	0.54	0.52	0.50	0.48	0.47	0.45	0.44	0.44	0.43					
		10.0	15.0	0.72	0.68	0.65	0.62	0.58	0.56	0.54	0.52	0.51	0.50	0.49	0.48	0.47	0.46	0.45			
		15.0	20.0	0.77	0.73	0.68	0.66	0.62	0.59	0.57	0.55	0.53	0.52	0.52	0.51	0.50	0.49	0.48	0.48	0.47	
		20.0	25.0	0.81	0.76	0.72	0.68	0.65	0.62	0.59	0.58	0.57	0.56	0.55	0.54	0.53	0.52	0.51	0.50	0.49	
0.3	8	1.0	2.5	0.42	0.40	0.38	0.37	0.34	0.32	0.31	0.30	0.30									12.9
		2.5	5.0	0.47	0.45	0.43	0.41	0.39	0.37	0.35	0.34	0.33	0.33	0.32							
		5.0	10.0	0.53	0.50	0.48	0.46	0.43	0.41	0.40	0.39	0.38	0.37	0.36	0.35	0.35					
		10.0	15.0	0.58	0.55	0.52	0.50	0.48	0.45	0.43	0.42	0.41	0.40	0.40	0.39	0.38	0.37	0.37			
		15.0	20.0	0.62	0.59	0.56	0.53	0.50	0.48	0.46	0.45	0.44	0.43	0.42	0.41	0.40	0.40	0.39	0.39	0.38	
		20.0	25.0	0.68	0.61	0.58	0.56	0.53	0.50	0.48	0.47	0.46	0.45	0.44	0.43	0.42	0.41	0.40	0.40	0.39	
	12	1.0	2.5	0.37	0.35	0.34	0.32	0.30	0.29	0.27	0.26	0.26									19.3
		2.5	5.0	0.41	0.39	0.38	0.36	0.34	0.32	0.31	0.30	0.29	0.29	0.28							
		5.0	10.0	0.46	0.43	0.41	0.40	0.38	0.36	0.35	0.34	0.33	0.32	0.31	0.31	0.31					
		10.0	15.0	0.51	0.48	0.45	0.43	0.41	0.39	0.38	0.37	0.36	0.35	0.34	0.34	0.33	0.33	0.33	0.33		
		15.0	20.0	0.55	0.52	0.49	0.47	0.44	0.42	0.40	0.39	0.38	0.37	0.36	0.35	0.35	0.34	0.34	0.34	0.33	
		20.0	25.0	0.57	0.54	0.51	0.49	0.46	0.43	0.42	0.41	0.40	0.39	0.38	0.37	0.36	0.36	0.35	0.35	0.35	
	20	1.0	2.5	0.30	0.29	0.28	0.27	0.26	0.24	0.24	0.23	0.23	0.23								32.2
		2.5	5.0	0.34	0.33	0.30	0.29	0.28	0.27	0.26	0.25	0.24	0.24	0.24							
		5.0	10.0	0.40	0.37	0.35	0.33	0.32	0.30	0.29	0.29	0.28	0.27	0.26	0.26	0.25					
		10.0	15.0	0.44	0.42	0.39	0.38	0.35	0.33	0.32	0.30	0.30	0.29	0.29	0.28	0.28	0.27	0.27			
		15.0	20.0	0.46	0.44	0.42	0.40	0.37	0.35	0.34	0.33	0.32	0.31	0.30	0.30	0.29	0.29	0.29	0.28	0.28	
		20.0	25.0	0.49	0.47	0.45	0.41	0.40	0.37	0.35	0.36	0.33	0.33	0.32	0.31	0.30	0.30	0.29	0.29	0.29	
0.2	0	1.0	2.5	1.02	0.96	0.93	0.88	0.81	0.78	0.74	0.71	0.69									0
	Natural convection	2.5	5.0	1.17	1.10	1.05	1.01	0.95	0.91	0.87	0.85	0.81	0.78	0.76	0.75						Natural convection
		5.0	10.0	1.29	1.23	1.17	1.12	1.06	1.01	0.99	0.95	0.93	0.90	0.87	0.85	0.83					
		10.0	15.0	1.39	1.30	1.26	1.23	1.15	1.09	1.04	1.02	1.00	0.97	0.95	0.93	0.91	0.90	0.89			
		15.0	20.0	1.45	1.39	1.34	1.29	1.22	1.17	1.12	1.08	1.05	1.03	1.01	0.99	0.97	0.96	0.95	0.94	0.93	
		20.0	25.0	1.52	1.43	1.38	1.33	1.26	1.21	1.17	1.13	1.10	1.07	1.05	1.02	1.01	1.00	0.99	0.98	0.97	
	2	1.0	2.5	0.67	0.63	0.59	0.57	0.55	0.52	0.51	0.50	0.49									3.2
		2.5	5.0	0.75	0.71	0.67	0.64	0.61	0.58	0.56	0.54	0.52	0.51	0.50							
		5.0	10.0	0.84	0.80	0.76	0.73	0.69	0.65	0.62	0.60	0.58	0.57	0.56	0.55	0.54					
		10.0	15.0	0.92	0.87	0.82	0.79	0.76	0.71	0.68	0.66	0.64	0.62	0.61	0.60	0.59	0.58	0.57			
		15.0	20.0	0.98	0.93	0.88	0.84	0.79	0.76	0.73	0.70	0.68	0.66	0.65	0.64	0.63	0.62	0.61	0.60	0.59	
		20.0	25.0	1.03	0.96	0.92	0.89	0.84	0.80	0.77	0.74	0.77	0.69	0.68	0.67	0.66	0.65	0.64	0.63	0.62	
			Degree C, Operating Temperature (t_{m_1})	93	121	149	177	204	260	316	371	427	482	538	593	649	704	760	816	871	

Note: Above resistances are in English Units °F, ft², hr, Btu.
Multiply by 0.1761 to obtain Metric Units K, m², W = R_{ms}.

Table B-13 Hot Service—Approximate Air Film Surface Resistance—Values of $\frac{1}{f}$ or (R_s) — English Units

∈ Emittance	Wind Velocity MPH	Insulation Resistance Min.	Insulation Resistance Max.	Operating Temperature (t_1) Degree F 200	250	300	350	400	500	600	700	800	900	1000	1100	1200	1300	1400	1500	1600	Wind Velocity km/hr
	4	1.0	2.5	0.55	0.52	0.49	0.47	0.44	0.42	0.40	0.39	0.38									6.4
		2.5	5.0	0.62	0.48	0.55	0.53	0.50	0.47	0.45	0.44	0.43	0.42	0.41							
		5.0	10.0	0.69	0.65	0.62	0.59	0.57	0.54	0.51	0.49	0.48	0.47	0.46	0.45	0.44					
		10.0	15.0	0.76	0.73	0.69	0.65	0.61	0.59	0.57	0.55	0.53	0.51	0.50	0.49	0.48	0.47	0.46			
		15.0	20.0	0.80	0.76	0.72	0.69	0.65	0.62	0.60	0.58	0.56	0.55	0.54	0.53	0.52	0.51	0.50	0.49	0.48	
		20.0	25.0	0.85	0.80	0.75	0.72	0.68	0.64	0.62	0.60	0.59	0.58	0.56	0.55	0.54	0.53	0.52	0.51	0.50	
	8	1.0	2.5	0.44	0.41	0.39	0.38	0.35	0.33	0.32	0.31	0.30									12.9
		2.5	5.0	0.50	0.47	0.44	0.42	0.40	0.38	0.35	0.34	0.34	0.34	0.33							
		5.0	10.0	0.56	0.52	0.50	0.48	0.45	0.42	0.41	0.40	0.39	0.38	0.37	0.36	0.36					
		10.0	15.0	0.62	0.58	0.53	0.52	0.49	0.47	0.45	0.44	0.43	0.42	0.41	0.40	0.39	0.38	0.38			
		15.0	20.0	0.66	0.61	0.59	0.56	0.52	0.50	0.48	0.47	0.45	0.44	0.43	0.42	0.41	0.40	0.40	0.40	0.39	
		20.0	25.0	0.68	0.64	0.62	0.59	0.55	0.52	0.50	0.49	0.48	0.46	0.45	0.44	0.43	0.42	0.42	0.41	0.40	
	12	1.0	2.5	0.38	0.36	0.34	0.32	0.30	0.29	0.28	0.27	0.27									19.3
		2.5	5.0	0.42	0.40	0.38	0.37	0.35	0.33	0.31	0.30	0.30	0.29	0.29							
		5.0	10.0	0.48	0.46	0.43	0.41	0.39	0.37	0.36	0.35	0.34	0.33	0.32	0.32	0.31					
		10.0	15.0	0.53	0.50	0.47	0.45	0.43	0.41	0.39	0.38	0.37	0.36	0.35	0.35	0.34	0.34	0.33			
		15.0	20.0	0.57	0.53	0.50	0.48	0.45	0.43	0.42	0.40	0.39	0.38	0.37	0.37	0.36	0.36	0.35	0.35	0.34	
		20.0	25.0	0.60	0.56	0.53	0.51	0.48	0.45	0.43	0.42	0.41	0.40	0.39	0.38	0.37	0.37	0.36	0.36	0.36	
	20	1.0	2.5	0.31	0.29	0.28	0.27	0.26	0.25	0.25	0.24	0.24									32.2
		2.5	5.0	0.35	0.34	0.32	0.31	0.29	0.28	0.27	0.26	0.25	0.25	0.24							
		5.0	10.0	0.41	0.38	0.36	0.34	0.33	0.31	0.30	0.29	0.28	0.28	0.27	0.27	0.26					
		10.0	15.0	0.46	0.43	0.41	0.39	0.37	0.34	0.32	0.31	0.30	0.30	0.29	0.29	0.28	0.28	0.27			
		15.0	20.0	0.48	0.46	0.44	0.42	0.39	0.36	0.35	0.34	0.33	0.32	0.31	0.31	0.30	0.30	0.29	0.29	0.28	
		20.0	25.0	0.50	0.48	0.46	0.44	0.41	0.38	0.36	0.35	0.34	0.33	0.33	0.32	0.32	0.31	0.30	0.30	0.29	
0.1	0 Natural convection	1.0	2.5	1.15	1.07	1.02	0.98	0.91	0.87	0.83	0.80	0.78									0 Natural convection
		2.5	5.0	1.30	1.21	1.17	1.12	1.05	1.00	0.95	0.92	0.89	0.87	0.85							
		5.0	10.0	1.47	1.36	1.30	1.24	1.17	1.12	1.07	1.04	1.01	0.98	0.96	0.94	0.92					
		10.0	15.0	1.57	1.49	1.43	1.35	1.27	1.21	1.17	1.12	1.10	1.07	1.05	1.03	1.01	0.99	0.97			
		15.0	20.0	1.65	1.57	1.51	1.46	1.35	1.27	1.22	1.19	1.15	1.13	1.10	1.08	1.06	1.05	1.04	1.02	1.01	
		20.0	25.0	1.71	1.62	1.57	1.51	1.43	1.35	1.29	1.25	1.21	1.19	1.16	1.14	1.12	1.10	1.08	1.07	1.06	
	2	1.0	2.5	0.71	0.66	0.63	0.61	0.57	0.54	0.52	0.51	0.50									3.2
		2.5	5.0	0.81	0.75	0.71	0.69	0.64	0.60	0.58	0.56	0.54	0.53	0.52							
		5.0	10.0	0.90	0.84	0.80	0.77	0.73	0.69	0.66	0.64	0.62	0.60	0.59	0.58	0.57					
		10.0	15.0	0.99	0.92	0.88	0.84	0.80	0.76	0.73	0.71	0.68	0.66	0.64	0.63	0.62	0.61	0.60			
		15.0	20.0	1.05	0.99	0.94	0.89	0.84	0.80	0.77	0.74	0.72	0.70	0.68	0.67	0.66	0.65	0.64	0.63	0.62	
		20.0	25.0	1.11	1.03	0.98	0.93	0.88	0.84	0.81	0.79	0.76	0.74	0.72	0.71	0.70	0.69	0.68	0.67	0.66	
	4	1.0	2.5	0.58	0.54	0.51	0.49	0.46	0.43	0.42	0.41	0.40									6.4
		2.5	5.0	0.65	0.61	0.58	0.55	0.52	0.50	0.48	0.46	0.45	0.44	0.43							
		5.0	10.0	0.74	0.69	0.65	0.62	0.59	0.56	0.54	0.52	0.51	0.49	0.48	0.47	0.46					
		10.0	15.0	0.80	0.75	0.71	0.67	0.64	0.61	0.59	0.57	0.55	0.54	0.53	0.52	0.51	0.50	0.49			
		15.0	20.0	0.85	0.80	0.76	0.73	0.69	0.66	0.63	0.61	0.59	0.58	0.56	0.55	0.54	0.53	0.52	0.51	0.50	
		20.0	25.0	0.90	0.84	0.80	0.76	0.72	0.68	0.65	0.63	0.61	0.60	0.59	0.58	0.57	0.56	0.55	0.54	0.53	
	8	1.0	2.5	0.45	0.42	0.40	0.39	0.36	0.34	0.33	0.32	0.31									12.9
		2.5	5.0	0.52	0.49	0.46	0.44	0.41	0.39	0.37	0.36	0.35	0.33	0.34							
		5.0	10.0	0.58	0.54	0.52	0.50	0.47	0.44	0.42	0.41	0.40	0.39	0.38	0.38	0.37					
		10.0	15.0	0.65	0.61	0.57	0.54	0.51	0.49	0.47	0.45	0.44	0.43	0.42	0.41	0.40	0.39	0.39			
		15.0	20.0	0.69	0.64	0.61	0.58	0.54	0.51	0.49	0.48	0.47	0.46	0.45	0.44	0.43	0.42	0.41	0.41	0.40	
		20.0	25.0	0.72	0.67	0.64	0.61	0.57	0.54	0.52	0.51	0.50	0.49	0.48	0.46	0.45	0.44	0.43	0.42	0.41	
	12	1.0	2.5	0.39	0.37	0.35	0.33	0.31	0.29	0.28	0.28	0.27									19.3
		2.5	5.0	0.44	0.42	0.40	0.38	0.36	0.34	0.32	0.31	0.30	0.30	0.29							
		5.0	10.0	0.50	0.48	0.45	0.43	0.40	0.38	0.37	0.36	0.35	0.34	0.33	0.33	0.32					
		10.0	15.0	0.55	0.52	0.49	0.47	0.45	0.43	0.41	0.39	0.38	0.37	0.36	0.36	0.35	0.34	0.34			
		15.0	20.0	0.60	0.56	0.53	0.51	0.48	0.45	0.43	0.42	0.41	0.40	0.39	0.38	0.37	0.36	0.36	0.35	0.35	
		20.0	25.0	0.63	0.58	0.56	0.53	0.50	0.48	0.46	0.44	0.42	0.41	0.40	0.39	0.38	0.38	0.37	0.37	0.36	
	20	1.0	2.5	0.32	0.30	0.29	0.27	0.26	0.25	0.24	0.24	0.24									32.2
		2.5	5.0	0.37	0.35	0.32	0.31	0.29	0.28	0.27	0.26	0.25	0.25	0.25							
		5.0	10.0	0.42	0.40	0.38	0.36	0.33	0.31	0.30	0.29	0.29	0.28	0.27	0.27	0.27					
		10.0	15.0	0.47	0.45	0.42	0.40	0.38	0.36	0.34	0.32	0.31	0.30	0.30	0.29	0.29	0.29	0.28			
		15.0	20.0	0.50	0.47	0.45	0.43	0.39	0.37	0.36	0.35	0.34	0.33	0.32	0.31	0.30	0.30	0.30	0.29	0.29	
		20.0	25.0	0.52	0.50	0.48	0.45	0.43	0.40	0.38	0.36	0.35	0.34	0.33	0.32	0.32	0.32	0.31	0.31	0.30	
				93	121	149	177	204	260	316	371	427	482	538	593	649	704	760	816	871	
				Degree C Operating Temperature (t_{m_1})																	

Note: Above resistances are in English Units °F, ft^2, hr, Btu.
Multiply by 0.1761 to obtain Metric Units K, m^2, W = R_{ms}.

Table B-13 Hot Service—Approximate Air Film Surface Resistance—Values of $\frac{1}{f}$ or (R_s) – English Units

∈ Emittance	Wind Velocity MPH	Insulation Resistance Min.	Max.	Operating Temperature (t_1) Degree F 200	250	300	350	400	500	600	700	800	900	1000	1100	1200	1300	1400	1500	1600	Wind Velocity km/hr
0.05	0	1.0	2.5	1.24	1.16	1.10	1.06	0.98	0.94	0.89	0.85	0.82									0
	Natural convection	2.5	5.0	1.37	1.31	1.22	1.18	1.10	1.05	1.00	0.96	0.94	0.92	0.91							Natural convection
		5.0	10.0	1.55	1.43	1.36	1.32	1.22	1.17	1.12	1.08	1.05	1.03	1.00	0.98	0.96					
		10.0	15.0	1.68	1.58	1.50	1.43	1.35	1.28	1.22	1.19	1.15	1.11	1.08	1.06	1.04	1.03	1.02			
		15.0	20.0	1.74	1.66	1.60	1.54	1.42	1.36	1.31	1.26	1.22	1.19	1.16	1.14	1.12	1.10	1.08	1.06	1.04	
		20.0	25.0	1.83	1.72	1.66	1.60	2.49	1.42	1.36	1.33	1.29	1.24	1.21	1.19	1.16	1.14	1.12	1.11	1.10	
	2	1.0	2.5	0.74	0.70	0.66	0.63	0.58	0.55	0.54	0.53	0.52									3.2
		2.5	5.0	0.84	0.78	0.74	0.71	0.65	0.62	0.60	0.58	0.56	0.55	0.54							
		5.0	10.0	0.94	0.88	0.83	0.80	0.75	0.71	0.68	0.65	0.63	0.62	0.61	0.60	0.59					
		10.0	15.0	1.02	0.96	0.91	0.88	0.82	0.77	0.74	0.72	0.70	0.68	0.66	0.65	0.64	0.63	0.62			
		15.0	20.0	1.09	1.01	0.97	0.94	0.88	0.83	0.79	0.77	0.75	0.73	0.71	0.70	0.69	0.68	0.67	0.66	0.65	
		20.0	25.0	1.15	1.07	1.01	0.98	0.92	0.88	0.84	0.81	0.78	0.76	0.74	0.72	0.71	0.70	0.69	0.68	0.67	
	4	1.0	2.5	0.60	0.55	0.52	0.50	0.46	0.44	0.42	0.41	0.41									6.4
		2.5	5.0	0.67	0.63	0.60	0.57	0.53	0.50	0.48	0.47	0.46	0.45	0.44							
		5.0	10.0	0.76	0.71	0.67	0.64	0.61	0.57	0.55	0.53	0.51	0.50	0.49	0.48	0.47					
		10.0	15.0	0.82	0.78	0.74	0.71	0.66	0.63	0.61	0.59	0.57	0.55	0.54	0.53	0.52	0.51	0.50			
		15.0	20.0	0.87	0.82	0.79	0.76	0.71	0.67	0.65	0.62	0.61	0.60	0.59	0.57	0.56	0.55	0.54	0.53	0.52	
		20.0	25.0	0.93	0.86	0.82	0.78	0.75	0.71	0.67	0.65	0.63	0.61	0.60	0.59	0.58	0.57	0.56	0.55	0.54	
0.05	8	1.0	2.5	0.46	0.43	0.41	0.39	0.37	0.35	0.34	0.33	0.32									12.9
		2.5	5.0	0.53	0.50	0.47	0.44	0.42	0.40	0.38	0.37	0.36	0.36	0.35							
		5.0	10.0	0.59	0.56	0.53	0.51	0.47	0.45	0.43	0.41	0.40	0.39	0.38	0.38	0.38					
		10.0	15.0	0.66	0.61	0.58	0.56	0.52	0.50	0.48	0.46	0.44	0.45	0.42	0.42	0.41	0.41	0.40			
		15.0	20.0	0.70	0.65	0.62	0.59	0.56	0.53	0.51	0.49	0.48	0.47	0.46	0.45	0.44	0.43	0.42	0.42	0.41	
		20.0	25.0	0.74	0.69	0.65	0.63	0.58	0.55	0.53	0.51	0.50	0.49	0.48	0.47	0.46	0.45	0.44	0.43	0.42	
	12	1.0	2.5	0.40	0.37	0.36	0.34	0.32	0.30	0.29	0.29	0.28									19.3
		2.5	5.0	0.45	0.42	0.40	0.39	0.37	0.35	0.33	0.32	0.31	0.31	0.30							
		5.0	10.0	0.51	0.47	0.44	0.42	0.41	0.39	0.37	0.36	0.35	0.34	0.33	0.33	0.33					
		10.0	15.0	0.56	0.53	0.49	0.46	0.44	0.42	0.41	0.40	0.39	0.38	0.37	0.36	0.36	0.35	0.35			
		15.0	20.0	0.61	0.56	0.53	0.50	0.46	0.44	0.43	0.42	0.41	0.30	0.39	0.38	0.38	0.27	0.27	0.36	0.36	
		20.0	25.0	0.64	0.60	0.56	0.53	0.49	0.46	0.44	0.43	0.42	0.42	0.41	0.40	0.39	0.39	0.38	0.38	0.37	
	20	1.0	2.5	0.33	0.31	0.29	0.28	0.26	0.25	0.24	0.24	0.24									32.2
		2.5	5.0	0.37	0.35	0.33	0.32	0.30	0.28	0.27	0.26	0.25	0.25	0.25							
		5.0	10.0	0.43	0.41	0.39	0.37	0.34	0.32	0.31	0.30	0.29	0.28	0.28	0.28	0.27					
		10.0	15.0	0.48	0.45	0.42	0.41	0.38	0.36	0.34	0.33	0.32	0.31	0.31	0.30	0.29	0.29	0.29			
		15.0	20.0	0.51	0.48	0.46	0.44	0.40	0.38	0.36	0.35	0.34	0.33	0.32	0.32	0.31	0.31	0.30	0.30	0.29	
		20.0	25.0	0.53	0.51	0.49	0.47	0.44	0.41	0.39	0.37	0.36	0.35	0.34	0.33	0.32	0.32	0.31	0.31	0.31	
			Degree C, Operating Temperature (t_{m_1})	93	121	149	177	204	260	316	371	427	482	538	593	649	704	760	816	871	

Note: Above resistances are in English Units °F, ft², hr, Btu.
Multiply by 0.1761 to obtain Metric Units K, m², W = R_{ms}.

Table B-14 Low temperature service – Approximate air film surface resistance values of $\frac{1}{f}$ or (R_s) – English Units
Heat transfer from air to lower temperature surface (cold service)

Air Temp. °F	Temp. Diff. Between Air & Surface °F	∈ = 1.0										∈ = 0.8										Temp. Diff. Between Air & Surface °C	Air Temp. °C
		Air Movement – MPH										Air Movement – MPH											
		0	1	2	3	4	5	10	15	20	25	0	1	2	3	4	5	10	15	20	25		
0	1	1.01	0.89	0.81	0.76	0.71	0.78	0.57	0.50	0.46	0.42	1.19	1.01	0.91	0.84	0.79	0.74	0.61	0.53	0.49	0.45	0.6	−17.8
	2	0.98	0.83	0.75	0.70	0.65	0.61	0.51	0.44	0.40	0.37	1.13	0.94	0.83	0.77	0.71	0.67	0.54	0.47	0.43	0.39	1.1	
	3	0.95	0.80	0.71	0.66	0.61	0.58	0.47	0.40	0.37	0.34	1.08	0.89	0.79	0.72	0.67	0.63	0.51	0.44	0.39	0.36	1.6	
	4	0.92	0.77	0.68	0.63	0.59	0.55	0.45	0.39	0.35	0.33	1.05	0.87	0.75	0.69	0.64	0.60	0.48	0.41	0.37	0.34	2.2	
	5	0.91	0.75	0.67	0.61	0.57	0.54	0.43	0.38	0.34	0.31	1.03	0.84	0.73	0.67	0.62	0.58	0.45	0.40	0.36	0.33	2.8	
	6	0.89	0.73	0.65	0.60	0.55	0.52	0.42	0.37	0.33	0.30	1.00	0.82	0.71	0.65	0.60	0.56	0.44	0.38	0.34	0.31	3.3	
	7	0.88	0.72	0.64	0.58	0.54	0.51	0.41	0.35	0.32	0.29	0.99	0.80	0.70	0.63	0.58	0.54	0.43	0.37	0.33	0.30	3.8	
	8	0.87	0.71	0.63	0.57	0.53	0.50	0.40	0.34	0.31	0.28	0.98	0.78	0.69	0.62	0.57	0.53	0.42	0.36	0.32	0.29	4.4	
	9	0.86	0.70	0.62	0.57	0.52	0.49	0.39	0.34	0.30	0.28	0.97	0.77	0.67	0.61	0.56	0.52	0.41	0.35	0.32	0.29	5.0	
	10	0.85	0.69	0.61	0.56	0.51	0.48	0.38	0.33	0.29	0.27	0.95	0.76	0.66	0.60	0.55	0.51	0.40	0.35	0.31	0.28	5.6	
10	1	0.96	0.86	0.78	0.73	0.68	0.65	0.55	0.49	0.45	0.41	1.13	0.97	0.88	0.81	0.76	0.72	0.60	0.52	0.48	0.44	0.6	−12.2
	2	0.94	0.80	0.73	0.67	0.63	0.60	0.50	0.44	0.40	0.37	1.09	0.91	0.81	0.75	0.70	0.65	0.53	0.47	0.42	0.39	1.1	
	3	0.91	0.77	0.69	0.64	0.60	0.56	0.46	0.41	0.37	0.34	1.04	0.86	0.77	0.70	0.65	0.61	0.50	0.43	0.39	0.36	1.6	
	4	0.89	0.75	0.67	0.62	0.57	0.54	0.44	0.39	0.35	0.32	1.01	0.83	0.74	0.67	0.62	0.59	0.47	0.41	0.37	0.34	2.2	
	5	0.87	0.73	0.65	0.60	0.56	0.52	0.43	0.37	0.34	0.31	1.00	0.81	0.72	0.65	0.61	0.57	0.45	0.40	0.35	0.32	2.8	
	6	0.86	0.71	0.63	0.58	0.54	0.51	0.41	0.36	0.32	0.30	0.97	0.79	0.69	0.63	0.59	0.55	0.44	0.38	0.34	0.31	3.3	
	7	0.85	0.70	0.62	0.57	0.53	0.50	0.40	0.35	0.32	0.29	0.95	0.78	0.68	0.62	0.57	0.54	0.43	0.37	0.33	0.30	3.8	
	8	0.84	0.69	0.62	0.56	0.52	0.49	0.39	0.34	0.31	0.28	0.95	0.76	0.68	0.61	0.56	0.52	0.42	0.36	0.32	0.29	4.4	
	9	0.83	0.68	0.60	0.55	0.51	0.48	0.39	0.33	0.30	0.28	0.93	0.75	0.66	0.60	0.55	0.51	0.41	0.35	0.31	0.29	5.0	
	10	0.82	0.67	0.59	0.54	0.50	0.47	0.38	0.33	0.29	0.27	0.91	0.74	0.65	0.59	0.54	0.50	0.40	0.34	0.30	0.28	5.6	
20	1	0.95	0.83	0.76	0.71	0.67	0.64	0.54	0.48	0.43	0.41	1.10	0.94	0.85	0.79	0.75	0.71	0.59	0.52	0.47	0.43	0.6	− 6.7
	2	0.90	0.78	0.70	0.65	0.61	0.58	0.48	0.43	0.39	0.36	1.04	0.88	0.78	0.73	0.68	0.64	0.52	0.46	0.41	0.38	1.1	
	3	0.87	0.74	0.67	0.62	0.58	0.55	0.45	0.40	0.36	0.33	1.00	0.84	0.74	0.68	0.64	0.60	0.49	0.42	0.38	0.35	1.6	
	4	0.85	0.72	0.65	0.60	0.56	0.53	0.43	0.38	0.34	0.32	0.98	0.81	0.72	0.66	0.61	0.57	0.47	0.40	0.36	0.34	2.2	
	5	0.83	0.70	0.63	0.58	0.54	0.51	0.42	0.37	0.33	0.30	0.96	0.79	0.70	0.64	0.59	0.55	0.45	0.39	0.35	0.32	2.8	
	6	0.82	0.69	0.61	0.57	0.53	0.50	0.40	0.35	0.32	0.29	0.94	0.77	0.68	0.62	0.57	0.54	0.43	0.37	0.34	0.31	3.3	
	7	0.81	0.67	0.60	0.55	0.52	0.49	0.39	0.34	0.31	0.28	0.93	0.66	0.66	0.61	0.56	0.52	0.42	0.36	0.33	0.30	3.8	
	8	0.81	0.67	0.59	0.55	0.51	0.48	0.38	0.33	0.30	0.28	0.91	0.74	0.65	0.59	0.55	0.51	0.41	0.35	0.32	0.29	4.4	
	9	0.80	0.66	0.58	0.54	0.50	0.47	0.38	0.33	0.30	0.27	0.90	0.73	0.64	0.58	0.54	0.50	0.40	0.34	0.31	0.28	5.0	
	10	0.79	0.65	0.58	0.53	0.49	0.46	0.37	0.32	0.29	0.27	0.89	0.72	0.63	0.57	0.53	0.49	0.39	0.34	0.30	0.28	5.6	
30	1	0.90	0.78	0.73	0.69	0.65	0.62	0.52	0.47	0.43	0.40	1.05	0.91	0.83	0.77	0.73	0.69	0.57	0.51	0.46	0.43	0.6	− 1.1
	2	0.86	0.75	0.68	0.63	0.60	0.57	0.47	0.42	0.38	0.35	1.00	0.85	0.76	0.71	0.66	0.62	0.51	0.45	0.41	0.38	1.1	
	3	0.84	0.72	0.65	0.60	0.56	0.54	0.44	0.39	0.36	0.33	0.96	0.81	0.72	0.67	0.62	0.59	0.48	0.42	0.38	0.35	1.6	
	4	0.82	0.70	0.63	0.58	0.54	0.52	0.42	0.37	0.34	0.31	0.94	0.78	0.70	0.64	0.60	0.56	0.46	0.40	0.36	0.33	2.2	
	5	0.80	0.68	0.61	0.56	0.53	0.50	0.41	0.36	0.32	0.30	0.92	0.76	0.68	0.62	0.58	0.54	0.44	0.38	0.34	0.31	2.8	
	6	0.79	0.67	0.60	0.55	0.52	0.49	0.40	0.35	0.31	0.29	0.90	0.75	0.66	0.61	0.56	0.53	0.43	0.37	0.33	0.30	3.3	
	7	0.78	0.66	0.59	0.54	0.50	0.48	0.39	0.34	0.31	0.28	0.89	0.73	0.66	0.59	0.55	0.51	0.41	0.36	0.32	0.29	3.8	
	8	0.78	0.65	0.59	0.53	0.50	0.48	0.38	0.33	0.30	0.27	0.88	0.72	0.64	0.58	0.54	0.50	0.40	0.35	0.31	0.29	4.4	
	9	0.77	0.64	0.56	0.53	0.49	0.46	0.37	0.32	0.29	0.27	0.87	0.71	0.62	0.57	0.53	0.49	0.39	0.34	0.31	0.28	5.0	
	10	0.76	0.63	0.56	0.52	0.48	0.45	0.37	0.32	0.29	0.26	0.87	0.70	0.62	0.56	0.52	0.49	0.39	0.33	0.30	0.27	5.6	
40	1	0.86	0.76	0.70	0.66	0.63	0.60	0.51	0.46	0.42	0.39	1.01	0.88	0.80	0.75	0.70	0.67	0.56	0.50	0.45	0.42	0.6	4.4
	2	0.83	0.72	0.66	0.61	0.58	0.55	0.46	0.41	0.37	0.35	0.97	0.82	0.74	0.69	0.64	0.61	0.50	0.44	0.40	0.37	1.1	
	3	0.80	0.69	0.63	0.58	0.55	0.52	0.43	0.37	0.35	0.32	0.93	0.79	0.70	0.65	0.61	0.57	0.47	0.41	0.37	0.34	1.6	
	4	0.78	0.67	0.61	0.56	0.53	0.50	0.41	0.36	0.33	0.31	0.91	0.76	0.68	0.62	0.58	0.55	0.45	0.39	0.35	0.32	2.2	
	5	0.77	0.66	0.59	0.55	0.51	0.49	0.40	0.35	0.32	0.30	0.89	0.74	0.66	0.61	0.56	0.53	0.43	0.37	0.34	0.31	2.8	
	6	0.76	0.65	0.58	0.54	0.50	0.47	0.39	0.34	0.31	0.29	0.88	0.73	0.64	0.59	0.55	0.51	0.42	0.36	0.33	0.30	3.3	
	7	0.75	0.64	0.57	0.53	0.49	0.46	0.38	0.33	0.30	0.28	0.86	0.71	0.63	0.58	0.54	0.50	0.41	0.35	0.32	0.29	3.8	
	8	0.74	0.63	0.57	0.52	0.48	0.46	0.37	0.32	0.29	0.27	0.85	0.70	0.63	0.57	0.52	0.49	0.40	0.34	0.31	0.28	4.4	
	9	0.74	0.62	0.55	0.51	0.47	0.45	0.36	0.32	0.29	0.26	0.84	0.69	0.61	0.56	0.52	0.48	0.39	0.34	0.30	0.28	5.0	
	10	0.73	0.61	0.55	0.50	0.47	0.44	0.36	0.31	0.28	0.26	0.84	0.68	0.60	0.55	0.51	0.48	0.38	0.33	0.29	0.27	5.6	
50	1	0.82	0.73	0.68	0.64	0.61	0.58	0.49	0.44	0.41	0.38	0.96	0.84	0.77	0.72	0.68	0.65	0.54	0.48	0.44	0.41	0.6	10.0
	2	0.79	0.69	0.64	0.60	0.56	0.54	0.45	0.40	0.37	0.34	0.93	0.80	0.72	0.67	0.64	0.60	0.49	0.44	0.39	0.36	1.1	
	3	0.77	0.67	0.61	0.57	0.53	0.51	0.43	0.38	0.34	0.32	0.91	0.76	0.70	0.63	0.59	0.56	0.46	0.40	0.37	0.34	1.6	
	4	0.76	0.65	0.59	0.55	0.52	0.49	0.41	0.36	0.33	0.30	0.88	0.74	0.66	0.61	0.57	0.54	0.44	0.38	0.35	0.32	2.2	
	5	0.75	0.64	0.58	0.53	0.50	0.48	0.39	0.35	0.31	0.29	0.86	0.72	0.65	0.59	0.55	0.52	0.42	0.37	0.33	0.31	2.8	
	6	0.74	0.63	0.56	0.52	0.49	0.46	0.38	0.34	0.30	0.28	0.85	0.71	0.63	0.58	0.54	0.50	0.41	0.36	0.32	0.30	3.3	
	7	0.73	0.62	0.55	0.51	0.48	0.45	0.37	0.33	0.30	0.27	0.84	0.70	0.62	0.57	0.53	0.49	0.40	0.35	0.31	0.29	3.8	
	8	0.72	0.61	0.54	0.50	0.47	0.45	0.36	0.32	0.29	0.27	0.83	0.69	0.61	0.56	0.51	0.48	0.39	0.34	0.30	0.28	4.4	
	9	0.71	0.60	0.54	0.50	0.46	0.44	0.36	0.31	0.28	0.26	0.82	0.68	0.60	0.55	0.50	0.47	0.38	0.33	0.30	0.27	5.0	
	10	0.71	0.60	0.53	0.49	0.46	0.43	0.35	0.31	0.27	0.25	0.81	0.67	0.59	0.54	0.50	0.47	0.38	0.32	0.29	0.27	5.6	

Note: Above resistances are in English Units °F, ft^2, hr, Btu
Multiply by 0.1761 to obtain Metric Units K, m^2, W = R_{ms}

Air movement in miles per hour
Multiply by 1.609 to obtain kilometers per hour

Table B-14 Low temperature service — Approximate air film surface resistance values of $\frac{1}{f}$ or (R_s) — English Units

Heat transfer from air to lower temperature surface (cold service)

Air Temp. °F	Temp. Diff. Between Air & Surface °F	∈ = 0.6										∈ = 0.4										∈ = 0.2										Temp. Diff. Between Air & Surface °C	Air Temp. °C
		Air Movement — MPH										Air Movement — MPH										Air Movement — MPH											
		0	1	2	3	4	5	10	15	20	25	0	1	2	3	4	5	10	15	20	25	0	1	2	3	4	5	10	15	20	25		
0	1	1.43	1.17	1.04	0.95	0.89	0.83	0.67	0.58	0.52	0.48	1.75	1.39	1.20	1.08	1.00	0.92	0.73	0.62	0.56	0.51	2.32	1.72	1.45	1.23	1.15	1.06	0.81	0.68	0.61	0.55	0.6	17.8
	2	1.33	1.07	0.94	0.85	0.79	0.73	0.59	0.50	0.45	0.41	1.61	1.25	1.07	1.04	0.88	0.81	0.64	0.54	0.48	0.44	2.06	1.50	1.25	1.10	1.00	0.92	0.70	0.58	0.51	0.46	1.2	
	3	1.27	1.01	0.88	0.80	0.73	0.69	0.54	0.46	0.42	0.39	1.52	1.17	1.00	0.89	0.81	0.75	0.56	0.50	0.44	0.40	1.91	1.39	1.15	1.01	0.91	0.84	0.62	0.53	0.47	0.42	1.6	
	4	1.22	0.97	0.84	0.76	0.70	0.65	0.51	0.44	0.39	0.36	0.45	.111	0.94	0.84	0.77	0.71	0.55	0.47	0.41	0.37	1.80	1.30	1.08	0.95	0.86	0.79	0.59	0.50	0.44	0.39	2.2	
	5	1.19	0.94	0.81	0.73	0.67	0.62	0.49	0.42	0.37	0.34	1.41	1.07	0.91	0.81	0.74	0.68	0.53	0.44	0.39	0.36	1.73	1.24	1.03	0.91	0.81	0.75	0.56	0.47	0.42	0.37	2.8	
	6	1.16	0.92	0.79	0.71	0.65	0.61	0.47	0.40	0.36	0.33	1.38	1.04	0.88	0.78	0.71	0.66	0.51	0.43	0.38	0.34	1.69	1.20	0.99	0.87	0.78	0.72	0.54	0.45	0.40	0.36	3.3	
	7	1.14	0.89	0.77	0.69	0.63	0.59	0.46	0.39	0.35	0.32	1.34	1.01	0.85	0.76	0.69	0.64	0.49	0.41	0.36	0.33	1.64	1.16	0.96	0.84	0.76	0.70	0.52	0.44	0.38	0.34	3.8	
	8	1.12	0.87	0.76	0.67	0.62	0.57	0.45	0.38	0.34	0.31	1.31	0.99	0.85	0.74	0.67	0.62	0.47	0.40	0.35	0.32	1.59	1.13	0.95	0.82	0.73	0.67	0.51	0.42	0.37	0.33	4.4	
	9	1.11	0.86	0.73	0.66	0.60	0.56	0.44	0.37	0.33	0.30	1.29	0.97	0.81	0.73	0.66	0.61	0.46	0.39	0.34	0.31	1.55	1.11	0.91	0.80	0.72	0.66	0.49	0.41	0.36	0.33	5.0	
	10	1.10	0.85	0.72	0.65	0.59	0.55	0.43	0.36	0.32	0.29	1.28	0.95	0.79	0.71	0.64	0.59	0.45	0.38	0.33	0.31	1.49	1.10	0.85	0.79	0.70	0.65	0.48	0.40	0.35	0.32	5.6	
10	1	1.37	1.14	1.10	0.93	0.86	0.81	0.65	0.57	0.51	0.47	1.73	1.37	1.19	1.07	0.99	0.92	0.73	0.62	0.56	0.51	2.23	1.69	1.43	1.27	1.13	1.05	0.81	0.68	0.60	0.54	0.6	−12.2
	2	1.28	1.04	0.91	0.84	0.77	0.72	0.58	0.50	0.45	0.41	1.57	1.22	1.05	0.95	0.87	0.80	0.63	0.54	0.48	0.43	2.02	1.48	1.23	1.09	0.99	0.91	0.69	0.58	0.51	0.46	1.2	
	3	1.22	0.99	0.86	0.78	0.72	0.67	0.53	0.46	0.41	0.38	1.47	1.14	0.98	0.88	0.80	0.74	0.58	0.49	0.44	0.40	1.87	1.37	1.13	1.00	0.91	0.83	0.62	0.53	0.47	0.42	1.6	
	4	1.18	0.95	0.82	0.75	0.69	0.64	0.51	0.53	0.39	0.35	1.41	1.09	0.93	0.83	0.76	0.71	0.54	0.46	0.41	0.37	1.78	1.29	1.07	0.94	0.85	0.78	0.59	0.50	0.43	0.39	2.2	
	5	1.15	0.92	0.80	0.72	0.66	0.62	0.48	0.42	0.37	0.34	1.38	1.05	0.90	0.80	0.73	0.67	0.52	0.44	0.39	0.35	1.70	1.23	1.02	0.90	0.81	0.74	0.56	0.47	0.41	0.37	2.8	
	6	1.13	0.89	0.77	0.70	0.64	0.61	0.47	0.40	0.36	0.33	1.34	1.02	0.87	0.77	0.70	0.65	0.50	0.42	0.38	0.34	1.65	1.19	0.98	0.87	0.78	0.71	0.54	0.45	0.40	0.36	3.3	
	7	1.11	0.87	0.75	0.68	0.62	0.58	0.45	0.39	0.35	0.31	1.31	1.00	0.84	0.75	0.68	0.63	0.48	0.41	0.36	0.33	1.61	1.15	0.95	0.84	0.75	0.69	0.52	0.44	0.38	0.35	3.8	
	8	1.09	0.86	0.75	0.66	0.61	0.57	0.44	0.38	0.34	0.31	1.28	0.97	0.84	0.73	0.66	0.61	0.47	0.40	0.35	0.32	1.57	1.12	0.94	0.81	0.73	0.67	0.50	0.42	0.37	0.33	4.4	
	9	1.07	0.85	0.73	0.65	0.59	0.55	0.43	0.37	0.33	0.30	1.26	0.94	0.80	0.72	0.65	0.60	0.46	0.39	0.34	0.31	1.53	1.10	0.90	0.80	0.71	0.65	0.49	0.41	0.36	0.32	5.0	
	10	1.06	0.83	0.71	0.64	0.58	0.54	0.42	0.36	0.32	0.29	1.24	0.93	0.78	0.70	0.63	0.59	0.45	0.38	0.33	0.30	1.47	1.08	0.89	0.78	0.70	0.64	0.48	0.40	0.35	0.31	5.6	
20	1	1.31	1.10	0.98	0.90	0.84	0.79	0.64	0.56	0.50	0.47	1.66	1.33	1.16	1.04	0.97	0.90	0.71	0.61	0.55	0.50	2.22	1.66	1.41	1.25	1.24	1.04	0.80	0.68	0.60	0.54	0.6	− 6.7
	2	1.24	0.99	0.89	0.82	0.75	0.71	0.57	0.49	0.44	0.40	1.53	1.19	1.03	0.93	0.85	0.79	0.62	0.53	0.47	0.43	2.00	1.49	1.23	1.09	0.98	0.90	0.69	0.58	0.51	0.46	1.2	
	3	1.18	0.96	0.84	0.77	0.71	0.66	0.53	0.45	0.41	0.37	1.44	1.12	0.96	0.86	0.79	0.73	0.57	0.49	0.43	0.39	1.85	1.36	1.13	1.00	0.91	0.83	0.61	0.53	0.46	0.42	1.6	
	4	1.14	0.92	0.80	0.73	0.67	0.63	0.50	0.43	0.38	0.35	1.38	1.07	0.91	0.82	0.75	0.69	0.54	0.46	0.41	0.37	1.76	1.28	1.06	0.94	0.84	0.78	0.59	0.49	0.43	0.39	2.2	
	5	1.12	0.89	0.78	0.70	0.65	0.61	0.48	0.41	0.37	0.33	1.34	1.03	0.88	0.79	0.72	0.67	0.52	0.44	0.39	0.35	1.68	1.22	1.01	0.90	0.80	0.74	0.56	0.47	0.41	0.37	2.8	
	6	1.09	0.87	0.76	0.68	0.63	0.59	0.46	0.39	0.35	0.32	1.31	1.00	0.85	0.76	0.69	0.64	0.50	0.42	0.37	0.34	1.62	1.18	0.98	0.86	0.77	0.71	0.53	0.45	0.39	0.36	3.3	
	7	1.07	0.85	0.74	0.66	0.61	0.57	0.45	0.38	0.34	0.31	1.28	0.98	0.83	0.74	0.67	0.62	0.48	0.41	0.36	0.33	1.58	1.14	0.95	0.83	0.75	0.69	0.52	0.43	0.38	0.34	3.8	
	8	1.06	0.84	0.73	0.65	0.60	0.56	0.44	0.37	0.33	0.30	1.26	0.95	0.82	0.72	0.65	0.61	0.47	0.39	0.35	0.32	1.55	1.11	0.93	0.81	0.73	0.67	0.50	0.42	0.37	0.33	4.4	
	9	1.04	0.82	0.71	0.64	0.58	0.54	0.43	0.36	0.32	0.30	1.24	0.94	0.79	0.71	0.64	0.59	0.45	0.38	0.34	0.31	1.52	1.09	0.89	0.79	0.71	0.65	0.49	0.41	0.36	0.32	5.0	
	10	1.03	0.81	0.69	0.63	0.57	0.53	0.42	0.35	0.31	0.29	1.22	0.92	0.77	0.68	0.63	0.58	0.45	0.38	0.33	0.30	1.50	1.07	0.88	0.77	0.69	0.64	0.48	0.40	0.35	0.32	5.6	
30	1	1.27	1.07	0.95	0.88	0.82	0.77	0.63	0.55	0.50	0.46	1.62	1.30	1.14	1.00	0.95	0.89	0.70	0.61	0.54	0.50	2.17	1.64	1.39	1.24	1.13	1.03	0.79	0.67	0.60	0.54	0.6	− 1.1
	2	1.20	0.99	0.87	0.80	0.74	0.68	0.56	0.48	0.44	0.40	1.48	1.17	1.01	0.91	0.84	0.78	0.61	0.53	0.47	0.43	1.96	1.45	1.21	1.08	0.97	0.89	0.68	0.58	0.51	0.46	1.2	
	3	1.15	0.94	0.82	0.75	0.69	0.65	0.52	0.45	0.40	0.37	1.40	1.10	0.94	0.85	0.78	0.72	0.57	0.48	0.43	0.39	1.82	1.34	1.12	0.99	0.89	0.82	0.61	0.52	0.46	0.42	1.6	
	4	1.10	0.90	0.79	0.72	0.66	0.62	0.49	0.42	0.38	0.35	1.35	1.05	0.90	0.81	0.74	0.69	0.53	0.45	0.40	0.37	1.71	1.26	1.05	0.93	0.84	0.77	0.58	0.49	0.43	0.39	2.2	
	5	1.08	0.87	0.77	0.69	0.64	0.59	0.47	0.41	0.36	0.33	1.31	1.01	0.87	0.78	0.71	0.66	0.51	0.43	0.39	0.35	1.65	1.20	1.00	0.89	0.80	0.73	0.55	0.47	0.41	0.37	2.8	
	6	1.07	0.85	0.74	0.67	0.62	0.58	0.46	0.39	0.35	0.32	1.28	0.99	0.84	0.75	0.68	0.63	0.49	0.42	0.37	0.34	1.60	1.17	0.97	0.85	0.77	0.70	0.53	0.45	0.39	0.36	3.3	
	7	1.04	0.83	0.72	0.65	0.60	0.56	0.44	0.38	0.34	0.31	1.25	0.96	0.82	0.73	0.66	0.62	0.48	0.40	0.36	0.33	1.57	1.13	0.94	0.83	0.74	0.68	0.51	0.43	0.38	0.32	3.8	
	8	1.03	0.82	0.72	0.64	0.59	0.55	0.43	0.37	0.33	0.30	1.23	0.94	0.81	0.71	0.65	0.60	0.46	0.39	0.35	0.32	1.53	1.10	0.92	0.80	0.71	0.66	0.50	0.42	0.37	0.33	4.4	
	9	1.02	0.80	0.68	0.63	0.57	0.54	0.42	0.36	0.32	0.29	1.20	0.92	0.78	0.70	0.63	0.59	0.45	0.38	0.34	0.31	1.49	1.08	0.89	0.79	0.70	0.65	0.49	0.41	0.36	0.32	5.0	
	10	1.00	0.79	0.68	0.61	0.56	0.53	0.41	0.35	0.31	0.29	1.19	0.91	0.76	0.68	0.62	0.56	0.44	0.37	0.33	0.30	1.48	1.06	0.87	0.77	0.69	0.63	0.48	0.40	0.35	0.32	5.6	
40	1	1.22	1.03	0.93	0.86	0.80	0.75	0.62	0.54	0.49	0.45	1.56	1.27	1.11	1.00	0.93	0.87	0.70	0.60	0.54	0.49	2.13	1.62	1.37	1.22	1.11	1.02	0.79	0.67	0.59	0.53	0.6	4.4
	2	1.16	0.69	0.85	0.78	0.72	0.68	0.55	0.48	0.43	0.39	1.44	1.14	0.99	0.90	0.83	0.77	0.61	0.52	0.46	0.42	1.92	1.43	1.20	1.06	0.95	0.88	0.68	0.57	0.50	0.46	1.2	
	3	1.11	0.91	0.80	0.73	0.68	0.64	0.51	0.44	0.40	0.36	1.36	1.08	0.93	0.84	0.77	0.71	0.56	0.48	0.43	0.39	1.79	1.32	1.10	0.98	0.88	0.81	0.61	0.52	0.46	0.41	1.6	
	4	1.07	0.87	0.77	0.70	0.65	0.61	0.48	0.42	0.37	0.34	1.31	1.03	0.88	0.80	0.73	0.68	0.53	0.45	0.40	0.36	1.70	1.25	1.04	0.92	0.83	0.76	0.58	0.49	0.43	0.39	2.2	
	5	1.05	0.85	0.74	0.68	0.62	0.58	0.46	0.40	0.36	0.33	1.28	0.99	0.85	0.76	0.70	0.65	0.51	0.43	0.38	0.35	1.64	1.19	0.99	0.88	0.79	0.73	0.55	0.46	0.41	0.38	2.8	
	6	1.03	0.83	0.72	0.66	0.60	0.56	0.45	0.39	0.35	0.32	1.25	0.96	0.82	0.74	0.67	0.63	0.49	0.41	0.37	0.33	1.58	1.15	0.96	0.84	0.76	0.70	0.53	0.44	0.39	0.35	3.3	
	7	1.10	0.81	0.71	0.64	0.59	0.55	0.44	0.38	0.34	0.31	1.22	0.94	0.80	0.72	0.65	0.61	0.47	0.40	0.36	0.32	1.54	1.12	0.93	0.82	0.74	0.68	0.51	0.43	0.38	0.34	3.8	
	8	1.00	0.80	0.70	0.63	0.58	0.54	0.43	0.37	0.33	0.30	1.20	0.92	0.79	0.70	0.64	0.59	0.47	0.39	0.35	0.31	1.50	1.09	0.92	0.80	0.71	0.66	0.50	0.41	0.37	0.33	4.4	
	9	0.99	0.78	0.68	0.62	0.57	0.53	0.42	0.36	0.32	0.29	1.18	0.90	0.77	0.69	0.62	0.58	0.45	0.38	0.34	0.31	1.47	1.06	0.90	0.78	0.70	0.64	0.48	0.40	0.36	0.32	5.0	
	10	0.97	0.77	0.67	0.60	0.55	0.52	0.41	0.35	0.31	0.28	1.16	0.89	0.75	0.67	0.61	0.57	0.44	0.37	0.33	0.30	1.44	1.04	0.87	0.76	0.68	0.63	0.47	0.40	0.35	0.31	5.6	
50	1	1.18	1.00	0.90	0.83	0.78	0.74	0.61	0.53	0.48	0.44	1.49	1.22	1.08	0.98	0.91	0.85	0.67	0.59	0.53	0.48	2.04	1.56	1.33	1.19	1.09	1.00	0.76	0.66	0.59	0.53	0.6	10.0
	2	1.16	0.96	0.85	0.78	0.72	0.68	0.55	0.48	0.43	0.40	1.41	1.12	0.97	0.89	0.81	0.76	0.60	0.51	0.46	0.42	1.88	1.41	1.19	1.05	0.95	0.88	0.68	0.57	0.50	0.45	1.2	
	3	1.08	0.89	0.78	0.72	0.67	0.62	0.50	0.44	0.39	0.36	1.33	1.05	0.91	0.82	0.76	0.70	0.55	0.47	0.42	0.38	1.75	1.30	1.09	0.97	0.88	0.81	0.61	0.52	0.46	0.41	1.6	
	4	1.04	0.85	0.75	0.68	0.63	0.59	0.48	0.41	0.37	0.34	1.28	0.98	0.87	0.78	0.72	0.67	0.52	0.44	0.40	0.36	1.67	1.23	1.03	0.91	0.83	0.76	0.58	0.49	0.43	0.39	2.2	
	5	1.02	0.83	0.73	0.66	0.61	0.57	0.46	0.40	0.36	0.32	1.25	0.97	0.85	0.75	0.69	0.64	0.50	0.42	0.38	0.34	1.61	1.19	0.99	0.88	0.79	0.72	0.56	0.46	0.41	0.37	2.8	
	6	1.00	0.81	0.71	0.65	0.59	0.56	0.44	0.38	0.34	0.31	1.22	0.95	0.81	0.73	0.67	0.62	0.48	0.41	0.36	0.33	1.56	1.15	0.95	0.84	0.76	0.69	0.53	0.44	0.39	0.35	3.3	
	7	0.99	0.80	0.69	0.63	0.58	0.54	0.43	0.37	0.33	0.30	1.19	0.93	0.79	0.71	0.65	0.60	0.47	0.40	0.35	0.32	1.52	1.11	0.93	0.81	0.71	0.68	0.51	0.43	0.38	0.34	3.8	
	8	0.97	0.78	0.68	0.61	0.57	0.53	0.42	0.36	0.32	0.29	1.18	0.91	0.78	0.69	0.63	0.58	0.45	0.39	0.34	0.31	1.47	1.08	0.90	0.79	0.70	0.66	0.49	0.42	0.37	0.33	4.4	
	9	0.96	0.76	0.67	0.61	0.56	0.52	0.41	0.35	0.31	0.29	1.18	0.90	0.76	0.68	0.61	0.57	0.44	0.38	0.33	0.30	1.45	1.05	0.87	0.78	0.69	0.64	0.48	0.40	0.36	0.32	5.0	
	10	0.94	0.77	0.67	0.60	0.55	0.51	0.40	0.34	0.31	0.28	1.18	0.88	1.74	0.67	0.60	0.56	0.43	0.37	0.32	0.30	1.43	1.03	0.85	0.76	0.68	0.62	0.47	0.39	0.34	0.31	5.6	

Note: Above resistances are in English Units °F, ft^2, hr, Btu
Multiply by 0.1761 to obtain Metric Units K, m^2, W = R_{ms}

Air movement in miles per hour
Multiply by 1.609 to obtain kilometers per hour

Table B-14 Low temperature service — Approximate air film surface resistance values of $\frac{1}{f}$ or (R_s) — English Units
Heat transfer from air to lower temperature surface (cold service)

Air Temp. °F	Temp. Diff. Between Air & Surface °F	∈ = 1.0										∈ = 0.8										Temp. Diff. Between Air & Surface °C	Air Temp. °C
		Air Movement — MPH										Air Movement — MPH											
		0	1	2	3	4	5	10	15	20	25	0	1	2	3	4	5	10	15	20	25		
	1	0.79	0.70	0.65	0.62	0.59	0.56	0.48	0.43	0.40	0.37	0.93	0.81	0.75	0.70	0.66	0.63	0.53	0.47	0.43	0.40	0.6	
	2	0.75	0.66	0.61	0.57	0.54	0.52	0.44	0.39	0.36	0.33	0.88	0.76	0.69	0.64	0.60	0.57	0.48	0.42	0.38	0.36	1.2	
	3	0.74	0.64	0.59	0.55	0.52	0.49	0.41	0.37	0.34	0.31	0.86	0.73	0.66	0.62	0.58	0.55	0.45	0.40	0.36	0.33	1.6	
	4	0.73	0.63	0.57	0.53	0.50	0.48	0.40	0.35	0.32	0.30	0.85	0.71	0.64	0.60	0.56	0.53	0.43	0.38	0.34	0.32	2.2	
60	5	0.72	0.62	0.56	0.52	0.49	0.46	0.39	0.34	0.31	0.29	0.83	0.69	0.62	0.58	0.54	0.51	0.42	0.36	0.33	0.30	2.8	15.6
	6	0.71	0.61	0.55	0.51	0.48	0.45	0.37	0.33	0.30	0.28	0.82	0.68	0.61	0.56	0.52	0.49	0.40	0.35	0.32	0.29	3.3	
	7	0.70	0.60	0.54	0.50	0.47	0.44	0.36	0.32	0.29	0.27	0.81	0.67	0.60	0.55	0.51	0.48	0.39	0.34	0.31	0.28	3.8	
	8	0.69	0.59	0.53	0.49	0.46	0.43	0.36	0.31	0.28	0.26	0.80	0.66	0.59	0.54	0.50	0.47	0.38	0.33	0.30	0.28	4.4	
	9	0.68	0.58	0.52	0.48	0.47	0.43	0.35	0.31	0.28	0.26	0.79	0.65	0.58	0.53	0.49	0.46	0.38	0.32	0.29	0.27	5.0	
	10	0.68	0.58	0.52	0.48	0.44	0.42	0.35	0.30	0.27	0.25	0.79	0.65	0.57	0.53	0.49	0.46	0.37	0.31	0.29	0.26	5.6	
	1	0.75	0.68	0.63	0.60	0.57	0.54	0.46	0.42	0.39	0.37	0.89	0.79	0.72	0.68	0.64	0.61	0.51	0.47	0.43	0.40	0.6	
	2	0.73	0.65	0.59	0.56	0.53	0.50	0.43	0.38	0.35	0.33	0.85	0.74	0.68	0.63	0.59	0.56	0.47	0.42	0.38	0.35	1.2	
	3	0.71	0.62	0.57	0.53	0.50	0.48	0.40	0.36	0.33	0.31	0.83	0.71	0.65	0.60	0.56	0.53	0.44	0.39	0.35	0.33	1.6	
	4	0.69	0.61	0.55	0.52	0.49	0.46	0.39	0.34	0.31	0.29	0.81	0.69	0.62	0.58	0.54	0.51	0.42	0.37	0.34	0.31	2.2	
70	5	0.68	0.60	0.54	0.50	0.48	0.45	0.38	0.33	0.30	0.28	0.80	0.68	0.61	0.56	0.53	0.51	0.41	0.36	0.32	0.30	2.8	21.1
	6	0.68	0.59	0.53	0.49	0.46	0.44	0.37	0.32	0.29	0.27	0.79	0.67	0.60	0.55	0.51	0.48	0.39	0.35	0.31	0.29	3.3	
	7	0.67	0.58	0.53	0.48	0.45	0.43	0.36	0.31	0.28	0.26	0.78	0.65	0.58	0.54	0.50	0.47	0.38	0.34	0.30	0.28	3.8	
	8	0.66	0.57	0.52	0.48	0.45	0.42	0.35	0.31	0.28	0.26	0.77	0.64	0.57	0.53	0.49	0.46	0.38	0.33	0.30	0.27	4.4	
	9	0.66	0.56	0.51	0.47	0.44	0.42	0.34	0.30	0.27	0.25	0.76	0.64	0.56	0.52	0.48	0.45	0.37	0.32	0.29	0.27	5.0	
	10	0.66	0.56	0.50	0.46	0.43	0.41	0.34	0.30	0.27	0.25	0.75	0.63	0.56	0.51	0.48	0.45	0.36	0.32	0.28	0.26	5.6	
	1	0.72	0.65	0.61	0.57	0.55	0.53	0.45	0.41	0.38	0.36	0.85	0.76	0.70	0.66	0.62	0.59	0.51	0.45	0.42	0.39	0.6	
	2	0.70	0.62	0.57	0.56	0.51	0.49	0.42	0.37	0.34	0.32	0.82	0.71	0.65	0.61	0.57	0.55	0.48	0.41	0.37	0.35	1.2	
	3	0.68	0.60	0.55	0.52	0.49	0.47	0.40	0.35	0.32	0.30	0.80	0.69	0.52	0.58	0.55	0.53	0.43	0.38	0.35	0.32	1.6	
	4	0.67	0.58	0.53	0.50	0.47	0.45	0.38	0.34	0.31	0.29	0.78	0.67	0.60	0.56	0.53	0.50	0.41	0.36	0.33	0.31	2.2	
80	5	0.66	0.57	0.52	0.49	0.46	0.44	0.37	0.33	0.30	0.28	0.77	0.65	0.59	0.55	0.51	0.48	0.40	0.35	0.32	0.29	2.8	26.7
	6	0.65	0.56	0.51	0.48	0.45	0.43	0.36	0.32	0.29	0.27	0.76	0.64	0.58	0.54	0.50	0.47	0.39	0.34	0.31	0.28	3.3	
	7	0.64	0.56	0.50	0.47	0.44	0.42	0.35	0.31	0.28	0.26	0.75	0.63	0.57	0.53	0.49	0.46	0.38	0.33	0.30	0.27	3.8	
	8	0.64	0.55	0.50	0.46	0.43	0.41	0.34	0.30	0.27	0.25	0.74	0.62	0.56	0.52	0.48	0.45	0.37	0.32	0.29	0.27	4.4	
	9	0.63	0.54	0.49	0.46	0.43	0.41	0.34	0.30	0.27	0.25	0.73	0.62	0.55	0.51	0.47	0.45	0.36	0.32	0.29	0.26	5.0	
	10	0.63	0.54	0.48	0.45	0.42	0.40	0.33	0.29	0.26	0.24	0.73	0.62	0.55	0.50	0.47	0.44	0.36	0.31	0.28	0.26	5.6	
	1	0.69	0.62	0.59	0.56	0.53	0.51	0.44	0.40	0.37	0.35	0.82	0.73	0.68	0.64	0.61	0.58	0.49	0.45	0.41	0.38	0.6	
	2	0.67	0.60	0.55	0.52	0.50	0.48	0.41	0.37	0.34	0.32	0.80	0.69	0.63	0.59	0.56	0.53	0.45	0.40	0.37	0.34	1.2	
	3	0.65	0.58	0.53	0.50	0.48	0.46	0.39	0.35	0.32	0.30	0.79	0.67	0.61	0.57	0.53	0.51	0.42	0.38	0.38	0.32	1.6	
	4	0.64	0.56	0.52	0.49	0.46	0.44	0.37	0.33	0.30	0.28	0.75	0.65	0.59	0.55	0.51	0.49	0.41	0.36	0.33	0.30	2.2	
90	5	0.63	0.55	0.51	0.47	0.45	0.43	0.36	0.32	0.29	0.27	0.74	0.64	0.57	0.53	0.50	0.47	0.39	0.25	0.31	0.29	2.8	32.2
	6	0.63	0.55	0.50	0.47	0.44	0.42	0.35	0.31	0.28	0.26	0.73	0.62	0.56	0.52	0.49	0.46	0.38	0.33	0.30	0.28	3.3	
	7	0.62	0.54	0.49	0.46	0.43	0.41	0.34	0.30	0.28	0.26	0.72	0.61	0.55	0.51	0.48	0.45	0.37	0.33	0.30	0.27	3.8	
	8	0.61	0.53	0.49	0.45	0.42	0.40	0.34	0.30	0.27	0.25	0.71	0.60	0.55	0.50	0.47	0.44	0.36	0.32	0.29	0.27	4.4	
	9	0.61	0.53	0.48	0.45	0.42	0.40	0.33	0.29	0.27	0.25	0.70	0.60	0.54	0.50	0.46	0.44	0.36	0.31	0.28	0.26	5.0	
	10	0.61	0.53	0.47	0.44	0.41	0.39	0.33	0.29	0.26	0.24	0.69	0.59	0.53	0.49	0.46	0.43	0.35	0.31	0.28	0.26	5.6	
	1	0.66	0.60	0.57	0.54	0.52	0.49	0.43	0.39	0.37	0.34	0.78	0.70	0.65	0.61	0.59	0.56	0.48	0.43	0.40	0.37	0.6	
	2	0.64	0.57	0.53	0.51	0.48	0.46	0.39	0.36	0.33	0.31	0.76	0.66	0.61	0.57	0.54	0.52	0.44	0.39	0.36	0.33	1.2	
	3	0.63	0.56	0.52	0.49	0.46	0.44	0.38	0.34	0.31	0.29	0.74	0.65	0.59	0.55	0.52	0.49	0.42	0.37	0.34	0.31	1.6	
	4	0.62	0.55	0.50	0.47	0.45	0.43	0.36	0.32	0.30	0.28	0.72	0.63	0.57	0.53	0.50	0.48	0.38	0.35	0.32	0.30	2.2	
100	5	0.61	0.53	0.49	0.45	0.44	0.42	0.35	0.31	0.29	0.27	0.71	0.63	0.56	0.52	0.49	0.46	0.38	0.34	0.31	0.29	2.8	37.8
	6	0.60	0.53	0.48	0.45	0.43	0.41	0.34	0.30	0.28	0.26	0.71	0.60	0.55	0.51	0.48	0.45	0.37	0.33	0.30	0.28	3.3	
	7	0.60	0.52	0.48	0.44	0.42	0.40	0.34	0.30	0.27	0.25	0.70	0.60	0.54	0.50	0.47	0.44	0.46	0.32	0.29	0.27	3.8	
	8	0.59	0.52	0.47	0.44	0.41	0.39	0.33	0.29	0.27	0.25	0.69	0.59	0.53	0.49	0.46	0.43	0.36	0.31	0.28	0.26	4.4	
	9	0.59	0.52	0.47	0.44	0.41	0.39	0.33	0.29	0.26	0.24	0.69	0.59	0.53	0.49	0.46	0.43	0.36	0.31	0.28	0.26	5.0	
	10	0.59	0.51	0.46	0.43	0.40	0.38	0.32	0.28	0.25	0.24	0.68	0.58	0.52	0.48	0.45	0.42	0.35	0.30	0.27	0.25	5.6	

Note: Above resistances are in English Units °F, ft^2, hr, Btu
Multiply by 0.1761 to obtain Metric Units K, m^2, W = R_{ms}

Air movement in miles per hour
Multiply by 1.609 to obtain kilometers per hour

Table B-14 Low temperature service – Approximate air film surface resistance values of $\frac{1}{f}$ or (R_s) – English Units

Heat transfer from air to lower temperature surface (cold service)

Air Temp. °F	Temp. Diff. Between Air & Surface °F	∈ = 0.6										∈ = 0.4										∈ = 0.2										Temp. Diff. Between Air & Surface °C	Air Temp. °C
		Air Movement – MPH										Air Movement – MPH										Air Movement – MPH											
		0	1	2	3	4	5	10	15	20	25	0	1	2	3	4	5	10	15	20	25	0	1	2	3	4	5	10	15	20	25		
60	1	1.12	0.96	0.87	0.81	0.76	0.71	0.59	0.52	0.47	0.44	1.45	1.19	1.05	0.96	0.89	0.83	0.66	0.58	0.52	0.48	2.00	1.54	1.32	1.17	1.07	0.99	0.75	0.65	0.58	0.53	0.6	15.6
	2	1.08	0.92	0.82	0.76	0.70	0.66	0.56	0.47	0.42	0.39	1.35	1.09	0.94	0.86	0.79	0.74	0.59	0.51	0.45	0.41	1.85	1.39	1.16	1.04	0.94	0.87	0.67	0.57	0.50	0.45	1.2	
	3	1.03	0.86	0.76	0.70	0.65	0.61	0.50	0.43	0.39	0.35	1.30	1.03	0.89	0.81	0.74	0.69	0.55	0.47	0.42	0.38	1.75	1.28	1.07	0.96	0.87	0.80	0.61	0.52	0.45	0.41	1.6	
	4	1.01	0.83	0.73	0.67	0.62	0.58	0.47	0.41	0.37	0.34	1.25	0.96	0.85	0.77	0.70	0.66	0.51	0.44	0.39	0.36	1.64	1.22	1.02	0.91	0.82	0.75	0.57	0.48	0.43	0.38	2.2	
	5	0.99	0.81	0.71	0.65	0.60	0.56	0.45	0.39	0.35	0.32	1.22	0.95	0.83	0.74	0.68	0.63	0.49	0.42	0.38	0.34	1.59	1.16	0.97	0.86	0.78	0.72	0.56	0.46	0.41	0.37	2.8	
	6	0.97	0.79	0.69	0.63	0.58	0.55	0.44	0.38	0.34	0.31	1.19	0.93	0.80	0.72	0.66	0.61	0.48	0.41	0.36	0.33	1.54	1.12	0.94	0.83	0.75	0.69	0.52	0.44	0.39	0.35	3.3	
	337	0.96	0.77	0.68	0.62	0.57	0.53	0.43	0.37	0.33	0.30	1.16	0.91	0.78	0.69	0.64	0.59	0.46	0.39	0.35	0.32	1.49	1.10	0.91	0.81	0.71	0.67	0.51	0.43	0.37	0.34	3.8	
	8	0.94	0.76	0.67	0.60	0.56	0.52	0.41	0.36	0.32	0.29	1.15	0.89	0.77	0.68	0.62	0.58	0.45	0.38	0.34	0.31	1.47	1.06	0.88	0.78	0.70	0.65	0.49	0.41	0.36	0.33	4.4	
	9	0.93	0.75	0.66	0.59	0.55	0.51	0.40	0.35	0.31	0.28	1.12	0.87	0.74	0.67	0.61	0.56	0.44	0.37	0.33	0.30	1.43	1.04	0.86	0.77	0.69	0.63	0.48	0.40	0.35	0.32	5.0	
	10	0.92	0.74	0.66	0.58	0.54	0.50	0.40	0.36	0.30	0.28	1.11	0.85	0.72	0.65	0.60	0.55	0.43	0.36	0.32	0.29	1.41	1.02	0.85	0.75	0.68	0.62	0.47	0.39	0.34	0.31	5.6	
70	1	1.09	0.93	0.85	0.79	0.74	0.70	0.58	0.51	0.46	0.43	1.41	1.16	1.03	0.94	0.88	0.82	0.65	0.57	0.52	0.47	1.96	1.52	1.30	1.16	1.06	0.98	0.75	0.65	0.58	0.52	0.6	21.1
	2	1.04	0.91	0.80	0.74	0.69	0.65	0.53	0.46	0.42	0.38	1.31	1.06	0.94	0.85	0.78	0.72	0.58	0.50	0.45	0.41	1.82	1.37	1.15	1.03	0.93	0.86	0.67	0.56	0.50	0.45	1.2	
	3	1.00	0.83	0.74	0.69	0.64	0.60	0.49	0.43	0.38	0.35	1.26	1.01	0.88	0.79	0.73	0.68	0.54	0.46	0.41	0.38	1.69	1.26	1.06	0.95	0.86	0.79	0.61	0.51	0.45	0.41	1.6	
	4	0.97	0.80	0.71	0.65	0.61	0.57	0.46	0.40	0.36	0.33	1.22	0.94	0.83	0.76	0.69	0.65	0.51	0.44	0.39	0.36	1.61	1.20	1.01	0.89	0.81	0.75	0.57	0.48	0.42	0.38	2.2	
	5	0.95	0.78	0.69	0.63	0.59	0.55	0.45	0.39	0.35	0.32	1.18	1.92	0.82	0.73	0.67	0.62	0.49	0.42	0.37	0.34	1.56	1.15	0.96	0.85	0.77	0.71	0.54	0.46	0.40	0.36	2.8	
	6	0.94	0.77	0.68	0.62	0.57	0.53	0.43	0.37	0.33	0.31	1.15	0.91	0.78	0.70	0.65	0.60	0.47	0.40	0.36	0.33	1.51	1.11	0.93	0.83	0.75	0.68	0.52	0.44	0.39	0.35	3.3	
	7	0.93	0.76	0.66	0.61	0.56	0.52	0.42	0.36	0.33	0.30	1.13	0.89	0.76	0.69	0.63	0.59	0.46	0.39	0.35	0.32	1.47	1.09	0.90	0.80	0.72	0.67	0.51	0.42	0.37	0.34	3.8	
	8	0.91	0.74	0.66	0.59	0.55	0.51	0.41	0.35	0.32	0.29	1.11	0.87	0.75	0.67	0.61	0.57	0.44	0.38	0.34	0.31	1.45	1.05	0.88	0.77	0.70	0.65	0.49	0.41	0.36	0.33	4.4	
	9	0.89	0.73	0.65	0.58	0.53	0.50	0.40	0.34	0.31	0.28	1.10	0.86	0.73	0.66	0.60	0.56	0.43	0.37	0.33	0.30	1.41	1.03	0.86	0.76	0.68	0.63	0.48	0.40	0.35	0.32	5.0	
	10	0.88	0.72	0.65	0.57	0.53	0.49	0.39	0.34	0.30	0.28	1.09	0.84	0.71	0.64	0.59	0.55	0.42	0.36	0.32	0.29	1.39	1.01	0.84	0.74	0.67	0.62	0.47	0.39	0.34	0.31	5.6	
80	1	1.05	0.91	0.83	0.77	0.72	0.68	0.57	0.50	0.46	0.42	1.35	1.12	1.00	0.92	0.85	0.80	0.64	0.57	0.51	0.47	1.92	1.49	1.28	1.15	1.05	0.74	0.65	0.65	0.57	0.52	0.6	26.7
	2	1.03	0.87	0.78	0.73	0.67	0.63	0.56	0.45	0.41	0.38	1.28	1.04	0.91	0.83	0.77	0.72	0.58	0.50	0.44	0.41	1.79	1.35	1.14	1.02	0.93	0.86	0.66	0.56	0.49	0.45	1.2	
	3	0.96	0.81	0.73	0.67	0.62	0.58	0.48	0.42	0.38	0.35	1.22	0.98	0.85	0.77	0.72	0.67	0.53	0.46	0.41	0.37	1.66	1.19	1.05	0.93	0.85	0.78	0.60	0.51	0.45	0.41	1.6	
	4	0.94	0.78	0.69	0.64	0.59	0.56	0.45	0.39	0.35	0.33	1.18	0.92	0.82	0.74	0.69	0.63	0.50	0.43	0.38	0.35	1.59	1.17	0.99	0.88	0.80	0.74	0.57	0.48	0.42	0.38	2.2	
	5	0.92	0.76	0.67	0.62	0.58	0.54	0.44	0.38	0.34	0.31	1.15	0.91	0.80	0.71	0.66	0.61	0.48	0.41	0.37	0.34	1.51	1.14	0.95	0.85	0.77	0.70	0.55	0.45	0.40	0.36	2.8	
	6	0.91	0.75	0.66	0.61	0.56	0.53	0.42	0.37	0.33	0.30	1.12	0.89	0.77	0.69	0.64	0.59	0.47	0.40	0.36	0.32	1.47	1.10	0.92	0.81	0.74	0.68	0.52	0.44	0.38	0.35	3.3	
	7	0.90	0.73	0.66	0.60	0.55	0.51	0.41	0.36	0.32	0.29	1.10	0.87	0.75	0.67	0.62	0.58	0.45	0.39	0.34	0.31	1.45	1.06	0.88	0.79	0.70	0.66	0.50	0.42	0.37	0.34	3.8	
	8	0.89	0.72	0.65	0.58	0.54	0.50	0.40	0.35	0.31	0.28	1.08	0.86	0.74	0.66	0.60	0.56	0.44	0.38	0.33	0.30	1.41	1.04	0.87	0.77	0.68	0.64	0.49	0.41	0.36	0.33	4.4	
	9	0.88	0.71	0.64	0.57	0.53	0.49	0.39	0.34	0.30	0.28	1.07	0.84	0.72	0.65	0.59	0.55	0.43	0.37	0.32	0.30	1.38	1.01	0.85	0.76	0.67	0.62	0.48	0.40	0.35	0.32	5.0	
	10	0.86	0.70	0.63	0.56	0.52	0.48	0.39	0.33	0.29	0.27	1.06	0.83	0.64	0.64	0.58	0.54	0.42	0.37	0.31	0.29	1.37	1.00	0.83	0.74	0.66	0.61	0.46	0.39	0.34	0.31	5.6	
90	1	1.01	0.88	0.80	0.75	0.70	0.67	0.56	0.50	0.45	0.42	1.32	1.10	0.98	0.90	0.84	0.79	0.63	0.56	0.51	0.46	1.89	1.47	1.27	1.14	1.04	0.96	0.74	0.64	0.57	0.52	0.6	32.2
	2	0.97	0.82	0.74	0.69	0.64	0.61	0.54	0.44	0.40	0.37	1.24	1.01	0.89	0.81	0.75	0.71	0.57	0.49	0.44	0.40	1.71	1.31	1.12	1.00	0.91	0.84	0.65	0.55	0.49	0.44	1.2	
	3	0.93	0.79	0.71	0.65	0.61	0.57	0.47	0.41	0.41	0.34	1.18	0.96	0.84	0.76	0.71	0.66	0.53	0.45	0.41	0.37	1.62	1.23	1.04	0.93	0.84	0.78	0.60	0.51	0.45	0.41	1.6	
	4	0.90	0.76	0.68	0.62	0.58	0.55	0.45	0.39	0.35	0.32	1.12	0.92	0.80	0.73	0.67	0.63	0.50	0.43	0.38	0.35	1.54	1.16	0.98	0.87	0.78	0.73	0.56	0.47	0.42	0.38	2.2	
	5	0.89	0.74	0.66	0.61	0.56	0.53	0.43	0.38	0.34	0.31	1.11	0.89	0.78	0.70	0.65	0.60	0.48	0.41	0.37	0.33	1.49	1.11	0.94	0.84	0.76	0.70	0.54	0.45	0.40	0.36	2.8	
	6	0.87	0.73	0.64	0.59	0.55	0.52	0.42	0.36	0.33	0.30	1.09	0.87	0.75	0.69	0.63	0.58	0.46	0.39	0.35	0.32	1.45	1.08	0.91	0.81	0.74	0.67	0.51	0.43	0.38	0.35	3.3	
	7	0.86	0.71	0.63	0.58	0.54	0.50	0.41	0.35	0.32	0.29	1.07	0.85	0.73	0.66	0.61	0.57	0.45	0.38	0.34	0.31	1.41	1.05	0.88	0.78	0.69	0.65	0.50	0.42	0.37	0.33	3.8	
	8	0.85	0.70	0.63	0.56	0.52	0.49	0.39	0.34	0.31	0.28	1.05	0.83	0.67	0.65	0.59	0.55	0.44	0.37	0.33	0.30	1.38	1.02	0.87	0.76	0.68	0.63	0.49	0.41	0.36	0.32	4.4	
	9	0.84	0.69	0.62	0.56	0.52	0.48	0.39	0.34	0.30	0.28	1.04	0.82	0.71	0.64	0.58	0.54	0.43	0.36	0.32	0.30	1.36	1.00	0.84	0.75	0.67	0.62	0.47	0.40	0.35	0.32	5.0	
	10	0.84	0.68	0.62	0.55	0.51	0.47	0.38	0.33	0.29	0.27	1.03	0.81	0.69	0.63	0.57	0.53	0.42	0.36	0.31	0.29	1.24	0.98	0.82	0.73	0.66	0.61	0.46	0.39	0.34	0.31	5.6	
100	1	0.97	0.85	0.77	0.73	0.69	0.65	0.55	0.50	0.44	0.41	1.28	1.07	0.96	0.89	0.83	0.78	0.63	0.55	0.50	0.46	1.85	1.45	1.25	1.12	1.03	0.95	0.75	0.64	0.57	0.52	0.6	37.8
	2	0.92	0.79	0.71	0.70	0.62	0.59	0.49	0.43	0.39	0.36	1.23	0.99	0.87	0.80	0.74	0.79	0.56	0.48	0.43	0.40	1.70	1.30	1.10	0.99	0.90	0.83	0.65	0.55	0.49	0.43	1.2	
	3	0.90	0.76	0.69	0.64	0.59	0.56	0.46	0.40	0.37	0.36	1.14	0.94	0.82	0.73	0.69	0.65	0.52	0.45	0.40	0.37	1.59	1.21	1.02	0.92	0.83	0.77	0.59	0.50	0.45	0.40	1.6	
	4	0.88	0.74	0.66	0.61	0.57	0.54	0.44	0.38	0.35	0.32	1.11	0.90	0.78	0.71	0.66	0.62	0.49	0.42	0.37	0.35	1.52	1.15	0.97	0.86	0.78	0.73	0.56	0.47	0.42	0.38	2.2	
	5	0.86	0.72	0.64	0.59	0.55	0.52	0.42	0.37	0.33	0.31	1.08	0.87	0.76	0.69	0.64	0.59	0.47	0.41	0.36	0.33	1.46	1.10	0.93	0.83	0.75	0.69	0.53	0.44	0.40	0.36	2.8	
	6	0.85	0.71	0.63	0.59	0.54	0.51	0.41	0.36	0.32	0.30	1.06	0.85	0.74	0.67	0.62	0.58	0.46	0.39	0.35	0.32	1.42	1.07	0.90	0.80	0.72	0.67	0.51	0.43	0.38	0.35	3.3	
	7	0.84	0.70	0.62	0.57	0.52	0.49	0.40	0.35	0.31	0.29	1.04	0.83	0.72	0.65	0.60	0.56	0.44	0.38	0.34	0.31	1.39	1.04	0.87	0.77	0.70	0.65	0.49	0.42	0.37	0.33	3.8	
	8	0.83	0.68	0.61	0.55	0.51	0.48	0.39	0.34	0.31	0.28	1.03	0.82	0.72	0.64	0.59	0.55	0.43	0.37	0.33	0.30	1.36	1.01	0.86	0.75	0.68	0.63	0.48	0.40	0.36	0.33	4.4	
	9	0.83	0.68	0.60	0.55	0.51	0.48	0.38	0.33	0.30	0.28	1.02	0.81	0.70	0.63	0.58	0.54	0.42	0.36	0.32	0.29	1.34	0.99	0.83	0.74	0.67	0.62	0.47	0.39	0.35	0.32	5.0	
	10	0.81	0.67	0.59	0.54	0.50	0.47	0.38	0.33	0.28	0.27	1.00	0.79	0.68	0.62	0.56	0.53	0.41	0.35	0.31	0.29	1.16	0.97	0.81	0.72	0.65	0.60	0.46	0.39	0.34	0.31	5.6	

Note: Above resistances are in English Units °F, ft^2, hr, Btu
Multiply by 0.1761 to obtain Metric Units K, m^2, W = R_{ms}

Air movement in miles per hour
Multiply by 1.609 to obtain kilometers per hour

Table B-15 Thermal expansion of pipes in inches per 100 linear feet

Temp. °F	Cast Iron Pipe	Steel Pipe	Wrought Iron Pipe	Copper Pipe	Temp. °C
− 20	0	0	0	0	−28.9
0	0.127	0.145	0.152	0.204	−17.8
20	0.255	0.293	0.306	0.442	− 6.7
40	0.390	0.430	0.465	0.655	4.4
60	0.518	0.593	0.620	0.888	15.6
80	0.649	0.725	0.780	1.100	26.7
100	0.787	0.898	0.939	1.338	37.8
120	0.926	1.055	1.110	1.570	48.9
140	1.051	1.209	1.265	1.794	60.0
160	1.200	1.368	1.427	2.008	71.1
180	1.345	1.528	1.597	2.255	82.2
200	1.495	1.691	1.778	2.500	93.3
220	1.634	1.852	1.936	2.720	104
240	1.780	2.020	2.110	2.960	116
260	1.931	2.183	2.279	3.189	127
280	2.085	2.350	2.465	3.422	138
300	2.233	2.519	2.630	3.665	149
320	2.395	2.690	2.800	3.900	160
340	2.543	2.862	2.988	4.145	171
360	2.700	3.029	3.175	4.380	182
380	2.859	3.211	3.350	4.628	193
400	3.008	3.375	3.521	4.870	204
420	3.182	3.566	3.720	5.118	216
440	3.345	3.740	3.900	5.358	227
460	3.511	3.929	4.096	5.612	238
480	3.683	4.100	4.280	5.855	249
500	3.847	4.296	4.477	6.110	260
520	4.020	4.487	4.677	6.352	271
540	4.190	4.670	4.866	6.614	282
560	4.365	4.860	5.057	6.850	293
580	4.541	5.051	5.268	7.123	304
600	4.725	5.247	5.455	7.388	316
620	4.896	5.437	5.660	7.636	327
640	5.082	5.627	5.850	7.893	338
660	5.260	5.831	6.067	8.153	349
680	5.442	6.020	6.260	8.400	360
700	5.269	6.229	6.481	8.676	371
720	5.808	6.425	6.673	8.912	382
740	6.006	6.635	6.899	9.203	393
760	6.200	6.833	7.100	9.460	404
780	6.389	7.046	7.314	9.736	416
800	6.587	7.250	7.508	9.992	427
820	6.779	7.464	7.757	10.272	438
840	6.970	7.662	7.952	10.512	449
860	7.176	7.888	8.195	10.814	460
880	7.375	8.098	8.400	11.175	471
900	7.579	8.313	8.639	11.360	482
920	7.795	8.545	8.867	11.625	493
940	7.989	8.755	9.089	11.911	504
960	8.200	8.975	9.300	12.180	516
980	8.406	9.916	9.547	12.473	527
1000	8.617	9.421	9.776	12.747	538

To obtain the amount of expansion between any two temperatures, take the proportionate difference between the values given for those temperatures.
To convert to Metric Units multiply by 8.2025 to obtain thermal expansion of pipes in millimeters per 10 metres in length.

Table B-16 Linear coefficients of expansion for one degree

Substance	Coefficient, n Celius	Fahrenheit
Metals and Alloys		
Aluminum, wrought	.0000231	.0000128
Brass	.0000188	.0000104
Brass wire	.0000193	.0000107
Bronze	.0000181	.0000101
Copper	.0000168	.0000093
German Silver	.0000183	.0000102
Gold	.0000150	.0000083
Iron, cast, gray	.0000106	.0000059
Iron, wrought	.0000120	.0000067
Iron, wire	.0000124	.0000069
Lead	.0000286	.0000159
Nickel	.0000126	.0000070
Platinum	.0000090	.0000050
Platinum-Iridium, 15% Ir.	.0000081	.0000045
Silver	.0000192	.0000107
Steel, cast	.0000110	.0000061
Steel, hard	.0000132	.0000073
Steel, medium	.0000120	.0000067
Steel, soft	.0000110	.0000061
Tin	.0000210	.0000117
Zinc, rolled	.0000311	.0000173
Miscellaneous Solids		
Glass	.0000085	.0000047
Graphite	.0000079	.0000044
Gutta-percha	.0005980	.0003322
Paraffin	.0002785	.0001547
Porcelain	.0000036	.0000020
Stone and Masonry		
Ashlar masonry	.0000063	.0000035
Brick masonry	.0000055	.0000031
Cement, Portland	.0000107	.0000059
Concrete	.0000143	.0000079
Concrete, masonry	.0000120	.0000067
Granite	.0000084	.0000047
Limestone	.0000080	.0000044
Marble	.0000100	.0000056
Plaster	.0000166	.0000092
Rubble masonry	.0000063	.0000035
Sandstone	.0000110	.0000061
Slate	.0000104	.0000058
Timber		
Fir (parallel to fiber)	.0000037	.0000021
Maple (parallel to fiber)	.0000064	.0000036
Oak (parallel to fiber)	.0000049	.0000027
Pine (parallel to fiber)	.0000054	.0000030
Fir (perpendicular to fiber)	.000058	.000032
Maple (perpendicular to fiber)	.000048	.000027
Oak (perpendicular to fiber)	.000054	.000030
Pine (perpendicular to fiber)	.000034	.000019
Liquid Substances	Volumetric Expan.	
Alcohol	.00104	.00058
Acid, nitric	.00110	.00061
Acid, sulphuric	.00063	.00035
Mercury	.00018	.00010
Oil, turpentine	.00090	.00050

Expansion of Water, Maximum Density = 1

C°	Volume	C°	Volume	C°	Volume	C°	Volume	C°	Volume	C°	Volume
0	1.000126	10	1.000257	30	1.004234	50	1.011877	70	1.022384	90	1.035829
4	1.000000	20	1.001732	40	1.007627	60	1.016954	80	1.029003	100	1.043116

Table B-17 Radiating area of flanged fittings

English Units

Including accompanying flanges in square feet and in equivalent length of same size pipe standard weight fittings

	Flanged couplings		90° ells		Long radius ells		Tees		Crosses	
Pipe size in.	Area, sq ft	Pipe length, ft	Area, sq ft	Pipe length, ft	Area, sq ft	Pipe length, ft	Area, sq ft	Pipe length, ft	Area, sq ft	Pipe length, ft
1	.32	.83	.79	2.31	.89	2.59	1.24	3.59	1.62	4.72
1¼	.38	.88	.96	2.20	1.08	2.49	1.48	3.40	1.94	4.47
1½	.48	.95	1.19	2.35	1.34	2.68	1.82	3.64	2.38	4.78
2	.67	1.08	1.65	2.65	1.84	2.96	2.54	4.08	3.32	5.34
2½	.84	1.12	2.09	2.78	2.32	3.08	3.21	4.26	4.19	5.56
3	.95	1.03	2.38	2.60	2.68	2.93	3.66	3.99	4.77	5.70
3½	1.12	1.07	2.98	2.85	3.28	3.13	4.48	4.28	5.83	5.56
4	1.34	1.14	3.53	2.90	3.96	3.36	5.41	4.59	7.03	5.97
4½	1.47	1.13	3.95	3.01	4.43	3.38	6.07	4.63	7.87	6.01
5	1.62	1.11	4.44	3.05	5.00	3.43	6.81	4.67	8.82	6.06
6	1.82	1.05	5.13	2.95	5.99	3.45	7.84	4.53	10.08	5.81
7	2.17	1.05	6.17	3.09	7.38	3.70	9.37	4.69	12.00	6.01
8	2.41	1.07	6.98	3.09	8.56	3.79	10.55	4.67	13.44	5.96
9	3.00	1.19	8.71	3.46	10.57	4.20	13.18	5.23	16.78	6.66
10	3.43	1.22	10.18	3.61	12.35	4.38	15.41	5.47	19.58	6.95
12	4.41	1.32	13.08	3.92	16.35	4.90	19.67	5.89	24.87	7.45
14	5.39	1.47	16.38	4.47	20.17	5.47	24.81	6.78	31.48	8.60
16	6.69	1.60	20.17	4.82	25.41	6.07	30.32	7.23	38.34	9.15

Table B-17a Radiating area of flanged fittings

Metric Units

Including accompanying flanges in square feet and in equivalent length of same size pipe standard weight fittings

	Flanged couplings		90° ells		Long radius ells		Tees		Crosses	
Pipe size in.	Area, m^2	Pipe length, metres	Area, m^2	Pipe length, metres	Area, m^2	Pipe length, metres	Area, m^2	Pipe length, metres	Area, m^2	Pipe length, metres
1	.029	.252	.073	.704	.083	.789	.115	1.094	.150	1.438
1¼	.035	.268	.089	.670	.100	.759	.137	1.036	.180	1.362
1½	.044	.290	.111	.716	.124	.816	.169	1.109	.221	1.457
2	.062	.329	.153	.808	.170	.774	.236	1.243	.308	1.627
2½	.078	.341	.194	.847	.215	.939	.298	1.298	.389	1.695
3	.088	.313	.221	.792	.248	.893	.340	1.216	.443	1.737
3½	.104	.326	.276	.868	.305	.954	.416	1.305	.541	1.695
4	.124	.347	.327	.884	.367	1.024	.503	1.399	.653	1.820
4½	.137	.344	.367	.917	.411	1.030	.563	1.411	.731	1.832
5	.150	.338	.412	.930	.465	1.045	.632	1.023	.819	1.847
6	.169	.320	.477	.899	.556	1.051	.728	1.381	.936	1.771
7	.201	.320	.573	.941	.686	1.127	.870	1.429	1.115	1.831
8	.223	.326	.648	.941	.792	1.155	.980	1.423	1.249	1.817
9	.279	.362	.809	1.055	.981	1.280	1.224	1.594	1.559	2.029
10	.319	.371	.946	1.100	1.147	1.335	1.432	1.667	1.819	2.118
12	.410	.402	1.215	1.195	1.519	1.493	1.827	1.795	2.310	2.270
14	.501	.448	1.522	1.362	1.874	1.667	2.305	2.067	2.924	2.621
16	.621	.487	1.873	1.469	2.361	1.830	2.817	2.204	3.562	2.789

Table B-18 Dimensions of insulation for NPS pipe in inches (Dec. and Fract.) – English Units

PIPE		INSULATION DIMENSIONS																	
		Inner diameter				Outer diameter for nominal thickness of:													
Nom. size	Outer dia	Diameter		Tolerance		1		1 1/2		2		2 1/2		3		3 1/2		4	
inches	inches	Dec."	Fract."	Minus	Plus	Dec."	Fract."	Dec."	Fract."	Dec."	Fract."	Dec."	Fract."	Dec."	Fract."	Dec."	Fract."	Dec."	Fract."
1/2	0.840	0.86	55/64	0	1/16	2.88	2 7/8	4.00	4	5.00	5	6.62	6 5/8	7.62	7 5/8	8.62	8 5/8	9.62	9 5/8
3/4	1.050	1.07	1 1/64	0	1/16	2.88	2 7/8	4.00	4	5.00	5	6.62	6 5/8	7.62	7 5/8	8.62	8 5/8	9.62	9 5/8
1	1.315	1.33	1 21/64	0	1/16	3.50	3 1/2	4.50	4 1/2	5.56	5 9/16	6.62	6 5/8	7.62	7 5/8	8.62	8 5/8	9.62	9 5/8
1 1/4	1.660	1.68	1 11/64	0	1/16	3.50	3 1/2	5.00	5	5.56	5 9/16	6.62	6 5/8	7.62	7 5/8	8.62	8 5/8	9.62	9 5/8
1 1/2	1.900	1.92	1 59/64	0	1/16	4.00	4	5.00	5	6.62	6 5/8	7.62	7 5/8	8.62	8 5/8	9.62	9 5/8	10.75	10 3/4
2	2.375	2.41	2 13/32	0	3/32	4.50	4 1/2	5.56	5 9/16	6.62	6 5/8	7.62	7 5/8	8.62	8 5/8	9.62	9 5/8	10.75	10 3/4
2 1/2	2.875	2.91	2 29/32	0	3/32	5.00	5	6.62	6 5/8	7.62	7 5/8	8.62	8 5/8	9.62	9 5/8	10.75	10 3/4	11.75	11 3/4
3	3.500	3.53	3 17/32	0	3/32	5.56	5 9/16	6.62	6 5/8	7.62	7 5/8	8.62	8 5/8	9.62	9 5/8	10.75	10 3/4	11.75	11 3/4
3 1/2	4.000	4.03	4 1/32	1/32	3/32	6.62	6 5/8	7.62	7 5/8	8.62	8 5/8	9.62	9 5/8	10.75	10 3/4	11.75	11 3/4	12.75	12 3/4
4	4.500	4.53	4 17/32	1/32	3/32	6.62	6 5/8	7.62	7 5/8	8.62	8 5/8	9.62	9 5/8	10.75	10 3/4	11.75	11 3/4	12.75	12 3/4
4 1/2	5.006	5.03	5 1/32	1/32	3/32	7.62	7 5/8	8.62	8 5/8	9.62	9 5/8	10.75	10 3/4	11.75	11 3/4	12.75	12 3/4	14.00	14
5	5.563	5.64	5 41/64	1/32	3/32	7.62	7 5/8	8.62	8 5/8	9.62	9 5/8	10.75	10 3/4	11.75	11 3/4	12.75	12 3/4	14.00	14
6	6.625	6.70	6 45/64	1/32	3/32	8.62	8 5/8	9.62	9 5/8	10.75	10 3/4	11.75	11 3/4	12.75	12 3/4	14.00	14	15.00	15
7	7.625	7.70	7 45/64	1/32	3/32			10.75	10 3/4	11.75	11 3/4	12.75	12 3/4	14.00	14	15.00	15	16.00	16
8	8.625	8.70	8 45/64	1/32	3/32			11.75	11 3/4	12.75	12 3/4	14.00	14	15.00	15	16.00	16	17.00	17
9	9.625	9.70	9 45/64	1/32	3/32			12.75	12 3/4	14.00	14	15.00	15	16.00	16	17.00	17	18.00	18
10	10.750	10.83	10 53/64	1/32	3/32			14.00	14	15.00	15	16.00	16	17.00	17	18.00	18	19.00	19
11	11.750	11.83	11 53/64	1/32	3/32			15.00	15	16.00	16	17.00	17	18.00	18	19.00	19	20.00	20
12	12.750	12.84	12 27/32	1/16	3/32			16.00	16	17.00	17	18.00	18	19.00	19	20.00	20	21.00	21
14[a]	14.000	14.09	14 3/32	1/16[b]	5/32[b]			17.00[a]	17[a]	18.00[a]	18[a]	19.00[a]	19[a]	20.00[a]	20[a]	21.00[a]	21[a]	22.00[a]	22[a]

[a]Sizes 16 through 36 in 1 inch increments
[b]Sizes 16 through 36 same tolerance as for 14 in NPS pipe, fraction conversions to nearest 64th of an inch

Table B-18a Dimensions of insulation for NPS pipe – Metric Units

Pipe		Insulation Dimensions – Millimeters									
Nom.	Outer	Inner Diameter			Outer Diameter for Nominal Thickness in Millimeters						
size	dia	Dia	Tolerance		1" nom.	1 1/2" nom.	2" nom.	2 1/2" nom.	3" nom.	3 1/2" nom.	4" nom.
inches	mm	mm	Minus mm	Plus mm	25 mm	38 mm	51 mm	64 mm	76 mm	89 mm	102 mm
1/2	21.3	22	0	1.6	73.15	101.6	127.0	168.1	193.5	218.9	244.3
3/4	26.7	27	0	1.6	73.15	101.6	127.0	168.1	193.5	218.9	244.3
1	33.4	34	0	1.6	88.9	114.3	141.2	168.1	193.5	218.9	244.3
1 1/4	42.2	43	0	1.6	88.9	127.0	141.2	168.1	193.5	218.9	244.3
1 1/2	48.3	49	0	1.6	101.6	127.0	168.1	193.5	218.9	244.3	273.1
2	60.3	61	0	2.4	114.3	141.2	168.1	193.5	218.9	244.3	273.1
2 1/2	73.0	74	0	2.4	127.0	168.1	193.5	218.9	244.3	273.1	298.5
3	88.9	90	0	2.4	141.2	168.1	193.5	218.9	244.3	273.1	298.5
3 1/4	101.6	102	0.8	2.4	168.1	193.5	218.9	244.3	273.1	298.5	323.5
4	114.3	115	0.8	2.4	168.1	193.5	218.9	244.3	273.1	298.5	323.5
4 1/2	127.0	128	0.8	2.4	193.5	218.9	244.3	273.1	298.5	323.9	355.6
5	141.3	143	0.8	2.4	193.5	218.9	244.3	273.1	298.5	323.9	355.6
6	168.3	180	0.8	2.4	218.9	244.3	273.1	298.5	323.9	355.6	381.0
7	193.7	196	0.8	2.4		273.1	298.5	323.9	355.6	381.0	406.4
8	219.1	221	0.8	2.4		298.5	323.9	355.6	381.0	406.4	431.8
9	244.5	246	0.8	2.4		323.9	355.6	381.0	406.4	431.8	457.2
10	278.0	275	0.8	2.4		355.6	381.0	406.4	431.8	457.2	482.6
11	298.4	300	0.8	2.4		381.0	406.4	431.8	457.2	482.6	508.0
12	323.8	326	1.6	2.4		406.4	431.8	457.2	482.6	508.0	533.4
14★	355.6	358	1.6	4.0		431.8	457.2	482.6	508.0	533.4	558.8

★Larger sizes 355.6 mm through 914.4 mm in 25.4 increments.

Table B-19 Approximate wall thickness of insulation for NPS pipes in inches (Dec. and Fract.) – English Units

PIPE		APPROXIMATE WALL THICKNESS													
		Nominal thickness													
Nom. size	Outer dia	1		1 1/2		2		2 1/2		3		3 1/2		4	
inches	inches	Dec."	Fract."	Dec."	Fract."	Dec."	Fract."	Dec."	Fract."	Dec."	Fract."	Dec."	Fract."	Dec."	Fract."
1/2	0.840	1.01	1 1/64	1.57	1 9/16	2.07	2 1/16	2.88	2 7/8	3.38	3 3/8	3.88	3 7/8	4.38	4 3/8
3/4	1.050	0.90	29/32	1.46	1 15/32	1.96	1 31/32	2.78	2 25/32	3.28	3 11/32	3.78	3 25/32	4.28	4 11/32
1	1.315	1.08	1 5/64	1.58	1 37/64	2.12	2 1/8	2.64	2 41/64	3.14	3 9/64	3.64	3 41/64	4.14	4 9/64
1 1/4	1.660	0.91	29/32	1.66	1 21/32	1.94	1 15/16	2.47	2 15/32	2.97	2 31/32	3.47	3 15/32	3.97	3 31/32
1 1/2	1.900	1.04	1 3/64	1.54	1 35/64	2.35	2 23/64	2.85	2 27/32	3.35	3 23/64	3.85	3 27/32	4.42	4 27/64
2	2.375	1.04	1 3/64	1.58	1 37/64	2.10	2 7/64	2.60	2 39/64	3.10	3 7/64	3.60	3 39/64	4.17	4 11/64
2 1/2	2.875	1.04	1 3/64	1.86	1 55/64	2.36	2 23/64	2.86	2 55/64	3.36	3 23/64	3.92	3 59/64	4.42	4 27/64
3	3.500	1.02	1 1/64	1.54	1 35/64	2.04	2 3/64	2.54	2 35/64	3.04	3 3/64	3.61	3 39/64	4.11	4 7/64
3 1/2	4.000	1.30	1 19/64	1.80	1 51/64	2.30	2 19/64	2.80	2 51/64	3.36	3 23/64	3.86	3 55/64	4.36	4 3/64
4	4.500	1.04	1 3/64	1.54	1 35/64	2.04	2 3/64	2.54	2 35/64	3.11	3 7/64	3.61	3 39/64	4.11	4 7/64
4 1/2	5.000	1.30	1 19/64	1.80	1 51/64	2.30	2 19/64	2.86	2 55/64	3.36	3 23/64	3.86	3 55/64	4.48	4 31/64
5	5.563	0.99	1	1.49	1 1/2	1.99	2	2.56	2 9/16	3.06	3 1/16	3.56	3 9/16	4.18	4 3/16
6	6.625	0.96	31/32	1.46	1 15/32	2.02	2 1/64	2.52	2 33/64	3.02	3 1/64	3.65	3 21/32	4.15	4 5/32
7	7.625			1.52	1 33/64	2.02	2 1/64	2.52	2 33/64	3.15	3 5/32	3.65	3 21/32	4.15	4 5/32
8	8.625			1.52	1 33/64	2.02	2 1/64	2.65	2 21/32	3.15	3 5/32	3.65	3 21/32	4.15	4 5/32
9	9.625			1.52	1 33/64	2.15	2 5/32	2.65	2 21/32	3.15	3 5/32	3.65	3 21/32	4.15	4 5/32
10	10.750			1.58	1 37/64	2.08	2 5/64	2.58	2 37/64	3.08	3 5/64	3.58	3 37/64	4.08	4 5/64
11	11.750			1.58	1 37/64	2.08	2 5/64	2.58	2 37/64	3.08	3 5/64	3.58	3 37/64	4.08	4 5/64
12	12.750			1.58	1 37/64	2.08	2 5/64	2.58	2 37/64	3.08	3 5/64	3.58	3 37/64	4.08	4 5/64
14[a]	14.000[a]			1.46	1 15/32	1.96	1 31/32	2.46	2 15/32	2.96	2 31/32	3.46	3 15/32	3.96	3 31/32

[a]Sizes 14" through 36" in 1 inch increments
Fraction conversions to nearest 64th of an inch

Table B-19a Approximate wall thickness of insulation for NPS pipe – Metric Units

Pipe		Insulation Dimensions									
Nom. size inches	Outer dia mm	Inner Diameter			Approximate Wall Thickness in Millimeters						
		Dia mm	Tolerance		1" nom.	1 1/2" nom.	2" nom.	2 1/2" nom.	3" nom.	3 1/2" nom.	4" nom.
			Minus mm	Plus mm	25 mm	38 mm	51 mm	64 mm	76 mm	89 mm	102 mm
1/2	21.3	22	0	1.6	26	40	53	73	86	99	111
3/4	26.7	27	0	1.6	23	37	50	71	83	96	109
1	33.4	34	0	1.6	27	40	54	67	80	92	105
1 1/4	42.2	43	0	1.6	23	42	49	63	75	88	101
1 1/2	48.3	49	0	1.6	26	39	60	72	85	98	112
2	60.3	61	0	2.4	26	40	53	66	79	91	106
2 1/2	73.0	74	0	2.4	26	47	60	73	85	100	112
3	88.9	90	0	2.4	26	39	52	65	77	92	104
3 1/2	101.6	102	0.8	2.4	33	46	58	71	85	98	111
4	114.3	115	0.8	2.4	26	39	52	65	79	92	104
4 1/2	127.0	128	0.8	2.4	33	46	58	73	85	98	114
5	141.3	143	0.8	2.4	25	38	51	65	78	90	106
6	168.3	180	0.8	2.4	24	37	51	64	77	93	105
7	193.7	196	0.8	2.4		39	51	64	80	93	105
8	219.1	221	0.8	2.4		39	51	67	80	93	105
9	244.5	246	0.8	2.4		39	55	67	80	93	105
10	278.0	275	0.8	2.4		40	53	66	78	91	104
11	298.4	300	0.8	2.4		40	53	66	78	91	104
12	323.8	326	1.6	2.4		40	53	66	78	91	104
14	355.6	358	1.6	4.0		37	50	62	75	88	101

*Larger sizes 355.6 mm through 914.4 mm in 25.4 increments.

Table B-20 Dimensions of insulation for tubes in inches (Dec. and Fract.) — English Units

TUBE		INSULATION																	
		Inner diameter				Outer diameter for nominal thickness of:													
Nom. size	Outer dia	Diameter		Tolerance		1		1 1/2		2		2 1/2		3		3 1/2		4	
inches	inches	Dec."	Fract."	Minus	Plus	Dec."	Fract."	Dec."	Fract."	Dec."	Fract."	Dec."	Fract."	Dec."	Fract."	Dec."	Fract."	Dec."	Fract."
3/8	0.500	0.52	33/64	0	1/16	2.38	2 7/8	3.50	3 1/2	4.50	4 1/2	5.56	5 9/16	6.62	6 5/8	7.62	7 5/8	8.62	8 5/8
1/2	0.625	0.64	41/64	0	1/16	2.88	2 7/8	3.50	3 1/2	4.50	4 1/2	5.56	5 9/16	6.62	6 5/8	7.62	7 5/8	8.62	8 5/8
3/4	0.875	0.89	57/64	0	1/16	2.88	2 7/8	4.00	4	5.00	5	6.62	6 5/8	7.62	7 5/8	8.62	8 5/8	9.62	9 5/8
1	1.125	1.14	1 9/64	0	1/16	2.88	2 7/8	4.00	4	5.00	5	6.62	6 5/8	7.62	7 5/8	8.62	8 5/8	9.62	9 5/8
1 1/4	1.375	1.39	1 25/64	0	1/16	3.50	3 1/2	4.50	4 1/2	5.56	5 9/16	6.62	6 5/8	7.62	7 5/8	8.62	8 5/8	9.62	9 5/8
1 1/2	1.625	1.64	1 41/64	0	1/16	3.50	3 1/2	4.50	4 1/2	5.56	5 9/16	6.62	6 5/8	7.62	7 5/8	8.62	8 5/8	9.62	9 5/8
2	2.125	2.16	2 5/32	0	1/16	4.00	4	5.00	5	6.62	6 5/8	7.62	7 5/8	8.62	8 5/8	9.62	9 5/8	10.75	10 3/4
2 1/2	2.625	2.66	2 21/32	0	1/16	4.50	4 1/2	5.56	5 9/16	6.62	6 5/8	7.62	7 5/8	8.62	8 5/8	9.62	9 5/8	10.75	10 3/4
3	3.125	3.16	3 5/32	0	1/16	5.00	5	6.62	6 5/8	7.62	7 5/8	8.62	8 5/8	9.62	9 5/8	10.75	10 3/4	11.75	11 3/4
3 1/2	3.625	3.66	3 21/32	0	1/16	5.56	5 9/16	6.62	6 5/8	7.62	7 5/8	8.62	8 5/8	9.62	9 5/8	10.75	10 3/4	11.75	11 3/4
4	4.125	4.16	4 5/32	1/32	3/32	6.62	6 5/8	7.62	7 5/8	8.62	8 5/8	9.62	9 5/8	10.75	10 3/4	11.75	11 3/4	12.75	12 3/4
5	5.125	5.16	5 5/32	1/32	3/32	7.62	7 5/8	8.62	8 5/8	9.62	9 5/8	10.75	10 3/4	11.75	11 3/4	12.75	12 3/4	14.00	14
6	6.125	6.20	6 13/64	1/32	3/32	8.62	8 5/8	9.62	9 5/8	10.75	10 3/4	11.75	11 3/4	12.75	12 3/4	14.00	14	15.00	15

Fraction conversions to nearest 64th of an inch

Table B-20a Dimensions of insulation for tubes — Metric Units

Tube		Insulation Dimensions — Millimetres									
		Inner Diameter			Outer Diameter for Nominal Thicknesses in Millimetres						
Nom. size	Outer dia	Dia	Tolerance		1" nom.	1 1/2" nom.	2" nom.	2 1/2" nom.	3" nom.	3 1/2" nom.	4" nom.
inches	mm	mm	Minus mm	Plus mm	25 mm	38 mm	51 mm	64 mm	76 mm	89 mm	102 mm
3/8	12.7	13	0	1.6	60.5	88.9	114.3	141.2	168.1	193.5	218.9
1/2	15.9	16	0	1.6	73.2	88.9	114.3	141.2	168.1	193.5	218.9
3/4	22.2	23	0	1.6	73.2	101.6	127.0	168.1	193.5	218.9	244.3
1	28.6	29	0	1.6	73.2	101.6	127.0	168.1	193.5	218.9	244.3
1 1/4	34.9	35	0	1.6	88.9	114.3	141.2	168.1	193.5	218.9	244.3
1 1/2	41.3	42	0	1.6	88.9	114.3	141.2	168.1	193.5	218.9	244.3
2	54.0	55	0	1.6	101.6	127.0	168.1	193.5	218.9	244.3	273.1
2 1/2	66.7	68	0	1.6	114.3	141.2	168.1	193.5	218.9	244.3	273.1
3	79.4	80	0	1.6	127.0	168.1	193.5	218.9	244.3	273.1	298.5
3 1/2	92.1	93	0	1.6	141.2	168.1	193.5	218.9	244.3	273.1	298.5
4	104.8	106	0.8	2.4	168.1	193.5	218.9	244.3	273.1	298.5	323.9
5	130.2	131	0.8	2.4	193.5	218.9	244.3	273.1	298.5	323.9	355.6
6	155.6	157	0.8	2.4	218.9	244.3	273.1	298.5	323.9	355.6	381.0

Table B-21 Approximate wall thickness of insulation for tubes in inches (Dec. and Fract.) — English Units

TUBE		APPROXIMATE WALL THICKNESS													
		Nominal thickness													
Nom. size inches	Outer dia inches	1		1 1/2		2		2 1/2		3		3 1/2		4	
		Dec.″	Fract.″	Dec.″	Fract.″	Dec.″	Fract.″	Dec.″	Fract.″	Dec.″	Fract.″	Dec.″	Fract.″	Dec.″	Fract.″
3/8	0.500	0.93	15/16	1.49	1 31/64	1.99	2	2.52	2 33/64	3.05	3 3/64				
1/2	0.625	1.12	1 1/8	1.43	1 7/16	1.93	1 15/16	2.46	2 15/32	2.99	3				
3/4	0.875	1.00	1	1.56	1 9/16	2.06	1 1/16	2.86	2 55/64	3.36	3 23/64	3.87	3 7/8	4.37	4 3/8
1	1.125	0.87	7/8	1.43	1 7/16	1.93	1 15/16	2.74	2 47/64	3.24	3 15/64	3.74	3 47/64	4.24	4 15/64
1 1/4	1.375	1.06	1 1/16	1.56	1 9/16	2.08	1 5/64	2.62	2 5/8	3.12	3 1/8	3.62	3 5/8	4.12	4 1/8
1 1/2	1.625	0.93	15/16	1.43	1 7/16	1.96	1 31/32	2.49	2 1/2	2.99	3	3.49	3 1/2	3.99	4
2	2.125	0.93	15/16	1.43	1 7/16	2.23	2 15/64	2.73	2 47/64	3.23	3 15/64	3.73	3 47/64	4.30	4 19/64
2 1/2	2.625	0.92	59/64	1.45	1 29/64	1.98	1 63/64	2.48	2 31/64	2.98	2 63/64	3.48	3 31/64	4.04	4 3/64
3	3.125	0.92	59/64	1.73	1 47/64	2.23	2 15/64	2.73	2 47/64	3.23	3 15/64	3.80	3 51/64	4.30	4 19/64
3 1/2	3.625	0.95	61/64	1.48	1 31/64	1.98	1 63/64	2.48	2 31/64	2.98	2 63/64	3.54	3 35/64	4.04	4 3/64
4	4.125	1.23	1 15/64	1.73	1 47/64	2.23	2 15/64	2.73	2 47/64	3.30	3 19/64	3.80	3 51/64	4.30	4 19/64
5	5.125	1.23	1 15/64	1.73	1 47/64	2.23	2 15/64	2.80	2 51/64	3.30	3 19/64	3.80	3 51/64	4.42	4 19/64
6	6.125	1.21	1 7/32	1.71	1 23/32	2.28	2 9/32	2.78	2 25/32	3.28	3 9/32	3.90	3 29/32	4.40	4 7/16

Fraction conversions to nearest 64th of an inch

Table B-21a Approximate wall thicknesses of insulation for tubes — Metric Units

Tube		Insulation Dimensions — Millimetres									
Nom. size inches	Outer dia mm	Inner Diameter			Approximate Wall Thickness in Millimetres						
		Dia mm	Tolerance		1″ nom.	1 1/2″ nom.	2″ nom.	2 1/2″ nom.	3″ nom.	3 1/2″ nom.	4″ nom.
			Minus mm	Plus mm	25 mm	38 mm	51 mm	64 mm	76 mm	89 mm	102 mm
3/8	12.7	13	0	1.6	24	38	51	64	77	90	103
1/2	15.9	16	0	1.6	28	36	49	62	76	89	102
3/4	22.2	23	0	1.6	25	40	52	73	85	98	111
1	28.6	29	0	1.6	22	36	49	70	82	95	108
1 1/4	34.9	35	0	1.6	27	40	53	67	79	92	105
1 1/2	41.3	42	0	1.6	24	36	50	63	76	89	101
2	54.0	55	0	1.6	23	36	57	69	82	95	109
2 1/2	66.7	68	0	1.6	23	37	50	63	76	88	103
3	79.4	80	0	1.6	23	44	57	69	82	97	109
3 1/2	92.1	93	0	1.6	24	38	50	63	76	90	103
4	104.8	106	0.8	2.4	31	44	57	69	84	97	109
5	130.2	131	0.8	2.4	31	44	57	71	84	97	112
6	155.6	157	0.8	2.4	31	43	58	71	83	99	112

Table B-22 Digest of simplified thickness standards – English Units

SIZE	COPPER TUBING Nominal 1" Thick	Nominal 1" Thick	Nominal 1½" Thick	Nominal 2" Thick	NPS PIPE Nominal 2½" Thick	Nominal 3" Thick	Nominal 4" Thick	Nominal 4½" Thick
¼"		⅞"		1 15/16" ¼" N 1" I.L. 2" N 1" O.L.				
⅜"	⅞"	1⅛"		2 3/16" ⅜" N 1" I.L. 2½" N 1" O.L.		3" ⅜" N 1" I.L. 2½" N 1½" O.L.	4" ⅜" N 1" I.L. 2½" N 1" I.L. 4½" N 1½" O.L.	4½" ⅜" N 1" I.L. 2½" N 1" I.L. 4½" N 2" O.L.
½"	1⅛"	1"	1 9/16"	2 1/16"	2⅞" ½" N 1" I.L. 2½" N 1½" O.L.	3⅜" ½" N 1½" I.L. 3½" N 1½" O.L.	4⅜" ½" N 2" I.L. 4½" N 2" O.L.	4½" ½" N 1" I.L. 2½" N 1½" I.L. 6" N 1½" O.L.
¾"	1"	29/32"	1 15/32"	1 31/32"	2 25/32" ¾" N 1" I.L. 2½" N 1½" O.L.	3 9/32" ¾" N 1½" I.L. 3½" N 1½" O.L.	4 9/32" ¾" N 2" I.L. 4½" N 2" O.L.	4 9/32" ¾" N 1" I.L. 2½" N 1½" I.L. 6" N 1½" O.L.
1"	1 3/16"	1 3/32"	1 19/32"	2⅛"	2 21/32" 1" N 1" I.L. 3" N 1½" O.L.	3 5/32" 1" N 1½" I.L. 4" N 1½" O.L.	4 5/32" 1" N 2" I.L. 5" N 2" O.L.	4 5/32" 1" N 1" I.L. 3" N 1½" I.L. 6" N 1½" O.L.
1¼"	1 1/16"	29/32"	1 21/32"	1 15/16"	2 15/32" 1¼" N 1" I.L. 3" N 1½" O.L.	2 21/32" 1¼" N 1" I.L. 3" N 2" O.L.	3 31/32" 1¼" N 2" I.L. 5" N 2" O.L.	4½" 1¼" N 1" I.L. 3" N 1½" I.L. 6" N 2" O.L.
1½"	1 3/16"	1 1/32"	1 17/32"	2 5/16"	2 27/32" 1½" N 1" I.L. 3½" N 1½" O.L.	3 11/32" 1½" N 1½" I.L. 4½" N 1½" O.L.	4 11/32" 1½" N 2" I.L. 6" N 2" O.L.	4 21/32" 1½" N 2" I.L. 6" N 2½" O.L.
2"	1 3/16"	1 1/16"	1 19/32"	2⅛"	2⅝" 2" N 1" I.L. 4" N 1½" O.L.	3⅛" 2" N 1½" I.L. 5" N 1½" O.L.	4 5/32" 2" N 2" I.L. 6" N 2" O.L.	4 21/32" 2" N 2" I.L. 6" N 2½" O.L.
2½"	1 3/16"	1 1/16"	1⅞"	2 11/32"	2⅞" 2½" N 1" I.L. 4½" N 1½" O.L.	3⅜" 2½" N 1½" I.L. 6" N 1½" O.L.	4⅜" 2½" N 2" I.L. 7" N 2" O.L.	4 29/32" 2½" N 2" I.L. 7" N 2½" O.L.
3"	1 7/32"	1 1/32"	1 9/16"	2 1/16"	2 9/16" 3" N 1" I.L. 5" N 1½" O.L.	3 1/16" 3" N 1½" I.L. 6" N 1½" O.L.	4 3/32" 3" N 2" I.L. 7" N 2" O.L.	4 31/32" 3½" N 2" I.L. 8" N 2½" O.L.
3½"		1 9/32"	1 13/16"	2 9/32"	2 25/32" 3½" N 1" I.L. 6" N 1½" O.L.	3 11/32" 3½" N 1½" I.L. 7" N 1½" O.L.	4 5/16" 3½" N 2" I.L. 8" N 2" O.L.	4 29/32" 3½" N 1" I.L. 6" N 2" I.L. 10" N 1½" O.L.
4"		1 1/16"	1 9/16"	2 1/16"	2 9/16"	3 3/32"	4 3/32" 4" N 2" I.L. 8" N 2" O.L.	4¾" 4" N 2" I.L. 8" N 2½" O.L.
4½"		1 9/32"	1 13/16"	2 5/16"	2 13/16" 4½" N 1" I.L. 7" N 1½" O.L.	3 11/32" 4½" N 1½" I.L. 8" N 1½" O.L.	3 27/32" 4½" N 1½" I.L. 8" N 2" O.L.	4½" 4½" N 1½" I.L. 8" N 2½" O.L.
5"		1 1/32"	1 17/32"	2 1/32"	2 9/16" 5" N 1" I.L. 7" N 1½" O.L.	3 1/16" 5" N 1½" I.L. 8" N 1½" O.L.	4 7/32" 5" N 1½" I.L. 8" N 2½" O.L.	4 3/16" 5" N 2" I.L. 9" N 2" O.L.
6"		1"	1½"	2 1/32"	2 17/32"	3 1/32"	4⅛" 6" N 2" I.L. 10" N 2" O.L.	4⅝" 6" N 2" I.L. 10" N 2½" O.L.

DEFINITIONS FOR SIMPLIFIED THICKNESS CHARTS

Prime Sizes

Single layer construction.

Assemblies

All other thicknesses, assembled from "Prime Sizes."

Nominal Thicknesses

Term commonly used to group actual thicknesses, under nearest regular thickness headings. For example (see Chart above left)—take 2" pipe size: Nominal 1" thick is actually 1 1/16" thick, and Nominal 3" thick is actually 3⅛" thick, etc.

Table B-22 (continued) – English Units

SIZE	NPS PIPE						
	Nominal 1½" Thick	Nominal 2" Thick	Nominal 2½" Thick	Nominal 3" Thick	Nominal 3½" Thick	Nominal 4" Thick	Nominal 4½" Thick
7"	1 17/32"	2 1/32"	2 9/16"	3⅛" 7" N 1½" I.L. 10" N 1½" O.L.	3⅝" 7" N 1½" I.L. 10" N 2" O.L.	4⅛" 7" N 2" I.L. 11" N 2" O.L.	4 21/32" 7" N 2½" I.L. 12" N 2" O.L.
8"	1 17/32"	2 1/32"	2 11/16"	3⅛"	3⅝" 8" N 1½" I.L. 11" N 2" O.L.	4⅛" 8" N 2" I.L. 12" N 2" O.L.	4½" 8" N 2" I.L. 12" N 2½" O.L.
9"	1 17/32"	2 5/32"	*2½"	3⅛" 9" N 1½" I.L. 12" N 1½" O.L.	3⅝" 9" N 1½" I.L. 12" N 2" O.L.	4 5/32" 9" N 2" I.L. 14" N 2" O.L.	4 21/32" 9" N 2" I.L. 14" N 2½" O.L.
10"	1 19/32"	2 3/32"	2 19/32"	3 3/32"	3 19/32" 10" N 1½" I.L. 14" N 2" O.L.	4 3/32" 10" N 2" I.L. 15" N 2" O.L.	4 19/32" 10" N 2½" I.L. 16" N 2" O.L.
11"	1 19/32"	2 3/32"	*2½"	3 3/32" 11" N 1½" I.L. 15" N 1½" O.L.	3 19/32" 11" N 1½" I.L. 15" N 2" O.L.	4 3/32" 11" N 2" I.L. 16" N 2" O.L.	4 9/32" 11" N 2" I.L. 16" N 2½" O.L.
12"	1 19/32"	2 3/32"	2⅝"	3 3/32"	3 19/32" 12" N 1½" I.L. 16" N 2" O.L.	4 3/32" 12" N 2" I.L. 17" N 2" O.L.	4½" 12" N 2" I.L. 17" N 2½" O.L.
14"	1½"	2"	2½"	3"	3½" 14" N 1½" I.L. 17" N 2" O.L.	4" 14" N 2" I.L. 18" N 2" O.L.	4½" 15" N 2" I.L. 19" N 2½" O.L.
15"	1½"	2"	*2½"	3" 15" N 1½" I.L. 18" N 1½" O.L.	3½" 15" N 1½" I.L. 18" N 2" O.L.	4" 15" N 1½" I.L. 18" N 2½" O.L.	4½" 15" N 1½" I.L. 18" N 3" O.L.
16"	1½"	2"	2½"	3"	3½" 16" N 1½" I.L. 19" N 2" O.L.	4" 16" N 2" I.L. 20" N 2" O.L.	4½" 16" N 2" I.L. 20" N 2½" O.L.
17"	1½"	2"	2½"	3" 17" N 1½" I.L. 20" N 1½" O.L.	3½" 17" N 1½" I.L. 20" N 2" O.L.	4" 17" N 2" I.L. 21" N 2" O.L.	4½" 17" N 2½" I.L. 22" N 2" O.L.
18"	1½"	2"	2½"	3"	3½" 18" N 2" I.L. 22" N 1½" O.L.	4"	4½" 18" N 2" I.L. 24" N 2½" O.L.
19"	1½"	2"	2½"	3" 19" N 1½" I.L. 22" N 1½" O.L.	3½" 19" N 1½" I.L. 22" N 2" O.L.	4" 19" N 1½" I.L. 22" N 2½" O.L.	4½" 19" N 1½" I.L. 22" N 3" O.L.
20"	1½"	2"	2½"	3"	3½" 20" N 2" I.L. 24" N 1½" O.L.	4" 20" N 2" I.L. 24" N 2" O.L.	4½" 20" N 2" I.L. 24" N 2½" O.L.
22"	1½"	2"	2½"	3"			
24"	1½"	2"	2½"	3"			

NOTES:

1. Prime Sizes and Thicknesses—Large Red Figures.
2. Assembled Sizes and Thicknesses—Large Black Figures.
3. Small Red Figures, shown immediately under each assembled thickness, are pipe sizes of layers used to make each assembled thickness. Thicknesses of such layers are indicated. For example: Pipe Size 12" and Nominal 3½" thick is actually 3 19/32" thick and is made by combining 12" N 1½" as the inner layer (I.L.) inside and 16" N 2" as the outer layer (O.L.).
4. **Sectional Insulation:**
   * "Precision Molded" Calcium Silicate—All the above sizes and thicknesses are furnished in sectional form except 9", 11" and 15" in 2½" thicknesses.

Reproduced with permission, from Heat Insulation Manual, Pabco Industrial Product Division, Fiberboard Corp.

Table B-23 Dimensions for mitered segments for short radius ells based on pipe insulation actual thickness – English Units

NOMINAL INSULATION THICKNESSES

Pipe Size	1"				1½"				2"				2½"			
	A	B	C	D	A	B	C	D	A	B	C	D	A	B	C	D
3"	3	3 3/32	⅛	2.22	—	—	—	—	—	—	—	—	—	—	—	—
3½"	3	3⅝	⅛	2.00	—	—	—	—	—	—	—	—	—	—	—	—
4"	3	3 15/16	5/16	1.66	2	6½	⅛	1.66	—	—	—	—	—	—	—	—
4½"	4	3 9/32	9/32	1.50	3	4 23/32	1/16	1.50	—	—	—	—	—	—	—	—
5"	4	3½	7/16	1.33	4	3 23/32	¼	1.33	3	5 9/32	1/16	1.33	—	—	—	—
6"	6	2 23/32	7/16	1.11	6	2 27/32	5/16	1.11	4	4½	¼	1.11	3	6⅜	1/16	1.11
7"	—	—	—	—	6	3¼	7/16	1.00	6	3⅜	5/16	1.00	4	5 5/16	¼	1.00
8"	—	—	—	—	6	3 21/32	9/16	.88	6	3 13/16	7/16	.88	6	3 31/32	¼	.88
9"	—	—	—	—	8	3 1/32	½	.75	8	3 3/32	⅜	.75	8	3¼	5/16	.75
10"	—	—	—	—	8	3⅜	19/32	.66	8	3 15/32	½	.66	8	3 9/16	13/32	.66
11"	—	—	—	—	8	3⅝	11/16	.66	8	3¾	19/32	.66	8	3 13/16	½	.66
12"	—	—	—	—	8	3 15/16	13/16	.58	8	4 1/32	11/16	.58	8	4⅛	19/32	.58
14"	—	—	—	—	8	4 7/16	1 3/16	.50	8	4 9/16	1 1/16	.50	8	4 11/16	15/16	.50
16"	—	—	—	—	8	5	1¼	.41	8	5⅛	1 3/16	.41	8	5 3/16	1⅛	.41
18"	—	—	—	—	8	5⅝	1½	.33	8	5 11/16	1⅜	.33	8	5¾	1¼	.33
20"	—	—	—	—	8	6⅛	1⅝	.28	8	6¼	1⅝	.28	8	6⅜	1½	.28
22"	—	—	—	—	8	6¾	1⅞	.28	8	6⅞	1¾	.28	8	7	1⅝	.28
24"	—	—	—	—	8	7⅜	2⅛	.25	8	7⅜	2	.25	8	7½	1⅞	.25

Pipe Size	3"				3½"				4"				4½"				5"			
	A	B	C	D	A	B	C	D	A	B	C	D	A	B	C	D	A	B	C	D
8"	3	8 5/16	7/32	.88	—	—	—	—	—	—	—	—	—	—	—	—	—	—	—	—
9"	6	4 15/32	¼	.75	6	4⅝	5/32	.75	—	—	—	—	—	—	—	—	—	—	—	—
10"	8	3 21/32	9/32	.66	8	3¾	7/32	.66	8	3⅞	⅛	.66	—	—	—	—	—	—	—	—
11"	8	3 29/32	⅜	.66	8	4	9/32	.66	8	4⅛	3/16	.66	8	4 7/32	⅛	.66	—	—	—	—
12"	8	4 7/32	½	.58	8	4 5/16	13/32	.58	8	4 7/16	5/16	.58	8	4 17/32	3/16	.58	8	4⅝	⅛	.58
14"	8	4 13/16	13/16	.50	8	4 15/16	11/16	.50	8	5 1/16	9/16	.50	8	5 3/16	7/16	.50	—	—	—	—
16"	8	5¼	1	.41	8	5 5/16	15/16	.41	—	—	—	—	—	—	—	—	—	—	—	—
18"	8	5⅞	1¼	.37																
20"	8	6½	1⅜	.29																
22"	8	7	1⅝	.29																
24"	8	7⅝	1¾	.20																

A = Number of Miters

B = Greater Dimension of outside face of pipe insulation

C = Lesser or throat dimension on outside face of pipe insulation

D = Number of short radius ells available from a linear foot of pipe insulation, based on alternating the cuts and allowing for saw kerf and waste

Dimensions are to nearest 1/32"

Reproduced with permission, from Heat Insulation Manual, Pabco Industrial Product Division, Fiberboard Corp.

Table B-24 Dimensions for mitered segments for long radius ells based on pipe insulation actual thickness – English Units

NOMINAL INSULATION THICKNESSES

Pipe Size	1″				1½″				2″				2½″				3″			
	A	B	C	D	A	B	C	D	A	B	C	D	A	B	C	D	A	B	C	D
2″	4	2	5/16	2.33	4	2 7/32	3/32	2.33	—	—	—	—	—	—	—	—	—	—	—	—
2½″	4	2 13/32	15/32	2.00	4	2 23/32	5/32	2.00	—	—	—	—	—	—	—	—	—	—	—	—
3″	4	2 13/16	21/32	1.66	4	3	15/32	1.66	3	4 5/16	11/32	1.66	3	4 13/32	3/32	1.66	—	—	—	—
3½″	4	3 9/32	¾	1.41	4	3 15/32	9/16	1.41	4	3 21/32	⅜	1.41	4	3 27/32	3/16	1.41	—	—	—	—
4″	6	2 7/16	11/16	1.20	6	2 17/32	9/16	1.20	6	2 11/16	7/16	1.20	6	2 13/16	5/16	1.20	4	4 9/16	¼	1.20
4½″	6	2¾	25/32	1.03	6	2⅞	⅝	1.03	6	3	½	1.03	6	3⅛	⅜	1.03	6	3¼	¼	1.03
5″	6	2 15/16	31/32	.93	6	3 1/16	27/32	.93	6	3 3/16	11/16	.93	6	3 11/32	9/16	.93	6	3 15/32	7/16	.93
6″	6	3½	1¼	.75	6	3⅝	1⅛	.75	6	3¾	1	.75	6	3⅞	⅞	.75	6	4	¾	.75
7″	—	—	—	—	6	4⅛	1 11/32	.66	6	4¼	1 7/32	.66	6	4⅜	1 3/32	.66	6	4½	15/16	.66
8″	—	—	—	—	8	3½	1 7/32	.58	8	3 19/32	1⅛	.58	8	3 23/32	1	.58	8	3 13/16	29/32	.58
9″	—	—	—	—	8	3⅞	1 13/32	.53	8	4	1 9/32	.53	8	4 1/16	1 7/32	.53	8	4 3/16	1 3/32	.53
10″	—	—	—	—	8	4 9/32	1 9/16	.50	8	4 13/32	1 15/32	.50	8	4½	1⅜	.50	8	4 19/32	1 9/32	.50
11″	—	—	—	—	8	4 11/16	1¾	.43	8	4 25/32	1 21/32	.43	8	4⅞	1 19/32	.43	8	5	1 15/32	.43
12″	—	—	—	—	8	5 1/16	1 31/32	.41	8	5 5/32	1⅞	.41	8	5 9/32	1¾	.41	8	5⅜	1 21/32	.41
14″	—	—	—	—	8	5¾	2 7/16	.33	8	5⅞	2 11/32	.33	8	5 31/32	2¼	.33	8	6 1/16	2 5/32	.33
16″	—	—	—	—	8	6 17/32	2 27/32	.29	8	6 21/32	2¾	.29	8	6¾	2⅝	.29	8	6⅞	2 9/16	.29
18″	—	—	—	—	8	7 11/32	3 7/32	.25	8	7 7/16	3⅛	.25	8	7½	3 1/32	.25	8	7⅝	3	.25
20″	—	—	—	—	8	8⅛	3⅝	.20	8	8¼	3½	.20	8	8⅜	3½	.20	8	8⅜	3⅜	.20
22″	—	—	—	—	8	8⅞	4	.20	8	9	4	.20	8	9⅛	3⅞	.20	8	9¼	3¾	.20
24″	—	—	—	—	10	7¾	3½	.166	10	7⅞	3½	.166	10	7⅞	3⅜	.166	10	8	3¼	.166

Pipe Size	3½″				4″				4½″				5″			
	A	B	C	D	A	B	C	D	A	B	C	D	A	B	C	D
2″	—	—	—	—	—	—	—	—	—	—	—	—	—	—	—	—
2½″	—	—	—	—	—	—	—	—	—	—	—	—	—	—	—	—
3″	—	—	—	—	—	—	—	—	—	—	—	—	—	—	—	—
3½″	—	—	—	—	—	—	—	—	—	—	—	—	—	—	—	—
4″	—	—	—	—	—	—	—	—	—	—	—	—	—	—	—	—
4½″	6	3 7/16	3/32	1.03	—	—	—	—	—	—	—	—	—	—	—	—
5″	6	3⅝	5/16	.93	4	5⅝	9/32	.93	—	—	—	—	—	—	—	—
6″	6	4 3/16	17/32	.75	4	6 9/16	19/32	.75	4	6¾	13/32	.75	4	6 31/32	7/32	.75
7″	6	4 11/16	13/16	.66	6	4 25/32	11/16	.66	6	4 15/16	17/32	.66	6	5 1/16	13/32	.66
8″	8	3 29/32	13/16	.58	8	4	23/32	.58	8	4 3/32	19/32	.58	8	4 3/16	½	.58
9″	8	4 9/32	1	.53	8	4 13/32	29/32	.53	8	4 15/32	13/16	.53	8	4 19/32	23/32	.53
10″	8	4 11/16	1 3/16	.50	8	4 27/32	1 3/32	.50	8	4 29/32	1	.50	8	4 31/32	29/32	.50
11″	8	5 3/32	1⅜	.43	8	5 3/16	1 9/32	.43	8	5 9/32	1 3/16	.43	8	5⅜	1 3/32	.43
12″	8	5 15/32	1 9/16	.41	8	5 9/16	1 15/32	.41	8	5 21/32	1⅜	.41	8	5¾	1 9/32	.41
14″	8	6 5/32	2 1/16	.33	8	6¼	1 31/32	.33	8	6 11/32	1 27/32	.33	—	—	—	—
16″	8	6 15/16	2 7/16	.29	—	—	—	—	—	—	—	—	—	—	—	—
18″	—	—	—	—	—	—	—	—	—	—	—	—	—	—	—	—

A = Number of Miters

B = Greater Dimension on outside face of pipe insulation

C = Lesser or throat dimension on outside face of pipe insulation

D—Number of long radius ells available from a linear foot of pipe insulation based on alternating the cuts and allowing for saw kerf and waste.

Dimensions are to nearest 1/32″

Dimensions based on formula—Chord length $= 2\,r\,\mathrm{Sin}\,\frac{A^\circ}{2}$

Where r = radius to outside of insulation (where segments are circular graphically) A° = 90° Divided by number of segments used.

Reproduced with permission, from Heat Insulation Manual, Pabco Industrial Product Division, Fiberboard Corp.

Table B-25 Cement (in pounds) – English Units

Required for Standard Screwed or Welded Fittings – for 1" thickness

Pipe Size	90° Ells			45° Ells			Valves and Tees			Crosses			Pipe Diameter	
Nom. NPS	Calcium Silicate	Finishing Cement	Insulating Cement	Calcium Silicate	Finishing Cement	Insulating Cement	Calcium Silicate	Finishing Cement	Insulating Cement	Calcium Silicate	Finishing Cement	Insulating Cement	Inches	mm
1/2"	0.24	0.42	1.10	0.12	0.21	0.55	0.30	0.52	1.37	0.32	0.56	1.46	0.840	21.3
3/4"	0.28	0.50	1.33	0.14	0.25	0.66	0.36	0.62	1.66	0.37	0.66	1.77	1.050	26.7
1"	0.36	0.63	1.66	0.18	0.32	0.83	0.45	0.78	2.07	0.47	0.84	2.20	1.315	33.4
1 1/4"	0.54	0.94	2.50	0.27	0.47	1.25	0.67	1.17	3.12	0.71	1.25	3.33	1.660	42.2
1 1/2"	0.71	1.25	3.33	0.36	0.63	1.66	0.89	1.56	4.16	0.95	1.66	4.43	1.900	48.3
2"	0.83	1.50	4.40	0.42	0.75	2.20	1.04	1.87	5.50	1.10	2.00	5.85	2.375	60.3
2 1/2"	0.95	1.65	5.75	0.47	0.83	2.87	1.18	2.06	7.18	1.26	2.20	7.65	2.875	73.0
3"	1.43	2.50	6.60	0.71	1.25	3.30	1.79	3.12	8.25	1.89	3.33	8.77	3.500	88.9
3 1/2"	1.66	2.92	7.75	0.83	1.46	3.87	2.08	3.65	9.68	2.20	3.88	10.30	4.000	101.6
4"	1.86	3.25	8.60	0.93	1.62	4.30	2.32	4.07	10.75	2.48	4.32	11.40	4.500	114.3
4 1/2"	2.14	3.75	10.00	1.17	1.82	5.00	2.68	4.70	12.50	2.85	5.00	13.30	5.000	127.0
5"	2.42	4.20	11.30	1.21	2.10	5.65	3.04	5.25	14.10	3.22	5.58	15.00	5.568	141.8
6"	3.00	5.25	14.00	1.50	2.62	7.00	3.76	6.56	17.50	4.00	6.98	18.60	6.625	168.3
7"	3.86	6.75	18.00	1.93	3.37	9.00	4.83	8.45	22.50	5.14	8.98	24.00	7.625	193.7
8"	4.15	7.20	19.30	2.08	3.60	9.65	5.18	9.00	24.10	5.51	9.57	25.70	8.625	219.1
9"	5.30	9.30	24.60	2.65	4.65	12.30	6.65	11.60	30.75	7.03	12.36	32.70	9.625	244.5
10"	6.14	10.70	28.60	3.08	5.35	14.30	7.67	13.30	35.75	8.15	14.23	38.00	10.750	273.0
11"	7.00	12.25	32.50	3.50	6.12	16.25	8.75	15.20	40.60	9.35	16.29	43.20	11.750	298.4
12"	7.86	13.80	36.60	3.94	6.90	18.30	9.80	17.20	45.75	10.48	18.35	48.70	12.750	323.8
14"	8.70	15.20	40.50	4.35	7.60	20.25	10.88	19.05	50.55	11.60	20.30	54.00	14.000	355.6
16"	10.45	18.30	48.50	5.23	9.15	24.25	12.92	22.65	60.20	13.92	24.40	64.80	16.000	431.8
17"	11.30	19.80	52.50	5.65	9.90	26.25	14.00	24.50	65.20	15.10	26.40	70.20	17.000	431.8
18"	12.10	21.20	56.30	6.05	10.60	28.15	15.10	26.40	70.03	16.28	28.50	75.60	18.000	457.2

Reproduced with permission, from Heat Insulation Manual, Pabco Industrial Product Division, Fiberboard Corp.
To obtain metric units multiply pounds, listed above, by 0.4536 to obtain kg's.

Table B-26 Areas of standard weight flanged fittings – English Units

Area–square feet per fitting (including single flanges)
150# steel fittings

Pipe Size	Flanged Coupling	90 Deg. Ell	Long Radius Ell	Tee	Cross	Pipe Diameter	
Nom. NPS						inches	mm
1"	.320	.795	.892	1.235	1.622	1.315	33.4
1 1/4"	.383	.957	1.084	1.481	1.943	1.660	42.2
1 1/2"	.477	1.174	1.337	1.815	2.38	1.900	48.3
2"	.672	1.65	1.84	2.54	3.32	2.375	60.3
2 1/2"	.841	2.09	2.32	3.21	4.19	2.875	73.0
3"	.945	2.38	2.68	3.66	4.77	3.000	88.0
3 1/2"	1.122	2.98	3.28	4.48	5.83	4.000	101.6
4"	1.344	3.53	3.96	5.41	7.03	4.500	114.3
4 1/2"	1.474	3.95	4.43	6.07	7.87	5.000	127.0
5"	1.622	4.44	5.00	6.81	8.82	5.568	141.3
6"	1.82	5.13	5.99	7.84	10.08	6.625	168.3
7"	2.17	6.17	7.38	9.37	12.00	7.625	193.7
8"	2.41	6.98	8.56	10.55	13.44	8.625	219.1
9"	3.00	8.71	10.57	13.18	16.78	9.625	244.5
10"	3.43	10.18	12.35	15.41	19.58	10.750	273.0
12"	4.41	13.08	16.35	19.67	24.87	12.750	323.8
14"	5.39	16.38	20.17	24.81	31.48	14.000	355.6
15"	6.18	18.50	22.92	27.91	35.48	15.000	381.0
16"	6.69	20.17	25.41	30.32	38.34	16.000	406.4

Reproduced with permission, from Heat Insulation Manual, Pabco Industrial Product Division, Fiberboard Corp.
To obtain metric units multiply area in square feet, listed above, to 0.0929 to obtain square meters.

Equations B-1

Methods of Determining Angle of Cut for Mitered Flat Block Insulation for Cylindrical Vessels

English Units

Where cylindrical vessels are to be insulated with flat blocks, in mitered form, the outside maximum width of the mitered block should be in accordance with the following table:

Diameter of Vessel	Maximum Outside Width of Miter
14″ to 24″	3″
25″ to 30″	$3^1/_2$″
31″ to 36″	4″
37″ to 42″	$4^1/_2$″
43″ to 48″	5″
49″ to 54″	$5^1/_2$″
55″ and larger	6″

To determine the angle of cut for the mitered blocks:

1. Add twice the thickness of the block to be used to the diameter of the vessel. Multiply this outside diameter of the insulation by 3.1416 to find the outside circumference of the insulation.
2. Divide circumference of insulation by width of miter from table above to determine number of miters, if this number is not whole, increase to next largest number. Multiply by two.
3. Divide 360° by above result, which will give angle of cut for miter.

Example:

A 48″ diameter vessel is to be insulated with 2″ thick flat blocks cut into lags or mitered form. Outside diameter of insulation = 48″ + 2″ + 2″ = 52″. Circumference of outside of insulation = 52″ × 3.1416 = 163.36″

From the above table the outside maximum width of miter for a 48″ diameter vessel is 5″.

$$\frac{163.36''}{5} = 32.67 \text{ pieces}$$

Increase to 33 pieces.

Outside width of miter =

$$\frac{163.36}{33} = 4.95'' = \text{approximately } 4\frac{61}{64}''$$

Angle of cut on each side of miter =

$$\frac{360^\circ}{2 \times 33 \text{ pcs.}} = \frac{360^\circ}{66} = 5.45^\circ$$

NOTE: Sectional pipe insulation, where available, may be used more economically for 18″ and smaller diameter vessels. Curved equipment blocks are available to insure snug fit for equipment up to 54″ in diameter.

Equations B-2

English or Metric Units

FORMULAE FOR AREAS ....

Area of circle = (3.1416) r^2

Area of ellipse = (3.1416) a b where a = $^1/_2$

Major axis and b = $^1/_2$ Minor axis

Area of equal circle = Area of square × 1.273

Area of right triangle = $^1/_2$ base times the altitude

Area of equilateral triangle = .43301 a^2 where a = Length of one side

Area of parallelogram = b^h where b = Length of greater side times perpendicular height between the two greater sides.

Areas of Bumped Heads of Vessels

Depth of Rise		Area
$\frac{D}{2}$	=	1.571 D^2
$\frac{D}{3}$	=	1.24 D^2
$\frac{D}{4}$	=	1.084 D^2

Note: D = Outside diameter of vessel.

Formulae For Surface and Volumes of Solids

Surface of sphere = (3.1416) d^2

Volume of sphere = 4/3 (3.1416) r^3 or Dia.^3 × .5236

Surface of right circular cylinder = 3.1416 × 2rh (where h = height) + 2 (3.1416) r^2

Volume of right circular cylinder = (3.1416) r^2 h

Surface of cone = (3.1416)d × $^1/_2$ slant height + (3.1416)r^2

Volume of cone = $^1/_3$ h (3.1416) r^2 (where h = Altitude)

Surface of frustrum of a cone = $^1/_2$ product of sum of circumference of two ends by the slant height plus areas of two ends.

Volume of frustrum of a cone = Sum of squares of diameters of the two ends plus the product of the two diameters times 0.7854; then the result multiplied by the altitude with the final result divided by 3.

Surface of a pyramid = Perimeter of base times one-half the slant height plus the area of the base.

Volume of a pyramid = $^1/_3$ area of base times the altitude or perpendicular height.

Surface of frustrum of a pyramid = Multiply the sum of perimeters of the two ends times the slant height of the frustrum and divide by two plus the area of the two ends.

To determine the volume of frustrum of a pyramid find the height that the pyramid would be if the top were put on, and then compute the volume of the completed pyramid and the volume of the part added; subtract the two volumes and the remainder would be the volume of the frustrum.

Table B-27 Surface areas of pipe insulation

English Units

Based on NPS pipe and ASTM dimensional standard/pipe insulation—square feet per linear foot													Pipe Diameter	
Nom. pipe size	Bare	Nominal insulation thickness — inches 1	1½	2	2½	3	3½	4	4½	5	5½	6	Inches	mm
1/8	0.106	0.62	0.92	1.18	1.46	1.73	2.00	2.26					0.405	10.3
1/4	0.141	0.75	1.05	1.30	1.46	1.73	2.00	2.26					0.504	12.8
3/8	0.177	0.75	1.05	1.30	1.46	1.73	2.00	2.26					0.695	17.1
1/2	0.220	0.75	1.05	1.30	1.73	2.00	2.25	2.52					0.840	21.3
3/4	0.275	0.75	1.05	1.30	1.73	2.00	2.25	2.52					1.050	26.7
1	0.344	0.92	1.18	1.46	1.74	2.00	2.26	2.52					1.315	33.4
1 1/4	0.435	0.92	1.31	1.46	1.74	2.00	2.26	2.52					1.660	42.2
1 1/2	0.498	1.05	1.31	1.74	2.00	2.26	2.52	2.81					1.900	48.3
2	0.622	1.18	1.46	1.73	2.00	2.26	2.52	2.81					2.375	60.3
2 1/2	0.753	1.31	1.73	2.00	2.26	2.52	2.81	3.08					2.875	73.0
3	0.917	1.46	1.73	2.00	2.26	2.52	2.81	3.08	3.34				3.500	88.9
3 1/2	1.047	1.73	2.00	2.26	2.52	2.81	3.08	3.34	3.67				4.000	101.6
4	1.178	1.73	2.00	2.26	2.52	2.81	3.08	3.34	3.67				4.500	114.3
4 1/2	1.309	2.00	2.26	2.52	2.81	3.08	3.34	3.67	3.67				5.000	127.0
5	1.456	2.00	2.26	2.52	2.81	3.08	3.34	3.67	3.93	4.19			5.563	141.3
6	1.734	2.26	2.52	2.81	3.08	3.34	3.67	3.93	4.18	4.45			6.625	168.3
7	1.996		2.81	3.08	3.34	3.67	3.93	4.19	4.45	4.71	4.97		7.625	198.7
8	2.258		3.08	3.34	3.67	3.93	4.19	4.45	4.71	4.97	5.24		8.625	219.1
9	2.520		3.34	3.67	3.93	4.19	4.45	4.71	4.97	5.24	5.50		9.625	244.5
10	2.814		3.67	3.93	4.19	4.45	4.71	4.97	5.24	5.50	5.76		10.750	273.0
11	3.076		3.93	4.19	4.45	4.71	4.97	5.24	5.50	5.76	6.02		11.750	298.4
12	3.338		4.19	4.45	4.71	4.97	5.24	5.50	5.76	6.02	6.28	6.54	12.750	323.8
14	3.665		4.45	4.71	4.97	5.24	5.50	5.76	6.02	6.28	6.54	6.81	14.750	855.6
16	4.189		4.97	5.24	5.50	5.76	6.02	6.28	6.54	6.81	7.07	7.33	16.000	406.4
18	4.712		5.50	5.76	6.02	6.28	6.54	6.81	7.07	7.33	7.59	7.85	18.000	459.2
20	5.236		6.02	6.28	6.54	6.81	7.07	7.33	7.59	7.85	8.12	8.38	20.000	508.0
22	5.759		6.54	6.81	7.07	7.33	7.59	7.85	8.12	8.38	8.64	8.90	22.000	558.8
24	6.283		7.07	7.33	7.59	7.85	8.12	8.38	8.64	8.90	9.16	9.42	24.000	609.5
26	6.807		7.59	7.85	8.12	8.38	8.64	8.90	9.16	9.42	9.69	9.95	26.000	6.604
28	7.331		8.12	8.38	8.64	8.90	9.16	9.42	9.69	9.95	10.21	10.47	28.000	711.2
30	7.854		8.64	8.90	9.16	9.42	9.69	9.95	10.21	10.47	10.73	11.00	30.000	762.0
Nominal insulation thickness — mm			25	38	51	64	76	89	102	127	140	152		

Table B-27a Surface areas of NPS pipe insulation

Based on NPS pipe and ASTM standard C-585 square metres per linear meter

Metric Units

Pipe			Nominal Insulation Thickness										
Nom Size Inches	Dia mm	Bare Area	1″ nom 25 mm	1½″ nom 38 mm	2″ nom 51 mm	2½″ nom 64 mm	3″ nom 76 mm	3½″ nom 89 mm	4″ nom 102 mm	4½″ nom 114 mm	5″ nom 127 mm	5½″ nom 140 mm	6″ nom 152 mm
1/8	10.3	0.032	0.189	0.280	0.360	0.445	0.527	0.610	0.689				
1/4	13.7	0.043	0.229	0.320	0.399	0.445	0.527	0.610	0.689				
3/8	17.1	0.054	0.229	0.320	0.399	0.445	0.527	0.610	0.689				
1/2	21.3	0.068	0.229	0.320	0.399	0.527	0.610	0.689	0.768				
3/4	26.7	0.084	0.229	0.320	0.399	0.527	0.610	0.689	0.768				
1	33.4	0.105	0.280	0.360	0.445	0.527	0.610	0.689	0.768				
1 1/4	42.2	0.133	0.280	0.399	0.445	0.527	0.610	0.689	0.768				
1 1/2	48.3	0.152	0.320	0.399	0.527	0.610	0.689	0.768	0.857				
2	60.3	0.190	0.360	0.445	0.527	0.610	0.689	0.768	0.857				
2 1/2	73.0	0.230	0.399	0.527	0.610	0.689	0.768	0.857	0.939				
3	88.9	0.210	0.445	0.527	0.610	0.689	0.768	0.857	0.939	1.018			
3 1/2	101.6	0.319	0.527	0.610	0.689	0.768	0.857	0.939	1.018	1.119			
4	114.3	0.359	0.527	0.610	0.689	0.768	0.857	0.939	1.018	1.119			
4 1/2	127.0	0.399	0.610	0.689	0.768	0.857	0.939	1.018	1.119	1.119			
5	141.3	0.444	0.610	0.689	0.768	0.857	0.939	1.018	1.119	1.198	1.277		
6	168.3	0.529	0.689	0.768	0.857	0.939	1.018	1.119	1.198	1.277	1.356		
7	193.7	0.608		0.857	0.939	1.018	1.119	1.198	1.277	1.356	1.436	1.515	
8	219.1	0.688		0.939	1.018	1.119	1.198	1.277	1.356	1.436	1.515	1.600	
9	244.3	0.768		1.018	1.119	1.198	1.277	1.356	1.436	1.515	1.600	1.676	
10	273.0	0.858		1.119	1.198	1.277	1.356	1.436	1.515	1.600	1.676	1.760	
11	298.4	0.938		1.198	1.277	1.356	1.436	1.515	1.600	1.676	1.760	1.834	
12	323.8	1.017		1.277	1.356	1.436	1.515	1.600	1.676	1.760	1.834	1.914	1.993
14	355.6	1.117		1.356	1.436	1.515	1.600	1.676	1.760	1.834	1.194	1.993	2.076
16	406.4	1.277		1.515	1.600	1.676	1.760	1.834	1.914	1.993	2.076	2.155	2.234
18	457.2	1.436		1.676	1.760	1.834	1.914	1.993	2.076	2.155	2.234	2.313	2.393
20	508.0	1.596		1.834	1.914	1.993	2.076	2.155	2.234	2.313	2.393	2.475	2.554
22	558.6	1.755		1.993	2.076	2.155	2.234	2.313	2.393	2.475	2.554	2.634	2.713
24	609.6	1.915		2.155	2.234	2.313	2.393	2.475	2.554	2.634	2.713	2.792	2.871
26	660.4	2.074		2.313	2.393	2.475	2.554	2.634	2.713	2.792	2.871	2.954	3.033
28	711.2	2.234		2.475	2.554	2.634	2.713	2.792	2.871	2.954	3.033	3.112	3.191
30	762.0	2.394		2.634	2.713	2.792	2.871	2.954	3.033	3.112	3.191	3.271	3.353

Table 27b Surface areas of tube insulation

Based on tube size insulation in accordance with ASTM standard C-585 square feet per linear foot

English Units

Tube				Nominal Insulation Thickness – Inches						
Nom Size Inches	Dia inches	Dia mm	Bare Area	1″ nom	1½″ nom	2″ nom	2½″ nom	3″ nom	3½″ nom	4″ nom
3/8	0.500	12.7	0.131	0.62	0.92	1.18	1.46	1.73	2.00	2.26
1/2	0.625	15.9	0.164	0.75	0.92	1.18	1.46	1.73	2.00	2.26
3/4	0.875	22.2	0.229	0.75	1.05	1.31	1.73	2.00	2.26	2.52
1	1.125	28.6	0.295	0.75	1.05	1.31	1.73	2.00	2.26	2.52
1 1/4	1.375	34.9	0.360	0.92	1.18	1.46	1.73	2.00	2.26	2.52
1 1/2	1.625	41.3	0.425	0.92	1.18	1.46	1.73	2.00	2.26	2.52
2	2.125	54.0	0.556	1.05	1.31	1.73	2.00	2.26	2.52	2.81
2 1/2	2.625	66.7	0.687	1.18	1.46	1.73	2.00	2.26	2.52	2.81
3	3.125	79.4	0.818	1.31	1.73	2.00	2.26	2.52	2.81	3.08
3 1/2	3.265	92.1	0.949	1.46	1.73	2.00	2.26	2.52	2.81	3.08
4	4.125	104.8	1.079	1.73	2.00	2.26	2.52	2.81	3.08	3.34
5	5.125	130.2	1.342	2.00	2.26	2.52	2.81	3.08	3.34	3.67
6	6.125	155.6	1.603	2.26	2.52	2.81	3.08	3.34	3.67	3.93

Table 27c Surface area of tube insulation

Based on tube size insulation in accordance with ASTM C-585 square metres per linear metre

Metric Units

Tube			Nominal Insulation Thickness						
Nom Size Inches	Dia mm	Bare Area	1″ nom 25 mm	1½″ nom 38 mm	2″ nom 51 mm	2½″ nom 64 mm	3″ nom 76 mm	3½″ nom 89 mm	4″ nom 102 mm
3/8	12.7	0.040	0.189	0.280	0.360	0.445	0.527	0.610	0.689
1/2	15.9	0.050	0.229	0.280	0.360	0.445	0.527	0.610	0.689
3/4	22.2	0.070	0.070	0.320	0.399	0.527	0.610	0.689	0.768
1	23.6	0.090	0.229	0.320	0.399	0.527	0.610	0.689	0.768
1 1/4	34.9	0.110	0.280	0.360	0.445	0.527	0.610	0.689	0.768
1 1/2	41.3	0.130	0.280	0.360	0.445	0.527	0.610	0.689	0.768
2	54.0	0.169	0.320	0.399	0.527	0.610	0.689	0.768	0.857
2 1/2	66.7	0.209	0.360	0.445	0.527	0.610	0.689	0.768	0.857
3	79.4	0.249	0.399	0.527	0.610	0.689	0.768	0.857	0.939
3 1/2	92.1	0.289	0.445	0.527	0.610	0.689	0.768	0.857	0.939
4	104.8	0.329	0.527	0.610	0.689	0.768	0.857	0.939	1.018
5	130.2	0.409	0.610	0.689	0.768	0.857	0.939	1.018	1.112
6	155.6	0.489	0.689	0.768	0.857	0.939	1.018	1.112	1.198

Table B-28 Outside surface area of prefabricated fitting covers — English Units

Fitting cover	Line pres psi	Nom pipe size inch	Outside surface area in square feet													
			Nominal insulation thickness—inches													
			1	1½	2	2½	3	3½	4	4½	5	5½	6	6½	7	7½
Line Flange	150	½	0.9	1.4	1.8	2.1	2.3	2.8	3.4	4.0	4.7	5.4	6.0	6.8	7.6	8.5
"	"	¾	0.9	1.4	2.1	2.1	2.3	2.9	3.4	4.0	4.7	5.4	6.1	6.8	7.6	8.5
"	"	1	1.2	1.8	2.4	2.5	2.8	3.3	3.9	4.6	5.3	6.0	6.7	7.5	8.4	9.2
"	"	1¼	1.6	2.1	2.7	3.0	3.3	3.9	4.6	5.3	6.0	6.8	7.5	8.3	9.2	10.1
"	"	1½	1.6	2.1	2.7	3.0	3.3	3.9	4.6	5.3	6.0	6.8	7.5	8.3	9.2	10.1
"	"	2	1.9	2.6	3.1	3.9	3.9	4.5	5.1	5.8	6.6	7.3	8.2	9.1	10.0	10.9
"	"	2½		3.0	3.5	3.8	4.8	4.9	5.6	6.3	7.0	7.8	8.7	9.6	10.5	11.5
"	"	3		3.0	3.6	4.5	4.8	4.9	5.6	6.4	7.1	8.0	8.8	9.7	10.7	11.6
"	"	3½		3.3	3.9	4.3	5.2	5.3	6.0	6.7	7.5	8.3	9.2	10.1	11.1	12.1
"	"	4		3.8	4.5	4.8	5.8	5.9	6.7	7.5	8.3	9.2	10.1	11.1	12.1	13.2
"	"	5		4.3	5.0	5.3	6.2	6.4	7.1	8.0	8.8	9.8	10.7	11.7	12.8	13.9
"	"	6		4.6	5.3	5.6	6.7	6.7	7.5	8.4	9.3	10.2	11.2	12.2	13.3	14.4
"	"	8		5.4	6.1	6.4	7.5	7.6	8.4	9.3	10.2	10.8	11.8	12.9	14.0	15.1
"	"	10		6.2	7.0	7.3	8.4	8.5	9.5	10.4	11.4	12.5	13.6	14.7	15.8	17.0
"	"	12		7.7	8.6	9.0	9.4	10.6	11.3	12.4	13.5	14.6	15.7	16.9	18.2	19.5
"	"	14		9.2	10.2	10.6	11.2	12.6	14.1	14.3	15.4	16.6	19.3	19.3	20.5	21.9
"	"	16		10.9	11.9	12.3	12.8	13.9	15.1	17.4	17.5	18.8	20.2	21.6	23.0	24.5
"	"	18		11.1	12.2	12.6	13.1	14.2	16.5	16.6	18.5	19.2	20.2	21.9	23.4	24.8
"	"	20		13.2	14.4	14.8	15.2	16.5	17.8	20.3	20.5	21.9	23.4	24.9	26.4	28.0
"	"	24		15.6	16.8	17.3	17.8	19.4	20.5	23.3	23.4	24.9	26.5	28.1	29.7	31.4
"	300	½	1.2	1.7	2.6	2.4	2.7	3.2	3.8	4.5	5.1	5.8	6.5	7.3	8.1	9.3
"	"	¾	1.6	2.2	2.7	3.0	3.2	3.8	4.6	5.2	5.9	6.7	7.4	8.2	9.1	10.4
"	"	1	1.6	2.2	2.7	3.0	3.2	3.8	4.6	5.2	5.9	6.7	7.4	8.2	9.1	10.4
"	"	1¼	1.6	2.2	2.7	3.5	3.5	3.9	4.7	5.3	6.0	6.8	7.5	8.4	9.3	10.4
"	"	1½	1.9	2.3	3.1	3.5	4.3	4.4	5.0	5.7	6.5	7.2	8.0	8.9	9.8	10.8
"	"	2	2.0	2.7	3.3	4.1	4.3	4.7	5.3	6.0	6.7	7.5	8.4	9.3	10.2	11.2
"	"	2½		3.1	3.7	4.0	5.0	5.1	5.8	6.6	7.3	8.1	9.0	9.9	10.9	11.9
"	"	3		3.7	4.3	4.8	5.7	5.8	6.6	7.4	8.1	9.1	10.0	11.0	11.9	13.0
"	"	3½		4.1	4.8	5.2	5.5	7.0	7.1	7.8	8.7	9.6	10.5	11.5	12.6	13.6
"	"	4		4.7	5.4	5.8	6.1	7.7	7.8	8.6	9.5	10.5	11.5	12.5	13.6	14.7
"	"	5		5.3	5.9	6.2	6.6	8.2	8.3	9.2	10.1	11.1	12.1	13.2	14.3	15.5
"	"	6		5.8	6.3	6.7	7.2	8.7	8.8	9.7	10.7	11.7	12.8	13.9	15.0	16.2
"	"	8		6.6	7.5	7.8	8.3	9.9	11.0	11.1	12.6	13.2	14.3	15.4	16.6	17.9
"	"	10		8.6	9.5	10.0	10.4	11.4	13.4	13.5	14.7	15.8	17.1	18.3	19.6	21.0
"	"	12		10.4	11.4	11.9	12.3	13.4	14.6	16.8	17.0	18.2	19.5	20.9	22.3	23.7
"	"	14		11.8	12.9	13.4	13.9	15.1	16.3	17.5	20.0	20.1	21.5	22.9	24.3	25.8
"	"	16		14.2	15.4	15.9	16.4	17.7	19.0	20.4	21.7	22.4	22.7	24.9	26.4	28.0
"	"	18		15.4	16.2	17.2	17.7	19.0	20.4	21.8	23.2	26.0	26.3	27.8	29.5	31.1
"	"	20		17.9	19.2	19.7	20.3	21.7	23.2	24.7	26.3	27.9	30.9	31.2	32.9	34.5
"	"	24		22.1	23.6	24.2	24.8	26.4	28.0	29.7	31.4	33.1	34.9	38.3	38.6	40.5
"	400	½	1.3	1.8	2.5	2.5	2.8	3.3	3.9	4.6	5.3	6.0	6.7	7.5	8.3	9.5
"	"	¾	1.7	2.3	2.9	3.1	3.4	4.0	4.7	5.4	6.1	6.8	7.6	8.4	9.4	10.2
"	"	1	1.7	2.3	2.9	3.1	3.4	4.0	4.7	5.4	6.1	6.8	7.6	8.4	9.4	10.2
"	"	1¼	1.7	2.3	2.9	3.4	3.4	4.0	4.8	5.5	6.2	6.8	7.7	8.6	10.2	10.4
"	"	1½	2.4	2.7	3.2	3.6	4.1	4.6	5.3	6.0	6.7	7.5	8.3	9.2	10.2	11.1
"	"	2	2.4	2.7	3.3	4.1	4.1	4.7	5.4	6.1	6.9	7.6	8.5	9.4	10.3	11.3
"	"	2½		3.2	3.7	4.1	5.1	5.2	5.9	6.7	7.4	8.2	9.1	10.0	10.9	12.0
"	"	3		3.7	4.3	4.8	5.8	5.9	6.7	7.5	8.3	9.2	10.1	11.1	12.1	13.8
"	"	3½		4.2	5.0	5.4	5.7	6.7	7.3	8.0	8.9	9.8	10.8	11.8	12.8	13.9
"	"	4		4.7	5.6	6.0	6.3	7.5	8.0	8.8	9.8	10.7	11.7	12.8	13.9	15.0
"	"	5		5.4	6.1	6.5	6.9	8.5	8.6	9.5	10.5	11.5	12.5	13.7	14.7	15.9
"	"	6		5.8	6.1	7.0	7.5	9.0	9.2	10.1	10.5	12.3	13.2	14.3	15.0	16.7
"	"	8		6.9	7.8	8.2	8.6	10.3	10.4	11.5	12.5	13.6	14.7	15.9	17.1	18.4
"	"	10		8.8	9.7	10.2	10.7	11.7	13.7	13.9	15.0	16.2	17.4	18.7	19.7	21.3
"	"	12		10.9	12.2	12.5	12.9	14.0	15.2	17.5	17.7	18.9	20.2	21.7	23.1	24.5
"	"	14		12.4	13.5	14.0	14.5	15.7	17.0	18.2	20.7	20.9	22.3	23.7	25.2	26.8

To convert above listed units in square feet to square meters multiply by 0.0929.

Table B-28 Outside surface area of prefabricated fitting covers (continued) – English Units

Fitting cover	Line pres psi	Nom pipe size inch	Outside surface area in square feet — Nominal insulation thickness–inches: 1	1½	2	2½	3	3½	4	4½	5	5½	6	6½	7	7½
Line Flange	400	16		14.7	15.8	16.4	17.0	18.3	19.6	20.9	22.3	25.1	25.3	26.9	28.4	30.1
"	"	18		15.9	17.1	17.7	18.3	19.6	21.0	22.4	23.8	26.7	26.9	28.5	30.1	31.8
"	"	20		18.6	20.0	20.5	21.1	22.6	24.1	25.6	27.1	29.8	31.9	32.1	33.9	35.7
"	"	24		23.0	24.5	25.1	25.8	27.3	29.0	30.7	32.4	34.1	35.9	39.4	39.7	41.6
"	600	½	1.3	1.8	2.5	2.5	2.8	3.3	3.9	4.6	5.3	6.0	6.7	7.5	8.3	9.5
"	"	¾	1.7	2.3	2.8	3.1	3.4	4.0	4.7	5.4	6.1	6.8	7.6	8.4	9.7	10.4
"	"	1	1.7	2.3	2.8	3.1	3.4	4.0	4.7	5.4	6.1	6.8	7.6	8.4	9.7	10.4
"	"	1¼	1.7	2.3	2.8	3.6	3.6	4.0	4.8	5.5	6.2	6.8	7.7	8.6	10.2	11.1
"	"	1½	2.4	2.7	3.2	3.6	4.1	4.6	5.3	6.0	6.8	7.5	8.3	9.2	10.2	11.1
"	"	2	2.4	2.7	3.3	4.1	4.1	4.7	5.4	6.1	6.9	7.6	8.5	9.4	10.3	11.3
"	"	2½		3.2	3.7	4.7	4.7	5.2	5.9	6.7	7.4	8.2	9.1	10.0	11.0	12.0
"	"	3		3.7	4.3	4.8	5.8	5.9	6.7	7.5	8.3	9.2	10.1	11.1	12.1	13.1
"	"	3½		4.2	5.0	5.4	6.3	6.5	7.3	8.0	8.9	9.8	10.8	11.8	12.8	13.9
"	"	4		5.1	5.7	6.1	6.5	8.1	8.2	9.1	10.0	11.0	12.0	13.0	14.1	15.3
"	"	5		6.9	7.7	8.2	8.7	9.6	10.5	11.4	12.4	13.5	14.6	15.7	16.9	18.2
"	"	6		6.9	7.7	8.2	8.7	9.6	11.1	11.4	12.4	13.5	14.6	15.7	16.9	18.2
"	"	8		8.8	9.7	10.2	10.6	10.7	11.7	12.8	14.8	15.0	16.2	17.2	20.0	21.4
"	"	10		10.8	11.9	12.5	13.0	14.3	15.6	16.6	19.0	19.2	20.5	21.9	23.3	24.8
"	"	12		12.0	13.1	13.7	14.2	15.3	16.6	17.8	20.2	20.4	21.8	23.2	24.7	26.2
"	"	14		13.7	14.9	15.4	16.0	17.3	18.5	19.9	21.3	23.8	24.1	25.6	27.2	28.8
"	"	16		16.3	17.5	18.1	18.7	20.1	21.5	22.9	24.4	25.9	28.9	29.1	30.7	32.4
"	"	18		19.0	20.4	21.1	21.7	23.1	24.6	26.2	27.7	29.3	31.0	34.3	34.5	36.3
"	"	20		21.1	22.5	23.1	23.8	25.3	26.9	28.5	30.1	31.8	33.6	36.9	37.2	39.0
"	"	24		26.6	28.2	28.9	29.7	31.4	33.1	34.9	36.7	38.6	40.5	42.4	46.2	46.5
"	900	½	1.8	2.4	2.9	3.2	4.0	4.2	4.9	5.6	6.3	7.2	7.9	8.7	9.6	10.6
"	"	¾	1.8	2.5	3.0	3.3	4.4	4.4	5.0	5.7	6.4	7.2	7.9	8.8	9.7	11.1
"	"	1	2.3	2.9	3.6	3.9	4.9	5.8	5.8	6.5	7.3	8.1	9.0	9.9	10.9	11.9
"	"	1¼	2.3	2.9	3.6	3.9	4.9	5.8	5.8	6.5	7.3	8.1	9.0	9.9	10.9	11.9
"	"	1½		3.5	4.0	4.4	5.4	6.2	6.3	7.1	7.9	8.7	9.6	10.6	11.6	12.6
"	"	2		4.0	4.6	5.2	5.6	6.9	7.1	7.8	8.8	9.7	10.7	11.7	12.6	13.8
"	"	2½		4.5	5.4	5.8	6.2	7.0	7.5	8.7	9.5	10.5	11.5	12.5	13.6	14.7
"	"	3		4.5	5.4	5.8	6.2	7.6	7.7	8.7	9.5	10.5	11.5	12.5	13.6	14.7
"	"	3½		5.4	5.8	6.7	7.2	7.7	8.7	9.5	10.3	11.5	12.5	13.6	14.6	15.7
"	"	4		5.9	6.7	7.2	7.6	8.5	10.2	10.3	11.3	12.3	13.4	14.6	15.7	16.9
"	"	5		7.2	8.1	8.6	9.1	10.4	11.0	12.9	14.1	15.0	16.2	17.4	18.7	20.0
"	"	6		8.0	8.7	9.2	9.7	10.7	11.6	13.7	13.9	15.0	16.2	17.4	18.7	20.0
"	"	8		10.8	11.8	12.3	12.9	14.1	15.2	16.4	17.7	20.0	20.3	21.7	23.1	24.6
"	"	10		13.0	14.1	14.7	15.3	16.5	17.7	19.1	20.4	21.7	24.5	24.7	26.2	27.8
"	"	12		14.6	15.7	16.6	16.9	18.2	19.5	20.9	22.3	23.7	26.6	26.8	28.4	30.0
"	"	14		16.7	18.0	18.6	19.3	20.7	22.0	23.5	25.0	26.6	28.2	29.8	31.5	33.2
"	"	16		18.3	19.6	20.3	20.9	22.4	23.9	25.8	27.3	28.5	30.2	33.3	33.6	35.3
"	"	18		22.4	23.8	24.6	25.3	26.9	28.5	30.0	31.8	33.5	35.3	37.1	40.6	40.9
"	"	20		25.8	27.4	28.2	28.9	30.6	32.3	34.1	35.9	37.7	39.6	41.5	43.5	47.3
"	"	24		36.5	38.4	39.4	40.3	42.3	44.3	46.3	48.4	50.6	52.8	55.1	57.1	59.6
"	1500	½	1.8	2.4	2.9	3.2	3.5	4.2	4.9	5.6	6.3	7.1	7.9	8.7	9.6	10.6
"	"	¾	1.8	2.5	2.9	3.3	4.2	4.2	5.0	5.7	6.4	7.2	8.0	8.8	9.7	11.1
"	"	1	2.3	2.9	3.6	3.9	4.9	5.8	5.8	6.5	7.3	8.1	9.0	9.9	10.9	11.9
"	"	1¼	2.3	2.9	3.6	3.9	4.9	5.8	5.8	6.5	7.3	8.1	9.0	9.9	10.9	11.9
"	"	1½		3.5	4.0	4.4	5.4	6.2	6.3	7.1	7.9	8.7	9.6	10.6	11.6	12.6
"	"	2		4.1	4.6	5.2	5.6	7.0	7.1	7.8	8.9	9.7	10.7	11.7	12.6	13.8
"	"	2½		4.6	5.4	5.8	6.2	6.9	8.5	8.7	9.5	10.5	11.5	12.5	13.6	14.7
"	"	3		5.6	6.3	6.8	7.2	8.1	9.8	10.0	10.9	11.9	12.7	13.8	15.0	16.2
"	"	3½		6.3	6.8	7.7	8.1	8.6	10.0	11.9	11.9	12.7	13.8	15.0	16.2	17.2
"	"	4		7.0	7.7	8.2	8.7	9.8	10.7	12.5	12.7	13.7	15.0	16.2	17.2	18.6
"	"	5		8.9	10.5	10.5	10.9	12.0	13.0	14.1	16.3	16.6	17.8	19.1	20.4	21.8
"	"	6		9.8	10.7	11.2	11.9	12.8	14.0	15.2	16.4	17.7	18.9	20.3	21.6	23.1

To convert above listed units in square feet to square meters multiply by 0.0929.

Table B-28 (continued) – English Units

Fitting cover	Line pres psi	Nom pipe size inch	Outside surface area in square feet													
			Nominal insulation thickness—inches													
			1	1½	2	2½	3	3½	4	4½	5	5½	6	6½	7	7½
Line Flange	1500	8		12.5	13.6	14.2	14.8	16.1	17.3	18.6	19.9	22.5	22.7	24.5	25.7	27.1
"	"	10		16.6	17.8	18.6	19.3	20.7	22.0	23.5	25.0	26.4	28.1	31.0	31.4	33.1
"	"	12		21.4	22.9	23.7	24.4	26.0	27.6	29.2	30.9	32.6	34.3	36.1	38.0	41.5
"	"	14		25.6	27.2	28.1	28.9	30.6	32.3	34.1	36.0	37.8	39.7	41.6	43.7	45.7
"	"	16		29.9	31.7	32.6	33.6	35.4	37.2	39.1	41.0	43.0	45.1	47.2	49.3	51.4
"	"	18		34.6	36.5	37.0	40.4	42.3	44.4	46.4	48.5	50.2	52.9	55.1	57.4	60.0
"	"	20		39.9	41.9	43.0	44.0	46.1	48.2	50.4	52.6	54.8	57.1	59.4	61.8	64.2
"	"	24		52.6	54.8	56.1	57.2	59.6	62.1	64.5	67.0	69.5	72.1	74.7	77.3	79.2
Screwed Globe Valve		½	1.2	1.2	1.6	2.1	2.6	3.3	3.9	4.7	5.5	6.5	7.4	8.4	9.4	10.5
"		¾	1.3	1.3	1.7	2.3	2.8	3.4	4.1	4.8	5.7	6.7	7.7	8.6	9.7	10.8
"		1	1.5	2.1	2.7	3.0	3.6	4.8	5.2	6.0	7.0	8.0	9.0	10.0	11.0	12.4
"		1¼	1.9	1.9	2.5	3.0	3.6	4.8	5.2	6.0	7.0	8.0	9.0	10.0	11.0	12.4
"		1½	2.5	2.9	3.1	3.4	4.1	5.1	5.6	6.6	7.6	8.6	9.7	10.6	11.9	13.4
"		2	2.6	2.9	3.2	3.8	4.6	5.4	6.2	7.3	8.2	9.3	10.4	11.2	12.8	14.2
"		2½	3.7	3.7	4.4	5.2	6.0	7.0	8.0	9.1	10.2	11.4	12.6	13.9	15.2	15.9
"		3	3.9	3.9	4.7	5.6	6.4	7.4	8.4	9.5	10.6	11.8	13.1	14.4	15.7	17.0
"		3½														
"		4														
"		5														
"		6														
"		8														
"		10														
"		12														
"		14														
"		16														
"		18														
"		20														
"		24														
Flanged Globe Valve	150	½														
"	"	¾														
"	"	1														
"	"	1¼														
"	"	1½														
"	"	2	5.6	5.6	6.2	6.7	7.7	8.8	9.9	11.0	12.1	12.3	14.5	17.0	17.3	19.4
"	"	2½		6.8	8.0	8.0	9.5	10.3	11.4	12.6	13.8	15.1	16.5	18.0	19.4	21.0
"	"	3		7.2	8.1	8.8	9.6	10.8	11.8	13.1	14.3	15.7	17.1	18.5	20.0	21.0
"	"	3½		8.1	8.8	9.6	10.8	13.1	14.3	15.7	17.1	18.5	20.0	21.0	22.3	24.4
"	"	4		9.9	10.8	11.5	12.7	14.0	15.2	16.2	17.9	19.2	21.2	22.3	24.4	26.0
"	"	5		11.9	12.6	13.4	14.7	16.0	17.5	18.9	20.4	22.2	23.9	25.5	27.9	28.6
"	"	6		12.8	13.6	14.3	15.6	16.9	18.6	20.1	21.6	23.2	24.9	26.7	28.5	30.9
"	"	8		16.8	18.8	19.7	20.7	22.3	24.0	25.7	27.5	29.4	31.3	33.2	35.2	37.3
"	"	10		21.5	23.2	24.2	35.3	27.0	29.0	31.0	32.9	34.9	36.9	39.1	41.2	43.6
"	"	12		24.9	26.8	27.2	29.0	30.9	32.9	34.8	36.9	39.0	41.1	43.5	45.6	48.0
"	"	14														
"	"	16														
"	"	18														
"	"	20														
"	"	24														
"	300	½														
"	"	¾														
"	"	1	2.9	3.8	4.5	5.2	5.8	6.7	7.9	8.9	10.1	11.3	12.5	13.8	15.2	16.4
"	"	1¼	3.3	4.0	4.8	5.3	5.8	6.7	7.9	8.9	10.1	11.3	12.5	13.8	15.2	16.2

To convert above listed units in square feet to square meters multiply by 0.0929.

Table B-28 Outside surface area of prefabricated fitting covers (continued) – English Units

Fitting cover	Line pres psi	Nom pipe size inch	Outside surface area in square feet													
			Nominal insulation thickness–inches													
			1	1½	2	2½	3	3½	4	4½	5	5½	6	6½	7	7½
Flanged Globe Valve	300	1½	3.3	4.3	5.0	5.6	6.1	6.9	8.0	9.0	10.1	11.3	12.5	13.8	15.2	16.2
"	"	2		7.1	8.1	8.4	9.4	10.7	11.8	13.2	15.7	15.7	17.1	18.5	20.0	21.5
"	"	2½		8.3	9.4	10.3	10.6	12.2	13.4	14.7	16.1	17.5	19.0	20.5	22.0	23.7
"	"	3		9.4	10.8	11.2	12.2	13.5	14.8	16.2	17.6	19.1	20.7	22.1	23.8	25.6
"	"	3½		11.6	12.9	13.8	14.4	15.8	17.3	18.7	20.2	21.8	23.4	25.2	26.9	28.7
"	"	4		14.3	15.9	16.5	17.4	19.0	20.4	22.0	23.6	25.4	26.7	29.7	30.9	32.8
"	"	5		14.7	16.5	17.4	19.0	20.4	20.8	21.6	24.9	26.7	28.0	30.4	32.3	34.0
"	"	6		15.2	16.6	17.5	18.5	20.0	21.6	23.3	24.9	26.7	28.0	30.4	32.3	34.3
"	"	8		20.8	22.5	23.4	24.4	26.2	28.0	30.0	31.9	33.8	35.9	38.0	40.1	42.4
"	"	10		27.4	29.1	30.7	32.3	33.7	35.7	37.8	40.0	42.2	45.4	46.8	49.2	51.7
"	"	12		36.8	38.9	40.3	41.6	44.0	46.3	48.9	51.2	53.7	56.3	58.9	61.6	64.3
"	"	14														
"	"	16														
"	"	18														
"	"	20														
"	"	24														
"	400	½														
"	"	¾														
"	"	1														
"	"	1¼														
"	"	1½	6.5	6.8	7.6	7.8	8.5	9.7	10.9	11.5	12.6	13.9	15.2	16.6	18.0	19.5
"	"	2		7.9	8.8	9.6	10.6	11.6	13.0	14.3	15.6	17.0	18.5	20.0	21.6	23.2
"	"	2½		9.3	10.3	11.2	12.7	13.2	14.5	15.9	17.3	18.8	20.3	21.9	23.5	25.3
"	"	3		9.8	10.9	11.9	12.8	14.0	15.4	16.8	18.3	19.8	21.4	23.0	24.7	26.4
"	"	3½		11.9	12.8	14.0	15.4	16.8	17.3	18.3	19.8	21.4	23.0	24.7	26.4	28.9
"	"	4		13.8	15.1	14.8	16.5	17.1	18.7	19.8	21.9	23.6	25.3	27.0	28.9	30.9
"	"	5		15.8	17.1	18.1	19.1	21.0	22.7	25.1	26.2	27.9	29.8	31.9	33.9	35.9
"	"	6		20.4	22.2	23.3	24.4	26.1	27.9	29.8	31.7	33.9	36.4	38.0	40.2	42.4
"	"	8		25.0	26.9	28.0	28.1	31.2	33.2	35.3	37.4	39.5	41.8	43.5	46.0	48.2
"	"	10														
"	"	12														
"	"	14														
"	"	16														
"	"	18														
"	"	20														
"	"	24														
"	600	½														
"	"	¾														
"	"	1	3.3	4.3	5.2	5.9	6.6	7.6	8.8	9.8	11.1	12.4	13.7	15.0	16.5	18.0
"	"	1¼	3.9	4.9	5.4	6.6	7.2	7.6	9.4	10.5	11.6	12.9	14.1	15.5	16.9	18.3
"	"	1½		6.9	7.9	8.5	9.4	10.5	11.7	12.8	14.0	15.5	16.8	18.3	19.7	21.2
"	"	2		7.7	8.7	9.4	10.3	11.5	12.7	14.0	15.3	16.7	18.2	19.7	21.2	22.8
"	"	2½		9.9	11.1	11.7	12.6	13.9	15.2	16.7	18.0	19.5	21.1	22.7	24.4	26.1
"	"	3		10.2	11.7	12.5	13.3	14.6	16.0	17.5	18.9	20.5	22.1	23.7	25.4	27.2
"	"	3½		11.7	12.5	13.6	14.6	16.0	17.5	18.9	20.5	22.1	23.7	25.4	27.2	29.3
"	"	4		12.9	14.2	15.3	16.0	17.5	19.1	20.6	22.2	23.9	25.5	27.4	29.3	31.1
"	"	5		18.0	19.5	20.6	21.6	23.3	25.0	26.7	28.7	30.6	32.6	34.5	36.6	38.7
"	"	6		20.2	22.0	23.1	24.1	25.8	27.6	29.6	31.6	33.6	35.6	37.7	39.9	42.0
"	"	8		30.0	32.0	33.3	34.5	36.6	38.8	41.9	43.2	45.7	47.9	50.8	52.8	55.4
"	"	10														
"	"	12														
"	"	14														
"	"	16														
"	"	18														
"	"	20														
"	"	24														

To convert above listed units in square feet to square meters multiply by 0.0929.

Table B-28 (continued) – English Units

Fitting cover	Line pres psi	Nom pipe size inch	Outside surface area in square feet													
			Nominal insulation thickness–inches													
			1	1½	2	2½	3	3½	4	4½	5	5½	6	6½	7	7½
Flanged Globe Valve	900	½														
"	"	¾														
"	"	1														
"	"	1¼														
"	"	1½		9.7	10.8	11.9	12.7	14.0	15.3	16.8	18.3	19.8	21.4	23.0	24.7	26.6
"	"	2		13.7	15.1	16.1	17.1	18.6	20.3	22.0	23.7	25.4	27.1	29.1	31.0	33.0
"	"	2½		15.7	16.9	18.0	19.0	19.7	22.4	24.1	25.9	27.6	29.6	32.1	34.1	36.2
"	"	3		15.7	16.9	18.0	19.0	19.9	22.6	24.5	26.3	27.9	30.3	33.3	34.7	36.8
"	"	3½		16.9	18.0	19.0	19.9	22.6	23.2	25.9	27.9	29.6	31.8	33.8	35.0	37.4
"	"	4		17.5	19.1	20.2	21.3	23.1	24.8	26.6	28.4	30.4	32.4	34.4	36.5	38.6
"	"	5		21.0	22.8	24.0	24.6	27.1	29.0	30.9	32.1	35.1	37.3	39.4	41.7	44.0
"	"	6		25.4	27.3	28.5	29.8	31.7	33.9	36.1	38.2	40.5	42.7	45.1	48.2	50.1
"	"	8		35.4	37.4	38.9	40.6	42.9	45.4	45.9	50.3	53.0	56.2	58.1	60.9	63.6
"	"	10														
"	"	12														
"	"	14														
"	"	16														
"	"	18														
"	"	20														
"	"	24														
"	1500	½														
"	"	¾														
"	"	1														
"	"	1¼														
"	"	1½														
"	"	2		13.1	14.5	15.4	16.3	17.9	19.5	21.1	22.7	24.4	26.1	28.0	29.4	31.9
"	"	2½		14.2	15.7	16.6	17.6	19.1	20.8	22.5	24.1	25.9	27.7	29.6	31.6	33.6
"	"	3		19.4	21.1	22.4	23.6	25.5	27.2	29.0	31.0	33.1	35.7	37.3	39.4	41.5
"	"	3½		22.4	23.8	25.5	25.9	27.2	31.0	33.1	35.2	37.3	39.4	41.5	43.8	44.0
"	"	4		24.6	26.4	27.6	28.9	30.8	33.0	35.0	37.2	39.5	41.6	43.8	46.2	49.6
"	"	5		31.3	32.4	34.8	36.3	38.7	40.9	43.2	45.6	48.2	50.6	53.3	55.8	58.6
"	"	6		34.6	36.8	38.5	40.0	42.3	44.6	47.3	49.8	52.3	55.0	57.5	60.3	63.0
"	"	8														
"	"	10														
"	"	12														
"	"	14														
"	"	16														
"	"	18														
"	"	20														
"	"	24														
Screwed Gate Valve		½														
"		¾														
"		1														
"		1¼														
"		1½														
"		2	5.1	5.1	6.3	7.5	8.9	10.5	12.1	13.7	15.5	17.4	19.4	21.3	23.7	26.0
"		2½	6.2	6.2	7.5	8.8	10.4	12.0	13.6	15.4	17.3	19.3	21.4	23.6	26.0	28.3
"		3	7.3	7.3	8.6	10.2	11.8	13.4	15.2	17.0	19.1	21.2	23.3	25.7	28.0	30.6
"		3½	7.6	7.6	8.9	10.5	12.1	13.8	15.6	17.5	19.6	21.6	23.8	26.1	28.5	31.1
"		4	8.6	8.6	10.2	11.7	13.4	15.2	17.0	19.0	24.1	23.3	25.6	28.0	30.6	33.2
"		5		11.2	13.6	14.5	15.2	19.0	19.3	23.3	25.6	28.0	30.6	33.2	36.5	39.4
"		6		13.2	15.4	17.3	19.4	21.5	23.7	26.1	28.5	31.1	33.8	36.5	39.4	42.4

To convert above listed units in square feet to square meters multiply by 0.0929.

Table B-28 Outside surface area of prefabricated fitting covers (continued) – English Units

Fitting cover	Line pres psi	Nom pipe size inch	Outside surface area in square feet — Nominal insulation thickness—inches: 1	1½	2	2½	3	3½	4	4½	5	5½	6	6½	7	7½
Screwed Gate Valve		8		17.6	19.7	21.7	24.1	26.5	29.0	31.6	34.3	37.1	40.0	43.0	46.2	49.5
"		10		20.9	23.2	25.5	28.0	30.6	33.2	36.1	38.9	42.0	45.1	48.3	51.5	55.0
"		12														
"		14														
"		16														
"		18														
"		20														
"		24														
Flanged Gate Valve	150	½														
"	"	¾														
"	"	1	4.1	4.4	4.8	5.3	6.6	8.5	9.3	10.4	12.0	12.6	13.9	15.2	16.5	18.0
"	"	1¼	4.3	4.6	5.3	5.6	6.9	9.0	10.4	11.2	12.4	13.6	15.0	16.3	17.7	19.2
"	"	1½	4.6	4.9	5.7	6.3	7.4	10.1	11.3	12.4	13.6	14.9	16.3	12.7	19.2	20.6
"	"	2		5.0	6.1	7.2	8.6	10.2	11.8	13.4	15.4	17.4	19.5	21.8	24.1	26.3
"	"	2½		5.6	7.0	8.1	9.7	11.3	13.0	14.8	16.5	18.8	20.6	23.4	25.7	28.3
"	"	3		7.0	8.1	9.5	11.1	12.7	14.5	16.5	18.6	20.8	23.0	25.4	28.0	30.6
"	"	3½		7.3	8.4	9.6	11.2	12.9	14.7	16.5	18.7	20.5	23.2	25.6	27.6	30.6
"	"	4		8.4	9.7	11.0	12.5	14.3	16.2	18.3	20.5	22.8	25.2	27.7	30.3	33.0
"	"	5		10.0	11.5	12.9	15.3	16.3	17.2	20.5	22.8	25.2	27.8	30.4	33.1	36.0
"	"	6		13.0	13.7	16.0	17.9	19.9	22.2	24.6	27.1	29.7	32.0	35.3	38.2	41.3
"	"	8		16.7	18.3	20.3	21.8	24.1	26.1	29.2	32.0	34.8	37.3	40.8	44.0	49.1
"	"	10		19.8	21.5	23.4	25.4	27.5	30.2	32.9	35.9	38.9	41.9	45.3	48.6	52.1
"	"	12		25.5	27.4	30.2	31.7	33.9	36.6	39.6	42.8	46.2	49.5	53.1	55.2	60.1
"	"	14		26.7	31.8	34.0	36.3	39.2	41.9	43.8	47.2	50.7	54.4	58.0	61.8	63.4
"	"	16		39.8	42.8	45.4	48.0	50.8	53.7	56.5	59.8	64.1	68.1	72.5	76.5	81.1
"	"	18		41.2	43.7	46.4	49.0	51.8	54.7	57.7	61.3	65.1	69.3	73.5	77.9	82.3
"	"	20		48.1	50.9	53.4	56.7	59.5	62.4	65.9	6.97	73.9	78.3	82.8	88.4	89.3
"	"	24		60.0	62.5	67.2	69.3	72.6	76.1	79.5	83.7	88.5	93.5	98.3	103.2	108.6
"	300	½														
"	"	¾														
"	"	1	4.1	5.2	6.1	6.6	6.9	8.5	9.3	10.4	12.0	12.6	13.9	15.2	16.5	18.0
"	"	1¼	4.6	5.6	6.6	7.3	7.8	9.0	10.4	11.2	12.4	13.6	15.0	16.3	17.7	19.2
"	"	1½	6.8	6.8	7.6	8.2	9.0	10.1	11.3	12.4	13.6	14.9	16.3	17.7	19.2	20.6
"	"	2		7.2	8.1	8.7	9.6	10.7	11.8	13.0	14.3	16.0	17.1	18.5	20.0	21.5
"	"	2½		8.2	9.1	10.0	10.7	11.8	13.1	14.3	15.7	17.1	18.5	20.0	21.5	23.2
"	"	3		9.7	11.0	11.8	12.5	13.8	15.1	16.5	18.3	19.7	21.4	22.9	24.6	26.5
"	"	3½		9.8	11.2	11.8	12.5	13.8	15.1	16.5	18.3	19.7	21.4	22.9	24.6	26.5
"	"	4		12.2	13.4	14.2	14.9	16.4	17.9	19.3	20.9	22.4	24.1	25.9	27.6	29.4
"	"	5		14.7	16.0	16.6	17.8	19.2	20.8	22.4	24.0	25.8	27.6	29.4	31.3	33.2
"	"	6		16.8	17.9	19.2	19.8	21.3	23.0	24.6	26.4	28.3	30.1	32.0	34.0	36.1
"	"	8		20.8	22.6	23.4	24.4	26.2	28.0	29.9	31.8	33.7	35.7	37.8	40.0	42.1
"	"	10		27.1	28.7	29.8	31.2	33.2	35.2	37.3	39.3	41.7	43.9	46.1	48.2	50.8
"	"	12		35.1	37.2	38.5	39.6	42.0	43.9	46.4	48.8	51.1	53.5	56.1	58.7	61.4
"	"	14		40.0	48.6	50.2	51.3	54.1	56.6	59.2	62.0	64.7	67.6	69.6	73.3	75.8
"	"	16		55.0	57.5	59.2	60.7	63.5	66.2	69.3	72.1	75.0	78.3	81.3	84.3	87.6
"	"	18		60.1	63.1	64.3	66.2	68.9	72.0	74.9	78.1	81.0	84.2	87.3	90.8	94.0
"	"	20		71.8	74.8	76.5	78.4	81.6	84.1	87.9	91.2	94.5	98.1	101.5	105.0	108.4
"	"	24		93.6	96.9	98.9	100.9	104.3	107.9	111.5	114.2	119.0	122.8	126.6	130.5	134.5
"	400	½														
"	"	¾														
"	"	1	4.6	5.7	6.7	7.3	7.8	9.0	10.1	11.2	12.4	13.7	15.0	16.3	17.7	19.2
"	"	1¼	4.7	5.9	6.9	7.5	8.1	9.3	10.4	11.6	12.8	14.0	15.4	16.7	18.2	19.7

To convert above listed units in square feet to square meters multiply by 0.0929.

Table B-28 (continued) – English Units

Fitting cover	Line pres psi	Nom pipe size inch	Outside surface area in square feet													
			Nominal insulation thickness–inches													
			1	1½	2	2½	3	3½	4	4½	5	5½	6	6½	7	7½
Flanged Gate Valve	400	1½	4.9	6.0	6.9	7.6	8.1	9.4	10.4	11.6	12.8	14.0	15.4	16.7	18.2	19.6
"	"	2		8.1	9.2	10.1	10.7	11.9	13.1	14.4	15.8	17.2	18.6	20.1	21.6	23.3
"	"	2½		9.6	11.0	11.0	12.4	13.7	15.0	16.5	17.8	19.3	20.9	22.4	24.1	25.9
"	"	3		10.3	11.8	12.5	13.3	14.6	16.0	17.5	18.9	21.0	22.1	23.7	25.4	27.2
"	"	3½		13.8	15.1	16.1	17.0	18.5	20.2	21.8	23.4	25.2	26.9	28.8	30.8	32.7
"	"	4		13.8	15.1	16.1	17.0	18.5	20.2	21.8	23.4	25.2	26.9	28.8	30.8	32.7
"	"	5		18.0	19.6	20.7	21.7	23.4	25.1	26.9	28.8	30.8	32.8	34.7	36.8	38.9
"	"	6		18.6	20.2	21.3	22.4	24.0	25.8	27.5	29.5	31.5	33.4	35.4	37.6	39.7
"	"	8		25.3	27.2	28.2	29.4	31.4	33.5	35.6	37.7	39.8	42.1	44.3	46.9	49.1
"	"	10		31.0	33.1	34.4	35.7	37.8	40.0	42.2	44.5	47.0	49.3	51.7	54.2	56.8
"	"	12		37.0	41.0	42.4	43.8	46.1	49.4	51.0	53.5	56.1	58.7	61.3	64.1	66.9
"	"	14		45.3	47.8	49.3	51.7	54.3	56.9	59.6	62.3	65.1	67.9	70.8	73.7	86.6
"	"	16		57.4	60.1	61.8	63.5	66.3	69.3	72.2	75.2	78.3	81.3	84.6	87.7	91.1
"	"	18		63.0	65.8	67.5	69.3	72.2	75.2	78.2	81.3	84.9	87.7	91.1	94.4	97.8
"	"	20		74.1	78.2	80.0	81.9	85.1	88.3	91.7	95.0	98.5	101.9	105.4	109.0	112.6
"	"	24		96.6	100.0	102.0	104.2	107.7	111.4	115.1	118.9	122.4	126.6	129.5	134.5	138.5
"	600	½														
"	"	¾														
"	"	1	4.0	5.1	5.9	6.6	7.2	8.1	9.4	10.4	11.6	12.8	14.0	15.4	16.7	18.8
"	"	1¼	5.1	6.3	7.4	8.0	8.6	10.0	11.1	12.3	13.5	14.8	16.2	17.8	19.1	20.6
"	"	1½		7.3	8.3	8.9	9.7	10.9	12.0	13.2	14.5	15.9	17.3	18.7	20.3	21.8
"	"	2		10.5	11.7	12.5	13.2	14.5	15.9	17.4	18.4	20.3	21.8	23.5	25.3	27.0
"	"	2½		11.2	12.5	13.5	14.1	15.4	16.9	18.4	19.9	21.4	23.0	24.8	26.6	28.3
"	"	3		12.9	14.2	15.0	15.9	17.4	18.9	20.5	22.0	23.6	25.4	27.2	29.0	30.9
"	"	3½		16.2	18.0	19.3	20.7	22.9	25.0	27.2	29.7	32.1	34.7	37.4	40.1	42.9
"	"	4		18.3	19.1	21.1	22.1	23.8	25.6	27.3	29.3	31.3	33.3	35.3	37.4	39.5
"	"	5		21.4	23.3	24.4	25.5	27.4	29.2	31.2	33.3	35.4	37.5	39.7	41.9	44.2
"	"	6		25.9	27.8	28.8	30.2	32.2	34.3	36.5	38.6	40.9	43.1	45.4	48.0	50.3
"	"	8		34.1	36.4	37.8	39.2	41.5	43.9	46.3	48.7	51.3	53.7	56.3	59.0	6.17
"	"	10		41.6	44.0	45.4	46.9	49.5	51.8	54.4	57.0	59.7	62.4	65.2	68.0	70.9
"	"	12		48.8	51.6	52.9	54.4	57.1	59.8	62.5	65.3	68.1	71.1	74.0	77.0	80.1
"	"	14		57.6	60.4	62.0	63.5	66.5	69.4	72.4	75.4	78.4	81.5	84.7	88.0	91.2
"	"	16		66.0	69.0	70.7	72.4	75.4	78.5	81.6	84.8	88.1	91.4	94.7	98.1	101.5
"	"	18		79.1	82.3	84.2	86.1	89.4	92.8	96.2	99.6	103.1	106.7	110.4	114.0	117.8
"	"	20		90.5	93.9	95.9	98.0	101.6	105.2	108.8	112.5	116.2	119.2	123.9	127.8	131.8
"	"	24		120.8	124.8	127.2	129.6	133.7	137.8	141.9	146.2	150.4	154.9	159.2	161.7	168.2
"	900	½														
"	"	¾														
"	"	1		8.7	9.7	10.7	11.4	12.6	13.9	15.2	16.7	18.2	19.6	21.2	22.7	24.5
"	"	1¼		9.3	10.4	11.4	12.2	13.5	14.9	16.2	17.8	19.3	20.8	22.4	24.0	25.8
"	"	1½		9.9	10.9	12.0	12.9	14.2	15.5	17.0	18.5	20.0	21.6	23.3	24.9	26.7
"	"	2		13.3	14.7	15.6	16.5	18.0	19.7	21.3	23.0	24.6	26.4	28.3	30.2	32.2
"	"	2½		16.7	18.2	19.2	20.3	22.0	23.7	25.5	27.2	29.1	31.1	33.1	35.2	37.2
"	"	3		17.1	18.7	19.2	20.8	22.5	24.3	26.1	27.8	29.8	31.8	33.8	35.9	38.0
"	"	3½		17.8	19.4	20.5	21.6	23.3	25.0	26.7	28.6	30.6	32.5	34.5	36.6	38.7
"	"	4		20.6	21.4	22.5	23.5	25.3	27.1	28.9	30.9	31.9	34.9	37.0	39.2	41.3
"	"	5		26.7	28.7	29.7	31.1	33.2	35.3	37.4	39.6	41.9	44.1	46.7	48.9	51.4
"	"	6		31.1	33.2	34.6	35.9	38.0	40.3	42.6	44.9	47.4	49.7	52.2	54.8	57.4
"	"	8		38.2	40.5	41.9	43.3	45.7	48.2	50.8	53.2	55.8	59.5	61.2	64.0	66.8
"	"	10		47.6	50.2	51.8	53.6	56.1	58.8	61.5	64.4	67.2	70.2	73.1	76.2	79.3
"	"	12		59.1	61.7	63.6	65.3	68.3	71.2	74.3	77.4	81.0	83.8	87.0	90.4	93.7
"	"	14		69.5	72.6	74.5	76.4	79.6	82.9	86.2	89.5	92.9	96.3	99.9	103.4	107.1
"	"	16		81.7	85.1	87.1	89.1	92.6	96.0	99.6	103.1	106.8	110.4	114.2	118.0	121.9
"	"	18		95.9	99.5	101.7	104.0	107.6	111.4	115.2	119.0	122.9	126.9	130.9	134.9	139.0
"	"	20														
"	"	24														

To convert above listed units in square feet to square meters multiply by 0.0929.

Table B-28 Outside surface area of prefabricated fitting covers (continued) – English Units

Fitting cover	Line pres psi	Nom pipe size inch	Outside surface area in square feet													
			Nominal insulation thickness–inches													
			1	1½	2	2½	3	3½	4	4½	5	5½	6	6½	7	7½
Flanged Gate Valve	1500	½														
"	"	¾														
"	"	1		7.9	8.9	9.5	10.6	11.8	13.0	14.3	15.7	17.2	18.6	20.1	21.7	23.3
"	"	1¼		9.3	10.5	11.5	12.3	13.6	14.9	16.3	17.9	19.4	20.9	22.5	24.1	25.9
"	"	1½		9.7	10.9	12.0	12.8	14.1	15.5	16.9	18.4	20.0	21.6	23.2	24.9	26.7
"	"	2		13.1	14.4	14.6	16.3	17.4	19.4	21.0	22.6	24.3	26.0	27.9	29.8	31.7
"	"	2½		16.4	17.8	18.8	19.9	21.5	23.2	25.0	26.7	28.6	30.6	32.5	34.5	36.6
"	"	3		19.5	21.1	22.4	23.6	25.3	27.2	29.1	31.0	33.1	35.2	37.3	39.5	41.7
"	"	3½		21.8	23.6	24.9	26.0	27.9	29.8	31.8	33.9	36.0	38.2	40.4	42.7	44.9
"	"	4		20.2	21.8	25.1	28.6	30.6	32.7	34.8	36.9	39.1	41.4	43.7	46.0	48.6
"	"	5		35.2	37.4	39.0	40.5	42.8	45.2	47.6	50.2	52.9	55.4	58.0	60.8	63.6
"	"	6		40.2	43.0	44.0	45.5	48.0	50.5	53.3	55.8	58.5	61.2	64.0	66.9	69.9
"	"	8		51.7	54.5	56.3	57.9	60.7	63.5	66.4	69.4	72.4	75.4	78.6	81.8	85.0
"	"	10		68.3	71.5	73.6	75.6	78.8	82.1	85.5	88.9	92.3	95.8	99.4	103.0	106.7
"	"	12		81.3	86.7	89.1	91.3	94.9	98.5	102.1	105.9	109.0	113.5	117.4	121.4	125.3
"	"	14		96.7	100.4	103.0	105.5	109.3	113.3	117.2	121.3	125.3	129.5	133.6	137.9	142.2
"	"	16		116.0	120.0	122.8	125.4	129.6	133.8	138.0	142.4	146.9	151.2	155.7	160.2	164.9
"	"	18														
"	"	20														
"	"	24														
Welding Tee		½														
"		¾	.7	1.1	1.6	2.6	3.3	4.0	4.9	5.9	6.8	7.9	9.3			
"		1	.9	1.3	1.8	2.6	3.3	4.0	4.9	5.9	6.8	7.9	9.3			
"		1¼	.9	1.3	1.8	2.6	3.3	4.0	4.9	5.9	6.8	7.9	9.3			
"		1½	1.1	1.6	2.5	3.2	4.0	4.8	5.8	6.8	7.8	9.3	10.5			
"		2	1.3	1.8	2.5	3.2	4.0	4.8	5.8	6.8	7.8	9.3	10.5	11.3	11.7	12.9
"		2½	1.5	2.5	3.2	3.9	4.7	5.7	6.7	7.8	9.2	10.4	11.7	13.1	14.6	16.1
"		3	1.8	2.5	3.2	3.9	4.7	5.7	6.7	7.8	9.2	10.4	11.7	13.1	14.6	16.1
"		3½	2.3	3.0	3.8	4.6	5.6	6.6	7.7	9.1	10.3	11.6	13.0	14.5	16.0	17.6
"		4	2.4	3.0	3.8	4.6	5.6	6.6	7.7	9.1	10.3	11.6	13.0	14.5	16.0	17.6
"		5	3.0	3.7	4.4	5.3	5.8	6.7	7.8	9.1	10.3	11.6	13.0	14.5	16.0	17.6
"		6	3.7	4.4	5.3	6.2	7.1	8.3	9.3	10.5	11.7	12.9	14.4	15.8	17.3	18.8
"		8		6.2	7.1	8.3	9.3	10.5	11.7	13.0	14.4	15.8	17.3	18.8	20.4	22.1
"		10		9.2	10.3	11.5	12.8	14.1	15.6	17.1	18.6	20.2	21.9	23.7	25.5	27.4
"		12		11.5	12.8	14.1	15.6	17.1	18.6	20.2	21.9	23.7	25.5	27.4	29.3	31.3
"		14		12.8	14.1	15.6	17.1	18.6	20.2	21.9	23.7	25.5	27.4	29.3	31.3	33.4
"		16		15.6	17.1	18.6	20.2	21.9	23.7	25.5	27.4	29.3	31.3	33.4	35.5	37.7
"		18		18.6	20.2	21.9	23.7	25.5	27.3	29.3	31.3	33.4	35.5	37.7	39.9	42.3
"		20		21.9	23.7	25.5	27.3	29.3	31.3	33.4	35.5	37.7	39.9	42.3	44.7	47.1
"		24		29.3	31.3	33.4	35.5	37.7	39.9	42.3	44.7	47.1	49.6	52.2	54.9	57.6
Screwed Tee		½	1.0	1.5	1.8	2.2	2.7	3.3	3.9	4.6	5.3	6.0	7.2	7.7	8.5	9.5
"		¾	1.1	1.6	2.6	3.2	4.0	4.8	5.9	6.9	8.0	9.3	10.5	11.8	13.2	14.7
"		1	1.3	1.8	2.6	3.2	4.0	4.8	5.9	6.9	8.0	9.3	10.5	11.8	13.2	14.7
"		1¼	1.6	2.6	3.3	4.0	4.8	5.9	6.9	8.0	9.3	10.5	11.8	13.2	14.7	16.2
"		1½	1.6	2.6	3.3	4.0	4.8	5.9	6.9	8.0	9.3	10.5	11.8	13.2	14.7	16.2
"		2	1.6	2.6	3.3	4.0	4.8	5.9	6.9	8.0	9.3	10.5	11.8	13.2	14.7	16.2
"		2½	2.5	3.2	3.9	4.8	5.8	6.8	7.8	9.2	10.4	11.7	13.1	14.6	16.2	17.7
"		3	3.1	3.9	4.8	5.7	6.7	7.8	9.1	10.4	11.7	13.0	14.5	16.1	17.7	19.4
"		3½	3.1	3.9	4.8	5.7	6.7	7.8	9.1	10.4	11.7	13.0	14.5	16.1	17.7	19.4
"		4	3.1	3.9	4.8	5.7	6.7	7.8	9.1	10.4	11.7	13.0	14.5	16.1	17.7	19.4
"		5		4.6	5.3	6.0	6.9	8.0	9.1	10.4	11.7	13.0	14.5	16.1	17.7	19.4
"		6		5.1	6.1	7.0	7.9	8.7	9.6	10.6	11.7	13.0	14.5	16.1	17.7	19.4
"		8		7.5	8.7	9.3	10.6	11.2	12.7	13.4	15.0	15.6	17.3	18.1	20.0	20.8
"		10		10.4	11.4	12.5	13.6	14.7	15.9	17.1	18.4	19.7	21.1	22.5	23.9	25.4
"		12		11.7	13.6	14.7	15.9	17.1	18.4	19.7	21.1	22.5	23.9	25.4	26.9	28.5
"		14														

To convert above listed units in square feet to square meters multiply by 0.0929.

Table B-28 (continued) – English Units

Fitting cover	Line pres psi	Nom pipe size inch	Outside surface area in square feet — Nominal insulation thickness–inches: 1	1½	2	2½	3	3½	4	4½	5	5½	6	6½	7	7½
Screwed Tee		16														
″		18														
″		20														
″		24														
Flanged Tee	150	½														
″	″	¾														
″	″	1	3.0	3.8	5.2	4.9	5.3	6.0	6.8	7.7	8.5	10.3	11.2	12.2	13.4	14.6
″	″	1¼	3.8	4.7	5.2	6.2	6.6	7.5	8.6	9.5	10.5	11.6	12.6	13.8	15.0	16.2
″	″	1½	3.9	4.9	5.9	6.1	6.6	7.5	8.6	9.5	10.5	11.6	12.6	13.8	15.0	16.2
″	″	2	4.8	5.9	6.9	8.2	8.5	9.5	10.5	11.6	12.7	13.8	15.1	16.3	17.6	18.9
″	″	2½		7.2	8.1	8.6	10.1	10.4	11.4	12.5	13.6	14.9	16.1	18.2	18.7	20.0
″	″	3		7.6	8.6	10.3	10.9	11.0	12.1	13.2	14.3	16.1	17.4	18.7	20.1	21.4
″	″	3½		8.5	9.6	10.4	11.9	12.1	13.2	14.3	15.6	16.9	18.2	19.6	20.9	22.4
″	″	4		9.8	11.2	11.9	13.3	13.6	14.9	16.1	17.4	18.8	20.2	21.6	24.1	24.6
″	″	5		12.2	13.5	14.2	16.0	17.1	17.9	18.9	20.0	21.8	23.3	24.8	26.5	28.1
″	″	6		13.4	14.7	15.5	17.4	17.5	18.1	19.8	21.8	23.3	24.8	26.5	28.1	29.8
″	″	8		16.1	17.5	18.5	20.8	21.0	22.5	24.0	25.6	27.2	28.9	30.6	32.4	34.2
″	″	10		20.5	22.0	22.9	25.2	25.4	27.0	28.8	30.5	32.2	34.1	35.9	37.8	39.9
″	″	12		25.2	26.9	27.9	28.8	32.1	32.3	34.1	36.0	37.9	40.0	41.8	43.8	46.7
″	″	14		31.2	33.1	34.1	35.2	37.1	40.9	41.1	43.3	45.3	47.4	49.6	51.9	54.1
″	″	16		34.7	38.7	39.8	40.9	43.1	45.0	49.1	49.4	51.6	53.9	56.2	58.6	61.0
″	″	18		40.3	42.4	43.6	44.9	47.8	51.4	51.7	54.4	56.7	59.5	62.0	64.1	66.6
″	″	20		48.2	50.3	51.5	52.7	55.1	57.4	62.1	62.3	64.9	67.4	70.0	72.7	75.4
″	″	24		62.9	65.4	66.9	68.3	71.0	73.7	79.0	79.2	82.1	85.0	87.9	90.9	93.9
″	300	½														
″	″	¾														
″	″	1														
″	″	1¼														
″	″	1½														
″	″	2	5.5	6.7	7.8	9.4	9.4	10.5	11.7	12.9	14.2	15.5	16.9	18.4	19.9	21.5
″	″	2½		7.9	9.0	9.8	11.6	11.7	13.0	14.5	15.6	17.1	18.6	20.0	21.6	23.2
″	″	3		9.1	10.5	11.5	13.2	13.6	14.9	16.3	17.8	19.2	20.8	22.4	24.1	25.8
″	″	3½		11.2	12.5	13.3	14.0	16.5	16.9	18.3	19.7	21.4	23.1	24.7	26.7	28.4
″	″	4		13.3	14.1	14.9	15.8	18.1	18.2	20.4	22.0	23.7	25.3	27.2	29.0	31.0
″	″	5		14.5	15.9	16.8	17.7	20.5	20.9	22.5	24.2	25.9	27.9	29.6	31.4	33.5
″	″	6		16.3	17.7	18.7	19.4	21.1	24.2	26.0	27.9	29.8	31.7	33.7	35.6	37.7
″	″	8		20.1	22.0	23.0	24.1	27.2	27.6	29.6	31.5	33.4	35.9	37.5	39.8	41.8
″	″	10		26.4	28.3	29.4	30.7	32.8	36.3	36.8	39.0	41.1	43.4	45.8	48.0	50.3
″	″	12		32.8	34.9	36.2	37.5	39.5	41.8	46.1	46.7	48.8	51.3	53.8	56.4	58.8
″	″	14		39.9	41.5	43.0	43.8	46.6	48.6	50.3	56.4	57.6	59.5	62.2	65.0	67.8
″	″	16		47.4	50.0	51.6	53.0	55.6	58.2	60.3	63.8	68.6	69.2	72.3	75.3	77.5
″	″	18		55.0	57.4	59.0	60.6	63.5	66.3	69.0	72.2	77.5	79.6	82.7	85.8	89.2
″	″	20		63.0	65.8	67.4	69.4	72.0	75.2	78.3	81.4	84.4	90.5	91.3	94.3	98.2
″	″	24		79.0	81.6	84.0	85.2	88.5	91.8	95.2	98.5	102.0	105.0	112.0	113.0	116.7
″	400	½	2.9	3.8	4.5	4.6	5.3	6.1	6.9	7.9	8.8	9.7	10.6	11.6	12.8	13.9
″	″	¾	3.2	4.1	4.9	6.1	6.4	6.6	7.6	8.6	9.4	10.5	11.4	12.5	13.7	14.9
″	″	1	4.1	5.2	5.3	6.6	7.1	8.0	9.2	10.1	11.2	12.3	13.3	14.5	15.8	17.1
″	″	1¼	5.5	5.7	6.2	7.4	7.6	8.3	9.5	10.5	11.6	12.7	13.7	15.0	16.3	17.6
″	″	1½	5.5	6.1	7.2	7.7	9.4	9.5	10.7	11.6	12.7	13.7	15.0	16.3	17.6	18.8
″	″	2	5.7	6.9	8.0	9.5	9.7	10.5	11.6	12.7	13.9	15.1	16.4	17.8	19.1	20.5
″	″	2½		8.7	9.8	10.6	12.5	12.7	13.9	15.1	16.3	17.7	19.2	20.6	22.0	23.5
″	″	3		10.2	11.4	12.4	14.2	14.3	15.7	17.0	19.4	19.8	21.3	22.8	24.3	25.9
″	″	3½		11.6	13.1	13.9	14.6	17.1	17.3	18.7	20.2	21.6	23.2	23.7	26.3	28.0
″	″	4		13.3	14.7	15.5	16.6	18.7	19.2	20.6	22.1	23.7	25.3	27.0	28.7	30.4
″	″	5		15.3	16.7	17.6	18.4	21.3	21.5	23.0	24.6	26.2	28.0	29.7	31.5	33.3
″	″	6		17.1	18.5	19.5	20.5	23.3	23.5	25.2	27.5	28.5	30.3	32.1	33.9	35.8

To convert above listed units in square feet to square meters multiply by 0.0929.

Table B-28 Outside surface area of prefabricated fitting covers (continued) – English Units

Fitting cover	Line pres psi	Nom pipe size inch	Outside surface area in square feet — Nominal insulation thickness–inches: 1	1½	2	2½	3	3½	4	4½	5	5½	6	6½	7	7½
Flanged Tee	400	8		21.7	23.5	24.4	25.4	28.6	28.7	30.5	32.3	34.2	35.1	38.1	39.1	42.1
"	"	10		27.3	29.6	30.7	31.9	33.9	38.1	37.7	39.8	41.8	44.0	45.3	48.3	50.5
"	"	12		35.0	37.1	38.3	39.5	41.6	43.8	47.1	48.3	50.4	52.7	55.1	57.5	60.0
"	"	14		40.4	42.6	43.9	45.1	47.4	49.8	52.0	56.5	56.8	58.2	61.7	64.3	66.9
"	"	16		48.2	50.5	52.1	52.3	55.7	57.2	60.7	63.2	68.1	68.5	71.2	73.9	75.7
"	"	18		53.4	56.7	58.1	59.6	62.1	64.7	67.4	70.0	75.2	75.5	77.3	81.2	84.1
"	"	20		63.6	66.2	67.7	69.3	72.0	74.9	77.7	78.5	83.5	89.1	89.5	92.5	95.6
"	"	24		82.3	85.3	87.0	88.8	91.9	95.1	98.2	101.5	104.7	108.1	114.4	114.8	118.3
"	600	½														
"	"	¾														
"	"	1	4.1	5.2	5.3	6.6	7.1	8.0	9.2	10.1	11.2	12.3	13.3	14.5	15.8	17.0
"	"	1¼	4.3	5.3	6.2	7.2	7.6	8.1	9.5	10.5	11.6	12.7	13.8	15.0	15.8	17.6
"	"	1½	5.2	6.3	7.2	7.8	9.3	9.5	10.5	11.6	12.7	13.7	15.0	16.3	18.0	18.8
"	"	2	5.7	6.9	8.0	9.5	9.6	10.5	11.6	12.8	13.9	15.1	16.4	17.8	19.1	21.5
"	"	2½		8.8	9.8	10.6	12.5	12.7	13.9	15.2	16.3	17.8	19.2	20.6	22.0	23.5
"	"	3		10.1	11.3	12.3	14.2	14.4	15.7	17.0	18.4	19.7	21.3	22.9	24.3	26.0
"	"	3½		11.6	13.1	13.9	14.6	17.1	17.3	17.7	20.2	21.6	23.2	24.7	26.4	28.0
"	"	4		13.8	15.0	15.9	17.0	19.8	20.0	21.4	23.0	24.7	26.2	28.0	29.7	31.5
"	"	5		19.2	20.6	22.0	22.9	24.5	26.1	29.4	29.7	31.6	33.4	35.3	37.2	39.2
"	"	6		20.6	21.8	22.9	23.9	25.6	28.8	29.0	30.9	32.8	34.7	36.6	38.6	40.7
"	"	8		27.0	28.9	29.9	31.0	33.0	35.1	38.8	39.1	41.1	42.3	45.3	47.8	49.8
"	"	10		35.1	37.3	38.7	39.7	42.1	44.3	46.6	51.1	51.4	53.6	56.0	58.5	61.3
"	"	12		39.7	41.9	43.2	44.5	46.8	49.1	51.5	56.0	56.2	58.8	61.3	63.9	66.5
"	"	14		45.2	47.5	48.9	50.3	53.1	55.3	57.8	60.1	64.9	65.3	68.0	73.6	76.4
"	"	16		59.7	59.9	60.1	60.2	62.8	65.5	68.2	71.0	73.7	79.1	79.5	83.4	85.4
"	"	18		65.0	67.8	69.5	71.1	74.9	76.8	79.8	83.7	85.7	88.9	94.6	95.1	98.2
"	"	20		73.9	76.8	78.6	80.4	83.4	86.5	89.6	92.7	95.0	99.2	105.4	105.8	109.2
"	"	24		96.4	99.7	101.8	103.8	106.3	110.8	114.3	117.8	121.4	125.1	128.8	135.9	136.3
"	900	½														
"	"	¾														
"	"	1	5.6	6.8	7.8	8.5	10.2	10.4	11.8	12.2	13.8	15.0	16.3	17.6	19.0	20.4
"	"	1¼	5.9	7.0	8.2	8.5	10.6	10.8	12.0	12.6	14.8	15.6	16.9	18.3	19.6	21.1
"	"	1½	8.3	8.4	9.4	10.0	11.0	13.3	13.4	14.5	15.8	17.0	18.5	19.8	21.3	22.7
"	"	2		10.9	12.1	13.2	14.1	16.5	16.7	18.2	19.7	21.1	22.7	24.3	25.8	27.6
"	"	2½		13.2	14.4	14.6	16.2	17.6	20.4	20.6	22.3	23.6	25.4	27.0	28.1	30.6
"	"	3		13.9	14.6	14.7	16.3	17.6	20.4	20.8	22.4	23.9	25.4	27.2	28.4	30.8
"	"	3½		14.7	15.6	16.3	17.2	19.2	22.1	22.6	23.4	25.7	27.8	29.6	31.4	32.0
"	"	4		16.0	17.5	18.5	18.8	21.0	24.0	24.2	25.9	27.5	29.7	31.2	33.0	34.8
"	"	5		20.9	22.6	23.7	24.9	26.6	28.3	31.8	33.8	35.7	37.7	39.7	41.8	43.9
"	"	6		23.1	25.2	26.3	27.5	29.2	31.0	34.8	35.1	37.1	39.0	41.1	43.3	45.4
"	"	8		33.0	35.1	36.4	38.1	39.9	42.0	44.3	46.5	51.0	52.2	54.6	57.1	59.6
"	"	10		44.1	43.4	44.8	46.2	48.6	50.9	53.5	55.8	58.3	63.1	63.5	66.1	68.8
"	"	12		49.0	51.6	53.1	54.6	57.3	59.8	62.4	65.1	67.8	73.0	73.4	76.2	79.2
"	"	14		56.0	58.9	60.8	62.3	66.0	67.8	70.6	73.4	76.4	79.3	84.9	85.3	88.4
"	"	16		64.3	67.0	68.8	70.6	73.5	76.4	79.4	82.4	85.5	88.6	94.6	94.8	98.2
"	"	18		76.2	79.2	81.2	83.1	86.2	89.4	92.6	95.8	97.8	102.6	106.0	112.5	113.0
"	"	20		88.7	92.0	94.1	96.1	99.5	102.9	106.3	109.9	113.4	117.1	120.6	124.3	131.3
"	"	24		124.2	128.1	130.6	133.0	136.9	141.0	145.0	149.2	153.3	157.5	161.8	166.0	170.4
"	1500	½														
"	"	¾														
"	"	1	5.6	6.8	7.9	8.5	10.2	10.4	11.8	12.2	13.8	15.0	16.3	17.6	19.0	20.4
"	"	1¼		7.1	8.2	8.5	10.6	10.8	12.0	12.6	14.8	15.6	16.9	18.3	19.6	21.1
"	"	1½		8.4	9.4	10.0	11.0	13.3	13.4	14.5	15.8	17.0	17.5	19.8	21.3	22.7
"	"	2		10.9	12.1	13.2	14.1	16.5	16.7	18.2	19.7	21.1	22.7	24.3	25.8	27.6
"	"	2½		12.7	14.4	14.6	16.2	17.6	20.3	21.0	22.4	23.7	25.3	27.0	28.0	30.5
"	"	3		15.4	16.9	17.9	18.8	20.4	23.5	23.8	25.4	27.1	28.8	30.7	32.6	34.5

To convert above listed units in square feet to square meters multiply by 0.0929.

Table B-28 (continued) – English Units

Fitting cover	Line pres psi	Nom pipe size inch	Outside surface area in square feet — Nominal insulation thickness–inches: 1	1½	2	2½	3	3½	4	4½	5	5½	6	6½	7	7½
Flanged Tee	1500	3½		17.3	18.6	19.2	20.7	22.8	25.0	26.4	27.3	29.3	32.7	33.3	35.0	38.0
"	"	4		19.7	21.3	22.4	23.6	25.4	26.3	30.4	30.5	32.6	34.6	36.6	38.6	40.7
"	"	5		26.3	28.3	29.7	31.0	33.0	34.9	37.0	41.2	41.5	43.7	45.9	48.2	50.6
"	"	6		29.1	31.1	32.5	33.8	35.7	38.0	40.2	44.0	44.7	47.0	49.3	51.7	54.3
"	"	8		38.2	40.4	42.0	43.5	45.2	48.2	50.6	53.2	58.1	58.4	60.9	63.5	66.3
"	"	10		52.0	53.9	55.7	58.2	60.9	63.9	66.5	69.3	72.2	75.2	80.1	81.2	84.2
"	"	12		67.0	70.9	72.8	74.6	77.8	80.9	84.1	87.4	90.6	94.4	97.3	100.6	107.1
"	"	14		81.6	85.4	87.1	89.7	93.1	96.6	100.2	103.7	107.2	110.9	114.5	118.3	122.0
"	"	16		97.3	101.2	103.4	104.7	106.4	110.1	113.8	117.6	121.4	125.3	129.2	133.2	137.2
"	"	18		115.2	119.1	121.8	124.5	128.4	132.5	136.6	140.8	144.9	149.2	153.3	157.8	162.2
"	"	20		133.5	138.4	140.6	143.4	147.8	152.1	156.4	160.9	165.4	170.0	174.5	179.1	183.8
"	"	24		179.0	183.9	187.2	190.4	195.3	200.4	205.4	210.6	215.7	220.9	226.1	231.3	235.4
Welding Ell		½	.3	.5	.9	1.6	2.1	2.8	3.5	4.3	5.2	6.2	7.2			
90°		¾	.3	.5	.9	1.6	2.2	2.8	3.5	4.3	5.2	6.2	7.2			
"		1	.4	.7	1.0	1.6	2.2	2.8	3.5	4.3	5.2	6.2	7.2			
"		1¼	.4	.8	1.0	1.6	2.2	2.8	3.5	4.3	5.2	6.2	7.2			
"		1½	.5	.8	1.5	2.0	2.6	3.3	4.1	5.0	5.9	7.1	8.3			
"		2	.8	1.1	1.5	2.0	2.6	3.3	4.1	5.0	5.9	7.1	8.3	9.3	10.6	12.7
"		2½	1.0	1.6	2.0	2.6	3.3	4.1	4.7	5.9	7.0	8.2	9.4	10.6	11.9	13.3
"		3	1.2	1.6	2.0	2.6	3.3	4.1	4.7	5.9	7.0	8.2	9.4	10.6	11.9	13.3
"		3½	1.8	2.2	2.7	3.2	3.9	4.7	5.7	6.8	7.8	9.0	10.2	11.6	12.9	14.3
"		4	1.8	2.2	2.7	3.2	3.9	4.7	5.7	6.8	7.8	9.0	10.2	11.6	12.9	14.3
"		5	2.4	2.9	3.6	4.4	4.9	5.8	6.5	7.8	8.4	9.7	10.9	12.5	14.1	15.3
"		6	3.3	3.8	4.4	5.2	5.9	7.1	7.7	8.5	9.4	10.4	11.8	13.5	15.0	16.6
"		8		4.8	5.4	6.2	7.0	7.8	8.6	9.1	10.3	11.8	12.6	14.7	14.9	15.5
"		10		6.9	7.7	8.5	9.4	10.3	11.5	11.5	13.5	14.7	15.7	17.0	18.3	19.6
"		12		11.6	12.6	13.8	14.9	16.5	17.5	18.7	20.2	21.5	22.6	24.6	26.0	27.6
"		14		13.8	15.2	16.4	18.0	19.2	20.4	21.8	23.3	25.0	26.3	27.9	29.7	31.7
"		16		17.8	19.5	20.7	22.6	23.5	25.0	26.6	28.3	30.0	31.8	33.5	36.3	37.1
"		18		22.1	23.4	24.9	26.5	28.3	30.0	31.7	33.0	35.3	37.0	38.9	40.7	42.8
"		20		26.1	27.6	29.4	31.1	32.9	34.8	36.6	38.4	40.3	42.0	44.3	46.6	48.0
"		24		36.2	38.3	40.4	42.3	44.0	46.2	48.7	51.1	53.2	55.3	57.6	60.2	62.8
Screwed Ell		½	.7	1.3	1.3	1.8	2.1	2.7	3.2	3.8	4.4	5.1	5.9	6.7	7.5	8.3
90°		¾	.8	1.3	1.7	2.1	2.7	3.2	3.8	4.4	5.1	6.0	6.7	7.5	8.3	8.7
"		1	1.0	1.3	1.8	2.1	2.7	3.2	3.8	4.4	5.1	6.0	6.7	7.5	8.3	8.7
"		1¼	1.1	1.9	2.5	3.1	3.7	4.6	5.4	6.3	6.4	6.7	6.7	7.5	8.3	8.7
"		1½	1.1	1.9	2.5	3.1	3.7	4.6	5.4	6.3	6.4	6.7	6.7	7.5	8.3	8.7
"		2	1.3	1.9	2.5	3.1	3.7	4.6	5.4	6.3	6.7	6.7	7.5	8.3	8.7	10.2
"		2½	1.9	2.5	3.1	3.7	4.6	5.1	5.4	6.3	6.4	6.7	7.5	8.3	8.7	10.2
"		3	1.9	2.5	3.1	3.7	4.6	5.1	5.5	6.3	6.7	7.5	8.3	8.7	10.2	11.2
"		3½	2.5	2.6	3.7	4.6	5.1	5.4	6.3	6.7	7.5	8.3	8.7	10.2	11.2	12.2
"		4	2.6	3.2	3.8	4.4	5.1	6.0	6.7	7.5	8.3	8.7	10.2	11.2	12.2	13.2
"		5		3.8	4.4	5.1	6.0	6.7	7.5	8.3	8.7	10.2	11.2	12.2	13.2	14.4
"		6		4.4	5.1	6.0	6.7	7.5	8.3	8.7	10.2	11.2	12.2	13.2	14.4	15.6
"		8		6.0	6.7	7.5	8.3	8.7	10.2	11.2	12.2	13.2	14.4	15.6	16.7	17.9
"		10		8.3	8.7	10.2	11.2	12.2	13.2	14.4	15.6	16.7	17.9	19.3	20.5	21.9
"		12		10.2	11.2	12.2	13.2	14.4	15.6	16.7	17.9	19.3	20.5	21.9	23.3	24.8
"		14														
"		16														
"		18														
"		20														
"		24														
"		½	.7	.9	1.0	1.4	1.7	2.1	2.5	3.0	3.6	4.1	4.8	5.4	6.0	6.7
45°		¾	.8	1.0	1.4	1.7	2.1	2.5	3.0	3.6	4.1	4.8	5.4	6.0	6.7	7.4
"		1	.8	1.0	1.4	1.7	2.1	2.5	3.0	3.6	4.1	4.8	5.4	6.0	6.7	7.4
"		1¼	.9	1.4	2.2	2.3	2.7	3.4	4.0	4.6	5.5	6.3	7.1	8.0	8.7	9.8

To convert above listed units in square feet to square meters multiply by 0.0929.

Table B-28 Outside surface area of prefabricated fitting covers (continued) – English Units

Fitting cover	Line pres psi	Nom pipe size inch	Outside surface area in square feet													
			Nominal insulation thickness–inches													
			1	1½	2	2½	3	3½	4	4½	5	5½	6	6½	7	7½
Screwed Ell 45°		1½	.9	1.4	2.2	2.3	2.7	3.4	4.0	4.6	5.5	6.3	7.1	8.0	8.7	9.8
		2	1.3	1.6	2.3	2.5	2.9	3.6	4.3	4.9	5.7	6.6	7.5	8.3	9.2	9.1
"		2½	1.4	1.8	2.4	2.7	3.2	3.9	4.5	5.3	6.0	6.8	7.7	8.6	9.6	10.5
"		3	1.7	2.3	2.6	3.2	3.7	4.3	5.2	5.8	6.7	7.4	8.4	9.4	10.3	11.4
"		3½	1.7	2.3	2.6	3.2	3.7	4.3	4.8	5.7	6.7	7.4	8.4	9.4	10.3	11.4
"		4	2.3	2.8	3.3	3.8	4.3	4.8	5.7	6.4	7.1	7.5	8.7	9.5	10.4	11.6
"		5		3.3	3.8	4.3	5.7	6.4	6.4	7.1	7.5	8.7	9.5	10.4	11.4	12.4
"		6		3.8	4.3	5.2	5.8	6.4	7.1	7.9	8.8	9.7	10.0	11.4	12.4	13.3
"		8		5.0	5.7	6.5	7.3	8.1	9.0							
"		10		7.7	8.1	8.9	10.1	11.2	12.3							
"		12		8.9	10.1	10.9	12.2	13.5	14.8							
"		14														
"		16														
"		18														
"		20														
"		24														
Flanged Ell	150	½														
Short Rad. 90°	"	¾														
"	"	1	2.4	3.2	3.8	4.3	4.8	5.7	6.4	8.5	9.4	10.8	12.2	13.7	15.5	17.2
"	"	1¼	3.3	3.9	4.3	5.0	6.0	7.2	8.6	9.9	11.2	12.6	14.3	16.0	17.7	19.3
"	"	1½	3.3	3.9	4.3	5.0	6.0	7.2	8.6	9.9	11.2	12.6	14.3	16.0	17.7	19.3
"	"	2	4.7	4.3	5.0	6.0	7.2	8.0	8.4	10.0	11.4	12.8	15.3	16.4	18.3	19.7
"	"	2½		5.9	5.9	6.3	7.2	8.0	8.9	10.2	11.6	13.1	15.6	16.9	18.7	20.3
"	"	3		5.9	5.9	6.7	7.2	8.0	8.9	10.4	11.8	13.3	15.9	17.7	19.0	20.9
"	"	3½		6.7	7.6	8.4	9.0	10.0	11.7	12.7	14.2	15.9	17.5	19.4	21.0	23.3
"	"	4		7.9	9.0	9.7	10.3	11.3	11.7	14.1	15.8	17.4	19.3	21.1	23.4	25.4
"	"	5		8.7	10.0	10.6	11.3	13.4	14.0	15.0	17.0	18.8	20.1	22.1	23.7	25.8
"	"	6		11.4	12.5	13.2	14.1	15.3	16.5	17.8	19.4	21.6	23.5	25.7	28.0	30.3
"	"	8		11.6	12.9	13.8	14.8	16.2	17.8	19.4	21.2	23.3	24.6	27.6	29.8	32.2
"	"	10		15.4	17.5	18.4	19.3	20.6	22.1	24.2	25.7	27.3	29.0	30.6	32.7	35.1
"	"	12		20.1	21.5	22.5	23.6	24.9	26.5	28.6	30.3	32.7	35.1	37.7	40.5	43.4
"	"	14		25.4	27.0	28.1	29.2	30.9	32.7	34.5	36.3	38.2	40.1	42.0	44.4	47.4
"	"	16		31.0	32.7	33.9	35.1	37.0	38.9	40.8	42.8	44.9	46.9	49.0	51.2	53.4
"	"	18		33.4	35.8	37.1	39.3	40.2	42.2	44.3	46.3	48.4	50.6	52.8	56.0	57.4
"	"	20		39.6	41.6	42.8	44.1	46.2	48.3	50.5	52.6	55.0	57.2	59.5	62.0	64.4
"	"	24		53.0	55.2	56.8	58.3	60.7	62.5	65.6	68.2	70.8	73.4	76.0	78.8	81.5
Flanged Ell	150	½														
Long Rad. 90°	"	¾														
"	"	1	2.4	3.2	3.8	4.3	4.8	5.7	6.4	8.5	9.4	10.8	12.2	13.8	15.5	17.2
"	"	1¼	3.1	3.9	4.4	5.0	6.0	7.2	8.6	10.0	11.2	12.6	14.3	16.0	17.7	19.3
"	"	1½	3.1	3.9	4.5	5.0	6.0	7.2	8.6	10.0	11.2	12.6	14.3	16.0	17.7	19.3
"	"	2	5.7	5.7	6.5	7.0	7.8	9.0	10.7	12.1	13.6	15.4	17.0	18.8	20.3	22.2
"	"	2½		5.7	6.5	7.0	7.8	9.0	10.7	12.1	13.6	15.4	17.0	18.8	20.3	23.2
"	"	3		6.0	6.5	7.0	7.8	9.0	10.7	12.1	13.6	15.4	17.0	18.0	20.3	23.2
"	"	3½		6.7	7.6	8.5	9.0	10.0	11.7	12.7	14.2	15.9	17.5	19.4	21.0	23.3
"	"	4		7.9	9.0	9.7	10.3	11.3	11.7	14.1	15.9	17.5	19.3	21.2	23.4	25.4
"	"	5		8.7	10.0	10.6	11.3	13.4	14.0	15.0	17.0	18.8	20.1	22.1	23.7	25.8
"	"	6		11.4	12.5	13.2	14.1	15.3	16.5	17.8	19.5	21.6	23.5	25.7	28.0	30.3
"	"	8		11.6	12.9	13.8	14.8	16.2	17.8	19.4	21.2	23.3	24.6	27.6	29.8	32.2
"	"	10		15.4	17.5	18.4	19.3	20.6	22.1	24.2	25.7	27.3	29.0	30.6	32.7	35.1
"	"	12		20.1	21.5	22.5	23.6	24.9	26.5	28.6	30.3	32.7	35.1	37.7	40.5	43.4
"	"	14		25.4	27.0	28.1	29.2	30.9	32.7	34.5	36.3	38.2	40.1	42.0	44.4	47.4
"	"	16		31.0	32.7	33.9	35.1	36.9	38.9	40.8	42.8	44.9	46.9	49.0	51.2	53.4
"	"	18		33.4	35.8	37.1	39.3	40.2	42.2	44.3	46.3	48.4	50.6	52.8	56.0	57.4
"	"	20		40.0	41.6	42.8	44.1	46.2	48.3	50.5	52.6	55.0	57.2	59.5	62.0	64.4
"	"	24		53.0	55.2	56.8	58.3	60.7	62.5	65.6	68.2	70.8	73.4	76.0	78.8	81.5

To convert above listed units in square feet to square meters multiply by 0.0929.

Table B-28 (continued) — English Units

Fitting cover	Line pres psi	Nom pipe size inch	Outside surface area in square feet													
			Nominal insulation thickness—inches													
			1	1½	2	2½	3	3½	4	4½	5	5½	6	6½	7	7½
Flanged Ell 45°	150	½														
"	"	¾														
"	"	1	1.2	1.6	1.9	2.1	2.4	2.8	3.2	4.2	4.7	5.4	6.1	6.9	7.7	8.6
"	"	1¼	1.3	1.6	2.0	2.2	2.5	3.0	3.6	4.3	4.9	5.6	6.3	7.1	8.0	8.8
"	"	1½	1.5	1.9	2.2	2.4	2.6	3.1	3.8	4.5	5.1	5.7	6.5	7.4	8.1	9.0
"	"	2	1.5	1.9	2.2	2.4	2.6	3.1	3.8	4.5	5.1	5.7	6.5	7.4	8.1	9.0
"	"	2½		1.9	2.2	2.4	2.6	3.1	3.8	4.5	5.1	5.7	6.5	7.4	8.1	9.0
"	"	3		2.6	2.9	3.3	3.6	4.0	4.4	5.2	5.9	6.6	7.9	8.7	9.5	10.4
"	"	3½		3.4	3.8	4.2	4.5	5.0	5.8	6.4	7.1	8.0	8.8	9.7	10.5	11.6
"	"	4		3.9	4.5	4.8	5.2	5.7	5.8	7.0	7.9	8.7	9.7	10.6	11.7	12.7
"	"	5		4.3	5.0	5.3	5.7	6.7	7.0	7.5	8.5	9.4	10.1	11.1	11.8	12.9
"	"	6		5.7	6.2	6.6	7.0	7.6	8.2	8.9	9.8	10.8	11.8	12.9	14.0	15.2
"	"	8		5.8	6.4	6.9	7.4	8.1	8.9	9.7	10.6	11.6	12.3	13.8	14.9	16.1
"	"	10		7.7	8.8	9.2	9.6	10.3	11.1	12.1	12.9	13.7	14.5	15.3	16.7	17.6
"	"	12		10.1	10.7	11.2	11.8	12.5	13.3	14.4	15.2	16.3	17.6	18.8	20.3	21.7
"	"	14		12.7	13.5	14.0	14.5	15.4	16.3	17.2	18.2	19.1	20.0	21.0	22.2	23.7
"	"	16		15.6	16.4	17.0	17.5	18.5	19.4	20.4	21.4	22.4	23.4	24.5	25.6	26.7
"	"	18		16.7	17.9	18.5	19.4	20.1	21.1	22.1	23.1	24.2	25.3	26.4	28.0	28.7
"	"	20		19.8	20.8	21.4	22.1	23.1	24.2	25.2	26.3	27.5	28.6	29.8	31.0	32.2
"	"	24		26.5	27.6	28.4	29.2	30.4	31.3	32.8	34.1	35.4	36.7	38.0	39.4	40.8

To convert above listed units in square feet to square meters multiply by 0.0929.

Table B-29 Square feet of insulation on duct work, per linear foot — English Units

1" Insulation				2" Insulation				1" Insulation				2" Insulation			
Semi perim"	Sq ft insul	Semi perim"	Sq ft insul	Semi perim"	Sq ft insul	Semi perim"	Sq ft insul	Semi perim"	Sq ft insul	Semi perim"	Sq ft insul	Semi perim"	Sq ft insul	Semi perim"	Sq ft insul
8	1.67	65	11.17	8	2.00	65	11.50	37	6.50	94	16.00	37	6.83	94	16.34
9	1.83	66	11.34	9	2.17	66	11.66	38	6.66	95	16.17	38	7.00	95	16.50
10	2.00	67	11.50	10	2.34	67	11.83	39	6.83	96	16.34	39	7.17	96	16.66
11	2.17	68	11.66	11	2.50	68	12.00	40	7.00	97	16.50	40	7.34	97	16.83
12	2.34	69	11.83	12	2.67	69	12.17	41	7.17	98	16.66	41	7.50	98	17.00
13	2.50	70	12.00	13	2.84	70	12.34	42	7.34	99	16.83	42	7.66	99	17.17
14	2.66	71	12.17	14	3.00	71	12.50	43	7.50	100	17.00	43	7.83	100	17.34
15	2.83	72	12.34	15	3.17	72	12.66	44	7.66	101	17.17	44	8.00	101	17.50
16	3.00	73	12.50	16	3.34	73	12.83	45	7.83	102	17.34	45	8.17	102	17.66
17	3.17	74	12.66	17	3.50	74	13.00	46	8.00	103	17.50	46	8.34	103	17.83
18	3.34	75	12.83	18	3.66	75	13.17	47	8.17	104	17.66	47	8.50	104	18.00
19	3.50	76	13.00	19	3.83	76	13.34	48	8.34	105	17.83	48	8.66	105	18.17
20	3.66	77	13.17	20	4.00	77	13.50	49	8.50	106	18.00	49	8.83	106	18.34
21	3.83	78	13.34	21	4.17	78	13.66	50	8.66	107	18.17	50	9.00	107	18.50
22	4.00	79	13.50	22	4.34	79	13.83	51	8.83	108	18.34	51	9.17	108	18.66
23	4.17	80	13.66	23	4.50	80	14.00	52	9.00	109	18.50	52	9.34	109	18.83
24	4.34	81	13.83	24	4.66	81	14.17	53	9.17	110	18.66	53	9.50	110	19.00
25	4.50	82	14.00	25	4.83	82	14.34	54	9.34	111	18.83	54	9.66	111	19.17
26	4.66	83	14.17	26	5.00	83	14.50	55	9.50	112	19.00	55	9.83	112	19.34
27	4.83	84	14.34	27	5.17	84	14.66	56	9.66	113	19.17	56	10.00	113	19.50
28	5.00	85	14.50	28	5.34	85	14.83	57	9.83	114	19.34	57	10.17	114	19.66
29	5.17	86	14.66	29	5.50	86	15.00	58	10.00	115	19.50	58	10.34	115	19.83
30	5.34	87	14.83	30	5.66	87	15.17	59	10.17	116	19.66	59	10.50	116	20.00
31	5.50	88	15.00	31	5.83	88	15.34	60	10.34	117	19.83	60	10.66	117	20.17
32	5.66	89	15.17	32	6.00	89	15.50	61	10.50	118	20.00	61	10.83	118	20.34
33	5.83	90	15.34	33	6.17	90	15.66	62	10.66	119	20.17	62	11.00	119	20.50
34	6.00	91	15.50	34	6.34	91	15.83	63	10.83	120	20.34	63	11.17	120	20.66
35	6.17	92	15.66	35	6.50	92	16.00	64	11.00			64	11.34		
36	6.34	93	15.83	36	6.66	93	16.17								

Semi-perimeter is the sum of the two adjacent sides. For example: A duct 4" x 6" has a semi-perimeter of 10 inches.
Metric Units: To convert inches to metres multiply by 0.0254.
To convert square ft per linear ft to square metres per linear metre multiply by 0.304815.

Table B-30 Tank areas in square feet — English Units

Where R = D

Tank diameter ft in.	Dished head area ft²	Flat head area ft²	Area, ft² per L.F. of cyl.	"H" Dimension"	Tank diameter ft in.	Dished head area ft²	Flat head area ft²	Area, ft² per L.F. of cyl.	"H" Dimension"
1'-0"	.84	.79	3.14	1.63	5'-0"	21.10	19.64	15.71	8.04
1-1	.99	.92	3.40	1.75	5-1	21.74	20.30	15.97	8.17
1-2	1.14	1.07	3.66	1.88	5-2	22.48	20.97	16.23	8.30
1-3	1.32	1.23	3.93	2.01	5-3	23.17	21.65	16.49	8.44
1-4	1.50	1.40	4.19	2.15	5-4	23.93	22.34	16.76	8.57
1-5	1.69	1.58	4.45	2.28	5-5	24.68	23.05	17.02	8.70
1-6	1.89	1.77	4.71	2.41	5-6	25.50	23.76	17.28	8.84
1-7	2.11	1.97	4.97	2.54	5-7	26.18	24.48	17.54	8.98
1-8	2.34	2.18	5.23	2.68	5-8	27.00	25.22	17.80	9.11
1-9	2.58	2.41	5.50	2.81	5-9	27.82	25.97	18.06	9.25
1-10	2.87	2.64	5.76	2.95	5-10	28.64	26.73	18.33	9.38
1-11	3.09	2.89	6.02	3.08	5-11	29.49	27.49	18.59	9.52
2'-0"	3.37	3.14	6.28	3.22	6'-0"	30.20	28.27	18.85	9.64
2-1	3.65	3.41	6.54	3.35	6-1	31.13	29.07	19.11	9.78
2-2	3.95	3.69	6.81	3.48	6-2	32.03	29.87	19.37	9.92
2-3	4.25	3.98	7.07	3.61	6-3	32.98	30.68	19.64	10.06
2-4	4.57	4.28	7.33	3.75	6-4	33.79	31.49	19.90	10.20
2-5	4.91	4.51	7.59	3.88	6-5	34.69	32.29	20.16	10.33
2-6	5.21	4.91	7.85	4.02	6-6	35.60	33.17	20.42	10.47
2-7	5.61	5.24	8.11	4.15	6-7	36.49	34.04	20.68	10.60
2-8	5.98	5.58	8.37	4.28	6-8	37.43	34.91	20.95	10.73
2-9	6.36	5.94	8.64	4.42	6-9	38.37	35.79	21.21	10.86
2-10	6.74	6.31	8.90	4.55	6-10	39.28	36.67	21.47	10.99
2-11	7.15	6.68	9.17	4.68	6-11	40.29	37.50	21.73	11.12
3'-0"	7.56	7.07	9.43	4.82	7'-0"	41.30	38.49	21.99	11.25
3-1	8.00	7.47	9.69	4.96	7-1	42.20	39.41	22.25	11.39
3-2	8.43	7.88	9.95	5.09	7-2	43.21	40.34	22.51	11.52
3-3	8.87	8.30	10.21	5.22	7-3	44.25	41.28	22.78	11.66
3-4	9.32	8.73	10.47	5.35	7-4	45.26	42.24	23.04	11.80
3-5	9.81	9.17	10.73	5.48	7-5	46.33	43.20	23.30	11.93
3-6	10.30	9.62	11.00	5.62	7-6	47.50	44.16	23.59	12.06
3-7	10.77	10.01	11.26	5.75	7-7	48.37	45.16	23.82	12.19
3-8	11.32	10.56	11.52	5.89	7-8	49.46	46.16	24.09	12.32
3-9	11.82	11.05	11.79	6.02	7-9	50.56	47.17	24.35	12.46
3-10	12.34	11.54	12.04	6.16	7-10	51.63	48.19	24.61	12.60
3-11	12.92	12.05	12.30	6.30	7-11	52.76	49.22	24.82	12.73
4'-0"	13.45	12.57	12.57	6.44	8'-0"	53.85	50.27	25.13	12.86
4-1	14.03	13.09	12.83	6.57	8-1	54.96	51.32	25.39	13.00
4-2	14.60	13.47	13.09	6.70	8-2	56.15	52.38	25.66	13.13
4-3	15.19	14.18	13.35	6.83	8-3	57.31	53.46	25.92	13.27
4-4	15.78	14.73	13.61	6.96	8-4	58.45	54.54	26.18	13.40
4-5	16.39	15.32	13.88	7.09	8-5	59.64	55.64	26.44	13.53
4-6	17.05	15.90	14.14	7.22	8-6	60.65	56.75	26.70	13.66
4-7	17.64	16.50	14.40	7.36	8-7	61.96	57.89	26.97	13.80
4-8	18.32	17.11	14.66	7.50	8-8	63.24	58.99	27.23	13.93
4-9	18.95	17.72	14.92	7.63	8-9	64.34	60.13	27.49	14.06
4-10	19.63	18.35	15.18	7.77	8-10	65.60	61.28	27.75	14.20
4-11	20.35	18.99	15.44	7.91	8-11	66.88	62.45	28.01	14.33

Metric Units: To convert above listed dimensions and areas:

Dia. — Change to inches, multiply by 0.0254 to obtain metres.

"H" Dimension is in inches, multiply by 0.0254 to obtain metres.

Dished Head, and Flat Head Area in square feet, multiply by 0.0929 to obtain square metres.

Area in ft² per L.F. multiply by 0.3048 to obtain metres per linear metre.

Table B-30 Tank areas in square feet — English Units

Where R = D

Tank diameter ft in.	Dished head area ft^2	Flat head area ft^2	Area, ft^2 per L.F. of cyl.	"H" Dimension"
9'-0"	68.10	63.62	28.27	14.46
9-1	69.33	64.80	28.54	14.59
9-2	70.65	66.00	28.80	14.72
9-3	71.97	67.20	29.06	14.86
9-4	73.22	68.42	29.32	15.00
9-5	74.61	69.64	29.58	15.13
9-6	76.00	70.88	29.85	15.26
9-7	77.14	72.13	30.11	15.40
9-8	78.64	73.39	30.37	15.54
9-9	79.96	74.66	30.63	15.67
9-10	81.37	75.94	30.89	15.80
9-11	82.81	77.93	31.15	15.95
10'-0"	84.15	78.54	31.42	16.08
10-1	85.54	79.85	31.68	16.21
10-2	86.96	81.18	31.94	16.34
10-3	88.37	82.52	32.20	16.48
10-4	89.78	83.86	32.46	16.61
10-5	91.26	85.22	32.73	16.74
10-6	92.71	86.59	32.99	16.87
10-7	94.15	87.97	33.25	17.00
10-8	95.68	89.35	33.51	17.14
10-9	97.22	90.76	33.77	17.27
10-10	98.69	92.11	34.03	17.40
10-11	100.26	93.60	34.30	17.53
11'-0"	101.89	95.03	34.56	17.68
11-1		96.48	34.82	
11-2		97.93	35.08	
11-3		99.40	35.34	
11-4		100.88	35.60	
11-5		102.37	35.87	
11-6	111.22	103.87	36.13	18.47
11-7		105.38	36.39	
11-8		106.90	36.65	
11-9		108.43	36.91	
11-10		109.98	37.18	
11-11		111.53	37.44	
12'-0"	121.18	113.10	37.70	19.31
12-1		114.67	37.96	
12-2		116.26	38.22	
12-3		117.86	38.48	
12-4		119.47	38.75	

Tank diameter ft in.	Dished head area ft^2	Flat head area ft^2	Area, ft^2 per L.F. of cyl.	"H" Dimension"
12-5		121.09	39.01	
12-6	131.49	122.72	39.27	20.09
12-7		124.36	39.53	
12-8		126.01	39.79	
12-9		127.67	40.06	
12-10		129.35	40.32	
12-11		131.04	40.58	
13'-0"	142.22	132.73	40.84	20.91
13-1		134.44	41.10	
13-2		136.16	41.36	
13-3		137.89	41.63	
13-4		139.63	41.89	
13-5		141.38	42.15	
13-6	153.36	143.14	42.41	21.72
13-7		144.97	42.67	
13-8		146.69	42.94	
13-9		148.36	43.20	
13-10		150.30	43.46	
13-11		152.11	43.72	
14'-0"	164.94	153.94	43.98	22.53
14-6	176.93	165.13	45.55	23.31
15-0	189.34	176.31	47.12	24.13
15-6	202.18	188.69	48.70	24.94
16-0	215.43	201.06	50.27	25.75
16-6		213.83	51.84	
17-0		226.98	53.41	
17-6		240.53	54.98	
18-0		254.47	56.55	
18-6		268.80	58.12	
19-0		283.53	59.69	
19-6		298.65	61.26	
20-0		314.16	62.83	
21-0		346.36	65.97	
22-0		380.13	69.11	
23-0		415.58	72.26	
24-0		450.79	75.40	
25-0		490.86	78.54	

"H" Constant for each 1" increase in diameter = .01116666 ft.
Area of dished head = 2π DH
Area of flat head = $.7854D^2$

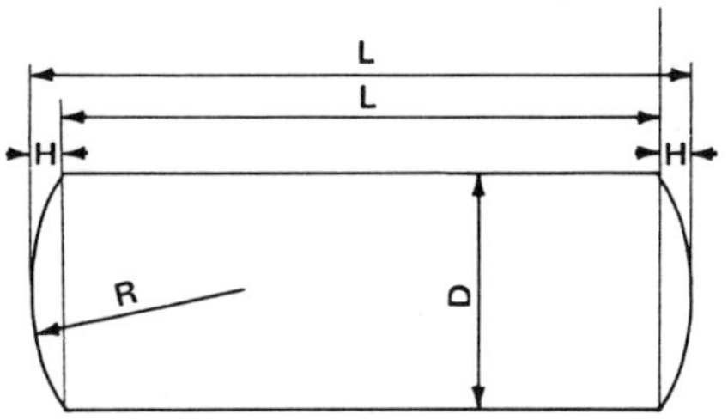

Metric Units: To convert above listed dimensions and areas:

Dia. — Change to inches, multiply by 0.0254 to obtain metres.

"H" Dimension is in inches, multiply by 0.0254 to obtain metres.

Dished Head, and Flat Head Area in square feet, multiply by 0.0929 to obtain square metres.

Area in ft^2 per L.F. multiply by 0.3048 to obtain metres per linear metre.

Table B-31 Capacities—cylinders and spheres – English Units

Dia. ft.	Cu. ft. per foot of cylinder	Gallons per foot of cylinder	Sphere surface in sq. ft.	Sphere volume in cu. ft.	dia. ft.	Cu. ft. per foot of cylinder	Gallons per foot of cylinder	Sphere surface in sq. ft.	Sphere volume in cu. ft.
½	.1963	1.4688	.7854	.0654	25½	510.71	3820.3	2042.8	8682.0
1	.7854	5.8752	3.1416	.5236	26	530.93	3971.6	2123.7	9202.8
1½	1.7671	13.219	7.0686	1.7671	26½	551.55	4125.8	2206.2	9744.0
2	3.1416	23.501	12.566	4.1888	27	572.56	4283.0	2290.2	10306
2½	4.9087	36.720	19.635	8.1812	27½	593.96	4443.1	2375.8	10889
3	7.0686	52.877	28.274	14.137	28	615.75	4606.1	2463.0	11494
3½	9.6211	71.971	38.485	22.449	28½	637.94	4772.1	2551.8	12121
4	12.566	94.033	50.265	33.510	29	660.52	4941.0	2642.1	12770
4½	15.904	118.97	63.617	47.713	29½	683.49	5112.9	2734.0	13442
5	19.635	146.88	78.540	65.450	30	706.86	5287.7	2827.4	14137
5½	23.758	177.72	95.033	87.114	30½	730.62	5465.4	2922.5	14856
6	28.274	211.51	113.10	113.10	31	754.77	5646.1	3019.1	15599
6½	33.183	248.23	132.73	143.79	31½	779.31	5829.7	3117.2	16366
7	38.485	287.88	153.94	179.59	32	804.25	6016.2	3217.0	17157
7½	44.179	330.48	176.71	220.89	32½	829.58	6205.7	3318.3	17974
8	50.265	376.01	201.06	268.08	33	855.30	6398.1	3421.2	18817
8½	56.745	424.48	226.98	321.56	33½	881.41	6593.4	3525.7	19685
9	63.617	475.89	254.47	381.70	34	907.92	6791.7	3631.7	20580
9½	70.882	530.24	283.53	448.92	34½	934.82	6992.9	3739.3	21501
10	78.540	587.52	314.16	523.60	35	962.11	7197.1	3848.5	22449
10½	86.590	747.74	346.36	606.13	35½	989.80	7404.2	3959.2	23425
11	95.033	710.90	380.13	696.91	36	1017.9	7614.2	4071.5	24429
11½	103.87	776.99	415.48	796.33	36½	1046.3	7827.2	4185.4	25461
12	113.10	846.03	452.39	904.78	37	1075.2	8043.1	4300.8	26522
12½	122.72	918.00	490.87	1022.7	37½	1104.5	8262.0	4417.9	27612
13	132.73	992.91	530.93	1150.3	38	1134.1	8483.8	4536.5	28731
13½	143.14	1070.8	572.56	1288.2	38½	1164.2	8708.5	4656.6	29880
14	153.94	1151.5	615.75	1436.8	39	1194.6	8936.2	4778.4	31059
14½	165.13	1235.3	660.52	1596.3	39½	1225.4	9166.8	4901.7	32269
15	176.71	1321.9	706.86	1767.1	40	1256.6	9400.3	5026.5	33510
15½	188.69	1411.5	754.77	1949.8	40½	1288.2	9636.8	5153.0	34783
16	201.06	1504.0	804.25	2144.7	41	1320.3	9876.2	5281.0	36087
16½	213.82	1599.5	855.30	2352.1	41½	1352.7	10119	5410.6	37423
17	226.98	1697.9	907.92	2572.4	42	1385.4	10364	5541.8	38792
17½	240.53	1799.3	962.11	2806.2	42½	1418.6	10612	5674.5	40194
18	254.47	1903.6	1017.9	3053.6	43	1452.2	10863	5808.8	41630
18½	268.80	2010.8	1075.2	3315.2	43½	1486.2	11117	5944.7	43099
19	283.53	2120.9	1134.1	3591.4	44	1520.5	11374	6082.1	44602
19½	298.65	2234.0	1194.6	3882.4	44½	1555.3	11634	6221.1	46140
20	314.16	2350.1	1256.6	4188.8	45	1590.4	11897	6361.7	47713
20½	330.06	2469.0	1320.3	4510.9	45½	1626.0	12163	6503.9	49321
21	346.36	2591.0	1385.4	4849.0	46	1661.9	12432	6647.6	50965
21½	363.05	2715.8	1452.2	5203.7	46½	1698.2	12704	6792.9	52645
22	380.13	2843.6	1520.5	5575.3	47	1734.9	12978	6939.8	54362
22½	397.61	2974.3	1590.4	5964.1	47½	1772.1	13256	7088.2	56115
23	415.48	3108.0	1661.9	6370.6	48	1809.6	13536	7238.2	57906
23½	433.74	3244.6	1734.9	6795.2	48½	1847.5	13820	7389.8	59734
24	452.39	3384.1	1809.6	7238.2	49	1885.7	14106	7543.0	61601
24½	471.44	3526.6	1885.7	7700.1	49½	1924.4	14396	7697.7	63506
25	490.87	3672.0	1963.5	8181.2	50	1963.5	14688	7854.0	65450

Metric Units: To convert above listed dimensions and areas:

Dia. in feet, multiply by 0.3048 to obtain metres.

Cu. ft. per foot of cylinder, multiply by 0.0929 to obtain m^3 per linear metres.

Gallons per foot of cylinder, multiply by 12.4186 to obtain liters per linear metres.

Surface Area in sq. ft., multiply by 0.0929 to obtain square metres.

Volume in cubic ft., multiply by 0.02832 to obtain cubic metres or multiply by 28.32 to obtain liters.

Table B-32 Volumes of pipe insulation — English Units

Nom. NPS Pipe size	CUBIC FEET PER LINEAR FOOT — Nominal Insulation Thickness — Inches: ½	1	1½	2	2½	3	3½	4	4½	5	5½	6	6½	7	7½	8	8½	9
½	0.013	0.040	0.083	0.133	0.231	0.313	0.400	0.498										
¾	0.010	0.038	0.082	0.130	0.228	0.310	0.398	0.496										
1	0.018	0.056	0.100	0.154	0.225	0.307	0.394	0.493										
1¼	0.029	0.051	0.122	0.148	0.219	0.302	0.388	0.487										
1½	0.027	0.071	0.120	0.218	0.300	0.388	0.486	0.611										
2	0.038	0.082	0.137	0.208	0.289	0.377	0.474	0.600										
2½	0.044	0.093	0.191	0.273	0.360	0.458	0.583	0.709										
3	0.044	0.098	0.169	0.251	0.338	0.437	0.562	0.688	0.818	1.003	1.162	1.331						
3½	0.049	0.148	0.229	0.317	0.414	0.540	0.665	0.796	0.892	0.982	1.140	1.309						
4	0.055	0.126	0.208	0.294	0.393	0.518	0.643	0.774	0.960	1.118	1.288	1.468						
6	0.082	0.169	0.268	0.393	0.518	0.649	0.834	0.993	1.162	1.342	1.533	1.734	1.948	2.171	2.405	2.651	2.908	3.174
8		0.223	0.349	0.480	0.666	0.823	0.993	1.173	1.363	1.565	1.778	2.002	2.237	2.482	2.738	3.005	3.283	3.573
10		0.257	0.442	0.600	0.769	0.949	1.140	1.342	1.554	1.778	2.013	2.258	2.514	2.782	3.060	3.349	3.649	3.960
12		0.343	0.513	0.693	0.883	1.086	1.298	1.522	1.757	2.003	2.258	2.525	2.803	3.093	3.393	3.703	4.025	4.356
14		0.328	0.508	0.698	0.900	1.113	1.337	1.571	1.817	2.073	2.340	2.618	2.908	3.207	3.518	3.840	4.173	4.516
16		0.363	0.573	0.786	1.009	1.243	1.489	1.745	2.013	2.291	2.580	2.88	3.191	3.513	3.845	4.189	4.543	4.909
18		0.414	0.638	0.873	1.118	1.374	1.642	1.920	2.209	2.509	2.820	3.142	3.474	3.818	4.173	4.538	4.914	5.302
20		0.458	0.703	0.960	1.228	1.505	1.794	2.094	2.405	2.728	3.060	3.403	3.758	4.123	4.500	4.887	5.285	5.694
24		0.546	0.834	1.134	1.445	1.768	2.100	2.443	2.798	3.163	3.540	3.927	4.325	4.734	5.154	5.585	6.027	6.480
30		0.677	1.031	1.397	1.773	2.160	2.558	2.968	3.388	3,818	4.259	4.713	5.176	5.651	6.136	6.633	7.139	7.658
36		0.808	1.222	1.658	2.100	2.553	3.016	3.491	3.976	4.473	4.980	5.498	6.027	6.567	7.118	7.679	8.253	8.836

Metric Units: To convert cubic feet per linear foot to cubic metres per linear metre multiply by 0.0929.

Table B-33 Areas, sizes and capacities of standard NPS pipe (All dimensions and weights are nominal) English Units

Nom. Size in.	Diameter, in. External	Diameter, in. Internal	Thickness in.	Circumference in. External	Circumference in. Internal	Transverse areas, sq in. External	Transverse areas, sq in. Internal	External surface areas sq ft/lin ft of pipe	Length of pipe containing cu ft	wt of water per foot, lb
1/8	0.405	0.269	0.068	1.272	0.845	0.129	0.057	0.1060	2533.775	0.025
1/4	0.540	0.364	0.088	1.696	1.144	0.229	0.104	0.1414	1383.789	0.045
3/8	0.675	0.493	0.091	2.121	1.549	0.358	0.191	0.1767	754.360	0.083
1/2	0.840	0.622	0.109	2.639	1.954	0.554	0.304	0.220	473.906	0.132
3/4	1.050	0.824	0.113	3.299	2.589	0.866	0.533	0.275	270.034	0.231
1	1.315	1.049	0.133	4.131	3.296	1.358	0.861	0.344	166.618	0.375
1¼	1.660	1.380	0.140	5.215	4.335	2.164	1.495	0.435	96.275	0.65
1½	1.900	1.610	0.145	5.969	5.058	2.835	2.036	0.498	70.733	0.88
2	2.375	2.067	0.154	7.461	6.494	4.430	3.355	0.622	42.913	1.45
2½	2.875	2.469	0.203	9.032	7.757	6.492	4.788	0.753	30.077	2.07
3	3.500	3.068	0.216	10.996	9.638	9.621	7.393	0.917	19.479	3.20
3½	4.000	3.548	0.226	12.566	11.146	12.566	9.886	1.047	14.565	4.29
4	4.500	4.026	0.237	14.137	12.648	15.904	12.730	1.178	11.312	5.50
4½	5.000	4.506	0.247	15.708	14.156	19.635	15.947	1.3009	9.030	6.91
5	5.563	5.047	0.258	17.477	15.856	24.306	20.006	1.4586	7.198	8.67
6	6.625	6.065	0.280	20.813	19.054	34.472	28.891	1.7384	4.984	12.51
7	7.625	7.023	0.301	23.955	22.063	45.664	38.738	1.996	3.717	16.80
8	8.625	8.071	0.277	27.096	25.356	58.426	51.161	2.2350	2.815	22.18
8	8.625	7.981	0.322	27.096	25.073	58.426	50.027	2.2058	2.878	21.70
9	9.625	8.941	0.342	30.238	28.089	72.760	62.786	2.5220	2.294	27.20
10	10.750	10.192	0.279	33.772	32.019	90.763	81.685	2.8174	1.765	35.37
10	10.750	10.136	0.307	33.772	31.843	90.763	80.691	2.8104	1.826	34.20
10	10.750	10.020	0.365	33.772	31.479	90.763	78.855	2.8104	1.826	34.20
11	11.750	11.000	0.375	36.914	34.558	108.434	95.033	3.076	1.515	41.20
12	12.750	12.090	0.330	40.055	37.892	127.676	114.800	3.338	1.254	49.70
12	12.750	12.000	0.375	40.055	37.699	127.676	113.097	3.338	1.273	49.00

Note: Conversions to metric units not given, as the above is specific information regarding standard NPS pipe which is based on English Units.

Table B-34 Gallons capacity per foot in height of rectangular tanks — English Units

Width of tank	Length of tank																				
ft in.	ft 2	ft in. 2 6	ft 3	ft in. 3 6	ft 4	ft in. 4 6	ft 5	ft in. 5 6	ft 6	ft in. 6 6	ft 7	ft in. 7 6	ft 8	ft in. 8 6	ft 9	ft in. 9 6	ft 10	ft in. 10 6	ft 11	ft in. 11 6	ft. 12
2	29.92	37.40	44.88	52.36	59.84	67.32	74.81	82.29	89.77	97.25	104.73	112.21	119.69	127.17	134.65	142.13	149.61	157.09	164.57	172.05	179.53
2 6	...	46.75	56.10	65.45	74.80	84.16	93.51	102.86	112.21	121.56	130.91	140.26	149.61	158.96	168.31	177.66	187.01	196.36	205.71	215.06	224.41
3	...	...	67.32	78.54	89.77	100.99	112.21	123.43	134.65	145.87	157.09	168.31	179.53	190.75	202.97	213.19	224.41	235.63	246.86	258.07	269.30
3 6	...	...	...	91.64	104.73	117.82	130.91	144.00	157.09	170.18	183.27	196.36	209.45	222.54	235.63	248.73	261.82	274.90	288.00	301.09	314.18
4	...	...	...	...	119.69	134.65	149.61	164.57	179.53	194.49	209.45	224.41	239.37	254.34	269.30	284.26	299.22	314.18	329.14	344.10	359.06
4 6	...	...	...	...	...	151.48	168.31	185.14	201.97	218.80	235.63	252.47	269.30	286.13	302.96	319.79	336.62	353.45	370.28	387.11	403.94
5	...	...	...	...	...	...	187.01	205.71	224.41	243.11	261.82	280.52	299.22	317.92	336.62	355.32	374.03	392.72	411.43	430.13	448.83
5 6	...	...	...	...	...	...	...	226.28	246.86	267.43	288.00	308.57	329.14	349.71	370.28	390.85	411.43	432.00	452.57	473.14	493.71
6	...	...	...	...	...	...	...	...	269.30	291.74	314.18	336.62	359.06	381.50	403.94	426.39	448.83	471.27	493.71	516.15	538.59
6 6	...	...	...	...	...	...	...	...	...	316.05	340.36	364.67	388.98	413.30	437.60	461.92	486.23	510.54	534.85	559.16	583.47
7	...	...	...	...	...	...	...	...	...	...	366.54	392.72	418.91	445.09	471.27	497.45	523.64	549.81	575.99	602.18	628.36
7 6	...	...	...	...	...	...	...	...	...	...	...	420.78	448.83	476.88	504.93	532.98	561.04	589.08	617.14	645.19	673.24
8	...	...	...	...	...	...	...	...	...	...	...	...	478.75	508.67	538.59	568.51	598.44	628.36	658.28	688.20	718.12
8 6	...	...	...	...	...	...	...	...	...	...	...	...	...	540.46	572.25	604.05	635.84	667.63	699.42	731.21	763.00
9	...	...	...	...	...	...	...	...	...	...	...	...	...	...	605.92	639.58	673.25	706.90	740.56	774.23	807.89
9 6	...	...	...	...	...	...	...	...	...	...	...	...	...	...	...	675.11	710.65	746.17	781.71	817.24	852.77
10	...	...	...	...	...	...	...	...	...	...	...	...	...	...	...	...	748.05	785.45	822.86	860.26	897.66
10 6	...	...	...	...	...	...	...	...	...	...	...	...	...	...	...	...	...	824.73	864.00	903.26	942.56
11	...	...	...	...	...	...	...	...	...	...	...	...	...	...	...	...	...	...	905.14	946.27	987.43
11 6	...	...	...	...	...	...	...	...	...	...	...	...	...	...	...	...	...	...	...	989.29	1032.3
12	...	...	...	...	...	...	...	...	...	...	...	...	...	...	...	...	...	...	...	...	1077.2

To figure the capacity of any rectangular tank, multiply the width, by the length, by the height, to obtain the cubic foot capacity, then multiply this result by 7.48 and the answer is the capacity in gallons.

Metric Units: Dimension in feet multiplied by 0.3048 gives dimension in metres.
Gallons capacity per foot in height multiplied by 12.4186 gives capacity in liters per metre in height.

Table B-35 Temperature and pressure of saturated steam — English Units

Abs press lbs/sq in.	Ga press lbs/sq in.	Temperature °F	Total heat Btu/lb	Abs press lbs/sq in.	Ga press lbs/sq in.	Temperature °F	Total heat Btu/lb
1		101.74	1106	190	175.3	377.51	1198
2		126.08	1116	200	185.3	381.79	1198
3		141.48	1123	210	195.3	385.90	1199
4		152.97	1127	220	205.3	389.86	1200
5		162.24	1131	230	215.3	393.68	1200
6		170.06	1134	240	225.3	397.37	1201
7		176.85	1137	250	235.3	400.95	1201
8		182.86	1139	260	245.3	404.42	1202
9		188.28	1141	270	255.3	407.78	1202
10		193.21	1143	280	265.3	411.05	1202
12		201.96	1147	290	275.3	414.23	1203
14		209.56	1150	300	285.3	417.33	1203
14.7	0	212.00	1150.4	320	305.3	423.29	1203
15	.3	213.03	1151	340	325.3	428.97	1204
20	5.3	227.96	1156	360	345.3	434.40	1204
25	10.3	240.07	1161	380	365.3	439.60	1204
30	15.3	250.33	1164	400	385.3	444.59	1205
35	20.3	259.28	1167	420	405.3	449.39	1205
40	25.3	267.25	1170	440	425.3	454.02	1205
45	30.3	274.44	1172	460	445.3	458.50	1205
50	35.3	281.01	1174	480	465.3	462.82	1205
55	40.3	287.07	1176	500	485.3	467.01	1204
60	45.3	292.71	1178	600	585.3	486.21	1203
65	50.3	297.97	1179	700	685.3	503.10	1201
70	55.3	302.92	1181	800	785.3	518.23	1199
75	60.3	307.60	1182	900	885.3	531.98	1195
80	65.3	312.03	1183	1000	985.3	544.61	1192
85	70.3	316.25	1184	1100	1085.3	556.31	1188
90	75.3	320.27	1185	1200	1185.3	567.22	1183
95	80.3	324.12	1186	1300	1285.3	577.46	1179
100	85.3	327.81	1187	1400	1385.3	587.10	1173
105	90.3	331.36	1188	1500	1485.3	596.23	1168
110	95.3	334.77	1189	1600	1585.3	604.90	1162
115	100.3	338.07	1190	1700	1685.3	613.15	1156
120	105.3	341.25	1190	1800	1785.3	621.03	1149
125	110.3	344.33	1191	1900	1885.3	628.58	1142
130	115.3	347.32	1192	2000	1985.3	635.82	1135
135	120.3	350.21	1192	2200	2185.3	649.46	1119
140	125.3	353.02	1193	2400	2385.3	662.12	1101
145	130.3	355.76	1194	2600	2585.3	673.94	1080
150	135.3	358.42	1194	2800	2785.3	684.99	1055
160	145.3	363.53	1195	3000	2985.3	695.36	1020
170	155.3	368.41	1196	3200	3185.3	705.11	934
180	165.3	373.06	1197	3206.2	3191.5	705.40	903

Table B-35a Properties of saturated and superheated steam

Note: p = absolute pressure.
ts = temperature in saturation state.
t = temperature in superheated state.
v = specific volume — m^3/kg.
u = specific internal energy — kJ/kg.
h = specific enthalpy — kJ/kg.
s = specific entropy — kJ/kg°K.
g (suffix) = property of saturated vapour.

Metric Units

p 10^5 N/m²	ts °C.		t °C.	50	100	150	200	250	300	400	500	p lbf/in²	ts °F.
0		u = h − RT	v										
			u	2446	2517	2589	2662	2737	2812	2969	3132		
			h	2595	2689	2784	2880	2978	3077	3280	3489		
			s										
0.006112	0.01	vg 206.1	v	243.9	281.7	319.5	357.3	395.0	432.8	508.3	583.8	0.0885	32
		ug 2375	u	2446	2517	2589	2662	2737	2812	2969	3132		
		hg 2501	h	2595	2689	2784	2880	2978	3077	3280	3489		
		sg 9.155	s	9.468	9.739	9.978	10.193	10.390	10.571	10.897	11.187		
0.01	7.0	vg 129.2	v	149.1	172.2	195.3	218.4	241.4	264.5	310.7	356.8	0.142	44
		ug 2385	u	2446	2517	2589	2662	2737	2812	2969	3132		
		hg 2514	h	2595	2689	2784	2880	2978	3077	3280	3489		
		sg 8.974	s	9.241	9.512	9.751	9.966	10.163	10.344	10.670	10.960		
0.05	32.9	vg 28.20	v	29.78	34.42	39.04	43.66	48.28	52.90	62.13	71.36	0.725	91
		ug 2420	u	2445	2516	2589	2662	2737	2812	2969	3132		
		hg 2561	h	2594	2688	2784	2880	2978	3077	3280	3489		
		sg 8.394	s	8.496	8.768	9.008	9.223	9.420	9.601	9.927	10.217		
0.1	45.8	vg 14.67	v	14.87	17.20	19.51	21.83	24.14	26.45	31.06	35.68	1.45	114
		ug 2437	u	2443	2516	2588	2662	2736	2812	2969	3132		
		hg 2584	h	2592	2688	2783	2880	2977	3077	3280	3489		
		sg 8.149	s	8.173	8.447	8.688	8.903	9.100	9.281	9.607	9.897		
0.5	81.3	vg 3.239	v		3.420	3.890	4.356	4.821	5.284	6.209	7.134	7.25	178
		ug 2483	u		2512	2585	2660	2735	2812	2969	3132		
		hg 2645	h		2683	2780	2878	2976	3076	3279	3489		
		sg 7.593	s		7.694	7.940	8.158	8.355	8.537	8.864	9.154		
0.75	91.8	vg 2.217	v		2.271	2.588	2.901	3.211	3.521	4.138	4.755	10.875	197
		ug 2496	u		2510	2585	2659	2734	2811	2969	3132		
		hg 2662	h		2680	2779	2877	2975	3075	3279	3489		
		sg 7.456	s		7.500	7.750	7.969	8.167	8.349	8.676	8.967		
1.0	99.6	vg 1.694	v		1.696	1.937	2.173	2.406	2.639	3.103	3.565	14.50	211
		ug 2506	u		2506	2583	2659	2734	2811	2968	3131		
		hg 2675	h		2676	2777	2876	2975	3075	3278	3488		
		sg 7.359	s		7.360	7.614	7.834	8.033	8.215	8.543	8.834		
1.01325	100.0	vg 1.673	v			1.912	2.145	2.375	2.604	3.062	3.519	14.69	212
		ug 2506	u			2583	2659	2734	2811	2968	3131		
		hg 2676	h			2777	2876	2975	3075	3278	3488		
		sg 7.355	s			7.608	7.828	8.027	8.209	8.537	8.828		
1.5	111.4	vg 1.159	v			1.286	1.445	1.601	1.757	2.067	2.376	21.75	232
		ug 2519	u			2580	2656	2733	2809	2967	3131		
		hg 2693	h			2773	2873	2973	3073	3277	3488		
		sg 7.223	s			7.420	7.643	7.843	8.027	8.355	8.646		
2.0	120.2	vg 0.8856	v			0.9602	1.081	1.199	1.316	1.549	1.781	29.00	248
		ug 2530	u			2578	2655	2731	2809	2967	3131		
		hg 2707	h			2770	2871	2971	3072	3277	3487		
		sg 7.127	s			7.280	7.507	7.708	7.892	8.221	8.513		
3.0	133.5	vg 0.6057	v			0.6342	0.7166	0.7965	0.8754	1.031	1.187	43.50	272
		ug 2544	u			2572	2651	2729	2807	2966	3130		
		hg 2725	h			2762	2866	2968	3070	3275	3486		
		sg 6.993	s			7.078	7.312	7.517	7.702	8.032	8.324		
4.0	143.6	vg 0.4623	v			0.4710	0.5345	0.5953	0.6549	0.7725	0.8893	58.00	290
		ug 2554	u			2565	2648	2727	2805	2965	3129		
		hg 2739	h			2753	2862	2965	3067	3274	3485		
		sg 6.897	s			6.929	7.172	7.379	7.566	7.898	8.191		
			t °F.	122	212	302	392	482	572	752	932		

Table B-35a Properties of saturated and superheated steam (continued)

Note: p = absolute pressure.
ts = temperature in saturation state.
t = temperature in superheated state.
v = specific volume — m^3/kg.
u = specific internal energy — kJ/kg.
h = specific enthalpy — kJ/kg.
s = specific entropy — kJ/kg°K.
g (suffix) = property of saturated vapour.

Metric Units

p 10^5 N/m²	ts °C.		t °C	200	250	300	350	400	450	500	600	p lbf/in²	ts °F.
5	151.8	vg 0.3748	v	0.4252	0.4745	0.5226	0.5701	0.6172	0.6641	0.7108	0.8040	72.5	305
		ug 2562	u	2644	2725	2804	2883	2963	3045	3129	3300		
		hg 2749	h	2857	2962	3065	3168	3272	3377	3484	3702		
		sg 6.822	s	7.060	7.271	7.460	7.633	7.793	7.944	8.807	8.351		
6	158.8	vg 0.3156	v	0.3522	0.3940	0.4344	0.4743	0.5136	0.5528	0.5919	0.6697	87.0	318
		ug 2568	u	2640	2722	2801	2881	2962	3044	3128	3299		
		hg 2757	h	2851	2958	3062	3166	3270	3376	3483	3701		
		sg 6.761	s	6.968	7.182	7.373	7.546	7.707	7.858	8.001	8.267		
7	165.0	vg 0.2728	v	0.3001	0.3364	0.3714	0.4058	0.4397	0.4734	0.5069	0.5737	101.5	319
		ug 2573	u	2636	2720	2800	2880	2961	3043	3127	3298		
		hg 2764	h	2846	2955	3060	3164	3269	3374	3482	3700		
		sg 6.709	s	6.888	7.106	7.298	7.473	7.634	7.786	7.929	8.195		
8	170.4	vg 0.2403	v	0.2610	0.2933	0.3242	0.3544	0.3842	0.4138	0.4432	0.5018	116.0	339
		ug 2577	u	2631	2716	2798	2878	2960	3042	3126	3298		
		hg 2769	h	2840	2951	3057	3162	3267	3373	3481	3699		
		sg 6.663	s	6.817	7.040	7.233	7.409	7.571	7.723	7.866	8.132		
9	179.4	vg 0.2149	v	0.2305	0.2597	0.2874	0.3144	0.3410	0.3674	0.3937	0.4458	130.5	348
		ug 2581	u	2628	2714	2796	2877	2959	3041	3126	3298		
		hg 2774	h	2835	2948	3055	3160	3266	3372	3480	3699		
		sg 6.623	s	6.753	6.980	7.176	7.352	7.515	7.667	7.811	8.077		
10	179.9	vg 0.1944	v	0.2061	0.2328	0.2580	0.2825	0.3065	0.3303	0.3540	0.4010	145.0	356
		ug 2584	u	2623	2711	2794	2875	2957	3040	3124	3297		
		hg 2778	h	2829	2944	3052	3158	3264	3370	3478	3698		
		sg 6.586	s	6.695	6.926	7.124	7.301	7.464	7.617	7.761	8.028		
15	198.3	vg 0.1317	v	0.1324	0.1520	0.1697	0.1865	0.2029	0.2191	0.2351	0.2667	217.5	389
		ug 2595	u	2597	2697	2784	2868	2952	3035	3120	3294		
		hg 2792	h	2796	2925	3039	3148	3256	3364	3473	3694		
		sg 6.445	s	6.452	6.711	6.919	7.102	7.268	7.423	7.569	7.838		
20	212.4	vg 0.0996	v		0.1115	0.1255	0.1386	0.1511	0.1634	0.1756	0.1995	290.0	416
		ug 2600	u		2681	2774	2861	2946	3030	3116	3291		
		hg 2799	h		2904	3025	3138	3248	3357	3467	3690		
		sg 6.340	s		6.547	6.768	6.957	7.126	7.283	7.431	7.701		
30	233.8	vg 0.0666	v		0.0706	0.0812	0.0905	0.0993	0.1078	0.1161	0.1324	435.0	453
		ug 2603	u		2646	2751	2845	2933	3020	3108	3285		
		hg 2803	h		2858	2995	3117	3231	3343	3456	3682		
		sg 6.186	s		6.289	6.541	6.744	6.921	7.082	7.233	7.507		
40	250.3	vg 0.0498	v			0.0588	0.0664	0.0733	0.0800	0.0864	0.0988	580.0	483
		ug 2602	u			2728	2828	2921	3010	3099	3279		
		hg 2801	h			2963	3094	3214	3330	3445	3674		
		sg 6.070	s			6.394	6.594	6.769	6.935	7.089	7.368		
50	263.9	vg 0.0394	v			0.0453	0.0519	0.0578	0.0632	0.0685	0.0786	725.0	507
		ug 2497	u			2700	2810	2907	3000	3090	3273		
		hg 2794	h			2927	3070	3196	3316	3433	3666		
		sg 5.973	s			6.212	6.451	6.646	6.818	6.975	7.258		
60	275.6	vg 0.0324	v			0.0362	0.0422	0.0473	0.0521	0.0566	0.0652	870.0	528
		ug 2590	u			2670	2792	2893	2988	3081	3266		
		hg 2784	h			2887	3045	3177	3301	3421	3657		
		sg 5.890	s			6.071	6.336	6.541	6.719	6.879	7.166		
70	285.8	vg 0.0274	v			0.0295	0.0352	0.0399	0.0441	0.0481	0.0556	1015.0	546
		ug 2581	u			2634	2772	2879	2978	3073	3260		
		hg 2772	h			2841	3018	3158	3287	3410	3649		
		sg 5.814	s			5.934	6.231	6.448	6.632	6.796	7.088		
			t °F.	392	482	572	662	752	842	932	1112		

Table B-35a Properties of saturated and superheated steam (continued)

Note: p = absolute pressure.
ts = temperature in saturation state.
t = temperature in superheated state.
v = specific volume — m^3/kg.
u = specific internal energy — kJ/kg.
h = specific enthalpy — kJ/kg.
s = specific entropy — kJ/kg°K.
g (suffix) = property of saturated vapour.

Metric Units

p $10^5 N/m^2$	ts °C.		t °C	350	375	400	425	450	500	600	700	p lbf/in²	ts °F.
80	295.0	vg 0.02352	v x 10^2	2.994	3.220	3.428	3.625	3.812	4.170	4.839	5.476	1160	563
		hg 2758	h	2990	3067	3139	3207	3272	3398	3641	3881		
		sg 5.744	s	6.133	6.255	6.364	6.463	6.555	6.723	7.019	7.279		
90	303.3	vg 0.02048	v x 10^2	2.578	2.794	2.991	3.173	3.346	3.673	4.279	4.852	1450	592
		hg 2743	h	2959	3042	3118	3189	3256	3385	3633	3874		
		sg 5.679	s	6.039	6.171	6.286	6.390	6.848	6.657	6.958	7.220		
100	311.0	vg 0.01802	v x 10^2	2.241	2.453	2.639	2.812	2.972	3.275	3.831	4.353	1450	592
		hg 2725	h	2926	3017	3097	3172	3241	3373	3624	3868		
		sg 5.615	s	5.947	6.091	6.213	6.321	6.419	6.596	6.902	7.166		
110	318.0	vg 0.01598	v x 10^2	1.960	2.169	2.350	2.514	2.666	2.949	3.465	3.945	1595	604
		hg 2705	h	2889	2989	3075	3153	3225	3360	3616	3862		
		sg 5.553	s	5.856	6.014	6.143	6.257	6.358	6.539	6.850	7.117		
120	324.6	vg 0.01426	v x 10^2	1.719	1.931	2.107	2.265	2.410	2.677	3.159	3.605	1740	616
		hg 2685	h	2849	2960	3052	3134	3209	3348	3607	3856		
		sg 5.493	s	5.762	5.937	6.076	6.195	6.301	6.487	6.802	7.072		
130	330.8	vg 0.01278	v x 10^2	1.509	1.726	1.901	2.053	2.193	2.447	2.901	3.318	1885	627
		hg 2662	h	2804	2929	3028	3114	3192	3335	3599	3850		
		sg 5.433	s	5.664	5.862	6.011	6.136	6.242	6.437	6.758	7.030		
140	336.6	vg 0.01149	v x 10^2	1.321	1.548	1.722	1.872	2.006	2.250	2.679	3.071	2030	638
		hg 2638	h	2753	2896	3003	3093	3175	3322	3590	3843		
		sg 5.373	s	5.559	5.784	5.946	6.079	6.193	6.390	6.716	6.991		
150	342.1	vg 0.01035	v x 10^2	1.146	1.391	1.566	1.714	1.844	2.078	2.487	2.857	2175	648
		hg 2611	h	2693	2861	2977	3073	3157	3309	3581	3837		
		sg 5.312	s	5.443	5.707	5.883	6.023	6.142	6.345	6.677	6.954		
160	347.3	vg 0.00932	v x 10^2	0.976	1.248	1.427	1.573	1.702	1.928	2.319	2.670	2320	657
		hg 2582	h	2617	2821	2949	3051	3139	3295	3573	3831		
		sg 5.248	s	5.304	5.626	5.820	5.968	6.093	6.301	6.639	6.919		
170	352.3	vg 0.00838	v x 10^2		1.117	1.303	1.449	1.576	1.796	2.171	2.506	2465	666
		hg 2548	h		2778	2920	3028	3121	3281	3564	3825		
		sg 5181	s		5.541	5.756	5.914	6.044	6.260	6.603	6.886		
180	357.0	vg 0.00751	v x 10^2		0.997	1.191	1.338	1.463	1.678	2.039	2.359	2610	675
		hg 2510	h		2729	2888	3004	3102	3268	3555	3818		
		sg 5.108	s		5.449	5.691	5.861	5.997	6.219	6.569	6.855		
190	361.4	vg 0.00668	v x 10^2		0.882	1.089	1.238	1.362	1.572	1.921	2.228	2755	683
		hg 2466	h		2674	2855	2980	3082	3254	3546	3812		
		sg 5.027	s		5.348	5.625	5.807	5.950	6.180	6.536	6.825		
200	365.7	vg 0.00585	v x 10^2		0.768	0.995	1.147	1.270	1.477	1.815	2.110	2900	690
		hg 2411	h		2605	2819	2955	3062	3239	3537	3806		
		sg 4.928	s		5.228	5.556	5.753	5.904	6.142	6.505	6.796		
210	369.8	vg 0.00498	v x 10^2		0.650	0.908	1.064	1.187	1.390	1.719	2.003	3045	698
		hg 2336	h		2500	2781	2928	3041	3225	3528	3799		
		sg 4.803	s		5.050	5.484	5.699	5.859	6.105	6.474	6.768		
220	373.7	vg 0.00368	v x 10^2		0.450	0.825	0.987	1.111	1.312	1.632	1.906	3190	705
		hg 2178	h		2300	2738	2900	3020	3210	3519	3793		
		sg 4.552	s		4.725	5.409	5.645	5.813	6.068	6.444	6.742		
221.2	374.15	vc 0.00317	v x 10^2	0.163	0.351	0.816	0.978	1.103	1.303	1.622	1.895	3207	706
		hc 2084	h	1637	2139	2733	2896	3017	3208	3518	3792		
		sc 4.406	s	3.708	4.490	5.398	5.638	5.807	6.064	6.441	6.739		
			t °F.	662	707	752	797	842	939	1112	1292		

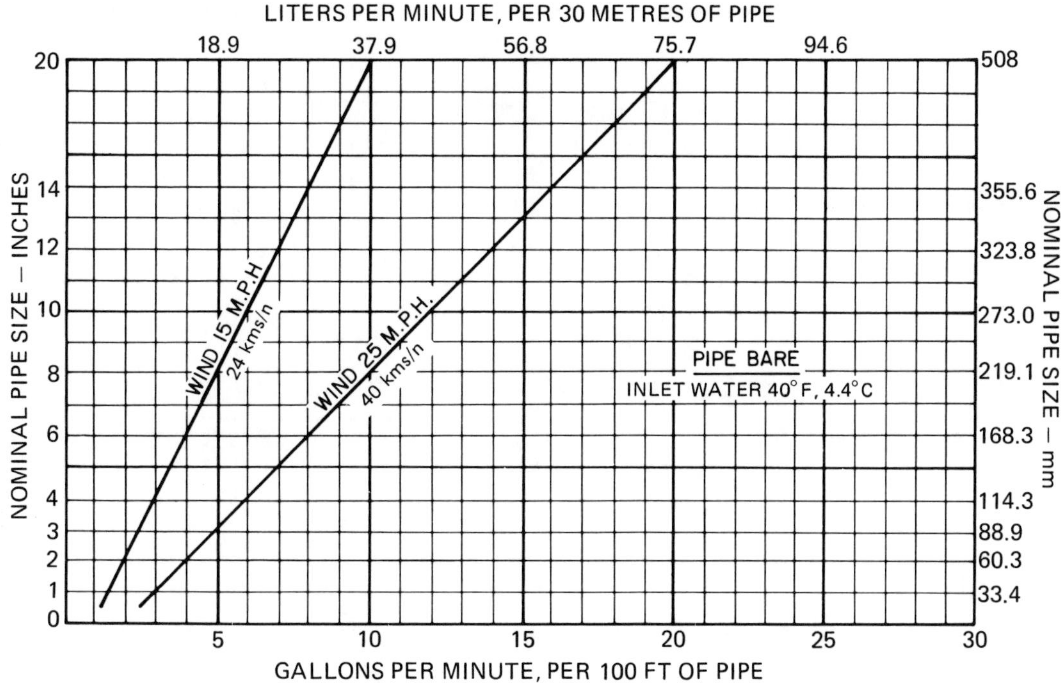

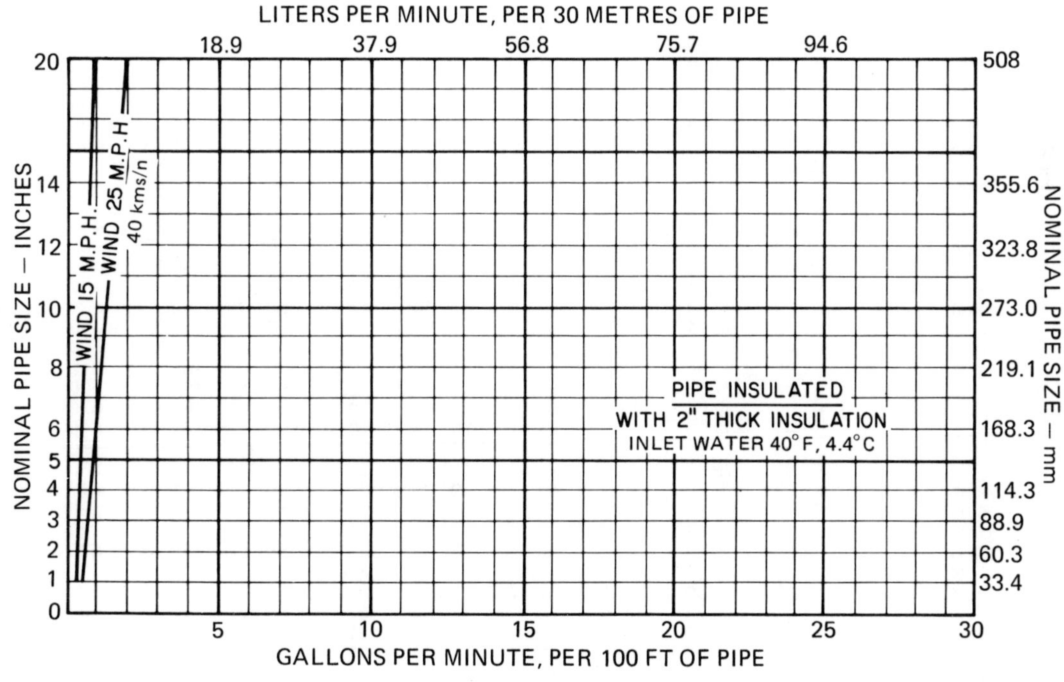

Chart B-11 Flow required to prevent freezing of water lines −20°F (−28.9°C) ambient air temperature

Table B-36 Percent relative humidity above which condensation will occur on surface of equipment or pipe not insulated

English and Metric Units (Eng. Units → ; Metric Units ←)

Equip or pipe temp °F	Ambient air temperature °F 40	45	50	55	60	65	70	75	80	85	90	95	100	Equip or pipe temp °C
35	82	68	57	48	39	33	28	24	20	17	15	13	11	1.7
40		82	68	57	48	40	33	28	24	20	17	15	14	4.4
45			82	69	57	48	41	34	29	25	21	18	16	7.2
50				82	69	57	49	41	35	30	26	22	19	10.0
55					83	70	59	50	42	36	31	27	23	12.8
60						84	70	60	51	43	37	32	27	15.6
65							84	71	60	52	44	38	33	18.3
70								84	71	61	52	45	38	21.1
75									84	72	62	53	46	23.9
80										85	73	63	54	26.7
85											85	73	63	29.4
90												86	74	32.2
95													86	35.0
Ambient air temperature °C	4.4	7.2	10.0	12.8	15.6	18.3	21.1	23.9	26.7	29.4	32.2	35.0	37.8	

Table B-37 Percent relative humidity above which condensation will occur on surface insulated with 1/4" thickness of insulation-mastic

English and Metric Units (Eng. Units → ; Metric Units ←)

Equip or pipe temp °F	Ambient air temperature °F 40	45	50	55	60	65	70	75	80	85	90	95	100	Equip or pipe temp °C
35	88	81	71	67	62	58	54	49	48	46	43	40	38	1.7
40		86	81	74	68	63	59	55	52	49	47	43	41	4.4
45			91	82	75	69	64	60	57	54	50	48	45	7.2
50				91	82	75	69	65	61	58	55	51	48	10.0
55					91	83	77	71	66	62	59	54	52	12.8
60						91	83	78	72	67	63	59	56	15.6
65							92	84	78	73	68	64	60	18.3
70								92	84	79	73	68	64	21.1
75									92	86	79	74	69	23.9
80										92	86	79	77	26.7
85											92	86	80	29.4
90												92	87	32.2
95													92	35.0
Ambient air temperature °C	4.4	7.2	10.0	12.8	15.6	18.3	21.1	23.9	26.7	29.4	32.2	35.0	37.8	

Table B-38 Weight of water and heat units per pound at different temperatures

English Units

Temp. deg F	Lb per cu ft	Btu per lb	Temp. deg F	Lb per cu ft	Btu per lb	Temp. deg F	Lb per cu ft	Btu per lb	Temp. deg F	Lb per cu ft	Btu per lb	Temp. deg F	Lb per cu ft	Btu per lb
32	62.41	0.	79	62.22	47.00	127	61.60	94.87	175	60.68	142.86	280	58.04	248.95
33	62.41	1.01	80	62.21	48.00	128	61.58	95.87	176	60.66	143.86	290	57.74	259.20
34	62.42	2.01	81	62.20	49.00	129	61.56	96.86	177	60.64	144.86	300	57.41	269.48
35	62.42	3.02	82	62.19	50.00	130	61.55	97.86	178	60.62	145.86	310	57.08	279.80
36	62.42	4.03	83	62.18	51.00	131	61.53	98.86	179	60.60	146.87	320	56.75	290.17
37	62.42	5.03	84	62.17	52.00	132	61.51	99.86	180	60.57	147.87	330	56.40	300.59
38	62.42	6.04	85	62.16	53.00	133	61.50	100.86	181	60.55	148.87	340	56.02	311.05
39	62.42	7.04	86	62.15	54.00	134	61.48	101.85	182	60.53	149.87	350	55.65	321.55
40	62.42	8.05	87	62.14	55.00	135	61.46	102.85	183	60.51	150.87	360	55.25	332.10
41	62.42	9.05	88	62.13	56.00	136	61.44	103.85	184	60.49	151.87	370	54.85	342.71
42	62.42	10.05	89	62.12	57.00	137	61.43	104.85	185	60.46	152.87	380	54.47	353.39
43	62.42	11.05	91	62.10	58.99	138	61.41	105.84	186	60.44	153.88	390	54.05	364.14
44	62.42	12.05	92	62.08	59.98	139	61.39	106.84	187	60.42	154.88	400	53.62	374.96
45	62.42	13.05	93	62.07	60.98	140	61.37	107.84	188	60.40	155.88	410	53.19	385.86
46	62.41	14.06	94	62.06	61.97	141	61.36	108.84	189	60.37	156.89	420	52.74	396.84
47	62.41	15.06	95	62.05	62.96	142	61.34	109.84	190	60.35	157.89	430	52.33	407.91
48	62.41	16.06	96	62.04	63.96	143	61.32	110.84	191	60.33	158.90	440	51.87	419.07
49	62.41	17.06	97	62.02	64.95	144	61.30	111.84	192	60.30	159.90	450	51.28	430.3
50	62.40	18.06	98	62.01	65.94	145	61.28	112.84	193	60.28	160.90	460	51.02	441.7
51	62.40	19.06	99	62.00	66.94	146	61.26	113.84	194	60.26	161.91	470	50.51	453.2
52	62.40	20.06	100	61.99	67.93	147	61.25	114.84	195	60.23	162.91	480	50.00	465.0
53	62.39	21.06	101	61.98	68.92	148	61.23	115.84	196	60.21	163.92	490	49.50	477.0
54	62.39	22.06	102	61.96	69.02	150	61.19	117.84	197	60.19	164.92	500	48.78	489.1
55	62.38	23.06	103	61.95	70.91	151	61.17	118.85	198	60.16	165.93	510	48.31	501.6
56	62.38	24.05	104	61.94	71.91	152	61.15	119.85	199	60.14	166.93	520	47.62	514.2
57	62.38	25.05	105	61.93	72.91	153	61.13	120.85	200	60.11	167.94	530	46.95	527.0
58	62.37	26.05	106	61.91	73.90	154	61.11	121.85	201	60.09	168.95	540	46.30	540.0
59	62.37	27.05	107	61.90	74.90	155	61.09	122.85	202	60.07	169.95	550	45.66	553.2
60	62.36	28.05	108	61.89	75.90	156	61.07	123.85	203	60.04	170.96	560	44.84	566.7
61	62.35	29.05	109	61.87	76.89	157	61.05	124.85	204	60.02	171.96	570	44.05	580.4
62	62.35	30.05	110	61.86	77.89	158	61.03	125.85	205	59.99	172.97	580	43.29	594.4
63	62.34	31.05	111	61.84	78.89	159	61.01	126.85	206	59.97	173.97	590	42.37	608.7
64	62.34	32.04	112	61.83	79.89	160	60.99	127.85	207	59.95	174.98	600	41.49	623.2
65	62.33	33.04	113	61.81	80.89	161	60.97	128.85	208	59.92	175.98	610	40.49	638.0
66	62.32	34.04	114	61.80	81.89	162	60.95	129.85	209	59.90	176.99	620	39.37	653.4
67	62.32	35.04	115	61.78	82.89	163	60.93	130.85	210	59.87	177.99	630	38.31	669.5
68	62.31	36.03	116	61.77	83.88	164	60.91	131.85	211	59.85	179.00	640	37.17	686.6
69	62.30	37.03	117	61.75	84.88	165	60.89	132.85	212	59.82	180.00	650	35.97	705.2
70	62.30	38.03	118	61.74	85.88	166	60.87	133.85	214	59.81	182.02	660	34.48	725.3
71	62.29	39.03	119	61.72	86.88	167	60.85	134.85	216	59.77	184.03	670	32.89	747.5
72	62.28	40.02	120	61.71	87.88	168	60.83	135.85	218	59.70	186.04	680	31.06	772.6
73	62.27	41.02	121	61.69	88.88	169	60.81	136.85	220	59.67	188.06	690	28.82	803.0
74	62.26	42.02	122	61.68	89.88	170	60.79	137.85	230	59.42	198.15	700	25.34	846.8
75	62.25	43.01	123	61.66	90.88	171	60.77	138.85	240	59.17	208.26	706	19.16	935.0
76	62.25	44.01	124	61.64	91.88	172	60.75	139.85	250	58.89	218.39			
77	62.24	45.01	125	61.63	92.87	173	60.73	140.85	260	58.62	228.55			
78	62.23	46.00	126	61.61	93.87	174	60.71	141.85	270	58.34	238.74			

Metric Units: To convert temperature use Table B-2.
To convert lb per cu ft multiply by 16.0185 to obtain kg/m^3.
To convert Btu per lb multiply by 2326 to obtain joules per kilogram.

Table B-39 Conductivities – English and Metric Units

Material			Mean Temp. °F	Conductivity k Btu in/ft²hr°F	Density kg/m³	Mean Temp. °C	Conductivity W/m K
			BUILDING CONSTRUCTION MATERIALS				
Asphalt, street	132#	Density	68	5.28	2116	20.0	0.7614
Beaver board							
Cane fiber	13.8#	Density	75	0.33	221	23.9	0.0476
Spruce fiber	31#	Density	75	1.97	497	23.9	0.2541
Cane fiber board	(Celotex)						
	13.8#	Density	75	0.33	221	23.9	0.0476
			90	0.34			0.0490
Ebonite	74#	Density	68	0.41	1185	20.0	0.0591
Glass							
Flint			59	4.16		15.0	0.5999
Plate			68	5.55		20.0	0.6003
Soda	161#	Density	68	4.94	2579	20.0	0.7123
Quartz			212	13.27		100.0	1.9135
Gypsum block	42.7#	Density	32	1.69	684	0.0	0.2437
			68	1.86		20.0	0.2682
			86	1.94		30.0	0.2797
Gypsum board							
Gypsum board							
covered with paper	62#	Density	70	1.44	993	21.1	0.2076
	1/2" Thick 53.5#	Density	68	2.60 (U)	857	20.0	14.7634 (U)
Linoleum			32	1.21		0.0	0.1745
			68	1.29		20.0	0.1860
			75	1.36		23.9	0.1961
Masonite	20#	Density	75	0.33	320	23.9	0.0476
Plaster and lath							
Metal lath and plaster			70	4.4 (U)		21.1	24.984 (U)
Wood lath and plaster			70	2.6 (U)		21.1	14.763 (U)
Plaster board							
Covered with paper	61#	Density			977		
3/8" Thick			70	3.73 (U)		21.1	21.1799 (U)
1/2" Thick			70	2.83 (U)		21.1	16.0695 (U)
Plaster, gypsum	52.4#	Density	68	1.77	839	20.0	0.2552
	46.2#	Density	86	2.32	740	30.0	0.3343
Procelain			329	11.30		165.0	1.6295
Quartz			212	13.27		100.0	1.9135
Rubber	68.6#	Density	86	1.22	1099	30.0	0.1759
Shingles							
Asbestos	65#	Density	75	6.0 (U)	1041	23.9	34.0696 (U)
Asphalt	70#	Density	75	6.5 (U)	1121	23.9	36.9087 (U)
Slate	201#	Density	75	10.37 (U)	3220	23.9	58.8835 (U)
Wood			75	1.28 (U)		23.9	7.2682 (U)
Slate							
Across cleavage			50	9.15 to 10.45		10.0	1.319 to 1.506
Along cleavage			50	10.00 to 18.9		10.0	1.442 to 2.725
Strawboard	43#	Density	86	0.50	689	30.0	0.721
Textan							
Rubber composition	81#	Density	86	1.17	1297	30.0	0.1687
Wood pulp board	43#	Density	86	0.49	689	30.0	0.0707
Wood felt	20.6#	Density	86	0.37	330	30.0	0.0534
Wood fiber board	19.8#	Density	75	0.33	317	23.9	0.0476
	11.9#	Density	86	0.30	191	20.0	0.0433
	28.5#	Density	75	0.50	457	23.9	0.0721

Note: Where values under the conductivity K column are marked (U) they are conductance values (not conductivity) and are given in the following units:
English: Btu/ft²hr°F
Metric: W/m²K

Table B-39 Conductivities (continued) – English and Metric Units

Material			Mean Temp. °F	Conductivity k Btu in/ft²hr°F	Density kg/m³	Mean Temp. °C	Conductivity W/m K
			MASONRY MATERIALS				
Brick							
Low density			70	5.0		21.1	0.7214
High density			70	9.2		21.1	1.3266
Red building, soft burned			600	4.3		316.0	0.6201
			800	4.6		427.0	0.6633
			1000	5.9		538.0	0.7066
Red building, hard burned			600	7.4		316.0	1.0671
			800	8.2		427.0	1.1824
			1000	9.0		538.0	1.2973
Brick–slag	87.4#	Density	59	87.4	1400	15.0	12.6031
Cement plaster			70	8.0		21.1	1.1536
Cinder block							
4 x 8 x 16–Solid			40	1.00 (U)		4.4	5.6783 (U)
8 x 8 x 16–With standard hollow spaces			40	0.58 (U)		4.4	3.2934 (U)
12 x 8 x 16–With standard hollow spaces			40	0.53 (U)		4.4	3.0094 (U)
Concrete							
Typical			40	12.00		4.4	1.7304
Concrete Block							
8 x 8 x 16–Sand and gravel aggregate (hollow)			40	0.9 (U)		4.4	5.1104 (U)
8 x 8 x 16–Limestone aggregate (hollow)			40	0.86 (U)		4.4	4.8833 (U)
12 x 8 x 16–Sand and gravel aggregate (hollow)			40	0.78 (U)		4.4	4.4290 (U)
Solid			88	8.2		31.1	1.1824
Concrete, cellulated							
	40#	Density	75	1.06	640	23.9	0.1528
	50#	Density	75	1.44	800	23.9	0.2076
	60#	Density	75	1.80	561	23.9	0.2596
	70#	Density	75	2.18	1121	23.9	0.3144
Concrete, cinder							
1:2:2.75 Ratio	104#	Density	75	4.63	1666	23.9	0.6676
1:2.75:4.5 Ratio	99#	Density	75	4.30	1585	23.9	0.6201
1:3.5:5.5 Ratio	92#	Density	75	3.73	1474	23.9	0.5379
Concrete–cork filled							
1 Portland: 2 Sand: 3 Granulated cork	79#	Density	185	1.79	1265	85.0	0.2581
Concrete gypsum							
87.5% Gypsum, 12.5% Wood chips	51#	Density	74	1.66	816	23.3	0.2394
Concrete, Haydite							
1:2:2.75 Ratio	80#	Density	75	4.15	1291	23.9	0.5984
1:2.75:4.5 Ratio	75#	Density	75	3.78	1201	23.9	0.5451
1:3.5:5.5 Ratio	72#	Density	75	3.67	1153	23.9	0.5292
1:8 Ratio	67#	Density	75	2.90	1073	23.9	0.4182
Concrete, limestone							
1:2:2.7 Ratio	135#	Density	75	11.2	2162	23.9	1.6150
1:2.75:4.5 Ratio	138#	Density	75	12.0	2210	23.9	1.7304
1:3.5:5.5 Ratio	136#	Density	75	11.5	2178	23.9	1.6583
Concrete, sand and gravel							
1:2:2.75 Ratio	145#	Density	75	13.1	2322	23.9	1.8890
1:2.75:4.5 Ratio	146#	Density	75	12.9	2338	23.9	1.8602
1:3.5:5.5 Ratio	145#	Density	75	13.2	2322	23.9	1.9034
Dolomite, compact			70	13.6 to 16.3		21.1	1.96 to 2.35
Domont brick (Terracotta)	113#	Density	196	4.62	1610	91.1	0.6662
Glagstone							
Across cleavage			70	12.8		21.1	1.8458
Along cleavage			70	18.4		21.1	2.6533
Freestone, sandstone			70	6.1		21.1	0.8796
Glass block			100	0.46 (U)		37.8	2.6120 (U)
			200	0.49 (U)		93.3	2.7823 (U)
			300	0.53 (U)		149.0	3.0095 (U)
			400	0.56 (U)		204.0	2.1798 (U)
			500	0.60 (U)		260.0	3.4070 (U)
Granite			70	15.0 to 22.0		21.1	2.16 to 3.17
Gravel							
Fine (0.16" to 0.35")	91#	Density	185	1.63	1457	85.0	0.2350
Dry Stone (1" to 3")	115#	Density	32	2.34	1642	0.0	0.3374
			68	2.58		20.0	0.3720
			104	2.83		40.0	0.4081

Note: Where values under the conductivity K column are marked (U) they are conductance values (not conductivity) and are given in the following units:
English: Btu/ft²hr°F
Metric: W/m²K

Table B-39 Conductivities (continued) – English and Metric Units

Material			Mean Temp. °F	Conductivity k Btu in/ft² hr °F	Density kg/m³	Mean Temp. °C	Conductivity W/m K
			MASONRY MATERIALS (continued)				
Gypsum Board							
Gypsum board							
covered with paper		62# Density	70	1.44	993	21.1	0.2076
	1/2″ Thick	53.5# Density	68	2.6	856	20.0	14.7634 (U)
Gypsum plaster							
		52.4# Density	68	1.77	839	20.0	0.2552
		46.2# Density	86	2.32	740	20.0	0.3345
Gypsum tile			50	.46		10.0	0.0663
3 x 3 x 16		67# Density	40	.50 (U)	1073	4.4	0.088 (U)
Gypsum tile			50	.46		10.0	0.0663
Haydite block							
8 x 8 x 16		67# Density	40	.50 (U)	1073	4.4	0.088 (U)
8 x 12		77# Density	40	.46 (U)	1233	4.4	0.081 (U)
Insulux Glass Block			100	.46 (U)		37.8	0.081 (U)
			200	.49 (U)		93.3	0.086 (U)
			300	.53 (U)		149.0	0.093 (U)
			400	.57 (U)		204.0	0.100 (U)
			500	.50 (U)		26.0	0.088 (U)
Lime							
Hard			50	25.57		10.0	3.687
Limestone							
			32	4.8		0.0	0.692
			59	4.9		15.0	0.707
			68	5.1		20.0	0.735
			77	5.2		25.0	0.750
Marble			86	14.5 to 19.9		80.0	2.09 to 2.74
Millstone		78.5# Density	177	3.36	1257	80.6	0.485
Mortar		107# Density	191	2.24	1714	88.3	0.323
		117# Density	191	3.71	1874	88.3	0.535
Onyx			86	16.14		30.0	2.327
Red Brick			200	5.4		98.3	0.779
			400	5.7		204.0	0.822
			600	6.1		316.0	0.880
			800	6.5		427.0	0.937
			1000	8.6		538.0	1.240
Sand							
Fine (less than .08″–Dry)		96# Density	32	2.10	1538	0	0.303
			68	2.26	1538	20.0	0.326
Fine–common moisture		98# Density	68	8.60	1570	20.0	1.240
Sandstone							
Fresh cut–natural gray		141# Density	50	10.72	2259	10.0	1.546
			68	11.62		30.0	1.676
			104	12.75		40.0	1.839
Slate							
Across cleavage			50	9.15 to 10.45		10.0	1.31 to 1.51
Along cleavage			50	16.00 to 18.9		10.0	2.31 to 2.73
Soapstone		171# Density	158	23.22	2739	70.0	3.398
Stucco			50	12.00		10.0	1.730
Terra Cotta		112# Density	196	4.62	1794	91.1	0.666
Terrazzo			50	12.00		10.0	1.730
Tile, clay hollow							
4″			50	1.00 (U)		10.0	0.176 (U)
6″			50	0.64 (U)		10.0	0.113 (U)
8″			50	0.60 (U)		10.0	0.106 (U)
10″			50	0.58 (U)		10.0	0.102 (U)
12″			50	0.40 (U)		10.0	0.070 (U)
Tile, gypsum							
4″ hollow			50	0.46 (U)		10.0	0.081 (U)

Note: Where values under the conductivity K column are marked (U) they are conductance values (not conductivity) and are given in the following units:
English: Btu/ft² hr °F
Metric: W/m² K

Table B-39 Conductivities (continued) – English and Metric Units

Material			Mean Temp. °F	Conductivity k Btu in/ft^{2}hr°F	Density kg/m^3	Mean Temp. °C	Conductivity W/m K
			METALS				
Aluminum		168 to			2690 to		
		170# Density	–290	1396.0	2723	–178.9	201.30
			32	1396.0		0.0	201.30
			210	1430.0		100.0	206.21
			570	1597.0		300.0	230.29
			930	1855.0		500.0	267.49
Antimony		413# Density	32	128.35	6615	0.0	18.51
			212	115.0		100.0	16.58
Brass	Yellow Brass		32	592.5		0.0	85.44
			212	738.0		100.0	106.42
	Red Brass		32	714.0		0.0	102.95
			212	820.0		100.0	118.24
Bronze			68	410.0		20.0	59.12
			210	492.0		98.9	70.94
Copper			32	2190.0		0.0	315.80
			212	2324.0		100.0	335.12
			390	2574.0		200.0	371.17
Gold			–420	1048.0		–251.1	151.12
			32	2160.0		0.0	311.47
			390	2145.0		200.0	309.31
Iron							
Cast			86	432.5		20.0	62.37
Iron wrought		492# Density	65	417.0	7881	18.3	60.13
			212	412.0		100.0	59.41
Lead			–297	313.8		–182.8	45.25
			10.4	276.3		– 12.0	39.84
			32	244.5		0.0	35.26
			64.4	241.0		18.0	34.75
			210	233.0		98.8	33.60
Magnesium		108# Density	210	1089.0	1729	98.8	157.03
Mercury			32	50.2		0.0	7.24
			212	62.8		100.0	9.06
Nickel (99%)			–256	374.5		–160.0	54.00
			50	403.0		10.0	58.18
			930	331.0		500.0	47.73
			1650	306.0		900.0	44.12
Platinum			212	485.0		100.0	69.94
Silver			–256	2900.0		–160.0	418.18
			32	3135.0		0.0	452.07
			212	2880.0		100.0	415.30
			644	2920.0		340.0	421.06
Steel							
Less 0.1% carbon			210	379.0		100.0	54.65
			570	347.0		300.0	50.03
			1110	258.0		600.0	37.20
			1650	234.0		900.0	33.74
Less than 0.6% carbon			210	290.0		100.0	41.82
			1110	234.0		600.0	33.74
			1650	202.0		900.0	29.13
Approximately 1.5% carbon			210	258.0		100.0	37.20
			570	249.0		300.0	35.91
			1110	234.0		600.0	33.74
			1650	202.0		900.0	29.13
Steel chromium			86	213.0 to 291.0		30.0	30.71 to 41.96
Steel, puddled			59	319		15.0	46.0
Steel Wool							
No. 2 Size Fiber		4.74# Density	132	0.63	76	55.6	0.0908
		6.3# Density	132	0.61	101	55.6	0.0897
		9.48# Density	132	0.55	152	55.6	0.0793
			32	443.0		0.0	63.88
Tin			212	413.0		100.0	59.55

Note: Where values under the conductivity K column are marked (U) they are conductance values (not conductivity) and are given in the following units:
English: Btu/ft^{2}hr°F
Metric: W/m^{2}K

Table B-39 Conductivities (continued) — English and Metric Units

Material				Mean Temp. °F	Conductivity k Btu in/ft^{2}hr°F	Density kg/m^3	Mean Temp. °C	Conductivity W/m K
MISCELLANEOUS								
Air								
(No heat transfer by radiation or convection)				70	0.175		21.1	0.0252
Air spaces and aluminum foil spacers—vertical								
1 1/2" space divided by								
Aluminum foil (bright both sides)				50	0.23 (U)		10.0	1.3060 (U)
3/4" space divided by								
Aluminum foil (bright both sides)				50	0.31 (U)		10.0	1.7603 (U)
2 1/4" space divided by two curtains								
Aluminum foil (bright both sides)				50	0.15 (U)		10.0	0.8517 (U)
3" space divided by three curtains								
Aluminum foil (bright both sides)				50	0.11 (U)		10.0	0.6246 (U)
3 3/4" space divided by four curtains								
Aluminum foil (bright both sides)				50	0.09 (U)		10.0	0.5110 (U)
Air spaces and aluminum foil spacers								
3 5/8" faced both sides with aluminum foil								
Vertical (heat flow across)				50	0.56 (U)		10.0	3.1798 (U)
Horizontal (heat flow up)				50	0.94 (U)		10.0	5.3376 (U)
Horizontal (heat flow down)				50	0.41 (U)		10.0	2.3281 (U)
Air spaces with ordinary building materials								
3 5/8" face with material of emissivities = .83								
Vertical (heat flow across)				50	0.17 (U)		10.0	6.6436 (U)
Horizontal (heat flow up)				50	1.32 (U)		10.0	7.4953 (U)
Horizontal (heat flow down)				50	0.94 (U)		10.0	5.3376 (U)
Celluloid, white				86	1.46		30.0	0.2105
Chalk				70	6.48		21.1	0.9344
Charcoal		11.85#	Density	32	0.41	189	0.0	0.0591
				104	0.46		40.0	0.0663
				176	0.51		80.0	0.0735
Clay	Dried			50	3.60		10.0	0.5192
	Wet			50	16.09		10.0	2.3202
Clinkers, from boilers		46.8#	Density	32	1.05	750	0.0	0.1151
				68	1.13		10.0	0.1629
Coal dust	Dry	62.4#	Density	32	0.97	999	0.0	0.1399
				68	1.05		10.0	0.1151
Lamp black		12.05#	Density	132	0.22	193	55.6	0.0317
				316	0.27		156.0	0.0389
				441	0.32		230.0	0.0461
Leather		62#	Density	50	1.10	993	10.0	0.1586
Linen				50	0.61		10.0	0.0879
Paraffin		55#	Density	86	1.60	861	30.0	0.2307
Peat Moss	Dry	11.8#	Density	32	0.33	189	0.0	0.0475
				68	0.34		20.0	0.0490
	Damp	12.17#	Density	68	0.57	195	20.0	0.0822
Plaster of Paris								
Powder				50	7.55		10.0	1.0887
Set				50	2.04		10.0	0.2942
Rubber	Hard	74.3#	Density	99	1.11	1190	37.2	0.1600
	Soft	68.6#	Density	86	1.22	1098	30.0	0.1759
	Sponge	14#	Density	50	1.38	224	10.0	0.1989
Sawdust								
Various, dry		12#	Density	90	0.41	192	32.2	0.0591
Pine, loose, dry		3.6#	Density	166	0.57	57	74.4	0.0822
Silk Fibers		9.2#	Density	32	0.32	147	0.0	0.0189
				122	0.38		50.0	0.0548
				212	0.42		100.0	0.0606
Soil	Dry			50	0.96		10.0	0.1384
	Including stones— Normal dampness			32	3.47		0.0	0.5004
				68	3.63		20.0	0.5234
				158	4.03		70.0	0.5811
	Wet			50	4.64		10.0	0.6691
Vacuum								
Silvered vacuum jacket								
Residual air pressure 0.001 MM of Hg				77	0.0042 (U)		25.0	0.0238 (U)

Note: Where values under the conductivity K column are marked (U) they are conductance values (not conductivity) and are given in the following units:
English: Btu/ft^{2}hr°F
Metric: W/m^{2}K

Table B-39 Conductivities (continued) – English and Metric Units

Material			Mean Temp. °F	Conductivity k Btu in/ft²hr°F	Density kg/m³	Mean Temp. °C	Conductivity W/m K
			WOOD				
Balsa							
Across grain	20.6#	Density	86	0.59	330	30.0	0.0851
	7.05#	Density	86	0.32	113	30.0	0.0461
California Redwood (across grain)							
0% Moisture	22#	Density	75	0.66	352	23.9	0.0952
8% Moisture	22#	Density	75	0.70	352	23.9	0.1009
16% Moisture	22#	Density	75	0.74	352	23.9	0.1067
0% Moisture	28#	Density	75	0.70	448	23.9	0.1009
8% Moisture	28#	Density	75	0.75	448	23.9	0.1008
16% Moisture	28#	Density	75	0.80	448	23.9	0.1154
Cypress (across grain)							
0% Moisture	22#	Density	75	0.67	352	23.9	0.0966
8% Moisture	22#	Density	75	0.71	352	23.9	0.1024
16% Moisture	22#	Density	75	0.79	352	23.9	0.1139
0% Moisture	32#	Density	75	0.79	512	23.9	0.1139
8% Moisture	32#	Density	75	0.84	512	23.9	0.1211
16% Moisture	32#	Density	75	0.90	512	23.9	0.1298
Elm–Soft (across grain)							
0% Moisture	28#	Density	75	0.73	449	23.9	0.1052
8% Moisture	28#	Density	75	0.77	449	23.9	0.1110
16% Moisture	28#	Density	75	0.81	449	23.9	0.1168
0% Moisture	34#	Density	75	0.88	544	23.9	0.1269
8% Moisture	34#	Density	75	0.93	544	23.9	0.1341
16% Moisture	34#	Density	75	0.97	544	23.9	0.1399
Fir (across grain)							
0% Moisture	26#	Density	75	0.61	416	23.9	0.0880
8% Moisture	26#	Density	75	0.66	416	23.9	0.0952
16% Moisture	26#	Density	75	0.76	416	23.9	0.1096
0% Moisture	34#	Density	75	0.67	544	23.9	0.0966
8% Moisture	34#	Density	75	0.75	544	23.9	0.1082
16% Moisture	34#	Density	75	0.82	544	23.9	0.1182
Hemlock, Eastern (across grain)							
0% Moisture	22#	Density	75	0.60	352	23.9	0.0865
8% Moisture	22#	Density	75	0.63	352	23.9	0.0908
16% Moisture	22#	Density	75	0.67	352	23.9	0.0966
0% Moisture	30#	Density	75	0.76	480	23.9	0.1096
8% Moisture	30#	Density	75	0.81	480	23.9	0.1108
16% Moisture	30#	Density	75	0.85	480	23.9	0.1226
Hemlock, West Coast (across grain)							
0% Moisture	22#	Density	75	0.68	352	23.9	0.0980
8% Moisture	22#	Density	75	0.73	352	23.9	0.1053
16% Moisture	22#	Density	75	0.78	352	23.9	0.1125
0% Moisture	30#	Density	75	0.79	480	23.9	0.1139
8% Moisture	30#	Density	75	0.85	480	23.9	0.1226
16% Moisture	30#	Density	75	0.91	480	23.9	0.1312
Mahogany (across grain)	34#	Density	86	0.90	544	30.0	0.1298
Maple, Hard							
Across Grain	45#	Density	127	1.26	720	52.8	0.1817
Along Grain	45#	Density	127	3.02	720	52.8	0.4355
0% Moisture	40#	Density (Across Grain)	75	1.01	640	23.9	0.1456
8% Moisture	40#	Density (Across Grain)	75	1.08	640	23.9	0.1557
16% Moisture	40#	Density (Across Grain)	75	1.15	640	23.9	0.1658
0% Moisture	46#	Density (Across Grain)	75	1.05	736	23.9	0.1514
8% Moisture	46#	Density (Across Grain)	75	1.13	736	23.9	0.1629
16% Moisture	46#	Density (Across Grain)	75	1.21	736	23.9	0.1745
Maple, Soft (Across grain)							
0% Moisture	36#	Density	75	0.89	576	23.9	0.1283
8% Moisture	36#	Density	75	0.96	576	23.9	0.1384
16% Moisture	36#	Density	75	1.01	576	23.9	0.1456
0% Moisture	42#	Density	75	0.95	672	23.9	0.1370
8% Moisture	42#	Density	75	1.02	672	23.9	0.1408
16% Moisture	42#	Density	75	1.09	672	23.9	0.1572

Note: Where values under the conductivity K column are marked (U) they are conductance values (not conductivity) and are given in the following units:
English: Btu/ft²hr°F
Metric: W/m²K

Table B-39 Conductivities (continued) — English and Metric Units

Material		Mean Temp. °F	Conductivity k Btu in/ft² hr° F	Density kg/m³	Mean Temp. °C	Conductivity W/m K
		WOOD (continued)				
Oak						
Across grain	51# Density	32	1.38	816	0.0	0.1990
		59	1.46		15.0	0.2105
Along grain	51# Density	54	2.42	816	12.2	0.3490
		60	2.50		15.6	0.3605
		120	2.99		48.9	0.4312
0% Moisture	38# Density (Across Grain)	75	0.98	608	23.9	0.1413
8% Moisture	38# Density (Across Grain)	75	1.03	608	23.9	0.1485
16% Moisture	38# Density (Across Grain)	75	1.07	608	23.9	0.1543
0% Moisture	48# Density (Across Grain)	75	1.18	769	23.9	0.1702
8% Moisture	48# Density (Across Grain)	75	1.24	769	23.9	0.1783
16% Moisture	48# Density (Across Grain)	75	1.29	769	23.9	0.1860
Pine, Norway (across grain)						
0% Moisture	22# Density	75	0.62	352	23.9	0.0894
8% Moisture	22# Density	75	0.68	352	23.9	0.0981
16% Moisture	22# Density	75	0.74	352	23.9	0.1067
0% Moisture	32# Density	75	0.74	512	23.9	0.1067
8% Moisture	32# Density	75	0.83	512	23.9	0.1197
16% Moisture	32# Density	75	0.92	512	23.9	0.1327
Pine, Sugar (across grain)						
0% Moisture	22# Density	75	0.54	352	23.9	0.0779
8% Moisture	22# Density	75	0.59	352	23.9	0.0851
16% Moisture	22# Density	75	0.65	352	23.9	0.0937
0% Moisture	30# Density	75	0.64	480	23.9	0.0923
8% Moisture	30# Density	75	0.71	480	23.9	0.1024
16% Moisture	30# Density	75	0.78	480	23.9	0.1125
Pine, White						
Across Grain	28# Density	167	0.74	448	75.0	0.1067
Along Grain	28# Density	133	1.78	448	56.1	0.2567
Across Grain	34# Density	86	0.80	544	30.0	0.1154
Pine, Yellow, Long Leaf (across grain)						
0% Moisture	30# Density	75	0.76	480	23.9	0.1096
8% Moisture	30# Density	75	0.83	480	23.9	0.1197
16% Moisture	30# Density	75	0.89	480	23.9	0.1283
0% Moisture	40# Density	75	0.86	640	23.9	0.1240
8% Moisture	40# Density	75	0.95	640	23.9	0.1370
16% Moisture	40# Density	75	1.03	640	23.9	0.1485
Pine, Yellow, Short Leaf (across grain)						
0% Moisture	26# Density	75	0.74	416	23.9	0.1067
8% Moisture	26# Density	75	0.79	416	23.9	0.1139
16% Moisture	26# Density	75	0.84	416	23.9	0.1211
0% Moisture	30# Density	75	0.91	480	23.9	0.1312
8% Moisture	30# Density	75	0.97	480	23.9	0.1399
16% Moisture	30# Density	75	1.04	480	23.9	0.1500
Sawdust						
Various, dry	12# Density	90	0.41	192	32.2	0.0591
Pine, loose, dry	3.6# Density	166	0.57	58	74.4	0.0822
Shavings—Planer						
Red Wood Bark	3# Density	90	0.31	48	32.2	0.0447
Red Wood Bark	5# Density	75	0.26	80	23.9	0.0375
Various	8.75# Density	86	0.41	140	30.0	0.0591
Beech and Birch	13.2# Density	90	0.36	211	32.2	0.0519
Teak Wood						
Across grain	40.5# Density	32	1.13	649	0.0	0.1629
Across grain	40.5# Density	59	1.21	649	15.0	0.1745
Across grain	40.5# Density	122	1.38	649	50.0	0.1990
Along grain	40.5# Density	32	2.59	649	0.0	0.3735
Along grain	40.5# Density	59	2.67	649	15.0	0.3850
Along grain	40.5# Density	122	2.75	649	50.0	0.3966

Note: Where values under the conductivity K column are marked (U) they are conductance values (not conductivity) and are given in the following units:
English: Btu/ft² hr° F
Metric: W/m² K

Table B-40 Radiation emittance table

English and Metric Units

Metals	Surface Temp °F	Surface Temp °C	Total Normal Emittance ∈	Metals	Surface Temp °F	Surface Temp °C	Total Normal Emittance ∈
Aluminum				Lead			
Highly polished	440-1070	227-577	0.039-0.057	Pure	260-440	127-227	0.06-0.08
Polished	100-1000	38-538	0.04-0.06	Gray, oxidized	75	24	0.28
Rough plate	78	26	0.055-0.070	Oxidized @ 390F	390	200	0.63
Oxidized @ 1110F	390-1110	139-599	0.11-0.19				
Roofing surface			0.216	Magnesium			
Oxide	530-1520	277-827	0.63-0.26	Polished	100-1000	38-538	0.07-0.22
Foil	212	100	0.087				
				Monel metal			
Bismuth	175	80	0.34	Washed, abrasive soap	75	24	0.17
				Repeated heating	450-1610	232-877	0.46-0.65
Brass							
Highly polished	497-710	260-377	0.03-0.04	Nickel and alloys			
Polished	100	38	0.05	Electrolytic, polished	74	23	0.05
Rolled plate, natural	72	22	0.06	Electroplated, not			
Rolled, coarse emeried	72	22	0.20	polished	68	20	0.11
Oxidized @ 1110F	390-1110	199-399	0.61-0.59	Wire	368-1844	185-1006	0.10-0.19
Dull plate	120-660	54-350	0.22	Oxidized @ 1110F	390-1110	200-599	0.37-0.48
				Oxide	1200-2290	654-1255	0.59-0.86
Chromium	100-1000	38-538	0.08-0.26	Nickel copper, polished	212	100	0.06
Polished	100-500	38-260	0.06-0.08	Nickel silver, polished	212	100	0.14
Polished	Solar		0.05	Nickelin, gray oxide	70	21	0.26
				Nichrome wire, bright	120-1830	50-1000	0.65-0.79
Copper				Nichrome wire, oxidized	120-930	50-500	0.95-0.98
Electrolytic, polished	176	80	0.02	Chrome-nickel			.36-.97
Comm'l plate, polished	66	19	0.030				
Heated @ 1110F	390-1110	199-599	0.57-0.57	Platinum, polished	440-2960	227-1485	0.05-0.17
Thick oxide coating	77	25	0.78				
Cuprous oxide	1470-2010	800-1100	0.66-0.54	Silver, pure, polished	440-1160	227-627	0.02-0.03
Everdur, dull	200	93	0.11	Stainless steels			
				Type 316, cleaned	75	24	0.28
Gold				316, repeated heating	450-1600	232-871	0.57-0.66
Highly polished	440-1160	227-627	0.02-0.40	304, 42 hrs @ 980F	420-980	215-527	0.62-0.73
Polished	100	38	0.06	310, furnace service	420-980	215-527	0.90-0.97
Iron and Steel				Tin, bright	76	25	0.04-0.06
Pure iron, polished	350-1800	177-980	0.05-0.37				
Wrought iron, polished	100-480	38-250	0.28	Tungsten			
Cast iron, polished			0.21	Filament	100-1000	38-538	0.03-0.08
Smooth oxidized iron	260-980	127-527	0.78-0.82	Filament	2000-5000	1093-2760	0.19-0.34
Strongly oxidized iron	100-480	38-250	0.95				
Steel, polished	100-1000	38-538	0.07-0.14	Zinc			
Steel, polished	Solar		0.045	Pure, polished	440-620	227-327	0.05
Steel, rolled sheet	70	21	0.657	Galv. iron, bright	82	28	0.23
Steel, rough plate	100-700	38-370	0.94-0.97	Galv. gray oxidized	75	24	0.28
Smooth sheet iron	1650-1900	900-1040	0.55-0.60	Galv. iron, dirty	2500	1371	0.90
Plate steel, rusted	67	20	0.69	Galv. iron, dirty	Solar		0.90
Steel, oxidized	100-1000	38-538	0.79-0.79	Galv. iron	Solar		0.54

Table 41 Surface emittances

English and Metric Units

Refractories	Surface °F	Temp. °C	Total Normal emittance ∈
Alumina Refractory	800	427	0.45
	1200	649	0.37
	1600	871	0.31
	2000	1093	0.28
	2400	1315	0.34
	2700	1482	0.37
Al_2O_3 Bonded Refractory	1500	816	0.47
Carbon Refractory	1500	816	0.97
Chrome Refractory	1500	816	0.97
Fosterlite Refractory	1500	816	0.95
Fused Castable Refractory	1500	816	0.51
Graphite Refractory	1500	816	0.97
Gypsum	100	38	0.91
Kaolin Insulating Brick	800	427	0.80
	1200	649	0.61
	1600	871	0.49
	2000	1093	0.48
	2400	1315	0.50
	2550	1398	0.59
Magnesium Oxide Refractory	800	427	0.56
98% MgO	1200	649	0.38
	1600	871	0.33
	2000	1093	0.32
	2400	1315	0.35
Magnesite	1500	816	0.48
Mullite, Converted	1500	816	0.51
Mullite, Synthetic	1500	816	0.51
Silica Refractory	1500	816	0.76
Silica, Translucent, 3/16" thick over Kaolin	800	427	0.98
	1200	649	0.80
	1600	871	0.70
	2000	1093	0.68
	2400	1315	0.67
Silicon Carbide Refractory	1000	538	0.95
(Crystolon)	1200	649	0.93
	1600	871	0.92
	2000	1093	0.90
	2400	1815	0.88
Silicon—Nitride Refractory	1500	816	0.93
Superduty Fireclay Refractory	1500	816	0.54
Zircon Refractory	1500	816	0.53
Zirconia Refractory	800	427	0.74
98% Zirconia	1200	649	0.44
	1600	871	0.33
	2000	1093	0.31
	2400	1315	0.25

Table 41 (continued)

English and Metric Units

Miscellaneous materials	Surface °F	Temp. °C	Total Normal emittance ∈
Aluminum paints	212	100	0.27-0.67
26% Al. 27% lacquer	212	100	0.3
Asbestos board	100	38	0.96
Asbestos paper	100-700	38-371	0.93-0.95
Asbestos cloth	200	93	0.90
Asphalt pavement	Solar	Solar	0.93
Brick			
Glazed	Solar	Solar	0.75
Red, rough	Solar	Solar	0.70
Silica	2500	1371	0.84
Refractory	200-1000	93-538	0.92-0.97
Carbon black	70-700	21-371	0.95
Concrete	2500		0.63
Concrete	Solar	Solar	0.65
Glass	72	22	0.937
Gypsum	70	21	0.903
Ice	32	0	0.96-0.99
Marble, polished	70	21	0.931
Mica	200	93	0.84
Oak, planed	70	21	0.895
Oil film	68	20	0.27-0.82
Paint, black	200-600	93-316	0.92-0.95
Black	100	538-38	0.97
Black	Solar	Solar	0.90
Green	200-600	93-316	0.93-0.90
Green	100	538-38	0.80
Green	Solar	Solar	0.50
White	200-600	93-316	0.92-0.84
White	100	538-38	0.68
White	Solar	Solar	0.30
Aluminum	212	100	0.27-0.67
Oil paint			0.92-0.944
Paper, white	70	21	0.924-0.944
Paper, black roofing	100	38	0.95
Procelain, glazed	70	21	0.924
Refractories	1500	816	0.65-0.90
Roofing paper	100	38	0.91
Rubber			
Hard, gloss	70	21	0.945
Soft, rough	76	24	0.859
Water	70	21	0.95-0.96
Wood	100	38	0.83-0.92

Table B-41a Observed emittances at atmospheric temperatures

Oxide surfaces	Observed range of emittances	Probable value for oxide forms on smooth metal	Oxide surfaces	Observed range of emittances	Probable value for oxide forms on smooth metal
Aluminum Oxide	0.22–0.40	0.30	Alumel (Oxidized)		0.87
Beryllium Oxide	0.07–0.37	0.35	Cast Iron (Oxidized)		0.70
Cerium Oxide	0.58–0.80		80 Ni 20 Cr (Oxidized)		0.90
Chromium Oxide	0.60–0.80	0.70	60 Ni 24 Fe 16 Cr (Oxidized)		0.83
Cobalt Oxide		0.75	55 Fe 37.5 Cr 7.5 Al (Oxidized)		0.78
Columbium Oxide	0.55–0.71	0.70	70 Fe 23 Cr 5 Al 2 Co (Oxidized)		0.75
Copper Oxide	0.60–0.80	0.70	Constantan (SS Co. 45 Ni) Oxidized)		0.84
Iron Oxide	0.63–0.98	0.70	Carbon Steel (Oxidized)		0.80
Magnesium Oxide	0.10–0.43	0.20	Stainless Steel (18–8) (Oxidized)		0.85
Nickel Oxide	0.85–0.96	0.90			
Thorium Oxide	0.20–0.57	0.59			
Tin Oxide	0.32–0.60				
Titanium Oxide		0.50			
Uranium Oxide		0.39			
Vanadium Oxide		0.70			
Yttrium Oxide		0.60			
Zirconium Oxide	0.18–0.43	0.40			

Table B-42 Radiation emittance table, weather-barriers and finishes used on thermal insulations

Weather-Barrier or Surface Finish	Conditions	Emittance ∈ (At Surface Temp of Approx 100 F)
Aluminum	Polished	0.03 to 0.06
	Gray-Dull	0.06 to 0.09*
	Oxidized	0.10 to 0.12*
Aluminum Paint	New	0.20 to 0.30
	After weathering	0.40 to 0.70
Asbestos Paper	Clean	0.90 to 0.94
Asphalt Asbestos Felts		0.93 to 0.96
Asphalt Mastics		0.90 to 0.95
Galvanized Steel	New–bright	0.06 to 0.10
	Dull	0.20 to 0.30
Paints	White–clean	0.55 to 0.70
	Green–clean	0.65 to 0.80
	Gray–clean	0.80 to 0.90
	Balck–clean	0.90 to 0.95
Painted Canvas	Color as painted	Will be approx. the same as ∈ for color of paint used
PVA Mastics	White–clean	0.60 to 0.70
	Green–clean	0.70 to 0.80
	Gray–medium–clean	0.85 to 0.90
	Black	0.85 to 0.95
Roofing Felts		0.90 to 0.95
Stainless Steel	Polished	0.22 to 0.26
	No. 4 mill finish	0.35 to 0.40
	Oxidized	0.80 to 0.85

*0.7 to 0.9 for solar heat in visable range

Table B-43 Specific heats and weights of materials – English Units

Material	Temperature °F	Mean Specific Heat	Weight lbs/cu ft
Aluminum		0.215	168
Asbestos	32-212	0.20	150
Bakelite	32-212	0.3-0.4	
Brass, yellow	32-212	0.088	534
Brass, red	32-212	0.09	534
Bronze	32-212	0.014	509-554
Brick	32-212	0.20-0.22	125-143
Carbon	32-212	0.165	139
Chalk	32-212	0.215	143
Charcoal	32-212	0.20	25
Cinders	32-212	0.18	
Coal	32-212	0.24-0.3	81-94
Concrete	32-212	0.156	137
Cork	32-212	0.485	
Coke	32-212	0.203	75
Copper	32-212	0.094	556
Glass	32-212	0.12-0.19	162
Graphite	32-212	0.201	135
Gold		0.031	1205
Granite	32-212	0.195	168
Gypsum	32-212	0.259	155
Humus (soil)	32-212	0.44	76-100
Ice		0.465	56
Ice	32	0.487	64
Iron, cast		0.130	442
Iron, wrought	32-212	0.110	485
Iron, wrought	32-572	0.122	485
Iron, at high temperatures	1382-1832	0.213	485
Lead		0.031	710
Limestone		0.217	155-162
Marble	32-212	0.210	170
Mercury	32-212	0.033	850
Masonry, brick	32-212	0.20-0.22	
Nickel		0.109	537
Oil, machine	0.	0.400	
Porcelain	32-212	0.22	
Quartz	32-212	0.17-0.28	165
Sand	32.212	0.195	100-125
Sandstone	32-212	0.22	143
Silver		0.056	655
Silica		0.191	
Steel, mild		0.116	485
Steel, high carbon		0.117	485
Stone, average		0.200	150
Tin		0.056	459
Water		1.000	62.4
Wood, fir		0.650	25-32
Wood, oak		0.570	42-54
Wood, pine		0.67	27-42
Zinc		0.095	440

Table B-44 Velocity of flow of water – English Units

In feet per minute, through pipes of various sizes, for varying quantities of flow

Gallons per Minute	$\frac{3}{4}$ in.	1 in.	$1\frac{1}{4}$ in.	$1\frac{1}{2}$ in.	2 in.	$2\frac{1}{2}$ in.	3 in.	$3\frac{1}{2}$ in.
5	218	$122\frac{1}{2}$	$78\frac{1}{2}$	$54\frac{1}{2}$	$30\frac{1}{2}$	$19\frac{1}{2}$	$13\frac{1}{2}$	$7\frac{2}{3}$
10	436	245	157	109	61	38	27	$15\frac{1}{3}$
15	653	$367\frac{1}{2}$	$235\frac{1}{2}$	$163\frac{1}{2}$	$91\frac{1}{2}$	$58\frac{1}{2}$	$40\frac{1}{2}$	23
20	872	490	314	218	122	78	54	$30\frac{2}{3}$
25	1090	$612\frac{1}{2}$	$392\frac{1}{2}$	$272\frac{1}{2}$	$152\frac{1}{2}$	$97\frac{1}{2}$	$67\frac{1}{2}$	$38\frac{1}{3}$
30	....	735	451	327	183	117	81	46
35	....	$857\frac{1}{2}$	$549\frac{1}{2}$	$381\frac{1}{2}$	$213\frac{1}{2}$	$136\frac{1}{2}$	$94\frac{1}{2}$	$53\frac{2}{3}$
40	....	980	628	436	244	156	108	$61\frac{1}{3}$
45	....	$1102\frac{1}{2}$	$706\frac{1}{2}$	$490\frac{1}{2}$	$274\frac{1}{2}$	$175\frac{1}{2}$	$121\frac{1}{2}$	69
50	....	.....	785	545	305	195	135	$76\frac{2}{3}$
75	....	.....	$1177\frac{1}{2}$	$817\frac{1}{2}$	$457\frac{1}{2}$	$292\frac{1}{2}$	$202\frac{1}{2}$	115
100	....	.....	.....	1090	610	380	270	$153\frac{1}{3}$
125	....	.....	.....	.....	$762\frac{1}{2}$	$487\frac{1}{2}$	$337\frac{1}{2}$	$191\frac{2}{3}$
150	....	.....	.....	.....	915	585	405	230
175	....	.....	.....	.....	$1067\frac{1}{2}$	$682\frac{1}{2}$	$472\frac{1}{2}$	$268\frac{1}{3}$
200	....	.....	.....	.....	1220	780	540	$306\frac{2}{3}$

Table B-45 Decimals of a foot – English Units

0″	.0000	1″	.0833	2″	.166667	3″	.2500
1/16″	.0052	1-1/16″	.0885	2-1/16″	.171875	3-1/16″	.2552
1/8″	.0104	1-1/8″	.09375	2-1/8″	.1771	3-1/8″	.2604
3/16″	.015625	1-3/16″	.0990	2-3/16″	.1823	3-3/16″	.265625
1/4″	.0208	1-1/4″	.1042	2-1/4″	.1875	3-1/4″	.2708
5/16″	.0260	1-5/16″	.109375	2-5/16″	.1927	3-5/16″	.2760
3/8″	.03125	1-3/8″	.1146	2-3/8″	.1979	3-3/8″	.28125
7/16″	.0365	1-7/16″	.1198	2-7/16″	.203125	3-7/16″	.2865
1/2″	.0417	1-1/2″	.1250	2-1/2″	.2083	3-1/2″	.2917
9/16″	.046875	1-9/16″	.1302	2-9/16″	.2135	3-9/16″	.296875
5/8″	.0521	1-5/8″	.1354	2-5/8″	.21875	3-5/8″	.3021
11/16″	.0573	1-11/16″	.140625	2-11/16″	.2240	3-11/16″	.3075
3/4″	.0625	1-3/4″	.1458	2-3/4″	.2292	3-3/4″	.3125
13/16″	.0677	1-13/16″	.1510	2-13/16″	.234375	3-13/16″	.3177
7/8″	.0729	1-7/8″	.15625	2-7/8″	.2396	3-7/8″	.3229
15/16″	.078125	1-15/16″	.1615	2-15/16″	.2448	3-15/16″	.328125
4″	.3333	5″	.416667	6″	.5000	7″	.5833
4-1/16″	.3385	5-1/16″	.421875	6-1/16″	.5052	7-1/16″	.5885
4-1/8″	.34375	5-1/8″	.4271	6-1/8″	.5104	7-1/8″	.59375
4-3/16″	.3490	5-3/16″	.4323	6-3/16″	.515625	7-3/16″	.5990
4-1/4″	.3542	5-1/4″	.4375	6-1/4″	.5208	7-1/4″	.6042
4-5/16″	.359375	5-5/16″	.4427	6-5/16″	.5260	7-5/16″	.609375
4-3/8″	.3646	5-3/8″	.4479	6-3/8″	.53125	7-3/8″	.6146
4-7/16″	.3698	5-7/16″	.453125	6-7/16″	.5365	7-7/16″	.6198
4-1/2″	.3750	5-1/2″	.4583	6-1/2″	.5417	7-1/2″	.6250
4-9/16″	.3802	5-9/16″	.4635	6-9/16″	.546875	7-9/16″	.6302
4-5/8″	.3854	5-5/8″	.46875	6-5/8″	.5521	7-5/8″	.6354
4-11/16″	.390625	5-11/16″	.4740	6-11/16″	.5573	7-11/16″	.640625
4-3/4″	.3958	5-3/4″	.4792	6-3/4″	.5625	7-3/4″	.6458
4-13/16″	.4010	5-13/16″	.484375	6-13/16″	.5677	7-13/16″	.6510
4-7/8″	.40625	5-7/8″	.4896	6-7/8″	.5729	7-7/8″	.65625
4-15/16″	.4115	5-15/16″	.4948	6-15/16″	.578125	7-15/16″	.6615
8″	.666667	9″	.7500	10″	.8333	11″	.916667
8-1/16″	.671875	9-1/16″	.7552	10-1/16″	.8385	11-1/16″	.921875
8-1/8″	.6771	9-1/8″	.7604	10-1/8″	.84375	11-1/8″	.9271
8-3/16″	.6823	9-3/16″	.765625	10-3/16″	.8490	11-3/16″	.9323
8-1/4″	.6875	9-1/4″	.7708	10-1/4″	.8542	11-1/4″	.9375
8-5/16″	.6927	9-5/16″	.7760	10-5/16″	.859375	11-5/16″	.9427
8-3/8″	.6979	9-3/8″	.78125	10-3/8″	.8646	11-3/8″	.9479
8-7/16″	.703125	9-7/16″	.7865	10-7/16″	.8698	11-7/16″	.953125
8-1/2″	.7083	9-1/2″	.7917	10-1/2″	.8750	11-1/2″	.9583
8-9/16″	.7135	9-9/16″	.796875	10-9/16″	.8802	11-9/16″	.9635
8-5/8″	.71875	9-5/8″	.8021	10-5/8″	.8854	11-5/8″	.96875
8-11/16″	.7240	9-11/16″	.8073	10-11/16″	.890625	11-11/16″	.9740
8-3/4″	.7292	9-3/4″	.8125	10-3/4″	.8958	11-3/4″	.9792
8-13/16″	.734375	9-13/16″	.8177	10-13/16″	.9010	11-13/16″	.984375
8-7/8″	.7396	9-7/8″	.8229	10-7/8″	.90625	11-7/8″	.9896
8-15/16″	.7448	9-15/16″	.828125	10-15/16″	.9115	11-15/16″	.9948

Reproduced with permission, from Heat Insulation Manual, Papco Industrial Product Div., Fiberboard Corp.

Table B-46

Decimal Equivalents

1/64	.015625	33/64	.515625
1/32	.03125	17/32	.53125
3/64	.046875	35/64	.546875
1/16	.0625	9/16	.5625
5/64	.078125	37/64	.578125
3/32	.09375	19/32	.59375
7/64	.109375	39/64	.609375
1/8	.125	5/8	.625
9/64	.140625	41/64	.640625
5/32	.15625	21/32	.65625
11/64	.171875	43/64	.671875
3/16	.1875	11/16	.6875
13/64	.203125	45/64	.703125
7/32	.21875	23/32	.71875
15/64	.234375	47/64	.734375
1/4	.25	3/4	.75
17/64	.265625	49/64	.765625
9/32	.28125	25/32	.78125
19/64	.296875	51/64	.796875
5/16	.3125	13/16	.8125
21/64	.328125	53/64	.828125
11/32	.34375	27/32	.84375
23/64	.359375	55/64	.859375
3/8	.375	7/8	.875
25/64	.390625	57/64	.890625
13/32	.40625	29/32	.90625
27/64	.421875	59/64	.921875
7/16	.4375	15/16	.9375
29/64	.453125	61/64	.953125
15/32	.46875	31/32	.96875
31/64	.484375	63/64	.984375
1/2	.5	1	1.0

Reproduced with permission from Heat Insulation Manual, Papco Industrial Product Div., Fiberboard Corp.

Table B-47 Table of gauges and weights for wire, bands and flat sheet metal

English Units

Material	Gauge	Diameter or Thickness	Linear Feet Per Pound
Galvanized or B. A. Wire	W & M		
12 Gauge	W & M	0.1055" Dia.	33.30
14 Gauge	W & M	0.0800" Dia.	58.82
16 Gauge	W & M	0.0625" Dia.	95.23
18 Gauge	W & M	0.0475" Dia.	166.66
Soft Copper Wire	B & S		
12 Gauge	B & S	0.0808" Dia.	52.63
14 Gauge	B & S	0.0640" Dia.	83.33
16 Gauge	B & S	0.0508" Dia.	142.85
18 Gauge	B & S	0.0403" Dia.	250.00
Galvanized Steel Bands			
1/2" Wide	–	.015" thk.	39.26
1/2" Wide	–	.020" thk.	29.45
3/4" Wide	–	.020" thk.	19.63
1 1/4" Wide	–	.035" thk.	6.73
Stainless Steel Bands 18-8 Chrome-Nickel			
1/2" Wide	–	.015" thk	39.26
1/2" Wide	–	.020" thk.	29.45
3/4" Wide	–	.015" thk.	26.18
3/4" Wide	–	.020" thk.	19.63

Material	Gauge	Thickness	Weight in Lbs Per Sq Ft
Flat Black Sheet Metal			
18 Gauge	U. S. Standard	.05"	2.00
20 Gauge	U. S. Standard	.0375"	1.50
22 Gauge	U. S. Standard	.03125"	1.25
24 Gauge	U. S. Standard	.025"	1.00
26 Gauge	U. S. Standard	.01825"	0.75
28 Gauge	U. S. Standard	.015625"	0.625
30 Gauge	U. S. Standard	.0125"	0.50
Flat Galvanized Sheet Metal			
18 Gauge	U. S. Standard	.0540"	2.1563
20 Gauge	U. S. Standard	.0415"	1.6563
22 Gauge	U. S. Standard	.03535"	1.4063
24 Gauge	U. S. Standard	.0290"	1.1563
26 Gauge	U. S. Standard	.02275"	.9063
28 Gauge	U. S. Standard	.019625"	.7813
30 Gauge	U. S. Standard	.0165"	.6563
Aluminum Sheet			
16 Gauge	B & S	.051	.716
18 Gauge	B & S	.040	.568
20 Gauge	B & S	.032	.450
22 Gauge	B & S	.025	.357
24 Gauge	B & S	.020	.283
26 Gauge	B & S	.016	.225
28 Gauge	B & S	.012	.178
Stainless Steel Sheet			
18 Gauge	U. S. Standard	.050	2.10
20 Gauge	U. S. Standard	.037	1.58
22 Gauge	U. S. Standard	.031	1.31
24 Gauge	U. S. Standard	.025	1.05
26 Gauge	U. S. Standard	.018	.788
28 Gauge	U. S. Standard	.015	.66
30 Gauge	U. S. Standard	.013	.53
32 Gauge	U. S. Standard	.010	.43

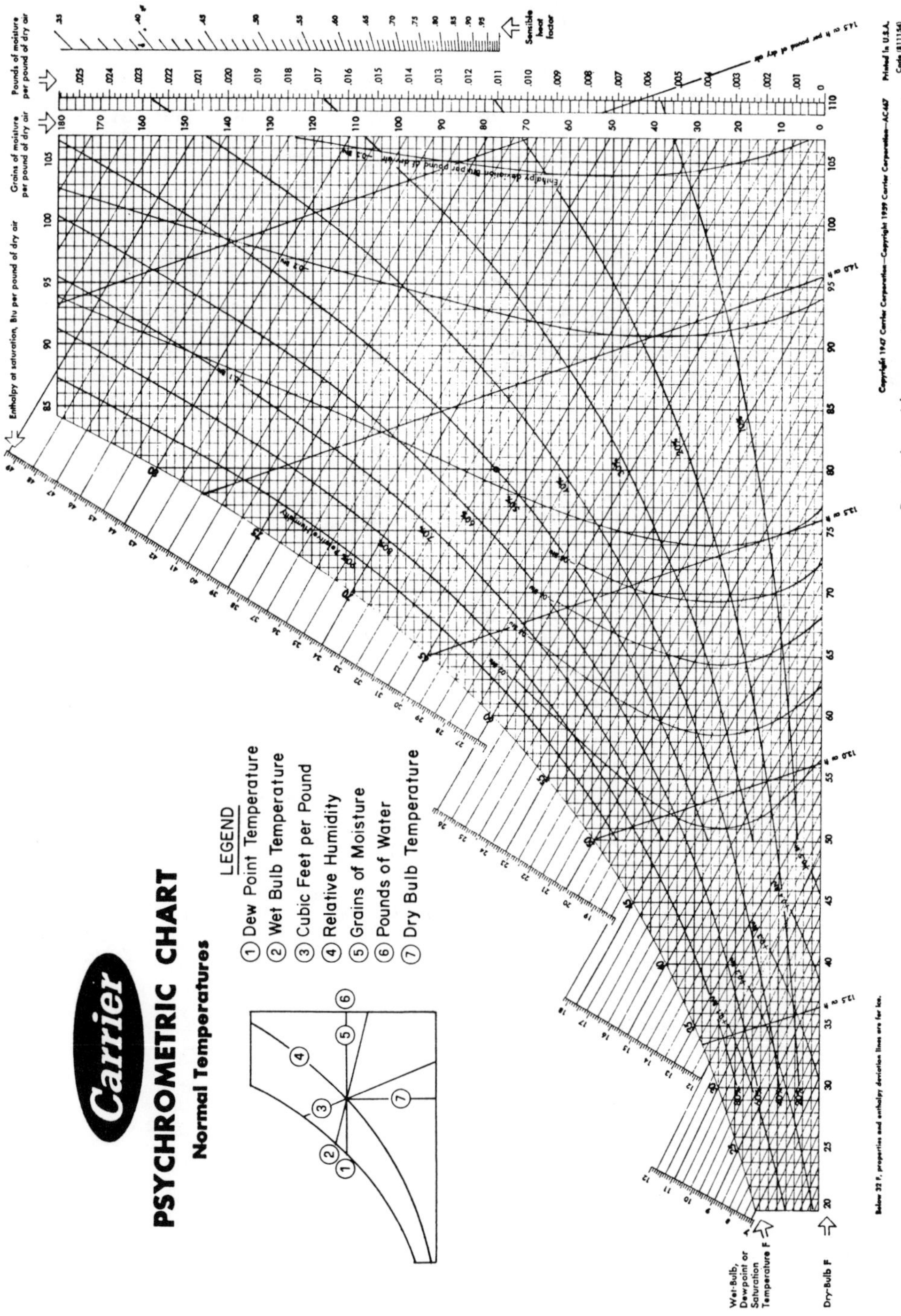

Reproduced by permission of Carrier Corporation

Chart B-12

Psychrometric Chart — English Units

Table B-48 Table of days between two dates

Day Mo.	Jan.	Feb.	March	April	May	June	July	Aug.	Sept.	Oct.	Nov.	Dec.
1	1	32	60	91	121	152	182	213	244	274	305	335
2	2	33	61	92	122	153	183	214	245	275	306	336
3	3	34	62	93	123	154	184	215	246	276	307	337
4	4	35	63	94	124	155	185	216	247	277	308	338
5	5	36	64	95	125	156	186	217	248	278	309	339
6	6	37	65	96	126	157	187	218	249	279	310	340
7	7	38	66	97	127	158	188	219	250	280	311	341
8	8	39	67	98	128	159	189	220	251	281	312	342
9	9	40	68	99	129	160	190	221	252	282	313	343
10	10	41	69	100	130	161	191	222	253	283	314	344
11	11	42	70	101	131	162	192	223	254	284	315	345
12	12	43	71	102	132	163	193	224	255	285	316	346
13	13	44	72	103	133	164	194	225	256	286	317	347
14	14	45	73	104	134	165	195	226	257	287	318	348
15	15	46	74	105	135	166	196	227	258	288	319	349
16	16	47	75	106	136	167	197	228	259	289	320	350
17	17	48	76	107	137	168	198	229	260	290	321	351
18	18	49	77	108	138	169	199	230	261	291	322	352
19	19	50	78	109	139	170	200	231	262	292	323	353
20	20	51	79	110	140	171	201	232	263	293	324	354
21	21	52	80	111	141	172	202	233	264	294	325	355
22	22	53	81	112	142	173	203	234	265	295	326	356
23	23	54	82	113	143	174	204	235	266	296	327	357
24	24	55	83	114	144	175	205	236	267	297	328	358
25	25	56	84	115	145	176	206	237	268	298	329	359
26	26	57	85	116	146	177	207	238	269	299	330	360
27	27	58	86	117	147	178	208	239	270	300	331	361
28	28	59	87	118	148	179	209	240	271	301	332	362
29	29	...	88	119	149	180	210	241	272	302	333	363
30	30	...	89	120	150	181	211	242	273	303	334	364
31	31	...	90	...	151	...	212	243	...	304	...	365

For leap year, one day must be added to each number of days after February 28.

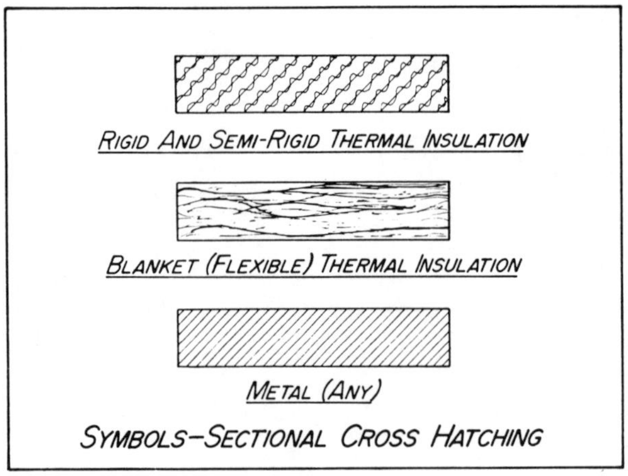

Fig. B-1.

Table B-49

HEAT LOSS "Q" ($Q = \frac{\Delta t}{R + R_s}$)

Q in Btu, sq ft, hr when $R + R_s$ is in English Units and Δt is °F.
Q in Wm² when $R + R_s$ is in Metric Units and Δt is °C or °K.

VALUE	Δt TEMPERATURE DIFFERENCE (Pipe or Equipment Surface Temperature Minus Ambient Air Temperature)											
$R+R_s$	100	200	300	400	500	600	700	800	900	1000	1100	1200
.1	1000	2000	3000	4000	5000	6000	7000	8000	9000	10000	11000	12000
.2	500	1000	1500	2000	2500	3000	3500	4000	4500	5000	5500	6000
.3	333	667	1000	1333	1667	2000	2333	2667	3000	3333	3667	4000
.4	250	500	750	1000	1250	1500	1750	2000	2250	2500	2750	3000
.5	200	400	600	800	1000	1200	1400	1600	1800	2000	2200	2400
.6	166	333	500	666	833	1000	1166	1333	1500	1666	1833	2000
.7	143	286	428	571	714	857	1000	1142	1285	1428	1571	1714
.8	125	250	375	500	625	750	875	1000	1125	1250	1375	1500
.9	111	222	333	444	556	667	778	889	1000	1111	1222	1333
1.0	100	200	300	400	500	600	700	800	900	1000	1100	1200
1.1	91	182	273	364	455	545	636	727	818	909	1000	1090
1.2	83	166	249	332	415	500	583	666	750	833	917	1000
1.3	77	154	231	308	383	461	588	615	692	769	846	923
1.4	71	142	213	284	355	426	500	571	642	714	785	857
1.5	66	133	200	266	333	400	467	533	600	666	733	800
1.6	63	125	188	250	313	375	438	500	563	625	687	750
1.7	59	118	176	235	294	353	412	470	529	588	697	706
1.8	56	111	166	222	277	333	388	444	500	555	611	666
1.9	53	105	157	210	263	315	368	421	473	526	578	631
2.0	50	100	150	200	250	300	350	400	450	500	550	600
2.1	48	95	142	190	238	286	333	380	428	476	524	571
2.2	45	91	136	182	227	272	318	364	409	454	500	545
2.3	43	87	130	174	217	261	304	347	391	434	478	522
2.4	42	83	125	166	208	250	291	333	375	416	458	500
2.5	40	80	120	160	200	240	280	320	360	400	440	480
2.6	38	77	115	154	192	231	269	308	346	384	423	462
2.7	37	74	111	148	185	222	259	296	333	370	407	444
2.8	35	71	107	142	178	214	250	285	321	357	392	428
2.9	34	69	103	137	172	206	241	275	310	344	379	413
3.0	33	67	100	133	167	200	233	267	300	333	367	400
3.1	32	65	97	129	161	193	225	258	290	322	354	387
3.2	31	62	94	125	156	187	218	250	281	312	343	375
3.3	30	61	91	121	151	181	212	242	272	303	333	363
3.4	29	59	88	118	147	176	205	235	264	294	323	352
3.5	29	57	86	114	143	171	200	228	257	285	314	343
3.6	28	55	83	111	138	167	194	222	250	278	306	333
3.7	27	54	81	108	135	162	189	216	243	270	297	324
3.8	26	53	79	105	131	157	184	210	236	263	289	315
3.9	26	51	77	102	128	153	179	205	230	256	282	307
4.0	25	50	75	100	125	150	175	200	225	250	275	300
4.1	24	48	73	98	121	146	171	195	219	243	268	293
4.2	24	47	71	95	119	142	166	190	214	238	261	285
4.3	23	46	70	93	116	139	162	186	209	232	256	279
4.4	23	45	68	91	114	136	159	182	204	227	250	272
4.5	22	44	66	89	111	133	156	178	200	222	244	267
4.6	22	43	65	87	109	130	152	174	196	217	239	260
4.7	21	43	64	85	106	127	149	170	191	213	234	255
4.8	21	42	62	83	104	125	146	167	188	208	229	250
4.9	20	41	61	81	102	122	142	163	184	204	224	245
5.0	20	40	60	80	100	120	140	160	180	200	220	240
5.1	20	39	59	78	98	118	137	157	176	196	216	235
5.2	19	38	58	77	96	115	134	154	173	192	211	231
5.3	19	38	57	75	94	113	132	150	170	189	207	226
5.4	19	37	56	74	93	111	130	148	167	185	203	222
5.5	18	36	55	73	91	109	127	145	164	181	200	218
5.6	18	36	54	71	89	107	125	142	160	179	196	214
5.7	18	35	53	70	88	105	122	140	158	175	192	210
5.8	17	34	52	69	86	103	120	138	155	172	188	207
5.9	17	34	51	68	85	102	119	136	153	169	186	203
6.0	17	33	50	67	83	100	117	133	150	167	183	200
6.1	16	33	49	66	82	98	115	131	148	164	180	197
6.2	16	32	48	65	81	97	113	129	145	161	177	193
6.3	16	32	48	63	79	95	111	127	142	159	174	190
6.4	16	31	47	63	78	94	109	125	140	156	172	187
6.5	15	31	46	62	77	92	108	123	138	154	169	185
6.6	15	30	45	61	76	91	106	121	136	152	167	182
6.7	15	30	45	60	75	90	104	119	134	150	164	179
6.8	15	29	44	59	74	88	102	118	132	147	162	176
6.9	14	29	43	58	72	87	101	116	130	145	159	174
7.0	14	29	43	57	71	86	100	114	129	143	157	171

VALUE	Δt TEMPERATURE DIFFERENCE (Pipe or Equipment Surface Temperature Minus Ambient Air Temperature)											
$R+R_s$	100	200	300	400	500	600	700	800	900	1000	1100	1200
7.1	14.0	28.2	42.2	56.3	70.4	84.5	98.6	112.7	126.8	140.8	154.9	169.0
7.2	13.9	27.7	41.6	55.5	69.4	83.3	97.2	111.1	125.0	138.8	152.8	166.6
7.3	13.7	27.4	41.1	54.8	68.5	82.2	95.9	109.6	123.3	137.0	150.7	164.4
7.4	13.5	27.0	40.5	54.1	67.6	81.1	94.6	108.1	121.6	135.1	148.6	162.2
7.5	13.3	26.6	39.9	53.3	66.7	80.0	93.3	106.6	120.0	133.3	146.6	160.0
7.6	13.2	26.3	39.4	52.6	65.8	78.9	92.1	105.2	118.4	131.6	144.7	157.9
7.7	13.0	26.0	39.0	51.9	64.9	77.9	90.9	104.0	116.9	129.9	142.9	155.8
7.8	12.8	25.6	38.5	51.3	64.1	76.9	89.7	102.6	115.4	128.2	141.0	153.8
7.9	12.7	25.3	38.0	50.6	63.3	75.9	88.6	101.3	118.9	126.6	139.2	151.9
8.0	12.5	25.0	37.5	50.0	62.5	75.0	87.5	100.0	112.5	125.5	137.5	150.0
8.1	12.3	24.7	37.0	49.4	61.7	74.1	86.4	98.8	111.1	123.5	135.8	148.1
8.2	12.2	24.4	36.6	48.8	60.9	73.2	85.4	97.6	109.8	123.0	134.1	146.3
8.3	12.0	24.1	36.1	48.1	60.2	72.3	84.3	96.3	108.4	120.5	132.5	144.6
8.4	11.9	23.8	35.7	47.6	59.5	71.4	83.3	95.2	107.1	119.0	131.1	142.9
8.5	11.8	23.5	35.3	47.0	58.8	70.9	82.3	94.1	105.8	117.6	129.4	141.2
8.6	11.6	23.3	34.9	46.5	58.1	69.8	82.0	93.0	104.7	116.3	127.9	139.5
8.7	11.5	23.0	34.5	46.0	57.5	69.0	80.5	92.0	103.4	114.9	126.4	137.9
8.8	11.4	22.7	34.1	45.5	56.8	68.2	79.5	90.9	102.2	113.6	125.0	136.4
8.9	11.2	22.4	33.7	44.9	56.2	67.4	78.7	89.9	101.1	112.4	123.6	134.8
9.0	11.1	22.2	33.3	44.4	55.5	66.7	77.8	88.9	100.0	111.1	122.2	133.3
9.1	11.0	22.0	33.0	44.0	55.0	66.0	77.0	88.0	99.0	110.0	120.9	131.9
9.2	10.9	21.7	32.6	43.4	54.3	65.2	76.1	87.0	97.8	108.7	119.6	130.4
9.3	10.8	21.5	32.3	43.0	53.8	64.5	75.3	86.0	96.8	107.5	118.3	129.0
9.4	10.6	21.3	31.9	42.6	53.2	63.8	74.5	85.1	95.7	106.4	117.0	127.7
9.5	10.5	21.0	31.6	42.1	52.6	63.2	73.7	84.2	94.7	105.3	115.8	126.3
9.6	10.4	20.8	31.3	41.7	52.1	62.5	72.9	83.3	93.8	104.2	114.6	125.0
9.7	10.3	20.6	30.9	41.2	51.5	61.9	72.2	82.5	92.8	103.1	113.4	123.7
9.8	10.2	20.4	30.6	40.8	51.0	61.2	71.4	81.6	91.8	102.0	112.2	122.4
9.9	10.1	20.2	30.3	40.4	50.5	60.6	70.7	80.8	90.9	101.0	111.1	121.2
10.0	10.0	20.0	30.0	40.0	50.0	60.0	70.0	80.0	90.0	100.0	110.0	120.0
10.2	9.8	19.6	29.4	39.2	49.0	58.8	66.6	78.4	88.2	98.0	107.8	117.6
10.4	9.6	19.2	28.8	38.5	48.1	57.7	67.3	76.9	86.5	96.2	105.8	115.4
10.6	9.4	18.9	28.3	37.7	47.2	56.6	66.0	75.5	84.9	94.3	103.8	113.2
10.8	9.2	18.5	27.8	37.0	46.3	55.6	64.8	74.1	83.3	92.6	101.9	111.1
11.0	9.1	18.2	27.3	36.4	45.5	54.5	63.6	72.7	81.8	90.9	100.0	109.0
11.2	8.9	17.9	26.8	35.7	44.6	53.6	62.5	71.4	80.4	89.3	98.2	107.1
11.4	8.8	17.5	26.3	35.0	43.9	52.6	61.4	70.2	78.9	87.7	96.5	105.3
11.6	8.6	17.2	25.9	34.5	43.1	51.7	60.3	69.0	77.6	86.2	94.8	108.4
11.8	8.5	16.9	25.4	33.9	42.4	50.8	59.3	67.8	76.3	84.7	93.2	101.7
12.0	8.3	16.6	24.9	33.2	41.5	50.0	58.3	66.6	75.0	83.3	91.7	100.0
12.2	8.2	16.4	24.6	32.8	41.0	49.2	57.4	65.6	73.8	81.7	90.2	98.4
12.4	8.1	16.1	24.2	32.3	40.3	48.4	56.5	64.5	72.6	80.6	88.7	96.8
12.6	7.9	15.9	23.8	31.7	39.7	47.6	55.6	63.5	71.4	79.4	87.3	95.2
12.8	7.8	15.6	23.4	31.3	39.1	46.9	54.7	62.5	70.3	78.1	85.9	93.8
13.0	7.7	15.4	23.1	30.8	38.5	46.1	53.8	61.5	69.2	76.9	84.6	92.3
13.2	7.6	15.2	22.7	30.3	37.9	45.5	53.0	60.6	68.2	75.8	83.3	90.9
13.4	7.5	14.9	22.4	29.9	37.3	44.8	52.2	59.7	67.2	74.6	82.1	89.6
13.6	7.4	14.7	22.0	29.4	36.8	44.1	51.5	58.8	66.2	73.5	80.1	88.2
13.8	7.2	14.5	21.7	29.0	36.2	43.4	50.7	58.0	65.2	72.5	79.7	87.0
14.0	7.1	14.2	21.3	28.4	35.5	42.6	50.0	57.1	64.2	71.4	78.5	85.7
14.2	7.0	14.1	21.1	28.2	35.2	42.2	49.3	53.3	63.4	70.4	77.5	84.5
14.4	6.9	13.9	20.8	27.8	34.7	41.7	48.6	55.6	62.5	69.4	76.4	83.3
14.6	6.8	13.7	20.5	27.4	34.2	41.0	47.9	54.8	61.6	68.5	75.3	82.2
14.8	6.7	13.5	20.3	27.0	33.8	40.5	47.3	54.1	60.8	67.6	74.3	81.1
15.0	6.6	13.3	20.0	26.6	33.3	40.0	46.7	53.3	60.0	66.6	73.3	80.0
15.2	6.6	13.2	19.7	26.3	32.8	39.5	46.1	52.6	59.2	65.7	72.4	78.9
15.4	6.5	13.0	19.5	25.9	32.5	39.0	45.4	51.9	58.4	64.9	71.4	77.9
15.6	6.4	12.8	19.2	25.6	32.0	38.5	44.9	51.3	57.7	64.1	70.5	76.9
15.8	6.3	12.7	19.0	25.3	31.6	38.0	44.3	50.6	57.0	63.2	69.6	75.9
16.0	6.3	12.5	18.8	25.0	31.3	37.5	43.8	50.0	56.3	62.5	68.7	75.0
16.2	6.2	12.3	18.5	24.7	30.9	37.0	43.2	49.4	55.6	61.7	67.9	74.1
16.4	6.1	12.2	18.3	24.3	30.5	36.6	42.7	48.8	54.9	61.0	67.0	73.1
16.6	6.0	12.0	18.0	24.1	30.1	36.1	42.2	48.2	54.2	60.2	66.3	72.3
16.8	6.0	11.9	17.8	23.8	29.8	35.7	41.7	47.6	53.6	59.5	65.5	71.4
17.0	5.9	11.8	17.6	23.5	29.4	35.3	41.2	47.0	52.9	58.8	64.7	70.6
17.2	5.8	11.6	17.4	23.3	29.0	34.9	40.6	46.5	52.3	58.1	64.0	69.8
17.4	5.7	11.5	17.2	23.0	28.7	34.5	40.2	46.0	51.7	57.5	63.2	69.0
17.6	5.7	11.4	17.0	22.7	28.4	34.1	40.0	45.5	51.1	56.8	62.5	68.2
17.8	5.6	11.2	16.8	22.5	28.0	33.7	39.3	44.9	50.6	56.2	61.8	67.4
18.0	5.6	11.1	16.6	22.2	27.7	33.3	38.8	44.4	50.0	55.5	61.1	66.6

TABLE B-49 (Sheet 1 of 2)

Table B-49 (Continued)

$$\text{HEAT LOSS ``Q'' } \left(Q = \frac{\Delta t}{R + R_s}\right)$$

Q in Btu, sq ft, hr when $R + R_s$ is in English Units and Δt is °F.
Q in Wm² when $R + R_s$ is in Metric Units and Δt is °C or °K.

VALUE	Δt TEMPERATURE DIFFERENCE (Pipe or Equipment Surface Temperature Minus Ambient Air Temperature)											
$R+R_s$	100	200	300	400	500	600	700	800	900	1000	1100	1200
18.2	5.5	11.0	16.5	22.0	27.5	33.0	38.5	44.0	49.5	54.9	60.4	65.9
18.4	5.4	10.9	16.3	21.7	27.2	32.6	38.0	43.5	48.9	54.3	60.0	65.2
18.6	5.4	10.8	16.1	21.5	26.9	32.3	37.6	43.0	48.4	53.8	59.1	64.5
18.8	5.3	10.6	15.9	21.2	26.6	31.9	37.2	42.6	47.9	53.2	58.5	63.8
19 0	5.3	10.5	15.7	21.0	26.3	31.5	36.8	42.1	47.3	52.6	57.3	62.5
19.2	5.2	10.4	15.6	20.8	26.0	31.3	36.5	41.7	46.9	52.1	57.3	62.5
19.4	5.1	10.3	15.5	20.6	25.8	30.9	36.1	41.2	46.4	51.5	56.7	61.8
19.6	5.1	10.2	15.3	20.4	25.5	30.6	35.7	40.8	45.9	51.0	56.1	61.2
19.8	5.1	10.1	15.2	20.2	25.3	30.3	35.4	40.4	45.5	50.5	55.6	60.6
20.0	5.0	10.0	15.0	20.0	25.0	30.0	35.0	40.0	45.0	50.0	55.0	60.0
20.5	4.9	9.8	14.6	19.5	24.4	29.3	34.1	39.0	43.9	48.8	53.7	58.5
21.0	4.8	9.5	14.2	19.0	23.8	28.6	33.3	38.0	42.8	47.6	52.4	57.1
21.5	4.7	9.3	13.9	18.6	23.3	27.9	32.6	37.2	41.8	46.5	51.2	55.8
22.0	4.5	9.1	13.6	18.2	22.7	27.2	31.8	36.4	40.9	45.4	50.0	54.5
22.5	4.4	8.8	13.3	17.8	22.2	26.7	31.1	35.6	40.0	44.4	48.9	53.3
23.0	4.3	8.7	13.0	17.4	21.7	26.1	30.4	34.7	39.1	43.4	47.8	52.2
23.5	4.3	8.5	12.8	17.0	21.2	25.5	29.8	34.0	38.3	42.6	46.8	51.1
24.0	4.2	8.3	12.5	16.6	20.8	25.0	29.1	33.3	37.5	41.6	45.0	50.0
24.5	4.1	8.2	12.2	16.3	20.4	24.5	28.6	32.7	36.7	40.8	44.9	49.0
25.0	4.0	8.0	12.0	16.0	20.0	24.0	28.0	32.0	36.0	40.0	44.0	48.0
26	3.8	7.7	11.5	15.4	19.2	23.1	26.9	30.8	34.6	38.4	42.3	46.2
27	3.7	7.4	11.1	14.8	18.5	22.2	25.9	29.6	33.3	37.0	40.7	44.4
28	3.5	7.1	10.7	14.2	17.8	21.4	25.0	28.5	32.1	35.7	39.2	42.8
29	3.4	6.9	10.3	13.7	17.2	20.6	24.1	27.5	31.0	34.4	37.9	41.3
30	3.3	6.7	10.0	13.3	16.7	20.0	23.3	26.7	30.0	33.3	36.7	40.0
31	3.2	6.5	9.7	12.9	16.1	19.3	22.5	25.8	29.0	32.2	35.4	38.7
32	3.1	6.2	9.4	12.3	15.6	18.7	21.8	25.0	28.1	31.2	34.3	37.5
33	3.0	6.1	9.1	12.1	15.1	18.1	21.2	24.2	27.2	30.3	33.3	36.3
34	2.9	5.9	8.8	11.8	14.7	17.6	20.5	23.5	26.4	29.4	32.3	35.2
35	2.9	5.7	8.6	11.4	14.3	17.1	20.0	22.8	25.7	28.5	31.4	34.3
36	2.8	5.5	8.3	11.1	13.8	16.7	19.4	22.2	25.0	27.8	30.6	33.3
37	2.7	5.4	8.1	10.8	13.5	16.2	18.9	21.6	24.3	27.0	29.7	32.4
38	2.6	5.3	7.9	10.5	13.1	15.7	18.4	21.0	23.6	26.3	28.9	31.5
39	2.6	5.1	7.7	10.2	12.8	15.3	17.9	20.5	23.0	25.6	28.2	30.7
40	2.5	5.0	7.5	10.0	12.5	15.0	17.5	20.0	22.5	25.0	27.5	30.0
41	2.4	4.8	7.3	9.8	12.1	14.6	17.1	19.5	21.9	24.3	26.8	29.3
42	2.4	4.7	7.1	9.5	11.9	14.2	16.6	19.0	21.4	23.8	26.1	28.5
43	2.3	4.6	7.0	9.3	11.6	13.9	16.2	18.6	20.9	23.2	25.6	27.6
44	2.3	4.5	6.8	9.1	11.4	13.6	15.9	18.2	20.4	22.7	25.0	27.2
45	2.2	4.4	6.6	8.9	11.1	13.3	15.6	17.8	20.0	22.2	24.4	26.7
46	2.2	4.3	6.5	8.7	10.9	13.0	15.2	17.4	19.6	21.7	23.9	26.0
47	2.1	4.3	6.4	8.5	10.6	12.7	14.9	17.0	19.1	21.3	23.4	25.5
48	2.1	4.2	6.2	8.3	10.4	12.5	14.6	16.7	18.8	20.8	22.9	25.0
49	2.0	4.1	6.1	8.1	10.2	12.2	14.2	16.3	18.4	20.4	22.4	24.5
50	2.0	4.0	6.0	8.0	10.0	12.0	14.0	16.0	18.0	20.0	22.0	24.0
51	2.0	3.9	5.9	7.8	9.8	11.8	13.7	15.7	17.6	19.6	21.6	23.5
52	1.9	3.8	5.8	7.7	9.6	11.5	13.4	15.4	17.3	19.2	21.1	23.1
53	1.9	3.8	5.7	7.5	9.4	11.3	13.2	15.0	17.0	18.9	20.7	22.6
54	1.9	3.7	5.6	7.4	9.3	11.1	13.0	14.8	16.7	18.5	20.3	22.2
55	1.8	3.6	5.5	7.3	9.1	10.9	12.7	14.5	16.4	18.1	20.0	21.8
56	1.8	3.6	5.4	7.1	8.9	10.7	12.5	14.2	16.0	17.9	19.6	21.4
57	1.8	3.5	5.3	7.0	8.8	10.5	12.2	14.0	15.8	17.5	19.2	21.0
58	1.7	3.4	5.2	6.9	8.6	10.3	12.0	13.8	15.5	17.2	18.8	20.7
59	1.7	3.4	5.1	6.8	8.5	10.2	11.9	13.6	15.3	16.9	18.6	20.3
60	1.7	3.3	5.0	6.7	8.3	10.0	11.7	13.3	15.0	16.7	18.3	20.0
61	1.6	3.3	4.9	6.6	8.2	9.8	11.5	13.1	14.8	16.4	18.0	19.7
62	1.6	3.2	4.8	6.5	8.1	9.7	11.3	12.9	14.5	16.1	17.7	19.3
63	1.6	3.2	4.8	6.3	7.9	9.5	11.1	12.7	14.2	15.9	17.4	19.0
64	1.6	3.1	4.7	6.3	7.8	9.4	10.9	12.5	14.0	15.6	17.2	18.7
65	1.5	3.1	4.6	6.2	7.7	9.2	10.8	12.3	13.8	15.4	16.9	18.5
66	1.5	3.0	4.5	6.1	7.6	9.1	10.6	12.1	13.6	115.2	16.7	18.2
67	1.5	3.0	4.5	6.0	7.5	9.0	10.4	11.9	13.4	15.0	16.4	17.9
68	1.5	2.9	4.4	5.9	7.4	8.8	10.2	11.8	13.2	14.7	16.2	17.6
69	1.4	2.9	4.3	5.8	7.2	8.7	10.1	1.6	13.0	14.5	15.9	17.3
70	1.4	2.9	4.3	5.7	7.1	8.6	10.0	11.4	12.9	14.3	15.7	17.1

VALUE	Δt TEMPERATURE DIFFERENCE (Pipe or Equipment Surface Temperature Minus Ambient Air Temperature)											
$R+R_s$	100	200	300	400	500	600	700	800	900	1000	1100	1200
71	1.40	2.82	4.22	5.63	7.04	8.45	9.86	11.27	12.68	14.08	15.49	16.90
72	1.39	2.77	4.16	5.55	6.94	8.33	9.72	11.11	12.50	13.88	15.28	16.66
73	1.37	2.74	4.11	5.48	6.85	8.22	9.59	10.96	12.33	13.70	15.07	16.44
74	1.35	2.70	4.05	5.41	6.76	8.11	9.46	10.81	12.16	13.51	14.86	16.22
75	1.33	2.66	3.99	5.33	6.67	8.00	9.33	10.66	12.00	13.33	14.66	16.00
76	1.32	2.63	3.94	5.26	6.58	7.89	9.21	10.52	11.84	13.16	14.47	15.79
77	1.30	2.60	3.90	5.19	6.49	7.79	9.09	10.40	11.69	12.99	14.29	15.58
78	1.28	2.56	3.85	5.13	6.41	7.69	8.97	10.26	11.54	12.82	14.10	15.38
79	1.27	2.53	3.80	5.06	6.33	7.59	8.86	10.13	11.39	12.66	13.92	15.19
80	1.25	2.50	3.75	5.00	6.25	7.50	8.75	10.00	11.25	12.55	13.75	15.00
82	1.22	2.44	3.66	4.88	6.09	7.32	8.54	9.76	10.98	12.20	13.41	14.63
84	1.19	2.38	3.57	4.76	5.95	7.14	8.33	9.52	10.71	11.90	13.11	14.29
86	1.16	2.33	3.49	4.65	5.81	6.98	8.20	9.30	10.47	11.63	12.79	13.95
88	1.14	2.27	3.41	4.55	5.68	6.82	7.95	9.09	10.22	11.36	12.50	13.64
90	1.11	2.22	3.33	4.44	5.56	6.67	7.78	8.89	10.00	11.11	12.22	13.33
92	1.09	2.17	3.26	4.34	5.43	6.52	7.61	8.70	9.78	10.87	11.96	13.04
94	1.06	2.13	3.19	4.26	5.32	6.38	7.45	8.51	9.57	10.64	11.70	12.77
96	1.04	2.08	3.13	4.17	5.21	6.25	7.29	8.33	9.38	10.42	11.46	12.50
98	1.02	2.04	3.06	4.08	5.10	6.12	7.14	8.16	9.18	10.20	11.22	12.24
100	1.00	2.00	3.00	4.00	5.00	6.00	7.00	8.00	9.00	10.00	11.00	12.00

TABLE B-49 (Sheet 2 of 2)

Table B-50

CAPITAL INVESTMENT FOR EQUIPMENT TO PRODUCE HEAT (OR STEAM)

Dollar Investment in Equipment to Produce Heat (or Steam) to Point of Use

$ per 1000 Btu/hr →		10.00	15.00	20.00	25.00	30.00	35.00	40.00	45.00	50.00	55.00	60.00
$ per 1 lb steam/hr →		12.50	18.75	25.00	31.25	37.50	43.75	50.00	56.25	62.50	68.75	75.00
$ per 1000 W(hr) →		34.12	51.18	64.24	85.30	102.36	119.42	136.48	153.55	170.61	187.67	204.73
↓Btu/hr	W (per hr)↓											
10	2.93	$0.10	$0.15	$0.20	$0.25	$0.30	$0.35	$0.40	$0.45	$0.50	$0.55	$0.60
20	5.86	0.20	0.30	0.40	0.50	0.60	0.70	0.80	0.90	1.00	1.10	1.20
30	8.79	0.30	0.45	0.60	0.75	0.90	1.05	1.20	1.35	1.50	1.65	1.80
40	11.72	0.40	0.60	0.80	1.00	1.20	1.40	1.60	1.80	2.00	2.20	2.40
50	14.65	0.50	0.75	1.00	1.25	1.50	1.75	2.00	2.25	2.50	2.75	3.00
60	17.58	0.60	0.90	1.20	1.50	1.80	2.10	2.40	2.70	3.00	3.30	3.60
70	20.51	0.70	1.05	1.40	1.75	2.10	2.45	2.80	3.15	3.50	3.85	4.20
80	23.45	0.80	1.20	1.60	2.00	2.40	2.80	3.20	3.60	4.00	4.40	4.80
90	26.37	0.90	1.35	1.80	2.25	2.70	3.15	3.60	4.05	4.50	4.95	5.40
100	29.30	1.00	1.50	2.00	2.50	3.00	3.50	4.00	4.50	5.00	5.50	6.00
110	32.24	1.10	1.65	2.20	2.75	3.30	3.85	4.40	4.95	5.50	6.05	6.60
120	35.16	1.20	1.80	2.40	3.00	3.60	4.20	4.80	5.40	6.00	6.60	7.20
130	38.10	1.30	1.95	2.60	3.25	3.90	4.55	5.20	5.85	6.50	7.15	7.80
140	41.02	1.40	2.10	2.80	3.50	4.20	4.90	5.60	6.30	7.00	7.70	8.40
150	43.90	1.50	2.25	3.00	3.75	4.50	5.25	6.00	6.75	7.50	8.25	9.00
160	46.89	1.60	2.40	3.20	4.00	4.80	5.60	6.40	7.20	8.00	8.80	9.60
170	49.82	1.70	2.55	3.40	4.25	5.10	5.95	6.80	7.65	8.50	9.35	10.20
180	52.75	1.80	2.70	3.60	4.50	5.40	6.30	7.20	8.10	9.00	9.90	10.80
190	55.66	1.90	2.85	3.80	4.75	5.70	6.65	7.60	8.55	9.50	10.45	11.40
200	58.61	2.00	3.00	4.00	5.00	6.00	7.00	8.00	9.00	10.00	11.00	12.00
210	61.54	2.10	3.15	4.20	5.25	6.30	7.35	8.40	9.45	10.50	11.55	12.60
220	64.47	2.20	3.30	4.40	5.50	6.60	7.70	8.80	9.90	11.00	12.10	13.20
230	67.41	2.30	3.45	4.60	5.75	6.90	8.05	9.20	10.35	11.50	12.65	13.80
240	70.34	2.40	3.60	4.80	6.00	7.20	8.40	9.60	10.80	12.00	13.20	14.40
250	73.27	2.50	3.75	5.00	6.25	7.50	8.75	10.00	11.25	12.50	13.75	15.00
260	76.20	2.60	3.90	5.20	6.50	7.80	9.10	10.40	11.70	13.00	14.30	15.60
270	79.13	2.70	4.05	5.40	6.75	8.10	9.45	10.80	12.15	13.50	14.85	16.20
280	82.06	2.80	4.20	5.60	7.00	8.40	9.80	11.20	12.60	14.00	15.40	16.80
290	84.99	2.90	4.35	5.80	7.25	8.70	10.15	11.60	13.05	14.50	15.95	17.40
300	87.92	3.00	4.50	6.00	7.50	9.00	10.50	12.00	13.50	15.00	16.50	18.00
310	90.85	3.10	4.65	6.20	7.75	9.30	10.85	12.40	13.95	15.50	17.05	18.60
320	93.78	3.20	4.80	6.40	8.00	9.60	11.20	12.80	14.40	16.00	17.60	19.20
330	96.71	3.30	4.95	6.60	8.25	9.90	11.55	13.20	15.30	17.00	19.70	19.80
340	99.64	3.40	5.10	6.80	8.50	10.20	11.90	13.60	15.30	17.00	19.70	20.40
350	102.57	3.50	5.25	7.00	8.75	10.50	12.25	14.00	15.75	17.50	20.25	21.00
360	105.50	3.60	5.40	7.20	9.00	10.80	12.60	14.40	16.20	18.00	20.80	21.60
370	108.44	3.70	5.55	7.40	9.25	11.10	12.95	14.80	16.65	18.50	21.35	22.20
380	111.37	3.80	5.70	7.60	9.50	11.40	13.30	15.20	17.10	19.00	21.90	22.80
390	114.30	3.90	5.85	7.80	9.75	11.70	13.65	15.60	17.55	19.50	22.45	23.40
400	117.23	4.00	6.00	8.00	10.00	12.00	14.00	16.00	18.00	20.00	22.00	24.00
410	120.16	4.10	6.15	8.20	10.25	12.30	14.35	16.40	18.45	20.50	22.55	24.60
420	123.09	4.20	6.30	8.40	10.50	12.60	14.70	16.80	18.90	21.00	23.10	25.20
430	126.02	4.30	6.45	8.60	10.75	12.90	15.09	17.20	19.35	21.50	23.65	25.80
440	128.95	4.40	6.60	8.80	11.00	13.20	15.40	17.60	19.80	22.00	24.20	26.40
450	131.88	4.50	6.75	9.00	11.25	13.50	15.75	18.00	20.25	22.50	24.75	27.00
460	134.81	4.60	6.90	9.20	11.50	13.80	16.10	18.40	20.70	23.00	25.30	27.60
470	137.74	4.70	7.05	9.40	11.75	14.10	16.45	18.80	21.15	23.50	25.85	28.20
480	140.67	4.80	7.20	9.60	12.00	14.40	16.80	19.20	21.60	24.00	26.40	28.80
490	143.65	4.90	7.35	9.80	12.25	14.70	17.15	19.60	22.05	24.50	26.95	29.40
500	146.53	5.00	7.50	10.00	12.50	15.00	17.50	20.00	22.50	25.00	27.50	30.00
510	149.47	5.10	7.65	10.20	12.75	15.30	17.85	20.40	22.95	25.50	28.05	30.60
520	152.40	5.20	7.80	10.40	13.00	15.60	18.20	20.80	23.40	26.00	28.60	31.20
530	155.33	5.30	7.95	10.60	13.25	15.90	18.55	21.20	23.85	26.50	29.15	31.80
540	158.26	5.40	8.10	10.80	13.50	16.20	18.90	21.60	24.30	27.00	29.70	32.40
550	161.19	5.50	8.25	11.00	13.75	16.50	19.25	22.00	24.75	27.50	30.25	33.00
560	164.12	5.60	8.40	11.20	14.00	16.80	19.60	22.40	25.20	28.00	30.80	33.60
570	167.05	5.70	8.55	11.40	14.25	17.10	19.95	22.80	25.65	28.50	31.35	34.20
580	169.98	5.80	8.70	11.60	14.50	17.40	20.30	23.20	26.10	29.00	31.90	34.80
590	172.91	5.90	8.85	11.80	14.75	17.70	20.65	23.60	26.55	29.50	32.45	35.40
600	175.84	6.00	9.00	12.00	15.00	18.00	21.00	24.00	27.00	30.00	33.00	36.00
610	178.77	6.10	9.15	12.20	15.25	18.30	21.35	24.40	27.45	30.50	33.55	36.60
620	181.70	6.20	9.30	12.40	15.50	18.60	21.70	24.80	27.90	31.00	34.10	37.20
630	184.63	6.30	9.45	12.60	15.75	18.90	22.05	25.20	28.35	31.50	34.65	37.80
640	187.51	6.40	9.60	12.80	16.00	19.20	22.40	25.60	28.80	32.00	35.20	38.40
650	190.50	6.50	9.75	13.00	16.25	19.50	22.75	26.00	29.25	32.50	35.75	39.00
660	193.43	6.60	9.90	13.20	16.50	19.80	23.10	26.40	29.70	33.00	36.30	39.60
670	196.36	6.70	10.05	13.40	16.75	20.10	23.45	26.80	30.15	33.50	36.85	40.20
680	199.29	6.80	10.20	13.60	17.00	20.40	23.80	27.20	30.60	34.00	37.40	40.80
690	202.22	6.90	10.35	13.80	17.25	20.70	24.15	27.60	31.05	34.50	37.95	41.40
700	205.15	7.00	10.50	14.00	17.50	21.00	24.50	28.00	31.50	35.00	38.50	42.00
710	208.08	7.10	10.65	14.20	17.75	21.30	24.85	28.40	31.95	35.50	39.05	42.60
720	211.01	7.20	10.80	14.40	18.00	21.60	25.20	28.80	32.40	36.00	39.60	43.20
730	213.94	7.30	10.95	14.60	18.25	21.90	25.55	29.20	32.85	36.50	40.15	43.80
740	216.87	7.40	11.10	14.80	18.50	22.20	25.90	29.60	33.30	37.00	40.70	44.40
750	219.80	7.50	11.25	15.00	18.75	22.50	26.25	30.00	33.75	37.50	41.25	45.00
760	222.73	7.60	11.40	15.20	19.00	22.80	26.60	30.40	34.20	38.00	41.80	45.60
770	225.66	7.70	11.55	15.40	19.25	23.10	26.95	30.80	34.65	38.50	42.35	46.20
780	228.60	7.80	11.70	15.60	19.50	23.40	27.30	31.20	35.10	39.00	42.90	46.80
790	231.53	7.90	11.85	15.80	19.75	28.50	27.65	31.60	35.55	39.50	43.45	47.40
800	234.46	8.00	12.00	16.00	20.00	24.00	28.00	32.00	36.00	40.00	44.00	48.00
↑Btu	Watts↑											

Energy Per Hour

Dollar Capital Investment

TABLE B-50 (Sheet 1 of 4)

Table B-50 (Continued)

CAPITAL INVESTMENT FOR EQUIPMENT TO PRODUCE HEAT (OR STEAM)

Dollar Investment in Equipment to Produce Heat (or Steam) to Point of Use

$ per 1000 Btu/hr →		10.00	15.00	20.00	25.00	30.00	35.00	40.00	45.00	50.00	55.00	60.00
$ per 1 lb steam/hr →		**12.50**	**18.75**	**25.00**	**31.25**	**37.50**	**43.75**	**50.00**	**56.25**	**62.50**	**68.75**	**75.00**
$ per 1000 W(hr) →		**34.12**	**51.18**	**64.24**	**85.30**	**102.36**	**119.42**	**136.48**	**153.55**	**170.61**	**187.67**	**204.73**
↓Btu/hr	**W (per hr)↓**											
810	**237.49**	$8.10	$12.15	$16.20	$20.25	$24.30	$28.35	$32.40	$36.45	$40.50	$44.55	$48.60
820	**240.31**	8.20	12.30	16.40	20.50	24.60	28.70	32.80	36.90	41.00	45.10	49.20
830	**243.25**	8.30	12.45	16.60	20.75	24.90	29.05	33.20	37.35	41.50	45.65	49.80
840	**246.18**	8.40	12.60	16.80	21.00	25.20	29.40	33.60	37.80	42.00	46.20	50.40
850	**249.11**	8.50	12.75	17.00	21.25	25.50	29.75	34.00	38.35	42.50	46.75	51.00
860	**252.04**	8.60	12.90	17.20	21.50	25.80	30.10	34.40	38.70	43.00	47.30	51.60
870	**254.97**	8.70	13.05	17.40	21.75	26.10	30.45	34.80	39.15	43.50	47.85	52.20
880	**257.90**	8.80	13.20	17.60	22.00	26.40	30.80	35.20	39.60	44.00	48.40	52.80
890	**260.83**	8.90	13.35	17.80	22.25	26.70	31.15	35.60	40.05	44.50	48.95	53.40
900	**263.76**	9.00	13.50	18.00	22.50	27.00	31.50	36.00	40.50	45.00	49.50	54.00
910	**266.69**	9.10	13.65	18.20	22.75	27.30	31.85	36.40	40.95	45.50	50.05	54.60
920	**269.63**	9.20	13.80	18.40	23.00	27.60	32.20	36.80	41.40	46.00	50.60	55.20
930	**272.56**	9.30	13.95	18.60	23.25	27.90	32.55	37.20	41.85	46.50	51.15	55.80
940	**275.49**	9.40	14.10	18.80	23.50	28.20	32.90	37.60	42.30	47.00	51.70	56.40
950	**278.42**	9.50	14.25	19.00	23.75	28.50	33.25	38.00	42.75	47.50	52.25	57.00
960	**281.35**	9.60	14.40	19.20	24.00	28.80	33.60	38.40	43.20	48.00	52.80	57.60
970	**284.28**	9.70	14.55	19.40	24.25	29.10	33.95	38.80	43.65	48.50	53.35	58.20
980	**287.20**	9.80	14.70	19.60	24.50	29.40	34.30	39.20	44.10	49.00	53.90	58.80
990	**290.14**	9.90	14.85	19.80	24.75	29.40	34.65	39.60	44.55	49.50	54.45	59.40
1000	**293.07**	10.00	15.00	20.00	25.00	30.00	35.00	40.00	45.00	50.00	55.00	60.00
1100	**322.4**	11.00	16.50	22.00	27.50	33.00	38.50	44.00	49.50	55.00	60.50	66.00
1200	**351.6**	12.00	18.00	24.00	30.00	36.00	42.00	48.00	54.00	60.00	66.00	72.00
1300	**381.0**	13.00	19.50	26.00	32.50	39.00	45.50	52.00	58.50	65.00	71.50	78.00
1400	**410.2**	14.00	21.00	28.00	35.00	42.00	49.00	56.00	63.00	70.00	77.00	84.00
1500	**439.0**	15.00	23.50	30.00	37.50	45.00	52.50	60.00	67.50	75.00	82.50	90.00
1600	**468.9**	16.00	24.00	32.00	40.00	48.00	56.00	64.00	72.00	80.00	88.00	96.00
1700	**498.2**	17.00	25.50	34.00	42.50	51.00	59.50	68.00	76.50	85.00	93.50	102.00
1800	**527.5**	18.00	27.00	36.00	45.00	54.00	63.00	72.00	81.00	90.00	99.00	108.00
1900	**555.6**	19.00	28.50	38.50	47.50	57.00	66.50	76.00	85.50	95.00	104.50	114.00
2000	**586.1**	20.00	30.00	40.00	50.00	60.00	70.00	80.00	90.00	100.00	110.00	120.00
2100	**615.4**	21.00	31.50	42.00	52.50	63.00	73.50	84.00	94.50	105.00	115.50	126.00
2200	**644.7**	22.00	33.00	44.00	55.00	66.00	77.00	88.00	99.00	110.00	121.00	132.00
2300	**674.1**	23.00	34.50	46.00	57.50	69.00	80.50	92.00	103.50	115.00	126.50	138.00
2400	**703.4**	24.00	36.00	48.00	60.00	72.00	84.00	96.00	108.00	120.00	132.00	144.00
2500	**732.7**	25.00	37.50	50.00	62.50	75.00	87.50	100.00	112.50	125.00	137.50	150.00
2600	**762.0**	26.00	39.00	52.00	65.00	78.00	91.00	104.00	117.00	130.00	143.00	156.00
2700	**791.3**	27.00	40.50	54.00	67.50	81.00	94.50	108.00	121.50	135.00	148.50	162.00
2800	**820.6**	28.00	42.00	56.00	70.00	84.00	98.00	112.00	126.00	140.00	154.00	168.00
2900	**849.9**	29.00	43.50	58.00	72.50	87.00	101.50	116.00	130.50	145.00	159.50	174.00
3000	**879.2**	30.00	45.00	60.00	75.00	90.00	105.00	120.00	135.00	150.00	165.00	180.00
3100	**908.3**	31.00	46.50	62.00	77.50	93.00	108.50	124.00	139.50	155.00	170.50	186.00
3200	**937.8**	32.00	48.00	64.00	80.00	96.00	112.00	128.00	144.00	160.00	176.00	192.00
3300	**967.1**	33.00	49.50	66.00	82.50	99.00	115.50	132.00	148.50	165.00	181.50	198.00
3400	**996.4**	34.00	51.00	68.00	85.00	102.00	119.00	136.00	153.00	170.00	197.00	204.00
3500	**1025.7**	35.00	52.50	70.00	87.50	105.00	122.50	140.00	157.50	175.00	202.50	210.00
3600	**1055.0**	36.00	54.00	72.00	90.00	108.00	126.00	144.00	162.00	180.00	208.00	216.00
3700	**1084.4**	37.00	55.55	74.00	92.50	111.00	129.50	148.00	166.50	185.00	213.50	222.00
3800	**1113.7**	38.00	57.00	76.00	95.00	114.00	133.00	152.00	171.00	190.00	219.00	228.00
3900	**1143.0**	39.00	58.50	78.00	97.50	117.00	136.50	156.00	175.50	195.00	224.50	234.00
4000	**1172.3**	40.00	60.00	80.00	100.00	120.00	140.00	160.00	180.00	200.00	220.00	240.00
4100	**1201.6**	41.00	61.50	82.00	102.50	123.00	143.50	164.00	184.50	205.00	245.50	246.00
4200	**1230.9**	42.00	63.00	84.00	105.00	126.00	147.00	168.00	189.00	210.00	231.00	252.00
4300	**1260.2**	43.00	64.50	86.00	107.50	129.00	150.50	172.00	193.50	215.00	236.50	258.00
4400	**1289.5**	44.00	66.00	88.00	110.00	132.00	154.00	176.00	198.00	220.00	242.00	264.00
4500	**1318.8**	45.00	67.50	90.00	112.50	135.00	157.50	180.00	202.50	225.00	247.50	270.00
4600	**1348.1**	46.00	69.00	92.00	115.00	138.00	161.00	184.00	207.00	230.00	253.00	276.00
4700	**1377.4**	47.00	70.50	94.00	117.50	141.00	164.50	188.00	211.50	235.00	258.50	282.00
4800	**1406.7**	48.00	72.00	96.00	120.00	144.00	168.00	192.00	216.00	240.00	264.00	288.00
4900	**1436.5**	49.00	73.50	98.00	122.50	147.00	171.50	196.00	220.50	245.00	269.50	294.00
5000	**1465.3**	50.00	75.00	100.00	125.00	150.00	175.00	200.00	225.00	250.00	275.00	300.00
5100	**1494.7**	51.00	76.50	102.00	127.50	153.00	178.50	204.00	229.50	255.00	280.50	306.00
5200	**1524.0**	52.00	78.00	104.00	130.00	156.00	182.00	208.00	234.00	260.00	286.00	312.00
5300	**1553.3**	53.00	79.50	106.00	132.50	159.00	185.50	212.00	238.50	265.00	291.50	318.00
5400	**1582.6**	54.00	81.00	108.00	135.00	162.00	189.00	216.00	243.00	270.00	297.00	324.00
5500	**1611.9**	55.00	82.50	110.00	137.50	165.00	192.50	220.00	247.50	275.00	302.50	330.00
5600	**1641.2**	56.00	84.00	112.00	140.00	168.00	196.00	224.00	252.00	280.00	308.00	336.00
5700	**1670.5**	57.00	85.50	114.00	142.50	171.00	199.50	228.00	256.50	285.00	313.50	342.00
5800	**1699.0**	58.00	87.00	116.00	145.00	174.00	203.00	232.00	261.00	290.00	319.00	348.00
5900	**1729.1**	59.00	88.50	118.00	147.50	177.00	206.50	236.00	265.50	295.00	324.50	354.00
6000	**1758.4**	60.00	90.00	120.00	150.00	180.00	210.00	240.00	270.00	300.00	330.00	360.00
6100	**1787.7**	61.00	91.50	122.00	152.50	183.00	213.50	244.00	274.50	305.00	335.50	366.00
6200	**1817.0**	62.00	93.00	124.00	155.00	186.00	217.00	248.00	279.00	310.00	341.00	372.00
6300	**1846.3**	63.00	94.50	126.00	157.50	189.00	220.50	252.00	283.50	315.00	346.50	378.00
6400	**1875.7**	64.00	96.00	128.00	160.00	192.00	224.00	256.00	288.00	320.00	352.00	384.00
6500	**1905.0**	65.00	97.50	130.00	162.50	195.00	227.50	260.00	292.50	325.00	357.50	390.00
6600	**1934.3**	66.00	99.00	132.00	165.00	198.00	231.00	264.00	297.00	330.00	363.00	396.00
6700	**1963.6**	67.00	100.50	134.00	167.50	201.00	234.50	268.00	301.50	335.00	368.50	402.00
6800	**1992.9**	68.00	102.00	136.00	170.00	204.00	238.00	272.00	306.00	340.00	374.00	408.00
6900	**2022.2**	69.00	103.50	138.00	172.50	207.00	241.50	276.00	310.50	345.50	379.50	414.00
7000	**2051.5**	70.00	105.00	140.00	175.00	210.00	245.00	280.00	315.00	350.00	385.00	420.00
↑Btu	**Watts↑**											

Energy Per Hour

Dollar Capital Investment

TABLE B-50 (Sheet 2 of 4)

Table B-50 (Continued)

CAPITAL INVESTMENT FOR EQUIPMENT TO PRODUCE HEAT (OR STEAM)

Dollar Investment in Equipment to Produce Heat (or Steam) to Point of Use

$ per 1000 Btu/hr →		10.00	15.00	20.00	25.00	30.00	35.00	40.00	45.00	50.00	55.00	60.00
$ per 1 lb steam/hr →		12.50	18.75	25.00	31.25	37.50	43.75	50.00	56.25	62.50	68.75	75.00
$ per 1000 W(hr) →		34.12	51.18	64.24	85.30	102.36	119.42	136.48	153.55	170.61	187.67	204.73
↓Btu/hr	W (per hr)↓											
7100	2080.8	$71.00	$106.50	$142.00	$177.50	$213.00	$248.50	$284.00	$319.50	$355.00	$390.50	$426.00
7200	2110.1	72.00	108.00	144.00	180.00	216.00	252.00	288.00	324.00	360.00	396.00	432.00
7300	2139.4	73.00	109.50	146.50	182.50	219.00	255.50	292.00	328.50	365.00	401.50	438.00
7400	2168.7	74.00	111.00	148.00	185.00	222.00	259.00	296.00	333.00	370.00	407.00	444.00
7500	2198.0	75.00	112.50	150.00	187.50	225.00	362.50	300.00	337.50	375.00	412.50	450.00
7600	2227.3	76.00	114.00	152.00	190.00	228.00	266.00	304.00	342.00	380.00	418.00	456.00
7700	2256.6	77.00	115.50	154.00	192.50	231.00	269.50	308.00	346.50	385.00	423.50	462.00
7800	2286.0	78.00	117.00	156.00	195.00	234.00	273.00	312.00	351.00	390.00	429.00	468.00
7900	2315.3	79.00	118.50	158.00	197.50	237.00	276.50	316.00	355.50	395.00	434.50	474.00
8000	2344.6	80.00	120.00	160.00	200.00	240.00	280.00	320.00	360.00	400.00	440.00	480.00
8100	2373.9	81.00	121.50	162.00	202.50	243.00	283.50	324.00	364.50	405.00	445.50	486.00
8200	2403.1	82.00	123.00	164.00	205.00	246.00	287.00	328.00	369.00	410.00	451.00	492.00
8300	2432.5	83.00	124.50	166.00	207.50	249.00	290.50	332.00	373.50	415.00	456.50	498.00
8400	2461.8	84.00	126.00	168.00	210.00	252.00	294.00	336.00	378.00	420.00	462.00	504.00
8500	2491.1	85.00	127.50	170.00	212.50	255.00	297.50	340.00	383.50	425.00	467.50	510.00
8600	2520.4	86.00	129.00	172.00	215.00	258.00	301.00	344.00	387.00	430.00	473.00	516.00
8700	2549.7	87.00	130.50	174.00	217.50	261.00	304.50	348.00	391.50	435.00	478.50	522.00
8800	2579.0	88.00	132.00	176.00	220.00	264.00	308.00	352.00	396.00	440.00	484.00	528.00
8900	2608.3	89.00	133.50	178.00	222.50	267.00	311.50	356.00	460.50	445.00	489.50	534.00
9000	2637.6	90.00	135.00	180.00	225.00	270.00	315.00	360.00	405.00	450.00	495.00	540.00
9100	2666.9	91.00	136.50	182.00	227.50	273.00	318.50	364.00	409.50	455.00	500.50	546.00
5200	2696.3	92.00	138.00	184.00	230.00	276.00	322.00	368.00	414.00	460.00	506.00	552.00
9300	2725.6	93.00	139.50	186.00	232.50	279.00	325.50	372.00	418.50	465.00	511.50	558.00
9400	2754.9	94.00	141.00	188.00	235.00	282.00	329.00	376.00	423.00	470.00	517.00	564.00
9500	2784.2	95.00	142.50	190.00	237.50	285.00	332.50	380.00	427.50	475.00	522.50	570.00
9600	2813.5	96.00	144.00	192.00	240.00	288.00	336.00	384.00	432.00	480.00	528.00	576.00
9700	2842.8	97.00	145.50	194.00	242.50	291.00	339.50	388.00	436.50	485.00	533.50	582.00
9800	2872.0	98.00	147.00	196.00	245.00	294.00	343.00	392.00	441.00	490.00	539.00	588.00
9900	2901.4	99.00	148.50	198.00	247.50	297.00	346.50	394.00	445.50	495.00	544.50	594.00
10000	2930.7	100.00	150.00	200.00	250.00	300.00	350.00	400.00	450.00	500.00	550.00	600.00
11000	3224	110.00	165.00	220.00	275.00	330.00	385.00	440.00	495.00	550.00	605.00	660.00
12000	3516	120.00	180.00	240.00	300.00	360.00	420.00	480.00	540.00	600.00	660.00	720.00
13000	3810	130.00	195.00	260.00	325.00	390.00	455.00	520.00	585.00	650.00	715.00	780.00
14000	4102	140.00	210.00	280.00	350.00	420.00	490.00	560.00	630.00	700.00	770.00	840.00
15000	4390	150.00	235.00	300.00	375.00	450.00	525.00	600.00	675.00	750.00	825.00	900.00
16000	4689	160.00	240.00	320.00	400.00	480.00	560.00	640.00	720.00	800.00	880.00	960.00
17000	4982	170.00	255.00	340.00	425.00	510.00	595.00	680.00	765.00	850.00	935.00	1020.00
18000	5275	180.00	270.00	360.00	450.00	540.00	630.00	720.00	810.00	900.00	990.00	1080.00
19000	5556	190.00	285.00	380.00	425.00	570.00	665.00	760.00	855.00	950.00	1045.00	1140.00
20000	5861	200.00	300.00	400.00	500.00	600.00	700.00	800.00	900.00	1000.00	1100.00	1200.00
21000	6154	210.00	315.00	420.00	525.00	630.00	735.00	840.00	945.00	1050.00	1155.00	1260.00
22000	6447	220.00	330.00	440.00	550.00	660.00	770.00	880.00	990.00	1100.00	1210.00	1320.00
23000	6741	230.00	345.00	460.00	575.00	690.00	805.00	920.00	1035.00	1150.00	1265.00	1380.00
24000	7034	240.00	360.00	480.00	600.00	720.00	840.00	960.00	1080.00	1200.00	1320.00	1440.00
25000	7327	250.00	375.00	500.00	625.00	750.00	875.00	1000.00	1125.00	1250.00	1375.00	1500.00
26000	7620	260.00	390.00	520.00	650.00	780.00	910.00	1040.00	1170.00	1300.00	1430.00	1560.00
27000	7913	270.00	405.00	540.00	675.00	810.00	945.00	1080.00	1215.00	1350.00	1485.00	1620.00
28000	8206	280.00	420.00	560.00	700.00	840.00	980.00	1120.00	1260.00	1400.00	1540.00	1680.00
29000	8499	290.00	435.00	580.00	725.00	870.00	1015.00	1160.00	1305.00	1450.00	1595.00	1740.00
30000	8792	300.00	450.00	600.00	750.00	900.00	1050.00	1200.00	1350.00	1500.00	1650.00	1800.00
31000	9083	310.00	465.00	620.00	775.00	930.00	1085.00	1240.00	1395.00	1550.00	1705.00	1860.00
32000	9378	320.00	480.00	640.00	800.00	960.00	1120.00	1280.00	1440.00	1600.00	1760.00	1920.00
33000	9671	330.00	495.00	660.00	825.00	990.00	1155.00	1320.00	1485.00	1650.00	1815.00	1980.00
34000	9964	340.00	510.00	680.00	850.00	1020.00	1190.00	1360.00	1530.00	1700.00	1870.00	2040.00
35000	10257	350.00	525.00	700.00	875.00	1050.00	1225.00	1400.00	1575.00	1750.00	1925.00	2100.00
36000	10550	360.00	540.00	720.00	900.00	1080.00	1260.00	1440.00	1620.00	1800.00	1980.00	2160.00
37000	10844	370.00	555.00	740.00	925.00	1110.00	1295.00	1480.00	1665.00	1850.00	2035.00	2220.00
38000	11137	380.00	570.00	760.00	950.00	1140.00	1330.00	1520.00	1710.00	1900.00	2090.00	2280.00
39000	11430	390.00	585.00	780.00	915.00	1170.00	1365.00	1560.00	1755.00	1950.00	2145.00	2340.00
40000	11723	400.00	600.00	800.00	1000.00	1200.00	1400.00	1600.00	1800.00	2000.00	2200.00	2400.00
41000	12016	410.00	615.00	820.00	1025.00	1230.00	1435.00	1640.00	1845.00	2050.00	2255.00	2460.00
42000	12309	420.00	630.00	840.00	1050.00	1260.00	1470.00	1680.00	1890.00	2100.00	2310.00	2520.00
43000	12602	430.00	645.00	860.00	1075.00	1290.00	1505.00	1720.00	1935.00	2150.00	2365.00	2580.00
44000	12895	440.00	660.00	880.00	1100.00	1320.00	1540.00	1760.00	1980.00	2200.00	2420.00	2640.00
45000	13188	450.00	675.00	900.00	1125.00	1350.00	1575.00	1800.00	2025.00	2250.00	2475.00	2700.00
46000	13481	460.00	690.00	920.00	1150.00	1380.00	1610.00	1840.00	2070.00	2300.00	2530.00	2760.00
47000	13774	470.00	705.00	940.00	1175.00	1410.00	1645.00	1880.00	2115.00	2350.00	2585.00	2820.00
48000	14067	480.00	720.00	960.00	1200.00	1440.00	1680.00	1920.00	2160.00	2400.00	2640.00	2880.00
49000	14365	490.00	735.00	980.00	1235.00	1470.00	1715.00	1960.00	2205.00	2450.00	2695.00	2940.00
50000	14653	500.00	750.00	1000.00	1250.00	1500.00	1750.00	2000.00	2250.00	2500.00	2750.00	3000.00
51000	14947	510.00	765.00	1020.00	1275.00	1530.00	1785.00	2040.00	2295.00	2550.00	2805.00	3060.00
52000	15240	520.00	780.00	1040.00	1300.00	1560.00	1820.00	2080.00	2340.00	2600.00	2860.00	3120.00
53000	15533	530.00	795.00	1060.00	1325.00	1590.00	1855.00	2120.00	2385.00	2650.00	2915.00	3180.00
54000	15826	540.00	810.00	1080.00	1350.00	1620.00	1890.00	2160.00	2430.00	2700.00	2970.00	3240.00
55000	16119	550.00	825.00	1100.00	1375.00	1650.00	1925.00	2200.00	2475.00	2750.00	3025.00	3300.00
56000	16412	560.00	840.00	1120.00	1400.00	1680.00	1960.00	2240.00	2520.00	2800.00	3080.00	3360.00
57000	16705	570.00	855.00	1140.00	1425.00	1710.00	1995.00	2280.00	2565.00	2854.00	3135.00	3420.00
58000	16990	580.00	870.00	1160.00	1450.00	1740.00	2030.00	2320.00	2610.00	2900.00	3190.00	3480.00
59000	17291	590.00	885.00	1180.00	1475.00	1770.00	2065.00	2360.00	2655.00	2950.00	3245.00	3540.00
60000	17584	600.00	900.00	1200.00	1500.00	1800.00	2100.00	2400.00	2700.00	3000.00	3300.00	3600.00
↑Btu	Watts↑											
Energy Per Hour												

Dollar Capital Investment

TABLE B-50 (Sheet 3 of 4)

Table B-50 (Continued)

CAPITAL INVESTMENT FOR EQUIPMENT TO PRODUCE HEAT (OR STEAM)

Dollar Investment in Equipment to Produce Heat (or Steam) to Point of Use

$ per 1000 Btu/hr →		10.00	15.00	20.00	25.00	30.00	35.00	40.00	45.00	50.00	55.00	60.00
$ per 1 lb steam/hr →		12.50	18.75	25.00	31.25	37.50	43.75	50.00	56.25	62.50	68.75	75.00
$ per 1000 W(hr) →		34.12	51.18	64.24	85.30	102.36	119.42	136.48	153.55	170.61	187.67	204.73
↓Btu/hr	W (per hr)↓											
61000	17877	$610.00	$915.00	$1220.00	$1525.00	$1830.00	$2135.00	$2440.00	$2745.00	$3050.00	$3355.00	$3660.00
62000	18170	620.00	930.00	1240.00	1550.00	1860.00	2170.00	2480.00	2700.00	3100.00	3410.00	3720.00
63000	18463	630.00	945.00	1260.00	1575.00	1890.00	2205.00	2520.00	2835.00	3150.00	3465.00	3780.00
64000	18757	640.00	960.00	1280.00	1600.00	1920.00	2240.00	2560.00	2880.00	3200.00	3520.00	3840.00
65000	19050	650.00	975.00	1300.00	1625.00	1950.00	2275.00	2600.00	2925.00	3250.00	3575.00	3900.00
66000	19343	660.00	990.00	1320.00	1650.00	1980.00	2310.00	2640.00	2970.00	3300.00	3630.00	3960.00
67000	19636	670.00	1005.00	1340.00	1675.00	2010.00	2345.00	2680.00	3015.00	3350.00	3685.00	4020.00
68000	19929	680.00	1020.00	1360.00	1700.00	2040.00	2380.00	2720.00	3060.00	3400.00	3740.00	4080.00
69000	20222	690.00	1035.00	1380.00	1725.00	2070.00	2410.00	3760.00	3105.00	3450.00	3795.00	4140.00
70000	20515	700.00	1050.00	1400.00	1750.00	2100.00	2450.00	3800.00	3150.00	3500.00	3850.00	4200.00
71000	20808	710.00	1065.00	1420.00	1775.00	2130.00	2485.00	2840.00	3195.00	3550.00	3905.00	4260.00
72000	21101	720.00	1080.00	1440.00	1800.00	2160.00	2520.00	2880.00	3240.00	3600.00	3960.00	4320.00
73000	21394	730.00	1095.00	1460.00	1825.00	2190.00	2555.00	2920.00	3285.00	3650.00	4015.00	4380.00
74000	21687	740.00	1110.00	1480.00	1850.00	2220.00	2590.00	2960.00	3330.00	3700.00	4070.00	4440.00
75000	21980	750.00	1125.00	1500.00	1875.00	2250.00	2625.00	3000.00	3375.00	3750.00	4125.00	4500.00
76000	22273	760.00	1140.00	1520.00	1900.00	2280.00	2660.00	3040.00	3420.00	3800.00	4180.00	4560.00
77000	22566	770.00	1155.00	1540.00	1925.00	2310.00	2695.00	3080.00	3465.00	3850.00	4235.00	4620.00
78000	22860	780.00	1170.00	1560.00	1950.00	2340.00	2730.00	3120.00	3510.00	3900.00	4290.00	4680.00
79000	23153	790.00	1185.00	1580.00	1975.00	2370.00	2765.00	3160.00	3555.00	3950.00	4345.00	4740.00
80000	23446	800.00	1200.00	1600.00	2000.00	2400.00	2800.00	3200.00	3600.00	4000.00	4400.00	4800.00
81000	23739	810.00	1215.00	1620.00	2025.00	2430.00	2835.00	3240.00	3645.00	4050.00	4455.00	4860.00
82000	24031	820.00	1230.00	1640.00	2050.00	2460.00	2870.00	3280.00	3690.00	4100.00	4510.00	4920.00
83000	24325	830.00	1245.00	1660.00	2075.00	2490.00	2905.00	3320.00	3735.00	4150.00	4565.00	4980.00
84000	24618	840.00	1260.00	1680.00	2100.00	2520.00	2940.00	3360.00	3780.00	4200.00	4620.00	5040.00
85000	24911	850.00	1275.00	1700.00	2125.00	2550.00	2975.00	3400.00	3835.00	4250.00	4675.00	5100.00
86000	25204	860.00	1290.00	1720.00	2150.00	2580.00	3010.00	3440.00	3870.00	4300.00	4730.00	5160.00
87000	25497	870.00	1305.00	1740.00	2175.00	2610.00	3045.00	3480.00	3915.00	4350.00	4785.00	5220.00
88000	25790	880.00	1320.00	1760.00	2200.00	2640.00	3080.00	3520.00	3960.00	4400.00	4840.00	5280.00
89000	26083	890.00	1335.00	1780.00	2225.00	2670.00	3115.00	3560.00	4005.00	4450.00	4895.00	5340.00
90000	26376	900.00	1350.00	1800.00	2250.00	2700.00	3150.00	3600.00	4050.00	4500.00	4950.00	5400.00
91000	26669	910.00	1365.00	1820.00	2275.00	2730.00	3185.00	3640.00	4095.00	4550.00	5005.00	5460.00
92000	26963	920.00	1380.00	1840.00	2300.00	2760.00	3220.00	3680.00	4140.00	4600.00	5060.00	5520.00
93000	27256	930.00	1395.00	1860.00	2325.00	2790.00	3255.00	3720.00	4185.00	4650.00	5115.00	5580.00
94000	27549	940.00	1410.00	1880.00	2350.00	2820.00	3290.00	3760.00	4230.00	4700.00	5170.00	5640.00
95000	27842	950.00	1425.00	1900.00	2375.00	2850.00	3325.00	3800.00	4275.00	4750.00	5225.00	5700.00
96000	28135	960.00	1440.00	1920.00	2400.00	2880.00	3360.00	3840.00	4320.00	4800.00	5280.00	5760.00
97000	28428	970.00	1455.00	1940.00	2425.00	2910.00	3395.00	3880.00	4365.00	4850.00	5335.00	5820.00
98000	28720	980.00	1470.00	1960.00	2450.00	2940.00	3430.00	3920.00	4410.00	4900.00	5390.00	5880.00
99000	29014	990.00	1485.00	1980.00	2475.00	2970.00	3465.00	3960.00	4455.00	4950.00	5445.00	5940.00
100000	29307	1000.00	1500.00	2000.00	2500.00	3000.00	3500.00	4000.00	4500.00	5000.00	5500.00	6000.00
↑Btu	Watts↑											

Energy Per Hour

Dollar Capital Investment

TABLE B-50 (Sheet 4 of 4)

Table B-51

DOLLAR COST OF ENERGY PER YEAR

BASED ON 8760 HRS/YR

$ Per Million Btu	$ Per M lbs Steam	HEAT LOSS IN WATTS PER HOUR										$ Per Million Watts
		2.9	5.0	8.8	11.7	14.7	17.6	20.5	23.4	26.3	29.3	
0.16	**0.20**	$0.01	$0.02	$0.04	$0.06	$0.07	$0.08	$0.10	$0.11	$0.13	$0.14	**0.57**
0.32	**0.40**	0.03	0.06	0.08	0.11	0.14	0.17	0.20	0.22	0.25	0.28	**1.09**
0.48	**0.60**	0.04	0.07	0.13	0.17	0.21	0.25	0.29	0.33	0.38	0.42	**1.64**
0.64	**0.80**	0.06	0.11	0.17	0.22	0.28	0.34	0.39	0.45	0.50	0.56	**2.18**
0.80	**1.00**	0.07	0.14	0.21	0.28	0.35	0.42	0.49	0.56	0.63	0.70	**2.73**
0.96	**1.20**	0.08	0.17	0.25	0.34	0.42	0.50	0.59	0.67	0.76	0.84	**3.27**
1.12	**1.40**	0.10	0.20	0.29	0.39	0.49	0.59	0.69	0.78	0.88	0.98	**3.82**
1.28	**1.60**	0.11	0.22	0.34	0.45	0.56	0.67	0.78	0.90	1.09	1.12	**4.37**
1.44	**1.80**	0.13	0.24	0.38	0.50	0.63	0.76	0.88	1.01	1.14	1.26	**4.91**
1.60	**2.00**	0.14	0.28	0.42	0.56	0.70	0.84	0.98	1.12	1.26	1.40	**5.46**
1.76	**2.20**	0.15	0.31	0.46	0.62	0.77	0.93	1.08	1.23	1.39	1.54	**6.01**
1.92	**2.40**	0.17	0.33	0.50	0.67	0.84	1.01	1.18	1.35	1.51	1.68	**6.55**
2.08	**2.60**	0.18	0.36	0.55	0.73	0.91	1.09	1.28	1.46	1.64	1.82	**7.09**
2.24	**2.80**	0.20	0.39	0.59	0.78	0.98	1.17	1.37	1.57	1.76	1.96	**7.64**
2.40	**3.00**	0.21	0.42	0.63	0.84	1.05	1.26	1.47	1.68	1.90	2.10	**8.19**
2.56	**3.20**	0.22	0.45	0.67	0.90	1.12	1.34	1.57	1.80	2.02	2.24	**8.73**
2.72	**3.40**	0.24	0.48	0.70	0.95	1.19	1.41	1.67	1.91	2.14	2.38	**9.28**
2.88	**3.60**	0.25	0.50	0.76	1.01	1.26	1.51	1.77	2.01	2.27	2.52	**9.82**
3.04	**3.80**	0.27	0.53	0.80	1.07	1.33	1.60	1.86	2.13	2.40	2.66	**10.37**
3.20	**4.00**	0.28	0.56	0.84	1.12	1.40	1.68	1.96	2.24	2.52	2.80	**10.92**
3.36	**4.20**	0.29	0.59	0.88	1.18	1.47	1.77	2.06	2.35	2.65	2.94	**11.45**
3.52	**4.40**	0.31	0.61	0.92	1.23	1.54	1.85	2.16	2.47	2.77	3.08	**12.01**
3.68	**4.60**	0.32	0.64	0.97	1.29	1.61	1.93	2.26	2.58	2.90	3.22	**12.56**
3.84	**4.80**	0.34	0.67	1.01	1.35	1.68	2.02	2.35	2.70	3.03	3.36	**13.10**
4.00	**5.00**	0.35	0.70	1.05	1.40	1.75	2.10	2.45	2.80	3.15	3.50	**13.65**
4.16	**5.20**	0.36	0.73	1.09	1.46	1.82	2.19	2.55	2.92	3.28	3.64	**14.20**
4.32	**5.40**	0.38	0.76	1.14	1.51	1.89	2.27	2.65	3.03	3.40	3.78	**14.74**
4.48	**5.60**	0.39	0.78	1.18	1.57	1.96	2.35	2.75	3.14	3.53	3.92	**15.29**
4.64	**5.80**	0.41	0.82	1.22	1.63	2.03	2.44	2.85	3.25	3.66	4.06	**15.84**
4.80	**6.00**	0.42	0.84	1.26	1.68	2.10	2.52	2.94	3.36	3.78	4.20	**16.38**
4.96	**6.20**	0.43	0.87	1.30	1.74	2.17	2.61	3.04	3.48	3.91	4.34	**16.92**
5.12	**6.40**	0.45	0.90	1.35	1.80	2.24	2.69	3.14	3.60	4.04	4.48	**17.47**
5.28	**6.60**	0.46	0.92	1.39	1.85	2.31	2.77	3.24	3.70	4.16	4.62	**18.02**
5.44	**6.80**	0.48	0.95	1.41	1.91	2.38	2.82	3.34	3.81	4.29	4.76	**18.57**
5.60	**7.00**	0.49	0.98	1.47	1.96	2.45	2.94	3.44	3.92	4.42	4.90	**19.11**
		10	**20**	**30**	**40**	**50**	**60**	**70**	**80**	**90**	**100**	
		HEAT LOSS IN Btu PER HOUR										

Table B-51 (Continued)

DOLLAR COST OF ENERGY PER YEAR

BASED ON 8760 HRS/YR

$ Per Million Btu	$ Per M lbs Steam	HEAT LOSS IN WATTS PER HOUR										$ Per Million Watts
		32.2	35.2	38.1	41.0	44.0	46.9	49.8	52.7	55.7	58.6	
0.16	**0.20**	$0.15	$0.17	$0.18	$0.19	$0.21	$0.22	$0.24	$0.25	$0.27	$0.28	**0.57**
0.32	**0.40**	0.31	0.34	0.36	0.39	0.42	0.45	0.48	0.50	0.53	0.56	**1.09**
0.48	**0.60**	0.46	0.50	0.55	0.59	0.63	0.68	0.71	0.76	0.80	0.84	**1.64**
0.64	**0.80**	0.62	0.67	0.73	0.78	0.84	0.90	0.95	1.00	1.06	1.12	**2.18**
0.80	**1.00**	0.77	0.84	0.91	0.98	1.05	1.12	1.19	1.26	1.33	1.40	**2.73**
0.96	**1.20**	0.93	1.01	1.09	1.17	1.26	1.36	1.43	1.51	1.60	1.68	**3.27**
1.12	**1.40**	1.08	1.18	1.28	1.37	1.47	1.57	1.67	1.77	1.86	1.96	**3.82**
1.28	**1.60**	1.22	1.34	1.46	1.57	1.68	1.79	1.91	2.02	2.13	2.24	**4.37**
1.44	**1.80**	1.37	1.51	1.64	1.77	1.89	2.02	2.14	2.27	2.40	2.52	**4.91**
1.60	**2.00**	1.54	1.68	1.82	1.96	2.10	2.24	2.38	2.52	2.67	2.80	**5.46**
1.76	**2.20**	1.70	1.85	2.00	2.16	2.31	2.47	2.62	2.78	2.92	3.08	**6.01**
1.92	**2.40**	1.85	2.01	2.18	2.35	2.52	2.71	2.86	3.03	3.20	3.36	**6.55**
2.08	**2.60**	2.00	2.18	2.37	2.55	2.73	2.92	3.10	3.28	3.46	3.64	**7.09**
2.24	**2.80**	2.15	2.35	2.55	2.75	2.94	3.14	3.34	3.53	3.73	3.92	**7.64**
2.40	**3.00**	2.31	2.52	2.73	2.94	3.15	3.36	3.57	3.78	3.99	4.20	**8.19**
2.56	**3.20**	2.45	2.69	2.91	3.14	3.36	3.58	3.81	4.04	4.26	4.50	**8.73**
2.72	**3.40**	2.62	2.86	3.10	3.34	3.57	3.81	4.05	4.28	4.53	4.77	**9.28**
2.88	**3.60**	2.73	3.03	3.28	3.53	3.78	4.04	4.29	4.54	4.79	5.05	**9.82**
3.04	**3.80**	2.93	3.20	3.46	3.73	3.99	4.26	4.53	4.79	5.06	5.33	**10.37**
3.20	**4.00**	3.08	3.36	3.64	3.92	4.20	4.49	4.77	5.05	5.33	5.61	**10.92**
3.36	**4.20**	3.24	3.53	3.82	4.12	4.42	4.70	5.00	5.29	5.59	5.89	**11.45**
3.52	**4.40**	3.39	3.70	4.01	4.32	4.63	4.93	5.24	5.55	5.86	6.17	**12.01**
3.68	**4.60**	3.54	3.87	4.19	4.51	4.83	5.16	5.48	5.80	6.12	6.45	**12.56**
3.84	**4.80**	3.70	4.03	4.37	4.71	5.05	5.42	5.72	6.06	6.32	6.74	**13.10**
4.00	**5.00**	3.85	4.20	4.56	4.91	5.26	5.61	5.96	6.31	6.66	7.01	**13.65**
4.16	**5.20**	4.01	4.37	4.74	5.10	5.47	5.83	6.20	6.56	6.92	7.29	**14.20**
4.32	**5.40**	4.16	4.54	4.92	5.30	5.68	6.05	6.43	6.81	7.19	7.57	**14.74**
4.48	**5.60**	4.32	4.71	5.10	5.49	5.89	6.28	6.67	7.06	7.46	7.85	**15.29**
4.64	**5.80**	4.47	4.88	5.28	5.69	6.10	6.50	6.91	7.32	7.72	8.13	**15.84**
4.80	**6.00**	4.62	5.05	5.47	5.89	6.31	6.73	7.15	7.57	7.99	8.41	**16.38**
4.96	**6.20**	4.78	5.21	5.65	6.08	6.52	6.95	7.39	7.82	8.26	8.69	**16.92**
5.12	**6.40**	4.89	5.38	5.83	6.28	6.74	7.17	7.62	8.07	8.52	8.99	**17.47**
5.28	**6.60**	5.09	5.55	6.01	6.48	6.94	7.40	7.86	8.33	8.78	9.25	**18.02**
5.44	**6.80**	5.24	5.71	6.20	6.67	7.05	7.62	8.10	8.58	9.05	9.53	**18.57**
5.60	**7.00**	5.40	5.89	6.38	6.87	7.36	7.85	8.34	8.83	9.32	9.81	**19.11**
		110	**120**	**130**	**140**	**150**	**160**	**170**	**180**	**190**	**200**	
		HEAT LOSS IN Btu PER HOUR										

Table B-51 (Continued)

DOLLAR COST OF ENERGY PER YEAR

BASED ON 8760 HRS/YR

$ Per Million Btu	$ Per M lbs Steam	HEAT LOSS IN WATTS PER HOUR										$ Per Million Watts
		61.5	64.5	67.4	70.3	73.3	76.2	79.1	82.0	85.0	87.9	
0.16	**0.20**	$0.30	$0.31	$0.32	$0.34	$0.35	$0.36	$0.38	$0.39	$0.41	$0.42	**0.57**
0.32	**0.40**	0.59	0.62	0.64	0.67	0.70	0.73	0.76	0.78	0.81	0.84	**1.09**
0.48	**0.60**	0.89	0.93	0.97	1.01	1.05	1.09	1.14	1.18	1.22	1.26	**1.64**
0.64	**0.80**	1.18	1.22	1.29	1.35	1.40	1.46	1.51	1.57	1.63	1.68	**2.18**
0.80	**1.00**	1.48	1.54	1.61	1.68	1.75	1.82	1.89	1.96	2.03	2.10	**2.73**
0.96	**1.20**	1.77	1.85	1.93	2.02	2.10	2.19	2.27	2.35	2.44	2.52	**3.27**
1.12	**1.40**	2.07	2.15	2.26	2.35	2.45	2.55	2.65	2.75	2.85	2.94	**3.82**
1.28	**1.60**	2.36	2.47	2.58	2.69	2.80	2.92	3.03	3.14	3.23	3.36	**4.37**
1.44	**1.80**	2.66	2.78	2.90	3.03	3.15	3.28	3.41	3.53	3.66	3.78	**4.91**
1.60	**2.00**	2.96	3.08	3.22	3.36	3.50	3.64	3.78	3.92	4.06	4.20	**5.46**
1.76	**2.20**	3.23	3.39	3.54	3.70	3.85	4.01	4.16	4.32	4.47	4.63	**6.01**
1.92	**2.40**	3.54	3.70	3.87	4.04	4.20	4.37	4.54	4.71	4.88	5.05	**6.55**
2.08	**2.60**	3.84	4.01	4.19	4.37	4.56	4.74	4.92	5.10	5.28	5.47	**7.09**
2.24	**2.80**	4.14	4.31	4.51	4.70	4.91	5.10	5.36	5.49	5.69	5.89	**7.64**
2.40	**3.00**	4.43	4.63	4.84	5.05	5.26	5.47	5.68	5.89	6.10	6.31	**8.19**
2.56	**3.20**	4.73	4.89	5.16	5.38	5.61	5.83	6.05	6.28	6.50	6.73	**8.73**
2.72	**3.40**	5.03	5.24	5.48	5.72	5.96	6.20	6.43	6.67	6.91	7.04	**9.28**
2.88	**3.60**	5.32	5.46	5.80	6.04	6.31	6.56	6.81	7.06	7.32	7.57	**9.82**
3.04	**3.80**	5.62	5.86	6.13	6.39	6.66	6.92	7.19	7.46	7.72	7.99	**10.37**
3.20	**4.00**	5.91	6.17	6.45	6.73	7.01	7.29	7.57	7.85	8.13	8.41	**10.42**
3.36	**4.20**	6.21	6.48	6.77	7.06	7.36	7.65	7.95	8.24	8.54	8.83	**11.45**
3.52	**4.40**	6.47	6.78	7.09	7.20	7.71	8.02	8.33	8.63	8.94	9.25	**12.01**
3.68	**4.60**	6.80	7.08	7.41	7.73	8.06	8.38	8.70	9.03	9.35	9.67	**12.56**
3.84	**4.80**	7.10	7.40	7.74	8.07	8.41	8.75	9.08	9.42	9.75	10.09	**13.10**
4.00	**5.00**	7.39	7.71	8.06	8.41	8.76	9.11	9.46	9.81	10.16	10.51	**13.65**
4.16	**5.20**	7.68	8.02	8.38	8.74	9.11	9.47	9.84	10.20	10.57	10.93	**14.20**
4.32	**5.40**	7.98	8.33	8.70	9.08	9.46	9.84	10.22	10.60	10.97	11.35	**14.74**
4.48	**5.60**	8.28	8.63	9.03	9.41	9.81	10.20	10.59	10.99	11.38	11.77	**15.29**
4.64	**5.80**	8.57	8.94	9.35	9.76	10.16	10.57	10.97	11.38	11.79	12.19	**15.84**
4.80	**6.00**	8.87	9.25	9.67	10.09	10.51	10.92	11.35	11.77	12.19	12.61	**16.38**
4.96	**6.20**	9.17	9.56	9.99	10.44	10.86	11.30	11.73	12.16	12.60	13.03	**16.92**
5.12	**6.40**	9.46	9.79	10.31	10.76	11.21	11.66	12.10	12.55	13.00	13.45	**17.47**
5.28	**6.60**	9.75	10.17	10.64	11.10	11.56	12.03	12.49	12.95	13.41	13.88	**18.02**
5.44	**6.80**	10.05	10.48	10.96	11.44	11.91	12.39	12.90	13.34	13.82	14.10	**18.57**
5.60	**7.00**	10.35	10.79	11.28	11.77	12.26	12.75	13.25	13.74	14.23	14.72	**19.11**
		210	**220**	**230**	**240**	**250**	**260**	**270**	**280**	**290**	**300**	
		HEAT LOSS IN Btu PER HOUR										

Table B-51 (Continued)

DOLLAR COST OF ENERGY PER YEAR

BASED ON 8760 HRS/YR

$ Per Million Btu	$ Per M lbs Steam	HEAT LOSS IN WATTS PER HOUR 90.8	93.8	96.7	99.6	102.6	105.5	108.4	111.3	114.3	117.2	$ Per Million Watts
0.16	**0.20**	$0.43	$0.45	$0.46	$0.48	$0.49	$0.50	$0.52	$0.53	$0.55	$0.56	**0.57**
0.32	**0.40**	0.87	0.90	0.93	0.95	0.98	1.01	1.04	1.07	1.09	1.12	**1.09**
0.48	**0.60**	1.30	1.36	1.39	1.43	1.47	1.51	1.56	1.60	1.64	1.68	**1.64**
0.64	**0.80**	1.74	1.79	1.85	1.91	1.96	2.02	2.07	2.13	2.19	2.24	**2.18**
0.80	**1.00**	2.17	2.24	2.31	2.38	2.45	2.52	2.59	2.66	2.73	2.80	**2.73**
0.96	**1.20**	2.61	2.71	2.77	2.86	2.94	3.03	3.11	3.20	3.27	3.36	**3.27**
1.12	**1.40**	3.04	3.14	3.24	3.34	3.43	3.53	3.63	3.73	3.83	3.92	**3.82**
1.28	**1.60**	3.48	3.58	3.70	3.81	3.92	4.04	4.15	4.26	4.37	4.50	**4.37**
1.44	**1.80**	3.91	4.04	4.16	4.29	4.42	4.54	4.67	4.79	4.92	5.05	**4.91**
1.60	**2.00**	4.35	4.49	4.63	4.77	4.91	5.05	5.19	5.33	5.47	5.61	**5.46**
1.76	**2.20**	4.78	4.93	5.09	5.24	5.40	5.55	5.70	5.86	6.01	6.17	**6.01**
1.92	**2.40**	5.21	5.42	5.55	5.71	5.89	6.06	6.22	6.39	6.55	6.74	**6.55**
2.08	**2.60**	5.65	5.83	6.01	6.20	6.38	6.56	6.74	6.92	7.11	7.28	**7.09**
2.24	**2.80**	6.08	6.28	6.48	6.67	6.87	7.06	7.26	7.46	7.65	7.85	**7.64**
2.40	**3.00**	6.51	6.72	6.94	7.15	7.36	7.57	7.78	7.99	8.20	8.41	**8.19**
2.56	**3.20**	6.95	7.17	7.40	7.62	7.85	8.07	8.30	8.52	8.75	8.99	**8.73**
2.72	**3.40**	7.39	7.62	7.86	8.10	8.34	8.58	8.82	9.05	9.29	9.53	**9.28**
2.88	**3.60**	7.82	8.07	8.32	8.57	8.83	9.08	9.33	9.59	9.84	10.09	**9.82**
3.04	**3.80**	8.25	8.52	8.79	9.05	9.32	9.59	9.85	10.12	10.39	10.65	**10.37**
3.20	**4.00**	8.69	8.97	9.25	9.52	9.81	10.09	10.37	10.65	10.93	11.21	**10.92**
3.36	**4.20**	9.12	9.40	9.71	10.01	10.30	10.59	10.89	11.18	11.48	11.77	**11.45**
3.52	**4.40**	9.56	9.86	10.08	10.43	10.79	11.10	11.41	11.72	12.03	12.33	**12.01**
3.68	**4.60**	9.99	10.31	10.64	10.96	11.28	11.61	11.93	12.25	12.57	12.89	**12.56**
3.84	**4.80**	10.43	10.84	11.10	11.44	11.77	12.11	12.95	12.78	13.09	13.47	**13.10**
4.00	**5.00**	10.86	11.21	11.56	11.91	12.26	12.61	12.96	13.32	13.67	14.01	**13.65**
4.16	**5.20**	11.30	11.66	12.03	12.39	12.75	13.12	13.48	13.75	14.21	14.57	**14.20**
4.32	**5.40**	11.73	12.11	12.49	12.87	13.25	13.62	14.00	14.38	14.76	15.14	**14.74**
4.48	**5.60**	12.17	12.56	12.95	13.34	13.74	14.13	14.52	14.91	15.31	15.70	**15.29**
4.64	**5.80**	12.60	13.01	13.30	13.82	14.23	14.63	15.04	15.45	15.85	16.26	**15.84**
4.80	**6.00**	13.03	13.45	13.87	14.30	14.71	15.14	15.56	15.98	16.40	16.82	**16.38**
4.96	**6.20**	13.47	13.90	14.34	14.77	15.21	15.64	16.08	16.51	16.95	17.38	**16.92**
5.12	**6.40**	13.90	14.34	14.80	15.25	15.70	16.15	16.60	17.04	17.49	17.98	**17.47**
5.28	**6.60**	14.34	14.80	15.26	15.73	16.19	16.65	17.11	17.58	18.03	18.50	**18.02**
5.44	**6.80**	14.77	15.25	15.73	16.20	16.68	17.16	17.63	18.10	18.59	19.06	**18.57**
5.60	**7.00**	15.20	15.70	16.19	16.68	17.17	17.66	18.15	18.64	19.13	19.62	**19.11**
		310	**320**	**330**	**340**	**350**	**360**	**370**	**380**	**390**	**400**	

HEAT LOSS IN Btu PER HOUR

Table B-51 (Continued)

DOLLAR COST OF ENERGY PER YEAR

BASED ON 8760 HRS/YR

$ Per Million Btu	$ Per M lbs Steam	HEAT LOSS IN WATTS PER HOUR										$ Per Million Watts
		120.1	123.1	125.9	128.9	131.9	134.8	137.7	140.6	143.6	146.5	
0.16	0.20	$0.57	$0.59	$0.60	$0.62	$0.63	$0.64	$0.66	$0.67	$0.69	$0.70	0.57
0.32	0.40	1.15	1.18	1.21	1.22	1.26	1.29	1.32	1.35	1.38	1.40	1.09
0.48	0.60	1.72	1.77	1.81	1.85	1.89	1.93	1.97	2.02	2.06	2.10	1.64
0.64	0.80	2.30	2.37	2.41	2.45	2.52	2.58	2.64	2.69	2.75	2.80	2.18
0.80	$.00	2.87	2.96	3.01	3.08	3.15	3.22	3.29	3.36	3.43	3.50	2.73
0.96	1.20	3.45	3.55	3.61	3.70	3.78	3.87	3.95	4.03	4.12	4.20	3.27
1.12	1.40	4.02	4.14	4.21	4.31	4.42	4.51	4.61	4.71	4.81	4.91	3.82
1.28	1.60	4.60	4.73	4.82	4.93	5.05	5.16	5.27	5.38	5.49	5.61	4.37
1.44	1.80	5.17	5.32	5.42	5.55	5.68	5.80	5.93	6.05	6.18	6.31	4.91
1.60	2.00	5.75	5.91	6.03	6.16	6.30	6.45	6.59	6.73	6.87	7.01	5.46
1.76	2.20	6.32	6.46	6.63	6.79	6.94	7.09	7.25	7.40	7.55	7.71	6.01
1.92	2.40	6.90	7.10	7.22	7.40	7.57	7.74	7.91	8.07	8.24	8.41	6.55
2.08	2.60	7.47	7.69	7.83	8.01	8.20	8.38	8.56	8.75	8.93	9.11	7.09
2.24	2.80	8.05	8.28	8.44	8.63	8.82	9.03	9.22	9.41	9.62	9.81	7.64
2.40	3.00	8.61	8.87	9.04	9.25	9.46	9.67	9.88	10.09	10.30	10.51	8.19
2.56	3.20	9.19	9.46	9.64	9.79	10.09	10.32	10.54	10.76	10.99	11.21	8.73
2.72	3.40	9.76	10.06	10.25	10.48	10.72	10.96	11.20	11.44	11.68	11.91	9.28
2.88	3.60	10.34	10.65	10.85	11.10	11.35	11.61	11.84	12.12	12.36	12.61	9.82
3.04	3.80	10.92	11.24	11.45	11.72	11.98	12.25	12.52	12.78	13.05	13.32	10.37
3.20	4.00	11.49	11.83	12.05	12.33	12.60	12.89	13.17	13.46	13.73	14.02	10.92
3.36	4.20	12.07	12.42	12.66	12.95	13.25	13.54	13.83	14.13	14.42	14.72	11.45
3.52	4.40	12.64	12.93	13.26	13.57	13.87	14.18	14.49	14.80	15.10	15.42	12.01
3.68	4.60	13.21	13.60	13.86	14.16	14.51	14.83	15.15	15.47	15.80	16.12	12.56
3.84	4.80	13.79	14.20	14.44	14.80	15.14	15.47	15.81	16.15	16.48	16.82	13.65
4.00	5.00	14.36	14.79	15.07	15.42	15.77	16.12	16.47	16.82	17.16	17.52	13.65
41.6	5.20	14.94	15.38	15.67	16.03	16.40	16.76	17.13	17.49	17.86	18.22	14.20
4.32	5.40	15.52	15.97	16.28	16.65	17.03	17.41	17.79	18.16	18.54	18.92	14.74
4.48	5.60	16.09	16.56	16.88	17.27	17.64	18.05	18.45	18.84	19.23	19.62	15.29
4.64	5.80	16.67	17.15	17.48	17.88	18.29	18.70	19.10	19.51	19.92	20.32	15.84
4.80	6.00	17.24	17.74	18.08	18.50	18.92	19.34	19.76	20.18	20.60	21.02	16.38
4.96	6.20	17.81	18.33	18.68	19.12	19.55	19.99	20.42	20.85	21.29	21.72	16.92
5.12	6.40	18.38	18.92	19.28	19.58	20.18	20.63	21.08	21.53	21.98	22.43	17.47
5.28	6.60	18.96	19.52	19.89	20.35	20.81	21.28	21.74	22.20	22.66	23.13	18.02
5.44	6.80	19.53	20.11	20.49	20.96	21.44	21.92	22.40	22.86	23.35	23.83	18.57
5.60	7.00	20.11	20.70	21.09	21.58	22.08	22.57	23.06	23.55	24.04	24.53	19.11
		410	420	430	440	450	460	470	480	490	500	
		HEAT LOSS IN Btu PER HOUR										

TABLE B-51 (Sheet 5 of 28)

Table B-51 (Continued)

DOLLAR COST OF ENERGY PER YEAR

BASED ON 8760 HRS/YR

$ Per Million Btu	$ Per M lbs Steam	HEAT LOSS IN WATTS PER HOUR 149.4	152.4	155.3	158.2	161.2	164.1	167.0	167.9	172.9	175.8	$ Per Million Watts
0.16	0.20	$0.71	$0.73	$0.74	$0.76	$0.77	$0.78	$0.80	$0.81	$0.83	$0.84	0.57
0.32	0.40	1.43	1.46	1.49	1.51	1.54	1.57	1.60	1.62	1.65	1.68	1.09
0.48	0.60	2.14	2.19	2.23	2.27	2.31	2.35	2.40	2.44	2.48	2.53	1.64
0.64	0.80	2.86	2.92	2.97	3.03	3.08	3.14	3.20	3.23	3.31	3.36	2.18
0.80	1.00	3.57	3.64	3.71	3.98	3.85	3.92	3.99	4.06	4.13	4.20	2.73
0.96	1.20	4.28	4.37	4.46	4.54	4.63	4.71	4.79	4.88	4.96	5.05	3.27
1.12	1.40	5.00	5.10	5.20	5.30	5.40	5.49	5.59	5.69	5.79	5.89	3.82
1.28	1.60	5.71	5.83	5.94	6.05	6.17	6.28	6.39	6.46	6.61	6.73	4.37
1.44	1.80	6.43	6.56	6.69	6.81	6.94	7.06	7.19	7.32	7.44	7.57	4.91
1.60	2.00	7.14	7.29	7.43	7.57	7.71	7.85	7.99	8.13	8.27	8.41	5.46
1.76	2.20	7.86	8.02	8.17	8.33	8.48	8.63	8.79	8.94	9.10	9.25	6.01
1.92	2.40	8.58	8.75	8.91	9.08	9.25	9.42	9.59	9.75	9.92	10.09	6.55
2.08	2.60	9.29	9.47	9.66	9.84	10.02	10.20	10.39	10.57	10.75	10.93	7.09
2.24	2.80	10.01	10.20	10.40	10.60	10.79	10.99	11.18	11.38	11.57	11.77	7.64
2.40	3.00	10.72	10.92	11.14	11.35	11.56	11.77	11.98	12.19	12.40	12.61	8.19
2.56	3.20	11.42	11.66	11.89	12.11	12.33	12.55	12.78	13.01	13.23	13.46	8.73
2.72	3.40	12.06	12.39	12.63	12.90	13.11	13.34	13.58	13.82	14.06	14.30	9.28
2.82	3.60	12.87	13.12	13.37	13.62	13.88	14.13	14.38	14.63	14.88	15.14	9.82
3.04	3.80	13.58	13.85	14.11	14.38	14.65	14.91	15.18	15.45	15.71	15.98	10.37
3.20	4.00	14.30	14.58	14.86	15.13	15.42	15.70	15.98	16.26	16.54	16.82	10.92
3.36	4.20	15.01	15.31	15.60	15.89	16.18	16.48	16.77	17.07	17.37	17.66	11.45
3.52	4.40	15.73	16.03	16.34	16.65	16.96	17.27	17.58	17.88	18.19	18.50	12.01
3.68	4.60	16.44	16.76	17.09	17.41	17.73	18.03	18.38	18.70	19.02	19.34	12.56
3.84	4.80	17.15	17.49	17.83	18.16	18.50	18.83	19.17	19.51	19.85	20.18	13.10
4.00	5.00	17.87	18.22	18.57	18.92	19.27	19.62	19.97	20.32	20.67	21.01	13.65
4.16	5.20	18.58	18.95	19.31	19.78	20.04	20.41	20.77	21.13	21.50	21.87	14.20
4.32	5.40	19.30	19.68	20.06	20.44	20.81	21.19	21.57	21.95	22.33	22.71	14.74
4.48	5.60	20.01	20.41	20.80	21.18	21.58	21.98	22.37	22.76	23.15	23.55	15.29
4.64	5.80	20.73	21.14	21.54	21.95	22.36	22.76	23.16	23.58	23.98	24.39	15.84
4.80	6.00	21.44	21.84	22.29	22.71	23.13	23.55	23.97	24.39	24.81	25.23	16.38
4.96	6.20	22.16	22.59	23.03	23.46	23.90	24.33	24.77	25.20	25.64	26.07	16.92
5.12	6.40	22.83	23.32	23.77	24.22	24.67	25.11	25.56	25.86	26.46	26.91	17.42
5.28	6.60	23.59	24.05	24.51	24.98	25.44	25.80	26.36	26.82	27.29	27.75	18.02
5.44	6.80	24.30	24.78	25.26	25.80	26.21	26.69	27.16	27.64	18.12	28.19	18.57
5.60	7.00	25.02	25.51	26.00	26.49	26.98	27.47	27.96	28.45	28.94	29.43	19.11
		510	520	530	540	550	560	570	580	590	600	
		HEAT LOSS IN Btu PER HOUR										

Table B-51 (Continued)

DOLLAR COST OF ENERGY PER YEAR

BASED ON 8760 HRS/YR

$ Per Million Btu	$ Per M lbs Steam	HEAT LOSS IN WATTS PER HOUR 178.7	181.7	184.6	187.5	190.5	193.4	196.3	199.2	202.2	205.1	$ Per Million Watts
0.16	**0.20**	$0.86	$0.87	$0.88	$0.90	$0.91	$0.93	$0.94	$0.95	$0.97	$0.98	**0.57**
0.32	**0.40**	1.71	1.74	1.77	1.79	1.82	1.85	1.88	1.91	1.93	1.96	**1.09**
0.48	**0.60**	2.56	2.61	2.65	2.71	2.73	2.78	2.82	2.86	2.90	2.94	**1.64**
0.64	**0.80**	3.42	3.48	3.53	3.58	3.64	3.70	3.76	3.81	3.90	3.92	**2.18**
0.80	**1.00**	4.27	4.35	4.42	4.49	4.56	4.63	4.70	4.77	4.84	4.91	**2.73**
0.96	**1.20**	5.13	5.21	5.30	5.42	5.47	5.55	5.63	5.72	5.80	5.89	**3.27**
1.12	**1.40**	5.98	6.08	6.18	6.28	6.38	6.48	6.57	6.67	6.77	6.87	**3.82**
1.28	**1.60**	6.84	6.95	7.06	7.17	7.29	7.40	7.51	7.62	7.74	7.85	**4.37**
1.44	**1.80**	7.69	7.82	7.95	8.07	8.20	8.33	8.45	8.58	8.70	8.83	**4.91**
1.60	**2.00**	8.55	8.69	8.83	8.97	9.11	9.25	9.39	9.52	9.67	9.81	**5.46**
1.76	**2.20**	9.40	9.56	9.71	9.87	10.02	10.18	10.33	10.48	10.64	10.79	**6.01**
1.92	**2.40**	10.26	10.43	10.60	10.84	10.93	11.10	11.27	11.43	11.60	11.77	**6.55**
2.08	**2.60**	11.11	11.30	11.48	11.60	11.84	12.02	12.21	12.39	12.57	12.75	**7.09**
2.24	**2.80**	11.97	12.17	12.36	12.55	12.75	12.95	13.15	13.34	13.54	13.74	**7.64**
2.40	**3.00**	12.82	13.03	13.24	13.45	13.66	13.88	14.09	14.30	14.51	14.72	**8.19**
2.56	**3.20**	13.68	13.90	14.13	14.34	14.58	14.80	15.03	15.25	15.47	15.70	**8.73**
2.72	**3.40**	14.53	14.77	15.01	15.25	15.49	15.73	15.96	16.20	16.44	16.68	**9.28**
2.82	**3.60**	15.39	15.64	15.89	16.15	16.40	16.65	16.90	17.16	17.41	17.66	**9.82**
3.04	**3.80**	16.24	16.51	16.78	17.04	17.31	17.58	17.84	18.11	18.38	18.64	**10.37**
3.20	**4.00**	17.10	17.38	17.66	17.94	18.22	18.50	18.78	19.04	19.34	19.62	**10.92**
3.36	**4.20**	17.95	18.24	18.54	18.80	19.13	19.43	19.72	20.01	20.31	20.60	**11.45**
3.52	**4.40**	18.81	19.12	19.43	19.74	20.04	20.16	20.66	20.97	21.28	21.58	**12.01**
3.68	**4.60**	19.66	19.99	20.31	20.62	20.95	21.28	21.60	21.92	22.24	22.56	**12.56**
3.84	**4.80**	20.52	20.86	21.19	21.69	21.87	22.20	22.54	22.87	23.21	23.54	**13.10**
4.00	**5.00**	21.37	21.72	22.07	22.42	22.78	23.13	23.47	23.83	24.18	24.53	**13.65**
4.16	**5.20**	22.23	22.59	22.96	23.32	23.69	24.05	24.42	24.78	25.14	25.50	**14.20**
4.32	**5.40**	23.08	23.46	23.84	24.22	24.60	24.98	25.35	25.73	26.11	26.48	**14.74**
4.48	**5.60**	23.94	24.33	24.72	25.12	25.51	25.90	26.29	26.69	27.08	27.47	**15.29**
4.64	**5.80**	24.79	25.20	25.61	26.01	26.42	26.59	26.83	27.23	27.64	28.45	**15.83**
4.80	**6.00**	25.65	26.06	26.49	26.91	27.33	27.75	28.17	28.59	29.01	29.43	**16.38**
4.96	**6.20**	26.50	26.94	27.37	27.81	28.24	28.68	29.11	29.55	29.98	30.40	**16.92**
5.12	**6.40**	27.36	27.81	28.26	28.67	29.15	29.60	30.05	30.49	30.95	31.40	**17.42**
5.28	**6.60**	28.21	28.68	29.14	29.60	30.06	30.53	30.99	31.45	31.91	32.38	**18.02**
5.44	**6.80**	29.06	29.54	30.02	30.49	30.98	31.45	31.92	32.40	32.88	33.36	**18.57**
5.60	**7.00**	29.92	30.41	30.91	31.40	31.89	32.38	32.87	33.36	33.85	34.34	**19.11**
		610	**620**	**630**	**640**	**650**	**660**	**670**	**680**	**690**	**700**	
		HEAT LOSS IN Btu PER HOUR										

Table B-51 (Continued)

DOLLAR COST OF ENERGY PER YEAR

BASED ON 8760 HRS/YR

$ Per Million Btu	$ Per M lbs Steam	HEAT LOSS IN WATTS PER HOUR										$ Per Million Watts
		208.0	211.0	213.9	216.8	219.8	222.7	225.6	228.5	231.5	234.4	
0.16	0.20	$1.00	$1.01	$1.02	$1.04	$1.05	$1.07	$1.08	$1.09	$1.11	$1.12	0.57
0.32	0.40	1.99	2.02	2.05	2.07	2.10	2.13	2.16	2.19	2.21	2.24	1.09
0.48	0.60	2.99	3.03	3.07	3.11	3.15	3.20	3.24	3.28	3.32	3.36	1.64
0.64	0.80	3.98	4.04	4.09	4.15	4.20	4.26	4.32	4.37	4.43	4.50	2.18
0.80	1.00	4.98	5.05	5.12	5.19	5.26	5.33	5.40	5.47	5.54	5.61	2.73
0.96	1.20	5.97	6.06	6.14	6.22	6.31	6.39	6.47	6.55	6.64	6.74	3.27
1.12	1.40	6.97	7.06	7.16	7.26	7.35	7.47	7.55	7.65	7.75	7.85	3.82
1.28	1.60	7.96	8.07	8.18	8.30	8.41	8.52	8.63	8.75	8.86	8.99	4.37
1.44	1.80	8.96	9.08	9.21	9.33	9.46	9.59	9.71	9.84	9.97	10.09	4.91
1.60	2.00	9.95	10.09	10.23	10.37	10.51	10.65	10.79	10.93	11.07	11.21	5.46
1.76	2.20	10.95	11.10	11.25	11.41	11.56	11.72	11.87	12.03	12.18	12.33	6.01
1.92	2.40	11.94	12.11	12.28	12.45	12.61	12.78	12.95	13.09	13.29	13.48	6.55
2.08	2.60	12.94	13.12	13.30	13.48	13.67	13.85	14.03	14.21	14.39	14.58	7.09
2.24	2.80	13.93	14.12	14.32	14.52	14.72	14.91	15.11	15.31	15.50	15.70	7.64
2.40	3.00	14.73	15.14	15.35	15.56	15.77	15.98	16.19	16.40	16.61	16.82	8.19
2.56	3.20	15.92	16.14	16.37	16.59	16.82	17.04	17.26	17.49	17.72	17.98	8.73
2.72	3.40	16.92	17.16	17.40	17.63	17.87	18.11	18.35	18.59	18.82	19.06	9.28
2.82	3.60	17.91	18.16	18.42	18.67	18.92	19.17	19.43	19.68	19.93	20.18	9.82
3.04	3.80	18.91	19.17	19.44	19.71	19.97	20.24	20.51	20.77	21.03	21.30	10.37
3.20	4.00	19.90	20.18	20.46	20.74	21.02	21.30	21.58	21.86	22.15	22.42	10.92
3.36	4.20	20.90	21.07	21.49	21.78	22.08	22.37	22.66	22.96	23.25	23.55	11.45
3.52	4.40	21.89	22.20	22.51	22.82	23.13	23.43	23.74	24.05	24.36	24.67	12.01
3.68	4.60	22.89	23.21	23.53	23.86	24.18	24.50	24.82	25.14	24.47	25.79	12.56
3.84	4.80	23.88	24.22	24.56	24.89	25.23	25.56	25.90	26.18	26.57	26.95	13.10
4.00	5.00	24.88	25.23	25.58	25.93	26.28	26.63	26.98	27.33	27.68	28.03	13.65
4.16	5.20	25.87	26.23	26.60	26.97	27.33	27.70	28.06	28.42	28.79	29.15	14.20
4.32	5.40	26.87	27.25	27.63	28.00	28.33	28.76	29.13	29.52	29.90	30.27	14.74
4.48	5.60	27.86	28.26	28.65	29.04	29.43	29.83	30.22	30.61	31.00	31.40	15.29
4.64	5.80	28.86	29.27	29.67	30.08	30.48	30.89	31.30	31.70	32.11	32.52	15.84
4.80	6.00	29.85	30.27	30.70	31.12	31.53	31.96	32.38	32.80	33.22	33.64	16.38
4.96	6.20	30.85	31.28	31.72	32.15	32.59	33.02	33.46	33.89	34.33	34.76	16.92
5.12	6.40	31.84	32.30	32.74	33.19	33.63	34.09	34.54	34.98	35.43	35.96	17.42
5.28	6.60	32.84	33.30	33.76	34.23	34.69	35.15	35.61	36.07	36.54	37.00	18.02
5.44	6.80	33.83	34.31	34.79	35.26	35.74	36.21	36.69	37.17	37.65	38.12	18.57
5.60	7.00	34.83	35.32	35.81	36.30	36.79	37.24	37.77	38.26	38.75	39.24	19.11
		710	720	730	740	750	760	770	780	790	800	
		HEAT LOSS IN Btu PER HOUR										

Table B-51 (Continued)

DOLLAR COST OF ENERGY PER YEAR

BASED ON 8760 HRS/YR

$ Per Million Btu	$ Per M lbs Steam	HEAT LOSS IN WATTS PER HOUR										$ Per Million Watts
		237.3	240.3	243.2	246.1	249.0	252.0	254.9	257.8	260.8	263.7	
0.16	0.20	$1.14	$1.15	$1.16	$1.18	$1.19	$1.21	$1.22	$1.23	$1.24	$1.26	0.57
0.32	0.40	2.27	2.30	2.33	2.35	2.38	2.41	2.44	2.45	2.49	2.52	1.09
0.48	0.60	3.41	3.45	3.49	3.55	3.57	3.62	3.65	3.70	3.74	3.78	1.64
0.64	0.80	4.54	4.60	4.65	4.73	4.76	4.82	4.88	4.93	4.99	5.05	2.18
0.80	1.00	5.68	5.75	5.81	5.91	5.96	6.03	6.10	6.16	6.24	6.31	2.73
0.96	1.20	6.81	6.90	6.99	7.06	7.15	7.22	7.32	7.40	7.48	7.57	3.27
1.12	1.40	7.95	8.04	8.14	8.24	8.34	8.44	8.54	8.63	7.73	8.82	3.82
1.28	1.60	9.08	9.19	9.31	9.41	9.53	9.64	9.75	9.86	9.97	10.09	4.37
1.44	1.80	10.21	10.34	10.47	10.64	10.72	10.84	10.97	11.10	11.23	11.35	4.91
1.60	2.00	11.35	11.49	11.63	11.82	11.91	12.05	12.19	12.33	12.47	12.60	5.46
1.76	2.20	12.49	12.64	12.80	12.93	13.11	13.26	13.41	13.57	13.72	13.87	6.01
1.92	2.40	13.62	13.79	13.96	14.13	14.29	14.44	14.63	14.80	14.96	15.14	6.55
2.08	2.60	14.76	14.94	15.12	15.30	15.49	15.67	15.85	16.03	16.22	16.40	7.09
2.24	2.80	15.89	16.09	16.28	16.48	16.68	16.88	17.07	17.27	17.46	17.64	7.64
2.40	3.00	17.02	17.23	17.45	17.66	17.87	18.08	18.29	18.50	18.71	18.92	8.19
2.56	3.20	18.16	18.38	18.61	18.92	19.06	19.29	19.51	19.73	19.96	20.18	8.73
2.72	3.40	19.30	19.53	19.77	20.01	20.25	20.49	20.73	20.97	21.21	21.44	9.28
2.88	3.60	20.44	20.69	20.93	21.29	21.44	21.70	21.94	22.20	22.45	22.71	9.82
3.04	3.80	21.57	21.83	22.10	22.47	22.64	22.90	23.17	23.43	23.70	23.97	10.37
3.20	4.00	22.70	22.99	23.27	23.65	23.82	24.11	24.39	24.67	24.95	25.20	10.92
3.36	4.20	23.84	24.13	24.42	24.84	25.02	25.31	25.60	25.90	26.20	26.49	11.45
3.52	4.40	24.98	25.28	25.59	25.86	26.22	26.52	26.83	27.14	27.44	27.73	12.01
3.68	4.60	26.11	26.43	26.76	27.08	27.40	27.72	28.05	28.33	28.69	29.00	12.56
3.84	4.80	27.25	27.58	27.92	28.26	28.59	28.89	29.27	29.60	29.94	30.38	13.10
4.00	5.00	28.38	28.73	29.08	29.59	29.78	30.13	30.48	30.83	31.19	31.54	13.65
4.16	5.20	29.52	29.88	30.25	30.61	30.98	31.34	31.70	32.07	32.43	32.80	14.20
4.32	5.40	30.65	31.03	31.40	31.79	32.17	32.55	32.92	33.30	33.68	34.05	14.74
4.48	5.60	31.78	32.18	32.57	32.97	33.35	33.75	34.14	34.53	34.93	35.28	15.29
4.64	5.80	32.92	33.33	33.74	34.30	34.14	34.55	34.96	35.36	35.17	36.58	15.84
4.80	6.00	34.05	34.47	34.90	35.32	35.74	36.16	36.58	37.00	37.42	37.84	16.38
4.96	6.20	35.19	35.63	36.06	36.50	36.93	37.37	37.80	38.23	38.67	39.10	16.92
5.12	6.40	36.33	36.77	37.22	37.67	38.12	38.57	39.02	39.46	39.92	40.36	17.45
5.28	6.60	37.46	37.93	38.39	38.85	39.31	39.78	40.24	40.70	41.76	41.63	18.02
5.44	6.80	38.60	39.08	39.55	40.03	40.51	40.98	41.46	41.93	42.41	42.89	18.57
5.60	7.00	39.73	40.23	40.71	41.20	41.70	42.19	42.68	43.16	43.65	44.15	19.11
		810	820	830	840	850	860	870	880	890	900	
		HEAT LOSS IN Btu PER HOUR										

Table B-51 (Continued)

DOLLAR COST OF ENERGY PER YEAR

BASED ON 8760 HRS/YR

$ Per Million Btu	$ Per M lbs Steam	HEAT LOSS IN WATTS PER HOUR 266.6	269.6	272.5	275.4	278.4	281.3	284.2	287.1	290.1	293.0	$ Per Million Watts
0.16	**0.20**	$1.28	$1.29	$1.30	$1.32	$1.33	$1.35	$1.36	$1.37	$1.39	$1.40	**0.57**
0.32	**0.40**	2.55	2.58	2.61	2.64	2.66	2.69	2.72	2.75	2.78	2.80	**1.09**
0.48	**0.60**	3.83	3.87	3.91	3.95	3.99	4.04	4.08	4.12	4.16	4.21	**1.64**
0.64	**0.80**	5.10	5.16	5.21	5.27	5.32	5.38	5.44	5.49	5.55	5.61	**2.18**
0.80	**1.00**	6.38	6.45	6.51	6.58	6.66	6.73	6.80	6.87	6.94	7.01	**2.73**
0.96	**1.20**	7.65	7.74	7.82	7.91	7.99	8.08	8.16	8.24	8.33	8.41	**3.27**
1.12	**1.40**	8.93	9.03	9.12	9.22	9.32	9.42	9.52	9.62	9.71	9.81	**3.82**
1.28	**1.60**	10.20	10.32	10.43	10.54	10.65	10.76	10.88	10.99	11.10	11.21	**4.37**
1.44	**1.80**	11.48	11.61	117.3	11.84	11.98	12.12	12.24	12.36	12.49	12.61	**4.91**
1.60	**2.00**	12.75	12.89	13.03	13.18	13.32	13.46	13.60	13.73	13.88	14.02	**5.46**
1.76	**2.20**	14.03	14.18	14.34	14.49	14.65	14.80	14.95	15.11	15.26	15.42	**6.01**
1.92	**2.40**	15.31	15.47	15.64	15.81	15.98	16.15	16.31	16.48	16.65	16.82	**6.55**
2.08	**2.60**	16.58	16.76	16.95	17.13	17.31	17.49	17.67	17.86	18.04	18.22	**7.09**
2.24	**2.80**	17.86	18.05	18.24	18.45	18.64	18.84	19.03	19.23	19.42	19.62	**7.64**
2.40	**3.00**	19.13	19.34	19.55	19.76	19.97	20.18	20.39	20.60	20.81	21.02	**8.19**
2.56	**3.20**	20.41	20.63	20.86	21.08	21.30	21.53	21.75	21.98	22.20	22.43	**8.73**
2.72	**3.40**	21.68	21.92	22.16	22.40	22.64	22.86	23.11	23.35	23.59	23.83	**9.28**
2.88	**3.60**	22.96	23.21	23.46	23.68	23.97	24.24	24.47	24.72	24.98	25.23	**9.82**
3.04	**3.80**	24.23	24.50	24.77	25.03	25.29	25.57	25.83	26.09	26.36	26.63	**10.37**
3.20	**4.00**	25.51	25.79	26.07	26.35	26.63	26.90	27.19	27.47	27.75	28.03	**10.92**
3.36	**4.20**	26.78	27.08	27.37	27.67	27.96	28.26	28.55	28.85	29.14	29.43	**11.45**
3.52	**4.40**	28.06	28.37	28.68	28.99	29.29	29.60	29.91	30.22	30.52	30.84	**12.01**
3.68	**4.60**	29.34	29.66	29.98	30.30	30.62	30.95	31.27	31.59	31.99	32.24	**12.56**
3.84	**4.80**	30.61	30.95	31.28	31.62	31.95	32.29	32.63	32.97	33.30	33.64	**13.10**
4.00	**5.00**	31.88	32.24	32.59	32.94	33.29	33.63	33.99	34.34	34.69	35.04	**13.65**
4.16	**5.20**	33.16	33.53	33.89	34.25	34.62	34.98	35.35	35.71	36.08	36.44	**14.20**
4.32	**5.40**	34.44	34.81	35.19	35.57	35.95	36.33	36.71	37.08	37.46	37.84	**14.74**
4.48	**5.60**	35.71	36.10	36.50	36.89	37.28	37.67	38.07	38.26	38.85	39.24	**15.29**
4.64	**5.80**	36.98	37.39	37.80	38.21	38.61	39.02	39.43	39.83	40.24	40.65	**15.84**
4.80	**6.00**	38.26	38.68	39.10	39.53	39.95	40.37	40.79	41.20	41.62	42.05	**16.38**
4.96	**6.20**	39.53	39.97	40.41	40.84	41.28	41.71	42.14	42.58	43.02	43.45	**16.92**
5.12	**6.40**	40.81	41.27	41.71	42.16	42.61	43.06	43.50	43.95	44.40	44.95	**17.47**
5.28	**6.60**	42.09	42.55	43.02	43.48	43.94	44.40	44.87	45.33	45.79	46.25	**18.02**
5.44	**6.80**	43.37	43.84	44.32	44.80	45.27	45.73	46.22	46.70	47.18	47.65	**18.57**
5.60	**7.00**	44.64	45.13	45.62	46.11	46.60	47.09	47.58	48.07	48.56	49.06	**19.11**
		910	**920**	**930**	**940**	**950**	**960**	**970**	**980**	**990**	**1000**	
		HEAT LOSS IN Btu PER HOUR										

Table B-51 (Continued)

DOLLAR COST OF ENERGY PER YEAR

BASED ON 8760 HRS/YR

$ Per Million Btu	$ Per M lbs Steam	HEAT LOSS IN WATTS PER HOUR										$ Per Million Watts
		293.0	322.3	351.6	380.9	410.2	439.5	468.8	498.1	527.4	556.7	
0.16	0.20	$1.40	$1.54	$1.68	$1.82	$1.96	$2.10	$2.24	$2.38	$2.52	$2.66	0.57
0.32	0.40	2.80	3.08	3.36	3.64	3.92	4.21	4.49	4.77	5.05	5.33	1.09
0.48	0.60	4.21	4.63	5.05	5.47	5.89	6.31	6.78	7.15	7.57	7.99	1.64
0.64	0.80	5.61	6.17	6.73	7.29	7.85	8.41	8.96	9.53	10.09	10.65	2.18
0.80	1.00	7.01	7.71	8.41	9.11	9.81	10.51	11.21	11.91	12.61	13.32	2.73
0.96	1.20	8.41	9.25	10.09	10.93	11.77	12.61	13.55	14.30	15.14	15.98	3.27
1.12	1.40	9.81	10.79	11.77	12.76	13.74	14.72	15.70	16.68	17.66	18.64	3.82
1.28	1.60	11.21	12.23	13.46	14.58	15.70	16.82	17.92	19.06	20.18	21.30	4.37
1.44	1.80	12.61	13.68	15.14	16.40	17.66	18.92	20.18	21.44	22.71	23.97	4.91
1.60	2.00	14.02	15.42	16.82	18.22	19.62	21.02	22.43	23.83	25.23	26.63	5.46
1.76	2.20	15.42	16.96	18.50	20.04	21.59	23.13	24.67	26.21	27.75	29.29	6.01
1.92	2.40	16.82	18.50	20.18	21.87	23.55	25.23	27.11	28.59	30.28	31.96	6.55
2.08	2.60	18.22	20.04	21.87	23.69	25.51	27.33	29.15	30.98	32.80	34.62	7.09
2.24	2.80	19.62	21.58	23.55	25.51	27.47	29.43	31.40	33.365	35.32	37.28	7.64
2.40	3.00	21.02	23.13	25.23	27.33	29.43	31.54	33.64	35.74	38.84	39.95	8.19
2.56	3.20	22.43	24.47	26.92	29.15	31.39	33.64	35.84	38.12	40.37	42.61	8.73
2.72	3.40	23.83	26.21	28.59	30.98	33.36	35.74	38.12	40.51	42.89	45.27	9.28
2.88	3.60	25.23	27.35	30.27	32.80	35.32	37.84	40.37	42.89	45.41	47.93	9.82
3.04	3.80	26.63	29.29	31.96	34.62	37.28	39.95	42.61	45.27	47.93	50.60	10.37
3.20	4.00	28.03	30.84	33.64	36.44	39.24	42.05	44.85	47.65	50.46	53.26	10.92
3.36	4.20	29.43	32.38	35.32	38.26	41.21	44.15	47.01	50.04	52.98	55.92	11.45
3.52	4.40	30.84	33.92	37.00	40.08	43.17	46.25	49.34	52.42	55.50	58.59	12.01
3.68	4.60	32.24	35.46	38.68	41.91	45.13	48.36	51.56	54.80	58.03	61.25	12.56
3.84	4.80	33.64	37.00	40.36	43.73	47.10	50.46	54.22	57.18	60.55	63.90	13.10
4.00	5.00	35.04	38.54	42.05	45.55	49.06	52.56	56.06	59.57	63.07	66.58	13.65
4.16	5.20	36.44	40.09	43.73	47.37	51.02	54.66	58.31	61.95	65.60	69.24	14.20
4.32	5.40	37.84	41.63	45.41	49.19	52.98	56.77	60.55	64.33	68.12	71.90	14.74
4.48	5.60	39.24	43.17	47.09	51.02	54.94	58.87	62.79	66.72	70.64	74.56	15.29
4.64	5.80	40.65	44.71	48.78	52.84	56.91	60.97	65.03	69.10	73.16	77.23	15.84
4.80	6.00	42.05	46.25	50.46	54.66	58.87	63.07	67.27	71.48	75.69	79.89	16.38
4.96	6.20	43.45	47.80	52.14	56.48	60.83	65.17	69.52	73.86	78.21	82.55	16.92
5.12	6.40	44.95	48.94	53.82	58.31	62.78	67.28	71.68	76.24	80.74	85.22	17.47
5.28	6.60	46.25	50.88	55.50	60.13	64.75	69.38	74.01	78.63	83.26	87.88	18.02
5.44	6.80	47.65	52.40	51.19	61.95	66.72	70.48	76.25	81.01	85.78	90.54	18.57
5.60	7.00	49.06	53.96	58.87	63.77	68.68	73.58	78.49	83.40	88.30	93.21	19.11
		1000	1100	1200	1300	1400	1500	1600	1700	1800	1900	
		HEAT LOSS IN Btu PER HOUR										

Table B-51 (Continued)

DOLLAR COST OF ENERGY PER YEAR

BASED ON 8760 HRS/YR

$ Per Million Btu	$ Per M lbs Steam	HEAT LOSS IN WATTS PER HOUR										$ Per Million Watts
		586.0	615.3	644.6	673.9	703.2	732.5	761.8	791.1	820.4	849.7	
0.16	0.20	$2.80	$2.96	$3.08	$3.22	$3.36	$3.50	$3.64	$3.78	$3.92	$4.07	0.57
0.32	0.40	5.61	5.91	6.17	6.45	6.73	7.01	7.29	7.57	7.85	8.13	1.09
0.48	0.60	8.41	8.87	9.92	9.67	10.09	10.51	10.93	11.35	11.77	12.19	1.64
0.64	0.80	11.12	11.83	12.23	12.90	13.46	14.02	14.58	15.14	15.70	16.26	2.18
0.80	1.00	14.02	14.79	15.42	16.12	16.82	17.52	18.22	18.92	19.62	20.32	2.73
0.96	1.20	16.82	17.74	18.50	19.34	20.18	21.02	21.87	22.71	23.55	24.39	3.27
1.12	1.40	19.62	20.70	21.58	22.57	23.54	24.53	25.51	26.49	27.47	28.45	3.82
1.28	1.60	22.43	23.66	24.67	25.79	26.92	28.03	29.15	30.27	31.39	32.32	4.37
1.44	1.80	25.23	26.62	27.75	29.01	30.27	31.54	32.80	34.06	35.32	36.58	4.91
1.60	2.00	28.03	29.57	30.84	32.24	33.64	35.04	36.44	37.84	39.24	40.65	5.46
1.76	2.20	30.84	32.33	33.92	35.46	37.02	38.54	40.08	41.63	43.17	44.71	6.01
1.92	2.40	33.64	35.49	37.00	38.68	40.36	42.05	43.73	45.41	47.10	48.78	6.55
2.08	2.60	36.44	38.45	40.09	41.91	43.73	45.55	47.37	49.20	51.02	52.84	7.09
2.24	2.80	39.24	41.40	43.17	45.13	47.09	49.06	51.02	53.58	54.94	56.91	7.64
2.40	3.00	42.05	44.36	46.25	48.36	50.46	52.56	54.66	56.77	58.87	60.97	8.19
2.56	3.20	44.95	47.31	48.94	51.58	53.82	56.06	58.31	60.55	62.78	65.03	8.73
2.72	3.40	47.65	50.28	52.42	54.80	57.19	59.57	61.95	64.33	66.72	69.10	9.28
2.88	3.60	50.46	53.23	54.60	58.26	60.45	60.07	65.60	68.12	70.64	73.16	9.82
3.04	3.80	53.26	56.19	58.59	61.25	63.91	66.58	69.24	71.90	74.57	77.23	10.37
3.20	4.00	56.06	59.15	61.67	60.47	67.28	70.08	72.88	75.69	78.49	81.29	10.92
3.36	4.20	58.87	62.11	64.75	67.70	70.64	73.58	76.53	79.47	82.41	85.36	11.45
3.52	4.40	61.67	64.66	67.84	70.92	72.00	77.09	80.17	83.26	86.34	89.42	12.01
3.68	4.60	64.47	68.02	70.81	74.15	77.37	80.59	83.82	87.04	90.26	93.49	12.56
3.84	4.80	67.38	70.98	74.00	77.37	80.83	84.10	87.46	90.82	94.19	97.54	13.10
4.00	5.00	70.08	73.93	77.09	80.59	84.10	87.60	91.10	94.61	98.11	100.16	13.65
4.16	5.20	72.88	76.89	80.17	83.82	87.46	91.10	94.75	98.39	102.04	105.68	14.20
4.32	5.40	75.69	79.85	83.25	87.04	90.82	94.61	98.39	102.18	105.96	109.75	14.74
4.48	5.60	78.49	82.81	86.34	90.26	94.08	98.11	102.04	105.96	109.89	113.81	15.29
4.64	5.80	81.29	85.76	89.42	93.49	97.55	101.62	105.68	109.75	113.81	117.86	15.84
4.80	6.00	84.10	88.72	92.50	96.71	100.92	105.12	109.22	113.53	117.73	121.94	16.38
4.96	6.20	86.90	91.70	95.59	99.93	104.38	108.62	112.97	117.31	121.66	126.00	16.92
5.12	6.40	89.90	94.62	97.88	103.17	107.65	112.13	116.61	121.09	125.57	130.07	17.47
5.28	6.60	92.51	97.59	101.76	106.38	111.07	115.63	120.26	124.88	129.50	134.13	18.02
5.44	6.80	95.31	100.55	104.84	109.61	114.37	119.14	123.90	129.00	133.43	138.20	18.57
5.60	7.00	98.11	103.51	107.92	112.83	117.73	122.64	127.55	132.45	137.36	142.26	19.11
		2000	2100	2200	2300	2400	2500	2600	2700	2800	2900	
		HEAT LOSS IN Btu PER HOUR										

Table B-51 (Continued)

DOLLAR COST OF ENERGY PER YEAR

BASED ON 8760 HRS/YR

$ Per Million Btu	$ Per M lbs Steam	HEAT LOSS IN WATTS PER HOUR 879.0	908.3	937.6	966.9	996.2	1025.5	1054.8	1084.1	1113.4	1142.7	$ Per Million Watts
0.16	0.20	$4.20	$4.35	$4.49	$4.63	$4.77	$4.91	$5.05	$5.19	$5.33	$5.47	0.57
0.32	0.40	8.41	8.69	8.96	9.25	9.53	9.81	10.09	10.37	10.65	10.93	1.09
0.48	0.60	12.61	13.04	13.55	13.88	14.30	14.72	15.14	15.56	15.98	16.40	1.64
0.64	0.80	16.82	17.38	17.92	18.50	19.06	19.62	20.18	20.74	21.30	21.86	2.18
0.80	1.00	21.02	21.73	22.43	23.13	23.83	24.53	25.23	25.93	26.63	27.33	2.73
0.96	1.20	25.23	26.07	27.11	27.75	28.59	29.43	30.28	31.12	31.96	33.73	3.27
1.12	1.40	29.43	30.42	31.40	32.38	33.36	34.34	35.32	36.30	37.28	38.26	3.82
1.28	1.60	33.64	34.76	35.84	37.00	38.20	39.24	40.36	41.49	42.61	43.73	4.37
1.44	1.80	37.84	39.11	40.37	41.63	42.89	44.15	45.41	46.67	47.93	49.20	4.91
1.60	2.00	42.05	43.45	44.85	46.25	47.65	49.06	50.46	51.86	53.26	54.66	5.46
1.76	2.20	46.25	47.80	49.34	50.88	52.42	53.96	55.50	57.05	58.59	60.13	6.01
1.92	2.40	50.46	52.14	54.21	55.50	57.18	58.87	60.55	62.23	63.90	65.46	6.55
2.08	2.60	54.66	56.48	58.31	60.13	61.95	63.77	65.60	67.42	69.24	71.06	7.09
2.24	2.80	58.87	60.83	62.79	64.76	66.72	68.68	70.64	72.60	74.56	76.53	7.64
2.40	3.00	63.07	65.17	67.27	69.38	71.48	73.58	75.69	77.79	79.89	81.99	8.19
2.56	3.20	67.28	69.52	71.68	74.00	76.24	78.49	80.74	82.98	85.22	87.46	8.83
2.72	3.40	70.48	73.86	76.25	78.63	81.01	83.95	85.78	88.16	90.54	92.93	9.28
2.88	3.60	75.68	78.21	80.73	83.26	85.78	88.30	90.82	93.35	95.87	98.39	9.82
3.04	3.80	79.89	82.55	85.22	87.88	90.54	93.21	95.87	98.53	101.20	103.86	10.37
3.20	4.00	84.10	86.90	89.70	92.50	95.21	98.11	100.91	103.72	106.52	109.32	10.92
3.36	4.20	88.30	91.24	94.02	97.13	100.07	103.02	105.86	108.00	111.85	114.79	11.45
3.52	4.40	92.51	95.59	98.67	100.76	104.84	107.92	111.01	114.09	117.16	120.26	12.01
3.68	4.60	96.71	99.93	103.12	106.38	109.61	112.83	116.05	119.28	122.50	125.72	12.56
3.84	4.80	100.91	104.28	108.43	111.00	114.37	117.74	121.10	124.46	127.80	130.91	13.10
4.00	5.00	105.06	108.62	112.13	115.63	119.14	122.64	126.14	129.65	133.16	136.66	13.65
4.16	5.20	109.32	112.97	116.61	120.26	123.90	127.55	131.19	134.03	138.48	142.12	14.20
4.32	5.40	113.53	117.31	121.10	124.88	128.67	132.45	136.24	140.02	143.80	147.59	14.74
4.48	5.60	117.74	121.66	125.58	129.51	133.43	137.36	141.28	145.20	149.13	153.06	15.29
4.64	5.80	121.94	126.00	130.07	132.98	138.20	142.26	146.33	150.39	154.46	158.52	15.84
4.80	6.00	126.14	130.35	134.53	138.76	142.96	147.17	151.37	155.58	159.79	163.99	16.38
4.96	6.20	130.35	134.69	139.04	143.38	147.74	152.07	156.52	160.76	165.11	169.45	16.92
5.12	6.40	134.56	139.04	143.36	148.00	152.48	156.98	161.47	165.95	170.43	174.91	17.47
5.28	6.60	138.76	143.38	148.01	152.63	157.26	161.89	166.51	171.14	175.76	180.39	18.02
5.44	6.80	140.96	147.74	152.49	157.26	162.02	166.79	171.56	176.32	181.09	185.85	18.57
5.60	7.00	147.17	152.07	156.98	161.89	166.79	171.70	176.60	181.51	186.42	191.32	19.11
		3000	3100	3200	3300	3400	3500	3600	3700	3800	3900	
		HEAT LOSS IN Btu PER HOUR										

Table B-51 (Continued)

DOLLAR COST OF ENERGY PER YEAR

BASED ON 8760 HRS/YR

$ Per Million Btu	$ Per M lbs Steam	HEAT LOSS IN WATTS PER HOUR										$ Per Million Watts
		1172.0	1201.3	1230.6	1259.9	1289.2	1318.5	1347.8	1377.1	1406.4	1435.7	
0.16	0.20	$5.61	$5.75	$5.91	$6.03	$6.17	$6.31	$6.45	$6.59	$6.73	$6.87	0.57
0.32	0.40	11.21	11.49	11.83	12.05	12.23	12.61	12.90	13.18	13.46	13.74	1.09
0.48	0.60	16.82	17.40	17.74	18.08	18.50	18.92	19.34	19.76	20.18	20.60	1.64
0.64	0.80	22.42	22.98	23.66	24.11	24.47	25.23	25.79	26.35	26.91	27.45	2.18
0.80	1.00	28.03	28.73	29.57	30.13	30.84	31.54	32.23	32.94	33.64	34.34	2.73
0.96	1.20	33.64	34.48	35.49	36.11	37.00	37.84	38.68	39.53	40.36	41.21	3.27
1.12	1.40	39.24	40.22	41.40	42.19	43.17	44.15	45.13	46.11	47.09	48.08	3.82
1.28	1.60	44.95	45.97	47.31	48.22	49.34	50.46	51.58	52.70	53.82	54.94	4.37
1.44	1.80	50.46	51.72	53.23	54.24	55.50	56.77	58.03	59.29	60.55	61.81	4.91
1.60	2.00	56.06	57.47	59.15	60.27	61.61	63.02	64.47	65.88	67.28	68.67	5.46
1.76	2.20	61.67	63.21	64.66	66.30	67.85	69.38	70.92	72.46	74.01	75.55	6.01
1.92	2.40	67.38	68.96	70.98	72.22	74.00	75.68	77.37	79.05	80.73	82.41	6.55
2.08	2.60	72.88	74.71	76.89	78.35	80.17	81.99	83.82	85.64	87.46	89.28	7.09
2.24	2.80	78.49	80.45	82.81	84.38	86.34	88.20	90.26	92.23	94.19	96.15	7.64
2.40	3.00	84.10	86.20	88.72	90.40	92.50	94.61	96.71	98.81	100.92	103.02	8.19
2.56	3.20	89.90	91.94	94.62	96.43	97.88	100.91	103.17	105.40	107.65	109.88	8.73
2.72	3.40	95.31	97.69	100.55	102.46	104.84	107.22	109.61	111.99	114.37	116.75	9.28
2.88	3.60	100.92	103.44	106.46	108.48	111.01	113.53	116.05	118.38	121.19	123.62	9.82
3.04	3.80	106.52	109.18	112.38	114.51	117.17	119.84	122.50	125.16	127.83	130.49	10.37
3.20	4.00	112.13	114.93	118.30	120.54	123.34	126.03	128.94	131.75	134.56	137.34	10.92
3.36	4.20	117.73	120.68	124.21	126.56	129.51	132.45	135.39	138.34	141.28	144.23	11.45
3.52	4.40	123.34	126.42	129.32	132.59	135.68	138.66	141.84	144.93	148.01	151.09	12.01
3.68	4.60	128.95	132.17	136.04	138.62	141.64	145.07	148.30	151.51	154.74	157.96	12.56
3.84	4.80	134.75	137.92	141.95	144.45	148.00	151.36	154.74	158.10	161.46	164.83	13.10
4.00	5.00	140.16	143.66	147.87	150.67	154.18	157.68	161.18	164.69	168.19	171.70	13.65
4.16	5.20	145.76	149.41	153.78	156.70	160.34	163.99	167.63	171.25	174.92	178.56	14.20
4.32	5.40	151.37	155.16	159.70	162.76	166.51	170.29	174.08	177.86	181.65	185.43	14.74
4.48	5.60	156.95	160.90	165.61	168.72	172.67	176.40	180.52	184.45	188.37	192.30	15.29
4.64	5.80	162.59	166.65	171.52	174.78	178.84	182.91	186.97	191.04	195.10	199.17	15.84
4.80	6.00	168.19	172.40	177.45	180.81	185.01	189.22	193.42	197.64	201.83	206.04	16.38
4.96	6.20	173.80	178.14	183.36	186.83	191.18	195.52	199.87	204.21	208.56	212.90	16.92
5.12	6.40	179.81	183.87	189.25	192.86	195.76	201.82	206.34	210.80	215.30	219.76	17.47
5.28	6.60	185.01	189.64	195.19	198.89	203.51	208.14	212.76	217.39	222.01	226.64	18.02
5.44	6.80	190.06	195.38	201.10	204.91	209.68	214.44	219.21	223.98	228.64	233.51	18.57
5.60	7.00	196.22	201.13	207.02	210.94	215.84	220.75	225.66	230.56	235.47	240.37	19.11
		4000	4100	4200	4300	4400	4500	4600	4700	4800	4900	
		HEAT LOSS IN Btu PER HOUR										

Table B-51 (Continued)

DOLLAR COST OF ENERGY PER YEAR

BASED ON 8760 HRS/YR

$ Per Million Btu	$ Per M lbs Steam	HEAT LOSS IN WATTS PER HOUR 1465.0	1494.3	1523.6	1552.9	1582.2	1611.5	1640.8	1670.1	1699.4	1728.7	$ Per Million Watts
0.16	0.20	$7.01	$7.15	$7.29	$7.43	$7.57	$7.71	$7.85	$7.99	$8.13	$8.27	0.57
0.32	0.40	14.02	14.30	14.58	14.86	15.14	15.41	15.70	15.98	16.26	16.54	1.09
0.48	0.60	21.02	21.44	21.87	22.29	22.71	23.13	23.55	23.91	24.39	24.81	1.64
0.64	0.80	28.03	28.59	29.15	29.71	30.27	30.84	31.39	31.96	32.32	33.08	2.18
0.80	1.00	35.04	35.74	36.44	37.14	39.84	38.54	39.24	39.95	40.65	41.35	2.73
0.96	1.20	42.05	42.89	43.73	44.57	45.41	46.25	47.10	47.93	48.78	49.62	3.27
1.12	1.40	49.06	50.04	51.02	52.00	52.98	53.96	54.94	55.92	56.91	57.89	3.82
1.28	1.60	56.06	57.08	58.31	59.43	60.55	61.67	62.78	63.91	64.64	66.16	4.37
1.44	1.80	63.07	64.33	65.60	66.86	68.19	69.38	70.64	71.90	73.16	74.43	4.91
1.60	2.00	70.08	71.48	72.88	74.29	75.69	77.09	78.49	79.89	81.29	82.69	5.46
1.76	2.20	77.09	78.63	80.17	81.71	83.26	84.80	86.34	87.88	89.42	90.96	6.01
1.92	2.40	84.10	85.78	87.46	89.14	90.82	92.51	94.19	95.87	97.54	99.23	6.55
2.08	2.60	91.10	92.93	94.75	96.57	98.39	100.21	102.04	103.86	105.68	107.50	7.09
2.24	2.80	98.11	100.07	102.04	104.00	105.96	107.92	109.89	111.85	113.81	115.77	7.64
2.40	3.00	105.12	107.22	109.22	111.43	113.53	115.63	117.73	119.84	121.94	124.04	8.19
2.56	3.20	112.13	114.17	116.61	118.86	121.10	123.34	125.57	127.83	130.07	132.31	8.73
2.72	3.40	119.14	120.63	123.90	126.28	129.00	131.05	133.43	135.82	138.20	140.58	9.28
2.88	3.60	126.14	128.67	131.19	133.71	136.24	138.76	141.28	143.80	146.33	148.85	9.82
3.04	3.80	133.15	135.82	138.48	141.14	143.80	146.47	149.13	151.79	154.46	157.20	10.37
3.20	4.00	140.16	142.96	145.76	148.57	151.37	154.18	156.98	159.78	162.58	165.39	10.92
3.36	4.20	147.17	150.11	153.05	156.00	158.94	161.88	164.83	167.77	170.72	173.66	11.45
3.52	4.40	154.18	157.26	160.33	163.43	166.51	169.59	172.68	175.76	178.84	181.93	12.01
3.68	4.60	161.18	164.41	167.63	170.86	174.08	177.30	180.33	183.75	186.97	190.20	12.56
3.84	4.80	168.19	171.55	174.92	178.28	181.65	185.01	188.38	191.74	195.08	198.47	13.10
4.00	5.00	175.20	178.70	182.21	185.71	189.22	192.72	196.22	199.73	203.23	206.74	13.65
4.16	5.20	182.21	185.85	189.50	193.14	197.78	200.43	204.07	207.72	211.36	215.01	14.20
4.32	5.40	189.22	193.00	196.78	200.57	204.35	208.14	211.92	215.71	219.49	223.27	14.74
4.48	5.60	196.22	200.15	204.07	208.00	211.82	215.85	219.77	223.70	227.62	231.54	15.29
4.64	5.80	203.23	207.30	211.36	215.43	219.49	223.56	227.62	231.68	235.75	239.81	15.84
4.80	6.00	210.24	214.45	218.45	222.85	227.06	231.26	235.47	239.67	243.88	248.08	16.38
4.96	6.20	217.25	221.59	225.94	230.28	234.62	238.97	243.32	247.66	252.01	256.35	16.92
5.12	6.40	224.26	228.34	233.22	237.71	242.19	246.82	251.14	255.65	258.56	264.62	17.47
5.28	6.60	231.26	235.89	240.51	245.14	249.77	254.39	258.02	263.64	268.27	272.89	18.02
5.44	6.80	238.27	243.04	247.80	252.57	258.00	262.10	266.86	271.63	276.40	281.16	18.57
5.60	7.00	245.28	250.19	255.09	260.00	264.90	269.81	274.71	279.62	284.52	289.43	19.11
		5000	5100	5200	5300	5400	5500	5600	5700	5800	5900	

HEAT LOSS IN Btu PER HOUR

Table B-51 (Continued)

DOLLAR COST OF ENERGY PER YEAR

BASED ON 8760 HRS/YR

$ Per Million Btu	$ Per M lbs Steam	HEAT LOSS IN WATTS PER HOUR										$ Per Million Watts
		1758.0	1787.3	1816.6	1845.9	1875.2	1904.5	1933.8	1963.1	1992.4	2021.7	
0.16	0.20	$8.41	$8.55	$8.69	$8.83	$8.96	$9.11	$9.25	$9.39	$9.53	$9.67	0.57
0.32	0.40	16.82	17.10	17.38	17.66	17.92	18.21	18.50	18.78	19.06	19.34	1.09
0.48	0.60	25.23	25.65	26.07	26.49	27.11	27.33	27.75	28.17	28.59	29.01	1.64
0.64	0.80	33.64	34.20	34.76	35.32	35.84	36.44	37.00	37.56	38.12	38.98	2.18
0.80	1.00	42.05	42.75	43.45	44.15	44.85	45.55	46.25	46.95	47.65	48.36	2.73
0.96	1.20	50.46	51.30	52.14	52.98	54.21	54.66	55.50	56.34	57.18	58.03	3.27
1.12	1.40	58.87	59.85	60.83	61.81	62.79	63.77	64.76	65.74	66.72	67.70	3.82
1.28	1.60	67.28	68.40	69.52	70.64	71.68	72.88	74.00	75.13	76.24	77.37	4.37
1.44	1.80	75.68	76.95	78.21	79.47	80.73	81.99	83.25	84.52	85.78	87.04	4.91
1.60	2.00	84.10	85.50	86.90	88.30	89.70	91.10	92.50	93.91	95.21	96.71	5.46
1.76	2.20	92.51	94.05	95.59	97.13	98.67	100.21	101.76	103.30	103.30	106.38	6.01
1.92	2.40	100.91	102.60	104.28	105.96	108.43	109.33	111.00	112.69	112.69	116.05	6.55
2.08	2.60	109.32	111.15	112.97	114.79	116.01	118.44	120.26	122.08	122.08	125.72	7.09
2.24	2.80	117.74	119.70	121.66	123.62	125.58	127.55	129.51	131.47	131.47	135.39	7.64
2.40	3.00	126.14	128.25	130.35	132.45	134.53	136.66	138.76	140.86	140.86	145.07	8.19
2.56	3.20	134.56	136.80	139.04	141.28	143.36	145.77	148.00	150.25	150.25	154.74	8.73
2.72	3.40	142.96	145.35	147.73	150.11	152.49	154.88	157.26	159.64	159.64	164.41	9.28
2.88	3.60	151.36	153.90	156.42	158.94	161.46	163.99	166.51	169.03	169.03	174.08	9.82
3.04	3.80	159.78	162.45	165.11	167.77	170.43	173.10	175.76	178.42	178.42	183.75	10.37
3.20	4.00	168.19	171.00	173.80	176.60	179.41	182.21	185.01	187.81	187.81	193.42	10.92
3.36	4.20	176.60	179.55	182.49	185.43	188.04	191.32	194.26	197.21	197.21	203.09	11.45
3.52	4.40	185.01	188.09	191.18	194.26	197.34	200.43	201.51	206.60	206.60	212.76	12.01
3.68	4.60	193.42	196.64	199.87	203.09	206.24	209.54	212.76	215.99	216.00	222.43	12.56
3.84	4.80	201.82	205.19	208.56	211.92	216.86	218.65	222.01	225.38	225.38	232.11	13.10
4.00	5.00	210.12	213.74	217.25	220.75	224.26	227.76	231.26	234.77	234.77	241.78	13.65
4.16	5.20	218.65	222.29	225.94	229.58	233.22	236.87	240.51	244.16	247.80	251.45	14.20
4.32	5.40	227.06	230.89	234.63	238.42	242.20	245.98	249.77	253.55	257.33	261.12	14.74
4.48	5.60	235.47	239.39	243.32	247.24	251.17	255.09	259.02	262.94	266.86	270.80	15.29
4.64	5.80	243.88	247.94	252.01	256.07	260.14	264.20	268.27	272.33	276.40	280.46	15.84
4.80	6.00	252.29	256.49	260.70	264.90	269.07	273.31	277.52	281.72	285.93	290.13	16.38
4.96	6.20	260.70	265.04	269.39	273.73	278.08	282.42	286.77	291.11	295.46	299.80	16.92
5.12	6.40	269.12	273.59	278.08	282.56	286.72	291.53	296.00	300.50	304.96	309.47	17.47
5.28	6.60	277.52	282.14	186.77	291.39	296.02	300.64	305.26	309.89	314.52	319.14	18.02
5.44	6.80	281.93	290.69	295.46	300.22	304.99	309.75	314.52	319.28	324.05	328.82	18.57
5.60	7.00	294.34	299.24	304.15	309.05	313.96	318.86	323.77	328.68	333.58	338.49	19.11
		6000	6100	6200	6300	6400	6500	6600	6700	6800	6900	
		HEAT LOSS IN Btu PER HOUR										

Table B-51 (Continued)

DOLLAR COST OF ENERGY PER YEAR

BASED ON 8760 HRS/YR

$ Per Million Btu	$ Per M lbs Steam	HEAT LOSS IN WATTS PER HOUR										$ Per Million Watts
		2051.0	2080.3	2109.6	2138.9	2168.2	2197.5	2226.8	2256.1	2285.4	2314.7	
0.16	0.20	$9.81	$9.95	$10.09	$10.23	$10.37	$10.51	$10.65	$10.79	$10.93	$11.07	0.57
0.32	0.40	19.62	19.90	20.18	20.46	20.74	21.02	21.30	21.59	21.86	22.15	1.09
0.48	0.60	29.43	29.85	30.28	30.70	31.12	31.54	31.96	32.38	32.80	33.22	1.64
0.64	0.80	39.24	39.81	40.37	40.93	41.49	42.05	42.61	43.17	43.73	44.29	2.18
0.80	1.00	49.06	49.76	50.46	51.16	51.86	52.56	53.26	53.96	54.66	55.36	2.73
0.96	1.20	58.87	59.71	60.55	61.39	62.23	63.07	63.90	64.75	65.46	66.44	3.27
1.12	1.40	68.68	69.66	70.64	71.62	72.60	73.58	74.56	75.55	76.53	77.51	3.82
1.28	1.60	78.49	79.61	80.74	81.85	82.98	84.10	85.22	86.34	87.46	88.58	4.37
1.44	1.80	88.30	89.56	90.82	92.09	93.35	94.61	95.86	97.13	98.39	99.64	4.91
1.60	2.00	98.11	99.51	100.92	102.32	103.72	105.12	106.52	107.92	109.32	110.73	5.46
1.76	2.20	107.92	109.47	111.01	112.55	114.09	115.63	117.16	118.72	120.26	121.80	6.01
1.92	2.40	117.36	119.42	121.10	122.78	124.46	126.14	127.80	129.51	130.91	132.87	6.55
2.08	2.60	127.55	129.37	131.19	133.01	134.83	136.66	138.48	140.30	142.12	143.94	7.09
2.24	2.80	137.36	139.32	141.28	143.24	145.20	147.17	149.13	151.09	153.06	155.02	7.64
2.40	3.00	147.17	149.27	151.37	153.47	155.58	157.68	159.79	161.89	163.99	166.09	8.19
2.56	3.20	156.98	159.22	161.47	163.71	165.95	168.19	170.43	172.58	174.91	177.16	8.73
2.72	3.40	166.79	169.17	171.56	173.94	176.32	178.70	181.09	183.47	185.85	188.24	9.28
2.88	3.60	176.60	179.12	181.64	184.17	186.69	189.22	191.74	194.26	196.78	199.31	9.82
3.04	3.80	186.41	189.08	191.74	194.40	197.07	199.73	202.39	205.05	207.72	210.33	10.37
3.20	4.00	196.22	199.03	201.83	204.63	207.44	210.24	213.04	215.85	218.65	221.45	10.92
3.36	4.20	206.04	208.98	211.72	214.87	217.81	220.75	223.70	226.64	229.58	232.53	11.45
3.52	4.40	215.85	218.93	222.02	225.10	228.18	231.26	234.33	237.43	240.51	243.60	12.01
3.68	4.60	225.66	228.88	232.10	235.33	238.55	241.78	245.00	248.22	251.45	254.67	12.56
3.84	4.80	235.47	238.83	242.21	245.56	248.93	252.29	255.61	259.02	261.81	265.74	13.10
4.00	5.00	245.28	248.78	252.29	255.79	259.30	262.80	266.32	269.81	273.31	276.92	13.65
4.16	5.20	255.09	258.74	262.38	266.02	269.67	273.31	276.95	280.60	284.24	287.88	14.20
4.32	5.40	264.80	268.69	272.47	276.26	280.04	283.22	287.61	291.39	295.18	298.96	14.74
4.48	5.60	274.71	278.64	282.56	286.49	290.41	294.34	298.26	302.19	306.11	310.03	15.29
4.64	5.80	284.53	288.59	292.65	296.72	300.78	304.85	308.91	312.58	317.04	321.11	15.84
4.80	6.00	294.34	298.54	302.74	306.95	311.15	315.36	319.57	323.77	327.97	332.18	16.38
4.96	6.20	304.05	308.49	312.83	317.18	321.53	325.87	330.22	334.56	338.91	343.25	16.92
5.12	6.40	313.95	318.44	322.95	327.41	331.90	336.38	340.86	345.35	349.82	354.32	17.47
5.28	6.60	323.77	328.40	333.02	337.64	342.27	346.90	351.52	356.15	360.77	365.40	18.02
5.44	6.80	333.58	338.35	343.11	347.88	352.64	357.41	362.17	366.94	371.70	376.47	18.57
5.60	7.00	343.39	348.30	353.20	358.11	363.01	367.92	372.42	377.73	382.64	387.54	19.11
		7000	7100	7200	7300	7400	7500	7600	7700	7800	7900	
		HEAT LOSS IN Btu PER HOUR										

Table B-51 (Continued)

DOLLAR COST OF ENERGY PER YEAR

BASED ON 8760 HRS/YR

$ Per Million Btu	$ Per M lbs Steam	HEAT LOSS IN WATTS PER HOUR 2344.0	2373.3	2402.6	2431.9	2461.2	2490.5	2519.8	2549.1	2578.4	2607.7	$ Per Million Watts
0.16	0.20	$11.21	$11.35	$11.49	$11.63	$11.83	$11.91	$12.05	$12.19	$12.34	$12.47	0.57
0.32	0.40	22.42	22.71	22.98	23.27	23.55	23.83	24.11	24.39	24.67	24.95	1.09
0.48	0.60	33.64	34.06	34.48	34.90	35.49	35.74	36.16	36.45	37.00	37.42	1.64
0.64	0.80	44.95	45.41	45.97	46.53	47.31	47.65	48.22	48.78	49.33	49.90	2.18
0.80	1.00	56.06	56.77	57.47	58.17	59.15	59.57	60.27	60.97	61.61	62.37	2.73
0.96	1.20	67.38	68.12	68.96	69.90	70.64	71.48	72.22	73.16	74.00	74.85	3.27
1.12	1.40	78.49	79.47	80.45	81.43	82.41	83.40	84.38	85.36	86.36	87.32	3.82
1.28	1.60	89.90	90.82	91.94	93.07	94.12	95.31	96.43	97.55	98.65	99.79	4.37
1.44	1.80	100.92	102.18	103.44	104.70	106.46	107.22	108.48	109.75	111.01	112.27	4.91
1.60	2.00	112.13	113.53	114.93	116.33	118.30	119.14	120.54	121.94	123.34	124.74	5.46
1.76	2.20	123.34	124.88	126.42	127.97	129.32	131.10	132.59	134.13	135.68	137.22	6.01
1.92	2.40	134.75	136.24	137.92	139.60	141.28	142.96	144.45	146.33	148.00	149.69	6.55
2.08	2.60	145.76	147.59	149.41	151.23	153.06	154.88	156.70	158.52	160.34	162.17	7.09
2.24	2.80	156.98	156.94	160.90	168.87	164.83	166.79	168.75	170.72	172.67	174.64	8.19
2.40	3.00	168.19	170.29	172.40	174.50	176.60	178.70	180.81	182.91	185.01	187.11	8.19
2.56	3.20	179.81	181.65	183.87	186.13	188.38	190.62	192.86	195.10	197.30	199.59	8.73
2.72	3.40	190.62	193.00	195.38	197.77	200.15	202.53	204.91	207.30	209.68	212.06	9.28
2.88	3.60	201.83	204.35	206.88	209.40	212.93	214.45	229.02	231.68	234.34	224.54	9.82
3.04	3.80	213.04	215.71	218.37	221.03	224.76	226.36	229.02	231.68	234.34	237.01	10.37
3.20	4.00	224.26	227.06	229.86	232.67	236.59	238.27	241.08	243.88	246.69	249.49	10.92
3.36	4.20	235.47	238.41	241.36	244.30	248.42	250.18	253.13	256.07	259.02	262.00	11.45
3.52	4.40	246.68	249.77	252.84	255.93	258.65	262.19	265.18	268.27	271.35	274.43	12.01
3.68	4.60	257.90	261.12	264.34	267.57	270.80	274.01	277.24	280.47	283.28	286.91	12.56
3.84	4.80	269.50	272.47	275.84	279.20	282.56	285.93	288.90	292.65	296.00	299.38	13.10
4.00	5.00	280.32	283.82	287.33	290.83	295.94	297.84	301.34	304.85	308.35	311.86	13.65
4.16	5.20	291.53	295.18	298.82	302.47	306.11	309.75	313.40	317.04	320.69	324.33	14.20
4.32	5.40	302.74	306.53	310.31	314.10	317.88	321.67	325.52	329.24	333.02	336.80	14.74
4.48	5.60	313.95	317.88	321.80	325.73	329.76	333.58	337.50	341.43	345.34	349.28	15.29
4.64	5.80	325.17	329.24	333.30	337.37	341.43	345.49	349.56	353.62	357.69	361.75	15.84
4.80	6.00	336.38	340.59	344.79	349.00	353.20	357.41	361.61	365.82	370.02	374.23	16.38
4.96	6.20	347.60	351.94	356.29	360.63	364.98	369.32	373.67	378.01	382.36	386.70	16.92
5.12	6.40	359.62	363.30	367.74	372.27	376.75	381.24	385.73	390.21	394.69	399.18	17.47
5.28	6.60	370.02	374.65	379.27	383.90	388.52	393.15	397.77	402.40	407.02	411.65	18.02
5.44	6.80	381.23	386.00	390.77	395.53	400.29	405.06	409.83	414.60	419.36	424.12	18.57
5.60	7.00	392.45	397.35	402.26	407.17	412.72	416.98	421.88	426.79	431.69	436.60	19.11
		8000	8100	8200	8300	8400	8500	8600	8700	8800	8900	
		HEAT LOSS IN Btu PER HOUR										

Table B-51 (Continued)

DOLLAR COST OF ENERGY PER YEAR

BASED ON 8760 HRS/YR

$ Per Million Btu	$ Per M lbs Steam	HEAT LOSS IN WATTS PER HOUR										$ Per Million Watts
		2637.0	**2666.3**	**2695.6**	**2724.8**	**2754.2**	**2783.5**	**2812.8**	**2842.1**	**2871.4**	**2900.7**	
0.16	**0.20**	$12.61	$12.76	$12.90	$13.04	$13.18	$13.32	$13.46	$13.60	$13.74	$13.88	**0.57**
0.32	**0.40**	25.23	25.51	25.79	26.07	26.35	26.63	26.91	27.19	27.47	27.75	**1.09**
0.48	**0.60**	37.84	38.26	38.68	39.11	39.53	39.95	40.34	40.79	41.21	41.63	**1.64**
0.64	**0.80**	50.46	51.02	51.58	52.14	52.70	53.26	53.82	54.38	54.94	55.50	**2.18**
0.80	**1.00**	63.07	63.77	64.97	65.17	65.88	66.58	67.28	67.98	68.67	69.40	**2.73**
0.96	**1.20**	75.68	76.53	77.37	78.21	79.05	79.89	80.76	81.57	82.41	83.26	**3.27**
1.12	**1.40**	88.20	89.28	90.26	91.24	92.23	93.21	94.18	95.17	96.15	97.13	**3.82**
1.28	**1.60**	100.91	102.04	103.17	104.28	105.40	106.52	107.65	108.76	109.88	111.01	**4.37**
1.44	**1.80**	113.53	114.79	116.05	117.31	118.38	119.84	121.20	122.36	123.62	124.88	**4.91**
1.60	**2.00**	126.03	127.55	128.94	130.35	131.75	133.15	134.56	135.96	137.34	138.76	**5.46**
1.76	**2.20**	138.66	140.03	141.84	143.38	144.93	146.47	148.09	149.55	151.09	152.63	**6.01**
1.92	**2.40**	151.36	153.06	154.74	156.42	158.10	159.78	161.46	163.15	164.83	166.51	**6.55**
2.08	**2.60**	163.99	165.81	167.63	169.45	171.28	173.10	174.92	176.74	178.58	180.39	**7.09**
2.24	**2.80**	176.40	178.56	180.52	182.48	184.45	186.41	188.37	190.34	192.30	194.26	**7.64**
2.40	**3.00**	189.22	191.32	193.42	195.52	197.64	198.73	201.83	203.93	206.03	208.14	**8.19**
2.56	**3.20**	201.82	204.07	206.34	208.56	210.80	213.04	215.30	217.53	219.76	222.01	**8.73**
2.72	**3.40**	214.44	216.83	219.21	221.59	223.98	226.36	228.64	231.12	233.51	235.89	**9.28**
2.88	**3.60**	227.06	229.58	232.10	234.63	236.75	239.67	242.38	244.72	247.24	249.77	**9.82**
3.04	**3.80**	239.67	242.34	245.00	247.66	250.33	252.99	255.65	258.32	260.98	263.64	**10.37**
3.20	**4.00**	252.07	255.09	257.88	260.70	263.50	266.30	269.02	271.91	274.67	277.52	**10.92**
3.36	**4.20**	264.90	267.85	270.79	273.73	276.68	279.62	282.56	285.51	288.45	291.39	**11.45**
3.52	**4.40**	277.32	280.60	283.68	286.77	289.85	292.93	296.02	299.10	302.18	305.27	**12.01**
3.68	**4.60**	290.02	293.36	296.58	299.80	303.03	306.25	309.47	312.70	315.92	319.14	**12.56**
3.84	**4.80**	302.73	306.11	309.47	312.84	316.20	319.57	322.93	326.29	329.66	333.02	**13.10**
4.00	**5.00**	315.36	318.86	322.37	325.87	329.38	332.88	336.38	339.89	343.39	346.90	**13.65**
4.16	**5.20**	327.97	331.62	335.26	338.91	342.50	346.20	349.84	353.48	357.13	360.77	**14.20**
4.32	**5.40**	340.49	343.37	348.16	351.94	355.73	359.51	363.30	367.08	370.86	374.65	**14.74**
4.48	**5.60**	352.80	357.13	361.05	364.98	368.90	372.83	376.74	380.67	382.60	388.52	**15.29**
4.64	**5.80**	365.82	369.88	373.95	378.01	382.08	386.14	390.21	394.27	398.33	402.40	**15.84**
4.80	**6.00**	378.43	382.64	386.84	391.05	395.27	399.46	403.66	407.87	412.07	416.28	**16.38**
4.96	**6.20**	391.05	395.39	399.74	404.08	408.43	412.77	417.12	421.46	425.81	430.15	**16.92**
5.12	**6.40**	403.65	408.15	412.67	417.12	421.60	426.09	430.59	435.06	439.52	444.03	**17.47**
5.28	**6.60**	416.28	420.90	425.52	430.15	434.78	439.40	444.03	448.62	453.28	457.90	**18.02**
5.44	**6.80**	428.89	433.66	438.42	443.86	447.95	452.72	457.28	462.25	467.01	471.78	**18.57**
5.60	**7.00**	441.50	446.41	451.32	456.21	461.13	466.03	470.94	475.94	480.75	485.65	**19.11**
		9000	**9100**	**9200**	**9300**	**9400**	**9500**	**9600**	**9700**	**9800**	**9900**	
		HEAT LOSS IN Btu PER HOUR										

Table B-51 (Continued)

DOLLAR COST OF ENERGY PER YEAR

BASED ON 8760 HRS/YR

$ Per Million Btu	$ Per M lbs Steam	HEAT LOSS IN WATTS PER HOUR 2930	3223	3516	3809	4102	4395	4688	4981	5274	5567	$ Per Million Watts
0.16	**0.20**	$14.02	$15.42	$16.81	$18.22	$19.62	$21.02	$22.43	$23.83	$25.23	$26.63	**0.57**
0.32	**0.40**	28.03	30.84	33.64	36.44	39.24	42.05	44.85	48.65	50.46	53.26	**1.09**
0.48	**0.60**	42.05	46.25	50.46	54.66	58.87	63.07	67.77	71.48	75.69	79.89	**1.64**
0.64	**0.80**	56.06	61.67	67.28	72.88	78.48	84.10	89.60	95.30	100.92	106.52	**2.18**
0.80	**1.00**	70.08	77.09	84.10	91.10	98.11	105.12	112.13	119.14	126.14	133.15	**2.73**
0.96	**1.20**	84.10	92.50	100.91	109.32	117.74	126.14	135.54	142.96	151.38	159.78	**3.27**
1.12	**1.40**	98.11	107.92	117.73	127.55	137.36	147.17	156.98	166.79	176.60	186.41	**3.82**
1.28	**1.60**	112.13	122.34	134.56	145.76	156.96	168.20	179.20	190.60	201.84	213.04	**4.37**
1.44	**1.80**	126.14	136.76	151.37	163.99	176.60	189.22	201.83	214.44	227.05	239.67	**4.91**
1.60	**2.00**	140.16	154.18	168.20	182.20	196.22	210.24	224.26	238.2	252.29	266.30	**5.46**
1.76	**2.20**	154.18	169.59	185.01	200.43	215.85	231.26	246.68	262.10	277.52	292.93	**6.01**
1.92	**2.40**	168.19	185.00	201.82	218.65	235.48	252.28	271.08	285.92	302.76	319.56	**6.55**
2.08	**2.60**	182.21	200.43	218.65	236.87	255.09	273.31	291.53	309.75	327.97	346.19	**7.09**
2.24	**2.80**	196.22	215.84	235.46	255.09	274.72	294.34	313.96	333.58	353.20	372.82	**7.64**
2.40	**3.00**	210.24	231.26	252.29	273.31	294.34	315.36	336.38	357.41	378.43	399.46	**8.19**
2.56	**3.20**	224.26	244.68	269.12	291.53	313.92	336.40	358.40	381.20	403.68	426.08	**8.73**
2.72	**3.40**	238.27	262.10	285.93	309.75	333.58	357.41	381.23	405.06	428.89	452.72	**9.28**
2.88	**3.60**	252.29	273.52	302.74	327.98	353.20	378.41	403.66	428.89	454.10	479.34	**9.82**
3.04	**3.80**	266.30	292.93	319.56	346.20	372.83	399.46	426.09	452.72	479.35	505.98	**10.37**
3.20	**4.00**	280.32	308.36	336.40	364.40	392.44	420.48	448.52	476.54	504.58	532.60	**10.92**
3.36	**4.20**	294.33	323.77	353.20	382.64	412.07	441.50	470.09	500.37	529.80	559.24	**11.45**
3.52	**4.40**	308.35	339.19	370.02	400.83	431.69	462.53	493.36	524.20	555.04	585.87	**12.01**
3.68	**4.60**	322.37	354.60	386.84	419.08	451.31	382.55	515.59	548.03	580.26	612.50	**12.56**
3.84	**4.80**	336.38	370.00	403.64	437.30	470.96	504.56	542.16	571.84	605.52	639.02	**13.10**
4.00	**5.00**	350.40	385.44	420.48	455.52	490.56	525.60	560.64	595.68	630.72	665.76	**13.65**
4.16	**5.20**	364.42	400.86	437.30	4733.74	510.18	546.62	583.06	619.50	655.95	692.38	**14.20**
4.32	**5.40**	378.43	416.27	454.12	491.96	529.80	567.65	605.49	643.33	681.18	719.02	**14.74**
4.48	**5.60**	392.44	431.68	470.92	510.18	549.44	588.68	627.92	667.16	706.40	745.64	**15.29**
4.64	**5.80**	406.46	447.11	487.76	528.40	569.05	609.70	650.34	690.99	731.63	772.28	**15.84**
4.80	**6.00**	420.48	462.52	504.58	546.62	588.67	630.72	672.67	714.82	756.86	798.93	**16.38**
4.96	**6.20**	434.50	477.95	521.39	564.84	608.29	651.74	695.19	738.64	782.09	825.54	**16.92**
5.12	**6.40**	449.52	489.39	538.24	583.06	627.84	672.80	716.80	762.40	807.36	852.16	**17.47**
5.28	**6.60**	462.53	508.78	555.03	601.28	647.54	693.79	740.05	786.30	832.55	878.80	**18.02**
5.44	**6.80**	476.54	524.20	571.85	619.51	667.16	704.82	762.47	810.12	857.78	905.43	**18.57**
5.60	**7.00**	490.56	539.61	588.67	637.73	686.78	735.84	784.90	833.95	883.00	932.06	**19.11**
		10000	**11000**	**12000**	**13000**	**14000**	**15000**	**16000**	**17000**	**18000**	**19000**	
		HEAT LOSS IN Btu PER HOUR										

Table B-51 (Continued)

DOLLAR COST OF ENERGY PER YEAR

BASED ON 8760 HRS/YR

$ Per Million Btu	$ Per M lbs Steam	HEAT LOSS IN WATTS PER HOUR										$ Per Million Watts
		5860	6153	6446	6739	7032	7325	7618	7911	8204	8497	
0.16	0.20	$28.03	$29.57	$30.84	$32.24	$33.62	$35.04	$36.44	$37.84	$39.24	$40.65	0.57
0.32	0.40	56.06	59.14	61.67	64.48	67.28	70.08	72.88	75.68	78.48	81.30	1.09
0.48	0.60	84.10	88.72	92.50	96.71	100.91	105.12	109.32	113.53	117.74	121.94	1.64
0.64	0.80	112.12	118.28	122.34	128.96	134.56	140.16	145.76	151.36	156.96	162.60	2.18
0.80	1.00	140.16	147.87	154.18	161.18	168.20	175.20	182.20	189.22	196.22	203.23	2.73
0.96	1.20	168.20	177.44	185.00	193.42	201.82	210.24	218.65	227.06	235.48	243.88	3.27
1.12	1.40	196.22	207.02	215.84	225.66	235.46	245.28	255.09	264.90	274.72	284.52	3.82
1.28	1.60	224.26	236.56	246.68	257.92	269.12	280.32	291.53	302.74	313.92	323.20	4.37
1.44	1.80	252.29	266.16	277.52	290.13	302.74	315.36	327.98	340.59	353.20	365.82	4.91
1.60	2.00	280.32	295.74	308.36	322.36	336.40	350.40	364.40	378.43	392.44	406.46	5.46
1.76	2.20	308.35	323.31	339.19	354.60	370.02	385.44	400.83	416.28	431.69	447.11	6.01
1.92	2.40	336.38	354.88	370.00	386.84	403.64	420.48	437.30	454.12	470.96	487.76	6.55
2.08	2.60	364.42	384.46	400.86	419.08	437.30	455.52	473.74	491.96	510.18	528.40	7.09
2.24	2.80	392.44	414.03	431.68	451.31	470.92	490.56	510.18	535.80	549.44	569.05	7.64
2.40	3.00	420.48	443.61	462.52	483.55	504.58	525.60	546.62	567.65	588.67	609.70	8.19
2.56	3.20	449.52	473.12	489.39	515.84	538.24	560.64	583.06	605.48	627.84	650.34	8.73
2.72	3.40	476.54	502.75	524.20	548.03	571.85	595.68	619.51	643.33	667.16	690.99	9.28
2.88	3.60	504.58	532.32	546.04	580.26	604.48	630.72	655.96	681.18	706.40	731.64	9.82
3.04	3.80	532.60	561.90	585.86	612.50	639.13	665.76	692.39	719.02	745.65	772.28	10.37
3.20	4.00	560.64	591.48	616.72	644.72	672.80	700.80	728.80	756.86	784.88	812.92	10.92
3.36	4.20	588.67	621.05	647.54	676.96	706.40	735.84	765.27	794.70	824.14	853.57	11.45
3.52	4.40	616.70	646.62	678.38	704.20	720.04	770.88	801.66	832.55	863.38	894.22	12.01
3.68	4.60	644.74	680.20	708.21	741.45	773.68	805.92	838.16	870.39	902.63	934.87	12.56
3.83	4.80	673.76	709.76	740.00	773.68	807.28	840.96	874.60	908.24	941.92	975.42	13.10
4.00	5.00	700.80	739.34	770.88	805.92	840.96	876.00	911.04	946.08	981.12	1016.16	13.65
4.16	5.20	728.82	768.92	801.72	838.16	874.60	911.04	947.48	983.92	1020.36	1056.80	14.20
4.32	5.40	756.86	798.49	832.54	870.39	908.24	946.08	983.92	1021.77	1059.61	1097.45	14.74
4.48	5.60	784.88	828.06	863.36	902.62	940.84	981.12	1020.36	1059.60	1098.88	1138.10	15.29
4.64	5.80	812.93	857.64	894.22	934.87	975.52	1016.16	1056.81	1098.45	1138.10	1178.75	15.84
4.80	6.00	840.96	887.23	925.04	967.10	1009.16	1051.20	1092.24	1135.30	1177.34	1219.39	16.38
4.96	6.20	868.99	916.79	955.90	999.34	1043.78	1086.24	1129.69	1173.10	1216.58	1260.04	16.92
5.12	6.40	899.04	946.24	978.78	1031.68	1076.48	1121.28	1166.12	1210.94	1255.68	1300.68	17.45
5.28	6.60	925.06	975.93	1017.56	1063.81	1110.06	1156.32	1202.56	1248.83	1295.08	1341.33	18.02
5.44	6.80	953.08	1005.50	1048.40	1096.05	1143.70	1191.36	1239.02	1289.99	1334.32	1381.98	18.57
5.60	7.00	981.12	1035.08	1079.22	1128.29	1177.34	1226.40	1275.46	1324.51	1373.57	1422.62	19.11
		20000	21000	22000	23000	24000	25000	26000	27000	28000	29000	
		HEAT LOSS IN Btu PER HOUR										

Table B-51 (Continued)

DOLLAR COST OF ENERGY PER YEAR

BASED ON 8760 HRS/YR

$ Per Million Btu	$ Per M lbs Steam	HEAT LOSS IN WATTS PER HOUR										$ Per Million Watts
		8790	9083	9376	9669	9962	10255	10548	10841	11134	11427	
0.16	0.20	$42.02	$43.45	$44.85	$46.25	$47.65	$49.06	$50.46	$51.86	$53.26	$54.66	0.57
0.32	0.40	84.10	86.90	89.60	92.50	95.30	98.11	100.92	103.72	106.52	109.32	1.09
0.48	0.60	126.14	130.35	135.54	138.76	142.96	147.17	151.38	155.58	159.78	163.99	1.64
0.64	0.80	168.20	173.80	179.20	185.00	190.60	196.22	201.84	207.44	213.04	218.64	2.18
0.80	1.00	210.24	217.25	224.26	231.26	238.27	245.28	252.29	259.30	266.30	273.31	2.73
0.96	1.20	252.28	260.70	271.08	277.51	285.92	294.34	302.76	311.16	319.56	327.28	3.27
1.12	1.40	294.34	304.15	313.96	323.78	333.58	343.39	353.20	363.01	372.82	382.64	3.82
1.28	1.60	336.40	347.60	358.40	370.00	371.20	392.44	403.68	414.88	426.08	437.38	4.37
1.44	1.80	378.41	391.05	403.66	416.28	428.89	441.50	454.10	466.73	479.34	491.96	4.91
1.60	2.00	420.48	434.50	448.52	462.52	476.54	490.56	504.58	518.59	532.60	546.62	5.46
1.76	2.20	462.53	477.95	493.36	508.78	524.20	539.62	555.04	570.45	585.87	601.28	6.01
1.92	2.40	504.56	521.40	542.16	555.02	571.84	588.68	605.52	622.32	639.02	654.56	6.55
2.08	2.60	546.62	564.84	583.06	601.29	619.50	637.73	655.95	674.17	692.38	710.61	7.09
2.24	2.80	588.68	608.30	627.92	647.56	667.16	686.78	706.40	726.02	745.64	765.28	7.64
2.40	3.00	630.72	651.74	672.67	693.79	714.82	735.84	756.86	777.88	798.93	819.94	8.19
2.56	3.20	672.80	695.19	716.80	740.00	762.40	784.88	807.36	829.76	852.16	874.56	8.73
2.72	3.40	704.82	738.64	762.47	786.30	810.12	833.95	857.78	881.60	905.43	929.26	9.28
2.88	3.60	756.82	782.10	807.32	832.55	857.78	883.01	908.20	833.46	958.68	983.92	9.82
3.04	3.80	798.91	825.54	852.17	878.80	905.43	932.06	958.70	985.32	1011.96	1038.59	10.37
3.20	4.00	840.96	869.00	897.04	925.04	952.08	981.12	1009.16	1037.18	1065.20	1093.24	10.92
3.36	4.20	883.00	912.44	940.18	971.30	1000.74	1030.18	1058.61	1089.04	1118.48	1147.91	11.45
3.52	4.40	925.06	955.89	986.72	1007.56	1048.40	1079.23	1110.08	1140.90	1171.64	1202.56	12.01
3.68	4.60	967.10	999.34	1031.18	1063.81	1096.05	1128.29	1160.52	1192.76	1225.00	1257.23	12.56
3.84	4.80	1009.12	1042.80	1084.32	1110.04	1143.68	1177.36	1211.04	1244.64	1278.04	1309.12	13.10
4.00	5.00	1050.60	1086.24	1121.28	1156.32	1191.36	1226.40	1261.44	1296.48	1331.58	1366.56	13.65
4.16	5.20	1093.24	1129.69	1166.12	1202.57	1239.00	1275.46	1311.90	1348.34	1384.76	1421.22	14.20
4.32	5.40	1135.30	1173.14	1210.98	1248.83	1286.66	1324.51	1362.36	1400.20	1438.04	1475.88	14.74
4.48	5.60	1177.36	1216.60	1255.84	1295.12	1334.32	1373.56	1412.80	1452.04	1491.28	1530.56	15.29
4.64	5.80	1219.40	1260.04	1300.68	1329.77	1381.98	1422.62	1463.27	1503.92	1544.56	1585.21	15.84
4.80	6.00	1261.44	1303.49	1345.34	1387.58	1429.64	1471.68	1513.72	1555.76	1597.86	1639.87	16.38
4.96	6.20	1303.49	1346.94	1390.38	1433.84	1477.28	1520.74	1564.18	1607.64	1651.08	1694.53	16.92
5.12	6.40	1345.60	1390.39	1433.60	1480.00	1524.80	1569.76	1614.72	1659.52	1704.32	1749.12	17.47
5.28	6.60	1387.58	1433.84	1480.10	1526.34	1572.59	1618.85	1665.10	1711.35	1757.60	1803.86	18.02
5.44	6.80	1409.64	1477.28	1524.94	1572.60	1620.24	1667.90	1715.56	1763.20	1810.86	1858.52	18.57
5.60	7.00	1471.68	1520.74	1569.79	1618.85	1667.90	1716.96	1766.01	1815.07	1864.12	1913.18	19.11
		30000	31000	32000	33000	34000	35000	36000	37000	38000	39000	
		HEAT LOSS IN Btu PER HOUR										

Table B-51 (Continued)

DOLLAR COST OF ENERGY PER YEAR

BASED ON 8760 HRS/YR

$ Per Million Btu	$ Per M lbs Steam	HEAT LOSS IN WATTS PER HOUR										$ Per Million Watts
		11720	12013	12306	12599	12892	13185	13478	13771	14064	14357	
0.16	0.20	$56.06	$57.46	$59.14	$60.27	$61.67	$63.07	$64.48	$65.88	$67.28	$68.68	0.57
0.32	0.40	112.12	114.92	118.28	120.54	122.34	126.14	128.96	131.75	134.56	137.35	1.09
0.48	0.60	168.20	172.40	177.44	180.81	185.00	189.22	193.42	197.63	201.81	206.04	1.64
0.64	0.80	224.24	229.84	236.56	241.08	244.68	252.28	257.92	263.50	269.12	274.70	2.18
0.80	1.00	280.32	287.33	295.74	301.34	308.36	315.36	322.36	329.38	336.40	343.34	2.73
0.96	1.20	336.38	344.80	354.88	361.12	370.00	378.41	386.84	395.25	403.64	412.07	3.27
1.12	1.40	392.44	402.25	414.03	421.88	431.68	441.50	451.31	461.13	470.92	480.75	3.82
1.28	1.60	449.52	459.68	473.12	482.16	493.36	504.56	515.84	527.00	538.24	549.40	4.37
1.44	1.80	504.58	517.19	532.32	542.41	555.03	567.65	580.26	592.88	605.49	618.11	4.91
1.60	2.00	560.64	574.66	591.48	602.68	616.12	630.17	644.72	658.75	672.80	686.68	5.46
1.76	2.20	616.70	632.12	648.62	662.96	678.54	693.79	709.20	724.63	740.05	755.46	6.01
1.92	2.40	673.76	689.60	709.76	722.24	740.00	756.82	773.68	790.50	807.32	824.14	6.55
2.08	2.60	728.82	747.05	768.92	783.49	801.72	819.94	838.16	856.38	874.60	892.82	7.09
2.24	2.80	784.88	804.50	828.06	843.76	863.36	882.01	902.62	922.25	941.86	961.50	7.64
2.40	3.00	840.96	861.98	887.23	904.03	925.04	946.08	967.10	988.13	1009.16	1030.18	8.19
2.56	3.20	899.04	919.36	946.24	964.32	978.78	1009.12	1031.68	1054.00	1076.48	1098.80	8.73
2.72	3.40	953.08	976.92	1005.50	1024.57	1048.40	1072.22	1096.05	1119.88	1143.70	1167.53	9.28
2.88	3.60	1009.16	1034.38	1064.64	1084.82	1110.06	1135.30	1160.52	1183.75	1211.92	1236.21	9.82
3.04	3.80	1065.20	1091.84	1123.80	1145.11	1171.72	1198.37	1225.00	1251.63	1278.25	1304.89	10.37
3.20	4.00	1121.28	1149.32	1182.96	1205.39	1233.44	1260.34	1289.44	1317.50	1345.60	1373.36	10.92
3.36	4.20	1177.34	1206.78	1242.10	1265.64	1295.08	1324.51	1353.94	1383.38	1412.81	1442.25	11.45
3.52	4.40	1233.40	1264.24	1293.24	1325.91	1356.76	1386.58	1418.40	1449.25	1480.09	1510.92	12.01
3.68	4.60	1289.48	1321.71	1360.39	1386.18	1416.42	1450.66	1482.89	1515.13	1547.37	1579.60	12.56
3.84	4.80	1347.52	1379.20	1419.52	1444.48	1480.00	1513.64	1547.36	1581.00	1614.64	1648.28	13.10
4.00	5.00	1401.60	1436.64	1478.69	1506.72	1541.76	1576.80	1611.84	1646.88	1681.92	1716.96	13.65
4.16	5.20	1457.64	1494.10	1537.84	1566.99	1603.44	1639.87	1676.32	1712.50	1749.20	1785.64	14.20
4.32	5.40	1513.72	1551.57	1596.98	1627.58	1665.08	1702.94	1740.79	1778.63	1816.47	1854.32	14.74
4.48	5.60	1569.76	1609.00	1656.12	1687.52	1726.72	1764.02	1805.24	1844.50	1883.72	1923.00	15.29
4.64	5.80	1625.86	1666.50	1715.28	1747.79	1788.44	1829.09	1869.74	1910.38	1951.03	1991.67	15.84
4.80	6.00	1681.92	1723.97	1774.46	1808.06	1850.08	1892.16	1934.20	1976.35	2018.32	2060.35	16.38
4.96	6.20	1737.98	1781.43	1833.57	1868.33	1911.80	1955.23	1998.68	2042.13	2085.58	2129.03	16.92
5.12	6.40	1798.08	1838.72	1892.48	1928.64	1957.56	2018.24	2063.36	2108.00	2152.96	2197.60	17.45
5.28	6.60	1850.12	1896.36	1951.86	1988.87	2035.12	2081.38	2127.62	2173.88	2220.13	2266.39	18.02
5.44	6.80	1906.16	1953.83	2011.00	2049.14	2096.80	2144.44	2192.10	2239.75	2286.40	2335.06	18.57
5.60	7.00	1962.24	2011.30	2070.16	2109.41	2158.44	2207.52	2256.58	2305.63	2354.69	2403.74	19.11
		40000	41000	42000	43000	44000	45000	46000	47000	48000	49000	
		HEAT LOSS IN Btu PER HOUR										

Table B-51 (Continued)

DOLLAR COST OF ENERGY PER YEAR

BASED ON 8760 HRS/YR

$ Per Million Btu	$ Per M lbs Steam	HEAT LOSS IN WATTS PER HOUR										$ Per Million Watts
		14650	14943	15236	15529	15822	16115	16408	16701	16994	17287	
0.16	0.20	$70.08	$71.48	$72.88	$74.29	$75.68	$77.09	$78.48	$79.89	$81.30	$82.69	0.57
0.32	0.40	140.16	142.96	145.76	148.60	151.36	154.18	156.96	159.78	162.60	165.39	1.09
0.48	0.60	210.24	214.44	218.65	222.85	227.06	231.26	235.48	239.67	243.88	248.08	1.64
0.64	0.80	280.32	285.92	291.53	297.13	302.72	308.35	313.92	319.56	323.20	330.78	2.18
0.80	1.00	350.40	357.41	364.40	371.42	398.43	385.44	392.44	399.46	406.46	413.47	2.73
0.96	1.20	420.48	428.88	437.30	445.70	454.12	462.53	470.96	479.34	487.76	496.17	3.27
1.12	1.40	490.56	500.37	510.18	519.99	529.80	539.62	549.44	559.23	569.05	578.86	3.82
1.28	1.60	560.64	570.84	583.06	594.28	605.48	616.70	627.84	639.13	646.40	661.55	4.37
1.44	1.80	630.72	643.34	655.96	668.56	681.18	693.79	706.40	719.02	731.64	744.25	4.91
1.60	2.00	700.80	714.82	728.80	742.85	756.86	770.88	784.88	798.91	812.92	826.94	5.46
1.76	2.20	770.88	786.30	801.66	817.13	832.55	847.97	863.38	878.80	894.22	909.64	6.01
1.92	2.40	840.96	857.76	874.60	891.42	908.24	925.06	941.92	958.69	975.42	992.33	6.55
2.08	2.60	911.04	929.26	947.48	965.70	983.92	1002.13	1020.36	1038.58	1056.80	1075.03	7.09
2.24	2.80	981.12	1000.74	1020.36	1039.98	1059.60	1079.23	1098.88	1118.48	1138.10	1157.72	7.64
2.40	3.00	1051.20	1072.22	1092.24	1114.27	1135.30	1156.32	1177.34	1198.37	1219.39	1240.42	8.19
2.56	3.20	1121.28	1141.68	1166.12	1188.56	1210.96	1233.40	1255.68	1278.26	1300.68	1323.11	8.73
2.72	3.40	1191.36	1206.25	1239.02	1262.84	1289.99	1310.50	1334.32	1358.15	1381.98	1405.80	9.28
2.88	3.60	1261.44	1286.68	1311.92	1337.13	1362.36	1387.58	1412.89	1438.04	1463.28	1488.49	9.82
3.04	3.80	1331.52	1358.15	1384.78	1411.41	1438.04	1464.67	1491.30	1517.93	1544.56	1571.19	10.37
3.20	4.00	1401.60	1429.64	1457.60	1485.69	1513.72	1541.76	1569.76	1597.82	1625.84	1653.89	10.92
3.36	4.20	1471.68	1501.11	1530.54	1559.98	1589.40	1618.84	1648.28	1677.71	1707.15	1736.58	11.45
3.52	4.40	1541.76	1572.59	1603.32	1634.26	1665.10	1695.94	1726.76	1757.61	1788.44	1819.27	12.01
3.68	4.60	1611.84	1644.07	1676.32	1708.55	1740.78	1773.02	1803.26	1837.50	1869.74	1901.97	12.56
3.84	4.80	1681.92	1715.55	1749.20	1782.83	1816.48	1850.11	1883.84	1917.39	1950.84	1984.67	13.10
4.00	5.00	1752.00	1787.04	1822.08	1857.12	1892.16	1927.20	1962.24	1997.28	2030.32	2067.36	13.65
4.16	5.20	1822.08	1858.52	1894.96	1931.40	1977.84	2004.29	2040.72	2077.17	2113.60	2150.05	14.20
4.32	5.40	1892.16	1930.00	1967.84	2005.68	2043.59	2081.38	2119.22	2157.06	2194.90	2232.74	14.74
4.48	5.60	1962.24	2001.48	2040.72	2079.97	2118.20	2158.46	2197.76	2236.95	2276.20	2315.44	15.29
4.64	5.80	2032.32	2072.96	2113.60	2154.26	2194.90	2235.55	2276.20	2316.83	2357.50	2398.14	15.84
4.80	6.00	2102.40	2144.45	2184.48	2228.54	2270.60	2312.64	2354.68	2396.74	2438.78	2480.83	16.38
4.96	6.20	2172.48	2215.92	2259.38	2302.83	2346.20	2389.72	2433.16	2476.63	2520.08	2563.52	16.92
5.12	6.40	2242.56	2283.36	2332.24	2377.11	2421.88	2466.82	2511.36	2556.51	2585.60	2646.22	17.47
5.28	6.60	2312.64	2358.89	2405.12	2451.39	2497.65	2543.90	2580.16	2636.41	2682.66	2728.91	18.02
5.44	6.80	2382.72	2430.37	2478.04	2525.68	2579.98	2620.99	2668.64	2716.30	2763.96	2811.61	18.57
5.60	7.00	2452.80	2501.86	2550.92	2599.97	2649.02	2698.08	2747.14	2796.19	2845.24	2894.30	19.11
		50000	51000	52000	53000	54000	55000	56000	57000	58000	59000	
		HEAT LOSS IN Btu PER HOUR										

Table B-51 (Continued)

DOLLAR COST OF ENERGY PER YEAR

BASED ON 8760 HRS/YR

$ Per Million Btu	$ Per M lbs Steam	HEAT LOSS IN WATTS PER HOUR 17580	17873	18166	18459	18752	19045	19338	19631	19924	20217	$ Per Million Watts
0.16	0.20	$84.10	$85.50	$86.90	$88.30	$89.60	$91.10	$92.50	$93.91	$95.30	$96.71	0.57
0.32	0.40	168.20	170.99	173.80	176.80	179.20	182.21	185.00	187.81	190.60	193.42	1.09
0.48	0.60	252.28	256.49	260.70	264.90	271.08	273.31	277.51	281.72	285.92	290.13	1.64
0.64	0.80	336.40	341.99	347.60	353.20	358.40	364.42	370.00	375.63	381.20	389.84	2.18
0.80	1.00	420.48	427.48	434.50	441.50	448.52	455.52	462.52	469.54	476.54	483.55	2.73
0.96	1.20	504.56	512.99	521.40	529.80	542.16	546.62	555.02	563.44	571.84	580.26	3.27
1.12	1.40	588.68	598.48	608.30	618.11	627.92	637.73	647.56	657.35	667.16	676.97	3.82
1.28	1.60	672.80	683.98	695.19	706.41	716.80	728.83	740.00	751.26	762.40	773.68	4.37
1.44	1.80	756.82	769.48	782.10	794.71	807.32	819.94	832.55	845.16	857.78	870.39	4.91
1.60	2.00	840.96	854.98	869.00	883.01	897.04	911.04	925.04	939.07	952.08	967.10	5.46
1.76	2.20	925.06	940.47	955.89	971.31	986.72	1002.14	1017.56	1032.98	1048.40	1063.81	6.01
1.92	2.40	1009.12	1025.97	1042.80	1059.60	1084.32	1093.25	1110.04	1126.89	1143.68	1160.52	6.55
2.08	2.60	1093.24	1111.47	1129.69	1147.91	1160.12	1184.35	1202.57	1220.79	1239.00	1257.23	7.09
2.24	2.80	1177.36	1196.97	1216.60	1236.21	1255.84	1275.46	1295.12	1314.70	1334.32	1353.94	7.64
2.40	3.00	1261.44	1282.46	1303.49	1324.51	1345.34	1366.56	1387.58	1408.61	1429.64	1450.66	8.19
2.56	3.20	1345.60	1367.96	1390.39	1412.81	1433.60	1457.66	1480.00	1502.52	1524.80	1547.37	8.73
2.72	3.40	1429.64	1453.45	1477.28	1501.11	1524.94	1548.77	1572.60	1596.42	1620.24	1644.07	9.28
2.88	3.60	1513.64	1538.96	1564.20	1589.41	1614.64	1639.87	1665.10	1690.32	1715.56	1740.78	9.82
3.04	3.80	1597.82	1624.45	1651.08	1677.72	1704.34	1730.98	1757.60	1784.24	1810.86	1837.50	10.37
3.20	4.00	1681.92	1709.95	1738.00	1766.02	1794.08	1822.08	1850.08	1878.14	1904.16	1934.21	10.92
3.36	4.20	1766.00	1795.45	1824.88	1854.32	1880.36	1913.18	1942.60	1972.05	2001.48	2030.92	11.45
3.52	4.40	1850.12	1880.94	1911.78	1942.62	1973.44	2004.29	2015.12	2065.95	2096.80	2127.63	12.01
3.68	4.60	1934.20	1966.44	1998.68	2030.92	2062.36	2095.39	2127.63	2159.89	2192.10	2224.34	12.56
3.84	4.80	2018.24	2051.94	2085.60	2119.23	2168.64	2186.50	2220.08	2253.77	2287.36	2321.05	13.10
4.00	5.00	2101.20	2137.44	2172.48	2207.52	2242.56	2277.60	2312.64	2347.68	2382.72	2417.76	13.65
4.16	5.20	2186.48	2222.94	2259.38	2295.82	2332.24	2368.70	2405.14	2441.59	2478.00	2514.47	14.20
4.32	5.40	2270.60	2308.43	2346.28	2384.12	2421.96	2459.81	2497.65	2535.49	2573.32	2611.18	14.74
4.48	5.60	2354.72	2393.93	2433.20	2472.42	2511.68	550.91	2590.24	2629.40	2668.64	2707.89	15.29
4.64	5.80	2438.80	2479.43	2520.08	2560.72	2601.36	2642.02	2682.66	2723.31	2763.96	2804.60	15.84
4.80	6.00	2522.88	2564.93	2606.98	2649.02	2690.68	2733.12	2775.16	2817.22	2859.28	2901.31	16.38
4.96	6.20	2606.98	2650.43	2693.88	2737.32	3780.76	2824.22	2867.67	2911.12	2954.56	2998.02	16.92
5.12	6.40	2691.20	2735.92	2780.78	2825.62	2867.20	2915.33	2960.00	3005.03	3049.60	3094.73	17.45
5.28	6.60	2775.16	2821.42	2867.67	2913.93	2960.20	3006.43	3052.64	3098.94	3145.18	3191.44	18.02
5.44	6.80	2819.28	2906.92	2954.56	3002.23	3049.88	3097.54	3145.20	3192.84	3240.48	3288.15	18.57
5.60	7.00	2943.36	2992.42	3041.47	3090.52	3139.58	3188.64	3237.70	3286.75	3335.80	3384.86	19.11
		60000	61000	62000	63000	64000	65000	66000	67000	68000	69000	
		HEAT LOSS IN Btu PER HOUR										

Table B-51 (Continued)

DOLLAR COST OF ENERGY PER YEAR

BASED ON 8760 HRS/YR

$ Per Million Btu	$ Per M lbs Steam	HEAT LOSS IN WATTS PER HOUR										$ Per Million Watts
		20510	20803	21096	21389	21682	21975	22268	22561	22854	23147	
0.16	0.20	$98.11	$99.51	$100.92	$102.32	$103.72	$105.12	$106.52	$107.92	$109.32	$110.73	0.57
0.32	0.40	196.22	199.03	201.84	204.64	207.44	210.24	213.04	215.85	219.64	221.45	1.09
0.48	0.60	294.34	298.54	302.76	306.95	311.16	315.36	319.56	323.77	327.98	332.18	1.64
0.64	0.80	392.44	398.05	403.68	409.27	414.88	420.48	426.08	431.69	437.28	442.91	2.18
0.80	1.00	490.56	497.57	504.58	511.58	518.59	525.60	532.60	539.62	546.62	553.63	2.73
0.96	1.20	588.68	597.08	605.52	613.90	622.32	630.72	639.02	647.54	654.56	664.36	3.27
1.12	1.40	686.78	696.60	706.40	716.22	726.02	735.84	745.64	755.46	765.28	775.08	3.82
1.28	1.60	784.88	796.11	807.36	818.53	829.76	840.96	852.16	863.39	874.56	885.81	4.37
1.44	1.80	883.01	895.62	908.20	920.85	933.46	946.08	958.60	971.31	983.92	996.54	4.91
1.60	2.00	981.12	995.14	1009.16	1023.17	1037.18	1051.20	1065.20	1079.23	1093.24	1107.26	5.46
1.76	2.20	1079.23	1094.65	1110.08	1125.48	1140.90	1156.32	1171.64	1187.16	1202.56	1217.99	6.01
1.92	2.40	1177.36	1194.16	1211.04	1227.80	1244.64	1261.44	1278.04	1295.08	1309.12	1328.72	6.55
2.08	2.60	1275.46	1293.68	1311.90	1330.12	1348.34	1366.56	1384.76	1403.00	1421.22	1439.44	7.09
2.24	2.80	1373.56	1393.19	1412.80	1432.43	1452.04	1471.68	1491.28	1510.92	1530.56	1550.17	7.64
2.40	3.00	1471.68	1492.70	1513.72	1534.72	1555.76	1576.80	1597.86	1618.85	1639.87	1660.90	8.19
2.56	3.20	1569.76	1592.22	1614.72	1637.07	1659.52	1681.92	1704.32	1725.77	1749.12	1771.62	8.73
2.72	3.40	1667.90	1691.73	1715.56	1739.39	1763.20	1787.04	1810.86	1834.69	1858.52	1882.35	9.28
2.88	3.60	1766.02	1791.24	1816.40	1841.70	1866.92	1892.16	1917.36	1942.62	1967.84	1993.08	9.82
3.04	3.80	1864.12	1890.76	1917.40	1944.02	1970.65	1997.28	2023.92	2050.54	2077.17	2103.80	10.37
3.20	4.00	1962.24	1990.27	2018.32	2046.34	2074.36	2102.40	2130.40	2158.46	2186.48	2214.53	10.92
3.36	4.20	2060.36	2089.79	2117.22	2148.65	2178.08	2207.52	2236.96	2266.39	2295.82	2325.25	11.45
3.52	4.40	2158.46	2189.30	2220.16	2250.97	2281.80	2312.64	2343.28	2374.31	2405.12	2435.98	12.01
3.68	4.60	2256.58	2288.81	2321.04	2353.29	2385.52	2417.76	2450.00	2482.23	2514.47	2546.70	12.56
3.84	4.80	2354.72	2388.32	2422.008	2455.60	2489.28	2522.88	2556.08	2590.16	2618.24	2657.43	13.10
4.00	5.00	2452.80	2487.84	2522.88	2557.92	2592.96	2628.00	2663.16	2698.08	2733.12	2768.16	13.65
4.16	5.20	2550.92	2587.35	2623.80	2660.24	2696.68	2733.12	2769.52	2806.00	2842.44	2878.88	14.20
4.32	5.40	2648.02	2686.87	2724.72	2762.55	2800.40	2832.24	2876.08	2913.93	2951.77	2989.61	14.74
4.48	5.60	2747.12	2786.38	2825.60	2864.87	2904.08	2943.36	2982.56	3021.85	3061.12	3100.34	15.29
4.64	5.80	2845.25	2885.89	2926.54	2967.19	3007.84	3048.48	3089.12	3129.77	3170.42	3211.06	15.84
4.80	6.00	2943.36	2985.41	3027.44	3069.50	3111.52	3153.60	3195.72	3237.70	3279.74	3321.79	16.38
4.96	6.20	3040.47	3083.92	3128.36	3171.82	3215.27	3258.72	3302.16	3345.62	3389.07	3432.52	16.92
5.12	6.40	3139.52	3184.43	3229.52	3274.14	3319.04	3363.84	3408.64	3453.54	3498.24	3543.24	17.47
5.28	6.60	3237.70	3283.95	3330.20	3376.45	3422.70	3468.96	3515.20	3561.47	3607.72	3653.97	18.02
5.44	6.80	3335.80	3383.46	3431.12	3478.77	3526.40	3574.08	3621.72	3669.39	3717.04	3764.70	18.57
5.60	7.00	3433.92	3482.98	3532.02	3581.09	3630.14	3679.20	3724.24	3777.31	3826.37	3875.42	19.11
		70000	71000	72000	73000	74000	75000	76000	77000	78000	79000	
		HEAT LOSS IN Btu PER HOUR										

Table B-51 (Continued)

DOLLAR COST OF ENERGY PER YEAR

BASED ON 8760 HRS/YR

$ Per Million Btu	$ Per M lbs Steam	HEAT LOSS IN WATTS PER HOUR										$ Per Million Watts
		23440	23733	24026	24319	24612	24905	25198	25491	25784	26077	
0.16	0.20	$112.12	$113.53	$114.92	$116.33	$118.28	$119.14	$120.54	$121.94	$123.34	$124.74	0.57
0.32	0.40	224.24	227.06	229.84	232.67	235.47	238.27	241.08	243.87	246.68	249.48	1.09
0.48	0.60	336.38	340.59	344.80	349.00	354.88	357.41	361.63	364.50	370.00	374.23	1.64
0.64	0.80	449.52	454.12	459.68	465.33	473.12	476.54	482.16	487.76	493.26	498.97	2.18
0.80	1.00	560.64	567.65	574.66	581.66	591.48	595.68	602.68	609.70	616.12	623.71	2.73
0.96	1.20	673.76	681.18	689.60	698.00	706.40	714.82	722.24	731.64	740.00	748.45	3.27
1.12	1.40	784.88	794.71	804.50	814.33	824.14	833.95	843.76	853.57	863.36	873.20	3.82
1.28	1.60	899.04	908.24	919.36	930.66	940.24	953.09	964.32	975.51	986.52	997.94	4.37
1.44	1.80	1009.16	1021.77	1034.38	1047.00	1064.64	1072.22	1084.82	1097.45	1110.07	1122.68	4.91
1.60	2.00	1121.28	1135.30	1149.32	1163.33	1182.96	1191.36	1205.39	1219.39	1233.44	1247.42	5.46
1.76	2.20	1233.40	1248.83	1264.24	1279.66	1293.24	1310.96	1325.91	1341.33	1356.76	1372.16	6.01
1.92	2.40	1347.52	1362.35	1379.20	1395.99	1412.87	1429.63	1444.48	1463.27	1480.00	1496.91	6.55
2.08	2.60	1457.64	1475.88	1494.10	1512.32	1530.55	1548.77	1566.99	1585.21	1603.44	1621.65	7.09
2.24	2.80	1569.76	1589.41	1609.00	1628.66	1648.28	16667.90	1687.52	1707.15	1726.72	1746.39	7.64
2.40	3.00	1681.92	1702.94	1723.97	1744.99	1766.02	1787.04	1808.06	1829.09	1850.08	1871.14	8.19
2.56	3.20	1798.08	1816.47	1838.72	1861.32	1883.75	1906.18	1928.64	1951.03	1973.04	1995.88	8.73
2.72	3.40	1906.16	1930.00	1953.83	1977.66	2001.49	2025.31	2049.14	2072.97	2096.80	2120.62	9.28
2.88	3.60	2018.32	2043.53	2068.76	2093.99	2129.28	2144.45	2169.64	2194.91	2220.13	2245.36	9.82
3.04	3.80	2130.40	2157.06	2183.68	2210.32	2247.60	2263.58	2290.22	2316.84	2343.44	2370.11	10.37
3.20	4.00	2242.56	2270.59	2298.64	2326.66	2365.92	2382.72	2410.78	2438.78	2466.88	2494.85	10.92
3.36	4.20	2354.68	2384.12	2413.56	2442.99	2484.20	2501.86	2531.29	2560.72	2590.16	2619.59	11.45
3.52	4.40	2466.80	2497.65	2528.48	2559.32	2586.48	2621.92	2651.82	2682.66	2713.52	2744.33	12.01
3.68	4.60	2578.96	2611.18	2643.42	2675.65	2707.89	2740.13	2772.36	2804.60	2832.84	2869.07	12.56
3.84	4.80	2695.04	2724.71	2758.40	2791.99	2825.62	2859.26	2888.96	2926.54	2960.00	2993.82	13.10
4.00	5.00	2803.20	2838.24	2873.28	2908.32	2959.38	2978.40	3013.44	3048.48	3083.52	3118.56	13.65
4.16	5.20	2915.28	1951.77	2988.20	3024.65	3061.10	3097.53	3133.98	3170.42	3206.88	3243.30	14.20
4.32	5.40	3027.44	3065.30	3103.14	3140.98	3178.83	3216.67	3255.15	3292.36	3330.16	3368.04	14.74
4.48	5.60	3139.52	3178.83	3218.00	3257.32	3297.56	3335.81	3375.04	3414.30	3453.44	3492.79	15.29
4.64	5.80	3251.72	3292.35	3333.00	3373.65	3414.30	3454.94	3495.59	3536.24	3576.88	3617.53	15.84
4.80	6.00	3363.84	3405.89	3447.94	3489.98	3532.03	3574.08	3616.12	3658.18	3700.16	3742.27	16.38
4.96	6.20	3475.96	3519.42	3562.87	3606.32	3649.77	3693.22	3736.66	3780.12	3823.60	3867.01	16.92
5.12	6.40	3596.16	3632.95	3677.44	3722.65	3767.50	3812.35	3857.28	3902.05	3946.90	3991.76	17.45
5.28	6.60	3700.24	3746.47	3792.73	3838.98	3885.24	3931.49	3977.74	4023.99	4070.24	4116.50	18.02
5.44	6.80	3812.32	3860.00	3907.66	3955.32	4002.97	4050.62	4098.28	4145.93	4193.60	4241.24	18.57
5.60	7.00	3924.48	3973.54	4022.59	4071.65	4120.70	4169.76	4218.82	4267.87	4316.88	4365.98	19.11
		80000	81000	82000	83000	84000	85000	86000	87000	88000	89000	
		HEAT LOSS IN Btu PER HOUR										

Table B-51 (Continued)

DOLLAR COST OF ENERGY PER YEAR

BASED ON 8760 HRS/YR

$ Per Million Btu	$ Per M lbs Steam	HEAT LOSS IN WATTS PER HOUR										$ Per Million Watts
		26370	26663	26956	27248	27542	27835	28128	28421	28714	29007	
0.16	0.20	$126.14	$127.55	$128.96	$130.35	$131.75	$133.15	$134.56	$135.96	$137.35	$138.76	0.57
0.32	0.40	252.28	255.09	257.92	260.70	263.50	266.30	269.12	271.91	274.70	277.52	1.09
0.48	0.60	378.41	382.64	386.84	391.05	395.25	399.46	403.64	407.87	412.07	416.28	1.64
0.64	0.80	504.56	510.18	515.84	521.39	527.00	532.61	538.24	543.82	549.40	555.03	2.18
0.80	1.00	630.72	637.73	644.72	651.74	658.75	665.76	672.80	679.78	686.68	693.79	2.73
0.96	1.20	756.82	765.27	773.68	782.09	790.50	798.91	807.60	815.73	824.14	832.55	3.27
1.12	1.40	882.01	892.82	902.62	912.44	922.25	932.06	941.84	951.68	961.50	971.30	3.82
1.28	1.60	1009.12	1020.37	1031.68	1042.79	1054.00	1065.22	1076.48	1087.64	1098.80	1110.07	4.37
1.44	1.80	1135.30	1147.91	1160.52	1173.14	1183.75	1198.37	1211.98	1223.60	1236.21	1248.83	4.91
1.60	2.00	1260.34	1275.46	1289.44	1303.49	1317.50	1331.52	1345.60	1359.55	1373.36	1387.58	5.46
1.76	2.20	1386.58	1403.00	1418.40	1433.84	1449.25	1464.67	1480.09	1495.51	1510.92	1526.34	6.01
1.92	2.40	1513.64	1530.55	1547.36	1564.19	1581.00	1597.82	1614.64	1631.46	1648.28	1665.10	6.55
2.08	2.60	1639.87	1658.09	1676.32	1694.53	1712.75	1730.97	1749.20	1767.42	1785.64	1803.86	7.09
2.24	2.80	1764.02	1785.64	1805.24	1824.88	1844.50	1864.13	1883.72	1903.37	1923.00	1942.62	7.64
2.40	3.00	1892.16	1913.18	1934.20	1955.23	1976.35	1997.28	2008.32	2039.33	2060.35	2081.38	8.19
2.56	3.20	2018.24	2040.73	2063.36	2085.58	2108.00	2130.43	2152.96	2175.28	2197.60	2220.13	8.73
2.72	3.40	2144.44	2168.27	2192.10	2215.92	2239.75	2263.58	2286.40	2311.24	2335.06	2358.89	9.28
2.88	3.60	2270.60	2295.82	2321.04	2346.28	2367.50	2396.74	2423.84	2447.19	2472.42	2497.65	9.82
3.04	3.80	2396.74	2423.37	2450.00	2476.63	2503.26	2529.89	2556.50	2583.15	2609.78	2636.40	10.37
3.20	4.00	2520.68	2550.91	2578.88	2606.98	2635.00	2663.04	2690.20	2719.10	2746.72	2775.16	10.92
3.36	4.20	2649.02	2678.46	2707.88	2737.32	2766.76	2796.19	2825.62	2855.06	2884.50	2913.92	11.45
3.52	4.40	2773.16	2806.00	2836.80	2867.67	2898.50	2929.34	2960.18	2991.01	3021.84	3052.68	12.01
3.68	4.60	2900.21	2933.55	2965.78	2998.02	3030.26	3062.49	3094.74	3126.96	3159.20	3191.44	12.56
3.84	4.80	3027.28	3061.09	3094.72	3128.37	3162.00	3195.65	3229.28	3262.92	3296.56	3330.20	13.10
4.00	5.00	3153.60	3188.64	3223.68	3258.72	3293.76	3328.80	3363.84	3398.88	3433.92	3468.96	13.65
4.16	5.20	3279.74	3316.19	3352.64	3389.07	3425.00	3461.95	3498.40	3534.84	3571.28	3607.72	14.20
4.32	5.40	3404.89	3443.73	3481.58	3519.42	3557.26	3595.10	3632.95	3670.79	3708.64	3746.48	14.74
4.48	5.60	3528.04	3571.28	3610.48	3649.77	3689.00	3728.26	3767.44	3806.74	3826.00	3885.24	15.29
4.64	5.80	3658.18	3698.82	3739.48	3780.11	3820.76	3861.41	3902.06	3942.70	3983.34	4023.99	15.84
4.80	6.00	3784.32	3826.37	3868.40	3910.46	3952.70	3994.56	4036.64	4078.66	4120.70	4162.75	16.38
4.96	6.20	3910.46	3953.91	3997.36	4040.81	4084.26	4127.71	4171.16	4214.61	4258.06	4301.51	16.92
5.12	6.40	4036.48	4081.46	4126.72	4171.16	4216.00	4260.86	4305.92	4350.56	4395.20	4440.27	17.47
5.28	6.60	4162.75	4209.00	4255.24	4301.51	4347.76	4394.02	4440.26	4486.52	4532.78	4579.03	18.02
5.44	6.80	4288.88	4336.55	4384.20	4431.86	4479.50	4527.17	4572.80	4622.47	4670.12	4717.78	18.57
5.60	7.00	4415.04	4464.10	4513.16	4562.21	4611.26	4660.32	4709.38	4758.43	4807.49	4856.54	19.11
		90000	91000	92000	93000	94000	95000	96000	97000	98000	99000	
		HEAT LOSS IN Btu PER HOUR										

Table B-52 Water vapor transmission units

English and Metric Units

Water Vapor Transmission

Water Vapor Transmission

$$= \frac{\text{Weight Change in Grains}}{\text{Time x Area}}$$

In English Units:

In Unit Time
Water Vapor Transmission, per Hour

$$= \frac{\text{Weight Change in Grains per Hour}}{\text{Area in Square Feet}}$$

In Unit Time, Unit Area

WVT = Rate of water vapor transmission in number of grains per hour per square foot, or

= Grains per hour square foot

$$= \frac{G}{\text{Hr, Sq Ft}}$$

When G = Grains

In Metric Units:

Water vapor transmission is expressed in unit time of 24 hours. Thus:

WVT = Rate of vapor transmission in number of grams per hour per square meter

= Grams per 24 hours square meter

$$= \frac{g}{\text{24 hrs, sq m}}$$

When g = Grams

Conversion Between the Two Systems:

Grams per 24 hours, sq m x 0.0598
= Grains per hour, sq ft.

Grains per hour, sq ft x 16.7
= Grams per 24 hour, sq m

Permeance of a Material

Permeance

$$= \frac{\text{Water Vapor Transmission}}{\text{Vapor Pressure Difference Between the Two Sides of the Material}}$$

$$= \frac{WVT}{\Delta P}$$

When ΔP = Vapor pressure difference between the two sides of the material

In English Units:

$$\text{Permeance (Perms)} = \frac{G}{\Delta P, \text{Hr, Sq Ft}}$$

In Metric Units:

$$\text{Permeance (Metric Perms)} = \frac{g}{\Delta p, \text{Hr, Sq Ft}}$$

When Δp = Vapor pressure difference between the two sides of the material

Conversion Between the Two Systems:

Metric Perms x 1.52 = Perms
Perms x 0.659 = Metric Perms

Permeability of a Material

For thick materials permeability is given in perm-inches for *English Units* and in Metric Perm-Centimeters for *Metric Units*

Conversion Between the Two Systems:

Metric Perm-Centimeters x 0.598 = Perm-inches
Perm-inches x 1.67 = Metric Perm Centimeters

Note: Permeability and Permeance as used in this Manual do not follow the pattern used in defining other terms in the Manual ending in "ance" and "ility," but follow the definitions as given by ASTM.

Units of Measure: In this book the commonly used term English Unit has been used to designate the unit of measure now used in the United States, although the correct designation, very seldom used, is British Units. Thus English Units and British Units are the same. Presently an International System of Units (S.I.) is being adopted throughout the world. This is based on Metric Units with some very minor changes. Therefore, S.I. Units and Metric Units are the same. As the term S.I. Units is not well known in the United States at this time, the term Metric Units was used in this book.

Thus: English Units are the same as British Units.
Metric Units are the same as S.I. Units.

Table B-53 English to Metric to English Conversion Table

Physical or Thermal Property	English System Quantity	English System Units Symbol	Conversion Factors Multiply by →	Metric System Quantity	Metric System Units Symbol	Conversion Factor Multiply by →	English System Quantity	English System Units Symbol
Heat	Calorie	cal	4.1868	joule	J	0.238846	Calorie	cal
Heat	British Thermal Unit	Btu	1055.056	joule	J	0.0009478	British Thermal Unit	Btu
Heat	therm (100,000 Btu)	Btu x 10^5	105505600	joule	J	0.00000000948	therm	Btu x 10^5
Heat Capacity	Btu per degree F	Btu/°F	1899.0	joule per degree C	J/°C	0.00052659	Btu per degree F	Btu/°F
Heat Enthalpy	Btu per pound	Btu/lb	2326.0	joule per kilogram	J/kg	0.0004299	Btu per pound	Btu/lb
Specific Heat	Btu per pound per °F	Btu/lb, °F	4187.0	joule per kilogram °C	J/kg°C	0.000238846	Btu per pound °F	Btu/lb°F
Entropy	Btu per degree Rankine	Btu/°R	1899.0	joule per degree K	J/°K	0.00052659	Btu per °R	Btu/°R
Specific Entropy	Btu per pound per °R	Btu lb/°R	4187.0	joule per kilogram °K	J/kg°K	0.000238846	Btu per pound °R	Btu lb/°R
Latent Heat	Btu per pound	Btu/lb	2326.0	joule per kilogram	J/kg	0.0004299	Btu per pound	Btu/lb
Vol. Heat	Btu per cubic foot	Btu/ft^3	37260.0	joule per cubic metre	J/m^3	0.00002684	Btu per ft^3	Btu/ft^3
Spec. Heat, Vol.	Btu per cubic foot °F	Btu/ft^3, °F	67010.0	joule per cubic metre °C	J/m^{3}C	0.000014923	Btu per ft^3, °F	Btu/ft^3, °F
Spec. Heat, Wt.	Btu per pound per °F	Btu/lb, °F	1.163	watt hour per kilogram K	W, hr/kgK	0.859845	Btu per pound °F	Btu/lb, °F
Spec. Heat, Wt.	Btu per pound per °F	Btu/lb, °F	4186.8	joule per kilogram kelvin	J/kgK	0.0002388	Btu per pound °F	Btu/lb, °F
Heat Flow	Calorie per second	cal/s	4.184	watt	W	0.239006	calorie/sec	cal/s
Heat Flow	Calorie per minute	cal/min	0.06973	watt	W	14.34103	calorie/minute	cal/min
Heat Flow	Btu per second	Btu/s	1055.056	watt	W	0.0009478	Btu per second	Btu/s
Heat Flow	Btu per minute	Btu/min	17.58426	watt	W	0.056869	Btu per minute	Btu/min
Heat Flow	Btu per hour	Btu/hr	0.2930711	watt	W	3.141214	Btu per hour	Btu/hr
Heat/area, time	Btu per sq foot per hr	Btu/ft^2, hr	3.152481	watts per sq metre	W/m^2	0.31721	Btu per sq ft per hr	Btu/ft^2, hr
Heat/lin, time	Btu per lin foot per hr	Btu/lin, ft, hr	0.9609097	watts per lin metre	W/lin, m	1.04068	Btu per lin ft per hr	Btu/lin, ft, hr
Heat/area, time °	Btu per sq ft per hr °F	Btu/ft^2, hr°F	5.67826	watts per sq m per °K	W/m^{2}K	0.17611	Btu ft^2 per hr °F	Btu/ft^2, hr°F
Conductance	Btu per sq ft per hour °F	Btu/ft^{2}hr°F	5.678	watts per sq m per °K	W/m^{2}K	0.17611835	Btu per sq ft per °F	Btu/ft^{2}hr°F
Conductivity	Btu per inch/sq ft, hr °F	Btu in./ft^2 hr°F	0.1442	watts m per sq m per °K	W/mK	6.9348	Btu in. per ft^2, hr°F	Btu in./ft^{2}hr°F
Conductivity	Btu per ft/sq ft, hr°F	Btu/ft/ft^2, hr°F	1.731	watts m per sq m per °K	W/mK	0.5777	Btu ft per ft^2, hr°F	Btu ft/ft^2 hr°F
Resistance	Degree F, sq ft hr/Btu	°F, ft^2, hr/Btu	0.1761	°K, metre^2/watt	K m^{2}W	5.6786	Degree, F, sq ft hr/Btu	°F, ft^2, hr/Btu
MOISTURE								
Content	Grains per pound		1.428	gram per kilogram	g/kg	0.70028	Grains per pound	
Vapor trans.	Perm-inch		1.8	Perm-centimetre		0.555555	Perm-inch	
Permeability	Pound foot/hr pound force		0.00000862	Kilogram, m/Newton sec		116.009.8	lb, ft/hr, lb	
Perm (23°C)	Grain/ft^2, hr, inch of Hg		5.74525 E-11*	Kilogram/pascal, sec sq m		17406 E + 10	Grain/ft^2, hr, inch of Hg	
Perm-in (23°C)	Grain/ft^2, hr, in. of Hg, in.		1.45929 E-12*	Kilogram/pascal, sec metre		.68526 E + 11	Grain/ft^2, hr, inch Hg, inch	
Perm, British	English Units		0.659	Metric Per m		1.517	Per m, British	

Note: This table of conversions from English Units to Metric Units and back to English Units is not a complete English-Metric conversion table. The units listed are those units which are likely to be used in the thermal insulation technology. Complete tabulation is available as ASTM Standard for Metric Practice E380-76.

*As defined by ASTM. Test methods E-11 and E-12.

Table B-53 English to Metric to English Conversion Table (Continued)

Physical or Thermal Property	English System Quantity	English System Units Symbol	Conversion Factors Multiply by →	Metric System Quantity	Metric System Units Symbol	Conversion Factor Multiply by →	English System Quantity	English System Units Symbol
Mass/Length	pound, per inch	lb/in.	17.85797	kilogram/metre	kg/m	0.055997	pound per inch	lb/in
Mass/Length	pound per foot	lb/ft	1.488164	kilogram/metre	kg/m	0.671968	pound per foot	lb/ft
Mass/area	ounce per sq foot	oz/ft^2	0.3051517	kilogram per metre^2	kg/m^2	3.277058	ounce/sq ft	oz./ft^2
Mass/area	pound per sq foot	lb/ft^2	4.882428	kilogram per metre^2	kg/m^2	0.204816	pound/sq ft	lb/ft^2
Specific Vol.	cubic foot per pound	ft^3/lb	0.06243	cubic metre per kilogram	m^3/kg	16.01794	cubic ft/lb	ft^3/lb
Specific Wt.	pound per cubic inch	lb/in.^{3}	27680.0	kilogram per cubic metre	kg/m^3	0.00003613	pound/cubic inch	lb/in.^{3}
Specific Wt.	pound per cubic foot	lb/ft^3	1.602	kilogram per cubic metre	kg/m^3	0.6242197	pound/cubic foot	lb/ft^3
Specific Wt.	pound per gallon	lb/gal	9.978	kilogram per cubic metre	kg/m^3	0.1002205	pound/gal	lb/gal
Mass Flow	pound per hour	lb/hr	0.000126	kilogram per second	kg/s	7936.507	pound/hr	lb/hr
Mass Flow/Area	pound/inch^2 per hour	lb/in.^{2}, hr	0.1953	kilogram/sq metre, second	kg/m^{2}s	5.120327	pound/inch^2, hr	lb/in.^{2}, hr
Mass Flow/Area	pound/foot^2 per hour	lb/ft^2, hr	0.001356	kilogram/sq metre, second	kg/m^{2}s	737.463127	pound/ft^2, hr	lb/ft^2, hr
Vol. Flow Rate	cubic foot/second	ft^3/sec	0.02832	cubic metre, per second	m^3/s	35.31073	cubic ft/sec	ft^3/sec
Vol. Flow Rate	cubic foot/second	ft^3/sec	2.832	litre per second	ℓ/s	0.353107	cubic ft/sec	ft^3/sec
Vol. Flow Rate	cubic foot/minute	ft^3/min	0.0094719	cubic metre per second	m^2/s	2119.093	cubic ft/min	ft^3/min
Vol. Flow Rate	cubic foot/minute	ft^3/min	0.4719	litre per second	ℓ/s	2.119093	cubic ft/min	ft^3/min
Vol. Flow Rate	gallon/minute	gal/min.	0.00006309	cubic metres per second	m^3/s	15850.372	gallon/minute	gal/min
Vol. Flow Rate	gallon/minute	gal/min	0.06309	litre per second	ℓ/s	15.850372	gallon/minute	gal/min
Voi. Flow Rate	gallon/hour	gal/hr	0.000001052	cubic metres per second	m^1/s	950570.34	gallon/hour	gal/hr
Vol. Flow Rate	gallon/hour	gal/hr	0.001052	litre per second	ℓ/s	950.57034	gallon/hour	gal/hr
MOTION								
Velocity	feet per sec	ft/sec	0.3048	metre per second	m/s	3.28084	feet/sec	ft/sec
Velocity	feet per min	ft/min	0.00508	metre per second	m/s	196.85039	feet/min	ft/min
Velocity	miles per hour	mph	0.447	metre per second	m/s	2.237136	miles/hour	mph
Velocity	miles per hour	mph	1.609	kilometres per hour	kms/hr	0.621504	miles/hour	mph
Velocity	miles per hour	mph	0.8684	knots	knots	1.151543	miles/hour	mph
Velocity	knots	knot	0.5144	metre per second	m/s	1.9440	knots	knot
Accelerotion	feet per second^2	ft/sec^2	0.3048	metre per second^2	m/s^2	3.28084	feet/sec^2	ft/sec^2
Frequency	cycle per second	cy/sec	1.00	hertz		1.00	cycle/sec	cy/sec
Frequency	sq ft per hour	ft^2/hr	0.0000258	sq metre per second	m^2/s	38750.0	sq ft per hour	ft^2/hr
Mass/area	gal/100 ft^2	gal/100 ft^2	0.228	litres per metre^2	ℓ/m^2	64.3859	gal per 100 sq ft	gal/100 ft^2
Mass/Volume	pound/gal	lb/gal	0.0978	kg per litre	kg/ℓ	8.34724	pound/gal	lb/gal
Area/Volume	sq ft/gal	ft^2/gal	0.024462	sq metres per litre	m^2/litre	40.8795	sq ft per gal	ft^2/gal

Note: This table of conversions from English Units to Metric Units and back to English Units is not a complete English-Metric conversion table. The units listed are those units which are likely to be used in the thermal insulation technology. Complete tabulation is available as ASTM Standard for Metric Practice E380-76.

Table B-53 English to Metric to English Conversion Table (Continued)

Physical or Thermal Property	English System		Conversion Factors	Metric System		Conversion Factor	English System	
	Quantity	Units Symbol	Multiply by →	Quantity	Units Symbol	Multiply by →	Quantity	Units Symbol
Length	inch	"	25.4	millimeter	mm	0.03937	inch	"
Length	inch	"	0.0254	metre	m	39.37	inch	"
Length	foot	'	3.048	metre	m	0.32808	foot	'
Length	yard	yd.	0.9144	metre	m	1.09361	yard	yd.
Length	mile	mi.	1.6093	kilometres	km	0.62139	mile	mi.
Area	square inch	in.^{2}	645.16	square, millimetre	mm^2	0.00155	sq inch	in.^{2}
Area	square inch	in.^{2}	0.00064516	square metres	m^2	1549.148	sq inch	in.^{2}
Area	square foot	ft^2	0.0929	square metres	m^2	10.76426	sq ft	ft^2
Area	square yard	yd^2	0.8361	square metres	m^2	1.196	sq yd	yd^2
Area	square mile	mi^2	2589998.0	square metres	m^2	0.000000386	sq mi	mi^2
Area	square mile	mi^2	2.589998	square kilometres	km^2	0.386103	sq mi	mi^2
Area	acre	acre	4046.873	square metres	m^2	0.00024698	acre	acre
Area	acre	acre	0.004046	square kilometres	km^2	246.98	acre	acre
Volume	cubic inch	in.^{3}	16.39	cubic millimetre	mm^2	0.061013	cubic in.	in.^{3}
Volume	cubic inch	in.^{3}	0.00001639	cubic metres	m^3	61013.0	cubic in.	in.^{3}
Volume	cubic foot	ft^3	0.02832	cubic metres	m^3	35.31	cubic ft	ft^3
Volume	cubic foot	ft^3	28.32	litre	ℓ	0.03531	cubic ft	ft^3
Volume	cubic yard	yd^3	0.7646	cubic metre	m^3	1.30787	cubic yd	yd^3
Volume	cubic inch	in.^{3}	0.01639	litre	ℓ	61.0128	cubic in.	in.^{3}
Volume	pint (liquid)	pt.	0.0004732	cubic metre	m^2	2113.271	pint (liquid)	pt.
Volume	pint (liquid)	pt.	0.4732	litre	ℓ	2.113271	pint (liquid)	pt.
Volume	quart (liquid)	qt.	0.0009463	cubic metre	m^3	1056.688	quart (liquid)	qt.
Volume	quart (liquid)	qt.	0.9463529	litre	ℓ	1.056688	quart (liquid)	qt.
Volume	gallon (liquid)	gal.	0.0037854	cubic metre	m^3	264.1728	gallon (liquid)	gal.
Volume	gallon (liquid)	gal.	3.785412	litre	ℓ	0.2641728	gallon (liquid)	gal.
Mass	ounce	oz.	0.029349	kilogram	kg	35.274612	ounce	oz.
Mass	pound	lb.	0.4535924	kilogram	kg	2.204622	pound	lb.
Mass	ton	ton	1016.047	kilogram	kg	0.0009941	ton	ton
Mass	ton (short)	short ton	907.1847	kilogram	kg	0.0011023	ton (short)	short ton
Mass	ton (metric)	metric ton	1000.0	kilogram	kg	0.001	ton (metric)	metric ton
Mass	grain	grain	0.0000648	kilogram	kg	15432.36	grain	grain
Length Expan.	length, foot per° F	ft/° F	1.600	metre per ° K	m/° K	0.625	length, ft/° F	ft/° F

Note: This table of conversions from English Units to Metric Units and back to English Units is not a complete English-Metric conversion table. The units listed are those units which are likely to be used in the thermal insulation technology. Complete tabulation is available as ASTM Standard for Metric Practice E380-76.

Table B-53 English to Metric to English Conversion Table (Continued)

Physical or Thermal Property	English System Quantity	English System Units Symbol	Conversion Factors Multiply by →	Metric System Quantity	Metric System Units Symbol	Conversion Factor Multiply by →	English System Quantity	English System Units Symbol
Force	ounce force		0.2780139	Newton	N	3.596942	ounce force	
Force	pound force	lbf	4.448222	Newton	N	0.224808	pound force	lbf
Force	pound thrust/pound	lbf/lb	9.80665	Newton per kilogram	N/kg	0.1019716	pound thrust/lb	lbf/lb
Force	poundal		0.138255	Newton	N	7.233	poundal	
Force	dyne		0.00001	Newton		100000.0	dyne	
Force/length	pounds foot per foot	lbf/ft	14.5939	Newton per metre	N/m	0.068522	pound foot/foot	lbf/ft
Force/length	pounds foot per inch	lbf/in.	175.1268	Newton per metre	N/m	0.0057101	pound foot/inch	lbf/in.
Torque	pound force per foot		1.356	Newton metre	N/m	0.737463	pound force per ft	
Pressure	pound force per sq inch	lbf/in.^{2}	6895.0	Newton per square metre	N/m^2	0.000145	pound/square inch	psi
Pressure	pound force per sq inch	lbf/in.^{2}	6895.0	pascal	Pa	0.000145	pound/square inch	psi
Pressure	pound force per sq inch		703.1	kilogram per square metre	kgs/m^2	0.0014227	pound per sq ft	lb/ft^2
Pressure	pound force per sq ft	lbf/ft^2	47.88026	Newton/m^2 or pascal	Pa	0.020885	pound/square foot	lbf/ft^2
Pressure	inch H_2O (at 40° F)	in.H_2O	249.082	Newton/m^2 or pascal	Pa	0.0040147	in. of H_2O (40° F)	in.H_2O
Pressure	inch of mercury (at 40° F)	1″ Hg	3386.0	Newton/m^2 or pascal	Pa	0.000029	in. of Hg (40° F)	in. Hg
Pressure	foot of H_2O (at 40° F)	1′ H_2O	2988.98	Newton/m^2 or pascal	Pa	0.0033457	foot of H_2O (40° F)	ft H_2O
Pressure	torr (millimetre of Hg)	torr	133.32	Newton/m^2 or pascal	Pa	0.0075007	torr (mm of Hg)	torr
Pressure	atmosphere (standard)		101325.	Newton/m^2 or pascal	Pa	0.00000987	atmosphere, std.	
Press./Drop	in. of H_2O per 100 feet		8.170	Newton per cubic metre	N/m^{2}m	0.122309	inch H_2O/100 lin ft	
Press./Drop	foot H_2O per 100 feet		98.1	Newton per cubic metre	N/m^{2}m	0.0101938	foot H_2O/100 lin ft	
Dynamic Viscosity	pound force second/foot^2		47.88026	Newton, second/metre	Ns/m	0.020885	pound force sec/ft^2	
Dynamic Viscosity	pound force second/foot^2		47880.0	Centipoise	cP	0.0000209	pound force sec/ft^2	
Dynamic Viscosity	pound force hour/foot^2		172400.	Newton, second/metre	Ns/m	0.0000058	pound force hour/ft^2	
Energy*	foot pound force	ftlbf	1.355818	joule	J	0.737562	pound force hour/ft^2	
Energy	foot poundal		0.04214	joule	J	23.73042	foot poundal	
Energy	horsepower hour		2684.000	joule	J	0.000000372	horsepower hr	
Energy	kilowatt hour	kWh	3600000	joule	J	0.000000277	kilowatt hr	
Energy	watt per second	Ws	1.0000	joule	J	1.000	watt per second W/s	
Energy	calorie		4.1868	joule	J	0.238846	calorie	
Energy Flow	horsepower		745.7	watts	W	0.001341	horsepower	
Energy Flow	horsepower		0.7457	kilowatts	kW	1.341022	horsepower	
Energy Flow	kilocalorie-hour		1.163	watts	W	0.859845	kilocalorie/hr	

*Other than thermal energy

Note: This table of conversions from English Units to Metric Units and back to English Units is not a complete English-Metric conversion table. The units listed are those units which are likely to be used in the thermal insulation technology. Complete tabulation is available as ASTM Standard for Metric Practice E380-76.

Table B-54 Safety insulation thicknesses

Outer Covering: Stainless Steel Jacket Emissivity 0.40

Nom. Pipe Size	Operating Temperature °C																				
	60 to 79	80 to 99	100 to 119	120 to 139	140 to 159	160 to 179	180 to 199	200 to 219	220 to 239	240 to 259	260 to 279	280 to 299	300 to 319	320 to 339	340 to 359	360 to 379	380 to 399	400 to 419	420 to 439	440 to 459	460 to 479
½	1	1	1	1	1	1	1	1	1	1	1½	1½	1½	1½	1½	1½	1½	2	2	2	2
¾	1	1	1	1	1	1	1	1	1	1½	1½	1½	1½	1½	1½	2	2	2	2	2	2
1	1	1	1	1	1	1	1	1	1	1½	1½	1½	1½	1½	1½	2	2	2	2	2	2
1¼	1	1	1	1	1	1	1	1	1	1½	1½	1½	1½	1½	1½	2	2	2	2	2	2
1½	1	1	1	1	1	1	1	1	1	1½	1½	1½	1½	1½	2	2	2	2	2	2	2
2	1	1	1	1	1	1	1	1	1	1½	1½	1½	1½	2	2	2	2	2	2½	2½	2½
2½	1	1	1	1	1	1	1	1	1½	1½	1½	1½	1½	2	2	2	2	2	2½	2½	2½
3	1	1	1	1	1	1	1	1	1½	1½	1½	1½	1½	2	2	2	2½	2½	2½	2½	3
3½	1	1	1	1	1	1	1	1	1½	1½	1½	1½	1½	2	2	2	2½	2½	2½	2½	3
4	1	1	1	1	1	1	1	1	1½	1½	1½	1½	2	2	2	2	2½	2½	2½	3	3
6	1½	1½	1½	1½	1½	1½	1½	1½	1½	1½	1½	2	2	2	2½	2½	2½	2½	3	3	3
8	1½	1½	1½	1½	1½	1½	1½	1½	1½	1½	1½	2	2	2	2½	2½	2½	2½	3	3	3
10	1½	1½	1½	1½	1½	1½	1½	1½	1½	1½	1½	2	2	2	2½	2½	2½	3	3	3	3½
12	1½	1½	1½	1½	1½	1½	1½	1½	1½	1½	1½	2	2	2	2½	2½	2½	3	3	3	3½
14	1½	1½	1½	1½	1½	1½	1½	1½	1½	2	2	2	2	2½	2½	2½	3	3	3	3½	3½
16	1½	1½	1½	1½	1½	1½	1½	1½	1½	2	2	2	2	2½	2½	3	3	3	3½	3½	3½
18	1½	1½	1½	1½	1½	1½	1½	1½	1½	2	2	2	2½	2½	2½	3	3	3	3½	3½	3½
20	1½	1½	1½	1½	1½	1½	1½	1½	1½	2	2	2	2½	2½	2½	3	3	3	3½	3½	3½
22	1½	1½	1½	1½	1½	1½	1½	1½	1½	2	2	2	2½	2½	2½	3	3	3	3½	3½	3½
24	1½	1½	1½	1½	1½	1½	1½	1½	1½	2	2	2	2½	2½	2½	3	3	3	3½	3½	4
26	1½	1½	1½	1½	1½	1½	1½	1½	1½	2	2	2	2½	2½	2½	3	3	3	3½	3½	4
28	1½	1½	1½	1½	1½	1½	1½	1½	1½	2	2	2	2½	2½	2½	3	3	3	3½	3½	4
30	1½	1½	1½	1½	1½	1½	1½	1½	1½	2	2	2	2½	2½	2½	3	3	3½	3½	3½	4
36	1½	1½	1½	1½	1½	1½	1½	1½	1½	2	2	2	2½	2½	2½	3	3	3½	3½	3½	4
Equip. Flat	1½	1½	1½	1½	1½	1½	1½	1½	1½	2	2	2	2½	2½	3	3	3½	3½	3½	4	4
	140 to 175	176 to 211	212 to 247	248 to 283	284 to 319	320 to 355	356 to 391	392 to 427	428 to 463	464 to 499	500 to 535	536 to 571	572 to 607	608 to 643	644 to 679	680 to 715	716 to 751	752 to 787	788 to 823	824 to 859	860 to 895
	Operating Temperature °F																				

Based on 90°F (32°C) ambient air temperature and conductivity curves of calcium silicate or expanded silica insulation.

Table B-54 Safety insulation thicknesses (continued)

Outer Covering: Stainless Steel Jacket Emissivity 0.40

Nom. Pipe Size	Operating Temperature °C																			
	480 to 499	500 to 519	520 to 539	540 to 559	560 to 579	580 to 599	600 to 619	620 to 639	640 to 659	660 to 679	680 to 699	700 to 719	720 to 739	740 to 759	760 to 779	780 to 799	800 to 819	820 to 839	840 to 859	860 to 879
½	2	2	2½	2½	2½	2½	2½	2½	2½	2½	3	3	3	3	3	3½	3½	3½	3½	4
¾	2	2½	2½	2½	2½	2½	2½	3	3	3	3	3	3½	3½	3½	3½	3½	4	4	4
1	2½	2½	2½	2½	2½	3	3	3	3	3½	3½	3½	3½	3½	4	4	4	4	4½	4½
1¼	2½	2½	2½	2½	3	3	3	3	3	3½	3½	3½	3½	4	4	4	4½	4½	4½	4½
1½	2½	2½	2½	2½	3	3	3	3	3	3½	3½	3½	3½	4	4	4	4½	4½	4½	4½
2	2½	3	3	3	3½	3½	3½	3½	3½	4	4	4	4	4	4½	4½	4½	4½	5	5
2½	2½	3	3	3	3½	3½	3½	3½	3½	4	4	4	4	4	4½	4½	4½	4½	5	5
3	3	3	3	3½	3½	3½	3½	4	4	4	4	4	4½	4½	4½	4½	5	5	5	5½
3½	3	3	3	3½	3½	3½	3½	4	4	4	4	4	4½	4½	4½	4½	5	5½	5½	5½
4	3	3	3½	3½	3½	4	4	4	4	4½	4½	4½	4½	4½	5	5	5	5½	5½	5½
6	3½	3½	3½	3½	4	4	4	4½	4½	4½	5	5	5	5	5½	5½	5½	6	6	6
8	3½	3½	3½	4	4	4	4½	4½	4½	5	5	5	5½	5½	5½	6	6	6½	6½	6½
10	3½	3½	4	4	4	4½	4½	5	5	5	5½	5½	5½	6	6	6	6½	6½	7	7
12	4	4	4	4	4½	4½	4½	5	5	5	5½	5½	6	6	6	6½	6½	7	7	7
14	4	4	4	4½	4½	5	5	5	5½	5½	5½	6	6	6	6½	6½	7	7	7½	7½
16	4	4	4	4½	4½	5	5	5	5½	5½	5½	6	6	6½	6½	6½	7	7	7½	7½
18	4	4	4	4½	4½	5	5	5½	5½	5½	6	6	6	6½	6½	7	7	7½	7½	8
20	4	4	4½	4½	4½	5	5	5½	5½	5½	6	6	6½	6½	6½	7	7	7½	8	8
22	4	4	4½	4½	5	5	5	5½	5½	5½	6	6	6½	6½	6½	7	7	7½	8	8
24	4	4	4½	4½	5	5	5	5½	5½	6	6	6½	6½	6½	7	7	7½	8	8	8
26	4	4	4½	4½	5	5	5	5½	5½	6	6	6½	6½	6½	7	7	7½	8	8	8
28	4	4	4½	4½	5	5	5	5½	5½	6	6	6½	6½	6½	7	7	7½	8	8	8
30	4	4½	4½	4½	5	5	5½	5½	6	6	6	6½	6½	7	7	7½	7½	8	8	8
36	4	4½	4½	5	5	5	5½	5½	6	6	6½	6½	7	7	7½	7½	8	8	8½	8½
Equip. Flat	4½	4½	5	5	5½	6	6	6½	6½	7	7	7½	8	8	8½	9	9	9½	10	10½
	896 to 931	932 to 967	968 to 1003	1004 to 1039	1040 to 1075	1076 to 1111	1112 to 1147	1148 to 1183	1184 to 1219	1220 to 1255	1256 to 1291	1292 to 1327	1328 to 1363	1364 to 1399	1400 to 1435	1436 to 1471	1472 to 1507	1508 to 1543	1544 to 1579	1580 to 1616
	Operating Temperature °F																			

Based on 90°F (32°C) ambient air temperature and conductivity curves of calcium silicate or expanded silica insulation.

Table B-55 Safety insulation thicknesses

Outer Covering: Aluminum Jacket Emissivity 0.05

Nom. Pipe Size	Operating Temperature °C																				
	60 to 79	80 to 99	100 to 119	120 to 139	140 to 159	160 to 179	180 to 199	200 to 219	220 to 239	240 to 259	260 to 279	280 to 299	300 to 319	320 to 339	340 to 359	360 to 379	380 to 399	400 to 419	420 to 439	440 to 459	460 to 479
½	1	1	1	1	1	1	1	1½	1½	1½	2	2	2	2	2	2½	2½	2½	2½	2½	2½
¾	1	1	1	1	1	1	1	1½	1½	1½	2	2	2	2	2½	2½	2½	2½	2½	2½	3
1	1	1	1	1	1	1	1	1½	1½	1½	2	2	2	2	2½	2½	2½	2½	3	3	3
1¼	1	1	1	1	1	1	1½	1½	1½	1½	2	2	2	2	2½	2½	2½	3	3	3	3
1½	1	1	1	1	1	1	1½	1½	1½	2	2	2	2	2	2½	2½	2½	3	3	3	3½
2	1	1	1	1	1	1	1½	1½	1½	2	2	2	2½	2½	2½	3	3	3	3½	3½	3½
2½	1	1	1	1	1	1	1½	1½	1½	2	2	2	2½	2½	2½	3	3	3	3½	3½	3½
3	1	1	1	1	1	1	1½	1½	2	2	2	2½	2½	2½	3	3	3	3½	3½	3½	4
3½	1	1	1	1	1	1	1½	1½	2	2	2	2½	2½	2½	3	3	3	3½	3½	3½	4
4	1	1	1	1	1	1	1½	1½	2	2	2½	2½	2½	3	3	3	3½	3½	4	4	4
6	1½	1½	1½	1½	1½	1½	2	2	2	2½	2½	2½	3	3	3	3½	3½	4	4	4	4½
8	1½	1½	1½	1½	1½	1½	2	2	2	2½	2½	2½	3	3	3½	3½	3½	4	4	4½	4½
10	1½	1½	1½	1½	1½	1½	2	2	2	2½	2½	3	3	3	3½	3½	4	4	4½	4½	5
12	1½	1½	1½	1½	1½	1½	2	2	2	2½	2½	3	3	3½	3½	4	4	4½	4½	5	5
14	1½	1½	1½	1½	1½	1½	2	2	2½	2½	3	3	3	3½	3½	4	4	4½	4½	5	5
16	1½	1½	1½	1½	1½	1½	2	2	2½	2½	3	3	3½	3½	3½	4	4	4½	5	5	5½
18	1½	1½	1½	1½	1½	1½	2	2	2½	2½	3	3	3½	3½	4	4	4½	4½	5	5	5½
20	1½	1½	1½	1½	1½	1½	2	2	2½	2½	3	3	3½	3½	4	4	4½	4½	5	5	5½
22	1½	1½	1½	1½	1½	1½	2	2	2½	2½	3	3	3½	3½	4	4	4½	4½	5	5	5½
24	1½	1½	1½	1½	1½	1½	2	2	2½	2½	3	3	3½	3½	4	4	4½	4½	5	5½	5½
26	1½	1½	1½	1½	1½	1½	2	2	2½	2½	3	3	3½	3½	4	4	4½	4½	5	5½	5½
28	1½	1½	1½	1½	1½	1½	2	2	2½	2½	3	3	3½	3½	4	4	4½	4½	5	5½	5½
30	1½	1½	1½	1½	1½	1½	2	2	2½	2½	3	3½	3½	3½	4	4½	4½	5	5	5½	6
36	1½	1½	1½	1½	1½	1½	2	2	2½	2½	3	3½	3½	3½	4	4½	4½	5	5	5½	6
Equip. Flat	1½	1½	1½	1½	1½	1½	2	2	2½	3	3	3½	3½	4	4½	4½	5	5½	6	6	6½
	140 to 175	176 to 211	212 to 247	248 to 283	284 to 319	320 to 355	356 to 391	392 to 427	428 to 463	464 to 499	500 to 535	536 to 571	572 to 607	608 to 643	644 to 679	680 to 715	716 to 751	752 to 787	788 to 823	824 to 859	860 to 895
	Operating Temperature °F																				

Based on 90°F (32°C) ambient air temperature and conductivity curves of calcium silicate or expanded silica insulation.

Table B-55 Safety insulation thicknesses (continued)

Outer Covering: Aluminum Jacket Emissivity 0.05

Nom. Pipe Size	Operating Temperature °C																			
	480 to 499	500 to 519	520 to 539	540 to 559	560 to 579	580 to 599	600 to 619	620 to 639	640 to 659	660 to 679	680 to 699	700 to 719	720 to 739	740 to 759	760 to 779	780 to 799	800 to 819	820 to 839	840 to 859	860 to 879
½	2½	3	3	3	3½	3½	3½	3½	4	4	4	4	4½	4½	4½	4½	5	5	5	5½
¾	3	3	3	3½	3½	3½	4	4	4	4½	4½	4½	4½	5	5	5	5½	5½	5½	6
1	3	3½	3½	3½	3½	4	4	4½	4½	4½	4½	5	5	5	5½	5½	5½	6	6	6
1¼	3½	3½	3½	4	4	4	4½	4½	5	5	5	5	5½	5½	6	6	6	6	6½	6½
1½	3½	3½	3½	4	4	4	4½	4½	5	5	5	5	5½	5½	6	6	6½	6½	6½	7
2	3½	4	4	4	4½	4½	4½	5	5	5½	5½	5½	5½	6	6	6	6½	6½	7	7
2½	3½	4	4	4	4½	4½	4½	5	5	5½	5½	6	6	6½	6½	6½	7	7	7½	7½
3	4	4	4½	4½	4½	5	5	5½	5½	6	6	6	6	6½	6½	7	7	7½	7½	7½
3½	4	4	4½	4½	4½	5½	5½	5½	6	6	6½	6½	6½	7	7	7½	7½	8	8	8
4	4½	4½	4½	5	5	5½	5½	5½	6	6	6½	6½	6½	7	7	7½	7½	8	8	8
6	4½	5	5	5½	5½	5½	6	6	6½	6½	7	7	7½	7½	7½	8	8	8½	8½	8½
8	5	5	5	5½	6	6	6	6½	6½	7	7	7½	7½	8	8	8½	8½	9	9	9
10	5	5½	5½	6	6	6½	6½	6½	7	7½	7½	8	8	8½	8½	9	9	9½	9½	9½
12	5	5½	5½	6	6½	6½	7	7	7½	7½	8	8	8½	8½	9	9	9½	10	10	10
14	5½	5½	6	6	6½	7	7	7½	7½	8	8	8½	8½	9	9	9½	10	10	–	–
16	5½	6	6	6½	6½	7	7	7½	8	8	8½	8½	9	9	9½	9½	10	–	–	–
18	5½	6	6	6½	7	7	7½	7½	8	8	8½	9	9	9½	9½	10	–	–	–	–
20	5½	6	6½	6½	7	7	7½	8	8	8½	8½	9	9	9½	10	10	–	–	–	–
22	5½	6	6½	6½	7	7	7½	8	8	8½	8½	9	9	9½	10	–	–	–	–	–
24	6	6	6½	6½	7	7½	7½	8	8	8½	9	9½	(½	10	10	–	–	–	–	–
26	6	6	6½	6½	7	7½	7½	8	8	8½	9	9½	9½	10	–	–	–	–	–	–
28	6	6	6½	6½	7	7½	7½	8	8	8½	9	9½	9½	10	–	–	–	–	–	–
30	6	6½	6½	7	7	7½	8	8	8½	9	9	9½	10	10	–	–	–	–	–	–
36	6	6½	7	7	7½	8	8	8½	8½	9	9½	10	10	–	–	–	–	–	–	–
Equip. Flat	7	7½	7½	8	8½	9	9½	10	–	–	–	–	–	–	–	–	–	–	–	–
	896 to 931	932 to 967	968 to 1003	1004 to 1039	1040 to 1075	1076 to 1111	1112 to 1147	1148 to 1183	1184 to 1219	1220 to 1255	1256 to 1291	1292 to 1327	1328 to 1363	1364 to 1399	1400 to 1435	1436 to 1471	1472 to 1507	1508 to 1543	1544 to 1579	1580 to 1616
	Operating Temperature °F																			

Based on 90°F (32°C) ambient air temperature and conductivity curves of calcium silicate or expanded silica insulation.

Table B-56 Safety insulation thicknesses

Outer Covering: Mastic or Jacket With Emissivity 0.9

Nom. Pipe Size	Operating Temperature °C																				
	60 to 79	80 to 99	100 to 119	120 to 139	140 to 159	160 to 179	180 to 199	200 to 219	220 to 239	240 to 259	260 to 279	280 to 299	300 to 319	320 to 339	340 to 359	360 to 379	380 to 399	400 to 419	420 to 439	440 to 459	460 to 479
½	1	1	1	1	1	1	1	1	1	1	1	1	1	1	1	1½	1½	1½	1½	1½	1½
¾	1	1	1	1	1	1	1	1	1	1	1	1	1	1	1	1½	1½	1½	1½	1½	1½
1	1	1	1	1	1	1	1	1	1	1	1	1	1	1	1	1½	1½	1½	1½	1½	1½
1¼	1	1	1	1	1	1	1	1	1	1	1	1	1	1½	1½	1½	1½	1½	1½	1½	1½
1½	1	1	1	1	1	1	1	1	1	1	1	1	1	1½	1½	1½	1½	1½	1½	2	2
2	1	1	1	1	1	1	1	1	1	1	1	1	1	1½	1½	1½	1½	1½	1½	2	2
2½	1	1	1	1	1	1	1	1	1	1	1	1	1½	1½	1½	1½	1½	1½	1½	2	2
3	1	1	1	1	1	1	1	1	1	1	1	1½	1½	1½	1½	1½	1½	2	2	2	2
3½	1	1	1	1	1	1	1	1	1	1	1	1½	1½	1½	1½	1½	1½	2	2	2	2
4	1	1	1	1	1	1	1	1	1	1	1	1½	1½	1½	1½	1½	2	2	2	2	2
6	1½	1½	1½	1½	1½	1½	1½	1½	1½	1½	1½	1½	1½	1½	1½	2	2	2	2	2	2½
8	1½	1½	1½	1½	1½	1½	1½	1½	1½	1½	1½	1½	1½	1½	1½	2	2	2	2	2	2½
10	1½	1½	1½	1½	1½	1½	1½	1½	1½	1½	1½	1½	1½	1½	1½	2	2	2	2	2	2½
12	1½	1½	1½	1½	1½	1½	1½	1½	1½	1½	1½	1½	1½	1½	1½	2	2	2	2	2½	2½
14	1½	1½	1½	1½	1½	1½	1½	1½	1½	1½	1½	1½	1½	1½	2	2	2	2	2½	2½	2½
16	1½	1½	1½	1½	1½	1½	1½	1½	1½	1½	1½	1½	1½	1½	2	2	2	2	2½	2½	2½
18	1½	1½	1½	1½	1½	1½	1½	1½	1½	1½	1½	1½	1½	1½	2	2	2	2	2½	2½	2½
20	1½	1½	1½	1½	1½	1½	1½	1½	1½	1½	1½	1½	1½	1½	2	2	2	2	2½	2½	2½
22	1½	1½	1½	1½	1½	1½	1½	1½	1½	1½	1½	1½	1½	1½	2	2	2	2	2½	2½	2½
24	1½	1½	1½	1½	1½	1½	1½	1½	1½	1½	1½	1½	1½	1½	2	2	2	2½	2½	2½	2½
26	1½	1½	1½	1½	1½	1½	1½	1½	1½	1½	1½	1½	1½	1½	2	2	2	2½	2½	2½	2½
28	1½	1½	1½	1½	1½	1½	1½	1½	1½	1½	1½	1½	1½	1½	2	2	2	2½	2½	2½	2½
30	1½	1½	1½	1½	1½	1½	1½	1½	1½	1½	1½	1½	1½	1½	2	2	2	2½	2½	2½	2½
36	1½	1½	1½	1½	1½	1½	1½	1½	1½	1½	1½	1½	1½	1½	2	2	2	2½	2½	2½	2½
Equip. Flat	1½	1½	1½	1½	1½	1½	1½	1½	1½	1½	1½	1½	1½	2	2	2	2	2½	2½	2½	2½
	140 to 175	176 to 211	212 to 247	248 to 283	284 to 319	320 to 355	356 to 391	392 to 427	428 to 463	464 to 499	500 to 535	536 to 571	572 to 607	608 to 643	644 to 679	680 to 715	716 to 751	752 to 787	788 to 823	824 to 859	860 to 895
	Operating Temperature °F																				

Based on 90°F (32°C) ambient air temperature and conductivity curves of calcium silicate or expanded silica insulation.

Table B-56 Safety insulation thicknesses (continued)

Outer Covering: Mastic or Jacket With Emissivity 0.90

Nom. Pipe Size	Operating Temperature °C																			
	480 to 499	500 to 519	520 to 539	540 to 559	560 to 579	580 to 599	600 to 619	620 to 639	640 to 659	660 to 679	680 to 699	700 to 719	720 to 739	740 to 759	760 to 779	780 to 799	800 to 819	820 to 839	840 to 859	860 to 879
½	1½	1½	1½	2	2	2	2	2	2	2	2½	2½	2½	2½	2½	2½	2½	2½	2½	2½
¾	2	2	2	2	2	2	2	2½	2½	2½	2½	2½	2½	2½	2½	2½	2½	2½	3	3
1	2	2	2	2	2	2	2	2½	2½	2½	2½	2½	2½	2½	2½	3	3	3	3	3
1¼	2	2	2	2	2	2	2½	2½	2½	2½	2½	3	3	3	3	3	3	3	3	3
1½	2	2	2	2	2	2	2½	2½	2½	2½	2½	3	3	3	3	3	3	3	3	3½
2	2	2	2	2	2½	2½	2½	2½	2½	2½	3	3	3	3	3	3	3½	3½	3½	3½
2½	2	2	2	2	2½	2½	2½	2½	2½	2½	3	3	3	3	3	3	3½	3½	3½	3½
3	2	2	2½	2½	2½	2½	2½	3	3	3	3	3	3	3½	3½	3½	3½	4	4	4
3½	2	2	2½	2½	2½	2½	2½	3	3	3	3	3	3	3½	3½	3½	3½	4	4	4
4	2	2½	2½	2½	2½	2½	3	3	3	3	3	3½	3½	3½	3½	3½	4	4	4	4
6	2½	2½	2½	2½	3	3	3	3	3½	3½	3½	3½	3½	3½	4	4	4	4	4½	4½
8	2½	2½	2½	2½	3	3	3	3	3½	3½	3½	3½	4	4	4	4	4½	4½	4½	4½
10	2½	2½	2½	3	3	3	3	3½	3½	3½	3½	4	4	4	4	4½	4½	4½	5	5
12	2½	2½	2½	3	3	3	3½	3½	3½	3½	4	4	4	4	4½	4½	4½	5	5	5
14	2½	2½	3	3	3	3½	3½	3½	3½	4	4	4	4	4½	4½	4½	5	5	5	5½
16	2½	3	3	3	3	3½	3½	3½	4	4	4	4	4½	4½	4½	5	5	5	5½	5½
18	2½	3	3	3	3½	3½	3½	3½	4	4	4	4½	4½	4½	4½	5	5	5	5½	5½
20	2½	3	3	3	3½	3½	3½	3½	4	4	4	4½	4½	4½	5	5	5	5½	5½	5½
22	2½	3	3	3	3½	3½	3½	3½	4	4	4	4½	4½	4½	5	5	5	5½	5½	5½
24	3	3	3	3	3½	3½	3½	4	4	4	4	4½	4½	4½	5	5	5	5½	5½	5½
26	3	3	3	3	3½	3½	3½	4	4	4	4	4½	4½	4½	5	5	5	5½	5½	5½
28	3	3	3	3	3½	3½	3½	4	4	4	4½	4½	4½	5	5	5	5	5½	5½	5½
30	3	3	3	3	3½	3½	3½	4	4	4	4½	4½	4½	5	5	5	5½	5½	5½	5½
36	3	3	3	3½	3½	3½	3½	4	4	4½	4½	4½	4½	5	5	5	5½	5½	6	6
Equip. Flat	3	3	3	3½	3½	4	4	4	4½	4½	5	5	5	5½	5½	6	6	6½	6½	6½
	896 to 931	932 to 967	968 to 1003	1004 to 1039	1040 to 1075	1076 to 1111	1112 to 1147	1148 to 1183	1184 to 1219	1220 to 1255	1256 to 1291	1292 to 1327	1328 to 1363	1364 to 1399	1400 to 1435	1436 to 1471	1472 to 1507	1508 to 1543	1544 to 1579	1580 to 1616
	Operating Temperature °F																			

Based on 90°F (32°C) ambient air temperature and conductivity curves of calcium silicate or expanded silica insulation.

Appendix C — Listing of Charts Equations Examples Figures Tables

Listing of Charts

*See definition of water vapor transmission units, Appendix B, page 599.

Listing of Equations

*See definition of water vapor transmission units, Appendix B, page 599.

Listing of Equations (Cont'd.)

*See definition of water vapor transmission units, Appendix B, page 599.

Listing of Examples

*See definition of water vapor transmission units, Appendix B, page 599.

Listing of Examples (Cont'd.)

*See definition of water vapor transmission units, Appendix B, page 599.

Listing of Figures

*See definition of water vapor transmission units, Appendix B, page 599.

Listing of Figures (Cont'd.)

*See definition of water vapor transmission units, Appendix B, page 599.

Listing of Figures (Cont'd.)

*See definition of water vapor transmission units, Appendix B, page 599.

Listing of Tables

*See definition of water vapor transmission units, Appendix B, page 599.

Listing of Tables (Cont'd.)

*See definition of water vapor transmission units, Appendix B, page 599.

**Conversions to Metric provided following the table.

Listing of Tables (Cont'd.)

*See definition of water vapor transmission units, Appendix B, page 599.

**Conversions to Metric provided following the table.

Index

ADDITIONAL REFERENCES

THERMAL INSULATION SYSTEMS
National Aeronautics and Space Agency

THERMAL DESIGN OF BUILDINGS Tyler Stewart Rogers

ASTM STANDARDS—PART 14
Thermal Insulation: Acoustical Materials
Fire Test: Building Construction

ASTM RECOMMENDED DIMENSIONAL STANDARDS
Prefabrication and Field Fabrication of Thermal Insulation Fitting Covers for NPS Pipe, Vessel Lagging and Dished Head Segments.

PART OF ASTM RECOMMENDED PRACTICE
C-450-657

HOW TO DETERMINE ECONOMIC THICKNESS OF INSULATION
National Insulation Manufacturers Association
441 Lexington Avenue
New York, New York 10017

ASHRAE GUIDE AND DATA BOOK

ASHRAE HANDBOOK OF FUNDAMENTALS

782187